Encyclopedia of Applied Physics

SPONSORS

AMERICAN INSTITUTE OF PHYSICS
DEUTSCHE PHYSIKALISCHE GESELLSCHAFT
JAPAN SOCIETY OF APPLIED PHYSICS
PHYSICAL SOCIETY OF JAPAN

AMERICAN INSTITUTE OF PHYSICS

DEUTSCHE PHYSIKALISCHE GESELLSCHAFT

JAPAN SOCIETY OF APPLIED PHYSICS

PHYSICAL SOCIETY OF JAPAN

ENCYCLOPEDIA OF APPLIED PHYSICS

VOLUME 8
Inductors to
Magnetic Devices

Edited by

GEORGE L. TRIGG

Associate Editors

EDUARDO S. VERA

WALTER GREULICH

Managing Editor

EDMUND H. IMMERGUT

Assistant Managing Editor

LINDEN BYASS

George L. Trigg
275 Beaver Dam Road
Brookhaven, New York 11719

Walter Greulich
Ziegeleiweg 30
D-6943 Birkenau
Federal Republic of Germany

Eduardo S. Vera
NTT Interdisciplinary Research Laboratories
Nippon Telegraph and Telephone Corporation
Musashino-shi, Tokyo 180, Japan

Edmund H. Immergut
2 Sidney Place
Brooklyn, New York 11201

Linden Byass
2 Sidney Place
Brooklyn, New York 11201

Library of Congress Cataloging-in-Publication Data

Encyclopedia of applied physics.
(Revised for Vol. 8)

"Sponsors, American Institute of Physics...[et al]."
Includes bibliographical references and index.
1. Physics—Encyclopedias. 2. Engineering—Encyclopedias. I. Trigg, George L. II. Vera, Eduardo S. III. Greulich, Walter. IV. American Institute of Physics.
QC5.E543 1991 530′.05 91-8738
ISBN 1-56081-067-X (VCH Publishers)

British Library Cataloguing in Publication Data

Encyclopedia of applied physics.
Vol. 8
I. Trigg, George L. (George Lockwood) II. Vera, Eduardo S. III. Greulich, Walter
621
ISBN 1-56081-058-0 (set)
ISBN 1-56081-067-X (vol. 8)

Printed in the United States of America.

ISBN 1-56081-067-X (Volume 8) VCH Publishers, Inc.
ISBN 1-56081-058-0 (set) VCH Publishers, Inc.
ISBN 3-527-28127-4 (Volume 8) VCH Verlagsgesellschaft mbH
ISBN 3-527-26841-3 (set) VCH Verlagsgesellschaft mbH

Printing History:
10 9 8 7 6 5 4 3 2 1

Published jointly by:

VCH Publishers, Inc.
220 East 23rd Street
New York, NY 10010

VCH Verlagsgesellschaft mbH
P.O. Box 10 11 61
69451 Weinheim
Federal Republic of Germany

VCH Publishers (UK) Ltd.
8 Wellington Court
Cambridge CB1 1HZ
United Kingdom

ADVISORY BOARD

EDITORIAL CONSULTANTS

MAIN ENTRIES

05-A	Inductors	1
01-E	Information Science	21
12-D	Insulators and Semiconductors, Conductivity in	41
10-B	Insulators, Structure of	85
09-D	Intense Particle Beams	103
14-D	The Interaction of Solids with Particles and Radiation	117
15-B	Intercalation Compounds	133
02-A	Interferometers and Interferometry	157
16-C	Ion-Beam Modification of Materials	173
11-D	Ionic Conduction and Diffusion in Solids	193
20-D	Ionosphere	223
02-A	Laboratory Instrumentation	247
06-D	Laser Photochemistry	283
06-D	Laser Physics	299
06-C	Laser Technology	321
06-A	Lasers, Dye, Technology and Engineering	331
06-A	Laser, Free-Electron	353
06-A	Lasers, Gas	371
06-C	Lasers, Industrial Uses of	397
06-A	Lasers, Semiconductor	425
06-A	Lasers, Solid State	443
11-D	Lattice Vibrations, Statistics of	465
06-A	Light-Emitting Diodes	485
10-B	Liquid Crystals, Structure of	515
10-B	Liquids, Simple, Structure of	533
05-A	Machines, Electrical	561
13-A	Magnetic Devices	587

The subject matter in the *Encyclopedia of Applied Physics* is presented in approximately 500 individual articles, arranged alphabetically. The topics can be classified into 20 sections, similar to the AIP Physics and Astronomy Classification Scheme (PACS):

01	General Aspects: Mathematical, Computational, and Information Techniques
02	Measurement Science, General Devices and/or Methods
03	Nuclear and Elementary Particle Physics
04	Atomic and Molecular Physics
05	Electricity and Magnetism
06	Optics (classical and quantum)
07	Acoustics
08	Thermodynamics and Properties of Gases
09	Fluids and Plasma Physics
10	Condensed Matter A: Structure and Mechanical Properties
11	Condensed Matter B: Thermal, Acoustic, and Quantum Properties
12	Condensed Matter C: Electronic Properties
13	Condensed Matter D: Magnetic Properties
14	Condensed Matter E: Dielectrical and Optical Properties
15	Condensed Matter F: Surfaces and Interfaces
16	Materials Science
17	Physical Chemistry
18	Energy Research and Environmental Physics
19	Biophysics and Medical Physics
20	Geophysics, Meteorology, Space Physics, and Aeronautics

Each article has been assigned a code number consisting of two digits which denotes the section, and a letter which gives the type of article. There are six types: A=Devices, Equipment; B=Materials; C=Methods, Processes; D=Phenomena, Effects; E=Scientific or Technological Fields; F=Institutions, Companies, Societies and other organizations.

CONTRIBUTORS

M. A. Abkowitz, Xerox Webster Research Center, Webster, NY 14580, U.S.A.
Insulators and Semiconductors, Conductivity in

Robert S. Averback, Dept. of Materials Science and Engineering, Univ. of Illinois, 1304 W. Green St., Urbana, IL 61801, U.S.A.
Ion-Beam Modification of Materials

James Benford, HPM Division/7200, Physics International Co., 2700 Merced St., San Leandro, CA 94577, U.S.A.
Intense Particle Beams

A. V. Chadwick, Chemical Laboratory, Univ. of Kent, Canterbury, Kent CT2 7NH, UK
Ionic Conduction and Diffusion in Solids

M. George Craford, Optoelectronics Division, Hewlett-Packard Company, San Jose, CA 95131, U.S.A.
Light-Emitting Diodes

E. M. Conwell, Xerox Webster Research Center, Webster, NY 14580, U.S.A.
Insulators and Semiconductors, Conductivity in

Norman E. Cusack, School of Physics, University of East Anglia, Norwich NR4 7TJ, UK
Liquids, Simple, Structure of

M. S. Dresselhaus, Department of Electrical Engineering and Computer Science and Department of Physics, Massachusetts Institute of Technology, Cambridge, MA 02139, U.S.A.
Intercalation Compounds

G. Dresselhaus, Francis Bitter National Magnet Laboratory, Massachusetts Institute of Technology, Cambridge, MA 02139, U.S.A.
Intercalation Compounds

F. J. Duarte, Imaging Research Laboratories, Eastman Kodak Company, Rochester, NY 14650-1744, U.S.A.
Lasers, dye, technology and engineering

J. G. Eden, Everitt Laboratory, Univ. of Illinois, Urbana, IL 61801, U.S.A.
Lasers, Gas

Alv Egeland, Dept. of Physics, University of Oslo, N-0316 Oslo, Norway
Ionosphere

Paul Engelking, Dept. of Chemistry, Univ. of Oregon, Eugene, OR 97403, U.S.A.
Laser Photochemistry

T. Y. Fan, Lincoln Laboratory, MIT, P.O. Box 73, Lexington, MA 02173, U.S.A.
Lasers, Solid State

D. R. Foster, Imaging Research Laboratories, Eastman Kodak Company, Rochester, NY 14650-1744, U.S.A.
Lasers, Dye, Technology and Engineering

F. J. Friedlaender, School of Electrical Engineering, Purdue University, West Lafayette, IN 47907, U.S.A.
Machines, Electrical

J. Heremans, General Motors Research Laboratories, 30500 Mound Road, Warren, MI 48090, U.S.A.
Magnetic Devices

I. P. Ipatova, A. F. Ioffe Physico-Technical Institute, Russian Academy of Sciences, St. Petersburg 194021, Russia
Lattice Vibrations, Statistics of

P. L. Kelley, Electro-Optics Technology Center, Tufts University, Medford, MA 02155, U.S.A.
Laser Physics

J. D. Litster, Department of Physics, Massachusetts Institute of Technology, Cambridge, MA 02139, U.S.A.
Liquid Crystals, Structure of

A. A. Maradudin, Department of Physics, University of California, Irvine, CA 92717, U.S.A.
Lattice Vibrations, Statistics of

Nelson C. Maynard, Phillips Laboratory, Hanscom Air Force Base, MA 01731, U.S.A.
Ionosphere

Charles T. Meadow, Faculty of Library & Information Science, Univ. of Toronto, Toronto, Ontario, Canada M5S 1A1
Information Science

D. W. Novotny, Department of Electrical and Computer Engineering, University of Wisconsin, Madison, WI 53706, U.S.A.
Machines, Electrical

Nagaatsu Ogasawara, Department of Electronics Engineering, University of Electro-Communications, Tokyo 182, Japan
Lasers, Semiconductor

Dennis Pavlik, Science and Technology Center, Westinghouse Electric Corp., 1310 Beulah Road, Pittsburgh, PA 15235, U.S.A.
Inductors

C. Pellegrini, Department of Physics, University of California, Los Angeles, Los Angeles, CA 90024, U.S.A.
Lasers, Free Electron

David M. Roessler, Physics Dept., General Motors Research & Development Ctr., Warren, MI 48090-9055, U.S.A.
Lasers, Industrial Uses of

S. E. Schnatterly, Physics Department, University of Virginia, Charlottesville, VA 22901, U.S.A.
Interaction of Solids with Particles and Radiation

R. Shashidhar, Center for Biomolecular Science and Engineering, Naval Research Laboratory, Washington, DC 20375, U.S.A.
Liquid Crystals, Structure of

Ayman A. Shourbaji, Instrumentation and Controls Division, Oak Ridge National Laboratory, Oak Ridge, TN 37831-6003, U.S.A.
Laboratory Instrumentation

W. H. Steel, School of Mathematics, Physics, Computing and Electronics, Macquarie Univ., Sydney, Australia 2109
Interferometers and Interferometry

Frank M. Steranka, Optoelectronics Division, Hewlett-Packard Company, San Jose, CA 95131, U.S.A.
Light-Emitting Diodes

Hiroshi Takuma, The Institute for Laser Science, University of Electro-Communications, Tokyo 182, Japan
Laser Technology

Cronin B. Vining, Jet Propulsion Laboratory, M/S 277-212, California Institute of Technology, 4800 Oak Grove Drive, Pasadena, CA 91109, U.S.A.
Insulators, Structure of

S. A. Werner, Dept. of Physics & Astron., Univ. of Missouri-Columbia, Columbia, MO 65211, U.S.A.
Interferometers and Interferometry

J. J. Zayhowski, MIT Lincoln Laboratory, Lexington, MA 02173, U.S.A.
Laser Physics

INDUCTORS

DENNIS PAVLIK, *Science and Technology Center, Westinghouse Electric Corporation, Pittsburgh, Pennsylvania, U.S.A.*

	Introduction	1
1.	**General Principles and Historical Perspective**	2
1.1	The Concept of Inductance	2
1.2	Analytical Methods	3
1.3	Types of Inductors	4
2.	**Principles of Operation**	6
2.1	Fixed Inductors	6
2.1.1	Resonant Circuits	6
2.1.2	Chokes	6
2.1.3	Energy Storage	6
2.2	Adjustable Inductors	7
2.3	Variable Inductors	7
2.3.1	Moving Core	7
2.3.2	Saturable Reactors	8
2.3.3	Magnetic Switches	8
3.	**Engineering Aspects**	10
3.1	Electromagnetics	11
3.2	Dielectrics	11
3.3	Thermodynamics	11
3.4	Mechanics	11
4.	**Inductor Applications**	11
4.1	Power Reactors	12
4.2	Energy-Transfer Inductors	12
4.3	Resonant Charging for Particle Accelerators	13
4.4	Superconducting Magnetic Energy Storage	14
4.5	Magnetic Switches	15
4.6	Electronic Circuit Applications	16
4.6.1	Resonant Tuning	16
4.6.2	Electronic Filters	16
4.6.3	Current-Source Reactors	17
	Glossary	18
	Works Cited	18
	Further Reading	19

INTRODUCTION

When a current flows through an electrical circuit, it creates a magnetic field. Inductance is the quantity that relates the energy stored in a magnetic field to the current flowing through that circuit. It defines the relationship between the voltage and current needed to add or extract energy from the magnetic field. *Inductor* is the generic name for a class of devices used for adding inductance to an electrical circuit. It covers a wide range of devices, applications, and configurations. Inductors range in size from small fractional-turn helixes having masses of several grams to devices hundreds of cubic meters in volume having masses in excess of 50 000 kg. Inductors are used as components of tuned circuits, electronic filters, magnetic energy storage devices, and measurement apparatus.

Because inductors usually comprise a number of turns or coils of conductor on a form, they are sometimes referred to as inductance coils, electric coils, or simply coils. There are other names used to describe inductors for specific applications. These include choke, choke coil, energy-storage choke, reactor, inductometer, induction coil, magnetic switch, pulsactor, and saturable reactor. A number of these will be described in detail to illustrate the distinction among different types and classes of inductors.

Inductors are classified according to their geometric configuration and whether or not a magnetic material is used to enhance the inductive effect. Inductors without magnetic materials are called air-core inductors. Those that use magnetic materials to enhance the inductive effect are referred to as iron-core inductors.

Inductors are typed according to whether or not they provide a fixed, adjustable, or variable inductance. The different classes and types are represented schematically by four different symbols as shown in Fig. 1. Combinations of these basic symbols are used to represent more complex inductor arrangements.

1-56081-067-X/94/$5.00+.50

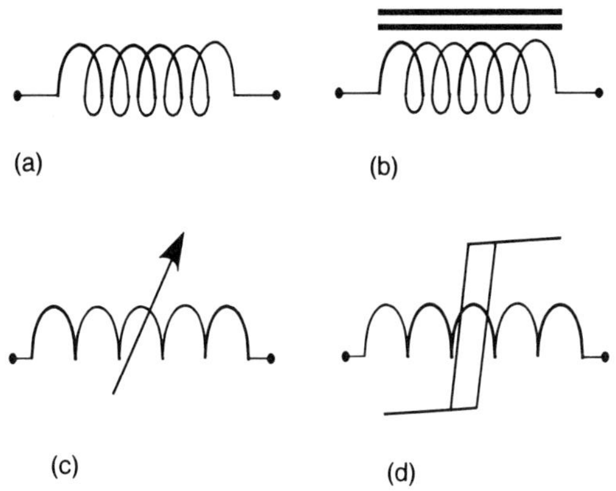

FIG. 1. Schematic symbols used to identify the different types of inductors: (a) air-core fixed inductor, (b) iron-core fixed inductor, (c) variable inductor, (d) saturating or nonlinear inductor.

The first section of this article explains the concept of inductance and the general principles and techniques used in its calculation. A historical perspective is taken to illustrate the development of different concepts and calculation procedures.

The class of air-core inductors is described and some formulas presented for the calculation of inductance. This is expanded to the more general class of linear inductors that can have iron cores but operate in the linear region. The operation of nonlinear and variable inductors is also presented.

The section on engineering aspects discusses how the electromagnetic system requirements are used in conjunction with the thermal and mechanical limitations of the available construction materials to synthesize a specific configuration. Examples of specific applications are presented in the last two sections.

1. GENERAL PRINCIPLES AND HISTORICAL PERSPECTIVE

1.1 The Concept of Inductance

In 1831 Faraday first showed that an electric current is induced in a closed circuit when the magnetic field in its vicinity is altered. This change in field could be caused by the movement of a permanent magnet in relation to the circuit or by a changing current in a neighboring electrical circuit. Lenz showed that the induced electromotive force (emf) or any current it generated in a closed circuit always acted to prevent the magnetic field linking it from changing. The amount of flux linking an electrical circuit is called the flux linkage λ. These two observations form Faraday's law:

$$\text{emf} = e = -\frac{d\lambda}{dt}. \tag{1}$$

Faraday (1839) reported that when a current flowing in a circuit was interrupted it produced an emf greater than that of the voltage source used to initiate the current. He showed that the inductive effect of a circuit on itself was part of the same phenomenon as that of a current affecting neighboring conductors. The inductive effect of a circuit on itself is called self-inductance. The inductive interaction between two or more distinct electric circuits is defined as the mutual inductance between the circuits (see TRANSFORMERS).

If the change in flux linkages is not due to a physical motion of conductors and the materials are linear, the induced emf can be related to changes in current by

$$\text{emf} = -L\frac{dI}{dt}, \tag{2a}$$

where L is the self-inductance of the circuit. For harmonic fields, Eq. (2a) becomes

$$V = j\omega LI = j(2\pi fL)I, \tag{2b}$$

where V is the magnitude of the counter emf, j is $\sqrt{(-1)}$ representing the 90° phase shift between voltage and current in phasor notation, ω is the angular frequency of the sinusoid equal to $2\pi f$, and I is the phasor current. The quantity $2\pi fL$ is known as the reactance of the inductive circuit element. The unit of inductance is a weber per ampere designated as the henry, after Joseph Henry, an American scientist and inventor. Assuming a constant inductance, Eqs. (1) and (2a) can be combined to give an equation for L in terms of λ and I:

$$L = \frac{d\lambda}{dI}, \tag{3a}$$

which, subject to linearity, is also

$$L = \lambda / I. \tag{3b}$$

The concept of flux linkages is often confusing and difficult to apply. For this reason,

Maxwell (1891) preferred to explore inductance in terms of energy and geometrical functions giving an alternate definition of inductance:

$$L=2W/I^2, \tag{4}$$

where W is the energy stored in the magnetic field as a result of the current I in the circuit. The stored magnetic energy can be related to the magnetic vector potential A. This is extremely useful in field calculations using numerical methods, such as the finite-element technique, as it allows for the direct calculation of inductance in very complicated arrangements and multicoil systems (see MAGNETOSTATICS).

Equations (2)–(4) give the same values of inductance and are useful when the surrounding media and materials are linear and there are no hysteresis, residual fluxes, or eddy-current effects. However, when these conditions are not met, using these equations interchangeably can result in serious errors. Rader and Litscher (1944) extensively treat the subject of inductance in the presence of nonlinear materials and eddy currents, giving examples showing when each formulation is appropriate.

The distinction between the formulas is illustrated in Fig. 2. Consider a nonlinear inductor operating at a current whose magnitude I is given by the length oa and total flux linkage is equal to length ab. The linear or flux-linkage inductance L_b from Eq. (3b) is ab/oa. However, from Eq. (4) the inductance

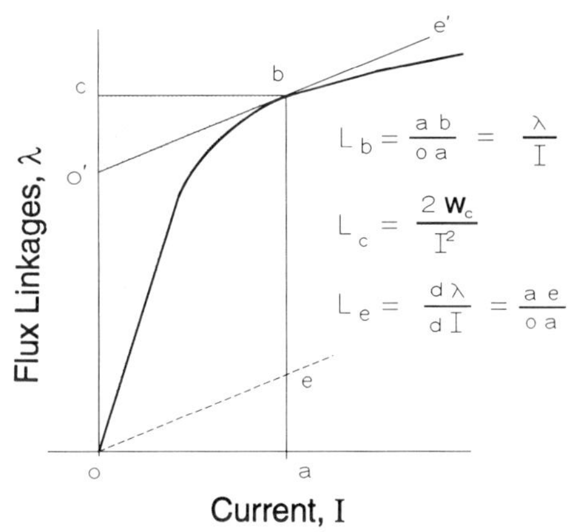

FIG. 2. Nonlinear magnetization curve illustrating how "inductance" differs depending on the selected definition and operating point.

L_c is equal to twice the total stored energy (shaded area obc) divided by the square of the current. At operating point b, the incremental inductance L_e given by Eq. (3a) is the slope of the line $o'e'$, which is just ae/oa. In nonlinear circuits $L_b > L_c > L_e$, and which one is appropriate to use depends on the circuit application.

In nonlinear devices such as saturable reactors and magnetic amplifiers the incremental inductance L_e is more appropriate. Magnetic switches are highly nonlinear, and their performance characteristics must be derived from a direct analysis of the B-H characteristics of the device (see related articles on MAGNETS and ENERGY CONVERSION).

1.2 Analytical Methods

The development of the concept of inductance was based on experimental observations. The defining equations are from this observational reference frame in terms of energy or voltage and current. Fundamentally, however, inductance is a measure of how an arrangement of conductors stores and exchanges electrical and magnetic energy. Therefore, it is a geometric quantity dependent on the physical arrangement of the currents and the permeability (inductive capacity) of the surrounding space.

The difficulty with magnetic materials arises because their inductive behavior is both nonlinear and hysteretic. When the surrounding space has materials that are conducting, currents are induced in them through mutual induction, and the observed inductance at the terminals is the resultant of the combined effects, making the calculation of inductance difficult.

Highly accurate formulas for inductance are available for cases where no magnetic materials, hysteresis, or eddy-current effects are present. Maxwell and his contemporaries developed exact solutions for the simple geometries comprising sheets or filaments of current. In all but the simplest cases, these formulas contained complicated inverse hyperbolic, trigonometric, and elliptic integrals. Analysis of these functions was difficult given the available tools. The approach taken was to represent the more difficult functions by rapidly converging series expansions that were then used to derive solutions for more complex geometries. Correction factors were developed to account for the fact that real inductors were made of finite-dimensioned wire with separation between turns. These complexities can best be illustrated by example.

Consider a circular loop of mean radius a, using a conductor with circular cross section of radius r. The inductance in henries with all dimensions in meters is given by

$$L=0.4\pi a[(1+r^2/8a^2)\ln(8a/r)(r^2/24a^2) - 1.75]\times 10^{-6}. \quad (5)$$

Lorenz developed a closed formula for the inductance of a finite-length right circular cylindrical solenoid, with a current-sheet winding (infinitesimal radial thickness), in terms of elliptic integrals, given by

$$L=(3.2\pi N^2a^3/3b^2) \times\{(2k^2-1)E/k^3 + (1-k^2)F/k^3-1\}\times 10^{-6}, \quad (6)$$

where a is the radius of the solenoid, b is its axial length, and F and E are complete elliptic integrals of the first and second kind, respectively, and modulus $k=[4a^2/(4a^2+b^2)]^{1/2}$.

A highly accurate formula for a solenoid with a finite radial build (thickness of winding including insulation) c, length b, and mean radius a, developed by Weinstein, is given by

$$L=0.4\pi aN^2(p+s)\times 10^{-6} \quad (7)$$

where $x=b/c$ and

$$p=\ln(8a/c)+\tfrac{1}{12}-\pi x/3+\ln(1+x^2) \times[1/(12x^2)-\tfrac{1}{2}]+(x^2/12)\ln(1+1/x^2) +\tfrac{2}{3}(x-1/x)\tan^{-1}(x) \quad (8)$$

$$s=(c^2/96a^2)\{[\ln(8a/c)-0.5\ln(1+x^2)] \times(1+3x^2)+3.45x^2+221/60-1.6\pi x^3 +3.2x^3\tan^{-1}(x)-\ln(1+x^2)/10x^2 +(x^4/2)\ln(1+1/x^2)\}. \quad (9)$$

Equation (7) loses accuracy as the solenoid approaches a current-sheet type winding ($x>25$) or as c approaches a. Although very accurate for squarish type sections ($x<10$) and $a>c$, the terms in Eqs. (8) and (9) consist of large positive and negative numbers that nearly cancel. These equations illustrate the two major difficulties in using series expansions:

1. The equations are generally only valid over a particular range; and
2. the individual terms in the series nearly cancel, requiring a high degree of accuracy in the separate terms for even moderate accuracy in the final answer.

Rosa and Grover (1911), working for the U. S. National Bureau of Standards, made what is perhaps the most comprehensive investigation of the various formulas available at the time. Later, Grover (1946) refined this work and produced a book of working formulas and tables that can be used to calculate the inductances for the important geometric arrangements. Grover's formulas employing only one or two terms give accuracies to greater than one part in 1000 over extremely wide ranges of geometric variation. Corrections for winding space and insulation are readily incorporated into the formulations for all of the important geometries.

From a geometric perspective, inductance is just N^2 times the permeance $\mathscr{P}$ of the magnetic path. If the magnetic path can be clearly defined, it is possible to derive reasonable approximations for inductance. In general, defining the magnetic path is difficult, and the magnetic-circuit approach offers no calculational advantage. Iron-core inductors are an exception, since one of the purposes of using a magnetic material is to provide a clearly defined magnetic path. Therefore, the magnetic-circuit approach provides a reasonably simple and accurate approximation for the inductance in many iron-core inductor problems.

Today numerical methods and high-speed computers make it possible to calculate inductances accurately for complex conductor arrangements including the effects of eddy currents, nonlinear magnetic materials, and hysteresis.

1.3 Types of Inductors

The three main parts of an inductor are the winding, the winding form, and the core. The winding usually comprises a number of turns of conductor wound about some axis of symmetry. The winding form is the structure used to support the winding. The region enclosed

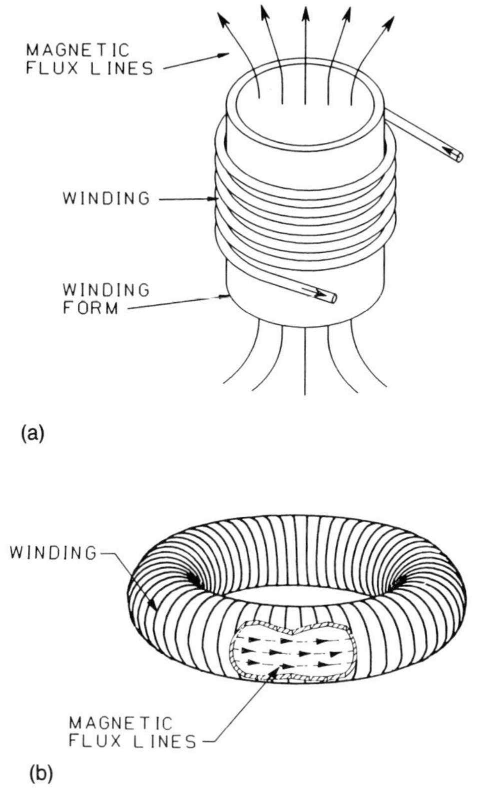

FIG. 3. The two most common types of air-core inductor configuration: (a) solenoid and (b) toroid.

by the winding in which the magnetic field is formed is called the core. Inductors with nonmagnetic core materials (wood, ceramic, plastic, air, oil, etc.) are called air-core inductors. If a magnetic material is used in the core to augment the inductive effect, the inductor is referred to as an iron-core inductor. The most common materials used in iron-core inductors are alloys of silicon steel. However, many different magnetic materials are used in the construction of inductors, especially at high frequencies, such as ferrites, powdered iron, cobalt-iron and nickel-iron alloys, and amorphous metals.

The two most common geometric forms of air-core inductors are solenoidal and toroidal. An inductor is solenoidal if the direction of the magnetic field is primarily straight within the body of the inductor and flux closure is external to the winding. Figure 3(a) shows a short solenoidal inductor of circular form. The winding form for a solenoid is generated by taking a closed planar figure (circle, rectangle, polygon) and extruding it along a straight line normal to its plane. If the length of the solenoid is short relative to its diameter, it is referred to as a pancake coil.

A toroidal inductor is one in which the direction of the magnetic field is predominantly a closed (generally circular) path so that flux closure is within the body of the winding. A toroidal winding form is generated by taking a closed planar surface and rotating it about a straight line in the same plane but external to the closed surface, as illustrated in Fig. 3(b). Because the winding closes on itself, toroidal inductors have a relatively small external magnetic field. In general, toroids have continuous smooth windings that approximate a current sheet over the surface of the winding form. It is not always possible or desirable to have a continuous winding over the surface of the form. An alternative to the smooth toroidal construction is to use discrete short solenoidal sections arranged in a generally circular pattern as illustrated by Fig. 4. This type of arrangement is known as a caged toroid and is used primarily in large energy-storage inductors and confining magnets used for fusion experiments.

In toroidal windings there are two components of electromagnetic forces acting on the windings. The first acts on the local cross section (minor radius) to expand the coil, as with any loop of wire. The second force acts to

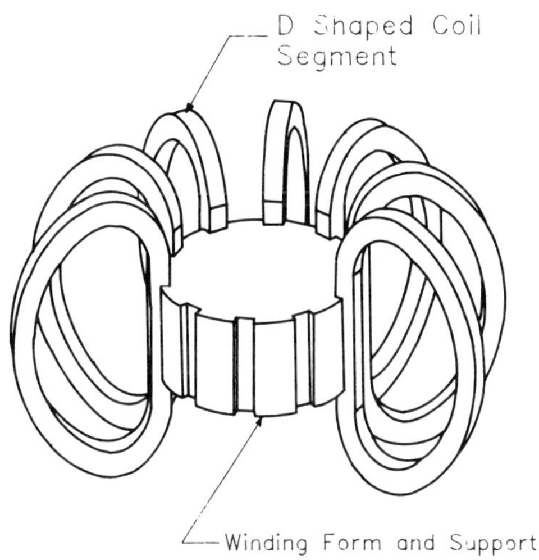

FIG. 4. Cage-type toroidal inductor with a 10-segment winding. The "D"-shaped coils provide a constant-tension winding that improves the inductor's ability to withstand electromagnetic forces.

compress the entire toroid inwardly toward the major axis. File *et al.* (1971) reported that a constant mechanical tension in the winding could be maintained by using a D-shaped cross section for the winding form. Shafranov (1973) later reported that this was also the optimum cross section for maximizing the inductance per given length of conductor for a toroidal arrangement. The D-shaped cross section is the shape taken by a sausage or water balloon resting on a flat surface. The internal expansive forces are like the local Lorentz forces in the coil, and the gravitational action flattening the bottom is analogous to the Lorentz forces acting to compress the toroid about the major axis.

The windings on iron-core inductors are generally either solenoidal or toroidal in form although alternative concepts are sometimes used. In addition to enhancing the inductive effect, the magnetic core also serves to contain the magnetic field within a controlled volume, limiting the "leakage flux." These inductors are generally named according to the shape of the magnetic core rather than winding type. A partial list of the more popular core configurations includes C-core, E-core, picture-frame, gapped-core, two-legged, three-legged, and racetrack.

Electrically conducting cores are sometimes used in variable-inductor high-frequency applications. Their operation is based upon a transformer action established between the main winding and eddy currents in the core, which act like a short-circuited secondary winding. As the core is inserted further into the winding, the magnetic coupling between winding and core increases and the inductance decreases, reaching a minimum when the core is fully inserted.

2. PRINCIPLES OF OPERATION

2.1 Fixed Inductors

These inductors provide a constant inductance over some range. There are many different configurations, but the essential feature is that the winding, core, and any air gaps remain fixed in position with respect to one another. Fixed inductors can be of the iron-core type, provided that their operating range is within the linear range of the core material. Many of these inductors use ferrites as the core material because of the high frequency (kHz to MHz) of operation.

2.1.1 Resonant Circuits. These are an important part of many electrical circuits and are the most common application for inductors. In these circuits, inductors and capacitors are selected to operate at a resonant frequency f_r given by

$$f_r = 1/2\pi\sqrt{(LC)}. \qquad (10)$$

2.1.2 Chokes. These are inductors designed to isolate low-frequency from high-frequency signals in mixed-frequency electrical circuits. They are designed to "choke off" the flow of ac from one portion of a circuit and allow dc or relatively low-frequency ac power to flow to some other portion of the circuit. The inductance needed to achieve a desired level of isolation is derived from Faraday's law as

$$L = V/2\pi f_c I \qquad (11)$$

where V is the voltage applied to the inductor and I is the maximum current that is to be allowed through the choke at the "cutoff" frequency f_c. Choke inductors come in a variety of geometric arrangements, but they are almost always of the iron-core type to maximize the inductance for minimum material and volume.

2.1.3 Energy-Storage Inductors. These inductors are designed to store energy in their magnetic fields (see STORAGE OF ENERGY). They are generally associated with power supplies of significant rating (several kVA to 1000 MVA) and are sometimes called inductive reactors or simply reactors. In the lower end of the energy ratings these devices are generally simple air-core solenoids. If stray fields are a problem, then an iron shield may be placed around the solenoid, or a toroidal arrangement may be selected having either a uniformly distributed or a cage-type winding construction (see Fig. 4).

In power systems these reactors are used

1. to supply energy for line charging in long transmission lines;
2. to limit fault currents through either series or neutral grounding;
3. to stabilize nonlinear loads in furnaces, welding, or other applications where arcs and nonlinear loads are present; and

4. to balance the current distribution between different parallel circuit branches.

They are also used in resonant power supplies for particle accelerator magnets (see Sec. 4.3), electromagnetic launchers, and some pulse-compression systems (see ACCELERATORS, LINEAR; CYCLOTRONS; SYNCHROTRONS).

These inductors are generally air-cored because of the high energy density achievable in nonmagnetic materials, equal to $B^2/2\mu$. Magnetic materials are often used for a portion of the magnetic flux path to concentrate the energy storage into small volumes of nonmagnetic material at a high flux density and to control the leakage flux. These reactors generally have solenoidal windings surrounded by an iron frame and are referred to as "solenoids in a box," or "picture frame" inductors as shown in Fig. 5(a). This type of inductor can still be classified as an air-core inductor as the volume surrounded by the winding is nonmagnetic. The air-core versus iron-core classification often becomes clouded as is the case with the inductor shown in Fig. 5(b). In this case, the volume surrounded by the winding as well as the outer legs consist of alternating magnetic and nonmagnetic regions. This arrangement is used to provide a linear inductance with high energy-storage capability and a well-defined magnetic path. The inductor in Fig. 5(c) uses two windings wound about the legs of a magnetic core and is clearly an iron-core-type inductor.

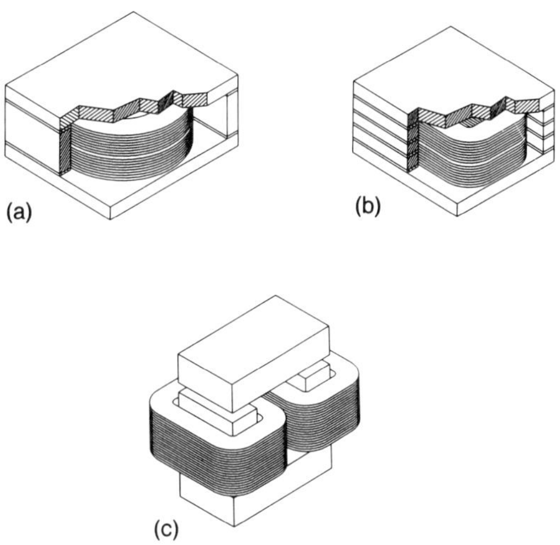

FIG. 5. Different construction methods used for large air-core and iron-core inductors. (a) Picture-frame construction with one large air gap in center of coil. (b) Picture-frame with continuous central leg and air gaps in the external legs. (c) Double-winding distributed air-gap construction.

The inductance required for a particular level of energy storage is found from Eq. (4). If the energy is to be stored in a region of constant permeability μ at a uniform flux density B, then the volume $\mathscr{V}$ needed is given by

$$\mathscr{V} = 2\mu W/B^2. \tag{12}$$

Equations (4) and (12) are used in combination with the characteristics of the power supply to define the optimum physical configuration for a particular application.

2.2 Adjustable Inductors

Adjustable inductors provide a variation in inductance through physical changes in the configuration of the winding arrangement. These inductors have either taps or sliding contacts that are used to alter the number of turns forming the inductor. They may also consist of several fixed inductors that are switched into various series or parallel combinations to provide a variation in inductance. In these inductors the variation in inductance is in discrete steps rather than a continuous variation.

2.3 Variable Inductors

Variable inductors provide an inductance that varies in a continuous manner over some range. This variation is achieved by

1. moving either all or part of the magnetic core with respect to the winding;
2. altering the permeability of the core material; or
3. moving one or more sections of the winding with respect to one another, thereby altering the mutual inductance between these sections and changing the inductance.

2.3.1 Moving Core. In this type of variable inductor a core is moved into or out of the center of a solenoidal coil. These devices are used in electronics applications where fine tuning of a resonant circuit is important. Typically the core is a low-loss ferrite material of cylindrical shape that moves on a threaded shaft. As the magnetic core is inserted further into the winding, the inductance increases over some range. In conducting-core inductors the inductance of the circuit decreases as the core is inserted further into the winding because of the eddy current generated in the conducting core material.

Moving-core inductors were formerly used to control the power flow to certain loads such as theater lights, where they operated in much the same manner as a rheostat without the attendant power loss. Solid-state devices have replaced variable inductors in most power delivery applications, but they are still widely used in communications circuits, using both magnetic- and conducting-core materials.

2.3.2 Saturable Reactors. These are devices that achieve a variable inductance by shifting the point on the magnetic saturation curve at which the inductor operates. This is accomplished by having an iron-core inductor with a control and load winding. The load winding is always divided into two separate windings as illustrated in Fig. 6. With this arrangement there is no ac voltage induced in the control winding; yet the control winding, shown on the central leg of the reactor, still functions to saturate the core. This can also be accomplished by using two separate magnetic cores, but the principle is the same.

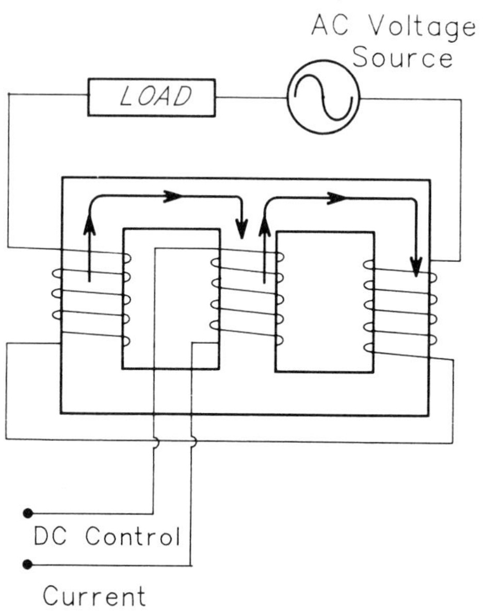

FIG. 6. Saturable reactor illustrating load and control winding arrangements. Note that the core and load windings are arranged so as to keep ac voltages from being induced in the dc control winding.

The inductance of the load winding is the incremental inductance determined by its operating point on the saturation curve (refer to Fig. 2). Shifting the operating point of the circuit is accomplished by varying a dc current through the control winding to cause either more or less saturation in the core. Saturable reactors were first introduced in the early 1900s, and their development proceeded along with that of vacuum tubes. Although saturable reactors tended to be heavy and bulky, they had many advantages over vacuum tubes. They are very simple, rugged, highly reliable devices capable of controlling large amounts of power. Saturable reactors can be designed to have essentially the same operating characteristics as silicon-controlled rectifiers (SCRs) and Triacs.

2.3.3 Magnetic Switches. These switches are saturable inductors capable of withstanding large voltages for short periods of time before saturating. Upon saturation the inductance of these devices decreases by two or three orders of magnitude. Magnetic switches are similar in operation to saturable reactors in that both depend on a large change in inductance caused by the saturation of a magnetic core material. Magnetic switches differ from saturable reactors in that they have a far greater swing in inductance, and it is the load circuit that drives the core to saturation, not a separate control winding. Schematically these devices are represented by the nonlinear inductor element shown in Fig. 1(d).

Melville (1951) first described the use of saturating reactors in pulse generators for radar modulators, nuclear particle acceleration, impulse testing in electrical apparatus, and auxiliary circuits in other power modulator applications. He coined the term "pulsactor" for these new devices that have since become known as magnetic switches. The principal advantages of these devices over electronic, spark-gap, or ignitron switching are that they have an indefinitely long and reliable life and can achieve high repetition rates, passive operation, and high efficiency. Their main disadvantage is that they are relatively heavy and difficult to operate in parallel because of triggering and timing problems.

Magnetic switches fell into relative obscurity until the 1980s, when the need developed for high-voltage (megavolt), high-power ($>10^{10}$ W), high-repetition-rate devices. Birx (1980) revisited the magnetic switch from the standpoint of these different application requirements taking advantage of new geometric configurations and better materials (amorphous metals; see METALLIC GLASSES).

A basic single-module magnetic switch circuit along with a magnetization curve and the voltage at several locations as functions of time are shown in Fig. 7. Operation of the switch is as follows: At time $t=0$ switch S_1 is closed and current begins to flow. Initially, the core is unmagnetized or is magnetized in the negative sense with respect to the applied voltage, and the inductance of the magnetic switch L_P is very large. As current begins to flow the flux in the magnetic core increases to the saturation flux density (B_s), and the impedance of the saturable reactor decreases rapidly. The inductance during the pulse is determined by the differential inductance as described in Sec. 1.1 and illustrated in Fig. 2. The initial inductance is the unsaturated value determined by the linear slope of the magnetization curve. The switch is driven so far into saturation that its saturated inductance can be calculated as if no iron core were present. The saturated inductance is typically 2000–3000 times less than the unsaturated inductance.

The time it takes for the switch to change from its "open" state to "closed" is given by

$$\int_0^{\tau_s} V_L(t)\, dt = NA\Delta B = NA[B_S - (-B_R)], \tag{13}$$

where V_L is the voltage across the magnetic switch, N is the number of turns on the core, A is the cross-sectional area of the core, ΔB is the change in flux density, and τ_s is the time to saturation. If the core is unpolarized, the flux swing will be from 0 to B_S. In general, however, magnetic switches are polarized so that the flux swing is from $-B_R$ to $+B_S$.

By using multiple stages in conjunction with capacitor storage, it is possible to get much higher changes in inductance as well as significant pulse compression compared to a single stage.

The Melville line is the classic example of a pulse-forming network using multiple magnetic switch modules, as shown in Fig. 8. Capacitor C_1 in Fig. 8 is charged through inductor L_0 at a rate roughly proportional to $\pi\sqrt{L_0C_1/2}$ for an ideal source. The charging continues until L_{P1} saturates. At this time C_1 will begin to discharge through L_{P1} and charge C_2. C_2 will charge much faster than C_1 because the saturated inductance of L_{P1} is designed to be much less than L_0. C_2 will then discharge into the load R once L_{P2} saturates. The saturated inductance of L_{P2} must be much less than that of L_{P1}. Figures 8(b) and 8(c) illustrate the voltage and current profiles at various points in the circuit. The voltage pulse is compressed in time but remains relatively constant in magnitude. The current pulse, on the other hand, is compressed in time but its amplitude is continually increasing from stage to stage. In an ideal system the energy transferred down the successive stages will remain constant with the magnitude of the current pulse increasing in direct proportion to the time compression it experiences. The energy

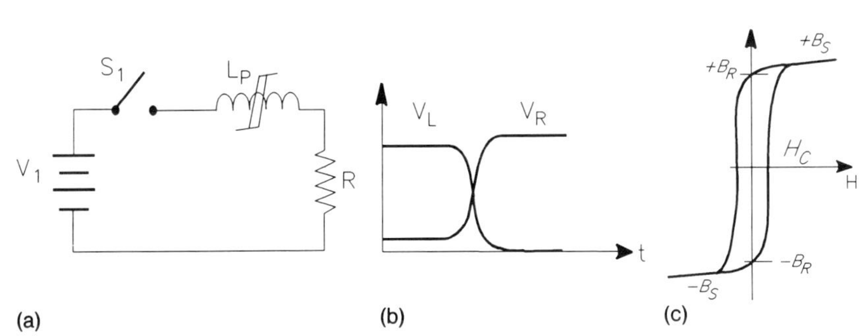

FIG. 7. Single-stage magnetic flux-compression circuit. (a) Schematic of the magnetic switch circuit. (b) Voltage wave forms across the magnetic switch and load as functions of time. (c) Magnetization curve of the switch core illustrating saturation and hysteresis.

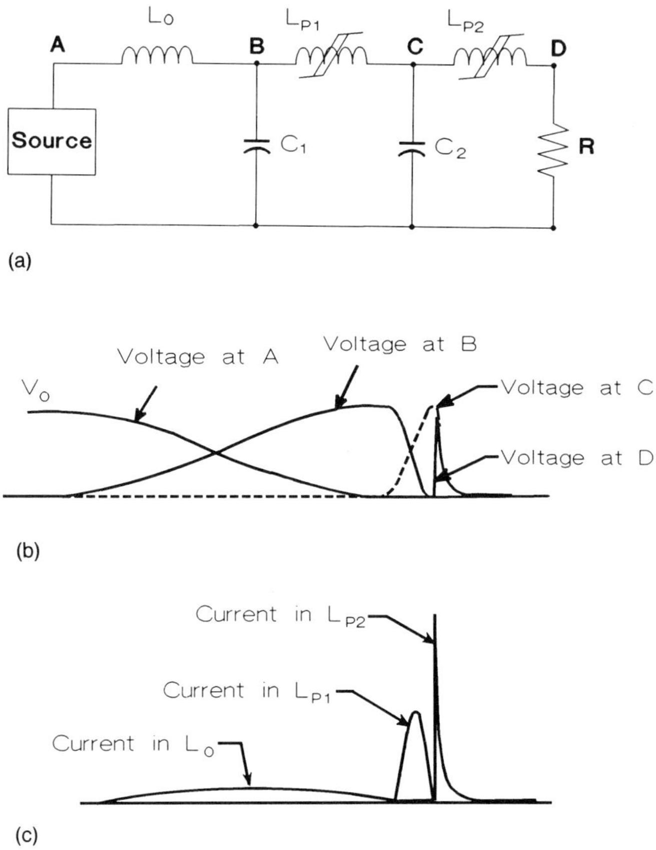

FIG. 8. Classic Melville line two-stage magnetic pulse compression circuit. (a) Schematic of compression line. (b) Voltage compression at selected points as a function of time. (c) Current through selected components illustrating compression of pulse width and increasing pulse magnitude.

lost per stage is typically less than 1% of the input energy.

The circuit described has only two magnetic switch/capacitor modules. In practice this arrangement can be extended to k elements where each C_n and L_{Pn} form a switching module. The gain g of an individual module is defined as the ratio of the capacitor's charge time to its discharge time. This is also equal to the square root of the ratio of the saturated inductance of module $n-1$ to that of module n. The overall gain G is equal to the product of the gains of each individual module.

For a given energy to be transferred, the minimum core volume is given by

$$\text{Minimum core volume} = (\pi/2)^2 \mu_{\text{sat}} W (g/\Delta B)^2 (1/sf). \tag{14}$$

It depends on the gain, flux swing ΔB, energy W to be transferred, saturation permeability μ_{sat}, and the ratio of magnetic material to total core volume, called the core-space factor, sf.

3. ENGINEERING ASPECTS

The first task in any design problem is to establish the performance requirements and constraints for the particular application. For inductors, these requirements include the inductance, operating voltage, frequency, duty cycle (continuous, intermittent, pulsed), allowable energy loss, and tolerances on each parameter. The constraints define the environment in which the inductor will operate. These include maximum operating temperature, ambient temperature, available and acceptable sources of cooling, limitations on stray magnetic fields, size, weight, and the thermophysical properties of the construction materials. Economic factors such as the relative cost of materials and labor as well as the operating costs act to constrain the design further.

The objective is to produce a design that utilizes the minimum amount of resources while meeting the performance specifications within the constraints. This requires input from a broad spectrum of engineering disciplines including electromagnetics, dielectrics, thermodynamics, and mechanics. The specific functional requirements determine the relative importance of the different disciplines, but each must be addressed to some degree. Making tradeoffs between the operating requirements and environmental constraints constitutes the bulk of the design effort as described in the following sections.

3.1 Electromagnetics

The electromagnetic analysis defines the inductor configuration best suited to the specific set of specifications. Tradeoffs of different configurations establish the best configuration for an application. The electromagnetic analysis determines the main field profile, leakage field, losses, and forces within the device and its surroundings. The analysis may be relatively simple or may involve sophisticated three-dimensional nonlinear magnetic-field analysis techniques.

3.2 Dielectrics

The insulation system is a critical part of the overall design. The spacings between conductors and structural members, winding form selection, and insulating materials are important design parameters. The requirements of the insulation system also influence the inductance and electromagnetic performance, in particular the leakage fields. The insulation designer must ensure that the insulating materials are compatible with the environment and perform over the life of the inductor. Transient performance must be carefully evaluated as inductors, by their very nature, can generate voltages far in excess of the normal operating levels when subjected to transients.

3.3 Thermodynamics

Losses of one type or another are generated in every inductor and must be managed. The thermodynamic analysis ensures that temperatures will not exceed the allowable operating values for the materials in the inductor or surrounding components. All aspects of cooling, loss management, and the thermodynamic properties of the various components must be addressed. The general progression of cooling schemes in order of increasing cost and effectiveness is from free convection to forced air to forced liquid and then to cryogenic methods.

The components in electronic circuits are generally thought of as having low loss. However, the loss density in these small devices is often quite high. For instance, some inductors in a household television set reach temperatures in excess of 100 °C. The coatings on the wire and the form itself must be designed to withstand these temperatures.

Thermal management plays a key role in the design of cryogenic inductors. In cryogenically cooled designs the thermal analysis is primarily directed toward heat leak from the outside into the cryostat. Internal losses are primarily due to movements of the conductors or structural components. Although these losses are small, even the slightest amount of heat can cause significant increases in temperature that can cause the inductor to become unstable, especially if it is superconducting.

3.4 Mechanics

Currents flowing through conductors generate forces and stresses that must be analyzed and supported. This requires an accurate knowledge of the magnetic-field profile and current distribution. In ac devices the forces are oscillatory and can cause unacceptable noise and/or mechanical fatigue after some period. Inductors have been designed that carry currents of several million amperes. Although transient in nature, the forces generated are huge, requiring massive structural supports. The mechanical analysis and design must ensure that the inductor will be self-supporting and structurally sound.

4. INDUCTOR APPLICATIONS

The diversity of names by which inductors are known only suggests the wide range of size, shape, and function to which these devices are applied. Some of the more novel

FIG. 9. An air-core power reactor used to provide a constant current for 100 μs during a switching operation is in the background, and an iron-core choke inductor used to limit ac current flow through a circuit is in the foreground.

aspects of inductor technology are presented to illustrate this diversity further.

4.1 Power Reactors

The loosely wrapped solenoidal air-core inductors shown in the background of Fig. 9 are typical of energy-storage reactors. These inductors have 94 turns in two parallel windings in a solenoidal arrangement 0.66 m long with a 0.7874-m outside diameter and 0.330-m inside diameter. If the current were uniformly distributed over the winding area, the inductance based on Eq. (7) would be 7.5 mH. However, the actual inductance, because of the looseness of the winding, is only 1.88 mH. This large variation between a theoretical uniformly distributed current assumption and reality illustrates the need for care in inductance calculations and why the correction factors for winding spacing must be considered in inductor design. These particular inductors are being used to provide a relatively flat current for a short period of time for a solid-state switching application. As the switch is opened, magnetic energy from the series-connected inductor is returned to the circuit. For time periods on the order of the switching time, 100 μs, the current remains essentially constant.

The relationship between the physical size of an inductor and its inductance can be misleading. For instance, the small iron-core inductor in the foreground of Fig. 9 is a 30-mH inductive choke, whose inductance is 15 times greater than that of the larger air-core reactors shown in the same figure. The volume of this inductor is 0.028 m^3 as compared to the 1.28 m^3 of the air-core reactors. This inductor is designed to limit the flow of ac current in a circuit.

4.2 Energy-Transfer Inductors

Inductors function as intermediate energy-storage and low-voltage to high-voltage load-matching devices in many pulsed-power applications. One such application is the hypervelocity electromagnetic macroparticle accelerator (or electromagnetic launcher). In

FIG. 10. Liquid-nitrogen–cooled, eight-segment caged-toroid energy-storage inductor used in a Westinghouse electromagnetic launcher power supply. The inductor will store 10 MJ of energy at 2 MA. Note the bracing structure needed to withstand the huge electromagnetically induced forces during operation.

this application, a low-voltage power supply is used to energize an inductor with high current over some relatively long period. At the appropriate time in the charging cycle, a commutating switch in series with the inductor and in parallel with the launcher is opened forcing current to flow into the high-voltage electromagnetic launcher. From Eq. (1) a sufficient voltage will develop across the inductor to keep the current flowing at its initial value. As the energy is extracted from the inductor the current will decay.

An inductor built for a 10-MJ particle-launcher system is shown in Fig. 10. It is a liquid-nitrogen–cooled toroidal cage-type inductor capable of storing 10 MJ of energy at 2 MA. Each of the eight coil segments measures 1.5 m high by 0.5 m across and is wound in the shape of a D to provide a constant-tension winding to maximize its energy-storage density. The Lorentz ($\mathbf{J}\times\mathbf{B}$) forces during operation are extremely high, which is the reason for the massive coil-support structure. The eight-section cage toroidal construction is a sufficiently close approximation to a pure toroid that the stray magnetic field is negligible.

4.3 Resonant Charging for Particle Accelerators

Synchrotron-type particle accelerators use powerful uniform-field electromagnets spaced periodically along the beam path to guide the particles along their equilibrium orbit. The field strength of these bending magnets must increase as the particles gain energy, and hence momentum, if they are to stay within their orbit. The field magnitude and rate of variation must be precisely controlled and balanced among all of the magnets for proper operation of the synchrotron.

White *et al.* (1956) introduced a resonant charging scheme that

1. reduced the cost and complexity of the magnet power supplies,
2. minimized the reactive power require-

ments from the electrical grid and,
3. could keep the magnets in synchronization.

The basic White circuit shown in Fig. 11 is designed with an inductor and capacitor resonant at the accelerator operating frequency and a dc power supply to pass the bias current. Energy is transferred back and forth between the magnet and inductor in precise synchronization with the particle acceleration. An ac source feeds power to the system via a secondary winding on the inductor to make up for losses. These resonant cells are connected in series to provide the synchronization among magnets. It is desirable to have all of the magnets connected in series; however, some accelerators require such a large number of magnets that this is not feasible.

The inductor is designed to store the energy of the magnets during the nonacceleration period and return it to the magnets as the particle accelerates. It also is used to provide makeup energy due to system losses through a secondary winding. Sasaki *et al.* (1989) designed a dual resonant power supply for the National Laboratory for High Energy Physics (KEK) in Japan that utilized a water-cooled air-core inductor comprising two pancake solenoidal coils enclosed in a picture-frame iron core as shown in Fig. 12. The secondary winding is directly interleaved with each primary turn to provide high magnetic coupling in this air-core inductor. Electrical losses are minimized by using a stranded cable transposed about a hollow conductor through which the cooling water circulates.

4.4 Superconducting Magnetic Energy Storage

Storage of large quantities of electrical power in a superconducting inductor for later recovery has been proposed by Boom and Peterson (1972) of the University of Wisconsin. In the superconducting magnetic energy storage (SMES) concept, electrical power from the ac grid is converted to dc, which is then used to energize a superconducting inductor. Since the inductor is superconducting, there are no losses and the current will circulate until extracted and converted back to ac power. There are significant refrigeration cost penalties associated with keeping the inductor at the low temperatures necessary to maintain the superconducting state. Studies have shown that efficiencies as great as 90% can be achieved by this method of energy storage as compared to 67% typical of hydroelectric pumped energy-storage systems. The greater efficiency is attributed to the ability of the inductor to store the electrical energy directly without intermediate energy conversion steps, as is the case with the pumped-storage concept.

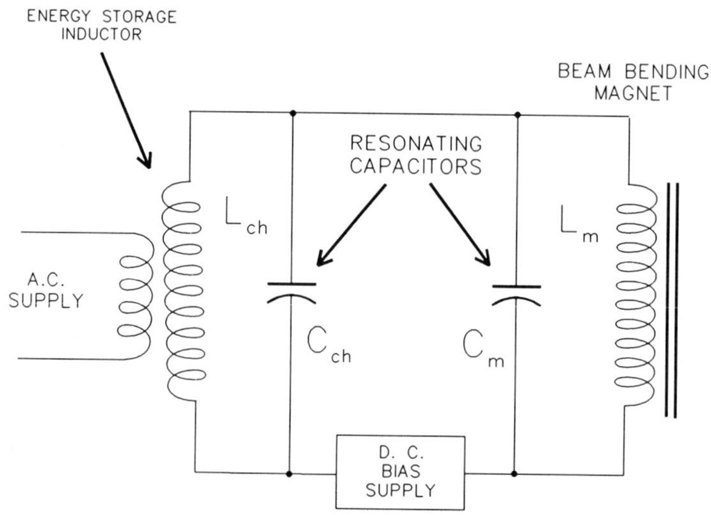

15p

FIG. 11. Circuit schematic of a single-module White circuit used for resonant charging of an accelerator magnet. Typically, these modules are connected in series to ensure that the same current flows in each magnet. Energy is shifted between magnet and inductor in precise synchronization with the beam acceleration profile.

FIG. 12. Energy-storage inductor used in the Japanese National Laboratory for High Energy Physics (KEK) rapid-cycling proton synchrotron. The inductor is of picture-frame construction with two water-cooled pancake coils enclosed by an iron core. It is 11.8 mH with a peak current of 2150 A.

A 19.8-TJ (5500 MW h) full-scale SMES is currently being considered. An artist's conception of the concept being proposed by a team headed by Ebasco Services, Chicago Bridge & Iron, Westinghouse Electric, and the University of Wisconsin is shown in Fig. 13. This solenoidal inductor is 872 m (2864 ft) in diameter with a winding height of over 27 m (89 ft). The entire structure is enclosed in a cryostat filled with liquid helium designed to keep the heat leak to a minimum and to contain the tremendous tensile stresses on the winding. The entire structure is buried in a trench to provide additional support and stability.

4.5 Magnetic Switches

Magnetic switches have a wide range of applications in all types of pulse-forming networks including radar modulators, pulse compression lines for free-electron lasers, particle-beam accelerators, and solid-state switching circuits. The best performance is achieved in high–repetition-rate, high-voltage, continuous-duty applications. They have been used extensively in neutral-particle beam experiments such as PBFA and Comet.

With a core mass in excess of 2000 kg, the magnetic switch employed in the Comet experiment at the Sandia National Laboratories is perhaps the largest magnetic switch built to date. It is a single-shot device that has been tested to 2.7 MV, and 270 kJ per pulse at a peak current of 900 kA. The switch was a single-turn coaxial arrangement which formed an integral part of the transmission line. Figure 14 shows the switch in various stages of construction.

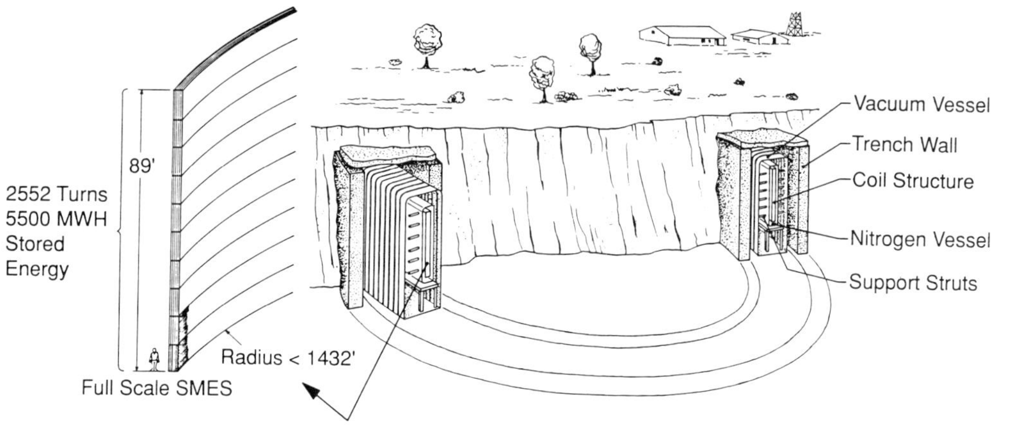

FIG. 13. Artist's conception of a full-scale superconducting magnetic energy-storage coil. The coil is over 870 m in diameter and stores 19.8×10^{12} J for later recovery by the power system.

4.6 Electronic Circuit Applications

Inductors in one form or another can be found in virtually every electronic circuit. Their uses in electronic circuits can be divided into three distinct functional categories: resonant tuning, filters, and current-source reactors.

4.6.1 Resonant Tuning. This type of tuning makes use of inductor-capacitor (*LC*) combinations to form a network selected to operate at a particular frequency as given by Eq. (10). These resonant networks provide regenerative feedback to an amplifier having nonlinear power characteristics to form a harmonic oscillator. Oscillator circuits are used in the generation, modulation, and demodulation of electromagnetic signals generally found in communications systems. Tuning to different frequencies is achieved by varying either the capacitance or inductance of the resonant combination. Variable inductors are used in circuits that may be susceptible to moisture or other contaminants because they are easier to seal against environmental factors than are variable capacitors.

In many applications, *LC* resonant elements have been replaced by circuits using piezoelectric crystals as the resonant elements because they are orders of magnitude more stable than inductor-capacitor combinations. Some integrated circuits have inductive elements built into them by the formation of a spiral conducting path. This arrangement, however, is generally avoided because of the relatively large area needed to construct inductors on the integrated circuit's substrate.

4.6.2 Electronic Filters. These make use of the variation in reactance as a function of frequency exhibited by inductors and capacitors to control which harmonics are either passed through or eliminated from a particular harmonic signal (see ELECTRONIC CIRCUITS). They are classified as either high-pass, low-pass, band-pass, or band-elimination depending on their function in the circuit. High-pass filters allow high-frequency signals through the system for further processing while rejecting frequencies below a particular value. The low-pass filter serves the opposite function. Band-pass filters select a band of frequencies for further processing, while band-elimination filters are used to reject a range of frequencies from the signal stream.

For inductors, the opposition to current flow, or inductive reactance, varies in direct proportion to the product of the inductance and frequency of the signal, as illustrated by Eq. (2b). In capacitors, the capacitive reactance varies inversely to the product of the capacitance and frequency. An inductor connected electrically in series with a load resistance represents a first-order low-pass filter because it presents a low series impedance to the lower frequencies of an applied harmonic signal. A second-order low-pass filter could use an inductor connected in series with the parallel combination of a capacitor and load. If the capacitor is placed in series with the parallel combination of an inductor and the

(a)

(b)

FIG. 14. Coaxial magnetic switch used in the Comet Project at Sandia National Laboratories at various stages of construction. (a) Magnetic core being wound from alternate layers of insulation and amorphous metal strip. (b) Finished magnetic switch ready for installation in the pulse compression line.

load, then a high-pass filter is obtained. Bandpass filters place an inductor and capacitor in series with the load, while band-elimination filters use a parallel combination of a capacitor and inductor in series with the load.

4.6.3 Current-Source Reactors. These reactors are used to control the magnitudes and rates of rise of current in electrical and electronic circuits. They are particularly useful in solid-state switching circuits used for power applications. Once triggered, solid-state devices can turn on in a relatively short space of time, generally 50 μs or less. If the current increases too rapidly, hot spots can be created that will thermally degrade the device. An inductor of appropriate size placed in series with the solid-state device will control the rate of rise of current through it to acceptable levels. Nonlinear inductors are often used for this application so that the series inductance decreases by several orders of magnitude after the solid-state device has been turned on. Many applications require that several solid-state devices operate in parallel. No matter how good the control system, not all devices will turn on at the same time. If this happens, all of the current will be flowing through only a few of the devices, resulting in an overload and premature failure of those devices. Eventually, the entire parallel combination will fail. By placing inductors in series with each device, the distribution of current can be controlled to ensure that the devices share the current load on an even basis.

GLOSSARY

Flux Linkage: The flux passing through a simple one-turn loop of wire. In a coil, the summation of the flux through each turn.

Henry: The unit of inductance. A circuit has 1 H of inductance if an emf of 1 V is induced when the current is changing at a rate of 1 A/s. Equivalent to a weber per ampere.

Hysteresis: The tendency for the field to persist in a magnetic material after the exciting field has been modified.

Permeability: A measure of the inductive capacity of a material given in units of henries per meter.

Phase Shift: The difference in phase angle between two sinusoidal waveforms.

Phasor notation: The technique of representing variables such as voltage, current, impedance, and resistance as complex numbers in the analysis of harmonic problems, having a magnitude and phase angle with respect to some reference.

Reactance: A complex constant of proportionality relating voltage to current, similar to resistance in Ohm's law. It is a function of frequency as well as inductance and capacitance.

List of Works Cited

Birx, D. L., *et al.*, (1980), "Basic Principles Governing the Design of Magnetic Switches," Lawrence Livermore Laboratory, UCID-18831, Berkeley, California, November 18, 1980.

Boom, R. W., Peterson, H. A. (1972), "Superconductive Energy Storage For Power Systems," *IEEE Trans. Magn.* **Mag-8**, 701–703.

Carter, G. W. (1967), *The Electromagnetic Field in Its Engineering Aspects,* 2d ed., New York: American Elsevier, Chaps. 8 and 9.

Faraday, M. (1839), *Experimental Researches in Electricity,* Vol. 1, Ser. 1, London: Bernard Quaritch.

File, J., Mills, R. G., Sheffield, G. V. (1971) "Large Superconducting Magnet Designs for Fusion Reactors," *IEEE Trans. Nucl. Sci.* **NS-18**, 277–282.

Grover, F. W. (1946), *Inductance Calculations, Working Formulas and Tables* (special printing by the Instrument Society of America with permission of Dover, New York, in 1982).

Maxwell, J. C. (1891), *A Treatise on Electricity and Magnetism,* 2nd ed., Vol. 2, Articles 693–706, London: Clarendon; reprint (1954), New York: Dover.

McLyman, W. T. (1978), *Transformer and Inductor Design Handbook,* New York: Marcel Dekker.

Melville, W. S. (1951), "The Use of Saturable Reactors as Discharge Devices for Pulse Generators," in: G. E. Williams (Ed.), *Proceedings of the Institution of Electrical Engineers (IEE)* (London), Vol. 98, Pt. 3 (Radio and Communications), No. 53, 185–207.

Rader, L. T., Litscher, E. C. (1944), "Some Aspects of Inductance When Iron Is Present," *Trans. Am. Inst. Elec. Eng. (AIEE)* **63**, 133–139.

Rosa, E. B., Grover, F. W. (1911), "Formulas and Tables for the Calculation of Mutual and Self-Inductance," U. S. Bureau of Standards Scientific Paper No. 169, Washington, DC: U.S. GPO, pp. 116–142.

Sasaki, H., Adachi, T., Someya, H., Sakai, I. (1989), "A DC Biased Rapid- cycling Magnet System Operating in a Dual Frequency Mode," in: A. Otter, J. Strathdee (Eds.), *Proceedings of the KAON PDS Magnet Design Workshop,* Vancouver, Univ. of British Columbia, B.C., pp. 33–40.

Shafranov, V. D. (1973), "Optimum Shape of a Toroidal Solenoid," *Sov. Phys. Tech. Phys.* **17**, 1433–1437.

Smith, S. (1985), *Magnetic Components Design and Applications,* New York: Van Nostrand Reinhold.

White, M. G., Shoemaker, F. C., O'Neill, G. K. (1956), "The 3 BEV High Intensity Proton Synchrotron," in: E. Regenstreif (Ed.), *Symposium du CERN sur les Accélérateurs de haute énergie et la Physique des Mesons, Genève, 11-23 Juin 1956*, Geneva, Switzerland: CERN, Vol. 1, pp. 525–529.

Further Reading

Information on design details and new applications and designs for inductors can generally be found in conference proceedings for applications to specific technology areas. The IEEE Pulsed Power Conference, the IEEE Power Modulator Symposium, the Symposium on Engineering Problems in Fusion Research, the *IEEE Transactions on Magnetics*, and the *IEEE Transactions on Nuclear Science* are good sources of related material for large inductor designs. The *Proceedings of the Institution of Electrical Engineers* (IEE) is also a good source of information. For accelerator applications there are numerous conferences sponsored by host laboratories throughout the world that can be good sources of information. For communications applications both the IEEE and IEE have societies that publish articles that may be of interest.

Carter (1967) provides a good physical insight into the electromagnetic field and treats the concept of inductance especially well. For anyone calculating inductances of standard geometries the Grover (1946) book is essential. Both Smith (1985) and McLyman (1978) do a thorough job of inductor design details for one wishing to build devices.

ICF

See FUSION, INERTIAL CONFINEMENT

IMPERFECTIONS IN CRYSTALS

See POINT DEFECTS AND OTHER IMPERFECTIONS IN CRYSTALS

INERTIAL CONFINEMENT FUSION

See FUSION, INERTIAL CONFINEMENT

INFORMATION SCIENCE

CHARLES T. MEADOW, *Faculty of Library & Information Science, University of Toronto, Toronto, Ontario, Canada*

	Introduction	21
1.	**General Aspects**	22
1.1	Nature of Information	22
1.2	Related Concepts	23
1.3	Representing Information	23
1.4	Information as a Commodity	24
2.	**Databases and Knowledge Bases**	25
2.1	General Aspects	25
2.2	Record Structure and Content	25
2.3	Representation of Attribute Values	25
2.4	Storage Structures and Indexes	26
2.5	Database Maintenance and Quality Control	26
2.6	Knowledge Bases	27
3.	**Information Systems**	28
3.1	General Aspects	28
3.2	Major Components	28
3.2.1	Sources and Acquisition Subsystems	28
3.2.2	Processing Subsystems	28
3.2.3	Communication and Dissemination Subsystems	28
3.3	Control of Information Flow	29
3.3.1	Control of Input	29
3.3.2	Selection of Output	29
3.3.3	Delivery of Output to Users	30
3.4	Examples of Information Systems	30
3.4.1	Database Search	30
3.4.2	A Typical Accounting System	30
3.4.3	The American Institute of Physics—An Organization as an Information System	30
4.	**Information Retrieval**	31
4.1	General Aspects and Models	31
4.2	Information Needs of Users	31
4.3	Query Languages	32
4.4	Intermediation—Assisting the User	32
4.5	Automatic Processing	33
4.6	Measurement and Evaluation of Retrieval Systems	33
5.	**Informetrics**	34
5.1	General Aspects	34
5.2	Language Use	34
5.3	Publication Measures	34
5.4	Usage Measures	35
6.	**Information Networks**	35
7.	**The Users of Information Systems**	36
8.	**Social and Commercial Aspects**	37
8.1	Use of Information Systems	37
8.2	Growth of the Information Economy	38
9.	**Societies and Journals in Information Science**	38
	Glossary	38
	Works Cited	39
	Further Reading	40

INTRODUCTION

Information science is a relatively new discipline, so new that the name is not widely known nor is there broad agreement as to what it means. Generally, it has grown out of studies of the design, use, and measurement of collections of information and of information systems, and, although rarely part of any formal definition, the assumption is often made that the systems studied are primarily computer systems. Computer science tends to concentrate on computer internals: design of hardware and algorithms, engineering of computer software, and the performance of the hardware or software components. Information science, on the other hand, tends to deal with systems as a whole, the interrelation of the systems with their users (indeed, even to include the users as *part of* the system), the

1-56081-067-X/94/$5.00+.50

effect of information systems on their environment, and the extent to which they accomplish their objectives.

As a formal discipline, one branch of information science grew out of library science when, in the early 1960s, library science began to turn its attention increasingly to new methods of organizing and retrieving information, at least partly by the application of new technology to these problems. Other aspects are found in the disciplines of computer science, linguistics, communication (the social science), and telecommunications engineering. In Europe, the expression *information science* is rare; *informatics* is more likely to be used. Today, there are several respected scholarly societies and journals in the field, listed in Sec. 9.

After World War II there was a large amount of scientific documentation that had been created on the Allied side and more captured from other countries. In the 1950s the new "cold war" began, which had an accelerating effect on the growth of science and technology. Government agencies that supported scientific work believed that a vast amount of information was not being used because it was not accessible, and they invested heavily in developing systems to handle this information. *Handle* meant to organize it—classify and index it—and to develop means to search, retrieve, and disseminate it.

Technologically, information science evolved along with several streams of developments in other disciplines. These were

1. computers, primarily the evolution of time-shared large computers or, latterly, networked small ones;
2. telecommunications, especially packet-switched data communications networks;
3. large databases that became necessary for many disciplines, for example, in recording information about publications in science, and that demanded the use of computer–computer communications for remote searching;
4. new concepts in the organization and representation of information for more effective retrieval;
5. algorithms for interpreting and searching natural language and for measuring the relationships among words or texts; and
6. a movement toward more direct use of computers by end users, requiring new appreciation of user behavior.

This article is organized as follows: Section 1 discusses the nature of information as a mathematical concept and as a commodity, and the representation of information and coping with its ambiguities. Section 2 discusses databases, the logical configurations in which information is stored, and their structure; maintaining quality of the content of a database; and knowledge bases that record information about how to process or interpret data. Section 3 contains an overview of modern information systems; the components and flow of information; and some examples. Section 4 presents an overview of information retrieval, which today means mostly retrieval of text or coded representations of text. Section 5 discusses informetrics, or the measurement of information phenomena. Section 6 discusses information networks, which have made access to information systems not only physically easier but economically attractive. Section 7 considers users of information systems, and characteristics of importance and how they are studied. Section 8 explores the impact of information systems on society as a whole and some of the social issues of current concern. Section 9 discusses societies and journals in information science.

1. GENERAL ASPECTS

1.1 Nature of Information

Information is ubiquitous, so much so that it is difficult to define it in a way that satisfies both scientific rigor and popular conceptions. In the context of transmitting messages composed of strings of characters, Shannon (Shannon and Weaver, 1959) defined the information content of a communication system as

$$H = -\sum p_i \log_2 p_i \,, \qquad (1)$$

where p_i is the probability of symbol i being selected for transmission, and summation is over the number of possible symbols. The units of H are *bits*. Shannon was a communications theorist, concerned with such problems as computation of channel capacity. He emphasized that his work was not intended to apply to semantic aspects of information or to the meaning of messages transmitted. He also intended that H measure the amount of information in a *system* that generates symbols with the given probabilities, *not* the information content of a *single message*. The probabilities refer to the choice of a letter or a word, not of a message. The frequency distributions of letters in English are fairly uniform over a variety of writings, as are the frequencies of the most commonly used words. But no one knows or can predict the probability of a given message if the source is free to send any combination of English words.

The popular meanings of information (there are many) center around the related concepts of novelty, dispelling of uncertainty, and the probability of occurrence of a message. A message contains information if its contents were not already known to the recipient (novelty). A message contains information if it clarifies an issue (dispels uncertainty). An unexpected message contains more information than an expected one. That a heavy favorite wins a race is not much information, but that a long shot won *is* very informative. It is, of course, impossible to know the *a priori* probability of occurrence of a news item, which is one reason why Shannon warned against computing H for semantic information.

Yet another set of definitions shows the influence of cybernetics. Here, information is a message that changes the state of a system receiving it. Or, information is what enables decisions to be made, i.e., a choice among alternatives. Making a choice changes the state of a system; hence these definitions are compatible with each other. For more information on the formal information theory as applied to coding and transmission of information, see TELECOMMUNICATIONS.

1.2 Related Concepts

Terms closely related to *information* include *data, knowledge, meaning, intelligence,* and *wisdom*. As may be expected, there are no universally accepted definitions for these terms. *Data,* in common usage, tends to mean a set of symbols whose meanings or values are not necessarily known or accepted, but whose structure is known. Hence, it is possible to read data or operate upon it with a program, but without some form of interpretation or evaluation we cannot gauge its significance. The structural aspect is important. If we have simply a set of numbers with no known internal structure, it hardly even qualifies as data. What is data to one person may be highly significant information to another. The subjective nature of the distinction between data and information is becoming widely accepted.

Information is data that has been understood and evaluated, or is meaningful. Thinking, interpreting, or processing converts data into information. *Knowledge,* in general usage, is a higher form of information. The term tends to be applied to bodies of information rather than individual messages. When we use this word, we often mean to imply information possessed or accepted by an authority or an authoritative community. The sentence "We have knowledge of the laws of bankruptcy" is, in common usage, a far stronger statement than "We have information about bankruptcy law." Information acquires the status of knowledge by virtue of general acceptance or acceptance by respected authority, such as by a dictionary or encyclopedia or the *Handbook of Physics and Chemistry*. In Sec. 2.6 we discuss a particular form of knowledge called a *knowledge base*.

Intelligence has a number of meanings, one of which is approximately what we have called *information*. *Wisdom* is a term not routinely used in scientific work but often used colloquially. It tends to have much the same connotation as knowledge, but now knowledge held or promulgated by a person or community of special distinction.

1.3 Representing Information

An aspect common to all the various definitions is that information is *not* the ink and paper or magnetic disks with which it is represented. It is not something tangible. Information is a consequence of the receipt of a message or sequence of messages that may be recorded in any medium, using any symbol set mutually understood by sender and receiver. The symbolic representation of information consumes space, matter, and time to transmit, but the information itself does not. However, the material and time requirements for storage and transmission of symbols are of critical importance in information-system design.

In scientific work, information is often represented by numbers or codes made up of letters and numbers. There is little of the concept of syntax associated with such representations, but there is some. Given a vector or array, we know that each element is normally the same kind of entity (temperatures, weights, distances, etc., all represented as integers, decimal numbers, or floating-point numbers). Computer programs can easily be written to "understand" such data structures. However, if information is represented in natural language, the syntax is highly complex and often ambiguous. There is no complete definition of the syntax of a natural language nor of the composition of a picture. It is relatively easy to search a numeric array for occurrence of a stored value or a function of a stored value, such as the point where a second difference changes sign. However, it is a far more uncertain procedure to search a body of natural language text to determine whether the concept of the effect of global warming is discussed therein, or to search a picture to see if it portrays a cow. Clearly, there are many ways to describe global warming or to express concern about it, and certainly many ways to picture a cow. The ambiguity or uncertainty that accompanies this kind of searching in collections of information, called *databases*, makes up a large part of the domain of interest of information science. As Sir Walter Scott (1828?) put it, "The end of uncertainty is the death of interest." For example, does "She read her paper on time" refer to the subject matter of the paper or "her" adherence to a schedule? Only context can determine the answer. The sentence "I really liked that" can mean what the words seem literally to say or the exact opposite, depending on tone of voice, if spoken, or the general sense of a written text—is sarcasm used elsewhere? How do we program a computer to recognize sarcasm?

To find information in a large collection of text requires that users understand the vagaries of the text and seek alternative ways of expressing concepts. They must anticipate how authors have represented concepts, not just how they, themselves, would represent them. This, in turn, may require training of users, classification or indexing of the text, and computer programs that can not only search for what the user has requested but help the user decide what to request. See Sec. 7 for a summary of human considerations in the design and use of databases.

Because it is often too expensive or even impossible to store a complete text, graphic, or audio recording, we must often make do with classification codes or other descriptors to convey information about content. In bibliographic work selection of such terms is called cataloguing and indexing. It is crucially important if the original cannot be stored and searched. Today, we are less dependent on indexing for text, still very much so for graphics and sound images.

1.4 Information as a Commodity

Information has been bought and sold, surely for as long as *information* has existed, probably beginning with tips on good hunting and fishing locations and practices. Economists and information scientists have long been fascinated with the fact that, since it is not really the information, but its representation, that is sold, the information remains with the seller after the sale, unlike any other commodity. Hence, scarcity resulting from depletion of stock is not a basic element in determining price. The kinds of information packages that bring the highest prices seem to be those for which interest or demand is high, but access is limited. It is often to the benefit of the purchaser to maintain the access limitation. Thus market research reports, stock market newsletters, and spy reports cost much. Purchasers do not necessarily want their competitors to have equal access. Textbooks and newspapers cost relatively little, partially because a purchaser normally does not object to others owning a copy of the work.

A trial was held in Ontario (Potter, 1988) in which an agent of a labor union was charged with conspiracy to steal a copy of the name and address list of employees of a hotel the union wanted to organize. The judge ruled that no theft could have occurred because the hotel would still have had the information for its own use. What the hotel was potentially deprived of was not information but *exclusivity of access* to that information. Essentially the same logic applies to cases such as the infamous Pentagon Papers episode (Sheehan, 1971), in which classified U.S. government documents were copied. The government was not deprived of the information but of exclusive control of it.

In the context of the cost of access to databases, when such services first became available to the public, in 1972, the basis for charging was almost always *connect time*, the amount of time the user's computer or communications terminal was connected to the database service's computer. Computers had limited numbers of communication ports, and when one of these was in use, it was denied to others. Then, gradually, charges were placed on records or portions of records retrieved, which of course means *copied*. The trend is to charge for a combination of work done and information transferred.

2. DATABASES AND KNOWLEDGE BASES

2.1 General Aspects

Databases are collections of data, with some commonality of content, capable of being searched or processed, and usually but not necessarily containing large amounts of data. A knowledge base is similar, but the implication is that it is used to guide some process, such as searching or interpreting a database. A database may be the set of news articles appearing in the *Washington Post* over a span of 10 years. A knowledge base may be the set of information and rules used to edit a news article from the point of view of English grammatical usage as interpreted by that paper's editors. The difference in concept may not be apparent to a casual observer.

2.2 Record Structure and Content

A *record*, as the term is used here, is a collection of data elements that concern a common *entity*. An entity is the subject of an element of information, such as a laboratory animal or a bank depositor. Information about an entity is represented as a set of *values* of *attributes* or *variables*. Examples of attributes are the *name* of a person, *time* of an event, and *amount* of a bank deposit.

Attribute values are represented as numbers, words, or arbitrary strings of any characters. Attributes also have names, and they have relationships to the entity described. Sometimes, the name adequately describes the relationship. If the entity is a person and the attribute is *date of birth*, that is reasonably self-explanatory. If, however, an attribute is called *pay*, it is not clear whether this is an hourly rate of pay or the amount of a forthcoming check. Therefore, the design of a record should include the list of attributes represented, the manner of representation (integer, character string, etc.) and the relationship of attribute to entity. The record itself then consists of a set of values of the attributes. The name of the attribute, its relationship, and nature of representation could be a part of the record, but they normally are not.

All attribute values for an entity can be stored together or stored separately and brought together when and as needed. In manual systems, those based on paper or other non-computer-readable records, all information related to a given entity is normally stored together, on a single sheet of paper or in a single file folder. In early data-processing systems records took the form of punched cards or were sequences of bits recorded contiguously on a magnetic tape. Today, the information about an entity may be physically distributed over several files on different computers, possibly located in different cities.

One subclass of records is *surrogate records*. A common example is the library catalog card. The card is not itself the book it describes, and it contains virtually none of the content of the book. However, it is descriptive of the book and it represents it in user searches. Search decisions are made using the surrogate, not necessarily the actual document. Gradually, surrogate records, once of vital importance, are becoming less so because it is becoming feasible in many cases to store the entire entity in the record, obviating the need for a surrogate.

2.3 Representation of Attribute Values

We have defined three levels of information representation in a database: the entities about which there is information, the attributes or aspects of an entity about which there will be information, and the representation of the attribute values. Database designers must decide what attributes to represent—whether, for example, a medical bibliographic database should contain only records about research and medical practice or also tutorials and articles for the layperson. Researchers will consider the latter to be clutter, and laypersons will not understand the former. Given a decision about entities, it must be determined what information should be recorded about them. Bibliographic records are fairly standard but typically do *not* contain information about the type of article (research, review, opinion, tutorial, *etc.*), nor do they contain citations used in the article. Finally, the representation of attributes can vary considerably. For example, the subject of an article can be described by a hierarchical classification code, subject headings, or a set of key words extracted from the text. Which is best, which best fits the needs and knowledge of users, must be determined by the designer.

2.4 Storage Structures and Indexes

Users often have no idea how records or individual attributes are stored in computer memory. This condition is called *transparency*: the users see through the physical organization to a *logical organization* or a way of envisioning data that may actually differ from the physical organization but be easier to work with. As an example, text created by a word processor is typically stored in segments of about one line in length, but the segments are not necessarily physically stored in the order in which they appear in the text. Each is likely to be stored in the next space available at the time of storage, wherever on a disk that happens to be. But the users are not required to be concerned about this aspect of storage. They visualize and work with the text as a series of lines in the order of their appearance on the printed page.

Physical organization of data has important effects on the speed, hence cost, of any operation using a database. To explain briefly, there must be some order to records or sets of attributes when stored in memory. Records are put in order on the basis of the value of an attribute, called a *sort key*. If records are so ordered, then it is possible to determine the approximate location of a record, given its key, and thereby to limit search time. Or, if records are stored wherever space is available, then each record will have a *pointer* to the location in memory of the next record in logical (sort key) order. The alternative to ordering is to require a search of every record in a database in order to find just one attribute value.

Suppose a file of employee information is ordered on *name* or *employee number*, but a search is to be done for a physicist who speaks French. Records in this file would be randomly ordered with respect to *job title* and *language spoken*. To avoid an exhaustive search through all employee records, indexes can be created for any attributes likely to be used for searching. An index is created for one attribute. Each index entry consists of the value of the attribute and a pointer to the location of a record containing that value. Then, to find a physicist, the value PHYSICIST is looked up in the job-title index and the set of pointers to records containing that value is retrieved. A similar search for the value FRENCH in an index for languages is then done. Then, the two sets of pointers are combined, retaining only those from both lists, the Boolean algebra AND operation. The user is unaware of the pointer set retrieval and processing—it is transparent. The user is told how many records are represented in the final set and may retrieve any of the indicated records from the employee database.

There are other organizational methods such as the increasingly popular *relational model*. This permits users to use a database as if it contained whatever subset of attributes and in whatever order is convenient for the user at the time of use. The user need not be concerned with whether indexes exist or in what actual order records are stored in the original database. A database management program creates or appears to create the desired structure as needed. This method vastly simplifies computer programming for users of the database, or the composition of search requests, but does add to the operational cost. The economics of modern computer systems tend to encourage this approach, which saves human costs, which continue to rise, at the expense of hardware costs, which continue to decline in terms of unit cost of memory or computational performance. For more information on data structures see COMPUTER DATABASES.

2.5 Database Maintenance and Quality Control

In any information-producing endeavor, aspects such as timeliness, accuracy, reliability and completeness of the corpus of information, or database, are important. It is rare that a database can simply be created and then remain in a constant state. Far more likely, new information will continue to be presented for inclusion, information in the database will change or must be deleted, and errors in existing records will be detected. This means that the managers or operators of any information system must create policies on how to select new information for inclusion, when to delete outdated records, and when and on what basis to make changes. It should be readily appreciated that if a spurious signal causes an error in computing the track of an aircraft in an air-traffic-control system, it must be quickly corrected. By contrast, an error in a bibliographic record would probably not be corrected until a subsequent update cycle, which might occur anywhere from a month to half a year after discovery of the error. This can put users at a disadvantage but the cost of faster change may be prohibitive.

Thus there are some very practical questions: How much error can a database or its users tolerate? What is it worth to users to have a database maintained free of error or, since that is often impossible, to keep them below some threshold? How can error rate or content be measured and monitored in any but the most simple situations? Information science does not have theoretical answers for such questions.

Another aspect of database maintenance is the logic, time, and cost of making changes. The logic is bound up with the physical storage structure (Korth and Silbershatz, 1986). The key elements to consider are two:

1. There is usually a trade-off between optimizing a file structure for speed of search and retrieval versus speed of making changes. A database search service will normally lean toward optimizing search time because that is the dominant operation and the major source of revenue.
2. There is a difference between item-by-item database changes and complete replacement of one version with a new database. The latter is called *reloading* and is costly but sometimes necessary. If the amount of change is great enough, reloading becomes necessary. Either item-by-item change or reloading require the database or a portion of it to be kept out of use for a time, which may vary from fractions of a second to hours.

2.6 Knowledge Bases

In the new field of *artificial intelligence*, or AI, (see ARTIFICIAL INTELLIGENCE), one branch is concerned with *expert systems*. Variously defined, expert systems tend to be information systems that perform tasks traditionally considered performable only by a human intelligence. Of course, there is no strict definition of such tasks and the question of how much human intelligence was required depends on the sophistication of both the performer of the task and the observer. Nonetheless, such systems exist and their number and capabilities increase steadily.

An expert system has to have knowledge in its field, or *domain*, of application approximately equal to that of a human expert. The range of expertise is quite limited. While expert systems have been built or postulated to play chess, diagnose a medical complaint, or search a database and select the best five records meeting the user's needs, no single present-day expert system could conceivably do all these tasks. Indeed, even within the domain of information retrieval, the expert systems tend to be restricted to a single subject area.

The information an expert system uses to perform its task is, in a structural sense, just another database. The difference is that the program using it must be able to interpret the information well enough to base its action thereon. In this case, the database is called a *knowledge base*, which generally consists of a set of data, or facts, and a set of rules (Davis and Lenat, 1982). The facts are used in interpreting the rules. A fact states, for example, that a pawn may move only one or two squares forward and two squares on its first move only, excluding capture situations. A rule tells which piece to move at a given stage in a game and where to move it. A discipline called *knowledge engineering* is developing that is devoted to interrogating human experts to extract from them the information needed for a knowledge base, and then putting this into a meaningful and useful form (Hart, 1992).

The field of expert systems is still quite new and its future uncertain. Some programs have been brilliantly successful—those that can play chess, for example. But there are computer scientists and others who feel there are limits to what can be done, partially on the basis of the fact that while a computer program can interpret information in a knowledge base, it cannot have had the full range of human experience that is often required for difficult decision-making tasks (Weizenbaum, 1976).

3. INFORMATION SYSTEMS

3.1 General Aspects

In a sense, almost anything we might call a *system* can be called an *information system*. A system of equations stores information about the variables represented. A lever might be considered a system that senses the force applied to it and the position of application and transmits information about a resultant force to be applied to a mass. But the more general meaning of *information system* is a set of components capable of acquiring, processing, storing, and transmitting information. The sources from which information is acquired or the destinations to which it is transmitted may be other information systems, people, or even animals. The components usually are assembled for a defined purpose and are interconnected according to a plan. An information system also includes the data or information stored within it. Take two identical computers, loaded with identical programs, but with different sets of data, and we might have quite different information systems.

In the sections to follow, we will consider the major components of typical information systems, how information flows among them, and how that flow is controlled. First, there is a question of the boundaries of a system. For example, are sources part of the system? Are destinations? Are operators (those humans who control the system)? If all these questions are answered in the negative, then the system is a collection of mechanical parts and of information. We often find that, for purposes of evaluation, this definition is too restrictive. To assess an air-traffic-control system as if its human controllers were not an integral part would yield only a moot evaluation. Unfortunately, published descriptions and evaluations of systems are not always clear on specifying what is and is not included.

3.2 Major Components

3.2.1 Sources and Acquisition Subsystems. *Sources* provide information and may control a system's content, reliability, timeliness, etc. The procedure or basis for selecting sources and for deciding what information to ask of or accept from them may also be a part of the system. For a library, sources are publishers, wholesalers, and used-book dealers. The selection decisions are not trivial, because they involve considerations of user needs, discounts offered, reliability in delivery, and the library's desire to spread its custom to assure good service in case of future need. For a newspaper, direct sources are its own reporters or press associations, but *their* sources are varied and cannot be completely controlled by a central office. Questions of reliability arise frequently and cannot be fully answered in most cases. A seismographic information system uses instruments that are part of the system, and hence whose reliability can be periodically verified. It receives information from other systems whose reliability may or may not be open to question.

3.2.2 Processing Subsystems. *Processing* can mean almost any form of transformation of data: conversion of codes, formatting for display, performing mathematical or statistical calculations, or simply changing the order of records (sorting). Information science is typically more concerned with the selection than the processing of data but there is no sharply defined boundary between these functions. Almost any information system performs some form of both.

3.2.3 Communication and Dissemination Subsystems. Modern computer-based information systems typically consist of one or more central computers, or servers, connected with remote users' personal computers. The nature of the connection can be a relatively permanent connection, as in *local-area networks* (LANs), or dial-up connections typical of *wide-area networks* (see NETWORKS, COMPUTER; TELECOMMUNICATIONS). A voice-telephone connection is an example of a dial-up connection—the link to the destination is established on demand and is maintained as long, but only as long, as the user wants it to be.

In information retrieval, *dissemination* has a special meaning not directly related to telecommunication. It is the process by which an information system decides to whom to send records and then performs the physical distribution. This has been an important part of science information systems. It requires the user, or potential recipient, to provide a *profile* or statement of the kind of information that the user would like to receive. Then new records are compared with this statement and those that match are marked for distribution to the person who submitted the profile. Many scientific organizations and individual scientists maintain profiles with the large scientific bibliographic database distribution services. This process is sometimes called *selective dissemination of information* or SDI (Meadow, 1992, pp. 145–146). It is analogous to the older practice of preparing a routing sheet and distributing a copy of a newly arrived journal within an office or laboratory to those who requested it.

3.3 Control of Information Flow

The essence of information science lies in the study of how to control, or what governs the control of, the flow of information from its origin to its final destination. While we can use the phrase *final destination,* it is not final in the sense of reaching a terminal point and going no further. Information generation and use tend to form a giant feedback loop. The recipient of information may be a decision maker who, or which, will then use this information for decision-making purposes over a long period of time. Thus the incoming information serves as a control signal governing outgoing information, possibly in a continuous feedback cycle. Or, information published in a scholarly journal, then retrieved through a retrieval service, may flow to an author and be used in that author's *new* paper, which enters the published literature and is, itself, available to be retrieved in an indefinite cycle.

3.3.1 Control of Input. Information or data can be said to be created when an event occurs and some form of record of the event is made in symbolic form. An event will concern one or more entities. The "event" need not be instantaneous; it could well be a condition of long duration. Colloquially, we say we record the *event,* but we do not and cannot actually do this. We record *information about the event,* that is, we transform a set of perceptions about the event into a set of symbols, and these symbols must be perceived by someone or something. The classical Berkeley question about whether a tree falling unobserved makes a noise may be restated to ask whether a set of symbols constitutes information if it is never perceived. By most modern technical definitions, the answer is *no,* unperceived symbols are not information. Hence the record of the event depends on the perception as well as on the event itself.

Examples of events are a political campaign reported by a journalist into a newspaper or television news system, and the automatic recording of temperatures in a laboratory by a programmed device. In order for either of these reports to be sent, there must previously have been

1. *a policy decision on what kind of events to report*—which, of all possible ones, are wanted in the database;
2. *a design for the record*—the determination of what attributes of an entity are to be recorded and in what manner the values are to be represented;
3. *a selection of events to report*—the determination of what actual events or entities to include. Given the policy, the reality may be different. A newspaper does *not* record all the news; a laboratory sampling procedure may miss a key event.

3.3.2 Selection of Output. Information is selected from a store, not only in information retrieval, but in management information systems and numeric data systems as well. When selection is to be done, there must be some form of language for expressing selection criteria. Most frequently, the manner of selecting is to define a subset of the database in terms of the values of attributes, of the form ATTRIBUTE **RELATION** VALUE, where ATTRIBUTE is the name of an attribute, and **RELATION** is an equivalence relation such as =, >, or CONTAINS (the character string BCD *contains* the string BC). Then Boolean algebraic expressions can be formed from these basic set-defining expressions, producing a final statement such as AUTHOR CONTAINS 'ITH' AND DATE > 1991. Almost any high-order computer-programming language can interpret such expressions (except for CONTAINS, which is relatively rare except in database systems).

3.3.3 Delivery of Output to Users. Computer software must, in any case, interpret such statements as directives for how to examine a record and to determine whether or not it matches selection criteria. The system must then present its information as electronic displays, printed output, or electronic messages to other computers or devices.

3.4 Examples of Information Systems

Here are brief descriptions of three kinds of information systems: a remote-access database search system, a typical accounting system, and a professional organization acting as an information service.

3.4.1 Database Search. Dialog Information Services, Inc., is an example of a service that acquires databases under license from other organizations, called *database producers*, who are analogous to publishers of books or journals. The producers acquire the raw information, put it into usable and standard form, possibly publish it in paper form, and then also distribute it to organizations such as Dialog that will redistribute it electronically to the ultimate user. Dialog provides the computer system, telecommunications for user access, computer software to maintain and search the databases, documentation, training, and user assistance.

The service is highly diversified, offering over 400 databases whose subject matter ranges from *Peterson's College Database*, a directory of educational institutions, to *Standard & Poor's Corporate Descriptions*, providing information about officers, subsidiaries, income, bond ratings, and the like on publicly held corporations, to *Searchable Physics Information Notices (SPIN)*, a bibliographic database describing publications in physics. Users pay for the service based on a combination of time connected to the central computer, the number and form of records retrieved, and in some cases the number of search terms used.

3.4.2 A Typical Accounting System. Every company and institution and almost every family has some kind of accounting system, however primitive, whose basic purpose is to inform management of the financial state of the organization. Clearly, a laboratory with many projects, supported by different sources under different terms, needs a much different system than does a family deriving almost all its income from a single salary and having no unusual expenditures. But the purposes are similar: to account for all monies handled, for tax and other official purposes, and to enable management to make financial plans and adjust them to changing conditions.

An accounting system normally requires the creation of a record for every transaction involving an expense or a commitment, or the receipt of funds. Contrary to the database search system model, all the data are generated internally to the using organization, not acquired *in toto* from an outside source. Controlling input, both assuring that every item is reported and that input values are correct, is a critical responsibility of the system or its managers.

Most accounting systems generate a series of standard reports: a listing of all income and expense items (so that they can be checked for accuracy and so that managers can see where the money is spent), a comparison of each subdivision's expenses compared with spending plans (budget), and a tally of the status of the bank account. But there is also a need for *ad hoc* reports, enabling any manager to request data suited to a particular problem, such as all expenses connected with a particular project, independent of organization, or a time series of expenses in some category to see what changes are taking place.

3.4.3 The American Institute of Physics—An Organization as an Information System. Many organizations, such as professional societies, travel agencies, and stock brokerages, are today recognized primarily as information agencies. They not only operate information systems, they *are* information systems. The American Institute of Physics (AIP) is a federation of professional societies concerned with physics. It is a primary publisher of physics literature, a secondary publisher (indexes and abstracts), a database distributor, and the provider of an online, interactive information service. It is an organization that observes the needs of its users for information; acquires information; selects, processes, and edits it; and distributes it to its users.

As a primary publisher AIP publishes scholarly journals of research, monographs, reference books, conference proceedings, and a newsletter reporting on Washington news of interest to physicists. As a secondary publisher it publishes *General Physics Advance Abstracts* and *Current Physics Index.* As a database producer AIP publishes *Searchable Physics Information Notices* (SPIN), a database derived from *Current Physics Index* that contains bibliographic information and abstracts of articles and other published items. SPIN is then distributed to several database distributors, including Dialog. As a provider of an electronic information service, AIP operates the Physics Information Network (PINET) whose services include electronic mail, society notices of various types, and access to bibliographic databases, including the abstracts of articles in this Encyclopedia. See PHYSICS LITERATURE AND PUBLICATIONS for more detail on publications.

4. INFORMATION RETRIEVAL

4.1 General Aspects and Models

A large part of information science is concerned with *information retrieval.* The purpose of an information storage and retrieval system, or information retrieval system (IRS), is to organize and store information in such a way that it can be retrieved and to perform the retrieval. Retrieval implies selecting a subset of records from the database, according to a specification provided by a user or inquirer. An IRS is an information system that provides

1. a physical storage facility for data or information;
2. an organizational scheme for the information, manifesting both the internal composition of records and the ordering of records and indexes to them; and
3. computer software that interprets queries presented in one or more of a variety of languages, natural or artifical, and carries out the search for records conforming to specifications provided in a formal statement of information need.

There may also be

4. an intermediary function whose purpose is to convert a user's informal statement of information need into the formal version that can be processed by the IRS. The intermediary may be a person trained for this type of task or a computer program or a combination of the two.

There may be

5. assistance in analyzing and evaluating retrieved information,

and there is almost always

6. a subsystem that compiles indexes to records and that gives some indication of content or location.

4.2 Information Needs of Users

Questions to an IRS originate with a user's awareness of a lack of information. Awareness of a lack does not mean that the need is clearly defined. Belkin (1980) called this an *anomalous state of knowledge,* or ASK. A user needing information would approach an information service, perhaps a library or a computer-based IRS, where, in one form or another, it is necessary to explain the *information need.* We distinguish between the ASK and the statement of information need in that the latter is a translation or mapping of the former. This mapping may not be accurate, for a variety of reasons to be discussed in Sec. 4.4.

The user's information need must then be expressed as a *query* or question phrased in the usually formal language of the IRS (see Sec. 4.3). Barring error in the use of language or in the spelling of terms, users usually manage to retrieve *some* records. Then it becomes necessary to evaluate these, or a sample of them, decide whether they are relevant or useful, change the query if necesary, and iterate. The evaluation of record content by the user is presumably based on that person's understanding of the extent to which the record satisfies the ASK. If the needs statement from which the query is derived does not represent the ASK well, then it is unlikely the records will be of great value. Misunderstandings occur easily if the user cannot or does not express his or her actual need. Then there is dissatisfaction with the outcome of the retrieval operation, which could only work from the *expressed* need, not the unexpressed need.

4.3 Query Languages

A query language is the set of symbols and rules for their use by which a user conveys to an information retrieval system what information is wanted and other specifications such as format and order of presentation. It is a form of special-purpose computer-programming language. (See COMPUTER PROGRAMMING LANGUAGES). Most query languages are artificial, using a limited vocabulary and syntax. Many users and system designers feel that, at least for inexperienced users, natural language would be preferable, but this would require that the computer be able to translate natural-language expressions into terms it could use to direct a search. This is being done to a limited extent. Systems that employ this approach do not require full natural-language understanding; they often treat the query as a set of individual words, without consideration of syntax (Salton and McGill, 1983; Croft and Thompson, 1987).

Most programming and query languages are *procedural.* They specify a procedure to be carried out by the issuance of commands, which consist of imperative verbs, objects of verbs, and sometimes prepositional phrases and adverbs. For example, SELECT TITLE FOR AUTHOR=BOHR? might specify that the user wants to see only titles of works represented in a database that were written by an author whose last name (in last-name-first style) begins with the letters B-O-H-R. The "?" asks for retrieval of any author name with these first four letters, followed by any number of others, whether a continuation of the last name or initials. The alternative to procedural language is nonprocedural language, in which the user specifies the characteristics of output, rather than how to produce it. A specification of the format of output would show what information is to appear on a page, and where placed on the page, rather than providing the computer program that formats the display.

The conventional method of entering commands into the computer is to type them on the keyboard. *Menus* and sets of *icons* have become popular recently as a fast means of entering a command. Their use is not an alternative to commands but is a means of entering them. The language in use and the manner in which the computer system interacts with the user in presenting options, receiving commands, and responding to them are major parts of a collection of input–output programs, languages, and display designs called the *user interface,* which is increasingly being recognized as a vital factor in the success of computer usage.

4.4 Intermediation—Assisting the User

The reasons why information retrieval is iterative are that

1. natural-language text is inherently ambiguous and
2. users are uncertain of what they really want.

One reason for the latter uncertainty is that the user has an incomplete mental image of the problem at hand. A second reason is uncertainty about how best to explain or represent the information need. A third is that, to some extent, what information is seen as needed depends upon what is available, and hence a complete and satisfactory query cannot be stated until the database has been explored relative to this user's need.

Given this triple uncertainty, we generally look to a combination of help from an *intermediary* and *feedback* from the system to achieve a desirable outcome. The intermediary converses with the user, the person having the information need, with the objective of learning the user's needs, priorities, evaluation criteria, and even budget so that a query can be formulated. The intermediary then helps the user interpret output and, if necessary, revises the query. The discourse between the two is not unlike that between physician and patient or lawyer and client—the layperson's expression must be translated into the technical terms of a profession. In recent years there have been attempts, with varying degrees of success, to have some aspects of the intermediary role performed by computer software (Marcus, 1991; Marchionini, 1992; Meadow, 1992, pp. 253–269). While progress has been made, it cannot be said that the human role has been replaced.

Feedback is typically provided in several ways. One form is for the system to tell the user what the results of a search were—for example, how many records were retrieved or that a proffered input has a value out of acceptable range. A second is provided by reading the set of records or data retrieved. This, in its raw state, requires that the user who requested the information examine the output, decide whether or not it meets the objective, and decide how to revise the request or commands to the system to improve the outcome. A modern variation of this is to have the system analyze the retrieved data and either help the user to decide what changes may be needed, or directly initiate the next iteration itself. This is an example of the application of artificial intelligence to information retrieval.

4.5 Automatic Processing

One of the areas of greatest accomplishment in information retrieval research has been in the assistance the system can provide to the user—for example, in helping select the search terms that will lead to satisfactory results. Users cannot generally anticipate all the ways authors have chosen to describe a subject. In chemistry, for example, there can be dozens of different ways of representing a compound: as a chemical name, a molecular formula, a ring structure, a popular name, or a trade name, with some of these having many variations. Thesauri can be manually compiled, and word association matrices can be automatically compiled. These show the frequency with which certain words co-occur with others, and this can be done at the level of a sentence, a paragraph, or the entire document. Thus, if a user is interested in the term DISK STORAGE, a word co-occurrence table might suggest other terms that frequently co-occur with either word or both, resulting in an expanded query that might include MAGNETIC DISK, CD-ROM, HARD DISK, FLOPPY DISK, DISK DRIVE, COMPUTER MEMORY etc. Often, even experts in a field will not take the trouble to anticipate all these synonyms, near synonyms, or synonyms in limited contexts.

A second class of assistance techniques is concerned with relationships between records or texts. If a user, by whatever means, has been able to find a few good records, then a useful next step is to ask the system to find any other records that are closely related to these. "Closely related" is taken to mean related in terms of the vocabulary used. A typical function for computing a relationship is the cosine coefficient (Salton and McGill, 1983, p. 121), defined as

$$\cos(R_j,R_k)=\frac{\Sigma\, t_{ij}t_{ik}}{\sqrt{\Sigma\, t_{ij}^2\, \Sigma\, t_{ik}^2}}, \qquad (2)$$

where R_j and R_k represent records, whether part of the database or a query, and t_{ij} is the frequency with which term i occurs in record j. The summations range from $i=1$ to n, the number of different terms in the combined documents. Using this measure a user can find a set of records, then ask for the subset most closely related to any given one.

In *relevance feedback* (Salton and McGill, 1983, pp. 140–145), a user retrieves some records and evaluates them. Then, these ratings are used as weights in performing a computation similar to $\cos(R_j,R_k)$, based not only on word co-occurrence but on the importance the particular user attached to a subset of the retrieved records. Other methods for computing relationships or making use of user judgements can be found in Salton and McGill (1983), pp. 199–256.

4.6 Measurement and Evaluation of Retrieval Systems

There is no measure of retrieval system performance that is at once well defined, used in research, and, perhaps the ultimate for a standard, accepted and used in trade. Evaluation is usually based on *relevance*, which can be a measure of *relatedness* between a query and a database record, or of *value* to the requestor. Unfortunately, users often do not have a clear objective in mind when using a retrieval system, leaving us with no clear measure of the degree of accomplishment. The measures most frequently used are *recall* and *precision*. Recall is the fraction of records of a database that are relevant to a query and that are retrieved by it. The difficulty of arriving at a value for recall is that, for a database of thousands to millions of records, it is virtually impossible to know which ones may be relevant but not retrieved. Precision measures the fraction of retrieved records that are relevant. This is a more practical measure, since a user or intermediary can be asked to make the necessary judgments. The difficulty is that a high value of precision is not necessarily the user's goal, although it is often assumed to be.

In practical use, for example, among commercial firms offering retrieval services, user surveys asking overall reaction to the service are more likely to be used and acted upon than measures of outcome or performance.

5. INFORMETRICS

5.1 General Aspects

Informetrics is a branch of information science that is "the study of the quantitative aspects of information in any form." (Tague-Sutcliffe, 1992). It includes patterns of language use, publication, and use of publications. It is related to, but broader than, *bibliometrics*, concerned with the measurement of various "aspects of the production, dissemination and use of recorded information" (Tague-Sutcliffe, 1992) and *scientometrics*, "the study of the quantitative aspects of science as a discipline or economic activity" (Tague-Sutcliffe, 1992). While these phenomena may seem, on the surface, to be almost random, they tend in fact to obey quite regular laws. This is a broad subject. We shall present only a few examples of the types of measures and their applications.

5.2 Language Use

The relationship of the frequency of occurrence of words in a text to the rank, by frequency, tends to follow a very smooth and mathematically simple curve. The relationship was first formalized by Condon (1928) but is commonly attributed to Zipf (1949). It states that $f=kr^{-1}$, where r is the rank, in descending order of word frequency f, and k is a constant, dependent on the sample size. This hyperbolic function is represented by a straight line, with slope -1, on a log-log scale. Most early studies of the phenomenon used "standard" English or French, but Whitehorn and Zipf (1943) also published some samples showing regular deviation from this geometric form, based upon Germanic language texts or English texts written by schizophrenic patients or by children. Mandelbrot (1953) modified it to

$$f=k(r+m)^B, \qquad (3)$$

which fits most distributions better. The parameters k, m, and B are attributes of the language sampled or the sample selected. So far, there is little practical application of this relationship in information science, but it may be possible to use the Mandelbrot parameters to characterize languages, including artificial ones, and their users.

Some more traditional measures are the *type-token ratio* and Yule's *characteristic* (Yule 1944). A word *type* is a generic word appearing in a text. A *token* is one specific instance of its appearance. THE is a type, usually the most frequent word in an English text. Any one occurrence of it is a token. The ratio of types to tokens is broadly characteristic of the language or its use by an author. It is the inverse of the mean word frequency. Yule's characteristic is defined as

$$K=(\sigma^2-M)/M^2\,, \qquad (4)$$

where σ^2 is the variance and M the mean of the word frequency distribution. Both these measures have been used in authorship studies to help determine the author of an anonymous text or to contest or verify that one person wrote several works attributed to a single person. Other authorship studies have used *discriminant analysis*, which involves finding relatively common words used with different frequencies by different authors (Mosteller and Wallace, 1963).

5.3 Publication Measures

A pressing problem for almost any library is to determine what scientific journals it should have in its collection to meet the demand of its users for maximum coverage of a discipline, while limiting its spending to meet budget limitations. Bradford's law (Kraft and Boyce, 1991) states that if journals in a discipline are ranked according to the number of articles on the subject in question, then organized into zones such that the journals in each zone contribute the same number of papers, then

$$N_k=N_1b^{k-1}, \qquad (5)$$

where N_k is the number of journals in the kth zone, N_1 the number in the first zone, and b a *scattering factor* descriptive of and specific to the subject field of the journals. Once some basic investigation has been done in a field, then Bradford's Law can predict the number of journals required for complete coverage or, given an actual list of journals on hand, can predict the cost of an additional increment of coverage or the saving by giving up an increment.

5.4 Usage Measures

We shall illustrate one form of usage measure, *citation* of one publication or work by another, indicating that one author has made use of another's work. A citation is an assertion by an author that the *citing* and *cited* works are related. If two works are cited in the same work, then the two are *co-cited*. If the one work is cited by two others, then we have an instance of *bibliographic coupling*. If the two were often co-cited, then the relationship is strong. A person doing a search of bibliographic files and recovering one title might want also to recover all works that are strongly coupled to the first. This information is not contained in conventional bibliographic files or library catalogs. There are *citation indexes* published in science, social science, arts and humanities, and law that provide such information. More information can be found in Kessler (1963), Small (1973), Garfield (1979), and White and McCain (1989).

6. INFORMATION NETWORKS

A network, in the most general sense, is a set of nodes connected by links. The nodes are entities capable of sending or receiving a message, or both. The links are communication channels. The nodes may be computers or people and the links may be telephone lines or a postal delivery system.

From the point of view of information science, information networks can be classified by content and by structure. In terms of content, one major type of network is primarily devoted to sending mail or messages or even complete databases. Generally, these are characterized by *not* having the data permanently stored in the system. Messages are created and sent as needed. They tend to be specifically addressed to recipients, although group addresses are allowed, and they are deleted from the system when delivery is complete. *Electronic-mail* systems are the prime example of these. These "e-mail" systems are the modern version of a telegraph service. Individual users enter their messages and address them. A network relays a message from one node to another until the destination is reached. There are many different methods of controlling the routing. A message is ultimately delivered to a local computer to which the recipient has access and is held there until read. Then the user can delete or retain the message, reply to it, or forward to another node. Bulletin boards and electronic conferencing systems are exceptions to the storage procedure. In these applications, messages may be entered into the system and then stored for viewing by anyone authorized to have access, but reading is generally done after the fashion of reading mail, i.e., by browsing to see what is available and asking for what seems interesting. There is not necessarily a separate copy provided to each reader. Deletion is done by a person responsible for maintaining the files. An extension of this concept allows a user at one node to operate a computer program at another node, obviating the need for transmitting the program or its input data from one machine to another.

The other major content type contains typically large databases, stored relatively permanently in the system, and not for specific users, but for anyone authorized to use the system (See Sec. 3.4.1). Because there are so many records, there must be some kind of content-based search or retrieval process. Permanence of storage is a relative term; records need not stay forever, but until superseded or deleted by an editor. They are not deleted solely because they are read. Again, they have no specific addressee.

Classification by structure or control is typified by two subtypes: *local-area networks* (LAN) and *wide-area networks* (WAN). A LAN consists of a set of computers, printers, and similar equipment, linked together by cable and usually existing within a relatively short distance of each other, limited to a few hundred meters. Wide-area networks may have the components cabled together but typically make extensive use of dial-up telephone links. For more information on networks see COMPUTER NETWORKS and TELECOMMUNICATIONS.

7. THE USERS OF INFORMATION SYSTEMS

Information scientists have come to realize that to study an information system that involves humans as users or operators requires that these human "components" be included in the study. Indeed, it might be said that the domain of information science is those systems that involve human behavior as well as mechanical operation.

Some of the earliest research involving the activity of retrieving information was concerned with how people search in a library (Taylor, 1968) and how cataloguers or indexers perceive the materials they work with, in particular, how consistent they are in their decisions (Lancaster, 1991). Clearly, in any searching situation principally dependent upon surrogate records, the consistency, even more than accuracy, of indexing is critical. How to achieve consistency among indexers or catalogers remains a problem in any discipline.

Searchers using an IRS may be broadly classified into three groups:

1. *professional searchers*, whose function is performing searches on behalf of others, an activity that requires skill in eliciting the information needs of others;
2. *professional users of information*, who are members of a profession or specialized discipline and need information for their professional work; they are also able to judge the value of retrieved information; and
3. *others*, which includes students typically below the graduate level who need information but generally lack the professional's skill at evaluating it.

In general, the second and third groups do not have skill or experience in the use of computer-based information systems, but that statement is gradually becoming less true as home and routine office use of computers spreads. In characterizing users, these three types are not enough. Users vary on knowledge of the mechanical aspects of the system they are using, the language or set of commands available, the content and structure of the database, and the subject matter of the field encompassing the database. Their purpose in searching also affects performance: is it on behalf of another person (implying that the searcher may not know the subject as well as, or the exact information goal of, the end user); is it to find a few good items, or a comprehensive set?

Research has found that these three groups perform differently when doing searches (Marchionini *et al.*, 1991; Cherry and Clinton, 1991), as is to be expected. But, since most information systems do not have significantly different modes of searching to offer users of different levels, the fit between user skill and system design is important. Systems do have differences in user language or methods of providing specifications to the computer. Research suggests that performance differences are at the level of understanding what needs to be done rather than the mechanics of how.

There is not an established set of measures of the performance of information systems or of user performance. This lack hampers research because it is difficult to compare one user group with another. We recognize that there are *performance* measures and *outcome* measures. The former include such variables as time required to perform a task, number of errors (invalid commands issued, for example), or number of commands required to perform a task as meaningful measures.

Studies have investigated relative speeds of work, errors made, commands used, and sequences of commands used. As one would expect, experienced users work faster and make fewer errors. They tend to use more of the commands available to them, analogous to an experienced writer using a larger vocabulary than a novice. Usage studies have tried to identify those commands or output displays that seem to be ambiguous or give trouble to users, and to see how different classes of users respond to an error or ambiguity. They try to determine which types of users prefer to issue commands to systems, and which may prefer to select functions from a menu.

Outcome measures are concerned with results obtained, not with the manner of obtaining them. Outcome measures usually are based on relevance (see Sec. 4.6), essentially measuring the extent to which a system retrieved information useful to or subject-related to that requested. Here is where user skill in evaluating information is involved. Again, as a broad generality, the professional intermediary is not able to judge the value of a record to a user and will instead concentrate on relatedness. The professional user is more likely to judge value, and users in the third category may simply take anything offerred as acceptable.

Some research has focused on *mental models*, essentially the mental image a user has of the system being used. If two users have different perceptions of how a system operates, they will use it differently. Research is becoming more intensive on how to record and compare mental models (Gentner and Stevens, 1983). Continuing research is needed to refine the measures and the laboratory procedures for taking them and to identify the characteristics of users that are of importance. Other aspects of the role of users in information systems have been discussed in Secs. 1.3, 2.2, 3.3, 4.2, and 4.4.

8. SOCIAL AND COMMERICAL ASPECTS

In science, education, and law, and to a lesser extent in business, we have reached the point where users of information have the expectation that all published information can be accessed by anyone who can use the appropriate information networks. If information is not published, which implies lack of an intent to make it public, then it is normally protected by special access restrictions, requiring not only a password, but also membership in some select group, such as those holding a military security clearance. We have not yet reached the long-predicted point where every family and office has a computer tied into the various information networks, but we move closer every year. Here are some current social issues.

8.1 Use of Information Systems

If all published information in a field is available, then teachers, supervisors, and clients expect their students, employees, and consultants to find and use it. Not knowing how is no excuse. This, in turn, creates a need for education and training in how to find and evaluate information, which is not yet covered in the educational programs of most fields. Law is a rare discipline in which knowledge of how to search the literature of the field is a part of professional education. Conceivably, we shall see the day when failure to find information becomes grounds for a charge of professional malpractice. One subfield of information science, overlapping with sociology, studies the patterns of adoption and use of new technology by individuals and groups: who takes the lead and why, who lags and why (Rogers, 1971).

The ease of transmission and copying of files creates concern over copyright protection, not only of the electronic files but also of their printed counterparts. Will readers buy a printed journal or a computer program if a copy can be made electronically at a fraction of the retail cost? The store of information is not depleted by a sale. Should this affect the price? How can the value of information be assessed? For a general review of information economics see Hall (1981).

Privacy and freedom of access to information and media are concepts that existed before electronic media, but have become of more intense interest because of the speed of computers and large amounts of information potentially available. The prospect of virtually all information about each of us being available to anyone—faceless agents of government, insurance or credit companies, prospective employers, or the like—can be daunting. It raises the spectre of "Big Brother," popularized by Orwell (1949). That these fears have not been realized may be partially due to the vigor with which privacy legislation was supported to prevent it (Rosenberg, 1986, Williams, 1988).

Access to media has three aspects:

1. the right to see information about one's self,
2. the right to see information in the hands of government if there is no current reason for maintaining secrecy, and
3. the right to get information *into* a medium.

Rights of access to stored information are now secured by law in many countries. Less well known is the issue of access to media, typified by the right of a political party to free television air time for its candidate in an election. Broadcast frequencies are universally allotted under government control and many governments have rules requiring equal treatment of political viewpoints. The issue is highly sensitive. A similar issue arises in science publishing, when proponents of certain unpopular views claim to be barred from presenting them in major professional publications.

8.2 Growth of the Information Economy

Bell (1976) has pointed out that employment in the primary sectors of the United States' economy, agriculture and extraction, has decreased sharply since about 1900, and "industry" decreased slightly, as a percent of U.S. gross national product. But services and information-related industries have filled the gap. This has led to the terms "post-industrial society" and "the information society." In about 1970 there was virtually no industry devoted to selling databases in electronic or magnetic form. Today, this is a thriving industry. Revenues have grown from about $300 million in 1977 to over $10 billion in 1990 (Information Hotline, 1992). These figures should be treated as approximate, as the definitions of the products vary considerably. Clearly, though, information as a commodity has established itself in world markets. Science information, directly exchanged among scientists, and also science information used in business decision making play a major role in this.

9. SOCIETIES AND JOURNALS IN INFORMATION SCIENCE

The principal professional societies in information science are the American Society for Information Science, the Canadian Association for Information Science, and the Institute of Information Scientists (U.K.). Closely related are the Special Interest Group on Information Retrieval of the Association for Computing Machinery and the Division of Chemical Information of the American Chemical Society. Each holds at least an annual conference.

The major journals are *Journal of the American Society for Information Science* (U.S.), *Information Processing and Management* (U.S.), *Canadian Journal of Information Science, Journal of Information Science* (U.K.), and *Acta Informatica* and *Informatik: Theorie und Praxis der Wissenschaftlich-Technischen Information* (Germany). Some journals closely akin include *Forum of the Association for Computing Machinery Special Interest Group on Research in Information Retrieval* and *Journal of Chemical Information and Computer Sciences* (U.S.), and *Journal of Documentation* and *International Journal of Man-Machine Communication* (U.K.).

GLOSSARY

Anomalous State of Knowledge (ASK): An anomaly in one's state of knowledge, or lack of knowledge, with respect to a problem faced. The start of the information retrieval process.

Attribute: An element of information about or a variable descriptive of an entity.

Data: A *datum* is a value of an attribute. A set of data implies not only a collection of such values but that there is a known organiztion to elements of the set.

Database: A collection of sets of data about entities; or a collection of records, where each is concerned with an entity; or a set of related files, all concerned with a common set of entities.

Entity: The individual about which information is or may be gathered, such as a person, a sensor, or a philosophical concept.

Expert System: An information system capable of, or intended to perform, tasks otherwise considered to be demanding of human intelligence for performance.

Information: (1) Data that is understood or deemed meaningful or valuable by a recipient. (2) $H = -\Sigma\, p_i \log_2 = p_i$, the sum of the product of the probability of a symbol's occurrence times the logarithm to the base 2 of the probability, over the set of possible symbols.

Information Retrieval: The process of organizing and storing information such that it can be retrieved, and performing the retrieval by selecting a subset of records from the database, according to a specification provided by a user or inquirer.

Informetrics: The study of the quantitative aspects of information in any form; including patterns of language use, the production, dissemination and use of recorded information, and the quantitative aspects of science as a discipline or economic activity.

Knowledge: In general usage, similar in meaning to *information*, but tending to be applied to bodies of information rather than individual messages. Often meant to imply information possessed or accepted by an authority or an authoritative community.

Query: A statement of a user's information need, expressed in the usually formal language of an information retrieval system.

Query Language: The set of symbols and rules for their use by which a user conveys to an information retrieval system what information is wanted, and other specifications such as format and order of presentation. It may be a special-purpose computer programming language.

Relevance: A measure used in the evaluation of information retrieval systems, which can be of the subject *relatedness* between a query and a database record, or of *value* of a retrieved record to the requestor.

List of Works Cited

Belkin, N. J. (1980), *Can. J. Info. Sci.* **5** 133–143.

Bell, D. (1976), *The Coming of Post-Industrial Society*, New York: Basic, pp. 134–137.

Cherry, J. M., Clinton, M. (1991), *Can. J. Info. Sci.* **16**, 2–22.

Condon, E. U. (1928), *Science* **68**, 300.

Croft, W. B., Thompson, R. H. (1987), *J. Am. Soc. Info. Sci.* **38**, 389–404.

Davis, R., Lenat, D. B. (1982), *Knowledge-Based Systems in Artifical Intelligence*, New York: McGraw-Hill.

Garfield, E. (1979), *Citation Indexing—Its Theory and Application in Science, Technology and Humanities*, New York: Wiley.

Gentner, D., Stevens, A. L. (Eds.) (1983), *Mental Models*, Hilldale, NJ: Lawrence Earlbaum.

Hall, K. (1981), *Info. Soc.* **1**, 143–166.

Hart, A. (1992), *Knowledge Acquisition for Expert Systems*, 2nd ed. New York: McGraw-Hill.

Information Hotline (1992) **24**, 1.

Kessler, M. M. (1963), *Am. Documentation* **14**, 10–25.

Korth, H. F., Silbershatz, A. (1986), *Database Systems Concepts*, New York: McGraw-Hill.

Kraft, D. H., Boyce, B. R. (1991), *Operations Research for Libraries and Information Agencies*, San Diego: Academic, p. 115.

Lancaster, F. W. (1991), *Indexing and Abstracting in Theory and Practice*, Champaign: University of Illinois Graduate School of Library and Information Science.

Mandelbrot, B. (1953), in: W. Jackson (Ed.), *Communication Theory*, London: Butterworths, pp. 486–502.

Marchionini, G. (1992), *J. Am. Soc. Info. Sci.* **43**, 156–163.

Marchionini, G., Meadow, C. T., Dwiggins, S., Lin, X., Wang, J., Yuan, W. (1991), *Can. J. Info. Sci.* **16**, 23–59.

Marcus, R. (1991), J.-M. Griffiths (Ed.) in: *Proceedings of the 54th ASIS Annual Meeting* Medford, NJ: Learned Information Inc., (Ed.) **28**, 49–59.

Meadow, C. T. (1992), *Text Information Retrieval Systems*, San Diego: Academic.

Mosteller, F., Wallace, D. L., (1963), *J. Am. Stat. Assoc.* **53**, 275–309.

Orwell, George (1949), *1984: Nineteen Eighty-Four, a Novel*, New York: Harcourt Brace.

Potter, R. B. (1988), *The Globe and Mail* (Toronto), 22 Sept., p. A7.

Rogers, E. M. (1971), *Diffusion of Innovations*, 2nd ed., New York: Free Press.

Rosenberg, R. S. (1986), *Computers and the Information Society*, New York: Wiley.

Salton, G., McGill, M. J. (1983), *Introduction to Information Retrieval*, New York: McGraw–Hill.

Scott, Sir W., *The Heart of Mid-Lothian*, place and publisher unknown. Republished (1965), New York: Holt, Rinehart and Winston, p. 11.

Shannon, C. E., Weaver, W. (1959), *The Mathematical Theory of Communication*, Urbana: University of Illinois Press, p. 19.

Sheehan, N. (1971), *The Pentagon Papers as Published by the New York Times*, New York: Quadrangle.

Small, H. (1973), "Cocitation in the scientific literature: a new measure of the relationship between two documents," *J. Am. Soc. Info. Sci.*, **24**, 265–269.

Tague-Sutcliffe, J. (1992), *Info. Processing Management* **28**, 1–3

Taylor, R. S. (1968), *College Res. Lib.* **29**, 178–194.

Weizenbaum, J. (1976), *Computer Power and Human Reason*, San Francisco: Freeman, pp. 202–227.

White, H. D., McCain, K. W. (1989), "Bibliometrics," in: M. Williams (Ed.), *Annual Review of Information Science and Technology*, Medford, NJ: Learned Information, Inc., Vol. 24, pp. 119–186.

Whitehorn, J. C., Zipf, G. K. (1943), *Arch. Neurol. Psych.* **49**, 831–851.

Williams, S. (1988), in: H. Dyer, G. Tseng (Eds.), *New Horizons for the Information Profession*, Los Angeles: Taylor Graham, pp. 212–225.

Yule, G. U. (1944), *The Statistical Study of Literary Vocabulary*, Cambridge: Cambridge University Press, p. 53.

Zipf, G. K. (1949), *Human Behavior and the Principle of Least Effort*, Reading: Addison-Wesley.

Further Reading

Bell, D. (1976), *The Coming of Post-Industrial Society,* New York: Basic, pp. 134–137.

Garfield, E. (1979), *Citation Indexing—Its Theory and Application in Science, Technology and Humanities,* New York: Wiley.

Hall, K. (1981), "The economic nature of information," *Info. Soc.* **1**, 143–166.

Losee, R. M., Jr. (1990), *The Science of Information, Measurements and Applications,* San Diego: Academic.

Meadow, C. T. (1992), *Text Information Retrieval Systems,* San Diego: Academic.

Mosco, V., Wasko, J. (Eds.) (1988), *The Political Economics of Information,* Madison, WI: University of Wisconsin Press.

Parsaye, K., Chignell, M., Khashafian, S., Wong, H. (1989), *Intelligent Databases, Object Oriented, Deductive Hypermedia Technologies,* New York: Wiley.

Porat, M. U. (1977), *Information Economy: Definition and Measurement,* OT Special Publication 77-12(11), Washington: U.S. Department of Commerce, Office of Telecommunications.

Rosenberg, R. S. (1986), *Computers and the Information Society,* New York: Wiley.

Salton, G. (1989), *Automatic Text Processing,* Reading, MA: Addison-Wesley.

Salton, G., McGill, M. J. (1983), *Introduction to Information Retrieval,* New York: McGraw-Hill.

Schiller, D. (1981), *Telematics and Government,* Norwood, NJ: Ablex.

Shatt, S., Fox, S. (1990), *Voice Data Telecommunications for Business,* Englewood Cliffs, NJ: Prentice-Hall.

Sherman, K. (1990), *Data Communication: A User's Guide,* 3rd ed., Englewood Cliffs, NJ: Prentice-Hall.

Stonebraker, M. (1988), *Readings in Database Systems,* San Mateo, CA: Morgan Kaufmann.

INFRARED SENSORS

See SENSORS, INFRARED

INFRARED SPECTROMETERS

See SPECTROMETERS, INFRARED

INSULATION MATERIALS

See THERMAL AND ELECTRICAL INSULATION MATERIALS

INSULATORS AND SEMICONDUCTORS, CONDUCTIVITY IN

E. M. CONWELL AND M. A. ABKOWITZ, *Xerox Webster Research Center, Webster, New York, U.S.A.*

	Introduction	42
1.	**Measurements**	44
1.1	Equilibrium Conductivity	44
1.2	Phenomenological Classification of Contact Behavior	44
1.2.1	Preparation of Contacts	45
1.3	Two- and Four-Probe Conductivity Measurements	45
1.4	Mobility Measurements	46
1.5	Measurement of Carrier Concentration	46
2.	**Band Structure**	47
2.1	What Determines Whether a Material is an Insulator, Conductor, or Semiconductor?	47
2.2	Hole Conduction	47
2.3	Energy, Momentum, and the Brillouin Zone	48
2.4	Band Structure of Column-IV and III-V Semiconductors	48
3.	**States Due to Defects in Crystalline Semiconductors**	51
3.1	The Hydrogen Model	51
3.2	Effects at High Impurity Concentration	52
4.	**Density of Current Carriers**	53
4.1	Intrinsic Carrier Concentration	53
4.2	Carrier Concentration in Doped Samples	54
4.2.1	Moderately Doped Samples	54
4.2.2	Heavily Doped Samples	56
4.3	Nonequilibrium Carrier Concentration: Photoconductivity	56
4.3.1	Intrinsic Photoconductivity	56
4.3.2	Extrinsic Photoconductivity	56
5.	**Formal Theory of Conductivity and the Hall Effect**	56
5.1	Effects of Electric and Magnetic Fields on the Carriers	56
5.2	The Relaxation Time	57
5.3	Calculation of dc Conductivity	57
5.4	ac Conductivity	58
5.5	The Hall Effect	58
5.5.1	Low Magnetic Fields	58
5.5.2	The Hall Effect for Two Types of Carriers	58
5.5.3	High Magnetic Fields	59
5.6	Magnetoresistance	59
5.7	Thermoelectric Effect	59
5.8	Galvanomagnetic and Thermomagnetic Effects	60
6.	**Relaxation Times and Mobilities**	60
6.1	Scattering Due to Lattice Vibrations	61
6.1.1	Acoustic and Optical Modes	61
6.1.2	Mobility Due to Acoustic-Mode Scattering	62
6.1.3	Mobility Due to Nonpolar Optical-Mode Scattering	62
6.1.4	Mobility Due to Intervalley Scattering	63
6.1.5	Mobility Due to Polar Optical-Mode Scattering	63
6.2	Impurity Scattering	64
6.3	Neutral-Impurity Scattering	64
6.4	Combination of Mobilities	65
6.5	Comparison with Experimental Data	65
7.	**High-Field Transport**	65
7.1	Hot Carriers	65
7.2	Mobility Changes in High Fields	66
7.3	Impact Ionization	67
7.4	The Gunn Effect	68
7.5	Negative Differential Conductivity	68
7.6	Acoustoelectric Effect	69

1-56081-067-X/94/$5.00+.50

8. **Impurity-Band Conduction** ... 70
8.1 Semiconductor Conductivity at Low Temperatures ... 70
8.2 Conduction Mechanisms at Very Low Temperatures ... 71
8.2.1 Variable-Range Hopping ... 72
8.3 Criterion for Localization ... 72
9. **Amorphous Semiconductors** .. 73
9.1 Atomic Structure ... 73
9.2 Electronic States ... 73
9.3 Native Defects ... 74
9.4 Conductivity Mechanisms ... 75
9.5 Doping ... 76
10. **Two-Dimensional and One-Dimensional Behavior** ... 76
10.1 Quantizing Magnetic Fields ... 77
10.1.1 Landau Levels ... 77
10.1.2 The Quantum Hall Effect ... 77
10.2 Transport in Inversion Layers ... 77
10.2.1 Effect of Electric Field on Semiconductor Surface ... 77
10.2.2 Quantization in the Inversion Layer ... 78
10.2.3 Transport in the Inversion Layer ... 78
10.2.4 Hot Electrons ... 80
10.2.5 Localization ... 80
10.2.6 Conductance Fluctuations ... 80
Glossary ... 81
Works Cited ... 82
Further Reading ... 83

INTRODUCTION

An ideal insulator would be one that does not conduct electricity, i.e., carries no current when a voltage is applied across it. In practice, all materials provide a measurable amount of electrical conduction. The effect of a voltage is to set up an electric field **E** that exerts a force on the electrons and ions within the material. In this article we will consider only electronic conduction. Ionic conduction is taken up in the article IONIC CONDUCTION AND DIFFUSION IN IONIC SOLIDS. Electronic conduction in a semiconductor or insulator differs from that in the usual metal in various ways, one being that current may be carried simultaneously by two types of carrier. One of these has the charge $-e$ (1.6×10^{-19} C) expected for an electron, while the other, called the hole, behaves as though it had a charge $+e$. The origin of the holes will be discussed in Sec. 2 when we take up band theory. If the carriers are free to move, the force due to the field results in a net motion, specifically a drift velocity $\mathbf{v}_d$, in the direction of the force. We define the resulting current density **j** as the amount of charge crossing a unit area (perpendicular to the flow) in unit time. Thus if only electrons were conducting we would have

$$\mathbf{j}=ne\mathbf{v}_d, \tag{1}$$

where n is the number of free electrons per unit volume. If there were also hole conduction, that would add a similar term to Eq. (1) for the hole contribution. The drift velocity is a function of the size of the electric field. At low fields $\mathbf{v}_d$ increases linearly with **E** and we may write

$$\mathbf{v}_d=\mu\mathbf{E}, \tag{2}$$

where μ, the mobility, defined by Eq. (2), is a measure of how well the carriers in a particular material respond to the electric field. The mobility becomes a function of **E** at high fields, as will be discussed in later sections. Defining the conductivity κ and resistivity ρ of a material by $\kappa=j/E=1/\rho$, we deduce from Eqs. (1) and (2) that

$$\kappa=ne\mu=1/\rho\ . \tag{3}$$

If there were also hole conduction, κ would have an additional term. The statement that κ and ρ are constants, independent of E, which is generally true at low enough fields, is Ohm's law. What determines the carrier concentration and μ, and how they vary with temperature, electric field intensity, etc., are the major subjects of this article.

Rather arbitrarily, we term a material an *insulator* if its conductivity is of the order of or less than 10^{-8} S m^{-1} ($10^{-10}\ \Omega^{-1}$ cm^{-1}). As shown in Table 1, with this definition the conductivity of insulators covers a range of some 8 orders of magnitude. Good metals, such as silver and copper, have conductivities 16 or more orders of magnitude higher. It is seen from Table 1 that the class of insulators includes a wide variety of materials, mostly organic. Semiconductors also have a wide range of conductivities; these lie between the extremes of metals and insulators just cited. It

Table 1. Representative conductivities.

Material	Conductivity (S/m)
Silver[a]	6.8×10^{7}
Copper[a]	5.8×10^{7}
Germanium (intrinsic)[b]	2.1
Silicon (intrinsic)[b]	3.2×10^{-4}
Nylon (polyamide)	10^{-5}–10^{-6}
Methyl methacrylates (acrylic plastic)	10^{-6}–10^{-7}
Gallium arsenide (intrinsic)[b]	2.4×10^{-7}
Aluminum nitride (undoped)[b]	10^{-9}–10^{-11}
Ceramics[c]	10^{-12}–10^{-14}
Amorphous selenium	10^{-14}
Corning 7911 Glass (96% silica)	$< 10^{-15}$
Paraffin[c]	3×10^{-17}

[a]Gray (1957).
[b]Madelung (1991).
[c]*Handbook of Chemistry and Physics* (1975).

is characteristic of semiconductors and insulators that the conductivity is quite dependent on impurities and defects, and varies strongly with temperature. The values given in Table 1 are those for nominally pure specimens at room temperature. As shown in Table 1, the class of semiconductors also includes a wide variety of materials. Important elemental semiconductors are silicon (see SILICON, CRYSTALLINE; SILICON, AMORPHOUS; SILICON, POLYCRYSTALLINE), extensively used for semiconductor electronic devices (see MICROELECTRONICS), and germanium (*q.v.*). Also important for applications are the compound semiconductors made up by combining an element from column III of the periodic chart [particularly aluminum (Al), gallium (Ga), or indium (In)] with one from column V [particularly phosphorus (P), arsenic (As), or antimony (Sb)]. Of these the most important is gallium arsenide (GaAs), used for electronic devices, light-emitting diodes (*q.v.*), and semiconductor lasers (see LASERS, SEMICONDUCTOR). Compound semiconductors made up by combining an element from column II of the periodic chart [particularly zinc (Zn), cadmium (Cd), or mercury (Hg)] with one from column VI [particularly sulfur (S), selenium (Se), or tellurium (Te)] are also of importance commercially for electroluminescence (*q.v.*), photodetectors (see PHOTODETECTOR DEVICES), and photoconductors. There are numerous other semiconductors—elemental, binary and ternary compounds, etc.

The sensitivity of properties of semiconducting materials to impurities has been taken advantage of by deliberate addition of chosen impurities, the process called doping, to modify conductivity values or to achieve spatial variation of properties. The effects of doping will be discussed extensively in what follows. Another property that significantly affects κ is the structure of the semiconductor sample. The sample may be crystalline, i.e., having a regular structure in which a pattern involving a small number of atoms is repeated identically throughout (see CRYSTALLINE STATE). More typically, the regular structure may occur over large regions of the sample but with the orientation, i.e., the axis, of the structure varying in a random way from one such region to another; such a sample is called polycrystalline. Alternatively, the sample may be amorphous; in the neighborhood of any particular atom the structure may be fairly close to what it is in the crystalline case (short-range order) but this pattern is not repeated, i.e., long-range order does not exist (see Sec. 9 and AMORPHOUS MATERIALS, STRUCTURE OF).

In what follows we first describe how conductivity is measured, emphasizing the role of contacts. What makes a material an insulator or semiconductor rather than a metal was understood early in the century; the explanation is taken up in the section on band structure. That section includes also descriptions of the conduction and valence bands of typical semiconductors. The structure of these bands underlies many features of the conduction process. It is followed by discussions of what determines carrier concentration and mobility in low and high electric fields in typical semiconductors. A major part of the article deals with conduction in column-IV and III-V semiconductors, on which a great deal of work has been done. This is followed by a section on amorphous semiconductors and one on conduction in cases where the carrier motion is restricted to a thin layer (two dimensional) or to a thin strip within the layer (one dimensional). Conduction through layers of different materials, or heterostructures, is taken up in HETEROSTRUCTURES AND SUPERLATTICES, SEMICONDUCTOR, and in the various articles on semiconductor devices.

1. MEASUREMENTS

1.1 Equilibrium Conductivity

Measurements of electrical conductivity are usually carried out by attaching contacts to the sample to connect it to a dc voltage source, as indicated in Fig. 1(a). To measure the conductivity, one first measures the conductance G, defined as the ratio of the current I through the sample to the voltage drop V across it. Conductance is the reciprocal of the resistance R. For a sample of simple geometry, as indicated in Fig. 1, from the definition of j we find $I=jA$, where A is the cross-sectional area. Also, $V=EL$, where L is the length of the sample. Thus the measured conductance yields the value of κ through

$$G=\frac{I}{V}=\frac{jA}{EL}=\kappa\frac{A}{L}=\frac{1}{\rho}\frac{A}{L}, \qquad (4)$$

from the definitions of κ and ρ given earlier.

For the measurements just described to yield a meaningful value of κ or ρ it is apparent that

1. the carrier concentration that exists in the sample in the absence of contacts, and of an applied field, must be retained under the conditions of the measurement, and
2. a uniform electric field must be sustained throughout the sample (with the possible exception of a very thin screening region at the contact–sample interface).

These conditions place severe requirements on the ability of the contacts to supply, or inject, charge, as well as on the sample. They are nevertheless satisfied with relative ease for metallic and high-conductivity semiconductor samples. For low-conductivity semiconductor samples and insulators they are much less likely to be satisfied. In those cases the current that flows in response to a voltage is mainly determined by the nature of the contacts.

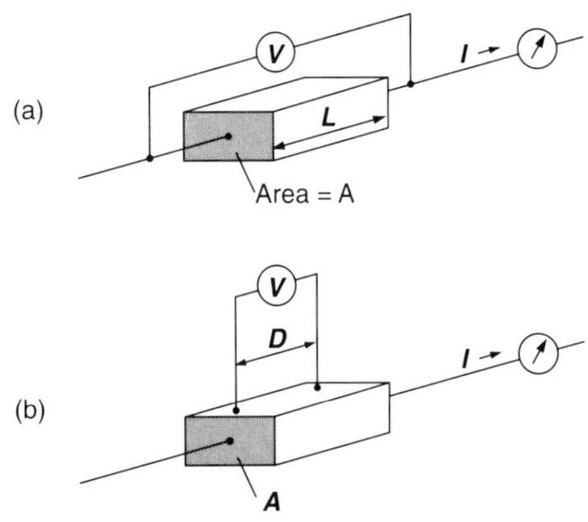

FIG. 1. Arrangements for measuring resistance or conductance. (a) Two-terminal method, (b) four-terminal method.

1.2 Phenomenological Classification of Contact Behavior

Contacts on an insulator or semiconductor may be operationally classified as Ohmic, finite injecting, or blocking. To understand these classifications it is necessary to review the elements of injection theory (see, for example, Lampert and Mark, 1970). We simplify the discussion by assuming that only one type of carrier is present. Consider the fate of a single excess carrier injected by a contact into a solid under the condition that the contact can act as an infinite reservoir of charge. The bulk dielectric relaxation time $\tau_r=\rho\epsilon$, where ϵ is the permittivity, may be defined as the time it takes for charges in the solid to rearrange themselves so as to neutralize any excess injected carrier. In a metal the relaxation is effectively instantaneous, while in a nearly perfect insulator τ_r is essentially infinite. Two regimes need to be considered, depending on the relative values of the injected carrier transit time (time to traverse the region between the contacts under the action of a field) and τ_r:

1. When the transit time is greater than the dielectric relaxation time, the carrier is neutralized before completing a transit, and the current at a given field is controlled in the usual way by the product of the resident thermal-equilibrium carrier population n and the mobility μ. In other words, Ohm's law is valid.
2. When the transit time is less than the dielectric relaxation time, the magnitude of the injected charge is limited only by the repulsion of the space charge already in the sample, including carriers that have been immobilized in traps. Such traps are quite common in high-resistivity materials.

In the simplest case of a trap-free specimen film configured as a plane parallel capacitor with capacitance C, under bias V the amount of charge injected per transit time to sustain this space-charge–limited current is about CV (Lampert and Mark, 1970). This injection rate can far exceed the rate required merely to sustain the equilibrium current. Thus, with increasing field, the onset of space-charge–limited injection occurs at the threshold field that just makes the transit time less than the dielectric relaxation time. The onset is marked by I increasing more rapidly than linearly with V, thus by Ohm's law no longer being valid. It is clear that the onset initiates a process that significantly increases the demand of the bulk on the interfacial source of injected charge. An Ohmic contact is generally defined as one fully capable of sustaining the demands of the bulk up to and including space-charge–limited conditions. A contact that is finite injecting imposes a supply limitation. The contact interface supply limitation can itself depend on field, temperature, and even the mobility of the solid. Thus there are circumstances in which it can be difficult experimentally to distinguish finite injecting from Ohmic contacts. A blocking contact is one that completely impedes injection. Usually a contact is blocking for only one sign of a carrier. Finally it is not uncommon for a given contact to evolve from Ohmic to finite injecting and then blocking with increasing field and for this entire process to be temperature dependent. For further discussion of the behavior of insulators see DIELECTRIC PROPERTIES OF INSULATORS.

1.2.1 Preparation of Contacts. Connection of specimens to measuring circuitry is made by contacts that are most commonly of the following types:

1. diffused;
2. painted (typically with metallic emulsions);
3. soldered, brazed, or spot welded;
4. alloyed;
5. evaporated or sputtered (commonly used on amorphous semiconductors, molecular crystals, and polymer films);
6. plated; or
7. pressure probe.

Diffused contacts are commonly used for semiconductors. A large dopant concentration is introduced by diffusion to bring their surfaces to a nearly metallic state. Metallization of the surface then produces a perfectly injecting electrical interface. It should be noted that contacts that work well on metals are relatively unpredictable on semiconductor surfaces because the interface typically undergoes uncontrolled compositional and thermal modification during the forming process. It is desirable to have current paths parallel to the surface of the specimen, which requires interelectrode spacing large compared to the specimen thickness. For further details on contacts and resistivity measurements, see Dunlap (1959).

1.3 Two- and Four-Probe Conductivity Measurements

The simplest arrangement for the measurement of dc conductivity is the terminal mode shown in Fig. 1(a). Here the voltage drop across the specimen and the current in series with the specimen are simultaneously recorded. It is necessary that the potential drops at the contacts to the specimen, i.e., the current contacts, be small compared to the drop across the specimen. In a typical configuration, the voltage source is an electronically regulated supply, sometimes buffered by a large storage capacitor. Voltage can be measured with an electrometer or electrostatic voltmeter, and current with an electrometer or picoammeter. When isolation or noise considerations are paramount, a shielded battery box still makes the best dc supply.

For high-resistivity materials, or cases where the contact potential drops are relatively large, measurements should be carried out in a four-terminal configuration, illustrated in Fig. 1(b). In this configuration the voltage drops across the contacts are unimportant because they are in series with much higher resistances in the voltage-measuring circuit. One such mode of measurement using a linear array of pin probes (commercially available) is commonly deployed for rapid routine evaluation of Si and Ge semiconductor specimens, or other materials with volume resistivities in the range 10^{-5} to about 10^{2} Ω m. Voltage measurements are taken across the innermost probes with both polarities of a constant-current source and averaged to eliminate extraneous thermal emf's or rectifying contact effects. If the probe spacing D is small compared to the smallest sample dimension and the contact area is negligible, the conductivity is given by $I/2\pi VD$. For high-resistivity materials, up to about 10^{9} Ω m, a particular four-probe configuration due to van der Pauw, which is commercially available, may be used (Dunlap, 1959). For $\rho > 10^9$ Ω m the two conditions specified in Sec. 1.1 are not likely to be satisfied. Values of $\rho > 10^9$ Ω m are probably qualitative at best.

1.4 Mobility Measurements

In addition to measuring conductivity or resistivity, it is possible in many materials to measure directly the mobility of the current carriers. The mobility may be obtained by measuring the transit time t_t over a known distance d of injected carriers in a known electric field. The drift velocity is $v_d = d/t_t$. Once v_d is measured, the mobility may be obtained from Eq. (2). There are basically two methods for measuring the mobility, one applicable to semiconductors, the other to insulators.

In the case of semiconductors the technique is applicable to a specimen in which one sign of carrier, either electrons or holes, is dominant under thermal-equilibrium conditions. This carrier is called the majority carrier. The technique relies on the fact that a suitable point contact applied to a semiconductor sample can inject or emit minority carriers into the sample for one sign of voltage relative to the sample, while it can collect, and thus detect, them for the reverse sign of voltage. The sample is fabricated as a uniform cylindrical rod and fitted with blocking contacts at each end. A dc voltage is applied between these contacts. Emitter and collector point contacts are located along the rod spaced a distance d apart. A pulse of minority carriers, typically of microsecond duration, is emitted into the sample at some time t_1 and drifts along the rod under the applied field, reaching the collector point contact at time t_2. The concentration of injected carriers is kept much less than the thermal-equilibrium carrier concentration to ensure that the transit occurs in a uniform field. The carrier mobility is then given by $\mu = d/E(t_2 - t_1)$, where t_1 and t_2 are determined by monitoring the voltage pulses induced across a sensing resistor in series with the specimen.

In the case of insulators, carriers are injected by means of light (see Sec. 9). The sample is prepared as a film of thickness d with semitransparent blocking contacts at the front and back through which the field $E = V/d$ is applied. Exposing the samples, through a contact, to a pulse of strongly absorbed light creates holes and electrons just below the contact. A sheet of either electrons or holes will be swept into the bulk, depending on the direction of the electric field, and through the sample to the other contact. To ensure that the transit occurs in a uniform field the amount of photoinjected charge must be kept small, below $0.1CV$ where C is the sample capacitance. The transiting carrier sheet induces a voltage across a series resistor that drops when the sheet exits through the other contact. The time between the creation of the sheet and its exit is the transit time t_t provided it is long compared to the duration of the light pulse, a condition that can usually be satisfied. Again, the mobility is given by $\mu = d/Et_t$.

If there are traps in the sample, μ measured as described above, called the drift mobility, will not agree with the conductivity mobility defined in Eq. (2). The conductivity mobility is a microscopic mobility, the mobility of an untrapped particle. For the case of shallow traps, where a carrier may be trapped and released from a trap many times during a transit, the drift mobility is the microscopic mobility times the fraction of time that the carrier is free. When the traps are not shallow, and there is a range of trap depths and thus a range of trapping times, the carriers straggle to the contact and it becomes impossible to define a transit time. Such dispersive transport is frequently seen for amorphous semiconductors and disordered materials.

1.5 Measurement of Carrier Concentration

Another important measurement for semiconductors is the Hall effect, from which the carrier concentration of samples with one type of carrier predominant may be determined. For good accuracy the sample should be long compared to the transverse dimensions. An electric field E_x applied along the long dimension, to be taken as x, sets up a current j_x. When a magnetic field is applied in a transverse direction, to be taken as z, it deflects the carriers in the y direction, whether toward $+y$ or $-y$ depending on their sign. During the initial transient phase a current flows in the y direction, causing charge to accumulate on the two faces perpendicular to y. This process continues until the field E_y due to this charge just balances the deflecting effect of the magnetic field. In the steady state the current again flows in the x direction. It is found that E_y is proportional to j_x and H_z, with the constant of proportionality being R_H, the Hall coefficient. The importance of the Hall effect lies in the fact that

$$R_H = r/ne \quad \text{or} \quad R_H = r/pe \,, \tag{5}$$

the proportionality constant r being of the order of unity, as will be shown in Sec. 5. Note that the sign of R_H is given by the sign of e. It is seen from Eqs. (3) and (5) that the product $R_H\kappa$ equals $r\mu$. It is usual to call the product $R_H\kappa$ the Hall mobility, μ_H.

2. BAND STRUCTURE

2.1 What Determines Whether a Material Is an Insulator, Conductor, or Semiconductor?

It is well known that the electrons in an atom or molecule occupy a set of discrete, well-separated energy levels (see ELECTRONIC STRUCTURE OF ATOMS AND MOLECULES). When atoms or molecules are brought close together to form a solid, electrons on different atoms or molecules interact and the energy levels spread into bands (see ELECTRONIC STRUCTURE OF SOLIDS). The deeper-level electrons, most tightly bound to the nuclei, interact least, forming low-lying narrow bands, well separated from each other. The electrons in the highest level (or levels) go into wide bands. These bands of allowed levels may be separated from each other, may overlap each other, or may overlap bands coming from levels ordinarily not filled in the atom.

It might be thought that, with the outermost electrons close to each other, they could move freely from one atom to another. All solids should then be conductors. The band picture given above can explain why this is not so. An electric field exerts a force on an electron, tending to increase its energy. This can only occur if there are unfilled levels nearby for the electron to move into. Thus if the uppermost band occupied by electrons were incompletely filled, or if it overlapped unfilled bands, the material would be a good conductor. This situation describes metals. It follows from these arguments that if the uppermost occupied band were completely filled, and did not overlap any unfilled bands, the material would not conduct. How then does one explain semiconductors? The band picture offers two possibilities. If the energy gap between the uppermost filled band, called the valence band, and the next higher unfilled band, called the conduction band, is small enough, electrons may be thermally excited into the unfilled band. This process will be discussed further in Sec. 4. This type of semiconductor is called an intrinsic semiconductor. The second possibility is that impurities or defects in the material give rise to energy levels in the energy gap between valence and conduction bands. Ordinarily electrons in these levels are bound or localized and cannot give rise to current. However, they can, at finite temperatures, produce unfilled bands in either of two ways. In one way an electron in a level close to the conduction band is excited into the conduction band by thermal energy. In the other way an electron from the valence band is thermally excited into an empty level close to the valence band. Both of these situations will be discussed further in Sec. 4. As a by-product of this discussion it is clear why some materials are insulators. An insulator is a material with a large gap between valence band and conduction band, typically greater than several electron volts. In materials with such large gaps it is not usual to find impurity levels close to the bands.

2.2 Hole Conduction

It has been found, e.g., by measurements of the Hall effect, that some metals and semiconductors act as though positive charges, rather than electrons, were the current carriers. Explanation of this apparent anomaly was another success of the band model. From the discussion in the last section it follows that semiconductors are characterized by a small fraction of filled states in the conduction band and/or a small fraction of unfilled states in the valence band. It is the latter states, called holes, that act like positive charges. This can be understood as follows. A filled band carries no current because every electron motion in a particular direction is canceled by an exactly opposite motion. When one electron is removed from the band the cancellation is no longer exact. In the presence of a field the net motion of all the remaining electrons is opposite to the field direction because of their negative charge. This results in the hole having a net motion in the field direction, like a positive charge. In fact, it can be shown that the behavior of all the remaining electrons in the band is equivalent to the behavior of a singly positively charged particle. The properties of this particle, the hole, can be deduced from the properties of the missing electron to which it corresponds. It is clear from its definition that the minimum energy for a hole occurs when it is at the top of the valence band.

2.3 Energy, Momentum, and the Brillouin Zone

The energy E of a free electron is given by $p^2/2m_0$, where p is its momentum and m_0 its mass. The electron may also be characterized by a wave vector k, which equals $p/\hbar$, $\hbar$ being Planck's constant, 6.63×10^{-34} J s, divided by 2π. An electron inside a crystal, which has a periodic arrangement of atoms, may still be characterized by a p value, or a k value. (Note, however, that it is no longer true that momentum is a constant as it is for a free electron.) However, unlike the free-electron case, for the electron in the crystal there is only a finite, although very closely spaced, set of allowed values of these quantities. That set is equal in number to the number of atoms in the crystal and is confined to a unit cell in k space. This unit cell is called the Brillouin zone (see ELECTRON STRUCTURE OF SOLIDS). The detailed boundaries, and thus the shape, of the Brillouin zone depend on the detailed geometry of the crystal lattice. For the diamond structure, possessed by Si and Ge, and the zinc-blende structure, which is the structure of GaAs, InSb, most of the other III-V compounds, and some of the II-VI compounds, the Brillouin zone is a truncated octahedron, shown in Fig. 2.

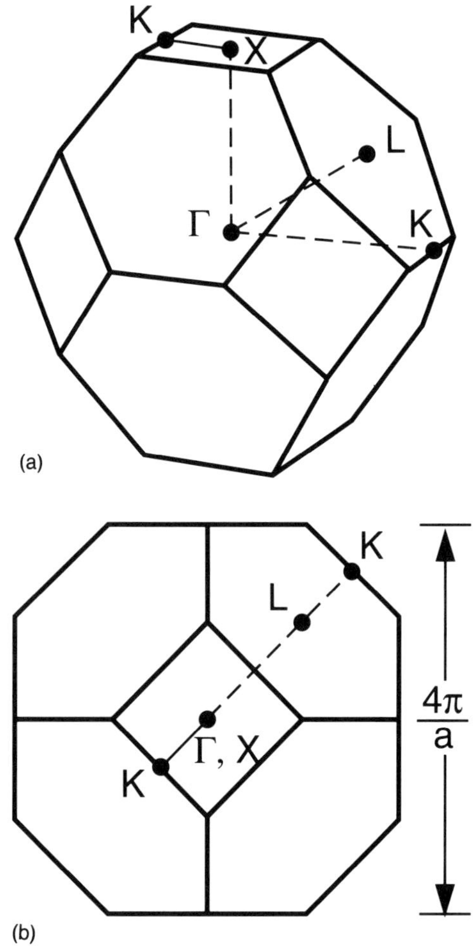

FIG. 2. The Brillouin zone and several symmetry points for the diamond and zinc-blende lattices. The view at the bottom is along a [110] (face diagonal) axis; that at the top has been tilted. The locations of the points shown are $\Gamma = (0,0,0)$; $X = (2\pi/a)(1,0,0)$; $L = (2\pi/a)(\frac{1}{2}, \frac{1}{2}, \frac{1}{2})$; $K = (2\pi/a)(\frac{3}{4}, \frac{3}{4}, 0)$, where a is the side of a cubic unit cell and the numbers in parentheses are the components of **k** along the cubic x, y, and z axes. (After Harrison, 1980).

2.4 Band Structure of Column-IV and III-V Semiconductors

Although many features of the band structure may be obtained from experiments, complete determination of the band structure for any particular material requires solution of the wave equation for that material (see ELECTRON STRUCTURE OF SOLIDS). The result of the band-structure calculation is a detailed relation between electron energy E and k. Because we have confined the possible **k** values to the Brillouin zone, there is for each **k** a set of E values, each corresponding to a different band. The significant energy bands for conduction are the valence and conduction bands, and in fact only the upper and lower edges of these bands, respectively, for low electric fields.

It is instructive to consider what some possible results of band-structure calculations could be for the column-IV and III-V zinc-blende semiconductors before going on to the actual results for these materials. In Fig. 3 we show some possible results for the band edges.

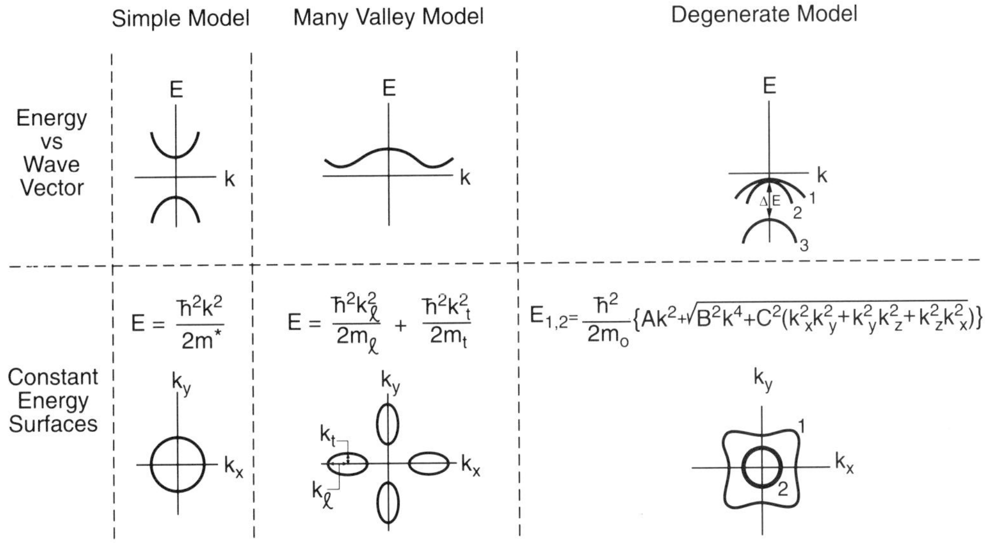

FIG. 3. *E* vs **k** and constant-energy surfaces for some possible semiconductor band structures. (Adapted from Conwell, 1958).

In the first column is the so-called simple model, where the energy measured from the edge of the band is a constant times k^2. E vs k for electrons is then a parabola as in the free-electron case and an inverted parabola for holes. To maintain the analogy between this model and the free-electron case, the proportionality constant between E and k^2 is taken as $\hbar^2/2m^*$, the m^* thus defined being called the effective mass of the carrier in the crystal. In the simple model the same parabola for E vs k, thus the same m^*, is obtained for all directions of **k**. This type of band structure could occur if the band edge were at the center of the Brillouin zone, the Γ point. Another way of representing this result is shown at the bottom of the first column. If all **k**'s belonging to the same E are connected, a single sphere is obtained, just like that for a free electron. As a result of this simple relation between E and **k**, for this model of the band structure it is possible for most purposes to treat the electron inside the crystal as a free electron, using, however, a different number for the mass.

When the minimum energy is not at the center of the zone there is a further complication. Symmetry requires that all equivalent **k**'s i.e., all **k**'s reachable from each other by crystal symmetry operations (rotations, reflections, or inversions that transform the crystal into itself), have the same energy. Thus the band edge or position of minimum energy must be multiple, and the electron (or hole) population must be divided among the various minima, or valleys. This band structure is therefore called the many-valley model. Within each valley, not far from the minimum energy, E can be represented as a quadratic function of the components of **k** measured from the minimum energy point. Thus the points in **k** space belonging to a particular energy are on a set of ellipsoids, one for each minimum. Actually, these are ellipsoids of revolution here, and so the energy can be written as a constant times k_l^2 plus another constant times k_t^2, where k_l and k_t are longitudinal and transverse components. In Fig. 3 the lower half of the second column illustrates the case for the minimum energy located at a point on a cube axis, in which case there are six equivalent valleys. This is the case for the conduction band of silicon. Because $m_l \neq m_t$, we must deal with two effective masses in this case. In later sections we will discuss how this type of band structure complicates transport.

A third possible type of band structure for the semiconductors we consider is the degenerate band structure shown in the third column of Fig. 3 for the case of the valence band. This is a set of three bands, one split off by an energy ΔE from the other two. This is the structure for the top of the valence band in Si, Ge, and the III-V compounds with zinc-blende structure. It is the result of the wave function at the top of the valence band having within each cell the character of an atomic p state, which is triply degenerate. The interaction of spin and orbital motion of the electron partially breaks up the degeneracy of the p states, leaving two states stuck together at $k=0$ and the third split off by a small amount ΔE (see ELECTRON STRUCTURE OF SOLIDS).

As is the case for the many-valley model, E vs $\mathbf{k}$ for the degenerate model is different in different directions. This results in the constant-energy surfaces being warped rather than spherical. The surfaces from the two upper bands belonging to a given energy are shown, again rather schematically, in the lower part of the third column of Fig. 3; their equation is given also. The values of A, B, and C are different for different materials. Because of the warping of the surfaces, the effective masses are functions of direction. The anisotropy is not very great, however, and for most purposes the holes in bands 1 and 2 may be described by constant masses. The band with the smaller curvature has the larger mass. Holes in this band are called heavy holes, while those in the other band are called light holes. At temperatures low enough that the thermal energy $kT \ll \Delta E$, only these two bands will be populated.

The calculated energy-band structure for GaAs is shown in Fig. 4. Note that many features of the band structure have been verified by other calculations and by experiments. The band structures of the other column-IV and the other III-V zinc-blende semiconductors are quite similar to that in Fig. 4. (For a tabulation of band structures and other data for all the column-IV and III-V compounds see Madelung, 1991). Plotted in this figure is E vs $\mathbf{k}$ of a number of bands for some high-symmetry directions in the Brillouin zone. Thus, referring to Fig. 2, we see that the curves from Γ to X represent the cube axis direction and those from Γ to L the cube diagonal direction, etc. Looking at the borders of the crosshatched region, which represent the lowest conduction band and the highest valence band, we see that the only case where the simple model comes close to being realized is the conduction band of GaAs, centered at Γ_6. The mass for that band is quite small, $\cong 0.06m_0$, where m_0 is the free-electron mass.

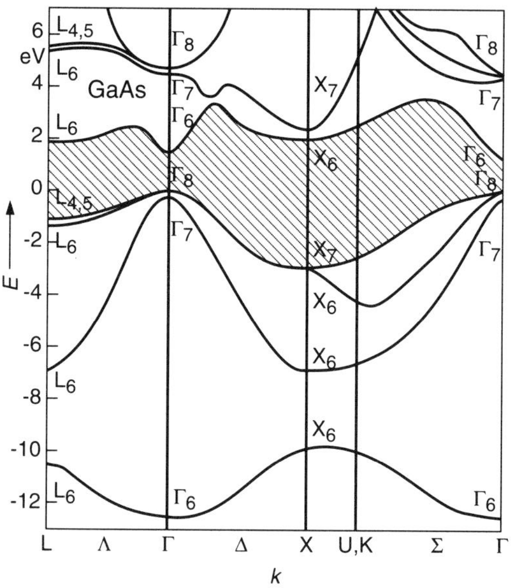

FIG. 4. Calculated band structure of GaAs. The crosshatched region represents the gap between the highest valence band and the lowest conduction band. (Adapted from Madelung, 1991.)

In the case of Si the conduction-band minimum is along the cube axis direction, actually close to the X point. Thus, as noted earlier, the conduction band of Si has six equivalent valleys, the situation depicted in Fig. 3. The ellipsoids are quite anisotropic, with $m_l = 0.92m_0$, $m_t = 0.19m_0$. The valence band of GaAs clearly fits the degenerate model of Fig. 3 with a spin-orbit splitting $\Delta E = 0.34$ eV. Light- and heavy-hole masses are well approximated by $0.08m_0$ and $0.05m_0$ (Madelung, 1991). The valence band of silicon also has the degenerate band structure, but the spin-orbit splitting is only 0.04 eV, so small that all three bands are populated at room temperature. Light- and heavy-hole masses are $0.15m_0$ and $0.54m_0$.

The band structures of semiconductors other than those in group IV and III-V compounds have not been studied as extensively. Regardless of the detailed lattice structure, band edges should still be described by the simple, many-valley, and degenerate models. Of course differences must be expected due to different symmetry when the lattice structure is not cubic.

3. STATES DUE TO DEFECTS IN CRYSTALLINE SEMICONDUCTORS

As indicated in the last section, impurities or defects give rise to energy levels in the forbidden gap that localize electrons. There is a great deal of information about such levels in silicon (see SEMICONDUCTORS, ELEMENTAL—ELECTRONIC PROPERTIES) and GaAs, for example (see SEMICONDUCTORS, COMPOUND—ELECTRONIC PROPERTIES). In this article we limit discussion to those impurities that are important for conduction. Specifically, we will discuss those that

1. provide energy levels close to the conduction- or valence-band edges, and
2. can be introduced into the semiconductor in relatively large concentrations.

The impurities that best satisfy these conditions are those that substitute for host atoms in the lattice and differ from the host in having one more or less valence electron. For Ge and Si these are the atoms of columns III and V of the periodic chart. For the III-V compounds such impurities include atoms from columns II, IV, and VI. As will be seen, the last electron in an impurity with one more electron than the atom it substitutes for is weakly bound to the impurity, i.e., occupies a level only a little below the conduction band. This last electron can easily be detached by thermal energy, going into the conduction band. Such an impurity is called a donor. By similar reasoning, an impurity with one less electron than the atom it substitutes for provides a level just above the valence band into which a valence electron can be excited by means of thermal energy, leaving a hole in the valence band. Such an impurity is called an acceptor. We show now that both donors and acceptors can be described by a simple intuitive picture that predicts reasonably well the energy levels they provide.

3.1 The Hydrogen Model

Consider as an example the case of a column-V donor atom, say P, substituting for a Si atom. P has five valence electrons. Four of these go into covalent bonds with the four nearest neighbors just as the electrons on a Si atom would. This results in a P^+ ion that attracts the fifth electron. This fifth electron uses wave functions from the conduction band. If the conduction-band structure were that of the simple model, the electron would move around the P^+ ion essentially as an electron in free space moves around the proton in a hydrogen atom, or more generally around a nucleus of charge Ze. The theory of such motion is well understood (see ELECTRONIC STRUCTURE OF ATOMS AND MOLECULES) It is found that the most probable distance a_H of the electron from the nucleus is given by $a_H = \hbar^2/Zm_0e^2$. The binding energy E_b of the electron to such a nucleus is $E_b = m_0e^4Z^2/2\hbar^2$. To utilize these results for the electron bound to a substitutional P^+ ion in Si it is necessary to take into account two important differences between the situations. Because the P^+ ion is embedded in a medium of permittivity ϵ, the attractive potential of the ion for the electron when they are separated by a distance r is reduced from e^2/r to $e^2/\epsilon r$. as though the nuclear charge $Z = 1/\epsilon$. The other difference arises from the band structure. In the case of the simple model (see Fig. 3) the free-electron mass m_0 may be replaced by m^*. For the simple model of the band structure, a_H then becomes

$$a_H^* = \varepsilon\hbar^2/m^*e^2, \quad (6)$$

and the binding energy becomes

$$E_b^* = m^*e^4/2\varepsilon^2\hbar^2. \quad (7)$$

E_b^* in this case represents the distance of the energy level below the conduction-band edge.

For typical semiconductors ε is large, ~12 to 16. This results in a_H^* being large and E_b^* quite small. For Si, for example, with $\varepsilon = 12$, by Eq. (6) the binding energy of the fifth electron to a P^+ ion would be $0.1m^*/m_0$ eV if the conduction-band structure were the simple model of Fig. 3. The measured value of this binding energy is 0.045 eV (Madelung, 1991), close to the value that would be obtained from Eq. (6) by inserting for m^* the geometric mean mass $(m_t^2m_l)^{1/3}$ for a conduction electron in Si. Thus the simple H model we have used gives a reasonably good value for the binding energy. A detailed calculation for the many-valley model is a great deal more complicated [see Altarelli and Bassani (1982)]. Equations (6) and (7) should be valid for the appropriate impurities in GaAs, where the simple model of the band structure does apply reasonably well for the conduction band. Some impurities for which the theory should be valid are column-IV impurities—C, Si, and Ge—substituting for a Ga atom, and column-VI impurities—S and Se—substituting for an As atom. With $\varepsilon=12.4$ for GaAs, and $m^*=0.06m_0$, Eq. (7) leads to $E_b^*=5.3$ meV. This is in good agreement with the observed value of 6 meV for all the impurities listed but S, for which the observed E_b^* is closer to 7 meV. Refinements in the theory bring the theoretical values close to the experimental values (see Altarelli and Bassani, 1982).

In the case of a substitutional impurity with one less electron than the atom it replaces, when the four covalent bonds with nearest neighbors are filled, it results in a net charge of $-e$ in the region. This negative charge attracts a hole, which can be treated just as the bound electron was in the previous case. The most probable distance of the hole from the impurity is given by Eq. (6), and the binding energy of the hole by Eq. (7), with m^* the hole effective mass. Note that in this case E_b^* equals the distance above the valence band of the electronic energy level provided by the acceptor. Again the actual calculation is more complicated because of the complications of the band structure, in this case the degenerate model for the semiconductors of most interest. The measured binding energies for holes arising from the substitution for Si atoms of column-III acceptors—B, Al, Ga, and In—range from 40 to 150 meV. For holes arising from column-II elements—Be, Mg, Zn, and Cd—substituting for Ga in GaAs, the measured binding energies range from 28 to 35 meV. The tendency of the binding energies to increase with increasing atomic number is due to the fact, not acknowledged in the hydrogen model, that the bound electron or hole does actually spend some time close to the impurity, where ε has less than its full value.

The energy required to release a bound electron or hole from a donor or acceptor, respectively, is also called the activation energy because it is the energy required to give rise to conductivity. The activation energies cited for hydrogen-model donors and acceptors are typical for such donors and acceptors in the other column-IV and III-V semiconductors (see Madelung, 1991). They are also the smallest activation energies; other impurities may also supply conduction-band electrons or valence-band holes, but greater energy is required to achieve this. The activation energies for the hydrogen-model impurities cited above are generally <25 meV, the thermal energy at room temperature, with the consequence that the bound electrons and holes will be mostly or entirely free at room temperature and even below. The detailed variation of carrier concentration with temperature to which they give rise will be taken up in the next section.

3.2 Effects at High Impurity Concentration

In this section so far we have considered the activation for a single isolated impurity. When there is a high concentration of impurities, so that the wave functions of bound electrons or holes begin to overlap, the activation energy decreases. In a process similar to that described earlier for band formation, as the impurities come closer the energy levels spread into bands. Band formation occurs first for excited electronic levels, which we have not considered so far in this section, because their wave functions extend further. When the hydrogen model is strictly applicable the excited levels are analogous to those of the H atom (see ELECTRONIC STRUCTURE OF ATOMS AND MOLECULES). Otherwise they are different in detail. Overlap of the higher excited levels to form bands, which then overlap (and mix with) the conduction band, clearly decreases the activation energy, one of a number of effects that do so. For impurity concentration large enough that the ground states of adjacent impurities overlap, an impurity band is formed. We can estimate the impurity concentration N_I at which this occurs by equating a_H^* of Eq. (6) to $N_I^{-1/3}$, a rough measure of the average distance between impurities. For the conduction band of GaAs this yields $N_I \simeq 10^{21}$ m^{-3} for formation of an impurity band. By this criterion a concentration several times as large would be required for impurity-band formation in Si or Ge. Such concentrations are not far below the maximum that can be incorporated into these semiconductor lattices for any impurity. The impurity band that results would be expected to have somewhat different properties from the bands we have been discussing because, apart from its being quite narrow at concentrations of 10^{21} or 10^{22} m^{-3}, the impurities are distributed randomly rather than on a regular lattice. Impurity-band conduction will be taken up in Sec. 8.

4. DENSITY OF CURRENT CARRIERS

A pure, perfect semiconductor crystal at the absolute zero of temperature would have no electrons in the conduction band and no holes in the valence band. The current carriers present at higher temperatures are excited both from the valence band and from donor and acceptor impurities by thermal energy. In what follows we will consider first the thermal-equilibrium carrier concentration in pure specimens, the so-called intrinsic carrier concentration, and then in doped or extrinsic specimens. The section will end with an introduction to nonequilibrium carrier concentrations. These lead to enhanced conductivity and other phenomena of great technological importance in semiconductors and insulators. The nonequilibrium carriers may be excited by light, in which case the enhanced conductivity is called photoconductivity. They may also be created by injection at contacts, as discussed in Sec. 1, and by ionizing radiation.

4.1 Intrinsic Carrier Concentration

The probability $f(E)$ of an electron occupying a state of energy E is given by the Fermi–Dirac distribution function (see STATISTICAL MECHANICS, QUANTUM)

$$f(E)=\{1+\exp[(E-E_F)/kT]\}^{-1}, \tag{8}$$

where E_F is the electrochemical potential, called the Fermi level in discussions of solids. The number of electrons per unit volume in the conduction band of a semiconductor is given by

$$n=\int f(E)N(E)dE, \tag{9}$$

where $N(E)dE$ is the number of available states per unit volume of the sample in the energy range dE at E, measured from the band edge, and the integration is taken over the conduction band. For the simple model of the band structure the states in the conduction band up to energy E occupy a sphere in k space of radius $k=[2m^*(E-E_c)/\hbar^2]^{1/2}$, where E_c is the energy of the conduction-band edge. From this, plus the fact (due to the uncertainty principle) that each state must occupy a volume of k space equal to $(2\pi)^3/$(volume of the sample), we deduce that

$$N(E)dE=(1/2\pi^2)(2m^*/\hbar^2)^{3/2} \times(E-E_c)^{1/2}dE. \tag{10}$$

This result may be formally generalized to other models by replacing m^* by $m^{(N)}$, called the density-of-states effective mass. For the many-valley model

$$m^{(N)}=g^{2/3}(m_t^2m_l)^{1/3}, \tag{11}$$

where g is the number of equivalent valleys.

We may now obtain n as a function of temperature by inserting Eqs. (8) and (10) in Eq. (9). Except for very heavily doped semiconductors, $E-E_F \gg kT$, and Eq. (9) may be approximated by the Maxwell–Boltzmann distribution

$$f(E)=\exp[(E_F-E)/kT]. \tag{12}$$

This allows the integration to be simplified by taking the limits from 0 to ∞. The result is

$$n=N_c\exp[-(E_c-E_F)/kT] \equiv 2(2\pi m_n^{(N)}kT/h^2)^{3/2}\exp[-(E_c-E_F)/kT], \tag{13}$$

which serves to define N_c, the effective density of states in the conduction band. Similar calculations may be made for holes in the valence band, starting from the probability $1-f(E)$ of finding a hole in the level at energy E. The result is that the hole concentration p varies with temperature according to

$$p=N_v\exp[-(E_F-E_v)/kT] \equiv 2(2\pi m_p^{(N)}kT/h^2)^{3/2}\exp[-(E_F-E_v)/kT], \tag{14}$$

which defines N_v, the effective density of states in the valence band. Because of the approximations made in the derivation, the expressions for n and p are accurate only when E_F is at least several kT from the respective band edge. As will be seen, this condition is always met for intrinsic samples. Multiplying together the expressions (13) and (14) we obtain the equilibrium value for the product np. For an intrinsic sample, in which the number of carriers is denoted n_i, $np=n_i^2$. We can use Eqs. (13) and (14) to obtain

$$n_i=p_i=(N_cN_v)^{1/2}e^{-E_g/2kT}. \tag{15}$$

Thus the intrinsic carrier concentration depends exponentially on the energy gap $E_g/$

$2kT$. Note that E_g is also somewhat temperature dependent, decreasing ~5% from 0 to 300 K for Si, for example. Using the fact that E_g for Si is 1.12 eV at 300 K (Madelung, 1991) and the effective masses given previously, we find that $n_i = 1 \times 10^{16}$ m^{-3} (1×10^{10} cm^{-3}) at 300 K for Si. For GaAs, with a gap of 1.4 eV at 300 K, n_i is only 2×10^{12} m^{-3} at that temperature. Particularly in the case of GaAs the intrinsic number of carriers is so small that carriers coming from impurities, which cannot be reduced to such a low level, must always be dominant.

We can use the equations derived in this section to determine the location of E_F in intrinsic material, E_F^i. To do this it is convenient to use as our reference point, i.e., the zero of energy, the top of the valence band. Equating n_i obtained from Eq. (13) to p_i obtained from Eq. (14) we find

$$E_F^i = (E_g/2) + \tfrac{3}{4}kT \ln(m_p^{(N)}/m_n^{(N)}). \qquad (16)$$

If $m_p^{(N)} = m_n^{(N)}$ the Fermi level for intrinsic material is at midgap, independent of temperature. Although this is generally not the case, E_F^i is usually not far from midgap because $m_p^{(N)}$ is usually relatively close in value to $m_n^{(N)}$.

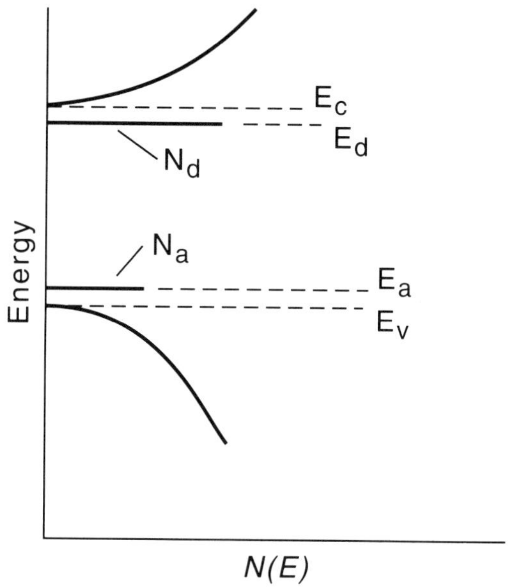

FIG. 5. Distribution of states in a lightly doped *n*-type crystalline semiconductor. The variation of $N(E)$ with E is given by Eq. (10).

4.2 Carrier Concentration in Doped Samples

4.2.1 Moderately Doped Samples. To illustrate the way in which carrier concentration varies with temperature in a doped sample, we will discuss the case of a donor-doped, or *n*-type, sample for specificity. The results apply to acceptor-doped samples with the obvious modifications. In the present section we discuss samples for which Maxwell–Boltzmann statistics, Eq. (12), are valid, typically samples with $N_d < 10^{22}$ m^{-3}. Actually, because it is not possible to control impurity content totally, the typical *n*-type sample with N_d donors/m^3 will also have some acceptors, present in concentration N_a/m^3. We will assume, as shown in Fig. 5, that the N_d donors all provide the same energy level E_d, while the N_a acceptors all provide the energy level E_a. In crystalline semiconductors that are not as well under control as column-IV and III-V semiconductors there may be different sets of donor and acceptor levels, or even a continuous range of levels across parts of the energy gap; the results we derive in this section will not apply in detail to such materials. As seen in Fig. 5, the energy required to raise an electron from a donor level to the conduction band, i.e., the electron binding energy, is $E_c - E_d$, while that required to raise it from the valence-band edge to the acceptor level, the hole binding energy, is $E_a - E_v$. Given N_d, N_a, N_c, N_v, and the various energy levels, it is in principle possible to determine the variation of n with T from Eq. (13), provided E_F is known as a function of T. A procedure for determining $E_F(T)$ will be given later. It is useful to precede this with a qualitative view of the situation.

Let n_d represent the number of electrons in donor levels. In the limit of low temperatures ($kT \ll E_c - E_d$), n is negligible and $n_d = N_d - N_a$, i.e., there are N_a empty donors because electrons on the donors can lower their energy by going into the empty acceptor levels. For a moderately doped sample with $N_a < N_d/2$, a typical situation, the majority of the donor levels are filled. Consequently the probability of an electron having energy E_d, $f(E_d)$, must be not much less than unity and, by Eq. (8), E_F must be within a few kT of E_d. With E_F close to E_d, the number of electrons in the conduction band, according to Eq. (13), is given by

$$n \simeq N_c \exp[-(E_c - E_d)/kT] \quad \text{for } kT \ll E_c - E_d. \qquad (17)$$

Note that relation (17) is frequently used for the experimental determination of $E_c - E_d$, or the activation energy. It is valid more generally than for the model of Fig. 5, requiring only that other donor and acceptor levels be more than several kT below E_d. What is required for its use is the observation of a range of temperatures in which ln n increases linearly with $1/T$.

When $E_c - E_d$ is small, as is true for many of the semiconductors of interest, the exponential increase in n with T predicted by Eq. (17) occurs over only a small temperature range. As the temperature is increased further, more electrons are excited to the conduction band. The conduction band is preferred over the donor levels in the competition for electrons because of its greater density of states. With more electrons in the conduction band E_F must move down; the total number of disposable electrons, $N_d - N_a$, is kept constant by reducing the number in the donor levels. There follows a temperature range in which, as E_F is moving down, there is substantially complete ionization of the donor levels, and $n \simeq N_d - N_a$. This temperature range starts as soon as E_F has fallen several kT below E_d, and is called the saturation range. With further increase in temperature E_F continues to drop until it reaches a position a few kT above midgap. When this occurs, holes begin to be produced, indicating the approach of intrinsic conductivity. Further increase in temperature brings E_F close to E_F of Eq. (16), thus essentially to the intrinsic condition.

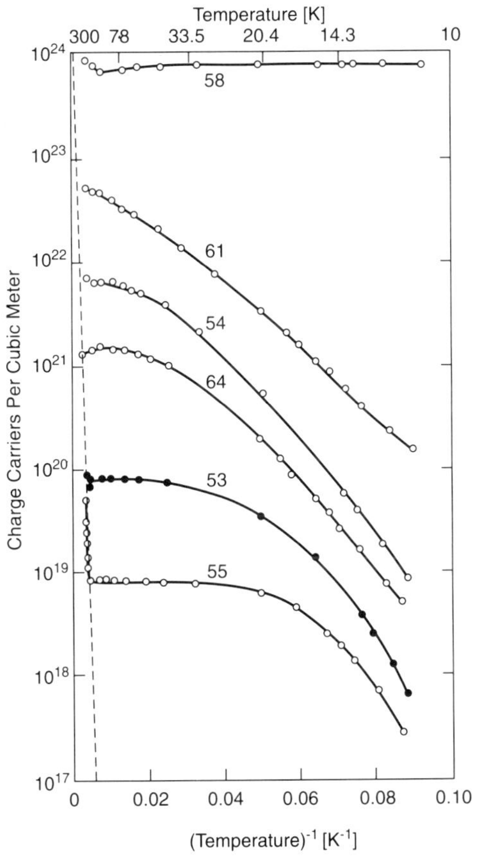

FIG. 6. Density of charge carriers vs the reciprocal of temperature for a set of As-doped Ge samples identified by numbers 55,53,... . The dashed line represents the density of intrinsic carriers. (Adapted from Debye and Conwell, 1954.)

Experimental data (obtained from Hall measurements) for n vs $1/T$ of a set of Ge samples doped with different amounts of As are shown in Fig. 6. In accordance with the above results it is seen that for all but the most heavily doped sample n is small and increases exponentially with $1/T$ at low temperatures. The increase becomes less rapid as T grows, and finally n reaches a saturation value, $N_d - N_a$, when essentially all the donors are ionized. With further increase in T the sample goes into the intrinsic range, n and p both increasing rapidly. For more heavily doped (or small-gap semiconductor) samples the intrinsic range may set in before saturation is achieved. The case of the most heavily doped sample, number 58, will be taken up in the next section.

A systematic procedure for finding E_F as a function of T is based on the requirement for electrical neutrality: the sum of the positive charges in the sample must equal the sum of the negative charges. The statement of this requirement for the situation of Fig. 5 is

$$p + p_a + N_d = n + n_d + N_a, \qquad (18)$$

where p_a is the number of holes in acceptor levels. From the definition of $f(E)$, $p_a = N_a[1 - f(E_a)]$ and $n_d = N_d f(E_d)$. With these expressions for p_a and N_a, and Eqs. (13) and (14), all quantities in Eq. (18) that depend on E_F are expressed in terms of E_F. Equation (18) may then be solved graphically for E_F for any set of N_d and N_a at any desired temperature (Shockley, 1950).

4.2.2 Heavily Doped Samples. For $N_d > 10^{24}$ m^{-3} it is found, as seen in the most heavily doped sample of Fig. 6, that $n \simeq N_d$ and essentially constant up to room temperature. (N_a should be negligible for such high doping.) This is the behavior of a metal. Material with such high doping is called degenerate because the carriers obey degenerate or Fermi–Dirac statistics, Eq. (8) rather than Eq. (12). The constancy of n with T means E_F must be constant with T in this case. This requires that E_F be at least several kT above the conduction-band edge. An exact description of the situation is not easily made because at such high concentration impurity levels overlap and mix with the conduction band. For further discussion see Sec. 8.

4.3 Nonequilibrium Carrier Concentration: Photoconductivity

4.3.1 Intrinsic Photoconductivity. In a semiconductor a photon of energy greater than the energy gap may raise an electron from the valence band to the conduction band, creating an electron and a hole (see Ryvkin, 1964). In some materials, particularly at low temperatures, the attraction of the electron and hole causes them to be bound to each other, the bound entity being called an exciton (see EXCITONS). In many materials, however, the electron and hole are free and may add to the conductivity. The number of electrons or holes generated per unit time per unit volume is given by the product of the number of photons (of energy greater than E_g) absorbed per unit time per unit volume and the probability β that an absorbed photon will create an electron–hole pair. β is called the quantum yield and in many cases has the value unity. For a thin sample and a wavelength of light not strongly absorbed, the number of photons absorbed per unit time per unit volume is the product of the number I incident on unit area of the surface per unit time and α, the fraction absorbed per unit distance the light travels in the sample. When the fraction of photons absorbed is not small it is necessary to allow for the decrease in the number as the light travels.

When a sample is illuminated with a light source of constant I it is found that the number of electrons and holes grows with time, at first rapidly and then more slowly, reaching a steady value or saturation. The rate of growth decreases because excess carriers tend to recombine. Saturation occurs when the recombination rate equals the generation rate. We can characterize the recombination rate by introducing average lifetimes τ_n and τ_p for electrons and holes, respectively. Using the generation rate $\beta\alpha I$ obtained above, we may write for the steady-state densities of electrons and holes $\beta\alpha I\tau_n$ and $\beta\alpha I\tau_p$, respectively. The lifetimes vary widely, typically ranging from 10^{-2} s to fractions of a microsecond.

Electrons and holes may recombine directly, i.e., an electron falls into a hole in the valence band, in which case $\tau_n = \tau_p$. The energy given up in this process, essentially E_g, may be in the form of heat (specifically phonons; see Sec. 6.1) or radiation. Radiative recombination is the process that results in light-emitting diodes (*q.v.*) or semiconductor lasers (see LASERS, SEMICONDUCTOR). The alternative to direct recombination is sequential capture of an electron and a hole, or vice versa, by a defect. Such defects are called recombination centers. When these centers mediate the recombination process, τ_n may be quite different from τ_p.

4.3.2 Extrinsic Photoconductivity. In a semiconductor with electrons in a donor level $E_c - E_d$ below the conduction band, a photon of energy $E_c - E_d$ or greater can raise electrons to the conduction band, resulting in a conductivity increase. Similarly, photons could excite electrons from the valence band into empty acceptor levels in the gap, resulting in an increase in hole concentration. The resulting increase in κ can be used for photodetection. (see PHOTODETECTOR DEVICES). The small amount of energy required to free an electron or a hole from a hydrogenic center in Ge or Si, for example, makes these and other semiconductors useful for infrared detectors.

5. FORMAL THEORY OF CONDUCTIVITY AND THE HALL EFFECT

5.1 Effects of Electric and Magnetic Fields on the Carriers

The force on an electron or hole due to an electric field **E** and a magnetic field **H** is given by

$$\mathbf{F}=e\mathbf{E}+e\mathbf{v}\times\mathbf{H}\,, \tag{19}$$

where e is negative for electrons, positive for holes, and $\mathbf{v}$ is the velocity of the electron or hole. The velocity of an electron or hole in a lattice is shown by band theory (see ELECTRONIC TRANSPORT IN SOLIDS) to be

$$\mathbf{v}=\nabla_p E=(1/\hbar)\nabla_k E\,. \tag{20}$$

For the simple model of band structure, with E given in Fig. 3, Eq. (20) leads to $\mathbf{v}=\mathbf{P}/m^*=\hbar\mathbf{k}/m^*$, the free-electron result with m_0 replaced by m^*. Thus $\mathbf{v}$ is parallel to $\mathbf{k}$. For the many-valley model that need not be the case. By definition of the ∇ operator, $\mathbf{v}$ must be perpendicular to the constant-energy surfaces. Only for $\mathbf{k}$ along one of the ellipsoid axes will $\mathbf{v}$ be parallel to $\mathbf{k}$.

As is expected by analogy with Newton's second law, the effect of a force on an electron or hole in a lattice is to change its momentum, i.e.,

$$\mathbf{F}=\frac{d\mathbf{P}}{dt}=\hbar\,\frac{d\mathbf{k}}{dt}\,. \tag{21}$$

5.2 The Relaxation Time

In the absence of an electric field the carriers are moving with random thermal velocities, but of course the net velocity vanishes when the velocities of all the carriers are summed up. In the presence of the field each carrier experiences the force $\mathbf{E}e$, which accelerates it and increases its velocity in the direction of the force. If the lattice were perfect the increase would continue indefinitely and the conductivity would be infinite. Of course the lattice is not perfect and the acceleration is interrupted on the average after a time τ when the carrier is scattered by, or collides with, a defect. We will assume, as turns out to be the case for most types of defect, that after each collision, i.e., after the time τ, the carrier has forgotten the acceleration it gained from the field and the acceleration process starts anew. We may therefore take the rate of loss of momentum due to collisions to be

$$\frac{d\mathbf{P}}{dt}=-\frac{\mathbf{P}}{\tau}\,. \tag{22}$$

Integration of this equation shows that excess momentum decays exponentially as $\exp(-t/\tau)$. τ is therefore called the momentum relaxation time. In the remainder of this section we will derive expressions for the conductivity, Hall effect, and magnetoresistance in terms of τ. Section 6 will be devoted to calculating τ for the most important scatterers: lattice atoms out of place as a result of their thermal vibrations, and foreign ions or atoms.

5.3 Calculation of dc Conductivity

Combining Eqs. (19) and (22) we obtain for the rate of momentum change of a single carrier

$$\frac{d\mathbf{P}}{dt}=e\mathbf{E}+e\mathbf{v}\times\mathbf{H}-\frac{\mathbf{P}}{\tau}\,. \tag{23}$$

Let us assume for the moment that all carriers have the same τ, i.e., that τ is not a function of the energy of the carrier. We may then sum Eq. (23) over all the carriers and obtain, for the simple model of the band structure,

$$\frac{d(m^*\mathbf{v}_d)}{dt}=e\mathbf{E}+e\mathbf{v}_d\times\mathbf{H}-\frac{m^*\mathbf{v}_d}{\tau}\,, \tag{24}$$

$\mathbf{v}_d$ being the drift velocity. For the steady state in a constant dc field $d(m^*\mathbf{v}_d)/dt=\mathbf{0}$. In the absence of a magnetic field Eq. (24) then gives

$$\mathbf{v}_d=(e/m^*)\tau\mathbf{E}\,. \tag{25}$$

With this equation and the definition, Eq. (2), of the mobility μ, we find for the simple model

$$\mu=(e/m^*)\tau\,, \tag{26}$$

and the dc conductivity is given by

$$\kappa_0=(ne^2/m^*)\tau\,. \tag{27}$$

Allowing for the variation of τ with energy E, we find with a more accurate theory (see, for example, Herring, 1955, or Conwell, 1982) that τ in Eqs. (25)–(27) is replaced by the weighted average $\langle E\tau\rangle/\langle E\rangle$.

For the many-valley model μ is a tensor. But the semiconductor crystals of greatest interest are cubic, and for low electric fields the mobility must be the same in all directions. To obtain the correct μ we must therefore average over the different valleys. The result is to replace $1/m^*$ in Eqs. (25)–(27) by an average inertial mass

$$1/m^{(I)}=\tfrac{1}{3}\,(2/m_t+1/m_l) \tag{28}$$

when the constant-energy surfaces are ellipsoids of revolution.

5.4 ac Conductivity

Consider now the case where the electric field applied to the sample is oscillatory, i.e., $\mathbf{E}=\mathbf{E}_0\, e^{i\omega t}$, where ω is the angular frequency. Such a field sets up an oscillatory current with the same frequency and thus a drift velocity $\mathbf{v}_d=\mathbf{v}_d^0\, e^{i\omega t}$. When these expressions for $\mathbf{E}$ and $\mathbf{v}_d$ are inserted into Eq. (24) with $\mathbf{H}=\mathbf{0}$, and the solution rationalized, we find the ac conductivity

$$\kappa(\omega)=\frac{ne^2\tau/m^*}{1+\omega^2\tau^2}-i\omega\,\frac{ne^2\tau^2/m^*}{1+\omega^2\tau^2}\,. \qquad (29)$$

The conductivity being complex reflects the fact that, at all frequencies for which $\omega\tau$ is not negligible compared to unity, the carriers can no longer follow the field. The current then consists of a part in phase with the applied field, the first term, and a part 90° out of phase, the second term. Comparing with Eq. (27) we see that the in-phase conductivity is κ_0 reduced by the factor $(1+\omega^2\tau^2)^{-1}$. By inserting the current $\kappa(\omega)\mathbf{E}_0e^{i\omega t}$ into Maxwell's equations it is found that $1/\omega$ times the imaginary part of $\kappa(\omega)$ represents a contribution to the permittivity ε. This free-carrier contribution to ε is negative because it is 180° out of phase with the displacement current. For n and τ large and $\omega\tau$ not too large, this contribution may be large enough to make ε negative. Again, when τ is allowed to be a function of E the quantities involving τ must be replaced by averages. For the many-valley model $1/m^*$ is replaced by $1/m^{(I)}$ defined in Eq. (28).

5.5 The Hall Effect

5.5.1 Low Magnetic Fields. From Eq. (24), using Eqs. (1) and (27), we find for the steady state of electrons with the simple model of the band structure in a magnetic field

$$\mathbf{E}=(\mathbf{j}/\kappa_0)-(1/ne)\mathbf{j}\times\mathbf{H}\,. \qquad (30)$$

If we assume the conditions of a Hall-effect experiment—$\mathbf{H}=H_z$, applied field $\mathbf{E_x}$, and current flowing only in the x direction—we find from Eq. (30) to terms linear in H that $E_x=j_x/\kappa_0$ and

$$E_y=(1/ne)j_xH_z,\quad \tau\neq\tau(E)\,. \qquad (31)$$

According to Eq. (31) the Hall constant R_{H}, defined in Sec. 1.5 as E_y/j_xH_z, is

$$R_{\mathrm{H}}=-1/n|e| \text{ for electrons},\quad \tau\neq\tau(E)\,; \qquad (32)$$

$$R_{\mathrm{H}}=1/p|e| \text{ for holes},\quad \tau\neq\tau(E)\,. \qquad (33)$$

If we allow τ to be a function of energy, the quantity $R_{\mathrm{H}}ne$ or $R_{\mathrm{H}}pe$ is no longer unity but is given by μ_{H}/μ, according to Sec. 1.5. Calculations (for example, Herring, 1955, or Conwell, 1982) lead to

$$\frac{\mu_{\mathrm{H}}}{\mu}=\frac{\langle E\tau^2\rangle\langle E\rangle}{\langle E\tau\rangle^2}\,B\,, \qquad (34)$$

where B depends on the shape of the constant-energy surfaces. Such a dependence is to be expected because a magnetic field, which cannot change the energy of a carrier, causes the carriers to rotate around the constant-energy surfaces. $B=1$ for spherical constant-energy surfaces and is generally of the order of unity. The other factor in μ_{H}/μ will be seen to lie between 1 and 2 for the important scattering mechanisms.

5.5.2 The Hall Effect for Two Types of Carriers. There are two situations in which we must deal with more than one type of carrier. In intrinsic or nearly intrinsic samples both electrons and holes are present. In a semiconductor with degenerate bands we must deal with sets of electrons or holes with different effective masses or, more generally, different effective-mass tensors. It is usually a good approximation to treat the different types as independent. When an electric field is applied without a magnetic field, j and κ are given by the sum of the contributions of each kind of carrier regardless of sign. When a magnetic field is applied in addition to the electric field, this is no longer the case. Holes drifting in an electric field are pushed by a magnetic field in the same direction as electrons because both v_d and e are opposite in sign to those of electrons. The resulting Hall field is then the difference of the two fields.

Quite generally, for two types of carriers, denoted by subscripts 1 and 2, it is found that

$$R_{\mathrm{H}}=\frac{n_1e_1\mu_1\mu_{\mathrm{H}1}+n_2e_2\mu_2\mu_{\mathrm{H}2}}{(n_1|e_1|\mu_1+n_2|e_2|\mu_2)^2}\,. \qquad (35)$$

If the two carriers are opposite in sign, the Hall voltage will go through zero when

$$n_1\mu_1\mu_{H1}=n_2\mu_2\mu_{H2} \ . \tag{36}$$

For the semiconductors we are discussing this will happen when p is somewhat greater than n because μ_n is generally larger than μ_p. If there are two carriers present with the same sign, as is generally the case for p-type column IV and III-V semiconductors, the ratio of R_H to the total carrier concentration may be increased by as much as a factor of 2 because μ of the light holes is much greater than μ of the heavy holes.

5.5.3 High Magnetic Fields. In the limit of high magnetic fields, although not so large that quantum effects occur, it is found theoretically that R_H goes to the value $1/ne$ or $1/pe$. Thus μ_H/μ is unity at high fields independent of the details of band structure and energy dependence of τ. For a sample with both electrons and holes, R_H in the high-field limit goes to $1/(n-p)e$.

As will be seen in Sec. 5.6, a high magnetic field is specified by the condition $\mu H \gg 1$. Thus, for example, the high-field condition is satisfied at lower values of **H** for light holes than for heavy holes. This can explain the fact that R_H for p-Ge, measured at a typical field (3000 Oe), was observed to decrease by a factor of 1.6 as T was decreased from 300 to 78 K. Clearly this decrease could not represent an increase in p with decreasing T. Nor, as will be seen, could it be accounted for by a change in $\tau(E)$. The explanation is that, with decreasing T, μ increases for both light and heavy holes. The light holes, with their higher μ, go into the high-**H** regime at a higher temperature than the heavy holes, decreasing their value of μ_H. As can be seen from Eq. (35) this decreases R_H considerably.

It can be concluded from this section that the carrier concentration deduced from measurements of the Hall effect is uncertain within about a factor of 2 unless one knows a great deal about the carrier mobilities and the detailed scattering mechanisms.

5.6 Magnetoresistance

It is readily seen from Eq. (30) that for the simple model with $\tau \neq \tau(E)$ there is no change in resistance in a magnetic field, i.e., no magnetoresistance. The resistivity may be calculated from the component of **E** along **j** divided by **j**, i.e., $\rho=\mathbf{j}\cdot\mathbf{E}/|j|^2$. According to Eq. (30) $\rho=1/\kappa_0$, independent of **H**. Magnetoresistance is a result of $\tau=\tau(E)$ or of nonspherical constant-energy surfaces.

To understand the situation for $\tau=\tau(E)$ we solve Eq. (30) for **j**, with the result

$$\mathbf{j}=\frac{\kappa_0\mathbf{E}}{1+\mu^2H^2}+\frac{\kappa_0\mu(\mathbf{E}\times\mathbf{H})}{1+\mu^2H^2} \ . \tag{37}$$

We have simplified the situation by assuming that $\mathbf{E} \perp \mathbf{H}$. For $\mathbf{E}=E_x$, $\mathbf{H}=H_z$, the first term on the right-hand side of Eq. (37) represents j_x, the second j_y. It is seen that

$$j_y=\mu H j_x=(He/m^*)\tau j_x \ . \tag{38}$$

This equation expresses the fact that j_y is produced by rotation of j_x in the magnetic field. The quantity He/m^*, called the cyclotron frequency, is the rate at which carriers rotate around the constant-energy surfaces in a magnetic field, while τ is the mean time to undergo a collision, which redirects the carrier and effectively interrupts the rotation. Because carriers with different τ have different j_y, it is not possible to find an electric field that will keep currents due to carriers with different τ's in the same direction. The net current is therefore reduced, making the apparent resistance of the medium larger.

Because the magnetic field dependence of **j** occurs in the form μH, a small magnetic field is defined by $\mu H \ll 1$. From Eq. (37) it is seen that for small μH, where $(1+\mu^2H^2)^{-1}$ may be expanded as a power series in μ^2H^2, $\Delta\rho$ increases as H^2. It is also seen that, in the limit $\mu H \gg 1$, ρ does not keep on increasing, but saturates.

For ellipsoidal constant-energy surfaces the magnetoresistance is a function of the orientation and shape of the constant-energy ellipsoids. As a result, measurements of magnetoresistance may make it possible to determine the location and parameters of the ellipsoids in a many-valley structure. For detailed calculations of magnetoresistance for the many-valley model see Herring (1955).

5.7 Thermoelectric Effect

A temperature gradient across a sample causes the Fermi energy to vary with position. This can give rise to a current even in the absence of an applied field. In an isolated sample the flow of charge results in an electric field that prevents further charge from flowing. Thermoelectric power, or thermopower for short, is defined as the ratio of the field developed to the temperature gradient. The thermopower Q is related to the Peltier coefficient Π, defined as the energy flux, measured relative to E_F, per unit current. The relation is $Q=\Pi/T$. It follows from this relation that, when one type of carrier is present,

$$Q=\mp\frac{k}{|e|}\left(\frac{|E_F-E_b|}{kT}+\frac{|\Delta E_T|}{kT}\right), \quad (39)$$

where the upper sign is for electrons, the lower for holes, $|E_F-E_b|$ is the distance of E_F from the conduction- or valence-band edge, and $|\Delta E_T|$ is a weighted average of the energy transported per carrier (across unit area in unit time) relative to the band edge. Specifically $\Delta E_T=\langle v^2(E-E_b)\tau\rangle/\langle v^2\tau\rangle$. The dependence of Q on the sign of the carriers means that, like the Hall effect, it can be used to determine whether the dominant carriers are electrons or holes. For a moderately doped semiconductor sample, Q varies a great deal with T because, as we have seen, E_F does. Q can be expressed in terms of the carrier concentration by using relation (13) or (14) to express E_F in terms of n or p, respectively. Other things being equal, Q is smaller for more heavily doped samples. In samples where electrons and holes are present in comparable numbers we have

$$Q=(Q_n\kappa_n+Q_p\kappa_p)/(\kappa_n+\kappa_p)\ . \quad (40)$$

5.8 Galvanomagnetic and Thermomagnetic Effects

These effects occur when a conductor carrying an electric current or thermal current, which may arise as a result of a thermal gradient, is placed in a magnetic field. In the isothermal situation discussed earlier in this section, introduction of a transverse magnetic field was found to give rise to the Hall effect and magnetoresistance. When thermal gradients are permitted, many additional galvanomagnetic effects occur. One of these is the Ettingshausen effect, the appearance in a transverse magnetic field of a temperature gradient perpendicular to both **H** and the current direction. Another is the Nernst effect in which, due to the transverse magnetic field, a temperature gradient appears parallel to the current direction. This by no means exhausts the possible galvanomagnetic effects. Among the thermomagnetic effects are magnetic field effects on thermal conductivity (analogous to magnetoresistance) and thermopower. The Righi–Leduc effect, a thermal analog of the Hall effect, is the appearance in a transverse magnetic field of a thermal gradient perpendicular to both **H** and the thermal current direction. In the Ettingshausen–Nernst effect the combination of thermal current and transverse magnetic field produces an electric field perpendicular to both. Many additional thermomagnetic effects could be specified. For a systematic listing of these effects see, for example, Conwell (1982).

6. RELAXATION TIMES AND MOBILITIES

Mobility in semiconductors is found to vary greatly with temperature and doping. In Fig. 7 are shown Hall mobility data versus temperature for the set of Ge samples whose carrier concentration variation with T was shown in Fig. 6. The least doped, thus purest, sample has the highest μ_H. Comparing it with the dashed line indicating a $T^{-3/2}$ dependence, we see that above ~100 K, μ_H for this

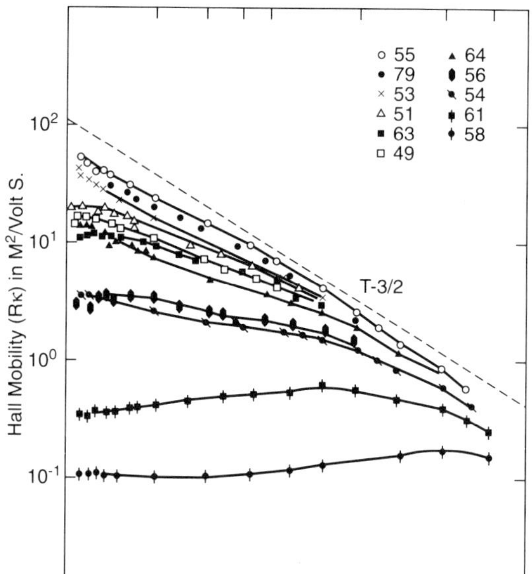

FIG. 7. Hall mobility vs temperature for the set of As-doped Ge samples for which n vs T is shown in Fig. 6. (After Debye and Conwell, 1954.)

sample decreases more rapidly with increasing T, while below ~100 K, the decrease is less rapid than $T^{-3/2}$. It will be shown shortly that the $T^{-3/2}$ dependence arises from scattering due to one kind of thermal vibration, the so-called acoustic modes. The steeper slope above 100 K indicates that another type of scattering, which can be attributed to another type of thermal vibration, is also operative. The progressively smaller mobility, and the less steep decrease with increasing temperature, seen with increasing doping will be shown to result from scattering by the doping ions.

In this section we discuss the momentum relaxation time τ defined in Eq. (22) for the most important scattering mechanisms. Also mobilities are given for these scattering mechanisms. At the end of the section we will present a more detailed comparison with experimental data.

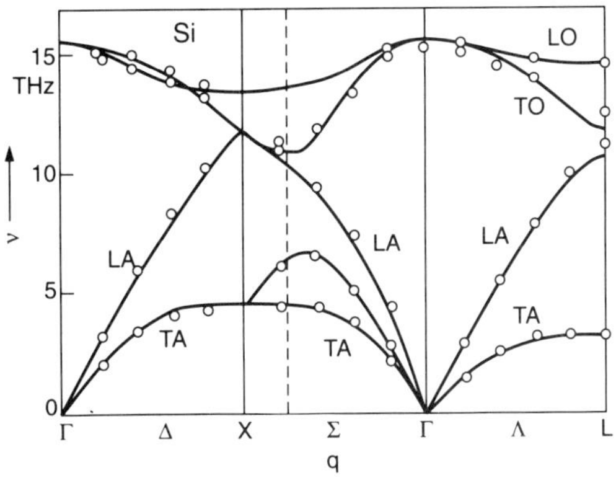

FIG. 8. Frequency ν vs wave vector **q** of normal modes of lattice vibration in Si for various directions in the Brillouin zone. (Adapted from Madelung, 1991.) See Fig. 2 for identification of the directions.

6.1 Scattering Due to Lattice Vibrations

6.1.1 Acoustic and Optical Modes. The random thermal motion of the lattice atoms can be described as the superposition of a set of normal modes, each mode describing the displacement of every atom in the sample in terms of a propagating wave of a single frequency (see PHONONS IN CRYSTAL LATTICES). Because of the periodicity of the lattice, there is associated with each wave, as with each electron wave function, a wave vector in the direction of propagation. This wave vector will be denoted by **q** to distinguish it from the electron wave vector **k**. By the same reasoning as employed earlier for the **k**'s, the **q**'s have a closely spaced set of discrete values and occupy the first Brillouin zone. Frequency ν of the normal modes in Si versus wave vector is shown in Fig. 8. It is seen that in most directions there are four branches. These are labeled L for longitudinal and T for transverse, indicating that the vibrations are parallel to **q** and perpendicular to **q**, respectively. Actually the vibrations are not purely longitudinal or transverse in an actual crystal, but the difference is usually neglected. It is also usual, at least in conductivity theory, to neglect the differences between ν vs q for a given branch in different directions.

The branches that start at $q=0$ (the Γ point) are called acoustic branches. The low-frequency modes of the LA branch are familiar as sound waves. The other branch is called transverse acoustic (TA). In the low-frequency acoustic vibrations the two atoms in the unit cell move almost together, i.e., with only a small phase difference between them. Another type of long-wavelength vibration is possible: the two atoms in the unit cell move opposite to each other, with the same motion repeated in the next cell slightly earlier or later depending on the propagation direction of the wave, i.e., in a slightly different phase. These vibrations can also be both longitudinal and transverse. They are called optical modes (TO or LO) because, in a lattice with ions of opposite sign, they produce a polarization that can interact with light waves.

For the long-wavelength acoustic waves ν and the wavelength λ are related according to $\nu\lambda=\text{const}$. This constant is identified as the phase velocity, which for the LA branch is the sound velocity. In contrast to the acoustic branches, for the two optical branches ν is quite large for long wavelengths, i.e., close to $q=0$. The reason is that the opposing motion of the two atoms in the unit cell gives rise to a large restoring force and therefore a high frequency of vibration even for very large wavelengths. This has the further consequence that ν is not strongly dependent on λ, i.e., the optical branches are relatively flat, because λ only determines the relative phase of adjacent unit cells.

The vibrational energies of each normal mode can only have certain discrete values differing by amounts $h\nu$. The energy steps or quanta are called phonons, as suggested by the analogy with light quanta or photons. For a crystal in thermal equilibrium the average number of phonons n_q in a normal mode having frequency ν_q is given by Planck's law:

$$n_q = 1/(e^{h\nu_q/kT} - 1) \ . \tag{40}$$

Thus around room temperature and below, the average number of optical phonons will be small because $h\nu$ for these phonons is larger than kT at room temperature.

6.1.2 Mobility Due to Acoustic-Mode Scattering. In the process of being scattered by the thermal vibrations, a carrier must emit or absorb a phonon. If the carrier wave vector changes from $\mathbf{k}$ to $\mathbf{k}'$ in the process, while the phonon wave vector is $\mathbf{q}$, conservation of momentum requires $\mathbf{k}' = \mathbf{k} \pm \mathbf{q}$. Energy must also be conserved, requiring $E_{\mathbf{k}'} = E_{\mathbf{k}} \pm h\nu_{\mathbf{q}}$, $h\nu_{\mathbf{q}}$ being the energy of the phonon with wave vector $\mathbf{q}$. The upper sign is required for absorption and the lower for emission. The probability of absorption or emission of a phonon, thus of scattering, is proportional to the number of phonons, given by Eq. (40). Let us consider first the case of long-λ, or low-q, acoustic phonons. For these phonons, at all but extremely low temperatures, $h\nu_q$ is small compared to kT; Eq. (40) then gives $n_q \propto kT/h\nu_{\mathbf{q}}$. The small energy of the phonons, coupled with the conservation of momentum and energy conditions, results in these acoustic phonons causing scattering only within a valley.

The probability of scattering by any given vibrational mode also depends on how strongly the carrier is coupled to that mode, which in turn depends on the strain caused by that mode. For acoustic waves the coupling is described by a deformation potential E_1, which is the shift in the band-edge energy per unit strain (for a cubic crystal the strain is the fractional change in volume of a small region of the crystal) due to the wave. Combining the various factors described above, we find for the relaxation time τ_{ac} due to long-λ acoustic-mode scattering of a carrier with the simple model of band structure

$$\tau_{ac} = \frac{\pi \hbar^4 \rho u_l^2}{2^{1/2} E_1^2 (m^*)^{3/2} kT} E^{-1/2} \ , \tag{41}$$

where ρ is the density of the material and u_l the sound velocity. (Note that TA waves do not scatter in this case.) It is seen that τ_{ac} depends inversely on $E^{1/2}$, or the velocity of the carrier, resulting in slow carriers having longer relaxation times or being scattered less.

Inserting Eq. (41) in Eq. (26), allowing for the averaging over carrier energy, thus over the Maxwell–Boltzmann distribution in this case, we obtain the mobility for long-λ acoustic-mode scattering:

$$\mu_{ac} = \frac{2(2\pi)^{1/2} e \hbar^4 \rho u_l^2}{3 E_1^2 (m^*)^{5/2} (kT)^{3/2}} \ . \tag{42}$$

Thus μ_{ac} decreases with increasing T as $T^{-3/2}$, $T^{-1/2}$ coming from the fact that $\langle E^{-1/2} \rangle \propto (kT)^{-1/2}$ for the Maxwell–Boltzmann distribution. The calculation of τ_{ac} and μ_{ac} for the many-valley model is more complicated because τ may not be constant over a constant-energy surface, and shear strains may also be important (see Herring and Vogt, 1956). However, the dependence on carrier energy and T remain the same.

In III-V and II-VI semiconductors, because they are somewhat polar and lack inversion symmetry, the strain of an acoustic wave produces also an electric polarization (see DIELECTRIC PROPERTIES OF INSULATORS). This causes scattering in addition to the scattering studied above, which may be called deformation-potential scattering. The additional scattering is called piezoelectric or piezoacoustic scattering. In contrast to τ_{ac} of Eq. (41), the relaxation time τ_{pe} for piezoelectric scattering is directly proportional to $E^{1/2}$. As a result, except for strongly piezoelectric material such as CdS or ZnO, it is expected to be important only for low-energy carriers, and thus for low temperatures only. It will be seen that, except for very pure samples, impurity scattering is the dominant scattering mechanism at low temperatures. Thus, in materials that are not strongly piezoelectric, piezoelectric scattering need only be considered at low temperatures in pure samples. For further discussion of piezoelectric scattering see Zawadzki (1982) and references therein.

6.1.3 Mobility Due to Nonpolar Optical-Mode Scattering. Optical-mode scattering in nonpolar crystals may be treated quite similarly to acoustic-mode scattering. For the coupling to the optical modes it is necessary to introduce a different deformation potential because the strain, in contrast to that due to an acoustic phonon, is not simply a change in linear dimensions or volume. In the case of a zinc-blende or diamond structure the strain is a displacement of the two sublattices, based on the two atoms in a unit cell, with respect to each other. The appropriate deformation potential is D, the shift of the band edge per unit relative displacement of the two sublattices. Another difference between optical- and acoustic-mode scattering is that, because of the larger size of the optical phonons, the exponential in Eq. (40) may not be expanded; n_q must be taken from Eq. (40). The relaxation time for optical-phonon scattering (all phonons assumed to have the same frequency ν_0) is given by

$$\tau_{op}=\frac{2^{3/2}\pi^2\hbar^3\rho\nu_0}{D^2(m^*)^{3/2}}\,[n_q(E+h\nu_0)^{1/2} + (n_q+1)(E-h\nu_0)^{1/2}]^{-1}\,. \tag{43}$$

The first term in the square brackets represents relaxation due to phonon absorption and the second relaxation due to phonon emission. The emission probability is proportional to n_q+1, n_q representing stimulated emission and 1 spontaneous emission. Emission is only possible, however, if $E>h\nu_0$. If $h\nu_0$ were negligible compared to E, as is the case for long-λ acoustic phonons, n_q and n_q+1 could be replaced by $kT/h\nu_q$ and τ_{op} would have the same dependence on E and T as τ_{ac}. Generalizing τ_{op} to the many-valley model is simple; it requires only replacing m^* by the geometric mean mass $(m_l m_t^2)^{1/3}$.

The mobility μ_{op} obtained after averaging τ_{op} over energy is usually larger than μ_{ac} because of the relatively small number of optical phonons. This same feature, however, results in μ_{op} decreasing more rapidly with increasing T as n_q increases, according to Eq. (40).

6.1.4 Mobility Due to Intervalley Scattering. Because of conservation of momentum, the **q** of a phonon that causes an intervalley transition will be approximately equal to the difference between the **k** vectors of the valley minima. This difference is usually of the order of the radius of the Brillouin zone, which means that the energy of the phonon required is comparable to the optical-phonon energy. The intervalley scattering may then be treated much like the optical-phonon scattering, leading to an expression for the intervalley scattering relaxation time τ_{iv} of the same form as τ_{op}. Of course, the deformation potential is different for this case, and high-energy acoustic phonons as well as optical phonons may be involved in the transition. Note that the energy in Eq. (41) or (43) must be measured from the bottom of the valley into which the carrier is scattered, which need not be the same energy as the bottom of the valley in which it originates. Also m^* of Eq. (42) must be replaced by the geometric mean mass.

6.1.5 Mobility Due to Polar Optical-Mode Scattering. Nonpolar optical scattering, discussed in Sec. 6.1.3, may exist in polar as well as nonpolar crystals. In polar crystals, however, the polarization of the crystal arising from the vibrations of the LO branch results in an additional interaction with the carriers, or scattering, the polar optical scattering. In the III-V semiconductors of most interest, GaAs and InSb, the polar optical scattering dominates over nonpolar scattering for electrons in the conduction band because the latter vanishes for spherically symmetric wave functions, which is more or less the situation for the conduction-band minimum at the Γ point. Polar optical-mode scattering is much more difficult to treat theoretically than nonpolar. A relaxation time does not exist unless $E\gg h\nu_0$, which requires temperatures well above room temperature. It is possible to get a qualitative idea of what is important by considering the mean free time between collisions, τ_{po}. For the simple model

$$\tau_{po}=(2m^*E)^{1/2}/2eE_0\{n_q\sinh^{-1}(E/h\nu_0)^{1/2} + (n_q+1)\sinh^{-1}[(E-h\nu_0)/h\nu_0]^{1/2}\}, \tag{44}$$

where $E_0=m^*e^2h\nu_0(1/\epsilon_\infty-1/\epsilon_0)/\hbar^2$, ϵ_∞ and ϵ_0 being the high-frequency and static permittivities, respectively. Note that Eq. (44) applies only for low carrier concentration because it neglects the screening of the electric fields due to the free carriers, which decreases the scattering effect.

Equation (44) may be used to predict the temperature dependence for polar optical scattering. For $\langle E\rangle\ll h\nu_0$ the energy dependence cancels out and the temperature dependence of μ_{po} is that due to $1/n_q$. In the other limit, $\langle E\rangle\gg h\nu_0$, the E dependence arising from the

$\sinh^{-1}$ terms becomes logarithmic and may be neglected. The combination of $E^{1/2}$ and n_q^{-1} dependence in this limit yields $\mu_{po} \propto T^{-1/2}$. For a detailed treatment of polar optical scattering see Zawadzki (1982) and references therein.

6.2 Impurity Scattering

The donors and acceptors that have given up an electron to the conduction band, or a hole to the valence band, are left charged. The charge results in an attraction or repulsion of the carriers, depending on the sign of the charge. In either case the carriers are deflected, thus scattered. There have been many calculations of the relaxation time for scattering by impurity ions, τ_I, varying in degree of sophistication and in which refinement of the process they concentrated on. A treatment that gives reasonably good agreement with experiment for not too heavily doped (nondegenerate) samples leads to

$$\tau_I = \frac{2^{1/2}}{\pi} \frac{(m^*)^{1/2}\epsilon^2}{e^4 N_I} \frac{E^{3/2}}{\ln(1+\xi) - \xi/(1+\xi)}, \tag{45}$$

where, for a sample with carrier concentration n, $\xi = (8\pi\epsilon m^* kT/ne^2h^2)E$, N_I being the number of ionized impurities per unit volume. The quantity ξ arises from the screening of the ions by the free carriers. For degenerate samples ξ is independent of T, although it still depends on E. Unless the sample has almost as many acceptors as donors (the reverse for a p-type sample), n may be taken equal to N_I. For moderately doped samples around room temperature $\xi \gg 1$, and Eq. (45) predicts that τ_I decreases as N_I increases, and that slow carriers are deflected or scattered more than fast carriers, both of these effects being expected. As the temperature decreases the terms involving ξ have more effect and τ_I increases less rapidly with E. The expression given above for ξ no longer represents the screening properly when many of the carriers have dropped out of the conduction band (or valence band), and n must be replaced by a quantity involving both the donor and acceptor concentrations. See, for example, Chattopadhyay and Queisser (1981), where also many of the improvements on the theory presented above are discussed.

Carrying out the averaging over carrier energy, we obtain from Eq. (45) the impurity mobility

$$\mu_I = \frac{2^{7/2}}{\pi^{3/2}} \frac{\epsilon^2 (kT)^{3/2}}{(m^*)^{1/2} e^3 N_I} \frac{1}{\ln(1+\bar{\xi}) - \bar{\xi}/(1+\bar{\xi})}, \tag{46}$$

where $\bar{\xi} = 24\pi\epsilon m^*(kT)^2/ne^2h^2$. In this form μ_I is valid for nondegenerate samples at temperatures where most of the carriers are still free. It increases with T less rapidly than $T^{3/2}$ because of the terms in $\bar{\xi}$. For degenerate samples $E \simeq E_F$, independent of temperature, and with ξ independent of T, as pointed out earlier, μ_I is independent of T. This explains the fact that the Hall mobility of sample 58 in Fig. 8, which is both degenerate and so heavily doped that impurity scattering dominates, is essentially independent of T.

6.3 Neutral-Impurity Scattering

As the temperature falls carriers return to the donors or acceptors and the number of neutral impurities increases. When neutral the impurities still scatter carriers. Because of the analogy, discussed in Sec. 3, between a neutral impurity and a hydrogen atom, the scattering has been treated similarly to scattering of a free electron by hydrogen (see ELECTRON SCATTERING BY ATOMS AND MOLECULES). The result is that at low temperatures ($T < 20$ K for Si and < 4 K for Ge) the relaxation time τ_N for scattering by a concentration N_N of neutral impurities is given by

$$\tau_N = (m^*/2)^{1/2} \pi (a_H^*)^2 N_N \times [(E/E_b^* + 0.0275)^{-1} + 10]^{-1} E^{-1/2}, \tag{47}$$

where a_H^* and E_b^* are defined in Eqs. (6) and (7), respectively (Zawadzki, 1982). Modifications for the many-valley model do not change τ_N greatly. Experimental verification of Eq. (47) is difficult because there are few free electrons at such low temperatures, and the effects of other scattering mechanisms must be subtracted. Nevertheless there are low-temperature data for Ge that indicate that μ_N

varies as $T^{-1/2}$, as predicted by Eq. (47). At higher temperatures μ_N is predicted to go to a constant value (Zawadzki, 1982).

6.4 Combination of Mobilities

In general, at any given temperature, more than one scattering mechanism is operative. To find the mobility one must first calculate an overall relaxation time τ by combining the τ's for the individual scattering mechanisms (which are essentially probabilities of scattering per unit time) according to $1/\tau = \Sigma_i(1/\tau_i)$. The resulting τ must then be averaged over energy to obtain μ. It is not unusual for this procedure to be simplified by calculating μ's for each scattering process and combining the reciprocals of these μ's to obtain the reciprocal of the overall μ. For the combination of acoustic-mode and ionized-impurity scattering the error due to this simplified procedure may be as much as 30%.

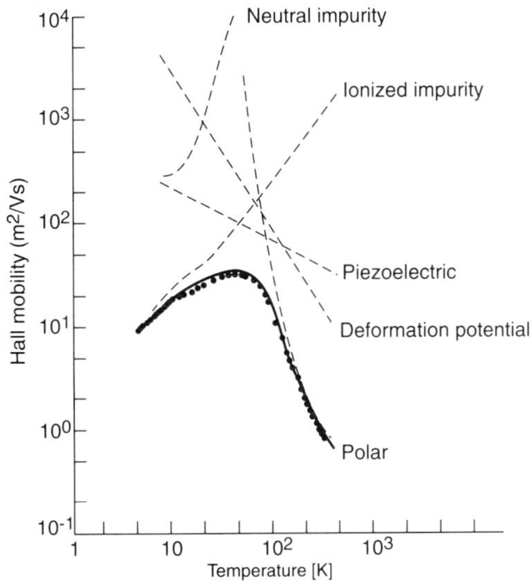

FIG. 9. Dots: The measured mobility of electrons in a sample of *n*-GaAs vs temperature. Dashed lines: Theoretical mobilities calculated for the various scattering process indicated. Solid line: The combined mobility calculated from all the scattering processes. (After Fletcher and Butcher, 1972.)

6.5 Comparison with Experimental Data

For well-characterized semiconductors, such as Si, Ge, and GaAs, the foregoing theory, with some refinements, agrees well with experiment for moderately doped and intrinsic samples. For *n*-Si around room temperature in very pure samples $\mu \propto T^{-2.42}$, resulting from an admixture of μ_{ac} ($\propto T^{-1.5}$) and μ_{iv} due to scattering between the [100] valleys. The latter has a much steeper T dependence arising from the large energy of the phonons required. Below 50 K the energy required for the intervalley process is too high to allow this process to occur, and in pure samples $\mu = \mu_{ac}$. For pure samples of *p*-Si and *p*-Ge around room temperature $\mu \propto T^{-2.33}$ from a mixture of μ_{ac} and nonpolar optical scattering. Plots of experimental and theoretical μ vs T for the cases just cited and for other intrinsic semiconductors may be found in Madelung (1991). Comparison of theory with experiment for a moderately doped *n*-type GaAs sample is shown in Fig. 9. It is seen that polar optical scattering dominates at high temperatures while impurity scattering dominates at low temperatures. In purer samples piezoelectric scattering may dominate at low temperatures.

7. HIGH-FIELD TRANSPORT

7.1 Hot Carriers

In this section we consider only nondegenerate samples, i.e., samples in which the carrier energies in the absence of an electric field are given by the Maxwell–Boltzmann distribution, Eq. (12). The effects to be described cannot be achieved with heavily doped samples. In the absence of an electric field the carriers interact with phonons, emitting and absorbing them, while the Maxwell–Boltzmann distribution of carrier energies and the Planck distribution of phonon numbers, Eq. (40), are maintained. Application of an electric field, which accelerates the carriers, increases their energy. This results in an increase in their rate of phonon emission, giving rise to the well-known Joule heating that accompanies a current. Achievement of a steady state, the situation for dc current, requires that the rate at which the electric field puts energy into the carriers equals the rate at which they lose energy through phonon emission. The rate of energy gain is $\mathbf{F} \cdot \mathbf{v}_d$, which, given Eqs. (19) and (2), equals $e\mu|\mathbf{E}|^2$. With $\langle dE/dt \rangle$ representing the average rate of energy loss, the steady-state condition may be written

$$e\mu|\mathbf{E}|^2 = \left\langle \frac{dE}{dt} \right\rangle . \quad (48)$$

In small electric fields, satisfying the condition (48) requires little change in the distribution, the essential change being the addition of a small drift term to the Maxwell–Boltzmann distribution. A high electric field may be defined as one for which this is no longer true. In semiconductors a steady state is always possible, but in high fields the drift term may not be small and the part of the distribution that is independent of the field direction no longer corresponds to an average energy of $\frac{3}{2}kT$, where T is the lattice temperature. It may still be the case that the latter part of the high-field distribution is reasonably described by a Maxwell–Boltzmann distribution, but the carrier temperature involved, denoted by T_e, is greater than the lattice temperature. For this reason the carriers are called "hot." Even when the high-field distribution is not Maxwellian it is usual to speak of a carrier temperature, which may be defined by equating the average carrier energy to $\frac{3}{2}kT_e$. It should be noted that to perform clear-cut experiments to measure hot-carrier effects it is necessary to apply the high electric field in pulses of low repetition rate to avoid raising the lattice temperature. When that is not possible it is necessary to consider the effects of a nonequilibrium phonon distribution along with the hot-electron effects.

From the preceding discussion it is clear that the higher the mobility of a material, the lower will be the magnitude of the electric field required to create hot carriers. High mobility means that the field can give the carriers energy at a high rate. It also suggests that the coupling to phonons is not strong, with the consequence that the rate of energy loss to phonons is not large.

Hot electrons may also be created by light. When a photon is absorbed by an electron, the electron energy is increased by the amount of the photon energy. Another method of creating hot electrons is through a potential step. A junction of two semiconductors with different band gaps (heterojunction) constitutes such a step. When an electric field is applied in the appropriate direction, electrons moving with thermal energy in the wide-band-gap region cross into the narrow–band-gap region. There they find themselves well above the conduction-band edge. In other words, the energy difference between the conduction-band edges in the two regions is converted into kinetic energy. In such a situation, before the carriers have undergone the collisions that set up a steady state their velocities exceed the steady-state drift velocity. This phenomenon is called velocity overshoot. It occurs also when, because of a sudden change in electric field intensity, the distribution of carriers has not yet reached its steady state. Velocity overshoot has been detected by electrical methods (Hess, 1990) and optical methods (Auston, 1990). It is of practical interest because it could be used to make faster devices. A potential step such as the heterojunction described above can also be used as a barrier, making possible the measurement of electron energies. This procedure could be called hot-electron spectroscopy (Hess, 1990).

In the remainder of this section we discuss some of the more striking hot-electron effects. Computer simulations of some of these effects are displayed in Hess (1990).

7.2 Mobility Changes in High Fields

Because the relaxation times for most scattering processes depend on carrier energy, as shown in Sec. 6, it is expected that the mobility will change in high electric fields. At fields where the carrier "temperature" is not much increased over the lattice temperature, called the warm-electron range, it is expected that μ in a field $\mathbf{E}$ depends on μ_O, the low-field (Ohmic) mobility, according to

$$\mu = \mu_O(1+\beta|\mathbf{E}|^2) , \quad (49)$$

where β is independent of field. The lowest-order deviation of μ from μ_O must be quadratic in field because μ should not change on reversal of the field duration. The behavior of μ predicted by relation (49) has been verified by many experiments, in which lattice temperature, impurity content, and crystal orientation have been varied (for data and references for the material in this section see Conwell, 1967).

We can predict how μ changes from the energy dependence of τ for the scattering mechanisms that operate. Thus, if acoustic or nonpolar optical or intervalley scattering is dominant, according to Eqs. (41) or (43) μ must decrease on electron heating. This has been verified for Ge and Si, with reasonably good agreement with theory. For n-Ge at 300 K the decrease in μ starts at $\sim 10^5$ V/m, while at 77 K, where the low-field mobility in highly purified samples is higher by an order of magnitude, it starts at $\sim 10^4$ V/m. For n-Si, with lower mobility, higher fields are required to reach the non-Ohmic region.

A by-product of the mobility changes in the many-valley model is that current is no longer parallel to **E**, as it must be at low fields because we are dealing with cubic crystals. The nonparallelism occurs for **E** in directions that are not axes of rotational symmetry. Although the current direction for a given valley is at an angle to the valley axes because of the mass variation in the valley, in low fields the components perpendicular to **E** cancel out. In high fields, valleys with **k** vectors of small-mass (m_t) electrons more closely aligned with **E** get hotter, and their contribution decreases relative to the other valleys. This has the result that the perpendicular contributions to **j** no longer cancel out. The effect is magnified by the fact that the hotter valleys lose carriers by intervalley scattering to the cooler valleys.

As the field increases in the non-Ohmic region, v_d continues to increase, but more and more gradually. For Ge, Si, and many other semiconductors at 300 K, v_d tends to saturate at high fields at a value of $\sim 10^5$ m/s. This value is important for devices, which are frequently operated at high fields, because it determines the transit time through the device. Interestingly, this drift velocity is high enough so that, in the very small devices that are now being investigated, a sizable fraction of the carriers can make the transit through the device without a collision. Such collisionless transport is called ballistic transport. It is of particular interest because it provides a way of making faster devices.

In many of the group-IV and III-V semiconductors, when the field is raised beyond the range where v_d tends to saturate, to close to 10^7 V/m for n-Ge or 1.2×10^4 V/m for InSb to cite two examples, the current begins to rise very steeply. It has been shown that this rise represents an increase in carrier concentration due to impact ionization of the lattice atoms by the hot carriers. This process will be discussed in Sec. 7.3. A different phenomenon, called the Gunn effect, occurs at high fields in GaAs, GaP, InAs, and some other materials. This will be taken up in Sec. 7.4.

The discussion so far in this section has focused on phenomena from room temperature down to ~ 78 K in relatively pure samples. We consider now the situation at low temperatures, typically 20 K and below, where impurity scattering is important and a sizable fraction of the carriers have been frozen out of the conduction and valence bands. According to Eq. (45) τ_I and therefore μ_I should increase when the carriers are heated. Thus in this range β of Eq. (49) should be positive. This was in fact found for Ge samples at 20 K. However, there is another reason for conductivity to increase with field in this temperature range: hot carriers can cause impact ionization of neutral donors or acceptors. Hall measurements made as a function of field were able to show that both effects occur at this temperature. The initial non-Ohmic behavior is due to increase in mobility, but once the carriers are hot, impact ionization dominates, with the rise in current with field becoming so steep at temperatures where most of the carriers are frozen out as to be called a breakdown. The field required for this type of breakdown is less than 10^3 V/m for n-Ge.

7.3 Impact Ionization

In the temperature range where the donors or acceptors are fully ionized or almost so, an electron or hole that has gained sufficient energy from the field may give up its energy to a valence electron, raising the latter into the conduction band and leaving a hole in the valence band. The threshold energy required for an electron to create a hole–electron pair in this way is readily found for a semiconductor with the simple model of the band structure by applying the conditions of conservation of energy and momentum. For electron and hole masses m_n^* and $m_p^* = \gamma m_n^*$, respectively, the threshold energy for pair creation by an electron is $E_g(2+\gamma)/(1+\gamma)$. Thus for equal electron and hole masses an electron must have a minimum energy $3E_g/2$ to create an electron–hole pair. An initiating hole would also require only $3E_g/2$ for pair creation if $m_p^* = m_n^*$. With the gap in silicon, for example, 1.1 eV, at 300 K the field must raise the energy of an average electron, 0.025 eV, by a factor close to 60 to achieve pair creation. The field required is over 10^7 V/m. For InSb, with a much smaller gap and higher mobility carriers, only 1.2×10^4 V/m is required.

At fields just above threshold, ionization is due to a relatively few "lucky" carriers that have, fortuitously, been scattered less than others. This leads to an ionization rate changing roughly with field as $\exp(-A/|\mathbf{E}|)$, where A is a constant. At still higher fields, where most of the carriers are raised to a higher temperature T_e, the ionization rate is expected to depend on field as $\exp(-B|\mathbf{E}|^2)$.

7.4 The Gunn Effect

In n-type GaAs and a number of other materials, instead of the near saturation at high fields with subsequent impact ionization of the lattice found in Si, for example, a different and unique phenomenon called the Gunn effect is observed. Application of a field results in little departure from Ohm's law until after attainment of a threshold field in the range $(2\text{–}4)\times10^5$ V/m at 300 K in GaAs. Beyond this field the current is found to vary strongly with time, its average value dropping below the threshold value. Many different types of time variation have been seen in different samples, the simplest being an oscillation between the value at threshold and a value as low as half the threshold. The period of the oscillations was found to be about equal to the transit time of an electron between the electrodes as calculated from the threshold current, the number of electrons, and the length of the sample. The oscillations are typically of microwave frequency and the Gunn effect is used for microwave generation.

The current oscillations arise from the mobility change that occurs on heating of the carriers. As seen in Fig. 9, polar optical scattering is dominant in GaAs at high temperatures. According to Sec. 5, this means that electrons that acquire $E>h\nu_0$, ~400 K in GaAs, are scattered progressively less as their energy increases. In a strongly polar insulator with no other important scattering mechanisms, this would lead to the electron energy growing without limit as the field increases, causing impact ionization, severe lattice heating, changes in chemical and mechanical properties, and, most important, breakdown of the insulating properties. In fact, this mechanism has been suggested as the reason for breakdown of polar insulators. In GaAs, however, when an electron reaches 0.3 eV it can be scattered, by a phonon, from the Γ minimum to one of the L valleys in the conduction band (see Fig. 4). At 0.46 eV it can also be scattered into one of the X valleys in the conduction band. The probability of these processes is quite high because the densities of states in the X and L valleys are much higher than that in the Γ valley. This intervalley scattering has been detected by picosecond optical measurements (Auston, 1990). The mobility of electrons in the L and X valleys is also considerably smaller than that of electrons in Γ. Similar mobility changes due to intervalley electron transfers of hot electrons occur in GaAs-GaP alloys, InP, CdTe, ZnSe, InSb, and GaSb-InSb alloys, which also show the Gunn effect. In InSb the effect can be observed only for a nanosecond after the beginning of current flow because, the energy gap being smaller than the energy separation between valleys, impact ionization dominates after a short time. Interestingly, the Gunn effect is found under uniaxial stress in p-Ge. Although p-Ge has the degenerate band structure, under appropriate stress it goes over into one with two sets of valleys appropriately separated.

The result of the intervalley transfers is that beyond the threshold field further increase in $\mathbf{E}$ causes a decrease in current. This is an example of negative differential conductivity. How this phenomenon leads to a current that oscillates in time is discussed in the next section.

7.5 Negative Differential Conductivity

Negative differential conductivity (NDC) is characterized by regions of negative slope in plots of current density versus field. There are two types of NDC, shown schematically in Fig. 10, distinguished by their behavior in circuits.

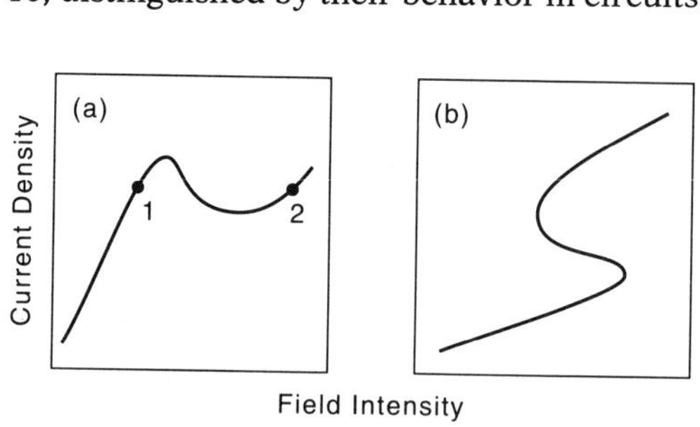

FIG. 10. The two types of negative differential conductivity. (a) The N-shaped curve representing voltage-controlled NDC, and (b) the S-shaped curve representing current-controlled NDC. (After Conwell, 1970.)

Unique electrical behavior results when a system operates in a region of NDC. In particular, a departure from equilibrium such as a fluctuation of field or of current will tend to grow rather than dissipate. In a sample with the voltage-controlled type of NDC, consider a small nonuniformity of the field at some point such that the field upstream (as regards current flow) is higher than the field downstream. This nonuniformity may be caused by a momentary deficiency of positive charge or accumulation of electrons, or by a permanent nonuniformity in impurity concentration or conductivity. If the differential conductivity of the sample is positive, a higher current will flow into the region than flows out, and any accumulated negative charge will be dissipated. If the differential conductivity is negative, however, a smaller current will flow into the region than out; more negative charge will accumulate and the original field discontinuity will be enhanced.

In a rectangular sample, as is ordinarily used, the equipotentials must be planes perpendicular to the sides of the sample, and the tendency will be to build up a plane of negative charge, an electron accumulation layer, at the discontinuity. If the initial nonuniformity or fluctuation had been of the opposite sign, there would be a tendency for a layer of positive charge, that is, an electron depletion layer, to grow. This is the type of instability that leads to the Gunn effect.

Imagine now a fluctuation in a sample with current-controlled NDC such that at some instant two adjacent lateral regions of filaments are carrying different currents. If the differential conductivity were positive, the filament with the higher current would have the higher field. A transverse current would then flow in such a direction as to equalize the currents. In a sample with NDC, however, the filament with the higher current would have the lower field, and transverse current would flow in the direction that increases the difference between the currents. Stability will be reached when the currents in the different filaments all correspond to the same value of electric-field intensity. Current-controlled NDC has been observed in Ge at low temperatures in situations where the carriers are hot enough to cause impact ionization of impurities, and in Ge and InSb at room temperature where the carriers are hot enough to ionize the lattice. In both these situations, once breakdown is achieved, the current can be carried at a lower voltage than was required for breakdown.

As has been seen, in the context of the Gunn effect, voltage-controlled NDC occurs in many materials because of mobility changes caused by the electric field. It also occurs in some semiconductors, notably CdS and high-resistivity GaAs, because the number of conduction electrons n decreases with increasing field. In CdS it is found that, beyond a threshold field $\sim 2\times10^4$ V/m, $n \propto E^{-m}$, where $m > 1$. This behavior has been attributed to "field quenching," in which the field serves to free holes from hole traps so that conduction electrons can combine with them. NDC in CdS is convenient for study because in the high-field regions E_g is decreased, making these regions opaque to light for which the low-field regions are transparent. It is found that with a blocking contact, which makes n at the cathode smaller than the value in the bulk, a high-voltage domain originates at the cathode at ~ 500 V and becomes larger with increasing voltage.

In GaAs exhibiting the Gunn effect a high-voltage region is also likely to form in the neighborhood of the cathode because of a small high-resistivity region there. The high-voltage region results from a dipole, with an accumulation of electrons in one region, a depletion in another. If the product of n and the sample length is large enough, the dipole will detach itself from the cathode and continue to grow until the upstream and downstream fields have a set of values such as 1 and 2 in Fig 10(a). The domain moves down the sample with the drift velocity because it is made of moving electrons. When it reaches the anode it exits from the sample, the current rises, and the process of building up another domain at the cathode begins. For further discussion and references see Conwell (1967, 1970).

7.6 Acoustoelectric Effect

In CdS, ZnO, and a number of other II-VI and III-V compounds for certain crystallographic directions, Ohm's law is obeyed only up to the field for which the drift velocity $v_d \simeq v_s$, the sound velocity. Note that this is not a high enough field for carrier heating. Further increase in **E** results either in saturation, i.e., a time-independent current that increases little with further increase in field, or in time-dependent behavior in which the current oscillates between the Ohmic value appropriate to the particular field and the lower, saturation value (Einspruch, 1965). In the latter case, probing the sample electrically led to the discovery that, similar to the Gunn effect, a high-resistance domain is generated at the cathode and moves down the sample. In this case, it moves with v_s. After it leaves at the anode a new domain is generated, etc. In the time-independent case, as in the Gunn effect, a high-resistance region is generated but the conditions for it to move are not satisfied.

A clue to the origin of these effects is that the directions for which they are observed are those for which there is piezoelectric coupling between an acoustic wave traveling in that direction and an electric field along the direction. An acoustic wave introduced from outside, through a transducer, for example, in such a direction produces an alternating electric field that bunches the conduction electrons or holes in its potential-energy minima. In the absence of an applied dc field the acoustic wave must drag the electron bunches along with it and experiences attenuation. If, however, a dc field is applied in this direction of sufficient magnitude to make the electrons drift with $v_d > v_s$, the electrons drag the wave, i.e., the wave experiences gain. Of course, there are acoustic waves, the phonons, present in these materials even when acoustic energy is not put in from the outside. Application of a dc field high enough to make $v_d > v_s$ in one of the piezoelectrically active directions results in amplifying acoustic waves that happen to be traveling in the field direction. The traveling domains have been found to be regions of high acoustic flux. The high resistance associated with these domains, or with stationary domains, is due to the large rate of momentum loss of the electrons in acoustic-wave generation. The acoustic-wave amplification we have been discussing may be usefully thought of as stimulated emission of phonons.

8. IMPURITY-BAND CONDUCTION

8.1 Semiconductor Conductivity at Low Temperatures

As discussed in Sec. 4, with decreasing temperature conductivity decreases rapidly in all but the most heavily doped (degenerate) semiconductor samples because electrons and holes fall back into the impurity levels. The decrease in n with T was shown in Fig. 6 for a set of n-Ge samples down to ~ 10 K. According to Eq. (17), the slope represents the activation energy, 0.01 eV for this case. A plot of $\ln\kappa$ vs $1/T$ for these samples would show the same decrease, with essentially the same activation energy, because the mobility has much weaker T dependence than n in this range. A set of p-Ge samples would give very similar data, with the same activation energy. As T is lowered further, carriers must continue to drop out of the conduction or valence band at the same rate, but the conductivity does not continue the rapid rate of decrease. This is shown in Fig. 11 where the logarithm of resistivity ρ is plotted versus $1/T$ for a set of p-Ge samples with acceptor concentrations ranging from 2×10^{22} to 7.3×10^{22} m^{-3}. From the minimum in ρ around 78 K, corresponding to all holes being in the valence band, ρ

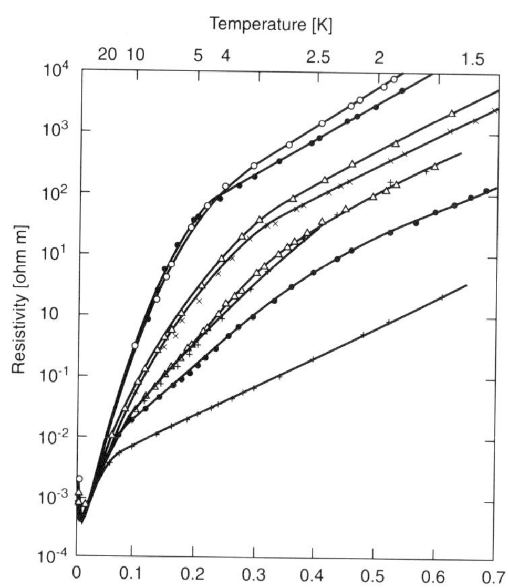

FIG. 11. Resistivity vs temperature for a set of p-Ge samples with acceptor concentrations ranging from 7.3×10^{22} m^{-3} for the lowest curve to 2×10^{22} m^{-3} for the uppermost curve. (After Fritzsche, 1955.)

rises exponentially for all samples, with a slope corresponding to the activation energy, 0.01 eV; this is a repetition of the data in Fig. 6 with a more compressed scale. However, beyond some temperature, ranging from 20 K for the lowest ρ (most doped) to 5 K for the highest ρ (least doped) sample, the rate of increase of resistivity with decreasing T is markedly smaller. Processes with lower activation energy are now dominating the conductivity.

The experimental curves of Fig. 11, from ~78 K on down, can be fitted by a sum of terms of the form

$$\rho=\rho_1 e^{\mathscr{E}_1/kT}+\rho_2 e^{\mathscr{E}_2/kT}+\rho_3 e^{\mathscr{E}_3/kT}, \quad (50)$$

where $\rho_1 \gg \rho_2 \gg \rho_3$ and $\mathscr{E}_1 > \mathscr{E}_2 > \mathscr{E}_3$. From the discussion above $\mathscr{E}_1 = 0.01$ eV. The second term is not seen for the lowest- and highest-resistivity samples but is seen in between. The third contribution, $\rho_3 \exp(\mathscr{E}_3/kT)$, is seen over the lowest temperature range for all samples.

8.2 Conduction Mechanisms at Very Low Temperatures

The $\mathscr{E}_2$ and $\mathscr{E}_3$ contributions can be understood in terms of the expected density of states. We will describe the situation in terms of electrons, but the description applies to holes with the obvious modifications. Figure 5 gives the density of states for a lightly doped semiconductor, Ge with $\sim 10^{20}$ donors/m^3, for example. At $\sim 10^{22}$ m^{-3}, as discussed in Sec. 3, the hydrogenic orbits of electrons bound on adjacent donors begin to overlap and the discrete level at E_d begins to spread into a band. The random arrangement of the impurities results in the overlap, and therefore the bandwidth, being somewhat greater than it would be for a regular arrangement. Nevertheless, the band is quite narrow at this concentration and the repulsion U_r for a second electron on a donor site is larger than the bandwidth B. In that situation the impurity band splits into two subbands, the upper band separated from the lower by an energy $\sim U_r - B$. In Fig. 12 the lower band is identified as the (a) band, the upper the (b) band. To explain the exponential temperature dependence of ρ in the (a) band, and other measured properties, it is necessary to assume that the electrons in this band are localized, one to

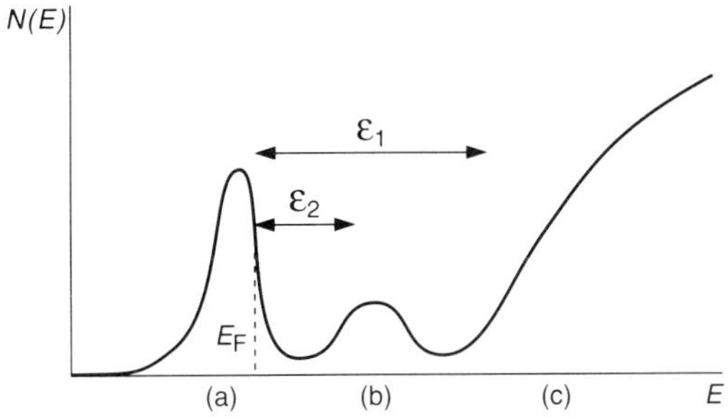

FIG. 12. Distribution in energy of the states in an n-type semiconductor showing the splitting of the impurity band into two bands (a) and (b). (c) is the conduction band. The position of the Fermi level is shown for the case $N_a < N_d/2$. (After Mott and Davis, 1971.)

a donor, for $N_d < 7 \times 10^{22}$ m^{-3}, the highest concentration in Fig. 11.

In the limit of very low temperatures the (a) band would be completely filled with one electron per site if there were no acceptors, while the higher bands would be empty. Actually there are always some acceptors present in an n-type sample, with the result that in the low-T limit E_F lies in the (a) band as shown in Fig. 12, and the band has a number of vacancies equal to N_a. It is this fact that makes conduction possible within the (a) band. Conduction takes place by an electron hopping from a neutral donor, i.e., a filled donor site, to a positively charged donor, or empty donor site. It is clear that the amount of conductivity due to this mechanism depends on N_a/N_d. For $N_a \ll N_d$, κ will increase as N_a increases. However, if N_a is close to N_d, κ would decrease with further increase in N_a because there would be fewer electrons to hop.

The randomness of donor and acceptor distributions results in electrons on different donors having somewhat different energies. To enable an electron on site i to hop to a donor on site j, where the energy level is higher by W_{ij}, the energy W_{ij} must be provided by a phonon. The hopping probability therefore depends on the abundance of phonons of energy W_{ij}, which at low temperatures ($kT \ll W_{ij}$) is given by $\exp(-W_{ij}/kT)$. It also depends strongly on the overlap of the electron wave functions in the initial and final states. This dependence, for two donors separated by a distance R_{ij}, is given by $\exp(-2R_{ij}/a_H)$, where a_H is the radius of the ground-state (hydrogenic) orbit. Thus the probability ν_{ij} of a transition per unit time between sites i and j may be written

$$\nu_{ij}=\nu_0 \exp(-2R_{ij}/a_{\rm H}-W_{ij}/kT), \qquad (51)$$

where ν_0 depends on the coupling to the phonons, given by the deformation potential (see Sec. 6.1) in this case, R_{ij}, and other factors. In calculating κ or ρ from Eq. (51) it is useful to distinguish between two ranges:

1. the "high-temperature" range where kT is large enough so that an electron can usually find a phonon of sufficient energy to jump to a nearest neighbor;
2. the "low-temperature" range where kT is small enough so that the electron will preferentially jump to a further site for which W_{ij} is smaller despite the decrease in $\exp(-2R_{ij}/a_{\rm H})$.

The latter hopping process is called variable-range hopping (VRH) because the average hopping distance increases with decreasing T.

8.2.1 Variable-Range Hopping. It is readily shown that the optimum range $(R_{ij})_m$ is $[9a_{\rm H}/8\pi N(E_{\rm F})kT]^{1/4}$, where $N(E_{\rm F})$ is the density of states at $E=E_{\rm F}$. When $(R_{ij})_m$ and the W_{ij} that corresponds are inserted in Eq. (51) we find the hopping rate

$$\nu_{ij}=\nu_0 \exp[(-T_0/T)^{1/4}], \qquad (52)$$

where $T_0=16/kN(E_{\rm F})a_{\rm H}^3$. The diffusion coefficient D that results from the hopping is given by $\langle \nu_{ij} R_{ij}^2\rangle$. From this the mobility may be obtained with the use of the Einstein relation $\mu=De/kT$. To calculate κ or ρ, n is taken as the number of electrons per unit volume within kT of $E_{\rm F}$ because essentially only those will be able to hop to an empty site by finding a phonon to assist them, or by emitting a phonon. This number is $2N(E_{\rm F})kT$, the factor 2 resulting from the fact that electrons of both spins may occupy each state.

In carrying out the averaging over R_{ij} and W_{ij} required to obtain D it is necessary to take into account not only the variations of these quantities but the fact that carriers may take different paths through the sample, avoiding those with particularly unfavorable R_{ij} or W_{ij}. For the high-temperature case, or nearest-neighbor hopping, the T dependence of κ will be dominated by the exponential in Eq. (51) with $\mathscr{E}_3$ representing a suitably averaged W_{ij}. For the "low-temperature" or VRH case the T dependence of κ will be essentially that shown in Eq. (52). The experimental T dependence seen in Fig. 11 in the lowest temperature range can clearly be well fitted by $\exp(\mathscr{E}_3/kT)$, consistent with nearest-neighbor hopping. Perhaps VRH can be seen at lower T's. Variable-range hopping will be found to fit the observations for amorphous semiconductors quite well.

8.3 Criterion for Localization

Conduction in the (b) band requires that an electron in (a) absorb a phonon of sufficient energy to make the transition to (b). Once the electron is in the (b) band, the conduction mechanism depends on whether the states are localized or extended. A criterion for determining this has been set up by Anderson (1958). He finds that if the actual energy spread ΔE of the states in a band is greater than $\sim$5 times the bandwidth B that there would be due to overlap if the (isolated) states all had the same energy, all the states in the band are localized. If $\Delta E/B<5$ the states in the outer edges of the band may still be localized. The condition $\Delta E>5B$ apparently applies to the (a) band. It need not apply to the (b) band. It is not unlikely that the extent of the higher-energy state is larger, which would make B larger and the ratio $\Delta E/B$ smaller. The available data suggest that the (b) states are extended in the doping range of Fig. 11. In that case (see Fig. 11) $\mathscr{E}_2=E_i-E_{\rm F}$, the energy required to get the electron into the level E_i in the (b) band.

At concentrations higher than those of the Fig. 11 samples, probably $>10^{24}$ m^{-3}, the (a) and (b) bands have spread out and strongly overlap each other and the conduction band. The states at $E_{\rm F}$ are now extended, and the sample is degenerate, as can be seen from the experimental data for sample 58 in Figs. 6 and 9. The states in the low-energy tail of the impurity band are still expected to be localized, however; these correspond to electrons that, because of the randomness of the impurity distribution, are in particularly deep potential wells.

9. AMORPHOUS SEMICONDUCTORS

9.1 Atomic Structure

Amorphous materials lack the lattice periodicity that characterizes the crystalline solid state. The departure of the atomic positions in an amorphous solid from a periodic array is often represented by a radial distribution function (RDF). The RDF displays a number of peaks at r values giving the average distance between nearest neighbors, next-nearest neighbors, and so on. The radial fluctuation in these interatomic distances is manifested in the peak width. The lower part of Fig. 13 compares the RDFs for an assembly of identical atoms arranged in (a) crystalline form, (b) a glass, and (c) the gas phase (instantaneous picture). The RDF for the first few atomic positions in the amorphous solid is similar to that of the crystal because short-range order in the solid state is controlled by the dictates of chemical bonding. Beyond the first few neighbor distances in the amorphous solid, however, the RDF becomes progressively featureless, thus more gaslike than crystallike. Experimentally the RDF is determined by a complex procedure (Zallen, 1983) from the real-space Fourier transform of the x-ray scattering interference function.

One important result of the application of RDFs to the analysis of covalent amorphous semiconductors is the demonstration of the general applicability of the continuous random network (CRN) model first introduced by Zachariasen (1932) as an idealized description of covalent glasses such as silica. In the CRN model it is assumed that each constituent atom (including any impurity atom) is able to optimize its own bonding coordination needs. The optimum coordination number Z of any atom in a CRN is given by the Mott $8-N$ rule, where N is the number of valence electrons. According to the $8-N$ rule, which is based on elementary chemical reasoning, when $N \geqslant 4$ the coordination is $Z=8-N$, while for $N<4$ the optimum coordination is N. Figure 13 (upper part) illustrates the two-dimensional CRN of a three-coordinated (i.e., each atom has three bonds) glass. For a given composition all configurations of the glass that allow the constituent atoms to satisfy their local bonding requirements are equally probable. The effect of the long-range disorder on the electronic structure is expressed in terms of a randomly varying disorder potential (Mott and Davis, 1979).

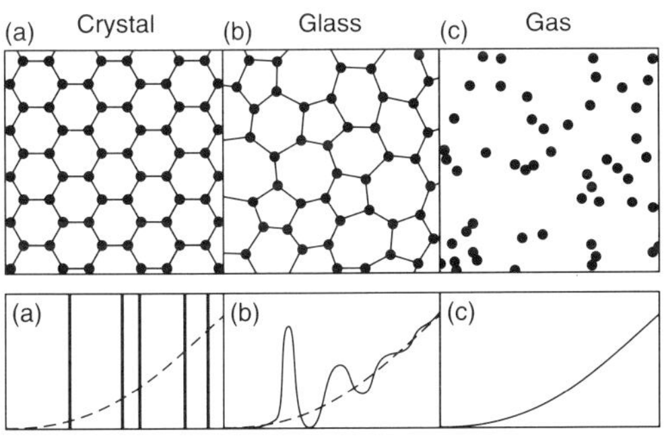

FIG. 13. Upper: Schematic representation of the atoms in (a) a covalent crystal, (b) a covalent glass network, and (c) a gas. Lower: Schematic represention of the corresponding RDFs (see text). (After Zallen, 1983.)

What, then, is the meaning of a structural defect in the CRN framework? In a crystal any atom not in a proper position in the lattice is a defect. In amorphous semiconductors, where CRN represents the ideal, the elementary defect is a miscoordinated atom, that is to say an atom with too few or too many bonds. Bonding miscoordination defects typically give rise to electronic states deep in the gap. These defects are now understood to play a major role in many of the key photoelectronic properties exploited in device applications (Street, 1991).

9.2 Electronic States

In the absence of lattice periodicity the vector **k** defined in Sec. 2.3 no longer has a meaning. For covalent-network amorphous solids whose crystalline counterparts have wide bands of order 5 eV or more, energy bands are still retained in the presence of disorder, but these must now contain localized as well as delocalized electronic states. Such disorder-modified bands can no longer be characterized by E-**k** dispersion relations such as that in Fig. 4, but must instead be described in terms of a density of states function $N(E)$. Important physical parameters like the carrier effective masses, defined in terms of the dispersion relations in the crystalline solid state, must be reformulated for amorphous semiconductors. The general problem of disorder-induced localization and its influence on the electronic band structure of amorphous covalent networks is very complex. It was elucidated by Anderson (1958), as discussed in Sec. 8. In hydrogenated amorphous silicon, *a*-Si:H, and in chalcogenide glasses, bandwidths are of order 5–10 eV, and the disorder estimated to be present, about 100–500 meV, is clearly too small to satisfy the Anderson criterion for complete localization. When the disorder potential is insufficient to induce complete localization of all states in the band, it can still act, as discussed earlier, to broaden the band edges into tails of localized states that extend into the gap. There is a significant difference in the mobility of carriers depending on whether they occupy extended or band-tail states. As a result, a demarcation energy separating extended states in the band from localized band-tail states was proposed (Mott, 1967). Because it was originally thought that the mobility should undergo a discontinuous change associated with the transition from band to hopping transport, this demarcation energy was said to define a mobility edge. It was suggested that for an amorphous semiconductor the appropriate analogy to the band gap of an intrinsic semiconductor crystal is the pseudogap bounded by the conduction- and valence-band mobility edges, denoted by E_C and E_V, respectively. The gap in an amorphous semiconductor is referred to as a pseudogap because it can be populated by a more or less continuous distribution of disorder-induced localized states (see Fig. 14). In addition to the band tails there are electronic states typically located deep in the gap, which arise from characteristic native defects associated with short-range disorder. Deep defects that can act as traps and recombination centers significantly influence the transport and photoelectronic properties of amorphous semiconductors and are believed to be involved in the mechanism of doping.

9.3 Native Defects

Every energy level of a defect can contain up to two electrons. For simplicity consider a defect that introduces a single level into the gap. That defect can exist in any of three charge states, corresponding to empty, singly, or doubly occupied, and its occupancy determines the position of the level it contributes to the gap relative to a mobility edge. The energies of the doubly occupied states differ from those of the singly occupied states because

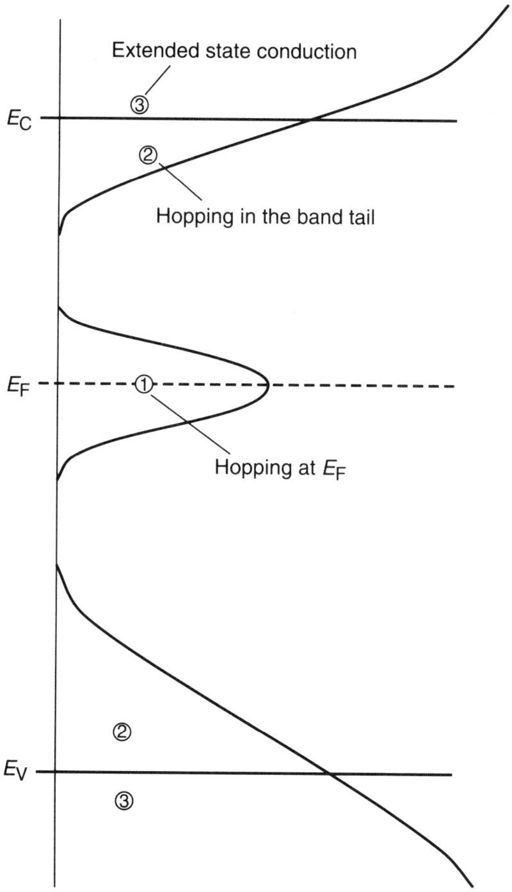

FIG. 14. Distribution in energy of the states, i.e., $N(E)$ vs E, in a prototypical undoped amorphous semiconductor.

1. the two electrons repel each other (Coulomb energy) and
2. addition of an electron can induce a bonding change or relaxation, which may either raise or lower the energy.

The algebraic sum of the two energies is called the correlation energy U. Both negative and positive correlation-energy defects exist and they have significantly different electronic properties. It can, for instance, be shown (Street, 1991) that for positive-U defects

1. the singly occupied, paramagnetic state has the minimum energy;
2. the Fermi level position is sensitive to dopants and there is minimal lattice relaxation.

On the other hand, for a negative-U defect

1. a large lattice relaxation is required for its formation;
2. the lowest energy state is diamagnetic; and
3. the Fermi level is constrained to lie between the defect energy levels.

The canonical example of a negative-U defect system is the valence alternation pair (VAP) that exists in chalcogenide glasses (Kastner *et al.*, 1976). *a*-Se, an elemental chalcogenide glass, provides the simplest example. According to the $8-N$ rule every Se atom should be bonded to two others, forming a one-dimensional network glass. A chain-terminating Se atom might be expected to have one unpaired or unbonded electron, called a dangling bond. Estimates based on viscosity data indicate that the number of such dangling bonds should exceed 10 ppm, which would be easily detectable by electron spin resonance (ESR). ESR is, however, not observed (Abkowitz, 1967), in *a*-Se or in any of the more complicated chalcogenide glasses. The neutral dangling-bond defect must be unstable against conversion to a diamagnetic species during glassy solidification. It has been proposed that this conversion process occurs by valence alternation and involves bond switching and electron transfer between chains (see AMORPHOUS MATERIALS, STRUCTURE OF).

9.4 Conductivity Mechanisms

It is clear from the preceding discussion that the analysis of conductivity in amorphous solids is greatly complicated by the broad energetic distribution and varied character of the states that can contribute channels for carrier motion. Instead of defining a single mobility for all states in the conduction band, or valence band, as done in Eqs. (2) and (26), it is useful to introduce a mobility $\mu(E)$ characteristic of the group of states in the range dE at E. The appropriate generalization of Eqs. (4) and (9) is then

$$\kappa=\int N(E)e\mu(E)f(E)dE\,. \tag{53}$$

Figure 14 is a schematic illustration of $N(E)$ vs E.

In Fig. 14 the Fermi level is shown contained in a submanifold of localized states near midgap. If the density of such states near the Fermi level is sufficiently high then, because their occupancy is heavily favored, these states will completely dominate the transport at low temperatures. For evaporated unhydrogenated amorphous Si, where the density of paramagnetic dangling-bond levels at the Fermi level is very large, it is found experimentally that $\ln\kappa\propto T^{-1/4}$ over a wide range of temperatures starting from the lowest (Street, 1991). As discussed in Sec. 8, this is the T dependence expected for VRH. In hydrogenated amorphous Si the dangling-bond defects from which the near-midgap states arise are in large measure eliminated by hydrogen complexation (i.e, passivation) and VRH is not observed.

At higher temperatures, nearest-neighbor hopping of carriers through band-tail states, regions 2 in Fig. 14, will contribute to the conductivity. When carriers of a single sign dominate, $1/\kappa$ vs T is activated, like the terms of Eq. (50). At still higher temperatures carrier transport is dominated by carriers that have been excited to states above the mobility edge E_C. κ may then be written

$$\kappa=\kappa_{E_C}\exp[-(E_C-E_F)/kT]\,, \tag{54}$$

where κ_{E_C} is the conductivity at the mobility edge and presumably also just above it. κ_{E_C} is called the minimum metallic conductivity. Mott and Davis (1979) have argued that for a three-dimensional case $\kappa_{E_C}=2.2e^2/Z^2\hbar a_E$, where Z is the coordination number and a_E the distance between localized states. It is difficult to verify this experimentally.

A reliable way of inferring the existence of localized states near band edges is via their influence on time-of-flight (TOF) drift mobility of injected carriers (Marshall, 1989). As discussed in Sec. 1.4, when the effective or drift mobility μ_d is limited by trapping it is reduced from the conductivity (free carrier) mobility μ_0 by the fraction of time that a carrier on average spends in traps, in this case the localized states adjacent to the mobility edge. When the injected carriers establish quasiequilibrium with trap and extended states in a small fraction of a transit time, the time of flight and the drift mobility are experimentally well defined. This condition is not satisfied, however, when there are deep traps with a range of depths, resulting in the carriers straggling to the contact. This behavior, called dispersive transport, is clearly exhibited by photoinjected holes in a-As_2Se_3. A successful statistical treatment of this type of transport has been carried out by Scher and Montroll (1975). On the basis of the Scher–Montroll treatment applied to time-of-flight and transient (bulk) photoconductivity data, a-As_2Se_3 is now generally believed to be characterized by a nearly exponential band tail, as suggested in Fig. 14.

9.5 Doping

It was thought earlier that the $8-N$ rule and the CRN model discussed in Sec. 9.1 would rule out the possibility of modifying the dc conductivity of an amorphous semiconductor by substitutional doping. In fact, a-Si:H exhibits doping phenomena (Spear and LeComber, 1975) with elements in groups V and III of the periodic table that are comparable to those in crystalline Si, though the underlying mechanism is believed to be quite different. Less widely appreciated and more subtle in its manifestations is a systematic doping effect in certain Se-rich chalcogenide glasses that is induced by both strong electropositive elements like the alkalis and electronegative substituents like the halogens. In both a-Si and a-Se (and related glasses) the underlying mechanism of doping is surprisingly similar and involves the interaction of dopant atoms with native defects.

The modification of conductivity by the doping of a-Si:H with P and B is dramatically illustrated in the data of Spear and LeComber (1975) shown in Fig. 15. Consider the case of P in a-Si:H. According to the $8-N$ rule P should be threefold coordinated (i.e., make three bonds) in its lowest-energy configuration. In fact the P doping efficiency in a-Si:H is very low, and it is generally believed that this is so because most of the substitutionally incorporated P is in threefold coordination, precisely as suggested by bonding and topology arguments. The smaller proportion of electronically active P dopant atoms that act as donors must be incorporated in fourfold coordination. The excess charge must somehow be compensated, and a large body of experimental evidence points to the dangling-bond defect D^- as the compensation center, with the Fermi level situated at an energy between P^+ and D^- levels. A complementary argument applies to the formation of ionized acceptors by B doping.

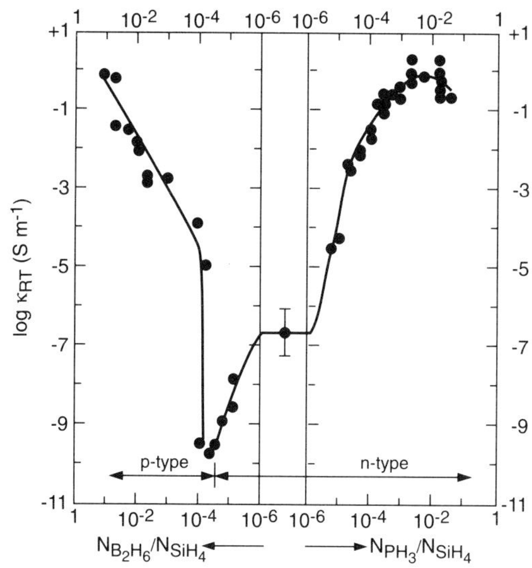

FIG. 15. Conductivity of a-Si:H at room temperature as a function of the dopant gases phosphine (PH_3) and borane (B_2H_6) introduced into silane (SiH_4), the deposition gas. (After Spear and LeComber, 1975.)

It should be noted that Mott (1976) suggested an analogous mechanism for the doping of a-Se by alkalis and halogens involving the native, negative-U, VAPs discussed earlier. In this case the VAPs, which normally exist in undoped material only as autocompensated defect pairs, would associate with some fraction of electropositive (e.g., Na) or electronegative (e.g., Cl) dopant atoms, thereby altering their relative populations while satisfying the requirement of overall charge neutrality. These predictions have been experimentally verified (Abkowitz, 1987). The doping effect of halogens on Se-rich chalcogenide alloys (e.g., Se:Te:As ternaries) is in fact systematically utilized in the design of the large-area photoconductors used in xerographic copiers and printers.

10. TWO-DIMENSIONAL AND ONE-DIMENSIONAL BEHAVIOR

The materials we have been discussing are all three dimensional (3D), in fact isotropic in most of their low-field properties. They can be made highly anisotropic, essentially 2D, by three different methods:

1. application of a high magnetic field;

2. confining the carriers in a thin layer at the surface; or
3. confining them in a thin layer of semiconductor between layers of different semiconductors, which may also be thin.

In the latter two cases further structuring the material to confine the carriers in one of the two dimensions can lead to quasi-one-dimensional behavior. This section is devoted to discussing the first two methods listed above; the third is the subject of HETEROSTRUCTURES, SEMICONDUCTOR.

10.1 Quantizing Magnetic Fields

10.1.1 Landau Levels. Solution of the wave equation has been carried out for a free electron in magnetic field. The same solution may be used for a carrier in the simple model of the band structure, with the replacement of m_0 by m^*. It is found that for a field **H** in the z direction the possible energy levels of the carrier are

$$E=(\hbar^2/2m^*)k_z^2+(\nu+\tfrac{1}{2})\hbar\omega_c, \qquad (55)$$

where ν takes on integer values 0, 1, 2, ..., and ω_c is the cyclotron frequency He/m^*, the rate at which carriers rotate around the constant-energy surfaces in a magnetic field (see Sec. 5.6). It is seen that the energy of motion in the z direction is unaffected by the field, as expected because, as is seen in Eq. (19), there is no component of force parallel to **H**. The energy of motion perpendicular to the field, however, is quantized in steps of $\hbar\omega_c$. The set of all levels with a given ν and arbitrary k_z is referred to as the νth Landau level. The effect of the field has been to cause all the states with different k_x and k_y to "condense" on the nearest Landau level.

If an electromagnetic field of frequency ω_c is applied together with the magnetic field, a carrier can absorb a photon $\hbar\omega_c$, making a transition to the next Landau level. The transition corresponds to the carrier going from the orbit in **k** space characterized by ν to the $\nu+1$ orbit. This phenomenon is called cyclotron resonance. Its realization requires well-defined Landau levels ($kT\ll\hbar\omega_c$) and well-defined orbits (scattering time $\tau>2\pi/\omega$, the time of one rotation). Cyclotron resonance experiments, with **H** applied in different directions, have been very important for determining the band structure of semiconductors. In the electromagnetic field with frequency ω_c transitions are also possible from the level ν to $\nu-1$. The interlevel transitions are seen as magnetic-field–produced structure in optical transitions between bands, called magneto-optical absorption.

Quantization of the energy levels, Eq. (55), is accompanied by quantization of the areas of the orbits in **k** space corresponding to the allowed levels (Ziman, 1972). The quantization conditions result in a sharp oscillatory structure in the electronic density of states $N(E)$. $N(E)$ will have a sharp peak whenever E is equal to the energy corresponding to an orbit with maximum or minimum area, called an extremal orbit. This is an important effect for metals and for degenerate semiconductors, where many properties depend on $N(E_F)$, the density of states at the Fermi energy. At a value of **H** for which E of an extremal orbit coincides with E_F, $N(E_F)$ will be large. Thus it is expected that the values of many properties of metals and semiconductors, among them magnetoresistance, will have an oscillatory behavior as **H** varies. Specifically, the properties depending on $N(E_F)$ are periodic in $1/H$ (Ziman, 1972). The resulting oscillations in magnetoresistance, called the Shubnikov–de Haas effect, have been seen at low temperatures in HgSe and a number of other degenerate semiconductors. Oscillations with a period of $1/H$ have also been seen in interband optical absorption and magnetic susceptibility (de Haas–van Alphen effect).

10.1.2 The Quantum Hall Effect. At low temperatures and high magnetic fields, in a situation where a high density of electrons is confined in two dimensions (see Sec. 10.2), unusual behavior of the Hall effect has been found. The ratio of j_x to E_y, which by Eq. (31) should be ne/H_z, is found to change discontinuously with H_z, showing plateaus. It has been found that the plateaus correspond to the filling of successive Landau levels as the field changes. This effect, with accompanying phenomena, is known as the quantum Hall effect (*q.v.*).

10.2 Transport in Inversion Layers

10.2.1 Effect of Electric Field on Semiconductor Surface. Early in the modern history of semiconductors (1948) it was discovered that application of an electric field perpendicular to the surface of a semiconductor could increase or decrease carrier concentration at the surface, depending on the polarity of the field. Because it is required that, in thermal equilibrium, the Fermi level must be the same throughout a sample, the effect of the field is to bend the bands at the surface. In Fig. 16 we show some of the possible situations that can occur for a *p*-type semiconductor. The same situations occur, of course, for an *n*-type semiconductor. As can be seen from Eq. (14), when E_F is closer to the valence-band edge at the surface than in the interior, the hole density is greater at the surface, making it an accumulation layer. Where E_F is further from E_V in the surface than in the bulk, the surface is depleted. When the bands are bent further still, so that at the surface E_F is closer to E_C than E_V, the surface becomes *n* type, constituting an inversion layer. We should note, however, that the band bending may not be entirely due to the field; the bands are usually bent at the surface in the absence of the field because the surface has different defects and different defect levels from the bulk, and in fact may have a much larger number of defect levels.

10.2.2 Quantization in the Inversion Layer. The case of the inversion layer is of particular interest. If the direction perpendicular to the surface is taken as *z*, electrons whose kinetic energy in the *z* direction lies below the bulk E_c are confined in a potential well. As is well known, if the well is deep enough, such electrons have quantized energy levels, the lowest level corresponding to a wave function with no nodes within the well, the next to a wave function with one node within the well, etc. (see QUANTUM MECHANICS). Some of these energy levels are indicated schematically in Fig. 16(c). The existence of these quantized levels was demonstrated by magnetoconductance measurements in which the Fermi level was swept through the Landau levels by varying the carrier concentration in the inversion layer as discussed in the last section. The characteristics of the Landau levels were those predicted for carriers confined to distances from the interface as small as 1–10 nm (Fowler, 1982). For such narrow wells the separation between levels is much larger than kT at 4 K and even somewhat higher temperatures, and the carriers occupy only the lowest quantum level at these temperatures. Of course, for each level there is a band of energies in the *x* and *y* directions, corresponding to all allowed wave vectors k_x and k_y, because motion parallel to the surface is unconfined. Thus 2D subbands are formed with electron energies, for the simple model of the band structure, given by the sum of E_n^0, the *n*th quantized energy level, and the kinetic energy $\hbar^2(k_x^2+k_y^2)/2m^*$.

10.2.3 Transport in the Inversion Layer. Transport in inversion layers can be conveniently studied in a MOSFET, a metal-oxide-semiconductor field-effect transistor (see TRANSISTORS, FIELD EFFECT). The MOSFET structure is shown in Fig. 17. The silicon dioxide (SiO_2) thickness at the gate varies from ~10 nm to 1 μm. When $V_g=0$, electrons in the source cannot get to the drain, even though there is an attractive field due to the gate voltage V_D, because the *p*-Si region separating source and drain is a barrier. Application of a positive gate voltage creates an *n*-type channel, i.e., an inversion layer, at the *p*-Si surface with the SiO_2, permitting the electrons to go through. The voltage V_{sub} also affects the field at the SiO_2 interface and thus the number of carriers in the inversion layer. Thus varying V_{sub} and V_g appropriately allows a variation of electron concentration in the inversion layer at constant surface field or alternatively a variation of surface field at constant electron concentration.

Most of the transport measurements made on bulk samples may be made on the surface. Mobility may be determined from the measured conductivity of the surface layer in the usual way, using Eq. (3), but it is difficult to determine accurately the carrier concentration in the inversion layer. Hall measurements may be made and Hall mobility obtained from them, but, as discussed in Sec. 5.5, the ratio μ_H/μ is uncertain within about a factor of 2.

Conductivity and Hall-mobility measurements have been made for different orientations of the surface. Among the simplest to interpret is the [100] orientation. For that orientation in Si four of the valleys have the light mass in the *z* direction, while the other two have the heavy mass in that direction. The field in the potential well results in the latter two valleys having lower energies than the other four.

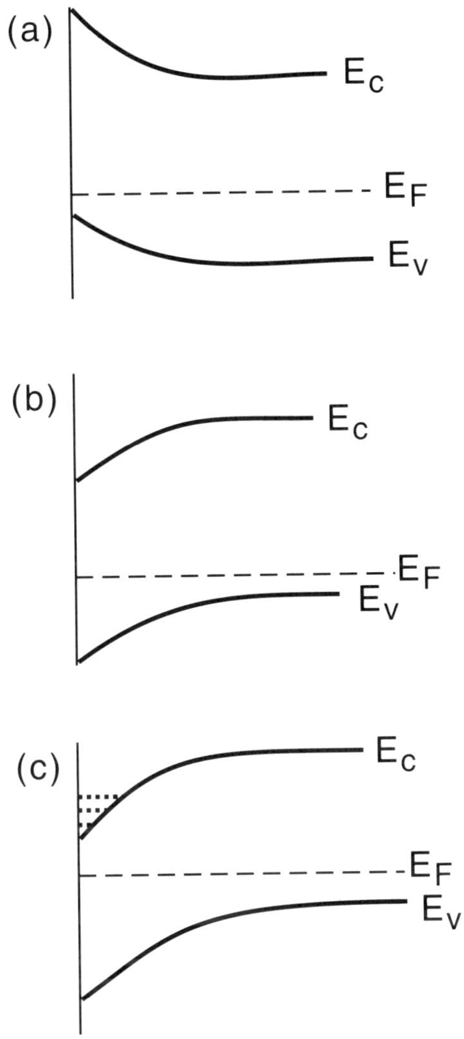

FIG. 16. Possible surface conditions for a *p*-type semiconductor: (a) accumulation layer; (b) depletion layer; (c) inversion layer. The dotted lines in (c) indicate, schematically, quantized energy levels.

The processes that scatter the carriers in the inversion layer are those discussed in Sec. 6, plus surface scattering. At low temperature, e.g., 4 K, the problem is relatively simple in that only the lowest quantized states are occupied and only impurity and surface scattering are significant. Impurity scattering can be due to either ions in the bulk or charges in the insulating SiO_2 layer. In practice, except for very heavy doping ($>10^{22}$ m^{-3}) in the bulk and for very clean oxides, the oxide ions are dominant. The samples in which impurity scattering is dominant at 4 K are likely to be degenerate, i.e., obey Fermi–Dirac statistics. For impurity scattering in a three-dimensional degenerate case we may use Eq. (45) for τ_I, replacing E by E_F. As the density of electrons in the inversion layer increases, E_F increases. Thus this equation predicts that at low temperatures impurity scattering gets less effective as the density of carriers in the inversion layer increases, which is in agreement with experiment. For detailed comparison with experiment, however, the Sec. 6 calculation of μ_I must be modified in a number of ways to describe better the inversion-layer situation (see Ando *et al.*, 1982).

As the carrier density increases and impurity scattering becomes less effective, surface roughness scattering comes to dominate. The surface roughness scattering is assumed to be due to random deviations from planarity of the Si–SiO_2 interface. It is characterized by two parameters, the rms deviation from surface planarity and the correlation length of the deviations. Information on these parameters may be obtained by transmission electron microscopy. The theory for this scattering process appears to be in reasonable agreement with experiment (Fowler, 1982).

At temperatures higher than 4 K phonon scattering is also important, specifically acoustic-phonon, nonpolar optical-phonon, and intervalley scattering. Extensive calculations are needed to take into account all the different phonons, and scattering within and between the different quantized bands and within and

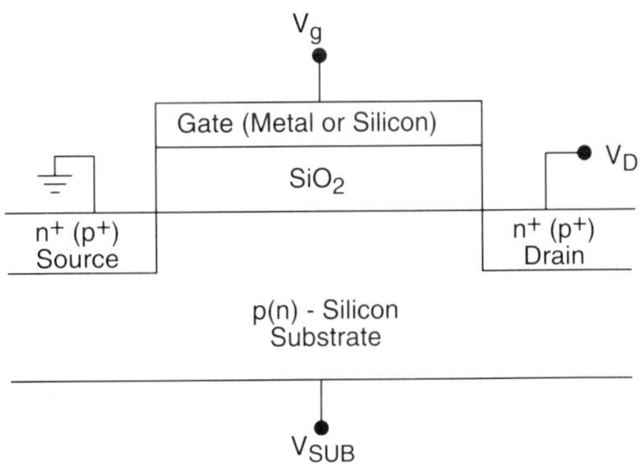

FIG. 17. Schematic MOSFET structure in cross section. The device may have a highly doped *n*-type (n^+) source and drain and a *p*-type substrate or a p^+ source and drain and an *n*-type substrate. (After Fowler, 1982.)

between valleys. For a review of the measurements and theoretical work see Ando *et al.* (1982).

10.2.4 Hot Electrons. Particularly at low temperatures, where the electron mobilities are high, the conductance is found to depend on the source–drain electric field. At 4 K mobility changes are observed at fields as low as a fraction of a V/cm. The mobility is found to rise initially when the electrons are heated by the field, as expected when the dominant scattering mechanism is ionized impurity scattering (see Sec. 7). The energy-loss mechanism that allows achievement of a steady state is acoustic-phonon scattering. For higher fields at 4 K impurity scattering becomes ineffective and optical-phonon scattering comes to dominate the scattering. In this field range mobility decreases with increasing field, as was described in Sec. 7 for hot carriers in the bulk. In the high-field limit drift velocity tends to saturate, as was also found for bulk *n*-Si. At 300 K, where impurity scattering is negligible, v_d vs **E** for the inversion layer is quite similar to that obtained for bulk *n*-Si, with no change up to $\sim 10^6$ V/m, beyond which v_d decreases with further increase in **E** and tends to saturate.

10.2.5 Localization. It is observed from ~ 2 to 50 K for samples with electron concentration in the inversion layer, n_{inv}, less than $10^{16}\ \text{m}^{-2}$ that the conductivity is activated, i.e., proportional to $\exp(-W/kT)$ with W a constant. W decreases monotonically with increasing n_{inv} until it vanishes (Fowler, 1982). It was early suggested that these phenomena indicate localization, attributed to potential fluctuations due to charges known to be in the oxide layer. When the fluctuations are of short range they produce a band tail similar to the impurity band (see Sec. 8), as is found also in amorphous semiconductors. Indeed, the data for samples showing this type of behavior agree well with the prediction [Eq. (54)] for transport in a band with a localized tail. k_{E_C} is found to be approximately equal to the value $0.1e^2/\hbar$ predicted for electrons propagating above the mobility edge in a two-dimensional case. The activation energy decreases as n_{inv} increases, raising E_{F}, and vanishes when E_{F} goes through the mobility edge. Further evidence that this model is correct is provided by the fact that as the temperature is lowered κ is no longer activated, but rather $\ln\kappa \propto T^{-1/3}$. This is the temperature dependence predicted for variable-range hopping (see Sec. 8) in two dimensions.

10.2.6 Conductance Fluctuations. More recently work in this area has focused on transport properties of localized electrons at temperatures below 4 K in very small Si metal-oxide field-effect transistors, patterned to have widths as small as 50 nm. Conduction processes expected to be important for strongly localized electrons when $T < 1$ K are VRH and tunneling. Devices appropriate for studies of hopping conductivity are long ones (because tunneling decreases exponentially with distance), with narrow Si accumulation layers. In a system with dimensionality d when VRH dominates $\ln \kappa \propto -(T/T_{0d})^{1/d+1}$, where T_{0d} is a characteristic temperature that depends on the density of states $N(E_{\text{F}})$ and d. It is found experimentally for these devices that, at temperatures $0.2\ \text{K} < T < 3$ K, VRH may be either 1D or 2D depending on the value of V_g, which affects $N(E_{\text{F}})$. (See ELECTRON STATES IN ZERO-, ONE-, AND TWO-DIMENSIONAL STRUCTURES for further discussion.)

A more striking effect in these samples is that at millikelvin temperatures conductance versus V_g has, superimposed on a general upward trend, a spiky behavior. The spikiness increases as T is lowered, the peaks being orders of magnitude above the background with widths at the top that are fractions of a millivolt. The spiky behavior results from the small size of the samples. As discussed in Sec. 8, an electron can hop only to states within a few kT of E_{F}. At 50 mK, for example, $2kT \simeq 10^{-5}$ eV. In a small device the number of states lying within this range is very small. Changing the gate voltage alters the energy of the electronic states in the channel relative to E_{F}. A small change in V_g may result in a large change in conductance G, for example, if it results in a particularly long hop being replaced by a short one. Further, each sample will have its own characteristic dependence of G on V_g because the locations of the levels, being determined by fluctuations in the oxide charge, are different in every sample.

Similar considerations apply to devices in which tunneling is dominant. In a short enough device an electron could tunnel directly from source to drain. Such short devices are difficult to make. In devices where the distance between source and drain is ⩾500 nm, resonant tunneling has been found of importance. In this process the electron tunnels from the source contact to a localized state in the channel and subsequently tunnels out of that state to the drain contact. For this process to occur the energy of the resonant localized state must equal, or be very close to, E_F in the source and drain contacts. Thus the situation is quite similar to that for VRH, and a similar spiky behavior is found in plots of G vs V_g (Fowler *et al.*, 1988).

GLOSSARY

Acceptor: Impurity that can accept an electron.

Accumulation Layer: Surface region in which the carrier density is greater than that in the bulk of the sample.

Activation Energy: When conductivity varies with temperature T as $\exp(-C/kT)$, C is called the activation energy.

Band Structure: The set of allowed energy E values of an electron in a crystal as a function of wave vector **k**. The kinds of band structure referred to are (1) simple model—E vs **k**, a parabola; (2) many valley—E vs **k** has a set of minima; (3) degenerate—E vs **k** consists of two bands, approximately parabolic, stuck together at $k=0$, and a third separated by a small energy interval.

Blocking Contact: A contact that completely cuts off the supply of charge to a sample.

Brillouin Zone: Unit cell in wave-vector or momentum space containing all the possible (discrete) wave-vector or momentum values.

Conduction Band: The lowest unfilled, or mostly unfilled, band of allowed energies.

Constant-Energy Surface: The surface or surfaces connecting points having the same energy in the Brillouin zone.

Dangling Bond: An unbonded electron. In amorphous silicon, for example, the paramagnetic defect state associated with a threefold-coordinated silicon atom.

Density of States: Number of states available for an electron per unit volume of a sample per unit energy range.

Density-of-States Effective Mass: The geometric mean effective mass for a given valley multiplied by the number of equivalent valleys to the 2/3 power.

Depletion Layer: Surface region in which the carrier density is smaller than that in the bulk of the sample.

Donor: Impurity that can give up an electron.

Doping: Addition of impurities that can increase conductivity of a semiconductor sample by adding electrons to the conduction band or holes to the valence band.

Effective Mass: A mass, which may be anisotropic, assigned to an electron or hole in a solid according to the energy–wave-vector relationship of the state it occupies.

Energy Gap E_g: The gap in energy between valence and conduction bands, empty of states in a sample without defects.

Fermi Energy E_F: The energy at which the probability of a state being occupied by an electron, under thermal equilibrium, is $\frac{1}{2}$.

Hall Effect: Application of a magnetic field perpendicular to the direction of current flow is found to give rise to an electric field perpendicular to both the magnetic field and the current. The Hall coefficient R_H, defined as the electric field developed per unit magnetic field per unit current density, is approximately equal to the reciprocal of the electron (hole) concentration times the charge on the electron.

Hall Mobility μ_H: The product of R_H (see **Hall Effect**) and conductivity, thus approximately equal to the mobility.

Hole: An unfilled state in a band more than half filled with electrons. The unfilled state acts as though it were filled with a charge equal in magnitude but opposite in sign to that of the electron.

Impurity Band: The group of states pulled out of the conduction (valence) band by the addition of donor (acceptor) impurities.

Inertial Mass: The reciprocal of the sum of reciprocals of the effective masses in the three principal-axis directions of the constant-energy surfaces for a given valley.

Injecting Contact: A contact able to supply charge to a semiconductor under the action of an electric field. An injecting contact that can be regarded as an infinite (i.e., nondepletable) charge reservoir, one able to fully satisfy the demands of the semiconductor even under space-charge–limited conditions, is called Ohmic.

Intrinsic Semiconductor: One in which carriers arise primarily from electrons that have crossed the gap into the conduction band, leaving behind holes in the valence band.

Inversion Layer: Surface region in which the current carriers have opposite sign to that of the carriers in the bulk.

Magnetoresistance: The change in resistance of a sample on application of a magnetic field.

Mobility μ: The net velocity in the electric field direction developed by an electron or hole due to acceleration by unit electric field.

Mobility Edge: The energy that separates localized from extended states near the band edge of an amorphous semiconductor.

Momentum Relaxation Time τ: The time for excess momentum, such as that created by an electric field, to decay to $1/e$ of its original value when the source of the excess is removed. τ is approximately equal to the mean free time of an electron or hole between scattering processes.

Ohm's Law: The statement that current increases linearly with applied voltage or electric field. This ceases to be true at high enough fields.

Phonon: A quantum of vibrational energy of one of the normal modes into which the random thermal motion of the lattice atoms may be decomposed.

Photoconductivity: An increase of electrical conductivity induced by the absorption of light.

Space-Charge–Limited Current (SCLC): The current demanded by a solid at fields that make the transit time of an excess injected carrier passing through the bulk to a collecting contact less than the time required to induce its local neutralization. Under these conditions the current is limited by the mutual repulsion of the transiting carriers.

Valence Band: The uppermost filled or nearly filled band of allowed energies.

Variable-Range Hopping: The transfer of an electron between localized sites at temperatures for which the thermal energy (required to make up the energy difference between sites) is so small that progressively larger hops are required with decreasing temperature for sites close enough in energy to the original to be found.

List of Works Cited

Abkowitz, M. A. (1967), *J. Chem. Phys.* **46**, 4537–4538.

Abkowitz, M. A. (1987), in: M. A. Kastner, G. A. Thomas, S. R. Ovshinsky (Eds.), *Disordered Semiconductors*, New York: Plenum, pp. 205–217.

Altarelli, M., Bassani, F. (1982), in: W. Paul (Ed.), *Handbook on Semiconductors*, Vol. 1, Amsterdam: North-Holland, Chap. 5.

Anderson, P. W. (1958), *Phys. Rev.* **109**, 1492–1505.

Ando, T., Fowler, A. B., Stern, F. (1982), *Rev. Mod. Phys.* **54**, 437–672.

Auston, D. H. (1990), *Phys. Today* **43** (2), 46–54.

Chattopadhyay, D., Queisser, H. J. (1981), *Rev. Mod. Phys.* **53**, 745–768.

Conwell, E. M. (1958), *Proc. IRE* **46**, 1281–1300.

Conwell, E. M. (1967), *High Field Transport in Semiconductors*, New York: Academic.

Conwell, E. M. (1970), *Phys. Today* **23** (6), 35–41.

Conwell, E. M. (1982), in: W. Paul (Ed.), *Handbook on Semiconductors*, Vol. 1, Amsterdam: North-Holland, Chap. 10.

Debye, P. P., Conwell, E. M. (1954), *Phys. Rev.* **93**, 693–706.

Dunlap, W. C. (1959), in: K. Lark-Horovitz, V. Johnson (Eds.), *Methods of Experimental Physics*, Vol. 6B, New York: Academic, pp. 32–70.

Einspruch, N. G. (1965), in: F. Seitz, D. Turnbull (Eds.), *Solid State Physics*, Vol. 17, New York: Academic, pp. 217–268.

Fletcher, K., Butcher, P. N. (1972), *J. Phys. C* **5**, 212–224.

Fowler, A. B. (1982), in: W. Paul (Ed.), *Handbook on Semiconductors*, Vol. 1, Amsterdam: North-Holland, Chap. 11B.

Fowler, A. B., Wainer, J. J., Webb, R. A. (1988), *IBM J. Res. Develop.* **32**, 372–383.

Fritzsche, H. (1955), *Phys. Rev.* **99**, 406–419.

Gray, Dwight E. (Ed.) (1957), *American Institute of Physics Handbook*, New York: McGraw-Hill.

Handbook of Chemistry and Physics (1975), Cleveland: CRC.

Harrison, W. A. (1980), *Electronic Structure and the Properties of Solids*, San Francisco: Freeman.

Herring, C. (1955), *Bell Syst. Tech. J.* **34**, 237–290.

Herring, C., Vogt, E. (1956), *Phys. Rev.* **101**, 944–961.

Hess, K. (1990), *Phys. Today* **43** (2), 34–42.

Kastner, M., Adler, D., Fritzsche, H. (1976),

Phys. Rev. Lett. **37**, 1504–1507.

Lampert, M. A., Mark, P. (1970), *Current Injection in Solids*, New York: Academic.

Madelung, O. (1991), "Semiconductors," in: R. Poerschke (Ed.), *Data in Science and Technology*, Berlin: Springer-Verlag.

Marshall, J. M. (1989), in: H. Fritzsche (Ed.), *Amorphous Silicon and Related Materials*, Vol. 1B, Singapore: World Scientific, pp. 799–824.

Mott, N. F. (1976), *Philos. Mag.* **34**, 1101–1108.

Mott, N. F., Davis, E. A. (1979), *Electronic Processes in Non-Crystalline Materials*, Oxford: Clarendon.

Ryvkin, S. M. (1964), *Photoelectric Effects in Semiconductors*, New York: Consultants Bureau.

Scher, H., Montroll, E. W. (1975), *Phys. Rev. B* **12**, 2455–2497.

Shockley, W. (1950), *Electrons and Holes in Semiconductors*, New York: Van Nostrand.

Spear, N. E., LeComber, P. G. (1975), *Solid State Commun.* **17**, 1193–1195.

Street, R. A. (1991), *Hydrogenated Amorphous Silicon*, Cambridge, U.K.: Cambridge University Press.

Zachariasen, W. H. (1932), *J. Am. Chem. Soc.* **54**, 3841–3851.

Zallen, R. (1983), *The Physics of Amorphous Semiconductors*, New York: Wiley.

Zawadzki, W. (1982), in: W. Paul (Ed.), *Handbook on Semiconductors*, Vol. 1, Amsterdam: North-Holland, Chap. 12.

Ziman, J. M. (1972), *Principles of the Theory of Solids*, Cambridge, U.K.: Cambridge University Press.

Further Reading

Beer, A. C. (1963), *Galvanomagnetic Effects in Semiconductors*, New York: Academic. Detailed treatment of theory and experiment for crystalline semiconductors.

Conwell, E. M. (1967), *High Field Transport in Semiconductors*, New York: Academic. Qualitative and quantitative discussion of hot electrons in crystalline semiconductors.

Lampert, M. A., Mark, P. (1970), *Current Injection in Solids*, New York: Academic. Standard text for physics of charge transfer across interfaces.

Mott, N. F., Davis, E. A. (1979), *Electronic Processes in Non-Crystalline Materials*, Oxford: Clarendon. Comprehensive description of electronic and optical properties of amorphous semiconductors.

Paul, W. (Ed.) (1982), *Handbook on Semiconductors*, Vol. 1, Amsterdam: North-Holland. General introduction to band structure and transport aimed at graduate students and young researchers in the field.

INSULATORS, STRUCTURE OF

CRONIN B. VINING, *Jet Propulsion Laboratory/California Institute of Technology, Pasadena, California, U.S.A.*

	Introduction	85
1.	**Properties of Insulators**	86
1.1	Dielectric Properties	86
1.1.1	The Complex Permittivity	86
1.1.2	Polarization Mechanisms	87
1.1.3	Typical Permittivity Behavior	88
1.1.4	Piezoelectricity	90
1.1.5	Pyroelectricity	91
1.1.6	Ferroelectricity	91
1.1.7	Lattice Dynamics and Soft Modes	92
1.1.8	Antiferroelectric and More Complex Order	92
1.2	Magnetism	93
1.2.1	Diamagnetism	93
1.2.2	Paramagnetism	93
1.2.3	Ferromagnetism	93
1.2.4	Antiferromagnetism and More Complex Order	94
1.3	Thermal Properties	94
2.	**Insulating Materials**	95
2.1	Elemental Insulators	95
2.2	Binary Compound Insulators	96
2.2.1	Covalent Insulators	96
2.2.2	Ionic Insulators	96
2.2.3	Binary Oxides and Other Binary Insulators	96
2.3	Multinary Compound Insulators	97
2.3.1	Rochelle Salts and Other Order/Disorder Systems	97
2.3.2	Perovskites and Displacive Transitions	98
2.3.3	Bronzes, Pyrochlores, and Nonmagnetic Oxides	99
2.3.4	Ferrites: Spinels and Garnets	100
2.3.5	Organics	100
3.	**Summary**	101
	Glossary	101
	Works Cited	102
	Further Reading	102

INTRODUCTION

Most of the substances around us, whether natural or synthetic, are electrical insulators. The distinction between insulators and metals has been recognized for centuries, and even children can tell the difference: metals are typically soft and malleable and can be polished to a high luster, while common insulators are hard, brittle, and often rather dull. Most electrical insulators are also thermal insulators. Hot beverages are served in electrical insulators, not metals, to retain the heat better and to prevent injury to the hand while drinking. These are among the more common properties, but insulators exhibit a broad range of characteristics that are perhaps less obvious, but just as important.

Insulators are needed whenever electrical current flow must be controlled, whether as a nerve impulse in a living organism, a bit of information in a digital computer, or the industrial transmission of electrical power. Certain insulators greatly enhance the performance of capacitors by increasing the amount of charge that can be maintained using a particular applied voltage. Other insulators respond to mechanical, thermal, magnetic, electrical, or optical stress in ways that make possible a wide variety of applications such as transducers, filters, motors, detectors, resonators, oscillators, imagers, memory devices, video displays, and many more. The tremendous variation of the properties of insulators plays a major role in the modern technological world, and it is important, therefore, to have some understanding of the relationship between their structure and their properties.

An insulator may be defined as any material that does not allow net movement of electric charge from one region of the material to any other region. The distinction between insulator and conductor, however, is relative and concerns both the conductivity of

1-56081-067-X/94/$5.00+.50

the surrounding medium and the duration of the observation. Every material conducts electricity to some extent, so that given enough time even the best electrical insulators will allow some charge motion. Among the best electrical insulators are diamonds, with electrical conductivities of 10^{-13} S m^{-1} or less. But the best metals, like silver, have conductivities of around 10^{8} S m^{-1}, more than 20 orders of magnitude larger. As a rough guideline, anything with a conductivity less than 10^{-3} S m^{-1} may be considered an insulator.

All materials are composed of atomic nuclei and enough electrons to maintain charge neutrality. An electron in such a material may be tightly bound, always remaining near a particular atomic nucleus (or group of nuclei), or it may be free to migrate from one region in the material to another. A material with a large number of free electrons is either a semiconductor or a metal. In an insulator, there is a negligible number of free carriers and a substantial activation energy for the creation of free carriers from the bound electrons.

Electrical insulation can result from a number of distinct mechanisms. In gases, the thermal activation energy is a result of the high ionization energy of isolated atoms. In liquids, glasses, and amorphous solids the thermal activation energy is a result of the atomic disorder that makes migration of carriers a relatively tortuous process. Crystalline solids sometimes have exactly enough electrons to fill entire bands of electronic states completely, leaving the next state available for free electrons at a significantly higher energy, called the band-gap energy. In plastic, polymeric, and liquid-crystal materials various combinations of ionization energies, disorder, and band filling may contribute to a substantial activation energy.

This binding of electrons in an insulator implies that the electrical conductivity, which is proportional to the number of free carriers, will be thermally activated:

$$\sigma=\sigma_o e^{-E_g/2kT}, \tag{1}$$

where E_g is the band-gap energy (or possibly some other, related activation energy). The division is somewhat arbitrary, but typically the term "electrical insulator" is reserved for a material with a band gap of at least 3 eV. The following sections are most appropriate for the properties and structure of homogeneous, crystalline insulators, but some other insulators will be briefly mentioned, and the concepts involved are of general utility. For more detailed information on amorphous materials, electronic structure of solids, polymers, liquid crystals, ceramics, and composite materials see the related articles in this encyclopedia. An overview of the important properties of insulators follows, including a brief discussion of the "soft mode" concept. The section on materials discusses the occurrence of insulating solids among the elements, binary compounds, and certain classes of more complex materials.

1. PROPERTIES OF INSULATORS

1.1 Dielectric Properties

The term "dielectric" emphasizes the response of an insulator to an applied field. The principal properties of interest in this regard are the permittivity (ϵ) and the loss tangent ($\tan\delta$), which characterize both the degree of polarization in an applied field and the losses that result from application of an alternating frequency field. A brief outline of dielectric properties is provided below, but for a more complete discussion see DIELECTRIC PROPERTIES OF INSULATORS.

1.1.1 The Complex Permittivity. When an electric field E is applied to a dielectric material, the charges within the material experience forces that tend to redistribute the charges. This redistribution is characterized by the polarization P, which represents the electric dipole moment induced in the material by the applied electric field. It is also convenient to define a third quantity, called the displacement field D, which represents the effects of the actual charges in the material and is given by

$$D=\epsilon_v E+P, \tag{2}$$

where ϵ_v is the permittivity of the vacuum.

In an ordinary dielectric, the polarization, the electric field, and the displacement field will all be proportional to each other. We define

$$D=\epsilon E=\epsilon_v\epsilon_r E, \tag{3}$$

where ϵ_r is the relative permittivity, a unitless quantity sometimes called the dielectric constant, although this term is more properly reserved for situations in which ϵ_r is actually independent of frequency. The polarization is now given by

$$P=\epsilon_{\mathrm{v}}(\epsilon_r-1)E. \tag{4}$$

The vacuum clearly has no polarization and $\epsilon_r^{\mathrm{vacuum}}=1$.

As an example, consider a capacitor in a simple charging circuit. The effect of inserting a dielectric material between the plates of the capacitor is to increase the capacitance (the charge the capacitor can hold) by the factor ϵ_r. If the applied voltage is held constant, eventually only a small leakage current will flow because of parasitic conduction through the dielectric material. If the applied voltage is sinusoidal in time, there will be a leakage current in phase with the applied voltage as well as a charging current that precedes the applied voltage by a 90° phase angle.

Thus, the total current through the capacitor (Fig. 1) is given by

$$i=j\omega\epsilon' E+\omega\epsilon'' E=\epsilon\frac{dE}{dt}, \tag{5}$$

where the first term represents the charging current and the second term represents the loss current. The permittivity is now considered to be a complex quantity with both a real and an imaginary component:

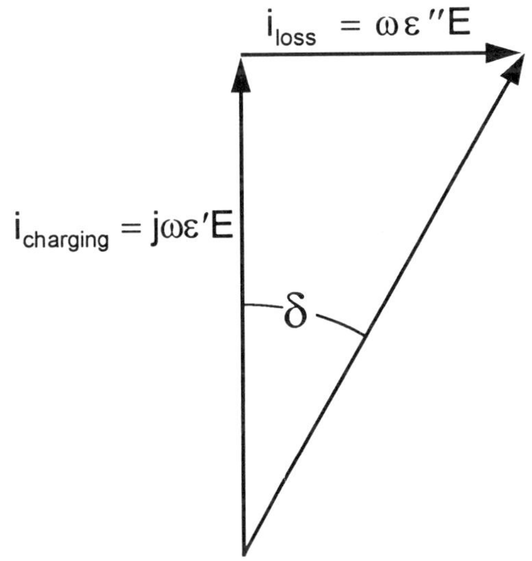

FIG. 1. Charging and loss currents, where tanδ is the loss tangent.

$$\epsilon=\epsilon'-j\epsilon'', \tag{6}$$

where the imaginary component is connected to the frequency-dependent electrical conductivity as

$$\sigma=\omega\epsilon''. \tag{7}$$

A useful quantity for characterizing an insulator is the so-called loss tangent (also called the dissipation factor), defined as the ratio of the loss current to the charging current:

$$\tan\delta=\frac{i_{\mathrm{loss}}}{i_{\mathrm{charging}}}=\frac{\epsilon''}{\epsilon'}. \tag{8}$$

The loss tangent is a material property related to the complex permittivity, which characterizes the electrical losses in an insulator.

1.1.2 Polarization Mechanisms. To understand the origin of polarization effects in materials it is best to begin with the smallest structural units involved, either the individual atoms or small groups of atoms. The electric dipole moment of a single atom (or small group) is taken to be proportional to the local electric field as

$$p_i=\alpha_i E_{\mathrm{local}}, \tag{9}$$

where α_i is the polarization factor. The polarization factor is a microscopic quantity and is convenient because it can be calculated from theory. However, the α_i's are connected to the local electric field, not to macroscopic quantities that may be measured in the laboratory: E, P, and D. In the simplest approximation, the local field is given by

$$E_{\mathrm{local}}=E+P/3\epsilon_{\mathrm{v}}. \tag{10}$$

Combining Eqs. (4), (9), and (10) leads to the Clausius-Mossotti equation, completing the required relationship between the microscopic polarization factors and the macroscopic relative permittivity:

$$\frac{\epsilon_r-1}{\epsilon_r+2}=\frac{1}{3\epsilon_{\mathrm{v}}}\sum_i N_i\alpha_i, \tag{11}$$

where N_i is the number density of dipoles with polarization factor α_i.

The polarization has its origins in the structure of the insulator and may have a number of contributing factors. Figure 2 summarizes the most important mechanisms for polarization in a solid. All materials are subject to *electronic polarization* (also called *atomic* or *optical polarization*) because in an applied electric field the atomic nucleus is displaced with respect to the surrounding electrons, inducing a dipole moment. In the absence of an applied field, the charges within an atom will tend to restore the symmetry of the atom and remove the dipole moment. An atom, therefore, may be thought of as a simple harmonic oscillator with a rather high natural frequency, typically in the ultraviolet range due to the small mass of the electrons. Using this description, and including a damping term to represent losses in the oscillator, the electronic polarization can be shown to have the form given in Fig. 3.

Ionic polarization results from the displacement of cations (positive ions) in the direction of the applied field and anions (negative ions) in the opposite direction, as indicated schematically in Fig. 2. These processes also may be thought of as resulting from simple harmonic oscillators, in this case involving the relative motion of cations and anions in the crystal. Ionic polarization, then, has the same functional form as electronic polarization, but since the motion here involves entire atoms, the natural frequencies of the oscillators are much lower and correspond to lattice vibration frequencies, typically in the infrared or visible region.

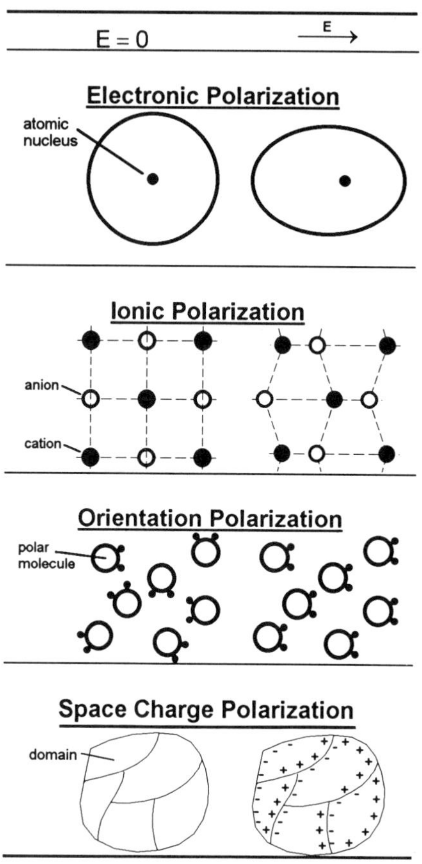

FIG. 2. Major polarization mechanisms. The displacement of the charges has been strongly exaggerated.

Orientation (or *dipole*) *polarization* occurs when a structural repeat unit or a polar domain with a permanent dipole moment (p_m) experiences a torque, tending to orient the dipole in the direction of the applied field. Thermal motion tends to randomize the orientation of dipoles. In the absence of an applied field a system of oriented dipoles will "relax" toward a random distribution of orientations with a characteristic time τ, called the relaxation time. Using this relaxation model, the orientation polarization can be shown to have the form shown in Fig. 3, with characteristic frequencies (τ^{-1}) ranging into the microwave region.

Space charge (or interfacial) polarization represents the contribution from charge carriers that normally are free to move through the solid, but become impeded by grain boundaries, electrodes, and the like as suggested in Fig. 2. In this case, a space charge eventually builds up, generating a field to counter the applied field until current ceases to flow.

1.1.3 Typical Permittivity Behavior. For the electronic, ionic, and orientational mechanisms three distinct frequency regimes may be identified: first, a low-frequency regime, where the applied field varies slowly enough for the polarization to track the applied field closely and the polarization factor is essentially a real constant; second, a high-frequency regime where the applied fields vary so rapidly that the polarization can not track the applied field at all and the polarization factor tends to zero; and finally, an intermediate regime near the characteristic frequency of the oscillators or relaxation mechanism, where both the real and imaginary components of the polarization factor vary relatively rapidly with frequency. It is only in this intermediate-frequency regime that α'' becomes of significant magnitude.

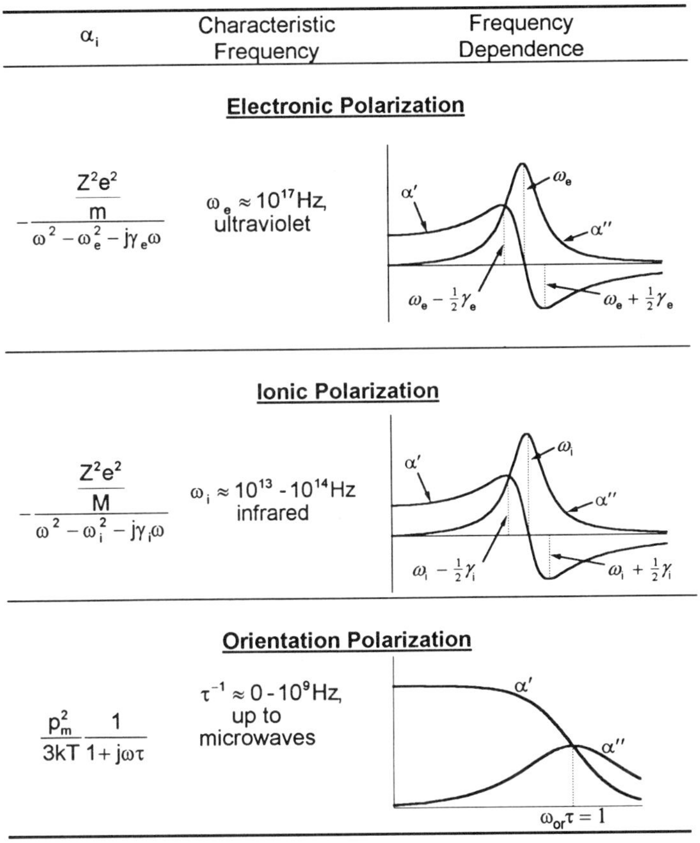

FIG. 3. Frequency dependence of the major polarization mechanisms.

Ordinarily the four polarization mechanisms may be considered as acting independently, in which case the total polarization factor is given by a sum of the four contributions:

$$\alpha_i = \alpha_{i,e} + \alpha_{i,i} + \alpha_{i,o} + \alpha_{i,s} \,, \tag{12}$$

where the second subscript indicates the contributions from the different mechanisms. Often, since the characteristic frequencies are so different for the different mechanisms, only one of the contributions will have a strong frequency dependence in any particular range of frequencies and the other mechanisms can be considered to provide a frequency-independent background.

The relative permittivity of a material can now be calculated using the sum of the individual contributions to the polarization factor and the Clausius-Mossotti relation. The relative permittivity typically has a frequency dependence similar to the frequency dependence of the contributing polarization factors, represented in Fig. 3. The real component of the relative permittivity is larger at low frequencies (usually called the static dielectric constant, ϵ_0) than at high frequencies (usually called the optical dielectric constant, ϵ_∞) and will vary rapidly only over a relatively narrow range around the characteristic frequency. At frequencies well beyond the ultraviolet, the relative permittivity would drop even further as even electronic polarization processes cannot track the applied field at these frequencies.

At frequencies below the ultraviolet, many materials exhibit a frequency-independent contribution to the polarization from electronic polarization processes and a frequency-dependent contribution to the polarization from ionic polarization processes. The relative permittivity in this case is given by

$$\epsilon_r(\omega) = \epsilon_\infty + \frac{\epsilon_\infty - \epsilon_0}{\omega^2/\omega_T^2 - 1} \,, \tag{13}$$

where

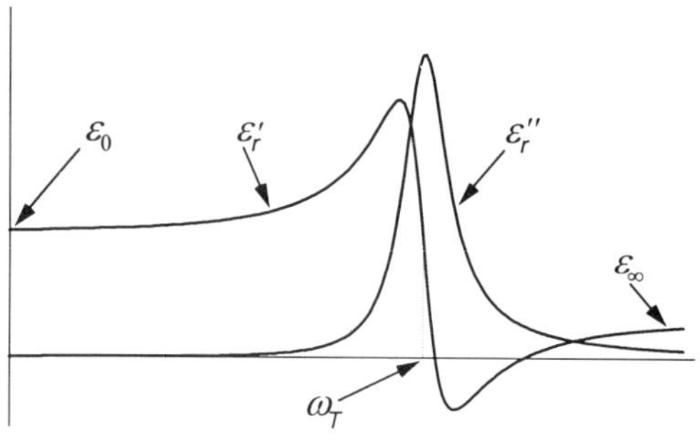

FIG. 4. Real and imaginary components of the relative permittivity typical of an ionic contribution to the polarization.

$$\omega_T^2=(\omega_i^2+j\gamma_i\omega)\left(\frac{\epsilon_\infty+2}{\epsilon_0+2}\right). \tag{14}$$

The real and imaginary components are shown in Fig. 4, where the change in sign of the real component just above ω_T is typical of an ionic contribution to the polarization.

Qualitatively different behavior characterizes the dipole orientation polarization mechanism, where the relative permittivity is given by

$$\epsilon_r(\omega)=\epsilon_\infty+(\epsilon_0-\epsilon_\infty)/(1+j\beta) \tag{15}$$

with

$$\beta=[(\epsilon_0+2)/(\epsilon_\infty+2)]\omega\tau\ . \tag{16}$$

Here $\epsilon_r'(\omega)$ decreases monotonically from ϵ_0 at low frequencies to ϵ_∞ at high frequencies, mimicking the frequency dependence of the orientation polarization factor shown in Fig. 3. In this case, a plot of $\epsilon_r''(\omega)$ as a function of $\epsilon_r'(\omega)$ has the particularly simple form of a circle, as indicated in Fig. 5. This type of graphical representation, called a Cole-Cole diagram, can be used to determine the relaxation time, τ. The procedure involves connecting a point $(\epsilon_r'(\omega),\ \epsilon_r''(\omega))$ at any particular frequency with either the high-frequency limit $(\epsilon_\infty,0)$ or the low-frequency limit $(\epsilon_0,0)$, as indicated by the straight dotted lines in Fig. 5. β may be estimated from the slope of either line and the relaxation time is then given by Eq. (16). Cole-Cole diagrams are useful in identifying relaxation-type polarization behavior, such as is typical of dipole orientation polarization, since other mechanisms appear more complex and qualitatively distinct when represented in this manner, as suggested in Fig. 6.

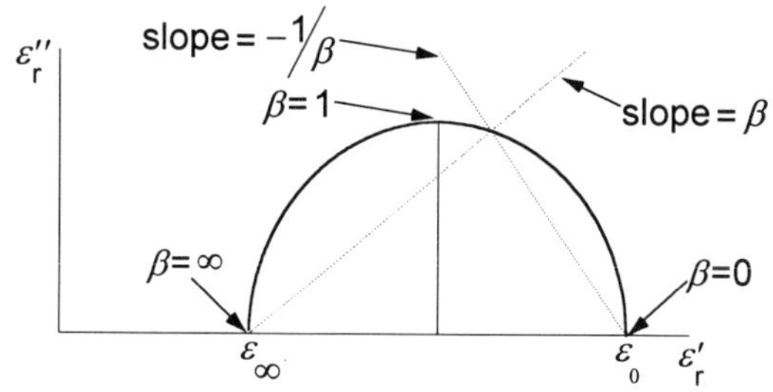

FIG. 5. Cole-Cole diagram for an ideal relaxation-time mechanism.

1.1.4 Piezoelectricity. Piezoelectric materials exhibit two characteristic effects: polarization in response to applied mechanical stress, and mechanical deformation in response to an applied electric field. Both effects are linear and change sign when the applied stress (mechanical or electrical) changes sign. A crystal with a center of symmetry cannot be piezoelectric, since mechanical stress alone will not disturb the balance among the dipole moments within a unit cell. Crystals with a center of symmetry will mechanically deform in an applied electric field, but the deformation will be proportional to the square of the applied electric field. This effect, also called electrostriction, is independent of the sign of the applied field and is usually quite small in magnitude.

Piezoelectricity is possible in any noncentrosymmetric crystal (which includes all the pyroelectric materials discussed below), as illustrated in Fig. 7. The Curie brothers (1880) discovered piezoelectricity in quartz, tourmaline $\{Na(Mg,Fe,Mn,Li,Al)_3Al_6[Si_6O_{18}]$-$(BO_3)_3(OH,F)_4\}$, and Rochelle salt. More recently (see PIEZOELECTRIC DEVICES) there has been great interest in organic materials such as polyvinylidene fluoride (PVDF).

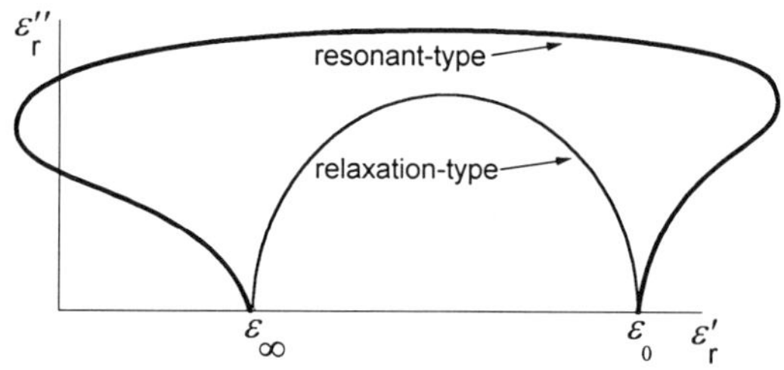

FIG. 6. Cole-Cole diagram illustrating qualitatively distinct behavior for different polarization mechanisms.

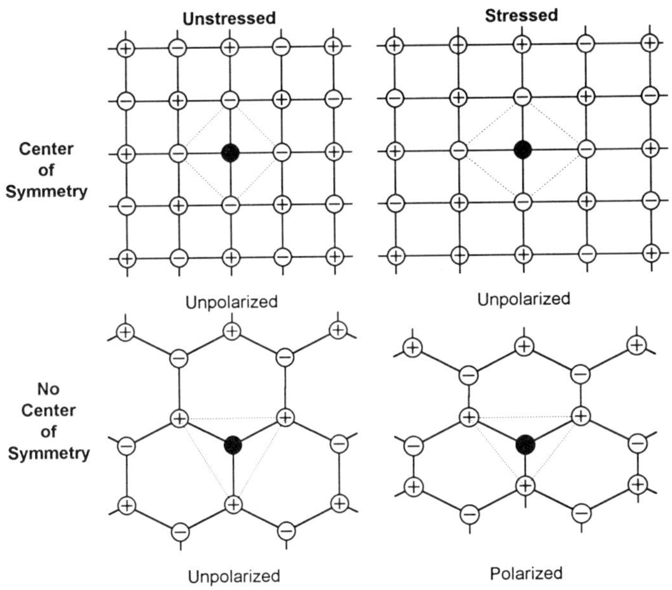

FIG. 7. Crystals with a center of symmetry are not polarized, even under stress. Piezoelectrics, without a center of symmetry, polarize under stress.

1.1.5 Pyroelectricity. Certain materials are called pyroelectric because they appear to generate a spontaneous polarization when heated (or cooled). In fact, polarization is always present internally in these materials, but is ordinarily masked by neutralizing charges collecting on the exposed surfaces. A net polarization is temporarily observed with temperature changes until the surface charges can reequilibrate with the temperature-dependent internal polarization. The observation of an internal polarization implies severe constraints on the symmetry of the crystal: there can be no mirror planes perpendicular to the axis of polarization and no rotational symmetry except along an axis parallel to the polarization axis.

1.1.6 Ferroelectricity. A material that is nonpyroelectric above a certain temperature T_C (the Curie temperature) and pyroelectric below T_C is called a ferroelectric. Just below T_C the spontaneous polarization is usually small and may be readily reversed in direction by an applied field opposing the original polarization, as shown by the characteristic hysteresis loop in Fig. 8. This hysteretic behavior in the polarization and the ability to reverse the polarization by an applied field is sometimes taken as the defining property of a ferroelectric. The term ferroelectric is in analogy to similar behavior observed in the magnetization of ferromagnetic materials.

The origin of ferroelectricity lies in the interaction between the local polarization and the macroscopic electric field. Solving Eqs. (9) and (10) for the total polarization yields

$$P=\frac{\Sigma_i N_i \alpha_i}{1-(1/3\epsilon_0)\Sigma_i N_i \alpha_i}E. \tag{17}$$

The orientational contribution to the polarization, $\alpha_{i,o}$, typically increases as $1/T$ as the temperature decreases, leading eventually to a zero in the denominator of Eq. (17) at some temperature, implying spontaneous polarization. Here, the local field produced by a dipole has a sufficiently strong effect on neighboring dipoles that all of the dipoles prefer to orient in the same direction, even in the absence of

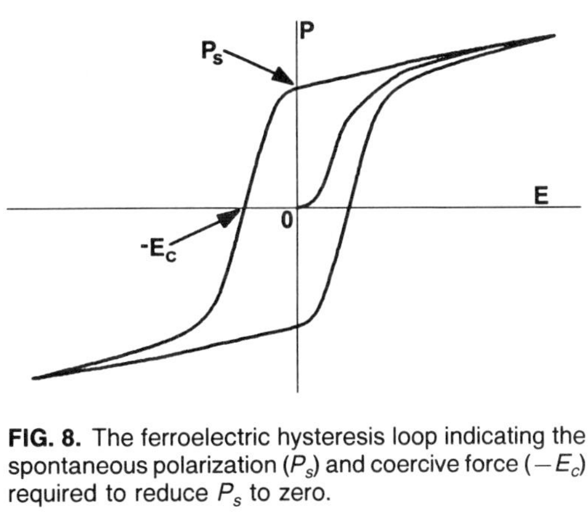

FIG. 8. The ferroelectric hysteresis loop indicating the spontaneous polarization (P_s) and coercive force ($-E_c$) required to reduce P_s to zero.

an applied field. This critical temperature is the Curie temperature T_C, and the dielectric susceptibility takes the form

$$\chi \equiv \epsilon_r - 1 = C/(T - T_C), \tag{18}$$

where C is a constant, again in close analogy to the magnetic susceptibility of a ferromagnet.

The dipole moments that can be induced by applied fields or stress above T_C become permanent below T_C. The mechanisms responsible for ferroelectricity vary from material to material and will be discussed in the materials section below. Generally, however, ferroelectricity results from a lowering of the symmetry of the crystal lattice, similar to that suggested in Fig. 7 except that an applied stress is no longer required to maintain the polarized state. Blinc and Žekš (1974) list over 300 known inorganic ferroelectric compounds (not counting solid solutions), and Fousek (1991) estimates that over 50 families of liquid-crystal ferroelectrics are now known. A few representative materials exhibiting a variety of properties are listed in Table 1.

1.1.7 Lattice Dynamics and Soft Modes. The permanent polarization in the low-temperature form of a ferroelectric represents a modification of the lattice compared to the more symmetric, high-temperature form of the material. The lattice modification can be examined with experiments such as x-ray or neutron diffraction that are sensitive to the atomic positions. Cochran (1959) and Anderson (1959) introduced the concept that ferroelectricity results from an instability of one of the normal lattice vibrational modes. The characteristic frequency of the so-called soft mode, referring to the unstable lattice vibration, decreases on approach to the Curie temperature, T_C, and the restoring force tending to return the displaced atoms to their normal positions approaches zero. The soft mode then freezes in and the atoms become permanently displaced, resulting in the ferroelectric state.

While a detailed discussion of soft modes is well beyond the scope of this article, it is important to emphasize the close connection between the stability of the lattice with respect to distortion and the occurrence of ferroelectricity. Much of the study of ferroelectrics has been directed toward a detailed understanding of the occurrence, symmetry characteristics, and energetics of soft modes. Blinc and Žekš (1974) provide a review of theoretical and experimental techniques in the study of soft modes. Articles in this encyclopedia on phonons and lattice vibrations may also be of interest in this regard.

1.1.8 Antiferroelectric and More Complex Order. While individual dipoles orient in the same direction in a ferroelectric (at least within a single domain), this is by no means the only possible arrangement of dipoles. The first soft-phonon mode to freeze into a static displacement may have a periodicity identical to the original lattice unit cell or a simple multiple of the unit cell, or may even be incommensurate with the original lattice. In an antiferroelectric material, for example, there are two sets of dipole moments, each in a regular arrangement and alternating in direction. In this case there is no net polarization as one set of dipoles precisely cancels the other. $PbZrO_3$, the first known antiferroelec-

Table 1. Typical ferroelectric (F) and antiferroelectric (A) materials.

Family name	Formula		T_C (K)	P (C/m^2)	at T (K)
TGS (triglycine sulfate)	$(C_2H_5NO_2)_3 \cdot H_2SO_4$	F	322	0.03	293
KDP (potassium dihydrogen phosphate)	KH_2PO_4	F	123	0.05	96
ADP (ammonium dihydrogen phosphate)	$NH_4 \cdot H_2PO_4$	A	148	—	
Perovskite	$NaNbO_3$	A	627	—	
Perovskite	$PbTiO_3$	F	763	>0.50	300
Rochelle salt	$NaKC_4H_4O_6 \cdot 4D_2O$	F	297[a] 255[b]	0.003	278
PVDF (polyvinylidene fluoride)	$(CH_2CF_2)_n$	F	≈370	0.1	300

[a]Upper T_C.
[b]Lower T_C.

tric, was identified by Kittel (1951), who also coined the term.

$Gd_2(MoO_4)_3$ has been found to be a ferroelectric, but the transition to the ferroelectric state is not associated with any observable dielectric anomaly and the substance is called an "improper" ferroelectric since the spontaneous polarization does not alter the symmetry of the crystal. Other and increasingly subtle types of dipole ordering are now recognized, in both homogeneous and inhomogeneous materials, and have been reviewed by Fousek (1991).

1.2 Magnetism

Magnetite (Fe_3O_4) may have been used as a compass by the Chinese as early as 2500 B.C., and magnetic materials continue to play an important role today as inductor cores, magnetic recording media, and permanent magnets for use in motors, generators, loudspeakers, and much more. Insulating magnetic materials are preferred to electrically conducting magnetic materials for high-frequency applications because the absence of induced eddy currents, which reduces power losses, makes them especially important for the manufacture of inductors.

The magnetic behavior of materials is no less varied than the electrical behavior described in the preceding sections. A more complete discussion is provided in the articles on magnetism and magnetic devices, but in broad terms the magnetic field H, the magnetic induction B, magnetic permeability μ, and magnetic susceptibility χ_m parallel the roles of the electric field E, the displacement field D, the permittivity ϵ, and the electric susceptibility χ already described. Caution must be exercised in pressing the analogies too far, however, because the physical origin of local magnetic-dipole moments is quite distinct from the origin of local electric-dipole moments, as will be described in the following sections.

1.2.1 Diamagnetism. Insulators composed of ions with all electronic shells filled are typically diamagnetic, and the total diamagnetic susceptibility may be reliably estimated from the contributions of the individual ions. The effect of an applied magnetic field, in this case, has been described as inducing a microscopic current in the orbital electrons about each atom. According to Lenz's law, the induced currents produce a weak magnetic field tending to cancel the applied magnetic field. A complete description of this behavior requires a quantum-mechanical treatment, but the effect is called diamagnetism. Under certain conditions superconducting materials exhibit "perfect diamagnetism," where the induced currents completely cancel the applied field and prevent the penetration of magnetic field to the interior of the superconductor. Diamagnetism arising from the orbital motion of electrons within an atom is about a million times weaker than perfect diamagnetism and is temperature independent. Although typical of most insulators, this weak diamagnetism is of little technological interest.

1.2.2 Paramagnetism. Paramagnetism refers to an induced magnetic field in the same direction as the applied field, which is to say, a positive magnetic susceptibility. Paramagnetism occurs in insulators composed of ions with partially filled $3d$ (Ni, Fe, Mn, etc.) or $4f$ (Sm, Gd, Dy, etc.) shells, representing first-row transition-metal ions and the lanthanides, respectively. The electrons occupying an incompletely filled shell experience electron-electron Coulomb interactions as well as an interaction between the electron spin and the orbital motion. These two interactions tend to keep the electron spins aligned, with the result that the ion has a permanent magnetic-dipole moment. The magnetic response of a solid composed of such ions will depend on temperature following Curie's law, analogous to Eq. (18), where the decrease in the susceptibility at high temperatures reflects the inability of an applied field to overcome the increased thermal agitation of the local moments.

1.2.3 Ferromagnetism. When in close proximity, the magnetic field produced by one localized magnetic moment may directly influence the orientation of the magnetic moment of a neighboring ion. In this case, the localized moments no longer behave entirely independently, and large numbers of moments may spontaneously orient in the same direction. This behavior is called ferromagnetism [Fig. 9(a)], and the temperature at which ferromagnetic order sets in will depend on the strength of the interaction between

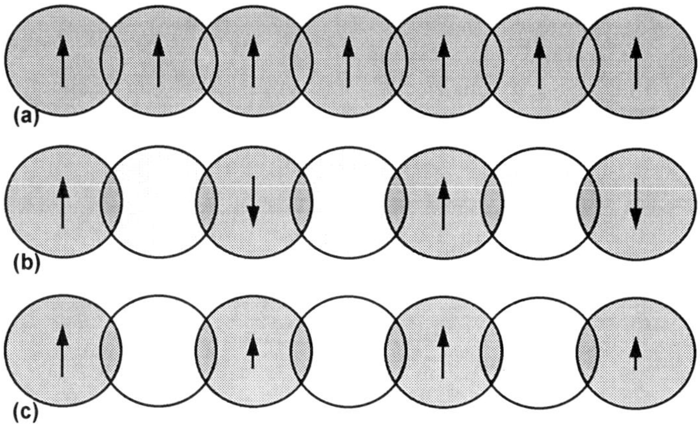

FIG. 9. (a) Direct exchange, local moments directly overlap leading to ferromagnetic order; (b) superexchange, local moments interact via an intermediate nonmagnetic ion leading to antiferromagnetic order; and (c) superexchange leading to ferrimagnetic order.

local moments. Above the ordering temperature, such materials exhibit paramagnetic behavior.

1.2.4 Antiferromagnetism and More Complex Order. In many crystal structures, local magnetic moments are separated by nonmagnetic ions, as illustrated in Fig. 9(b). Since the direct interaction between local moments decreases rapidly with increasing distance, the nonmagnetic ion can mediate a magnetic interaction that is stronger than the direct interaction. This type of interaction is called superexchange and is typically negative in sign, leading to the alternating spin-up, spin-down arrangement known as antiferromagnetism, as suggested in Fig. 9(b). An antiferromagnet exhibits no net magnetization, and the magnetic susceptibility, which peaks in a cusp at the antiferromagnetic ordering temperature, need not reach extremely large values. Complex crystal structures with more than one type of magnetic ion may exhibit a wide variety of magnetic ordering behavior with the magnetic moment varying in magnitude, as suggested by Fig. 9(c), or direction from one crystallographic site to another.

1.3 Thermal Properties

The heat capacity of insulators is remarkably well behaved, and even relatively simple models based on the statistical mechanics of phonons provides a quite accurate description. At high temperatures, each possible vibrational mode of an insulator is equally excited and the heat capacity approaches the temperature-independent Dulong-Petit value of $3k$ for each atom in the crystal. At low temperatures, the lattice vibrations decrease rapidly and the heat capacity varies as T^3 with relatively minor and well-understood deviations, except for materials with special considerations such as phase transitions.

Metals are typically good thermal conductors because of the ability of free electrons to conduct heat as well as electricity. The thermal conductivity of insulators is somewhat more complex, as discussed in more detail in THERMAL CONDUCTION IN SOLIDS, and can be either large or small depending on the material. Essentially, the heat carried by phonons in an insulator can be estimated from a knowledge of how fast and how far a phonon moves between scattering events. The thermal conductivity may be expressed as

$$\lambda=\tfrac{1}{3}Cv_sl\,, \qquad (19)$$

where C is the heat capacity per unit volume, v_s is the speed of sound, and l is the phonon mean free path. The speed of sound typically reflects the bond strength, and high–melting-point materials tend to have a high sound velocity and, therefore, high thermal conductivity. If there is only one atom per unit cell, all the phonon modes are acoustic modes with relatively high propagation velocities. As the crystal structure becomes more complex, the unit cell grows, and more phonon modes become optical in character with very low propagation velocity. Thus, the more complex the crystal structure, the lower the thermal conductivity is expected to be. Amorphous materials, such as fused silica, may be considered the limiting case in regard to structural complexity, and indeed these materials have among the lowest thermal conductivity values known.

In materials relatively free of defects, phonons may propagate a long distance before scattering, and in this case the long mean free path leads to a high thermal conductivity. Classic examples are diamond and aluminum nitride, which have thermal conductivity values greater than metals. Even if no other defects are present, the phonons themselves may be thought of as deviations from the ideal crystal structure and phonon-phonon scattering is often the limiting factor. At high temperatures, phonon-phonon scattering leads to a $1/T$ temperature dependence. At low temperatures, as phonon-phonon scattering decreases, the phonon mean free path becomes larger until the walls of the specimen prevent further propagation. Eventually, the thermal conductivity decreases as T^3 with decreasing temperature, reflecting the temperature dependence of the heat capacity discussed above.

2. INSULATING MATERIALS

Electrically insulating materials are far too common to enumerate in any complete way, but some appreciation of the trends and patterns may be achieved by a brief survey. Some of the more important materials will be discussed, as well as the connection between the structure of the materials and their properties.

2.1 Elemental Insulators

The rare gases, hydrogen, nitrogen, oxygen, and fluorine and chlorine are insulating gases under standard conditions and remain insulators even when they condense to liquids and finally solids at sufficiently low temperatures. Bromine is an insulating liquid and iodine and astatine are insulating solids under standard conditions. The common, stable form of selenium has infinite helical chains and is, like tellurium, semiconducting, but Se_8 polymorphs are red insulators similar to solid S_8, which is a very poor electrical conductor. Arsenic is a semimetal, and the electrical conductivity of phosphorus varies greatly among the four known allotropes, but the volatility and reactivity of these elements severely limit their use as solid insulators. Boron, the only refractory nonmetallic element other than carbon, occurs in two distinct crystal structures (the α and β forms), both of which are semiconductors with band gaps around 1.5 to 2.0 eV.

Carbon, in the diamond crystal structure shown in Fig. 10, is expected to become increasingly important since the development of diamondlike thin films. The diamond crystal structure may be described as two identical, interpenetrating face-centered-cubic (fcc) lattices. The second fcc lattice is displaced from the first by a translation along a diagonal of the cube by a distance of $\frac{1}{4}$ of the length of the diagonal. With its tightly bound electrons and strong covalent bonds, diamond is uniquely suited for many applications. The electrical properties of diamond are dominated by the very large band gap of 5.4 eV, which leads to electrical conductivity values as low as 10^{-13} S m^{-1}, although doping can increase this value to 10^{0} S m^{-1} or more. The strong bonds and small mass of carbon lead to sound velocity and thermal conductivity values among the highest known. For more information the reader is referred to DIAMOND AND DIAMONDLIKE CARBON.

Other forms of carbon are not electrical insulators. Graphite, for example, consists of two-dimensional sheets of hexagonally bonded carbon with only weak bonding between the sheets. Graphite is typically a fairly good electrical conductor, with conductivity values on the order of 10^5 S m^{-1}. Thus, by changing the crystal structure from the diamond arrangement to the graphite arrangement the

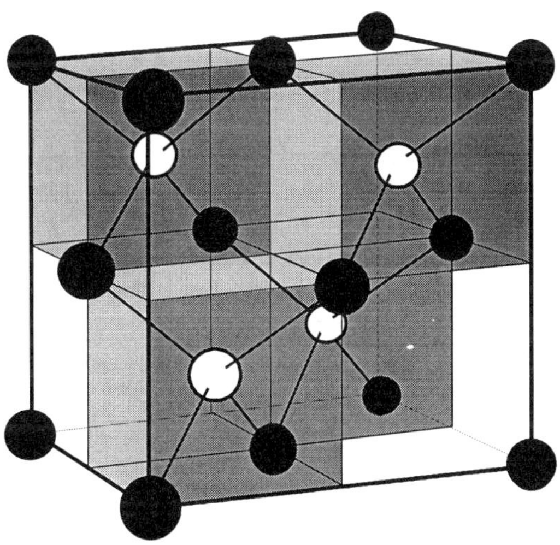

FIG. 10. The sphalerite (or zinc-blende) crystal structure. Also, the diamond crystal structure when the solid and open circles represent the same atom.

Table 2. Selected covalent insulators.

Structure type	Compound
Diamond	C
Sphalerite	BAs, BP, BeTe, ZnSe, BeSe, ZnS, CuBr, BN, BeS, ZnO, MgO
Wurtzite	BP, AgI, SiC, ZnS, BN, GaN, AlN, ZnO, BeO
Hexagonal	α-Si_3N_4, β-Si_3N_4, SiO_2

conductivity can increase by 18 orders of magnitude, which emphasizes the role of structure in determining the electrical properties of solids.

All of the remaining elements are semiconductors, semimetals, or true metals. Thus, among the elements, solid electrical insulators are relatively rare.

2.2 Binary Compound Insulators

2.2.1 Covalent Insulators. Many covalently bonded compounds are semiconductors, but some important insulators found in this group are listed in Table 2. The first group includes those compounds that share the same basic structure as diamond, except that the two interpenetrating sublattices shown in Fig. 10 are occupied by different atoms. This is the sphalerite (or zinc-blende) structure type, and Table 2 lists a number of common insulators with this structure. The sphalerite structure is symmetric about each atom in the unit cell and therefore is neither pyroelectric nor piezoelectric.

Other covalent insulators form in the wurtzite structure, which can be considered as two interpenetrating close-packed hexagonal lattices. As in the sphalerite structures, each atom is tetrahedrally coordinated with four atoms of the opposite type. BN can be found in the zinc-blende or wurtzite forms, as well as in a graphitelike hexagonal structure, all of which are good insulators. SiC forms in a large number of closely related structures composed of sphalerite and wurtzite layers, and although it has interesting semiconducting properties, all SiC polytypes may be considered insulators if sufficiently pure. The ease with which the covalent bond is polarized generally increases with atomic number, which can lead to somewhat larger static dielectric constants than expected for a nonpolarizable bond, but qualitatively the permittivity behaves very much as expected on the basis of the ionic polarization mechanism described in Figs. 2, 3, and 4.

2.2.2 Ionic Insulators. The ionic crystal structures for sodium chloride (NaCl) and cesium chloride (CsCl) are shown in Fig. 11. Under ordinary conditions, all of the alkali halides form in the NaCl-type structure, except CsCl, CsBr, and CsI, which form in the CsCl-type structure. Both are high-symmetry structures, exhibit no piezoelectric or pyroelectric effects, and have dielectric behavior in good accord with the ionic polarization mechanism described above. In Table 3 note the distinctly greater polarizability of the CsCl-type compounds compared to the trends among the NaCl-type as indicated by the relatively larger static dielectric constants ϵ_0 of the former compounds. This indicates that even closely related crystal structures with similar bonding can exhibit distinct properties.

2.2.3 Binary Oxides and Other Binary Insulators. Although both insulators and conductors can be found in almost any family of materials, many binary chalcogenides (especially oxides and sulfides) and pnictides

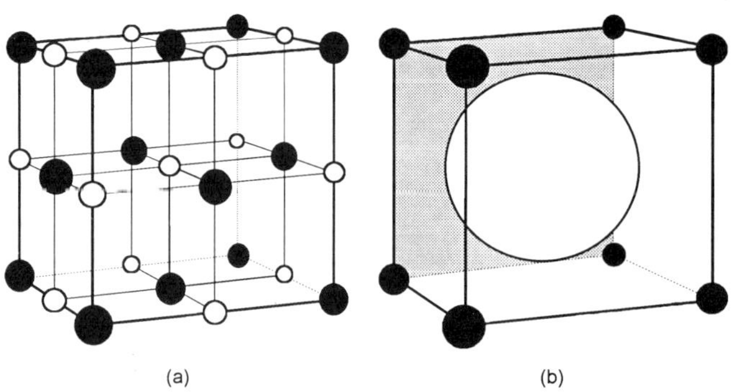

FIG. 11. (a) Sodium chloride and (b) CsCl crystal structure types.

(including nitrides and phosphides) and practically all halides are typically good electrical insulators. These compounds are neither purely covalent nor purely ionic, and while many are insulators, they tend to have relatively high symmetry crystal structures and fairly ordinary dielectric properties.

Most of the binary oxides are insulators, and materials such as alumina (Al_2O_3), magnesia (MgO), and beryllia (BeO) see widespread use for their dielectric properties. 3*d* transition-metal oxides such as Fe_3O_4 and CrO_2 are typically magnetic and exhibit the characteristic phenomena of saturation and hysteresis so important for magnetic recording applications. Many other 3*d* oxides such as MnO, FeO, and CrO are antiferromagnets, and in most cases the superexchange mechanism is responsible for the magnetic order.

Practically all the binary carbides and most of the borides not listed above, however, have at most relatively small band-gap energies and are semiconductors, and many, in fact, are metallic compounds. Only a handful of binary compounds (such as FeS, Bi_2S_3, WO_3 and possibly a few others) are ferroelectrics. Indeed, the scarcity of ferroelectric behavior among binary compounds left the impression for many years that ferroelectricity was rather rare, an impression later dispelled by studies of more complex materials. Finally, SiO_2 in its various forms deserves special attention and is discussed in dedicated articles on GLASSES and QUARTZ.

Table 3. Alkali halide ionic insulators in the NaCl structure type, except where indicated (after Knox and Teegarden, 1968).

Compound	ϵ_0	ϵ_∞	$\hbar\omega_T/k$ (K)
LiF	9.01	1.96	442
NaF	5.05	1.74	354
KF	5.46	1.85	274
RbF	6.48	1.96	224
CsF	—	2.16	125
LiCl	11.95	2.78	276
NaCl	5.90	2.34	245
KCl	4.84	2.19	215
RbCl	4.92	2.19	183
CsCl[a]	7.2	2.42	151
LiBr	13.25	3.17	229
NaBr	6.28	2.59	195
KBr	4.90	2.34	166
RbBr	4.86	2.34	139
CsBr[a]	6.67	2.42	114
LiI	16.85	3.80	—
NaI	7.28	2.93	167
KI	5.10	2.62	156
RbI	4.91	2.59	118
CsI[a]	6.59	2.62	95

[a]CsCl structure type.

2.3 Multinary Compound Insulators

As increasingly complex materials have come under investigation, new types of behavior have been observed and characterized. Often, these new behaviors have no precedent among the binary compounds or among the elements. As far as ordinary dielectric properties are concerned, polystyrene, polyethylene, Mylar, Teflon, and similar materials are not qualitatively superior to or even especially distinct from the classic binary compounds discussed above. Still, modern plastics and polymers are often vastly preferred in dielectric applications for cost, packaging, or other practical considerations.

Many of the more complex materials exhibit effects completely unobserved in the binary compounds. Table 4, for example, provides an overview of the ferroelectric materials now known to exist. Typically, as each class of materials has been developed, new theoretical concepts have been required as well.

2.3.1 Rochelle Salts and Other Order/Disorder Systems. The earliest known ferroelectrics were the Rochelle salts (also called Seignette salts), such as $NaKC_4H_4O_6 \cdot 4H_2O$ (Valasek, 1920, 1921). The orientation polarization mechanism is vividly demonstrated in these materials in that the hydrogen bonding is sufficiently weak to allow a type of hindered rotation of the water molecules. The orientation of one water molecule influences the preferred orientation of others nearby, and as the temperature is lowered through the Curie temperature, the water molecules tend to order with respect to each other, establishing a net polarization.

An interesting feature of the Rochelle salts is the presence of two critical temperatures, as indicated in Table 1. Above the upper transition temperature and below the lower temperature, two different sublattices of water mol-

Table 4. Overview of ferroelectric materials (modified from Fousek, 1991).

I. Inorganic compounds

	Nonhydrated	
Perovskites, $BaTiO_3$	K_2SO_4 type, K_2SeO_4	Barium fluoride type, $BaMnF_4$
$LiNbO_3$ type, $LiTaO_3$	Langbeinites, $(NH_4)_2Cd_2(SO_4)_3$	Halides, HCl
Manganites, $YMnO_3$	Sulfates, $(NH_4)_2SO_4$	Antimony sulfide iodine, SbSBr
Tungsten bronze type, $PbNb_2O_6$	Selenites, $LiH_3(SeO_3)_2$	Oxyfluorides, $Rb_3MoO_3F_3$
Pyrochlore type, Cd_2Nb	Nitrites, $NaNO_2$	Boracites, $Mg_3B_7O_{13}Cl$
Layer structure oxides, $Bi_4Ti_3O_{12}$	Nitrates, KNO_3	Iodates, NH_4IO_3
Molybdates, $Gd_2(MoO_4)_{32}O_7$	Thiophosphates, $Sn_2P_2S_6$	Oxyiodates, Sb_5O_7I
Vanadates, $Ca_3(VO_4)_2$	Potassium dihydrogen phosphate (KDP) type, KH_2PO_4	
Other double oxides, $SbNbO_4$		
	Hydrated	
Colemanite, $Ca_2B_6O_{11} \cdot 5H_2O$	Potassium cyanides, $K_4Mn(CN)_6 \cdot 3H_2O$	Alums, $NH_4Fe(SO_4)_2 \cdot 12H_2O$
Sulfates, $NaNH_4SO_4 \cdot 2H_2O$	Uranospinite, $H_2(VO_2)_2(AsO_4)_2 \cdot 8H_2O$	

II. Organic-inorganic compounds

Nonhydrated	
Triglycine sulfate (TGS), $(CH_2NH_2COOH)_3 \cdot H_2SO_4$	Sarcosine salt, $(CH_3NHCH_2COOH)_3CaCl_2$
Betaines, $(CH_3)_3NCH_2COO \cdot H_3AsO_4$	Tetramethyl ammonium salt, $N(CH_3)_4 \cdot HgCl_3$
Chloracetates, $(CH_2ClCOO)_2HNH_4$	Propionate, $(CH_3CH_2COO)_6Ca_2Sr$
Alkyl ammonium halogenantimonates, $(CH_3NH_3)_3Sb_2Br_9$	
Hydrated	
Rochelle salts (tartrates), $NaKC_4H_4O_64H_2O$	Formates, $Cu(HCOO)_2 \cdot 4H_2O$
Triglycine sulfate (TGS), $(CH_2NH_2COOH)_2 \cdot MnCl_2 \cdot 2H_2O$	Alums, $CH_3NH_3Al(SO_4)_2 \cdot 12H_2O$
Betaines, $(CH_3)_3NCH_2COO \cdot CaCl_2 \cdot 2H_2O$	Guanidinium salts, $C(NH_2)_3Al(SO_4)_2 \cdot 6H_2O$

III. Organics

Thioruea, $SC(NH_2)_2$	Tanane, $C_9H_{18}NO$	Benzil, $C_{14}H_{10}O_2$	Phenanthrene, $C_{14}H_{10}$
PVDF type, $(CH_2CF_2)_n$ and copolymers with $CHF{=}CF_2$, $CF_2{=}CF_2$			
Liquid crystals, $\approx$50 distinct homologues known			
Langmuir-Blodgett films: ω-tricosenoic acid $(CH_2{=}CHC_{21}H_{42}COOH)$ + docosylamine $(C_{22}H_{45}NH_2)$			

ecules are equally polarized, but in opposite directions resulting in no net polarization. Between the two transition temperatures the two sublattices are still polarized in opposite directions, but the polarizations do not exactly cancel. Fairly detailed models, which may be described as a competition between two distinct relaxation times rather than the single relaxation time assumed in Fig. 3, have been developed to describe the Rochelle salts and their heavy water analogs (Blinc and Žekš, 1974).

Materials systems with related order-disorder transitions generally involve hydrogen bonding. Examples include the heavily studied potassium dihydrogen phosphate (KH_2PO_4 or KDP) type system, sulfates, selenites, nitrites, triglycine sulfate (TGS), formates, colemanite, alums, and more. These systems have been rather extensively studied and are reasonably well described by relaxation-time, order/disorder models.

2.3.2 Perovskites and Displacive Transitions. Figure 12 shows the typical perovskite crystal structure in the high-temperature, cubic phase. The oxygen atoms in the centers of the cube faces have two titanium nearest neighbors (atoms) along an axis perpendicular to the cube face, but there are no titanium atoms in the cube faces at all. Although the overall lattice is cubic, the oxygen atoms do not experience full cubic symmetry

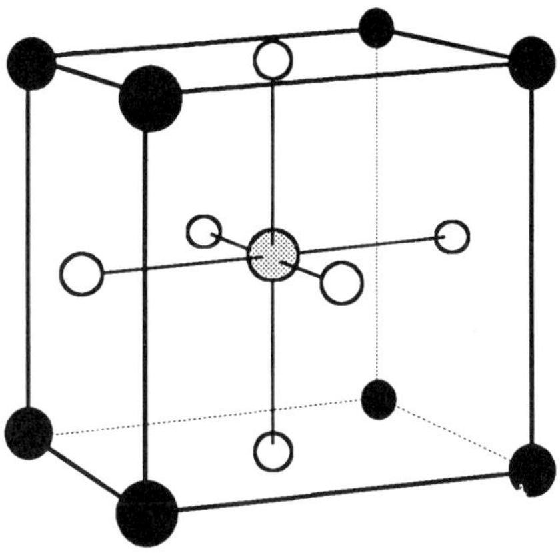

FIG. 12. The perovskite crystal structure for $BaTiO_3$. Ba atoms are at the cube corners, a Ti atom is at the cube center, and O atoms are at the centers of the cube faces. Note that each oxygen atom has only two titanium nearest neighbors, which is the origin of the ferroelectric behavior instability.

and can displace within the plane much more easily than in a direction perpendicular to the plane.

It is this asymmetric movement of the oxygen atoms that leads to an instability in the lattice, a rather clear example of a soft-mode instability. And indeed, upon cooling the unit cell distorts to a tetragonal structure and a net polarization develops along the direction of a cube edge, as shown in Fig. 13. On further cooling, the unit cell distorts again, this time along a diagonal in the face of the cube, and finally at still lower temperatures a rhombohedral distortion sets in, with a net polarization along a full cube diagonal.

The perovskites are called displacive ferroelectrics, since the mechanism involves a permanent displacement of atoms. Above the Curie temperature the resonant frequency (from Fig. 3) goes to zero as

$$\omega_i^2 \propto T - T_C , \tag{20}$$

a typical soft-mode signature.

2.3.3 Bronzes, Pyrochlores, and Nonmagnetic Oxides. The metal oxides (niobates, manganites, tungsten bronzes, pyrochlores, molybdates, vanadates, and some others) listed in Table 4 undergo displacive transitions upon cooling and exhibit resonant behavior in the dielectric properties characteristic of an ionic soft mode, as with the perovskites. The theory of these ferroelectric materials can be said to be reasonably mature. Important for some time as commercial ceramics (see Table 5), many of these oxides continue to be the subject of research in the form of thin films, composites, and layered structures (see MATERIAL PROPERTIES OF CERAMICS).

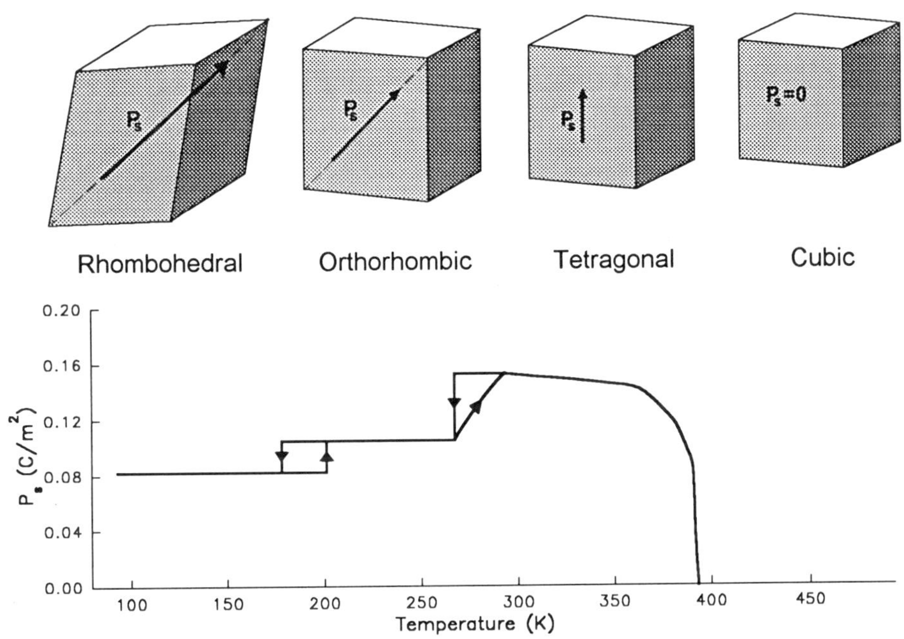

FIG. 13. Projection of the spontaneous polarization of $BaTiO_3$ onto one edge of the unit cell (modified from Merz, 1949).

2.3.4 Ferrites: Spinels and Garnets. The term ferrite is typically used to refer to ferrimagnetic oxides. These materials are hard, brittle, and poor electrical conductors widely used in inductors, particularly at high frequencies. An important family of ferrites are the spinels with the chemical formula MFe_2O_4, where the iron is trivalent and M denotes one of the divalent ions Mg^{2+}, Mn^{2+}, Fe^{2+}, Co^{2+}, Ni^{2+}, Cu^{2+}, Zn^{2+}, or Cd^{2+}. Spinels have a cubic crystal structure containing 8 chemical formula units with 32 oxygen sites per unit cell. The metal ions occupy one of two types of sites within the unit cell, designated A and B, and are coordinated by either four or six nearest-neighbor oxygen ions, respectively. Thus, there are two types of magnetic moments, corresponding to the Fe^{3+} ions or the divalent metal ions, and ferrimagnetic behavior due to the superexchange mechanism dominates the magnetic properties. There are two structural variants, called normal and inverse spinels, depending on whether the divalent ions occupy A sites or B sites, respectively, and real materials are typically mixtures of both structural variants.

The garnets, with chemical formula $3M_2O_3 \cdot 2Fe_2O_3 \cdot 3Fe_2O_3$, have 160 atoms per unit cell and three distinct types of crystallographic sites occupied by magnetic ions. In these materials M represents one of the trivalent rare-earth elements; perhaps the best known example is yttrium iron garnet, abbreviated YIG, which is notable for particularly low magnetic losses at high frequencies.

Table 5. Oxides often used in insulating ceramics (modified from von Hippel, 1954).

Kaolinite	$Al_2(OH)_4Si_2O_5$
Quartz	SiO_2
Feldspar	$K_xNa_{1-x}[Al \cdot Si_3]O_8$
Mullite	$Al_6Si_2O_{13}$
Cordierite	$Mg_2Al_2Si_5O_{15}$
Alumina (corundum)	Al_2O_3
Steatite (talc)	$Mg_3 \cdot (OH)_2 \cdot Si_4O_{10}$
Clinoenstatite	$MgSiO_3$
Forsterite	Mg_2SiO_4
Zircon	$ZrSiO_4$
Spodumene	$Li_2Al_2Si_4O_{12}$
Wollastonite	$CaSiO_3$
Rutile	TiO_2
Barium titanate	$BaTiO_3$
Sodium tantalate	$NaTaO_3$
Sodium niobate	$NaNbO_3$
Cadmium niobate	$Cd_2Nb_2O_7$
Magnetite	Fe_3O_4
Spinel	$MgAl_2O_4$
Ferrite spinel	$Ni_xZn_{1-x}Al_2O_4$

2.3.5 Organics. Plastics, polymers, and organic materials of many types have become increasingly important for a variety of applications, from packaging to textiles (see POLYMERS, MATERIAL PROPERTIES OF). Organic chemistry techniques applied to ferroelectric materials have led to the development of novel structures with unique properties. A few examples follow.

Polyvinylidene fluoride (PVDF) is an organic polymer that can be prepared from solvents in the form of thin films. As prepared, PVDF crystallizes in a nonpolar α phase, which upon stretching transforms to a ferroelectric β phase. Mixtures of PVDF and trifluoroethylene (and other copolymers) will crystallize directly in the ferroelectric phase without further mechanical treatment, and the Curie temperature can be adjusted in the range of 330 K to 400 K. PVDF and related materials may be inexpensively prepared in thin layers ($< 10\ \mu m$) and have therefore received considerable attention for various piezoelectric and pyroelectric applications, particularly those requiring rapid response times (Gerhard-Multhaupt *et al.*, 1987).

Liquid crystals represent a second family of organic insulators with particularly interesting ferroelectric properties. Liquid-crystal ferroelectrics typically consist of relatively long, rodlike molecules which "crystallize" in regular layers, as suggested in Fig. 14. These layered phases, called smectic phases, come in a variety of types enumerated alphabetically, as in SmA for the smectic A-type arrangement and so on. The arrangement depicted in Fig. 14 is the SmC arrangement, indicating that the long axes of the molecules are not quite in the direction normal to the planes, but are tilted somewhat. Within the layers, however, the molecules have no long-range order and behave essentially as a liquid, but the molecules remain essentially parallel and from plane to plane behave much like a crystal.

If the molecules in a smectic-C type phase are also chiral, meaning that the molecule has distinct right-handed and left-handed versions, then the liquid crystal will exhibit ferroelectric properties. Liquid-crystal ferroelectrics have become increasingly important for their potential in low-cost flat-panel displays,

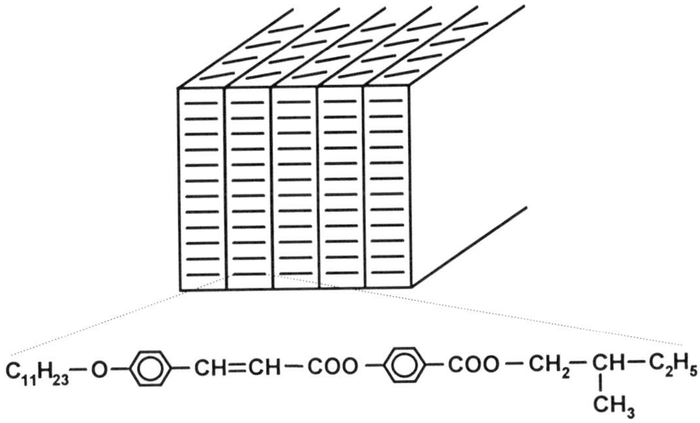

FIG. 14. Typical ferroelectric liquid crystal molecule, shown in the smectic-C arrangement.

and most of the new ferroelectric materials identified recently are liquid crystals (Fousek, 1991). For a more detailed discussion of liquid crystals, see LIQUID CRYSTALS, STRUCTURE OF.

Finally, the Langmuir-Blodgett technique is another powerful method for preparing organic thin-film structures with well-defined symmetry (Roberts, 1985). Typically, molecules with a hydrophilic head and a hydrophobic tail are dissolved in an organic solvent and spread on a water surface, and the solvent is evaporated. Monomolecular layers of the film-forming material can then be collected on a solid substrate by repeated passes through the water surface. A pyroelectric material can be produced by depositing alternating layers of fatty acids and amines, as shown in Fig. 15.

3. SUMMARY

The list of insulating materials has not been exhausted here. Rubbers, elastomers, plastics, polyvinyl chloride (PVC), gases, liquids, most minerals, glasses, and more have been nearly excluded. But the basic polarization mechanisms described (electronic, ionic, orientational, and space charges) provide a sound foundation for understanding the major insulating properties of a wide variety of materials. The principles of crystal structure and symmetry in allowing or forbidding various kinds of electric dipole ordering will also play a key role, regardless of the material or application in question.

The theoretical understanding of the classical insulating materials is quite mature and provides a firm foundation for further developments. The advent of liquid-crystal ferroelectrics and the ability to design and manufacture new structures nearly at will is sure to have an impact on future sensor, imaging, memory, and video technologies.

GLOSSARY

Critical Temperature: Typically a cusp or other sharp feature in the temperature dependence of an observable property, representing an abrupt change in the order among physical objects such as ions, electric dipoles, or magnetic dipoles.

Dielectric: A material that internally redistributes charges in response to an applied electric field.

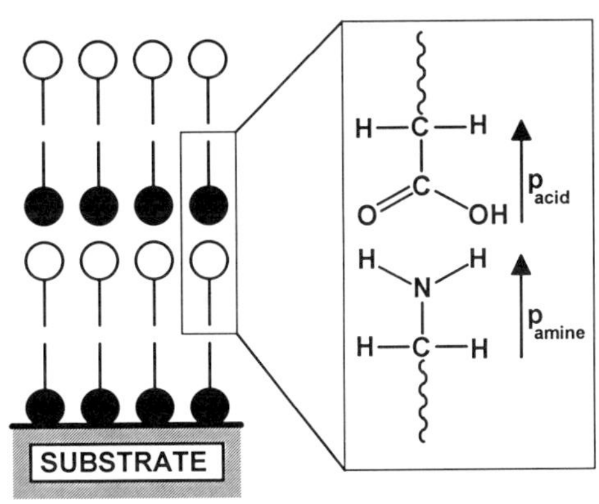

FIG. 15. Langmuir-Blodgett film of alternate acid/amine layers to produce a pyroelectric (redrawn from Roberts, 1985).

Phonon: A single lattice vibration mode representing physical displacement of atoms or ions about their equilibrium positions.

Polarization: The electric-dipole moment induced in a dielectric by an applied electric field.

Soft Mode: An unstable lattice vibration requiring little or no energy to displace atoms or ions from their equilibrium positions.

List of Works Cited

Anderson, P. W. (1959), in: G. I. Skanavi (Ed.) *Fizika Dielektrikov*, Moscow: Akad. Nauk SSSR, p. 290.

Blinc, R., Žekš, B. (1974), *Soft Modes in Ferroelectrics and Antiferroelectrics*, New York: North-Holland.

Curie, J., Curie, P. (1880), *C. R. Acad. Sci.* **91**, 294.

Cochran, W. (1959), *Phys. Rev. Lett.* **3**, 412–414.

Fousek, J. (1991), *Ferroelectrics* **113**, 3–20.

Gerhard-Multhaupt, R., Gross, B., Sessler, G. M. (1987), in: G. M. Sessler (Ed.), *Electrets*, Berlin: Springer, pp. 383–420.

Kittel, C., (1951), *Phys. Rev.* **82**, 729–732.

Knox, R. S., Teegarden, K. J. (1968), in: W. B. Fowler (Ed.), *Physics of Color Centers*, New York: Academic, p. 625–626.

Merz, W. J. (1949), *Phys. Rev.* **76**, 1221–1225.

Roberts, G. G. (1985), *Adv. Phys.* **34**, 475–512.

von Hippel, A. (1954), *Dielectric Materials and Applications*, New York: Wiley.

Valasek, V. (1920), *Phys. Rev.* **15**, 537–538.

Valasek, V. (1921), *Phys. Rev.* **17**, 475–481.

Further Reading

Blinc, R., Žekš, B. (1974), *Soft Modes in Ferroelectrics and Antiferroelectrics*, New York: North-Holland. A thorough discussion of lattice dynamics and soft modes in insulators.

Flügge, S. (1956), *Encyclopedia of Physics*, Vol. 16, *Dielectrics*, Berlin: Springer-Verlag. Classic presentation including detailed case studies of some of the most important ferroelectric systems.

Robert, P. (1988), *Electrical and Magnetic Properties of Materials*, Norwood, MA: Artech House. An introduction to the properties of materials with an emphasis on engineering applications.

von Hippel, A. (1954), *Dielectric Materials and Applications*, New York: Wiley. Somewhat dated, but an outstanding discussion of the basic concepts important for dielectric applications.

Wells, A. F. (1987), *Structural Inorganic Chemistry*, Oxford: Clarendon Press. A comprehensive discussion of the crystallographic aspects of inorganic materials, most of which are electrical insulators.

INTEGRATED CIRCUITS

See VLSI—INTEGRATED CIRCUITS

INTEGRATED FERROELECTRICS

See THIN-FILM INTEGRATED FERROELECTRICS

INTENSE PARTICLE BEAMS

JAMES BENFORD, *Physics International Company, San Leandro, California, U.S.A.*

	Introduction	103
1.	**Methods of Generation**	104
1.1	Acceleration Techniques	104
1.2	Electron Emitters	106
1.3	Accelerating Diodes	107
1.4	Ion-Beam Diodes	108
2.	**Propagating Intense Beams**	109
2.1	Beams in Gas and Plasma	109
2.2	Beams in Drift Tubes	110
3.	**Applications of Intense Particle Beams**	110
3.1	Inertial Fusion	111
3.2	High-Power Microwaves	112
3.3	X-Ray Production	115
3.4	Industrial Applications	116
	Glossary	116
	Works Cited	116
	Further Reading	116

INTRODUCTION

Particle beams are characterized as intense if their self-fields are sufficiently strong to dominate their behavior. Such beams can be controlled with guiding magnetic fields, but they can also propagate freely in a variety of ways. They are characterized by the strength parameter ν/γ. The fundamental scale for intense beams is the Alfvén current

$$I_A = \frac{4\pi\epsilon_0 m_0 c^2}{e}\,\beta\gamma, \tag{1}$$

where m_0 is the electron mass and the relativistic factor $\gamma^2=(1-\beta^2)^{-1}$; $\beta=v/c$, and v is the particle velocity. In practical units

$$I_A = 17\beta\gamma \text{ kA}. \tag{2}$$

The current of a beam can be written in terms of the Alfvén current as

$$I/I_A = \nu/\gamma, \tag{3}$$

where ν is the Budker parameter, defined as the product of the number N of beam particles per unit length and the classical electron radius:

$$\nu = Ne^2/4\pi\epsilon_0 m_0 c^2. \tag{4}$$

For $\nu/\gamma \ll 1$, electron trajectories are roughly sinusoidal and propagation is possible. Such a propagating beam has a characteristic "betatron wavelength" that is the characteristic oscillation length of the electron about the axis as it propagates. The betatron wavelength for a beam of radius a is

$$\lambda_\beta = a\gamma/\nu. \tag{5}$$

As beam intensity is increased, a point is reached where the Larmor radius of beam electrons in the magnetic field of the current is about equal to the beam radius. Beam electrons turn completely around and flow stops. This limit occurs at $I \sim I_A$ for a uniform radial current distribution. So when $\nu/\gamma \approx 1$, self-fields strongly influence beam motion.

Intense particle beams were first developed in the 1960s when pulsed electrical systems producing tens of kiloamperes of electron current at megavolt energies succeeded in propagating beams at about the Alfvén current. Since that time, vigorous efforts have produced beams with megampere currents at several megavolts with typical pulse duration of 100 ns. Electrical beam impedance is low, $<100\ \Omega$, and can be $<1\ \Omega$. The large energy per pulse, as high as a megajoule, gives this technology many applications from nuclear effects simulation to inertial-confinement fusion, high-power microwave generation, and x-ray production. In this article we first discuss methods of intense-beam acceleration, then electron- and ion-beam emission, followed by beam propagation and, finally, the applications of intense beams.

1-56081-067-X/94/$5.00+.50

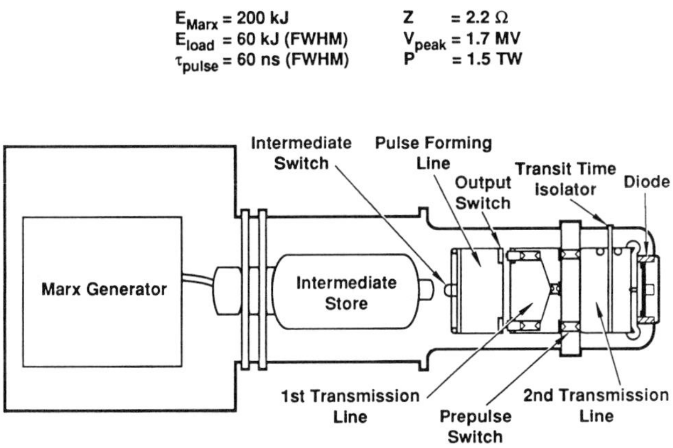

FIG. 1. Schematic of a typical pulsed-power machine used for generation of intense particle beams. The machine is the KALIF generator in Karlsruhe, Germany.

1. METHODS OF GENERATION

1.1 Acceleration Techniques

Intense beams can be produced by a variety of means, usually involving a voltage-multiplying Marx generator that pulse-charges a liquid dielectric intermediate line in a time scale of about a microsecond (Fig. 1). The Marx generator works by charging capacitors in parallel to a high voltage (~100 kV), switching them into a series arrangement, and discharging them into a load (see Miller, 1982). Firing a gas- or liquid-insulated intermediate switch then allows the voltage pulse from the intermediate line to be applied to the pulse-forming line. This line determines the output pulse duration: when the output switch fires, a wave propagates along the pulse-forming line, reflects at the end, and returns. Therefore the output pulse is twice the transit time of the line, typically 100 ns. There is a prepulse switch in some machines that fires early in the pulse, preventing any early voltage transients from reaching the output. Finally, the pulse passes through an insulator and enters a vacuum diode consisting of electrodes from which particles, usually electrons or protons, are accelerated across the potential difference and extracted as an intense beam. Intense electron beams were first generated in the 1960s, and intense ion beams appeared in the 1970s. Figure 2 shows the historical development of intense-beam generators. Machines approaching 100 TW have been built, consisting of tens of modules, each of a few terawatts each.

An important variant of the Marx generator–pulse-forming line–diode approach to beam generation is the linear induction accelerator (LIA). This technique is essentially a transformer distributed in vacuum and allows substantial downsizing of the pulsed power system. The LIA is in fact a series of one-to-one pulse transformers, each threaded by an electron beam or a solid cathode. In the latter mode it is called a voltage adder (Fig. 3). Each of the *N* transformer sections generates an

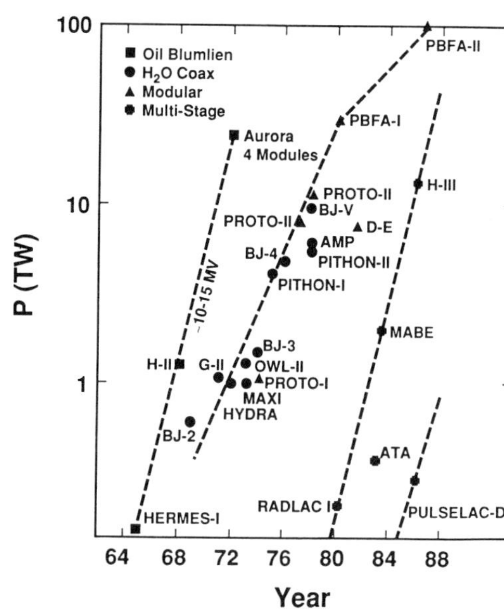

FIG. 2. Evolution of pulsed-power drivers of intense particle beams in the United States. The Blumlein is a type of pulse-forming line. Multistage devices accelerate beams incrementally; the LIA is of this type.

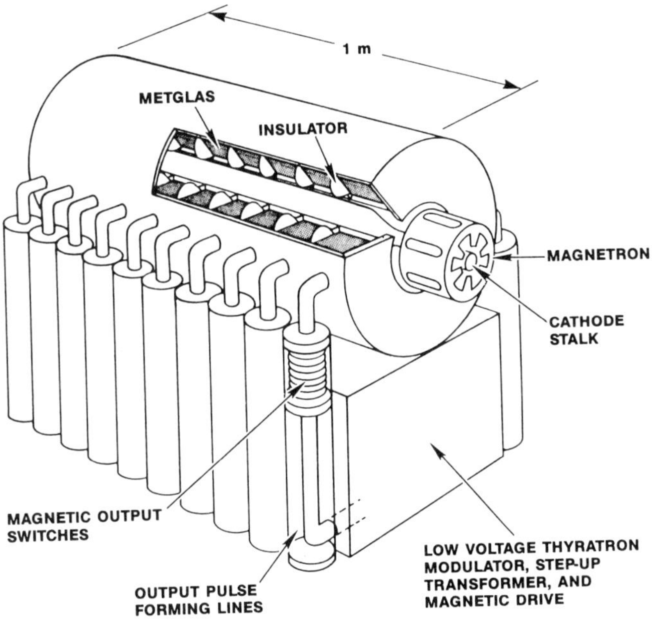

FIG. 3. Compact version of voltage adder. Metglas is a ferrite material produced by Allied Corp. Figure courtesy Physics International Company.

increment of voltage, V, on the cathode or beam so that the peak voltage NV occurs at the tip of the cathode or on the exiting electron beam. Its principal virtue is that the required accelerating voltage is distributed among a number of pulse-forming subsystems so that the full voltage appears only on the beam or at a beam-emitting diode. LIAs are useful for short pulses, < 100 ns. Greater output voltage can be obtained by adding more cavities since the system is fundamentally modular. While diode machines are limited to voltages of a few MeV, LIAs have reached 50 MeV. For example, Hermes III is a very large voltage adder built from modules (Fig. 4). This is in contrast to the diode machines, with which much higher currents (megamperes) can be generated. The disadvantages of the LIA are

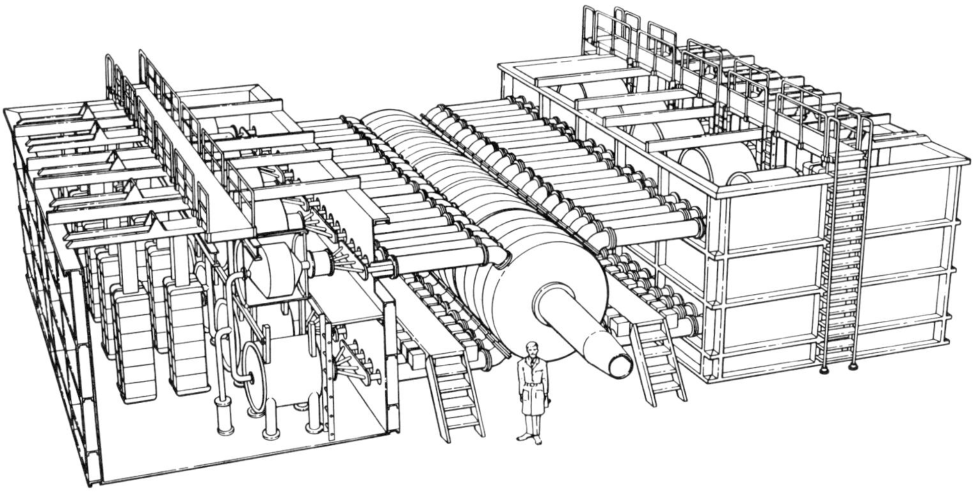

FIG. 4. Hermes III, a large-scale voltage adder (20 MV, 200 kA, 40 ns).

that it is expensive and complex compared to the diode machines.

Table 1 shows representative high-power beam-generation machines as of 1992. Most of these devices can generate intense particle beams for a wide variety of applications. For a discussion of these machines and those in Fig. 2, see Smith (1991).

1.2 Electron Emitters

The high currents needed make the electron source a vital factor in intense-particle-beam generators. We cover beam emission here; extraction and propagation are treated in Sec. 2. The most common electron sources are based on the following emitting phenomena, in order of decreasing emission current density: explosive emission, nonexplosive field emission, thermionic emission, and photoemission. In addition to the capacity for emitting at high current density, the attractiveness of each option depends on the requirements for "quality" of emission, as measured by spreads in the electron velocities and energies, ruggedness, and lifetime.

The most commonly used electron sources in present pulsed devices are based on the process of explosive emission. Applied average electric fields in excess of roughly 100 kV/cm at the emitting surface are further enhanced by the presence of naturally occurring micropoints. The enhanced fields lead to the flow of large field-emission currents through these micropoints, which rapidly heat and explode to form plasma flares within a few nanoseconds. Individual flares then expand and merge in 5–20 ns to form a uniform electron emitter capable of providing current densities of 10^7–10^8 A/m^2. The microscopic points are regenerated from one shot to the next, so that these sources can be reused. Explosive emission sources can be made from a wide variety of materials, including aluminum, graphite, stainless steel, and cloth emitters such as velvet. They produce the highest current densities of all options, are quite rugged, can tolerate poor vacuum conditions (the operationally limiting factor is arcing at $\sim 10^{-2}$ Pa), and have reasonable lifetimes. Undesirable features of explosive emission sources include the spotty character of the emission process at early times before the flares merge and the expansion of the dense plasma from which the electrons are drawn. This expansion causes the emitting surface to move with time, a phenomenon known as gap closure, which over the course of a long pulse will short out the accelerating voltage and terminate electron emission.

Field-emission sources use macroscopic points or grooves on a surface to get enhanced electric fields and produce nonexplosive field emission. Arrays of tungsten needles, grooved graphite, and polymer-metallic matrices are examples of emitters of this type; common velvet and cloth also make excellent field emitters. Current densities ranging from 10^6 to 10^7 A/cm^2 are achievable. Their advantages include operating at field levels below those required for explosive emission and avoidance of expanding plasma at the emitting surface. A disadvantage of these sources relative to explosive emission sources is the lower peak current density achievable; source lifetime is low.

Thermionic emission is based on heating an emitting surface to allow electrons to overcome the work function and escape the surface. No expanding surface plasma is required, and long-pulse to continuous electron beams can be produced. The most common thermionic emitter types are pure materials such as tungsten and LaB_6; oxide cathodes

Table 1. Representative intense beam generators, from Smith (1991).

	Pulse Power	Pulse energy	Duration (ns)	Voltage	Current
		Pulse Line Machines			
PBFA-II	70 TW	2.8 MJ	40	12 MV	5.8 MA
Double Eagle	7 TW	0.5 MJ	75	2 MV	3.5 MA
KALIF	1.5 TW	60 kJ	60	1.7 MV	1.3 MA
		Linear Induction Accelerators			
ATA	0.25 TW	10 kJ	40	50 MV	5 kA
Hermes III	16 TW	0.64 MJ	40	20 MV	0.8 MA

such as BaO; dispenser cathodes (tungsten impregnated with a mix of oxides such as BaO:CaO:Al_2O_3); cermet cathodes; and thorium-based cathodes. Thermionic emitters have the advantages of producing current densities up to 10^6 A/m^2 for a long or continuous pulses (lower than that for explosive and field emitters), and have good uniformity, long lifetime, and high beam quality. Disadvantages are the need for a cathode heater and susceptibility to poisoning at pressures above about 10^{-5} Pa.

In photoemission sources, the work-function energy for electron emission from materials such as GaAs and Cs_3Sb is supplied by laser illumination. These sources have the useful feature that electron emission can be turned on and off coincident with the laser pulse. Modulating the laser has the effect of producing a beam that is effectively bunched. Furthermore, because the electron emission is "cold," with very little excess energy over that required for emission, random thermal motions of the electrons imparted in thermionic sources can be eliminated. Current densities as high as 2×10^6 A/m^2 have been produced. The disadvantages of photoemitters are the possibility of plasma formation at the surface if the laser intensity is sufficiently high, the complication of adding a laser to the system, and the requirement for vacuums as low as 10^{-8} Pa.

Ion beams are produced by extraction of ions from cathode plasma; see Sec. 1.4.

1.3 Accelerating Diodes

The design of accelerating structures to produce electron or ion beams of high quality usually involves computer simulation in order to shape properly the electrodes and magnetic fields (if a field is used). See Septier (1983). Nevertheless, the basic features of diodes are illustrated by two idealized models. First is the Child-Langmuir diode in parallel planar geometry, which applies when the self-magnetic-field of the beam is small. The second is the pinched-flow model, parapotential flow, valid when the self-magnetic-field of the beam strongly influences the particle trajectories in the accelerating region. At nonrelativistic voltages below 500 kV, provided we can ignore the self-magnetic-field of the beam, the beam current density J depends on the diode voltage V according to the Child-Langmuir law. For an anode–cathode gap d, the law takes the form (in commonly used units)

$$J=(2.33\ \text{kA MV}^{-3/2})V^{3/2}/d^2. \tag{6}$$

This expression is also based on the assumption that the cathode can supply enough current to be space-charge limited, in which case the electric field at the cathode is zero and the current is a maximum. The relativistically correct version of the Child-Langmuir law is complex, but for 500 kV and above the following approximate expression is quite accurate:

$$J=\frac{(2.71\ \text{kA})}{d^2}\left[\left(1+\frac{V}{0.511\ \text{MV}}\right)^{1/2}-0.847\right]^2. \tag{7}$$

Note the identical scaling with the anode–cathode gap spacing in both cases. The voltage scaling is quite different, however, with the familiar $V^{3/2}$ scaling at low voltages, while at high voltages $J\sim V$.

In real diodes of the Child-Langmuir type, several phenomena left out of this analysis can be important. First, the azimuthal self-magnetic-field of the beam causes electron beams to pinch at high currents due to the force $F_r=v_zB_\theta$. The critical beam current above which beam pinching occurs is estimated by equating the Larmor radius for an electron at the edge of the beam with the anode–cathode gap:

$$I_c=8.5(r_c/d)\gamma\ \text{kA}, \tag{8}$$

where r_c is the cathode radius. Pinching leads to a very different flow pattern, the parapotential flow (Fig. 5). For $I>I_c$ electron flow is derived from a model in which electrons move along conical equipotentials, for which the parapotential current is,

$$I_P=I_c\ln[\gamma+(\gamma^2-1)^{1/2}]. \tag{9}$$

In order to extract the electron beam, the anode must be a screen or foil thin enough (i.e., much less than one electron range) that the electrons can pass through without undue angular scattering. The foil becomes a stress point in the design, since it will have a limited lifetime. Also, beam heating of the anode can create an anode plasma, which contributes to gap closure. Finally, the conducting anode

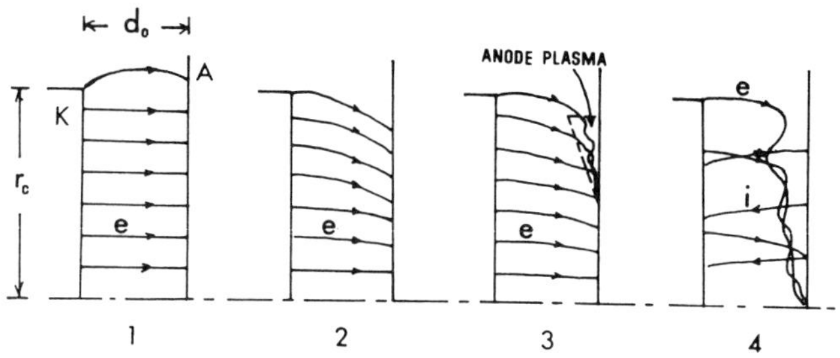

FIG. 5. Evolution of charge flow in an intense pinched electron-beam diode. (1) Child-Langmuir electron flow at low voltage; (2) electron flow at high voltage; (3) anode-plasma–induced collapsing electron ring; (4) electron and ion trajectories in pinched diode (parapotential flow).

will locally short-circuit the radial electric field of the beam, which disturbs the beam electron trajectories and introduces further spread in the transverse velocities of the electrons. Gap closure has the effect of making the diode gap a time-dependent quantity: $d=d_0 - v_p t$, where closure velocities are typically greater than 100 m/μs.

A means of avoiding the problems associated with foil or mesh anodes is to use a foil-less diode.* An example of such a diode is shown in Fig. 6. A strong magnetic field guides the electrons, preventing them from striking the wall so that they can be extracted down a drift tube. Foil-less diodes produce hollow electron beams, since emission at the outer edge of the cathode screens the smaller-radius portion of the cathode from the anode potential. The exact relation between current and voltage in these diodes depends on the details of the diode configuration. See Humphries (1990).

Just as with Child-Langmuir diodes, plasma expansion from the cathode is a problem with foil-less diodes. Plasma expansion both along and across the magnetic field can occur. Expansion along the field is faster, by roughly an order of magnitude, than cross-field expansion.

1.4 Ion-Beam Diodes

Intense ion beams at powers greater than a terawatt can be generated by megavolt-level voltages applied between two electrodes in vacuum if the electron flow in the diode is suppressed by using magnetic insulation. A magnetic field is supplied perpendicular to the electron path and to the electrodes so that it prevents electrons from crossing the gap. An anode plasma provides the ions that, being less affected by the magnetic field because of their greater mass, are accelerated by the electric field toward the cathode. The magnetic field in an ion diode serves two functions: it insulates the diode so that ions can be extracted, and it also traps electrons. The accumulated space charge of the electrons causes enhancement of the ion flux beyond the Child-Langmuir limit.

There are several ion diode types, depending upon how the field is generated and what geometry is chosen. One of the simplest types is the pinch reflex diode, which uses no static magnetic field. The flow develops as in Fig. 5. The electrons at first flow straight across the gap, then begin to pinch. They heat the anode, which begins to emit ions. Their space charge cancels that of the electrons, altering their flow. The result is the complex trajectories shown in the fourth part of Fig. 5. Since electrons move radially inward, the scale length for them is r_c. The ions cross directly, so their scale is d. This has the net result that the ion current is enhanced by roughly d/r_c

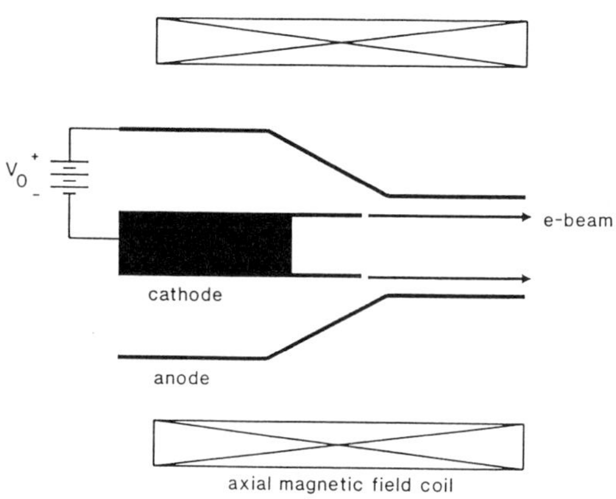

FIG. 6. Schematic of a foil-less diode.

*Diodes of this type are also commonly called *magnetically insulated diodes* in the Soviet literature.

relative to the Child-Langmuir diode. Therefore for large-aspect-ratio diodes ($r_c/d > 10$), large ion currents (~ 100 kA) can occur for megavolt applied voltages.

The pinch reflex diode has been largely superseded by applied-B-field diodes in which an external coil provides the insulating field, i.e., the magnetic field that must be applied to keep electrons drawn into the gap by the voltage from striking the anode. The two most common geometries are shown in Fig. 7. In Fig. 7(a), both anode and cathode are "barrel shaped," i.e., a structure symmetric about the center line. The ions cross the cathode and are focused in the center of the diode. The extractor diode, Fig. 7(b), is used in order to separate the target from the diode.

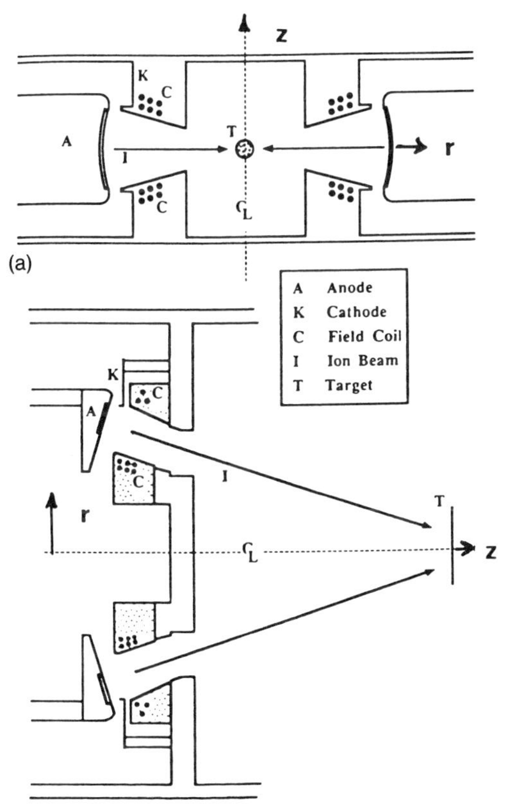

FIG. 7. Two forms of ion diodes used in inertial-confinement fusion studies. Above: barrel diode; below: extractor diode.

2. PROPAGATING INTENSE BEAMS

2.1 Beams in Gas and Plasma

Intense particle beams have self-fields sufficiently strong to dominate their behavior. Once a beam leaves the accelerator, its ability to travel a distance depends on canceling in some way the repulsive electrostatic force due to the charge of the beam electrons and the attractive magnetic force that arises from the beam current. In vacuum, the electrostatic force dominates, and the beam rapidly expands radially. Electron repulsion can be countered by supplying neutralizing ions from preexisting plasmas, by direct ionization of gas by beam electrons or by laser beam, or by applying a longitudinal magnetic guide field. In principle, fractional space-charge neutralization sufficient to provide force balance is possible, but in practice propagating beams very quickly become space-charge neutralized. When electrostatic fields have been eliminated by one or more of the above means, the magnetic force determines particle orbits.

There are three ways of exceeding the Alfvén limiting current. The naturally occurring method is "current neutralization," which arises from the tendency of plasma or any conductor to resist sudden changes in magnetic field. A fast-rising beam current induces an opposing axial electric field that drives a partially canceling return current in the plasma. The resulting net current (beam minus plasma current) determines electron orbits, so that ν/γ is replaced by $(\nu/\gamma)_{\text{net}}$. For highly conducting plasma, the net current can be reduced to such a low value that beam electrons are not contained and expand to the boundaries of the chamber.

The second method of propagation is to employ a radially nonuniform current distribution. For example, a beam with most of the current flowing in a shell near the edge will propagate with $\nu/\gamma > 1$ by turning beam electrons only in the outer shell. Alternatively, in the Bennett distribution, the current density is peaked on the axis. It can be shown that ν/γ is proportional to the ratio of the magnetic field energy within the beam to its kinetic energy. Nonuniform beam distributions reduce the magnetic energy, increasing the allowable current that can be propagated.

The third method, application of a longitudinal magnetic guide field, will prevent electrons from turning in their self-field, allowing propagation. The energy density of the field must be at least as great as that of the beam.

In recent years, experiments have demonstrated an electrostatic propagation aid. A laser ionizes a channel through a background gas. When the head of an electron beam passes, the electrons are expelled, leaving a background of ions. The remainder of the beam has its charge neutralized by the ions, so the beam propagates along the channel.

For further discussion of the above phenomena, see Humphries (1990).

2.2 Beams in Drift Tubes

A case of interest in many accelerators and beam-transport systems is the propagation of beams through metal pipes called drift tubes, usually in vacuum. Primary issues are space-charge depression of the energy of the beam electrons, limiting currents, beam equilibria, and beam quality.

Space-charge depression is a reduction of kinetic energy of the beam electrons by the self-electric-field arising from the beam space charge. Compare this to the ideal case of a low-current beam in which space-charge effects are negligible. In this case, the potential within the drift tube will be the same as the anode potential of the walls, and the kinetic energy of a beam electron will be given by the full diode voltage: $E_K = (\gamma - 1)mc^2 = eV_0$.

For an intense beam with significant space charge, the potential within the beam will be reduced from the anode potential by an amount determined by the beam space-charge density. Far from the diode region, the potential in the drift tube will be solely a function of radius, so that space-charge depression introduces shear (or radial variation) into the axial drift velocity of the electrons. As the beam current increases, the electrons are increasingly slowed down, until at some point a maximum current, known as the space-charge-limiting current, is achieved. One commonly occurring case is a thin annular beam drifting along a strong guiding magnetic field, for which the space-charge-limiting (SCL) current is

$$I_{SCL} = (8.5\ \text{kA})\, \frac{(\gamma^{2/3} - 1)^{3/2}}{1 + r_0/a}, \qquad (10)$$

where a is the radius of the beam and r_0 is the radius of the drift tube wall. Note that I_{SCL} scales as $V_0^{3/2}$ at low voltages, and as V_0 at large voltages, similar to the Child-Langmuir diode.

For example, if a beam of 5 kA is injected at 425 keV with $a = 1.9$ cm and $r_0 = 2.4$ cm, then $I_{SCL} = 12.6$ kA. The kinetic energy at injection is 340 keV, and the electrostatic potential energy is 85 keV, which is 20% of the total injected energy.

Electric and magnetic forces between the pipe wall and the beam can provide equilibrium in some cases. There is an induced image charge on the conducting wall. If the length of the beam pulse is shorter than the time it takes for magnetic field to penetrate the metal, there can be no magnetic beam-generated field outside the pipe, and therefore a return current exists on the wall equal and opposite to that of the beam. The image charges on the wall attract the beam and the return current repels it. If the beam is in the center of a cylindrical pipe, the image charges and currents are symmetric and no net force occurs on the beam. If the beam is off-center in the pipe, the beam shifts toward the outer wall and the repulsive force increases. An equilibrium ensues in which the force from the image charge balances the centripetal force. However, the beam can be destabilized if it encounters too abrupt a bend.

We have only touched on the large subject of beam instabilities. Many of the applications use instabilities to extract energy from beams in useful forms. For further discussion of beam instabilities and the above phenomena, see Humphries (1990) and Lawson (1988).

3. APPLICATIONS OF INTENSE PARTICLE BEAMS

The principal applications of intense particle beams are discussed here. The most studied applications are the topics described below. There is much interest today in industrial processes that could involve intense beams, but none of these has received sufficient attention in the research community to date.

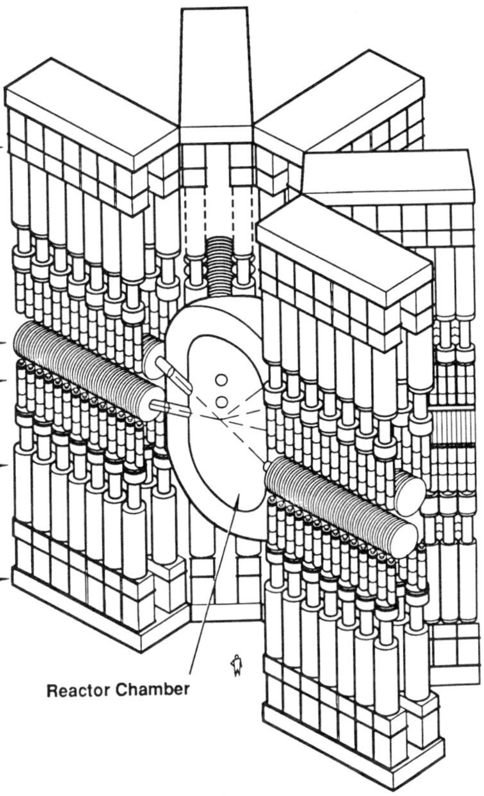

FIG. 8. LIBRA concept for a light-ion fusion reactor (30-MV, 4-MJ, 9-ns driver; 16 modules, 28 cells/module). A voltage adder injects beams into a reactor chamber. Note scale of human figure. PFL is pulse-forming line.

3.1 Inertial Fusion

Inertial fusion, as distinct from magnetically confined fusion techniques, centers around the rapid heating and compression of a small deuterium–tritium target at solid densities. The technique is called inertial-confinement fusion (ICF) because inertia of the target prevents its disassembly before fusion can occur (see FUSION, INERTIAL CONFINEMENT). Lasers and particle beams have been suggested for the driver. There are two approaches involving intense particle beams: one uses light ions and the other heavy ions.

The challenge of inertial fusion with particle beams is great. The particle-beam driver must deliver about 4 MJ of energy to the target, cost less than $100 per joule delivered, operate at an efficiency of at least 10%, avoid preheat of the fuel, and focus the beam to a 1-cm-diam target at about 100 TW/cm^2 with good uniformity. Any nonexpendable portion of the reactor must be at least 3–5 m from the pellet and must be built to survive the explosion. It would operate repetitively, at a few shots per second, in an environment with nuclear products.

The deuterium–tritium fuel must be compressed to approximately a thousand times its liquid density, and a small region in the center must reach a temperature of 4 keV. Ablatively driven spherical implosions compress and heat the fuel, driving a high-gain target. The key element is the ion-beam-producing diode, which is the juncture between the electrical energy produced by the pulsed power system and the transport system that carries the beam through the reactor cavity to the target. Heavy-ion fusion and laser fusion both have relative isolation because the transport methods to the target are ballistic and optical focusing, respectively, which place far fewer constraints on the total system than does light-ion focusing. The greatest challenge of light-ion fusion is achieving 100-TW/cm^2 power densities. An aid to focusing is that light ions at the approximately 30 MeV required are nonrelativistic, so that the voltage can be ramped upward to bunch the ion beam as it transits to the pellet. This technique also is used in heavy-ion fusion.

Many electrical technologies will be used in developing light-ion fusion, ranging from present-day, single-module, TW proton accelerators to repetitive, multimodule, multibeam, few-hundred-TW, multimegajoule lithium-ion accelerators, which will be needed to develop the final reactor chamber. The approach now being considered for the final reactor is the voltage-adder technique extended to much higher voltages, while keeping the current high enough for magnetic insulation. A conceptual view of such an accelerator is shown in Fig. 8.

Current concepts for an ion-beam diode are challenging. For example, the laboratory microfusion facility shown in Fig. 9 is conceived as 12 Hermes-III scale modules, each producing a 40-ns ion beam in which energy ramps from 18 to 20 MeV and current ramps from 460 to 720 kA. In addition, there are 24 accelerators delivering an ion pulse that ranges from 26 to 32 MeV and current that ranges from 800 to 1200 kA. Each beam bunches during its 4-m drift to the target, after which the pulse duration is 15 ns. The 36 lithium beams deposit 20 MJ onto a 2-cm-diam target. Extraction ion diodes are used with ballistic transport followed by focusing using magnetic lenses (Fig. 10). The challenges of efficient, low-cost, repetitive operation are immense.

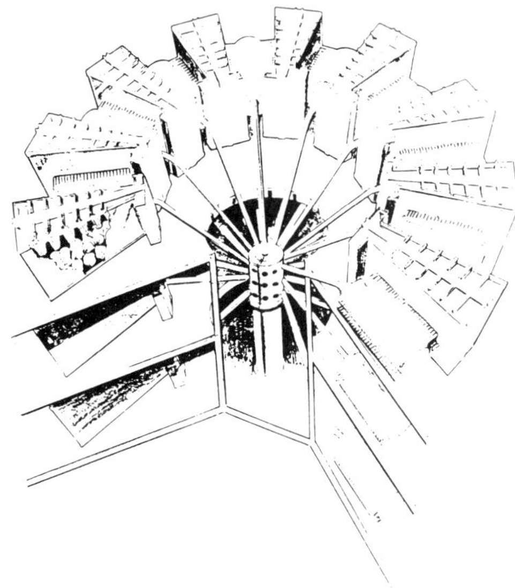

FIG. 9. Concept for a laboratory microfusion facility employing Hermes III–type technology.

3.2 High-Power Microwaves

High-power microwaves (HPM) is one of the most vigorous areas for applications of intense particle beams. Applications for high-power microwaves vary from particle accelerators to modification of the atmosphere, fusion heating, laser pumping, and electronic warfare. They generally require peak powers varying from 100 MW to 10 GW and average powers from 1 kW to as much as 1 MW radiated power. The more intense powers are required for large-scale applications such as modification of the atmosphere. For example, microwaves at a power level of 1 GW can be beamed into the lower ionosphere around 70 km to enhance local ionization. The moving microwave beam quickly paints an ionizing region, and a low-frequency ($<$ 100-MHz) signal is bounced over the horizon, allowing long-range radar detection. Another suggestion is that microwaves can break up atmospheric contaminants such as chlorofluorocarbon (CFC), which is causing destruction of the ozone. Of course, the power required to clean the atmosphere is enormous. For a survey of HPM applications, see Benford and Swegle (1992).

Particle acceleration using rf fields has long been an application for conventional microwave devices. To achieve gradients of 100 MV/m requires raising the power or frequency. Typical values are $>$ 100-MW sources operating above 10 GHz. Hundreds of such sources are necessary in order to drive next-generation accelerators.

The use of high voltages to produce relativistic electron beams has had several profound consequences for microwave generation. Very high power levels have been achieved. Figure 11 shows the peak power generated by a representative sample of high-power sources

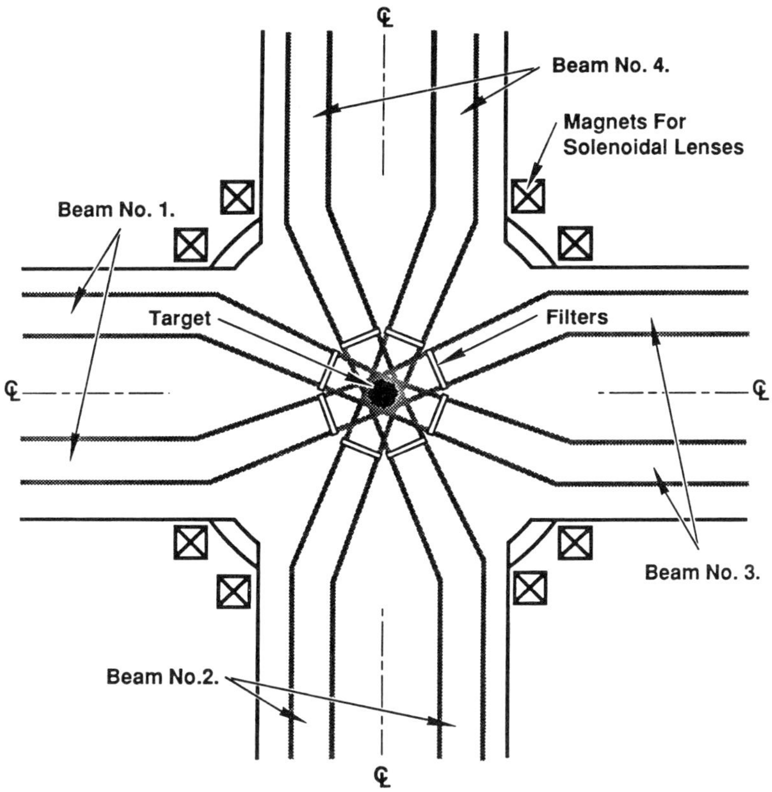

FIG. 10. Transport and focusing of multiple beams with ion lenses, filters, and overlap of beams on target.

as a function of the frequency. The factor P/λ^2 is a measure of the power density that can be produced. It appears that source power falls off roughly as f^{-2} at high frequencies; however, no clear trend emerges for the lower frequencies, below 10 GHz.

The limit on HPM source power is not known. Electrical pulses with power up to about 10 TW are available from a single pulse generator, and one can buy 1-TW generators commercially for a few million dollars. At a moderate extraction efficiency of 10%, one could therefore expect a peak power of 100 GW. Such a power will probably be achieved within the next decade. This will nevertheless require substantial improvements in the understanding of the limitations of specific devices and the overcoming of breakdown limits

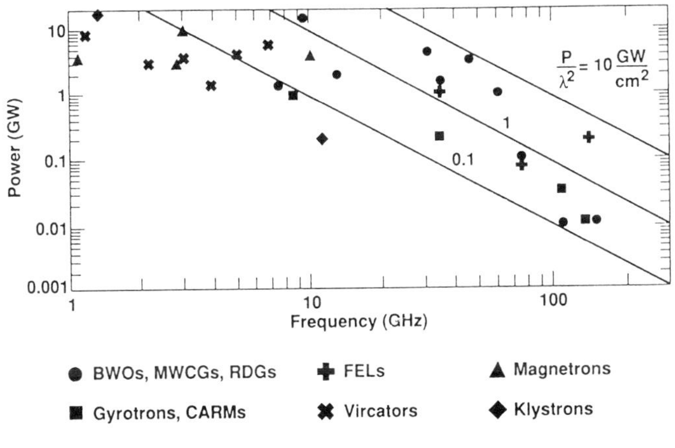

FIG. 11. Peak powers of pulsed high-power microwave sources.

Table 2. Classification of HPM sources.

	Slow wave	Fast wave
O type	Backward-wave oscillator (BWO) Traveling-wave tube (TWT) Surface-wave oscillator Relativistic diffraction generator Orotron Flimatron Multiwave Cherenkov generator Dielectric Cherenkov maser Plasma Cherenkov maser Relativistic klystron	Free-electron laser (ubitron) Optical klystron Gyrotron Gyro-BWO Gyro-TWT Cyclotron autoresonant maser Gyroklystron
M type	Relativistic magnetron Crossed-field amplifier Magnetically insulated line oscillator	Rippled-field magnetron
Space charge	Vircator Reflex triode	

and source-specific problems such as spurious mode generation.

Microwave sources can be classified into three basic types, as shown in Table 2:

1. O type: Devices in which the electrons drift parallel to an externally imposed magnetic field. This field helps to guide the intense beams used in HPM sources and, in some instances, plays an essential role in the microwave-generation process.
2. M type: Devices in which the electrons drift perpendicular to crossed electric and magnetic fields.
3. Space charge: Devices, possibly involving an applied magnetic field, in which the interaction producing microwaves is intrinsically traceable to intense space-charge interaction.

A consequence of the use of intense beams for HPM is the introduction of devices based explicitly on relativistic effects, such as the gyrotron. A second is stronger energy-, as opposed to velocity-, dependent tuning of the output frequencies in devices such as the free-electron laser (FEL) and cyclotron autoresonant maser (CARM). A third is the development of devices such as the virtual-cathode oscillator and the relativistic klystron that depend fundamentally on the very high beam currents that accompany high-voltage operation. Finally, a number of HPM sources are descendants of conventional microwave sources such as the magnetron, the backward-wave oscillator, and the traveling-wave tube. Nevertheless, in these devices, relativistic operation has brought increased power through the use of higher operating currents and stronger beam–field couplings within the interaction region.

An example of a space-charge device is the virtual-cathode oscillator, or vircator. An electron beam is generated in a diode with a thin foil anode that allows the beam to pass through to the drift space beyond. The maximum beam current that will pass through the drift region is the space-charge-limiting current I_{SCL}. One can ensure that the beam current exceeds this value by injecting the beam through a transmitting anode into a drift space of sufficiently large radius. In this event, a strong potential barrier will form at the head of the beam, from which a significant number of the beam electrons will be reflected back toward the anode. The reflection point for the electrons is known as the virtual cathode. Essentially, the excess space charge in the beam is rejected from the drift space. This situation is intrinsically different from other microwave-generation mechanisms. Instead of starting with a well defined initial beam state in which perturbations grow from small amplitudes at early times, in virtual-cathode formation there is a very fundamental breakdown of the beam state with the result that no beam equilibrium is possible. The position of the virtual cathode oscillates axially, shedding charge and behaving like a

FIG. 12. Commercial vircator with tunable end wall and two extraction waveguides. Figure courtesy of Physics International Company.

relaxation oscillator. The motion of the space charge and potential well of the virtual cathode produces radiation at a frequency that is proportional to the square root of the beam current (of the order of the beam plasma frequency). The bandwidth of the vircator is quite large, though, and if the virtual cathode is formed in a resonant cavity, it will interact with the cavity modes to oscillate at a frequency tuned to a cavity mode. Figure 12 shows an advanced version of the vircator that is commercially available.

3.3 X-Ray Production

Intense beams allow production of very large fluences of x rays. At high voltages the efficiency of x-ray production can be large; for example, at 10 MeV x rays can be produced at about 10% efficiency in an optimized converter. Such converters contain a high–atomic-number material with a thickness of about a quarter of the electron range. Because the beam is intense, additional low–atomic-number material may be necessary to contain the impulse generated by the deposition of the beam. Efficiency is roughly proportional to voltage in this range, and therefore power will be proportional to the product of current and voltage squared. However, relativistic narrowing of the radiation cone enhances the radiated power in the forward direction so that power on target is more sensitive to voltage. At energies above 5 MeV, x-radiation power scales roughly as $V^{2.8}$. For example, at 10 MeV and 300 kA, x radiation can be generated at 0.3 TW and produce radiated energies of order 10 kJ. From such converters the mean x-ray energy is about a third of the energy of the accelerating potential of the electrons.

In order to produce much lower-energy x rays, pulsed-power devices are used to drive linear z pinches that, when their collapse stagnates on axis, generate copious amounts ($\sim$10 kJ) of low-temperature x rays with energy of order of 1 keV.

3.4 Industrial Applications

Intense particle beams have been developed for a variety of applications, usually involving defense, such as nuclear x-ray simulation and directed-energy weapons. Nonmilitary applications of particle beams are at an early stage of development. Examples are radiation treatment of food and waste, x-ray lithography, and powering of industrial lasers. These applications have a common thread of requiring high average power as well as high peak power. This is because they are performing high-volume processing. Continuous operation at high repetition rates is required. The ratio of peak to average power, usually referred to as the duty factor, must be large in order to make a system economically viable. Intense particle beams have not been developed with average power above about 100 kW, which is a threshold value for an industrial process. Therefore successful industrial applications of intense particle beams will have two basic requirements. The first is that the process must itself require high peak power in order for intense beams to be applicable. The second, the high average power of any industrial process, will stress the technology of intense particle beams that have been developed for single-shot work or low average powers. The future utility of intense particle beams in the emerging area of industrial processing lies in the ability to develop high–average-power, high–peak-power devices.

GLOSSARY

Bennett Beam Equilibrium: Model in which particle density is peaked on the axis.

Image Charge: Fictional charge used to describe electric fields in a region by replacement of the actual problem with charges external to the region of interest.

Larmor Radius: Radius of the circular orbit of an electron in a magnetic field.

Linear Induction Accelerator: Means of generating particle beams using vacuum pulse transformers based on ferrite magnetic cores.

Marx Generator: Means of producing high voltages by charging a group of capacitors in parallel and then firing switches to connect them; they then discharge in series.

Return Current: Conduction current on conducting surfaces near the beam that allows the current to return to its point of origin.

Space-Charge Fields: Electric fields due to distribution of electrons.

Voltage Adder: A linear induction accelerator with a metal cathode threading it, allowing peak voltage to appear only at its tip.

List of Works Cited

Benford, J., Swegle, J. (1992), *High Power Microwaves*, Norwood, MA: Artech House.

Humphries, S. (1990), *Charged Particle Beams*, New York: Wiley.

Lawson, J. D. (1988), *The Physics of Charged Particle Beams*, Oxford: Clarendon.

Miller, R. B. (1982), *An Introduction to the Physics of Intense Charged Particle Beams*, New York: Plenum.

Septier, A., (Ed.) (1983), *Applied Charged Particle Optics, Part C: Very-High-Density Beams*, New York: Academic.

Smith, I. (1991), "Pulsed Power in the United States," in: *Proceedings of the 8th IEEE International Pulsed Power Conference*, IEEE 91-CH52-8, p. 15.

Further Reading

Granatstein, V. L., Alexeff, I. (Eds.) (1987), *High Power Microwave Sources*, Norwood, MA: Artech House.

Humphries, S. (1986), *Physics of Charged Particle Acceleration*, New York: Wiley.

VanDevender, J. P. (Ed.) (1992), Special Section on Pulsed Power Technology, *Proc. IEEE* **80**, 931–1018.

Velurde, G., Martinez, J. (Eds.) (1991), *Nuclear Fusion by Inertial Confinement*, New York: CRC.

Yonas, G., Toepfer, A. (1978), "Relativistic Electron Beam Produced Plasmas," in: M. N. Hirsch and H. J. Oakam (Eds.), *Gaseous Electronics: Electrical Discharge*, New York: Academic, Vol. 1, Chap. 6, p. 399.

INTERACTION OF SOLIDS WITH PARTICLES AND RADIATION

S. E. SCHNATTERLY *Physics Department, University of Virginia, Charlottesville, Virginia, U.S.A.*

	Introduction	117
1.	**Particles**	117
1.1	Electrons	117
1.1.1	Low Energies	117
1.1.2	Intermediate Energies	118
1.1.2.1	Examples of Spectra for Various Solids	119
1.1.3	Radiation Damage	122
1.2	Ions	124
1.3	Neutrons	126
2.	**Electromagnetic Radiation**	126
2.1	Soft X Rays	126
2.2	Hard X Rays	128
2.3	γ Rays	128
	Glossary	130
	Works Cited	130
	Further Reading	131

INTRODUCTION

The interactions between solids and both energetic particles and ionizing radiation produce a wide range of phenomena and lead to numerous applications. These interactions form the basis of most detectors of particles and radiation. They also allow spectroscopies of the electronic structure of solids to be developed. In addition, various properties of the solid target can be altered, either by radiation damage or by the implantation of incident particles. In this article we describe and illustrate a number of these phenomena and applications. In Sec. 1 particles are discussed, including electrons, ions, and neutrons. In Sec. 2 electromagnetic radiation is discussed, including soft x rays, hard x rays, and γ rays.

1. PARTICLES

Several elementary particles and many ionized atoms can be formed into beams and directed onto a solid target. At high energies the interesting aspects of the collisions that occur have to do with the properties of elementary particles and the nuclei in the solid. Here we deal with lower energies where characteristics of the solid or the modification of its properties are the primary focus.

1.1 Electrons

Electrons are bound to the atoms out of which every solid is made. The outermost, or valence, electrons create the cohesive forces that bind the solid together (see COHESIVE ENERGY). These electrons are held within the solid by a binding energy of typically several electron volts (eV). An electron volt is the energy an electron gains or loses in passing through a potential difference of one volt. Valence electrons do not, however, all have the same total energy. They exist in distinct states having a range of energies. This range is referred to as the valence-band width, and varies from less than 1 eV to several eV in different solids. When an additional electron is injected into a solid the extent of its interaction with valence and core (very tightly bound) electrons depend on its own kinetic energy.

1.1.1 Low Energies. An electron in a solid with energy in the range of a few to several thousand eV interacts strongly with the valence electrons. Its electric field causes valence electrons to be excited to higher energy states. The energy gained by the valence electrons is lost by the moving electron, causing it to slow down rapidly. In a metal this rapid slowing down continues all the way to the Fermi energy, which is the lowest energy the extra electron can have. In an insulator or semiconductor the strong coupling to the va-

1-56081-067-X/94/$5.00+.50

lence electrons continues until the kinetic energy of the electron is equal to the band gap measured from the bottom of the conduction band. Below this point it does not have enough energy to excite an electron across the band gap and can therefore no longer lose energy by this process.

Transitions like these in which one electron loses energy while another gains energy are called Auger excitations. In this energy range Auger collisions occur so frequently that they can be thought of as causing a viscous friction force. The average distance an electron moves between Auger collisions is called the mean free path and depends strongly on the kinetic energy of the electron. The minimum value of the mean free path is typically 3–4 Å and occurs for a kinetic energy of about 30 eV. The mean free path is also sometimes referred to as the escape depth, since in photoemission experiments the important information comes from electrons excited by ultraviolet photons that emerge from the solid without losing energy.

1.1.2 Intermediate Energies. As the kinetic energy of the electron rises above 30 eV, the mean free path increases, and continues to do so to higher energies. The strength of the coupling between the fast electron and those in the solid decreases, and inelastic collisions become less frequent. It then becomes possible to produce sample materials in the form of thin films such that only one collision occurs as the fast electron passes through the sample. This means that the energy and momentum lost by the fast electron are transferred entirely to a single excitation in the solid, making it possible to demonstrate the existence of and study the dynamics of electronic excitations in solids by this means.

As the incident electron moves rapidly through the solid, the electrons in the material feel a pulse of electric field as it moves past. When an electric field is applied to a solid, its response can be characterized by a dielectric function $\epsilon(q,\omega)$, Here q is the wave vector of the applied electric field, equal to $2\pi/\lambda$, where λ is its wavelength and ω is its frequency. When the fast electron scatters in the solid, creating such an electric field, in order to conserve energy and momentum it must give up energy $\hbar q$, where $\hbar$ is Planck's constant divided by 2π. The relation between the scattering cross section and the dielectric function is (Schnatterly, 1977)

$$\frac{d^2\sigma}{d\Omega\, d\omega}=\frac{\hbar\gamma^2}{(\pi e a)^2 n q^2}\,\mathrm{Im}\left(\frac{-1}{\epsilon(q,\omega)}\right), \tag{1}$$

where n is the number density of electrons in the sample, a is the Bohr radius, and γ is the relativistic mass enhancement factor. Since the dielectric function has roughly the same magnitude for different momentum transfers q, this shows that the cross section falls off approximately as $1/q^2$.

The above equation for the scattering cross section is of great importance for our understanding of the dielectric properties of solids. It says that we can determine $\epsilon(q,\omega)$ by measuring the scattering of fast electrons through a thin film of a given material. Actually, what is directly measured is not the dielectric function itself, but the imaginary part of its inverse. The real part of the inverse dielectric function can be obtained from the imaginary part using a Kramers–Kronig integral relation. This requires knowing the imaginary part over a wide energy range. In the limit that q becomes small, the dielectric function obtained in this way becomes equal to that measured optically. To evaluate the total cross section for scattering through all angles and losing all possible energies, the above equation must be integrated with respect to Ω and ω. Using a simple statistical model of atoms to do so results in (Schnatterly, 1977)

$$\sigma=8\pi\gamma^2 Z^{1/2}\left[k^2 \ln\left(\frac{Z^{3/2}}{(\Delta E/2E)^2(ka)^2}\right)\right]^{-1}. \tag{2}$$

This total inelastic cross section allows mean free paths to be estimated as a function of atomic number Z and electron beam energy E with forward momentum $\hbar k$.

In addition to the inelastic scattering processes described above, elastic scattering of the fast electron can occur from the atoms in the solid, with no loss of energy and no excitation of electrons in the material. The elastic cross section remains nearly constant with q out to a few Å^{-1}, while the inelastic cross section starts at a larger value and decreases rapidly with increasing q.

An important property of the dielectric function is that it obeys an important integral relation known as the oscillator strength sum rule (see Wooten, 1972):

$$\int \epsilon_2(q,\omega)\omega\, d\omega = \frac{\pi}{2}\omega_p^2. \quad (3)$$

Here $\omega_p^2 = 4\pi ne^2/m$ is called the plasma frequency of the electrons with number density n. Since the dielectric function is defined in terms of the polarizability per unit volume of the solid (see **Glossary**), the oscillator strength is proportional to the number of electrons per unit volume. This sum rule is a great help in making sure that measured cross sections or dielectric functions have the correct overall magnitude.

1.1.2.1 Examples of Spectra for Various Solids. Figure 1 shows inelastic electron scattering data for LiF, an ionic insulator with a very large band gap. LiF is sometimes regarded as the simplest of the alkali halides, since Li and F have fewer electrons than the other alkali-metal or halogen atoms. Figure 1 shows scattering power plotted against excitation energy in eV for four values of q in both the (100) and (110) directions. The lowest energy peak, at about 14 eV, is due to an exciton, which is an electron in the conduction band bound to a hole, or missing electron, in the valence band. An exciton is similar to a hydrogen atom with the hole playing the role of the proton. The excitation spectrum of an exciton is similar to that of a hydrogen atom but with a different energy scale for each solid. The dominant feature in the spectra shown in Fig. 1 is a strong peak near 25 eV, which is due to a plasma excitation. This is a collective motion of all the valence electrons in LiF. As q is increased the plasma peak loses strength relative to the other features in the spectrum and splits into two partially resolved peaks. This splitting is due to the interaction with the periodic field of the ions of the crystal lattice. The other features in the spectra are due to interband transitions that also change shape and position with q.

The change in position and shape of these features with q provides dynamical information about the electronic properties of LiF. For example, note that the exciton energy increases as q is increased. Figure 2 shows a graph of this exciton energy plotted against q for both the (100) and (110) directions. The curvature of this graph for small q gives an effective mass for the exciton. In other words, if you could reach into the solid and push on the exciton, how much inertia would it have? Since mass is a measure of the inertia of an object, the greater the mass (i.e., the less the curvature shown in Fig. 2), the greater the inertia of the exciton.

This question has a more fascinating answer than one might at first expect. In classical physics, when two particles are bound together, their total inertia is unaffected. This is a fundamental conclusion coming directly from Newton's laws of motion. Here we have an electron and a hole bound together, but it turns out that their inertia is affected by the binding. Figure 3 shows a measurement of E_p, the exciton peak energy, versus the square

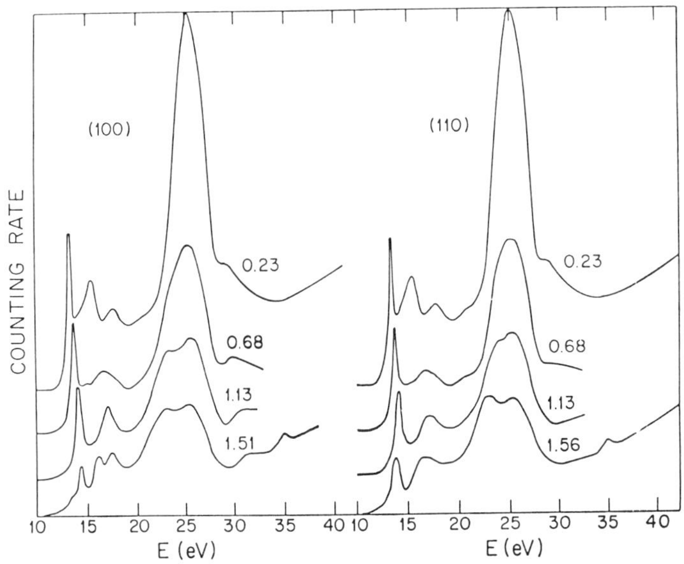

FIG. 1. Inelastic electron-scattering spectra for LiF. Momentum transfers in the (100) and (110) directions are given in Å^{-1} (Schnatterly, 1977).

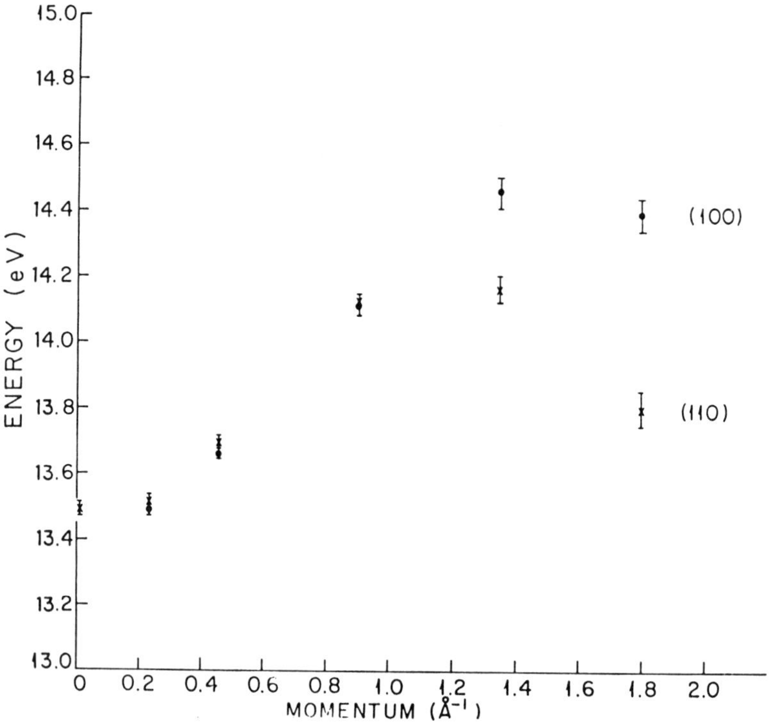

FIG. 2. Energy of the LiF valence exciton vs q for the (100) and (110) directions (Schnatterly, 1977).

of q for NaCl. The slope of this line gives a value for the mass of the exciton in this material. Plotted on the same graph is the energy to create an unbound electron and hole, E_g. The slope of this line is greater, indicating that the unbound electron–hole pair in NaCl is lighter than the bound pair. This is a quantum-mechanical effect due both to the wave nature of matter on the atomic scale, and to the fact that these electrons and holes are moving in a lattice with which they interact rather than in free space. This example stands in contrast to the other way that the mass of a system depends on its binding, namely special relativity. In that case the total mass of the composite system is decreased; here it is increased, and the effect is due to quantum mechanics, not relativity.

Figure 4 shows the real (dashed line) and imaginary (solid line) parts of the dielectric function for LiF plotted versus energy. These results were obtained from electron-scattering measurements, as described above, using a small momentum transfer q, and so they should agree with the results of optical experiments. In the 10–20-eV range are the exciton and valence excitations discussed above. At

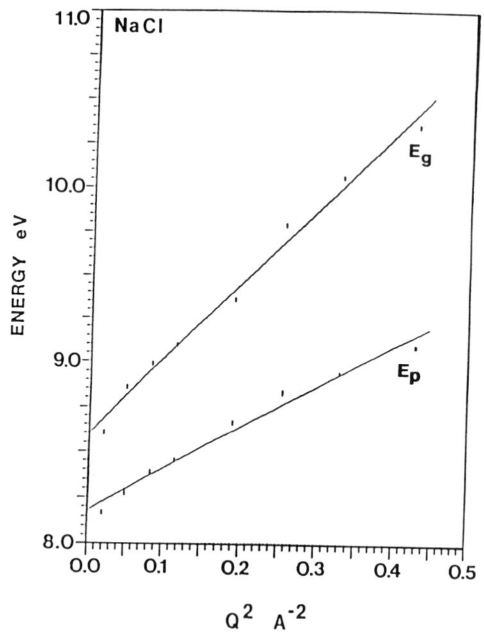

FIG. 3. E_g, the energy of an unbound electron–hole pair at threshold, and E_p, the energy of a bound electron–hole pair for NaCl vs q^2 (Schnatterly *et al.* (1988).

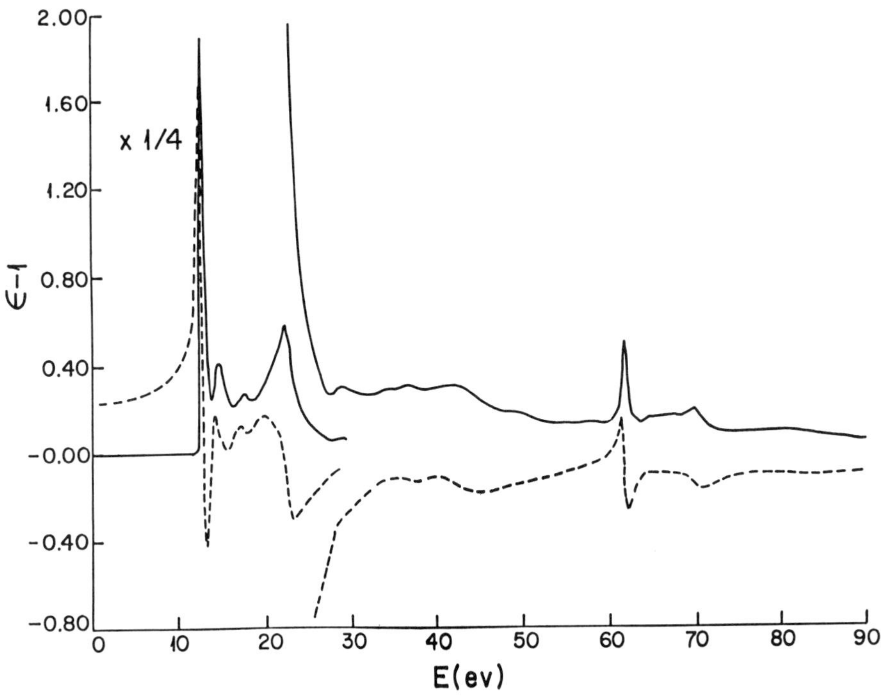

FIG. 4. $\epsilon-1$ for small q vs E for LiF. The dashed line shows the real part and the solid line the imaginary part of the dielectric function (Schnatterly, 1977).

much higher energies, in the 60–70-eV range, is another set of features. These are due to the excitation of an electron out of the 1*s* core level of Li. The 1*s* level is the most tightly bound state on any atom. These electrons have a small-radius state and do not interact significantly with the other electrons in LiF. Core levels such as this play an important role in other spectroscopies of solids such as XPS (x-ray photoemission spectroscopy) and Auger experiments (see AUGER SPECTROSCOPY).

Simple metals such as Na and Al were the first kinds of materials studied by inelastic electron scattering. This is partly because some metals are easy to prepare in thin-film form by vacuum evaporation, and partly because the measured spectra were dominated by the plasma excitation, and so appeared simple to interpret. In simple metals the plasma peak stands out even more strongly than that for LiF above. Figure 5 shows an example for Al. The shoulder at 1.5 eV is an interband transition; at 7 eV is a surface plasma excitation; at 15 eV, and stronger by a factor of more than 100, is the bulk plasma excitation; at 30, 45, and 60 eV are multiples of this dominant feature in which electrons create two, three, and four plasma excitations. Beginning at 72 eV are features due to the 2*p* core of Al. All these excitations were measured with q small compared with the Fermi wave vector of Al.

The plasma excitation is a direct consequence of the strong tendency of electrons in solids to screen out applied electric fields. For metals this screening is complete for macroscopic fields at zero frequency. One of the important boundary conditions used in electrostatics is that there can be no electric field inside a metal. The highest frequency that the electrons in a metal can follow is the plasma frequency. Above that frequency metals become much more transparent to light; the reflectivity and absorption coefficient both drop sharply.

As the wavelength of an applied field decreases (the wave vector increases), the screening decreases even below the plasma frequency. It is in this regime that we need to understand the screening effect in order to develop quantitative models of the interactions between electrons as they move about inside a solid. The most direct way to study this experimentally is to measure the q dependence of the plasma excitation in simple met-

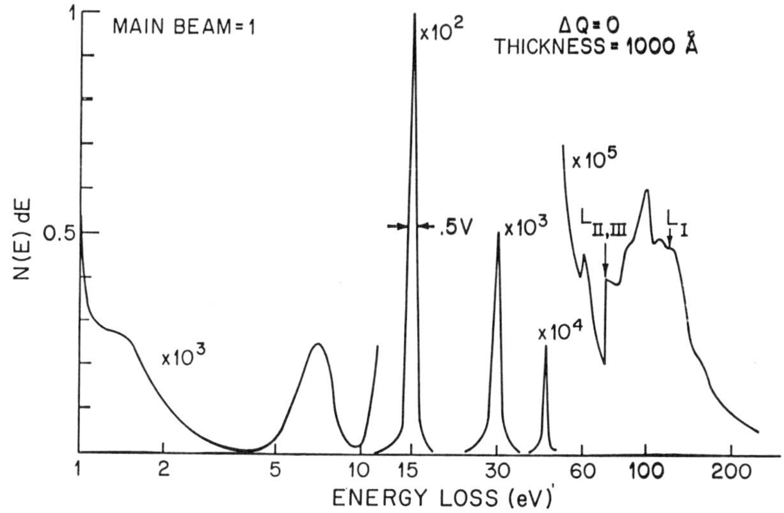

FIG. 5. Excitations in Al measured by inelastic electron scattering (Schnatterly, 1977).

als. As q is increased the plasma energy and the width of the peak both increase. Figure 6 shows the plasma energy of Al as a function of q. Figure 7 shows the width of the plasma peak in Al versus q. The peak width increases very rapidly for q greater than a cutoff wave vector q_c. In the simplest model, below this value the width is expected to be zero. Understanding the shape of this excitation in simple metals over a wide range of q values is a fundamental problem in condensed-matter physics. Much effort has been expended in developing models for this purpose (see Mahan, 1990).

Intermediate in terms of electronic properties between simple metals such as Al and good insulators like the alkali halides are the semiconductors. Excitons and interband transitions are much weaker in typical semiconductors than for good insulators. Plasma excitations in semiconductors often have widths for small q that are intermediate between those for metals and insulators. All atoms have core levels, and so semiconductors exhibit core spectra that often reveal information about the chemical state of the atom involved. Figure 8 shows the $2p$ core threshold for both Si and SiO_2. The threshold energy is shifted up in the oxide compared with the semiconductor and the shape is markedly different.

1.1.3 Radiation Damage. Electrons with energies below roughly 100 keV damage some materials rapidly and others not at all. In this energy range damage is caused by creation of electron–hole pairs in the solid. The number of electron–hole pairs created is roughly equal to the ionizing energy deposited divided by $2.8E_g$, where E_g is the band gap of the material. An electron excited out of the valence band of an insulator is a broken chemical bond. Whether the bond is reformed before a nearby atom moves from its normal lattice site is a question with no simple answer. Generally speaking, metals and covalent semiconductors do not damage rapidly, if at all, by

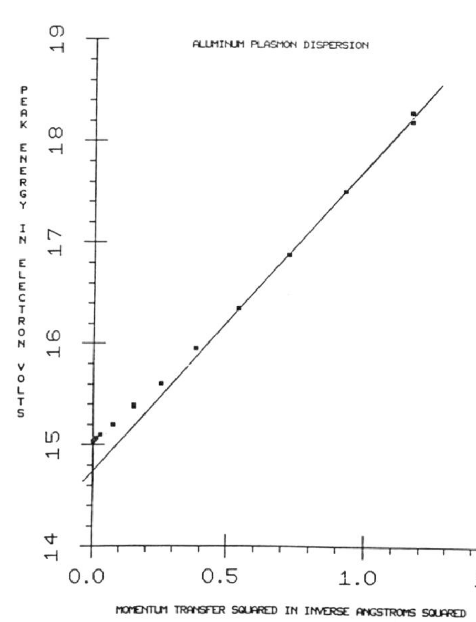

FIG. 6. Energy of the Al plasmon vs q^2.

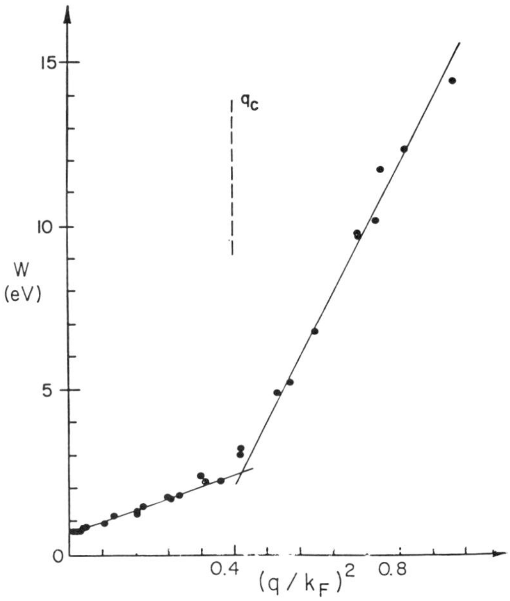

FIG. 7. Width of the Al plasmon vs q^2. q_c is the value of q below which the width should be zero according to the random-phase approximation (Schnatterly, 1977).

this mechanism. Ionic and organic materials are much more sensitive to damage by electron–hole pair excitation. In some materials two electron–hole pairs must be excited very near each other before damage occurs.

As the energy of an electron beam is increased, eventually the electrons have enough momentum to dislodge an atom from its lattice site by direct collision. Since the nucleus must be hit head-on, the cross section for this kind of damage is small. When an electron with energy E strikes an initially stationary atom of mass M, the maximum kinetic energy that can be transferred to the atom $T_{\max}$ is

$$T_{\max}=2(E+2mc^2)E/Mc^2 \,. \qquad (4)$$

For each atom in a solid there is a threshold value of energy transferred to the atom, T_d, that will result in its being displaced from its normal lattice position. Values of T_d vary from 4 to 25 eV, with the value for a particular atom being typically about twice the bond energy for that atom. For example, for Si, T_d=13 eV, which corresponds to 145 keV for the incident electron-beam energy.

Radiation damage effects in alkali halides are representative of those in insulators generally and have been extensively studied. The most readily produced atomic defect caused by electron–hole pair excitation in alkali halides is a negative-ion vacancy, which then traps an electron to produce a color center called an *F* center. The *F* center is an example on the atomic sale of a particle in a box. It is a direct consequence of the wave nature of matter that, as the box is made smaller, the energy of the electron in it increases. This also means that the energy difference between the ground and first excited states of the electron increases as the size of the box is reduced. The color associated with color centers comes from light being absorbed as the electron makes a transition between these two states. Changing the lattice constant of the crystal then changes the wavelength of light that is absorbed. This variation is described by the Mollwo–Ivey law: $\lambda_p=703d^{1.84}$ where λ_p is the wavelength of the peak of the absorption band in Ångström units, and d is the lattice constant of the alkali halide in ångströms.

As the temperature is lowered, the *F*-center absorption peak shifts to shorter wavelengths and the width of the band decreases. The width is due to the interaction of the electron with the vibrations of the surrounding atoms, and the shift is due primarily to thermal contraction as the lattice constant becomes smaller at lower temperatures. The magnetic moment of the trapped electron allows *F* centers to be studied by electron paramagnetic resonance. *F* centers are a model system in solid-state physics, which have stimulated

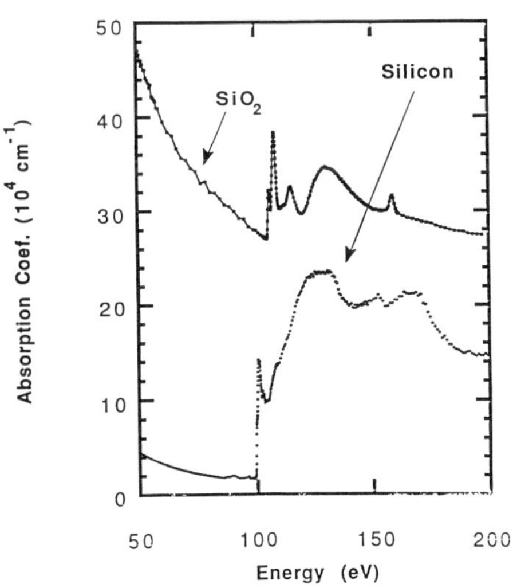

FIG. 8. Si 2*p* core threshold region for Si and SiO_2 (Husk *et al.*, 1991).

the development of new experimental techniques and theoretical concepts (see Fowler, 1968).

F centers interact with each other, and a kind of *F*-center chemistry occurs. Two *F* centers bound together form a new entity originally called the *M* center, which is the analog of the hydrogen molecule. Three and four *F* centers in nearest-neighbor positions also form stable aggregates. Higher concentrations of *F* centers are required to form these aggregate centers, which of course requires a greater dose of ionizing radiation. Eventually enough halogen atoms leave the crystal that alkali-metal particles are formed within the material. These particles also give rise to interesting optical effects since they scatter incident light strongly and in a manner that depends on wavelength.

Perhaps the most common material in the Earth's crust is SiO_2. It is also important in the semiconductor industry, since it is used to form insulating layers in Si integrated circuits. Exposure to ionizing radiation causes the formation of oxygen vacancies, creating the E_1' center, which absorbs light in the ultraviolet region of the spectrum. Defects such as this trap charge and cause inhomogeneities that interfere with the operation of high-density integrated circuits. (see Verweij and Wolters, 1983.)

Radiation damage produces defects in a number of insulators that can trap charge carriers and hold them for very long times, even millions of years. As such a material is heated, the trap lifetimes become rapidly shorter as a result of the Boltzmann factor, which describes the probability of thermally ionizing the trap. By observing the light emitted as the crystal is heated, the total number of defects caused by radiation can be determined. This technique is called thermoluminescence, and can be used to determine how long ago certain minerals were at elevated temperatures. The rate at which radiation produces traps can be measured for a given material. Then if the rate of cosmic-ray incidence or natural radioactivity is known, the material's thermoluminescent output indicates how long it has been subjected to the radiation at ambient temperatures. This is useful in geology (age since mineral formation) and anthropology (e.g., age of pottery).

Electrons with sufficiently high energies lose energy in solids primarily through bremsstrahlung. This is a scattering process in which the electron comes very close to the nucleus of an atom and is strongly accelerated. According to electromagnetic theory any accelerated charge must radiate. The radiation in this case is in the form of x rays. For example, in Pb, above about 10 MeV, the electron inelastic-scattering cross section is dominated by bremsstrahlung and is proportional to the energy to well above 1 GeV. This means that the energy lost by an energetic electron in a certain path length is proportional to the electron's energy, thus providing a method of measuring that energy.

As energetic electrons pass through Pb, or any other material, they generate x rays that propagate out in all directions from the path of the electron beam. This is called an electromagnetic shower. The x rays are scattered and eventually absorbed, creating electron–hole pairs. The number of pairs is proportional to the energy deposited in the material. Semiconductors allow the number of pairs to be measured by the current they generate in an electric field; phosphors (insulators doped with optically active charge traps) allow the number to be measured by the amount of light generated. In either case these interactions allow the energy of the electron to be determined through a variety of different experimental arrangements.

1.2 Ions

Ions incident on a solid surface cause many different physical and chemical processes to occur that are of both scientific and technological interest (see ION-BEAM MODIFICATION OF MATERIALS). Light ions of keV or more energy undergo a series of energy-loss collisions with both atoms and electrons, eventually stopping hundreds to thousands of atomic layers from the surface. Even for a monoenergetic beam of incident ions, there will be a distribution of penetration depths, approximately Gaussian in shape, due to the statistical nature of the collisions. This provides a means of doping the target material near the surface. Ion implantation, in contrast to chemical diffusion, allows a greater range of composition and concentration of dopant atoms because of the nonequilibrium nature of the process. The *p-n* junctions in many integrated circuits are formed using ion implantation.

Heavy ions with keV or more energy can create a collision cascade in which many hundreds of atoms in the target are displaced from their regular lattice sites. For example, 10-keV Bi ions incident on Si create amorphous tubes in which thousands of Si atoms along the flight path of a Bi ion have completely lost their local crystalline arrangement. These tube diameters are approximately 50 Å, and when the incident dose is sufficient for them to overlap, the entire surface region is amorphous.

Another process, important for surface studies, is sputtering (*q.v.*). Incident ions, often 3-keV Ar ions, deliver most of their energy very near the surface. Surface atoms, through multiple collisions with other target atoms, can be ejected into the vacuum. This provides a means of cleaning oxides and other contaminants from a surface. By using a mass spectrometer to identify the ejected ions, sputtering can be used to determine target composition as a function of depth. For some materials electron–hole pair excitation alone can cause sputtering without the need for direct momentum transfer from an incident ion.

In all of the above processes radiation damage plays a role. Often this is an undesirable aspect of the process, and it is desirable to anneal the damage away. Some fascinating things have been learned in studying these phenomena in semiconductors. If a heat pulse lasts less than about 1 ns and it contains sufficient energy, the result is to convert crystalline material to an amorphous structure. The heat pulse could come from a passing heavy ion, in which case it might last 10^{-11} s, or from a very short laser pulse. The physical origin of the lattice energy does not seem to matter. Longer heat pulses, usually produced by lasers, in the ns to μs regime, can produce local melting, followed by rapid crystal growth. Annealing can also be accomplished by the more conventional method of heating the entire target to an appropriate temperature.

In metals phenomena similar to those in semiconductors occur. Bulk radiation damage can be monitored by measuring the electrical resistivity of the metal. A metal with no structural disorder will have zero resistivity at a temperature of 0 K. A metal irradiated at low temperature and then warmed up while its resistivity is monitored reveals the temperatures at which various kinds of radiation damage anneal. Surprisingly, even for "hard" metals (e.g., Pt) annealing begins at very low temperatures, typically 10 K. X-ray diffraction and EXAFS (extended x-ray absorption fine structure) can also be used to monitor radiation damage.

For gas ions (e.g., H, He, Ar) incident on metals in the keV range an important phenomenon occurs at high dose levels. After the incident ions have been neutralized in the metal and have come to rest, they can usually diffuse around at room temperature. If the concentrations are high enough, they will aggregate to form bubbles. This often occurs for incident doses in the $(10^{16}–10^{17})/\text{cm}^2$ range. Bubbles cause blistering, which is an important source of degradation of thermonuclear fusion reactors. The densities of gas atoms or molecules in a bubble are comparable to those of the liquid, and yet the molecules are at room temperature, and so the pressures are very high. The bubbles near the surface eventually burst, causing a pock-marked appearance under a microscope and degraded vacuum and strength characteristics.

For incident ions in the MeV range deeper penetrations occur. For crystalline targets the depth of penetration is highly dependent on the target orientation relative to the incident beam direction. This is due to the occurrence in most crystal lattices of open regions within each unit cell, which line up with each other in particular directions in the lattice. If an incident ion beam is collimated and aimed along one of these directions, much of the beam penetrates considerably further than if the beam direction is changed by only 1° or 2°. This effect, referred to as channeling, can be observed by monitoring the fraction of backscattered ions as a function of target orientation. Channeling disappears completely for amorphous materials, and so it provides a means of monitoring the crystallinity of the surface of a target during experimentation or processing.

Energetic ions incident on thin targets will penetrate through and emerge from the other side, having lost some energy. The emerging ions often have lost a number of electrons and are in a highly excited state. Many excited-state lifetimes can be measured by monitoring the intensity of emitted light as a function of distance from the thin target. Experiments based on this technique are referred to as beam-foil spectroscopy. Another fascinating aspect of this kind of experiment is that electrons often emerge from the target having the same velocity as the ions, but not bound to them. These are referred to as convoy electrons. Convoy electrons are largely an atomic phenomenon, since their velocities are much greater than the Fermi velocity and they also occur in gases.

1.3 Neutrons

Neutrons interact with atoms in solids in a completely different manner than ions or electrons, since they have no charge. The cross sections for neutrons to scatter from atoms are about eight orders of magnitude smaller than those for keV ions or electrons, and so the mean free paths are that much longer. Thermal neutrons are available from research nuclear reactors in collimated beams that can be directed onto a sample. They have wavelengths of a few ångströms, comparable with the lattice constants of typical crystals. This means that thermal neutrons can be used to study crystal structures by diffraction, just as can x rays of a similar wavelength. A major difference is that the thermal-neutron scattering cross section varies irregularly from nucleus to nucleus in the periodic table, whereas the x-ray cross section varies as Z^2, where Z is the atomic number. This means that light elements are difficult to observe with x rays, while they may even dominate when observed with neutrons. Thermal neutrons also cause no radiation damage since they carry so little energy and momentum. Thermal neutrons are also used to study the vibrational energies of atoms in solids. For this they are uniquely suited, having both the correct momentum and energy for the task.

In addition to scattering directly from nuclei, neutrons possess a magnetic moment, which allows them to scatter from the magnetic moments of electrons in magnetic solids. In particular, antiferromagnetic materials, which have no net magnetization, could only be understood when thermal-neutron sources became available to study them after World War II.

Nuclear reactors also produce neutrons in the MeV energy range, which cause radiation damage by direct displacement of atoms from their regular lattice sites. Both point defects and collision cascades, in which a large number of atoms in a region are displaced, can occur. Samples can be irradiated by lowering them into the core region of a reactor and leaving them for a specific period of time. An interesting variation on this process is transmutation doping, in which the neutrons transmute one chemical element into another. For example, ^{30}Si, which has a natural abundance of about 3%, can be converted to P by absorbing a thermal neutron. This turns the Si into an n-type conductor with nearly perfect random placement of the P atoms and no introduced disorder. High-quality power diodes are doped in this way (see POWER ELECTRONICS).

2. ELECTROMAGNETIC RADIATION

The electromagnetic spectrum covers a wide range of wavelengths from radio waves to γ rays. The longer-wavelength regions are discussed in other articles in this Encyclopedia. Here we understand radiation to mean ionizing radiation, and begin with the soft–x-ray region.

2.1 Soft X Rays

X rays used for diffraction studies of solids have energies of 10 keV or more. Soft x rays have energies in the range of 20–30 eV to about 1000 eV. In this region of the spectrum man-made diffraction gratings can be used to resolve different wavelengths, but they must be used at grazing incidence to achieve reasonable reflectivities. Soft x rays are absorbed strongly by all solids, penetrating only 0.1–1 μm depending on wavelength.

Soft–x-ray absorption spectra exhibit sharp and detailed structure above the threshold energy for exciting an electron from a core level into empty states in the conduction band of the material being studied. Core spectra have sharp features in this energy range because these core levels have long lifetimes. At higher energies core states have shorter lifetimes, and in agreement with the energy uncertainty principle their spectra exhibit more broadening.

Core absorption spectra such as those shown in Fig. 8 reveal information about states available to electrons in the immediate vicinity of the atom being excited, in this case Si. As the figure shows, these states change significantly when the chemical state of the atom is altered. Spectra like those shown in Fig. 8 can be measured using soft x rays produced by a synchrotron, as well as by inelastic electron scattering, as was the case for the results shown in the figure.

Soft–x-ray emission spectra consist of soft x rays emitted by a sample after a core state has been ionized. An empty core state can be filled in two primary ways: A valence electron makes a transition into the core state while another valence electron is excited to higher energies. This is called an Auger transition, and typically dominates the process in this energy range. The second possibility is that a valence electron makes a transition into the core state and a soft x ray is emitted. The energy distribution of soft x rays then reveals the energy distribution of valence electrons since the core-state energy is fixed.

Figure 9 shows an example of soft–x-ray emission spectra for crystalline and amorphous Si. This is the inverse transition to that shown in Fig. 8. The energy position of the emission band is determined by the core-state binding energy, and thus will be different for

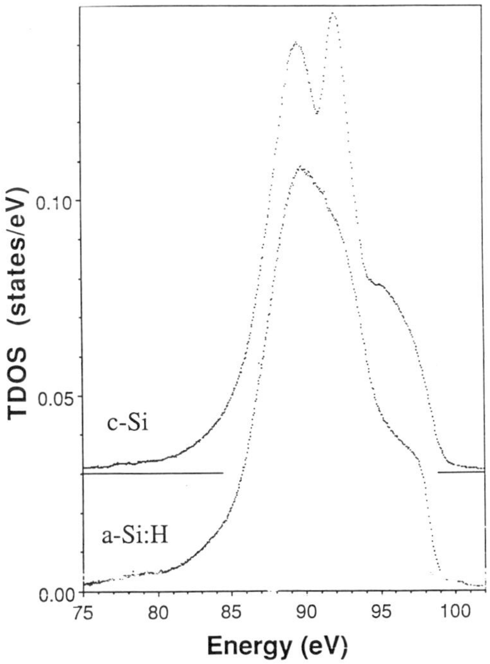

FIG. 9. Soft–x-ray emission of crystalline and amorphous Si due to transitions from the valence band to the 2*p* core level. The crystalline Si spectrum has been displaced upward by 0.03 units (Bruhwiler and Schnatterly, 1989).

each kind of atom. The shape of the spectrum is determined by the electronic structure of the valence band of the material. It is clear from Fig. 9 that crystalline Si has more structure than amorphous Si. The loss of long-range order in amorphous materials blurs the valence states of electrons, smearing out the structure in the density of states.

Measurements such as these can be used as a check on our understanding of the valence electronic states in a material. This emission band in Si is due to transitions from valence states to the 2*p* core level. The dipole selection rule, which governs transitions such as these, requires that only *s*-like and *d*-like states in the valence band can contribute to the spectrum. The Si valence band is made up of 3*s* and 3*p* states, with a small contribution from 3*d* states. Since the 3*p* states are suppressed in this spectrum, it allows us to observe the smaller contribution of the 3*d* states, which is not possible in other spectroscopies such as photoemission, where the 3*p* states dominate.

Theoretical calculations of the shape of the Si emission spectrum of crystalline Si agree quite well with the result shown in Fig. 9. Calculations for amorphous Si have not yet been carried out.

Figure 9 also shows that some emission intensity remains at energies below the emission band. This is caused by a radiative–Auger transition in which both processes occur: A soft x ray is emitted and a valence electron is excited to higher energies, reducing the energy available to the soft x ray.

Figure 10 shows the emission spectrum of BN. This material has a layered structure, and its valence electrons have states either in the planes of the layers (σ states) or perpendicular to those planes (π states). This causes the emission spectrum to be anisotropic, allowing it to be separated into σ and π components as shown. The dashed line shows the radiative–Auger contribution for the π states. Also shown is a peak at about 192 eV in the π spectrum that is not part of the valence band. This is a core exciton, which for some materials contributes to the emission. Figure 9 shows a weak feature near 100 eV in the case of amorphous Si that is also due to a core exciton.

2.2 Hard X Rays

Higher-energy x rays are produced by both bremsstrahlung and transitions to core states. They are absorbed by matter predominantly by electron–hole pair creation. Structural studies in condensed-matter physics are greatly aided by x-ray diffraction, in which x rays scatter elastically from atoms, revealing their relative positions within a material.

Inelastic scattering of x rays in matter also occurs. In this case the incident x ray gives some of its energy and momentum to a single electron, emerging with a lower energy and longer wavelength. This process is known as Compton scattering and played an important role earlier in this century in verifying the reality of photons as the quanta of electromagnetic radiation. Compton showed that if λ_1 is the wavelength of the incident x ray, and λ_2 the wavelength of the scattered x ray,

$$\lambda_2-\lambda_1=(h/mc)(1-\cos\theta) , \tag{5}$$

where h is Planck's constant, m the mass of the electron, and c the speed of light. The combination h/mc is called the Compton wavelength of the electron and has the value 0.00243 nm.

Compton's formula above applies to electrons that have zero energy, either kinetic or potential, before interacting with the x ray. Electrons in solids have both. The interaction with the nucleus and the other atoms creates a binding energy, which can be very large for core states and is different for different atoms. Valence electrons have a range of kinetic energies as described above. This means that Compton's formula must be modified to include the effects of the initial energy and momentum of the electron. Measurements of inelastic x-ray scattering from solids then provide a means of probing these energies and momenta.

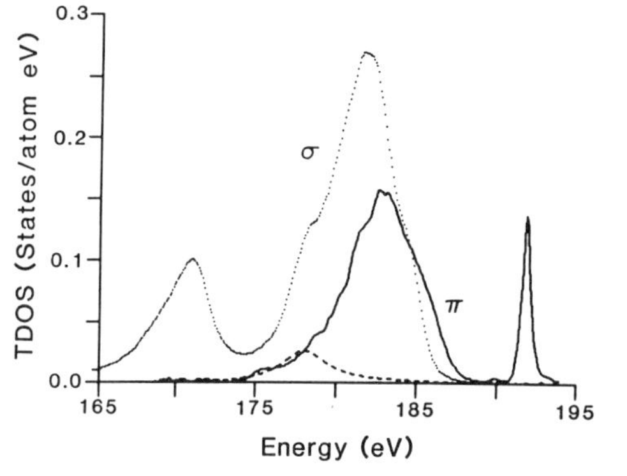

FIG. 10. Soft-x-ray emission spectrum of BN due to transitions from the valence band to the *B* 1s level (Mansour and Schnatterly, 1987).

The cross section for inelastic x-ray scattering from a solid has the same energy and momentum dependence as that for inelastic electron scattering, that is, it is proportional to the imaginary part of the inverse dielectric function. The coefficients multiplying the inverse dielectric function are completely different from those in Eq. (1) since the interaction of x rays with matter is different from that of charged particles. The size of the cross section is smaller than that for electrons by about 10 orders of magnitude. This means that macroscopic samples can be used as targets rather than the thin films used in electron scattering.

In principle, all the kinds of information described above obtainable about solids with electron scattering can also be found using x-ray scattering. In the past the counting rates in the measurements have been much lower, and the energy resolution obtainable has been worse, than that for electron scattering. As synchrotrons become more and more intense x-ray sources, this comparison will improve.

One interesting application of inelastic x-ray scattering derives from the fact that different atoms can be distinguished from one another because of their distinct electron energy levels. This makes it possible to scan an object, e.g., a package, to determine whether a given substance is present inside. With very sensitive x-ray detectors and a raster scanning source, the total radiation exposure is very low, and so this procedure can even be applied to human beings.

2.3 γ Rays

Higher-energy electromagnetic radiation is produced by nuclear transitions and collisions between elementary particles. As the energy increases, absorption of radiation by electron–hole pair creation decreases rapidly. Compton scattering becomes dominant typically above 0.1–1 MeV. At still higher energies, pair creation dominates the interaction of radiation with matter. At these energies, a photon near the nucleus of an atom can create an electron–positron pair. This requires a minimum of about 1 MeV, and the probability increases rapidly above that energy. For these and higher energies electromagnetic showers are produced by incident γ rays similar to those produced by energetic electrons.

Electron–hole pair creation, Compton scattering, and electron–positron pair creation are the three mechanisms by which γ rays interact with matter. These mechanisms underlie the production of detectors of γ rays, which can not only monitor their presence but also record their energies. In these detectors the following things must occur: first, the conversion of the γ-ray energy to kinetic energy of electrons by one of the above three mechanisms; second, the production of electron–hole pairs or excited molecular states by these electrons; and third, the collection and measurement of the charge carriers, or the light produced in the detector material.

Each γ ray then produces a pulse of either charge or light that can be measured. If the entire energy of the γ ray is deposited in the detector, then the size of the pulse is a measure of its energy.

Two such detectors will be described here briefly. NaI(Tl) is a scintillator material that has been used for years to detect x rays and γ rays. This consists of a crystal of NaI in which a small fraction of the Na atoms have been replaced with Tl atoms. The Tl impurities are optically active and can capture electrons and holes producing light. The cross section for stopping γ rays in a solid increases with atomic number, which is the reason the alkali iodide is used. These detectors are quite useful up to energies of several MeV. An important characteristic of these detectors is how much of the γ ray's energy is needed to produce each detected photon. For NaI(Tl) this is about 100 eV.

Ge is a semiconductor that is also widely used to detect γ rays. In pure Ge at low temperature no free carriers of charge exist within the material. Each of the four outer electrons of the Ge atom are contained in a valence band that cannot carry current. Above the top of the valence band is a gap region with no allowed electron states, followed by a conduction band. Only by exciting electrons into the conduction band can charge be carried through the Ge under the influence of an electric field. The gap region in Ge is about 0.7 eV wide.

A very pure crystal of Ge can be used as a γ-ray detector in the following way. An electric field is placed across the crystal, and the resulting current measured with an external circuit. Typically this is done at a temperature of 77 K, the boiling point of liquid nitrogen, so as to reduce the number of electrons thermally excited across the gap. When a γ-ray deposits energy in the Ge, many electron–hole pairs are created causing a pulse of current, which is measured. The average energy needed per created electron–hole pair is only about 3 eV.

The creation and detection of electron–hole pairs are statistical processes. This means that a beam of monoenergetic γ rays will produce pulses with a range of sizes. The energy resolution of the detector is determined by the width of this distribution of pulse sizes. This width increases approximately as the square root of the number of pairs detected. The γ-ray energy increases linearly with the number of pairs. This means that the relative resolution (energy divided by the uncertainty in energy) improves as the number of detected pairs increases. In this regard, Ge is a better detector than NaI(Tl), since more detectable events are created in Ge by the same energy γ ray than in NaI(Tl). In a typical detector, for example, a 1-MeV γ ray can be detected with a pulse width of about 1.6 keV.

Energy-sensitive detectors such as these are used in numerous applications involving radioactivity. For example, a company producing a radioactive source of ^{201}Tl, which is used in medical imaging, must specify not only how much activity there is from this isotope, but also how much is present due to impurities. Using a Ge detector, it is possible to detect impurity radionuclides down to the 0.001% activity level.

An important method used to identify the various chemical elements present in a sample of material is neutron-activation analysis. The sample is placed in a nuclear reactor, where it is subjected to a thermal neutron flux of up to 10^{14} cm^{-2} s^{-1} for perhaps a few days. The neutrons react with the nuclei present in the sample, changing them to new nuclides, many of which will be unstable and will emit radiation. γ-ray emitters are particularly valuable in this application because the cross sections for their production vary by large factors between radionuclides, making it easier to distinguish them. An important prerequisite for this to work well is either a low neutron cross section for the matrix material and a high cross section for the trace element of interest, or a large difference in half-lives of the produced radionuclides. Some elements, in particular the light ones, cannot be detected by neutron-activation analysis, since their nuclei do not become active under neutron exposure. This is an advantage in looking for trace elements in organic and geological materials, since the activation of the matrix (i.e., H, C, N, O, Si, P, S, and Ca) is negligible. For example, rare-earth elements can be detected in rock samples at the level of nanograms per gram. A Ge detector is a good choice for this application.

GLOSSARY

Auger Transition: A relaxation process in an atom, molecule, or solid in which one electron moves to an empty state with lower energy and another takes the energy gained in being excited to a higher energy state.

Band Gap: An energy region for which no electron states exist within a solid. The band gap most widely referred to is that just above the valence band.

Bremsstrahlung: The production of electromagnetic radiation, especially x rays, by the acceleration of an incident electron by the electric field of an atomic nucleus.

Channeling: Deep penetration of a beam of incident ions into a crystal when the direction of the beam is aligned along certain crystallographic axes.

Collision Cascade: The displacement of hundreds of atoms in a crystal by a single incident heavy ion with sufficient energy.

Compton Scattering: The scattering of an x ray by an electron with negligible initial energy.

Core Electron: An electron tightly bound to an atom in a small orbit so that in a molecule or solid it does not interact strongly with electrons on other atoms.

Dielectric Function: A coefficient describing the electric polarization of a solid, usually defined by $\epsilon E = E + 4\pi P$, where ϵ is the dielectric function, E the electric field, and P the induced dipole moment per unit volume. ϵ varies with q and ω where q is the wave vector and ω the frequency of the applied field.

Electron Volt: The energy an electron gains or loses upon passing through a potential difference of 1 V.

Exciton: An electron and a hole bound to each other in a semiconductor or insulator.

F Center: A negative-ion vacancy in an ionic solid that has trapped an electron.

Hole: A missing electron in a distinct state in an atom or solid. The hole behaves like a positively charged particle.

Ion Implantation: The process of adding atoms to a solid by exposing it to a beam of energetic ions.

Kramers–Kronig Relation: An integral equation relating the real and imaginary parts of the dielectric function. This allows the complex dielectric function to be evaluated if some measured property (reflectivity, inelastic electron scattering cross section) is known over a wide energy range.

Mean Free Path: Average distance between collisions for a particle moving in a medium (e.g., an electron in a solid).

Neutron-Activation Analysis: A test procedure for identifying chemical elements in a sample by exposing it to thermal neutrons in a reactor. This converts many elements to a radioactive form that can then be identified by their emissions.

Plasma Excitation: An excitation in solid in which a large number of electrons, e.g., all the valence electrons, move coherently together at a characteristic frequency (the plasma frequency) in response to an applied field.

Sputtering: The removal of surface atoms from a solid by bombarding with energetic ions, often Ar.

Thermoluminescence: Emission of light by a material as it is being gradually heated. The increasing temperature causes electrons and holes to be released from traps; as they recombine light is emitted.

Transmutation Doping: A method of doping Si in which one of the Si isotopes is transmuted to P by exposure to thermal neutrons in a reactor.

Valence Electrons: The outermost electrons on an atom, which form the chemical bonds that hold molecules and solids together.

List of Works Cited

Bruhwiler, P. A., Schnatterly, S. E. (1989), *Phys. Rev. B* **39**, 12 649–12 654.

Fowler, W. B. (1968), *Physics of Color Centers*, New York: Academic, Eq. (2-2).

Husk, D. E., Tarrio, C., Benitez, E., Schnatterly, S. E. (1991), *J. Appl. Phys.* **70**, 3338–3344.

Mahan, G. D. (1990), *Many Particle Physics*, 2nd ed., New York: Plenum, Chap. 5.

Mansour, A., Schnatterly, S. E. (1987), *Phys. Rev. B* **36**, 9234–9240.

Schnatterly, S. E. (1977) in: H. Ehrenreich, F. Seitz, D. Turnbull (Eds.), *Solid State Physics*, New York: Academic, Vol. 34.

Schnatterly, S. E., Tarrio, C., Cafolla, A. A. (1988), *Excitons in Confined Systems*, Springer Proceedings in Physics Vol. 25, p. 68.

Verweij, J. F., Wolters, D. R. (1983), *Insulating Films on Semiconductors*, Amsterdam: North-Holland.

Wooten, F. (1972), *Optical Properties of Solids*, New York, Academic, Chap. 6.

Further Reading

Debertin, K., Helmer, R. G. (1988), *Gamma and X-ray Spectrometry with Semiconductor Detectors*, Amsterdam: North-Holland.

Johnson, R. A., Orlov, R. N. (1986), *Physics of Radiation Effects in Crystals*, Amsterdam: North-Holland.

Schnatterly, S. E. (1977), "Inelastic Electron Scattering in Solids," in: H. Ehrenreich, F. Seitz, D. Turnbull (Eds.) *Solid State Physics*, Vol. 34, New York: Academic.

Williams, J. S., Poate, J. M. (Eds.) (1984), *Ion Implantation and Beam Processing*, New York: Academic.

INTERCALATION COMPOUNDS

M. S. DRESSELHAUS, *Department of Electrical Engineering and Computer Science and Department of Physics, Massachusetts Institute of Technology, Cambridge, Massachusetts, U.S.A.*
G. DRESSELHAUS, *Francis Bitter National Magnet Laboratory, Massachusetts Institute of Technology, Cambridge, Massachusetts, U.S.A.*

Introduction 133
1. **Structural Organization** 133
1.1 Layer Stacking Arrangements . 134
1.2 Requirements for Intercalation 136
1.3 Classification According to Rigidity 138
1.4 Classification of the Intercalate Layers 139
2. **Structural Characterization** .. 140
2.1 *c*-Axis Ordering 140
2.2 In-Plane Ordering 143
2.3 Interlayer Correlations........ 145
2.4 Intercalate Phase Transitions.. 145
3. **Electronic Properties** 146
3.1 Charge-Transfer Phenomena .. 146
3.2 Transport Properties........ 148
3.3 Electronic Structure 148
4. **Magnetic Intercalation Compounds**........ 148
5. **Superconducting Intercalation Compounds** 151
Acknowledgment 154
Glossary........ 154
Works Cited 154
Further Reading 155

INTRODUCTION

Intercalation compounds in general are formed by the insertion of atomic or molecular layers of a guest chemical species, called the intercalate, between layers of a host material. Intercalation compounds have received a great deal of attention from the physics, chemistry, and materials science communities because the intercalation process leads to the synthesis of new classes of materials, modifying the properties of both the host material and the guest species. In this article, the structure and properties of intercalation compounds are reviewed. These materials are of special interest because of their potential both for practical applications and for the study of novel physical phenomena. Shown in Fig. 1 are several schematic examples of layered intercalation compounds, with the basic structural unit of the host material ranging from a thin monolayer to a thick multilayer unit. In each case, the addition of the intercalant has little effect on the basic structural unit of the host material.

1. STRUCTURAL ORGANIZATION

To allow insertion of entire layers of guest species, the host material is usually a layered material, with very strong in-plane bonding and very weak interplanar bonding. There are many examples of host materials that support intercalation (see Fig. 1), including graphite (Dresselhaus, 1984; Dresselhaus and Dresselhaus, 1981; Solin, 1982, 1984; Kamimura, 1987), which has been studied most extensively; transition-metal dichalcogenides (Gamble and Geballe, 1976; Liang, 1987); some silicates and metal chlorides (Gamble and Geballe, 1974); some clays (Wada *et al.*, 1987; Newman, 1987); and some polymers and gels (Aldebert *et al.*, 1982). Intercalation modifies the properties of the host materials, giving rise to materials that are more conducting, more anisotropic, magnetic, or capable of storing large concentrations of hydrogen, to give some examples. Correspondingly, intercalation results in the modification of the intercalate, often changing the chemical reactivity drastically because of the bonding between the intercalate and the host material.

1-56081-067-X/94/$5.00+.50

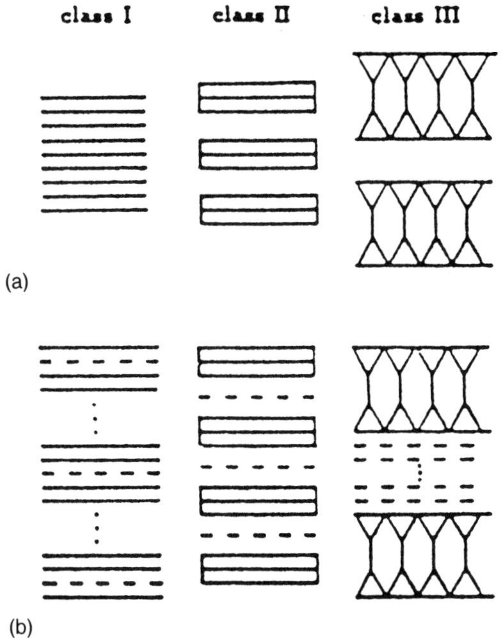

FIG. 1. Schematic classification of (a) pristine layered host materials and (b) the intercalated forms of (a). In (b) the dashed (solid) lines represent guest (host) layers. Three classes of intercalation compounds are shown, class I having the lowest layer rigidity and class III having the greatest rigidity for the host material (Solin, 1984).

1.1 Layer Stacking Arrangements

In discussing the structural organization of intercalation compounds it is necessary to comment on the structure of the host material and of the intercalate prior to intercalation, the effect of intercalation on these structures, and the structure of the entire intercalation compound. In this subsection, the layer stacking arrangements are discussed. First, host materials have their own specific structures prior to intercalation. In general, the basic structural unit of the host material is unaltered in the intercalation process. For example, the graphite host material is a layered compound with *ABAB* stacking and four atoms per unit cell (see GRAPHITE). After an intercalate layer has been inserted, the same intralayer atomic arrangements persist, and the same interlayer correlations between adjacent host layers are maintained within the basic structural unit of the intercalation compound, but the host-layer stacking sequence across basic structural units will in general be different if an intercalate layer is introduced between two layers of the host material.

Likewise the transition-metal dichalcogenide host material shown in Fig. 2 consists of a trilayer containing two chalcogenide monolayers *X* between which a transition-metal monolayer is sandwiched. This strongly bound trilayer unit maintains its structure as an intercalate layer is introduced in the van der Waals gap between the trilayer sandwiches, as indicated in Fig. 2. Similarly, the large, strongly bonded silicate basic structural unit also remains unaltered as the intercalate is inserted between silicate basic structural units as shown in Fig. 3. The intercalate may enter every available van der Waals gap (or intercalate gallery) between basic structural units of the host material in the case of a saturated intercalation compound with the maximum intercalate concentration. For more dilute intercalation compounds, only a fraction of the available van der Waals gaps are occupied by intercalate guest species. For those basic structural units of the host material connected by unoccupied van der Waals gaps, the stacking sequence remains un-

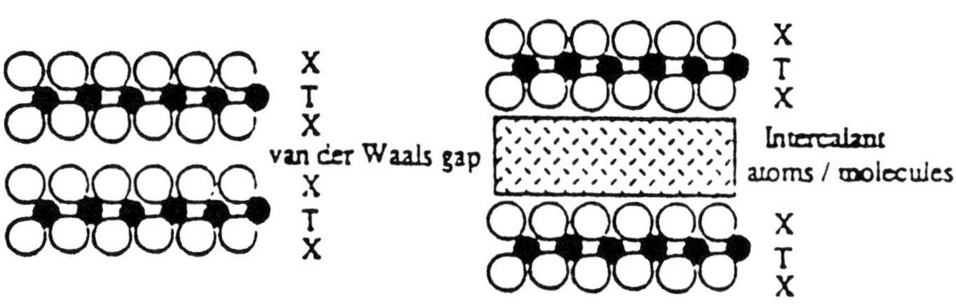

FIG. 2. Schematic representation of the trilayer transition-metal dichalcogenide TX_2 sandwiches and the interlayer van der Waals gap before and after addition of an intercalation complex (Liang, 1987) between layers of TX_2, where *T* denotes the transition-metal layer (usually from groups IV, V, and VI in the periodic table) and *X* is the chalcogenide layer (i.e., S, Se, or Te).

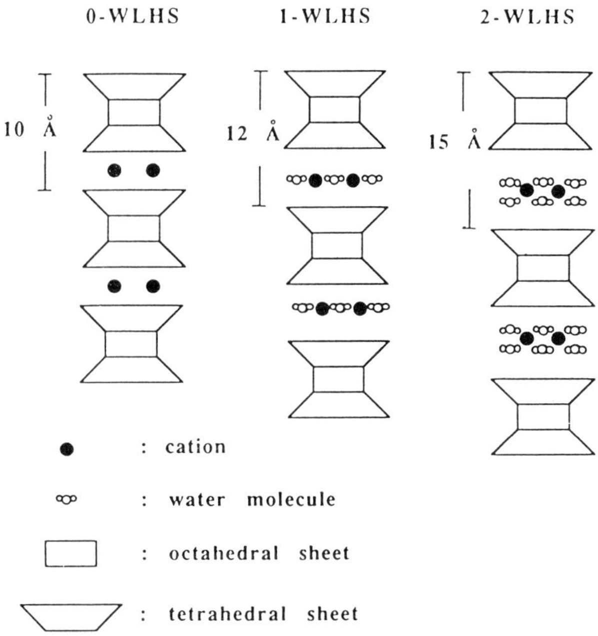

FIG. 3. Schematic representation of the vermiculite clay structure for zero-, one-, and two-water-layer hydration states (WLHS). Typical values for the *c*-axis repeat distance are indicated.

changed relative to the bulk host material prior to intercalation. The structure of the intercalate layer and of the resulting intercalation compound is more complex and is discussed in Sec. 2.2.

A special feature of intercalation compounds is the staging phenomenon, by which a constant number of layers of a host material are sandwiched between sequential intercalate layers, as shown in Fig. 4. The staging phenomenon thus gives rise to a periodic array of intercalate sheets [a one-dimensional (1D) superlattice normal to the layer planes] that can independently undergo 2D order–disorder and structural phase transitions within the layer planes (see Sec. 2.4). Staging allows control of the separation between sequential intercalate layers, thereby providing a means for control of the magnitude of the intercalate interplanar interactions. This control is exploited in studies of a variety of 2D phenomena including quasi-2D magnetism. The two layers of the host material between which the intercalated layer is sandwiched are called the *bounding* layers, while the layers of the host material with host layers on either side are called *interior* layers. Of all the host materials, graphite forms intercalation compounds (GICs) exhibiting the highest degree of staging fidelity.

Whereas intercalation normally refers to the insertion of complete atomic layers of guest species into a layered host material, the term intercalation is also used to denote the insertion of lines of atoms into host materials. For example, the high-T_c superconductor $YBa_2Cu_3O_7$ has chains of oxygen and copper (Cu2) atoms where the stoichiometry is CuO, and planes of the Cu and O species (Cu1) where the stoichiometry is CuO_2. The insertion of oxygen atoms into the chain planes, which contain available sites for O atoms, has also been called intercalation. It should be noted, however, that in the highly anisotropic high-T_c superconductors of the bismuth-based oxide series Bi-Sr-Ca-Cu-O [$Bi_2Sr_2CuO_x$ (2:2:0:1), $Bi_2Sr_2CaCu_2O_x$ (2:2:1:2), and $Bi_2Sr_2Ca_2Cu_3O_x$ (2:2:2:3)], layered intercalation compounds have been synthesized with the indicated basic structural units for the host material, using planes of iodine for the intercalate (Xiang *et al.*, 1991).

In some cases, such as fullerenes, the molecules of the host material are bound to each other by weak van der Waals forces, allowing the guest species to enter interstitial tetrahedral or octahedral sites. In this case the host material has been ordered crystals of C_{60} or C_{70} molecules, and the intercalate species

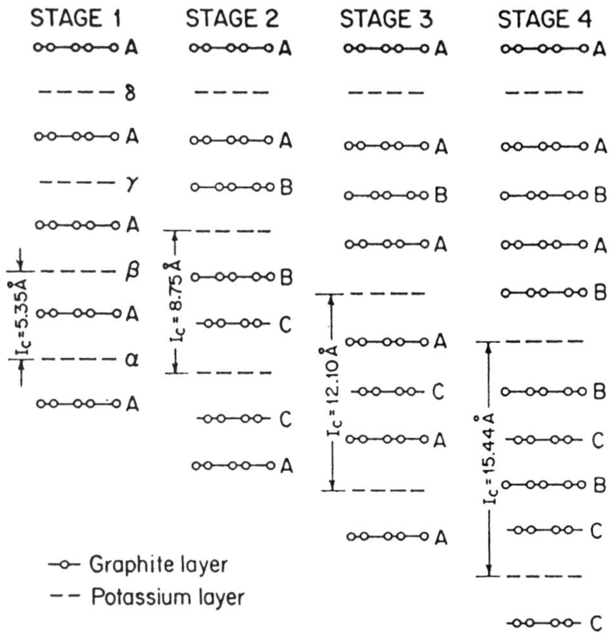

FIG. 4. Schematic diagram illustrating the staging phenomenon in certain layered intercalation compounds for stages $1 \geqslant n \geqslant 4$. In this figure the potassium intercalate layers are indicated by dashed lines and the host graphite layers by solid lines connecting open circles, and indicating schematically a projection of the carbon atom positions. For each stage, the distance I_c between adjacent intercalate layers is indicated.

have been various alkali metals (Na, K, Rb, Cs) and alkaline-earth metals (Ca, Ba) (see FULLERENES).

1.2 Requirements for Intercalation

The occurrence of the intercalation process in layered materials generally requires sufficient chemical activity to

1. transfer charge from the intercalate to the adjacent host layers in order to provide the electrostatic attraction between the intercalate and the host layers that is necessary for their mutual bonding, and
2. supply the elastic energy necessary to separate adjacent layers of the host material to accommodate the intercalate.

The insertion of isolated intercalate molecules into a layered host material introduces large local strains. These strains are greatly reduced by the insertion of a second similar intercalate molecule nearby. Thus both elastic forces and normal chemical bonding forces in solids give rise to an attractive in-plane intercalate-intercalate interaction, so that once nucleation of an intercalate layer is initiated, the subsequent growth of the intercalate layer is rapid, exploiting the enhanced in-plane diffusion coefficients of layered materials. In the case of intercalation compounds where the intercalate seeks specific sites (such as intercalated fullerenes) or specific atomic rows (such as certain cuprate compounds), the intercalation-induced change in lattice constant of the host material is quite small. In contrast, the repeat distance of a layered intercalation compound can be much larger than the basic structural unit of the host material, as discussed in Sec. 2.3.

A variety of chemical and structural considerations determine the facility with which a particular intercalate enters a given host material (Hérold, 1979). For example, graphite is a host material that is easily intercalated, forming intercalation compounds with several hundred chemical species. Although nature has synthesized intercalation compounds in the form of clays and minerals over geological time periods, and scientists have synthesized intercalation compounds for over 150 years (Schafhäutl, 1840), it is only recently that methods have been perfected to the point that one can prepare materials with specific structures and properties (Dresselhaus and Dresselhaus, 1981; Solin, 1982; Zabel and Solin, 1990). In the case of clay intercalation compounds, which are also layered materials, intercalation results from ion exchange between a guest ion and an ion in the host material. During this ion exchange process no charge transfer occurs.

A host material with very strong intraplanar binding and very weak interplanar binding-ing provides a necessary condition for intercalation in layered compounds but not a sufficient condition. For example, BN has essentially the same crystal structure and lattice constants as graphite (see Table 1). However, guest species enter into graphite easily but do not enter BN easily. The major difference between the graphite and BN host materials is the semimetallic electronic structure of graphite with a 40-meV band overlap, while BN is an insulator with a band gap of 4.5 eV. Therefore graphite has a relatively high density of states near the Fermi level E_F to facilitate the charge transfer that provides the electrostatic attraction between the intercalate layer and the two adjacent graphite bounding layers. On the other hand, the wide-gap insulator BN does not have available states near E_F to facilitate charge transfer, so that the weak electrostatic interaction may account for the poor intercalation properties of BN host materials.

For a given host material, some species can easily be intercalated and other species cannot. With the exception of the clays where intercalation occurs by ion exchange, a strong electrostatic interaction and a large amount of charge transfer favors intercalation, while the separation of two sequential layers of the host materials to accept the intercalate layer requires the expenditure of elastic energy. The competition between the electrostatic process and the elastic process determines which species will intercalate easily and for which stages the intercalation process is favorable. Thus for the graphite intercalation compounds (GICs) with molecular bromine (denoted by Br_2-GIC), the stage-1 compound is energetically unfavorable while the stability of higher-stage compounds is favorable. As another example, alkali-metal–GICs can usually be prepared for many stages, though the

Table 1. A comparison between the crystalline and electronic properties of graphite and boron nitride.

Property	Graphite	BN
In-plane lattice constant	2.46 Å	2.49 Å
c-axis lattice constant	3.35 Å	3.33 Å
c-axis layer stacking	*ABAB*	*ABAB*
Electronic structure	Semimetallic	Insulating
Band gap (overlap)	−0.040 eV	4.5 eV

formation of Na-GICs is energetically unfavorable except for high-stage ($n \geqslant 6$) compounds.

It is experimentally found that transition-metal dichalcogenide and fullerene host materials can be intercalated relatively easily with donor species, but not by acceptor intercalates. The intercalates that can be readily inserted into transition-metal dichalcogenides include metal atoms (such as alkali metals Li, Na, K, Rb, noble metals such as Ag, and polyvalent metals such as Ca, In, Sn, Pb), organic molecules (such as amines, pyridine, ammonia and hydrazine), and 3*d* transition metals (such as V, Cr, Fe, Co, Ni). Clays also permit a large number of species to be inserted into their van der Waals gap.

When an intercalation compound is energetically stable, it can often be prepared by a variety of techniques, such as vapor transport, liquid-phase growth, growth from solution, or electrochemical techniques (Hérold, 1979). Of these various possibilities, electrochemical techniques offer perhaps the greatest number of possibilities for preparing exotic intercalation compounds (Brec, 1987; Besenhard *et al.*, 1983), while vapor transport often results in materials with the greatest staging fidelity (Hérold, 1979).

While energetics determine whether or not an intercalation compound is formed, kinetic considerations are important for determining the time necessary to form a given compound. One of the most intriguing of the kinetic considerations concerns the transition from one stage to another as the intercalation process proceeds. The model due to Daumas and Hérold, which minimizes mass transport (see Fig. 5), provides a simple explanation for a staging transition from stage n to stage $n-1$ (Hérold, 1979), such as is commonly observed for graphite intercalation compounds. *In situ* x-ray (00*l*) diffraction or *in situ* Raman-scattering studies of the staging transition provide detailed information on the staging kinetics (Suematsu *et al.*, 1988). In such an *in situ* study, the intercalate species in the gas phase is brought into contact with the host material, which is held at constant temperature, and x-ray spectra are taken every few minutes to monitor the sample stage. The results presented in Fig. 6 for the K-GIC system show a stepwise dependence of the stage index on vapor pressure, with little evidence for multiple-stage coexistence.

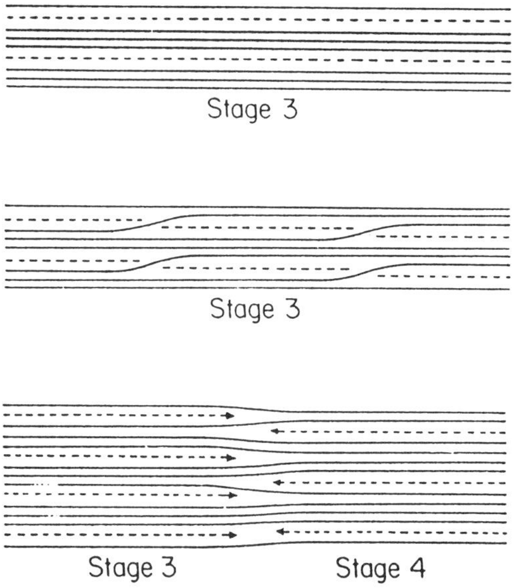

FIG. 5. (a) Single-domain model for a stage-3 compound with every third layer occupied by the intercalate species. (b) Daumas–Hérold domain model for a stage-3 compound with intercalate contained between all graphite planes. (c) Interchange of fourth- and third-stage regions as might occur during a stage transformation.

1.3 Classification According to Rigidity

The layered host materials that can be intercalated are conveniently classified according to their rigidity (Solin, 1984). For the class-I host materials, such as graphite and BN (see Fig. 1), the host material has a basic structural unit that is one monolayer thick (i.e., 3.35 Å for graphite), thereby giving rise to a relatively thin basic structural unit and to relatively flexible materials. A single isolated layer of the graphite structure is called a graphene layer. The basic structural unit of the host material for the transition-metal dichalcogenide intercalation compounds, which are examples of class-II compounds, consists of a trilayer (*XMX*), where *M* denotes the transition-metal layer flanked on either side by a chalcogen layer (see Fig. 2). This trilayer structural unit is intermediate between class I and class III with regard to layer rigidity. Other examples of host materials in the class-II category are the group III–VI (III: Ga, In; VI: S, Se) layered compound semiconductors, where the metal (*M*) and chalcogenide (*X*) layers are arranged in the sequence *XMMX* to form a four-layer slab with strong covalent bonding between each of the four layers and weak van der Waals bonding between these slabs (Julien *et al.*, 1986). Examples of the rigid class-III materials are the silicate clays, where the basic structural unit is typically ~10 Å thick, or thicker when the waters of hydration are included, as shown in Fig. 3 (Wada *et al.*, 1987).

In general, the less rigid the host material, the more likely that it can support the staging phenomenon, which depends also on the competition between the electrostatic and elastic interactions. Of the various host materials, graphite exhibits the highest degree of staging periodicity. Though some transition-metal dichalcogenide intercalation compounds exhibit evidence for staging, this occurs only for low-stage (i.e., $n=1,2$) compounds, and these compounds characteristically have low staging fidelity. Clay intercalation compounds (CICs), which show the largest layer rigidity, do not generally show any evidence for staging, except for stage-1 compounds. On the other hand, clay host materials permit the intercalation of multiple sequential intercalate layers (see Fig. 1), in contrast to the graphite and transition-metal dichalcogenide host materials, which only permit intercalation of a single intercalate unit (either a monolayer or a trilayer unit, as discussed in Sec. 1.4).

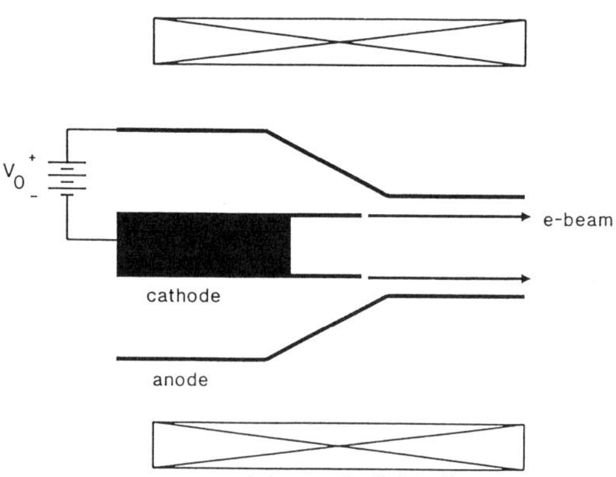

FIG. 6. *In situ* x-ray study of staging transitions (where T_G, the graphite temperature, is 400.1 °C). The kinetics of the staging process in K-GICs is obtained from the time dependence of (00*l*) reflections for transitions from a stage-7 compound to a stage-3 compound (Suematsu *et al.*, 1988).

1.4 Classification of the Intercalate Layers

There are several possible structural arrangements for the intercalate layers. The most common are binary compounds for which the intercalate is a single chemical species, such as an alkali metal, a halogen, or a metal chloride. In some cases, such as the alkali metals or halogens, the intercalate layer consists of a single atomic layer, sandwiched between the basic structural units of the host material. In other cases, such as transition-metal chlorides, the intercalate basic structural unit is a trilayer sandwich. For example, $FeCl_3$ forms a trilayer sandwich of Cl-Fe-Cl constituent layers. Not all compound intercalates form multilayer sandwiches; for example, $AuCl_3$ forms a monolayer intercalate.

A great deal of attention has also been given to the study of ternary intercalation compounds, which are formed by the insertion of a second intercalate species into a binary intercalation compound. Some common examples include the insertion of a second intercalate species into an alkali-metal graphite intercalation compound. Examples of intercalate species that intercalate into alkali-metal GICs include metal intercalates such as Ag and Tl, inorganics such as hydrogen or NH_3, and organic constituents such as benzene (C_6H_6) or tetrahydrofuran (THF) (Zabel and Solin, 1990). In general, the structural arrangement for these ternary intercalation compounds is that of the basic structural unit(s) of the host material, followed by an intercalate trilayer of *MNM* in which *M* refers to an alkali-metal layer and *N* refers to the new ternary constituent, subsequently added to the original binary intercalation compound. The classification of the ternary compounds is complicated by three issues. First, some of the ternary compounds, such as potassium hydride graphite intercalation compounds (KH-GICs), can also be prepared by direct intercalation; independent of whether KH-GICs are prepared sequentially or directly, they are classified as ternary compounds. (Here the intercalate contains two species, potassium and hydrogen.) Second, two-component metal alloy intercalates are said to form ternary intercalation compounds, even though the intercalate is inserted into the host material directly, and the intercalate occupies a single atomic layer as a random-alloy monolayer. The third complication in the ternary classification arises from the possibility of preparing bi-intercalation compounds where the two intercalate constituents are located in different van der Waals gaps or intercalate galleries. Further examples of the various types of binary and ternary systems are given later in this section.

For either binary or ternary intercalation compounds, the separation of two host layers d_s between which the intercalate is sandwiched can vary over wide ranges. In general d_s depends on the size of the guest species. The wide variation of d_s is most clearly seen in the GICs. One example of a GIC where d_s deviates only slightly from that of the unintercalated host material is the case of the Li donor intercalate in graphite where d_s=3.70 Å and $\tilde{c}_0$=3.35 Å is the thickness of the basic structural unit of a single graphene sheet. In the opposite limit, d_s can be greater than $3\tilde{c}_0$ for trilayered intercalates into graphite such as transition-metal chlorides (see Table 2) (Dresselhaus and Dresselhaus, 1981; Solin, 1982). Ternary compounds such as KHg-GICs tend to have large d_s values as indicated in Table 2.

The separation between specific intercalate layers can be further increased by the synthe-

Table 2. Intercalate sandwich thickness d_s for several graphite intercalation compounds (GICs) showing that the repeat distance I_c for a low-stage GIC can be varied over a wide range by proper choice of intercalate.

Intercalate	d_s (Å)
Li	3.706
Eu	4.87
K	5.35
Rb	5.65
Cs	5.94
Br_2	7.04
HNO_3	7.84
AsF_5	8.15
SbF_5	8.46
$MnCl_2$	9.30
$NiCl_2$	9.30
$MoCl_5$	9.31
$FeCl_3$	9.37
$SbCl_5$	9.42
$CrCl_3$	9.45
$MnCl_2$	9.48
$CoCl_2$	9.50
$FeCl_2$	9.51
$AlCl_3$	9.54
KHg	10.22

sis of bi-intercalation compounds (Zabel and Solin, 1990). Bi-intercalation compounds contain two different intercalants that assume special ordered arrangements. Layered bi-intercalation compounds with intercalants M_1 and M_2 can be prepared, forming the layered sequence $M_2CM_1CM_2$ where C denotes carbon (a single graphene sheet). One synthesis route for such bi-intercalation compounds can be the intercalation of species M_2 into a stage-2 M_1-GIC. For example, in a bi-intercalation metal chloride compound, the separation between sequential M_1 layers is close to 20 Å. Such compounds are of particular interest for the synthesis of quasi-2D magnetic systems, and in this connection the compound $CoCl_2$-$GaCl_3$-GIC has been synthesized and studied magnetically (see Sec. 4).

More generally, ternary intercalation compounds provide a mechanism whereby more than one intercalate species can be introduced into a host material (Solin and Zabel, 1988). As mentioned above, aside from the bi-intercalation compounds, there are basically two different categories of ternary compounds: intercalates introduced as alloys, and intercalates introduced sequentially. An example of an intercalate alloy system is K_xRb_{1-x}, which intercalates into graphite as a substitutional random alloy on each intercalate monolayer (Solin and Zabel, 1988). Such substitutional alloy systems offer very interesting possibilities for studying the physics of quasi two-dimensional random alloys. Intercalate alloys can also be introduced into nonlayered host materials, such as fullerenes. For the case where the two intercalate species have nearly the same size, the two intercalate species enter the fullerene lattice randomly in either octahedral or tetrahedral sites, while for cases like Na_xCs_{1-x}, where the two species have very different radii, the Na ions tend to favor the smaller tetrahedral sites and the larger Cs ions strongly favor the octahedral sites, so that an ordered compound of stoichiometry Na_2CsC_{60} can be prepared.

The more widely studied classes of ternary compounds comprise the sequentially intercalated compounds (Solin and Zabel, 1988), which allow the intercalation of species M_2 that could not itself be intercalated into the host material. In such cases, the prior intercalation of species M_1, often an alkali metal, permits subsequent intercalation of species M_2. Examples of such ternary compounds formed with M_1 as an alkali-metal species include as the M_2 species tetrahydrofuran (THF), an organic species; NH_3, an inorganic species; and the metallic species Hg and Tl, which form superconducting ternary GICs (see Sec. 5). In the case of the hydrogen ternary compounds, both chemisorbed and physisorbed compounds are possible, where the structure of the chemisorbed compound is similar to that for the ionic KHg-GICs (see Fig. 7), while the physisorbed compound can be prepared from stage-2 K-GICs and is important for high-density hydrogen storage applications. Because of the large differences in the amount of physisorption of the various isotopic hydrogen species (tritium > D_2 > H_2), the physisorbed hydrogenated K-GICs offer important possibilities for isotope separation applications (Terai and Takahashi, 1981).

In the above discussion of the intercalate layer, it was assumed that the guest species in the intercalate layer is the same as the guest species prior to intercalation. This is not always the case. For compounds such as the $SbCl_5$-GICs, the intercalation process results in a chemical reaction called disproportionation, giving rise to different species in the intercalate layer, as, for example, the species $SbCl_3$, $SbCl_6^-$, $SbCl_5$, and $SbCl_4^-$ relative to the parent intercalate species $SbCl_5$ prior to intercalation (Zabel and Solin, 1990).

2. STRUCTURAL CHARACTERIZATION

In Sec. 1, the general structural organization and classification of intercalation compounds was summarized. In this section, the structural ordering along the *c* axis normal to the layer planes and in the *a-b* layer planes is reviewed. Interlayer correlations and phase transitions in intercalation compounds are also discussed.

2.1 *c*-Axis Ordering

The dominant ordering along the *c* axis is the staging periodicity, discussed in Sec. 1.2, and characterized by the stage index *n* and the staging fidelity, denoted by the average length of the stage-*n* material (both in-plane and along the *c* axis), and by the amount of each minority stage *n*′, expressed as a percentage of the sample volume. The simplest method

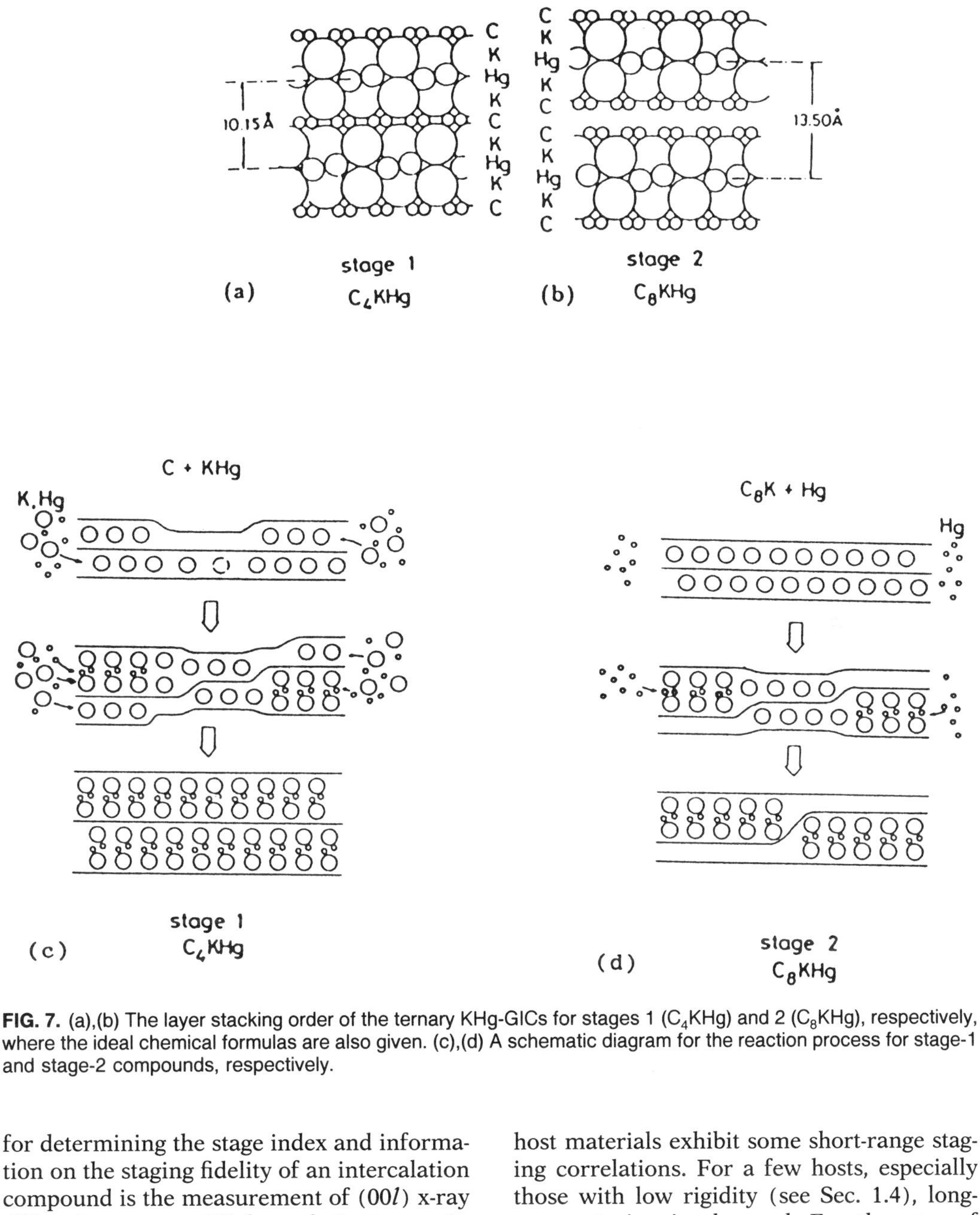

FIG. 7. (a),(b) The layer stacking order of the ternary KHg-GICs for stages 1 (C_4KHg) and 2 (C_8KHg), respectively, where the ideal chemical formulas are also given. (c),(d) A schematic diagram for the reaction process for stage-1 and stage-2 compounds, respectively.

for determining the stage index and information on the staging fidelity of an intercalation compound is the measurement of (00*l*) x-ray diffraction patterns. High-resolution transmission electron microscopy (TEM) studies of lattice fringe images provide a powerful technique for showing detailed aspects of the ordered lattice planes. Since the intercalate often has elements of higher atomic weight than the host material (e.g., carbon), the intercalate layer scatters the TEM electrons more effectively, thereby giving clear contrast between the intercalate and the light host materials. Such measurements show that many host materials exhibit some short-range staging correlations. For a few hosts, especially those with low rigidity (see Sec. 1.4), long-range staging is observed. For the case of graphite intercalation compounds, high-resolution TEM lattice fringe images show that a high degree of staging perfection (over a scale of ~100 Å) can be achieved under favorable circumstances (Zabel and Solin, 1990). For GICs, the *c*-axis repeat distance I_c between sequential intercalate layers follows the simple relation $I_c = d_s + (n-1)\tilde{c}_0$, where $\tilde{c}_0$ is the interlayer separation in pristine graphite, or the thickness of a single graphene sheet. The validity of this simple relation indicates that, to first order, GICs can be considered as a linear superposition of graphite and intercalate layers. Similar results are obtained in some cases for other host materials. Graphite intercalation compounds are, however, of particular interest with regard to staging because they can be synthesized with stages up to $n \sim 10$ and because of their relatively high degree of staging fidelity (Hérold, 1979; Zabel and Solin, 1990).

TEM studies also provide information about the coupling between the host material and intercalate at their interface. Some compounds such as $CuCl_2$-GICs show sharp interfaces with weak coupling between the intercalate and the graphite host, while K-GICs and F-GICs show very wavy fringes, indicating a strong interlayer coupling. In the case of F-GICs the coupling is so strong that the intercalate layers become puckered as covalent bonding is initiated between the fluorine intercalate and the graphite host material.

The basic structural units of the transition-metal dichalcogenide crystals TX_2 (see Fig. 2) also show *c*-axis ordering both before and after intercalation. The arrangements of atoms within a sheet and the stacking of the atomic TX_2 sheets are now described. Each sheet consists of atoms in a hexagonally close-packed network as shown in Fig. 8(a), which also identifies the three inequivalent close-packed sites *A*, *B*, and *C*. Two types of coordination structure are possible, and either one or both can form the basic structural units of the TX_2 crystal. In the *trigonal prismatic coordination*, the sandwich of *X-T-X* follows the ...*AbA*... sequence, as in Fig. 8(b), and in the octahedral coordination, the ...*AbC*... sequence, as in Fig. 8(c). Here capital letters are used to denote the chalcogen positions and lower-case letters for the metal atoms. In an ideal structure, the distance between chalcogen atoms across the sandwich is the same as that in the plane, which is also equal to the *a* and *b* lattice constants as well as the distance between metal atoms (in plane).

In addition to the two *intrasandwich* octahedral and trigonal prismatic coordination arrangements, there are many variations for *intersandwich* packings, and a large number of polytypes are known. The six most common polytypes for transition-metal chalcogenides are shown in Fig. 9. The nomenclature that identifies these polytypes specifies the number of trilayer sandwiches contained in the unit cell perpendicular to the plane of the layers, and the overall symmetry (trigonal, hexagonal, or rhombohedral) of the structure. Lower-case subscripts are sometimes further required to distinguish polytypes otherwise similarly labeled (e.g., $2H_a$ and $2H_b$). Thus the simplest polytype with octahedral coordination, labeled $1T$, has trigonal symmetry and a repeat distance, perpendicular to the layers, of one trilayer sandwich. The simplest trigonal prismatic coordination polytypes contain two trilayer sandwiches, and are designated $2H_a$ and $2H_b$ (see Fig. 9). The $2H_a$ structure is adopted by metallic group-V materials such as NbS_2 and stacks the metal atoms directly above each other along the *c* axis, while in the $2H_b$ structure, adopted by semiconducting group-VI materials such as MoS_2, the metal atoms are staggered. Octahedral interstitial sites between sandwiches are also shown in Fig. 9 in parentheses, and are the vacant sites into which intercalate atoms or molecules may enter. The crystallographic axes for the *T* and *H* structures are labeled by

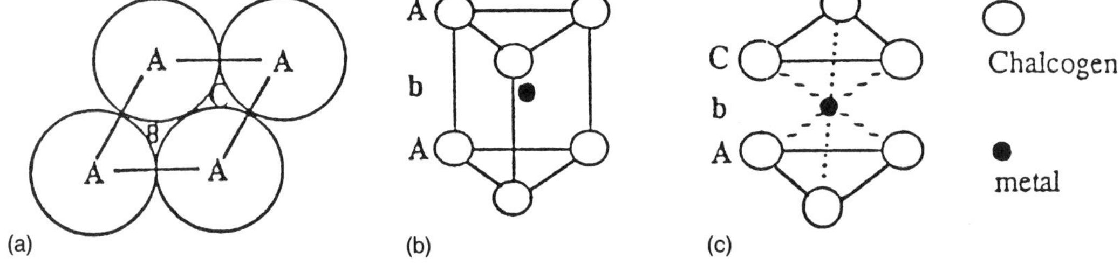

FIG. 8. The octahedral and trigonal prismatic structures of transition-metal dichalcogenides, TX_2. (a) *A*, *B*, and *C* represent the three nonequivalent positions for close-packed stacking; (b) the trigonal prismatic structure; and (c) the octahedral stacking arrangement. The capital letters designate chalcogen positions and the lower-case letters the metal positions.

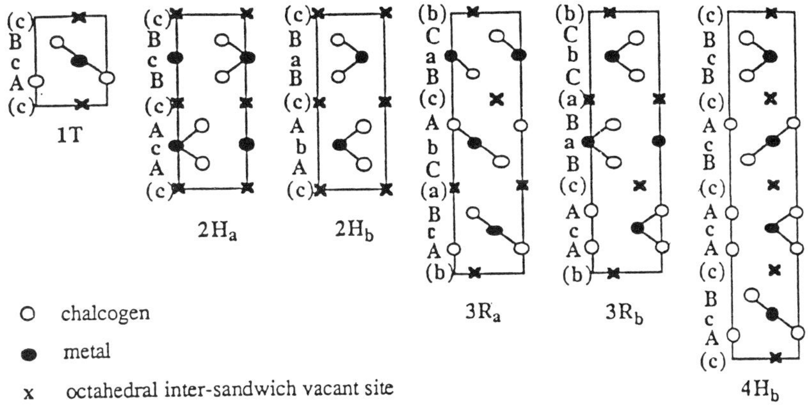

FIG. 9. Unit cells of simple polytypes shown with respect to their *c*-axis arrangement. The capital letters designate chalcogenide positions and the lower-case letters give the metal positions. Vacant octahedral sites for intercalation are indicated.

$a(x,y)$ parallel and $c(z)$ perpendicular to the layers. Typical values are 3–4 Å for the a lattice parameter, and 6 Å for the sandwich repeat distance along the c axis.

In addition to the octahedral interstitial sites shown in Fig. 9, there are also tetrahedrally coordinated intersandwich sites that lie immediately below and above the chalcogen atoms and are displaced from the symmetric plane between the trilayer sandwiches. These are shown as the "a" and "b" sites in Fig. 10. Octahedral interstitial sites are usually favored in the intercalation of alkali metals, Li, Na, etc., and the 3*d* transition metals, and the maximum intercalate uptake M gives a limiting stoichiometry of MTX_2. Tetrahedral site occupation is found in intercalation complexes formed with silver and copper, and with hydrazine; in this case the limiting stoichiometry is M_2TX_2. Li ions in VSe_2 can occupy both octahedral sites, forming $LiVSe_2$, and tetrahedral sites, as in Li_2VSe_2. Transition-metal dichalcogenide hosts are especially interesting for the possibility of introducing very large spacers between the TX_2 trilayers, as, for example, the insertion of *n*-octadecylamine into $TaSe_2$ leading to a separation of nearly 60 Å between sequential Ta layers in the layered structure of the intercalation compound.

2.2 In-Plane Ordering

Not only can intercalation compounds show long-range staging fidelity with regard to their *c*-axis ordering (1D superlattice), but intercalation compounds also show various in-plane orderings, which are of interest in their own right and bear importantly on the observed properties of this class of materials. The two major classes of in-plane structures are the commensurate and the incommensurate arrangements of the intercalate layers with respect to the adjacent host layers. In the commensurate structures (see Fig. 11), the intercalate atoms are in registry with the adjacent host, forming intercalate unit cells with areas that can be multiples of the areas of

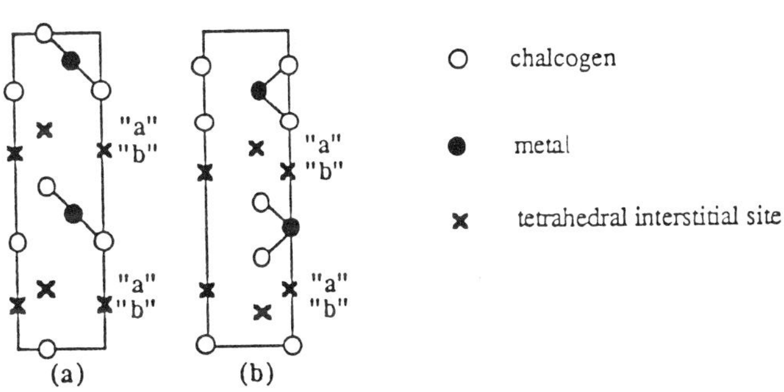

FIG. 10. Tetrahedral intersandwich sites *x* in the (a) 1*T* and (b) 2H_b TX_2 polytype structures.

the 2D unit cell of the host material. The following are common examples of commensurate structures that occur in GICs: $(\sqrt{3}\times\sqrt{3})R30°$, $(2\times2)R0°$, $(\sqrt{7}\times\sqrt{7})R19.1°$, and $(\sqrt{3}\times2)R(30°,0°)$, where the notation lists the lengths and angles of the intercalate basis vectors relative to those for the graphite lattice, and angular redundancies are not repeated. Commensurate structures are found in a wide variety of model GIC systems (see Table 3), including stage-1 binary donor compounds, such as C_6Li, C_6Eu, C_8K; coplanar molecular acceptor compounds, such as $C_{7n}Br_2$ for stages $n \geqslant 2$; triple-layer sandwich acceptors, such as the $SbCl_5$-GICs; and triple-layer sandwich donor ternary systems, such as KHg-GICs and KH-GICs (Zabel and Solin, 1990).

The formation of commensurate in-plane structures has a close analogy to the growth of semiconductor strained-layer superlattices by the molecular beam epitaxy (*q.v.*) (MBE) method or by metal-organochemical vapor deposition [see CVD (CHEMICAL VAPOR DEPOSITION)] (Dresselhaus, 1987). Because the intercalate layer in an intercalation compound is so thin (a single monolayer or a single trilayer sandwich), commensurate in-plane structures can be formed without the need for misfit dislocations to relieve local strains.

Table 3. Structure of common commensurate GICs.

Intercalate	Stage	Formula[a]	Structure[b]
Li	1	C_6Li	$(\sqrt{3}\times\sqrt{3})R30°$
Eu	1	C_6Eu	$(\sqrt{3}\times\sqrt{3})R30°$
K	1	C_8K	$(2 \times 2)R0°$
Br_2	$\geqslant 2$	$C_{7n}Br_2$	$(7 \times \sqrt{3})R(0°,30°)$
$SbCl_5$	$\geqslant 1$	$C_{14n}SbCl_5$	$(\sqrt{7}\times\sqrt{7})R19.1°$
KHg	1	C_4KHg	$(2 \times 2)R0°$
KH	1	C_4KH	$(2 \times 2)R0°$

[a]The nominal chemical formula for a stage-*n* compound.
[b]The notation for the 2D superlattice refers to the length scaling and the angle of rotation of the unit vectors for the intercalated superlattice relative to graphite.

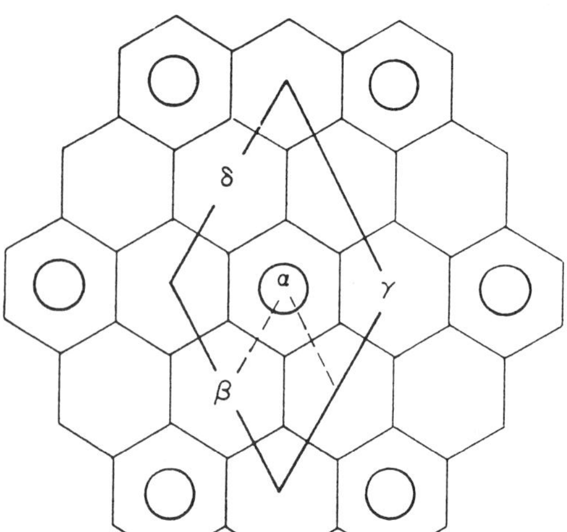

FIG. 11. Schematic diagram of an in-plane unit cell for the commensurate $(2 \times 2)R0°$ structure. Carbon atoms from the host material are located at the corners of the hexagons on one layer and the intercalate species are shown as the large open circles, which are located on the adjacent layer. In this diagram the intercalate and graphite layers are projected onto a single plane. For this structure there are four equivalent intercalate sites, labeled α, β, γ, δ. On a given intercalate layer only one of these sites is occupied. The dashed lines represent the in-plane unit cell for pristine graphite. For the first-stage C_8K and C_8Rb, the intercalate on consecutive layers is arranged on sequential α, β, γ, δ sites.

The lowest energy state for an intercalate atom is obtained by placing it over (or below) the center of a graphite hexagon, thereby achieving the greatest packing density (Dresselhaus and Dresselhaus, 1981; Hérold, 1979). A close-packed in-plane configuration for the intercalate maximizes its chemical bonding and minimizes the strain energy introduced into the adjacent graphite layers from insertion of the intercalate. When both registry with the graphite hexagons and close packing of the intercalate layers can be simultaneously satisfied on a long-range basis, commensurate intercalate layers are favored. Otherwise, the close-packed configuration of the intercalate layer implied by chemical binding considerations prevails and an incommensurate solid forms. In many cases, the in-plane intercalate lattice constant for incommensurate interlayer stacking is close to that of the parent material. Examples of intercalates forming incommensurate GICs are transition-metal chlorides and bromides (Vogel and Hérold, 1977; Hérold *et al.*, 1983), some of which have especially interesting magnetic properties, as discussed in Sec. 4.

Intercalate registration (the tendency for an intercalate atom to be located in a position of minimum energy with respect to the adjacent graphite layers) affects the structural ordering of GICs in many ways besides the formation of commensurate GICs, as discussed above. For the case of incommensurate GICs, local registration of the intercalate with respect to the graphite layers leads to *orientational* alignment of the intercalate, as evidenced by the spot patterns observed by ($hk0$) diffraction spectra for many incommensurate GICs. In addition, many incommensurate intercalates (e.g., transition-metal chlorides) tend to form intercalate islands ($\sim$100 Å in diameter) to minimize the number of intercalate atoms not in registry with the graphite host material (Dresselhaus, 1987).

2.3 Interlayer Correlations

Not only is there intercalate ordering within the intercalate layer, but there is also 3D stacking order of intercalate layers with respect to one another. A clear example of this phenomenon is observed in stage-1 alkali-metal–GICs (Dresselhaus and Dresselhaus, 1981). Because the intercalate unit cells can be much larger than that for the host, there will be several equivalent sites for the intercalate atoms in a commensurate GIC. For example, for a $(2\times2)R0°$ in-plane lattice (see Fig. 11), there will be four equivalent intercalate sites, labeled α, β, γ, δ. But once one of these sites is occupied by an intercalate atom (e.g., an α site), then the intercalate atoms on the same layer (within a single domain) will also tend to occupy the same type of site (an α site), thereby giving the maximum packing density for the intercalate (Dresselhaus and Dresselhaus, 1981). Because of the local strain introduced by placing an intercalate atom on site α, the intercalate atom on the adjacent intercalate layers will be less likely to be located on an α site than on a non-α site. This interlayer exclusion argument implies that when there is sufficient intercalate-intercalate interlayer correlation mediated by the intervening graphite layers, interlayer stacking order will result. A number of low-stage binary GICs and ternary GICs have been reported to exhibit interlayer stacking correlation effects (Solin and Zabel, 1988). Even when there is no long-range interplanar stacking order, there is often short-range interplanar correlation based on the interplanar intercalate exclusion argument given above. Because the interplanar intercalate-intercalate interaction energy is small, thermal excitation at room temperature is often sufficient to destroy interplanar intercalate correlation, thereby producing quasi-2D intercalate layers, uncorrelated in the c direction.

2.4 Intercalate Phase Transitions

In general, the host material and intercalate materials have different melting points. Thus unusual phase transitions can be observed, as one of the constituents melts while the other remains solid. Such behavior is, for example, observed in GICs, since graphite does not undergo structural phase transitions up to very high temperatures ($\sim$4000°C). Examples of such phase transitions are order–disorder transitions in the intercalate layer, as occur in the HNO_3-GIC system (Dresselhaus and Dresselhaus, 1981). The disordered intercalate phase often exhibits the characteristics of a lattice gas, whereby each intercalate atom is fluidlike in the intercalate layer, but has a high probability of being in registry with the adjacent graphite layers. Such effects are observed in high-resolution x-ray diffraction measurements (Zabel and Solin, 1990) of high-stage ($n>2$) alkali-metal GICs. Intercalate melting is a closely related phenomenon and has been observed for a variety of intercalates (Dresselhaus and Dresselhaus, 1981).

For some GICs, two or more intercalate orderings have approximately the same free energy, leading to multiphase coexistence, with domain walls that are either more dense (heavy walls) or less dense (light walls) than the ordered domains. An ordered array of domain walls may occur, giving rise to modulated structures, as occur in many high-stage ($n>2$) alkali-metal GICs (Zabel and Solin, 1990). Heavy and light walls also arise at the boundary between regions of different site occupation in a commensurate in-plane superlattice [e.g., the boundary between an α domain and a β domain in the same intercalate layer of a $(\sqrt{3}\times\sqrt{3})R30°$ superlattice]. These phenomena have been widely studied for adsorbed rare-gas phases on graphite (Zabel and Solin, 1990).

A particularly interesting example of a modulated in-plane structure in GICs is the stripe domain phase (Zabel and Solin, 1990), which has been observed in a stage-4 Br_2-GIC with nominal stoichiometry $C_{28}Br_2$. This intercalation compound has a commensurate $(7\times\sqrt{3})R(0°,30°)$ in-plane structure with

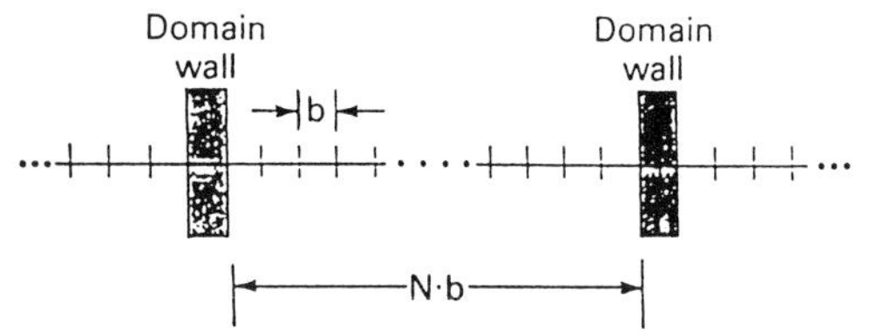

FIG. 12. Schematic diagram of a stripe domain phase. The intercalate lattice remains commensurate except in the domain wall where a phase slip occurs every N unit cells.

a rectangular unit cell containing four Br_2 molecules. The driving mechanism for this phase transition is the differential thermal expansion between the graphite basal plane, which has essentially zero thermal expansion, and that for a bromine layer, which has a normal thermal expansion coefficient. For the Br_2-GIC system, the coupling between the graphite and intercalate layers is sufficiently large so that the structure remains commensurate over a large temperature range. At a temperature $T_c = 69$ °C, a phase transition occurs whereby the graphite and bromine layers retain long-range registry along the $\sqrt{3}$ direction, but form a stripe domain phase along the seven-fold direction. As the temperature increases above T_c, the average bromine lattice constant along the seven-fold direction increases, though the graphite and bromine remain locally commensurate over a domain containing N unit cells (see Fig. 12). These domains are arranged in a periodic array with lattice constant $Nb+\tau$, where τ is the thickness of the domain wall, and $\tau = 2b/7$ where $b = 7a_0$ and a_0 is the graphite in-plane lattice constant. The domain wall is also called a misfit dislocation in the terminology of the strained-layer superlattice. This stripe domain phase is one example of a more general discommensuration phenomenon, which features local commensurate behavior but long-range incommensurate behavior. Discommensuration effects are also observed in high-stage alkali-metal–GICs (Zabel and Solin, 1990).

3. ELECTRONIC PROPERTIES

In accordance with the structural ordering discussed above is the periodicity in the charge distribution perpendicular to the layer planes, which arises from the charge transfer associated with the intercalation process. In general, charge transfer is the driving mechanism for intercalation and is responsible for the observed transport and electronic properties of intercalation compounds. In this section the concept of charge transfer is discussed and the transport and electronic properties that result from charge transfer are also reviewed.

3.1 Charge-Transfer Phenomena

To illustrate the charge-transfer concept in a general intercalation compound, let us consider the charge transfer from an alkali metal to a host material such as graphite in a graphite intercalation compound (see Fig. 13). Let us suppose that each alkali-metal atom contains one nearly free electron. If this electron is donated to the graphite layers, the alkali-metal layers become positively charged, thereby attracting the negative charges in the graphite layers. Thus the majority of the electrons transferred to the graphite will be located in the graphite bounding layers adjacent to the intercalate layers, as shown in Fig. 13. Lower concentrations of transferred charge will be available for the graphite interior layers that are surrounded only by graphite layers (Dresselhaus, 1984). Calculations of the charge distribution in the graphite layers

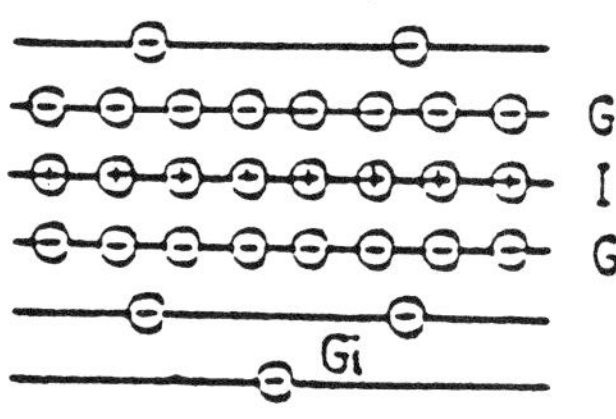

FIG. 13. Schematic representation of charge transfer in graphite intercalation compounds, showing the intercalate layer (I) for a donor compound transferring electrons to the graphite bounding (G_b) layer, with a small amount of charge also transferred to the graphite interior (G_i) layers.

show that the graphite bounding layers effectively screen the graphite interior layers from the intercalate layers with a power-law dependence of z^{-4} relative to the position of the intercalate layer. The charge-transfer concept for alkali-metal intercalation into other host materials is similar.

Since the carrier concentration in graphite is very low ($\sim 2\times 10^{-4}$ carrier/C atom at room temperature and $\sim 6\times 10^{-5}$ carrier/C atom at low temperature), the charge transferred to the graphite layers through intercalation causes a major increase in the mobile carrier concentration in the graphite layers. An estimate of three orders of magnitude increase in carrier concentration follows from assuming a charge transfer of one carrier per intercalate unit to the graphite layers and assuming an intercalate concentration one order of magnitude smaller than that for graphite. Since the mobility of pristine graphite is relatively high (13 000 $cm^2/V\,s$ at 300 K), intercalation compounds with very high in-plane electrical conductivity can be synthesized. On the other hand, the conductivity in the intercalate layers is expected to be low because of the low mobility of ions generally, and the low density and low mobility of electrons (or holes) in the intercalate layers. Thus electrical transport in the basal planes of GICs is mainly by graphitic π-band electrons and holes. Intercalation with donors raises the Fermi level so that conduction is predominantly by electrons, while intercalation with acceptors lowers the Fermi level, giving rise to conduction by holes (Dresselhaus and Dresselhaus, 1981).

Because of the high electron affinity of the transition-metal dichalcogenides and fullerenes, these materials lead only to donor intercalation compounds. Charge transfer in the donor transition-metal dichalcogenide compounds is especially interesting, since TX_2 compounds containing a group-IV transition metal, such as ZrS_2, are semiconducting, so that, for example, intercalation with an alkali metal can partially fill the d_{z^2} level, making the material conducting. On the other hand, if the transition metal T is a group-V element such as Nb, the TX_2 compound is metallic; however, if sufficient charge is transferred, a filling of the d_{z^2} band may occur, resulting in a semiconductor. These various possibilities are considered in Fig. 14.

The current that flows during the electrochemical intercalation process provides one direct measure of the charge transfer in intercalation compounds. In principle, magneto-oscillatory Fermi-surface experiments should also yield accurate values of the charge transfer. This has been demonstrated by very careful magneto-oscillatory measurements made on several well-characterized alkali-metal–GICs and a very few acceptor GIC materials. Because of the difficulty of these experiments, it is not likely with present techniques that Fermi-surface studies will become a widespread characterization tool for charge transfer in GICs.

Charge transfer in ternary GICs is of particular interest. For example, the addition of Hg or H to the first-stage compound C_8K reduces the electron concentration in the graphite layers by a back-donation process, while the Hg or H layers become negatively charged. This can be more easily understood by referring to Fig. 7 where the trilayer structure of the intercalate sandwich for a ternary compound (such as KHg-GIC) is shown. Here the Hg or H layers behave somewhat like an acceptor intercalate, though residing in an otherwise donor compound. It is believed that the Hg conduction band gains electrons but remains partially occupied in order to support

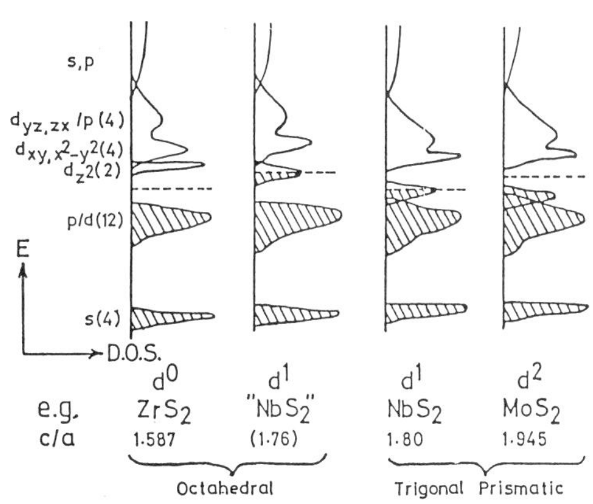

FIG. 14. Schematic density-of-states diagrams for groups IV, V, and VI transition-metal dichalcogenides. The numbers in parentheses following the atomic orbital notation indicate the number of states per formula unit that can be accommodated in each band. The compounds used as examples are taken from metals of the same row of the periodic table. NbS_2 does not actually crystallize in the octahedral form, and the second diagram represents an extrapolated hypothetical band diagram (Liang, 1987).

superconductivity in the KHg ternary compounds as discussed in Sec. 5.

3.2 Transport Properties

Because of the large increase in carrier concentration that often accompanies intercalation, large increases in the in-plane conductivity σ_a is often observed in intercalation compounds. For semiconducting group-IV or group-VI transition-metal dichalcogenides, the change in conductivity can be many orders of magnitude, since the material prior to intercalation has such a low electrical conductivity. For the GIC host material, the room-temperature in-plane conductivity can with suitable intercalates (e.g., AsF_5) become comparable to that of the best room-temperature conductors. Because of this potential for practical applications, most transport studies on intercalation compounds have been done on GICs. For most GICs, in-plane conductivity increases approaching an order of magnitude are frequently observed. Some exceptions to this general behavior are found in high-concentration F-GICs, where the intercalation process itself leads to disorder, the formation of covalent bonding, and a reduction in σ_a.

Not only is the electrical conductivity of layered intercalation compounds very high in the basal planes, but it is also highly anisotropic. The most anisotropic intercalation compounds are the GICs, a result in part of the anisotropy of the host material. The conductivity of pristine graphite in the *c* direction is between 3 and 4 orders of magnitude lower than that in the basal plane, depending on the temperature (Dresselhaus, 1987). Since intercalation with acceptors tends to introduce intercalate layers containing molecular species with very low electrical conductivity, the electrical anisotropy of graphite is increased by intercalation with acceptors, reaching values of σ_a/σ_c in the range 10^6–10^7 (McRae and Marêché, 1988). On the other hand, intercalation with alkali-metal donors often hybridizes the graphitic π electrons with intercalate electrons containing *s* and *p* character, thereby reducing the electrical anisotropy relative to pristine graphite, with anisotropy ratios as small as 10 for the case of stage-1 C_6Li (Dreselhaus and Dresselhaus, 1981). However, the ternary donor compounds (e.g., KHg-GICs and CsBi-GICs), show anisotropy ratios σ_a/σ_c comparable to the magnitudes previously reported for typical acceptor GICs (McRae and Marêché, 1988).

3.3 Electronic Structure

An extensive amount of work has been done in calculating the electronic structure of the various host materials supporting intercalation, such as graphite and the transition-metal dichalcogenides, and in some cases the electronic structure of the intercalation compounds has also been calculated (Zabel and Solin, 1992). From an experimental standpoint, photoemission has been extensively used for determinations of the electronic structure.

Although surface science techniques such as photoemission can, in principle, yield a great deal of valuable information about intercalation compounds, the near-surface sensitivity of these techniques causes them to emphasize the surface electronic structure rather than that typical of the bulk. This is a significant problem for intercalation compounds because of the large sizes of their *c*-axis unit cells. In addition, surface-science studies have thus far shown that the first few atomic layers of donor GICs have an excess intercalate concentration, while in acceptor GICs, the first few atomic layers exhibit a very low intercalate concentration (Dresselhaus, 1987). The cleavage of GICs at low temperature (e.g., ~77 K) has demonstrated the possibility of preparing surfaces retaining intercalate concentrations typical of the bulk for long enough times (minutes) to permit properties measurements to be carried out. Optical measurements, which probe many unit cells along the *c* axis, have also been widely applied to study interband transitions and phonon modes in various intercalation compounds. Because of the high degree of anisotropy in the structural properties, the electronic and optical properties are also highly anisotropic.

4. MAGNETIC INTERCALATION COMPOUNDS

A variety of magnetic materials has been prepared by the intercalation of magnetic species into various nonmagnetic host materials. In general, the intercalation of magnetic species into a host material results in a magnetically ordered intercalation compound below a magnetic ordering temperature T_c. By varying the separation between sequential magnetic layers through the staging mechanism, the interplanar magnetic coupling can be reduced in a controlled manner to very small values, thereby providing a convenient system for studying quasi-2D magnetism. Graphite intercalation compounds are of particular interest in the case of magnetic intercalation compounds because they provide novel systems for the study of quasi-2D magnetic behavior. The transition-metal dichalcogenides and clays also allow for the intercalation of layers of magnetic species in a nonmagnetic host material. The transition-metal dichalcogenides offer further possibilities insofar as by proper selection of the transition metal, the host material can itself be magnetically ordered; in this case, the intercalate can be either nonmagnetic or a different kind of magnetic layer. Some of the most interesting magnetic studies of 2D systems have so far been done on the graphite intercalation compounds because of their good staging characteristics.

Both donor (Eu) and acceptor ($FeCl_3$, $FeCl_2$, $CoCl_2$, $NiCl_2$, $MnCl_2$, $MoCl_5$) magnetic GICs have been synthesized and studied for their magnetic properties (Zabel and Solin, 1992). The magnetic ordering temperatures of the magnetic intercalation compounds are generally lower than those for their parent magnetic materials because of the smaller number of near neighbors imposed by the layer arrangement of the intercalation compounds. For the rare-earth donor GICs, the intercalate in-plane structure is commensurate with that of the adjacent graphite layers, forming a $(\sqrt{3}\times\sqrt{3})R30°$ superlattice structure, and thereby simplifying the treatment of the magnetic problem to some degree. Since the in-plane structures of incommensurate metal chloride acceptor GICs tend to be almost identical to those in the corresponding parent materials, the intraplanar magnetic exchange constants remain essentially unchanged upon intercalation, thereby providing a major simplification in the treatment of the magnetic problem in metal chloride acceptor GICs.

Of the various types of magnetic acceptor GICs, the quasi-2D *XY* systems (for which the spins are constrained to be in the layer planes) have been most widely studied, because of the special role these systems play in theoretical studies of critical phenomena. The absence of long-range order in 2D systems results in short-range correlations in the form of vortices. Close approximations to such magnetic phases can be observed in magnetic GICs over a limited range of temperatures and magnetic fields (Zabel and Solin, 1992).

Of the various magnetic acceptor GICs, the $CoCl_2$-GICs exhibit the largest *c*-axis magnetic anisotropy ($J_A/J=0.56$), so that these compounds are expected to provide the best approximation to a 2D *XY* model for GIC systems. Furthermore, it is possible to synthesize $CoCl_2$-GICs for a variety of stages, and thus to investigate the stage dependence of the observed phenomena. Moreover, the structural coherence lengths in the intercalate islands are large enough to contain a statistically meaningful number of spins ($>10^3$ spins/island), and yet small enough to permit a meaningful comparison with theoretical calculations for specific spin arrangements.

For all the acceptor 2D *XY* GIC systems that have been studied, two critical temperatures are identified (T_{cl} and T_{cu}), thereby delineating three temperature regimes, each exhibiting distinct magnetic behaviors: $T<T_{cl}$, $T_{cl}<T<T_{cu}$, and $T>T_{cu}$. At high temperature ($T\gg T_{cu}$), Curie–Weiss paramagnetic behavior is found, while at low temperature magnetically correlated states are observed (Zabel and Solin, 1992).

In the temperature range $T_{cl}<T<T_{cu}$ for the high-stage compounds, the spins are approximately confined to the basal plane and are arranged in a bound vortex configuration with strong in-plane spin-spin correlation but negligible interplanar correlation. Inelastic neutron-scattering studies provide strong evidence for the bound vortex state in the stage-2 $CoCl_2$-GIC. The absence of a remanent magnetization and the observation of a linear dependence of the magnetization on magnetic field for $T>T_{cu}$ identifies the phase above T_{cu} with paramagnetism.

Weak perturbations, such as the six-fold in-plane symmetry-breaking crystal field H_6 and the interplanar magnetic interaction J', become important at low temperatures and give rise to a 2D–3D crossover transition at T_{cl}. The spontaneous magnetization due to the coupling of the spins to H_6 gives rise to ferromagnetically ordered in-plane domains below a temperature of T_{cl}. The antiferromagnetic interplanar exchange interaction couples these ferromagnetic planar domains and results in an antiferromagnetic stacking arrangement of these domains, as implied by neutron scattering experiments (Zabel and Solin, 1992).

Magnetic phase transitions can be identified through the different field and temperature dependences that are found for the various properties such as the magnetic susceptibility, magnetization, and the magnetic contribution to the heat capacity. The most thoroughly studied magnetic donor GIC at the present time is the stage-1 europium compound, C_6Eu, for which the crystal structure is commensurate with the adjacent graphite layers. Each Eu atom has six nearest-neighbor Eu atoms at a distance of 4.31 Å within the layer plane, and six next-nearest-neighbor Eu atoms lie at a distance of 5.47 Å in *adjacent planes.* The configuration of the Eu^{2+} ion is $^8S_{7/2}$, which arises from a $(4f)^7$ electronic configuration, giving a total spin of 7/2 with no orbital angular momentum. Because the nearest-neighbor exchange interaction is antiferromagnetic, the spins form an antiferromagnetic frustrated triangular spin lattice. This spin system can easily be modified by introducing weak interactions, so that a variety of magnetic phases can be realized by small variations of the temperature and magnetic field. Field-dependent resistivity measurements (see Fig. 15) can be used to identify magnetic phase transitions, and the nature of

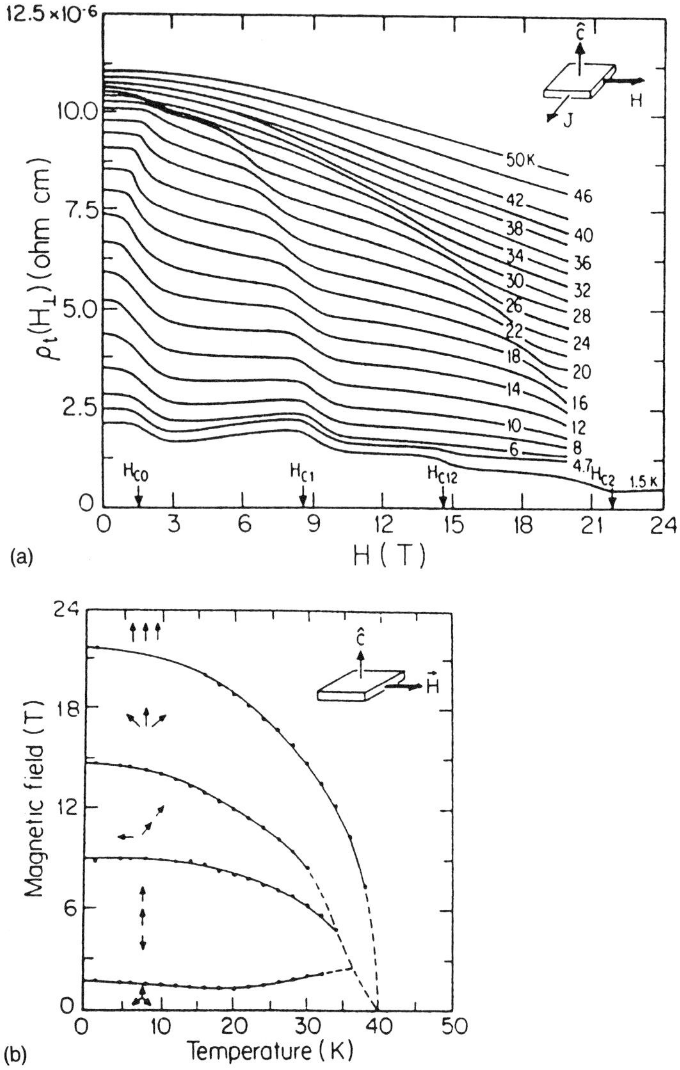

FIG. 15. (a) Anomalies in the field-dependent resistivity can be used to map out the magnetic phase diagram of a magnetic intercalation compound shown in (b). Specific results are shown here for the stage-1 donor compound C_6Eu. The orientations of three spins within the magnetic unit cell identify each of the magnetic phases. Ferromagnetic spin alignment along the magnetic field direction is achieved in the high–magnetic-field phase.

these magnetic phases is determined by magnetic measurements. On the basis of field-dependent resistivity measurements, a magnetic phase diagram has been developed for the low-temperature magnetic phases of C_6Eu (see Fig. 15).

The magnetic properties of donor GICs are distinctly different from those of the acceptor compounds because of differences relating to the basic magnetic coupling mechanism that occurs both in the pristine magnetic donor intercalates (which are metals) and in the intercalation compound (where the metallic behavior is associated predominantly with the graphite π bands). In the magnetic donor compounds, the mechanism for the magnetic exchange is through the conduction-band electrons, whereas the behavior of the magnetic acceptor GICs is dominated by the coupling between magnetic dipoles through induced dipoles on the nonmagnetic layers (i.e., by the superexchange mechanism).

5. SUPERCONDUCTING INTERCALATION COMPOUNDS

Superconductivity in intercalation compounds was first observed by Hannay *et al.* (1965), who identified superconducting transitions in the stage-1 heavy alkali-metal compounds with the intercalates K, Rb, and Cs in the graphite host material. The discovery of superconductivity in these materials attracted a great deal of interest insofar as neither the intercalate (an alkali metal) nor the host material (graphite) is by itself superconducting. Superconductivity in C_8K is also interesting with regard to its large anisotropy (the upper critical field H_{c2} is an order of magnitude larger for $H \perp c$ axis than for $H \parallel c$ axis), and for the unusual effect that for $H \parallel c$ axis, the intercalation compound exhibits type-I superconducting behavior (with no vortex formation), while for $H \perp c$ axis, the same sample can exhibit type-II behavior and vortex formation. Because of the low value for the superconducting transition temperature for C_8K (T_c=0.14 K), only a limited study of this system has been carried out (see Table 4). Other GICs with intercalates that are themselves superconducting (such as the stage-1 C_4KHg or the stage-2 C_8KHg) have T_c values in the range 1–2 K. Under pressure the alkali-metal intercalants enter the intercalate galleries at higher concentrations and give rise to T_c values up to 5 K (Zabel and Solin, 1992). The intercalation of fullerenes with alkali metals also leads to superconductivity (see FULLERENES), but with much higher transition temperatures (up to 33 K), because of the very much higher density of states near the Fermi level that can be achieved in this highly molecular solid.

Superconductivity is also observed in transition-metal dichalcogenides, but in this case the studies in the intercalation compounds have been focused in a different direction, since selected transition-metal dichalcogenides (such as TaS_2) are already superconducting prior to intercalation. In this case intercalation has been used to enhance the anisotropy of the superconducting phenomena.

Anisotropic superconductivity has been observed in a variety of layered materials (Dresselhaus and Chaiken, 1987), including deliberately structured materials, transition-metal dichalcogenides intercalated with large organic molecules, and graphite intercalation compounds. Because of the long range of the superconducting coherence distance ξ, the observation of 2D superconductivity focuses on samples with superlattice repeat distances I_c large compared with the superconducting coherence distance (in the *c*-axis direction) to

Table 4. Fundamental parameters[a] characterizing the superconductivity of C_8K (Zabel and Solin, 1992).

Parameter	Experimental value
T_c (mK)	128–198
Θ_D (K)	234.8
$N(0)V$	0.14
H_c (Oe)	10–13
$\kappa_\perp (T_c)$	0.78
$\kappa_\parallel (T_c)$	0.22
$\kappa_\perp / \kappa_\parallel$	3.55
$\lambda_\perp$ (0) (Å)	2300
$\lambda_\parallel$ (0) (Å)	8100
$\lambda_\perp / \lambda_\parallel$	0.28
$\xi_\perp$ (0) (Å)	10000
$\xi_\parallel$ (0) (Å)	2900
$\xi_\perp / \xi_\parallel$	3.45

[a] Θ_D is the Debye temperature, κ is the Landau–Ginzburg parameter, λ is the London penetration depth, ξ is the Ginzburg–Landau coherence length, and the subscripts $\parallel$ and $\perp$ refer to the directions parallel and perpendicular to the *c* axis, respectively.

achieve the condition of 2D superconductivity, $l_c > \xi_c$, where ξ_c denotes the coherence distance in the c direction for an anisotropic superconductor. Molecular-beam epitaxy and magnetron sputtering offer great promise for quantitative studies of 2D superconductivity and the 2D–3D crossover because of the flexibility of these synthesis techniques for the generation and control of large superlattice repeat distances in deliberately structured superconductors (Dresselhaus, 1987).

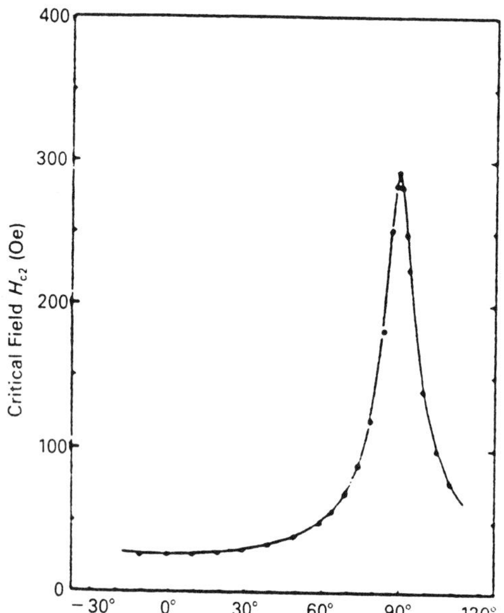

(a)

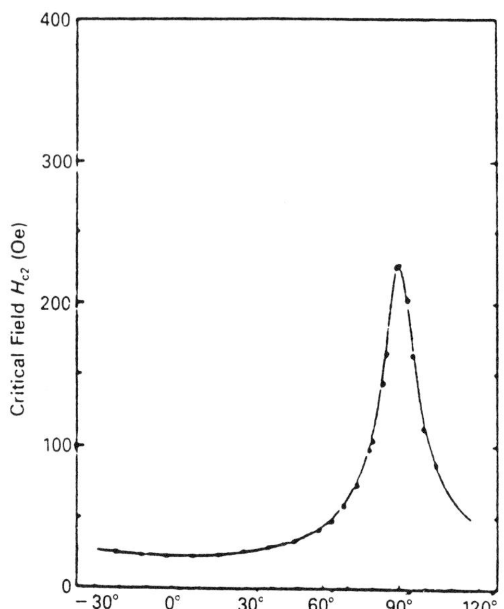

(b)

FIG. 16. Angular dependence of the upper critical field H_{c2} of the stage-1 ternary compounds (a) C_4KHg and (b) C_4RbHg. The curve in each case represents the best fit of the effective-mass model to the experimental points (Iye and Tanuma, 1984).

The earliest reports of superconductivity in layered transition-metal dichalcogenides intercalation compounds focused on the increase in T_c that occurred through intercalation with large organic molecules (Gamble and Geballe, 1976). The effect of lower dimensionality was observed in this work, especially in systems such as TaS_2 intercalated with *n*-octadecylamine, where a separation of nearly 60 Å was achieved between sequential Ta layers. This work was historically important in stimulating theoretical calculations of 2D superconductivity.

Most useful for studies of the 3D–2D superconducting transition have been the synthetic superlattice structures prepared by molecular beam epitaxy, because of the dimensional control that can be exercised in preparing the multilayers. Because of the abrupt interfaces and the possibility of preparing ordered sequences of superconducting monolayers, GICs provide a complementary tool for studying highly anisotropic superconducting materials. There is no known acceptor GIC that has shown evidence for superconductivity.

Critical-field studies for anisotropic superconductors have attracted considerable interest because of their similarity to the high-T_c superconducting cuprates, which are also layered compounds with anisotropic superconducting properties. For the intercalation compounds the angular dependence of the critical field is well approximated by the anisotropic Ginzburg–Landau formula (Zabel and Solin, 1992),

$$H_{c2}(\theta) = (\Phi_0/2\pi\xi_\perp^2) \times (\cos^2\theta + \epsilon^2 \sin^2\theta)^{-1/2}, \tag{1}$$

in which $\Phi_0 = hc/2e = 2.07 \times 10^{-7}$ G cm^2 is the elementary flux quantum and θ is the angle between the magnetic field and the c axis. Here the parameter ϵ characterizes the anisotropic behavior and is given by

$$\epsilon^2 = m_\perp / m_\parallel = (H_{c2\perp}/H_{c2\parallel})^2 = (\xi_\parallel/\xi_\perp)^2, \tag{2}$$

where ξ denotes the coherence length and m denotes the effective mass in the $\parallel$ and $\perp$

directions. We denote by $H_{c2\perp}$ the upper critical field for $\theta=0°$ whereby the diamagnetic screening currents lie in the layer planes. Similarly we denote by $H_{c2\parallel}$ the upper critical field for $\theta=90°$ whereby the diamagnetic screening currents must traverse different layer planes. These relations have been used to obtain qualitative fits to the experimental $H_{c2}(\theta)$ curves shown in Fig. 16. From these fits, values for the parameters $H_{c2\parallel}$, $H_{c2\perp}$, $\xi_{\parallel}$, and $\xi_{\perp}$ have been deduced. Specific values of these parameters are given in Table 5 for stage-1 KHg-GICs in comparison with values for pristine intercalate KHg, which is also superconducting, but shows no anisotropy. It is also of interest to note that intercalation increases the T_c value of a superconducting intercalate species (Zabel and Solin, 1992). For comparison a number of fundamental parameters characterizing the superconducting compound C_8K are listed in Table 4.

Also of interest is the temperature dependence of the critical fields $H_{c2}(T)$. Measurements of $H_{c2}(T)$ on GICs have been interpreted in terms of the anisotropic Ginzburg–Landau theory, which provides a qualitative fit to the data near T_c. Assuming an elliptical fluxoid to describe the anisotropy in the coherence length, the anisotropic Ginzburg–Landau theory yields the following expressions for the temperature dependence for the critical fields parallel and perpendicular to the c axis:

$$H_{c2\parallel}(T)=\frac{\Phi_0}{2\pi\xi_{\parallel}\,\xi_{\perp}}\left(1-\frac{T}{T_c}\right),$$
$$H_{c2\perp}(T)=\frac{\Phi_0}{2\pi\xi_{\perp}^2}\left(1-\frac{T}{T_c}\right). \quad (3)$$

Characteristic of the Ginzburg–Landau theory is a linear T dependence of H_{c2} near T_c for both $H_{c2\perp}$ and $H_{c2\parallel}$. The anisotropy in H_{c2} comes in through the prefactor coefficients in Eq. (3). Critical-field measurements on stage-2 KHg-GICs for both $H\parallel c$ axis and $H\perp c$ axis are well approximated by a linear T dependence near T_c in accordance with Eq. (3). Fits to these data yield values for the parameters $\xi_{\parallel}$ and $\xi_{\perp}$, consistent with values obtained from fits to the angular dependence of $H_{c2}(\theta)$ discussed above.

The standard BCS model for superconductivity with only occupied graphite π bands cannot explain the observed transition temperatures because the π-band density of states is too small. On the other hand, a model with only heavy-mass intercalate-related bands cannot explain the large observed anisotropy in the superconducting properties of GICs. Therefore the model for superconductivity in GICs must include consideration of both the graphite π-band electrons and the occupied heavy-mass intercalate bands, showing that it is necessary to have at least two types of occupied bands or strong hybridization between them to produce superconductivity in GICs.

Table 5. Comparison of the superconducting parameters of stage-1 KHg-GICs and pristine KHg (Dresselhaus and Chaiken, 1987).

Parameter	Stage-1 KHg-GIC	Pristine KHg
I_c (Å)	10.22	—
T_c (K)	1.53	0.98
$H_{c2\perp}$ (Oe)	83	244
$H_{c2\parallel}$ (Oe)	850	244
$\xi_{\perp}$ (0) (Å)	1960	1135
$\xi_{\parallel}(0)$ (Å)	192	1135
$\xi_{\perp}/\xi_{\parallel}$	10.2	1

The theoretical models predict both the temperature dependence and anisotropy of H_{c2} for GICs. The models are generally consistent with the observed data. The models explain many of the detailed properties of superconducting GICs, including the absence of superconductivity in acceptors that lack occupied heavy-mass intercalate levels. The models show that the value of the superconducting transition temperature is more sensitive to changes in the π-band electronic density of states than to the corresponding changes in the heavy-mass band.

One key to gaining a detailed understanding of superconductivity has been through in-depth investigations of the tunneling characteristics in the normal and superconducting states, which in turn yield direct information on the density of states. The layered nature of many intercalation compounds makes them amenable for tunneling measurements. Such experiments have also been performed on intercalated fullerenes, thereby yielding a direct measurement of the temperature dependence of the superconducting band gap in these relatively high-T_c superconductors (see FULLERENES).

Since the critical fields for superconducting GICs are easily achieved in the laboratory, GICs provide model systems of interest for the study of anisotropic superconductors. Such studies have been of interest with regard to high-T_c superconductors for which the magnetic fields required to study the corresponding effects are difficult to achieve experimentally.

Acknowledgment

The preparation of this manuscript was supported by the National Science Foundation Grant No. 92-01878DMR.

GLOSSARY

Basal Plane: The plane normal to the main symmetry axis, usually called the *c* axis.

BCS Theory: Microscopic theory of superconductivity based on the pairing of electrons that are coupled by the electron–phonon interaction.

Bi-intercalation: The intercalation of a layered binary intercalation compound with a second intercalate species to form an intercalation compound that is ordered with respect to each of the two intercalate species.

Binary Compound: An intercalation compound for which the intercalate is a single atomic or molecular species.

Chalcogen: Element from column VI of the periodic table: O, S, Se, Te.

Clays: Compounds in a class of layered aluminosilicates such as vermiculite, which intercalate water between layers and take up other metal ions by ion exchange interactions.

Commensurate: Intercalation compound in which atoms in the intercalate layer and in the host material are in registry.

Covalent Bonding: A localized bond between two atoms to which the valence electrons from each atom contribute.

Discommensuration: A modulation in the structure of two materials with different lattice constants at an interface to achieve regions where the two structures are locally in registry, but separated by boundaries called misfit dislocations where the material with the smaller lattice constant has an extra row of atoms.

Disproportionation: Formation of chemical species in the intercalate galleries with proportions of atomic constituents different from that prior to intercalation.

Electron Affinity: Energy required to take an electron from the top of the valence band (in a semiconductor) to the lowest ionization state.

Ginzburg–Landau theory: Macroscopic phenomenological theory of superconductivity.

Graphene layer: An isolated single layer of hexagonally arranged carbon atoms in a honeycomb structure such as found in graphite.

Intercalate Gallery: van der Waals gap between basic structural units of a layered host material in an intercalation compound.

Ionic Bonding: The electrostatic bonding between two atoms caused by the transfer of charge from the donor atom to the acceptor atom.

Octahedral Sites: Sites with six nearest neighbors in the cubically symmetric locations $(\pm a,0,0)$, $(0,\pm a,0)$, $(0,0,\pm a)$.

Polytypes: The ensemble of ordered arrangements possible for a given basic structural unit.

Stage: The ordered periodic superlattice arrangement of the intercalate in an intercalation compound, whereby the intercalate layers are separated by a fixed number of host layers.

Stage Index: Integer specifying number of host layers between consecutive intercalate layers.

Ternary Compound: An intercalation compound into which two distinct intercalates are introduced.

Tetrahedral Sites: Sites in the lattice with nearest neighbors at the corners of a regular tetrahedron.

Transition-Metal Ion: Metal ions containing unfilled *d* shells (where *d* electrons denote electrons with angular momentum of $2\hbar$).

van der Waals Bonding: Weak bonding between two atoms, such as might arise from π bonding between electrons, falling off rapidly with distance.

Vortex: A flow of charge or an arrangement of spins with circulation about a singular point or line.

List of Works Cited

Aldebert, P., Baffier, N., Legendre, J. J., Livage, J. (1982), *Rev. Chim. Minérale* **19**, 485–495.

Besenhard, J. O., Wudy, E., Moewald, H., Nickl, J. J., Biberacher, W., Foag, W. (1983), *Synth. Met.* **7**, 185–192.

Brec, R. (1987), in: M. S. Dresselhaus (Ed.), *Intercalation in Layered Materials*, NATO Advanced Study Institute, Series B: Physics, Vol. 148, New York: Plenum, pp. 93–134.

Dresselhaus, M. S. (1984), *Phys. Today* **37** (3), 60–68.

Dresselhaus, M. S. (Ed.) (1987), *Intercalation in Layered Materials*, NATO Advanced Study Institute, Series B: Physics, Vol. 148, New York: Plenum.

Dresselhaus, G., Chaiken, A. (1987), in: M. S. Dresselhaus (Ed.), *Intercalation in Layered Materials*, NATO Advanced Study Institute, Series B: Physics, Vol. 148, New York: Plenum, pp. 387–406.

Dresselhaus, M. S., Dresselhaus, G. (1981), *Adv. Phys.* **30**, 139–326.

Gamble, F. R., Geballe, T. H. (1976), in: N. B. Hannay (Ed.), *Treatise on Solid State Chemistry*, Vol. 3, New York: Plenum, pp. 89–166.

Hannay, N. B., Geballe, T. H., Matthias, B. T., Andres, K., Schmidt, P., MacNair, D. (1965), *Phys. Rev. Lett.* **14**, 225–226.

Hérold, A. (1979), in: F. Lévy (Ed.), *Physics and Chemistry of Materials with Layered Structures*, New York: Reidel, Vol. 6, pp. 323–421.

Hérold, A., Guérard, D., Lagrange, P. (1983), *Synth. Met.* **7 & 8**, Proceedings of the Third International Conference on Intercalation Compounds of Graphite.

Iye, Y., Tanuma, S. (1984), in: S. Tanuma, H. Kamimura (Eds.), *Graphite Intercalation Compounds*, Summary report of the Special Distinguished Research Project supported by the Ministry of Education, Japan, pp. 253–267. See also *Phys. Rev. B* **25**, 4583–4592 (1982).

Julien, C., Hatzikraniotis, E., Paraskevopoulos, K., Chevy, A., Balkanski, M. (1986), *Solid State Ionics* **18–19**, 859–864.

Kamimura, H. (1987), *Phys. Today* **40** (12), 64–71.

Liang, W. Y. (1987), in: M. S. Dresselhaus (Ed.), *Intercalation in Layered Materials*, NATO Advanced Study Institute, Series B: Physics, New York: Plenum, Vol. 148, pp. 31–74.

McRae, E., Marêché, J. F. (1988), *J. Mater. Res.* **3**, 75–86.

Newman, A. C. D. (Ed.) (1987), *Chemistry of Clays and Clay Minerals*, New York: Wiley.

Schafhäutl, C. (1840), *J. Prakt. Chem.* **21**, 129–157.

Solin, S. A. (1982), *Adv. Chem. Phys.* **49**, 455–532.

Solin, S. A. (1984), *J. Molecular Catal.* **27**, 293–303.

Solin, S. A., Zabel, H. (1988), *Adv. Phys.* **37**, 87–254.

Suematsu, H., Suda, K., Metoki, N. (1988), *Synth. Met.* **23**, 7–12.

Terai, T., Takahashi, Y. (1981), *J. Nucl. Sci. Tech.* **18**, 643–648.

Vogel, F. L., Hérold, A. (Eds.) (1977), *Proceedings of the Conference on Intercalation Compounds of Graphite*, Lausanne: Elsevier Sequoia, Vol. 31.

Wada, N., Suzuki, M., Hines, D. R., Koga, K., Nishihara, H. (1987), *J. Mater. Res.* **2**, 864–870.

Xiang, X. D., Zettl, A., Vareka, W. A., Corkill, J. L., Barbcc, T. W., III, Cohen, M. L. (1991), *Phys. Rev. B* **43**, 11 496–11 499.

Zabel, H., Solin, S. A. (Eds.) (1990), *Graphite Intercalation Compounds I: Stucture and Dynamics*, Springer Series in Materials Science, Vol. 14, Berlin: Springer.

Zabel, H., Solin, S. A. (Eds.) (1992), *Graphite Intercalation Compounds II: Transport and Electronic Properties*, Springer Series in Materials Science, Vol. 18, Berlin: Springer.

Further Reading

Dresselhaus, M. S. (1984), *Phys. Today* **37** (3), 60–68.

Dresselhaus, M. S. (Ed.) (1987), *Intercalation in Layered Materials*, NATO Advanced Studies Institute, Series B: Physics, Vol. 148, New York: Plenum.

Dresselhaus, M. S., Dresselhaus, G. (1981), "Intercalation compounds of graphite," *Adv. Phys.* **30**, 139–326.

Gamble, F. R., Geballe, T. H. (1976), in: N. B. Hannay (Ed.), *Treatise on Solid State Chemistry*, New York: Plenum, Vol. 3, pp. 89–166.

Hérold, A. (1979), in: F. Lévy (Ed.), *Physics and Chemistry of Materials with Layered Structures*, New York: Reidel, Vol. 6, pp. 323–421.

Kamimura, H. (1987), *Phys. Today* **40** (12), 64–71.

Solin, S. A. (1982), *Adv. Chem. Phys.* **49**, 455–532.

Solin, S. A., Zabel, H. (1988), *Adv. Phys.* **37**, 87–254.

Zabel, H., Solin, S. A. (Eds.) (1990), *Graphite Intercalation Compounds I: Structure and Dynamics*, Springer Series in Materials Science, Vol. 14, Berlin: Springer.

Zabel, H., Solin, S. A. (Eds.) (1992), *Graphite Intercalation Compounds II: Transport and Electronic Properties*, Springer Series in Materials Science, Vol. 18, Berlin: Springer.

INTERFEROMETERS AND INTERFEROMETRY

W. H. Steel, *School of Mathematics, Physics, Computing and Electronics, Macquarie University, Sydney, Australia*
Samuel A. Werner, *Department of Physics and Astronomy, University of Missouri–Columbia, Columbia, Missouri, U.S.A.*

	Introduction	157
1.	**General Aspects**	158
1.1	Theory	158
1.2	Polarization Effects	158
1.3	Types of Measurement	159
1.3.1	Source Spectra	159
1.3.2	Imaging by Interference	159
1.3.3	Phase at One Point	159
1.3.4	Phase Across a Plane	160
1.4	Classifications	160
2.	**Division of Amplitude**	160
2.1	Michelson	160
2.2	Mach–Zehnder	161
2.3	Cyclic or Sagnac	162
2.4	Fizeau	162
3.	**Common-Path Interferometers**	162
3.1	Lateral Shear	162
3.1.1	Interference Microscopes	163
3.2	Radial Shear	163
3.3	Point Diffraction	163
3.4	Testing Lasers	163
4.	**Multiple-Beam Interferometers**	163
4.1	Fabry–Pérot	163
4.1.1	Spherical Surfaces	164
5.	**Division of Wave Front**	164
5.1	Interferometers for Astronomy	164
6.	**Guided Waves**	164
7.	**Hologram and Speckle Interferometry**	164
7.1	Hologram Interferometry	164
7.2	Speckle Interferometry	165
8.	**X-Ray Interferometers**	165
8.1	General Principles	165
8.2	Index of Refraction and Scattering Factors	165
8.3	Absolute Measurement of Lattice Parameter	166
9.	**Neutron Interferometers**	167
9.1	General Principles	167
9.2	Gravitationally Induced Quantum Interference	168
9.3	Neutron Sagnac Effect	169
10.	**Electron Interferometry**	169
11.	**Atom-Beam Interferometers**	170
12.	**Future Trends**	170
12.1	Optical Interferometers	170
12.2	X-Ray and Matter-Wave Interferometers	171
	Acknowledgment	171
	Glossary	171
	Works Cited	172
	Further Reading	172

INTRODUCTION

An interferometer measures by using wave interference, the patterns of reinforcement and cancellation formed when two or more coherent waves meet. Interference has been important in the development of physics, and its use for measurement owes much to Michelson, who in 1907 was awarded the Nobel prize for "his optical instruments of precision and the spectroscopic and metrological investigations he executed therewith." Interferometers were used first with visible light, and these are commonly called *optical interferometers* to distinguish them from the newer types using other electromagnetic radiation or acoustic or matter waves.

Interferometers employing electromagnetic radiation with wavelengths extending from the radio-frequency range (kilometers) all the way down to x rays (ångströms) have been constructed and successfully employed in a wide variety of physical measurements, such as determination of the precise size and shape

1-56081-067-X/94/$5.00+.50

of objects, the smoothness of their surfaces, their densities, and the distance between objects. Interferometry based upon de Broglie matter waves of electrons and neutrons is now well established. In addition to fundamental physics experiments, such as the observation of gravitationally induced quantum interference and the topological Aharonov–Bohm effects, experiments analogous to a number of the classical optical interferometry experiments have now been carried out. Examples are the observation of the phase shift due to rotation of the interferometer (Sagnac effect) and the phase shift due to the motion of matter within one leg of the interferometer (Fizeau effect). Matter-wave interferometry using atoms is a very new field, with some potential technological applications. Although the basic theory underpinning interferometry with photons (Maxwell's equations) and with massive particles (Schrödinger's equation) is different, the concepts coming directly from the principle of linear superposition of waves, such as the idea of an index of refraction, are central to both fields.

1. GENERAL ASPECTS

1.1 Theory

Interference is traditionally explained in terms of the addition of waves, yet this treatment has become appropriate only with the use of lasers and monoenergetic matter waves. The radiation from classical light sources is a random superposition of many waves, and any interference depends on the correlation of such radiation at different times and different locations. Beams of radiation take different paths to the point where interference is observed. Their coherence depends on the time difference or *delay* between their arrival and the separation of the images of this point as seen from the source along the different paths. The lateral component of this separation is the *shear*. This is measured either as the distance between the images or as their angular separation (*angular shear).*

The power or *intensity* measured in two-beam interference of radiation of frequency ν has the form

$$I=I_1+I_2+2(I_1I_2)^{1/2}|\gamma| \cos(\arg \gamma+2\pi\nu\tau):$$

a background, the sum of intensities I_1 and I_2 of each beam, and a cosine *interferogram,* in which γ is the (*complex) degree of coherence* between beams that can have left the source in different directions and have a delay τ between their arrival times. The maxima and minima of visible interference patterns are called *fringes,* a word now applied to any interferogram.

The fringe contrast or *visibility* for beams of equal intensity is $|\gamma|$, with a maximum of 1 when the beams are fully coherent. A laser source can provide coherent beams under a variety of conditions, but coherence with a source of finite size or spectral bandwidth requires an interferometer that is specially corrected. If the bandwidth is finite, the delay must be small; if the source has a finite size, the shear must be small. In both cases the interferogram is a Fourier transform. As a function of the delay, it is the transform of the source spectrum; as a function of the shear, it is the two-dimensional transform of the intensity distribution. The shear of an interferometer varies along the paths, and the fringes with an extended source have high visibility only where the shear is small; they are said to be *localized* there. A longitudinal image separation also reduces coherence, but at a smaller rate.

Viewed from the point of interference, the two images of the source can also have separations. These determine the form of the fringe pattern formed across a plane. With a lateral separation the interfering beams arrive with a relative *tilt* and give almost straight fringes; with a longitudinal separation they give concentric circles.

These two sets of images, plus the delay, describe the main properties of an interferometer (Hansen, 1955). In simple interferometers, the two longitudinal image separations and the delay are connected; if one is changed, the others change also.

1.2 Polarization Effects

When polarization is taken into account, for maximum visibility the beams must start in the same state of polarization and interfere in the same state. In addition, for natural (unpolarized) light, the delay should be the same for all polarizations. Interferometers made in one plane are usually free of polarization effects, but three-dimensional layouts can have problems as can those made of optical fibers.

1.3 Types of Measurement

A two-beam interferometer can measure both the spectral and spatial distributions of a radiation source through their Fourier transforms. In the second case, this is equivalent to forming an image.

But these are not the only applications. More common is the use of an interferometer to study media. For this, measurements of phase only are required. The visibility need only be sufficient to detect this phase and any convenient source can be used. Now lasers are usually chosen, since they give high visibility without careful matching of the beams. These measurements yield the various quantities on which the delay, and hence the phase, depend.

1.3.1 Source Spectra. An interferometer can accept more light than a prism or grating spectrometer, which requires narrow slits to resolve spectral detail. If an interferogram is recorded as the delay is varied, its Fourier transform is the spectrum of the source. Two-beam interferometers are used for Fourier spectroscopy in the infrared (Connes, 1961; Vanasse and Sakai, 1967) in regions where detector noise is predominant. The signal-to-noise ratio of the measurement is then improved if all parts of the spectrum are recorded simultaneously rather than in sequence, each being coded so that it can be separated out later. Here the Fourier transform provides the coding and Fourier spectrometers have this further *multiplex advantage* over other instruments. In the visible and ultraviolet, where the limit is photon noise, this advantage does not exist. Multiple-beam instruments, such as the Fabry–Pérot interferometer, which interfere many beams with a series of delays, are then used. A measurement is the sum of many Fourier components and approximates the direct spectra rather than its transform.

1.3.2 Imaging by Interference. With the wavelengths used for radio astronomy about 10^4 times visual wavelengths, radio telescopes need to be this factor larger than visible-light telescopes to have the same resolving power. This is impractical for full dishes and interferometers are used widely. The image is built up from Fourier components from relatively small antennas at different spacings, that is, with shears between them. Signals are taken either simultaneously from fixed arrays or sequentially from antennas that are moved between measurements. With modern precise time measurement, the signals from two antennas can be recorded separately and correlated later. This has made possible interferometers that span different continents, and soon antennas in space will be included.

With visible light, resolving power beyond that of telescopes is needed to measure stellar diameters, and Michelson used his *stellar interferometer* for this. It had two beams of diameter small enough to reduce the degradation caused by atmospheric turbulence, and their separation was increased until the fringes vanished. Others found his technique too difficult, and instead the *intensity interferometer* was developed by Hanbury Brown and Twiss. This was very insensitive to atmospheric turbulence but could measure only bright stars. It was a unique, historic instrument (Hanbury Brown, 1974); now modern techniques (Davis and Tango, 1979) have allowed the development of several new stellar interferometers that either follow Michelson and use small diameters, with shears now up to 0.6 km, or combine telescopes of larger diameter and overcome the seeing by a technique called *speckle interferometry* (Labeyrie, 1976) but different from the speckle interferometry described later.

1.3.3 Phase at One Point. It is convenient to divide phase measurements into those made at one point for different settings of the interferometer and those that form an interference pattern across a plane. The time of passage is the product of the distance travelled and the refractive index of the medium, divided by the speed of light, and, if the phase of the interference is measured at two settings of the interferometer, these quantities can be derived. The speed of light is of interest in a rotating frame of reference; otherwise it is now a defined quantity.

Length measurement is of technical importance. Under both the earlier definition of the meter in terms of a specified wavelength and the 1983 definition that fixes the speed of light, an interferometer is needed to convert the definition to a practical length scale. Precise phase measurement involves optical phase shifters such as a moving grating, an acoustic-wave cell, or a rotating wave plate in polarized light. But visual techniques of estimating fringe position and indirect methods for finding the whole number of fringes, once used because counting was not reliable, have been largely replaced by photoelectric methods. The interferometers used have precisions ranging from 1 part in 10^{11} for standardization to parts in 10^7 for *lambdameters* that measure laser wavelengths. Simpler and more robust interferometers, made possible by electronic techniques and laser sources, are now becoming workshop equipment, where they provide a "micrometer with a one-wavelength pitch." A widely used instrument uses a Zeeman-split laser, which produces two circularly polarized beams of different frequencies, and this allows heterodyne detection.

Heterodyne interferometry can also measure flow speeds. The frequency difference is produced by reflecting one beam from particles in the moving fluid.

Length measurements in air must be corrected for the refractive index, which is either measured separately by another interferometer or derived from the atmospheric composition, temperature, and pressure. For electromagnetic radiation, measurements of refractive index relate to the density of a medium, but the effective refractive index for electron or neutron waves depends on other quantities, which can now be measured by interferometry.

1.3.4 Phase Across a Plane. If an interferogram is recorded across a plane, it gives a picture of the phase variations in two dimensions. Such measurements are used for surface topography, optical testing (Malacara, 1978, 1991), interference microscopy, and measurements of refractive index variations in wind tunnels and plasmas. These measurements used to be made visually and often this is still sufficient.

Detector arrays have made possible photoelectric measurement. When interference fringes are irregularly spaced, new techniques were required to realize the accuracy these made possible. In one method, called *temporal heterodyning* or *phase stepping*, the path difference between the beams is changed in several finite steps and an interferogram is recorded at each step. This gives enough data at each array point to calculate the phase there, independently of intensity variations (K. Creath, in Malacara, 1991). Alternatively, in *spatial heterodyning*, a considerable tilt is introduced to produce fringes that are close enough for accurate interpolation. The resulting interferogram is filtered to remove the effect of the tilt.

1.4 Classifications

Interferometers are traditionally described as *two beam* or *multiple beam*, and with having *division of amplitude* or *division of wavefront*. In division of amplitude, the beams are separated and recombined by *beam splitters*, which reduce the intensity. In division of wavefront, they leave the source in different directions with no loss. The most common beam splitter is a partially reflecting coating of metal or dielectric on an inclined plate or the diagonal of a cube; metal films have absorption losses but their properties vary less with polarization, wavelength, or angle. Polarizing components and diffraction gratings can also separate beams; Bragg diffraction in a crystal can separate x rays or matter waves. What is essentially the same interferometer can have a very different appearance if a different form of beam splitter is used.

Most interferometers are examples of a few standard types, which are described below with their uses.

2. DIVISION OF AMPLITUDE

2.1 Michelson

The important interferometers with division of amplitude are shown in Fig. 1. The Michelson interferometer is the most common, and it is usually used in collimated light, the Twyman–Green form. The same beam splitter separates the beams and later recombines them. Half the radiation is lost back to the source and, with lasers, some form of isolation may be needed so that this will not affect their operation. The delay can be varied by moving one mirror, and this instrument is commonly used for both length measurement and Fourier spectroscopy. The plane mirrors are often replaced by retroreflectors, since these are less affected by errors of the tracks on which they are moved. This interferometer also tests optical surfaces and components,

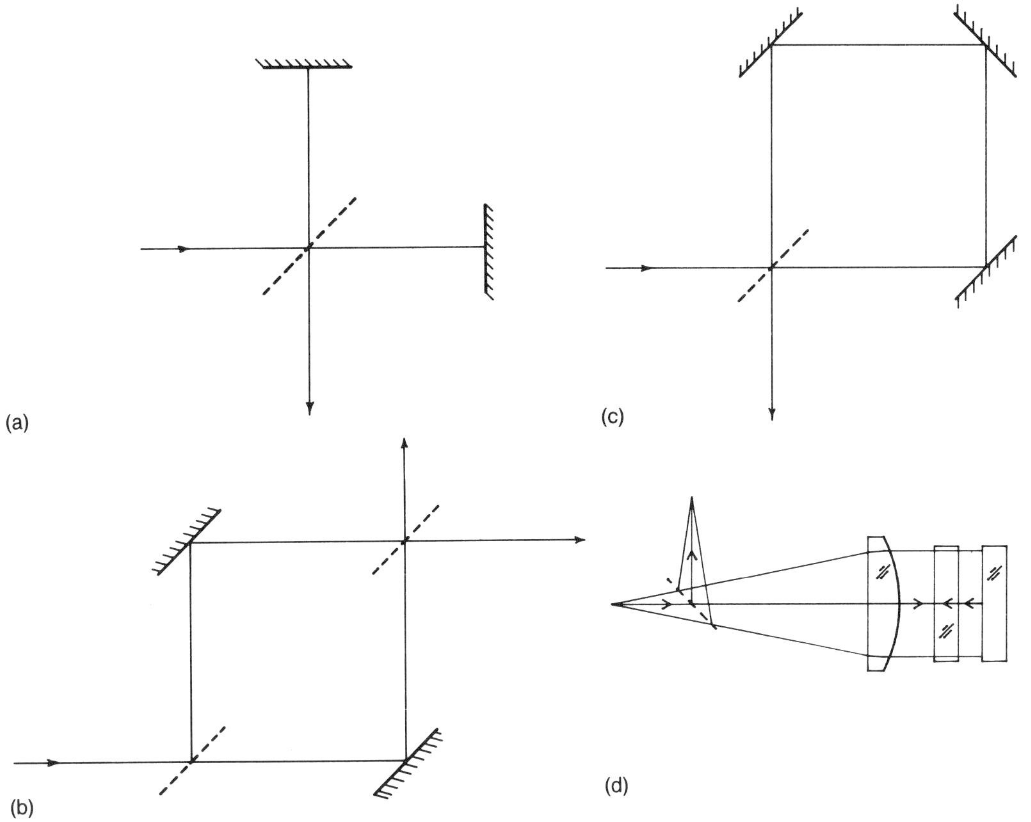

FIG. 1. Layouts of the main interferometers with division of amplitude. Beam splitters are represented by broken lines, mirrors by full lines shaded at the back. They are (a) Michelson, (b) Mach–Zehnder, (c) cyclic or Sagnac, (d) Fizeau. From Steel (1983), by permission of Cambridge University Press.

which replace the mirror in one arm. An optical system must be followed by a suitable reflector to return the light; the test is in *double passage.* Combined with a microscope, the Michelson interferometer becomes the *Linnik interference microscope* for studying the microtopography of surfaces.

Figure 2 shows the two types of fringes this interferometer can form, although not simultaneously. With a source at infinity, *Fizeau fringes* are localized at the ends of the arms and it is these that are used for optical testing. But if the beam is limited in size at the mirrors, so that the effective source is now there, another set of circular *Haidinger fringes* is seen at infinity. As the longitudinal image separation increases, either because of the introduction of an optical system for testing or because a mirror is moved back to change the delay, these fringes become smaller. At the same time the visibility of the Fizeau interferogram decreases. The inner fringe shows the largest source size that will give good visibility of either Fizeau fringes or an integrated interferogram. Any interferometer has similar independent, or *alternative,* sets of fringes, one of which can be used to choose a source size suitable for obtaining the other set with good visibility (Hansen, 1955; Steel, 1983).

For length measurement, the use of lasers avoids the source-size problem, but it remains in Fourier spectroscopy. However, there are more complex *field-widened* interferometers that can be used, in which the delay can be changed without changing the image separation.

2.2 Mach–Zehnder

When it is undesirable to measure in double passage, the Mach–Zehnder interferometer, Fig. 1(b), can be used. Both outputs are now accessible, and the shear and tilt are adjustable rather than the delay. The Mach–

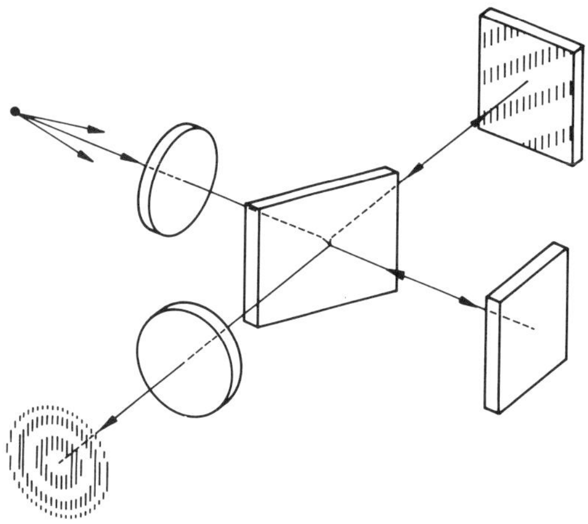

FIG. 2. The two independent sets of interference fringes in a Michelson interferometer with collimated light (Twyman–Green interferometer). Each set can be seen only when the beam is limited to be smaller than the spacing of the fringes of the other set. From Steel (1983), by permission of Cambridge University Press.

Zehnder interferometer measures variations of refractive index in wind tunnels and plasmas and is a common form for x-ray and matter-wave interferometers. The Jamin interferometer for measuring refractive indices of gases is a variant.

2.3 Cyclic or Sagnac

In the cyclic interferometer, Fig. 1(c), the two beams go around the same path in opposite directions. It is used to measure the speed of light in a rotating system (Sagnac, 1913) and, if sensitive enough, can act as a gyroscope. The sensitivity depends on the area of the circuit and can be increased by sending the beams around many times. This is most easily done in a fiber interferometer.

2.4 Fizeau

The Fizeau interferometer, Fig. 1(d), was originally used to test optical flats without placing two in contact and risking damage. In this interferometer the surface of the reference flat is also the beam splitter, and a second beam splitter is needed to separate the entering and leaving light. A modified form tests spherical surfaces. With a thermal source, the two surfaces being compared need to be close, but with a laser source, optical components can replace the flat for testing. The Fizeau is now taking over from the Michelson in this field and is the usual instrument used with phase stepping.

3. COMMON-PATH INTERFEROMETERS

The Fizeau interferometer is less affected by outside disturbances than the Michelson because the beams follow much the same path and are not widely separated. These advantages are taken further in *common-path* interferometers. These are chiefly for measuring phase variations, and, instead of the beam through the specimen being compared with a separate *reference* beam, the two beams follow the same path but pass through different regions of the specimen.

3.1 Lateral Shear

In a lateral-shear interferometer, the interference is between two images of the same beam, separated laterally. An early form, based on the Mach–Zehnder, is shown in Fig. 3. A laser is now commonly used, and, since correction of delay is not then needed, simpler forms are possible.

A lateral-shear interferogram does not show the phase distribution directly but differences between phases at points the shear apart. If the shear is small, this is approximately the phase gradient. But, for optical testing, a direct picture of the wave front, as given by

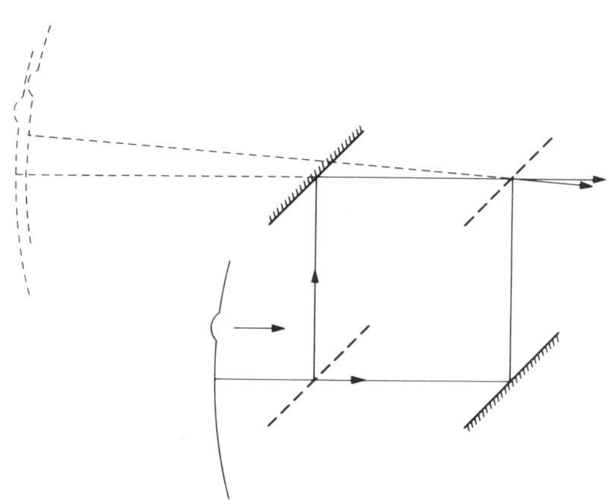

FIG. 3. Early form of an adjustable lateral-shear interferometer with equal optical paths. From Steel (1983), by permission of Cambridge University Press.

interferometers with a separate reference beam, is usually preferred.

3.1.1 Interference Microscopes. Interference is an alternative to phase-contrast microscopy for making transparent objects visible. While some interference microscopes have a separate reference beam, most use a lateral shear. A compensating shear is commonly used, removing the limit on source size but leaving a shear at the object.

3.2 Radial Shear

With a *radial shear* the two interfering beams have different magnifications. A form is shown in Fig. 4. The wave fronts from symmetrical optical systems follow a regular pattern, with little variation over their central region, and this can then give a reference wave if the ratio of magnifications, the *shear ratio,* is large. If, however, the ratio is close to unity, the interferometer can cover a larger range and the results can be corrected for the known shear.

Other shears can be used, such as rotational, inverting, or folding shears, each with its own application.

3.3 Point Diffraction

This simple interferometer, first described by Linnik and rediscovered by Smartt (Smartt and Steel, 1975), is shown in Fig. 5. It interferes an attenuated wave that has passed

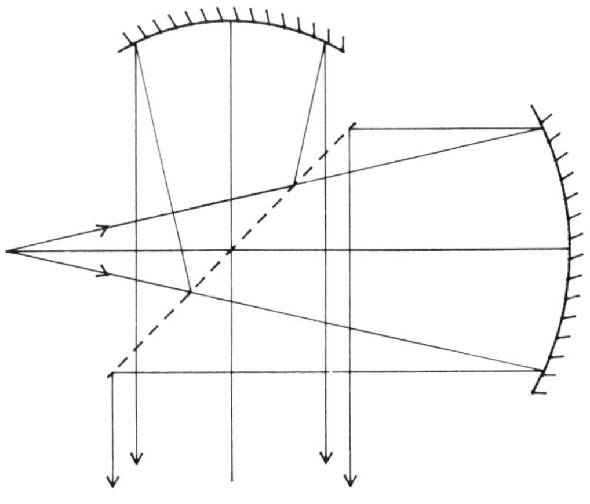

FIG. 4. A simple radial-shear interferometer, for use with a laser source (paths not equal).

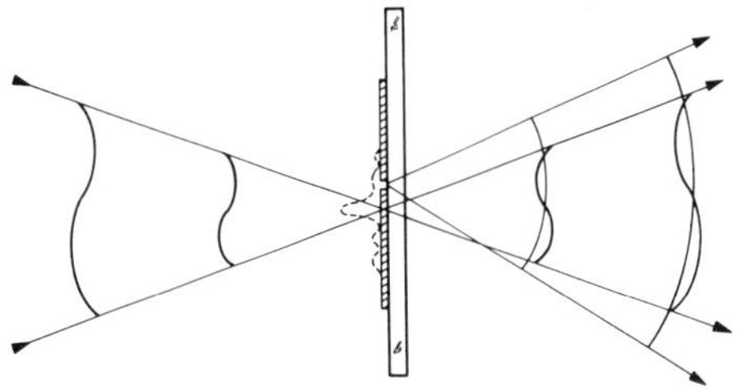

FIG. 5. Point-diffraction or Smartt interferometer. From Steel (1983), by permission of Cambridge University Press.

through an absorbing film with an approximately spherical wave diffracted by a pinhole in the film.

3.4 Testing Lasers

Common-path interferometers have the further application of testing laser outputs, where no separate reference beam is available. They measure either the coherence or the phase across the wave front.

4. MULTIPLE-BEAM INTERFEROMETERS

4.1 Fabry–Pérot

The Fabry–Pérot interferometer is the multiple-beam analog of the Michelson interferometer and has similar uses. It consist of two plates with reflecting coatings on their adjacent parallel surfaces. Successive beams have increasing delays, plus a small reduction in amplitude, and the fringes approach a series of spikes. Their shape is described by the *Airy function* and their sharpness by the *finesse,* the fringe spacing divided by their width. The finesse depends on the reflectance of the coatings and the flatness of the surfaces. It is reduced by lack of parallelism of the plates, and, if the separation is to be varied, interferometric control of parallelism is desirable (Ramsay, 1962).

An interference filter is a Fabry–Pérot interferometer with a half-wavelength separation of the reflecting surfaces, made of deposited layers. For narrower bandwidths, Fabry–Pérot interferometers with larger spacings are used. These can be tunable.

4.1.1 Spherical Surfaces. Another form of Fabry–Pérot interferometer has two concave spherical reflectors separated by their common radius. This can accept more light for spectroscopy at high resolving powers. Systems with spherical mirrors have become important as resonant cavities for lasers.

5. DIVISION OF WAVE FRONT

Because the beams in a wave-front–division interferometer leave the source in different directions, there is always a shear. An early interferometer of this type, the *Rayleigh interferometer,* is an alternative to the Jamin for measuring refractive indices, but its shear is a drawback. A shear does not matter with a laser source and it is required for the interference measurements of images. The main use of division of wave front is therefore in astronomy.

5.1 Interferometers for Astronomy

To avoid degradation of the image by the atmosphere, the first stellar interferometers limited the beams used by a telescope by two small apertures. Michelson increased the shear possible by mounting mirrors on a frame in front of the telescope to collect the two beams. These could be separated by more than the telescope diameter. In the intensity interferometer, the collecting mirrors could be given different separations by moving them around on a circular track. Now some modern stellar interferometers have collecting mirrors at fixed separations.

Radio interferometers commonly have a similar array of fixed antennas, sometimes along two baselines at right angles. But they may also use antennas that are moved along a track to change the shear.

6. GUIDED WAVES

In the interferometers already described, the radiation propagates in free space. But interferometers can have all or part of the beam paths in waveguides. Microwave interferometers, built up of these, are used for studies of plasmas, for which the useful wavelengths range from millimeters to the submillimeter infrared. With the advances in the technology of producing optical fibers for communication, particularly single-mode fibers, *guided-wave* interferometers are used also in the visible and near infrared. Their properties depend on the delay only, and the beams are split and recombined by standard fiber couplers. Polarization is a problem; it is difficult to make optical fibers free from strain, which causes birefringence. Even in a strain-free fiber, strain is introduced by bending. One solution is to use single-polarization fibers, which transmit one polarization only. Fiber interferometer sensors have become a rapidly advancing field of research. As well as measuring rotation, they can measure currents by sensing the magnetic field produced, and other quantities that change the optical path in a fiber: temperature, pressure, and mechanical stress, for example.

7. HOLOGRAM AND SPECKLE INTERFEROMETRY

A hologram is itself an interferogram, recorded on a material such as a photographic plate. Its ability to reconstruct the wave that produced it brings in the extra factor of time. Hologram interferometry can show how an object has changed by interfering a wave reflected by it some time ago, reconstructed from a hologram, with one reflected from it now. It allows interferometry of rough or irregular objects since only a wave from the same object will cancel out the surface detail and show only the changes.

7.1 Hologram Interferometry

Hologram interferometry can store a wave from a specimen and interfere it at any time with another wave from the same specimen to give *live fringes;* an example is shown in Fig. 6. Reconstruction of two holograms stored on the same plate gives *frozen fringes,* a method needed for rapid changes, which are recorded by a double-pulse laser. A third technique, *time-averaged fringes,* makes a hologram of a vibrating surface, which reconstructs to show the stationary nodal regions.

FIG. 6. Real-time holographic interferogram of an electric lamp, showing the change of the refractive index of the air when the filament is heated. From Steel (1983), by permission of Cambridge University Press.

7.2 Speckle Interferometry

Two single speckle interferograms of a rough surface illuminated by a laser can be correlated photographically or by television and electronics to give a *speckle interferogram* that similarly shows any changes. The conditions of illumination and recording can be arranged so that the in-plane and normal displacements of a surface can be measured separately.

Speckle images with no reference beam can also show displacements. Two images recorded on the same photographic plate produce Young's fringes when illuminated by a laser; this is *speckle photography.* The sequence of techniques for measuring displacements of rough surfaces, becoming simpler but less sensitive, is hologram interferometry, speckle interferometry, speckle photography, white-light speckle, and finally moiré.

8. X-RAY INTERFEROMETERS

8.1 General Principles

Diffraction effects at wavelengths on the order of 1 Å have been known since Max von Laue's demonstration of x-ray diffraction in 1912. Bonse and Hart (1965) demonstrated that interference between well-separated, coherent x-ray beams could be obtained in an interferometer, precisely machined from a perfect crystal of pure silicon. Such a device is shown schematically in Fig. 7. It is topologically identical to an optical Mach–Zehnder interferometer (see Sec. 2.2). Interferometers of up to about 10 cm in linear dimension have been constructed. Dimensional tolerances of better than about 1 μm in the machining of the device with diamond cutting tools must be maintained. Careful chemical etching of the work-damaged surface is required before the device can be used as an interferometer. In operation, temperature gradients and microphonic vibrations must be very carefully controlled.

The interferometer, in its simplest form, is a monolithic crystal in which three identical silicon slabs have been cut perpendicular to a set of strongly (x-ray) reflecting lattice planes, typically the (220) planes. A collimated, nominally monoenergetic x-ray beam is directed from the source to the first slab of the interferometer where it is coherently split by Bragg diffraction near point A. The two resulting beams are split by the second crystal slab, near points B and C. Two of these beams overlap, mix, and interfere in the third crystal slab near point D. The outgoing beams labelled O and G are coherent linear superpositions of the internal beams I and II that have passed through the interferometer. The device shown in Fig. 7 is called an LLL interferometer, since it consists of three transmission Laue geometry diffraction slabs. Other geometries, utilizing Bragg reflection crystals, and skew-symmetric arrangements have been successfully constructed. Interferometers consisting of two thick crystals, topologically similar to the Rayleigh optical interferometer, have been used in neutron interferometry.

The dynamical theory of diffraction is required for a detailed understanding of the beam splitting, reflection, and recombination in the three Si crystals of the interferometer.

8.2 Index of Refraction and Scattering Factors

If a semitransparent slab of material of thickness D is introduced in leg II of the interferometer, a phase difference $\Delta\Phi$ of the two beams traversing the interferometer will result, given by

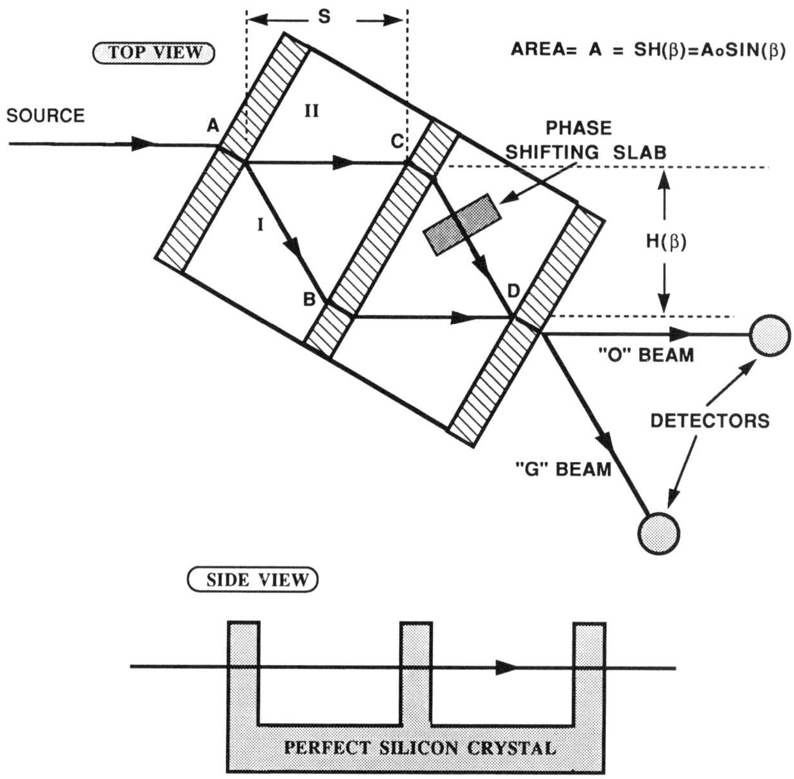

FIG. 7. Schematic diagram of a Bonse–Hart perfect silicon crystal interferometer.

$$\Delta\Phi = 2\pi(1-n)D/\lambda,$$

where λ is the x-ray wavelength, and n is the index of refraction of the material. In the x-ray region of the electromagnetic spectrum, the index of refraction differs from unity by a small number

$$\delta = 1-n = \lambda^2 r_e N f/2\pi,$$

which is typically of order 10^{-5}. Here r_e is the classical electron radius, N is the atom density, and f is the real part of the forward scattering amplitude of the atoms in the slab. Under conditions in which the x-ray wavelength is not close to an atomic absorption edge, the scattering amplitude f will be equal to the number of electrons per atom, Z. However, if the x-rays are tuned to wavelengths in the vicinity of an absorption edge, f will differ from Z and will display the fine structure due to the local arrangement of atoms in the slab material seen in the absorptive (imaginary) part of f and commonly referred to as extended x-ray–absorption fine structure (EXAFS). X-ray interferometry provides a very precise method to measure f in the region of an absorption edge.

8.3 Absolute Measurement of Lattice Parameter

The Si crystals in the x-ray interferometer act as diffraction gratings. Thus, if the third crystal is mechanically translated on the scale of fractions of an ångström, x-ray interference fringes will by generated. This difficult task can be accomplished by machining a weak joint between the second and third crystals of the LLL interferometer that acts as an elastic spring, and smoothly translating the third crystal by elastic deformation, as first demonstrated by Hart (1968). Bonse and TeKaat (1968) showed that a completely separated third crystal could be aligned, positioned, and translated by elastic deformation of spring elements while still maintaining fringe visibility. Deslattes (1969) showed that a fully developed scanning "engine" equipped with Fabry–Pérot optics and a separated interferometer could yield simultaneous x-ray and optical interference signals of comparable spatial resolution. A mirror attached to the third Si crystal is used as a back-reflecting optical element in a Michelson (or Fabry–Pérot) interferometer using a highly stabilized laser source. As the third crystal is translated, the rapidly shifting x-ray fringes are counted relative to the shift of the optical fringes. This then provides a direct measurement of the Si lattice spacing, calibrated to the standard optical laser wavelength. It can thereby be referred to the SI definition of the meter. Hart called this idea the "ångström ruler." This standardization can then be used to obtain Avogadro's constant. The precision of this technique is currently about 0.1 ppm.

9. NEUTRON INTERFEROMETERS

9.1 General Principles

According to the de Broglie hypothesis, under certain conditions moving particles such as electrons or neutrons will display wave properties, where their wavelength λ is inversely proportional to their momentum p,

$$\lambda = h/p,$$

and h is Planck's constant. For a thermal neutron of energy $E=25$ meV, velocity $v=2185$ m/s, the wavelength is 1.81 Å. Thus, thermal neutrons, like x-rays, will exhibit diffraction from single crystals in which they interact with the periodic array of nuclei. Ten years after the invention of the perfect Si-crystal x-ray interferometer, this device was also shown to work for neutrons by Rauch *et al.* (1974). This demonstration opened up an interesting and important field of fundamental quantum-mechanical investigations using the Bonse–Hart interferometer. Unlike x-rays, thermal neutrons suffer essentially zero absorption in passing through a silicon crystal. Thus, the thickness of the crystal slabs in the interferometer is typically several mm. As the phase difference $\Delta\Phi$ due to the introduction of a region of optical potential V in one leg of the interferometer is varied, the intensities in the two outgoing beams will oscillate sinusoidally according to

$$I_O = a + b\cos(\Delta\Phi),$$

and

$$I_G = c - b\cos(\Delta\Phi).$$

The constants a, b, and c are dependent upon the incident neutron flux and other instrumental parameters. These formulas will be recognized as being identical in form to the intensity distribution introduced in Sec. 1.1. The index of refraction of a medium of neutron optical potential V is given by the ratio of the wave vector K inside the medium to the wave vector $k=2\pi/\lambda$ in free space. By conservation of energy this is easily seen to be

$$n = K/k = \sqrt{1 - V/E} \approx 1 - V/2E.$$

In terms of the neutron scattering amplitude b, the neutron mass m, and the atom density N, one has

$$V = 2\pi\hbar^2 bN/m,$$

which is typically 10^{-4} to 10^{-5} meV. The phase difference induced by the potential extending over a distance D is then given by

$$\Delta\Phi = \pi VD/\lambda E = \lambda NbD.$$

Thus, a measurement of the phase shift $\Delta\Phi$ due to the presence of a medium in one leg of the interferometer provides a precise method for measuring the neutron-nuclear scattering amplitude (length), provided the atom density N is known. Conversely, if b is known, the density N can be determined. This is analogous to measuring the density of a medium, such as a gas, by optical interferometry. The interferograms observed in the O beam and the G beams are 180° out of phase with each other, thus showing the conservation of neutron flux. (In the x-ray case, because of absorption in the silicon and the details of the dynamical diffraction process, the interferograms in the two outgoing beams are in phase with each other.) Neutron interferometry experiments impacting fundamental quantum physics are (among others)

1. gravitationally induced quantum interference,
2. change of sign of the neutron Fermion wave function Ψ during a 2π precession in a magnetic field,
3. measurement of the longitudinal coherence length of a neutron beam,
4. observation of the effect of the Earth's rotation upon the quantum-mechanical

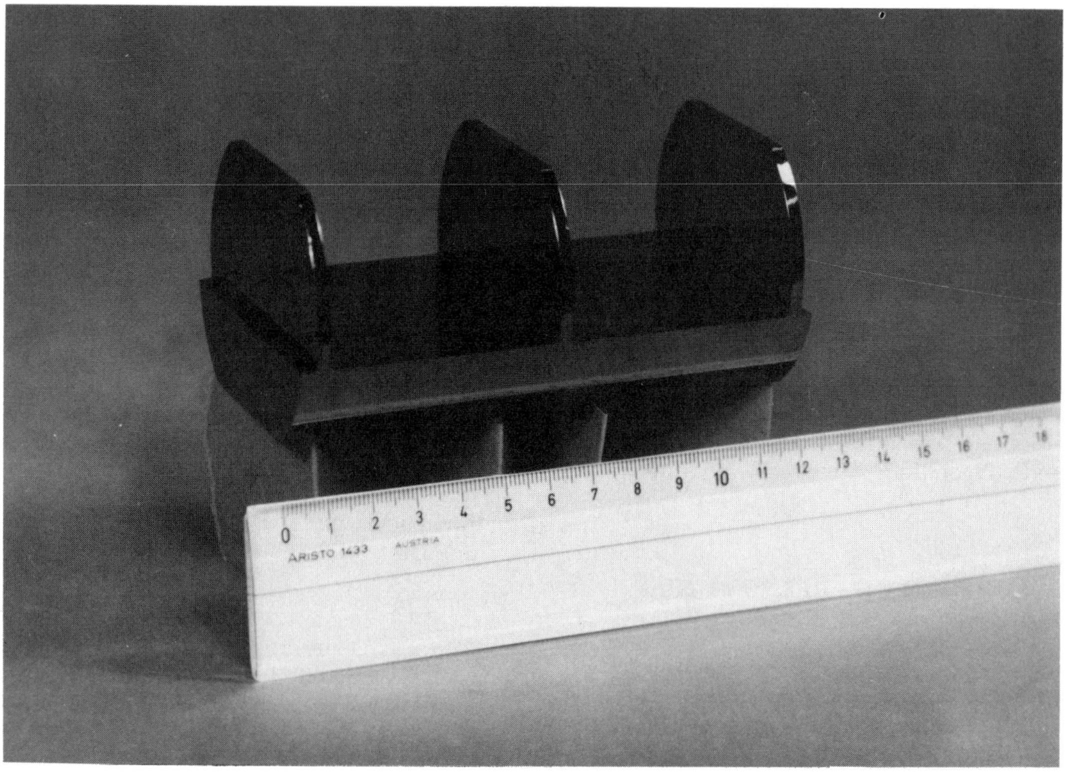

FIG. 8. Photograph of a neutron interferometer at the University of Missouri–Columbia.

phase of the neutron (Sagnac effect),

5. observation of the topological phase shift due to diffraction of a neutron around a line of electric charge (Aharonov–Casher effect), and
6. observation of the scalar Aharonov–Bohm effect.

We discuss two of these experiments in the following sections. A photograph of a silicon crystal neutron interferometer used at the University of Missouri–Columbia is shown in Fig. 8.

9.2 Gravitationally Induced Quantum Interference

In most phenomena of interest in terrestrial physics, gravity and quantum mechanics do not simultaneously play an important role. In a neutron interferometry experiment carried out by Colella *et al.* (1975), it was shown that quantum interference can be induced by gravity. This experiment is commonly referred to as the COW effect (for Colella, Overhauser, and Werner) (see also Staudenmann *et al.*, 1980). Suppose that the interferometer shown in Fig. 7 is oriented such that the plane of the parallelogram ABDC is vertical. Then the beam on path II is higher above the surface of the Earth on ray CD than on ray AB on path I; that is, it experiences a higher gravitational potential. The Newtonian Hamiltonian affecting the neutron's motion is

$$H = p^2/2m_i + m_g g H(\beta),$$

where m_i is the neutron's inertial mass and m_g is its gravitational mass, g is the acceleration due to gravity, and $H(\beta) = H_0 \sin(\beta)$ is the height difference between rays AB and CD as a function of the tilt angle β of the interferometer about the incident beam line AB. The gravitational field can be viewed as creating a region of variable index of refraction $n(\mathbf{r})$ for the neutron wave. This then gives rise to a phase difference for the neutron traversing path II with respect to path I in the interferometer, given by the formula

$$\Delta\Phi_{\text{gravity}} = 2\pi m_i m_g g \lambda A_0 \sin(\beta)/h^2,$$

where $A_0 = H_0 S$ is the area of the parallelogram enclosed by the neutron beam trajectories within the interferometer (see Fig. 7),

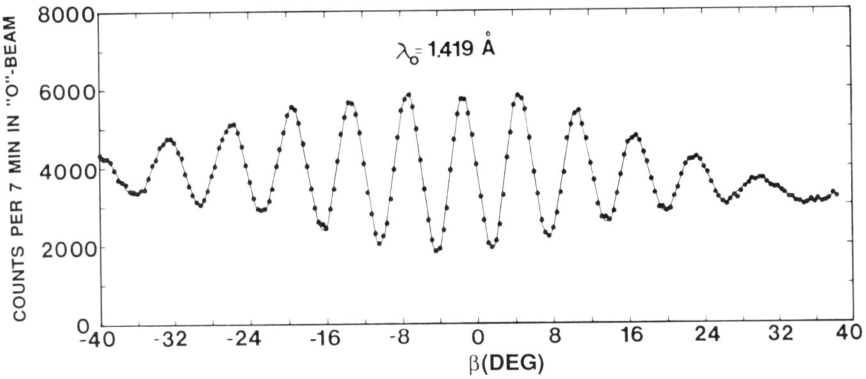

FIG. 9. Gravitationally induced quantum interference observed by neutron interferometry.

and the tilt angle $\beta=0$ when both paths I and II lie in the horizontal plane. Thus, tipping the interferometer through an angle β induces a phase shift in the neutron wave traversing path II relative to path I. An example of the resulting interferogram seen in the O detector is shown in Fig. 9 for incident neutrons of wavelength $\lambda=1.419$ Å (see Staudenmann *et al.*, 1980). The result of this experiment can be viewed as extending the principle of equivalence ($m_i=m_g$) into the quantum regime. The current limits of error in the most recent version of this experiment are about 1/10%.

9.3 Neutron Sagnac Effect

In 1925 Michelson *et al.* carried out a remarkable experiment designed to detect the effect of the Earth's rotation on the speed of light. An experiment demonstrating that angular rotation could be detected by optical interferometry had been carried out earlier by Sagnac (1913). The physics underlying this effect is the basis of the ring-laser gyroscope used for navigation on most large commercial and military aircraft (see Sec. 2.3). Using a neutron interferometer Werner *et al.* (1979) showed that the same effect exists for matter waves. In addition to the kinetic energy and gravitational potential energy terms in the Hamiltonian for a neutron moving through their interferometer, there is a term $-\mathbf{\Omega}\cdot\mathbf{L}$, where $\mathbf{\Omega}$ is the angular frequency of rotation of the Earth, and $\mathbf{L}=\mathbf{r}\times\mathbf{p}$ is the angular momentum of the neutron's motion about the center of the Earth. According to Hamilton's equations, the canonical momentum of the neutron is then

$$\mathbf{p}=m\mathbf{v}+m\mathbf{r}\times\mathbf{\Omega}.$$

Calculating the line integral of this momentum around the interferometer paths gives two terms: one due to gravity as already discussed above, and another due to the rotation. The second term is the quantum-mechanical analog of the Sagnac effect. The predicted phase shift is

$$\Delta\Phi_{\text{Sagnac}}=(2m/\hbar)\mathbf{\Omega}\cdot\mathbf{A},$$

where $\mathbf{A}$ is the normal area of the interferometer. Werner *et al.* observed this phase shift in an experiment with a vertically directed beam (for which the gravity effect is zero) having the expected symmetry and of magnitude 91°, in agreement with the above formula to within about 3%. The analogy with the optical Sagnac effect can be readily obtained, if one notes that the neutron's momentum is $mv=\hbar k$; thus, $m/\hbar=k/v$, where k is the neutron wave vector. Replacing the velocity v with the speed of light c, one obtains the formula given in standard optics textbooks for the Sagnac effect. It should be pointed out that since v/c is of order 10^{-5}, the neutron interferometer is very sensitive to rotation. Thus, an atom-beam interferometer (discussed below) will also be very sensitive to the Sagnac effect.

10. ELECTRON INTERFEROMETRY

The realization of electron interferometry was first provided by Marton (1948). The primary motivation was to observe the topological Aharonov–Bohm (AB) effect. A subsequent definitive experiment was carried out by Chambers (1960) and by Fowler *et al.* (1961). The conceptual idea of the experiment is shown in Fig. 10. A beam of electrons is allowed to diffract around a solenoid of essentially infinite extent, external to which the magnetic field **B** is zero, and thus the electron passes in a region of space where the classical Lorentz force is zero. Nevertheless, AB (1959) predicted that there should be a phase shift due to the existence of a magnetic vector potential **A** in the region external to the solenoid, since the canonical momentum of the electron in an electromagnetic field is given by

$$\mathbf{p}=m\mathbf{v}+e\mathbf{A}/c,$$

where **v** is the electron's velocity, e its charge, and c the velocity of light. External to a solenoid, **A** is a solenoidal vector field. The phase shift of the electron traversing the solenoid on the left with respect to the right is given by the line integral of the canonical momentum around the solenoid, namely

$$\Delta\Phi_{AB}=(1/\hbar)\oint \mathbf{p}\cdot d\mathbf{r}=e\phi/\hbar,$$

where $\hbar$ is $h/2\pi$, and ϕ is the magnetic flux contained within the solenoid. This phase shift was first observed by Chambers (1960) and more recently in an electron microscope by Tonomura (1988) for tiny toroidal coils with a superconducting overlayer.

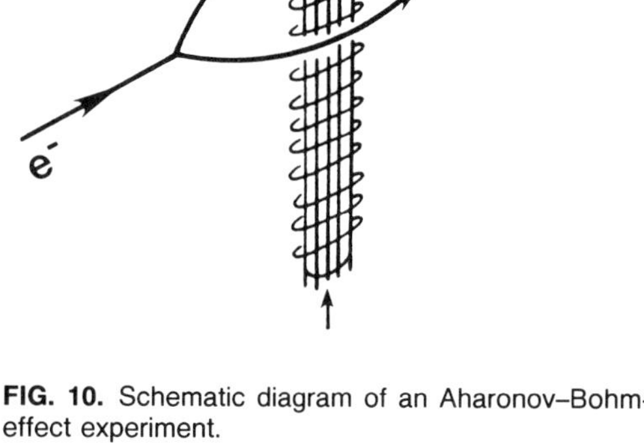

FIG. 10. Schematic diagram of an Aharonov–Bohm-effect experiment.

11. ATOM-BEAM INTERFEROMETERS

With the recent successes in cooling atoms to very low temperature by the technique of laser trapping, there has been a substantial interest in constructing atom-beam interferometers (see Chu, 1992). Like other particles (neutrons and electrons) of mass m and velocity v, atoms display de Broglie wave characteristics. The experimental challenge has been to create the necessary optical elements for beam splitters, mirrors, and mixers that the silicon crystals provide so beautifully in neutron interferometry. Standing light waves have been shown to diffract atoms. The first atom-beam interferometer, using wave-front division by diffraction of He atoms from a 2-μm slit in a gold foil, was realized by Carnal and Mlynek (1991). The resulting transversely coherent beam impinges on a microfabricated transmission structure, consisting of two 1-μm-wide slits with separation of 8 μm. The interference pattern of this double-slit "Young's type" experiment was observed with a 2-μm-wide scanning slit. Beam division by diffraction of Na atoms from a free-standing grating with period 0.4 μm has also been realized in a 3-grating interferometer by Keith *et al.* (1991).

The experimental difficulty in atom-beam interferometry is to obtain larger beam separations than are currently available. This problem is apparent when one notes that the angle between two beams from a diffraction grating is of order λ/d, where d is the grating spacing. Consequently, for a de Broglie wavelength of $\lambda=0.2$ Å, and a grating spacing of $d=0.2$ μm, we find an angle $\delta\theta=10^{-4}$ radians. Thus, the possibilities for manipulating the beams separately pose severe limitations. Nevertheless, if this difficulty can be ameliorated, the potential for constructing very sensitive gravity gradiometers (using the COW effect) and Sagnac gyroscopes would be quite attractive with atom-beam devices, especially in view of their portability and source intensity.

12. FUTURE TRENDS

12.1 Optical Interferometers

The new fields of optics are affecting interferometers. There are interferometers that react with the laser source and others based on nonlinear effects such as second-harmonic generation or phase conjugation. The development of *squeezed states* gives the possibility of more precise measurement of phase.

Applications are increasing. Lasers and electronics are simplifying interferometers sufficiently for them to be a common tool for workshop metrology or chemical spectroscopy, and fiber sensors have wide possibilities. The most ambitious application proposed is the detection of gravity waves, which will require an interferometer capable of detecting 1 part in 10^{21}. While this is beyond present systems, it is not considered to be an impossible goal.

12.2 X-Ray and Matter-Wave Interferometers

It is likely that Bonse–Hart x-ray interferometers will continue to be used to measure the forward scattering amplitude of atoms with increasing precision in the region of absorption edges. Experiments of the Deslattes type, in which optical and x-ray interferometry are combined to set x-ray and γ-ray energies (wavelengths) on the optical scale and to measure the Avogadro constant, will be carried on with increasing precision.

The importance of neutron interferometry in elucidating fundamental quantum-mechanical principles and the interactions of nature is continuing to attract interest. The detailed and precise study of topological Aharonov–Bohm-type interference is expected to yield new and deep understanding of electrodynamics and quantum physics. There are plans in several laboratories to carry out a Wheeler delayed-choice experiment in which the "which slit?" equation is most clearly displayed. There are theoretical reasons in connection with the conformal factor in the Lorentz transformations to pursue carrying out a Michelson–Morley experiment with neutrons. The challenge will be to construct a Michelson-type interferometer that works in the ångström region of wavelengths. There several ideas for doing this, but it is not clear whether they will yield sufficient intensity with thermal neutrons. Recent progress with ultracold-neutron interferometers (using ruled gratings) in this connection provides new possibilities.

It is difficult to predict the future in atom-beam interferometry.

Acknowledgment

The work of one of us (S.A.W.) was supported in part by National Science Foundation grant No. PHY 9024608.

GLOSSARY

Aharonov–Bohm Effects: Topological quantum-mechanical interference effects in which matter waves are phase shifted because of the presence of a potential, but for which there is no classical force on the particle (electron or neutron).

Beam Splitter: A component that divides a beam of radiation into two or more beams of the same wave form but reduced intensity.

Bonse–Hart Interferometer: A device consisting of three or more perfect silicon crystal plates machined from a monolithic silicon ingot in which x-rays or neutron beams are split and recombined by Bragg diffraction in the crystal plates.

Coherence: A complex quantity that describes the relationship between two radiation fields separated in space and time. Its magnitude denotes the amount of correlation between the fluctuations of their amplitudes and its phase denotes any effective phase difference between them.

COW Effect: See Gravitationally Induced Quantum Interference.

Degree of Coherence: The coherence normalized so that 1 denotes full amplitude correlation.

Gravitationally Induced Quantum Interference: A phenomenon observed in neutron interferometry in which the Earth's gravity causes a quantum phase shift. This experiment is sometimes called the COW effect after Colella, Overhauser, and Werner, who first observed it.

Index of Refraction: A quantity (usually wavelength dependent) that gives the ratio of the wave vector of the electromagnetic wave (or de Broglie matter wave) inside a material to that in vacuum.

Interferogram: The varying part of an interference pattern, without the uniform background.

Shear: A lateral separation of two images of the same wave front.

Speckle: The random pattern of light and dark obtained when radiation from a small source is reflected from a rough surface or refracted by a medium with random variations of refractive index.

Speckle Interference: Speckle patterns contain information on the wave that produced them and can be correlated to give the equivalent of an interferogram. The speckle images in a telescope, due to atmosphere turbulence, can be processed to find the true image.

List of Works Cited

Aharonov, Y., Bohm, D. (1959), *Phys. Rev.* **115**, 485–491.

Bonse, U., Hart, M. (1965), *Appl. Phys. Lett.* **6**, 155–158.

Bonse, U., TeKaat, E. (1968), *Z. Phys.* **214**, 16–21.

Carnal, O., Mlynek, J. (1991), *Phys. Rev. Lett.* **66**, 2689–2692.

Chambers, R. G. (1960), *Phys. Rev. Lett.* **5**, 3–5.

Chu, S. (1992), *Sci. Am.* **267** (2), 71–76.

Colella, R., Overhauser, A. W., Werner, S. A. (1975) *Phys. Rev. Lett.* **34**, 1472–1473.

Connes, J. (1961) *Rev. Opt.* **40**, 45–79, 116–140, 171–190, 231–265.

Davis, J., Tango, W. (Eds.) (1979), *High Angular Resolution Stellar Interferometry*, IAU Symposium 50, Sydney: Univ. of Sydney Astronomy Department.

Deslattes, R. D. (1969), *Appl. Phys. Lett.* **15**, 386–389.

Fowler, H. A., Marton, L., Simpson, J. A., Suddeth, J. A. (1961), *J. Appl. Phys.* **32**, 1153–1155.

Hanbury Brown, R. (1974), *The Intensity Interferometer, Its Applications to Astronomy*, London: Taylor and Francis.

Hansen, G. (1959), *Optik* **12**, 5–16.

Hart, M. (1968), *Brit. J. Appl. Phys.* **1**, 1405–1408.

Keith, D. W., Ekstrom, C. R., Turchette, Q. A., Pritchard, D. E. (1991), *Phys. Rev. Lett.* **6**, 2693–2696.

Labeyrie, A. (1976), in: E. Wolf (Ed.), *Progress in Optics*, Vol. 14, Amsterdam: North-Holland, pp. 47–87.

Malacara, D. (Ed.) (1978, 1991), *Optical Shop Testing*, New York: Wiley.

Marton, L. (1948), *J. Appl. Phys.* **19**, 863.

Michelson, A. A., Gale, H. G., Pearson, F. (1925), *Astron. J.* **61**, 140–142.

Ramsay, J. V. (1962), *Appl. Opt.* **1**, 411–413.

Rauch, R., Treimer, W., Bonse, U. (1974), *Phys. Lett.* **47A**, 369–372.

Sagnac, M. G. (1913), *C. R. Acad. Sci.* **157**, 708–712.

Smartt, R. N., Steel, W. H. (1975), *Jpn. J. Appl. Phys.* **14**, Suppl. 14-1, 351–356.

Staudenmann, J.-L., Werner, S. A., Colella, R., Overhauser, A. W., (1980) *Phys. Rev.* **A21**, 1419–1438.

Steel, W. H. (1983), *Interferometry*, 2d ed., Cambridge, U.K.: Cambridge University Press.

Tonomura, A. (1988), *Physica* **B151**, 206–213.

Vanasse, G. A., Sakai, H. (1967), in: E. Wolf (Ed.), *Progress in Optics*, Vol. 6, Amsterdam: North-Holland, pp. 349–393.

Werner, S. A., Staudenmann, J.-L., Colella, R. (1979), *Phys. Rev. Lett.* **42**, 1103–1106.

Further Reading

Badurek, G., Rauch, H., Zeilinger, A. (Eds.) (1988), "Matter Wave Interferometry," *Physica* **B151**; Proceedings of a workshop.

Bonse, U., Graeff, W. (1977), "X-Ray and Neutron Interferometry" in: H.-J. Queisser (Ed.), *X-Ray Optics*, Topics in Applied Physics Vol. 22, Berlin: Springer-Verlag, Chap. 4, pp. 93–140.

Hariharan, P. (1990), *Rep. Prog. Phys.* **54**, 339–390. An up-to-date review.

Hariharan, P. (1991), *Selected Papers on Interferometry*, SPIE Milestone Series Vol. MS 28, Bellingham, WA: SPIE. A collection of the historic papers on optical interferometry.

Klein, A. G., Werner, S. A. (1983), "Neutron Optics," *Rep. Prog. Phys.* **46**, 259–335.

Post, E. J. (1967), *Rev. Mod. Phys.* **39**, 475–493. A review of Sagnac-type experiments.

Steel, W. H. (1983), *Interferometry*, 2d ed., Cambridge U.K.: Cambridge University Press.

Werner, S. A. (1980), "Neutron Interferometry," *Phys. Today* **33** (12), 2–8.

ION-BEAM MODIFICATION OF MATERIALS

ROBERT S. AVERBACK, *Department of Materials Science and Engineering, University of Illinois at Urbana-Champaign, Urbana, Illinois, U.S.A.*

	Introduction	173
1.	**Depth Distributions of Implanted Ions**	
1.1	Electronic Stopping Power	174
1.2	Nuclear Stopping Power	175
1.3	Implantation Range Distribution	176
1.3.1	Transport Theory	176
1.3.2	Computer Simulation	177
1.4	Channeling	178
2.	**Implantation Damage**	179
2.1	Displacement Processes	179
2.2	Cascade Damage	179
2.3	Thermal Spikes	179
2.4	Applications of Ion Implantation	180
2.4.1	Doping of Semiconductor Devices	180
2.4.2	Buried Layers	181
2.4.3	Tribological Applications	182
3.	**Ion-Beam Mixing**	183
3.1	Mechanisms of Ion-Beam Mixing	183
3.1.1	Recoil Implantation	183
3.1.2	Cascade Mixing	183
3.1.3	Thermal-Spike Mixing	184
3.1.4	Phase Formation During Ion-Beam Mixing	184
3.2	Ion-Beam–Assisted Deposition	184
3.2.1	Dopant Incorporation	186
3.2.2	Improved Adhesion	186
4.	**Implantation at Elevated Temperature**	186
4.1	Theory of Radiation-Enhanced Diffusion	186
4.2	Examples of Radiation-Enhanced Diffusion	187
5.	**Summary**	189
	Glossary	189
	Works Cited	190
	Further Reading	191

INTRODUCTION

The possibility of using ion beams for materials modification was first explored nearly 50 years ago. However, not until the early 1970s, when the electronics industry began their frenetic drive toward minaturization of integrated circuits, did ion implantation become a wide-scale commercial process. It was also about 50 years ago when studies of radiation damage in metals and ceramics were initiated. These early investigations were part of the research and development program on nuclear weapons, but they were soon expanded to include materials problems in fission and fusion reactors. These two fields of ion implantation and radiation damage have now become integrated in the far more comprehensive field of ion-beam modifications of materials. No longer is ion-beam processing of materials restricted to the implantation of electrically active impurities into Si for semiconductor applications, but rather ion beams are being employed as a basic processing step in the fabrication of many advanced materials. For example, ion implantation is being used to provide corrosion and oxidation resistance in metals; to create nonequilibrium surface phases with special properties; to improve the quality, epitaxy, and adhesion of vapor-deposited thin films; and to clean or pattern surfaces. One of the most celebrated recent applications of ion-beam modifications of materials has derived from the field of tribology, where it was shown that the implantation of nitrogen into titanium hip joint replacements can extend the wear resistance of these prostheses beyond their required service lifetime. Indeed, the field of ion–solid interactions has moved from the workbench of nuclear physicists to the design tables of materials scientists.

Several features of ion-beam processing underlie its rapid growth in recent years:

1-56081-067-X/94/$5.00+.50

1. The concentration depth profile of an implanted species can be tailored to the materials requirements, whereas those obtained by the more conventional means of in-diffusing surface films are dictated by the diffusion equation.
2. Any ion species can be implanted into the solid. Because ion implantation is a high-energy process, thermodynamic and kinetic constraints are minimized. Consequently, nonequilibrium phases can be produced.
3. Ion implantation is intrinsically a low-temperature processing step, and so high-temperature anneals, which can cause interdiffusion in complex devices or phase transformations of metastable systems, can be avoided.
4. Ion irradiation can be employed to create structural defects in the surface region. Although defects are generally undesirable for implantation of dopants in semiconductors, they often have beneficial effects for other applications. Defects serve as nucleation sites for epitaxial growth, they modify the index of refraction for waveguide applications, and in some metals they lead to amorphization and concomitant improvement in tribology properties.
5. Ion implantation is conducive to lateral patterning using masks or focused ion beams. With its natural control of ion depth penetration, three-dimensional patterning becomes possible with submicron resolution.

Although ion implantation is often associated with the simple addition of foreign atoms into a substrate, doping is only one of many changes that occur in the material. As ions slow down in a target, they initiate cascades of energetic collisions between lattice atoms, as illustrated in Fig. 1. These collisions lead to the ejection of atoms from the surface (sputtering), the creation of Frenkel pairs (vacancies and interstitial atoms) in the interior of the crystal, and the relocation of many target atoms from one lattice site to another. These various processes, and their relative frequencies of occurrence, are indicated in Fig. 1. In this article, an elementary discussion of the physical principles underlying ion-beam modifications of materials is presented, and several examples are cited to illustrate the versatility of this new technology.

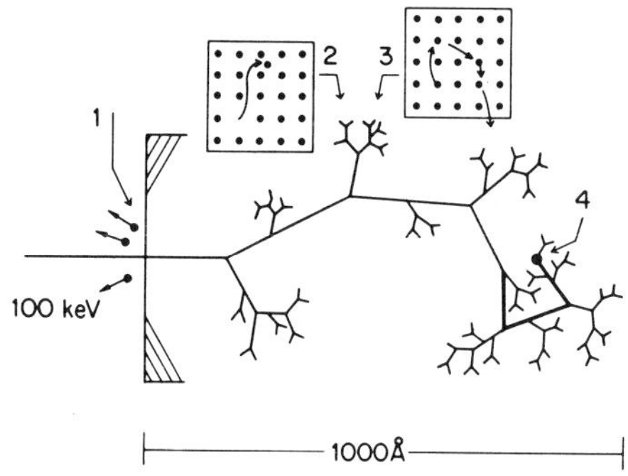

Summary of Events

Process	Relative Occurrence
1. Sputtered Atoms	~3
2. Frenkel Pairs	~500
3. Relocated Atoms	~5000
4. Implanted Atoms	1

FIG. 1. Schematic view of the different processes that occur and their frequency as an ion penetrates a solid and comes to rest. The values are typical for an ion implantation energy of approximately 100 keV.

1. DEPTH DISTRIBUTIONS OF IMPLANTED IONS

One of the primary advantages of ion-beam processing of materials is that the depth distribution of the implanted ions can be controlled with high precision, and with the use of a mask or a focused ion beam, lateral patterning of the substrate is also possible. It is now feasible to predict these implantation distributions accurately, greatly facilitating the design of new structures. In this section, the theory of the stopping of ions in materials and the methods for calculating implantation profiles are presented. In the next section, defect production during ion implantation is discussed, and several illustrative examples of the use of ion implantation in materials are provided.

1.1 Electronic Stopping Power

As energetic ions penetrate a solid, they lose energy by colliding with atomic nuclei and electrons. Theories of electronic stopping in the energy regime most relevant to ion-beam modifications of materials were developed by Firsov (1959) and Lindhard *et al.* (1963). The Firsov model, which is the easier to visualize, assumes that the ion velocity is slow relative to the electron orbital velocities, so that electrons behave "adiabatically," i.e., they are able fully to readjust their electronic states relative to the projectile position. Consequently, electrons are free to exchange between the projectile and target atom during collisions. Since the ion must perform work in accelerating target electrons up to its own velocity during an electron exchange, but without recapturing that energy when electrons leave, the ion loses energy. The energy loss is thus determined by calculating the electron currents flowing between the moving ion and the substrate atoms. Within the Thomas–Fermi model of an atom, the stopping power is given by the expression

$$\frac{dE}{dx} = -NS_e(E) = -N\{(0.336\ \text{eV}^{1/2}\ \text{Å}^2)[Z_1+Z_2] \times [E/A_1]^{1/2}\}, \tag{1}$$

where $S_e(E)$ is the so-called electronic stopping cross section and has units of energy-area, Z_1 and Z_2 are the atomic numbers of the implanted ion and target atoms, respectively, A_1 is the atomic mass of the ion, N is the atomic density, E is the projectile energy, and x is the distance along the ion trajectory. Equation (1) shows that the electronic stopping power increases with increasing ion velocity and atomic number; the theory of Lindhard and Scharff yields similar results. It is interesting to note that because the stopping power depends linearly on the velocity of the ion, enormous electronic stopping powers can be obtained with heavy, GeV ions—energy losses exceeding 5 keV/Å. This incredibly high power density has recently been exploited for materials modification applications. In high-temperature superconductors, for example, it was shown that heavy GeV particles produce long, highly damaged tracks, over 20 μm in length and approximately 40 Å in diameter, and that these tracks are very effective in pinning magnetic flux lines. Thus, GeV particle irradiation can be employed to enhance the critical current densities of high-temperature superconductors (Roas *et al.* 1990). GeV irradiations are also being used to produce high-precision filters. For this application, long, porous tubes of uniform diameter are created by selective etching of the damage tracks (Spohr, 1990).

1.2 Nuclear Stopping Power

The loss of projectile energy due to the nuclear collisions is described by the expression

$$\frac{dE}{dx} = -NS_n(E) = -N\int_0^{T_{max}} T\frac{d\sigma(E,T)}{dT}\,dT. \tag{2}$$

The nuclear stopping cross section $S_n(E)$ is defined by the right side of Eq. (2) (it also has units of energy-area), and $d\sigma(E,T)$ is the differential cross section for an ion of energy E to transfer an energy between T and $T+dT$ to a target atom. The cross section is governed by the appropriate interatomic potential, which for ion implantation is a screened Coulomb potential. Historically, Thomas–Fermi potentials have been employed for this purpose, although recently the more accurate Universal potential (Ziegler *et al.* 1985) has been preferred. The maximum energy transfer in Eq. (2), T_{max}, occurs in a head-on collision; it is given by

$$T_{max} = 4\frac{A_1A_2}{(A_1+A_2)^2}E. \tag{3}$$

Equation (3) shows that the maximum energy transfer is largest when the ion and substrate atoms have similar masses. For this reason, energetic electrons, even those with energies in excess of 1 MeV, cause little structural damage in materials. Lindhard and co-workers (1963) showed that by employing reduced units, the nuclear stopping of all ion–solid combinations could be described by a single universal stopping function,

$$s_n(\epsilon) = -d\epsilon/d\rho, \tag{4}$$

where the reduced energy ϵ is

$$\epsilon = \left\{\frac{A_2E}{A_1+A_2}\right\}\left\{\frac{Z_1Z_2e^2}{a}\right\}^{-1} \tag{4a}$$

and the reduced length ρ is

$$\rho = N4\pi a^2A_1A_2x/(A_1+A_2)^2\,. \tag{4b}$$

The screening length a is

$$a = 0.8853a_0/(Z_1^{2/3}+Z_2^{2/3})^{1/3}\,. \tag{4c}$$

where a_0 is the Bohr radius. The stopping function is shown in Fig. 2. Stopping powers for all ion–solid combinations can be determined from this one curve using the expression

$$\frac{dE}{dx} = -NS_n(E) = -N\left\{\frac{\pi a^2 T_m}{\epsilon}\right\} s_n(\epsilon). \quad (5)$$

Also shown in reduced units in Fig. 2 is the electronic stopping power. [$s_e(\epsilon)$ is not strictly a universal function, but it does not vary strongly with atomic number.] It is noteworthy that the total stopping power $s_{tot}(\epsilon) = s_n(\epsilon) + s_e(\epsilon)$ is rather constant over the range $0.2 < \epsilon < 2$, which is the range of energies of most implantations. The maximum nuclear stopping occurs at $\epsilon = 0.44$, and the electronic stopping becomes greater than the nuclear stopping at $\epsilon = 4$. This scaling with reduced energy units thus provides a natural way to compare implantations for all ion–target combinations. Table 1 lists the implantation energy at which $\epsilon = 1$ for various ion–solid combinations.

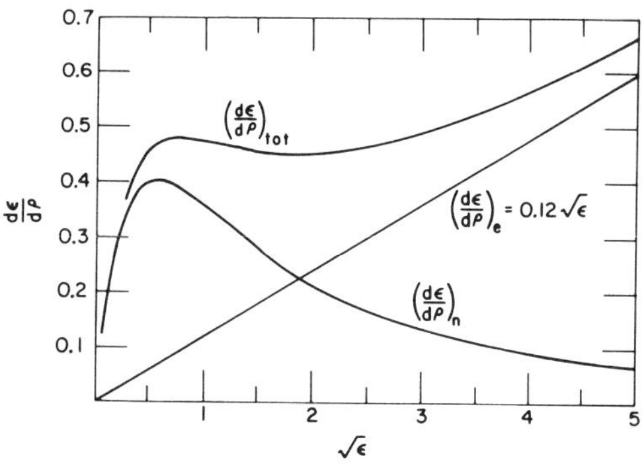

FIG. 2. Energy dependence of the nuclear and electronic stopping powers in reduced units. The nuclear curve is a universal function, but the electronic stopping depends somewhat on the ion–target combination. At low energies, most energy is dissipated through collisions with substrate atoms, but at high energies, most is lost by electronic excitation. (After Schiøtt, 1970.)

1.3 Implantation Range Distribution

1.3.1 Transport Theory. The distance that an ion travels before coming to rest within a solid, i.e., the path length, is given by the expression

$$L(E) = \int_E^0 \frac{dE}{dE/dx} = \frac{1}{N}\int_E^0 \frac{dE}{S_n(E) + S_e(E)}. \quad (6)$$

The path length, however, is usually of less interest for ion implantation than is the ion range, i.e., the distance between an ion's point of entry at the surface and its final location in the solid. Since the ranges of ions vary because of their different points of entry in the solid, lattice vibrations, and the discrete nature of collisions, a distribution of ranges is generated by an implantation, even for a single ion type, energy, and angle of incidence. The range distribution is comprised of longitudinal (normal to the surface) and transverse (parallel to the surface) components. Although the longitudinal component, or projected range R_P, and its variance are generally of primary concern for ion implantation, the transverse straggling ($\langle R_T^2 \rangle / \langle R_P^2 \rangle$) is of interest when implantations are performed through a mask, since it determines the minimum line width in circuit patterns (see, e.g., Tamura, 1991).

The range distributions of particles can be determined by solving the appropriate Boltzmann transport equation. Although the formalism is straightforward, solving the integro-differential equations is unwieldy. A more tractable procedure has been to calculate the moments of the distribution and then to construct the range distribution from an expansion of the various moments. The Edgeworth expansion has been found useful for this purpose (see Sigmund, 1969; Winterbon, 1975). The first two moments of the Edgeworth expansion, the mean range and its variance, determine a Gaussian distribution, and they often provide sufficient accuracy. The third moment becomes important for low-energy implantations since these distribu-

Table 1. Projectile energies (in keV) for which $\epsilon = 1$.

Projectile	Target Be	Si	Ge	Ag	Au
H	0.256	1.16	3.32	5.47	10.8
C	2.44	11.2	25.1	39.0	72.1
Al	18.1	37.1	69.2	101.3	176
Ni	89.3	147	219	297	185
Au	1017	1347	1535	1810	2331

tions are skewed toward the surface. A convenient expression for the mean range is (Schiøtt, 1970)

$$R_P = C_1(\mu)A_2 \frac{(Z_1^{2/3}+Z_2^{2/3})^{1/2}}{Z_1 Z_2} E\mu\text{g/cm}^2\ \text{keV}, \quad (7)$$

where $\mu = A_2/A_1$ and $C_1(\mu) \approx 0.62 + 0.171 \log\mu$. The rule of thumb (Davies, 1992)

$$\langle R_P^2\rangle^{1/2}/\langle R_P\rangle = 0.35[2(A_1A_2)^{1/2}/(A_1+A_2)] \quad (8)$$

can be used for the variance. Figure 3 compares the range distributions generated from a Gaussian distribution, using Eqs. (7) and (8), with those determined more accurately by computer simulation, a method now described.

1.3.2 Computer Simulation. With the vast improvements in personal computers in recent years and the development of efficient codes, simulation has become the primary means of calculating range distributions. These codes also yield the depth distributions of defects and electronic excitations that are produced by the ions. The procedure employed by simulation programs is to follow the path of an implanted ion by calculating the scattering integrals and electronic losses of the ion as it collides with target atoms, one by one. Distributions are generated by compiling the histories of many trajectories, with each having a different point of entry in the target. The trajectories of the recoil atoms can also be obtained in a similar manner, although at much greater computer overhead due to their greater number. Only marginal improvement in the accuracy of the implantation profile is obtained by this procedure, but for calculations of sputtering, defect production, and ion-beam mixing (see below), the inclusion of recoil atoms is necessary. The most widely used simulation code, TRIM, was developed by Biersack and Haggmark (1980). It is distinguished by an ingenious scheme to reduce the computational effort in calculating the scattering integral. Additional computer economy is achieved in TRIM by the assumption that the target material is amorphous. This reduces the lengthy process of locating the ion's next scattering partner in a lattice to a simple application of a random number generator. Because of this inherent randomness, TRIM is a Monte Carlo simulation and is most useful for providing statistical averages. Accurate concentration distributions can presently be obtained running TRIM on a personal computer in under 1 h (see Zeigler *et al.*, 1985, for details). For some applications, the crystalline structure of the target has a large influence on the ion trajectories. In these cases, a lattice must be included in the simulation model, as in the code MARLOWE (Robinson and Torrens, 1974). Since these codes must search the lattice for the successive collision partners of the ion and recoil atoms, they are accordingly slower than TRIM.

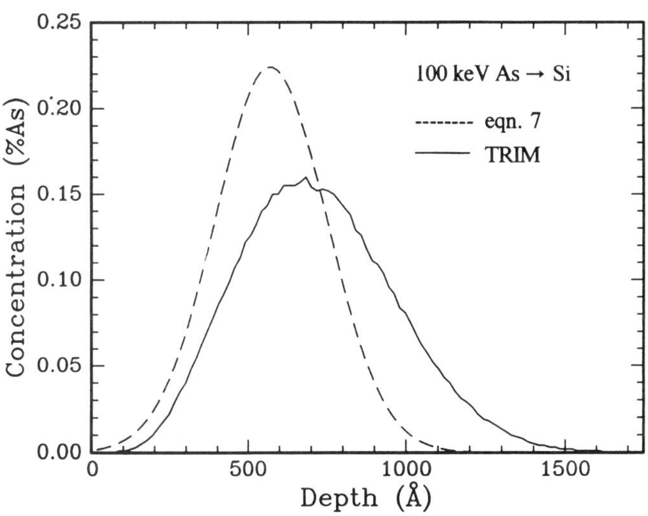

FIG. 3. Range distribution for the implantation of Si with 100-keV As calculated by TRIM and by Eqs. (7) and (8).

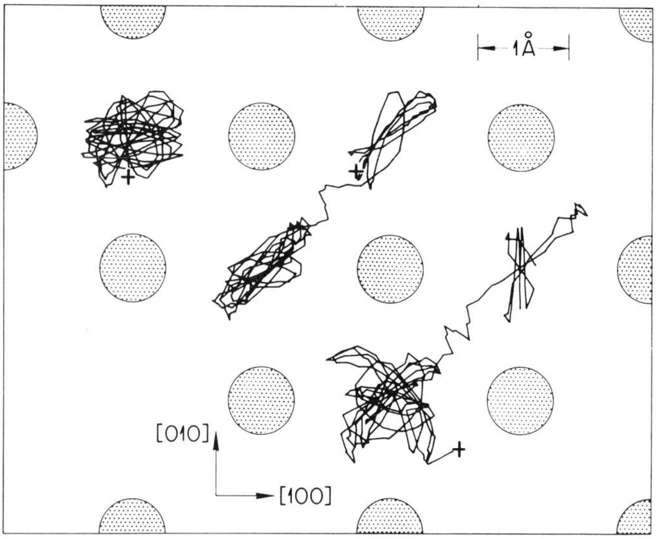

FIG. 4. Computer simulations showing the oscillatory motion of channeled ions in the (100) plane perpendicular to the ⟨001⟩ channeling direction. Radius of shaded circles is the distance of closest approach. (After Robinson and Oen, 1963.)

1.4 Channeling

Years ago, Stark (1912) predicted that the crystal lattice should influence the trajectory of ions in a crystalline solid, and by the early 1960s several experimenters began noticing long tails in their ion range distributions. But it was not until 1963 that the phenomenon of "channeling" was clearly elucidated by Robinson and Oen (1963), using computer simulation. In their simulations, which employed a lattice model, Robinson and Oen observed that some Cu projectiles had extremely long trajectories in a Cu lattice, particularly when the initial ion velocity was aligned with a major crystal axis. By following these special trajectories, they revealed that ions could travel through the relatively open spaces of a lattice as a result of a steering effect of the atomic rows. Figure 4 shows the transverse motion of a 1-keV Cu ion as it channels in a ⟨100⟩ direction of Cu. As the ion nears an atomic row, the repulsive interatomic potential nudges the ion back toward the center of the channel. The ion thus oscillates in two dimensions as it penetrates into the solid. Experimental evidence for channeling is shown in Fig. 5 (Piercy *et al.*, 1963), where the ranges of 40-keV Kr^+ in single-crystalline Al are shown for different incident ion directions. (Actual range distributions are the slopes of the curves in Fig. 5 as a function of depth.) The range in amorphous Al_2O_3 is shown for comparison. In the absence of channeling, the ranges of 40-keV Kr in Al and Al_2O_3 should be nearly the same. In many cases, the beam and substrate need only be aligned to within 3°–10° for channeling to occur, so that some fraction of the implanted ions can scatter into a channeling direction, independent of the initial orientation. This propensity for channeling poses serious problems for ion implantation applications where long tails in the range distribution are undesirable, such as shallow junction devices. Accordingly, much effort has been expended in finding crystallographic directions in Si, and other semiconductors, where channeling effects are minimized.

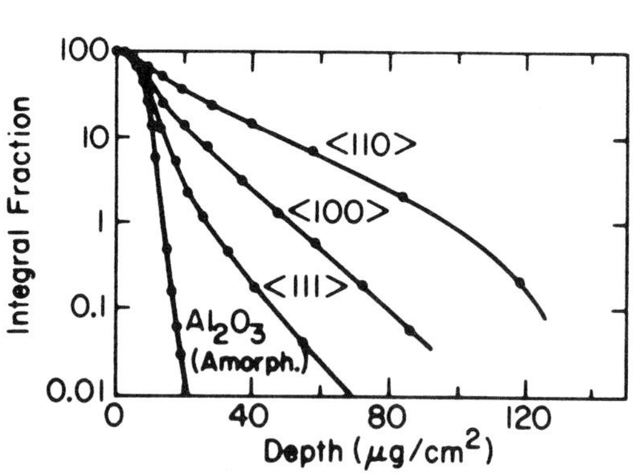

FIG. 5. Measurements of the range of 40-keV Kr in Al when the implantation beam is aligned with different crystallographic directions. Here the number of ions implanted deeper than a given depth is shown. For comparison, the range in amorphous Al_2O_3 is shown (after Piercy *et al.*, 1963).

2. IMPLANTATION DAMAGE

2.1 Displacement Processes

If a target atom recoils with an energy greater than the displacement energy E_d, it leaves its lattice site and creates a vacancy. The recoil atom may then fall into an interstitial site, or, if its energy is sufficiently high, it may displace additional atoms before coming to rest. Displacement energies in metals and semiconductors range from 5 to 40 eV (see Table 2). These values of E_d are surprisingly low in metals and derive from an efficient means of transporting energy and mass in a crystalline lattice by linear replacement collision sequences. Vineyard and co-workers (Gibson *et al.*, 1960) revealed, using molecular-dynamics computer simulations, that defects are created by series of collisions whereby one atom collides with its neighbor and replaces it in a linear sequence, as shown in Fig. 6.

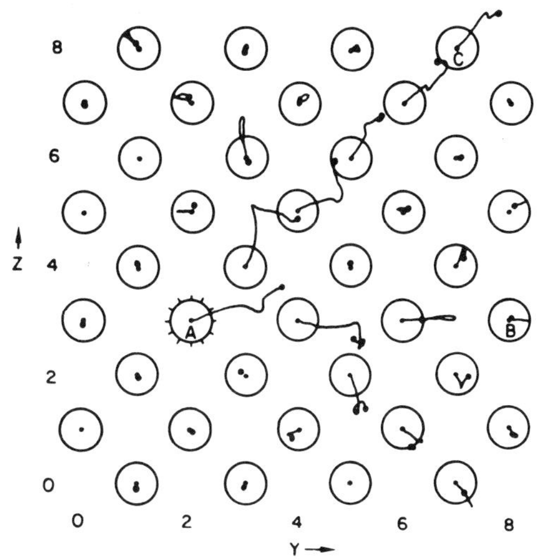

FIG. 6. Computer simulations illustrating replacement collision sequences in Cu (after Gibson *et al.*, 1960).

2.2 Cascade Damage

High-energy particles can produce recoils with energies that are much greater than the minimum energy required for displacements. In this case, both the primary and the recoil atom leave the lattice site and produce additional displacements at other sites. If the energies of the secondary recoil atoms are sufficiently high, they too continue the multiplication process of displacing atoms, thus creating a displacement cascade. Since the cross section for atomic collisions increases rapidly with decreasing energy, because of the screened Coulomb interaction between atoms, a projectile will produce recoils closer and closer together as it slows down. The energies of the recoil atoms are usually low; therefore they too do not travel far before undergoing additional collisions in the localized vicinity of their creation (see Fig. 1). Consequently, defects produced in a displacement cascade are spatially correlated and often form defect clusters. This has been demonstrated experimentally using field ion microscopy, as shown in Fig. 7. In some semiconductor systems and intermetallic compounds, the high concentration of defects can lead to small amorphous zones.

Table 2. Threshold displacement energies for several elements or compounds.

Material	Energy (eV)
Al	16
Si	13
Cu	21
Ge	14
GaAs	10
Ni	23
Mo	33
Au	35

The number of Frenkel pairs that are produced in a cascade, $\nu_{FP}(T)$, can be estimated from the "modified Kinchin-Pease" expression (1955),

$$\nu_{FP}(T)=0.8E_D(T)/2E_d, \tag{9}$$

where $E_D(T)$, the damage energy, is the total cascade energy T less the energy lost to electronic excitations. [The original Kinchin-Pease expression was $\nu_{FP}(T)=T/2E_d$.] Computer simulations, such as TRIM, are convenient for obtaining both the depth distribution and total number of defects produced by ion implantation.

2.3 Thermal Spikes

Although some of the kinetic energy of the implanted ion is lost to electronic excitations and some is stored as potential energy in the Frenkel pairs, most of the ion energy is dissipated in the lattice as heat. For 100-keV As

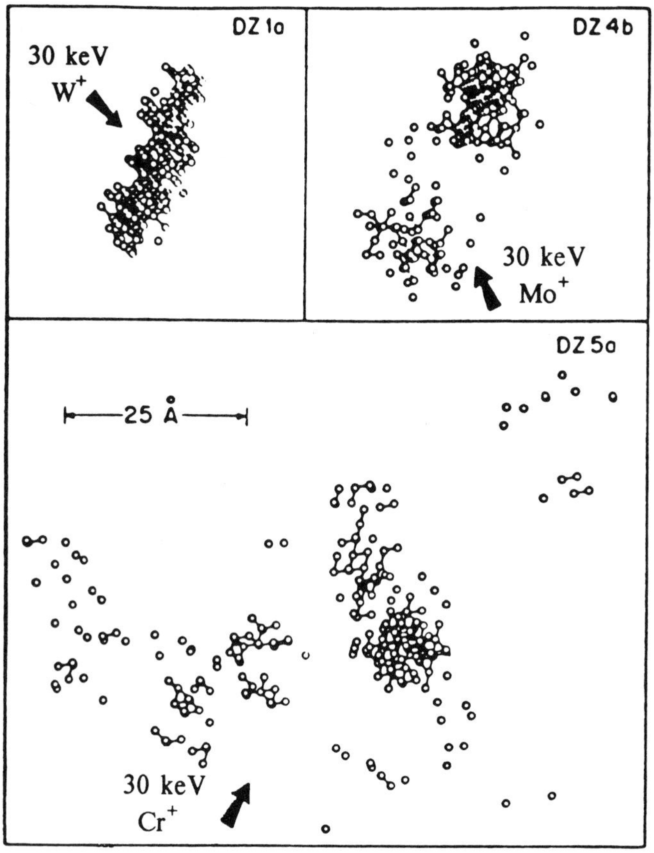

FIG. 7. Distribution of vacancies produced in displacement cascades of W. As the projectile mass increases, the density of defects also increases because of the increasing value of $S_n(E)$. Arrows indicate the initial directions of the incoming ions. (After Wei *et al.*, 1981.)

implantation of Ge, for example, 40 keV of energy is lost to electronic excitation, $\approx$2 keV is stored in defect energy, and the remainder, 58 keV, is converted to heat. The localization of such a large amount of energy in the small volume of a cascade can result in energy densities of several eV/atom. A long-standing question in the field of atomic collisions in solids has been whether this large localization of energy persists sufficiently long for thermal processes to evolve. The most vivid evidence for thermal spikes has derived from molecular-dynamics computer simulations. Figure 8, for example, shows the collision dynamics of a 10-keV Au ion impacting a Au surface (Ghaly and Averback, 1993). Clearly illustrated in Fig. 8(a), which shows a snapshot of the atoms in the vicinity of a cascade early in the event, are a shock wave emanating from the cascade center, replacement collision sequences, and a highly disordered cascade core. Shown in Fig. 8(b), which is a snapshot of the atoms some time later, is a well-defined local melt zone in the near surface; the melt is found to persist for several picoseconds. In materials with higher melting temperatures and smaller atomic numbers, the effects of thermal spikes are smaller.

2.4 Applications of Ion Implantation

The number of applications of ion implantation has grown enormously in the past decade, and they cannot be categorized here. A few examples, however, serve to illustrate the diversity of materials problems for which ion implantation has been successful.

2.4.1 Doping of Semiconductor Devices. The use of implantation for the doping of semiconductors has now been an industrial process for over 20 years, and a number of different devices include implantation as a processing step (see e.g., Ryssel and Ruge, 1986). As outlined above, implantation of semiconductors causes damage as well as increasing the dopant concentration, and damage generally degrades the electrical proper-

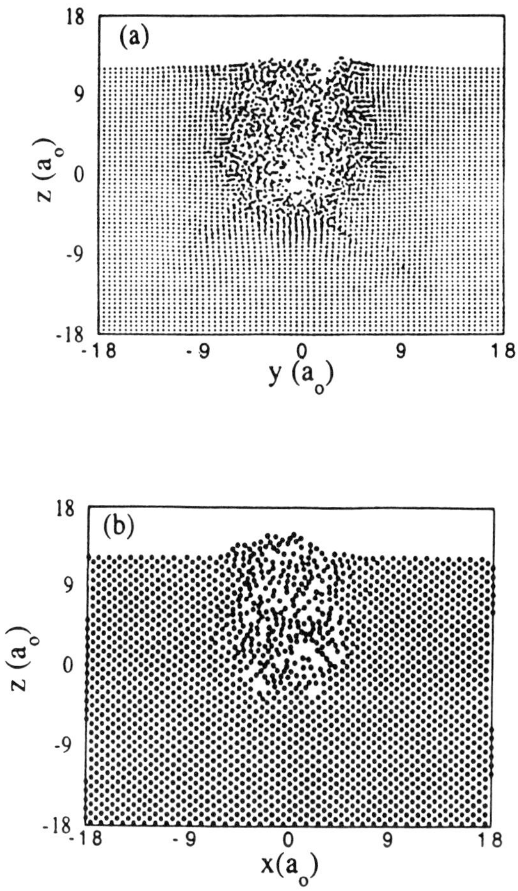

FIG. 8. Snapshots of atomic positions at two instants of time, 1.0 and 9.8 ps after a 10-keV Au atom impinges on a Au substrate. For clarity, only atoms in a cross-sectional slab containing four (100) lattice planes are shown. (a) A shock wave propagating from the cascade core, trails of replacement sequences, and atomic disorder are observed early in the event, while (b) local melting is observed sometime later. (After Ghaly and Averback, 1993.)

ties of the device. Moreover, the dopants do not always lie on substitutional sites in the lattice when initially implanted, and therefore they may not be electrically active. Consequently, annealing treatments are necessary to remove the damage and to activate the impurities. In Si, the damage can consist of defect clusters and dense dislocation networks. This type of damage is very resistant to annealing and typically requires annealing temperatures of 950 °C. Such high-temperature anneals are unpractical for most devices because of problems of interdiffusion. In many cases, however, the irradiation renders the implanted region amorphous. This can be advantageous because crystallization occurs at a relatively low annealing temperature, less than 650 °C in Si. The crystallization begins at the amorphous–crystal interface located at the end of the implanted ion range, and it advances toward the surface by a process called solid-phase epitaxy (SPE). SPE leaves very few defects in the reformed crystal, while dopants find lattice sites and become activated. Rates of regrowth on various cystallographic planes in Si are illustrated in Fig. 9.

2.4.2 Buried Layers. One of the advantages of ion implantation over diffusion doping is that the peak impurity concentration can be located beneath the surface. An interesting application of this capability is the production of planar waveguides in $KNbO_3$ and $LiNbO_3$ using He implantation (Townsend, 1987). Because He ions are very light, their stopping in a material occurs primarily via electronic excitation until they reach the end of their range, where the nuclear stopping becomes larger (see Fig. 2). Consequently, the damage produced by He implantation is mostly

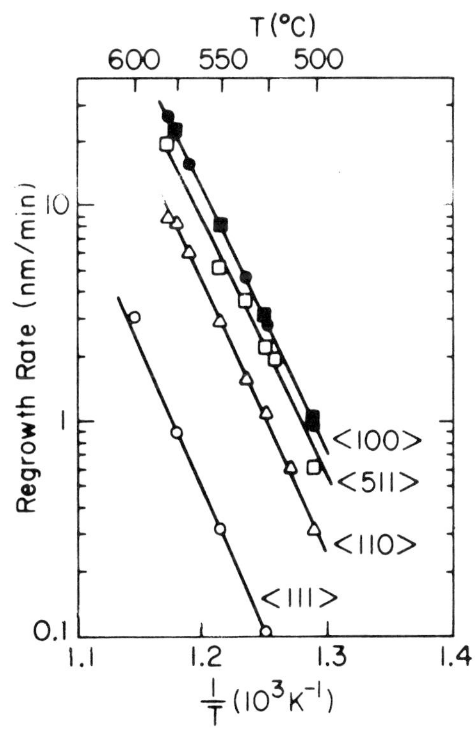

FIG. 9. The SPE regrowth rate at various amorphous-crystalline Si interfaces as a function of temperature (after Csepregi *et al.*, 1978).

confined to a narrow layer at a depth about equal to the projected range. This subsurface damage layer in $KNbO_3$ and $LiNbO_3$, in fact, becomes amorphous after sufficiently prolonged irradiation. Since the densities of these amorphous glasses are less than those of their corresponding crystalline phases, their indices of refraction are smaller. The buried amorphous layers thus create optical barriers or waveguides, with the thickness of the guiding region dictated by the energy of the He ion.

Buried layers are of interest for a variety of other applications as well. Co, for example, is implanted into Si to form a buried silicide layer, $CoSi_2$, a process called mesoepitaxy. These buried conductive layers serve as permeable base regions in transistor devices and interconnects for high-speed integrated circuits (Mantl, 1992). Oxygen is also used for implantation in Si to form buried insulator layers of SiO_2. These layers isolate the device from the silicon substrate. Consequently, they are useful for radiation hardening since the insulating layer prevents electron-hole pairs that are created by ionizing radiation anywhere in the thick Si substrate from entering the thin, active, device region at the substrate surface.

2.4.3 Tribological Applications. Materials selected for particular applications on account of such properties as strength, density, etc., often do not possess similarly attractive surface properties, such as oxidation resistance, wear resistance, or toughness. Ion implantation can be used to modify these surface properties without affecting the bulk material. Moreover, implantation has the advantages over surface plating and in-diffusion in that any material can be treated, any ion can be implanted, and discrete interfaces between the surface alloy and bulk material, which are susceptible to spalling, are avoided. One success of this method is illustrated in Fig. 10; it shows the improvement in sliding friction and wear resistance in 304 stainless steel resulting from implantation with Ti ions. The improved properties were attributed to the formation of a thin amorphous layer at the surface of the steel. Nitrogen implantation has been found effective for these purposes in a number of other metals, and because of its lower cost to implant, it is more commonly used than Ti (Rehn *et al.*, 1987).

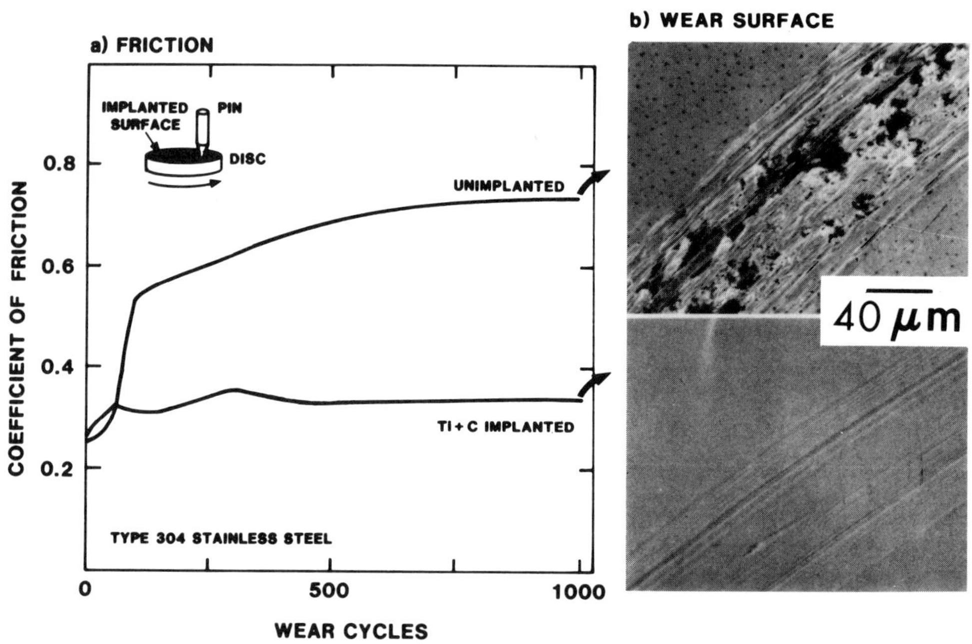

FIG. 10. Improvement of wear resistance in 304 stainless steel due to implantation with Ti and C. Doses of 2×10^{17} cm^{-2} 90–180-keV Ti and 2×10^{17} cm^{-2} 50-keV C were employed. (a) Coefficient of friction versus time (wear cycles); (b) scanning electron micrograph of the wear tracks. (After Follstaedt *et al.*, 1989.)

3. ION-BEAM MIXING

The large number of ions set into motion in a displacement cascade gives rise to a "mixing" of atoms in the solid. Three distinct mechanisms of mixing can be distinguished: recoil implantation, cascade mixing, and thermal spike mixing; these are schematically illustrated in Fig. 11.

3.1 Mechanisms of Ion-Beam Mixing

3.1.1 Recoil Implantation. The first target atom impacted by an implanted ion necessarily recoils in the forward direction, deeper into the solid. This process of recoil implantation can be important during ion implantation. For example, oxygen can be driven from a thin surface layer deeper into a target and degrade electrical properties of thin devices. Or, the stoichiometry of compound targets can be altered by preferential recoil implantation of one of the components. This is most significant in targets like InP or $LiNbO_3$ where the atomic numbers (and collision cross sections) of the components are substantially different. For a thin surface layer, the distribution of recoil implantation atoms can be approximated by the expression

$$S(x)=N^{2/3}\int dT\,\frac{d\sigma(E,T)}{dT}\,F(x-x',\theta), \qquad (10)$$

where $F(x-x',\theta)$, the range distribution of a recoil atom, can be obtained from Eq. (7) with a correction for scattering angle θ relative to the surface normal.

3.1.2 Cascade Mixing. After several collisions, the momenta of the implanted ion and recoil atoms become randomized so that the relocation of most displaced atoms is a stochastic process. Calculations of the mixing due to the random relocation of atoms is far more difficult than those for recoil mixing as they require solving the Boltzmann transport equation (Sigmund and Gras-Marti, 1981). Accurate results, however, can be obtained more simply using computer simulations based on TRIM. The parameter Q is often employed for quantifying ion-beam mixing; it is defined as

$$Q=\langle\Delta x^2\rangle/2\phi F_D, \qquad (11)$$

where $\langle\Delta x^2\rangle$ is the variance of the mixing profile, ϕ is the ion fluence, and F_D is the deposited damage energy per unit length normal to the surface. The parameter Q is a measure of mixing per unit damage energy per unit volume and thus is normalized for all ion–solid combinations. Calculated values for the mixing parameter are in the range $Q=1$–7 $\text{Å}^5/\text{eV}$. Table 3 shows measured values of Q. It is observed that the mixing parameter is much larger than the values calculated by cascade

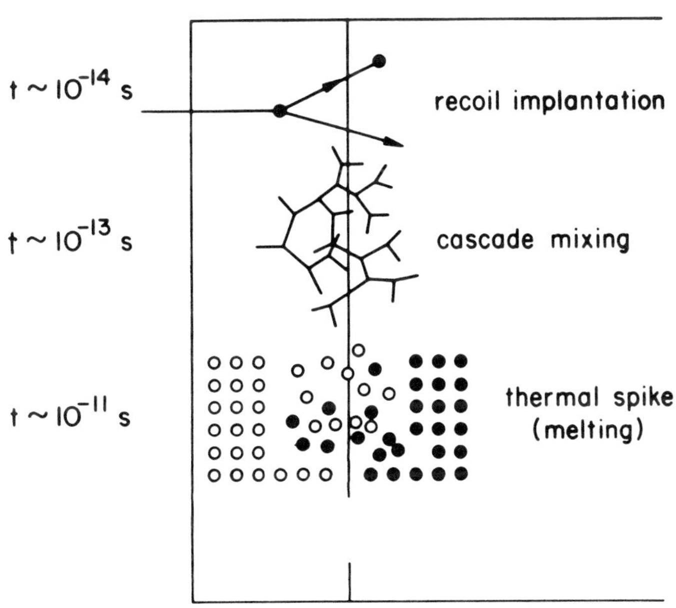

FIG. 11. Schematic illustration of the three contributions to ion-beam mixing: recoil implantation, cascade mixing, and thermal spike mixing.

Table 3. Mixing parameter Q in several metals.

Metal	Q (Å^5/eV)
Al	20
Ti	4
Ni	7
Cu	25
Ag	75
Pd	12
Au	120
Pt	19

mixing or recoil implantation for many metals and semiconductors.

3.1.3 Thermal-Spike Mixing. The possibility of local melting in a cascade provides another contribution to atomic mixing. Calculations of thermal-spike mixing have been performed using the equation

$$\langle \Delta x^2 \rangle = 2 \int \frac{d\sigma(E,T)}{dT}\, dT \int NV(T,t) D(T,t)\, dt, \tag{12}$$

where $V(T,t)$ and $D(T,t)$ are the instantaneous volume and diffusion coefficient of the hot cascade. (The outer integral sums over all the possible recoils that an ion makes with target atoms.) Since the diffusion coefficient of liquids is much larger than that of solids, most of the thermal-spike mixing takes place in the melt. Uncertainties in the cooling rates and appropriate diffusion coefficients make Eq. (12) difficult to evaluate accurately, but it nevertheless shows that mixing in the thermal spike is large in metals that have low melting temperatures and high energy densities (high atomic number). Values of the mixing parameter listed in Table 3 are consistent with these predictions. Johnson *et al.* (1985) have shown that because the characteristic energy of thermal-spike mixing is some tenths of electron volts, in contrast to the much higher energies of the implanted ion and recoil and cascade mixing, thermodynamics can sometimes play a role. This is a fundamentally new concept in the theory of ion–solid interactions, and it has helped significantly in understanding many details of ion-beam modifications of materials.

Calculations of the contributions to ion-beam mixing from thermal spikes cannot be obtained from simulation programs like TRIM, since these codes assume independent, two-body collisions and low densities of moving atoms. For thermal phenomena, however, the collective motion of atoms is vital. Recent studies of ion-beam mixing using molecular-dynamics computer simulations are beginning to clarify this process in metals (see Averback *et al.*, 1991), but the role of thermal spikes in ion-beam mixing in semiconductors and insulators is still poorly understood.

3.1.4 Phase Formation During Ion-Beam Mixing. The large number of atoms relocated during ion irradiation makes it possible to fabricate surface alloys and compounds far more efficiently than by direct ion implantation. The production of a 100-Å contact layer of Ni_2Si by direct implantation of Ni into Si requires an ion dose of over 2×10^{16} cm^{-2}. In comparison, a dose of only $\approx 1\times10^{15}$ cm^{-2} Xe ions is needed to mix a Ni film with a Si layer to produce a film of similar thickness. This technique has been applied to produce Ohmic contacts on semiconductor devices as well as surface alloys on metals for tribology applications.

An example of the use of ion-beam mixing to create metastable phases with specialized properties is illustrated by the mixing of a Cu/Bi diffusion couple. Cu and Bi are virtually immiscible in the solid state, so that CuBi alloys cannot be produced by conventional methods. Irradiation at low temperature, however, causes their intermixing and the formation of an amorphous phase as shown by the diffraction patterns in Fig. 12. The amorphous Cu-Bi phase has the interesting property of being superconducting at low temperatures, even though neither Cu nor Bi is itself a superconductor. On warming, however, the constituent metals precipitate and the superconductivity is lost.

3.2 Ion-Beam–Assisted Deposition

A variation of the ion-beam mixing process of growing technological significance is ion-beam–assisted deposition (IBAD), for which a low-energy ion beam is focused onto the specimen surface during the growth of a thin film. IBAD is employed for two distinct purposes:

1. creating a mixed interfacial layer between the substrate and the thin film to improve film adhesion;
2. improving the quality, stoichiometry, or

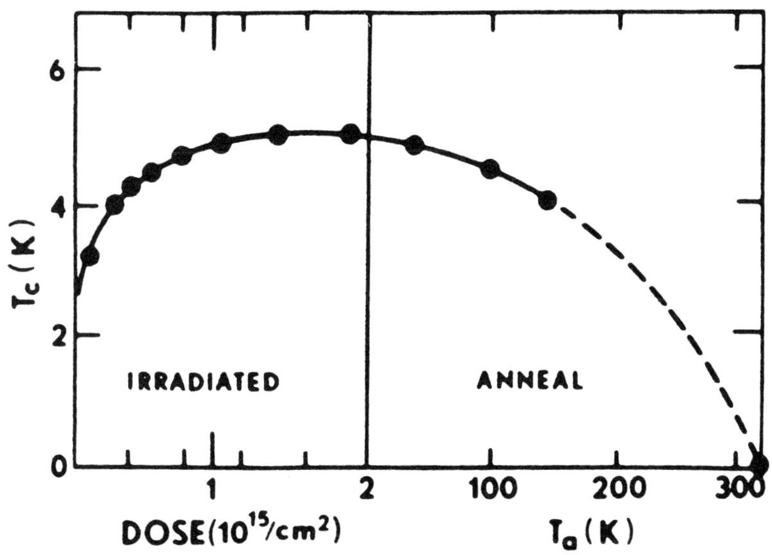

(a)

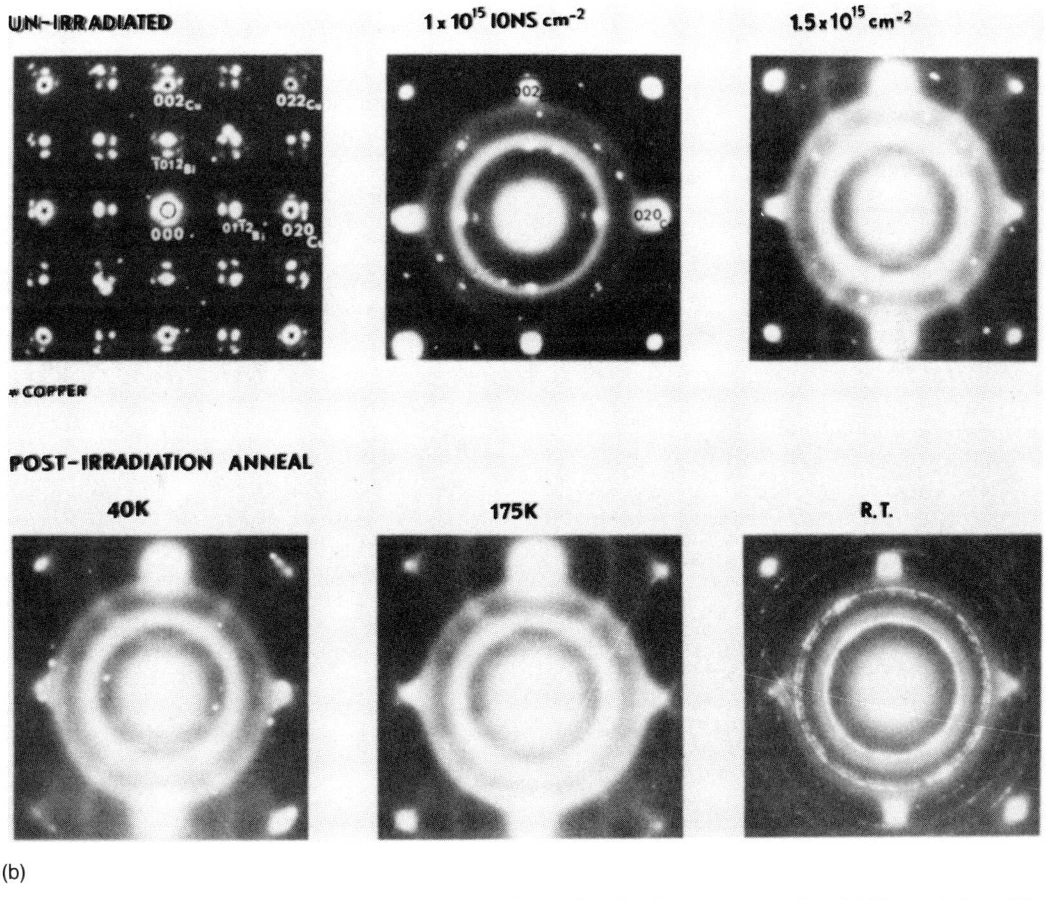

(b)

FIG. 12. Ion-beam mixing of Cu/Bi produces an amorphous alloy that is superconducting. (a) The variation of the critical temperature with 380-keV Kr fluence and annealing temperature; (b) electron diffraction patterns illustrating the amorphization during irradiation and subsequent crystallization during annealing. (After Averback *et al.*, 1985.)

epitaxy of the growing thin film.

A variation of the second process is direct ion-beam deposition (IBD), for which a high-intensity, low-energy ion-beam source is employed but without simultaneous vapor deposition. Two examples illustrate these purposes.

3.2.1 Dopant Incorporation. The synthesis of devices by thin-film deposition, e.g., molecular-beam epitaxy (MBE), is often limited by the difficulty of incorporating electrically active impurities in the growing film. As, Sb, In, and Ga are difficult to incorporate into Si, while Zn, Cd, Mn, S, and Te are difficult to incorporate into GaAs. The use of accelerated dopant beams during growth is proving to be a practical solution to this problem (Greene *et al.*, 1989). Shown in Fig. 13 is the incorporation probability of In in a growing Si film, plotted as a function of inverse temperature, for various energies of the In beam (Hasan *et al.*, 1989). The incorporation rate is increased over five orders of magnitude when the energy of the In beam is increased from thermal energies to over 50 eV. In addition, the incorporation was near unity for beam energies over 200 eV for growth temperatures up to 850 °C. These investigations also showed that the In atoms were located on substitutional sites and that they were electrically active for concentrations as high as 2×10^{18} cm^{-3}, which is well above the solubility limit. No evidence of radiation damage was observed in these experiments, and carrier (hole) mobilities were higher than those obtained using implantation and high-temperature annealing (Hasan *et al.*, 1989).

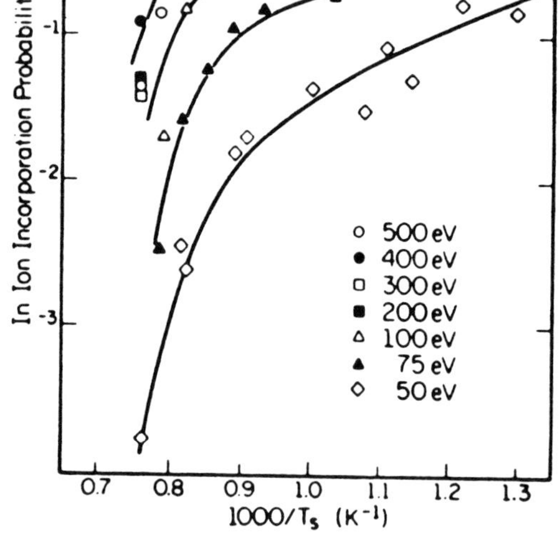

FIG. 13. The incorporation probability of In ions in MBE-grown Si as a function of the In energy and substrate temperature (after Hasan *et al.*, 1989).

3.3.2 Improved Adhesion. The fabrication of Schottky contacts that have good thermal stability, interface integrity, and adhesive strength on semiconductor devices poses several problems. In principle, Au should make an excellent Schottky barrier on GaAs; however, conventional deposition of Au on GaAs does not provide adequate bonding. High-temperature anneals improve the bonding, but they also cause interdiffusion and surface roughening. Good adhesion strength has been obtained, however, by directing a 300-eV Ar beam onto the GaAs surface during the deposition of Au, as shown in Fig. 14 (Kellock *et al.*, 1991). Here, the peel strength of the Au film, i.e., the force needed to peel the film from the substrate, is shown as a function of the Ar:Au arrival rate for several substrate temperatures. Far better adhesion is observed in the film prepared by beam-assisted growth. These films were stable at least up to 250 °C; in addition, they were all Schottky type with an ideality factor of 1.3. The physical details of how the ion beam improves the bonding in this system is not yet clear; however, it was suggested that the Ar bombardment liberates As from the near surface and replaces it with Au, thus leading to the formation of a complex Au-Ga-As phase (Kellock *et al.*, 1990).

4. IMPLANTATION AT ELEVATED TEMPERATURE

4.1 Theory of Radiation-Enhanced Diffusion

It is well established that diffusion in solids involves the motion of point defects—vacancies and interstitials. This is expressed through the equation

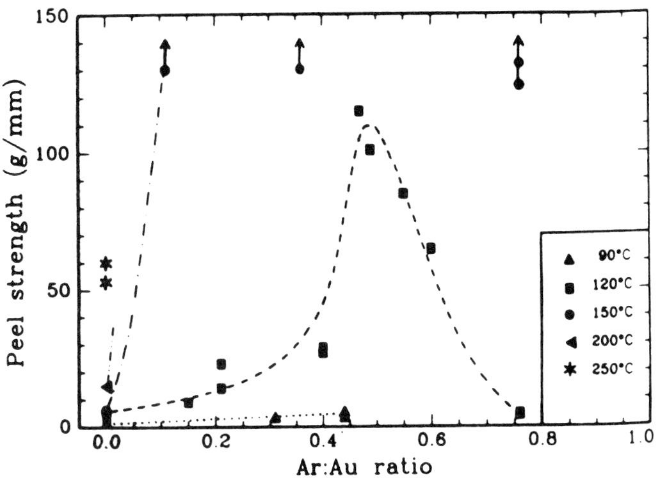

FIG. 14. The peel strength of a Au film vapor deposited on GaAs plotted as functions of the ratio of arrival of 300-eV Ar ions and Au atoms and temperature (after Kellock *et al.*, 1991).

$$D = D_i C_i + D_v C_v \qquad (13)$$

where D_j and C_j are diffusivities and concentrations of vacancies and interstitials. The diffusivities have the form $D_j = D_{0j} \exp\{\Delta H_j / kT\}$, and ΔH_j is the activation enthalpy for defect migration; D_{0j} is a preexponential, or entropic, factor. In absence of irradiation, the concentrations of defects refer to their values in thermal equilibrium. However, during implantation, the displacement processes described above introduce large supersaturations of point defects, both interstitials and vacancies, and according to Eq. (13), they enhance diffusion. The instantaneous concentration of defects can be calculated using chemical rate theory (see, e.g., Sizmann, 1978). The diffusion coefficient that derives from this theory is plotted in Fig. 15 as a function of inverse temperature. At low temperatures, the defects are nearly immobile so that the radiation-enhanced diffusion (RED) becomes smaller than ion-beam mixing. At very high temperatures, normal thermal diffusion prevails. At intermediate temperatures, however, diffusion is controlled by RED. The temperature dependence of RED varies depending on a number of factors: temperature, ion flux, and the perfection of the crystal. The important point, however, is that RED occurs at substantially lower temperatures than normal thermal diffusion. Moreover, it is distinguished from ion-beam mixing in that

1. it can be orders of magnitude larger than ion beam mixing;
2. it is a "near" equilibrium process. This means that defects tend to anneal out during the irradiation and that equilibrium, rather than metastable, phases are usually produced.

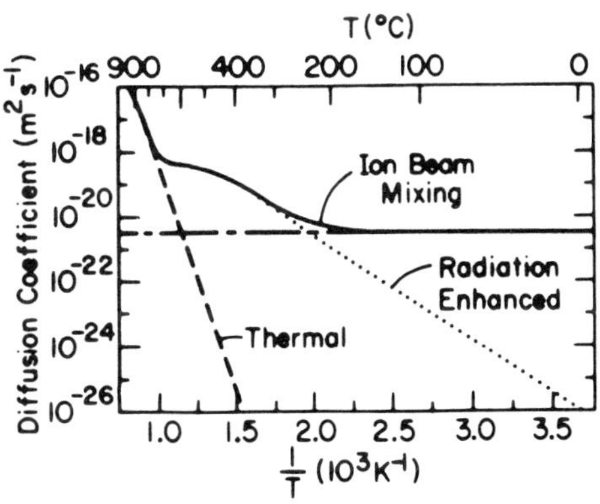

FIG. 15. Radiation-enhanced diffusion coefficient as a function of inverse temperature.

4.2 Examples of Radiation-Enhanced Diffusion

It may be expected from the current discussion that implantation at elevated temperatures may result in defect-free material and activated impurities in semiconductor devices. In some cases, doping of semiconductors is performed at elevated temperatures, but because of the enhanced diffusion of the impurity, precipitation, and the buildup of complex damage structures, this method is not widely used. Rather, rapid thermal annealing is employed to remove defects in implanted semiconductors. A more promising application of high-temperature implantation is the synthesis of buried $CoSi_2$ in Si discussed in Sec. 2.4.2. To produce good mesoepitaxial layers, it is imperative that the Si substrate remain crystalline during the Co implantation, otherwise the coherence between the silicide and Si substrate would be lost. Consequently, the Co is implanted at a temperature above which amorphous Si crystallizes, $\approx$300 °C. In addition, RED provides sufficient atomic mobility for the implanted Co to precipitate as an epitaxial layer of $CoSi_2$. The quality of these buried silicide layers is illustrated in Figs. 16 and 17. In Fig. 16(a), the cross-sectional transmission electron microscope image shows the flatness of the buried layer, while Fig. 16(b) reveals the atomic perfection at the interfaces. Figure 17 shows the effect of substrate temperature on crystal perfection. Here channeling of He ions is employed as a measure of the crystal quality (low yields implying little damage). For implantation below $\approx$300 °C the surface region becomes polycrystalline, even after high-temperature annealing. At too-high implantation temperatures, the crystal quality also diminishes since the $CoSi_2$ precipitates grow too rapidly and become three dimensional, rather than two-dimensional, flat films. Thus, careful control of RED is critical to this emerging technology.

A second example of RED is the formation of $NbSi_2$ using Ar irradiation at high temperatures. In Fig. 18 it is shown that $NbSi_2$ can be efficiently formed by ion-beam mixing at temperatures less than $\approx$300 °C. The formation of

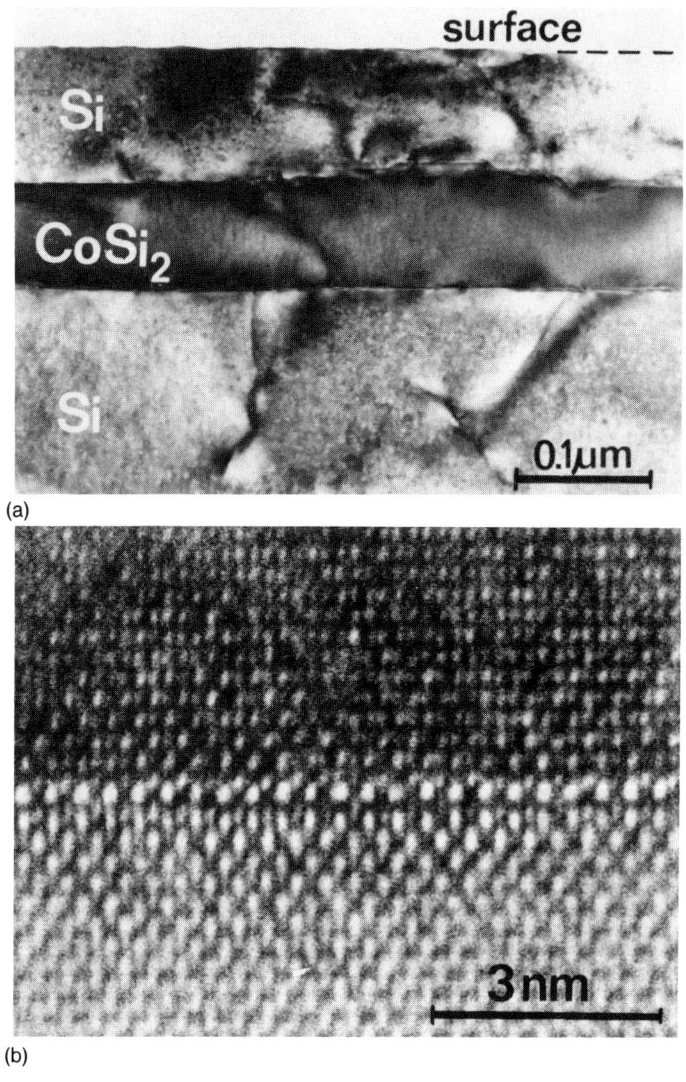

FIG. 16. Perfection of mesoepitaxial layer of $CoSi_2$ in Si. (a) Cross-sectional TEM image showing the flatness of the buried layer; (b) high-resolution TEM image showing the atomic perfection of the interfaces between the $CoSi_2$ and Si interfaces. (After Werner and Jaeger, 1993.)

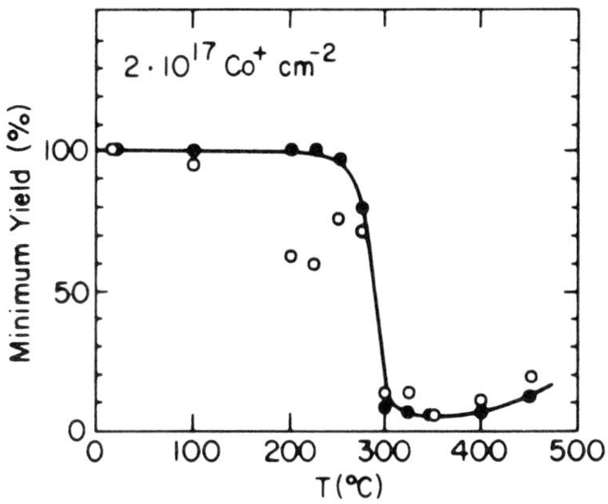

FIG. 17. The effect of implantation temperature on mesoepitaxial perfection. The quality of the film is characterized here by ion channeling measurements. A large yield of backscattered ions means that there is a large number of crystalline imperfections blocking the channels. (After Mantl, 1992.)

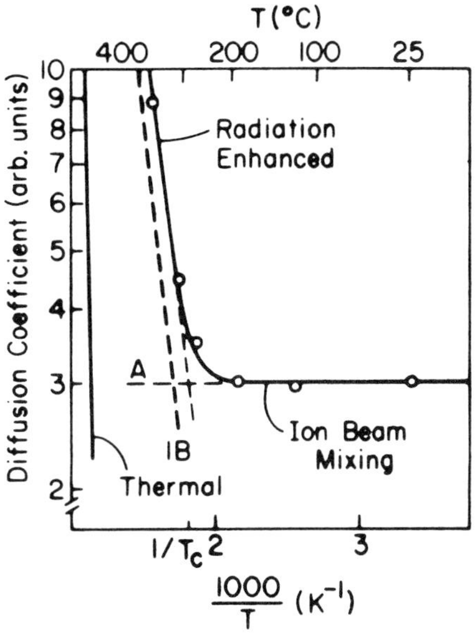

FIG. 18. Logarithmic plot of the intermixing of Nb and Si films to form $NbSi_2$ during Si irradiation as a function of inverse temperature. At low temperatures the intermixing is due to ion-beam mixing and is independent of temperature. At higher temperatures, the mixing is due to radiation-enhanced diffusion. Thermal diffusion occurs at much higher temperatures. (After Matteson *et al.*, 1979.)

$NbSi_2$ by thermal annealing, however, does not take place until $\approx$450 °C. Ion-beam mixing of Nb and Si at low temperatures requires much higher fluences, and it does not yield the $NbSi_2$ phase.

5. SUMMARY

In recent years, ion-beam modification of materials has become an important processing step for a number of advanced materials. This development has become possible after many years of fundamental research on the interaction of energetic particles with solids and defects in crystals. Computer simulation programs are now readily available for accurately calculating the depth distributions of implanted ions, even in complex substrates. These programs can also reliably calculate sputtering yields and defect production. The damage created by the implantation is also becoming better understood, and along with this better understanding, methods to eliminate it, or to exploit it, are being developed. In this article, the fundamental processes involved in ion-beam modification were presented. Several applications were described that illustrate how various aspects of implantation can be utilized. These ranged from the doping of semiconductors, to ion-beam–assisted growth of thin films, to the formation of metastable phases at metal and ceramic surfaces for wear and oxidation resistance. In this age of advanced materials, where processing must operate on the submicron level, ion-beam modification of materials is certain to play an ever increasing role.

GLOSSARY

Channeling: The guidance of implanted atoms by lattice atoms to travel in the open spaces of axial or planar channels in a crystalline target.

Compositional Disordering: The intermixing of compositionally modulated structures.

Damage Energy: The fraction of an implanted ion's energy that is utilized in displacing atoms from their lattice sites. It is approximately the energy of the implantation particle less that dissipated in electronic excitations.

Displacement Energy: The energy required to displace an atom from its lattice site and create a Frenkel pair.

Frenkel Pair: An interstitial atom and its associated vacancy.

Ion-Beam–Assisted Deposition: The deposition of a thin film onto a surface while simultaneously bombarding the surface with ions.

Ion-Beam Mixing: The athermal rearrangement of atoms during ion irradiation. It is often associated with the intermixing of contiguous thin-film layers by ion irradiation.

Mesoepitaxy: The formation of a subsurface, coherent layer by ion implantation.

Radiation-Enhanced Diffusion: The enhancement of diffusion in a material at elevated temperatures by using irradiation to produce point defects.

Range Distribution: The concentration profile of an implanted species as a function of distance beneath the surface.

Recoil Implantation: The forward recoil of substrate atoms deeper into a substrate during ion implantation.

Replacement Collision Sequence: A linear chain of collisions along the close-packed directions of a lattice whereby one atom collides with and replaces its neighbor.

Sputtering: The erosion of a surface during ion irradiation. The sputtering yield is the number of target atoms ejected from the surface per incident ion.

Stopping Power: The loss of energy of an implanted ion per unit distance along its trajectory in a material.

Straggling: A measure of the variance of a range distribution. The straggling normal to the surface is the ratio of the square root of the variance to the mean range. The transverse straggling is the ratio of the variances in directions parallel and perpendicular to the surface.

Thermal Spike: Local, transient heating of a material around the trajectory of an implanted ion.

TRIM (Transport of Ions in Matter): A popular computer code for generating range damage and ionization profiles for ion implantation.

Lists of Work Cited

Averback, R. S., Okamoto, P. R., Baily, A. C., Strizker, B. (1985), *Nucl. Instrum. Methods. B* **7/8**, 556–560.

Averback, R. S., Diaz de la Rubia, T., Hsieh, H., Benedek, R. (1991), *Nucl. Instrum. Methods B* **59/60**, 709–717.

Biersack, J. P., Haggmark, L. G. (1990), *Nucl. Instrum. Methods* **174**, 257–269.

Csepregi, L., Kennedy, E. F., Mayer, J. W., Sigmon, T. W. (1978), *J. Appl. Phys.* **49**, 3906–3911.

Davies, J. A. (1992), *MRS Bull.* **17**, 26–29.

Firsov, O. B. (1959), *Zh. Eksp. Teor. Fiz.* **36**, 1517–1523 [*Sov. Phys. JETP* **36**, 1076–1080].

Follstaedt, D. M., Knapp, J. A., Pope, L. E. (1989), in; L. E. Rehn, J. Greene, F.A. Smidt (Eds.), *Processing and Characterization of Materials Using Ion Beams*, MRS Symposium Proceedings Vol. 128, Pittsburgh: Materials Research Society.

Ghaly, M., Averback, R. S. (1993) (unpublished).

Gibson, J. B., Goland, A. N. Milgram, M., Vineyard, G. H. (1960), *Phys. Rev.* **120**, 1229–1253.

Greene, J. E., Barnett, S. A., Sundgren, J. E., Rockett, A. (1989), in: T. Itoh (Ed.), *Ion Assisted Film Growth*, Amsterdam, Elsevier, p. 101.

Hasan, M. A., Knall, J., Barnett, S. A., Sundgren, J. E., Markert, L. C., Rockett, A., Greene, J. E. (1989), *J. Appl. Phys.* **65**, 172–179.

Johnson, W. L., Chang, Y. T., Van Rossum, M., Nicolet, M.-A. (1985), *Nucl. Instrum. Methods B* **7/8**, 657–665.

Kellock, A. J., Baglin, J. E. E., Bardin, T. T., Pronko, J. G. (1991), *Nucl. Instrum. Methods B* **59/60**, 249–253.

Kinchin G. H., Pease, R. S. (1955), *Rep. Prog. Phys.* **18**, 1–51.

Lindhard, J. L., Scharff, M., Schiøtt, H. E. (1963), *K. Dan. Videnskab. Selskab., Mat. Fys. Medd.* **33**, No. 10.

Mantl, S. (1992), *Mat. Sci. Rep.* **8**, 1–95.

Matteson, S., Roth, J., Nicolet, M.-A. (1979), *Radiat. Eff.* **42**, 217–226.

Piercy, G. R., Brown, F., Davies, J. A., McCargo, M. (1963), *Phys. Rev. Lett.* **10**, 399–400.

Rehn, L. E., Picraux, S. T., Wiedersich, H. (1987), *Surface Alloying by Ion, Electron, and Laser Beams*, Metals Park, OH: American Society for Metals.

Roas, B., Hensel, B., Henke, S., Klaumunzer, S., Kabius, B., Watanabe, W., Saemann-Ischenko, G., Schultz, L., Urban, K. (1990), *Europhys. Lett.* **11**, 669–674.

Robinson, M. T., Oen, O. S. (1963), *Phys. Rev.* **132**, 2385–2398.

Robinson, M. T., Torrens, I. M. (1974), *Phys. Rev. B* **9**, 5008–5024.

Ryssel, H., Ruge, I. (1986), *Ion Implantation*, Chichester: Wiley, p. 316.

Schiøtt, H. E. (1970), *Radiat. Eff.* **6**, 107–113.

Sigmund, P. (1969), *Phys. Rev.* **184**, 383–416.

Sigmund, P., Gras Marti, A. (1981), *Nucl. Instrum. Methods* **182/183**, 25–41.

Sizmann, R. (1978), *J. Nucl. Mater.* **69/70**, 386–412.

Spohr, R. (1990), *Ion Tracks and Microtechnology: Principles and Applications*, Braunschweig: Vieweg.

Stark, J. (1912), *Phys. Z.* **13**, 973–977.

Tamura, M. (1991), *Mat. Sci. Rep.* **6**, 1–214.

Townsend, P. D. (1987), *Rep. Prog. Phys.* **50**, 501–558.

Wei, C. Y., Current, M. I., Seidmann, D. N. (1981), *Philos. Mag.* **A44**, 459–491.

Werner, P., Jaeger, W. J. (1993), *J. Appl. Phys.* (to be published).

Winterbon, K. B. (1975), *Ion Implantation Range and Energy Deposition Distributions*, Vol. 2, New York: IFI/Plenum

Ziegler, J. F., Biersack, J. P., Littmark, U. (1985), *The Stopping and Range of Ions in Solids*, Vol. 1, New York: Pergamon.

Further Reading

The following books provide both a background to the theory of ion–solid interactions and many detailed examples:

Auciello, O., Kelly, R. (Eds.) (1984), *Ion Beam Modifications of Materials, Ion Beam Modifications of Surfaces*, Amsterdam: Elsevier. Sputtering applications.

Itoh, T. (Ed.) (1984), *Ion Beam Assisted Deposition*, Amsterdam: Elsevier. Beam-assisted deposition.

Ryssel, H., Ruge, I. (1986), *Ion Implantation*, Chichester: Wiley. Doping of semiconductors.

Townsend, P. D., Kelly, J. C., Hartley, N. E. W. (1976), *Ion Implantation, Sputtering and their Applications*, London: Academic. Sputtering applications.

Several international conference series are dedicated to ion-beam modifications of materials: International Conference on Ion-Beam Modification of Materials, Ion-Beam Analysis, Atomic Collisions in Solids, Ion-Implantation Technology. All meet biannually and their proceedings are published in *Nuclear Instruments and Methods, Part B.* In addition, several symposia of the Materials Research Society focus on ion-beam modification of materials, and their proceedings are published in the MRS Symposium Proceedings.

ION AND ELECTRON SOURCES

See SOURCES, ION AND ELECTRON

ION SPECTROMETERS

See SPECTROMETERS, ELECTRON AND ION

IONIC CONDUCTION AND DIFFUSION IN SOLIDS

A. V. CHADWICK, *Chemical Laboratory, University of Kent, Canterbury, Kent, United Kingdom*

	Introduction: General Background to Diffusion in Solids.	
1.	**Theory of Solid-State Diffusion**	?
1.1	Phenomenological Theory of Diffusion	194
1.1.1	Classical Approach	194
1.1.2	Irreversible Thermodynamics	195
1.2	Atomistic Theory of Solid-State Diffusion	196
1.2.1	Introduction to Point Defects	196
1.2.2	Thermodynamics of Point Defects in a Monatomic Solid	197
1.2.3	Point-Defect Thermodynamics in an Ionic Solid	199
1.2.4	Point-Defect Concentrations in Solids	202
1.2.5	Mechanisms of Diffusion	202
1.2.6	Random-Walk Theory	203
1.2.7	Relations between Macroscopic and Microscopic Parameters	204
1.2.8	Correlation Effects in Diffusion	204
2.	**Experimental Methods of Studying Diffusion**	205
2.1	Methods of Measuring Diffusion Coefficients	205
2.1.1	Study of the Diffusion Penetration Profile	205
2.1.2	Measurement of the Atomic Jump Frequency	206
2.2	Ionic Conductivity Measurements	207
2.3	Methods of Identifying Diffusion Mechanisms	211
2.4	Computational Techniques	212
3.	**Diffusion in Different Solid Types**	213
3.1	Metals and Alloys	213
3.2	Molecular Solids	214
3.2.1	Rare-Gas Crystals	214
3.2.2	Plastic Crystals	215
3.2.3	Nonplastic Crystals	215
3.3	Covalent Solids	215
3.4	Ionic Solids	215
3.4.1	Normal Ionic Solids	216
3.4.2	Fast-Ion Conductors	217
4.	**Diffusion in an Applied Field**	218
5.	**Summary**	219
	Glossary	219
	Works Cited	220
	Further Reading	221

INTRODUCTION: GENERAL BACKGROUND TO DIFFUSION IN SOLIDS

The classical "textbook" picture of a solid is that the atoms in a solid stay located on their lattice sites, and the only motions that are present are the lattice vibrations. In fact, atoms in a real solid are continually changing their positions, and at temperatures close to the melting point an atom will change site about a million times per second. The study of these translational atomic movements, *diffusion*, in solids is a mature field of chemical physics with a long and rich history. In his pioneering studies of the electrical behavior of materials in the 1830s, Michael Faraday measured the conductivity of a number of ionic solids. It was almost a century ago that Sir William Roberts-Austen (a student of Thomas Graham, who formulated the laws of diffusion in gases) made the first diffusion measurement on a metallic system, gold in lead. Since those early days there has been a steady growth of interest in all aspects of diffusion in solids. From the viewpoint of basic science, the ability to model diffusion is a severe test of the theoretical concepts of

1-56081-067-X/94/$5.00+.50

physics, particularly the understanding of the interatomic potentials, and there will be a continual striving to predict accurately the diffusion parameters of solids. A vast number of technological processes involve diffusion phenomena, including the manufacture of semiconductor devices by doping with impurities, sintering, plastic deformation, and corrosion. It is therefore important that physical scientists understand the basic phenomena and theories of solid-state diffusion and ionic conductivity. The objective of this article is to present an overview of the current state of knowledge that emphasizes the principal features. Readers wishing more detail are referred to the extensive bibliography at the end of the chapter, which will guide them to several authoritative texts and articles on specific topics. The focus is on diffusion in crystalline solids where the basic phenomena and ideas are now well established. Although there have been extensive studies of amorphous materials and glasses, the theory of diffusion in these systems is still in a state of flux (Greaves, 1990).

The first section treats the basic theories of diffusion in solids, both the macroscopic, phenomenological aspects and the atomistic mechanisms. The next section covers the variety of experimental techniques. A special emphasis is given to the measurement of conductivity of ionic solids, as this has proved particularly fruitful in the development of many basic concepts of solid-state diffusion. Also included in this section is a brief treatment of the methods of computer simulation, a technique that is proving invaluable in understanding the details of the diffusion processes. The third section reviews the current knowledge of the diffusion processes in the different solid types. A final section briefly introduces the difficult, but important, topic of diffusion in an applied field, i.e., a gradient of temperature (*thermomigration*) or of electric field (*electromigration*).

1. THEORY OF SOLID-STATE DIFFUSION

1.1 Phenomenological Theory of Diffusion

1.1.1 Classical Approach. Translational diffusion in classical physics is governed by *Fick's laws*. In a one-dimensional system the flux of a given type of particle, i.e., the number passing through unit cross-sectional area in unit time, can be expressed as

$$J=-D\left(\frac{\partial c}{\partial x}\right), \tag{1}$$

where c is the number concentration of particles, $\partial c/\partial x$ is the gradient of the concentration of particles along the x direction, and D is the diffusion coefficient. A dimensional analysis shows that D has units of L^2T^{-1} and is usually reported as $cm^2\ s^{-1}$ or $m^2\ s^{-1}$. Equation (1) is *Fick's first law*. In three dimensions this can be written as

$$\mathbf{J}=-\mathbf{D}\cdot\nabla c, \tag{2}$$

where D is now a second-rank tensor.

Fick's first law is analogous to Fourier's law of heat flow and Ohm's law and is a specific statement of the more general law of physics that a flow is directly proportional to a driving force, the driving force being a gradient of "potential." In the form of Eqs. (1) and (2), Fick's first law is not particularly useful to the experimentalist, as the evaluation of D would involve the measurement of a flux and a concentration gradient, both of which could be time dependent. If the particles are conserved, then the equation of material balance will apply, and therefore a further condition must be obeyed, i.e.,

$$\frac{\partial J}{\partial x}=-\frac{\partial c}{\partial t}\,. \tag{3}$$

This continuity equation is simply derived by considering the net flow into and out of a small volume along the diffusion axis. Combining Eqs. (1) and (3) yields

$$\frac{\partial c}{\partial t}=\frac{\partial}{\partial x}\left(D\,\frac{\partial c}{\partial x}\right). \tag{4}$$

If D is independent of concentration, then Eq. (4) reduces to

$$\frac{\partial c}{\partial t}=D\,\frac{\partial^2 c}{\partial x^2}\,, \tag{5}$$

which is often referred to as *Fick's second law*.

Equation (5) is a second-order differential equation to which a variety of solutions can be derived to correspond with the different experimental situations, i.e., different boundary conditions. These can be found in the standard texts (e.g., Crank, 1975), and the only solution quoted here is that for the initial condition of an *infinitely* thin source on the surface of an *infinite* sample. This geometry is commonly used in tracer diffusion experiments, and the appropriate solution to Eq. (5) is

$$c(x,t)=\frac{A}{\sqrt{\pi Dt}}\exp\left(-\frac{x^2}{4Dt}\right). \tag{6}$$

Here $c(x,t)$ is the concentration of particles at a penetration depth x into the surface after time t, and A is the initial concentration of particles on unit area of the surface. Thus the penetration profile, the plot of $c(x,t)$ vs x, will have a Gaussian shape, and the diffusion coefficient can be determined from the plot of $\ln c(x,t)$ vs x^2; i.e., the slope is $-1/4Dt$.

External forces acting on the system will perturb the diffusion, and intuitively Eq. (1) may be modified by including an additional flux. This term is referred to as a *drift* and can be simply equated to the concentration of particles multiplied by an average drift velocity, $\bar{v}$, imparted by the force. Thus the resulting total flux can be expressed as

$$J=-D\left(\frac{\partial c}{\partial x}\right)+\bar{v}c. \tag{7}$$

Examples of external driving forces are gradients of electrical potential, temperature, and stress. For particles of charge q in a gradient of electric potential E $(=-d\Phi/dx)$, the drift velocity is given by

$$\bar{v}=\mu E, \tag{8}$$

where μ is the mobility of the charged particle, i.e., the drift velocity in unit potential gradient. If the system is in a steady state with no net flux, then from Eq. (7)

$$D\left(\frac{\partial c}{\partial x}\right)=\bar{v}c=\mu Ec. \tag{9}$$

Assuming thermodynamic equilibrium in the system, the concentration of particles will obey Boltzmann statistics, and hence

$$c(x)=a\exp\left(\frac{-q\Phi(x)}{kT}\right), \tag{10}$$

where a is a constant, k is the Boltzmann constant, and T is the temperature. Differentiating Eq. (10) with respect to x yields

$$\frac{\delta c}{\delta x}=\frac{-cq}{kT}\frac{d\Phi}{dx}=\frac{cqE}{kT}. \tag{11}$$

Combining Eqs. (9) and (11) gives the result

$$\mu/D=q/kT\,; \tag{12}$$

and since the conductivity σ is defined by

$$\sigma=cq\mu, \tag{13}$$

the final result relating D and σ is

$$\sigma=cq^2D/kT\,. \tag{14}$$

Equation (14) is the *Nernst–Einstein equation* and is important in the study of atomic transport in ionic solids, as it provides a quantitative link between diffusion and conductivity experiments. This is discussed later in greater detail.

The treatment of diffusion in a temperature gradient, where the driving force is dT/dx, is not so straightforward. In the format of Eq. (7) the flux is given in this case by

$$J=-D\left(\frac{\partial c}{\partial x}\right)-\frac{cQ^{*\prime}}{kT^2}\frac{dT}{dx}. \tag{15}$$

Here $Q^{*\prime}$ is the reduced heat of transport and has the units of energy. By convention $Q^{*\prime}$ is positive if the particle moves down the temperature gradient. The meaning of $Q^{*\prime}$ is further explained in the treatment given below.

1.1.2 Irreversible Thermodynamics. The treatment of diffusion theory presented above is based on empirical and intuitive approaches to the phenomenon. A rigorous approach to the problem is possible using *irreversible* (or *nonequilibrium*) *thermodynamics*. Only the basic ideas can be treated here, and for more detail the reader is referred to standard texts (de Groot and Mazur, 1962; Prigogine, 1967), while the special applications to solid-state diffusion are treated by Howard and Lidiard (1964).

In a system in which diffusion is occurring there will be local fluxes of atoms or species of type i, $\mathbf{J}_i$, and there will be a heat flux $\mathbf{J}_q$. The thermodynamics of irreversible processes relates these fluxes linearly to certain conjugate thermodynamic forces, $\mathbf{X}_i$ and $\mathbf{X}_q$, to give the relations

$$\mathbf{J}_i=\sum_{j=1}^{n} L_{ij}\mathbf{X}_j+L_{iq}\mathbf{X}_q \tag{16}$$

and

$$\mathbf{J}_q=\sum_{j=1}^{n} \mathbf{L}_{qj}\mathbf{X}_j+L_{qq}\mathbf{X}_q. \tag{17}$$

The L_{ij}'s are called the *phenomenological coefficients* and are independent of the forces (but will depend on concentration and temperature).

The conjugate forces are given by

$$\mathbf{X}_i=-T\nabla(\mu_i/T) \tag{18}$$

and

$$\mathbf{X}_q=T\nabla(1/T), \tag{19}$$

where μ_i is the chemical potential of species i.

The diagonal coefficients (L_{ii}) are simply related to atomic mobilities and thermal conductivities. The off-diagonal or cross coefficients express the coupling between the various fluxes and forces. For example, a heat flux will give rise to a diffusional flux and vice versa. An important point in the general theory is the *Onsager reciprocal relations* between the coefficients, i.e.,

$$L_{ij}=L_{ji} \tag{20}$$

and

$$L_{iq}=L_{qi}. \tag{21}$$

It is more convenient to use the *reduced heat flux* $\mathbf{J}'_q$ in which the flux of enthalpy due to the diffusion of species is subtracted from $\mathbf{J}_q$, i.e.,

$$\mathbf{J}'_q=\mathbf{J}_q-\sum_{i=1}^{n} \mathbf{J}_i h_i. \tag{22}$$

Here h_i is the partial specific enthalpy of species i. In choosing the reduced heat flux the conjugate thermodynamic force needs to be redefined as $\mathbf{X}'_i$, given by

$$\mathbf{X}'_j=-(\nabla\mu_i)_T. \tag{23}$$

Thus $\mathbf{X}'_i$ is the gradient of chemical potential produced by the variation of composition in the absence of a temperature gradient.

It is useful to define the *heat of transport* Q^*_i by the relation

$$L_{iq}=\sum_{j=1}^{n} Q^*_j L_{ij}. \tag{24}$$

Thus Eqs. (16) and (17) can be reexpressed as

$$\mathbf{J}_i=\sum_{j=1}^{n} L_{ij}(\mathbf{X}_j+Q^*_j\mathbf{X}_q) \tag{25}$$

and

$$\mathbf{J}_q=\sum_{i=1}^{n} Q^*_i\mathbf{J}_i+\left(L_{qq}-\sum_{j=1}^{n} L_{jq}Q^*_j\right)\mathbf{X}_q, \tag{26}$$

which in the absence of a temperature gradient reduces to

$$\mathbf{J}_q=\sum_{j=1}^{n} Q^*_j\mathbf{J}_j. \tag{27}$$

In the framework of reduced heat flux a *reduced heat of transport* $Q^{*\prime}_j$ is defined by

$$Q^{*\prime}_j=Q^*_j-h_i, \tag{28}$$

and then Eqs. (25) and (27) are

$$\mathbf{J}_i=\sum_{j=1}^{n} L_{ij}(\mathbf{X}'_j+Q^{*\prime}_j\mathbf{X}'_q) \tag{29}$$

and

$$\mathbf{J}'_q=\sum_{j=1}^{n} Q^{*\prime}_j\mathbf{J}_j. \tag{30}$$

Equation (30) provides a physical meaning of the reduced heat of transport as the flux of heat in an isothermal system due to the flux of matter.

It is important to note that the fluxes of all species in a system cannot be independent. Thus measuring the fluxes relative to the lattice frame at rest yields

$$\sum_{i=1}^{n} \mathbf{J}_i=0, \tag{31}$$

so that if there are n species in a system there are only $n-1$ *independent* heats of transport. This is important when considering diffusion due to point defects.

1.2 Atomistic Theory of Solid-State Diffusion

1.2.1 Introduction to Point Defects. All real crystals are imperfect, and they contain a variety of types of defects. These include *bulk defects,* such as grain boundaries and dislocations, which extend over many lattice spacings, *surfaces*, and *point defects*, the latter being at the atomic level. The mechanisms of diffusion in crystalline solids that move an atom from one lattice site to another, *lattice* or *bulk diffusion*, involve the point defects. Diffusion will occur along the other types of defect—in fact it should occur more readily because of the open nature of their structures; however, it will not be covered here. It is more difficult to treat dislocation and surface diffusion in a detailed analysis, as the atomic pathway is less well defined, although there has been some progress in recent years (Kaur and Gust, 1988; King, 1987). In addition, these defects are not thermodynamically inherent in crystals, and their concentrations, and hence the contribution to the total diffusion, will depend on the history of the sample. At high temperatures, lattice diffusion will usually dominate.

In a monatomic solid, there are four basic types of point defect, as illustrated in Fig. 1. In an absolutely pure crystal there may be *vacancies* [Fig. 1(a)], unoccupied lattice sites, and *interstitials* [Fig. 1(b)], atoms on interstitial positions in the lattice. These are sometimes referred to as *intrinsic* or *thermally activated* point defects as they will be present in any crystal above 0 K, arising from the fact that the energy of some atoms in the Boltzmann distribution will be sufficient to cause them to migrate from their normal lattice sites. A real crystal will also contain impurity atoms dissolved in the lattice. There are two types of these *extrinsic* or *impurity point defects: substitutional impurities* [Fig. 1(c)], where the impurity occupies a site normally filled by the host atom, and *interstitial impurities* [Fig. 1(d)], where the impurities occupy interstitial sites in the lattice.

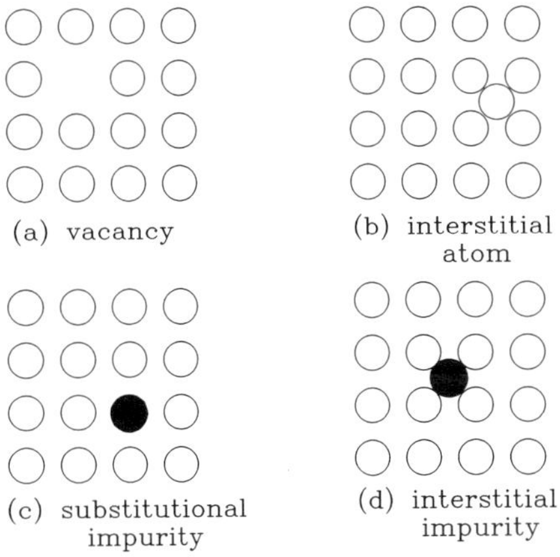

FIG. 1. Schematic representation of the basic point defects in monatomic solids.

In a real crystal, there will be interactions between the point defects, and this can lead to association of the simple defects to form more complex *aggregates* or *clusters*. Obvious examples are the *divacancy* (or vacancy pair), with two vacancies in adjacent lattice sites, and the *impurity–vacancy pair*, a substitutional impurity and vacancy on adjacent lattice sites. The concentration of a particular type of aggregate will depend on temperature, the concentration of the basic component defects, and their interaction energy. The last term will depend on the nature of the intermolecular forces, and the importance of aggregation, as will be seen later, depends very much on the type of solid under consideration.

1.2.2 Thermodynamics of Point Defects in a Monatomic Solid. The simplest case to deal with is a crystal containing vacancies in a single electronic state, and the treatment given below follows that developed in other texts (Howard and Lidiard, 1964; Chadwick and Glyde, 1977). The results for other types of defect will be described as appropriate.

The Gibbs free energy of a perfect crystal containing N atoms at a temperature T and uniform pressure P can be defined as $G_0(T,P)$. If n_v noninteracting, mobile vacancies are introduced into the crystal, the Gibbs free energy of this real crystal, $G(T,P)$, is given by

$$G(T,P)=G_0(T,P)+n_v g_v-kT\ \ln\Omega. \qquad (32)$$

Here g_v ($\equiv h_v-Ts_v$) is the Gibbs free energy of formation of a single vacancy, k is the Boltzmann constant, and Ω is the number of states of the system, i.e., the number of ways of arranging N atoms and n_v vacancies on $N+n_v$ lattice sites. Thus the final term in Eq. (32) is a *configurational entropy term*, given by

$$\ln\Omega=\ln\left(\frac{(N+n_v)!}{N!n_v!}\right). \qquad (33)$$

At thermal equilibrium $G(T,P)$ will be a minimum with respect to n_v, i.e.,

$$\left(\frac{\partial G}{\partial n_v}\right)_{T,P,N}=\mu_v=0, \qquad (34)$$

where μ_v is the *chemical potential of the vacancies*. The contributions to $G(T,P)$ from the vacancy free energy and the configurational entropy term are shown schematically in Fig. 2. Substituting from Eq. (33) into (32) and performing the differentiation (which requires the use of Stirling's theorem, i.e., $\ln N! = N \ln N - N$ for large values of N), yields

$$\mu_v = g_v + kT \ln\left(\frac{n_v}{N+n_v}\right) = 0 \qquad (35)$$

or the site fraction of vacancies* c_v as

$$c_v = \frac{n_v}{N+n_v} = \exp\left(\frac{-g_v}{kT}\right). \qquad (36)$$

Equation (36) shows that the vacancy concentration increases exponentially with temperature, reaching a maximum at the melting point. The physical process of vacancy formation may involve an atom *in* the surface layer jumping from its site to a new site *on* the surface and the subsequent migration of the vacancy so created into the bulk of the crystal. The exact reversal of the process leads to vacancy annihilation. In a crystal there is a dynamic equilibrium, the rates of vacancy formation and annihilation being equal. These rates are governed by the distances between

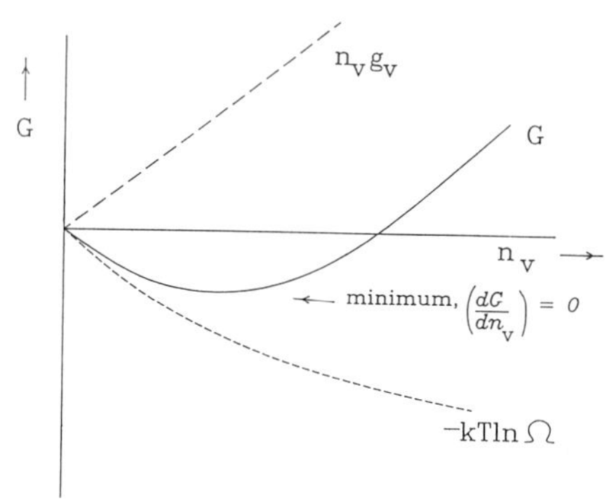

FIG. 2. Schematic representation of the vacancy contribution to the Gibbs free energy of a solid.

*In the treatment of point defects it is more convenient to use *site fractions*, as defined in Eq. (36), as the expression of concentration rather than the more conventional number concentration. Since $N \gg n_v$ the site fraction is *virtually* equal to the mole fraction of vacancies.

the surfaces (*sources* and *sinks* for vacancies) and the diffusion coefficient of the vacancies. Fortunately for the experimentalist, internal surfaces, such as grain boundaries and dislocations, act as sources and sinks so that vacancy concentrations are usually in thermal equilibrium. In fact, special conditions have to be employed to perturb the equilibrium; e.g., rapid cooling ($\sim 1000\ \mathrm{K\ s^{-1}}$) from near the melting point to low temperatures (where diffusion is very slow) will quench in high vacancy concentrations.

The derivation leading to Eq. (36) can easily be extended to a crystal containing n_i interstitial atoms to yield a site fraction (the fraction of interstitial sites in the crystal that are occupied by atoms) as

$$c_i = \frac{n_i}{\beta(N-n_i)} = \exp\left(\frac{-g_i}{kT}\right). \qquad (37)$$

Here g_i ($\equiv h_i - Ts_i$) is the Gibbs free energy to form an interstitial atom, and β is the number of interstitial sites associated with a normal lattice site, i.e., a purely geometric factor depending on the crystal structure of the material. It is clear from Eqs. (36) and (37) that both vacancies and interstitial atoms will be present in any real crystal above absolute zero, although their relative concentrations will depend on g_v and g_i. Simple arguments based on atom sizes and compressibility lead to the prediction that in close-packed structures $g_i > g_v$ and the interstitial concentration is negligible with respect to vacancy concentrations.

It is worth considering the extension of the derivation of Eq. (36) to the case of vacancy pairs. This yields an expression for the *site fraction of pairs* c_p as

$$c_p = \frac{z}{2} \exp\left(\frac{-(2g_v + \Delta g_{v2})}{kT}\right), \qquad (38)$$

where z is the coordination number of the lattice and Δg_{v2} is the binding energy of the pair. Written in this form Δg_{v2} is the free-energy change on bringing two vacancies from infinity to form a pair; i.e., if Δg_{v2} is negative, then divacancy formation is favorable. The term $z/2$ arises as there are $z/2$ ways of orienting a pair in the lattice. Combining Eqs. (36) and (38) yields

$$\frac{c_p}{c_v^2} = \frac{z}{2} \exp\left(\frac{-\Delta g_{v2}}{kT}\right), \qquad (39)$$

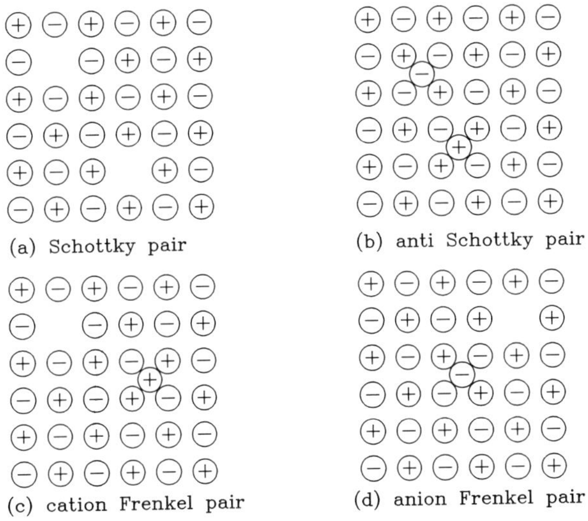

FIG. 3. Schematic representation of the point defects in a *MX* ionic crystal.

which can also be derived by considering the vacancy association reaction (vacancy + vacancy ⇆ divacancy) in terms of the mass action law. It should be noted that Eq. (38) will apply even when the vacancies are not in thermal equilibrium. Under normal circumstances, c_p will be expected to depend exponentially on temperature and to be significantly smaller than c_v.

Finally the perturbation on the system by the addition of impurities needs to be discussed. Assuming that the impurity concentration is low and that the impurity dissolves substitutionally, then the formation of impurity–vacancy pairs, i.e., an impurity and a vacancy on adjacent lattice sites, needs to be included in the treatment of the point-defect thermodynamics. In an analogous manner to the treatment of vacancy pairs, it can be shown that the site fraction of impurity–vacancy pairs, c_{Iv}, is given by

$$c_{Iv}=zc_I \exp\left(\frac{-(g_v+\Delta g_{Iv})}{kT}\right) \quad (40)$$

or

$$\frac{c_{Iv}}{c_I c_v}=z \exp\left(\frac{-\Delta g_{Iv}}{kT}\right), \quad (41)$$

where c_I is the site fraction of isolated impurity atoms and Δg_{Iv} is the binding energy of the pair. The total vacancy concentration $\overset{T}{c}_v$ $(=c_v+c_{Iv})$ is given by

$$\overset{T}{c}_v=\exp\left(\frac{g_v}{kT}\right)\left[1+zc_I \exp\left(\frac{-\Delta g_{Iv}}{kT}\right)\right]. \quad (42)$$

Thus if Δg_{Iv} is negative and there is a tendency to form these pairs, the site fraction of vacancies is increased in comparison to the pure metal.

1.2.3 Point-Defect Thermodynamics in an Ionic Solid. In an ionic solid the point-defect picture is more complicated than that in a monatomic solid. This arises from the fact that an isolated point defect, such as a vacancy or an interstitial ion, will have an *effective* charge with respect to the lattice, and any perturbation must maintain electrical neutrality and preserve the lattice structure. In a 1:1 crystal *MX* that is strongly ionic the point defects in a pure crystal will occur in pairs.* The common types of intrinsic defects are shown in Fig. 3. The *Schottky pair* [Fig. 3(a)] consists of a cation vacancy and an anion vacancy, the *cation Frenkel pair* [Fig. 3(c)] consists of a cation vacancy and a cation interstitial, and the *anion Frenkel pair* [Fig. 3(d)] consists of an anion vacancy and an anion interstitial. The *anti-Schottky pair* [Fig. 3(b)], consisting of a cation and an anion

*Only the case of a strongly ionic crystal, such as an alkali halide, is considered here. In semi-ionic solids, such as a transition-metal oxide, charge compensation on formation of a lattice defect may occur by creation of an electron or hole.

interstitial, is a theoretically possible type of defect pair although it is not found in real systems.

The concentration of point defects in an ionic crystal can be treated in a similar manner to that outlined in Sec. 1.2.2 for vacancies in a monatomic solid. Treatments for ionic crystals can be found in several standard texts (see, for example, Lidiard, 1957; Corish and Jacobs, 1973; Jacobs, 1983). The example taken here will be that of an MX crystal consisting of N_M cations and N_X anions and containing Schottky defects, n_{cv} cation vacancies, and n_{av} anion vacancies. The Gibbs free energy of the crystal can be expressed as

$$G(T,P)=G_0(T,P)n_{cv}g_{cv}+n_{av}g_{av}-kT\ \ln\Omega, \tag{43}$$

where g_{cv} $(\equiv h_{cv}-Ts_{cv})$ and g_{av} $(\equiv h_{av}-Ts_{av})$ are the Gibbs free energies of formation of cation vacancies and anion vacancies, respectively. In this case, with two sublattices, the configurational entropy is given by

$$kT\ \ln\Omega=kT\ \ln\left[\left(\frac{(N_M+n_{cv})!}{N_M!n_{cv}!}\right)\times\left(\frac{(N_X+n_{av})!}{N_X!n_{av}!}\right)\right]. \tag{44}$$

The minimization of $G(TP)$ with respect to the defect concentrations to find the equilibrium condition, i.e.,

$$\delta G=\sum_j\left(\frac{\partial G}{\partial n_j}\right)_{T,P,N}\delta n_j=0, \tag{45}$$

where n_j is the number of defects of type j, must be performed with the constraints that electrical neutrality is maintained and the number of sites on the two sublattices remains equal. If virtual chemical potentials of the defects are defined by

$$\mu_{cv}=g_{cv}+kT\ \ln\left(\frac{n_{cv}}{N_M+n_{cv}}\right) \tag{46}$$

and

$$\mu_{av}=g_{av}+kT\ \ln\left(\frac{n_{av}}{N_X+n_{av}}\right), \tag{46}$$

then the equilibrium condition can be rewritten as

$$\mu_{cv}+\mu_{av}=0, \tag{47}$$

which leads to

$$\frac{n_{cv}}{N_M+n_{cv}}\frac{n_{av}}{N_X+n_{av}}=c_{cv}c_{av}=\exp\left(\frac{-g_s}{kT}\right). \tag{48}$$

Here $g_s=g_{cv}+g_{av}$, the Gibbs free energy of formation of a Schottky pair, and c_{cv} and c_{av} are the site fractions of cation and anion vacancies, respectively.

Equation (48) has the form of a solubility product, and it is useful in ionic solids to consider defect equilibria in terms of quasi-chemical equilibria controlled by mass-action equations. Thus Schottky defect formation can be expressed as

$$\text{perfect lattice} \leftrightarrows \text{cation vacancy} + \text{anion vacancy} \tag{49}$$

controlled by the mass-action equation

$$c_{cv}c_{av}=K_S=\exp(-g_S/kT), \tag{50}$$

where K_S is the equilibrium constant for the Schottky-pair formation equilibrium. Clearly it follows that in a pure crystal N_M will be equal to N_X and c_{cv} will be equal to c_{av} $(=K_S^{1/2})$.

Expressions similar to Eq. (50) can be obtained for other types of defect; for example, for cation Frenkel pairs the expression is

$$c_{cv}c_{ci}=\exp(-g_{cF}/kT), \tag{51}$$

where c_{ci} is the concentration of cation interstitials and g_{cF} $(\equiv h_{cF}-s_{cF}T)$ is the Gibbs free energy of formation of cation Frenkel pairs. Clearly the type of point defect dominant in a particular crystal will depend on the relative magnitudes of their free energies of formation, which in turn depend on the ionic sizes and crystal structures. Among the simple materials the alkali halides contain Schottky pairs, the alkaline earth fluorides contain anion Frenkel pairs, and silver chloride and bromide contain cation Frenkel pairs.

A great deal of information on the point defects in an ionic crystal can be obtained by adding (or "doping") the crystal with aliovalent impurity. The additional charge introduced into the lattice has to be compensated, and this is usually achieved by perturbing the defect concentrations. Taking again the example of an MX crystal containing Schottky defects, the effect of adding substitutionally a small amount of a divalent cation, IX_2, is to introduce cation vacancies. The conditions of electrical neutrality, i.e.,

$$N_M+2n_I=N_X, \tag{52}$$

and site conservation, i.e.,

$$N_M+n_I+n_{cv}=N_X+n_{av}, \tag{53}$$

give the result

$$n_{cv}=n_I+n_{av} \tag{54}$$

and hence

$$c_{cv}=c_I+c_{av}. \tag{55}$$

Here n_I is the number of I^{2+} ions, and c_I is the corresponding site fraction. Combining Eq. (55) with (50) yields the result

$$c_{cv}=\tfrac{1}{2}[c_I+(c_I^2+4K_s)^{1/2}]. \tag{56}$$

This expression has two limits:

1. at low temperatures,
   $$c_I \gg 4K_S;\quad \text{then}\quad c_{cv}=c_I \tag{57}$$
   and c_{cv} is independent of temperature—the *extrinsic region*—and
2. at high temperatures,
   $$c_I \ll 4K_S;\ \text{then}\ c_{cv}=c_{av}=K_S^{1/2} =\exp(-g_S/2kT) \tag{58}$$
   and defects are thermally created—the *intrinsic region*.

Since the cation vacancy and the I^{2+} ion have opposite effective charges, they will tend to associate as a neutral pair on adjacent lattice sites (termed *impurity-vacancy pairs* or *I-V pairs*). The reaction can be written as

$$\text{cation vacancy}+I^{2+} \leftrightarrows I\text{-}V\ \text{pair} \tag{59}$$

and the mass-action relation as

$$\frac{c_{IV}}{c_I c_{cv}}=K_{IV}=\exp\left(\frac{-g_{IV}}{kT}\right). \tag{60}$$

Here c_{IV} is the site fraction of pairs, K_{IV} the equilibrium constant for association, and g_{IV} ($\equiv h_{IV}-Ts_{IV}$) the binding energy of the *I-V* pair.

The concentration of cation vacancies, c_{cv}, as a function of temperature in a *MX* crystal in which Schottky disorder dominates is shown schematically in Fig. 4. Three regions are shown: the intrinsic high-temperature region where the defects are thermally controlled, the extrinsic region at moderate temperatures where c_{cv} is equal to the divalent impurity concentration, and the association region where the cation vacancies form *I-V* pairs. As the temperature decreases a fourth region superimposes on the association region, the *precipitation region*, where the *I-V* pairs aggregate to large clusters and eventually form a separate phase, i.e., crystallites of IX_2. The plot is representative of the behavior of alkali halides, as they usually contain traces of divalent cations (e.g., calcium, magnesium, etc.) as natural impurities.

It is relatively straightforward to extend the above treatment of other types of defect, and examples can be found in several texts (Lidiard, 1957; Corish and Jacobs, 1973; Jacobs, 1983). The treatment can also be made more exact by replacing concentrations with

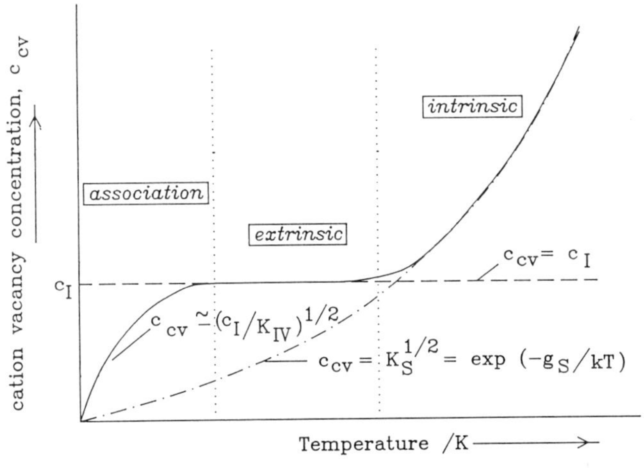

FIG. 4. Cation vacancy concentration in a *MX* crystal with Schottky disorder and doped with a divalent cation as a function of temperature.

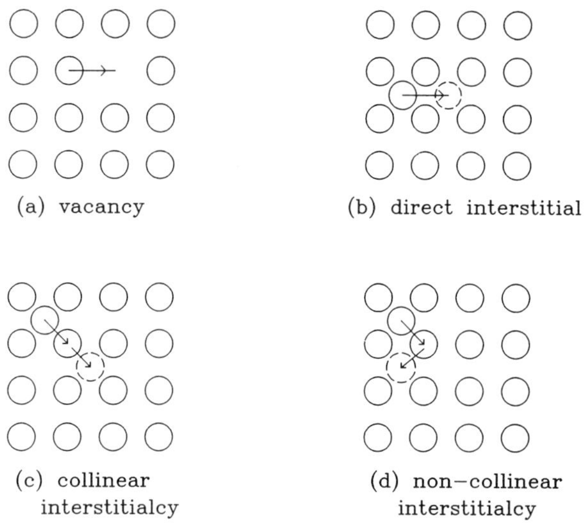

FIG. 5. Schematic representation of the diffusion mechanisms in a solid.

thermodynamic activities. The nonideal behavior arises from the long-range Coulombic interactions between the defects, which are usually treated in a similar manner to electrolyte solutions by the Debye–Hückel–Lidiard theory.

1.2.4 Point-Defect Concentrations in Solids. The process of taking an atom from the lattice to form a vacancy is similar to that of sublimation. If there is no relaxation of the lattice around the vacancy then it is possible to a first approximation (Jost, 1960) to equate the enthalpy of vacancy formation, h_v, with the latent heat of sublimation, L_s, i.e.,

$$h_v \simeq L_s. \tag{61}$$

Estimation of the entropy of vacancy formation is more difficult, as it arises from the change in vibrational frequency spectrum of the crystal on forming the vacancy. Calculations for solid argon yield the result

$$s_v \simeq 2k. \tag{62}$$

Combining these values into g_v and substituting into Eq. (36) gives an estimate of c_v as about 1×10^{-4} for solid argon at the melting point T_m. This is quite a good estimate, and for many types of solid $c_v(T_m)$ lies in the range 10^{-4} to 10^{-3}. In close-packed solids c_i will be orders of magnitude smaller than c_v since $g_i > g_v$.

A variety of techniques can be used to determine the point-defect concentrations in solids, and the reader is referred to specific texts (see, for example, Flynn, 1972; Crawford and Slifkin, 1972). In general, this can be experimentally difficult because of the low magnitude of the concentrations and is usually only possible at temperatures close to T_m. An exception is the case of ionic solids where the measurement of ionic conductivity, which will be described in Sec. 2.2, can provide accurate information in a basically simple experiment.

1.2.5 Mechanisms of Diffusion. In nearly all crystalline solids where the mechanism of diffusion has been firmly established the mechanism involves point defects. The earliest theories of solid-state diffusion proposed that at the atomic level two atoms on neighboring sites simply exchanged positions. It was soon realized that this mechanism, known as the *direct exchange mechanism*, would be energetically unfavorable in close-packed solids, as it would require considerable dilation of the lattice to allow the atoms to move past each other. An extension of this mechanism is the *ring mechanism* where a group of atoms move around in a circular fashion, each atom in the group taking the place of its neighbor in the direction of motion. Although this reduces the energy of the process, the ring mechanism does not operate in most systems.

The diffusion mechanisms involving point defects are shown schematically in Fig. 5. The simplest of these is the *vacancy mechanism*, shown in Fig. 5(a), in which an atom "jumps" into a vacancy on a neighboring site. This process is energetically favorable and is the dominant diffusion process in most close-packed crystal structures. For example, this is the mechanism that operates for self-diffusion in fcc metals. In systems where the dominant defects are interstitial atoms, the diffusion can occur via these defects by a number of possible processes. In the *direct interstitial mechanism*, shown in Fig. 5(b), the interstitial atom jumps directly from interstitial site to interstitial site. This is the mechanism by which small impurity atoms that occupy the interstitial sites in metals diffuse—for example, carbon in iron. In pure solids interstitials can diffuse by the *interstitialcy mechanism*, shown in Figs. 5(c) and 5(d), where an interstitial moves onto a neighboring normal lattice site with the displacement of the original occupant atom onto a new interstitial site. There are two variants of this mechanism, the *collinear interstitialcy mechanism* [Fig. 5(c)], where the motion of the two atoms is along the same direction, and the *noncollinear interstitialcy mechanism*, where the two atoms move along different directions. Proven examples of the interstitialcy mechanism can be found in the motion of silver ions in silver halides and fluoride ions in the alkaline earth fluorides.

Variations on the basic point-defect mechanisms described above are also possible, such as the diffusion of atoms via divacancies and the *dumbbell interstitial* (two atoms located around the same lattice site).

It is important to note the use of the term *jump* to describe the process of the atom's motion. This implies that the movement is rapid, occurring over the time scale of the period of a lattice vibration, and that the time the atom is in "flight" between its initial and final sites is very short compared to its average "residence" time on a lattice site.

1.2.6 Random-Walk Theory. The diffusion of atoms in a solid with no applied driving force is analogous to Brownian motion, as each atom can be pictured as making jumps from lattice site to lattice site, the direction of each jump being in a random direction relative to preceding jumps. Thus the process can be treated by the random-walk theory developed originally by Einstein and Smoluchowski.

If an atom makes n consecutive jumps, each described by the translational vector $\mathbf{r}_i$, then the total displacement vector between the initial and final sites of the atom is given by

$$\mathbf{R}_n=\mathbf{r}_1+\mathbf{r}_2+\mathbf{r}_3\cdots=\sum_{i=1}^{n}\mathbf{r}_i. \qquad (63)$$

It follows that the displacement squared is simply

$$\mathbf{R}_n^2=\sum_{i=1}^{n}\mathbf{r}_i^2+2\sum_{i=1}^{n-1}\sum_{j=i+1}^{n}\mathbf{r}_i\cdot\mathbf{r}_j. \qquad (64)$$

The average displacement squared of many atoms after n jumps is given by

$$\langle\mathbf{R}_n^2\rangle=\sum_{i=1}^{n}\langle\mathbf{r}_i^2\rangle+2\sum_{i=1}^{n-1}\sum_{j=i+1}^{n}\langle\mathbf{r}_i\cdot\mathbf{r}_j\rangle. \qquad (65)$$

If there is only one mechanism of diffusion operative then in a cubic crystalline solid the length of each jump step will be the same, r. In addition, if the jumps are *truly random* then the last term in Eq. (65) will be averaged out to zero. For a cubic solid $\langle\mathbf{r}_i\cdot\mathbf{r}_j\rangle$ will be replaced by $\langle r^2\cos\theta_{ij}\rangle$, where θ_{ij} is the angle between the ith and jth jumps. Since all angles between jumps i and j will be found, the equation can be written simply as

$$\langle\mathbf{R}_n^2\rangle=nr^2. \qquad (66)$$

Thus the average displacement, the root mean square displacement, is given by

$$\sqrt{\langle\mathbf{R}_n^2\rangle}=\sqrt{n}r. \qquad (67)$$

If the n jumps take place in a time t, then the jump frequency Γ is given by

$$\Gamma=n/t \qquad (68)$$

and

$$\langle\mathbf{R}_n^2\rangle=t\Gamma r^2. \qquad (69)$$

The Einstein equation relates the diffusion equation to the mean square displacement, and for a cubic solid this is

$$D=\langle\mathbf{R}_n^2\rangle/6t. \qquad (70)$$

Thus combining this equation with Eq. (67) yields

$$D=\Gamma r^2/6, \qquad (71)$$

which is the *Einstein–Smoluchowski equation.*

1.2.7 Relations between Macroscopic and Microscopic Parameters. It is the Einstein–Smoluchowski equation that allows the connection to be made between the measured diffusion coefficient in an experiment and the parameters involved in the atomistic mechanism of diffusion. Here the case of self-diffusion of atoms in a monatomic solid by a vacancy mechanism is taken as the example.

The jump frequency of an atom, Γ, will depend on the probability that a vacant site is adjacent to the atom (the product of the site fraction of vacancies, c_v, and the coordination number z) and the jump frequency w_0 of the vacancy:

$$\Gamma = zc_v w_0. \tag{72}$$

Absolute reaction-rate theory, developed for chemical kinetics (Glasstone *et al.*, 1941), can be used to obtain an expression for w_0 in the form

$$w_0 = \nu \exp(-\Delta g_v^m/kT). \tag{73}$$

Here ν is the vibrational frequency of the atom in the direction of the vacancy (usually taken as the Debye frequency) and Δg_v^m ($\equiv \Delta h_v^m - T\Delta s_v^m$) is the Gibbs free energy of vacancy migration, i.e., the height of the potential-energy barrier that the atom has to overcome in passing from its initial site into its final site. There has been debate about the validity of this approach, as it implies that the jump process would have to occur slowly enough for thermodynamic equilibrium to be maintained. Thus "dynamic theories" have been proposed; however, it has been shown (Glyde, 1967) that these lead to essentially the same form of result.

Combining Eqs. (72) and (73) and substituting into Eq. (71) with the expansion of c_v yield the final expression

$$D = \frac{\nu z r^2}{6} \exp\left(\frac{s_v + \Delta s_v^m}{k}\right) \exp\left(\frac{-(h_v + \Delta h_v^m)}{kT}\right). \tag{74}$$

In experiments the temperature dependence of the diffusion coefficient in a crystalline solid usually obeys the *Arrhenius equation*:

$$D = D_0 \exp(-Q/kT), \tag{75}$$

where D_0 is called the preexponential factor and Q is the activation energy. Hence by comparing Eqs. (74) and (75) the following identities between the experimental and microscopic parameters can be made:

$$D_0 = \frac{\nu z r^2}{6} \exp\left(\frac{s_v + \Delta s_v^m}{k}\right) \tag{76}$$

and

$$Q = h_v + \Delta h_v^m. \tag{77}$$

It is possible to make a rough estimate of Δh_v^m at the same level of approximation used to estimate h_v, i.e., Eq. (61). As an atom moves from its site the number of bonds it needs to "break" with its neighbors is similar to those broken in vacancy formation. Thus Δh_v^m will also be approximately equal to L_s, i.e.,

$$\Delta h_v^m = L_s \tag{78}$$

and therefore

$$Q \simeq 2L_s. \tag{79}$$

This prediction should be reasonable where there is no relaxation, either electronic or lattice—for example, in molecular solids. If it is assumed that $\Delta s_v^m \simeq s_v \simeq 2k$, then it is possible to estimate the value of D at T_m as 10^{-8} cm^2 s^{-1}, which is the value typically found experimentally for most solids.

1.2.8 Correlation Effects in Diffusion. In order to measure the diffusion coefficient it is necessary to "tag" atoms so that their motions can be followed; for example, they may be isotopically labeled as in a tracer experiment. If the motion of an individual atom that is tagged is followed, then it will be found that there can be correlations between successive diffusive jumps, depending on the mechanism. The simplest example that demonstrates correlation is that of diffusion in a pure solid via a vacancy mechanism. If the tagged atom makes a jump into a vacancy, then its *most probable next jump* is back to its original site, as there is a vacancy there ready to accept it. Thus the tracer jumps will be correlated although the jumps of the vacancy are random. For the tracer it is necessary to modify the diffusion coefficient to

$$D_t = fD_{\text{random}}. \tag{80}$$

Here D_t is the measured diffusion coefficient of the tracer, D_{random} is the diffusion coefficient given by random-walk expressions, and f is the correlation factor. Sometimes f is referred to as the *Bardeen–Herring correlation factor*, after the first workers (Bardeen and Herring, 1952) to note this correlation in the motion of tagged atoms. Since accurate values of f can be theoretically calculated for all the possible diffusion mechanisms in the different lattice structures, the determination of f provides an experimental route to the identification of diffusion mechanisms (LeClaire, 1970). Values of f are given in Table 1.

2. EXPERIMENTAL METHODS OF STUDYING DIFFUSION

A wide variety of methods can be employed in the investigation of diffusion. The focus here is on those that can provide direct, reliable information on the diffusion parameters and insights into the diffusion mechanisms. Of particular interest are clearly those methods that can yield accurate values of the diffusion coefficient. Thus techniques that provide indirect or limited diffusion information are not treated here. For example, the rate of plastic deformation (*creep*) under certain conditions is governed by atomic diffusion, and a study of the temperature dependence will yield good values of the activation energy for diffusion, Q, of the atoms. However, this technique, although useful, is limited in the detail it can provide and is not included in the discussion.

2.1 Methods of Measuring Diffusion Coefficients

2.1.1 Study of the Diffusion Penetration Profile. The most reliable method of measuring the diffusion coefficient is to determine the profile of labeled atoms that have been diffused from the surface of a sample in a known time at a constant temperature. In order to ensure that the true bulk diffusion coefficient is measured, the sample should be a single crystal with the minimum number of dislocations.

The majority of the accurate diffusion coefficients that are available have been measured by the *radiotracer serial sectioning technique*, or its variants, and it is applicable to a wide range of solid types. The full details can be found in a number of texts (see, for example, Bénière, 1983; Mundy and Rothman, 1983). The basic procedures of this technique are shown in Fig. 6. First, a very thin layer of radiotracer is applied to a flat surface of the sample by vacuum sublimation or by deposition from a solution. The tracer may be an isotope of the host crystal (e.g., ^{22}Na to measure self-diffusion in Na, ^{36}Cl to measure chloride ion self-diffusion in NaCl, ^{14}C to measure carbon diffusion in steel, etc.). The sample is then annealed at constant temperature T for a known time t. Following the anneal the sample is sectioned parallel to the face of the tracer deposit and the sections subjected to a radioactive assay. This allows the determination of the penetration profile and the evaluation of the tracer diffusion coefficient from Eq. (6), i.e., using a plot of ln(tracer concentration) versus (penetration)^2. Two advantages of this technique are its simplicity and the relative cheapness of the equipment. In addition, the determination of D_t *directly* from the penetration profile provides a good test that bulk diffusion is being observed; the diffusion plot should be linear, and D_t should be independent of annealing time. An obvious disadvantage is the destructive nature of the sectioning, while the technique is limited to systems where convenient radiotracers, in terms of half-life and radiological safety, are available.

The controlling parameters in this experiment are basically the anneal time and the thickness of the sections. Reasonable times for the anneal lie in the range 10^3 s (any shorter and heating–cooling corrections become large) to 10^6 s (longer times would generally be inconvenient). A variety of techniques can be used to section the sample. For bulk samples, of the order of 10 mm dimensions, the sections are generally taken off with a microtome (for reasonably hard materials,

Table 1. Values of the correlation factor f for self-diffusion in simple lattices.

Mechanism of diffusion	Lattice	f
Vacancy	fcc	0.7815
	bcc	0.727
	sc	0.653
	Honeycomb	0.3333
Divacancy	fcc	0.458
Direct interstitial	Independent of structure	1
Collinear interstitialcy	fcc	0.9643
	sc	0.8000

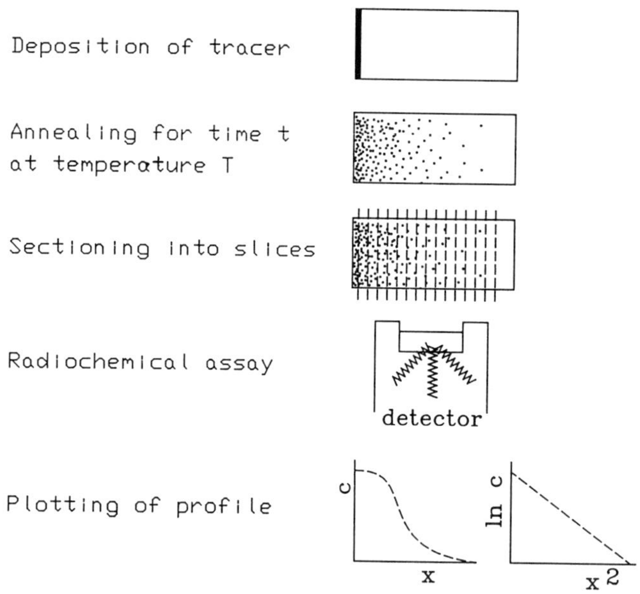

FIG. 6. Procedures of the radiotracer serial-sectioning diffusion experiment.

e.g., alkali halides), with a lathe (for metals), or by grinding (very hard materials), and the individual sections are usually in the 1- to 100-μm range. Using the criterion (Bénière, 1983) that the diffusion zone $[\sim 4(Dt)^{1/2}]$ should be divided into at least 13 sections over which the activity falls by a factor of 100 gives the useful range of D_t that can be studied with these types of sectioning as roughly 10^{-14} to 10^{-6} cm^2 s^{-1}. For smaller values of D_t, thinner sections need to be taken, and this is possible by sputtering (sections ~5 to 100 nm) and electrochemical etching (sections ~50 nm), putting the lower limit of D_t at around 10^{-17} cm^2 s^{-1}.

A very important variation of the technique is to use a stable isotope and secondary-ion mass spectroscopy (SIMS) to study the penetration profile (Macht and Naundorf, 1982; Lodding, 1987). This has proved particularly useful in the study of oxygen diffusion, as there is no radioactive isotope of this element and the diffusion is usually very slow. The sections taken in SIMS are in the range 1–100 nm allowing values of D_t as low as 10^{-19} cm^2 s^{-1} to be studied. In addition, very small samples, around 1 mm dimensions, can be used. However, the equipment required for a SIMS experiment is extremely expensive and requires a highly trained operator. Other means of determining the penetration profile of a tracer that are particularly useful in studying slow diffusion are the use of Rutherford backscattering and nuclear reaction analysis.

2.1.2 Measurement of the Atomic Jump Frequency. A wide variety of relaxation techniques can be used to deduce the jump frequency of atoms and hence allow the evaluation of D via the Einstein–Smoluchowski equation, i.e., Eq. (71). Generally these techniques rely on a theoretical model to relate the observed parameter to the microscopic atomic motion. This involves making some assumption about the nature of the *correlation function* (the function that expresses the probability that a particle originally at a position $\mathbf{r}=0$ at time $t=0$ is found at a position $\mathbf{r}=\mathbf{r}$ after a time t) and the *pair correlation function* (the probability that if an atom is at $\mathbf{r}=\mathbf{0}$ another atom will be found at $\mathbf{r}=\mathbf{r}$) (van Hove, 1954). Techniques that come into this category are dielectric relaxation spectroscopy, neutron scattering (Springer and Richter, 1987; Lechner, 1983), Mössbauer spectroscopy, and nuclear magnetic resonance (NMR) spectroscopy (Wolf, 1979; Strange, 1986; Kanert, 1982). The general advantages of these techniques are that they are nondestructive and rapid (the atoms only need to make a few jumps to give the effect and thus the observation time is short), and they are usually less demanding on sample requirements when compared to tracer methods (powders can be used in some cases). However, they often have limited applications to certain elements (e.g., Mössbauer spectroscopy can only be used for ^{57}Fe, ^{119}Sn, and a few other elements) and systems where diffusion is relatively fast, i.e., $D > 10^{-9}$ cm^2 s^{-1}. In addition, it is usually difficult to use these methods to give accurate values of D, because of deficiencies in the theoretical treatment of the correlation functions for an ensemble of atoms, although they will yield good values of Q. An exception to some of these current limitations is the NMR technique.

NMR diffusion experiments (see MAGNETIC RESONANCE, NUCLEAR) are best suited to systems where the NMR-active isotope is in high abundance and has nuclear spin s equal to $\frac{1}{2}$ (e.g., ^1H, ^{19}F) or a small quadrupolar moment (e.g., ^7Li, ^{23}Na). A reasonably large number of interesting systems fall into this category, particularly organic solids. There are a variety of NMR experimental procedures, but those in current common usage involve the measurement of a spin relaxation time in a pulsed spectrometer (Farrar and Becker, 1971; Fukushima and Roeder, 1981). The basis of the experiments involves disturbing the equilibrium spin population between a certain set of energy levels in the system by the application of a short burst (pulse) of radio-frequency power and then measuring the characteristic time required for the spins to return to equilibrium by monitoring the magnetization of the sample. For example, the spin–spin relaxation time T_1 is the time for the spins to equilibrate between the Zeeman levels whose separation is determined by the magnitude of the applied external magnetic field, $\mathbf{B}_0$. A spin in an excited state can only relax down to the lower level by losing a quantum of energy equal to the separation between the states, $\Delta E = \hbar\omega$, where ω is the Larmor precession frequency. This involves the spin interacting with oscillating magnetic fields in the system that fluctuate at a frequency equal to ω, i.e., it is a stimulated emission. In the system a spin will experience a *local field* due to dipole–dipole interactions with the neighboring spins. One source of oscillating magnetic field is the modulation of the local field caused by relative motion of the spins. Diffusive motion provides a fluctuating magnetic field whose frequency components are similar to those of a "noise spectrum." Analysis of this spectrum into its Fourier components yields the spectral density function, which depends on the atomic (spin) jump frequency Γ. The precise relationship between the measured relaxation time and Γ involves an assumption about the correlation function, and various models have been developed. However, it is clear that the relaxation will be most efficient (the relaxation time will be a minimum) when Γ is equal to ω. By measuring different relaxation times, i.e., probing different sets of energy levels for the spins, a wide range of Γ, and hence D, can be determined. Typically T_1 will monitor D in the range 10^{-11}–10^{-6} cm^2 s^{-1}; the spin–spin relaxation time T_2, the range 10^{-10}–10^{-6} cm^2 s^{-1}; the spin–lattice relaxation time in the rotating frame, $T_{1\rho}$, the range 10^{-14}–10^{-8} cm^2 s^{-1}; and the spin–lattice relaxation time in the local dipolar field, T_{1D}, the range 10^{-20}–10^{-12} cm^2 s^{-1}. Thus it is possible with one technique, using one spectrometer, to probe a very wide range of D.

Two further points concerning NMR methods are worthy of note. First, for cubic systems a precise theory relating relaxation times to Γ has been developed, tested, and shown to be reliable, and can be used to obtain accurate values of D. This is the *encounter model* as exploited by Wolf (1979) (see also Strange, 1986). Second, there is a further NMR technique, *pulsed-field gradient NMR*, that does not involve measurement of the relaxation time. In this experiment the spin is used to "label" atoms, like a tracer, and the time scale of the experiment is such that the atoms move macroscopic distances. Thus it is not subject to the criticisms outlined above and is particularly useful in the study of diffusion in liquids or solids where diffusion is rapid, i.e., $D > 10^{-9}$ cm^2 s^{-1} (Strange, 1986).

2.2 Ionic Conductivity Measurements

For materials that are ionic the measurement of the ionic conductivity provides a very powerful method of studying diffusion. As will be seen below, these experiments can yield not only information on the migration of the ions but also the concentration of point defects. This has been one of the major factors that has led to the current deep level of understanding of point defects and transport in ionic solids—a better understanding than in any other type of solid. Thus the measurement and interpretation of ionic conductivity has been described in a number of texts

(Lidiard, 1957; MacDonald, 1987; Chadwick, 1991; Corish and Jacobs, 1973; Jacobs, 1983) and is described here in some detail.

Samples for conductivity measurements are typically disks, about 100 mm^2 in cross-sectional area and 5 mm thick. Ideally the sample is a single crystal that has a known orientation, but this is not always available and it is common to use a compacted pellet. It is usually necessary to anneal the specimen close to the melting point prior to measurement to release any strains from the sample and to ensure a stable geometry. In the two-probe experiment most commonly in use, the large faces of the disk are placed in contact with the measuring electrodes in a cell with a controlled atmosphere and temperature. Many designs of conductivity cell have been published and the essential features of common systems are shown in Fig. 7. The key features of this design are that

1. the cell atmosphere can be controlled using either a purified inert gas stream or a dynamic vacuum (a very important feature when studying reactive materials like fluorides),
2. inert electrodes (usually graphite) are coated on the large faces of the sample to ensure good contact with the measuring electrodes and a true evaluation of σ from the measured conductance,
3. temperature is accurately measured by means of two thermocouples on each side of the sample, and
4. construction materials (high-purity silica, alumina, graphite, and platinum) that are involatile and unreactive to most materials are used in the hot zone.

This type of system can be used to make very precise conductivity measurements, and, provided the sample is thoroughly thermally equilibrated, the absolute accuracy is governed mainly by the accuracy of the thermocouples. The upper temperature limit of this cell, around 1000 °C, can be extended by replacing the silica components with alumina.

The conductivity is best measured by an ac technique to avoid polarization effects at the electrodes. Highly accurate ac conductivity–capacitance bridges, with errors around 1 part in 10^4, have been commercially available for many years. The earlier models were tedious to use as they were manually balanced and operated at a fixed, basic frequency, although separate signal generators and detectors could be attached to make frequency-dependence studies. Later generations of bridges were autobalancing, but still operated at fixed frequencies. The current generation of commercial systems are autobalancing, can rapidly scan a wide frequency range, and are readily interfaced to microcomputers to allow completely automated data collection. The evaluation from the raw data of the true "bulk" conductivity due to the motion of ions within the matrix of the lattice depends on the nature of the sample. The approach now in general use is that of complex-plane impedance analysis (Bauerle, 1969; Archer and Armstrong, 1981; MacDonald, 1987).

The application of a small, sinusoidally oscillating potential $\Delta E \sin \omega t$ with angular frequency ω ($=2\pi f$) across the electrodes will give rise to a current $\Delta i \sin(\omega t + \theta)$ that is

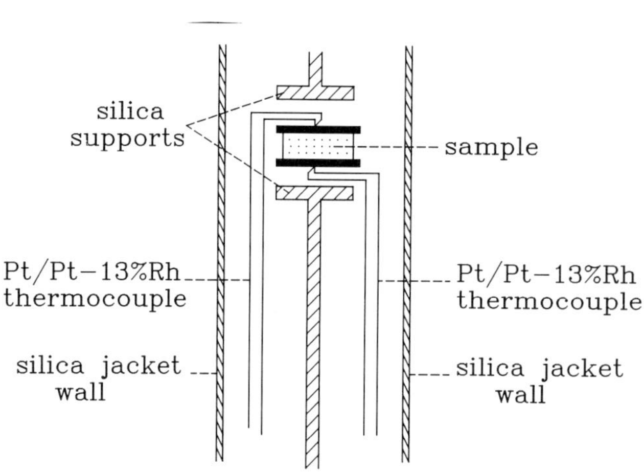

FIG. 7. Essential features of a cell for measuring the ionic conductivity of solids.

phase shifted by an angle θ. The impedance Z can be characterized by a combination of the ratio $\Delta E/\Delta i$ and the phase angle θ. It is convenient to represent Z at a given ω as a complex function and hence by a point on the complex impedance plane, where the y axis is the imaginary component, Z'', and the x axis is the real component, Z'. Thus $|Z|$ is the length of the line from the origin to the point and θ is the angle between this line and the x axis. The locus of Z as ω is varied will depend on the components in the circuit between the electrodes. The advantage of the complex-impedance representation is that a series combination of elements can be described by the vectorial addition of individual components. The situation is more complicated for a parallel combination of elements. For example, a simple combination of resistance R and capacitance C leads to a semicircle in the complex-impedance plot. For a parallel combination of elements it is often more useful to plot the admittance Y ($=1/Z$) in the complex plane, i.e., the imaginary part B (the susceptance) versus the real part G (the conductance). All the parameters are interrelated, and it is possible to present the basic measurement of $|Z|$, ω, and θ in a variety of formats; the interconversion is a built-in feature of some of the commercial impedance analyzers. The problems in many studies are first to find the equivalent circuit that corresponds to the measurements and then to identify the circuit elements with physical features or chemical processes in the electrode-sample arrangement.

The simplest situation is that for a single crystal of a normal ionic conductor such as an alkali halide. In this case the circuit can be represented as a resistance and capacitance in parallel. Provided the frequency is high enough to avoid electrode polarization, usually $\geqslant 1$ kHz, then the measured conductance G will be the true bulk conductance of the sample, and extremely accurate measurements can be taken (the individual data points can be accurate to $\pm 0.1\%$). For materials with high ionic conductivities and for polycrystalline pellets the bulk conductance has to be extracted from the complex impedance or admittance plot. The ideal complex impedance spectrum for a ceramic pellet would consist of three semicircles, as shown in Fig. 8. Each of the contributions to the electrical behavior of the bulk material, i.e., the interior of the grains, the grain boundaries, and the sample–electrode interface, is represented by an RC component, and these are connected in series. Thus, in principle, it is possible to extract the true bulk conductance and to analyze it in terms of defect structures and transport mechanisms. However, the accuracy is necessarily lower than that achievable for alkali halides, and the number of data points that can be feasibly collected over a given temperature range is less. The complex-impedance spectra for many real pellets are more complicated than Fig. 8, and a number of strategies have to be adopted to identify the bulk resistance (e.g., varying the sample geometry, electrode preparation, impurity concentration, etc.). Good examples can be found in the original paper by Bauerle (1969).

Typical conductivity data for an ionic crystal, in this case sodium chloride, are shown in Fig. 9. It should be noted that this is very

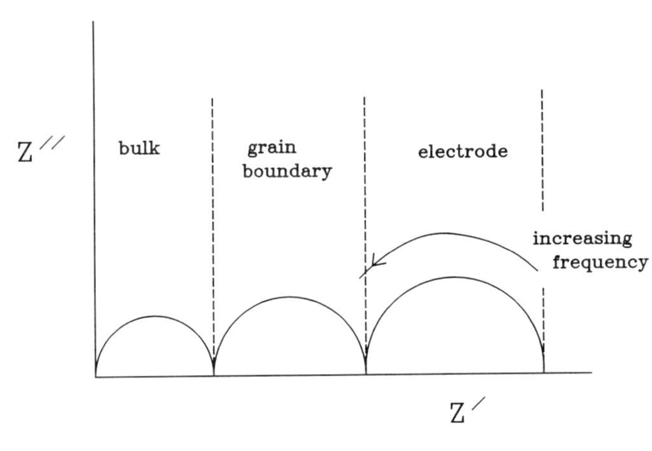

FIG. 8. Idealized complex-impedance spectrum for a ceramic ionic conductor.

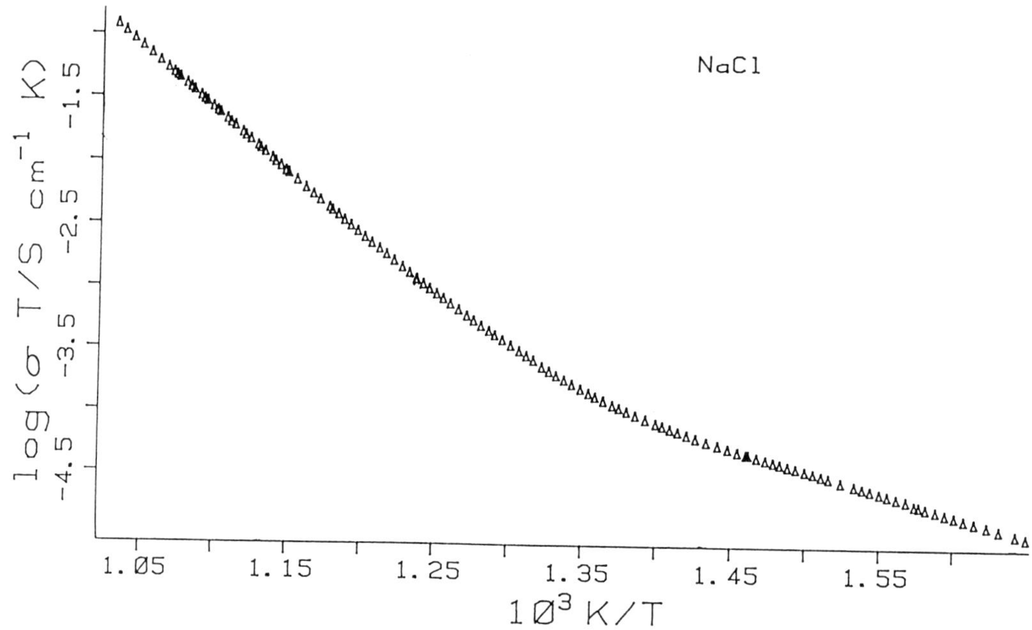

FIG. 9. Ionic conductivity data for a sodium chloride single crystal (after Hooton, 1988).

high-quality data with a high density of accurate points. The simplest first step in the analysis of data of this type is to determine the Arrhenius energies from the plot* of $\log(\sigma T)$ vs $1/T$. The next stage is to attempt to relate these to the microscopic conduction processes. The specific conductance of a substance is given by the relationship

$$\sigma = \sum_{\text{all } r} n_r |q_r| \mu_r, \tag{81}$$

where n_r is the number of conducting species of type r per unit volume, q_r and μ_r are its charge and mobility, respectively, and the sum is taken over all the types of mobile species. For a simple ionic crystal the species can be regarded either as the ions themselves or as the point defects by which they move. Thus values of n_r can be written down explicitly in terms of point-defect equilibria. Alkali halides will be considered as a simple example, and here the predominant point defects are Schottky pairs (a cation vacancy *cv* and an anion vacancy *av*) and the defect equilibrium is described by the equations in Sec. 1.2.2. It has already been mentioned that long-range defect interactions can be included by including activity coefficients γ_r, which can be calculated from the Debye–Hückel–Lidiard theory (Lidiard, 1957). This gives

$$\gamma_r = \exp\left(\frac{-q_r^2 \kappa}{2\epsilon kT(1+\kappa R)}\right), \tag{82}$$

where κ^{-1} is the Debye length, given by

$$\kappa^2 = 4\pi e^2 \sum_r \frac{c_r}{2a^3 \epsilon kT}. \tag{83}$$

Here, e is the protonic charge, ϵ is the permittivity, $2a^3$ is the volume per molecule of the salt, and R is the distance of closest approach between defects before the formation of a defect pair or complex. Thus it is possible to solve the mass-action equations using the constraints of charge and site balance and to write down an explicit expression for the concentration of any defect in terms of the thermodynamic parameters governing its formation and association with other defects and the corresponding activity coefficients.

The defect mobility μ_r is related to the defect jump frequency w_r by the expression

$$\mu_r = S_r |q_r| a_r^2 w_r / 6kT, \tag{84}$$

*In the older literature (and in some current papers) data are plotted as $\log\sigma$ vs $1/T$. However, this is not correct, as the mobility μ in Eq. (81) depends on $1/T$, and activation energies deduced from these plots will usually be slightly in error.

where S_r is the number of equivalent sites into which the defect can make a jump of distance a_r. The jump frequency w_r is given by Eq. (73). The mobilities can to be corrected for the drag experienced from the surrounding defect cloud by using the *Onsager–Pitts correction* (Lidiard, 1957).

Thus, given a model of the defect structure and transport mechanisms, it is possible to rewrite Eq. (81) with explicit expressions for the defect concentrations and mobilities. For an alkali halide crystal containing M^{2+} impurity this will contain nine parameters: the Schottky formation enthalpy and entropy, h_s and s_s; the migration enthalpies and entropies of the cation vacancy, $\Delta \mathscr{H}^m_{cv}$ and $\Delta \mathscr{S}^m_{cv}$, and anion vacancy, $\Delta \mathscr{H}^m_{av}$ and $\Delta \mathscr{S}^m_{av}$; the association enthalpy and entropy of the cation vacancy–M^{2+} complex, h_{IV} and s_{IV}; and the total M^{2+} concentration c_T. The modern procedure is to make a computer fit to the experimental data with these parameters as the variables in a nonlinear least-squares-fitting program. These programs will minimize the sum of the residual deviations squared, ϕ, i.e.,

$$\phi = \sum [\log(\sigma T)_{\text{calc}} - \log(\sigma T)_{\text{exptl}})]^2. \tag{85}$$

The Debye–Hückel–Lidiard activity coefficients and the Onsager–Pitts correction are incorporated in the expressions, as are the temperature dependences of ϵ and a, if they are available. The advantages of the least-squares computer fitting approach are that it makes the best use of accurate data and that the *whole* conductivity plot is analyzed to yield a full set of defect parameters, for both defect formation and migration. Clearly, precautions in the fitting have to be taken (Jacobs, 1983; Chadwick, 1991); and to get really reliable defect parameters, crystals with a range of dopant concentrations have to be studied, and often other diffusion data are used as guidelines.

However, it is not always possible to obtain high-quality data over a wide temperature regime. In these situations a less detailed approach is required. It is usually reasonable to assume that only one type of the defect pair is present, and simple theoretical models will normally be able to identify which of the defects is the more mobile. It is then possible to obtain estimates of the enthalpy of defect migration and formation. For example, the more mobile defect in NaCl is the cation vacancy, and the dominant impurities in these crystals are divalent cations. Thus in the low-temperature region of Fig. 9 the cation vacancy concentration is determined by the impurities and, if *I-V* pairs are neglected, c_{cv} is given simply by Eq. (57). Substituting this into Eq. (81) shows that the slope of the *extrinsic conductivity region* is therefore equal to $-\Delta \mathscr{H}^m_{cv}/k$. Similarly, the value of c_{cv} in the high-temperature *intrinsic conductivity region* is given by Eq. (58), and the slope here is equal to $-(h_s/2 + \Delta \mathscr{H}^m_{cv})/k$. Thus it is still possible to extract information on the defect-formation and -migration processes.

2.3 Methods of Identifying Diffusion Mechanisms

The ultimate aim of diffusion studies is to identify the nature of the atomic diffusion mechanism that is operative, i.e., vacancy, divacancy, interstitialcy, etc. Clearly, information on the nature of the dominant point defects in the solid will provide a starting point. For example, in close-packed metals the dominant defects are known to be vacancies, so that mechanisms based on interstitials can be ruled out. In less obvious situations one approach is to compare the experimentally determined diffusion parameters (predominantly activation energies but also activation volumes) with those obtained from theoretical calculations for the different mechanisms. In most cases this has been quite a sound approach, and with the advent of the modern computer-simulation procedures this is becoming even more reliable. For ionic solids the predictions from these calculations have been quite enlightening on some complex systems.

Direct experimental identification of the diffusion mechanism has usually relied on the determination of the correlation factor f. In principle, this can be determined from the study of the *isotope effect for diffusion* (LeClaire, 1970; Peterson, 1975), which involves the measurement of the differences in diffusion coefficient of two radiotracers, A and B, with different masses, e.g., ^{22}Na and ^{24}Na. However, there are only a few elements with convenient isotopes to use this approach, and the relations between f and D_A and D_B are not as simple as expected from the early work on this topic. The experiment will measure the product $f\Delta K$, where ΔK is the fraction of the kinetic energy of the jump that is carried by the tracer atom. In a jump many atoms around the tracer and defect will be involved; therefore, ΔK is not unity and is typically 0.8–0.9. In some well-understood systems where the diffusion mechanism and hence f are known, it is possible to work backward to determine ΔK; however, in an unknown system it can be hard to separate f and ΔK, and identification of the mechanism is not simple.

In ionic crystals the ionic conductivity arises from the movement of the effectively charged defects in the electric field. Since the diffusion of the defects is random, the diffusion coefficient calculated from the ionic conductivity D_σ via the Nernst–Einstein equation [Eq. (14)] will be the diffusion coefficient of the ions if they were moving randomly. Thus using Eq. (80) yields

$$f = D_t / D_\sigma, \tag{86}$$

and in principle the mechanism can be identified. However, the situation is often more complex. The ions may diffuse by more than one mechanism; e.g., the Ag^+ ions can diffuse by both vacancy- and interstitial-based mechanisms, and the ratio D_t/D_σ is not simply f. This ratio is still useful and is now referred to as the *Haven ratio*.

2.4 Computational Techniques

Computer simulation of defects and transport in solids is now well established, particularly for ionic materials, and the methodologies are described in a number of recent reviews (Catlow and Mackrodt, 1983; Catlow, 1987; Harding, 1990; Kehr and Binder, 1987). The simulation methods fall into three general types: *static lattice simulations, molecular dynamics* (MD) *methods*, and *Monte Carlo* (MC) *methods*.

Briefly, in the static lattice methods the atoms, with some appropriate mathematical representation of the interatomic potential, are placed on the sites of a representative portion of the lattice, and the energy of the lattice is calculated by pairwise summation of all the intermolecular forces. The calculation is then repeated with a defect in the center of the lattice to yield the energy of defect creation. The activation energy for diffusion can be calculated by placing the defect at the saddle point of the jump path—e.g., for the vacancy mechanism with the diffusing atom midway between the initial and final lattice sites. The success of the methods depends on the reliability of the interatomic potential and the ability to treat "edge" effects and relaxation in the system on defect creation. In ionic and semi-ionic solids the codes HADES and CASCADE have been thoroughly tested and shown to be remarkably successful for calculating defect formation and migration energies (Catlow and Mackrodt, 1983; Harding, 1990). These codes use the Mott–Littleton (1938) two-region procedure, where an inner region of ions ($\sim$100 ions) is treated explicitly and a large outer region is treated as an elastic and dielectric continuum. Fast minimization techniques are used to determine the minimum-energy configuration of the system. Quite sophisticated potentials, either empirical or *ab initio*, can be employed, and the effects of polarization of the ions have been incorporated using shell-model potentials. In essence the static lattice methods are calculations at 0 K and yield the internal energies of defect formation or migration. However, using the code SHEOL (Harding, 1990) the calculations for ionic solids can be extended to yield defect-free energies, enthalpies, and entropies. It should be noted that there are other methods of computing the defect entropies.

MD calculations (see MOLECULAR DYNAMICS) are often regarded as "experiments" (Bennett, 1984; Ciccotti and Hoover, 1986). An ensemble of atoms is arranged in the simulation box and the system allowed to evolve. Typically the box will contain 100 atoms with periodic boundary conditions to generate an infinite crystal. The atoms interact with a given potential, and they are started from coordinates close to lattice positions with trajectories and velocities corresponding to some given temperature. The system is allowed to "run," which means solving the classical equations of motion in a succession of small time steps (typically 10^{-15} s, which is less than the characteristic time of any motional process for the atoms, e.g., a lattice vibration) and continually updating the positions and velocities. Following an initial equilibration period the simulation should model all the properties, structural and dynamic, of the system. This method is useful for systems where there is rapid motion and a significant number of diffusive events can be observed in the time scale of the simulation. A limitation has been the demands on computing power due to the large memory requirement and therefore the necessity to use rather simplified potentials. With the developments in computers this limitation is not expected to be a major problem in the future.

MC methods (*q.v.*) (Kehr and Binder, 1987) are useful where high levels of disorder are present and a variety of types of jump of an atom may be possible in a system. The technique involves generating an ensemble of configurations by random moves and using an "acceptance criterion" to decide which of the configurations should be included in the ensemble. This leads to a picture of the most likely jump processes in the system.

3. DIFFUSION IN DIFFERENT SOLID TYPES

3.1 Metals and Alloys

Because of their technological importance there have been extensive studies of point defects and diffusion in metals (Brebec, 1983; Philibert, 1986). The most recent comprehensive and up-to-date collection of data can be found in the survey edited by Mehrer (1990). As expected for close-packed structures, the dominant disorder is single vacancies. Inevitably there will be vacancy aggregates (divacancies, trivacancies, etc.) present, but the concentrations are significantly smaller, and it is difficult to obtain reliable thermodynamic parameters of these defects. A variety of different techniques have been used to determine values of h_v in pure metals, and in the well-studied systems the values obtained from different experiments are consistent [see, for example, the comparison between the values from the positron-annihilation technique and quenching experiments (Eldrup, 1986)]. The values are considerably smaller than the simple prediction given in Sec. 1.2.3 that h_v would be $\sim L_S$; typically $h_v \sim (0.3\text{–}0.4)L_S$. The origin of this discrepancy is the atomic relaxation and electron charge redistribution that occurs on vacancy formation in a metal. Explicit calculations of defect parameters in metals are complicated by the complex electronic structure and the difficulty in describing the interatomic potential.

In the thoroughly studied fcc metals, like Al, Ag, Au, Cu, and Pt, the self-diffusion occurs predominantly via a vacancy mechanism. In cases where h_v and $\Delta \mathscr{H}_v^m$ can be determined separately, the sum of the two parameters agrees well with experimental values of Q, as illustrated in Table 2. The table also demonstrates the empirical observation that $\Delta \mathscr{H}_v^m \simeq 0.8 h_v$ for fcc metals. The Arrhenius plots of self-diffusion measurements for some of these metals exhibit slight curvature, the apparent activation energy increasing as T_m is approached. This can be seen in the data for Ag shown in Fig. 10. The most common interpretation of this effect is that there is a contribution to the diffusion from the migration via divacancies (Brebec, 1983; Philibert, 1986). Although the divacancy concentration is generally much lower than c_v, it is argued that they are more mobile than the single vacancies and hence have a significant effect on D as the temperature approaches T_m. It should be noted that an alternative interpretation is that single vacancies alone are responsible for self-diffusion but the defect enthalpies and entropies are slightly temperature dependent because of anharmonicity of the lattice vibrations (Mundy, 1992).

The self-diffusion in a number of bcc metals, notably β-Ti, β-Zr, β-Hf, and the rare-earth elements, reveals a strong curvature of the Arrhenius plots, and these have long been regarded as "anomalous" systems. A number of explanations have been proposed to explain the effect [see, for example, the recent reviews (Brebec, 1983; Philibert, 1986; Herzig and Kohler, 1987; Mundy, 1992)], including the presence of oxygen impurity, the presence of bulk defects, a contribution from divacancy diffusion, and temperature-dependent single-vacancy parameters. The contribution from divacancies has been a favored explanation, but recent work (Herzig and Kohler, 1987) related the self-diffusion to soft-phonon modes and provides a uniform explanation of the data in terms of single vacancies.

Table 2. Experimental values for the enthalpy of vacancy formation, h_v, enthalpy of vacancy migration, $\Delta \mathscr{H}_v^m$, and self-diffusion in fcc metals.

	Experimental enthalpies eV		
Metal	h_v	$\Delta \mathscr{H}_v^m$	Q
Cu	1.30	0.80	2.05
Ag	1.10	0.83	1.91
Au	0.98	0.82	1.81
Al	0.77	0.65	1.46

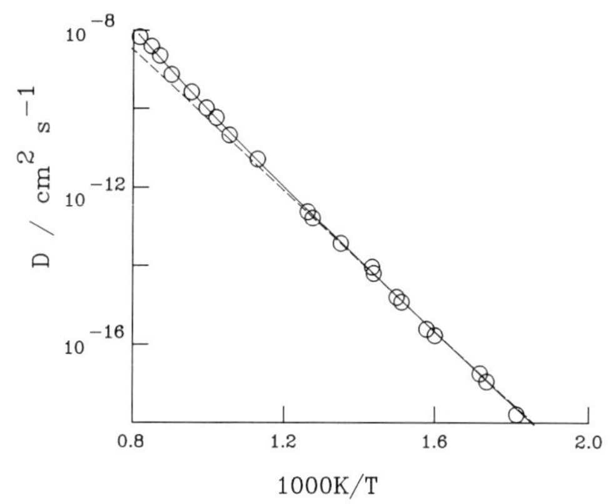

FIG. 10. Compiled self-diffusion coefficient of silver as a function of temperature (after Brebec, 1983).

The literature on metallic alloys is also very extensive, and a wide range of materials have been investigated (Brebec, 1983; Philibert, 1986). It is appropriate here to consider only the dilute alloys. The sites occupied in the host lattice depend mainly on the size of the impurity; atoms of a similar size will dissolve substitutionally, and small atoms can occupy interstitial sites (e.g., C, N, and O in bcc metals like α-Fe). Interstitial impurity diffusion in dilute alloys is consistent with the impurity migrating via the direct interstitial mechanism. In substitutional alloys the diffusion data are usually interpreted in terms of the *five-frequency model* (Brebec, 1983; Philibert, 1986). In this model the diffusion coefficients of the host impurity atoms are derived as functions of the defect concentrations and the five distinct types of jump of the vacancy:

1. vacancy adjacent to impurity exchanging with the impurity,
2. vacancy adjacent to impurity exchanging with host atom and remaining adjacent to the impurity,
3. vacancy adjacent to impurity exchanging with host atom and moving away from the impurity,
4. the reverse of **3**, and
5. vacancy exchanging with a host atom on sites not adjacent to the impurity.

There are some well-documented systems where the impurity exhibits anomalously fast diffusion in the alloy—for example, Cu, Pd, Au, Ni, and Ag in Pb, and Au in the alkali metals. In these cases the vacancy mechanism seems inappropriate and various forms of interstitial process have been suggested to explain the anomalies.

3.2 Molecular Solids

Molecular solids represent the largest group of materials, as they include all the organic and biological materials, which are virtually infinite in number given the facility of carbon to bond covalently with itself and a range of other elements (e.g., O, N, H, S, etc.). Information on the point defects in these materials is sparse, a result of both the experimental difficulties in studying these crystals (i.e., high vapor pressures, low purity) and the lack of technological interest in the systems as solids. The interest has been primarily of an academic nature, particularly in the case of the rare-earth gas solids where relative simplicity of the interatomic forces makes them ideal model systems.

It is convenient to divide molecular solids into three types:

1. *rare-gas crystals*,
2. *plastic crystals*—systems where the molecules are roughly spherical, undergo a rapid endospherical reorientation, and adopt highly symmetric crystal structures, usually fcc, bcc, and hcp (Sherwood, 1979), and
3. *nonplastic crystals*—the majority of organic crystals, where the molecular symmetry is low, the molecules are not reorienting, and the crystal structure is determined by packing factors.

In all three types the crystal structures are close packed and the expected dominant point defects are vacancies.

3.2.1 Rare-Gas Crystals. There have been a large number of theoretical estimates of the vacancy formation and migration parameters in rare-gas crystals (Chadwick and Glyde, 1977; Chadwick, 1983). The interatomic potentials employed range from Lennard–Jones potentials to sophisticated empirical potentials, and in most cases it is concluded that $h_v \simeq 1L_s$ and $\Delta \mathcal{H}_v^m \simeq 1L_s$, consistent with the simple predictions in Secs. 1.2.3 and 1.2.7. There have been several experimental studies of vacancy concentrations in rare-gas solids. The experiments are extremely difficult, and there is still some confusion over the experimental vacancy formation parameters. For example, the earlier experiments for solid krypton yielded $h_v \sim 0.7L_s$, whereas the value for neon was $1L_s$. The experiments for krypton were over a limited temperature range and the data analysis has been questioned, and there is a need for new studies on this system.

The diffusion data for rare-gas crystals from a number of sources, including radiotracers and NMR, suggest that these systems are well behaved (Chadwick and Glyde, 1977; Chadwick, 1983). The parameters are consistent with the simple theoretical predictions, i.e., Q is close to $2L_s$.

3.2.2 Plastic Crystals. In many respects the simpler plastic crystals can be considered as analogous to rare-gas crystals; the rapid molecular reorientation averages out the interactions between atoms on separate molecules, and these can be replaced by a single intermolecular potential acting between molecular centers of gravity. The simplest predictions (Chadwick, 1983) are therefore that $h_v \simeq 1L_s$ and $\Delta h_v^m \simeq 1L_s$. The same results have been obtained in the case of adamantane, where calculations have used atom–atom potentials (Meyer *et al.*, 1986). Reliable information on vacancy-formation parameters in plastic crystals is limited to a positron-annihilation study of two materials, adamantane and bicyclooctane, which yielded $h_v = 1L_s$ in both cases (Lightbody *et al.*, 1980).

There have been a large number of self-diffusion studies in plastic solids (Sherwood, 1979), as the motion is relatively rapid at moderate temperatures (D at T_m is typically 10^{-9}–10^{-8} cm^2 s^{-1} and T_m is less than 300 °C) and a wide number of diffusion techniques can be employed. NMR measurements for these materials are particularly appropriate, as D is in the required range and the organic plastic crystals usually contain a suitable nucleus (e.g., ^{1}H). In many cases the simple prediction of $Q \simeq 2L_s$ seems to be observed, but there are a number of systems where NMR methods yield lower values of Q (e.g., $Q \simeq 1L_s$ in cyclohexane and white phosphorus). This apparent anomaly is still unresolved.

3.2.3 Nonplastic Crystals. The diffusion in nonplastic solids is around 4 to 5 orders of magnitude *slower* than that in plastic crystals. The diffusion experiments are therefore difficult, and only a few systems, mainly simple aromatic ring systems like anthracene and naphthalene, have been studied. The general results are that $Q \simeq 2L_s$ and that the dominant point defects are single vacancies (Chadwick, 1983).

3.3 Covalent Solids

There has been considerable debate over the nature of the point defects in both elemental (e.g., C, Si, Ge) and compound covalent crystals. In essence the problems arise in assigning the electronic structure around the defect, e.g., what happens to the "dangling bonds" when an atom is removed to form a vacancy. Each broken bond contains one electron, which may occupy lower energy levels by forming new bonds across the vacancy with the dangling bonds of neighboring atoms. Self-diffusion and impurity diffusion have been very actively studied in these materials, particularly for systems where there are semiconductor applications, as the information is useful in device fabrication. In the elemental crystals the self-diffusion is extremely slow (D at T_m is typically 10^{-13}–10^{-12} cm^2 s^{-1}), and the mechanisms that are operative have not been fully resolved (Gosele and Tan, 1992; Leroy, 1990). It is now believed that both self-interstitials and vacancies are involved in the self-diffusion and most impurity diffusion processes in Si and GaAs, a suggestion made over 20 years ago (Seeger and Chik, 1968) that is now being substantiated by experiments.

3.4 Ionic Solids

Of all the solid types, the point defects and diffusion are best understood in ionic crystals. This is due to a combination of factors, some of which are evident from the foregoing sections. From the experimental viewpoint there are a number of good model systems, like the alkali and silver halides, that are readily available in the form of large single crystals. In addition, it is possible to control the defect concentrations precisely by means of doping with aliovalent impurities. However, the greatest advantage over other materials in defect and diffusion studies is that the technique of ionic conductivity provides a relatively simple and accurate means of determining defect-formation and -migration parameters. The combination of ionic conductivity and diffusion measurements also provides a means of identifying mechanisms via the Haven-ratio method. From the theoretical viewpoint the interionic potential can be well represented by fairly simple mathematical functions, and therefore computer simulation methods have proved highly successful.

In the last two decades it has been convenient to consider ionic solids as being of two types, *normal ionic conductors* and *fast-ion conductors*, and these will be discussed separately.

3.4.1 Normal Ionic Solids. Normal ionic conductors are materials like the alkali halides where the conductivity is well described by jump mechanisms of simple point defects and the ionic conductivity is around 10^{-3} S cm^{-1} at the melting point. Most of the detailed conductivity and diffusion work has focused on the simple systems like the metal halides, particularly those with the rock-salt and fluorite structures, because of the ease of experiment, although very similar behavior is expected for the higher-melting-point oxides. For a large number of systems there is now a wealth of experimental information, including the nature of the dominant point defects, the thermodynamic parameters for point-defect formation and migration, and the nature of the self-diffusion mechanisms (Lidiard, 1957; Corish and Jacobs, 1973; Chadwick, 1991; Jacobs, 1983). In many of these systems static lattice computer simulations have been extremely successful in calculating the defect parameters (Catlow, 1987).

The systems for which there is the best experimental information are the alkali halides, the silver halides, and the alkaline-earth fluorides. Some of these data are shown as an example in Table 3. Only the principal point-defect enthalpies are shown; however, in many cases the entropies and the parameters for defect association are available. The dominant defects in the alkali halides are Schottky pairs, and both ions diffuse by a vacancy mechanism. The diffusion coefficients of both ions are similar. For systems like NaCl and KCl there is a wealth of data, and the level of interpretation is now extremely sophisticated (see, for example, Hooton and Jacobs, 1988), including contributions to the ion transport from vacancy pairs and Frenkel defects (which are present in very low concentrations but are mobile enough to make a significant effect on the transport at high temperatures). In silver chloride and bromide the dominant point defects are cation Frenkel pairs, and not surprisingly the cation diffusion coefficient is about 1000 times higher than that for the anion. In the intrinsic region the silver ions move predominantly by interstitialcy mechanisms, both collinear and noncollinear, while in the extrinsic region of crystals doped with divalent cations (i.e., doped with excess cation

Table 3. Point-defect enthalpies (in eV) for simple ionic crystals.

Material	Point-defect enthalpy	Experimental value determined from ionic conductivity	Calculated value from computer simulation
NaCl	h_S	2.41	2.32
	Δh^m_{cv}	0.67	0.65
	Δh^m_{av}	0.77	0.71
KCl	h_S	2.50	2.50
	Δh^m_{cv}	0.67	0.71
	h^m_{av}	0.85	0.69
AgCl	h_{cF}	1.49	1.4
	Δh^m_{ci} (noncollinear)	0.10–0.12	
	Δh^m_{ci} (collinear)	0.002–0.004	
	Δh^m_{cv}	0.27–0.28	
CaF_2	h_{aF}	1.85	1.98
	Δh^m_{av}	0.52	0.40
	Δh^m_{ai} (noncollinear)	0.79	0.72
BaF_2	h_{aF}	1.91	1.98
	Δh^m_{av}	0.57	0.46
	Δh^m_{ai} (noncollinear)	0.76	0.72
MgO	h_S	5–7	7.5
	Δh^m_{cv}	2.2–2.6	2.18

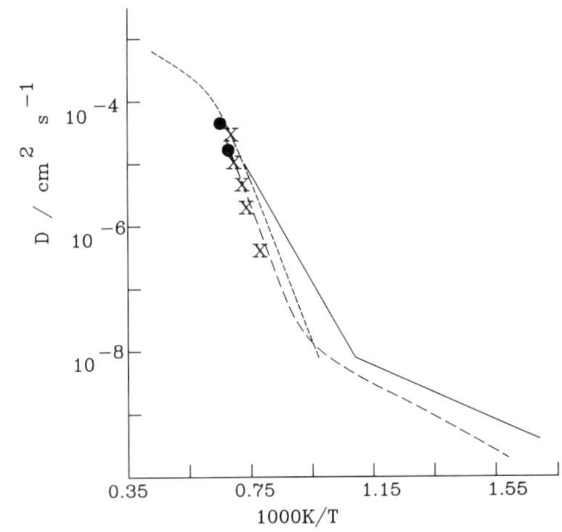

FIG. 11. Ionic transport data for Li_2O. A plot of the Li^+ diffusion coefficient from ionic conductivity (long-dashed curve), molecular-dynamics simulations (short-dashed curve), static lattice simulations of HADES and SHEOL (solid line), tracer (crosses), and neutron data (filled circles) (after Flack, 1990).

vacancies) the silver ions diffuse by a vacancy mechanism. The alkaline-earth fluorides at moderate temperature, up to about $0.8T_m$, behave as normal ionic crystals with anion Frenkel pairs as the dominant defects, and the fluoride ion diffuses predominantly by the vacancy mechanism, although there is a significant contribution from the noncollinear interstitialcy process (Jacobs, 1983).

The success of the computer modeling when applied to the point-defect parameters of ionic crystals is also seen in Table 3. In a recent study (Chadwick *et al.*, 1988; Flack, 1990) of lithium oxide, Li_2O (a material with the antifluorite structure), it was possible to compare experimental conductivity and diffusion data with a range of computer-simulation methods. The results are shown in Fig. 11 and the agreement is excellent. The successful testing of the computer calculations in simple systems means that they can now be used with confidence in predicting the behavior of more complex ionic materials.

3.4.2 Fast-Ion Conductors. There are a number of ionic solids with unusually high ionic conductivities, the magnitude of the conductivity approaching that found in molten salts and aqueous electrolyte solutions, and these are now variously referred to as *fast-ion conductors* or *superionic conductors* or *solid electrolytes*. Tremendous interest has been shown in these materials for technological applications where a purely ionic conductor is required, e.g., in solid-state batteries and fuel cells. There is an extensive literature on these materials, including books (Chandra, 1981; Laskar and Chandra, 1990; Stoneham, 1989), reviews (Schulz, 1982; Catlow, 1990; Chadwick, 1992), and a specialized journal (*Solid State Ionics*). Conductivity plots for a few fast-ion conductors are shown in Fig. 12. It can be seen that a wide range of solid types exhibit the phenomenon, i.e., fluorite-structured solids at high temperature (e.g., PbF_2), silver iodide and related compounds (e.g., $RbAg_4I_5$), and layered structures (e.g., β-aluminas). Clearly in terms of Eq. (81), in order for there to be a high ionic conductivity the point defect in the solid must have exceptionally high concentrations and/or exceptionally high mobilities.

The types of ion that show fast-ion conduction are usually small with low formal charges, e.g., Li^+, Na^+, F^-, H^+, and O^{2-}. It has now been realized that there is no unique origin to fast-ion conduction in the different systems, and therefore a variety of theoretical approaches have to be adopted. There are several ways of classifying fast-ion conductors, depending on their structure (open, tunnel, layered, glasses, polymers, etc.) or their applications. A very useful approach developed by Catlow (1990) and explored by others has been to use a hybrid classification based on structure and mechanism. Thus the types of class include solids that are fast-ion conductors as a result of a phase transition (like AgI), massively disordered and heavily doped materials (such as ZrO_2 stabilized in the cubic

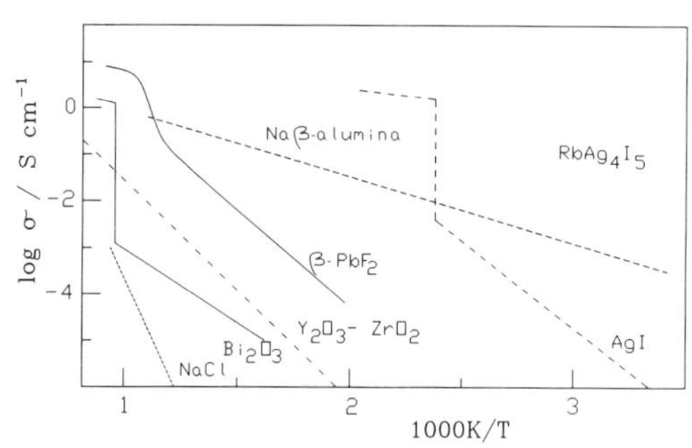

FIG. 12. Conductivity plots for several fast-ion conductors.

phase by impurity), and layered structures (like the β-aluminas).

There are still many problems to be resolved in fast-ion conductors, particularly the precise nature of the diffusion mechanisms. In many cases the diffusion is so rapid that many of the conventional experimental probes are inadequate. An additional problem is the lack of availability of good single crystals, and a measurement of the ionic conductivity is virtually the only transport technique that is applicable. Thus there is often a lack of precise information on which to base deductions concerning mechanisms. In much of the early work it was believed that the diffusion of the ions was liquidlike, with ions in continuous motion, and that defect-jump models would be inappropriate. However, a picture is now emerging, particularly from molecular-dynamics computer simulations (Gillan, 1989), that in most fast-ion conductors the mobile ion spends most of the time on a lattice site and a relatively short time "in transit" between sites. Molecular-dynamics simulations are particularly applicable to fast-ion conductors, as the diffusion is sufficiently rapid that a significant number of diffusive jumps can be observed in a computational time that is not prohibitive.

4. DIFFUSION IN AN APPLIED FIELD

The investigation of diffusion in an applied field, either an electric field (*electromigration*) or a temperature gradient (*thermomigration*), is difficult from both experimental and theoretical viewpoints (Allnatt and Chadwick, 1967; Wever, 1973; Hummel and Huntington, 1977; Gillan, 1983). However, the phenomena are extremely important technologically as they can lead to "unimixing" of a system. When there are large electric fields and current densities, as in a thin-film metal connection on a semiconductor device, electromigration can lead to a net movement of the metal in one direction and an eventual break in the connection. Similarly, a large temperature gradient, as is found in the walls of a furnace or in the fuel element of a nuclear reactor, can lead to segregation of one of the components of a material and ultimately mechanical or chemical breakdown. The academic interest in these phenomena lies in the fact that they provide a severe test of the atomistic models of diffusion in solids.

There have been few studies of electromigration. One method of study of impurity electromigration is to place a tracer between two slices of material and measure the drift of the tracer after the application of a large field across (and large current through) the sandwich. Self-electromigration is studied by placing markers along a specimen and noting their relative movement in the applied field. In metals the movement of the atoms can be considered as due to the *effective charge* of the atom interacting with the flow of electrons, termed the *electron wind effect*.

Thermomigration in pure solids can also be studied in marker-movement experiments, and these have been used to study metals. However, the experiments are extremely difficult to perform, as distortion of the samples complicates the interpretation. For impurity thermomigration it is possible to use the Soret effect, the steady concentration gradient of impurity developed in the temperature gradient. From Eq. (15) at steady state the flux will be zero and hence $Q^{*\prime}$ can be deduced from the concentration gradient, i.e.,

$$\frac{1}{c}\left(\frac{\delta c}{\delta x}\right)=\frac{Q^{*\prime}}{kT^2}\left(\frac{dT}{dx}\right) . \tag{87}$$

This experimental approach has been used to study thermomigration of impurities in a variety of types of solids. In ionic crystals the effect of thermomigration is to tend to separate the charged species. The incipient separation is opposed by an electric field, the *thermoelectric power*, and this provides a very convenient means of studying thermomigration (Allnatt and Chadwick, 1967; Gillan, 1983) and determining the heats of transport.

From the earlier description of the heat of transport, the heat carried from the initial site to the final site in an isothermal jump of an atom, it is clearly a complex parameter to evaluate, since at the atomic level the energy of the jump is spread over many atoms and $Q^{*\prime}$ depends on how this energy is redistributed after the jump. An early, simple theory, due to Wirtz (1943), predicted that $Q^{*\prime}$ would be equal to or less than the activation enthalpy of migration of the atom, Δh^m. Later this was simplified to $Q^{*\prime}=\Delta h^m$ in the LeClaire–Brinkman theory (LeClaire, 1954; Brinkman, 1954), i.e., the atom carries all the migration enthalpy from initial site to final site. In systems where there is simple vacancy diffusion, e.g., in alkali halide crystals, this seems to be reasonable as experiments show $Q^{*\prime} \sim \Delta h^m$. However, in cases where interstitial migration is known then this theory fails and does not even predict the correct sign for $Q^{*\prime}$; in these systems experimental values of $Q^{*\prime}$ are usually negative (Allnatt and Chadwick, 1967; Gillan, 1983). This probably reflects the highly cooperative nature of interstitial diffusion processes.

5. SUMMARY

The objective of this article has been to present the basics of the diffusion and ionic conductivity in solids to form a foundation on which to build by further reading. The nature of the different types of point-defect diffusion has been presented along with an introduction to defect thermodynamics. The situation in the different types of solid can be summarized as follows:

Metals: In pure metals the dominant defects are single vacancies. The defect physics of fcc metals is well understood as a result of reliable measurements of defect concentrations and diffusion. In bcc metals the dominant defects are also single vacancies, although there has been some debate concerning the role of divacancies in diffusion. The lattice positions occupied by impurities depends mainly on their size, small impurities (with respect to the host atoms) occupying interstitial sites and larger impurities dissolving substitutionally.

Covalent solids: The defects are less well understood in these systems because of the role of "dangling bonds" and the very low diffusion coefficients. The diffusion is complicated and appears to involve both vacancies and interstitial defects in some form.

Molecular solids: These have not been thoroughly studied, mainly because of the lack of technological applications. As the structures are close packed it is generally assumed that the dominant defects are single vacancies and diffusion is via a vacancy mechanism; these assumptions are upheld by the very limited amount of experimental data.

Ionic solids: These are extremely well understood in terms of their defect physics as the result of the ability to vary the defect concentrations deliberately, experimentally, by doping with aliovalent ions. In addition, ionic conductivity measurements provide an excellent probe of both the defect concentrations and mobilities.

The references given in the list of Further Reading will provide plenty of material for the interested reader to extend his or her knowledge and to consider the various processes where the defects and diffusion play a vital role, for example, in corrosion, catalysis, the photographic process, materials design, etc.

GLOSSARY

Correlation Factor: The degree of correlation between successive diffusive jumps of a labeled atom in a solid (more correctly termed the *Bardeen–Herring correlation factor*).

Divacancy: Two vacancies on neighboring lattice sites.

Doping: The deliberate addition of impurities to a solid.

Electromigration: Diffusion in an applied electric field.

Frenkel Defect: Defect in an ionic crystal consisting of a vacancy and an interstitial of one of the ions (hence two types are possible: *cation Frenkel defects* and *anion Frenkel defects*).

Fast-Ion Conductors: Solid with an unusually high ionic conductivity (sometimes referred to as *superionic conductors*).

Haven Ratio: For an ionic solid, the ratio of the tracer diffusion coefficient to the diffusion coefficient calculated from the ionic conductivity via the Nernst–Einstein equation.

Interstitial Atom: An atom on an interstitial lattice site in a crystal.

Interstitial Diffusion: The diffusion process in which an interstitial atom jumps from interstitial site to interstitial site.

Interstitialcy Diffusion: The diffusion process in which an interstitial atom jumps onto a normal lattice site as the atom on that site moves onto an interstitial site.

I-V Pairs: An impurity atom and a vacancy on neighboring lattice sites.

NMR: Nuclear magnetic resonance.

Plastic Crystal: A molecular solid with globular molecules that are undergoing rapid endospherical reorientation. The crystals have high symmetry (cubic or hexagonal lattices) and tend to be easily plastically deformed.

Point Defect: A defect in a crystal at the atomic level (see **Vacancy** and **Interstitial Atom**).

Schottky Defects: Point defect in an ionic solid consisting of cation and anion vacancies.

SIMS: Secondary-ion mass spectrometry.

Site Fraction: The concentration of a species in a crystal measured as the ratio of the number of molecules of the species to the total number of sites in the crystal the species could occupy.

Thermomigration: Diffusion in an applied temperature gradient.

Tracer Diffusion: The measurement of diffusion coefficients by monitoring the penetration of isotopic tracer atoms into a solid.

Vacancy: An unoccupied lattice site in a crystal.

Vacancy Diffusion: The diffusion process in which an atom moves by jumping into a neighboring vacancy.

List of Works Cited

Allnatt, A. R., Chadwick, A. V. (1967), *Chem. Rev.* **67**, 681–705.

Archer, W. I., Armstrong, R. D. (1981), in: H. R. Thirsk (Ed.), *Electrochemistry*, London: The Chemical Society, p. 157.

Bardeen, J., Herring, C. (1952), in: W. Shockley (Ed.), *Imperfections in Nearly Perfect Crystals*, New York: Wiley, p. 261.

Bauerle, J. E. (1969), *J. Phys. Chem. Solids* **30**, 2657–2670.

Bénière, F. (1983), in: F. Bénière, C. R. A. Catlow (Eds.), *Matter Transport in Solids*, NATO Advanced Study Institute, Series B, Physics, Vol. 97, New York: Plenum, p. 21.

Bennett, C. H. (1984), in: G. E. Murch, A. S. Nowick (Eds.), *Diffusion in Crystalline Solids*, New York: Academic, p. 73.

Brebec, G. (1983), in: F. Bénière, C. R. A. Catlow (Eds.), *Matter Transport in Solids*, NATO Advanced Study Institute, Series B, Physics, Vol. 97,New York: Plenum, p. 251.

Brinkman, J. A. (1954), *Phys. Rev.* **93**, 345.

Catlow, C. R. A. (1987), in: A. K. Cheetham, P. Day (Eds.), *Solid State Chemistry: Techniques*, Oxford: Clarendon, p. 231.

Catlow, C. R. A. (1990), *J. Chem. Soc. Faraday Trans.* **86**, 1167–1176.

Catlow, C. R. A., Mackrodt, W. C. (Eds.) (1983), *Computer Simulation of Solids*, Berlin: Springer.

Chadwick, A. V. (1983), in: F. Bénière, C. R. A. Catlow (Eds.), *Matter Transport in Solids*, NATO Advanced Study Institute, Series B, Physics, Vol. 97, New York: Plenum, p. 285.

Chadwick, A. V. (1991), *Philos. Mag. A* **64**, 983–998.

Chadwick, A. V. (1992), in: G. E. Murch (Ed.), *Unsolved Problems in Diffusion*, Zurich: Trans. Tech., p. 489.

Chadwick, A. V., Glyde, H. R. (1977), in: M. L. Klein, J. A. Venables (Eds.), *Rare Gas Solids*, New York: Academic, Vol. 2, Chap. 19.

Chadwick, A. V., Flack, K. W., Strange, J. H., Harding, J. H. (1988), *Solid State Ionics*, **28-30**, 185–188.

Chandra, S. (1981), *Superionic Solids*, Amsterdam: North-Holland.

Ciccotti, G., Hoover, W. G. (Eds.) (1986), *Molecular Dynamics Simulation of Statistical Mechanical Systems*, Amsterdam: North-Holland.

Corish, J., Jacobs, P. W. M. (1973), in: M. W. Roberts, J. M. Thomas (Eds.), *Surface and Defect Properties of Solids*, London: The Chemical Society, Vol. 2, p. 184.

Crank, J. (1975), *The Mathematics of Diffusion*, Oxford: Clarendon.

Crawford, J. H., Slifkin, L. M. (Eds.) (1972), *Point Defects in Solids*, New York: Plenum, Vols. 1–3.

de Groot, S. R., Mazur, P. (1962), *Non-Equilibrium Thermodynamics*, Amsterdam: North-Holland.

Eldrup, M. (1986), in: A. V. Chadwick, M. Terenzi (Eds.), *Defects in Solids—Modern Techniques*, NATO Advanced Study Institute, Series B, Physics, Vol. 147, New York: Plenum, p. 145.

Farrar, T. C., Becker, E. D. (1971), *Pulse and Fourier Transform NMR*, New York: Academic.

Flack, K. W. (1990), Ph.D. thesis, University of Kent (unpublished).

Flynn, C. P. (1972), *Point Defects and Diffusion*, Oxford: Clarendon.

Fukushima, E., Roeder, S. B. W. (1981), *Experimental Pulse NMR: A Nuts and Bolts Approach*, Reading, MA: Addison-Wesley.

Gillan, M. J. (1983), in: F. Bénière, C. R. A. Catlow (Eds.), *Matter Transport in Solids*, NATO Advanced Study Institute, Series B, Physics, Vol. 97, New York: Plenum, p. 227.

Gillan, M. J. (1989), in: A. M. Stoneham (Ed.), *Ionic Solids at High Temperatures*, Singapore: World Scientific.

Glasstone, S., Eyring, H., Laidler, K. J. (1941), *The Theory of Rate Processes*, New York: McGraw-Hill.

Glyde, H. R. (1967), *Rev. Mod. Phys.* **39**, 373–382.

Gosele, U., Tan, T. Y. (1992), in: G. E. Murch

(Ed.), *Unsolved Problems in Diffusion*, Zurich: Trans. Tech., p. 189.

Greaves, G. N. (1990), in: D. R. Uhlmann, N. Kreidl (Eds.), *Glass Science and Technology*, London: Academic, Vol. 4B, p. 1.

Harding, J. H. (1990), *Rep. Prog. Phys.* **53**, 1403–1466.

Herzig, C., Kohler, U. (1987), *Mater. Sci. Forum* **1–3**, 301–322.

Hooton, I. E. (1988), Ph.D. thesis, University of Western Ontario (unpublished).

Hooton, I. E., Jacobs, P. W. M. (1988), *Can. J. Chem.* **66**, 830–835.

van Hove, L. (1954), *Phys. Rev.* **95**, 249–262.

Howard, R. E., Lidiard, A. B. (1964), *Rep. Prog. Phys.* **27**, 161–240.

Hummel, R. E., Huntington, H. B. (Eds.) (1977), *Electro- and Thermotransport in Alloys*, New York: AIME.

Jacobs, P. W. M. (1983), in: F. Bénière, C. R. A. Catlow (Eds.), *Matter Transport in Solids*, NATO Advanced Study Institute, Series B, Physics, Vol. 97, New York: Plenum, p. 81.

Jost, W. (1960), *Diffusion in Solid, Liquids and Gases*, New York: Academic.

Kanert, O. (1982), *Phys. Rep.* **91**, 183–232.

Kaur, I., Gust, W. (1988), *Fundamentals of Boundary and Interface Boundary Diffusion*, Stuttgart: Zeigler.

Kehr, R., Binder, K. (1987), in: K. Binder (Ed.), *Applications of Monte Carlo Simulation and Statistical Physics*, Topics in Current Physics Vol. 36, Berlin: Springer, p. 181.

King, A. H. (1987), *Int. Mater. Rev.* **32**, 173–189.

Laskar, A. L., Chandra, S. (Eds.) (1990), *Supersonic Solids and Solid Electrolytes*, New York: Academic.

Lechner, R. (1983), in: F. Bénière, C. R. A. Catlow (Eds.), *Matter Transport in Solids*, NATO Advanced Study Institute, Series B, Physics, Vol. 97, New York: Plenum, p. 169.

LeClaire, A. D. (1954), *Phys. Rev.* **93**, 344.

LeClaire, A. D. (1970), in: H. Eyring, D. Henderson, W. Jost (Eds.), *Physical Chemistry—An Advanced Treatise*, New York: Academic, Vol. 10, p. 261.

Leroy, B. (1990), in: A. L. Laskar, G. Brebec, J. L. Boucquet, C. Monty (Eds.), *Diffusion in Materials*, NATO Advanced Study Institute, Series E, Applied Sciences, Vol. 179, Dordrecht: Kluwer, p. 525.

Lidiard, A. B. (1957), in S. Flügge (Ed.), *Handbuch der Physik*, Berlin: Springer, Vol. 20, p. 246.

Lightbody, D., Eldrup, M., Sherwood, J. N. (1980), *Chem. Phys. Lett.* **70**, 487–491.

Lodding, A. (1987), in: F. Adams, R. Gijbels, R. van Grieken (Eds.), *Inorganic Mass Spectroscopy*, New York: Wiley, p. 125.

MacDonald, J. R. (1987), *Impedance Spectroscopy*, New York: Wiley.

Macht, M. P., Naundorf, V. (1982), *J. Appl. Phys.* **53**, 7551–7557.

Meyer, M., Mahric, C., Ciccotti, G. (1986), *Mol. Phys.* **58**, 723–733.

Mehrer, H. (Ed.) (1990), *Diffusion in Solid Metals and Alloys*, Landolt-Bőrnstein, New Series Vol. III/26, Berlin: Springer.

Mott, N. F, Littleton, M. J. (1938), *Trans. Faraday Soc.* **34**, 485–499.

Mundy, J. N. (1992), in: G. E. Murch (Ed.), *Unsolved Problems in Diffusion*, Zurich: Trans. Tech., p. 1.

Mundy, J. N., Rothman, S. J. (1983), in: J. N. Mundy, S. J. Rothman, M. J. Fluss, L. C. Smedskjaer (Eds.), *Methods in Experimental Physics*, New York: Academic, Vol. 21, p. 1.

Peterson, N. L. (1975), in: A. S. Nowick, J. J. Burton (Eds.), *Diffusion in Solids, Recent Developments*, New York: Academic, p. 115.

Philibert, J. (1986), in: A. V. Chadwick, M. Terenzi (Eds.), *Defects in Solids—Modern Techniques*, NATO Advanced Study Institute, Series B, Physics, Vol. 147, New York: Plenum, p. 349.

Prigogine, I. (1967), *Introduction to the Thermodynamics of Irreversible Processes*, New York: Wiley-Interscience.

Schulz, H. (1982), *Annu. Rev. Mater. Sci.* **12**, 351–376.

Seeger, A., Chik, C. P. (1968), *Phys. Status Solidi* **29**, 455–542.

Sherwood, J. N. (Ed.) (1979), *The Plastically Crystalline State*, London: Wiley.

Springer, T., Richter, D. (1987), in: D. L. Price, K. Sköld (Eds.), *Methods of Experimental Physics*, Orlando, FL: Academic, Vol. 23B, p. 131.

Stoneham, A. M. (Ed.) (1989), *Ionic Solids at High Temperatures*, Singapore: World Scientific.

Strange, J. H. (1986), in: A. V. Chadwick, M. Terenzi (Eds.), *Defects in Solids—Modern Techniques*, NATO Advanced Study Institute, Series B, Physics, Vol. 147, New York: Plenum, p. 243.

Wever, H. (1973), *Elektro- und Thermotransport in Metallen*, Leipzig: Barth.

Wirtz, K. (1943), *Physik. Z.* **44**, 221–231 and 369–392.

Wolf, D. (1979), *Spin Temperature and Nuclear Spin Relaxation in Matter*, Oxford: Clarendon.

Further Reading

Textbooks Devoted to Diffusion in Solids

Adda, Y., Philibert, J. (1966), *La Diffusion dans les Solides*, Paris: Presses Universitaire de France, Vols. 1 and 2.

Flynn, C. P. (1972), *Point Defects and Diffusion*, Oxford: Clarendon Press.

Kofstad, P. (1972), *Non-Stoichiometry, Diffusion and Electrical Conduction in Binary Metal Oxides*, New York: Wiley.

Manning, J. R. (1968), *Diffusion Kinetics for Atoms in Crystals*, Princeton, NJ: Van Nostrand.

Philibert, J. (1985), *Atom Movements*, Paris: Les Editions de Physique: translation into English by S. J. Rothman (1991) with additional material.

Shewmon, P. G. (1963), *Diffusion in Solids*, New York: McGraw-Hill. 2nd ed. (1989), Warrendale, PA: The Metallurgical Society of AIME.

Edited Texts on Diffusion in Solids

Abromeit, C., Wollenberger, H. (Eds.) (1988), *Interstitials and Vacancies in Metals and Alloys*, Aedermansdorf, Switzerland: Trans. Tech.

Crawford, J. H., Slifkin, L. M. (Eds.) (1972), *Point Defects in Solids*, New York: Plenum, Vols. 1–3.

Dayananda, M., Murch, G. E. (Eds.) (1985), *Diffusion in Solids, Recent Developments*, Warrendale, PA: The Metallurgical Society of AIME.

Lodding, A., Lagerwall, T. A. (Eds.) (1970), *Atomic Transport in Solids and Liquids*, Tübingen, Germany: Zeitschrift für Naturforschung.

Murch, G. E. (Ed.) (1992), *Unsolved Problems in Diffusion*, Zurich: Trans. Tech.

Murch, G. E., Nowick, A. S. (Eds.) (1984), *Diffusion in Crystalline Solids*, New York: Academic.

Murch, G. E., Birnbaum, H. K., Cost, J. (Eds.) (1984), *Non-traditional Methods in Diffusion*, Warrendale, PA: The Metallurgical Society of AIME.

Nowick, A. S., Burton, J. J. (Eds.) (1975), *Diffusion in Solids, Recent Developments*, New York: Academic.

Shaw, D. (Ed.) (1973), *Atomic Diffusion in Semiconductors*, New York: Plenum.

Sherwood, J. N., Chadwick, A. V., Muir, W. M., Swinton, F. S. (Eds.) (1971), *Diffusion Processes*, London: Gordon and Breach, Vols. 1 and 2.

Proceedings of Conferences Devoted to Diffusion in Solids

Bénière, F., Catlow, C. R. A. (Eds.) (1983), *Matter Transport in Solids*, NATO Advanced Study Institute, Series B, Physics, Vol. 97, New York: Plenum.

Chadwick, A. V., Terenzi, M. (Eds.) (1986), *Defects in Solids—Modern Techniques*, NATO Advanced Study Institute, Series B, Physics, Vol. 147, New York: Plenum.

Kedves, F. J., Beke, D. L. (Eds.) (1983), *DIMETA-82, Diffusion in Metals and Alloys*, Aedermansdorf, Switzerland: Trans. Tech.

Kedves, F. J., Beke, D. L. (Eds.) (1990), *DIMETA-88, Diffusion in Metals and Alloys*, Aedermansdorf, Switzerland: Trans. Tech.

Laskar, A. L., Brebec, G., Bocquet, J. L., Monty, C. (Eds.) (1990), *Diffusion in Materials*, NATO Advanced Study Institute, Series E, Applied Sciences, Vol. 179, Dordrecht: Kluwer.

IONOSPHERE

ALV EGELAND, *University of Oslo, Oslo, Norway*
NELSON C. MAYNARD, *Phillips Laboratory, Hanscom Air Force Base, Massachusetts, U.S.A.*

	Introduction	223
1.	**Definition and Historical Preamble**	223
2.	**The Earth's Upper Atmosphere and Magnetic Field**	224
3.	**Ionization Sources**	228
4.	**The Ionospheric Layers**	229
4.1	The Topside Ionosphere (>400 km)	230
4.2	The F Region (>140 km)	230
4.3	The E Layer (90–140 km)	231
4.4	The D Region (50–90 km)	231
5.	**Global Distribution of Ionization**	232
6.	**Electric Fields, Conductivity, and Currents**	233
7.	**Aurora—The Polar Light**	236
8.	**Dynamics of the Ionosphere**	239
9.	**Radio Propagation Effects**	240
10.	**Planetary Ionospheres**	240
11.	**Questions for the Future**	241
	Glossary	242
	Works Cited	242
	Further Reading	243

INTRODUCTION

The formation, structure, and dynamics of the ionosphere—a layer of partially ionized gas extending from about 50 to beyond 1000 km—are the central topics of this article. Up to the start of the space age—around the 1960s—our main information about the ionosphere was obtained by the use of radio waves reflected from the different ionospheric layers. During the last three decades instrumented rockets and satellites have been widely used in the studies of the ionosphere, particularly the ionosphere poleward of approximately 60° latitude.

The polar ionosphere is much more complicated than the ionosphere at lower latitudes. While solar radiation with wavelengths below ~250 nm is the principal ionization source of the low-latitude ionosphere, energetic electrons and ions from the magnetosphere and solar wind significantly contribute to the production of the polar ionosphere. These charged particles are also the main source of the aurora, contributing to intense electric fields and currents within the ionosphere. The electric and magnetic fields control the motion of electrons and ions within the ionosphere and thereby influence the global distribution of ionization.

After a short historical preamble and a brief review of the Earth's neutral atmosphere and its magnetic field, we will concentrate on the production and the characteristics of the ionosphere. Physical and chemical processes as well as interactions between the different ionospheric regions and the magnetosphere will be discussed. A section on ionospheres of other planets together with a short summary of items that should have high priority during the coming decade conclude this article.

1. DEFINITION AND HISTORICAL PREAMBLE

The Earth's ionosphere extends from ~50 to over 1000 km and is a partially ionized gas that envelops the Earth, interfacing the upper atmosphere with the magnetosphere. Electrons in concentrations from 10^8 to greater than 10^{12} m^{-3} roam free from their parent ions and move with their thermal speed under the influence of the ambient electric (**E**) and magnetic (**B**) fields in an anisotropic manner relative to the geomagnetic field. As the atmo-

1-56081-067-X/94/$5.00+.50

spheric density decreases with increasing altitude, the motion of charged particles changes from collision-dominated to collisionless motion following **E**×**B** drift paths. The electrons carry currents and influence radio-wave propagation, an effect that historically provided the first clues of the ionosphere's existence.

When Appleton and co-workers in 1925 had demonstrated the existence of the Earth's ionosphere by the use of radio waves and obtained reliable measurements of its height, studies were begun of both the nature of the upper atmosphere and the ionospheric formation (cf. Ratcliffe, 1972). Naturally, one first studied the simpler ionospheric aspects and only later concentrated on the more complicated polar ionosphere. Actually, between 1925 and 1955 ground-based radio measurements led to a fairly complete understanding of the average characteristics of the ionosphere and the atmosphere between the heights of about 90 and 300 km. The name ionosphere for this region was coined by R. A. Watson-Watt in 1926.

Quite a bit was known about the "ionosphere" even before it was discovered by the radio technique. The First International Polar Year (1882–83) could be regarded as marking the beginning of modern ionospheric and auroral research. Studies of the Earth's magnetic field by Stewart (1882) suggested that the upper atmosphere acted as the armature of an "atmospheric dynamo." Thus, the current hypothesized by Stewart explains the regular daily variations (S_q; S is short for Sun and q for quiet) of the geomagnetic field observed on the ground at mid and low latitudes. Two important suggestions regarding the ionosphere were made around the turn of our century, namely:

1. A. E. Kennelly and O. Heaviside suggested the existence of a layer of ionized gas in the upper atmosphere to explain G. Marconi's transmission of radio signals from England to Newfoundland.
2. On the basis of data from a network of geomagnetic observatories in the auroral regions, Birkeland (1908) introduced and estimated the heights of "ionospheric currents" 25 years before the basic character of the ionosphere was actually established (cf. Ratcliffe, 1972; Chapman and Bartels, 1940).

Probably the most spectacular of nature's phenomena, the aurora (also called the northern light and/or the polar light) has stirred man's imagination, curiosity, and fear since the dawn of man. The aurora is the visible means by which we can observe the coupling of energy from the Sun via the magnetosphere to the ionosphere and atmosphere. These processes involve field-aligned currents (now referred to as Birkeland currents) and horizontal currents (the auroral electrojet) in the polar ionosphere. It was from studies of the aurora around the turn of the century that sunspots and charged particles were first tied to ionospheric dynamics by Birkeland (1908).

Man's interest in the ionosphere mainly developed because of the need for long-distance radio communications and out of general curiosity. A fairly detailed ionospheric knowledge would be desirable in improving and extending communication quality. Today, modern, *in situ* ionospheric research is mainly conducted through the use of sophisticated instruments on board rockets and satellites, closely coordinated with data from advanced ground-based equipment. Even active controlled experiments—e.g., using radio transmitters to heat the ionosphere from the ground and/or chemical and charged-particle injections from rockets and satellites—are today regularly performed. The use of instrumented rockets and satellites, beginning about 1955, at first mainly confirmed previous findings and ideas about the ionosphere. However, after about 1960 new information about the topside ionosphere (i.e., above about 400 km) as well as the lower region (i.e., below 90 km) was obtained. In addition, new ground-based techniques, such as partial-reflection, cross-modulation, coherent-scatter, and incoherent-scatter radars, have yielded a lot of new information about the ionosphere during the last 30 years.

2. THE EARTH'S UPPER ATMOSPHERE AND MAGNETIC FIELD

The Earth's upper atmosphere and its magnetic field strongly influence the structure and dynamics of the ionosphere. The atmospheric composition affects the density of charged particles while the geomagnetic field influences charged-particle transport. Only those aspects important to the ionosphere will be discussed here.

We define the upper atmosphere as the atmosphere above the stratopause, i.e., above ~50 km. This is also where ionization levels become significant. The atmosphere has no sharp upper limit, but here we will concentrate on the atmosphere below ~1000 km.

Density, composition, and temperature are the atmospheric properties of primary importance to the ionosphere because of their effects on collision frequency and recombination processes, respectively. The average density profiles for the major atmospheric constituents of an International Standard Atmosphere are sketched in Fig. 1 (see also ATMOSPHERIC PHYSICS; ATMOSPHERIC STRUCTURE). Typical temperature curves are shown in Fig. 2. The density curves represent idealized, year-round, mid-latitude conditions under mean solar activity. Large temporal and spatial variations occur, especially above 200 km. The primary gases up to ~120 km are molecular oxygen (≈20%) and nitrogen (78%), while atomic oxygen is very important above 120 km. Minor constituents such as water vapor (H_2O), nitric oxide (NO), hydroxyl (OH), and sodium (Na) are all important for photochemical processes in the upper atmosphere up to 100 km. Notice that density and pressure decrease rapidly (approximately exponentially) with height up to approximately 100 km. Above this height the decrease is significantly slower (see Fig. 1). The atmosphere is considered to rotate with the Earth (cf. *U.S. Standard Atmosphere*, 1976).

The energy budget for the upper atmosphere, whose main energy source is ultraviolet radiation from the Sun, is sketched in Fig. 3. Electromagnetic radiation from the Sun (in units of W m^{-2} nm^{-1}) outside the atmosphere and at the Earth's surface is shown in Fig. 4. The albedo is approximately 0.3, which means that ~30% of the radiation is reflected back into space. Roughly 45% of the energy is found within the visible spectrum, normally defined as from 400 to 700 nm, while approximately 10% of the energy is within the ultraviolet (uv) and x-ray bands. The total energy—integrated over all wavelengths—is called the solar constant because it is practically independent of solar activity; however, the uv segment of the spectrum varies significantly with solar activity. The solar constant outside the atmosphere is ≈1.370 kW m^{-2}. If averaged over the whole sphere, a realistic value at the Earth's surface of 250 W m^{-2} is obtained. The upper atmosphere acts as a filter because, as it becomes ionized, it absorbs practically all x rays and uv radiation from the Sun (i.e.,

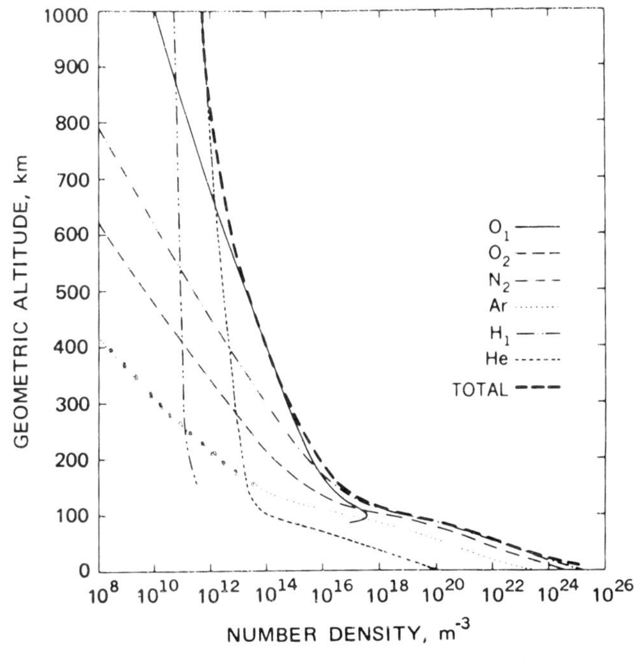

FIG. 1. Number densities of individual species and total number density as functions of altitude for the International Standard Atmosphere (from *U.S. Standard Atmosphere*, 1976).

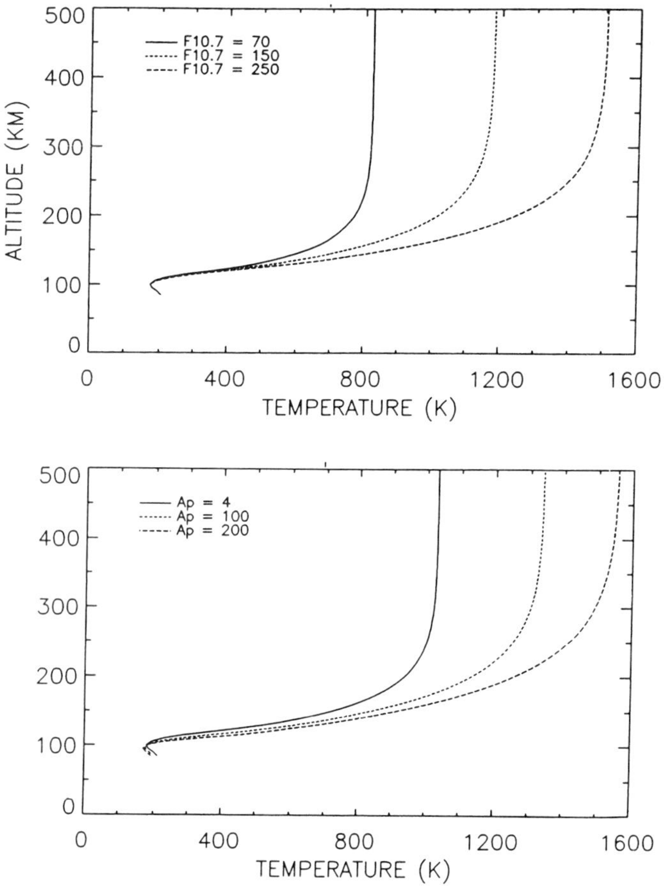

FIG. 2. Neutral temperature as a function of altitude. Top: The variation at 14 h LT and latitude 30° with solar activity, represented by the magnitude of the F10.7 flux (F10.7 is the electromagnetic radiation from the Sun at the wavelength 10.7 cm). Bottom: The variation at 2 h LT and latitude 65° with changes in magnetic storm activity represented by the global magnetic activity index A_p. The solar flux was held constant at the middle level of the top panel. The profiles are from the MSIS-90 model of Hedin (1991) [courtesy of F. Marcos, Geophysics Directorate, Philips Laboratory (USAF)].

solar radiation below ~250 nm, which we in the following will call the xuv band).

Notice that atmospheric waves generated close to the Earth contribute to the energy budget around 100 km. Also, as will be discussed later, energy from the magnetosphere, in the form of charged-particle precipitation and Joule heating, heats the upper atmosphere in the high-latitude regions. Neutral winds are also important for the energy balance. Winds and waves, as well as turbulence, affect the transport of energy to different regions of the ionosphere (see Fig. 3).

The temperature is variable and decreases in the mesosphere, reaching a minimum near 90 km. Above this height the temperature gradient is always positive (Fig. 2). A large fraction of the absorbed ultraviolet radiation from the Sun goes into heating of the upper atmosphere. Infrared radiation to outer space is a major energy loss (Fig. 3). The vertical temperature distribution, which results from the delicate balance of the various sources of energy and energy transport, varies considerably as it increases between 100 and 200 km. The temperature continues to increase up to approximately 500 km, where it approaches an asymptotic value (above about 500 km temperature no longer has a clear physical meaning). The top panel of Fig. 2 shows the variation of the dayside temperature with increasing solar radiation at 1400 LT (local time) and latitude 30°. The bottom panel shows the high-latitude, nightside temperature variability with magnetic activity at the middle level of solar flux, but at 0200 LT and latitude 65° in the auroral zone. Energy is deposited at high latitudes from the magnetosphere by energetic particle precipitation and field-aligned currents. Local heating affects both neutral temperature and composition, which in turn affect the ionization level (discussed in Secs. 4 and 5).

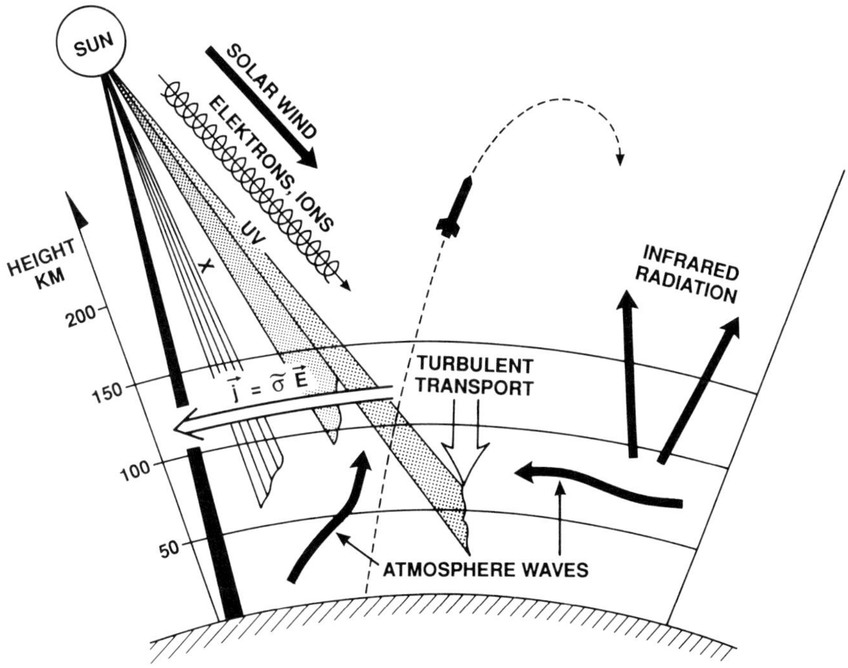

FIG. 3. A schematic representation of the energy budget of the upper atmosphere. The main energy source is electromagnetic and particle radiation from the Sun, but even atmospheric waves contribute to the energy budget around 100 km. The main energy sink is infrared radiation.

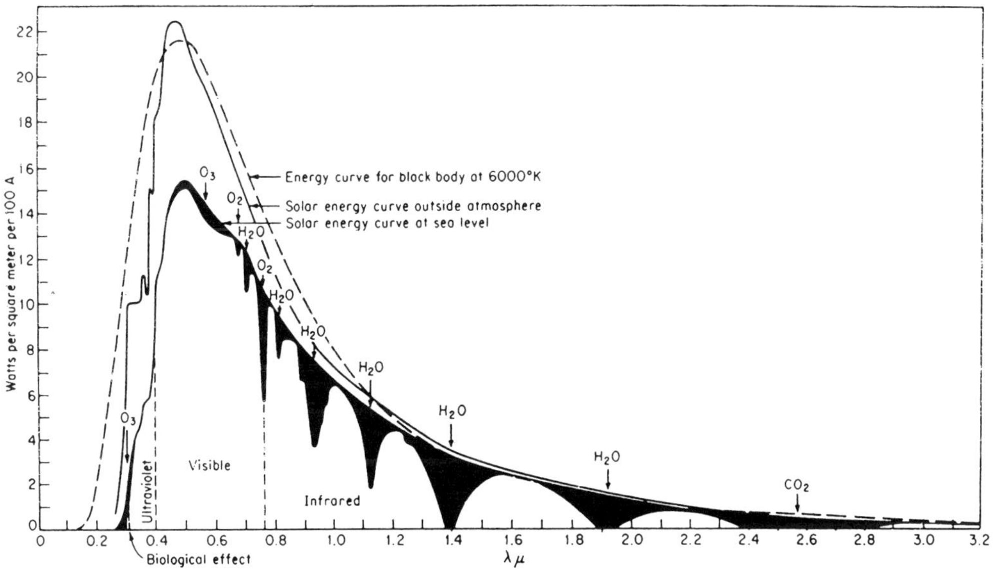

FIG. 4. Spectral energy curves of the solar energy outside the atmosphere and at sea level. Atmospheric absorption is indicated. For comparison the energy curve for a black body at 6000 K is also drawn. (From Egeland *et al.*, 1973.)

The solar xuv band significantly influences the upper atmosphere and changes its composition dramatically, particularly through photodissociation of atmospheric molecules (X_2) according to the formula

$$X_2+h\nu \rightarrow X+X, \qquad (1)$$

where $h\nu$ is a photon—within the xuv band—with energy sufficient to dissociate atmospheric molecules. Equation (1) explains why molecules dominate the region below ~150 km, whereas atoms (nitrogen and oxygen) are the primary particles above this altitude (Fig. 1).

Diffusive separation becomes increasingly important with increasing altitudes, with the heavier constituents experiencing more gravitational attraction and the lighter ones less. As a result, the light gases dominate the upper atmosphere. Thus two elements that are rare near the Earth [i.e., helium (He) and hydrogen (H)] become relatively more and more abundant at higher altitudes. Above 600 km, layers of He and H dominate (cf. Fig. 1) because of their low masses. More important for the physics of the upper ionosphere is the existence of a several-hundred-kilometer-thick layer of atomic oxygen (O) above about 250 km. Even between 120 and 250 km, O is a very important constituent.

Because the Earth's magnetic field controls charged-particle motion in the ionosphere, a few of its main characteristics will be given here. The basic shape of this field is that of a dipole—similar to that of a bar magnet. However, significant differences exist that cause the magnetic field to deviate from the dipole shape and require a 15th-order spherical harmonic expansion to describe it accurately in the ionosphere. The intensity of the geomagnetic field gradually increases from the magnetic equator to the pole, where its strength of roughly 60 000 nT is nearly twice that at the equator. The Earth's magnetic axis is offset from its geographic axis in the northern hemisphere by ~11.5° and in the southern hemisphere by nearly 15°. Thus the dipole axis is offset from the center of the Earth. Magnetic ore deposits cause small spatial perturbations. Strong, variable electric currents in the ionosphere and magnetosphere further perturb the dipole shape. Because these time-varying currents (especially those associated with the aurora; cf. Sec. 8) cause large temporal perturbations in the Earth's magnetic field, a magnetic compass is not always an accurate aid in the auroral zones.

The dipole field decreases with distance from the Earth's surface as R^{-3}. At a distance of more than ~4 Earth radii the magnetic field is significantly compressed on the dayside of the Earth, while it is stretched out into a tail on the nightside by the continuous interaction of the magnetic field with the solar wind (cf. Fig. 4 of GEOELECTROMAGNETISM; also MAGNETOSPHERE). The resulting paraboloid-shaped surface with its nose at local noon encloses the region called the Earth's magnetosphere; i.e., the region where charged particles have greater energies than thermal particles and where their motion is controlled mainly by the Earth's magnetic field. The position of the nose depends on the balance of the solar-wind pressure against the magnetic pressure of the Earth's magnetic field, which dominates over the plasma pressure in the outer magnetosphere. The standoff distance is typically at 10 Earth radii. The magnetic field lines that connect the magnetosphere to the ionosphere guide the motion of charged particles and thus provide the paths for the interchange of currents and energy. At polar latitudes, where the magnetic field lines are nearly vertical, ions and electrons can escape from the ionosphere in a process termed the polar wind.

3. IONIZATION SOURCES

The atmosphere above 50 km and extending out to the order of 1000 km is called the ionosphere, because in this region charged particles—free electrons and ions—exist in large enough numbers to influence the propagation of radio waves. Ions are created by attachment of free electrons, by collisions with energetic particles, or by photoionization. The latter two processes create free electrons. The density of charged particles in any particular region depends on the balance between production and loss processes. All of these processes are dependent on the upper atmosphere composition and density.

The solar xuv band is particularly important for producing the ionosphere—via the process called photoionization, which can be written symbolically

$$X+h\nu \rightarrow X^{+}+e_n, \qquad (2)$$

where X is an atmospheric gas, atom, or molecule, $h\nu$ is solar radiation within the xuv band, while $X^+ + e_n$ is an ion–electron pair. Only photons within the xuv band have sufficient energy to ionize the atmospheric gases. Thus the xuv radiation changes the upper atmosphere from electrically neutral to electrically conducting. X^+ is a positive ion while e_n is a free, low-energy electron. The closer the solar radiation reaches the Earth, the denser the atmosphere. The production of ion pairs increases until a large fraction of the radiation is absorbed. It is therefore a combination of atmospheric density and composition affecting solar xuv radiation absorption that determines the height of the ionosphere.

The density of electrons and ions is determined by the balance between production and loss. Electrons and ions are lost by recombination processes

$$X^+ + e_n \rightarrow X, \quad (3a)$$

$$YX^+ + e_n \rightarrow X + Y, \quad (3b)$$

in which X is an atom and XY a molecule. Electrons and ions may also be transported away and thus lost to a particular area. Because solar xuv radiations vary with time, season, and location, large temporal variations are found in the concentrations of ion pairs in the ionosphere. Other variations result from different recombination rates for different atmospheric constituents and from variations in the magnetospheric particle input. For instance, NO^+ recombines with electrons much more quickly than O_2^+, and so spatial changes in the relative abundance of these constituents create local variations in the ion or electron density.

At low and medium latitudes photoionization [Eq. (2)] is mainly responsible for the creation of charged-particle distributions. At higher latitudes precipitating energetic electrons and ions from the magnetosphere and from the solar wind (a stream of charged particles with an average speed of 400 km s^{-1} and a density of $\sim 10^7$ m^{-3} constantly blowing towards the Earth's magnetosphere) significantly contribute to the production of ionospheric electrons via the following processes:

$$X + e \rightarrow X^+ + e_n + e', \quad (4a)$$

$$X + i \rightarrow X^+ + e_n + i' \quad (4b)$$

$$(e' = e - 36 \text{ eV}, \quad i' = i - 36 \text{ eV}),$$

where e and i are energetic ($\geqslant$ 100 eV) electrons and ions, respectively. [One electron volt (1 eV) is the energy an electron gets when passing through a voltage drop of 1 V.] Each collision process of the type shown in Eq. (4) decreases the energy of an incoming particle by an average 36 eV. A 10-keV particle bombarding the atmosphere will on the average participate in 300 collisions and thus produce 300 ion pairs before it comes to a complete stop at or above $\sim$90-km altitude.

Notice that in the processes listed in Eqs. (2) and (4) the molecules and/or atoms will have lost an electron and thereby become positively charged. A gas that contains free electrons and positive ions is called a plasma. Because an ionospheric plasma contains roughly the same number density of electrons and ions, it is for all practical purposes electrically neutral.

In the polar regions (i.e., poleward of $\sim 60°$ magnetic latitude), the energy source from the solar wind and its interaction with the magnetosphere is of particular importance. The solar wind is indirectly the main generator of auroras, electric fields, and currents in the upper atmosphere (i.e., $\geqslant$100 km) (cf. Secs. 6 and 7).

4. THE IONOSPHERIC LAYERS

On the basis of the results of ionospheric studies by radio technique, it has been agreed to divide the ionosphere into different regions as illustrated in Table 1. The first to be identified was a layer of ionization near 100 km, the E layer, so named because of the electric field (i.e., the E field) of the reflected radio waves with which it was measured. Above the E region lies the F region, while the region underlying was named D.

A typical ionospheric electron density profile for mid-latitude, daytime conditions is shown in Fig. 5. The different regions reflect radio waves with different wavelengths. While short waves (i.e., $\simeq$100-m wavelengths equivalent to a frequency $\geqslant$3 MHz) are reflected from the F region, the longer radio waves (i.e., $\sim$1000 m) require less density for reflection and are reflected from the E region. These reflection differences allow the ionosondes to measure the ionization profile with height by

Table 1. Characteristics of the ionospheric regions.

Region	Height (km)	Electron density Day (m^{-3})	Electron density Night (m^{-3})	Production sources
D	50–90	10^9	10^8	xuv+cosmic rays
E	90–140	10^{11}	$<10^{10}$	xuv+auroral particles
F_1	140–200	3×10^{11}	10^{10}	uv+auroral particles
F_2	200–400	10^{12}	10^{11}	uv+transport from below

measuring the variation of travel-time delay of the reflected radio signal with frequency.

At higher frequencies or shorter wavelengths signals pass through and are not reflected. Thus the variable ionosphere has a pronounced influence on long-distance radio communication and navigation. By comparing the electron density given in Table 1 for the D and E regions with Fig. 1, it is clearly seen that only a tiny fraction of the atmospheric gases are ionized. The majority of the molecules remain neutral in the lower ionosphere.

4.1 The Topside Ionosphere (> 400 km)

The region above the peak density of the F layer is called the topside ionosphere. The density decreases slowly with increasing altitude, with the rate of decrease being a function of temperature. This region was called the topside because its density profile could not be determined by radio-wave sounding techniques from the ground. Its characteristics were first determined by *in situ* sounding from above with transmitters on satellites and by sounding rocket measurements. In the lower part O^+ and O are the major ion and neutral species, respectively. Above approximately 600 km the lighter atomic ions He^+ and H^+ can dominate. In the upper part of the topside ionosphere, the plasma is effectively fully ionized.

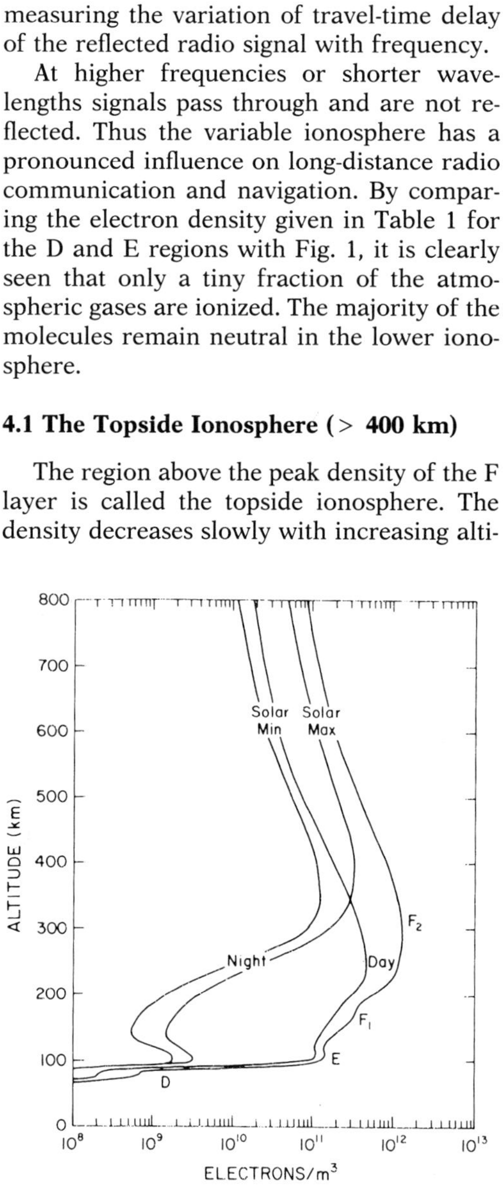

FIG. 5. Typical mid-latitude electron density profiles for solar maximum and minimum for both daytime and nighttime conditions. The different ionospheric layers are marked off in this figure (from Rees, 1989).

4.2 The F Region (> 140 km)

Above the E layer (i.e., >140 km up to a height arguable in the scientific community) lies the F region. In the following we will define the upper height as that where the light ions (helium and hydrogen) begin to dominate over the heavier ions, in particular oxygen (between 500 and 1000 km depending on conditions). This height will vary with temperature. Note that the F region overlaps the lower part of the topside ionosphere. Even if production of ionization peaks below 200 km, the maximum electron density is found at considerably greater (up to 350 km) heights. In 1933 it was found that the F region during daytime often has two maxima, thus splitting into F_1 and F_2 layers. Typical peak electron densities are 3×10^{11} and 10^{12} m^{-3} in the F_1 and F_2 layers, respectively, while the neutral density is approximately 10^{15} m^{-3}. The most important ions in the F region are O^+, NO^+, O_2^+, and N_2^+. The important F-layer ionization source is found in the xuv band between 10 and 90 nm. It should be pointed out that N_2^+ reacts quickly with O via the process

$$N_2^+ + O \rightarrow NO^+ + N. \quad (5)$$

In the F_1 region (normally only a ledge in the lower F region with peak density near 180 km) the following reaction is also important:

$$O^+ + O_2 \rightarrow O_2^+ + O. \qquad (6)$$

The conclusion is that NO^+ and O_2^+ together with O^+ are the most numerous ions below 200 km, while O^+ and the light ions dominate in the upper part of the F region (i.e., F_2 layer). Thus most of the ions in the F_2 layer are atomic.

The continuity equation for the F_2 layer can be written

$$\frac{dn_e}{dt} = q - \beta n_e + \mathbf{v} \cdot \nabla n_e, \qquad (7)$$

where q is the production rate of electrons, n_e is the electron density, and β is the recombination (attachment) coefficient. Note that the loss of electrons depends linearly on electron density. In addition, the last term in Eq. (7) shows that there are pronounced drift effects. Electrons and ions are also able to escape from the ionosphere along open magnetic fields at polar latitudes. This loss of plasma—called the polar wind—can have a significant effect in the polar ionosphere. Plasma flow from one hemisphere to the magnetic conjugate ionosphere is important at midlatitudes. The inner magnetosphere or plasmasphere is populated by outflow of plasma along magnetic field lines from the ionosphere at mid latitudes.

4.3 The E Layer (90–140 km)

The E layer was the first ionospheric layer to be recognized. Since the typical daytime electron density is 10^{11} m^{-3}, and the neutral density at 120 km is 10^{17} m^{-3}, the E region is weakly ionized, and collisions between charged particles and neutral species are important in determining the conductivity and ion motion. The chief ions are O_2^+ and NO^+. The dominant neutral species in this region are N_2 and O_2. In the upper part of the layer O^+ occurs in significant quantities. N_2^+ is also produced but rapidly lost through charge exchange via the reaction

$$N_2^+ + O_2 \rightarrow N_2 + O_2^+. \qquad (8)$$

In the lower part of the layer, NO^+ dominates.

Because NO has such a low ionization potential, Lyman-α radiation ($\lambda = 121.6$ nm) is the main ionization source. The predominant loss term is dissociative recombination, which takes the form [cf. also Eq. 3(b)]

$$O_2^+ + e_n \rightarrow O_2 \rightarrow O + O. \qquad (9)$$

[In process (9) both momentum and energy are conserved, which makes this process very fast.] The continuity equation for this layer can be written

$$\frac{dn_e}{dt} = q - \alpha\, n_e^2, \qquad (10)$$

where α is the recombination coefficient for all reactions of the type described by Eq. (9). Because of the high neutral density, drift effects of type shown by the gradient term in Eq. (7) can be neglected. More recent results based upon equilibrium between production and loss (i.e., no time variations) have given α values in good agreement with laboratory measurements. Also, rocket measurements have been very useful in testing Eq. (10).

Near 100 km a thin layer of enhanced ionization, known as sporadic E, sometimes exists. This layer can contain enough electrons to reflect high-frequency (HF) radio signals that would normally propagate through to the F region.

4.4 The D Region (50–90 km)

Because of the relatively high neutral density in this region, the low ion densities, and the dominance of collisions, the classical radio techniques used to study the E and F regions of the ionosphere are not applicable. Rocket techniques provide the only *in situ* measurements since this region is not accessable to satellites and balloons. Electron density does not show a marked peak, but rather a gradual increase with height (cf. Fig. 5).

The important ionization sources are listed in Table 1. X rays between 0.1- and 1-nm wavelengths are a principal source below 90 km. Solar protons from flares on the Sun also create D-region ionization. The bottom of the D region, where the electron density is very low (i.e., 10^8 to 10^9 m^{-3}), is produced by galactic cosmic rays. Thus there are hardly any diurnal variations in the lower D region during quiet conditions.

The primary ions are NO^+ and O_2^+. In addition, negative molecular ions and hydrated ions (water-vapor complexes) have been observed. One reaction producing negative ions may be

$$2O_2+e_n \rightarrow O_2^- + O_2. \tag{11}$$

Production (of particularly complicated positive and negative ions), recombination, and other loss processes in the D region are complicated and not fully understood.

Because the D region is important for the energy balance as well as a principal region for absorption of radio waves propagating over long distances, further studies by extending modern incoherent-scatter radar and lidar techniques to this region have just begun.

5. GLOBAL DISTRIBUTION OF IONIZATION

Detailed studies of the ionosphere by simultaneous, coordinated satellite, rocket, and ground observations have clearly demonstrated large temporal and spatial ionospheric variations. Before the space age, studies were limited mainly to the regular ionospheric variations; i.e., seasonal, diurnal, and solar-cycle responses. These variations are at least partially due to the rotation of the Earth with its geographic axis tilted relative to the Sun and the geomagnetic axis tilted relative to the geographic axis, and long-term activity variations on the Sun. Both the photoionization, the main source for maintaining the medium- and low-latitude ionosphere, and magnetospheric and solar-wind ionization contributions will also vary considerably and thereby add to the complexity of the global distribution.

As seen from Eq. (10), the electron density in the E region during stationary conditions is

$$n_e=(q/\alpha)^{1/2}=c_1(\cos\chi)^{1/2}, \tag{12}$$

because the ion production rate q is proportional to the cosine of the zenith angle (χ) of the Sun. Because of the regular diurnal variation alone, the electron density will thus vary as $\cos\chi$. During winter in the northern hemisphere, when the polar region is completely in darkness for a few months, photoionization is practically absent, and precipitating particles become the principal ionization source for the E region.

Because the lifetime of ions is much longer in the F regions than below, they can be transported considerable distances from their origin. The polar F-region densities vary greatly depending on the proximity to auroral particle and dayside uv sources. Just equatorward of the nocturnal auroral oval a minimum in electron density, called the electron trough, is very pronounced during the winter months. This trough has a limited latitudinal extent and it is very deep from 16, to 0600 MLT (magnetic local time). During this time, the F-region density can be as low as 10^9 m^{-3} (cf. Table 1).

To illustrate the global distribution of ionization and its seasonal variation, Plate 1 presents the local time-averaged global distribution of plasma density versus month of the year. These measurements are from the Defense Meteorological Satellite Program (DMSP) satellite orbiting at 830 km and are averaged over a 2-y period. The distribution at the topside ionosphere is representative of other F-region altitudes, although variation with altitude will occur. All p.m. local times are averaged together. The dayside low-latitude maximum, which stretches to higher latitudes in the summer hemisphere, is clearly evident, as is the depletion in the winter polar cap. The trough at subauroral latitudes does not appear because this latitude variation is lost by averaging. A polar view of a typical ionization distribution in the southern high-latitude regions at the altitude of F-region maximum is shown in Plate 2(a) along with a similar plot in Plate 2(b) for 12 h later. This is a simulation using the Utah State University ionospheric model (Schunk, 1988) that has been run over a 24-h period with constant inputs. The polar hole, ionization from auroral oval precipitation, and the mid-latitude trough are clearly seen.

An additional diurnal variation occurs at high latitudes because of the offset of the geomagnetic pole from the geographic pole. Electrodynamically driven effects are controlled by the geomagnetic field, while neutral atmospheric and solar ionization effects are dominated by the geographic axis. As a result the ionosphere is stirred daily with the rotation of the geomagnetic axis about the geographic axis in universal time (UT). Comparing Plate 2(a) with 2(b) illustrates the maximum differences from this effect as these times were chosen when the magnetic pole was toward and away from the Sun relative to the geographic pole. These variations from stirring are added to those from solar and magnetic activity to produce very dynamic behavior of the ionosphere in the auroral and polar cap regions.

6. ELECTRIC FIELDS, CONDUCTIVITY, AND CURRENTS

There are three principal sources of electric fields in the global electrical circuit: the thunderstorm-generated atmospheric electric field, the tidal-wind–generated S_q ionospheric electric field, and the magnetospheric convection electric field driven by the interaction of the solar wind with the magnetosphere. The last two are important to the upper ionosphere, while all three interact at the lower boundary. The coupling of these generators is dependent on the ionospheric conductivity.

The motion of charged particles in the ionosphere depends on electric and magnetic fields, on the ionization density, and on collisions. Thus conductivity is directionally dependent. It may be divided into three terms: σ_0 is the specific conductivity along the magnetic field **B**, σ_P is the Pedersen conductivity in the direction perpendicular to the magnetic field and along the perpendicular component of the electric field **E**, and σ_H is the Hall conductivity in the direction perpendicular to both the magnetic and electric fields. The profiles of the conductivities with altitude are shown in Fig. 6. Note that the specific conductivity scale is six orders of magnitude greater than the other two. For this reason the electric field parallel to the magnetic field is, to first approximation, zero, and the perpendicular electric field maps along magnetic field lines. The Hall conductivity typically peaks near 110 km while the Pedersen conductivity peaks near 130 km. Above 200 km the collision frequencies are so small that the charged-particle motion is controlled strictly by the electric and magnetic fields. The ion velocity at these altitudes can be expressed by

$$\mathbf{v}_i = (\mathbf{E} \times \mathbf{B})/B^2. \tag{13}$$

At high latitudes the magnetic field lines project out into the magnetosphere. The magnetospheric electric fields project along the magnetic field into the ionosphere and impose a potential pattern on the high-latitude regions. Electric fields at high latitudes are typically 20 to 60 mV m^{-1} and can exceed 200 mV m^{-1} during disturbed conditions. The ion flow is along the equipotentials, and this flow is also the principal driving force for neutral winds at F-region altitudes. When the interplanetary magnetic field has a southward component, the flow patterns form two convection cells with a negative potential minimum near 16 h and a positive potential maximum near 04 h. The shape of the patterns is dependent on the dawn–dusk component of the interplanetary magnetic field. Representative patterns are shown in Fig. 7. For positive (negative) B_y the BC pattern is applicable to the northern (southern) hemisphere while the DE pattern is applicable to the southern (northern) hemisphere. Typical values of the difference in potential between the maximum and minimum range from 30 to 150 kV depending on magnetic activity. In extreme cases

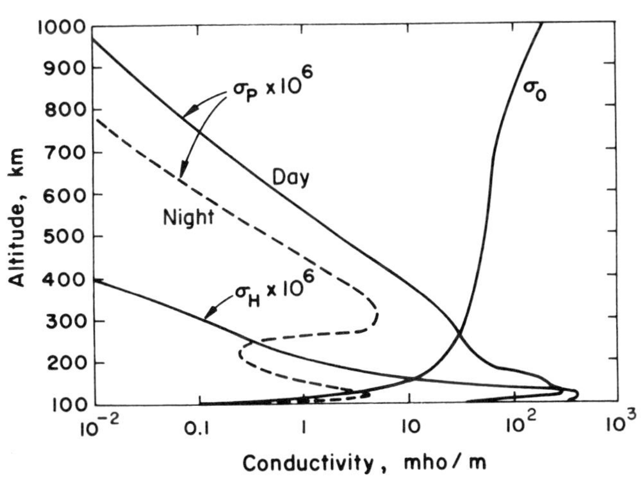

FIG. 6. Typical conductivity values for the mid-latitude daytime ionosphere. Notice the change of scale for the Hall and Pedersen conductivities. The dashed curve is a typical nighttime profile of the Pedersen conductivity also with the scale change (from Kelley, 1989; adapted from Johnson, 1961).

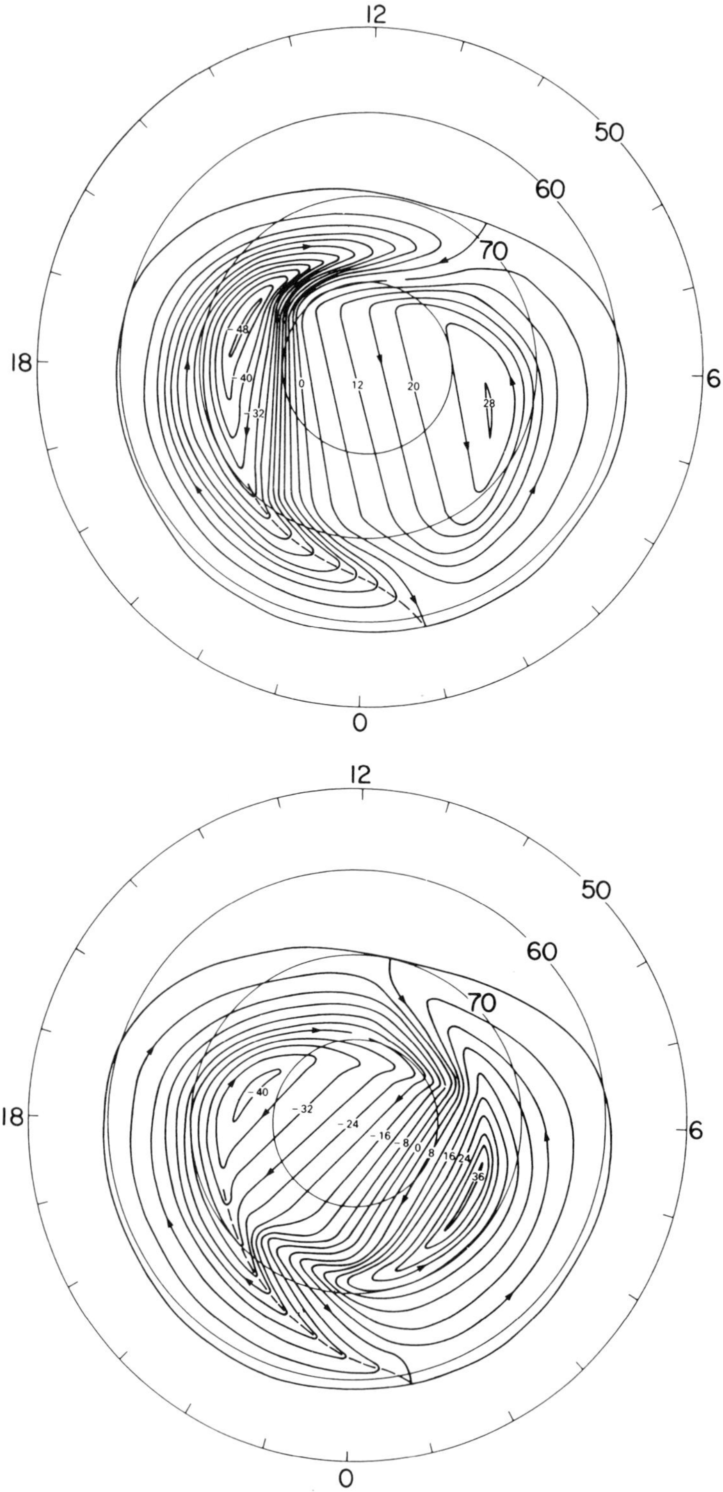

FIG. 7. Two convection patterns of Heppner and Maynard (1987) for a southward interplanetary magnetic field to illustrate the variation in the pattern with the east–west component (B_y) of the interplanetary magnetic field (IMF). The BC pattern (bottom) is for the northern hemisphere for positive B_y. The DE pattern (top) is for the northern hemisphere during negative B_y. For the southern hemisphere the same patterns apply with the opposite sign of the IMF B_y component. The contours represent equipotentials spaced at 4 kV and are representative of average activity conditions. The magnitude of the potential difference between the potential minimum on the evening side and the potential maximum on the morning side (76 kV in the above patterns) varies from $\sim$20 to over 200 kV with varying magnetic activity.

it may exceed 200 kV. When the interplanetary magnetic field component B_z is northward, the two-cell patterns become distorted and can evolve into more than two cells. The potential differences are generally very small and the electric fields are extremely variable, in some cases making it difficult to recognize a pattern (Heppner and Maynard, 1987).

Under steady-state conditions the electric fields from the magnetospheric generator are poleward of about 60° magnetic latitude. A shielding layer in the inner magnetosphere prevents penetration of the magnetospheric electric fields inside of about 4 Earth radii, which projects along the Earth's magnetic field to 60° latitude. As magnetic activity changes, the shielding must be reestablished. Magnetospheric electric fields temporarily penetrate closer to the Earth and project along the Earth's nearly dipolar magnetic field lines, which are equipotentials, to lower latitudes in the ionosphere. At middle latitudes the ionospheric electric field is a combination of these penetration electric fields and S_q dynamo electric fields generated by neutral wind and tidal motions of the neutral atmosphere, which through collisions drive ions across magnetic field lines. This ionospheric E-region tidal generator is driven by a complex combination of solar-heating–induced neutral winds and gravitational effects from the Sun and Moon. Electric fields are small, of the order of a few mV m^{-1}, with a total potential drop of about 10 kV.

At the lower edge of the D region the gas is only weakly ionized and the conductivity becomes isotropic. The atmospheric electric field is vertical and couples to **E** fields from the other two sources. The atmospheric vertical electric field in the lower D region is small since 90% of the 300-kV potential drop between the Earth and the ionosphere exists below 10 km and the electric field is inversely proportional to the conductivity. Under conditions where all electrons are tied to heavy negative ions, gravitationally driven and/or differential neutral-wind–driven ion motion can generate anomalously large (V m^{-1}) local electric fields.

The ionospheric electric fields drive horizontal currents in the ionosphere, which are observed in the ground magnetometer recordings. These are principally Hall currents in a layer around 110 km (Fig. 6). At midlatitudes these are strongest during the day when the conductivity peaks. At the magnetic equator the Hall current would be principally downward and is impeded from flowing because of the decreasing conductivity. A polarization electric field results, causing an apparent enhancement of conductivity (Cowling conductivity) and an enhanced east–west current called the equatorial electrojet. The magnitude of the electrojet current peaks with the conductivity near 1100 LT (see Chapman and Bartels, 1940).

In high-latitude regions ionization from solar xuv radiation is supplemented by ionization from the energetic particles producing the aurora (see Sec. 8). Within the auroral oval the conductivity due to auroral particle precipitation dominates, and the conductivity varies with magnetic activity. The Hall currents driven by the magnetospheric convection electric fields are large. They are referred to as auroral electrojets because of the primary currents being confined in altitude to the peak of the Hall conductivity profile. On the evening side the currents are eastward, while after midnight they are westward. The two electrojets overlap in the midnight region at the Harang discontinuity. The conductivity, electric fields, and currents are extremely variable. Current continuity under these spatially and temporally variable conditions requires field-aligned currents, or Birkeland currents, to flow.

Birkeland currents couple the ionosphere with the magnetosphere. The upward currents carried by the precipitating auroral electrons are balanced by downward currents, primarily carried by upward-flowing, low-energy electrons. The large-scale pattern of field-aligned currents (shown in Figure 8) can be divided at most local times into two principal regions. Region-2 currents couple to the inner magnetosphere, and currents flow into the ionosphere on the evening side and out of the ionosphere on the morning side. Region-1 currents couple to the magnetospheric boundary layers and flow out of the ionosphere on the evening side and into the ionosphere on the morning side. Near noon and midnight these patterns overlap, and three current layers are common (Iijima and Potemra, 1978; Potemra, 1990). Additional weaker current systems are observed poleward of the oval especially during periods when the interplanetary magnetic field has a northward component. Field-aligned currents are closed in the

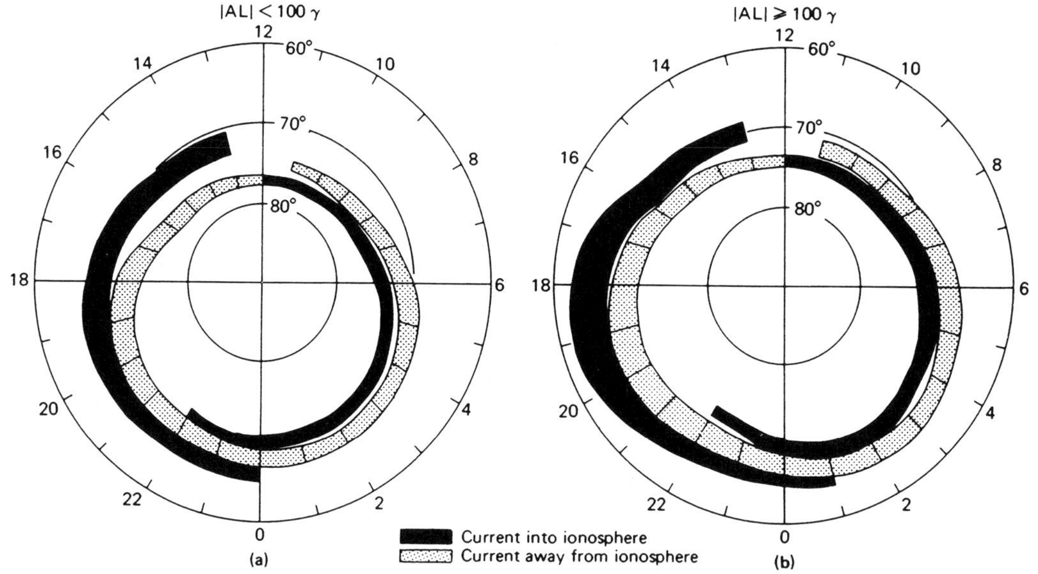

FIG. 8. A summary of the distributions and flow directions of large-scale field-aligned currents based on Triad data for weakly disturbed and active conditions (from Iijima and Potemra, 1978). AL is related to the maximum current intensity in the westward auroral electrojet and is the larger of the two components of the auroral activity index AE. The units are in nT (1 $\gamma = 1$ nT). Thus the left plot is for magnetically quiet conditions and the right plot is for active conditions.

ionosphere principally by Pedersen currents. In all these current regions the direction of the current is the average direction. Structured field-aligned currents in both directions exist side by side in all regions, a necessary consequence of the variability of the conductivity and the electric fields.

7. AURORA—THE POLAR LIGHT

The aurora, as illustrated in Plate 3, is certainly one of the most spectacular of nature's phenomena. In earlier times, inhabitants of northern Canada, Greenland, and the Nordic countries saw the northern lights as omens from the gods portending disaster, signs from deceased relatives, a battle of the gods, or a weather sign. For those interested in the history of the aurora, monographs by, e.g., Brekke and Egeland (1983) and Eather (1980) are available.

Those who admire the beauty of wonders in the dark sky may find nothing comparable to a night with magnificent auroral displays. Unfortunately, this impressive and imagination-inspiring display can only be seen regularly by people living near the polar region. At these latitudes, Aurora Borealis (i.e., in the north) and Aurora Australis (in the southern hemisphere) routinely appear in the so-called auroral zones. Consequently, auroras are far from most population centers. Even today, the area of the southern auroral zone (roughly around Antarctica) is inhabited only intermittently. In the northern auroral zone, which crosses Alaska, central Canada, northern Scandinavia, and northern Siberia, these majestic lights are a routine part of the daily life of these northern people.

The instantaneous distribution of aurora versus time and latitude is given by an oval-shaped belt of maximum activity called the auroral oval (cf. Fig. 9). These zones represent the average statistical occurrence of aurora and are approximately centered about the geomagnetic poles. Notice that both the location and width vary significantly from midnight to midday. They also vary with solar and magnetic disturbances as shown in Fig. 9. The oval expands equatorward during disturbed periods and contracts during quiet periods. Auroral emissions called polar cap arcs also appear poleward of the oval, but with markedly less frequency and intensity. The frequency of auroral occurrence, as well

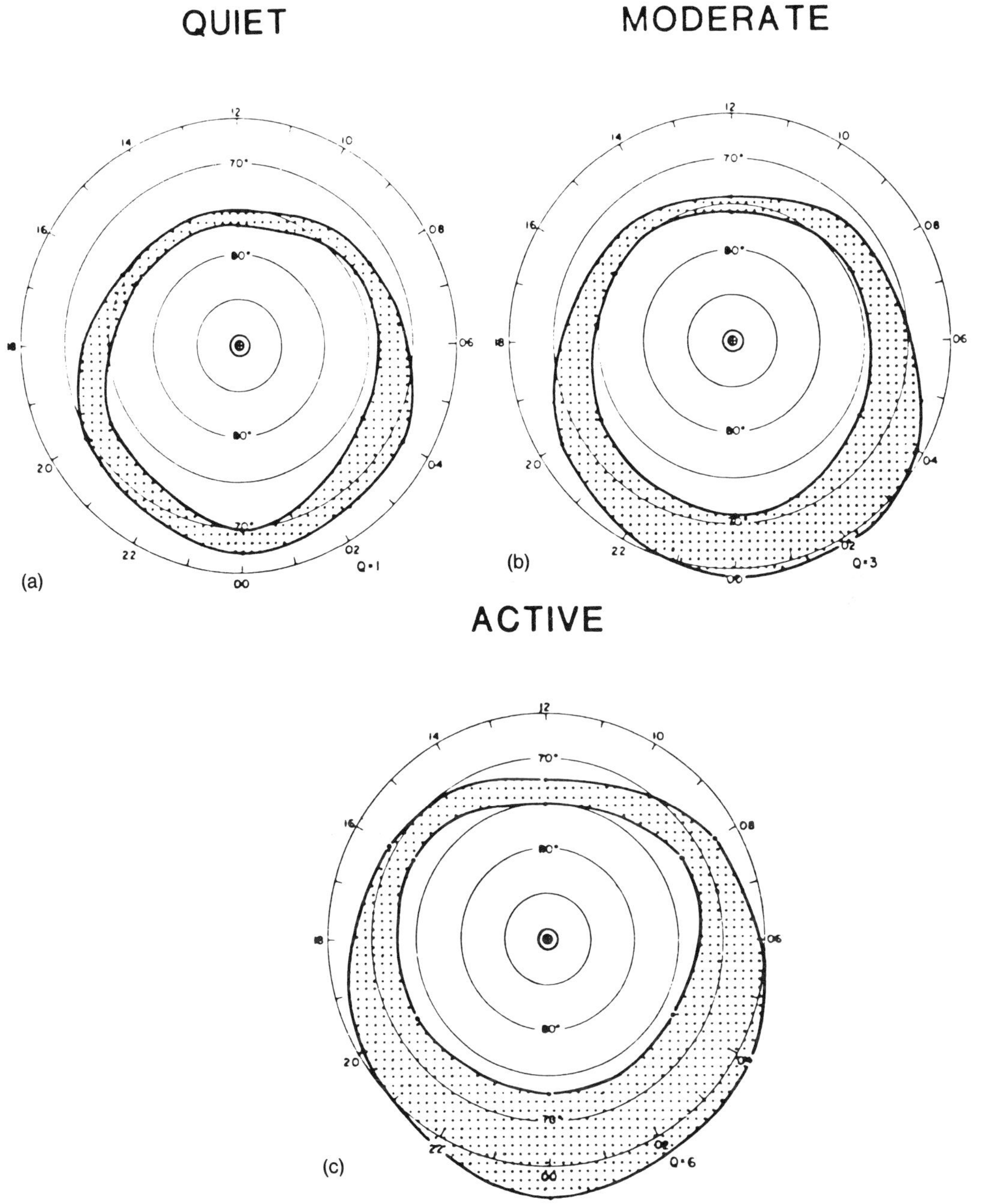

FIG. 9. Variation in the size of the auroral oval for different solar or magnetic activity; quiet, moderate, and disturbed conditions.

as its intensity and location, is clearly correlated with the degree of activity on the Sun. This is also borne out by the 27-d recurrence tendency (i.e., the rotation period of the Sun) as well as the 11-y sunspot cycle. However, auroral activity peaks one to two years after sunspot maximum. The oval width and location for quiet, moderate, and active conditions are illustrated in Fig. 9.

Auroral intensity can vary from being barely detectable among the stars to a luminance bright enough to veil the stars completely.

Thus the intensity of this polar light can vary over more than four orders of magnitude. At maximum intensity the illumination on the ground is comparable to moonlight, and ~100 times stronger than starlight. With a modern 35-mm camera, an f/1.4 lens, and a fast color film (e.g., 400 ASA), an exposure time of ~1 s is needed to photograph a bright aurora.

By watching the aurora, one may get the impression that a confusingly large number of different auroral forms and structures occur. However, by categorizing them into four primary forms—i.e., arcs and bands, auroral rays, spiral structures, and diffuse surfaces—the customary occurrence forms can be explained.

Of all characteristic properties of the northern light, none have been discussed and studied more than its height. On the basis of about 20 000 parallactic measurements between 1910 and 1940 (Störmer, 1955), it was proved that auroral arcs and bands, auroral glow, and surfaces lie predominantly within the height interval 95–150 km, whereas isolated rays or bundles of rays may have an altitude considerably higher (i.e., >200 km). The average height of dayside aurora is typically 150 km higher than nighttime aurora. A few auroral forms—called sunlit aurora—are found up to 1000 km. Today, the luminosity height profile may be measured *in situ* by instrumented rockets. A few tens of such flights have been carried out, showing that the height profile is directly related to the energy of the precipitating particles from the magnetosphere. The lower border near 100 km is excited by electrons of 5 to 15 keV. The peak emissions on the dayside, which typically are near 300 km, are from particles precipitating with energies of ten to a few hundred eV.

Although a weak aurora appears to the eye as white, increased intensity reveals the predominant green color together with red and blue. As illustrated in Plate 4, the optical spectrum of an aurora consists of a great number of spectral lines and bands—from ultraviolet to infrared wavelengths. These emissions are primarily due to a two-step process; i.e., precipitating energetic particles collide with the Earth's upper atmosphere, converting their kinetic energy into chemically excited states of atoms and molecules that in turn generate photons of wavelengths determined by the specific gas (i.e., oxygen, nitrogen, helium, etc.) and the relaxation processes. The auroral emissions of atomic lines and molecular bands can therefore be considered as the "fingerprints" of atmospheric constituents.

Symbolically, this can be written

$$N_2+(e \text{ and/or } i)\rightarrow(N_2^+)^*+(e' \text{ or } i')+e_n, \quad (14)$$

where e' $(i')=e$ $(i)-36$ eV and the asterisk denotes an excited state. Equation (14) is followed by radiation of the first negative bands:

$$(N_2^+)^*\rightarrow N_2^+ + h\nu \quad \text{(aurora at 391.4 and 427.8 nm).} \quad (15)$$

Because the probability of process (14) followed by (15) (spontaneous emission as excited species relax to a lower energy level) is nearly independent of energy of the primary particles, these blue auroral bands at 391.4 and/or 427.8 nm can be used to determine the net downward particle energy. Quantitative considerations show that one photon is emitted at $\lambda=391.4$ nm per 25 ion pairs produced.

The brightest auroral emission is the "green line" at 557.7 nm, which is due to atomic oxygen. But, as these emission photons are not emitted spontaneously as in process (15), it is a forbidden transition. Another commonly observed line—particularly in the polar cusp—is the red line at 630 nm, which is also emitted from excited oxygen. This is also a forbidden transition. Allowed or permitted transitions occur very rapidly (i.e., 10^{-7} s). Forbidden transitions, which violate the quantum-mechanical selection rules, still do occur, but only after a much longer time, i.e., ≈0.7 and 110 s for the green and the red lines, respectively. For the nearly 2-min lifetime of the red-line transition, collision deactivation (without light emission) is very likely for altitudes below about 200 km. Thus, the red 630-nm [OI] emission will always peak at altitudes above 200 km.

The intensity, height, and spectral content of the aurora are projections of magnetospheric processes onto the high-latitude ionosphere, much like the cathode-ray tube in a television set. The magnetic field lines to which the particles are tied trace out to the plasma sheet in the nightside and magnetospheric boundary layers and the cusp on the dayside. The aurora provides a continuous monitor of these activities. The advantage of using the aurora in the ionosphere as a monitor of these near-Earth processes over any conceivable system of *in situ* measurements is that the size over which the phenomena are occurring in the ionosphere differs in scale from that in the magnetosphere by a factor of perhaps as much as 10^6. Continuous high spatial and temporal resolution observations provide another advantage of studying ground measurements of aurora.

8. DYNAMICS OF THE IONOSPHERE

During recent years ionospheric variations, both small-scale irregularities and large-scale variations connected with specific solar events, such as solar flares—and more recently coronal mass ejections—have received special attention (cf. SOLAR RADIATION). Flares are the most dramatic of the solar events. Here we will briefly list in chronological order the most dramatic ionospheric effects of a large solar event.

Approximately 8 min after a flare occurs, the first effect in the Earth's ionosphere—the so-called sudden ionospheric disturbance (SID)—is observed on the sunlit portion of the Earth. The existing source is enhanced x rays from the Sun with a factor of 100 and even more increase in solar radiation between 0.1 and 1 nm. As a result, the electron density in the D region is significantly enhanced. Long-distance radio-wave propagation is seriously disturbed, while the propagation conditions for frequencies with longer wavelengths may be improved. The SID events normally last less than 1 h.

Major flares are normally connected with large fluxes of relativistic protons (typically 100 MeV, i.e., comparable to the energies of cosmic rays). Within one to a few hours these energetic solar particles reach the Earth's ionosphere poleward of typically 50° magnetic latitude. The result is a complete "blackout" of all long-distance radio communications at high latitudes, which often lasts for many days. These are called polar cap absorption (PCA) events, named for their effect upon radio-wave transmissions. The physical explanation is that these high energy particles cause a new ionospheric layer in or below the regular D region. Even with an electron density of only 10^8 to 10^9 m^{-3}, the absorption effects are so large because of the high collision frequency at these low altitudes. Again, the enhanced D region will improve long-wavelength transmissions that are reflected from the D layer.

The third class of ionospheric disturbance is called auroral absorption (AA) in radio terminology, or, more generally, magnetic substorm events. A major solar event will produce both a magnetic storm and substorm events (while a magnetic storm is a global phenomenon lasting days, substorms are episodic releases of magnetospheric energy into the ionosphere primarily to the auroral oval, lasting on the order of one hour). AA events occur not only from flares, but can also depend on other solar and solar-wind irregularities (not always well defined). As both x rays and the solar xuv radiation vary in phase with the solar activity, marked changes in electron densities and irregularities are also caused by these radiations. About 30 to 60 h after this solar activity, AA effects are observed particularly in the oval within ±5–6 h around magnetic midnight. The ionospheric disturbances from magnetic storms and substorms are primarily caused by auroral electrons; i.e., electrons between 0.1 and 100 keV. The occurrence of individual substorm events is normally irregular, and there may be several events in one disturbed night. Particle bombardment introduces local changes in E- and F-region ionization. The electric fields are generally enhanced and variable, and activity expands equatorward. Since auroral particles are the principal source of nighttime E- region ionization, changes in this region are large and rapid during the substorm. Transport increases in importance for determining F-region densities. As the substorm activity passes its peak, large electric fields, called subauroral ion drift (SAID) events, develop on the equatorward edge of the auroral region in the trough. In this region, also, the subauroral red arc or SAR arc is sometimes seen as a result of heating of the upper F region by energy coupled along magnetic field lines from the inner edge of the plasma sheet in the inner magnetosphere.

Solar abnormalities may exist for several solar rotation periods, and ionospheric disturbances may peak at 27-d intervals, i.e., when the same place on the Sun faces the Earth again. Not all magnetic storms are caused by solar flares. Recent research points to coronal mass ejections, which may come from flares or may be from other active regions on the Sun, as the carriers of the enhanced plasma in the solar wind that causes magnetic storms. The magnetic storm typically lasts a few days with enhanced substorm activity during the period.

9. RADIO PROPAGATION EFFECTS

For the same reasons that radio waves were important in the discovery of the ionosphere, the ionosphere greatly influences radiowave propagation. For the longer wavelengths that cannot penetrate the ionosphere, the reflection of these waves to a first approximation obeys Snell's law. Signals can propagate within the Earth–ionosphere cavity to great distances. Amateur radio operators quickly learn that different frequency bands propagate better depending on daytime or nighttime conditions. The E layer is normally greatly reduced at night because of higher neutral densities, shortening the recombination time, and signals that would be reflected by the E layer in the daytime pass through to the F region at night. This allows communication to greater distances while bypassing nearer regions, a process often referred to as "skip."

Reflection from a smooth layer is well behaved, while reflection from an irregular region can greatly distort the signal. During magnetic storms the high-latitude regions are disturbed, which in turn disrupts or greatly changes communication characteristics. Skip becomes irregular. In the case of a PCA event (cf. Sec. 8), the energetic solar protons create extra ionization in the D region. This results in communication blackouts for high frequencies.

The ionosphere also influences the propagation of VHF and UHF radio waves to and from satellites at frequencies that normally are not reflected. Irregularities in the ionization density create both phase and amplitude distortions in radio waves as they propagate through the ionosphere. Increasing the bit error rate can make communications marginal. These distortions are referred to as scintillations. Scintillations vary widely in space and time; however, two regions of high scintillation activity deserve specific mention. At low latitudes scintillations are especially prevalent in the evening hours. After sunset the E region disappears because of recombination, leaving a sharp gradient in the plasma density. This results in a more dense fluid over a less dense fluid, which is subject to a Rayleigh-Taylor instability. Bubbles of low-density plasma form and rise through the F region creating irregularities referred to as equatorial spread F. This causes major communications problems because of the resulting scintillations. Auroral precipitation, especially in the auroral oval, is also a major source of irregularities and therefore scintillations. In addition, the polar cap can also have significant amounts of irregularities. Patches of ionization are created in the auroral-zone F region near noon and are convected across the polar cap to the nightside by the electric fields. These patches are scources of scintillations.

10. PLANETARY IONOSPHERES

Since the flyby of the planet Venus in the 1960s by the Mariner probes, a wealth of information on planetary ionospheres has been gathered from other spacecraft by *in situ* and remote sounding techniques. The outer planets have been explored by Voyagers 1 and 2. Pioneer Venus orbited Venus for nearly a decade, providing extensive information about ionospheric structure and dynamics. During the coming decade planetary ionospheres will receive more and more attention. In this section we will only list a few important ionospheric facts about the planets.

Even though several of the basic physical and chemical processes (i.e., production, loss, and transport) that control planetary ionospheres may be similar to those on Earth, major differences occur in the composition and height distribution of the neutral atmospheres as compared to the Earth. One other important factor is the difference in the magnetic fields of the planets. The magnetic moments on the other inner planets, particularly Venus, are low, while the outer planets have large magnetic fields when compared to the Earth.

The ionospheres of the inner planets have several similarities to the Earth's E layer. The most abundant ion is O_2^+ even if there is practically no neutral O_2 in their upper atmospheres. This is due to photodissociation of CO_2, which is a major gas in the planets' atmospheres. Even the maximum plasma densities are found between 100 and 200 km. In a similar way to the terrestrial ionosphere, transport processes are important for determining electron and ion densities at greater altitudes. However, because their magnetic fields are so weak, magnetic conjugate effects are generally nonexistent. The solar wind interacts directly with the ionosphere on Venus, with the plasmapause and bow shock standoff distance being determined by the balance of the plasma pressure with the solar-wind pressure.

Both the densities and, especially, the vertical height distributions of the ionospheres are significantly different at the major outer planets. Taking Jupiter as a typical example, the major ion is H^+ with its peak density above 500 km. Also H_3^+ (H_2^+ reacts with H_2 to form H_3^+) is of special importance in Jupiter's ionosphere. The upper atmosphere of Jupiter consists mainly of molecular hydrogen, while H, He, and CH_4 are also important constituents. Again, both photoionization and solar-wind–ionospheric interactions are principal ionization sources. Because of Jupiter's large magnetic field, a significant ionospheric difference is expected from mid- and low-latitude to polar regions where energetic particles also become important. Particularly for the outer planets, information about background atmospheres is insufficient, and their chemical reactions are poorly understood. Furthermore, there is no direct information on electron and ion temperatures in the ionosphere.

11. QUESTIONS FOR THE FUTURE

Although the fundamental processes related to the large-scale formation and nature of the ionosphere are fairly well understood, many problems concerning structure and dynamics, energy balance, physical and chemical couplings, etc., still remain. Here we will only briefly mention some items that will have a high priority during the coming decade.

Coupling and transport of ionization between the auroral zones and regions further poleward and equatorward are not well mapped and explained. Both horizontal and particularly vertical transport (the so-called polar wind) from high latitudes need further experimental and theoretical attention. New incoherent-scatter radars for the central polar regions, where even the electron density profiles are poorly known, closely coordinated with simultaneous rocket and satellite measurements, should have priority.

It may surprise readers that the ionospheric layer closest to the Earth—i.e., the D region between 50 and 90 km—is least known and understood. Here the atmospheric pressure is large and collisions totally dominate so that the usual ground techniques for studying the ionosphere are not applicable. Because collision frequencies exceed the critical frequencies for the region, absorption dominates.

Progress in the next decade will be toward exploration of the lower ionosphere and understanding the complexities of this intricate coupling between regions. Global scale models like the Utah State University ionospheric model (Schunk, 1988; Schunk and Szuszczewicz, 1988) along with *in situ* satellite measurements and ground-based radar sounding of the ionosphere should greatly increase our understanding of processes in the upper ionosphere. However, data at lower altitudes are limited in scope and coverage by the inaccessibility of the D and E regions to satellite measurements. Sounding rockets have provided limited coverage, but cannot provide a large-scale picture of the dynamics of these regions necessary to address the complicated feedback mechanisms associated with geomagnetic storms and substorms. Models can simulate most instantaneous conditions, but have too many unconstrained variables to specify the time dependence accurately. NASA is planning a mission in the latter part of the decade called TIMED (thermosphere–ionosphere–mesosphere energetics and dynamics) to address some of these problems. TIMED will perform an exploratory study of the physical and chemical processes acting within and upon the coupled mesospheric, thermospheric, and ionospheric system between 50 and 180 km. The energetics of these regions, the global circulation and dynamics, the chemistry and composition, and the ionospheric structure and electrodynamics will be primary objectives of this mission. It will also begin to address possible anthropogenic changes in this region, questions raised because of global change that appears to be taking place in the atmosphere below.

GLOSSARY

Aurora: Optical emissions in the upper atmosphere (> 90 km) due to excitation of molecules and atoms by precipitating energetic particles from the magnetosphere and solar wind.

Auroral Oval: An oval-shaped belt around the geomagnetic pole in each hemisphere between about 60° and 80° geomagnetic latitude that is the location of the majority of the auroral emissions. The oval expands and contracts with changes in magnetic activity.

Conjugate Auroras/Disturbances: Simultaneous phenomena in both hemispheres at locations connected by the same geomagnetic field lines.

Convection: The global scale motion of plasma in the ionosphere driven by electric fields. The motion is generally sunward in the auroral zone and antisunward over the polar cap following lines of equipotentials in the electric-field pattern.

Ionosonde: A transmit and receive system for determining the structure of the ionosphere by measuring the reflection height of radio waves at various frequencies.

Ionosphere: A region of partially ionized gas between 50 and 1000 km where free electrons and ions coexist in an electrically neutral plasma. Subdivided into **D region:** ionospheric region between 50 and 90 km above the Earth; **E region:** ionospheric region between 90 and 140 km; **F region:** ionospheric region above 140 km; **topside ionosphere:** ionospheric region above the peak of the F region.

Magnetic Local Time: Local time defined in a magnetic latitude and longitude system centered on the Earth's geomagnetic axis.

Magnetic Storm: Large-scale disturbance of the magnetosphere and ionosphere caused by the arrival of an enhanced region of solar-wind density and velocity caused by a solar disturbance. The magnetic storm typically lasts 1–3 d from onset to recovery.

Magnetic Substorm: An episodic release of energy into the auroral region from the magnetosphere causing enhanced auroral activity and magnetic disturbances in the polar ionosphere lasting approximately 1 h.

Magnetosphere: The region between the ionosphere and the magnetopause, a boundary determined by the balance of the plasma pressure of the solar wind and the magnetic pressure of the Earth's magnetic field. The Earth's dipole magnetic field is pushed in on the dayside and stretched out into a tail on the nightside. The typical standoff distance of the magnetopause on the dayside is 10 Earth radii.

Plasma: An electrically neutral, partially or fully ionized gas containing free electrons and ions.

Polar Cap: The region poleward of the auroral oval.

Polar Ionosphere: The ionosphere poleward of about 60° magnetic latitude (encompassing the auroral oval and polar cap).

Polar Wind: The outflow of ionized particles from the polar ionosphere.

Solar Wind: Low-density supersonic plasma flowing out from the Sun which interacts with the Earth's magnetosphere.

Trough: A region of low electron density equatorward of the nighttime auroral oval.

Universal Time (UT): Time referenced to 0° longitude or the Greenwich meridian.

List of Works Cited

Birkeland, Kr. (1908), *Norwegian Aurora Polaris Expedition, 1902–3,* Christiana, Norway: Aschehoug, Vol. 1, pp. 1–315.

Brekke, A., Egeland, A. (1983), *The Northern Light*, Heidelberg: Springer.

Chapman, S., Bartels, J. (1940), *Geomagnetism*, Oxford: Clarendon.

Eather, R. H. (1980), *Majestic Lights*, Washington, DC: American Geophysical Union.

Egeland, A., Holter, Ø., Omholt, A. (1973), *Cosmical Geophysics*, Oslo, Norway: Universitetsforlaget, p. 30.

Hedin, A. E. (1991), "Extension of the MSIS Thermospheric Model into the Middle and Lower Atmosphere," *J. Geophys. Res.* **96**, 1159–1172.

Heppner, J. P., Maynard, N. C. (1987), "Empirical High Latitude Electric Field Models," *J. Geophys. Res.* **92**, 4467–4489.

Iijima, T., Potemra, T. A. (1978), "Large Scale Characteristics of Field Aligned Currents Associated with Substorms," *J. Geophys. Res.* **83**, 599–615.

Johnson, F. S. (Ed.) (1961), *Satellite Environment Handbook*, Stanford, CA: Stanford University Press.

Kelley, M. C. (1989), *The Earth's Ionosphere*, San Diego: Academic, p. 40.

Potemra, T. (1990), *Satellite Measurements of Birkeland Currents*, Birkeland Lecture No. 3, Oslo: The Norwegian Academy of Science and Letters.

Ratcliffe, J. A. (1972), "The Formation of the Ionosphere," *Geophys. Norveg.*, **29**, 13–27.

Rees, M. H. (1989), *Physics and Chemistry of the Upper Atmosphere*, Cambridge, U.K.: Cambridge University Press.

Schunk, R. W. (1988), "A Mathematical Model of the Middle and High Latitude Ionosphere," *Pure Appl. Geophys.* **127**, 255–303.

Schunk, R. W., Szuszczewicz, E. P. (1988), "First Principle and Empirical Modelling of the Global Scale Ionosphere," *Ann. Geophysicae* **6**, 19–30.

Stewart, B. (1882), "Terrestrial Magnetism," *Encyclopedia Brittanica*, 9th ed., Edinburgh: Adam Black.

Störmer, C. (1955), *The Polar Aurora*, Oxford: Clarendon.

U.S. Standard Atmosphere (1976), National Oceanographic and Atmospheric Administration, National Aeronautics and Space Administration, and U.S. Air Force, Washington, DC.

Further Reading

Akasofu, S.-I., Chapman, S. (1972), *Solar-Terrestrial Physics*, Oxford: Clarendon.

Banks, P. M., Kokarts, G. (1973), *Aeronomy*, New York: Academic.

Brekke, A., Egeland, A. (1983), *The Northern Light*, Heidelberg: Springer.

Chapman, S., Bartels, J. (1940), *Geomagnetism*, Oxford: Clarendon.

Eather, R. H. (1980), *Majestic Lights*, Washington, DC: American Geophysical Union.

Jursa, A. S. (Ed.) (1985), *Handbook of Geophysics and Space Environment*, Air Force Geophysics Laboratory, Air Force Systems Command, Springfield, VA: National Technical Information Service.

Heppner, J. P., Maynard, N. C. (1987), "Empirical High Latitude Electric Field Models," *J. Geophys. Res.* **92**, 4467–4489.

Kelley, M. C. (1989), *The Earth's Ionosphere*, San Diego: Academic.

Ratcliffe, J. A. (1972), *An Introduction to the Ionosphere and Magnetosphere*, Cambridge, U.K.: Cambridge University Press.

Rees, M. H. (1989), *Physics and Chemistry of the Upper Atmosphere*, Cambridge, U.K.: Cambridge University Press.

Rishbeth, H., Garriott, D. K. (1969), *Introduction to Ionospheric Physics*, Amsterdam: Academic Publishers Assn.

Vallance Jones, A. (1974), *Aurora*, Dordrecht, Holland: Reidel.

IRON AND STEEL

See STEEL

LABORATORY INSTRUMENTATION, CALIBRATION AND MAINTENANCE

See MAINTENANCE AND CALIBRATION OF LABORATORY INSTRUMENTATION

LANs

See NETWORKS, COMPUTER

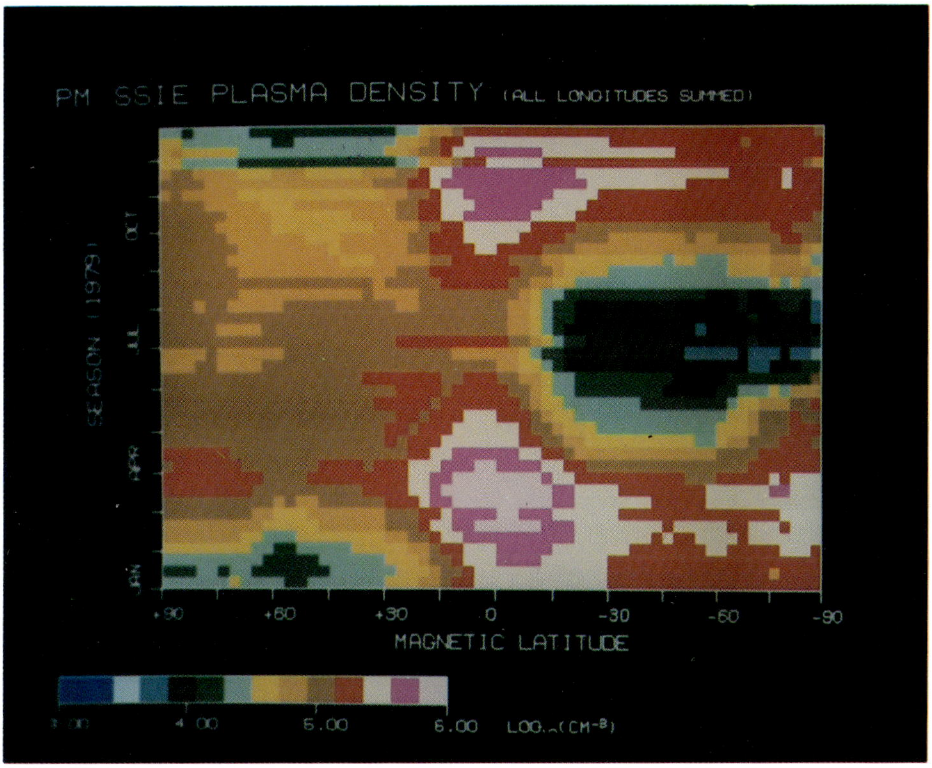

PLATE 1. An illustration of the global distribution of ionization and its seasonal variation at 830 km. This is an averaged distribution over all p.m. local times from the DMSP F2 and F4 satellites over a 2-y period (courtesy of F. J. Rich, Geophysics Directorate, Phillips Laboratory, USAF).

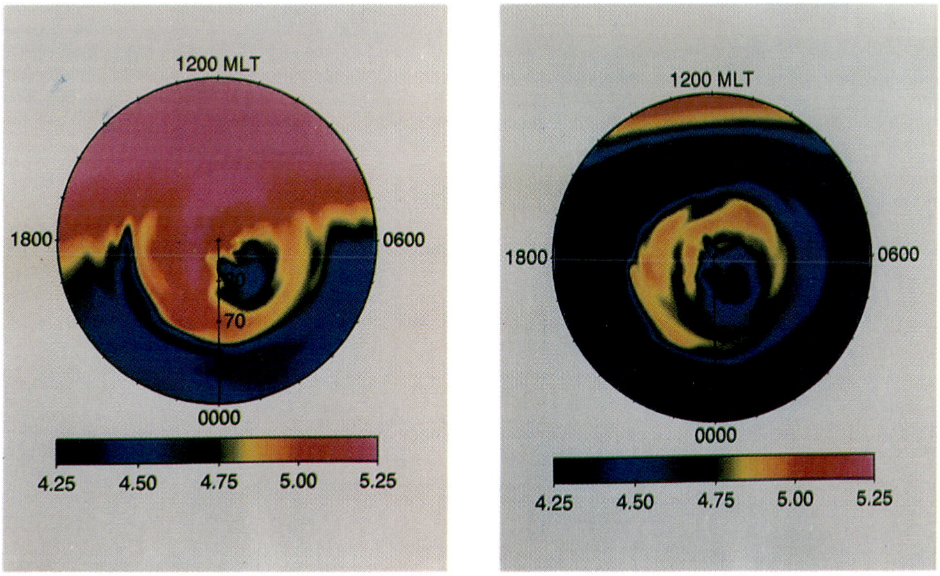

PLATE 2. Polar views of a typical ionization distribution for southern-hemisphere winter for low solar activity and quiet geomagnetic conditions. (a) For 0500 UT when the geomagnetic pole is tipped toward the Sun; (b) for 12 h later when the geomagnetic pole is tipped away from the Sun. These distributions are calculated for the altitude of the F-region peak using the Utah State University ionospheric model (courtesy of R. W. Schunk and J. J. Sojka, Utah State University).

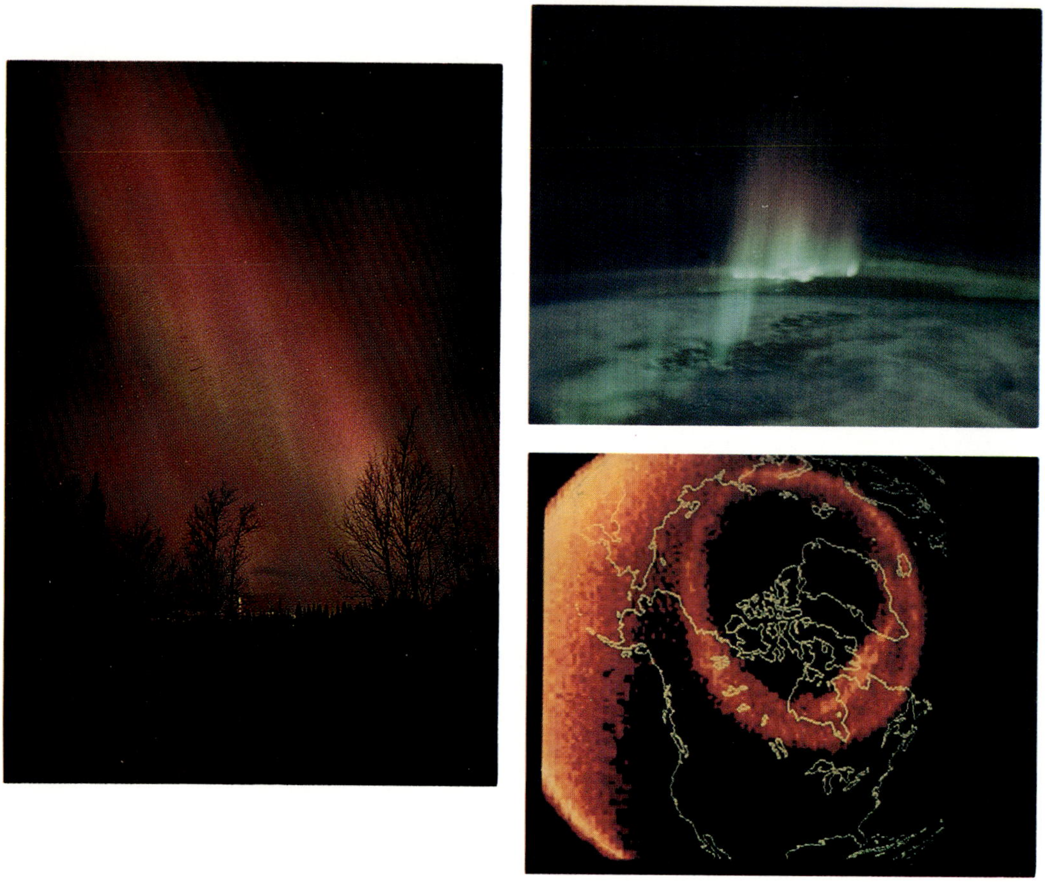

PLATE 3. Auroral pictures—(a) Rayed aurora band photographed in Northern Scandinavia. (b) A view of the aurora australis photographed from the flight deck of NASA Space Shuttle *Discovery* by one of its seven crew members on Mission STS-39. Note the vertical extent of the rays and the red color of the higher altitude portions. (Photo courtesy of the National Aeronautics and Space Administration). (c) The aurora as seen from deep space by the DE-1 spacecraft (from nearly 4 Earth radii) showing the sunlit dayside of the Earth and the auroral oval over the United States and Scandanavia (courtesy of L. A. Frank, University of Iowa).

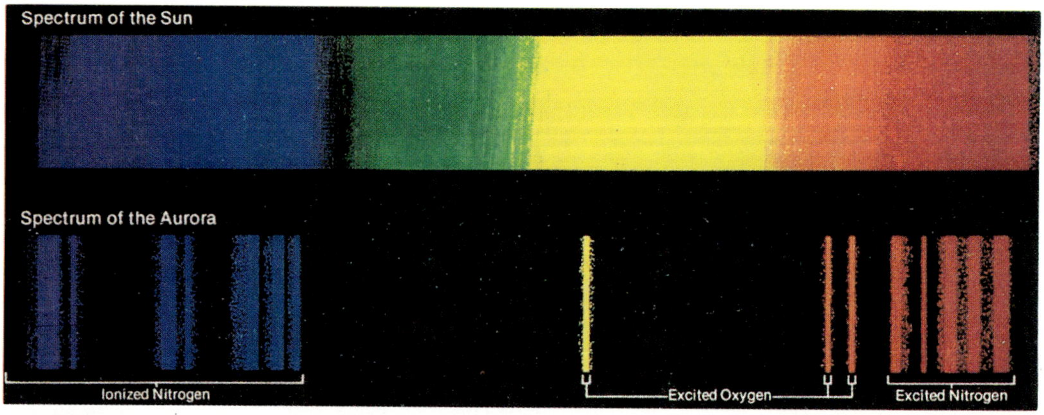

PLATE 4. The auroral spectrum, which consists of distinct lines and bands. For comparison the spectrum of the Sun is shown.

LABORATORY INSTRUMENTATION

AYMAN A. SHOURBAJI, *Instrumentation and Controls Division, Oak Ridge National Laboratory, Oak Ridge, Tennessee, U.S.A.*

	Introduction	247
1.	**Test and Measurement Instrumentation**	248
1.1	Digital Voltmeters	248
1.2	Oscilloscopes	250
1.2.1	Analog Oscilloscopes	252
1.2.2	Analog Storage Oscilloscopes	252
1.2.3	Digital Oscilloscopes	253
1.3	Function Generators	253
1.4	Recorders	256
1.4.1	Analog Recorders	256
1.4.2	Digital Recorders	258
1.4.2.1	Digital Wave Recorders	258
1.4.2.2	Programmable Recorders	258
1.4.3	*X-Y* Recorders	258
1.5	Spectrum Analysis	259
1.5.1	Fourier Analyzers	260
1.5.2	Spectrum Analyzers	261
1.6	Counters	262
1.7	Boxcar Integrators	265
1.8	Lock-In Amplifiers	266
1.9	Time-Domain Reflectometry	267
2.	**Data-Acquisition Instrumentation**	268
2.1	Data Loggers	268
2.2	PC-Based Data-Acquisition Systems	270
2.2.1	Software	271
2.3	Instrument Interfaces	272
3.	**Measurement Interface**	275
3.1	Digital-to-Analog Converters	275
3.2	Analog-to-Digital Converters	276
3.3	Signal Transmission	278
4.	**Conclusion**	279
	Glossary	279
	Works Cited	279
	Further Reading	279

INTRODUCTION

The main mission of every laboratory is to conduct scientific investigations to develop an aspect of a new technology or product or to improve an existing technology or product. The success of the experiments, which are an important part of the scientific investigative efforts, depends to a large extent on the availability of the appropriate test and measurement tools to quantify physical phenomena in complex measurements. Consequently, instrumentation users are faced with the ultimate responsibility of selecting the appropriate instrument for the scientific task in view of the diversified and rapidly changing technologies associated with the instrumentation field. The instrumentation industry, particularly, has undergone major changes that are directly related to the technlogical advances in electronics that evolved over the past five decades.

One major change was marked by the invention of the transistor in the 1940s. That was the beginning of semiconductor electronics. By replacing the vacuum tube, the transistor (solid state device) sent the instrument designer back to the drawing board, so to speak, and a new generation of instruments emerged. The newly developed instruments afforded better accuracy and improved stability, were more efficient and more compact, and offered faster response than their predecessors that were based on vacuum-tube technology. In the 1960s and the 1970s, electronics research developed solid-state devices capable of performing complex functions that included many discrete components such as chips, transistors, resistors, diodes, and capacitors integrated in a single silicon substrate. Today, these devices are known as integrated circuits (ICs). Further research and development in the IC field led to the microprocessor, which can be considered as a second turning point—if we consider the transistor as being the first—in the electronics and the instrumentation industries. Most modern test and measuring devices use one to three microproces-

1-56081-067-X/94/$5.00+.50

sors to add "intelligence" (i.e., capabilities that allow these devices to perform complex functions that previously required several instruments and off-line calculations). In addition, with microprocessors, test and measurement instruments combined the best of two worlds—the digital and the analog.

This article, in its scope and content, is intended for readers in a wide range of technical disciplines. The topics covered and their treatment, therefore, were chosen with that in mind. The article introduces a class of instrumentation that is most commonly used in the laboratory for conducting a variety of tests and performing a wide range of measurements as well as measurement systems for large-scale data monitoring and processing. The discussion includes principles of operation, and the underlying technologies associated with the latest development and functionality of these instruments. Also discussed are issues that confront instrument users when interfacing analog and digital devices for measurement applications necessitated by modern technologies.

It should be noted that all standards, communication protocols, and conventions cited in this article are based on those adopted in the United States—which may or may not be the same as those adopted by other countries.

1. TEST AND MEASUREMENT INSTRUMENTATION

1.1 Digital Voltmeters

The digital voltmeter (DVM), also known as the digital multimeter (DMM), is one of the most versatile instruments used in the laboratory. It is a multifunction instrument capable of measuring ac and dc electrical quantities such as volt, current, and resistance. The analog signal to be measured is introduced to the instrument, and it produces nearly instantaneously a direct visual digital display of the measured quantity. This type of instrument is available in analog form, but DVMs are now more commonly used than the analog counterpart. The DVM displays the readings on a liquid crystal display (LCD) or on a light-emitting diode (LED) device, which is an integral part of the meter. (See DISPLAY DEVICES for more details on digital display.)

In dc measurement, the dc quantity is an average value taken over some time interval T to smooth any fluctuations. The average value is computed by a simple averaging circuit in one of two ways. The first approach is to apply the signal to a true integrating circuit (true averaging) by using an operational amplifier with a capacitor in its feedback path (Fig. 1). In the second approach, the average value is obtained by using a low-pass filter circuit,

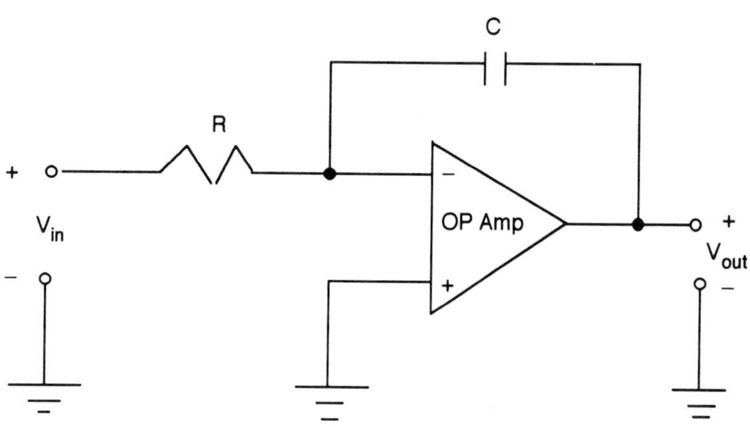

$$V_{out} = -\frac{1}{RC} \int V_{in}\,dt = -\frac{t}{RC} V_{in}$$

time constant (τ) in seconds = RC ; R in ohms, C in farads

FIG. 1. Active integrator using operational amplifier (op-amp). It is used in some voltmeters as true averaging circuits in root-mean-square measurements.

which is simply a resistance–capacitor (RC) network (Fig. 2). The RC network continuously smooths the signal fluctuations to produce, after approximately three to five time constants, an average value of the signal (RC averaging). Some other digital voltmeters consist simply of an analog-to-digital converter (ADC) (discussed in Sec. 3.1) and an integrated display device such as a LED or LCD. Some DVM designs employ other techniques to compute the average value for a dc signal, but they will not be discussed here; the reader can consult ELECTRICAL INSTRUMENTATION or Helfrick and Cooper (1990).

Ordinarily, ac measurement is given in terms of a root mean square (rms) value, also known as the effective value. Mathematically, the rms value of a voltage signal $v(t)$ is given by

$$V_{\mathrm{rms}}=\left(\frac{1}{T}\int_0^T v^2(t)\,dt\right)^{1/2}. \tag{1}$$

If $v(t)$ in Eq. (1) is a sine wave, that is, given by

$$v(t)=V_m \sin \omega t,\quad \omega=2\pi f, \tag{2}$$

where V_m is the sine-wave amplitude, ω is the radian frequency in radians per second, and f is the frequency in cycles per second (cps) or hertz (Hz), then the rms value (V_{rms}) can be computed from Eq. (1) to be $V_m/\sqrt{2}$.

DVMs, which provide rms measurements based on the concept expressed by Eq. (1), are known as *true mean square* (TMS) instruments. The underlying TMS concept can be simply outlined by

1. squaring the instantaneous value of the signal,
2. averaging the squared instantaneous value over the sampling time, and
3. taking the square root of the measured mean square value.

One type of TMS DVM squares the voltage of the applied signal by using a detector circuit with a square-law characteristic. The squared voltage signal is time averaged by using either a true averaging circuit or an RC averaging circuit. The square-root function can be accomplished by a proper readout scale calibration. Other types of TMS-type DVMs use detectors that compute rms value directly. When highly accurate rms measurements of nonsinusoidal signals are required (e.g., random signals), methods based on the heating power of the wave forms are used since heating power is directly proportional to the mean squared voltage.

Some voltmeters are calibrated to read the rms value of a sinusoidal signal, but they do not actually measure the rms value of the applied signal. These instruments usually consist of a simple rectifier followed by an equivalent low-pass smoothing filter such that a quantity proportional to either the mean absolute value (average rectified value) or peak value of the applied signal is measured. The mean absolute value for a sinusoidal voltage signal $v(t)$ is defined as

$$\overline{|v(t)|}=\frac{1}{T}\int_0^T |v(t)|\,dt, \tag{3}$$

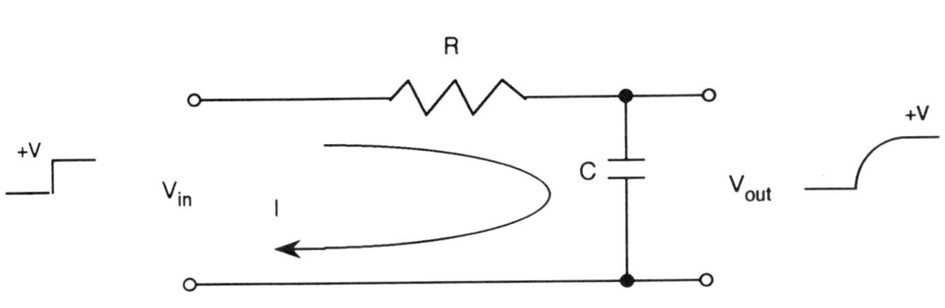

$V_{out}(t) = V_{in}(1 - e^{-t/RC})$

time constant (τ) in seconds = RC; R in ohms, C in farads

FIG. 2. Low-pass resistance–capacitor (RC) filter used in some voltmeters as RC averaging circuits in root-mean-square measurements.

where $v(t)$ is the absolute value and $\overline{|v(t)|}$ is the mean of the absolute value. For either a peak or mean absolute-value measurement, a readout in terms of the rms value of a sine wave can be obtained by a proper readout scale calibration. This calibration is easy to accomplish since the peak, mean absolute, and rms values of a sine wave have fixed relationships. For the sine wave in Eq. (2), the peak value is V_m, the mean absolute value is $\overline{|v(t)|}=(2/\pi)V_m$ and the rms value is $V_m/\sqrt{2}$. Some instrument designs take advantage of these relationships because peak- or mean absolute-value detection circuits are rather cheaper to build than true rms detection circuits. These simple instruments will produce inaccurate readings if they are used to measure complex periodic signals that are not sinusoidal.

DVMs are available in a wide range of capabilities to measure dc and ac voltages from <1 mV to >1000 V, currents from less than a milliampere to several amperes, and dc resistances from less than an ohm to up to several gigohms. Inaccuracy of the meters is generally determined by the degree of stability of the reference voltage used in the circuit and the inherent error of ± 1 unit in the least significant digit (LSD). Better accuracy can be realized with an increased number of display digits because of the higher resolution that can be attained. Digital voltmeters are available with up to $8\frac{1}{2}$ digits for bench-type versions and up to $4\frac{1}{2}$ digits for the portable-type (hand-held) versions. A meter with $4\frac{1}{2}$ digits can display four full digits ranging from 0 to 9, and one digit (leftmost) that displays only a 1 or 0. Therefore, this meter can display up to 19 999 counts of resolution, whereas a meter with $5\frac{1}{2}$ digits can display 199 999 counts of resolution, as shown in Table 1. Most recently, frequency measurement has been added to the DVM's measurement capability. Hand-held DVMs with built-in frequency counters that can measure from 1 Hz to >20 kHz are now commercially available.

1.2 Oscilloscopes

The oscilloscope, also commonly known as a scope, has been and is still considered to be one of the most versatile general-purpose electronic measuring instruments available for industrial applications as well as scientific experimentation. The scope is capable of capturing, displaying, and analyzing analog signals of various wave forms. The applied signal is displayed on a phosphorescent screen as an illuminated two-dimensional *x-y* trace. The signal magnitude is always displayed on the vertical *y* axis (vertical deflection), while the *x* axis (horizontal deflection) is used either as the time base or to display a second input signal to produce an *X-Y* plot of the two signals.

The most common use of an oscilloscope is to produce a plot of the input signal against time. The input signal is sensed and applied to the scope through a probe designed for that purpose. Most laboratory oscilloscopes have a 12.7-cm-diam (5-in.-diam) screen, but larger screens up to 53.3 cm (21 in.) are also available. Large-screen oscilloscopes usually cannot attain the high-frequency capability of the 12.7-cm (5-in.) types.

The modern oscilloscope, regardless of its sophistication, consists basically of the same building blocks, or components used in earlier models, as illustrated in Fig. 3. The cathode-ray tube (CRT) is the main output device that displays a trace of the signal wave form. The signal trace is produced by a focused narrow beam of electrons projected from an electron gun through a set of vertical and horizontal deflection plates. Application of voltages to these plates creates an electric field that deflects the electron beam, causing vertical and horizontal displacement of its focused beam

Table 1. Relationship between the number of display digits and the resolution.

Number of display digits	Number of counts of full scale	Resolution (%)
$3\frac{1}{2}$	1999	1 part in 1999=0.05
$4\frac{1}{2}$	19 999	1 part in 19 999=0.005
$5\frac{1}{2}$	199 999	1 part in 199 999=0.0005
$6\frac{1}{2}$	1 999 999	1 part in 1 999 999=0.000 05

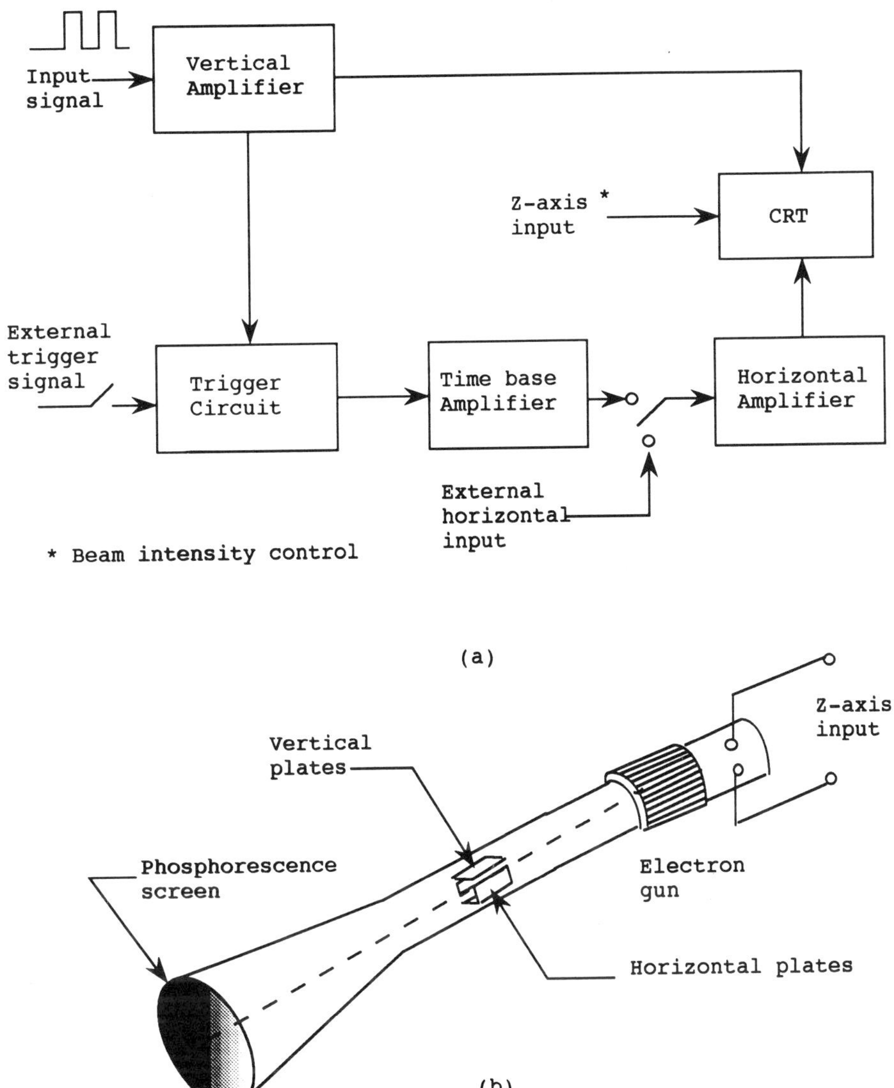

FIG. 3. Basic components of oscilloscopes: (a) block diagram; (b) cathode-ray tube basic structure.

point on the phosphorescent screen. With the proper design, the displacement can be made closely linear with the applied voltages. In standard CRTs, the phosphor is electrically refreshed before the phosphorescence decays to avoid blinking or flickering of the displayed trace. Some CRTs, storage CRTs, do not require continual refreshing to maintain the displayed trace. This type of CRT is discussed in Sec. 1.2.2.

The horizontal amplifier drives the horizontal plates with a ramp voltage, causing the focused spot to sweep from left to right at a constant speed. The trigger or synchronizing circuit provides the mechanism that defines the starting point of every successive sweep of the displayed trace. For proper synchronization of the sweep and the input signal applied to the vertical deflection plates, triggering of the sweep is initiated by energizing the trigger circuit from the leading edge of the input signal itself. External sources also can be used for sweep triggering.

Laboratory oscilloscopes may be classified in many ways. Usually, the distinctions are based either on frequency-response capability or on CRT characteristics.

1.2.1 Analog Oscilloscopes. An analog oscilloscope usually deploys an electrostatically deflected CRT with a standard refresh phosphor and real-time display circuitry. With this circuitry, the CRT beam deflection traces exactly the signal across the screen as the event occurs. For repetitive signals, the beam traces the signal pattern over and over, resulting in a bright wave form displayed on the screen.

Analog oscilloscopes can be classified into low- and high-frequency categories. The low-frequency oscilloscope typically has video-frequency capability ranges from dc to about 10 MHz (10-MHz bandwidth), and the high-frequency oscilloscope is capable of displaying single-shot signals with rise times as fast as 0.7 ns ($\sim$500 MHz).

Two of the most important considerations when evaluating a scope are the bandwidth and the rise time. The bandwidth is defined as the frequency range between dc and the maximum frequency at which the scope could display a signal with an attenuation of no more than 3 dB, and it determines the signal capturing performance that could be expected from the scope. The rise time is defined as the time it takes the signal to change from 10% to 90% of its final value, as shown in Fig. 4. The rise time determines the ability of the scope to deal with short events such as fast transients with short durations. The bandwidth and the rise time t_r are related by the approximation

$$\text{bandwidth} = 0.35/t_r. \quad (4)$$

It should also be realized that the rise time of a signal as observed on the screen is not just the rise time of the signal but also includes the rise time contributed by the scope and according to the approximate relationship

$$t_r(\text{observed}) = [t_r^2(\text{signal}) + t_r^2(\text{oscilloscope})]^{1/2}. \quad (5)$$

1.2.2 Analog Storage Oscilloscopes. Analog storage oscilloscopes can display signals of very low frequencies or low repetition rates, or even single fast events. They can temporarily store (for several minutes) wave forms in their display CRTs for extended visual analysis. The storage CRT, unlike the standard type, does not require continual refreshing to maintain a displayed trace. The storage capabilities of this type of CRT are attributed to the type of phosphor used with the screen. Phosphor types with different characteristics are classified according to their storage capabilities as short, medium, or long persistence, as indicated by Table 2.

Storage oscilloscopes are useful in applications where display measurements are difficult to observe on a standard oscilloscope CRT. Single-shot events, such as the wave

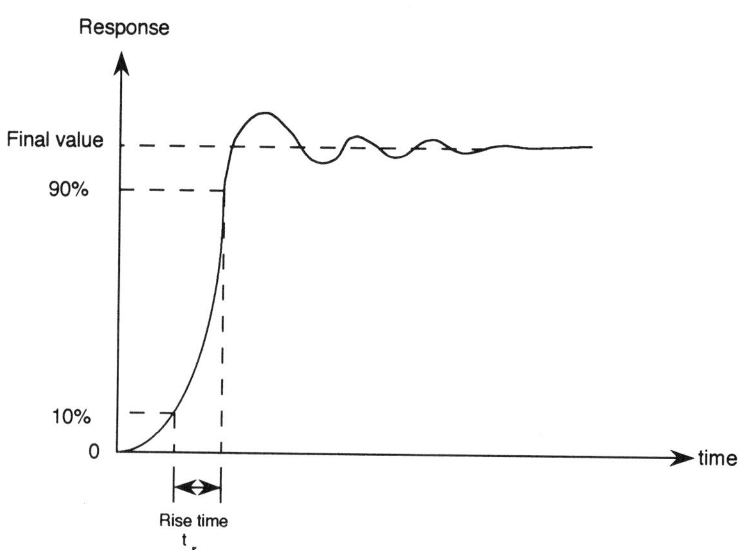

FIG. 4. Example of a typical oscillatory response of a transducer showing the rise time.

Table 2. Some of the commonly used phosphors.

Phosphor type	Color		Persistence	Display application
	Fluorescence[a]	Phosphorescence[a]		
P1	Yellow-green	Yellow-green	Medium (<2 s)	Scopes, radar
P2	Blue-green	Green	Medium	Scopes
P7	Blue-white	Yellow-green	Variable[b]	Radar, medical
P11	Blue-violet	Blue	Medium to short (<1 s)	Photographic recording
P19	Orange	Orange	Long (few min)	Radar

[a]Fluorescence is the light emitted as a beam strikes the phosphor, whereas phosphorescence is the light after an excitation beam is removed.
[b]Persistence can vary from short to long depending on filter color.

form resulting from some destructive testing or fast-occurring phenomena, are transient in nature and are quickly lost unless the wave form can be either photographed or stored on the CRT itself. Signals with very slow repetition rates are also difficult to view on standard oscilloscopes because of the phosphorescence decay. Comparison of two wave forms is often difficult when a standard oscilloscope is used since CRT refresh characteristics do require simultaneous testing of the two circuits. In storage oscilloscopes, one signal trace is stored for comparison with a second signal obtained later. Analog storage oscilloscopes have their own limitations because of the high cost of storage CRTs and the limited storage time depending on the type of phosphor used.

1.2.3 Digital Oscilloscopes. The digital oscilloscope is capable of measuring and displaying a repetitive wave form by sampling the input signal with an ADC at regular intervals and internally storing the values in solid-state memory. Nonvolatile memory can provide data storage without degradation or loss of data for up to five years with the power removed from the instrument. The stored data can be retrieved later to reconstruct the signal by a digital-to-analog converter (DAC) (discussed in Sec. 3.2) long after the signal is removed from the input. The reconstructed signal can be displayed on the screen as a series of dots, representing signal values, located closely enough to appear as a continuous trace. The number of these values on the screen determines the vertical resolution as dictated by the input ADC. A 10-bit ADC provides a full scale of 1024 discrete values that, when used to represent a single-shot signal, results in a resolution comparable to that of an analog storage oscilloscope.

The sampling rate for sampling a real-time signal needs to be carefully considered to reconstruct the signal accurately. As shown in Fig. 5, sampling at a rate less than the signal's own frequency gives false indication of a much slower signal. This phenomenon is known in sampling theory as *aliasing*. Theoretically, the sampling frequency f_s, also known as the Nyquist frequency, should be at least twice that of the input signal's highest-frequency component to ensure accurate results.

The number of ADCs used in the sampling process is an important consideration. When one ADC is shared by two or more input channels and one channel is in use, it can sample the input signal at full speed. When two channels are in use, the effective sample rate for each channel is reduced by 50%. The result is that a single-shot frequency performance can be reduced in half. Some manufacturers recently introduced a digital storage oscilloscope and DVM integrated into one compact hand-held meter. The scope is dual trace with 50-MHz bandwidth and sampling rate of 25 megasamples per second. The DVM can measure the true rms of ac or ac-plus-dc signals with 8-bit resolution.

1.3 Function Generators

The function generator instrument is commonly used in testing, calibration, and signal simulation applications. The generator can deliver a variety of wave forms whose frequencies are adjustable over a wide range with high accuracy. The only input to the instrument is the ac power required for the electronic circuitry. The output can be selected from a set of commonly used standard wave-

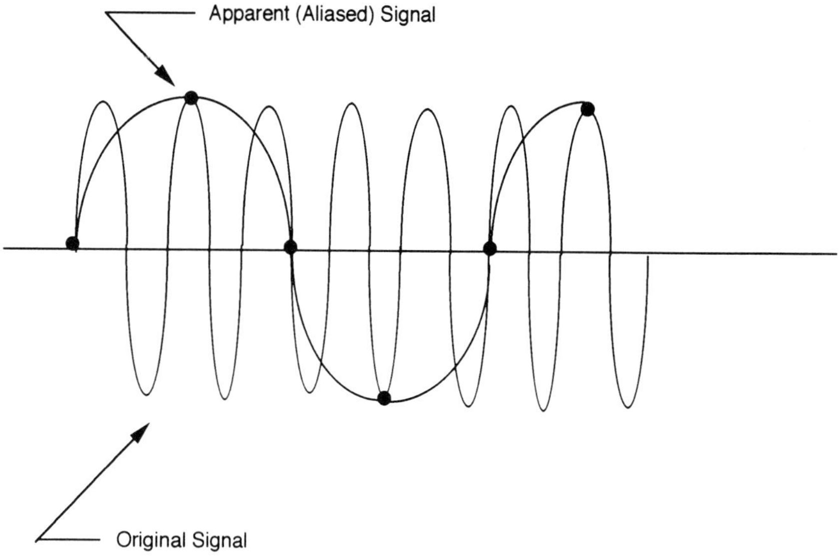

FIG. 5. Aliasing. A 16-kHz sine wave sampled at 20 kHz appears to be a 4-kHz sine wave.

forms such as the sine wave, haversine, triangular, square, and sawtooth (Fig. 6). Modern function generators can offer waveforms with frequencies adjustable from 1 μHz to 80 MHz. In addition to standard wave forms, arbitrary or user-defined waveforms (for special applications) can be created on a personal computer (PC) by using a specific software developed for this purpose and then downloaded to the generator by a standard instrument interface (IEEE-488/GPIB interface discussed in Sec. 2.3). Arbitrary signals such as video test signals and mechanical vibration can be used in simulating responses of some physical systems. When equipped with the IEEE-488 interface, the function generator can also provide remote control of various functions and allows the generator to be integrated with automatic testing systems.

Some function generators have a very useful feature for phase locking the generator with external signal sources. With phase lock, one function generator can be phase locked to a harmonic of the sine wave of a second function generator. By adjusting the phase and amplitude of the harmonics, many other wave forms can be generated as the result of summing the fundamental frequency from one generator with the harmonics of the other. The function generator can also be phase locked to a frequency standard. With such arrangements, the generated outputs have the standard source frequency accuracy and stability.

It is very difficult to give a unified description of the internal structure of all available generators since the structure varies from one

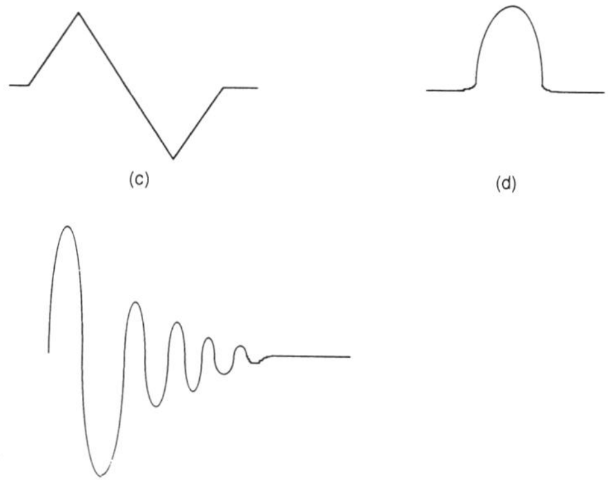

FIG. 6. Typical output wave forms produced by function generators: (a) sine wave; (b) square wave; (c) triangular wave; (d) haversine; (e) example of an arbitrary (user-defined) wave form.

instrument designer to another. However, the underlying concept can be illustrated by the structure of the function generator in Fig. 7 (Helfrick and Cooper, 1990).

This structure represents a function generator with three outputs (sine, triangular, and square waves) with a frequency range of 0.01 Hz to 100 kHz. The frequency can be adjusted either manually from the instrument front panel or from an external source. The frequency control circuit has the function of regulating two current sources.

The upper current source supplies a constant current i to the integrator A whose output voltage V_{out} increases linearly with time according to the relationship

$$V_{out} = -\frac{1}{C}\int i\,dt, \tag{6}$$

where C is the capacitance of the condenser in the feedback path of the operational amplifier K. The slope of V_{out} varies proportionally with the upper current. The integrator output is connected to a comparator whose state changes at a predetermined level on the positive slope of the integrator's output. This change of state results in cutting off the upper current supply to the integrator, and the lower current source is switched in.

The lower current source supplies a reverse current to the integrator, resulting in a linear decrease of its output with time. At a predetermined level on the negative slope of the output, the comparator switches to cut off the lower current source while simultaneously switching in the upper source once again. The switching action between the two current sources results in a triangular wave form at the output of the integrator. The switching action of the comparator causes its output to toggle between two fixed voltage levels, resulting in a square wave. The sine wave can be generated from the triangular signal by a diode-resistance network. Such a network can be designed to alter the slope of the triangular wave in a piecewise linear approximation fashion as its amplitude changes, with $<1\%$ distortion. Figure 8 illustrates the basic idea of piecewise-linear approximation of nonlinear functions by using the diode-resistance circuit. The accuracy of such an approximation is determined by the number of line segments used. The approximate curve is obtained by the summation of the individual line segments, whose "breakpoint" voltages and slopes are determined separately for each segment. The operational amplifier function is to sum up the segments by summing up the currents from the individual diode-resistance circuits at the summing junction. The feedback resistor R_f provides the scaling function.

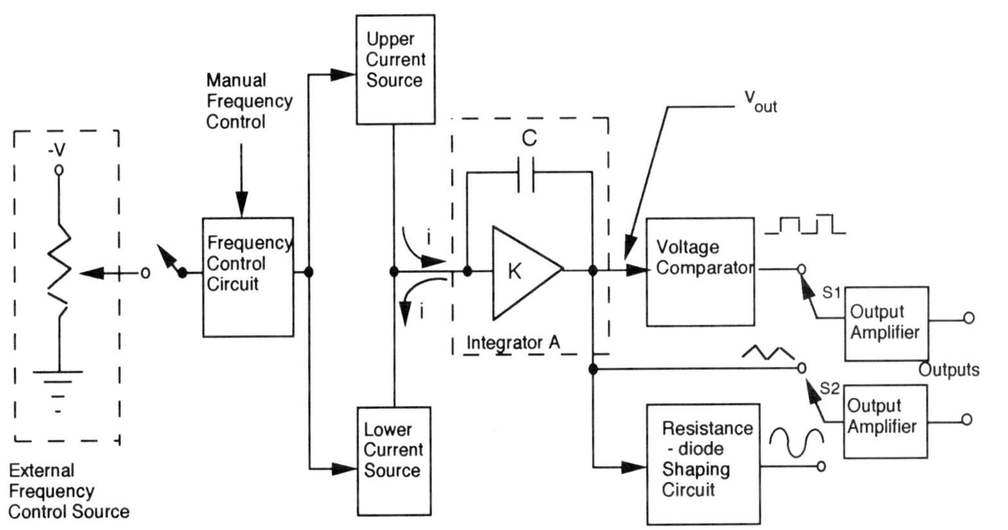

FIG. 7. Basic elements of a function generator providing three standard wave forms through two separate outputs. Some generators have only one output.

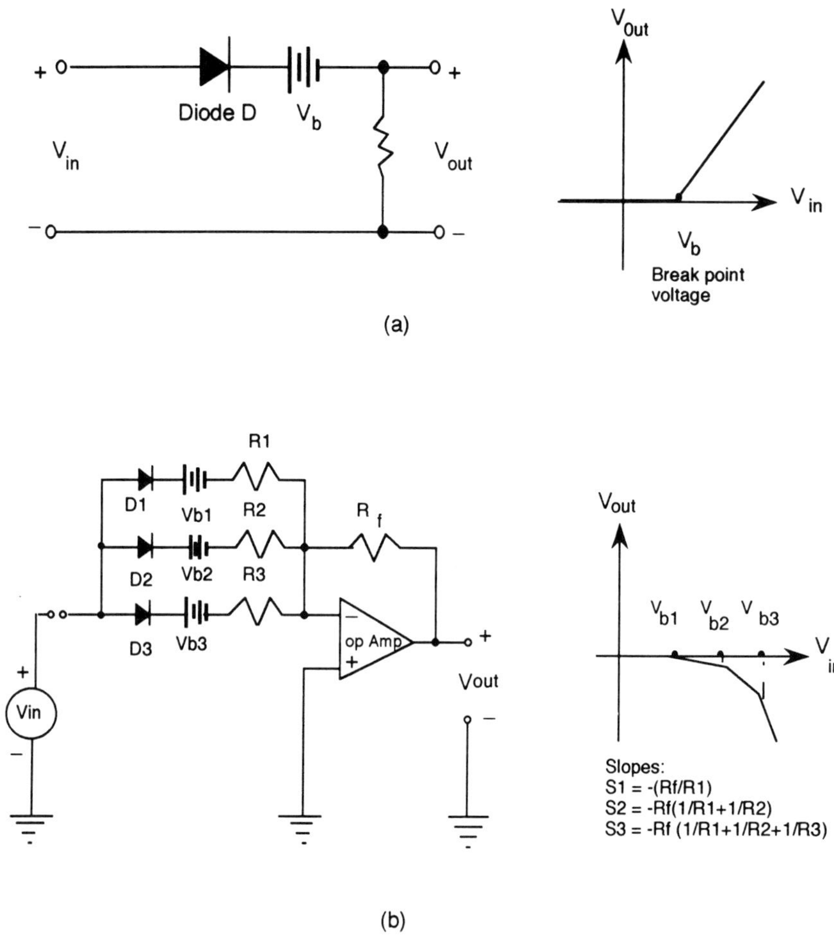

FIG. 8. Piecewise approximation: (a) simple line segment; (b) multiline segment approximation with summing amplifier.

1.4 Recorders

Recorders, in general, represent a fairly inexpensive class of recording devices primarily used for capturing and preserving data long after the signal has ceased. These recorders can be classified on the basis of their principle of operation and function. In this section, three conventional types of recorders that can be described as stand-alone instruments are discussed. The other forms of recorders require interfacing with some peripheral devices to provide the user with the level of man–machine interaction necessary for recalling and displaying the data. These types of recorders are discussed in the ELECTRICAL INSTRUMENTATION and RECORDING DEVICES and, therefore, will not be covered here.

1.4.1 Analog Recorders. The analog recorder receives analog (real-time) signals and produces a continuous time record or trace on a strip of calibrated paper (chart) traveling at a constant velocity under a motor-driven writing pen (Fig. 9). The widely known strip-chart recorder, commonly used in laboratory applications, has chart motion perpendicular to the pen deflection. The basic recorder has a single writing pen dedicated for a single input. The multipen version, usually up to six pens, produces traces of several signals simultaneously with different colors. Also available are multi-input recorders, each accommodating as many as 30 input signals and using a dot printing head instead of the writing pen. Circular-chart recorders are also available and are used primarily in industrial applications.

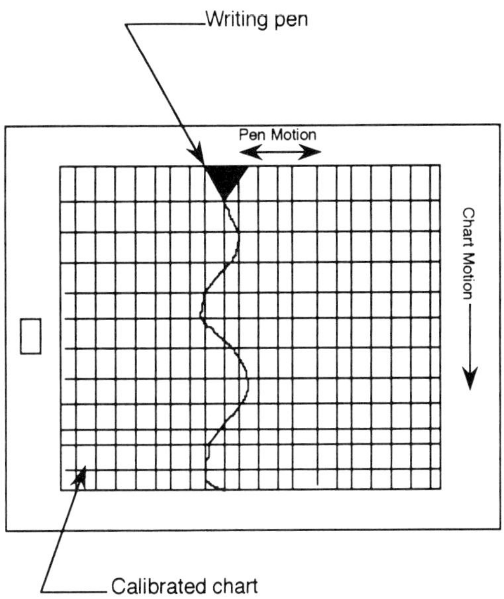

FIG. 9. Strip-chart recorder.

Some modern recorders use a pen-deflection mechanism based on improved versions of the moving-coil galvanometer type widely recognized as the D'Arsonval movement. More modern designs employ a digital servomechanism to drive the writing pen. Although servo systems vary in their design and hardware implementation, their principle of operation is very much the same. Figure 10 is a block diagram illustrating the principle of a closed-loop null-balance servo system used for driving the writing pen in recorders. The modern servomechanism consists basically of a dc motor and associated components (gears or linkages) to drive the pen and an encoder (position transducer) to detect and feed back the pen position. As a null-balance positioning system, the balance position is always being sought by the servo system as the input signal changes. The error produced is proportional to the signal change, causing the servo to drive the pen in the appropriate direction to offset the error. Scaling circuits are used to scale the input and the position signals appropriately before they are summed to produce the positional error.

With the advances in microprocessor technology, analog recorders are provided with advanced features unsurpassed by the earlier generation. Such features provide the analog recorder with improved flexibility useful for laboratory experimentation. One of the new powerful features is the programmability of input types, full-scale ranges, and chart speed without the need for operator programming skills. With this feature, the recorder can be programmed to record temperature by using a thermocouple or resistance-thermometer detector (RTD) in one application and can be easily reprogrammed to use the same recorder hardware with different input signals such as voltage or current in another application. Some computational capabilities are also now possible, such as computing the mean, minimum, and maximum values as well as having an integral digital readout and printout of the programmed setup. Although modern analog recorders are reliable, economical, and accurate (better than ±1%), their response is limited by the inherently slow pen mechanism. Typical frequency response is on the order of a few hundred hertz. Analog

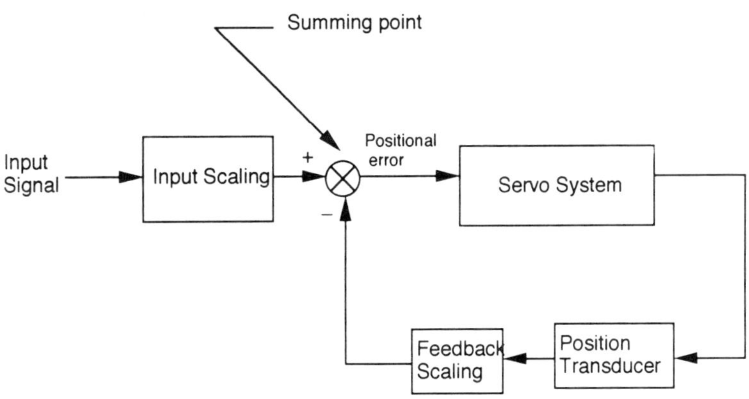

FIG. 10. Functional block diagram of a null-balance servo.

oscilloscopes (discussed in Sec. 1.2.1) also fall into the class of analog recording instruments.

1.4.2 Digital Recorders

1.4.2.1 Digital Wave Recorders. The development of high-speed ADCs (discussed in Sec. 3.1) together with the ready availability, small size, and reasonable cost of digital memories and associated digital-processing hardware has resulted in the development of the digital wave recorder.

The incoming analog voltage signal is digitized by an ADC with a resolution dependent on the digital output capability of the converter. An 8-bit ADC has a resolution of 1 part in 256 (0.4% accuracy), and a 10-bit ADC has a resolution of 1 part in 1024 (0.1% accuracy). The analog input can be sampled at adjustable rates (100 000 samples/s or more is possible). The sampled values are stored in a nonvolatile solid state memory for later processing of the data. The sampling rate and memory size must be selected to suit the duration and wave form of the physical event being recorded.

To view the wave form, stored data are read from memory through a DAC (discussed in Sec. 3.2), and the resultant signal is displayed on an integral CRT screen (discussed in Sec. 1.2). The data can be read from memory at a slow rate to allow the original event (which may have been rapidly changing, such as a one-shot transient) to be reproduced at a rate that is easy to view or even to record on one of the slower recording devices (such as a strip-chart or *X-Y* recorder). Data can also be read out from memory at a faster rate repetitively where, in such a case, a one-shot transient can appear as a continuous wave form that can be easily observed and analyzed. Digital memory can also be accessed directly (by bypassing the DAC) by a PC-based data acquisition system (DAS) (discussed in Sec. 2.3) for manipulation and processing of the recorded data.

Other developments in digital recording devices include visual trend display, where several channels of information are continuously and simultaneously updated, stored, and displayed on the CRT as bar charts or in pen-recorder-type traces.

1.4.2.2 Programmable Recorders. Combining digital technology, microcomputing, and analog recording capabilities has allowed a programmable class of recorders to become available. The result is a digital potentiometric recorder that is superior to the analog recorder. The programmability of the device provides the kind of flexibility and diversity required by the experimental nature of laboratory investigations. The sampled and stored data can be plotted against time or one channel against another with 0.1% full-scale accuracy. Additional capabilities attributed to the intelligence provided by on-board microprocessors are computing functions such as addition, subtraction, multiplication, division, averaging any two curves point by point, differentiation, integration, and normalization. In addition to these functions, the data for plotting can be offset, scaled, or expanded. Another form of digital recording device is the digital oscilloscope discussed in Sec. 1.2.2.

1.4.3 *X-Y* Recorders. The *X-Y* recorder, also known as the *X-Y* plotter, is another type of analog recorder, in which one variable can be plotted against a second variable in two-dimensional Cartesian coordinates (Fig. 11). The function plot is produced by simultaneously introducing two signals to the instrument, where each signal serves as an input to independent pen-driving mechanisms. The writing pen is driven in a two-dimensional plane by the two systems (*X* axis and *Y* axis) to produce the plot on chart paper. The most common chart paper sizes for this type of recorder are the standard 21.6×27.9 cm ($8\frac{1}{2}\times11$ in.) or 27.9×43.2 cm (11×17 in.). The chart paper in this case is held in place on a flat surface by an electrostatic field. Other instruments are available with roll chart paper with a variable speed selectable over a wide range to a maximum of 80 cm/s (31.5 in./s) (for example), with accuracy $>1\%$ of full scale.

Two common mechanisms are available for driving the writing pen: (1) the servo system, described earlier, mostly employed in modern *X-Y* recorder designs, and (2) the optical galvanometer. The optical galvanometer offers significant improvements by eliminating the friction and inertia problems associated with the conventional galvanometer. As illustrated by Fig. 12, a small fixed mirror fastened to a moving coil is positioned to reflect a beam of light (from a light source) on light-sensitive, moving chart paper. When the coil turns, the light beam deflects over the paper, causing a trace on the chart. The re-

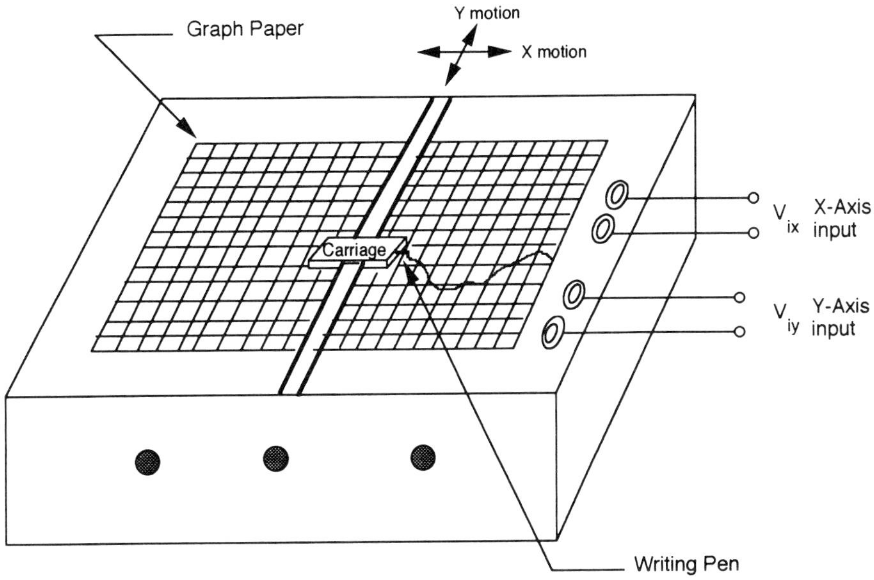

FIG. 11. *X-Y* recorder (plotter).

cording speed varies somewhat between self-balancing potentiometer types (earliest form of recording mechanism) and oscilloscopes (analog form). The frequency response of the structure shown is flat from dc to 100 Hz at full-scale deflections and nonlinearity of 1% of full scale (Doebelin, 1983).

Many *X-Y* recorders also can perform as strip-chart recorders by allowing for one of the axes to serve as a time base if so desired. The strip-chart recording function can be accomplished by introducing a voltage ramp to one of the axes and the signal to the other. The ramp can be generated by charging a capacitor with a constant current, thus causing the voltage to change linearly with time. The corresponding servo system will then cause its axis to translate at a constant speed. Most modern *X-Y* recorders offer the choice of a wide range of dc input voltages for each axis. As many as 19 different ranges and a sensitivity from as low as 0.1 mV/cm to 50 V/cm is not uncommon.

1.5 Spectrum Analysis

In many measurement applications, the analysis of signals in the frequency domain in terms of amplitude versus frequency is quite important and desirable; vibration analysis is a good example. The distribution of signal amplitude against frequency is known as the spectrum. Spectrum analysis and computation depend on signal behavior whether periodic or nonperiodic. Almost all physical measurements that we deal with can be classified as *periodic* or *nonperiodic*.

Fourier (the mathematician) showed that periodic signals (e.g., sine, square, triangular wave) expressed in the continuous-time domain can be represented by a sum of components at harmonic frequencies. If $x(t)$ is a periodic function [i.e., $x(t+T)=x(t)$, where T is the time period], then it can be approximated by Fourier transform by the summation

$$x(t)=\sum_{n=0}^{\infty} a_n \cos\omega_n t + b_n \sin\omega_n t, \tag{7}$$

where

$$a_n=\frac{1}{T}\int_0^T x(t)\cos\omega_n t\, dt,$$

$$b_n=\frac{1}{T}\int_0^T x(t)\sin\omega_n t\, dt,$$

$$\omega_n=n\omega_0,$$

ω_0=fundamental radian frequency,

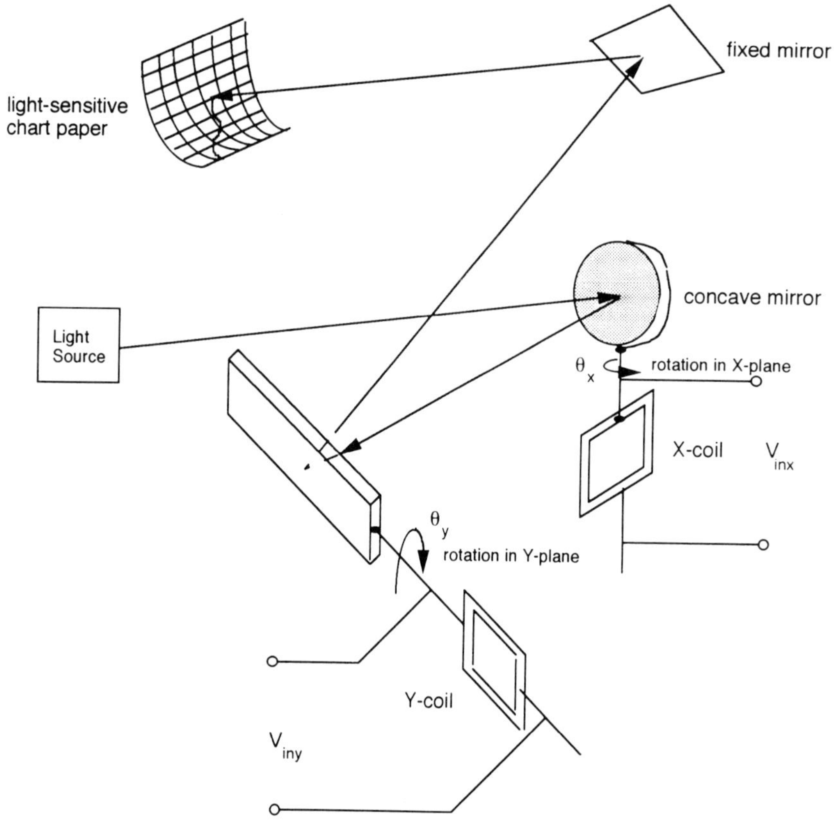

FIG. 12. A simplified structure of an optical galvanometric *X-Y* recorder.

which results in a discrete amplitude–frequency spectrum where frequency contents exist at discrete values only, as shown in Fig. 13(a).

With the advent of computer technology and its growth in the 1960s, a fast-Fourier-transform (FFT) algorithm [based on the discrete Fourier form (DFT)] was developed by Tuckey and Colley to compute the FFT by using digital computers [see FOURIER AND OTHER MATHEMATICAL TRANSFORMS or one of the textbooks on the subject (e.g., Brigham, 1974)]. The FFT algorithm is simply an efficient means of calculating the DFT $X(n)$ of a discrete signal $x(kT)$ (signal expressed in discrete-time domain) by

$$X(nf_s/N) = \sum_{k=0}^{N-1} x(kT_s)\, W^{nk}, \tag{8}$$

where $W = e^{-j(2\pi/N)}$ is a complex exponential weighting term, and the sampling frequency $f_s = 1/T_s$, which for sampling interval $T_s = 1$ can be written in the simpler form

$$X(n/N) = \sum_{k=0}^{N-1} x(k)\, W^{nk}. \tag{9}$$

For nonperiodic signals, typical of most physical measurements, the amplitude–frequency spectrum is continuous where frequency contents are distributed over some range depending on the signal, as shown in Fig. 13(b). To calculate the frequency spectrum of nonperiodic signals, another technique based on the concept of power spectrum density function is used, as will be discussed in Sec. 1.5.2.

1.5.1 Fourier Analyzers. The development of the computationally efficient FFT technique led to the development of the Fourier analyzer. FFT-based analyzers provide fast, high-resolution spectrum and network analysis with the information presented on a CRT. Unlike the spectrum analyzer discussed below (swept tune), the Fourier analyzer can measure dynamic signals (in real time) because they measure all frequencies simulta-

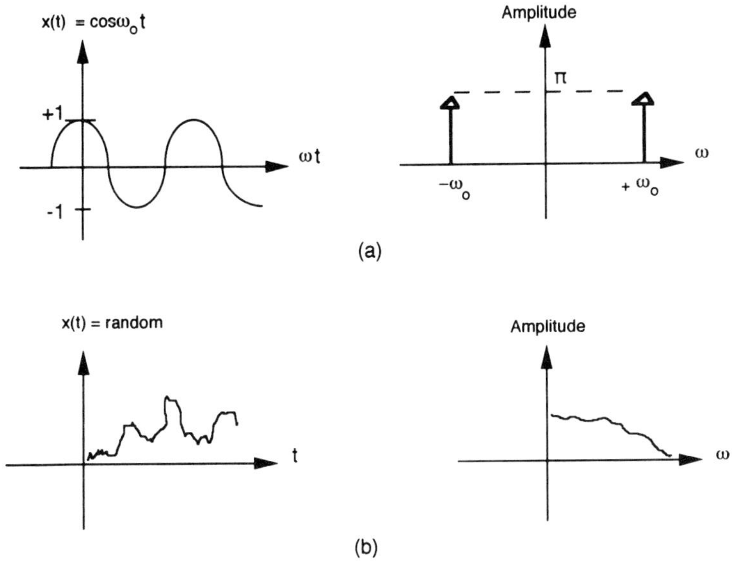

FIG. 13. Examples illustrating the frequency spectrum of periodic and nonperiodic functions: (a) periodic function; (b) nonperiodic function.

neously (not one at a time). Also, this analyzer provides both the magnitude and the phase information of each frequency component. The analyzer is useful for measuring signals from a few microhertz to about 100 kHz. In transient–wave-form analysis, signals can be sampled, digitized, and stored in internal memory or sent by one of several interfaces [e.g., general-purpose interface bus (GPIB), discussed in Sec. 2.3] to an external storage device. The stored information can be recalled and analyzed in time, frequency, and amplitude domains.

Some modern designs incorporate measurement of the statistical properties of signals as well as the joint properties of two or more signals. These measurements are made possible because of the availability of phase information afforded by the analyzer.

1.5.2 Spectrum Analyzers. For nonperiodic functions (signals), the spectrum is calculated on the basis of the power spectral density concept. For a complex, nonperiodic function $x(t)$, its power spectral density $S_{xx}(\omega)$ can be expressed as (Sydenham, 1982)

$$S_{xx}(\omega)=\frac{1}{2\pi}\lim_{T\to\infty}\frac{|X(j\omega)|^2}{2T}, \qquad (10)$$

where $X\ (j\omega)$ is the Fourier transform of $x(t)$ given by

$$X(j\omega)=\int_{-\infty}^{+\infty} x(t)e^{-j\omega t}dt. \qquad (11)$$

It is clear from Eq. (10) that the power spectral density is a real-valued function of ω and even, so that $S_{xx}(\omega)=S_{xx}(-\omega)$. The power P of the entire signal in the frequency domain is related to the power spectral density $S_{xx}(\omega)$ by Parseval's equation (Sydenham, 1982) as

$$P=\overline{x^2(t)}=\lim_{T\to\infty}\frac{1}{2\pi}\int_{-T}^{+T} x^2(t)dt$$

$$=\int_{-\infty}^{+\infty} S_{xx}(\omega)d\omega. \qquad (12)$$

Equation (12) shows that the spectrum of a signal at various frequencies is directly related to the signal power at these frequencies. Such a relationship provides the basic concept in spectrum analyzer design.

Design implementation of spectrum analyzers can be generally summarized in the following steps:

1. frequency filtering of the signal by a narrow-band–pass filter,
2. squaring the instantaneous value of the filtered signal,
3. averaging the squared instantaneous value over the sampling time, and
4. computing the rms value of the result.

As shown by Fig. 14, most modern spectrum analyzers perform the filtering function by a technique known as heterodyning, which al-

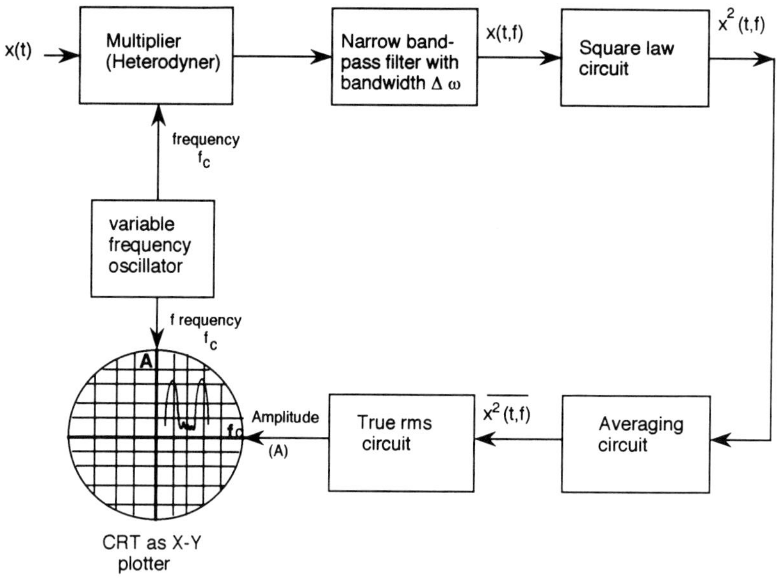

FIG. 14. Functional block diagram of a spectrum analyzer.

lows the use of a single fixed-band–pass filter throughout the entire range of frequencies. Without heterodyning, several filters of fixed bandwidth staggered in frequency would have to be used to cover the desired frequency range. The heterodyning operation consists of simply multiplying the signal with a sine wave whose frequency f_c can be accurately and quickly changed through a range of values. This multiplication process is equivalent to amplitude modulation that shifts any frequency content of the signal at, say, frequency f_s to f_c+f_s.

The filtered signal $x(t,f)$ is that part of $x(t)$ at frequency f. After squaring and averaging $x(t,f)$, the spectrum can be computed by using a calibrated TMS voltmeter circuit, as described in Sec. 1.1. The variable frequency from the oscillator provides the sweep signal for the synchronization with the x-axis sweep of the CRT to generate the amplitude spectrum on an oscilloscopelike display.

Spectrum analyzers can provide a broad frequency range from a few hertz to >300 GHz with a wide amplitude range from less than −100 to about +30 dBm. With the proper instrument interface protocol, the analyzers can be directly connected to plotters or printers to provide hard copies of the CRT displays.

1.6 Counters

A most convenient and widely used instrument for accurate measurement of frequency and time intervals is the electronic counter. Counters can perform a wide range of functions including high-accuracy frequency calibration of radio-frequency (rf) signals, frequency measurements on noisy signals, accurate propagation delay measurements, and phase shift in filter signals.

Modern counters employ the state of the art in digital technology and ICs by providing high-precision measurement and powerful analyzing capabilities—all in one compact, cost-effective instrument. With on-board microprocessors, counters can even provide some computational capabilities with built-in mathematical and statistical functions such as inversion, normalization, scaling, mean value, standard deviation, and minimum and maximum values. In addition to the integrated digital display of the measurement, the counter can have other outputs such as binary-coded decimal (BCD) (discussed below) and analog output for strip-chart recording. The versatility of the counter can also be extended by adding the IEEE-488 standard instrument interface bus for communication with other instruments and remote programmability of the counter.

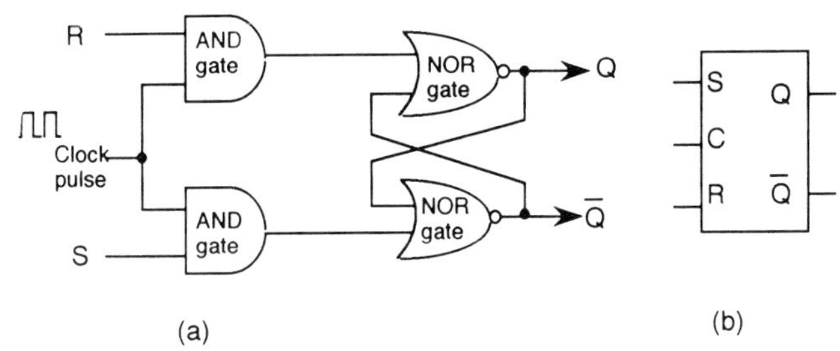

R	S	Q	$\bar{Q}$
0	0	Unchanged	
0	1	1	0
1	0	0	1
1	1	Undefined	

(c)

FIG. 15. Clocked SR flip-flop using NOR gates: (a) circuit diagram; (b) block diagram; (c) state table.

The most basic element of the counter is a digital device called the *flip-flop* (FF). The flip-flop, also called a bistable multivibrator, is an extremely versatile logic device that is used extensively in most digital circuitry including digital computers. The basic FF, known as the RS FF, can be formed by cross-connecting two NOR gates or two NAND gates (Fig. 15 shows the NOR configuration to illustrate the basic operation of the FF). The RS FF device has three inputs: S (SET), R (RESET), and CP (CLOCK PULSE). The S and R inputs are normally at logic 0 (also referred to in the digital field as logic LOW or 0 state). If the input applied to S is changed from logic 0 to logic 1 (referred to in the digital field as logic HIGH or 1 state), the Q output goes to logic 1 and the complemented (inverted) output $\bar{Q}$ goes to logic 0. Even if the S input returns to the 0 state, Q remains at the 1 state because of the internal feedback between the two NOR gates. This operation is called *setting the FF.* Similarly, when the input applied to R changes to the 1 state, Q goes to the 0 state (and $\bar{Q}=1$ state) and remains in that state. This operation is called *resetting the FF,* as indicated by the table of Fig. 15(c). Important to note is that the S and R inputs should not go to logic 1 simultaneously, because the output state in this case is not defined. In practice, the FF would be one component in a digital circuit performing a specific function. Therefore, the operation of the FF must be synchronized with the rest of the circuitry. Synchronization is accomplished by a CLOCK PULSE at the CP input. This pulse is usually a square pulse with

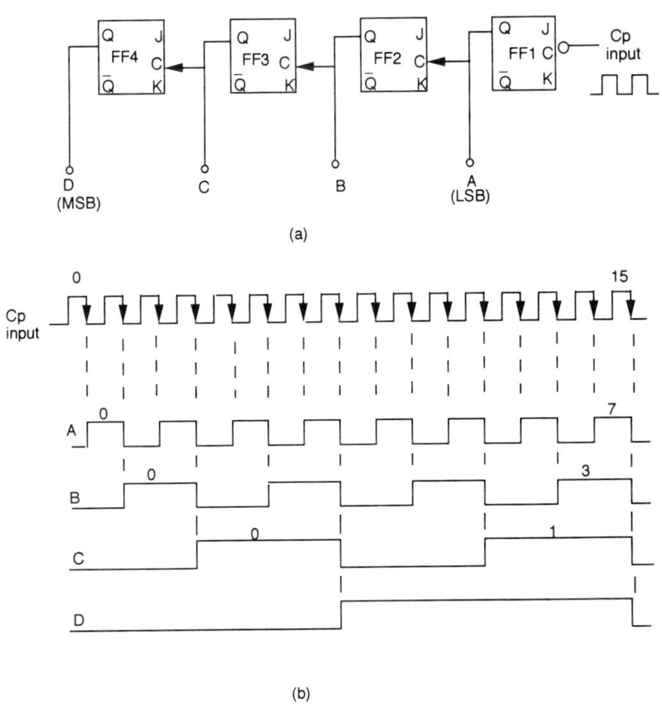

FIG. 16. Four flip-flops forming a 4-bit (MOD-16) binary counter: (a) interconnection diagram; (b) wave forms showing frequency division.

specific frequency. It allows the FFs to change states, on the basis of the present states of inputs R and S, only once for each clock pulse. Other FF forms exist, but their discussion is beyond the scope of this article. For further information on digital logic devices, refer to a textbook on the subject (e.g., Kohavi, 1978). A wide variety of counters (in their basic forms) are available as standard IC packages, and it is seldom necessary to construct a counter from individual FFs. Therefore, coverage of internal operation of counters will be limited to discussing external operating characteristics of currently available counters.

Flip-flops can be connected in various arrangements to function as binary counters that count input clock pulses or events. A 4-bit counter, for example, contains four FFs, one per bit, with the number of outputs equal to the number of bits (i.e., four). As shown by Fig. 16, the counter basically has one input, the clock pulse CP. It is assumed in this counter structure that the FFs are of a form known as JK, where J and K are the FF inputs, and that both J and K are kept at logic 1. In this configuration, the outputs of the JK FF change states on the high-to-low transition of the clock pulse, as indicated by the small circle at the input CP. Some counters trigger on low-to-high transition, in which case the input CP would not have the small circle. The states of the four FFs represent a binary number equal to the number of pulses that have been applied to the CP input. In the diagram, the A output represents the least significant bit (LSB) and D is the most significant bit (MSB) of the binary count. For example, after the fifth input pulse, the output DCBA=0101, which is the binary equivalent of five.

The 4-bit counter can count from 0000_2 (zero) to 1111_2 (fifteen), for a total of 16 distinct states. After the count reaches the maximum, the next input pulse causes the FF outputs to recycle to the 0000 count, and the counting sequence is then repeated with subsequent pulses. Generally, a counter with N FFs can count from 0 up to 2^N-1, for a total of 2^N distinct states.

The total number of states is known as the MOD number. The counting sequence can be cut short at any time by applying logic HIGH to a second input (RESET) that can be incorporated in the design. The decade counter is such an example, which counts from 0000 to 1001 (0 to 9) and then recycles to 0000. It has 10 distinct states and thus is referred to as a MOD-10 counter. It is also called a BCD counter because the FF outputs count through 10 in the common digital code known as the BCD code. The BCD is a weighted code where four bits are used to represent each decimal digit. The bits carry weights 8, 4, 2, 1 in the order from the MSB to the LSB, as shown in Table 3.

The counting capability of FFs can be demonstrated by examining the wave form in Fig. 16(a). It is evident that the frequency of the square wave (at CP) is divided by 2 successively by each FF. Therefore, cascading more FF units will produce division by 4, 8, 16,..., etc. In general, the wave form out of the MSB FF of a counter will divide the input frequency by the MOD number. For example, the MSB FF of a 4-bit counter has a frequency that is 1/16 of the clock input, as is evident by comparing waveform D to the clock pulse CP.

Counters can count up from zero to some maximum count and then reset to zero, and up/down counters can count in either direction. Some counters have a single CP input that is used for both count-up and count-down operations, whereas others have two separate CP inputs—one for counting up and another for counting down. For the latter type, only one input can be pulsed at one time to avoid

Table 3. The 8-4-2-1 binary-coded decimal (BCD) code representation of decimal numbers.

Decimal number	8-4-2-1 BCD code
0	0000
1	0001
2	0010
3	0011
4	0100
5	0101
6	0110
7	0111
8	1000
9	1001
10	0001 0000
11	0001 0001
12	0001 0010
⋮	⋮
19	0001 1001
20	0010 0000
⋮	⋮
99	1001 1001
100	0001 0000 0000
⋮	⋮

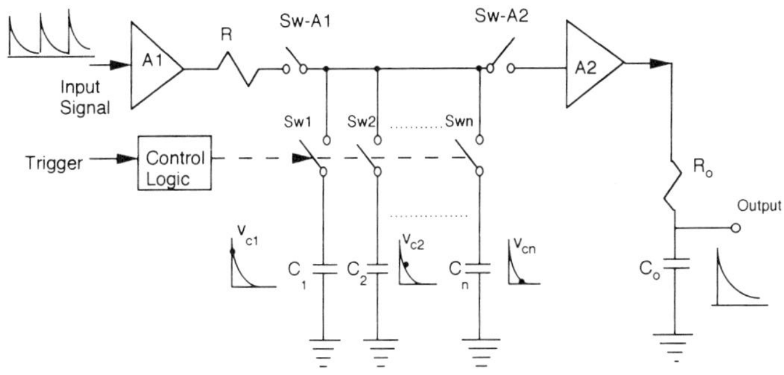

FIG. 17. Descriptive schematic diagram of a signal averager. Upon receiving a trigger, the control logic closes the switches one at a time in sequential order.

erratic operations. Some IC counters are designed to give the user the capability of varying the counting sequence through appropriate external connections. These types of counters are usually referred to as variable-MOD counters.

Commercially, counters are classified into frequency counters and general-purpose timer–counters. The frequency counter has a single input channel and sometimes an additional prescaled input for measuring high frequencies. In addition to frequency measurements, most frequency counters can also perform period measurements. The timer–counter version can perform time-interval measurements between two events in addition to frequency and period functions. For a given counter, a frequency can be used as a guide in deciding the suitability of frequency versus period measurement. For a 1-s sampling period, this frequency f_0 is given by (Doebelin, 1983)

$$f_0=(f_c)^{1/2}, \qquad (13)$$

where f_c is the frequency of the crystal oscillator ("clock" frequency).

For example, a 10-MHz clock-frequency counter has $f_0=3160$ Hz, above which frequency measurements should be used and below which period measurement should be used.

A more advanced version can provide pulse-width measurements; totalizing capabilities (pulse counting); for the timer/counter type, rise/fall time computation; and phase delay between two events.

1.7 Boxcar Integrators

In some measuring applications, it is desirable to trigger a process repeatedly, such that on repetition, the resulting signal will be identical each time and will appear to be superimposed upon itself. To enhance the signal-to-noise ratio (SNR) of the repetitive signal, an averaging process is performed to smooth out the signal. Two known analog techniques used for this purpose are the signal averager and the boxcar integrator.

The signal averager, as shown in Fig. 17, consists of a bank of parallel capacitors connected to the output amplifier A2 through switch SW-A2. The capacitors are charged by closing in a specific sequence a set of electronic switches known as field-effect transistors (FETs), which are solid-state devices with very high input impedance. The sequence is determined by the timing control circuit controlling the FETs in response to an external trigger signal. The sequence always begins at a fixed time following application of the trigger signal. Consequently, upon repetition, each capacitor circuit receives part of the applied signal through amplifier A1 with the same phase each time by charging the capacitors through the resistance R by a small amount each time. The result after a large number of repetitions is that each capacitor contains an averaged sum corresponding to a point on the signal trace. The information stored on the capacitors can be retrieved by allowing switches SW1–SWn to be closed one at a time while SW-A1 is opened and SW-A2 is closed. The output resistance–capacitance combination R_0C_0 is used to smooth the final response, which otherwise would look like changing steps because of the switching action from one channel to the next in succession.

The boxcar integrator, also known as the boxcar averager or detector, performs the same function as that of the signal averager, except with a rather simpler circuit. Instead of using a bank of many capacitors, the boxcar employs a single capacitor circuit, as shown in Fig. 18, to take only one sample during each

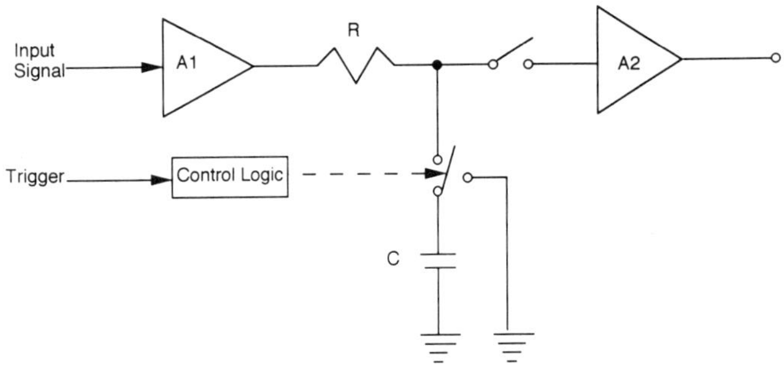

FIG. 18. A simplified schematic of a boxcar integrator.

signal occurrence and requires a trigger signal at a fixed time interval.

The trigger pulse is used to delay by a fixed time interval the closing of the switch. The capacitor in this case is clearly taking an average of a set of readings for one single point. The entire process is then repeated with different delay-time increments until all points are averaged. [See Sydenham (1982) for more details.]

1.8 Lock-In Amplifiers

The lock-in amplifier is used to recover weak ac signals corrupted with noise. One common application of lock-in amplifiers is spectroscopy, where the generated signal is typically weak. By amplitude-modulating signals at some reference or carrier frequency, the noise often can be reduced and dc drift problems avoided. The main principle of signal recovery is based on the synchronous detector (also known as the phase-sensitive detector or PSD) part of the circuit, which in essence functions as a demodulator. As shown by Fig. 19, the PSD circuit consists primarily of a mixer (multiplier) followed by a low-pass filter. The output of the mixer is the product of two signals x and y. The phase difference between the two inputs can be adjusted precisely by the phase shifter. Mixer circuits used in lock-in amplifiers must be capable of withstanding large amounts of noise without overloading. The square wave used by the mixer is supplied by a sine-to-square-wave converter (this could be simply a comparator—sensing zero crossing). If the square-wave frequency is precisely the same as that of the input signal and if it coincides in phase with its zero crossing, the result would be a type of full-wave rectification. The SNR of the noisy amplitude-modulated signal can be improved by reducing the bandwidth with a band-pass filter tuned to the carrier frequency. The signal can finally be recovered by a low-pass filter, which removes all components of the mixer output that have frequencies beyond the filter cutoff frequency. Some lock-in amplifiers are commercially available with built-in oscillators that can be used to supply excita-

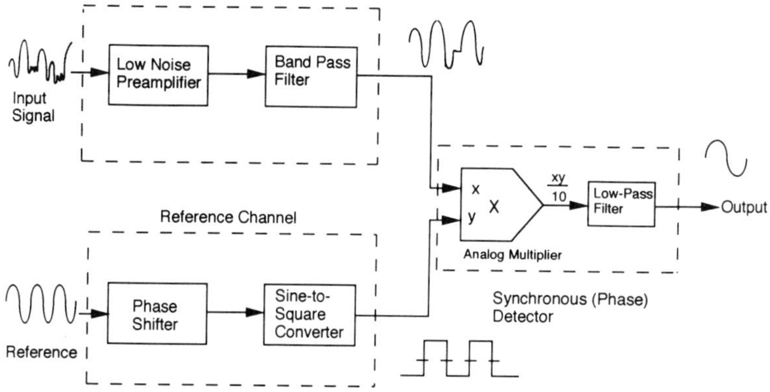

FIG. 19. Functional block diagram of a lock-in amplifier.

tion to an external system as well as the needed reference signal such as in strain gauge measurement (Vassos and Ewing, 1985; Sydenham, 1982).

1.9 Time-Domain Reflectometry

Time-domain reflectometry (TDR) is an echo technique used in transmission-line analysis to characterize and detect discontinuities in the line. This method also can identify the location and nature of the mismatch resulting from the discontinuity. This identification can be accomplished by applying a pulse of energy and monitoring the reflection of that energy by an oscilloscope. In the laboratory, the TDR is used primarily to test electrical cable (e.g., coaxial cable) and to check thermocouple characteristics in a thermopile.

The TDR system consists primarily of a step generator and an associated oscilloscope designed for this type of application, as shown by Fig. 20. The step generator produces a positive-going wave (incident wave) that is applied to the transmission system under test. The wave propagates through the transmission line at the velocity of propagation of the line. If the load impedance Z_L is equal to the characteristic impedance Z_0 of the line, then no wave reflection occurs, and the incident step voltage is displayed on the oscilloscope without change in its shape. If a mismatch exists at the load, part of the incident wave is reflected. The reflected wave is displayed on the oscilloscope's CRT as algebraically added to the incident wave, as depicted in Figs. 21 and 22 (Hewlett-Packard, 1988).

Analyzing the time delay, magnitude, and shape of the reflected wave form permits determination of location and nature of the mismatch whether the mismatch is at the load or somewhere along the transmission line.

Time-domain reflectometry is much more useful and advantageous than the classical standing-wave-ratio technique (frequency-domain measurement), which requires measurement at many frequencies to identify the mismatch—a rather tedious and time-consuming procedure. Also, with the technological advances in the optical field and the increased applications of fiber optics both in the laboratory and in industry, the optical TDR presents a powerful tool for testing the fiber. This device can analyze the reflected light energy to determine the existence and location of breaks in the fiber and losses at splices and connectors as well as the total loss of the system (Helfrick and Cooper, 1990).

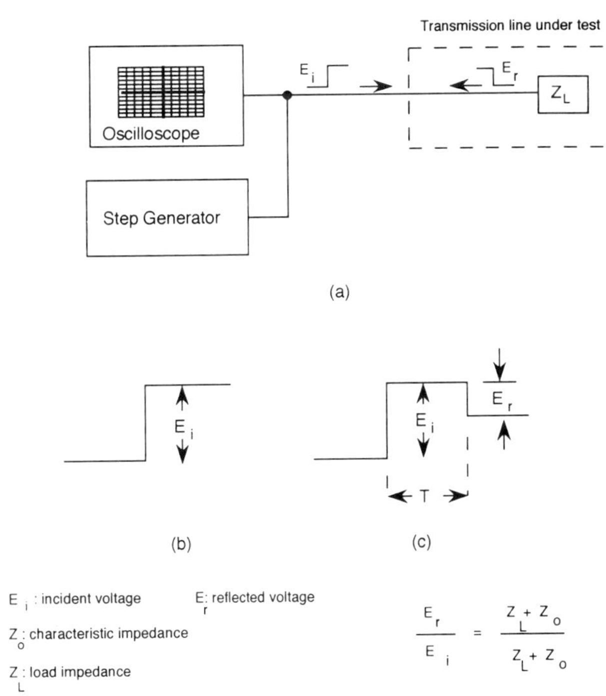

FIG. 20. Time-domain reflectometer system: (a) typical reflectometer measurement set-up; (b) oscilloscope display when E_r=0; (c) oscilloscope display when $E_r \neq 0$.

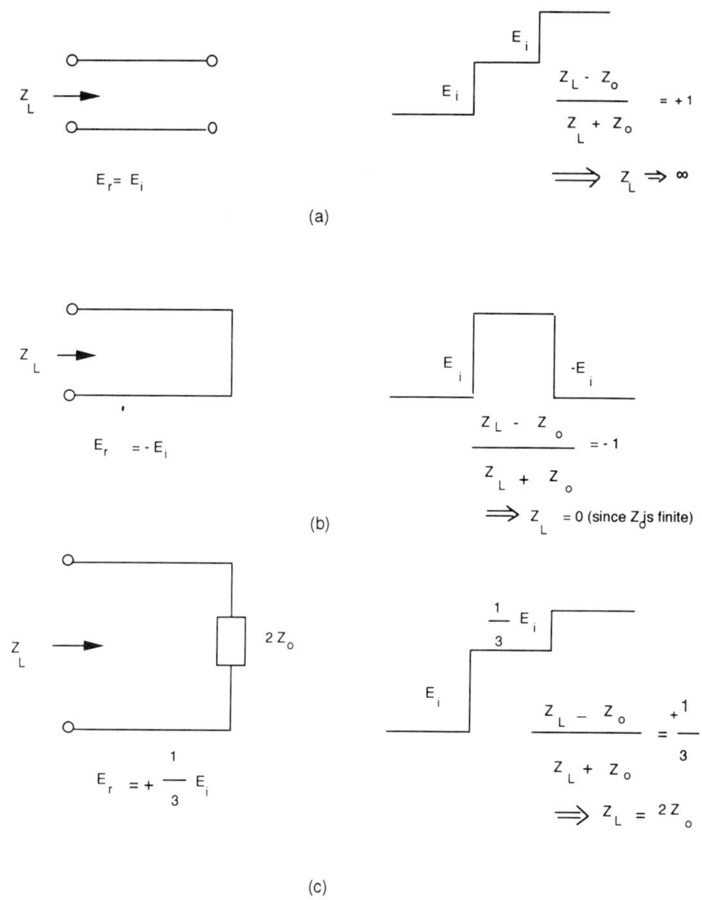

FIG. 21. Time-domain reflectometer displays for typical loads: (a) open-circuit termination; (b) short-circuit termination; (c) line terminated with $Z_L = 2Z$.

2. DATA ACQUISITION INSTRUMENTATION

In many measurement situations, one is faced with vast quantities of information—mostly in analog form—generated by transducers. Gathering and processing the measured data with the speed and accuracy demanded by the challenging and complex scientific development tasks necessitates a more complex, more sophisticated type of instrumentation. Data acquisition systems (*q.v.*) (DASs) are used for that purpose. A DAS may be as simple as a DVM or it may be complex enough to contain a large-scale computer as part of its hardware. Ordinarily, DASs are digital systems since, with today's technology, complex circuits are low cost, accurate, and relatively simple to implement and easy to interface with human operators. As depicted by Fig. 23, the DAS generally accepts a wide mix of analog signals representing physical parameters; performs signal conditioning functions (filtering, linearization, scaling, and square root); converts the signals into digital form; performs various computations; analyzes data; and interfaces with many external peripheral devices (e.g., CRTs, digital recorders, printers, and mass storage devices). The analog multiplexer is a high-speed scanning device allowing inputs to be switched sequentially one at a time for processing and converted to digital form by using the same ADC.

2.1 Data Loggers

The data logger is a data acquisition system whose architecture resembles very closely that shown in Fig. 23. It provides an automated method of performing measurements and recording data as well as performing some computations with high speeds and accuracy. These functions can be performed at precise,

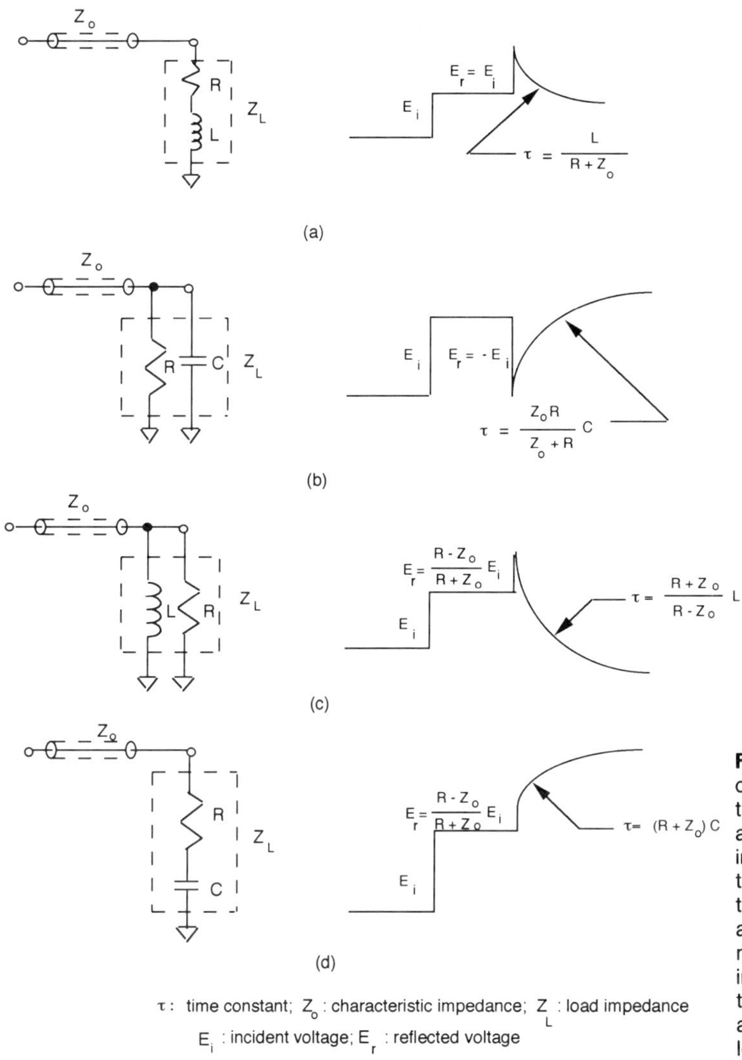

FIG. 22. Oscilloscope displays for complex load impedance Z_L: (a) transmission line terminated with a series resistance-inductance load impedance; (b) transmission line terminated with a parallel resistance-capacitance load impedance; (c) transmission line terminated with parallel resistance-inductance load impedance; (d) transmission line terminated with a series resistance-capacitance load impedance.

programmable intervals; once the logger is configured for the application, it will carry out the measurement precisely, unattended.

Data loggers offer good flexibility such as mixing signals representing various physical parameters and display each signal with the appropriate engineering units (pascals, liters per second, degrees Celsius or Fahrenheit). In addition, the same logger can be easily configured for different applications. Data loggers, however, have limitations compared with other DAS forms. Loggers are limited in their ability to provide histograms, trends, time function plots, and curve fits of data required for analyzing data. Although data loggers can interface with many external devices (output devices), they are still limited to some extent in their ability to interface directly with mass storage devices such as hard-disk drives. Some modern data-logger designs include microprocessors and two separate internal memories. One memory (RAM) for the system and user program can run without interference from a host, while the other memory can provide >100 kbytes of data storage. In addition, an RS-232 or IEEE-488 interface can be included for communication with a host computer for data transfer and remote control of the logger. With a 12-bit ADC, the logger can provide up to 62 500 readings/s; while with a 16-bit ADC, it can provide up to 50 000 readings/s.

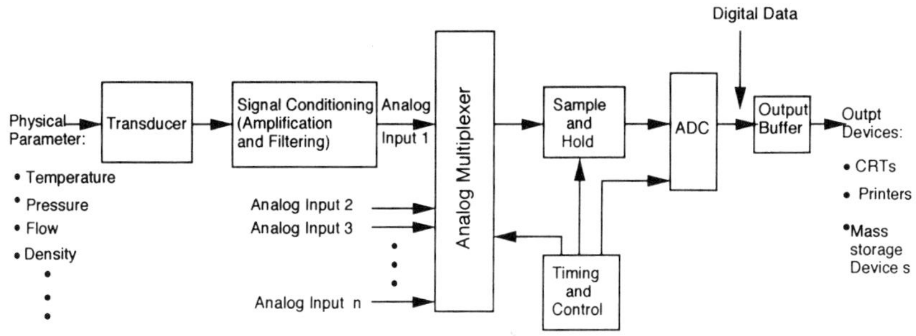

FIG. 23. Typical architecture of data acquisition system.

2.2 PC-Based Data Acquisition Systems

The high speed and low cost of personal computers (PCs) along with extensive hardware and software support made PCs an ideal choice for automated data acquisition systems in the laboratory and even in industrial applications. Most of today's PC-based DASs are based on either the IBM PC or the Apple Macintosh. As illustrated by Fig. 24, the basic elements of a PC-based DAS are listed below.

1. A PC with an operating system. The common operating system for the IBM PC is the PC-Disk Operating System (PC-DOS) and the Microsoft Disk Operating System (MS-DOS) for IBM-compatible PCs (commercially known as *clones*).
2. DAS boards for analog measurements and control. Analog input boards include an ADC, which is necessary to convert analog signals to digital form for further processing by the PC. Analog output boards also include a DAC, which is necessary to interface the DAS with the analog world. A host of other input/output (I/O) boards are available for practically any application. DAS boards are designed to be inserted directly into available expansion slots inside the PC chassis. In this case, the boards connect directly (through a standard connector, which is part of the board) into the PC bus. Some manufacturers offer standalone systems in which DAS boards can be inserted into a separate chassis, which in turn connects to the PC through an RS-232 or an IEEE-488 interface.
3. An instrument interface IEEE-488. This interface provides complete automated measurement and control capabilities of IEEE-type instruments such as DVMs.
4. Output devices. The PC can offer a wide choice of output devices from many differ-

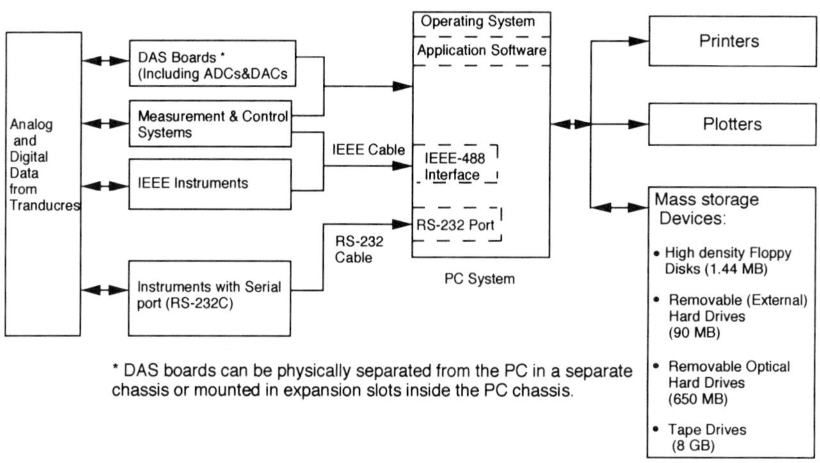

FIG. 24. Personal-computer-based data acquisition system structure.

ent manufacturers. In addition to the PC's own CRT, external devices such as printers, plotters, and many mass storage devices with various capacities are available from numerous manufacturing sources.

5. Application software. A wide range of software has been developed specifically for PC-based DASs for both the IBM PC and the Apple Macintosh at reasonable cost. Plug-in boards also are available that include the necessary hardware and resident software that can turn the PC into an oscilloscope, a DVM, a signal generator, a logic analyzer, etc. These simulated instruments, known as virtual instruments (VIs), can be controlled from the PC keyboard, which emulates the control knobs and buttons used with the actual instruments by means of software modules known as instrument drivers. These instrument drivers are essential building blocks in the construction of VIs. The drivers handle the details of controlling and communicating with specific instruments. With the instrument drivers, the user can create VIs by combining several instruments with data analysis, data presentation, and user interface code. A multitasking software development system for DOS called LabWindows is becoming the industry standard. LabWindows primarily serves three purposes. It provides an integrated development environment, a high-level interface to the instrumentation hardware, and the user interface to the virtual instruments. LabWindows also has extensive libraries of functions for data acquisition, data analysis, and data presentation along with tools for editing and debugging user programs written in BASIC or C (discussed in the following). PC-based DASs offer a high sampling rate [100 000 samples/s (i.e., 10-μs sample period)].

2.2.1 Software. Numerous programming languages have already been developed and exist in a variety of dialects (see COMPUTER PROGRAMMING LANGUAGES). Some of these languages were developed specifically for scientific applications such as those performed in the laboratory from customized PC-based DASs to data analysis routines.

BASIC (Beginner's All-purpose Symbolic Instruction Code) is an interpreting, high-level, unstructured, interactive, easy to learn and use language. It became popular among programmers because of its English-like commands. The difficulty of the original version of BASIC is the global internal structure, which does not lend itself to reusing or sharing parts of the program. Many PCs include improved versions of BASIC (e.g., Microsoft BASIC) by incorporating the BASIC interpreter in ROM, which is also available on disk. Another version of BASIC known as *QuickBASIC* was developed and became very popular for developing specialized DASs. Although QuickBASIC is an interpreting language, it is still considered fast enough in comparison with a compiled language such as FORTRAN.

Recently, a programming language called Visual Basic, for Windows, was introduced, primarily to improve users' interface development of virtual instrumentation as well as of general-purpose languages.

C is a block-structured, general-purpose programming language developed by Bell Laboratory. It has been closely associated with the UNIX system because it was developed on that system; also, UNIX and its software are written in C. However, the language is not tied to any one operating system or machine. Because of its portability, C can be used to compose programs that can run on a host of microcomputers with minimal alteration. C is quickly becoming the language of choice among professional application and system developers for the PC, other microbased systems, minicomputers, and mainframe computers. C can be used with LabWindows (for DOS) to develop customized PC-based ADSs. Recently, a programming language called C++ was introduced to improve users' ability to interface with virtual instrumentation as well as handling interfacing requirements of general-purpose languages.

Assembly generates the tightest code of any language and is considered to be the fastest to execute except for machine language. It is used to design programs such as the well-known spreadsheet software, Lotus 1-2-3. Assembly programs are generated from individual instructions represented by three- or four-letter mnemonics like MOV, JMP, and PUSH. When assembled, a program translates directly into machine code that the processor can read from memory and execute. Although the language is somewhat difficult to learn and use, once the programming is complete the results are hard to match.

2.3 Instrument Interfaces

Transmitting and capturing data among instruments in a measurement system requires some physical links or interface to integrate instruments from different manufacturers in a system. Also, the increased application of computers in test, measurement, and control in the instrumentation field requires some form of communication between instruments and computers. Several interface standards adopted by industry are discussed below (see also BUS SYSTEMS AND COMPUTER INTERFACES).

1. IEEE-488/GPIB interface: a standard communication bus adopted by the Institute of Electrical and Electronics Engineers (IEEE) in 1978. This interface was initially developed in 1972 by Hewlett-Packard (HP) as a computer–to–instrument-interface bus (HP-IB) to standardize interface configuration and operating protocol for all HP equipment supplied with digital control interfaces. The HP-IB was adopted by IEEE in 1975 and became IEEE Standard 488-1975 (Digital Interface for Programmable Instrumentation) and was later reissued as IEEE 488-1978 and 488-1987. The standard is also known as the general-purpose interface bus (GPIB). It defines the physical and data-link structure of a local area network for digital communications between a main controller (e.g., computer) and various pieces of instrumentation and laboratory equipment. Instruments (e.g., scopes, analyzers) equipped with this standard can be controlled remotely as part of automated test and measurement systems. The interface bus is available as a module (printed circuit board) containing all the necessary hardware and firmware to implement the specified communication protocol. The IEEE board includes a standard connector for interfacing with the IEEE instruments through a standard cable. The basic bus structure as it interfaces with the instruments is shown in Fig. 25. The diagram shows a typical computer-controlled measurement system in which the various instruments are connected by the IEEE-488 bus. The diagram clearly shows that all the instruments are connected in parallel to the 16 different bus lines. The various instruments can perform one or more of the following functions: (a)Talker, which can transfer data to other instruments by the data bus; (b) Listener, which can receive data from other instruments by the data bus; (c) Controller, which manages the operation of the bus system, primarily by sending addresses and commands over the data bus and sending control signals over the control bus. Only one Talker is allowed to put data on the data bus at a given time, but more than one Listener can receive data simultaneously. The Controller sends a Talk address out over the data bus while activating the appropriate control line. The instrument with that particular address is the only Talker that is activated, and all other Talkers are disabled. A similar procedure selects the one or more Listeners that can receive data from the active Talker.
2. VXIbus interface. The VXIbus interface is a modular instrument standard developed in 1987 to define a bus structure architecture that allows a wide range of instruments, interface cards, or computers from different manufacturers to coexist as modules in the same enclosure (cardcage). The VXIbus specification describes in detail the technical requirements of VXIbus-compatible components such as mainframes, backplanes, power supplies, and functional plug-in instrument and communication interface modules. The VXIbus standard is used in custom building powerful microcomputer-based DASs with distributed intelligence and extremely high data-transfer rates. The VXIbus instrument standard is an extension to the VMEbus that is open to all manufacturers and is compatible with present industry standards. The VMEbus architecture was developed primarily with focus on computer systems and with limited instrumentation capabilities. A new open architecture instrumentation bus, based on VMEbus, the Eurostandard (printed circuit board standard IEC 297-3), and other instrumentation standards such as IEEE-488.2, was created and became known as the VXIbus. A typical VXIbus (abbreviation for VMEbus eXtension for Instrumentation) system consists of a single frame [48.3-cm (19-in.) rack] with up to 13 modules plugged into numbered slots. A common configuration places into slot 0 a system resource module such as the VXIbus-

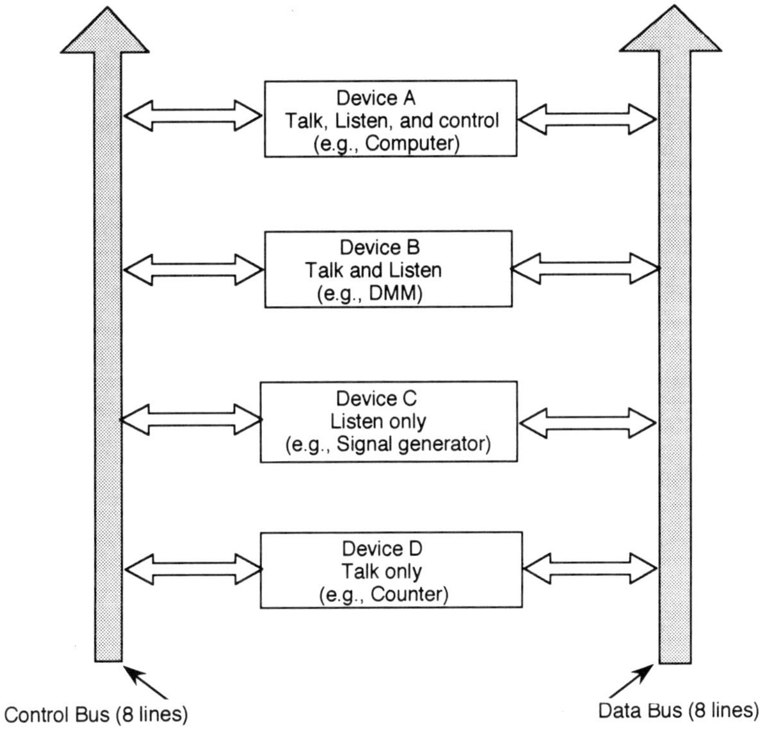

FIG. 25. Typical IEEE-488 bus system.

mandated timing generation or communication port such as IEEE-488 or RS-232. The other positions are general-purpose slots for the user to mix and match. Figure 26 shows some of the possible system configurations and illustrates the flexibility that is afforded by the VXIbus standard. [For more details, see *The System Engineer's Handbook*, a guide to building VMEbus and VXIbus systems (Black, 1992).]

3. RS-232C interface. The RS-232 serial (asynchronous) communication standard was developed initially by the Electronic Industries Association (EIA) to standardize the interface between modems and data terminal equipment (e.g., computers and termi-

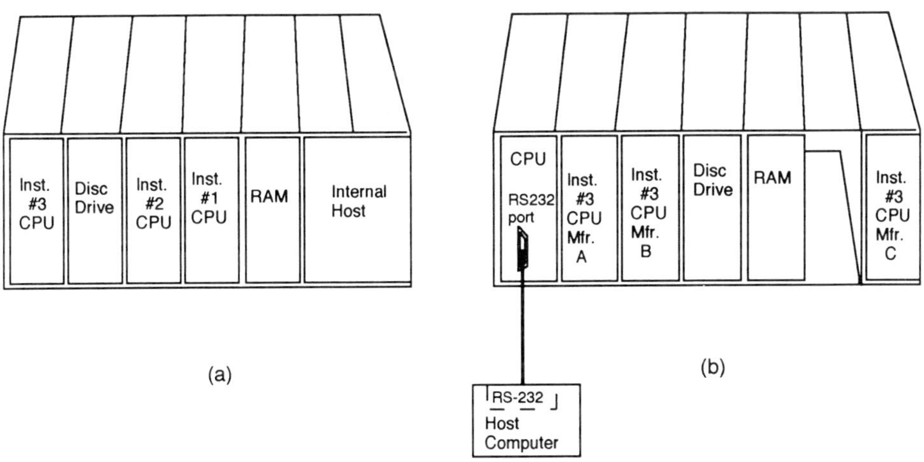

FIG. 26. Typical VXIbus system configurations: (a) stand-alone system; (b) multiple-CPM system.

nals) connected to them. The latest version, RS-232C, was adopted in 1969 and has become the serial interface standard when one microprocessor system communicates with another microprocessor system over long distances or with peripheral devices such as printers or CRT terminals. Serial transmission requires only two wires to carry all the necessary data, address, and control information. The information is transmitted one bit at a time, making serial transmission slower than parallel transmission. The RS-232C standard is typically used for baud rates up to 20 kb/s using signal lines up to 15 m (50 ft). For higher transmission rates over longer distances, EIA has adopted other standards such as the RS-422, RS-423, and RS-485, listed in Table 4 along with their electrical characteristics. Almost all modern instrumentation manufacturers offer serial interface as an option for their products to communicate with computers and other devices. An interface can be purchased as a card (printed circuit board), commonly known as a serial port, with all the necessary hardware and software (serial interface protocol). It is usually inserted into an open slot in the instrument or in the computer. The connection among devices equipped with RS-232C ports is through standard 25-wire cable that connects directly into a 25-pin connector mounted on the card.

4. CAMAC interface. The CAMAC is a modular instrumentation and digital interface system developed to address instrument incompatibility that results from employing instrumentation from various manufacturers in a given measurement system both in the laboratory and in industrial fields. An initialism for Computer-Automated Measurement And Control, the CAMAC is an ANSI/IEEE standard (Standard 583-1982) that is concerned with mechanical, electrical, and signal interfaces. As shown in Fig. 27, measurement devices are in modular units and are housed in a standard card enclosure (cage or crate) providing a common dc power required by these devices to function. Data input and output from the modules are handled by a specified bus called a data highway (Dataway) at the rear of the cage. The system initially has a 1-μs Dataway cycle time and a high address capability. The Dataway is a parallel bus with 24 read lines and 24 write lines so that words of 24 bits or less can be handled in a single data transfer. Additional levels of compatibility are achieved through the use of standardized parallel and serial highways that have been developed for use with the basic CAMAC system. The serial highway can be used to interconnect up to 62

Table 4. Communication Electronic Industries Association standards comparison.

	Interface			
Mode of operation	RS-232C Single-ended	RS-422 Single-ended	RS-423A Differential	RS-485 Differential
Number of drivers and receivers allowed on-line	1 driver, 1 receiver	1 driver, 10 receivers	1 driver, 10 receivers	32 drivers, 32 receivers
Maximum cable length [m (ft)]	15 (50)	1200 (4000)	1200 (4000)	1200 (4000)
Maximum common-mode voltage	±25 V	±6 V −0.25 V	±6 V	±12 V −7 V
Driver output	±5 V min ±15 V max	±2 V min	±3.6 V min ±6.0 V max	±1.5 V min
Driver load	3 to 7 KΩ	100 Ω min	450 Ω min	60 Ω min
Driver slew rate	30 V/μs max	NA[a]	Externally controlled	NA[a]
Driver output short-circuit current limit	500 mA to V_{cc} or ground	150 mA to ground	150 mA to ground	150 mA to ground
Receiver input resistance	3 or 7 KΩ	4 KΩ	4 KΩ	12 KΩ
Receiver sensitivity	±3 V	±200 mV	±200 mV	±200 mV

[a]NA=Not applicable.

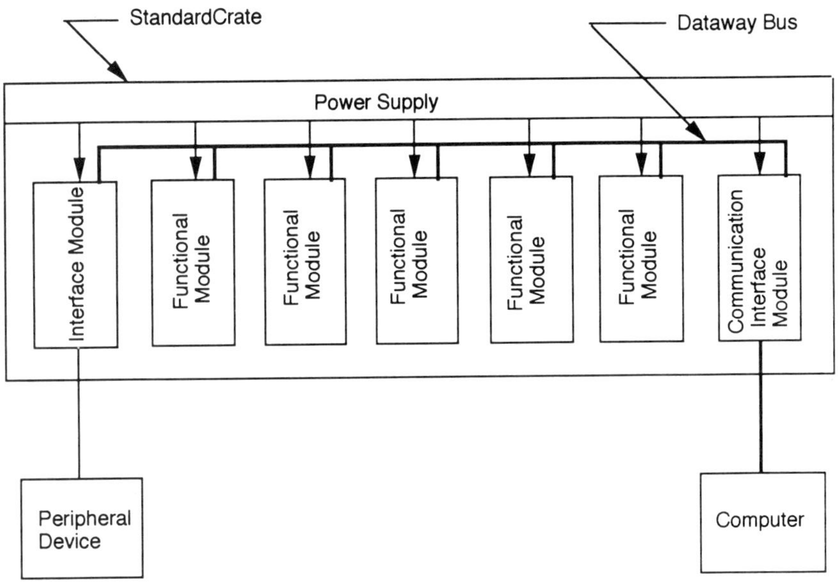

FIG. 27. Basic computer automated measurement and control (CAMAC) standard crate shown with a computer as the control device. A self-contained CAMAC crate system can be used with a CAMAC-type crate controller in place of the communication module.

crates, providing a widely distributed system, and the parallel highway can handle up to 7 crates. (See CAMAC Instrumentation and Interface Standards, IEEE 1982, or latest version for more details.)

3. MEASUREMENT INTERFACE

Every measurement application consists of a number of components that together form a system used to perform a measurement and processes the result. The application could be data storage, recording, display, controlling functions, or analysis. The components within the measurement system are most likely to be a mix of analog and digital devices. Such a mix in interdomain devices gives rise to two important factors that must be considered and well understood for the measurement system to perform its intended function in the most satisfactory manner. These factors are data conversion from analog to digital and vice versa and signal transmission within the system.

The coverage in this section of interdomain conversion and signal transmission is intended to be brief and to serve as an introduction of this subject to the reader. A complete and more comprehensive coverage is provided in ELECTRICAL INSTRUMENTATION.

3.1 Digital-to-Analog Converters

The digital-to-analog converter (DAC), also known as the D/A converter, is an electronic device that accepts digital information and transforms it into an analog voltage. The digital information is in the form of a binary number with some fixed number of digits. In the digital field, the binary number is called a binary word, and the digits are called bits. For example, the 8-bit word 10101011_2 would be a binary number having eight digits consisting of 0s and 1s.

The DAC converts the digital word into an analog voltage by scaling the analog output to be zero when all bits are 0 and some maximum value when all bits are 1, as illustrated by the simplified functional diagram of Fig. 28. This function can be mathematically represented by expressing the binary number as a fractional number according to the relationship

$$V_x \simeq V_r(a_1 2^{-1} + a_2 2^{-2} + \ldots + a_n 2^{-n}), \quad (14)$$

where V_x is the analog voltage output, V_r is some voltage reference in the DAC internal

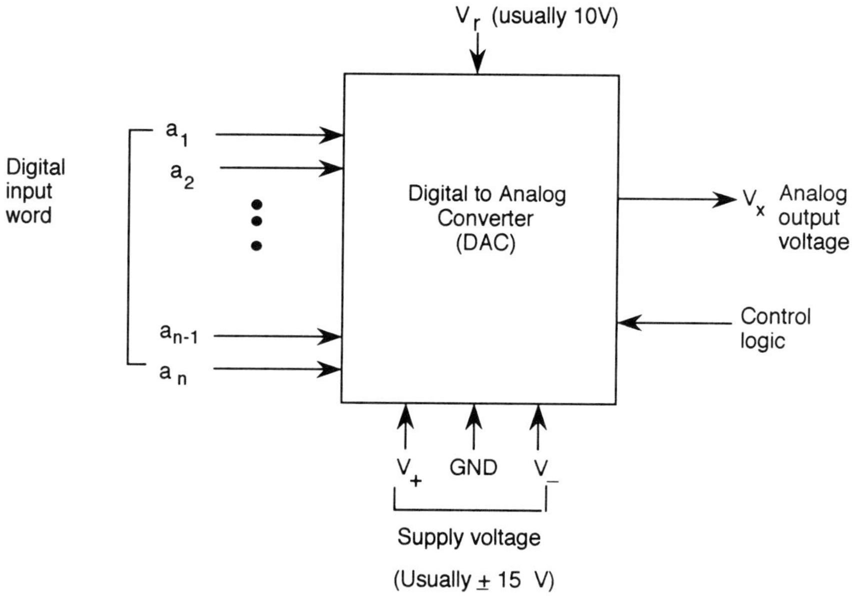

FIG. 28. Input/output diagram of an *n*-bit digital-to-analog converter.

circuit, and the coefficients $a_1,...,a_n$ represent the n-bit binary word. For a 10-bit DAC ($n=10$) with a 10-V reference, the output can have any of 1024 evenly spaced voltages, including 0 and 10.0 V, thus offering a resolution (1/1024) of slightly better than 0.1%.

DACs can be used in various applications other than for data conversion from digital to analog form. A DAC, for example, when driven by a binary counter, can produce a *staircase*. This staircase is useful in a variety of experiments. One form of DAC (based on a $R/2R$ resistor ladder network structure), when controlled with a BCD-coded input, can also be used as a digitally controlled voltage divider with high resolution. One such device is designed to provide an output between 0 and 1.999 times the input voltage, with a resolution of 0.001.

3.2 Analog-to-Digital Converters

The analog-to-digital converter (ADC), also known as the A/D converter, is an electronic circuit that accepts an analog voltage signal and transforms it into an encoded digital word, as illustrated by the simplified function diagram of Fig. 29. The relationship between the analog input voltage and the digital output sequence of the ADC is the same as that expressed by Eq. (14), where V_x is the analog input voltage, V_r is a reference voltage, and the coefficients $a_1,...,a_n$ represent the n-bit digital output (digital word). It is clear from Eq. (14) that the voltage, as expressed by the right-hand side, can change only by a finite step size given by

$$\Delta V = V_r 2^{-n}. \quad (15)$$

The reference voltage V_r for the ADC also represents the full-scale value of the input voltage that can be applied to the converter. It is for that reason that Eq. (15) often appears in publications as

$$\Delta V = V_{fs} 2^{-n}. \quad (16)$$

The step size ΔV thus constitutes the smallest change in the analog input voltage that the ADC can resolve. Because of this resolution, the approximate equality in Eq. (14) is used rather than the equality sign. It is also clear that the resolution increases with an increase in the number of bits of the digital output. Historically, the first generation of ADCs was developed with 4-bit digital output. With the increased demand for higher performance, ADCs with 8, 10, 12, 14, and 16 bits offering superior resolution capabilities were developed.

The ADC sampling of the analog signal for conversion gives rise to two important considerations that could adversely affect its perfor-

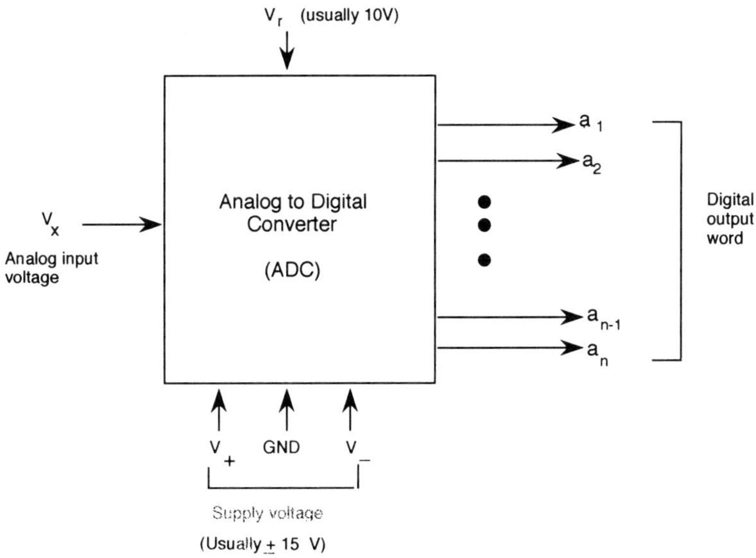

FIG. 29. Input/output diagram of an *n*-bit analog-to-digital converter.

mance if not properly addressed. The first is the sampling frequency, which must be at least twice the greatest frequency component of the signal to avoid the aliasing problem discussed in Sec. 1.2.3. Sampling the analog signal at a frequency less than that would lead to erroneous results when the sampled values are reconstructed. The second consideration concerns the finite time required for the ADC to complete its conversion. This time is known as aperture time, which is defined as the actual time the ADC requires to obtain an accurate reading of the analog signal. The aperture time must not be greater than the period of the greatest frequency component of the signal multiplied by the resolution. If the ADC is not fast enough to make a conversion within the aperture time for the given signal, then a circuit known as a sample-and-hold should be used. The sample-and-hold circuit, shown in its simplest form by Fig. 30, samples the signal by closing switch S1 long enough to charge capacitor C to the present signal value and hold that value for the time required by the ADC to make its conversion.

Many design methods exist for performing the ADC function. These methods vary in their techniques for realizing improvement in performance associated mainly with conversion speed and noise rejection. These designs resulted in the development of the following ADCs, which are listed for reference only because they are discussed in more details in ELECTRICAL INSTRUMENTATION:

1. successive approximation,
2. single-slope integrating,
3. dual-slope integrating,
4. servo,
5. ramp,

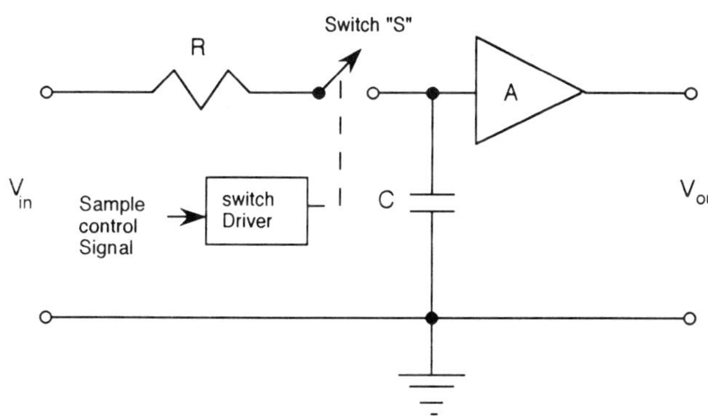

FIG. 30. Basic sample-and-hold circuit. The input is sampled periodically by closing switch S until capacitance C is charged to the new input value. The switch is then opened to hold the value to be delivered to the analog-to-digital converter through high–input-impedance amplifier A.

6. voltage-to-frequency,
7. parallel or flash,
8. cyclic, and
9. sigma-delta.

Until recently, the successive approximation and the dual-slope integrating-type ADCs have dominated because of their compromise in performance versus cost benefits. The dual-slope integrating-type ADC is not as fast as some of the others, but its inherent excellent noise immunity as well as its cost effectiveness due to design simplicity makes it ideal for many applications.

The most recent development in data-conversion technology is the sigma-delta ADC. It is considered state of the art and uses the latest development in digital technology to offer simplicity combined with powerful performance and resolution of up to 20 bits and high SNR.

3.3 Signal Transmission

Measurement starts with a transducer to quantify the physical measurement in terms of an electrical signal. Most transducers have analog outputs that need to be transmitted to other instrumentation for display, recording, analysis, or other purposes. It is crucial that the transmitting medium preserves signal integrity between the transmitting and the receiving instruments. Noise corruption from surrounding sources [e.g., electromagmetic interference (EMI)] and transmitting medium characteristics are two limitations that must be carefully considered.

The transmitting media can be classified in three categories: analog, digital, and optical. In industrial applications, the instruments within the same loop are separated by large distances, and the application environment is more noisy than the laboratory environment. Digital transmission methods [e.g., voltage-to-frequency (V/F), RS-232C, RS-422, RS-423, RS-425] offer a practical solution in industrial applications because of the inherent noise immunity and protection against signal degradation due to voltage drops along the transmitting media (wires). Analog transmission using a 4- to 20-mA current signal is also used for limited distances and when practical.

In laboratory applications, analog transmission of signals is the most practical and thus most commonly used method. Among the various signals used in analog data transmission (10–50 mA, 0–20 mA, 4–20 mA, 0–5 V, 0–10 V), the 4- to 20-mA current has been widely accepted as the industry standard signal because of the advantages that it has to offer. The 4- to 20-mA current is usually scaled to represent the signal range where the 4-mA current corresponds to the lowest range value, and the 20-mA current corresponds to the maximum range value. The 4-mA current, compared with the 0-mA current, value is recognized as the "live zero" because of the distinction it provides between loss of signal (electrical zero) due to physical interruption in the transmission medium and signal zero (representing the minimum range value). In addition to this advantage, current signals in general are less susceptible to noise pickup than voltage signals are, and degradation in signal due to voltage drops does not apply. The transmitting medium is typically a twisted pair of wires to reduce noise corruption further. Figure 31 illustrates a typical transmission configuration known as two-wire transmission. In this configuration, the 4- to 20-mA current is transmitted around a loop, allowing

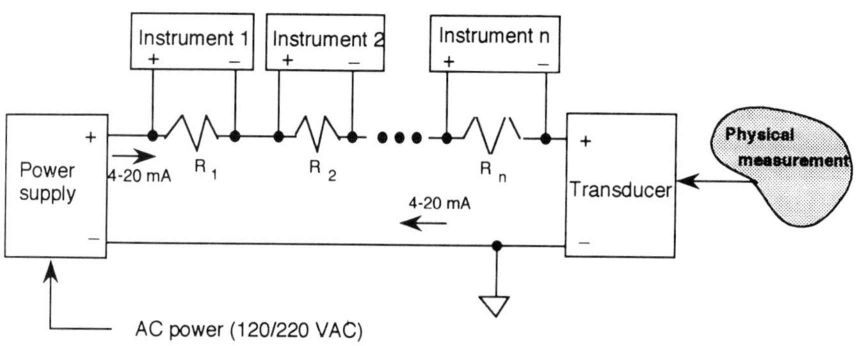

FIG. 31. Two-wire signal transmission configuration.

several instruments to share the same signal simultaneously. The signal is introduced to each instrument through an individual resistance R. The total number of instruments that can share the signal within the loop is limited by the sum of resistances around the loop. The maximum value of the total resistance in ohms is calculated by

$$R(\max)=R_1+R_2+\ldots+R_n$$
$$=(V_{\text{supply}}-12\ \text{V})/(20\ \text{mA}). \quad (17)$$

The transducer in the two-wire configuration is self-powered since its connection with the power supply serves to provide the dc power to and carry the signal from the transducer at the same time.

4. CONCLUSION

As was stated in the Introduction, this article is intended to acquaint readers within the scientific community with the kind of instruments and related technologies that are available to carry out scientific endeavors. In view of the rapidly changing technologies associated with the instrumentation industry, it is difficult to predict what will happen over the next few years. However, in observing the trend in the research and development in the electronics industry including the digital field and the their explosive markets, we can see that the basic ideas and theories will probably remain the same while details on device or instrument implementation will change. For example, in the area of instrument interface, our discussion of the VXIbus noted that the idea was not totally new, but it was an improvement over the VMEbus developed earlier. Another example is the future development in laboratory automation. PCs (such as the i386- and i486-based IBM-compatible PCs and the recently announced Pentium and the Macintosh) have invaded the workplace. Future development in this area is toward improving the processing power, networking [such as Ethernet and local area network (LAN)] capabilities, and data storage capacity. Workstations with 10^9 floating-point operations per second (Gflops) have evolved as a result of reduced instruction-set computer (RISC) processors.

GLOSSARY

ADC (Analog-to-Digital Converter): A device that changes analog voltage or current signal into a digitized number that can be digitally displayed or further analyzed by digital instruments.

Algorithm: A step-by-step procedure for solving a problem or performing a computation.

Analog device: A device that produces or processes analog signals.

Analog signal: A signal that varies continuously and can manage an infinity of values in any given range.

Aperture time: The time window, or time uncertainty, in making a measurement. For an analog-to-digital converter, it is the conversion time; for a sample-and-hold circuit, it is the signal averaging time during the sample-to-hold transition.

BCD (Binary Coded Decimal): A binary code used to represent decimal numbers in which each digit from 0 to 9 is represented by four bits weighted 8-4-2-1. Only 10 of the 16 possible states that are represented by the four bits are used.

Binary: A number representation system with a radix of 2.

Binary digit: In binary notation, either of the digits 0 or 1.

Bit: An abbreviation of "binary digit."

Bus: A group of wires used as a common path connecting various components within a microprocessor system (or computer) for transferring data by using control signals–for example, for interfacing the central processing unit of a digital computer to memory and inputs/outputs.

Byte: A group of adjacent binary digits operated on by the computer as a unit. The most common byte size is eight binary digits.

CRT (Cathode-Ray Tube): An electronic vacuum tube with a screen for graphical or alphanumeric visual display of information. Display is produced by means of proportionally deflected electron beams.

DAC (Digital-to-Analog Converter): A device that converts a digital signal, often from a computer, into a proportional analog signal (voltage or current).

Data Conversion: The process of changing the form of data representation. For example, converting analog signal to digital form and vice versa.

dB (Decibel): A logarithmic unit that expresses the magnitude of change in the level of a signal. The gain or attenuation of a device is commonly expressed in decibels defined in terms of output/input ratio as voltage (V): $dB_V = 20 \log V_{out}/V_{in}$; current ($I$): $dB_I = 20 \log I_{out}/I_{in}$; power ($P$): $dB_P = 10 \log P_{out}/P_{in}$. The decibel scale has the advantage of using small numbers to represent large ratios. Consecutive stages of amplification or attenuation lead to a sum (rather than a product) of individual gains.

dBm: A logarithmic measure in decibels of absolute signal level above or below some reference. For power measurement, m signifies a reference power of 1 mW. For example, 20 dBm is 100 mW. For other measurements (e.g., volts), the reference could be 1 V (dBV) or 1 mV (dBmV).

Digital Device: A device that produces or processes digital signals.

Digital Signal: A signal that varies in discrete steps and can manage only a finite number of different values in any given range.

EPROM (Erasable Programmable Read-Only Memory): A solid-state memory device that can be erased with ultraviolet light and reprogrammed by using equipment specialized for this purpose.

GPIB (General-Purpose Interface Bus): The IEEE-488 bus (HP-IB) designed by Hewlett-Packard in 1965 and adopted by the Institute of Electrical and Electronic Engineers in 1974.

LCD: Liquid crystal display.

LED: Light-emitting diode.

LSB (Least-Significant Bit): The rightmost binary digit in a digital word.

MSB (Most-Significant Bit): The leftmost binary digit in a digital word.

Peripherals: Auxiliary equipment that can be attached to and controlled by a central processing unit or a computer (e.g., CRTs, printers, magnetic tape transports).

Port: An electronic device that allows a computer to communicate with other external devices having communication capabilities.

Protocol: A formalized set of conventions including wiring arrangements necessary to establish and maintain communication between two devices.

RAM (Random Access Memory): This term, through common usage, refers to high-speed, volatile, solid-state (semiconductor) memory used to store and recall information. This type of memory can be read from or written to.

Refresh: Update information on the cathode-ray tube to maintain data on the screen.

Resolution: The magnitude of step changes distinguished by an instrument or device as the measured signal is continuously varied over the instrument range.

ROM (Read-Only Memory): A memory used for permanently stored computer programs and data.

Sampling Rate: The number of analog-to-digital conversions per second that are actually performed.

Sampling Theorem: A theorem by Nyquist that states that if a continuous bandwidth-limited signal contains no frequency components higher than clock frequency (f_c), then the original can be recovered without distortion if it is sampled at a rate of at least $2f_c$ samples per second.

Serial Communication: A form of communication that allows only one bit at a time to be sent or received between two communicating devices.

Serial Port: An electronic device that sends and receives data one bit at a time (also known as asynchronous communication port).

Spectrum: The spectrum of an analog signal is the amplitude versus frequency for all components within the signal.

Transducer: A primary device that converts physical measurements into low-level electrical signals.

Transmitter: A primary device that translates physical measurements into electrical signals, usually 4 to 20 mA.

Transmitter (Two-Wire): A device that is used to transmit data by a two-wire current loop. The loop usually has an external direct current power supply, and the transmitter acts as a variable resistor with respect to its input signal.

Volatility: An inability of memory to retain stored data during and after a power failure. An example of volatile memory is random access memory.

List of Works Cited

Black, J. (1992), *System Engineer's Handbook,* New York: Academic.

Brigham, O. E. (1974), *The Fast Fourier Transform,* Englewood Cliffs, NJ: Prentice-Hall.

Doebelin, E. O. (1983), *Measurement Systems: Application and Design*, New York: McGraw-Hill.

Helfrick, A. D., Cooper, W. D. (1990), *Modern Electronic Instrumentation and Measurement Techniques*, Englewood Cliffs, NJ: Prentice-Hall.

Hewlett-Packard (1988), *Fundamentals of Time-Domain Reflectometry*, Application Note No. 62.

Institute of Electrical and Electronics Engineers (1982), *CAMAC Instrumentation and Interface Standards*.

Kohavi, Z. (1978), *Switching and Finite Automata Theory*, 2nd ed., New York: McGraw-Hill.

Sydenham, P. H. (1982), *Handbook of Measurement Science*, New York: Wiley, Vol. 1.

Vassos, B. H., Ewing, G. W. (1985), *Analog and Digital Electronics for Scientists*, 3rd ed., NewYork: Wiley.

Further Reading

Bendat, J. S., Piersol, A. G. (1971), *Random Data: Analysis and Measurement Procedures*, New-York: Wiley.

Black, J. (1992), *System Engineer's Handbook*, New York: Academic.

Brigham, O. E. (1974), *The Fast Fourier Transform*, Englewood Cliffs, NJ: Prentice-Hall.

Kohavi, Z. (1978), *Switching and Finite Automata Theory*, 2nd ed., New York: McGraw-Hill.

Ouchi, G. I. (1987), *Personal Computers for Scientists*, Washington, DC: American Chemical Society.

Sydenham, P. H. (1982), *Handbook of Measurement Science*, New York: Wiley, Vol. 1.

VanDoren, A. H. (1982), *Data Acquisition Systems*, Reston, VA: Prentice-Hall.

LASERS, MEDICAL USE OF

See MEDICAL USE OF LASERS

LCDs

See DISPLAY TECHNOLOGY

LEDs

See LIGHT-EMITTING DIODES

LASER PHOTOCHEMISTRY

PAUL ENGELKING, *Department of Chemistry, University of Oregon, Eugene, Oregon, U.S.A.*

	Introduction	283
1.	**Photophysics**	284
1.1	Laser Light	284
1.2	Interaction with Matter	285
1.2.1	Linear Absorption	285
1.2.2	Multiphoton Absorption	285
1.2.3	Level Shifts	286
1.2.4	Hole Burning	286
2.	**Multiphoton Processes**	286
2.1	Electronic Multiphoton Absorption	286
2.1.1	Multiphoton Selection Rules	286
2.1.2	Multiphoton Ionization (MPI)	287
2.1.3	Multiphoton Dissociation (MPD)	288
2.2	Vibrational Multiphoton Absorption	289
2.2.1	Vibrational-State Ladders	289
2.2.2	Isolated Molecules	289
2.2.3	Effects of Collisions	290
3.	**Energy Partitioning**	290
3.1	Randomization of Energy in Molecules	290
3.2	Energy Localization	291
4.	**Chemical Changes**	292
4.1	Ionization and Detachment	292
4.2	Dissociation	292
4.3	Isomerization	294
4.4	Condensation and Addition	294
5.	**Applications**	294
5.1	Holography	294
5.2	Isotope Selection	294
5.3	Surface Modification	295
5.3.1	Etching	295
5.3.2	Deposition	295
5.4	Information Storage	295
5.5	Chemical Analysis	295
5.6	Biochemistry	295
	Glossary	296
	Works Cited	297
	Further Reading	298

INTRODUCTION

Photochemistry transforms matter with energy in the form of light; laser photochemistry transforms matter with light from lasers. While most chemistry emphasizes the production of bulk materials, much photochemistry aims at the production of specialty materials, the manufacture of structures, or the processing of materials to obtain, transfer, or store information. Laser photochemistry retains these further emphases.

Manmade photons are expensive chemical reagents; economics restricts photochemistry to high-value–added processes, or to where only a few photons selectively initiate a process. Photons are very effective at causing chemical changes, but current photon sources are inefficient. While a mole of visible photons contains energy valued at only a cent at today's electric rates, inefficiencies of conventional light sources can quickly push the cost of photons to over a dollar per mole. These economics generally make photochemistry unattractive for the production of bulk chemicals. Photochemistry is more useful for high-value products such as drugs (currently vitamin D is produced photochemically), insecticides, perfumes, and polymer resins. Established photochemical technologies are found in photography, coating (especially of thermoplastics, such as used in magnetic recording or adhesive tapes), photocuring of optical and dental cements, drug phototherapy, uv disinfection, sterilization, and surface cleaning, printing, lithography, and photoresist masking in semiconductor and printed-circuit manufacturing.

The capital costs associated with most laser sources—particularly those operating in the photochemical uv—can bring the cost of photons up to $1000 per mole. Production of exceptionally valuable materials, such as isotopes or isotopically labeled molecules, can

1-56081-067-X/94/$5.00+.50

benefit from laser photochemistry. More commonly, laser photochemical processes transfer or store information within materials. Lasers find applications in photographic rendering of computer-drawn images, laser marking, and photoresist masking. In stereolithography, laser photochemistry is used to realize complex three-dimensional plastic prototypes or mold models directly from computer-aided designs. The latter complements automated machining: solid material adds to an object held on a mandrel wherever a laser beam initiates curing of a polymer resin.

Lasers have found extensive application in the investigation of chemical processes themselves, where they have given us invaluable information about the details of chemical reactions. Photoinitiation is much more selective than thermal initiation of reactions and provides greater control in the investigation of chemical kinetics (now often called "chemical dynamics" to emphasize the higher precision afforded to prepare and investigate reactions of single quantum states). Lasers have moved into nearly every chemical kineticist's laboratory—if not to initiate reactions, then to analyze products. A whole field of "half reactions" has developed, in which a reaction intermediate, normally prepared by a precursor reaction or collision process, is instead prepared directly by a photoprocess. This allows the experimenter to prepare an intermediate of known chemical conformation, having a given energy, at a specified time.

The availability of short-duration (picosecond and femtosecond) pulses has shortened the time scale directly accessible for kinetic experiments in the time domain (Mukamel, 1990; Khundkar and Zewail, 1990). For example, nearly a dozen transient molecular states and conformers (different, relatively stable arrangements of atoms) have been identified by pump-probe laser experiments subsequent to the absorption of light by rhodopsin, the optical pigment in the retina (Birge, 1990).

Fundamental reaction dynamics also can be investigated in the frequency domain. For example, photoemissions from the dissociating states of the molecules methyl iodide and ozone have been observed. The structure of each intermediate, changing as it dissociates, is impressed upon the spectra as a series of resonances, while the dynamics and time scales of the reaction appear in the pattern of intensities and spectral widths of these features (Imre *et al.*, 1984). A laser provides a high flux of photons of a single frequency, making such investigation of weak emissions possible.

Information from the analysis of materials is becoming more in demand, and laser techniques have become more popular there. Laser spectroscopy is useful, but often lasers are used to prepare the sample by a mixture of physical and photochemical effects. Solid material may be introduced into the gas phase by laser ablation, for example. Photoionization of a sample may prepare ions for identification by mass spectrometry, or photodissociation may prepare atoms for atomic spectroscopy.

In much of photochemistry brought about by lasers, the fundamental uniqueness of laser radiation has not yet been used; lasers often represent specialized "light bulbs" that, in principle, could be replaced by thermal light sources. As such, much of this laser photochemistry is not qualitatively different from photochemistry conducted with weaker light sources, the subject of the article on PHOTOCHEMISTRY. The present article concentrates on the unique properties of laser radiation and focuses on the qualitative differences in the photophysics, energy partitioning, and chemistry that can appear when laser radiation is used.

1. PHOTOPHYSICS

1.1 Laser Light

The radiation field within a laser must stimulate emission in competition with spontaneous emission. This establishes a characteristic brightness of laser radiation.

Spontaneous emission can be viewed as emission stimulated by the random, zero-point fluctuations of the radiation field. The spectral density of these spontaneous fluctuations approximates that of the density of states per wavelength interval, $4\lambda^{-4}$. In terms of photon flux, or brightness, this is

$$\frac{d(\text{photons})}{dA\ d\Omega\ d\lambda\ dt} = \frac{c}{3\pi\lambda^4}, \tag{1}$$

or, in terms of flux power density M, solid angle Ω, and wavelength spread $\Delta\lambda$ of a laser beam,

$$M/\Omega\Delta\lambda = hc^2/3\pi\lambda^4. \tag{2}$$

Laser light differs from light from lower-brightness, thermal sources in one or more of these quantities:

- high power flux density,
- low angular spread,
- narrow frequency distribution.

These give laser light several advantages for photochemistry.

1.2 Interaction with Matter

1.2.1 Linear Absorption. Until intense laser sources of radiation became available, the fundamental photochemical reaction step, absorption, depended linearly upon intensity. At low power density, light interacts with matter by the absorption of one quantum of radiation at a time.

Even without entailing direct multiphoton processes, lasers may still significantly enhance rates of photoabsorption. At low light levels, using thermal light sources, the absorption of light is usually the rate-limiting step in a sequence of steps that compose the mechanism of a photochemical process. At high light levels, resonant absorption may become the fastest step in a photochemical mechanism.

While the line shapes of individual transitions and the spectral distribution of transition bands depend upon numerous details of the molecule and its interaction with the radiation field, the overall line strengths, or band intensities, can be related to simple perturbation expressions. The absorption cross section σ for transition to the final state $|f\rangle$, when integrated over frequency ν, is given theoretically by the leading, first-order perturbation of the initial state $|i\rangle$ in the electric dipole approximation,

$$\int \sigma_{fi}(\nu)d\nu = 4\pi^2\alpha\bar{\nu}\left|\left\langle f\left|\sum_e \mathbf{r}_e\cdot\mathbf{n}\right|i\right\rangle\right|^2. \tag{3}$$

The sum is taken over all charges e at distances $\mathbf{r}_e$; α is the fine-structure constant, and $\mathbf{n}$ is a unit polarization vector. The bar over the frequency indicates averaging.

Each electron contributes to the total integrated absorption cross section, averaged over orientations and summed over possible final states, one "electron oscillator," $\pi r_0 c \approx 0.03$ $cm^2\ s^{-1}$; r_0 is the classical electron radius. Individual electronic absorption bands typically contain the transition strength corresponding to 0.01–0.5 of an electron oscillator, while weaker vibrational transitions contain perhaps 10^{-6}–10^{-4} of an electron oscillator in each band.

1.2.2 Multiphoton Absorption. At the high power densities achievable with lasers (easily 10^{18} W/m^2 in focused pulses), light interacts with matter through the additional mechanism of direct multiphoton absorption.

The second-order perturbation correction gives the two-photon contribution to the absorption. An absorption cross section $\sigma^{(2)}N_1$ for a photon of frequency ν_2 is induced by an off-resonance monochromatic field of frequency ν_1 and photon flux N_1. When integrated over the frequency of the second photon, the second-order cross section can be related to the dipole matrix elements:

$$\int \sigma_{fi}^{(2)}(\nu_1,\nu_2)N_1(\nu_1)d\nu_2 = \frac{4\pi^2\alpha^2}{\nu_1\bar{\nu}_2}\left|\sum_m \frac{\nu_{fm}\nu_{mi}\langle r_{fm}\rangle\langle r_{mi}\rangle}{\nu_{mi}-\nu_1}\right|^2, \tag{4}$$

where

$$\langle r_{fm}\rangle = \left\langle f\left|\sum_e \mathbf{r}_e\cdot\mathbf{n}_2\right|m\right\rangle,$$
$$\langle r_{mi}\rangle = \left\langle m\left|\sum_e \mathbf{r}_e\cdot\mathbf{n}_1\right|i\right\rangle, \tag{5}$$

and the sum is taken over intermediate states $|m\rangle$ having frequencies ν_{mi} and ν_{fm} for transitions to the initial and final states. The energy for the overall transition comes from two photons; hence the two-photon resonance condition $\nu_{fi} = \nu_1 + \nu_2$. A special case often occurs when only one radiation frequency is used: then $\nu_1 = \nu_2$ and two-photon resonance is achieved when the energy of the transition corresponds to twice the frequency of the radiation field.

This integrated induced cross section is roughly of the order $\alpha(\pi c r_0)^2 N_1/(\bar{\nu}_1\nu_2)$. Making the two-photon absorption comparable to the one-photon absorption requires off-resonance fields on the order of

$$N_1 \approx \nu_1\nu_2/\alpha\pi c r_0 \tag{6}$$

or about 10^{21} photons $s^{-1}\ cm^{-2}$ (10^{17} $W\ m^{-2}\ s^{-1}$ for typical green light).

Expressions for higher-order, multiphoton cross sections are readily derived. Typical dipole-allowed molecular multiphoton absorption cross sections have been found on the orders of 10^{-58} m^4 s for two photons and 10^{-94} m^6 s^2 for three photons.

1.2.3 Level Shifts. High radiation powers can cause atomic and molecular resonant frequencies to vary, responding to the ac Stark effect (Carmichael and Walls, 1976; Li *et al.*, 1985).

In atoms and diatomic molecules, effects of high power can be rather striking, even at small laser powers. In sparse spectra, associated with atoms and small molecules, each individual resonance has a relatively large oscillator strength. Even moderate fields, tuned near resonance, interact strongly with an atom, which undergoes rapid excitation to, and stimulated emission from, excited states at a characteristic, power-dependent rate, the Rabi frequency. The fluorescence from such an interacting system has a characteristic "head and shoulders" appearance, best understood as radiation at the fundamental frequency undergoing amplitude modulation at the Rabi frequency (Mallow, 1969). Such a spectrum may be "smeared" to higher or lower wavelength by an ac Stark effect (Li *et al.*, 1985).

Large molecules, in contrast to small molecules, spread the oscillator strength of an electronic or vibrational transition over a large number of molecular "rovibronic" states. Especially when a number of initial states are populated at finite temperatures, transitions are heterogeneously broadened. Each individual state-to-state transition may have only a small fraction of the original oscillator strength; at any given frequency, many resonances with widely varying transition strengths are likely to overlap. Thus, power broadening and ac Stark shifts require much higher powers (and are less observable) in the multiphoton photochemistry of large polyatomic molecules. However, power-driven state "pulling," field-induced state mixing, and spontaneous and stimulated Raman effects often explain the excitation to higher states that would otherwise appear to be inaccessible.

1.2.4 Hole Burning. At high power densities, a general excitation of rotational and vibrational levels accompanies excitation and deexcitation of states by the radiation field. Absorbing states that interact resonantly with radiation become depleted as molecules enter and remain in "dark" states having weaker interactions with the radiation field. These optical-pumping effects can make resonance features in an absorption spectrum disappear at high power levels (Rothberg *et al.*, 1981).

Radiation at a particular frequency generally moves population out of states that interact strongly with that radiation. This depletion may appear as a "hole" in the absorption spectrum. "Hole burning" is the basis for Doppler-free Lamb-dip spectroscopy. Atoms or molecules having little or no velocity component along the axis of two counterpropagating beams are temporarily depleted from the population distribution of velocities. At exactly the frequency center of an absorption line, the absorption cross section acquires a power-dependent decrease or "dip." This technique is commonly used for laser frequency stabilization, such as the iodine-locked He-Ne laser. Recently, interest has shifted to permanent hole burning as a method of information storage in materials.

2. MULTIPHOTON PROCESSES

The most striking examples of new photophysics and photochemistry that occur when lasers are used in place of conventional light sources involve multiphoton absorption.

2.1 Electronic Multiphoton Absorption

Electronic multiphoton absorption, excitation to electronic states energetically inaccessable by one-photon absorption, was one of the first effects explored with lasers. Two-photon absorption was first reported for inorganic crystals containing europium ions (Kaiser and Garrett, 1961). Blue and ultraviolet fluorescence appeared in the interaction of red, ruby-laser light with organic compounds (Peticolas *et al.*, 1963; Peticolas and Reickhoff, 1963; Singh and Stoicheff, 1963; Weisz *et al.*, 1964). Others observed two-photon absorption directly (Hopfield *et al.*, 1963; Staginnus *et al.*, 1968).

2.1.1 Multiphoton Selection Rules. Using the symmetry of the absorbing molecule, and the electronic-state symmetry assignments, McClain (1974) derived explicit selection rules for two-photon electric-dipole interactions. Most notable in two-photon excitations, even-parity transitions appear for centrosymmetric molecules, normally forbidden in single-photon electric-dipole transitions. Light polarization characteristics may also affect absorption, even for collections of randomly oriented molecules, especially those possessing high molecular symmetry.

2.1.2 Multiphoton Ionization (MPI). One striking multiphoton effect, runaway electrical breakdown of air, often occurs at the focus of high-power, pulsed laser light. Initiated by direct multiphoton absorption, the resulting molecular excited states undergo additional single-photon absorption directly into the ionization continuum. In collisions with molecular and ionic excited states, slow electrons efficiently extract energy: slow electrons "superelastically" scatter to become fast electrons; cross sections for this process diverge for low electron-impact velocities. Thus initiated, high-pressure ionization rapidly transforms the initially transparent gas into an optically thick plasma, in which most of the absorbing power is derived from electronic transitions originating on excited states of molecules and ions, and terminating on states consisting of additional free electrons and ions. Taken to its ultimate, a totally ionized plasma produced by laser radiation is now the object of laser fusion research.

Some researchers began to use ionization as a tool to investigate weak multiphoton absorption. Two groups almost simultaneously, Johnson *et al.* (1975) and Petty *et al.* (1975), devised a simple technique. A tunable laser was focused into a proportional-counter cell containing an inert gas and a trace amount of the molecule being studied (Fig. 1). For two-photon spectroscopy, when the energy of two photons was resonant with an allowed, two-photon transition, the energy of a third photon would be sufficient to ionize the resulting excited molecules, as represented in Fig. 2(a). Since ionization usually has a relatively unstructured absorption spectrum, and the power levels employed to produce two-photon absorption saturate the one-photon ionization step, the overall rate of ionization depends predominantly upon the rate of the initial two-photon step. Monitoring the amount of ionization produced in the proportional-counter cell as a function of wavelength generates a "two-photon" ionization spectrum. One of the first molecules investigated this

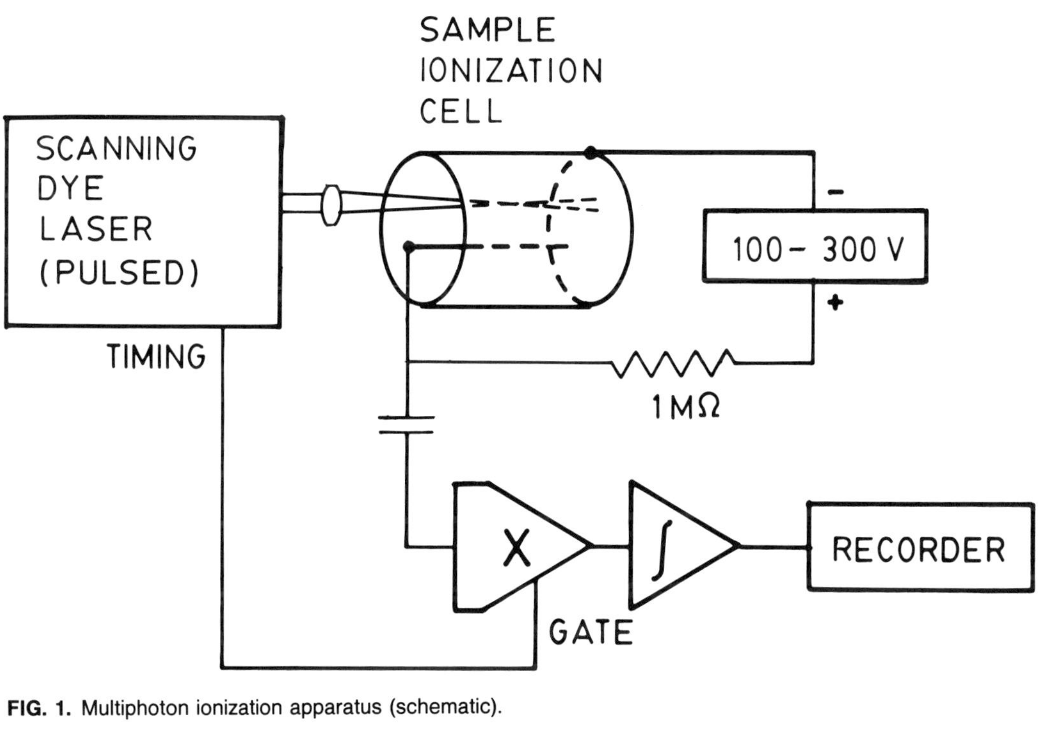

FIG. 1. Multiphoton ionization apparatus (schematic).

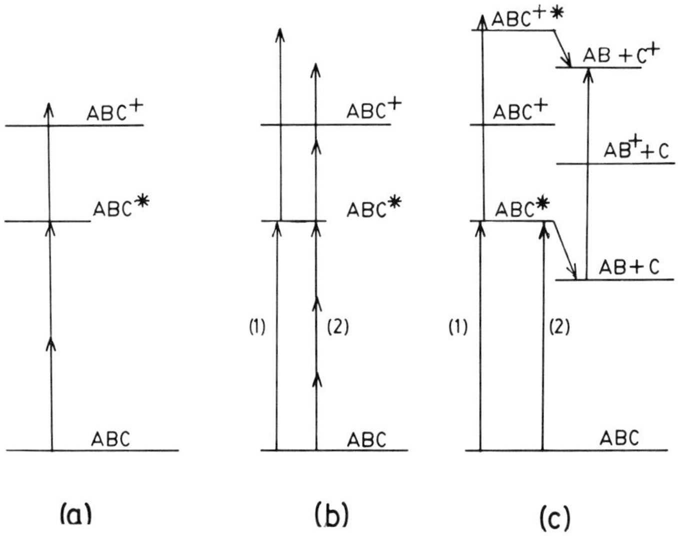

FIG. 2. Multiphoton ionization processes. (a) 2+1 multiphoton ionization occuring by two-photon absorption to a molecular excited state and subsequent one-photon ionization. (b) 1+1 and 3+2 multiphoton ionization processes. (c) Multiphoton ionization either (1) preceding dissociation of the ion, or (2) subsequent to dissociation of the neutral.

way was nitric oxide, NO; one of the first polyatomic molecules investigated was benzene, C_6H_6 (Johnson, 1975, 1976). Molecular states near the ionization limit (principally molecular Rydberg states) appear, states that are usually invisible because of the selection rules of one-photon spectroscopy.

In addition to this simple "2+1" process (two-photon absorption followed by one-photon ionization), numerous other *n*-photon combinations leading to absorption and ionization could, and did, occur, as exemplified in Fig. 2(b). Any of these processes are subsumed under the name "multiphoton ionization." Processes that depended upon resonant one-, two-, or many-photon intermediate processes were generally indicated by the additional term "resonant" or "selective."

2.1.3 Multiphoton Dissociation (MPD). With the recognition that intense visible or UV radiation rapidly excites molecules to high electronic states, culminating in ionization, a seductive argument can be made that saturated electronic absorption at high power densities would completely dominate "slower" processes involving nuclear displacements, such as molecular rearrangement and dissociation. Until the early 1980s, multiphoton dissociation was generally thought to compete ineffectively with multiphoton ionization; molecular fragmentation was thought to occur only after ionization of the parent molecule.

Studies of ions produced by multiphoton ionization of organic molecules suggested that rearrangement and dissociation could occur in the neutral molecule, before ionization, but poor understanding of the absorption spectra of the intermediate, neutral fragments did not allow definitive proof. Clear evidence that dissociation may precede one or more slow steps of absorption, and that it certainly precedes ionization, came from organometallic photochemistry. Multiphoton ionization of transition-metal compounds occurs by multiple molecular absorption events and rapid dissociation of the neutral molecule, followed by multiphoton ionization of naked, neutral transition-metal atoms (Gerrity *et al.*, 1980; Engelking, 1980; Leutwyler *et al.*, 1980).

In such cases, the MPI spectrum of a molecule actually expresses the atomic MPI spectrum of one or more of the molecule's constituent atoms, which have been ejected from the molecule by MPD. Subsequent work has shown the prevalence of MPD in the photochemistry of polyatomic organic molecules, such as benzene, where laser-induced fluorescence and resonance Raman spectroscopy show the presence of C_2, CH, and carbon atoms at high power levels.

Now we understand that the situation in a photochemical mechanism involving laser radiation is complicated by numerous competing pathways, and that photophysical and chemical steps can be intertwined, as represented in Fig. 2(c). In retrospect, the reason for this is clear: the electric fields within a molecule exceed those imposed by radiation, and any time an electron is displaced from the bonding regions between nuclei for any reason, nuclei are exposed to strong Coulomb repulsion. Two reviews in the same journal published within months of each other (Bernstein, 1982; Gedanken *et al.*, 1982) show the rapidity of this change of perspective.

2.2 Vibrational Multiphoton Absorption

2.2.1 Vibrational-State Ladders. Molecules couple to infrared (IR) radiation principally through the overall vibrational motion of molecular charge. Typically involving only a fraction of an electronic charge moving through only a fraction of a picometer, the resulting strength of a vibrational transition is commonly 10^{-4} of an electronic transition.

Low-energy vibrations of molecules usually take place in nearly harmonic potentials, containing "ladders" of nearly equally spaced vibrational energy levels. An IR laser frequency can almost coincide with a sequence of vibrational transitions corresponding to the intervals between levels of a particular normal mode, as shown in Fig. 3. At high laser powers, these nearly resonant intermediate states enhance multiphoton processes. Since vibrational spacing generally decreases with increasing mode energy, laser radiation that is resonant with molecular transitions at the bottom of a vibrational ladder becomes less resonant and couples less effectively with molecules already excited to higher energies, as shown in Fig. 3.

2.2.2 Isolated Molecules. The effectiveness of multiple resonances in assisting IR multiphoton transitions in isolated molecules depends to a large degree upon molecular size. For small molecules, such as diatomic molecules, with only one vibrational mode, anharmonicity quickly moves subsequent absorptions out of resonance. Except at extreme power densities, IR excitation of small molecules saturates lower transitions, but cannot give access to higher levels unless measures are taken to account for anharmonicity. In medium-sized molecules, it is easy to excite molecules rapidly to high vibrational levels, but difficult to drive the molecules to dissociation or rearrangement. In a process akin to hole burning, the energy ensemble of vibrating molecules acquires a bimodal, non-Boltzmann distribution: some low-energy molecules occupy vibrational and rotational states not effectively coupled to the radiation, while a high-energy group of molecules in a broader distribution of states interacts with the field. Bimodal population distributions are most often produced when molecular modes of vibration that occur at isolated frequencies are excited by IR laser radiation.

Large molecules, having a great number of vibrational modes, present many nearly coincident vibrations and overtones, even at moderate levels of excitation. Mixing of the numerous states reduces the interaction of any one state with the radiation field, giving rise to a dense cluster of states around each resonance. This is especially true at high energies. States that fill the gaps in a vibrational-state ladder in large molecules compensate for the anharmonicity that interrupts resonant multiphoton excitation to higher levels in small

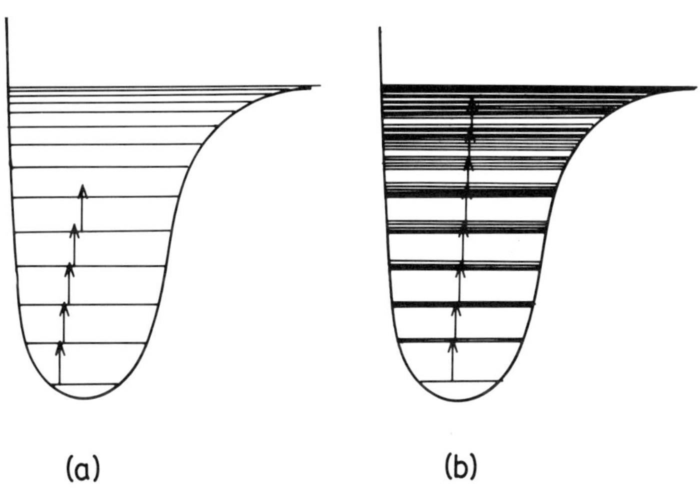

FIG. 3. IR multiphoton events with single-frequency radiation. (a) Resonant excitation drives small molecules up a ladder of nearly evenly spaced energy levels until anharmonicity moves states from further resonance. (b) Large molecules have high densities of states, providing near resonances all the way to dissociation.

molecules. With the increasing likelihood that each transition will be assisted by a resonance, excitation to dissociation becomes easier yet. Furthermore, with the dilution of resonances and the complication of the absorption spectrum, it is less likely that a narrow internal energy distribution will result, and the distribution of excitation in energy levels becomes more statistical, resembling a thermal Boltzmann distribution.

Vibrational transitions are weaker and more numerous than electronic transitions, making sequential resonant, or nearly resonant, transitions more important in the infrared and direct nonresonant multiphoton excitation less important in the infrared. Typically, one expects to need 10^{14} W m^{-2} or more to dissociate a diatomic molecule by near-resonant multiple transitions, and 10^{8}–10^{12} W m^{-2} for dissociation of a large molecule.

2.2.3 Effects of Collisions. If molecules are allowed to collide, the rates for IR ladder climbing increase. Colliding pairs of vibrationally excited molecules, which are no longer resonant with the radiation field, may "pool" energy into higher vibrational levels of one collision partner, leaving the other collision partner in lower vibrational levels. This brings the second molecule again in resonance with the exciting IR.

Collisional redistribution of vibrational energy is a statistical process: one expects to obtain a Boltzmann energy distribution. The persistence of a fraction of molecules in states that have transitions resonant with the radiation allows efficient heating. Power levels of 10^{4}–10^{7} W m^{-2} may lead to dissociation, rearrangement, or other thermally initiated chemistry.

IR heating of gases has proven to be effective in thermally initiating reactions for kinetic studies. This may be done either by directly coupling IR to the reactive molecules, or by indirectly heating the reactive molecules seeded in a bath of strongly absorbing molecules (Golden *et al.*, 1981).

3. ENERGY PARTITIONING

Chemists have, since the 1930s, known that certain bonds between atoms gave rise to characteristic frequencies in the IR, corresponding to bond stretches and bends. Thermal sources of radiation proved ineffective in breaking specific bonds, since energy could not be supplied fast enough to stretch a bond to breaking, in competition with intramolecular and intermolecular energy relaxation.

In recent years, the topic of mode-specific photochemistry has revived in the context of a better understanding of the energy redistribution that may occur inside molecules.

3.1 Randomization of Energy in Molecules

Energy flow in molecules falls between two extremes. In the "small-molecule" limit, radiation excites isolated states; the interaction with the radiation field is more important than the interaction between internal molecular states. In the "large-molecule" limit, radiation excites a dense band of internal states strongly mixed with the absorbing state; the interaction between numerous molecular states exceeds that with the radiation field. The difference is quite striking: small-molecule states typically fluoresce; large-molecule states typically retain the energy within the molecule.

The coupling between states within molecules is mediated by numerous mechanisms: vibrational mixing, Coriolis interactions, velocity and acceleration coupling of electronic and vibrational motion, and spin-orbit coupling.

The density of internal states is as important as the strength of interaction between them in determining the large-molecule limit. A single radiating molecular state interacting strongly with just a few molecular states produces simply a few, widely separated states. The large-molecule limit must be defined a bit more carefully by the two conditions

$$\Gamma \ll \gamma \tag{7}$$

and

$$\Delta E \ll \Gamma, \tag{8}$$

where Γ is the natural width of the radiative state, γ is the average strength of the interaction of the radiative state with other perturbing molecular states, and ΔE is the average spacing between perturbing states (the reciprocal of the density of states). The effect of various combinations of these parameters is outlined in Fig. 4.

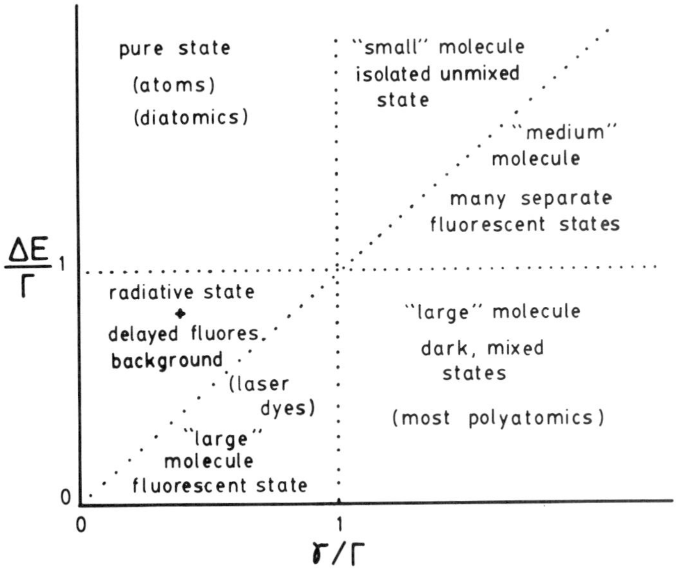

FIG. 4. Regimes of "large" and "small" molecule behavior. Here γ is the strength with which a radiative state interacts with other molecular nonradiative states; Γ is the strength of the radiative interaction; ΔE is the average spacing between mixing states.

Under the large-molecule conditions, where both relationships hold, the high density of perturbing states provides the molecule with its own thermodynamic bath. If a single radiative state couples with a large number of nonradiative bath states, the mixture of perturbed states will rapidly dephase into a mixture of incoherently radiating states, each with a small fraction of the original transition strength. It appears that the energy absorbed by the initial state has dissipated into the bath states.

The phenomenon of a high density of internal molecular vibrations acting as an intramolecular thermodynamic bath has been carefully investigated in the case of alkyl benzenes in which the length *n* of the "floppy" carbon chain attached to the rigid benzene ring is systematically increased:

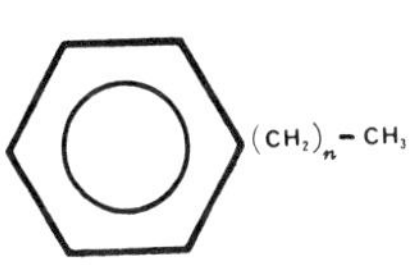

The electronic transition occurs primarily on the benzene ring, but couples with the molecular vibrations. As might be expected, the fluorescence quantum yields and the radiative lifetimes of molecules decrease as the chain length, and hence the density of vibrational states, increases (Smalley, 1982).

This ability of large molecules to transfer energy between electronic, vibrational, and spin-rotational manifolds of states has a profound effect upon the chemistry of molecules. In laser photochemistry, the rate at which this happens often competes with the rate at which subsequent photons are absorbed. In effect, the rate of intersystem crossing, internal conversion, and vibrational mixing defines the rate by which energy deposited in a given state, or mode, naturally "randomizes" within molecules, and it is against this rate that photoexcitation must compete.

In the condensed phase, vibrational relaxation typically occurs on the order of tens of femtoseconds to hundreds of picoseconds, as shown in laser pump-probe experiments (Elsaesser and Kaiser, 1991). One should be aware, however, that in simple liquids, such as liquid nitrogen or inert gases, vibrational relaxation times on the order of minutes have been found (Brueck and Osgood, 1976)!

3.2 Energy Localization

Recent work on molecules has shown the conditions for maintaining local concentrations of energy in spite of relatively high densities of vibrational states. In some molecules, at high levels of vibrational excitation the energy coupling between modes becomes weak. Under these conditions, it is possible to have vibrational energy localized primarily in one or more modes for many vibrational periods.

Examples of mode-specific chemistry include the work of Sinha *et al.* (1990) in which singly deuterated water was alternatively excited along the O—H bond or along the O—D bond. Upon reaction with hydrogen atoms,

$$H+HOD(4\nu_{OH}) \rightarrow \begin{cases} OD+H_2, & (9) \\ OH+HD, & (10) \end{cases}$$

reaction (9) is enhanced, compared to (10), when the O—H bond is excited to the fourth overtone of the OH stretch. Bar *et al.* (1991) found that photodissociation of the same molecule at 193 nm depended upon the mode in which one quantum of vibration was placed. In this case, HOD excited to $\nu_3=1$ produced a different mix of photodissociation products from that normally produced by dissociation of HOD ground state, while there was no significant change in the product distribution when HOD was excited in the vibrational mode $\nu_1=1$.

Not all molecules are suitable for mode-specific photochemistry, as pointed out theoretically in the work of Kellman *et al.* (1991) on the HOD system. For example, Marquardt and Quack (1991), working on CHX_3 theoretical models, point out that the energy initially localized in a wave packet on the CH stretch mixes with CH bends in 80 fs, and is randomly distributed in hundreds of femtoseconds.

4. CHEMICAL CHANGES

4.1 Ionization and Detachment

One of the more striking effects of laser photochemistry is the frequent appearance of ions, even when working with photon energies less than the molecular ionization potential. This ionization usually results from the processes, shown in Fig. 2, of direct multiphoton absorption, resonant multiphoton ionization involving several intermediate states, or ionization of chemical products resulting from rearrangements or elimination of groups of atoms.

An analogous process for negative ions is the production of neutral species by multiphoton photodetachment. In fact, one of the first structured multiphoton spectra observed was that of photodetachment of C_2^- (Lineberger and Patterson, 1972). Detachment is reviewed (along with ion dissociation) by Miller (1981).

At high pressures, collisional mechanisms allow molecules to pool energy, and "chemi-ionization" may result. This is thought to be one possible mechanism to account for multiphoton ionization by IR absorption, similar to the production of ionization in flames (Avouris *et al.*, 1980).

4.2 Dissociation

The major photochemical pathway is that of dissociation. Lasers have been instrumental in exploring photodissociation by their ability to initiate and probe the sequence of events as molecules come apart. For example, it has been possible to excite, and observe, the emission by molecules while they are dissociating. Imre *et al.* (1984) reviewed the seminal experiments, along with the accompanying advances in theory that allow these experiments to be understood.

Lasers have also provided insight into the distribution of products, in terms of energy release, angular distributions of products, and the internal rotation and vibrational states of the fragments (Leone, 1985).

One study looked at the details of predissociation in formaldehyde by carefully comparing the rotational state distribution expected from an RRKM (Rice-Ramsperger-Kassels-Marcus) statistical theory with the state populations measured for CO (Bamford *et al.*, 1985). The RRKM model assumes an equipartitioning of energy between all molecular modes. Although there are only four atoms present, the excited electronic state efficiently couples to the vibrational manifold associated with the electronic ground state, which then decomposes by elimination of hydrogen:

$$H_2CO+h\nu \rightarrow H_2CO(S_1) \rightarrow H_2CO(S_0,\text{vib.}) \rightarrow H_2+CO(\text{rot.}). \quad (11)$$

In liquid solutions, photodissociation may be followed immediately by recombination of species trapped by a solvent cage. The effect of this cage has been the object of much study since it was first observed, and the prototypical system here is the photodissociation of I_2.

This system is still not fully understood, but the use of picosecond and faster lasers has allowed us a better understanding than in the past. A recent review is provided by Harris *et al.* (1988). As currently understood, dissociation happens on the subpicosecond time scale, followed by recombination on the order of several picoseconds. Vibrational relaxation occurs within several hundred picoseconds in polyatomic solvents. The most difficult step, which is not entirely understood, is the I-atom escape from the solvent cage, which can take place in anywhere from tens of picoseconds to tens of nanoseconds, depending upon the solvent in a way that is not yet understood.

One area of investigation that would be nearly impossible without lasers is photodissociation of isolated molecular ions in molecular beams or ion traps (Moseley, 1985). Examples of these studies include work on aromatic molecular ions of benzene (Freiser and Beauchamp, 1975),

$$C_6H_6^+ + h\nu \ (2.5 \ \text{eV}) \rightarrow C_6H_6^{+*},$$

$$C_6H_6^{+*} + h\nu \ (1.5 \ \text{eV}) \rightarrow C_6H_5^+ + \text{H}, \qquad (12)$$

and toluene (Eyler, 1976),

$$C_7H_8^+ + h\nu \rightarrow C_7H_7^+ + \text{H}. \qquad (13)$$

In the first case, the benzene cation

requires sequential electronic absorption of two photons in order to surmount the barrier to dissociation. This is one example of large-molecule photoabsorption, as discussed above, followed by dissociation. Interestingly, the intermediate formed by the absorption of one photon is very long-lived (seconds) and appears to be a vibrationally hot ion, but in its electronic ground state. Relaxation of the initially produced, electronically excited state into a mixture of vibrational levels of the ground electronic state prevents the excited molecular ion from radiating, and prepares the ion for absorption of an additional quantum of radiation. The combined energy of two photons now exceeds the threshold for dissociation of about 4 eV.

The toluene ion,

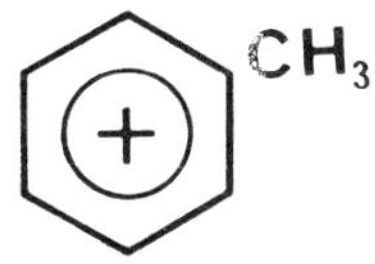

illuminated at almost the same energy, dissociates in a one-photon process. Direct elimination of a hydrogen from the methyl group to produce the benzyl cation

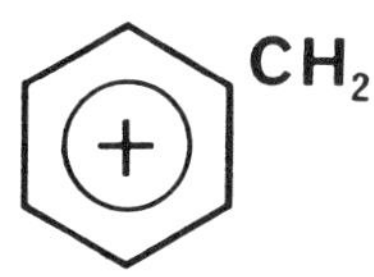

is energetically allowed when 2.3 eV is supplied by radiation. However, isomerization of the carbon atom backbone of the molecule, enlarging the ring to seven members,

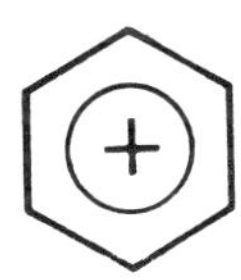

in concert with the elimination of a hydrogen, would be energetically possible with the addition of only 1.7 eV of energy from radiation. The near coincidence of the photothreshold (2.3 eV) with the predicted thermodynamic threshold for the production of benzyl appears to indicate that direct dissociation occurs without (at least initially) isomerization of the molecular ion fragment.

Dissociation is also the usual step initiating radical chain reactions, such as polymerization. Commercially important are reactions involving azides,

$$R\text{N}_3 + h\nu \rightarrow R\text{N:} + \text{N}_2, \qquad (14)$$

triazines, and electron-transfer initiators such as diaryliodonium and triarylsulfonium salts.

4.3 Isomerization

Isomerizations, the rearrangement of atoms within single molecules, are some of the cleanest photochemical transformations. Classical photochemistry has focussed intently on *cis-* to *trans-* isomerizations, and these are the object of laser photochemistry, too.

IR photochemistry of the isomers of 1,2-dichloroethene at 981 cm^{-1} results in a photostationary mixture in which the higher-energy *cis-* conformation dominates, in contrast to the result produced by simple heating (Ambartzumian *et al.*, 1976).

Photophysical and photochemical processes accompanying isomerizations of the visual pigment rhodopsin have been studied on time scales ranging from femtoseconds to seconds with lasers. The key transformation occurring there is the isomerization of the conjugated chain of retinal between conformations (a) and (b):

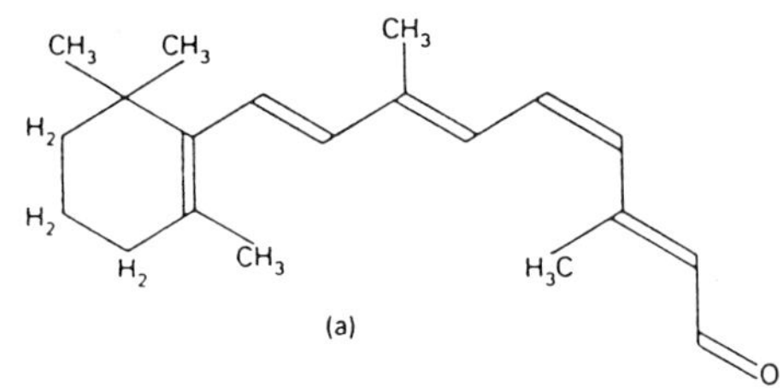

(a)

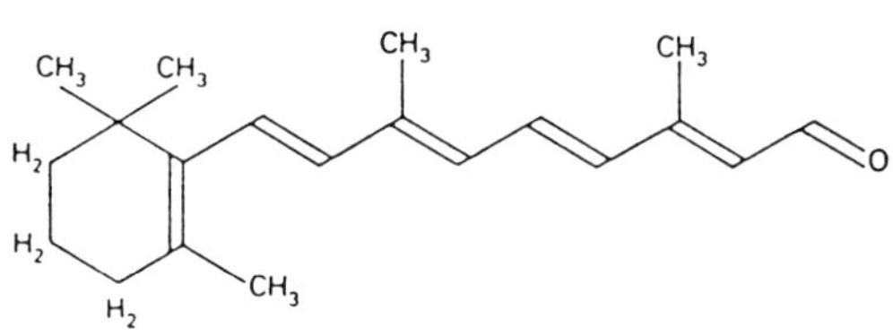

(b)

4.4 Condensation and Addition

A common photochemical process is the initiation of polymerization, the condensation of many monomer molecules into larger polymer molecules. While the usual mechanism is radical-initiated or radical-ion–initiated chain reaction, occasionally a direct addition of an excited reactant with another is causative.

One of the more striking effects observed upon passing laser light through an absorbing gas sample is that of photoinduced condensation, producing brilliant opalescence along the laser beam. Examples of this include alkali-metal vapors through which light from a dye laser is directed. In this configuration, a significant fraction of alkali-metal-atom excited states are produced, which then may act as nucleation centers for the formation of diatomic molecules and clusters. Several mechanisms appear to be operating; the most striking occurs when the radiation density is sufficient to cause ionization. Ion-atom potentials are always attractive, and they operate at long range, and this electrostatic force, in contrast to covalent bonding forces, does not saturate, allowing the growth of large clusters of molecules that otherwise might not even react.

5. APPLICATIONS

Their cost per mole makes laser photons an expensive reagent, economical in only high-value products. Uses that can be foreseen for laser photons include the production of drugs, the isolation of isotopes, processing of materials on or near surfaces, and the storage and retrieval of information.

5.1 Holography

Laser photochemistry proved important early in the field of holography (see HOLOGRAPHY, OPTICAL). Lasers provided the coherent light, and the interference patterns were recorded by the photochemistry of either silver halide photographic emulsions or organic photoresists. Holographic technology now provides nearly ghost-free commercial diffraction gratings (see MONOCHROMATORS).

5.2 Isotope Selection

High wavelength specificity of lasers, along with sufficient power to cause chemistry on the mole scale, led to interest in using lasers for isotope production. Some possible methods are reviewed by Letokhov (1983) and some of the concerns are raised by Krauss (1977). Numerous processes have become available for laser isotope separation (Cantrell *et al.*, 1979; Hartford, 1979; Jensen *et al.*, 1980; Letokhov, 1979; Letokhov and Moore, 1976, 1977; Zare, 1977). The application to the separation of uranium isotopes has been the focus of studies on both atomic U and its compounds (Dietz *et al.*, 1982).

5.3 Surface Modification

The most common industrial use of lasers in materials processing is for cutting or machining, applying the high energy densities of laser beams to burn or ablate materials. Secondary to this use, lasers have found their way into hardening of materials, thermal treatments, and exposure of photoresists.

5.3.1 Etching. One possibility for increasing the effectiveness of laser ablation techniques is in modification of a nonabsorptive material to make it absorptive. High–band-gap materials will develop color centers upon electron irradiation. At low power levels, lasers can be used to photobleach these color centers; at high power levels, these same color centers can be used to provide the initial absorption necessary to initiate ablation.

Lasers have been found to enhance greatly the rate at which a chemical process removes material from surfaces. Plasma etching and electrolytic etching are examples where lasers enhance the rate of reaction. One effect of lasers that is different from most other techniques is the ability to etch long channels in materials along the axis of light propagation with diameters narrower than the beam profile (Podlesnik and Gilgen, 1989).

5.3.2 Deposition. Laser decomposition of organometallic compounds over surfaces generates metal atoms and deposits metal layers at the adjacent surface (Osgood, 1983; Ehrlich and Tsao, 1983; Jones *et al.*, 1985). Direct dissociation of gaseous organometallic compounds occurs promptly, and the metal ions often possess significant translational energy, as observed in the Doppler profiles of optical transitions of these ions (Marinero and Jones, 1985). This would seriously complicate the ability to deposit patterns at surfaces, since metal atoms generated in the gas phase would tend to spread over the surface, away from the region of photodissociation. Dissociation has been found, however, to be enhanced at surfaces, which permits the writing of structures directly on the surface with dimensions on the order of 10^{-7} m (Wilson and Houle, 1985; Houle *et al.*, 1986).

The field of laser deposition writing is summarized in the book edited by Ibbs and Osgood (1989).

5.4 Information Storage

Many lasers in the information industry provide controlled light for exposure of optical recording media. Most photochemical processes are based upon photographic or photoresist technology, although currently information storage is dominated by photophysical techniques such as xerography (*q.v.*) or thermal melting, such as in the now familiar commercial CD and "laser" disks. Currently, the relative lack of cw uv lasers operating at wavelengths below 300 nm, compared to the ready availability of IR and visible lasers, limits the types of photochemical recording methods available to information technology.

5.5 Chemical Analysis

Lasers have been used to ablate surfaces, thereby generating gaseous ions, atoms, and molecular fragments from samples of surfaces and molecules at surfaces. One use to which multiphoton ionization has been applied is in the detection of molecules eluting from a gas chromatograph (Klimcak and Wessel, 1980) (see CHROMATOGRAPHY).

5.6 Biochemistry

Since a number of biological processes are photochemical, it has been natural to use light in the study of these. The photochemistry of rhodopsins, the pigments of the eye, has been extensively investigated with ever-shortening laser pulses. For a review, see Birge (1990). The major chemical mechanism of vision is isomerization of the long, conjugated-chain molecule (see Sec. 4.3), but numerous conformational changes occur in this molecule before and after this isomerization occurs.

One area of concern to molecular biologists is the rate at which proteins, especially enzymes, change configurations. Optical techniques have been used to study these changes. Miller (1991) reviews how transient optical gratings have provided information about the transformations occurring in hemoglobin with the addition and release of oxygen or CO, and the rate at which various conformations of the heme group relax.

GLOSSARY

Ablation: Removal of material from a surface caused by the impact of light, atoms, molecules, or ions. Generally, a thermal process involving local evaporation.

Bond-Specific Photochemistry: Photochemistry having products determined by a specific bond or combination of bonds set into vibrational motion.

Chemi-ionization: Ionization caused by a chemical reaction.

Conformation: The (possibly unstable) arrangement of atoms inside a molecule.

Conformers: Different, relatively stable, arrangements of atoms, each consisting of identical numbers and types of atoms. A dissociating or rearranging molecule may pass through different conformers. Conformers need not be capable of being physically or chemically isolated.

Hole Burning: Depletion of part of an inhomogeneous distribution of molecules, caused by the interaction of light with those molecules; selective photobleaching.

Internal Conversion: Mixing of the vibrational levels of an electronic state with the vibrational levels of another electronic state.

Intersystem Crossing: Mixing between states of different spin.

Isomers: Different molecular conformations, consisting of identical atoms, that are stable, or metastable long enough to be physically or chemically separated.

Lamb Dip: The reduction in absorption (hole burning) observed when two counterpropagating laser beams interact with the same atoms. Usually observed at the center of a Doppler-broadened transition.

"Large" Molecule: A molecule having a high density of states, forming numerous overlapping resonances that interact simultaneously with radiation of a single frequency.

Local Mode: Classically, relative motion of one atom, or a group of atoms, in a molecule. Quantum mechanically, an eigenfunction or superposition of degenerate, or nearly degenerate, eigenfunctions of the vibrational Hamiltonian with considerable amplitude on one or a few atoms.

Mechanism: Sequence of physical and chemical steps leading to a chemical change. Sometimes used to refer to the most unique or characteristic step of an overall mechanism.

Mode-Specific Photochemistry: Photochemistry having products determined by the vibrational mode set into motion.

Multiphoton Absorption: Direct multiphoton absorption occurs when two or more photons are absorbed at once, mediated by higher-order perturbation terms of the molecular Hamiltonian interaction with the radiation field. Sequential one-photon events are sometimes called by this name, but more properly may be called "multiple photon absorption."

Multiphoton Ionization (MPI): Ionization caused by multiphoton absorption, either directly into the ionization continuum, or by absorption to an intermediate resonant state, followed by photoionization. The latter is termed "resonant" or "selective."

Normal Mode: Classically, molecular vibration in which (nearly) all atoms move concertedly and periodically with the same frequency, but possibly not with the same phase. Quantum mechanically, eigenfunction of the vibrational Hamiltonian.

Organometallic: A compound containing organic carbon and metal atoms.

Photodetachment: Ejection of an electron from a negative ion by the effect of light.

Photodissociation: Fragmentation of a molecule into two or more pieces by the effect of light.

Photoresist: An organic resin that polymerizes under the influence of light.

Rabi Frequency: The rate at which a stimulated electronic transition would occur in response to a radiation field. A simple two-level system will alternately absorb and emit, toggling between excited and ground states at this frequency.

Rovibronic State: An individual state of an isolated molecule, including contributions from rotational, vibrational, and electronic motion.

RRKM: Rice-Ramsperger-Kassel-Marcus theory of unimolecular reactions, based upon the statistical flow of vibrational energy within a molecule. The rate at which fluctuations localize energy in critical, reactive bonds determines the rate of reaction.

"Small" Molecule: A molecule having a sparse density of states in which isolated resonances interact individually with a radiation field of a single frequency.

Spontaneous Raman Scattering: Inelastic scattering by matter of photons into a random final state.

Stimulated Raman Scattering: Inelastic scattering of photons stimulated by the radiation field interacting with matter.

Vibrational Relaxation: Loss of internal vibrational energy into other modes of motion.

List of Works Cited

Ambartzumian, R. V., Chekalin, N. V., Doljikov, V. S., Letokohov, V. S., Lokhman, V. N. (1976), *Opt. Commun.* **18**, 400–402.

Avouris, P., Chan, I. Y., Loy, M. T. (1980), *J. Chem. Phys.* **72**, 3522–3527.

Bamford, D. J., Filseth, S. V., Foltz, M. F., Hepburn, J. W., Moore, C. B. (1985), *J. Chem. Phys.* **82**, 3032–3041.

Bar, I., Cohen, Y., David, D., Arusi-Parpar, T., Rosenwaks, S., Valentini, J. J. (1991), *J. Chem. Phys.* **95**, 3341–3347.

Bernstein, R. B. (1982), *J. Phys. Chem.* **86**, 1178–1184.

Birge, R. R. (1990), *Annu. Rev. Phys. Chem.* **41**, 683–733.

Brueck, S. R. L., Osgood, R. M., Jr. (1976), *Chem. Phys. Lett.* **39**, 568–572.

Cantrell, C. D., Freund, S. M., Lyman, J. L. (1979), in: M. L. Stitch (Ed.), *Laser Handbook*, Vol. 3, Boca Raton, FL: CRC Press, pp. 485–576.

Carmichael, H. J., Walls, D. F. (1976), *J. Phys. B: Atom. Mol. Phys.* **9**, 1199-1219.

Dietz, T. G., Duncan, M. A., Smalley, R. E., Cox, D. M., Horsley, J. A., Kaldor, A. (1982), *J. Chem. Phys.* **77**, 4417–4426.

Ehrlich, D. J., Tsao, J. Y. (1983), *J. Vac. Sci. Technol. B* **1**, 969–984.

Elsaesser, T., Kaiser, W. (1991), *Annu. Rev. Phys. Chem.* **42**, 83–107.

Engelking, P. C. (1980), *Chem. Phys. Lett.* **74**, 207–210.

Eyler, J. R. (1976), *J. Am. Chem. Soc.* **98**, 6831–6834.

Freiser, B. S., Beauchamp, J. L. (1975), *Chem. Phys. Lett.* **35**, 35–40.

Gedanken, A., Robin, M. B., Kuebler, N. A. (1982), *J. Phys. Chem.* **86**, 4096-4107.

Gerrity, D. P., Rothberg, L. J., Vaida, V. (1980), *Chem. Phys. Lett.* **74**, 1–5.

Golden, D. M., Rossi, M. J., Baldwin, A. C., Barker, J. R. (1981), *Acc. Chem. Res.* **14**, 56–62.

Harris, A. L., Brown, J. K., Harris, C. B. (1988), *Annu. Rev. Phys. Chem.* **39**, 341-366.

Hartford, A., Jr. (1979), *AIChE Symp. Ser.* **75**, 44–49.

Hopfield, J. J., Warlock, J. M., Park, K. (1963), *Phys. Rev. Lett.* **11**, 414–417.

Houle, F. A., Wilson, R. J., Baum, T. H. (1986), *J. Vac. Sci. Technol. A* **4**, 2452–2458.

Ibbs, K. G., Osgood, R. M. (Eds.) (1989), *Laser Chemical Processing for Microelectronics*, Cambridge, U.K.: Cambridge University Press.

Imre, D., Kinsey, J. L., Sinha, A., Krenos, J. (1984), *J. Phys. Chem.* **88**, 3956–3964.

Jensen, R. J., Sullivan, J. A., Finch, F. T. (1980), *Separation Sci. Technol.* **15**, 509–532.

Johnson, P. M. (1975), *J. Chem. Phys.* **62**, 4562–4563

Johnson, P. M. (1976), *J. Chem. Phys.* **64**, 4143–4148.

Johnson, P. M., Berman, M. R., Zakheim, D. (1975), *J. Chem. Phys.* **62**, 2500–2502.

Jones, C. R., Houle, F. A., Kovac, C. A., Baum, T. H. (1985), *Appl. Phys. Lett.* **46**, 97–99.

Kaiser, W., Garrett, C. G. B. (1961), *Phys. Rev. Lett.* **7**, 229–231.

Kellman, M. E., Li, Z., Xiao, L. (1991), *Chem. Phys. Lett.* **181**, 141–150.

Khundkar, L. R., Zewail, A. H. (1990), *Annu. Rev. Phys. Chem.* **41**, 15–60.

Klimcak, C. M., Wessel, J. E. (1980), *Anal. Chem.* **52**, 1233–1239.

Kraus, A. S. (1977), *Science* **196**, 721–731.

Leone, S. (1985), *Science* **227**, 889–895.

Letokhov, V. S. (1979), *Nature* **277**, 605–610.

Letokhov, V. S. (1983), *Nonlinear Laser Chemistry*, Berlin: Springer-Verlag.

Letokhov, V. S., Moore, C. B. (1977), in: C. B. Moore, (Ed.), *Chemical and Biochemical Applications of Lasers*, Vol. 3, New York: Academic, pp. 1–165.

Leutwyler, S., Even, U., Jortner, J. (1980), *Chem. Phys. Lett.* **74**, 11–14.

Li, Leping, Yang, Bing-Xin, Johnson, P. M. (1985), *J. Opt. Soc. Am. B* **2**, 748-752.

Lineberger, W. C., Paterson, T. A. (1972), *Chem. Phys. Lett.* **13**, 40–44.

Mallow, B. R. (1969), *Phys. Rev.* **188**, 1969–1975.

Marinero, E. E., Jones, C. R. (1985), *J. Chem. Phys.* **82**, 1608–1609.

Marquardt, R., Quack, M. (1991), *J. Chem. Phys.* **95**, 4854–4876.

McClain, W. M. (1974), *Acc. Chem. Res.* **7**, 129–135.

Miller, R. J. D. (1991), *Annu. Rev. Phys. Chem.* **42**, 581–614.

Miller, T. M. (1981), *Adv. Elec. Electron. Phys.* **55**, 119–188.

Moseley, J. T. (1985), in: K. P. Lawley (Ed.), *Photodissociation and Photoionization*, New York: Wiley, pp. 245–298.

Mukamel, S. (1990), *Annu. Rev. Phys. Chem.* **41**, 647–681.

Osgood, R. M., Jr. (1983), *Annu. Rev. Phys. Chem.* **34**, 77–101.

Peticolas, W. L., Rieckhoff, K.E. (1963), *J. Chem. Phys.* **39**, 1347–1348.

Peticolas, W. L., Goldsborough, J. P., Rieckhoff, K. E. (1963), *Phys. Rev. Lett.* **10**, 43–45.

Petty, G., Tai, C., Dalby, F. W. (1975), *Phys. Rev. Lett.* **34**, 1207–1209.

Podlesnik, D. V., Gilgen H. H. (1989), in: K. G. Ibbs, R. M. Osgood (Eds.), *Laser Chemical Processing for Microelectronics*, Cambridge, U.K.: Cambridge University Press.

Rothberg, L. J., Gerrity, D. P., Vaida, V. (1981), *J. Chem. Phys.* **75**, 4403–4412.

Singh, S., Stoicheff, B. P. (1963), *J. Chem. Phys.* **38**, 2032–2033.

Sinha, A., Hsiao, M. C., Crim, F. F. (1990), *J. Chem. Phys.* **92**, 6333–6335.

Smalley, R. E. (1982), *J. Phys. Chem.* **86**, 3504–3512.

Staginnus, B., Frölich, D., Caps, T. (1968), *Rev. Sci. Instrum.* **39**, 1129–1130.

Weisz, S. Z., Zahlen, A. B., Gilreath, J., Jarnagin, R. C., Silver, M. (1964), *J. Chem. Phys.* **41**, 3491–3495.

Wilson, R. J., Houle, F. A. (1985), *Phys. Rev. Lett.* **55**, 2184–2187.

Zare, R. N. (1977), *Sci. Am.* **236** (2), 86–98.

Further Reading

Albertson, B. (1969), *Photochemical Processes*, Park Ridge, NJ: Noyes Data Corp.

Ambartzumian, R.V., Letokhov, V. S. (1977), *Acc. Chem. Res.* **10**, 61–67.

Andrews, D. L. (1986), *Lasers in Chemistry*, Berlin: Springer-Verlag.

Bauerle, D. (1986), *Chemical Processing with Lasers*, Berlin: Springer-Verlag.

Bloembergen, N., Yablonovich, E. (1978), *Phys. Today* **31** (5), 23–30.

Braun, A. M. (1991), *Photochemical Technology*, New York: Wiley.

Ibbs, K. G., Osgood, R. M. (Eds.) (1989), *Laser Chemical Processing for Microelectronics*, Cambridge, U.K.: Cambridge Univ. Press.

Letokhov, V. S. (1983), *Nonlinear Laser Chemistry*, Berlin: Springer-Verlag.

Luxon, J. T., Parker, D. E. (1987), *Lasers in Manufacturing*, Kempston, Bedford, U.K.: IFS.

Phillips, R. (1983), *Sources and Applications of Ultraviolet Radiation*, New York: Academic.

Ranney, M.W. (1975), *Irradiation in Chemical Processes: Recent Developments*, Park Ridge, NJ: Noyes Data.

Journals specializing in laser photochemistry include *Applied Physics B: Photophysics and Laser Chemistry* and *Laser Chemistry*.

LASER PHYSICS

P. L. Kelley, *Electro-Optics Technology Center, Tufts University, Medford, Massachusetts, U.S.A.*
J. J. Zayhowski, *MIT Lincoln Laboratory, Lexington, Massachusetts, U.S.A.*

	Introduction	299
1.	**Basic Theory of Operation**	299
1.1	Population Inversion and Stimulated Emission	299
1.2	Pumping and Relaxation Processes	300
1.3	Resonators and Cavity Modes	301
2.	**Important Characteristics of Laser Radiation**	302
2.1	Linewidth—Spectral Brightness	
2.2	Intensity and Directionality (Angular Confinement)—Spatial Brightness	302
2.3	Short Pulses—Temporal Brightness	302
3.	**Types of Lasers**	303
3.1	Solid-State Lasers	303
3.2	Gas Lasers	304
3.3	Dye Lasers	305
3.4	Semiconductor Lasers	307
3.5	uv and X-Ray Lasers	309
3.6	Free-Electron Lasers	310
4.	**Laser Dynamics**	310
4.1	Rate-Equation Model	310
4.2	Buildup from Noise	311
4.3	Threshold	311
4.4	Gain Saturation	312
4.5	Laser Efficiency	312
4.6	Multimode Operation	312
4.6.1	Spatial Hole Burning	313
4.6.2	Spectral Hole Burning	313
4.6.3	Single-Frequency Operation	313
5.	**Types of Pulsed Operation**	314
5.1	Long-Pulse Operation	314
5.1.1	Relaxation Oscillations	314
5.2	*Q*-Switched Operation	315
5.3	Gain-Switched Operation	315
5.4	Cavity-Dumped Operation	316
5.5	Mode-Locked Operation	316
5.5.1	Active Mode Locking	316
5.5.2	Passive Mode Locking	317
6.	**Control of Laser Output**	317
6.1	Frequency Tuning	317
6.2	Amplitude Modulation	318
7.	**Issues in Laser Design**	318
8.	**Frequency Conversion and Nonlinear Control of Laser Radiation**	318
	Further Reading	319

INTRODUCTION

The laser is a device that generates coherent, nearly single-wavelength (and -frequency), highly directional electromagnetic radiation somewhere in the wavelength range from submillimeter through ultraviolet and x-ray. The word *laser* is an acronym for "*l*ight *a*mplification by *s*timulated *e*mission of *r*adiation." The principle of operation of the laser is similar to that of the maser, which is somewhat arbitrarily defined as a device operating in the range from the radio or microwave region down to millimeter wavelengths. Since the first laser was operated in 1960, the laser has come to play an important role through its revolutionary impact on applied optical technology, including fiber optical communications and optical data storage.

1. BASIC THEORY OF OPERATION

1.1 Population Inversion and Stimulated Emission

Quantum theory shows that matter can exist in only certain allowed energy levels or states (see Quantum Mechanics). In thermal equilibrium, lower-energy states of matter are

1-56081-067-X/94/$5.00+.50

preferentially populated, with an occupation probability proportional to $e^{-E/kT}$, where E is the state energy, T the temperature, and k the Boltzmann constant. An excited state can decay spontaneously (i.e., with only zero-point electromagnetic radiation present) to a lower-energy state, emitting a quantum or wave packet of electromagnetic radiation (photon) with transition frequency $\nu=\Delta E/h$, where ΔE is the energy difference between the two states and h is Planck's constant. In the presence of radiation at frequency ν, a transition from the upper state to the lower state can be induced by the radiation, with the simultaneous emission of a photon in phase (coherent) with the stimulating radiation. This stimulated emission process is the reverse of the absorption process. If matter can be forced out of thermal equilibrium to a sufficient degree, so that the upper state has a higher population than the lower state (population inversion), then more stimulated emission than absorption occurs, leading to coherent growth (amplification or gain) of the electromagnetic wave at the transition frequency.

Laser transitions in the optical region are most often electric dipole in character. In the dipole approximation to the Hamiltonian, the transitions arise from a term of the form erE, where e is the electronic charge, r is the quantum-mechanical coordinate operator defined relative to the center of coordinates of the material system (such as an atom or molecule), and E is the electric field of the optical wave at the center of coordinates. The transition rate and gain cross section are proportional to the square of this interaction term. The transition matrix element of the coordinate operator between upper and lower laser levels ranges from about one hundredth of a Bohr radius ($\approx 0.5 \times 10^{-8}$ cm), for vibrational transitions in molecules and for local-field–induced transitions of rare earths in solids, to several hundreds of Bohr radii, for highly excited Rydberg atoms.

A laser generally consists of three components: an active medium with energy levels that can be selectively populated, a pump to produce population inversion between some of these energy levels, and (usually) a resonant electromagnetic cavity that contains the active medium and provides feedback to maintain the coherence of the electromagnetic field (see Fig. 1). In a continuously operating laser, coherent radiation will build up in the cavity to the level required to balance the stimulated emission and cavity losses (see Sec. 4.3). The system is then said to be lasing, and radiation is emitted in a direction defined by the cavity.

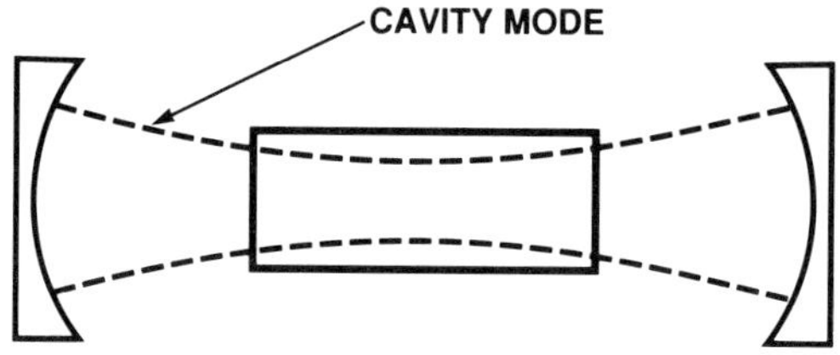

FIG. 1. Simplified schematic of a laser oscillator. The mirrors at the ends of the laser form an open resonator. Stable modes that consist of electromagnetic waves that travel back and forth in the resonator are amplified by the active laser material. In the radiative steady state, the gain due to amplification balances the loss due to intracavity absorption, mirror reflection losses, and diffraction beyond the edges of the mirrors. The pumping system is not shown, nor are ancillary intracavity elements that are often used for temporal (including frequency selection) and spatial control of the laser output. Usually, one of the mirrors is partially transmitting so that some of the highly directional radiation leaves the cavity through the mirror. The dashed lines are approximately characteristic of the transverse extent of the lowest-order transverse mode.

1.2 Pumping and Relaxation Processes

A material system can become excited and displaced from normal thermal equilibrium when driven by processes such as chemical reactions or under sufficiently strong external influence. External influences include electron beams and optical fields, which selectively excite energy levels of the material. Applied voltages can create electrical currents, also resulting in disequilibrium. Disequilibration, if by a sufficiently selective process, can result in population inversion and laser operation. This "pumping" can be carried out continuously, with single pulses, or with multiple pulses of excitation. The inversion and its duration depend on the relaxation rates for the different energy levels and the degrees of freedom of the system, as well as on the rate of stimulated emission. The energy-level scheme of the laser plays an important role in obtaining inversion; in Sec. 4.1 we will discuss the difference in the operation of three-level and four-level lasers.

Let us try to understand pumping and relaxation in an "ideal" four-level laser with the aid of Fig. 2. The pumping process, indicated by the upward arrow, is assumed to excite the system from the lowest energy level, denoted by g for ground state, to the highest level, denoted by e for excited state. Pumping might occur in a variety of ways, one of which could be through radiative excitation using light whose frequency coincides with the transition frequency between g and e. The state e is assumed to relax to the upper laser level, u. The population of the upper laser level is radiatively transferred, through either spontaneous or stimulated emission, to the lower laser level, l. Finally, the lower laser level can either relax to the ground state or absorb the laser radiation and repopulate the upper laser level.

Several conclusions concerning optimum operation can be made from this model. First, the relaxation rates from e to u and from l to g should be as rapid as possible in order to maintain the maximum population inversion between u and l. Second, the pumping rate between g and e should be sufficiently rapid to overcome the spontaneous emission from u to l. Third, the thermal equilibrium population of l should be as small as possible. Fourth, decay of e to any level other than u should be as slow as possible (for optical pumping, e can decay radiatively to g), and the nonradiative decay of u should be slow. For radiative pumping it is advantageous to have e be distinct from u and not to have radiative decay of u to g; for other types of pumping this advantage is not obvious.

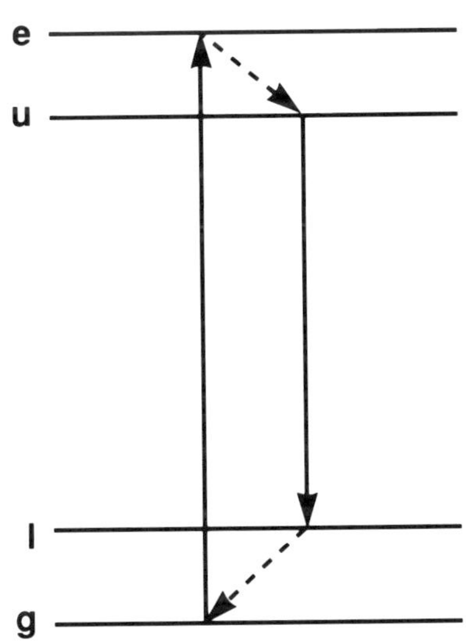

FIG. 2. Schematic representation of a four-level system. Population is pumped from g to e and laser operation occurs on the transition between u and l.

Not every practical laser satisfies this "ideal" model. The Cu-vapor laser violates the first conclusion listed in the preceding paragraph, since the lower laser level has a very slow decay rate and the laser self-terminates because of filling of this level, with concomitant reduction in the population inversion. Nevertheless, the Cu-vapor laser operates with fair efficiency as a powerful repetitively pulsed laser.

1.3 Resonators and Cavity Modes

An important aspect of the laser involves the design of resonators to accommodate the characteristics of the active medium and the diffractive properties of radiation, and at the same time meet requirements such as low angular beam divergence and high efficiency. The electromagnetic field in a resonator has well-defined modes that have patterns both transverse to and along the cavity axis. Waveguiding with index-of-refraction profiles or reflecting walls can be usefully employed in some cases (most notably in semiconductor and fiber lasers), not only to confine the radiation to the amplifying medium, but to force the laser to operate in a single transverse mode. More frequently, however, laser resonators are open in the sense that the transverse structure is defined only by axial mirrors or lenses (see Fig. 1). Open resonators formed with convergent optics ("stable" resonators) generally have the lowest diffraction losses, while planar resonators have higher losses, and resonators formed with divergent optics ("unstable" resonators) have the highest losses. Figure 3 shows a few of the lowest-order transverse-mode distributions for a stable resonator.

There is generally one transverse mode of a cavity that has the largest net gain (product of the amplifier gain and the transmittance of the remainder of the cavity). This is the transverse mode that oscillates first and is typically the lowest-order (TEM_{00}) mode (see Fig. 3). Lasers operating in the TEM_{00} mode have nearly optimal "spatial brightness" in the sense that the beam divergence is near the minimum value for the spot size of the laser on the output mirror; such a laser is said to

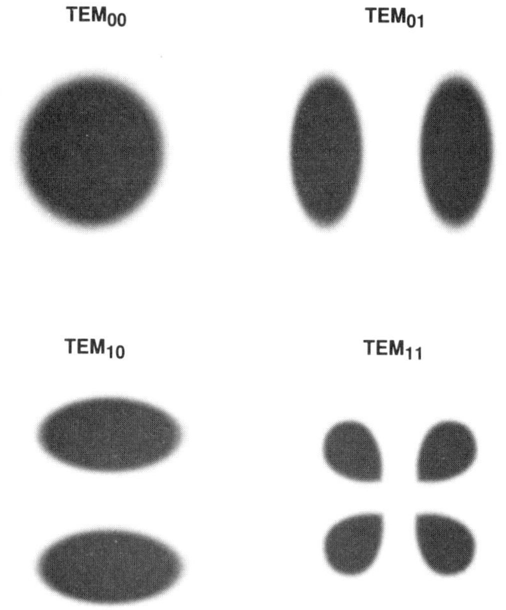

FIG. 3. Lowest-order transverse modes for a "stable" resonator with square symmetry. The TEM_{mn} notation indicates that the modes have nearly transverse electric and magnetic fields with m nodes vertically and n nodes horizontally.

have near-diffraction-limited output. The optimal resonator for a particular laser (e.g., "stable" or "unstable") is determined by the geometry of the gain medium, the desired cavity length, and the single-pass gain.

Cavity modes have a frequency periodicity that is determined primarily by the cavity length. The spacing between the axial (longitudinal) modes is the inverse of the round-trip time for radiation in the cavity. The gain in a laser is peaked at a transition frequency determined by the energy levels of the active medium, and laser operation tends to occur at the axial-mode frequency (or frequencies) closest to the gain peak.

2. IMPORTANT CHARACTERISTICS OF LASER RADIATION

2.1 Linewidth—Spectral Brightness

Lasers can have very narrow linewidths and, therefore, very high spectral brightness (power per unit spectral interval). Linewidths as narrow as a fraction of a hertz have been obtained (1-Hz stability corresponds to a fractional stability of about 2×10^{-15} for visible lasers).

The first step in achieving narrow-line operation is to design the laser to operate in a single longitudinal and transverse mode (see Sec. 4.6.3). Once single-mode operation has been obtained, the linewidth of the output is determined by technical and fundamental noise. Technical noise arises from sources that can be controlled, such as power-supply fluctuations, variations in the thermal environment, environmental vibrations, etc. Fundamental noise arises from sources that cannot be eliminated, such as spontaneous emission and fundamental thermal fluctuations. The random effect of noise causes the laser frequency to drift. The influence of noise can often be reduced by adjusting controllable parameters such as cavity lifetime. In addition, the frequency drift can be reduced by measuring the frequency of the laser and providing feedback to adjust the cavity mode frequency to compensate for the drift. The frequency can be measured in several ways: by using a stable Fabry-Pérot etalon, by comparing the laser frequency with a narrow resonance in an atom or molecule, or by comparing the frequency with a nearby frequency source.

2.2 Intensity and Directionality (Angular Confinement)—Spatial Brightness

As discussed in Sec. 1.3, lasers operating solely in the lowest-order transverse mode (TEM_{00}) have optimal spatial brightness and are described as operating in the diffraction limit. In this case, for propagation of initially collimated radiation in a nonconfining medium, the product $A\Omega$, where A is the laser output aperture and Ω is the solid angle into which the power is confined at long distances, has a minimum value of order λ^2, where $\lambda = c/\nu$ is the wavelength of the radiation and c is the velocity of light. This aperture–solid-angle product is invariant to transformations of the laser beam such as beam expansion. Because of this diffraction-limited character, single-mode laser beams can maintain high intensities over long distances and have a high degree of directionality (small solid angle).

2.3 Short Pulses—Temporal Brightness

Short pulses of laser radiation can be made in a variety of ways including gain switching, Q switching, mode locking, and nonlinear frequency broadening with pulse compression. These techniques are described in following sections. To obtain the shortest pulses, mode locking is used to obtain pulses in the picosecond (1 ps $= 10^{-12}$ s) range followed by frequency broadening with pulse compression to achieve pulses in the range of tens of femtoseconds (1 fs $= 10^{-15}$ s). Pulses as short as 6 fs have been obtained, and very high peak powers are possible.

High temporal brightness (high peak power) competes with high spectral brightness in the sense that the shortness of a pulse is limited by the bandwidth of the radiation. Said another way, for a fixed average power in a continuous train of bandwidth-limited pulses, peak power cannot be increased without reducing spectral brightness.

3. TYPES OF LASERS

A rich variety of physical systems has been exploited to produce laser radiation over a five-decade range of wavelengths. We can only briefly describe here some of the most significant of these. More details can be found in the articles on specific types of lasers.

3.1 Solid-State Lasers

(See LASERS, SOLID STATE). The first laser used a rod of ruby (single-crystal Al_2O_3 with the ion Cr^{3+} substituted for a small fraction of the Al^{3+} ions) as the active medium. In this laser, a xenon flashlamp pumps the chromium ions from their ground state to a broad band of states, from which they rapidly decay nonradiatively to a long-lived state at an energy of about 14 422 cm^{-1} (693.4 nm) above the ground state. (In laser physics, energy E and frequency ν are often given in terms of wave number $1/\lambda$, where λ is wavelength, $E=hc/\lambda$, $\nu=c/\lambda$, h is Planck's constant, and c is the speed of light in vacuum.) It is the narrow ($\approx$11 cm^{-1}) emission line from this level back to the ground state that gives rise to ruby laser emission. When lasing occurs, the emission narrows in frequency width to less than 1 cm^{-1} and leaves the ruby rod with an angular spread of a few milliradians.

The ruby laser is a three-level system (see Fig. 4 and Sec. 4.1), and therefore requires depopulation of the ground state by more than 50% to obtain population inversion. For this reason, pumping of the laser requires a high-intensity source, and continuous-wave (cw) operation of the system is difficult to achieve. In the free-running mode, as opposed to the Q-switched mode (see below and Sec. 5.2), the output radiation of the ruby laser fluctuates rapidly over a time of about 1 ms and, for a 1-cm-diam$\times$10-cm-long laser rod, is emitted in pulses of about 1 J.

The high threshold pump-power requirement of a three-level laser is greatly relaxed in a four-level system. In the four-level system, radiative transitions do not terminate on the ground state (see Fig. 2). If the final state is not significantly populated at the operating temperature, then population inversion can be maintained with only moderate pump power, and such lasers can usually be operated on a continuous basis.

One type of four-level solid-state laser is based solely on the electronic levels of ions in a crystal or other solid-state host, with the laser transition terminating on an excited electronic level of the ion. The rare-earth ions Nd^{3+}, Ho^{3+}, Er^{3+}, and Tm^{3+} are the most frequently used active ions in such lasers. Extensively developed lasers of this type are

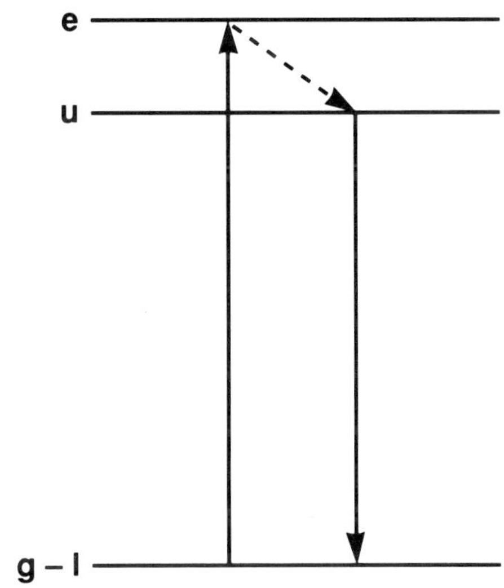

FIG. 4. Schematic representation of a three-level system. Note that g and l are now the same level.

the Nd-glass laser and the Nd:YAG (yttrium aluminum garnet, $Y_3Al_5O_{12}$) laser, both of which have their most important transitions in the vicinity of 1.06 μm. Nd:YAG lasers have produced in excess of 1 kW in continuous operation, several joules per pulse in low–repetition-rate (10-Hz) pulsed operation, and about 10 mJ per pulse in high–repetition-rate (1-kHz) pulsed operation. In order to increase efficiency and decrease heat loading in the solid-state laser medium, Nd:YAG lasers have been pumped with arrays of semiconductor diode lasers; up to 300 W of continuous output has been obtained. Large Nd-glass lasers have produced tens of joules per pulse at repetition rates of several hertz, and several kilojoules per pulse at very low repetition rates ($\approx$ 1 per hour). Rare-earth–doped glass-fiber lasers are of utility in optical communications; Er^{3+} fiber lasers have given output powers of the order of 100 mW and are in use as amplifiers in long-distance fiber transmission.

A second type of four-level solid-state laser involves fewer electronic levels of the active ion. Instead, laser transitions are employed that terminate on an excited vibrational level of the ion in the host lattice. The basic optical pumping–emission cycle is similar to that of a molecular dye laser (see Sec. 3.3). One important class of these solid-state lasers uses transition-metal ions such as Cr^{3+}, Ti^{3+}, Ni^{2+}, and Co^{2+}. As with the ruby (Al_2O_3) laser, the alexandrite ($BeAl_2O_4$) laser employs the Cr^{3+} ion. In the latter case, however, the laser transition can terminate on a variety of final vibrational states. As a result, this laser is tunable from 700 to 818 nm and has pulsed output energies similar to the ruby laser. The Ti:Al_2O_3 laser has even broader tunability, covering a range from 660 nm to beyond 1.1 μm. It has operated continuously at power levels up to 17 W with Ar-laser pumping, and pulsed at energies of hundreds of millijoules per pulse when pumped by doubled Nd:YAG laser radiation.

Lasers based on color centers in alkali halide crystals operate on a principle similar to dye lasers (see Sec. 3.3) and transition-metal lasers. Using different types of *F* centers in various alkali halides, wavelength coverage over a range from 0.82 to 3.3 μm can be obtained, with continuous output powers ranging from tens of milliwatts to over a watt. Stability of *F* centers can be a problem, and low-temperature storage is required for several of these lasers.

The energy stored in a pumped solid-state laser medium can be delivered as a giant pulse in a time much shorter than the spontaneous lifetime of the upper laser energy level through the use of *Q* switching (quality factor switching). This technique (see Sec. 5.2), which was first used in the ruby laser, has been applied to many laser systems. In the case of the Nd:YAG laser, *Q*-switched pulses are typically of the order of 10 ns in duration and can be obtained with the use of a saturable dye absorber, an electro-optic element, or an acousto-optic element as the switch in the laser cavity.

Because solid-state lasers have spectrally broad gain regions, many equally spaced longitudinal cavity modes can lie within the gain bandwidth when cavity lengths are of the order of tens of centimeters. As a result, mode locking (see Sec. 5.5) can be used to obtain a train of high-intensity, short pulses. Mode locking has been applied to many solid-state laser systems such as Nd:YAG (20-ps pulses) and Ti:Al_2O_3 (200-fs pulses). Mode locking has also been used in many other types of lasers (see, e.g., Sec. 3.3 on dye lasers).

3.2 Gas Lasers

A number of methods can be used to produce population inversion in gaseous media. Inversion can exist between some of the energy levels of the constituents in a gas discharge (electrical discharge in a gas) (see LASERS, GAS). The first such system, demonstrated not long after the announcement of the ruby laser, was the He-Ne laser, now a standard item in optics laboratories. This system makes use of a discharge in He at a pressure of about 1 Torr, with an admixture of Ne at about 0.1 Torr. The discharge excites He atoms to their first excited level, about 160 000 cm^{-1} above their ground state. This excitation is readily transferred by collisions to a Ne atomic level with nearly the same excitation energy (resonant transfer). These excited states decay radiatively to lower-energy Ne states, giving rise to continuous laser emission in the red at 15 820 cm^{-1} (632.8 nm) with an output power of about 10^{-2} W. Other transitions produce strong emission at 8 680 cm^{-1} (1.15 μm) and 2957 cm^{-1} (3.39 μm). In a pure Ne discharge, excitation would occur to many Ne levels, and population inversion would not occur as effectively.

An important gas-discharge laser is based on the energy levels of the argon ion (Ar^+). Through a complex series of steps, argon-ion–electron collisions in the discharge lead to population inversion and lasing at a number of frequencies near 20 500 cm^{-1} (488 nm). Continuous output at power levels of tens of watts in the blue-green make this device especially useful as a spectroscopic source in Raman scattering and for pumping continuously operating tunable dye lasers.

Other intense laser sources arise from atomic transitions of metal ions in a pulsed He discharge (the Cu and Cd vapor lasers, for example). Gas-discharge lasers have also been made to operate in the uv, but special problems arise in this frequency range; these are discussed below in Sec. 3.5.

Gas lasers with output at longer wavelengths make use of the vibration–rotational energy levels of molecules (see MOLECULAR SPECTROSCOPY). In addition to the electronic-state energy-level structure characteristic of atoms, there is vibration–rotational structure associated with the relative motion of the nuclei. The spacing of vibrational energy levels corresponds to frequencies in the infrared. It is in this region of the electromagnetic spectrum that molecular gas-discharge lasers are especially important.

The most efficient and powerful of the molecular gas-discharge lasers is the CO_2 laser. One version of this laser makes use of a dilute mixture of CO_2 in a N_2 discharge. The N_2 molecules are excited by collisions with electrons to their first excited vibrational state, from which the excitation is resonantly transferred by molecular collisions to excite CO_2 molecules preferentially to a particular vibrational state. These molecules in turn undergo radiative transitions to lower vibrational levels. The presence of the rotational structure gives rise to a cluster of many lines that can lase, grouped near frequencies of 944 cm^{-1} (10.6 μm) and 1042 cm^{-1} (9.6 μm). The electric-discharge CO_2 laser is quite efficient (better than 10% electrical power converted to laser power) and is capable of producing continuous output powers of greater than 1 kW. Other important molecular gas-discharge lasers make use of vibration–rotational or pure rotational transitions of H_2O, CO, and HCN, and produce emission at 78 and 119 μm, 5.3 μm, and 337 and 311 μm, respectively.

A number of other methods of generating laser radiation using molecular or atomic energy levels in gases have been devised. Powerful pulses of laser radiation at 1.3 μm from excited iodine atoms have been produced by flash photolysis (uv photodecomposition) of CH_3I (methyl iodide). In the gas-dynamic laser, a nonthermal distribution of molecular vibrational energy levels is produced by the rapid expansion of a hot gas through a nozzle. This method has produced continuous emission of tens of kilowatts at 10.6 μm from CO_2 gas. In the chemical laser, two reacting molecular species in a gas produce a product that is left in an excited vibration–rotation state and returns to the ground state radiatively. An example is the HF (DF) laser, which lases in the 2.5–3.5-μm region when H_2 (D_2) and F_2 gases combine chemically. Other types of lasers produced by excitation of gases are TEA (transverse electric discharge excited) lasers, *e*-beam (electron-beam excited) lasers, and uv-preionized electric-discharge lasers.

CO_2 laser radiation has been used to pump other gases, yielding far-infrared emission. If there is a coincidence between a CO_2 laser line and a vibration–rotational transition in another gas, an excited level can be populated directly or by collisional transfer of excitation. Pure rotational transitions can take place radiatively to an unpopulated level, producing far-infrared laser radiation. Such gases as NH_3 (output at 291 μm) and CH_3OH (output at 164 and 205.3 μm) have been made to lase by this method.

3.3 Dye Lasers

Many organic dyes, when illuminated with visible or uv radiation, fluoresce strongly at lower frequencies. This so-called Stokes shift of the fluorescence can be understood in terms of the change of equilibrium internuclear position with electronic excitation, the rapid vibrational relaxation within an electronic state, and the Franck–Condon principle (see MOLECULAR SPECTROSCOPY), which states that an electronic transition in a molecule takes place so rapidly that the nuclear coordinates can be regarded as nearly fixed. An optical pumping–fluorescence cycle for a molecular system is indicated schematically in Fig. 5(a). After absorption of a photon and nonradiative vibrational cascading to the lowest vibrational level, the excited electronic state can decay by fluorescence at a lower frequency, and the molecule returns to the ground state by nonradiative vibrational cascading (see ORGANIC DYE LASERS). Since the final state in the radiative transition process is unoccupied, the pumping requirements for such a laser system need not be too severe. Discrete vibrational structure is washed out in a liquid, but the general outline of the cycle indicated in Fig. 5(a) is still preserved for dye fluorescence in a liquid. This process is virtually identical to that occurring in tunable transition-metal solid-state lasers such as Ti:Al_2O_3 or Cr:$BeAl_2O_3$. In the solid, the substitutional-ion's coordinates with respect to neighboring host ions play the same role as the internuclear coordinates in the dye molecules.

A problem that arises with the use of dyes in a laser is the existence of long-lived electron spin-triplet states into which the usual excited electron spin-singlet state can relax nonradiatively. (For rhodamine 6G, a commonly used dye, this nonradiative relaxation time is $\approx 10^{-7}$ s.) This process interferes with laser action, but can be largely circumvented by using dye jets to circulate the dye solution through the laser cavity. Dye lasers can be flashlamp pumped or pumped with radiation from an Ar-ion, frequency-doubled Nd:YAG, or N_2 laser. Dye systems fluoresce over a wide band of frequencies from the near infrared through the visible, and are well adapted for use in a cw laser, with wavelength tuning over as much as 40 nm for a single dye. A battery of dyes placed in optical cavity structures gives tunable laser radiation over a range from roughly 10 000 cm^{-1} (1 μm) to 25 000 cm^{-1} (400 nm).

Because of the broad bandwidth of dye fluorescence lines and, in the case of pulsed dye lasers, the low Q of the cavity, the dye-laser emission line is fairly broad. It can, however, be greatly narrowed without much loss in output power through the use of a diffraction grating in place of one of the mirrors in the cavity. Even narrower linewidths may be obtained by using intracavity frequency-selective elements (etalons). Single-longitudinal-mode operation can be obtained in both pulsed and cw dye lasers. Dye-jet fluctuations normally limit the stability of cw dye lasers, although stabilities of hundreds of kilohertz are readily obtained and stabilities

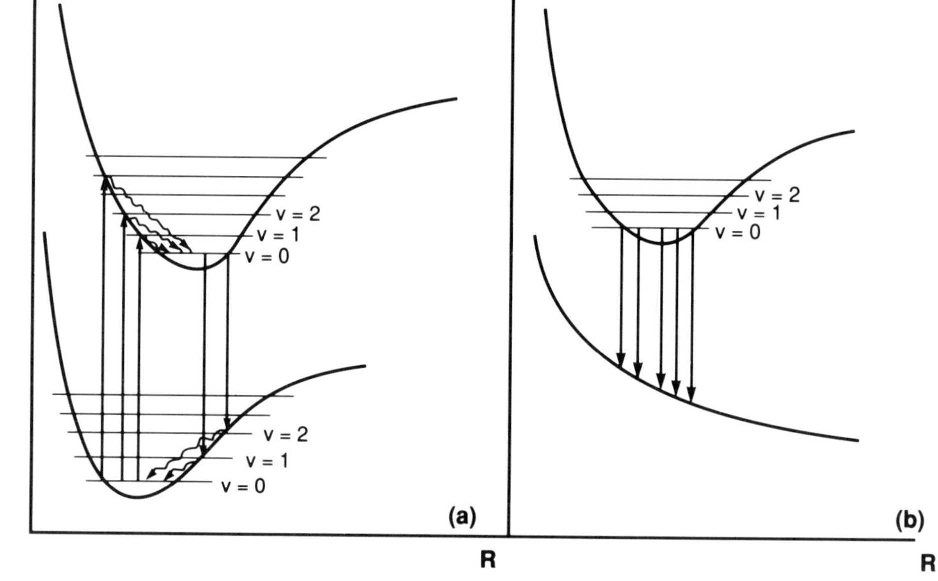

FIG. 5. Light emission from molecular systems: (a) bound–bound system; (b) bound–free system. The curves give the variation of electronic energy (vertical axis) as a function of internuclear coordinate (horizontal axis). The light horizontal lines indicate the location of the vibrational levels and are labeled by vibrational quantum number.

of a fraction of a hertz have been achieved. Thus the dye laser is an important source of radiation for spectroscopy.

Since the dye laser has a spectrally broad gain region and many equally spaced longitudinal cavity modes falling within the gain bandwidth, it is well suited for mode-locked operation (see Sec. 5.5). Using continuous mode-locked dye lasers, pulse trains having pulses of less than 10 fs duration have been obtained. These short pulses are used for studies of fast electronic processes in solids and organic molecules.

3.4 Semiconductor Lasers

In 1962, it was reported that a forward-biased semiconductor diode of GaAs radiated efficiently at about 11 800 cm^{-1} (850 nm). In the following year, a number of groups reported the observation of laser emission in this frequency region from suitably prepared diode structures. While these semiconductor devices are solids, they are differentiated from those lasers involving optically active ions in ionic hosts on account of their markedly different physical and technological characteristics (see LASERS, SEMICONDUCTOR).

To understand how semiconductor lasers function, it is necessary to consider the nature of the electronic energy states in a semiconductor (see the various articles on semiconductors). A periodic crystal has bands of allowed energy levels separated by forbidden energy gaps. In an intrinsic semiconductor at low temperatures, there are just enough electrons present to fill the uppermost occupied energy band (valence band), leaving the next higher band (conduction band) empty. In an *n*-type semiconductor, impurity atoms (donors) are present that contribute electrons to the conduction band; in a *p*-type semiconductor, there are impurity atoms (acceptors) present that can bind electrons, leaving behind missing electrons (holes) in the valence band.

Figure 6 shows a schematic of a *p-n* junction, fabricated by forming *p* and *n*-type semiconductor layers in intimate contact. When a voltage is applied in the forward direction, electrons are injected from the *n* region into the depletion region of the junction (a region of the junction about 1 μm thick). At the same time, holes are injected from the *p* region. As electrons drop into hole states they may emit radiation (electron–hole recombination radiation) at a frequency in the vicinity of the energy gap ($\nu = E_g/h$). When the injection current density is sufficiently high, the electron–hole pair density becomes sufficiently large that gain is induced.

A typical semiconductor diode-laser heterostructure is shown schematically in Fig. 7. The dielectric film serves to guide the current into a narrow stripe region in order to concentrate the electron–hole pair density. The x coordinate shown in Fig. 6 corresponds to the vertical direction in Fig. 7.

In the diode (or injection) laser, a cavity structure (typically a fraction of a millimeter in dimensions) is provided by plane-parallel, cleaved faces at right angles to the junction plane. Laser emission is perpendicular to the cleaved facets, once the diode injection current reaches a threshold value. Multimode behavior often occurs in the plane of the

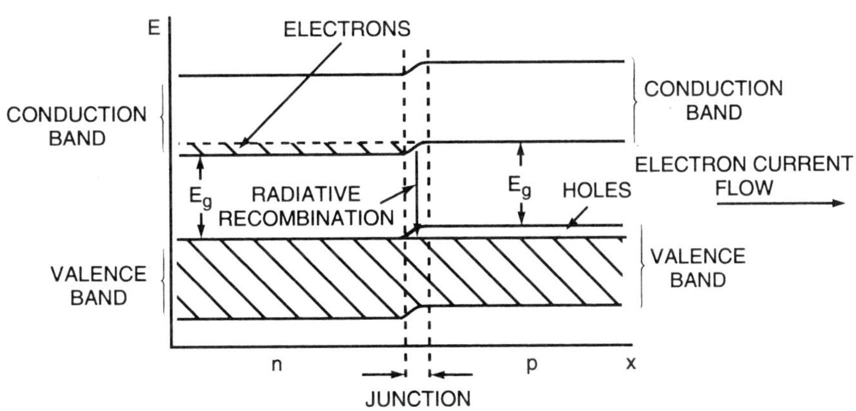

FIG. 6. Schematic showing the emission of radiation from a forward-biased light-emitting diode.

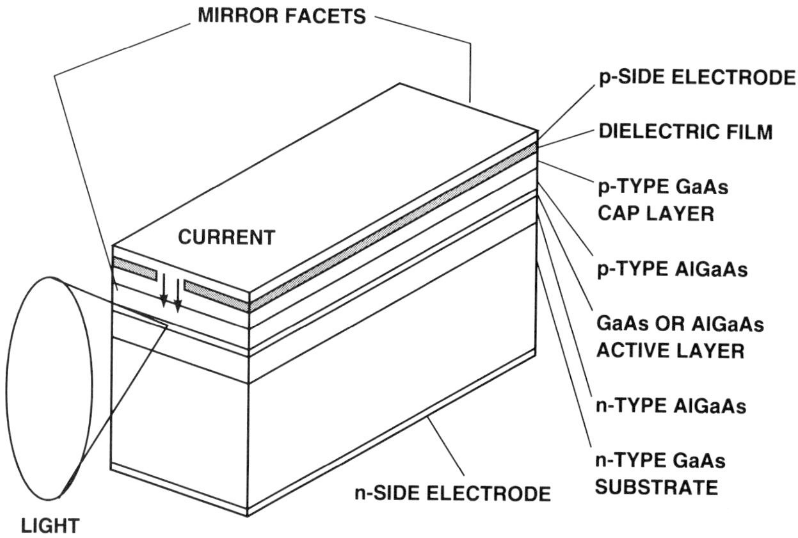

FIG. 7. Typical semiconductor diode-laser heterostructure. The material compositions shown produce lasers operating in the 750–850-nm wavelength range. The device length is typically 250 μm, the width 50–100 μm, and the height of the order of 50 μm. The epitaxial layers (i.e., excluding the substrate and the contacts) contribute only a few micrometers to the height. For thermal and electrical connection, the device is soldered to an electrically conducting heat sink, and an electrical lead is attached to the top side of the device. The width of the emitting region is 10–20 μm, while the height is approximately 0.5 μm; this causes the beam to diverge rapidly in the vertical direction.

junction. Since stimulated emission occurs in a narrow area near the junction (a few to several hundred micrometers in the plane of the junction and a fraction of a micrometer perpendicular to the junction), the angular spread of emitted radiation is fairly large, as expected from consideration of diffraction.

The technology of diode lasers has undergone considerable development, with the primary goals of achieving continuous room-temperature operation, low thresholds, high output powers, improved mode quality, wavelength diversity, and long lifetimes. Progress has included improvements in electrical and optical confinement and closer coupling of the gain region to the heat sink. Much of the improvement has involved advances in materials growth, including molecular-beam epitaxy (MBE) and metal-organic chemical vapor deposition (MOCVD). These growth techniques have resulted in material of exceptionally high quality as well as structures with quantum confinement of carriers (electrons and holes).

Quantum-well diode lasers have exceptional optical and electrical properties including very low laser thresholds. By reducing the volume of the active region through the use of a quantum well, the amount of current required to bleach the absorption of the semiconductor (the transparency current) is reduced, with a commensurate reduction in threshold. In addition, the quantum confinement of carriers in the active region increases their radiative gain cross section. In strained quantum-well devices the density of states in the valence band is reduced, leading to a further reduction in the transparency current and reducing parasitic effects such as Auger recombination, which is important at long wavelengths.

In 1969 (before the advent of quantum-well devices), room-temperature continuous operation was achieved in a GaAlAs double-heterostructure laser. This laser structure consists of a small region of GaAs sandwiched between *p*- and *n*-type layers of the wider-bandgap alloy $Al_xGa_{1-x}As$ ($x<1$). With further development, device lifetimes of tens of years were obtained, with output powers in the tens of milliwatts. These improvements opened up applications of considerable significance—in particular, fiber-optical communication and optical data storage. GaAlAs diode lasers have also been operated continuously at room temperature with output powers in the range of several watts. Linear arrays of diodes, in the form of multiple stripes in a 1-cm bar, have given a total output of over 50 W. In addition, electrical-to-optical power conversion efficiencies of greater than 50% have been obtained.

Continuous laser operation at room temperature has also been obtained in the quaternary alloy system InGaAsP. Because laser operation is much less sensitive to dislocation effects in this system, long-lifetime devices were readily achieved. When operating at 1.3 and 1.55 μm, these lasers are matched to low-loss (< 1 dB km^{-1}), low-dispersion fiber optics, and are currently used in high–data-rate, long-distance communication.

The compositionally tuned lead salt lasers ($Pb_x Sn_{1-x}Te$, $PbS_x Se_{1-x}$) operate at cryogenic temperatures. As with other semiconductor systems, changes in composition change E_g and, therefore, the frequency of laser emission. Furthermore, since E_g and hence the dielectric constant at near-bandgap wavelengths are sensitive functions of temperature in these small-gap semiconductor systems, cavity-mode frequencies and gain peaks are tunable, giving rise to temperature-tunable laser output. These tunable sources have been used extensively for high-resolution infrared spectroscopy in the 5–20-μm region.

In the visible region, using the quaternary alloy AlGaInP, continuous room-temperature operation has been obtained at 640 nm, while room-temperature pulsed operation has been observed at wavelengths as short as 603 nm; nitrogen-temperature continuous operation has been achieved at 583.6 nm, which is in the yellow region of the spectrum. Recently, operation in the blue-green region has been achieved in $Zn_{1-x}Cd_x Se/ZnS_y Se_{1-y}$ quantum-well heterostructure diodes. These lasers have been operated continuously at 77 K and pulsed at room temperature. Here again, advances in materials growth technology has been primarily responsible for progress in this material system.

In addition to injection pumping of semiconductor lasers, a number of other methods not requiring the fabrication of a junction have been used successfully. These include electron-beam pumping and optical pumping.

3.5 uv and X-Ray Lasers

To produce stimulated emission in the uv and x-ray regions, special problems must be addressed. Short lifetimes of inverted populations become important, since the spontaneous radiative lifetime varies with frequency as ν^{-3}. For example, the nitrogen gas-discharge laser, which radiates in the near uv (337 nm), can only operate in a pulsed mode, and must be pumped by a powerful intermittent source. Another problem is associated with the difficulty of devising resonant structures, since the reflectivity of materials becomes very small in the vacuum-uv and x-ray regions. A different conceptual scheme must frequently be used in these spectral regions, involving directionally amplified spontaneous emission from an inverted population (superradiance), where the directional amplification is achieved by the geometry of the pumped region.

The nitrogen and hydrogen discharge lasers make use of radiative transitions between two bound molecular electronic levels (bound-bound transitions) occuring at 337 nm and 160 nm, respectively. On the other hand, there are molecular systems (excimers) in which radiative transitions occur between a bound excited state and a free or very weakly bound ground state [bound–free transitions—see Fig. 5(b)]. Xe and Kr form the excited molecular states Xe_2^* and Kr_2^*, although the diatomic molecules are unstable in their ground states. At high pressure these gases, when pumped by powerful electron-beam sources, emit superradiantly at 172 and 145.7 nm, respectively. At somewhat longer wavelengths, rare-gas–halide excimers, such as ArF* at 248 nm, KrF* at 193 nm, and XeCl* at 308 nm, have been operated both by electron-beam pumping and by transversely excited discharge. These lasers can produce multijoule pulses at 100-Hz repetition rates and are commercially available.

X-ray lasers at wavelengths shorter than a few tens of nanometers present a more difficult challenge. Nevertheless, several groups have achieved laser operation in this region. A multijoule visible laser can directly pump the gain medium through inversion produced in a laser-generated plasma, or indirectly by using the x rays from a laser plasma to pump a separate x-ray laser medium. The first approach has been used to achieve operation near 21 nm in Se XXV, which has a $1s^2 2s^2 2p^6$ Ne-like electron configuration. Current research involves efforts to improve the very low efficiency of x-ray lasers as well as studies of potential new x-ray laser systems.

3.6 Free-Electron Lasers

Relativistic electrons traveling in a periodically alternating transverse magnetic field (wiggler) can be stimulated to give up radiation to a copropagating electromagnetic field of wavelength $\lambda = \lambda_w/2\gamma^2$, where λ_w is the wiggler period and γ is the ratio of the electron energy to its rest mass energy. This phenomenon can be pictured as stimulated inverse Compton scattering from the electromagnetic field of the wiggler seen in the electron rest frame (see LASERS, FREE-ELECTRON). To obtain efficient conversion over reasonable distances with high output power, the electron beam must be very monoenergetic, constitute a high current, and have very low angular divergence. Efficient operation also requires either the use of an electron storage ring or a tapered-period wiggler. Free-electron lasers have been operated at several wavelengths ranging from the visible to millimeter waves. While free-electron lasers do not have the wavelength restrictions imposed on other lasers and do not have the inhomogeneities characteristic of many laser media, they require electron beams of a current level and quality that are difficult to achieve simultaneously. Furthermore, the electron beam and wiggler requirements tend to become more demanding as wavelength decreases. These systems have the large sizes that are associated with relativistic electron-beam sources.

4. LASER DYNAMICS

4.1 Rate-Equation Model

Many of the properties of the laser can be determined from a rate-equation model for the population of the laser levels and the photon number in the laser cavity. The rate equations provide a simple and intuitive, yet accurate, picture of the behavior of many lasers. In the most simplified form, the increase in photon number within the laser cavity is balanced by the decrease in the population difference between the upper and lower laser levels. In addition, the population difference increases on account of the pumping process, while the photon number decreases because of absorption, diffraction of the beam out of the cavity, and transmission through the mirrors.

The rate-equation model can be derived as an approximation to the fundamental equations relating the electromagnetic field, the material polarization, and the populations. The validity of the rate equations requires that the material polarization can be accurately approximated by assuming that it instantaneously follows the field; this is a situation that applies to most lasers. In order to describe the problem in terms of total population and total photon number within the laser cavity, it is necessary that the gain of the laser be small during one pass through the cavity and that the laser operate in a single longitudinal mode (see Sec. 4.6.1).

As stated above, the number of photons q within the laser cavity is affected by two types of events: the emission of a photon by the gain medium ($\dot{q}_{\rm emm}$) and the escape of a photon from the cavity or absorption by unpumped transitions ($\dot{q}_{\rm dec}$). Photon emission can be either stimulated ($\dot{q}_{\rm stim}$) or spontaneous ($\dot{q}_{\rm spon}$). Once a laser is above threshold the stimulated emission rate is much greater than the spontaneous rate and, to first order, spontaneous emission can be ignored. We will return to the issue of spontaneous emission in the next section when we discuss the buildup of a laser from noise.

The stimulated emission rate is proportional to the number of photons within the cavity, the total population inversion N, and the probability per unit time B that a given photon will interact with a given inverted site. The interaction probability B is the product of the probability that a photon will pass within the gain cross section σ of a given inverted site as it traverses the laser cavity and the number of times the cavity is traversed. Mathematically, this reduces to $B=(\sigma/A)(c/l)=\sigma c/V$, where A is the cross-sectional area of the laser mode, l is the cavity length, c is the average speed of light in the cavity, and V is the volume of the lasing mode. The stimulated emission rate is, therefore, $\dot{q}_{\rm stim}=qN\sigma c/V$.

The escape of photons from the laser cavity and their absorption within the cavity are characterized by the cold-cavity lifetime τ_c (cavity lifetime in the absence of any inversion) and correspond to $\dot{q}_{\rm dec} = q/\tau_c$. The total rate of change of the number of photons within the laser cavity is thus

$$\dot{q}=\dot{q}_{\text{stim}}-\dot{q}_{\text{dec}}=qN\sigma c/V-q/\tau_c. \quad (1)$$

We will now derive the rate equations for the population inversion of both a four-level and a three-level laser. The energy-level diagrams of a four- and a three-level laser are shown in Figs. 2 and 4. In both cases, the pump excites the active medium from the ground level (with population N_g) to level e (N_e). It is then assumed that the excited state quickly decays to the upper laser level, level u (N_u), so that $N_e \approx 0$. Lasing occurs between levels u and l (N_l). The difference between a four- and a three-level laser is that for the three-level laser the lower laser level is also the ground state.

To derive the rate equation for the population inversion $N=N_u-N_l$, we start by considering the population of the upper laser level N_u. The population of the upper level is affected by pumping P_r, stimulated emission $\dot{N}_{u,\text{stim}}$, and spontaneous decay $\dot{N}_{u,\text{spon}}$. For most pumping schemes, the pump rate is proportional to the number of ions in the ground state and can be written as $P_r = W_pN_g$. The stimulated-emission process decreases the population of the upper laser level by one for every photon created, so that $\dot{N}_{u,\text{stim}}=-\dot{q}_{\text{stim}}$. Spontaneous decay is characterized by the spontaneous lifetime τ, corresponding to $\dot{N}_{u,\text{spon}}=-N_u/\tau$. The rate equation for the upper population level is thus

$$\begin{aligned}\dot{N}_u&=P_r+\dot{N}_{u,\text{stim}}+\dot{N}_{u,\text{spon}}\\&=W_pN_g-qN\sigma c/V-N_u/\tau.\end{aligned} \quad (2)$$

In an ideal four-level laser there is a very rapid decay of the lower laser level to the ground state, so that N_l and $\dot{N}_l$ are approximately equal to zero. Since the total number of active ions N_t is constant, $N_t \approx N_g + N_u$. The rate equations for a four-level laser are, therefore,

$$\dot{q}=(N\sigma c/V-1/\tau_c)q, \quad (3)$$

$$\dot{N}=W_p(N_t-N)-qN\sigma c/V-N/\tau. \quad (4)$$

Since, in an ideal three-level laser, the lower laser level is the ground state, $N_t \approx N_l + N_u$ and $\dot{N}_l \approx -\dot{N}_u$. As a result, the rate equations for a three-level system reduce to

$$\dot{q}=(N\sigma c/V-1/\tau_c)q, \quad (5)$$

$$\dot{N}=W_p(N_t-N)-2qN\sigma c/V-(N_t+N)/\tau. \quad (6)$$

The photon rate equations [Eqs. (3) and (5)] for the four-level and three-level lasers are the same. The rate equations for the population inversion are different, however. In particular, the stimulated emission term for a three-level laser is twice that of a four-level laser.

4.2 Buildup from Noise

In the photon rate equation derived in Sec. 4.1, the term corresponding to spontaneous emission was left out. Laser action is initiated by spontaneous emission, or noise. As a result, these rate equations cannot account for the onset of lasing, as is seen by setting $q = 0$ at time $t = 0$. When spontaneous emission is properly taken into account, the photon rate equation becomes

$$\dot{q}=(qN+N_u)\sigma c/V-q/\tau_c. \quad (7)$$

The net effect is as if there were initially one photon in the cavity. This one photon of noise stimulates optical transitions and initiates lasing.

4.3 Threshold

The threshold inversion required for lasing is derived by requiring that the photon rate equation have a nontrivial solution in steady state. In steady state $\dot{q} = 0$, resulting in the condition $qN\sigma c/V - q/\tau_c = 0$. Physically, this states that the number of photons leaving the cavity must be balanced by the number of photons generated though the stimulated emission process. The threshold inversion is, therefore,

$$N_{\text{thresh}}=V/\sigma c\tau_c. \quad (8)$$

The pump rate W_p required to reach threshold is obtained by setting $\dot{N} = 0$, $\dot{q} = 0$, and $N = N_{\text{thresh}}$. For a four-level laser,

$$W_{p,\text{thresh}}=N_{\text{thresh}}/(N_t-N_{\text{thresh}})\tau\approx V/N_t\sigma c\tau_c\tau, \quad (9)$$

where we have assumed that $N_{\text{thresh}} \ll N_t$. For a three-level laser

$$W_{p,\text{thresh}}=(N_t+N_{\text{thresh}})/(N_t-N_{\text{thresh}})\tau\approx 1/\tau. \quad (10)$$

For the same value of τ, the threshold pump rate for a four-level laser is smaller than the threshold pump rate for a three-level system by a factor of $V/N_t\sigma c\tau_c$, which is usually quite large. This is the basis of the superior performance of a four-level system over a three-level system in cw operation.

4.4 Gain Saturation

The gain g of an active medium is defined as the fractional change in optical intensity per unit length as a light beam passes through. From the discussion in Sec. 4.1, it follows that $g = N\sigma/V$. In the presence of a strong optical field the population inversion is reduced and the gain is saturated. The rate equation for the population inversion (of a three- or four-level gain medium) in steady state can be rewritten in the form

$$g=g_0/(1+I/I_s), \tag{11}$$

where g_0 is the unsaturated gain (the gain in the absence of an optical field), $I=h\nu qc/V$ is the optical intensity ($h\nu$ is the energy of one photon), and I_s is known as the saturation intensity.

The steady-state photon rate equation ($\dot{q} = 0$) predicts that above threshold ($q\neq 0$) the inversion density and the gain of a laser are clamped at their threshold values; the round-trip gain of the cavity is equal to the round-trip loss. With increased pumping, the gain remains fixed while the photon number and the output of the laser increase.

4.5 Laser Efficiency

The efficiency of a laser is often discussed in terms of the slope efficiency η_s. Slope efficiency is defined as the ratio of the change in output power to the change in pump power of a laser once it has reached threshold, and is determined by four factors: the pump efficiency η_p, the area efficiency η_a, the quantum efficiency η_q, and the output-coupling efficiency η_o. Mathematically,

$$\eta_s=\eta_p\eta_a\eta_q\eta_o. \tag{12}$$

The pump efficiency η_p is the ratio of the energy absorbed by the gain medium to the energy of the pump source. In an optically pumped laser, part of the incident optical energy may be reflected by the gain medium, and part may be transmitted. Both of these effects decrease the pump efficiency. The area efficiency η_a is a measure of how well the pumped volume is used by the oscillating mode. If the cross section of the pumped volume is much larger than the cross section of the lasing mode, only a small portion of the pumped volume contributes to the gain of the system, and the area efficiency will be low. The quantum efficiency η_q is simply the ratio of the energy of the photon created during lasing to the energy required to create one excitation. In an optically pumped system, the quantum efficiency is the ratio of the energy of a photon at the oscillating frequency to the energy of an absorbed pump photon. Finally, the output coupling efficiency η_o is the ratio of the output coupling to the total round-trip loss of the laser cavity.

The total efficiency η of a laser (power out divided by power in) is dependent on the slope efficiency and the laser threshold. Using the rate equation model, the slope efficiency is constant and the total efficiency is given by

$$\eta=\eta_s(1-P_{\text{thresh}}/P), \tag{13}$$

where P is the total pump power and P_{thresh} is the pump power required to reach threshold.

4.6 Multimode Operation

For most lasers, the frequency spacing between adjacent longitudinal cavity modes is much less than the gain bandwidth. As a result, lasers tend to oscillate at several frequencies simultaneously.

Although the above statement is true, the reasons are more subtle than they may initially seem. In the early days of lasers, it was believed that lasers with homogeneously broadened gain spectra should operate in a single longitudinal mode. The reasoning behind this can be understood from the rate equations. If we assume a uniform optical intensity within the laser cavity, the steady-state solution to the photon rate equation fixes the inversion density at its threshold value. The first cavity mode to lase (the one with the highest net gain) clamps the inversion density, and no other mode can reach threshold. The flaw in this reasoning lies in the assumption of uniform optical intensity.

Experimentally, lasers with both homogeneously and inhomogeneously broadened gain media tend to oscillate in several longitudinal modes as the result of spatial and spectral hole burning.

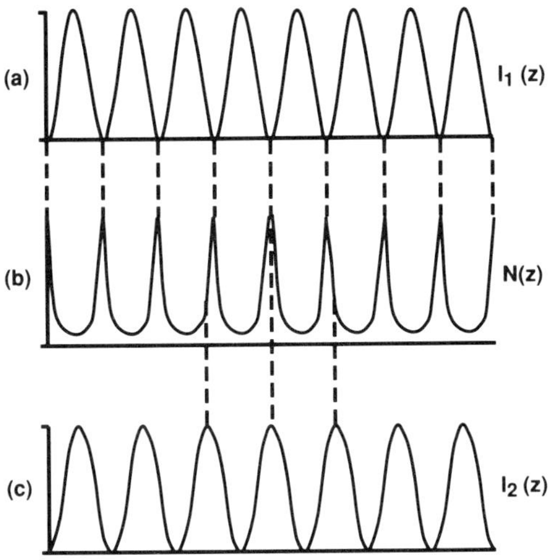

FIG. 8. Illustration of spatial hole burning, showing (a) the intensity profile of the first longitudinal mode to lase, $I_1(z)$; (b) the population inversion in the presence of the first oscillating mode, $N(z)$; and (c) the intensity profile of the second longitudinal mode to lase, $I_2(z)$.

4.6.1 Spatial Hole Burning. In standing-wave laser cavities, the coherent superposition of the optical fields traveling in opposite directions within the cavity results in a sinusoidal intensity distribution. At positions where the intensity distribution is at its maximum, there is strong gain saturation and the population inversion is depleted. However, at nulls in the optical field the oscillating mode is unable to deplete the inversion. As a result, the inversion density is no longer uniform, but has "holes" at the positions corresponding to the peaks in the optical intensity. This phenomena is known as spatial hole burning. The gain at the nulls in the optical field will continue to increase as the gain medium is pumped harder. Since other cavity modes have a different spatial profile than the first mode and can use the population inversion at these positions, this will lead to the onset of multimode operation. These ideas are illustrated in Fig. 8.

4.6.2 Spectral Hole Burning. Homogeneous gain broadening occurs when each excited state of the gain medium has exactly the same energy distribution. This is often the case in crystalline solid-state gain media at room temperature. In materials such as glasses, each excited ion sees a slightly different environment, resulting in a slightly different energy spectrum. The ensemble effect is inhomogeneous gain broadening. Another example of an inhomogeneous system is a gas, where each atom or molecule is moving at a slightly different speed and has its energy spectrum Doppler shifted by a different amount.

In an inhomogeneous system, only those excitations with gain at the lasing frequency are able to participate in the stimulated emission process. As a result, only those excitations become depleted, producing a gain spectrum that has a dip at the lasing frequency. This is known as spectral hole burning, and is illustrated in Fig. 9. Excitations that cannot contribute to the lasing process for the first mode can contribute to the onset of lasing for other modes, resulting in multimode oscillation.

4.6.3 Single-Frequency Operation. There are many techniques for obtaining single-frequency operation of a laser. Several of these involve introducing an element into the cavity such that the cavity sees a frequency-dependent loss, thereby decreasing the net gain bandwidth and selecting an individual longitudinal cavity mode. Examples of such elements are a prism, a grating, a Fabry-Pérot

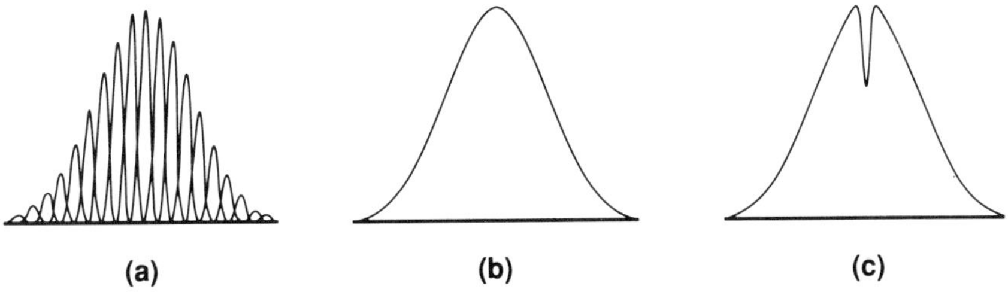

FIG. 9. Illustration of spectral hole burning, showing (a) several closely spaced homogeneously broadened spectra; (b) the inhomogeneously broadened spectrum resulting from the sum of several homogeneously broadened spectra; and (c) the inhomogeneously broadened spectrum with a spectral hole burnt in the center as a result of the saturation of one of the homogeneously broadened components.

etalon, and the combination of a birefringent filter and a polarizing element. In some cases, a cavity may require more than one device in order to obtain enough frequency selectivity to ensure single-frequency operation.

Single-mode operation has also been obtained by reducing the length of the cavity so that the longitudinal mode spacing is comparable to, or greater than, the gain bandwidth. This is most easily done with gas lasers, which have a narrow gain bandwidth, but has also been achieved in very short solid-state lasers.

Alternatively, single-mode operation can be achieved in a laser with a homogeneously broadened gain medium by reducing or eliminating the effect of spatial hole burning. A unidirectional ring cavity has a uniform optical intensity within the cavity, rather than the sinusoidal intensity distribution of a standing-wave cavity. Spatial hole burning is therefore eliminated, and such a laser may operate at a single frequency well above threshold. A variation of the same idea involves placing a quarter-wave plate on either side of the gain medium. As a result, the optical fields traveling in opposite directions in the gain medium are orthogonally polarized and do not interact coherently. The optical intensity within the gain medium is, therefore, uniform and there is no spatial hole burning.

In some gain media there is a large amount of energy diffusion. Energy diffusion moves some of the excited states away from the peaks in the population inversion, toward the minima. This smooths out the population inversion profile, reducing the effects of spatial hole burning. One important example of such a gain medium is a semiconductor.

Finally, the effects of spatial hole burning are reduced if the gain medium is located very close to a cavity mirror. At a mirror, the phase of the optical spatial intensity distribution is pinned at zero for all of the cavity modes. The phase difference between longitudinal modes increases gradually as one moves toward the center of the cavity. Quite close to the mirror, the peaks and nulls of the optical intensities for each of the modes occur in approximately the same place, and that portion of the gain that is not depleted by one mode will not be in a good position to contribute to any other.

5. TYPES OF PULSED OPERATION

5.1 Long-Pulse Operation

Long-pulse, or quasi-cw, operation refers to a pulsed laser with a pulse duration long enough for all relevant parameters within the system to come to their equilibrium values. Although the behavior of the system is cw-like at the end of the pulse, it will, in general, be quite different at the beginning of the pulse.

Let us consider a laser with a step-function pump source. The pump may quickly create a population inversion. It will take some time, however, for a lasing mode to build up from spontaneous emission. During this time the inversion density may greatly exceed threshold. The large inversion density eventually results in a large optical intensity, well in excess of the cw value. This optical intensity, in turn, drives the inversion density below threshold, substantially reducing the laser intensity. The entire process then starts again. For a single-mode laser, this often leads to regular spiking at the beginning of the pulse. The process is damped, however, and with time the intensity of the spikes decreases. Spiking eventually gives way to damped oscillations (known as relaxation oscillations) in the optical intensity and finally cw-like behavior.

Figure 10 shows the computer solutions to the rate equations for a laser under the conditions just described. In this computer simulation the spiking was heavily damped and cw-like behavior was quickly obtained. In a multimode laser the interaction between modes often leads to mode hopping, mode beating, and very irregular spiking, which may never damp out.

5.1.1 Relaxation Oscillations. Relaxation oscillations occur whenever the population inversion of a laser is disturbed from its equilibrium value. It is a result of the coupling between the population inversion and the photon density within the laser cavity, as described above. From the rate equations, it can be shown that for a single-mode laser a small disturbance in the inversion density results in damped oscillations with an oscillation frequency

$$\omega = [(N/N_{\text{thresh}} - 1)/\tau_c\tau]^{1/2} \tag{14}$$

and a damping constant

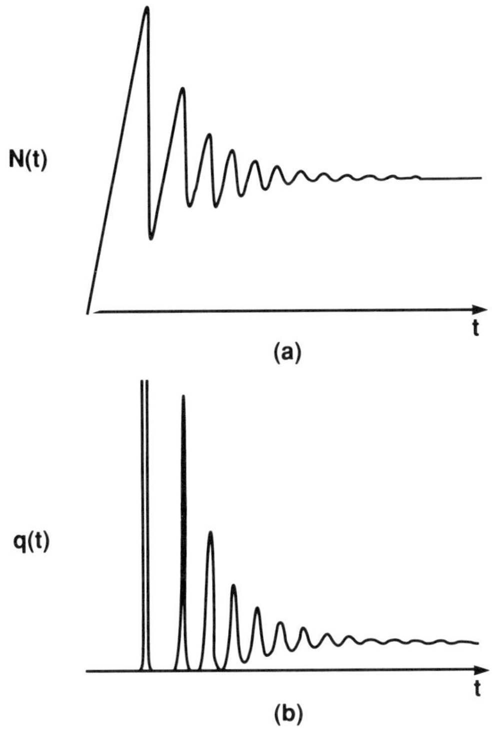

FIG. 10. Computer solutions to the rate equations for a laser with a step-function pump source, showing (a) the population inversion $N(t)$, and (b) the photon number $q(t)$.

$$t_0 = 2\tau N_{\text{thresh}}/N. \tag{15}$$

Note that if $1/t_0 > \omega$ the oscillations are overdamped and spiking will not occur. Although this condition is not satisfied in solid-state lasers, it is common in gas lasers.

5.2 *Q*-Switched Operation

For many applications it is desirable to obtain short, high-peak-power pulses from a laser. This can be achieved by creating a large population inversion and then quickly decreasing the cavity loss so that the inversion density is well in excess of its new threshold value. The large inversion density allows an intracavity optical field to develop rapidly. This field then depletes the population inversion and turns itself off. The cavity loss is subsequently increased to prevent the development of a second pulse. This technique is known as *Q* switching, since the quality factor, or *Q*, of the optical cavity is changed.

Q switching relies on the fact that the lifetime of the population inversion is much longer than the output pulse. The gain medium is therefore able to store energy, which can be quickly released in the form of an output pulse.

The length of the *Q*-switched output pulse is dependent on several factors. In order to obtain a rapid buildup of the optical pulse, it is desirable to have a large gain cross section and a large inversion population when the *Q* of the cavity is switched. The rapid buildup of the output pulse also relies on the presence of a strong intracavity optical field, which argues in favor of a high cavity *Q*. In order to obtain a short output pulse, however, the intracavity intensity must also decay quickly after the peak of the pulse. Since the *Q* of the cavity is constant for the duration of the pulse (it is difficult to change the *Q* of the cavity significantly during the duration of a short output pulse, although it can be changed in the relatively long pulse buildup time preceding the pulse), this would argue for a low cavity *Q*. Solutions to the rate equations show that the minimum-width output pulse is obtained when the total round-trip loss of the cavity during the output pulse is adjusted so that the initial inversion density is about three times the new threshold value.

Methods of *Q* switching a laser include the use of electro-optic shutters, mechanical devices, and acousto-optic *Q* switches. Passive *Q* switching can be obtained through the use of an intracavity saturable absorber with a long recovery time.

5.3 Gain-Switched Operation

Gain switching is another way to obtain short, high-peak-power pulses from a laser. The idea is to increase the pump power rapidly so that the population inversion of the laser is well in excess of the threshold value by the time the first output spike develops. The optical pulse then drives the population inversion below its threshold value. The pump power is subsequently reduced so that the inversion remains below threshold and only a single pulse is obtained.

The factors that are important for short gain-switched pulses are the same as for short *Q*-switched pulses.

5.4 Cavity-Dumped Operation

Cavity dumping allows the energy in a laser cavity to be output in a time comparable to the cavity round-trip time. The concept is to introduce rapidly (within a cavity round-trip time) a large output coupling (nearly 100%) into a cavity that previously had no output coupling. Methods of cavity dumping include the use of an electro-optic Pockels cell and polarizing beam splitter, and acousto-optic devices.

5.5 Mode-Locked Operation

Mode locking refers to the situation when the phases of several cavity modes are fixed (or locked) with respect to each other such that the electric fields add coherently and constructively for a short period of time. This allows the generation of a train of high-intensity, ultrashort pulses. To understand this, consider the case of $2n+1$ equally spaced longitudinal modes oscillating with the same amplitude E_0. Assume that the phases ϕ_m of the modes are locked according to $\phi_m-\phi_{m-1}=\phi$, where ϕ is a constant. The total electric field is the sum of all of these modes:

$$E(t)=\sum_{m=-n}^{n} E_0\exp\{i[(\omega_0-m\Delta\omega)t+m\phi]\}$$

$$=A(t')\exp(i\omega_0 t), \tag{16}$$

where ω_0 is the frequency of the center mode, $\Delta\omega$ is the frequency difference between two adjacent modes, $t' = t + \phi/\Delta\omega$, and

$$A(t')=E_0\frac{\sin[(2n+1)\Delta\omega t'/2]}{\sin[\Delta\omega t'/2]}. \tag{17}$$

Equation (16) shows that $E(t)$ can be represented in terms of a wave with a carrier frequency ω_0 whose amplitude $A(t')$ is time dependent. The intensity of this wave is given by $A^2(t')$, which consists of a train of pulses whose peak intensity is $A^2(0)=(2n+1)^2E_0^2$, pulse width is $\Delta\tau_p=2\pi/(2n+1)\Delta\omega$, and separation between pulses is $\tau_p=2\pi/\Delta\omega$. Since the total oscillating bandwidth $\Delta\nu_{\mathrm{osc}}$ is given by $(2n+1)\Delta\omega/2\pi$, the pulse width can also by written as $\Delta\tau_p=1/\Delta\nu_{\mathrm{osc}}$. Since $\Delta\omega=\pi c/l$, where l is the length of the laser cavity, the time between pulses is $\tau_p=2l/c$. In words, a mode-locked laser produces a train of output pulses whose pulse width is given by the inverse of the oscillating bandwidth and whose separation between pulses is equal to the round-trip time of the laser cavity. Figure 11 shows $A^2(t')$ for $n=5$.

In the above example, the phases of all of the modes were locked so that the output pulse had its minimum possible duration. Such a pulse is referred to as a transform-limited pulse, since its temporal profile is the Fourier transform of its spectral profile. This need not be the case—it is possible to obtain longer pulses, but not shorter pulses. Also, it is worth noting that, unlike the other pulsed schemes described in this section, a mode-locked laser is a cw device and there is phase coherence between pulses.

The above discussion tells us what mode locking is, but sheds little light on why or how it occurs. In general, mode locking will occur if the net gain for a mode-locked train of pulses is greater than the net gain for any other combination of cavity modes.

5.5.1 Active Mode Locking. Mode locking can be obtained actively or passively. Active mode locking can be broken into two categories, "AM mode locking" (produced using an amplitude modulator within the laser) and "FM mode locking" (produced using a frequency or phase modulator within the laser). In AM mode locking, the loss of some element in the laser cavity is modulated at the round-trip cavity frequency. Light circulating in the cavity will see less loss, and therefore more net gain, when it is incident at the loss element during the time of minimum loss.

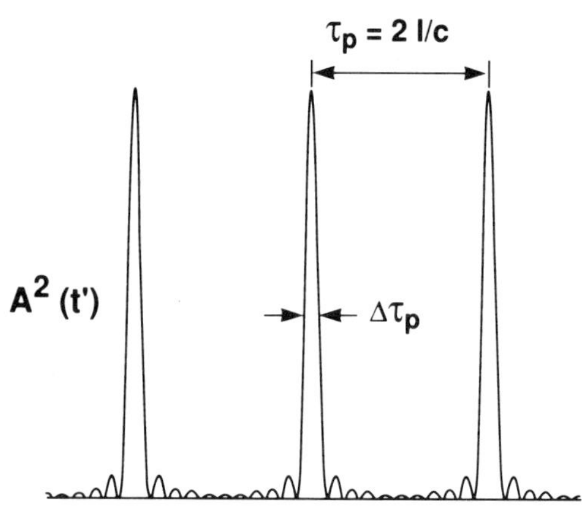

FIG. 11. Train of mode-locked pulses, made up of 11 modes of equal intensity.

This encourages short-pulsed operation and mode locking can be induced. The same result occurs if the gain of the cavity is modulated. Gain modulation through modulation of the pump source is known as synchronous pumping.

In FM mode locking, the optical length of the laser cavity (length or refractive index) is modulated at the round-trip cavity frequency. For simplicity, let us consider the case where one of the mirrors is moved sinusoidally along the direction of the cavity axis. Light incident on the mirror during its motion will be Doppler shifted. As a result, it will not reproduce itself after one round trip and will not produce a coherent oscillating mode. Light incident on the mirror at its turning points (maximum or minimum cavity length) sees a stationary mirror and will not experience a Doppler shift. The net result is that mode-locked pulses will tend to form. Modulation of the refractive index at some point in the cavity has the same effect.

5.5.2 Passive Mode Locking. Passive mode locking can occur when a laser cavity contains a nonlinear optical element, such as a saturable absorber. In this case, the more intense the light incident on the saturable absorber, the less the total absorption. The total loss of the cavity is therefore minimized by putting all of the energy into short pulses. This is essentially self-induced AM mode locking. A similar effect is obtained by putting a Kerr lens and an aperture within the cavity. Other techniques include the use of interferometric elements containing nonlinear media.

Passive mode locking must be initiated by the presence of a pulse within the cavity. If the optical intensity within the cavity is uniform in time, there is no loss or gain element that is modulated at the round-trip cavity frequency in order to induce mode locking. Noise, however, is capable of introducing a small amplitude modulation on the optical field. In some lasers, this small modulation is sufficient to start the mode-locking process. Such lasers are referred to as self-starting. In other systems a pulse (or AM modulation) must be intentionally introduced into the cavity to start the mode-locking process. Once started, however, mode locking can persist for long times.

The introduction of an appropriate nonlinear optical element into a laser cavity is not sufficient to guarantee mode locking. In order for passive mode locking to work, the relative phases of all of the longitudinal modes must remain constant. One effect that can destroy the phase relationship, and therefore prevent mode locking, is dispersion. Passively mode-locked lasers must be dispersion free if mode locking is to occur. To accomplish this, prisms are often introduced into the laser cavity to compensate the dispersion of other intracavity elements, such as the gain medium. In a properly compensated cavity, non-mode-locked operation can be unstable.

The other effect that can destroy the phase relationship between the spectral components of a mode-locked pulse train is spontaneous emission, or noise. The phase of the noise is unrelated to the phase of the oscillating mode. The net phase will be shifted when the two are combined.

6. CONTROL OF LASER OUTPUT

6.1 Frequency Tuning

Frequency tuning of a laser can occur in one of two ways. If the longitudinal mode spacing of the laser cavity is much less than the gain bandwidth, the cavity is capable of supporting several modes, each at a different frequency. A single frequency is then selected through the insertion of an element into the cavity such that the cavity sees a frequency-dependent loss, as discussed in Sec. 4.6.3. In most of the examples listed in that section, a small repositioning of the frequency-selective element would result in a new longitudinal mode (and hence a new operating frequency) being selected. The frequency-selective element is used to select one of the several cavity modes, and discrete tuning is obtained. Fast tuning can be obtained through the use of electro-optic frequency-dependent components.

The other way in which a laser can be tuned is to change the frequency of a given cavity mode, by changing the optical length of the cavity. Since the cavity length can be changed continuously, this leads to continuous tuning. This type of tuning is often limited by the free spectral range of the cavity. Once the cavity modes are shifted by a full free spectral range, an adjacent cavity mode is positioned at the frequency where the initial mode started. For the same reasons that the initial mode was originally favored, the adjacent mode is now favored, and the laser will have a tendency to mode-hop back to the original frequency.

6.2 Amplitude Modulation

The output power of a laser can be controlled by changing the pump power, the output coupling, or the intracavity loss. This type of amplitude modulation is usually limited to frequencies below the frequency of the relaxation oscillations. The relaxation frequency characterizes the response time of the cavity. Near the relaxation frequency there is resonant enhancement of the modulation response; above the relaxation frequency the response rolls off.

Methods used for direct amplitude modulation of a laser may have the side effect of introducing frequency modulation as well. For example, changing the pump power affects the thermal load on the gain medium, and therefore the temperature. This, in turn, affects the refractive index, changing the optical length of the cavity and the oscillating frequency. For amplitude modulation applications where frequency stability is critical, it is often better to modulate the laser power external to the cavity.

7. ISSUES IN LASER DESIGN

The design of a laser is dependent on many interdependent factors, including the requirements placed on the output beam (wavelength, spectral purity, tunability, divergence, polarization, power, and power stability), the operating environment (temperature, humidity, vibration, acceleration, and externally applied forces), and practical considerations (size, cost, and available power). There is an increasingly large number of gain media, cavity designs, and pump configurations that have been employed in lasers, and several texts have been written on the subject of laser design. No one design is well suited for all applications; every laser is optimized for operation at one point in the multidimensional parameter space outlined above.

A very important issue in the design of many lasers is the extraction of heat from the gain medium. In the process of pumping the gain medium, heat is generated. As the temperature of the gain medium changes, so too do its physical length and refractive index. Each of these contributes to a change in the optical length and resonant frequencies of the laser cavity. Nonuniform heating results in thermal lensing and internal stress. Thermal lensing changes the confocal parameters of the laser cavity and can destabilize an otherwise stable cavity (or vice versa). Internal stress leads to stress birefringence and, eventually, stress fracture.

Other issues that must be considered in high-power lasers are nonlinear optical effects and optical damage. The electrical field within the optical beam of a high-power laser can be large enough to damage optical components. This is particularly important in high-power pulsed lasers. At optical intensities below the optical damage level, deleterious nonlinear optical interactions can still degrade the performance of the laser and even destroy the device. One example is stimulated Brillouin (acoustic wave) scattering in fiber lasers. In this case, nonlinear interactions create acoustic waves that can blow off the ends of the fiber.

8. FREQUENCY CONVERSION AND NONLINEAR CONTROL OF LASER RADIATION

Nonlinear optical techniques can be used to extend the frequency coverage of lasers as well as to modify other characteristics of laser radiation. Frequency conversion is a very important adjunct, converting the output of practical lasers to regions where primary laser sources may not exist or may not be very practical. Harmonic generation, frequency mixing, optical parametric oscillation, and stimulated Raman scattering have been used for frequency conversion. Nonlinear processes have also been used to produce mode locking (see Sec. 5.5.2), to improve transverse beam quality (i.e., to produce output closer to the diffraction limit), and to dampen relaxation oscillations in pulsed lasers. The article OPTICS, NONLINEAR, describes the nonlinear processes in detail; here, we will only discuss their general significance for lasers. Finally, it should be mentioned that laser oscillators and amplifiers are nonlinear optical devices in that there is a reduction in optical gain caused by partial depletion of the population inversion by the laser radiation. This partial depletion, or saturation, stabilizes the laser output amplitude in cw lasers and plays a vital role in determining the operating characteristics of pulsed lasers, as we have seen in Sec. 5.

The most frequently used frequency-conversion techniques are second-harmonic and sum-frequency generation. Typically, the output of infrared lasers in the 1-μm region, such as Nd:YAG and Nd:glass, is doubled into the green. Depending on the application, shorter wavelengths may be obtained (for example, by summing the green radiation with the infrared). There are a variety of uses for this short-wavelength output, including pumping short-wavelength lasers and optical parametric oscillators (OPOs). For second-harmonic generation to the green, conversion efficiencies as high as 75% have been reported, and average powers of greater than 50 W have been obtained.

Difference-frequency generation and OPOs have been used as tunable sources of radiation for various spectroscopic applications, primarily in the infrared. Continuous-output difference-frequency generation, while producing very low average power with low efficiency, has been used for ultrahigh-resolution molecular spectroscopy in the mid-infrared. OPOs are useful sources of tunable pulsed output with high peak power and high efficiency. Stimulated Raman scattering has been used to generate large pulse energies with essentially unity quantum efficiency at a variety of wavelengths. However, because energy is deposited in the Raman process, moving Raman media are often required to obtain high–average-power conversion.

The optical Kerr effect, which is the change in refractive index proportional to the optical intensity, can cause short pulses, typically 0.1–10 ps in duration, to acquire a frequency sweep, or chirp. By sending the chirped pulse through a frequency-dispersive delay line, which might be a grating or prism pair, even shorter pulses, typically 5–50 fs, can result.

Nonlinear phase conjugation can reverse the phase distortion acquired from the active medium in a laser (for example, an optically imperfect laser crystal or a laser material that has large thermal gradients due to high–average-power operation). On a second pass through the distorting medium, the reversed phase distortion is canceled. Most often, backward stimulated Brillouin scattering is used to reverse the phase; the small acoustic frequency shift produces an optical wave that remains within the gain bandwidth of most lasers and has linear propagation characteristics that match the input to the Brillouin cell.

Harmonic generation has been used to moderate the amplitude of the relaxation oscillations that often occur in long-pulse lasers. Relaxation oscillations (see Sec. 5.1.1) involve the flow of energy back and forth between the population inversion and the radiation field. Conversion to the second harmonic can clip the high initial peaks in the laser power, an action that reduces the amount of depletion in the population inversion, thereby damping the oscillations.

Further Reading

Popular discussions of lasers and their applications and history are given in the following:

Hecht, J., (1985), *Laser Pioneers*, rev. ed., San Diego, CA: Academic.

Hecht, J., Teresi, R. (1982), *Laser, Supertool of the 1980's*, New Haven: Ticknor and Fields.

A general discussion of laser science and technology, accessible to anyone with a background of electromagnetic theory, basic quantum mechanics, and calculus, can be found in any of the following:

Eastham, D. A. (1986), *Atomic Physics of Lasers*, London: Taylor and Francis.

Kaminow, I. P., Siegman, A. E. (1973), *Laser Devices and Applications*, New York: IEEE.

Shimoda, K. (1986), *Introduction to Laser Physics*, 2nd ed., New York: Springer.

Siegman, A. J. (1986), *Lasers*, Mill Valley, CA: University Science.

Svelto, O. (1989), *Principles of Lasers*, 3rd ed., New York: Plenum.

Yariv, A. (1988), *Quantum Electronics*, 3rd ed., New York: Wiley.

Detailed theoretical discussion of lasers is contained in the following:

Haken, H. (1984), *Laser Theory*, New York: Springer.

Sargent, M., III, Scully, M. O., Lamb, W. E., Jr. (1974), *Laser Physics*, Reading, MA: Addison-Wesley.

The underlying physics of laser resonators and its application to the design of many types of lasers in use today is the topic of the following:

Hall, D. R., Jackson, P. E. (1989), *The Physics and Technology of Laser Resonators*, Bristol, U.K.: Adam Hilger.

Specific types of lasers are discussed in the following:

Brown, D. C. (1981), *High-Peak-Power Nd:Glass Laser Systems*, New York: Springer.

Casey, H. C., Jr., Panish, M. B. (1978), *Heterostructure Lasers*, Parts A and B, New York: Academic.

Duarte, F. J., Hillman, L. W. (1990), *Dye Laser Principles With Applications*, San Diego, CA: Academic.

Garrett, C. G. B. (1967), *Gas Lasers*, New York: McGraw-Hill.

Kaminskii, A. A. (1981), *Laser Crystals*, New York: Springer.

Koechner, W. (1988), *Solid-State Laser Engineering*, 2nd ed., New York: Springer.

Mollenauer, L. F., White J. C. (Eds.) (1988), *Tunable Lasers*, New York: Springer.

Schäfer, F. P. (Ed.) (1973), *Dye Lasers*, New York: Springer.

Willett, C. S. (1974), *Introduction to Gas Lasers: Population Inversion Mechanisms*, Oxford: Pergamon.

A concise source of data in tabular and graphical form is provided for workers in the areas of laser research and development in the following series:

Weber, M. J. (Ed.) (1982), *Handbook of Laser Science and Technology*, Vols. I and II; (1986), *ibid.*, Vols. III and IV; (1987), *ibid.*, Vol. V; (1991), *ibid.*, Suppl. 1, Boca Raton, FL: CRC.

LASER TECHNOLOGY

HIROSHI TAKUMA, *The Institute for Laser Science, University of Electro-Communications, Tokyo, Japan*

	Introduction	321
1.	**Measurements of Basic Characteristics of Lasers**	321
1.1	Mode of Oscillation and Beam Divergence	321
1.2	Output Power	322
1.3	Temporal Brightness	322
1.4	Spectral Brightness	323
1.4.1	Grating Spectrometer and Fabry–Pérot Interferometer	323
1.4.2	Scanning Confocal Fabry–Pérot Optical Spectrum Analyzers	323
1.4.3	Homodyne Detector and Electronic Spectrum Analyzer	323
1.4.4	Autocorrelation Measurements	
2.	**Wavelength Measurements**	323
2.1	Moving Michelson-Type Wavelength Meter	324
2.2	Direct Frequency Comparison with Microwave Frequency Standard	324
2.3	Fizeau Wavelength Meter	325
3.	**Measurement of the Characteristic Parameters of Laser Media**	325
4.	**Stabilization of the Laser Output**	327
4.1	Amplitude Stabilization	327
4.2	Frequency Stabilization	327
5.	**Squeezing**	328
6.	**Manipulation of Laser Beams**	328
6.1	*Q* Switching	328
6.2	Modulation and Switching of Laser Beams	329
6.3	Optical Harmonic Generation	329
	Works Cited	330
	Further Reading	330

INTRODUCTION

Laser technology deals with operation of lasers and the characterization and the use of their output beams. Although lasers are used in various kinds of systems that are important in various practical applications, only basic techniques are presented here. For more detailed descriptions on various kinds of lasers and application systems, the reader is advised to consult appropriate other articles of this Encyclopedia and the references listed at the end of the present article.

1. MEASUREMENT OF BASIC CHARACTERISTICS OF LASERS

1.1 Mode of Oscillation and Beam Divergence

Normally a laser is designed to operate in its fundamental TEM_{00} transverse mode so that an output beam having Gaussian intensity distribution is generated, although sometimes a higher-mode or mixed-mode output may be sought for some very specialized applications. In practice some amount of higher modes is mixed with the fundamental mode, and the beam pattern shows some deviation from the pure Gaussian distribution. Precise measurement of the beam pattern can be made by observing the intensity distribution of the cross section of the laser beam with a charge-coupled-device (CCD) camera and analyzing the two-dimensional pattern with a computer.

The Gaussian beam behaves just like a part of a spherical wave at a long distance from the beam waist, which is the point where the spot size of the beam has its minimum value. A beam waist corresponds to the focal point in geometrical optics and has a nonzero size because of diffraction of the light wave. The spot size of a Gaussian beam can be defined as

1-56081-067-X/94/$5.00+.50

the radius of the contour of the spot where the intensity is $1/e^2$ of the peak value. When the beam is focused by a lens of focal length f, the minimum spot size, which shall be designated as d, is given by the relationship $d=f\theta$, where θ is the half-angle beam divergence in radians. For a Gaussian beam, θ is given by the equation

$$\theta=[(\lambda/\pi w_1)^2+w_1^2/R_1^2]^{1/2}, \tag{1}$$

where λ, w_1, and R_1 are the wavelength, the beam spot size, and the radius of curvature of the laser beam at the input face of the lens, respectively. In practice, it is convenient to make the second term negligibly small by making the input laser beam nearly parallel using a beam expander, which is an inverted Galilean telescope. In a practical laser beam the measured beam divergence is always larger than that given by (1) because of the presence of the higher modes.

In above discussion, we assumed axial symmetry of the laser beam. However, in some kinds of lasers, the active part of the laser medium strongly deviates from axial symmetry. The output beam of such lasers may not be axially symmetric. The typical example of such is the semiconductor diode laser: the active part of it is very thin in the direction perpendicular to the *pn* junction layer. The output of a diode laser reflects such a peculiar structure, showing an asymmetric diffraction pattern. The typical beam divergence of a diffraction-limited, single-mode diode laser is 10° in the direction parallel to the junction layer, and 30°–40° in the perpendicular direction. Because even such a large angular spread is due to diffraction, a single-mode diode laser output can be converted into a parallel axially symmetric beam by the use of properly designed optics.

1.2 Output Power

The absolute value of the average output power is normally obtained by a calorimeter. Let the laser beam or a well-defined part of it be absorbed by a body, the heat capacity of which is well known, for a known duration of time. The average power is determined by measuring the temperature rise of the body due to the absorbed laser energy. A more conventional method is to use a commercially available calibrated bolometer. In carrying out the measurement, care must be taken that the effective area of the detector is large enough to collect all the input beam, the absorption rate of the detector is well known, and the intensity of the laser-beam is not strong enough to burn the detector surface or to cause any nonlinear optical effect (which scatters part of the energy of the laser beam out of the detector as photons of a different wavelength).

In a pulsed laser, the output energy per pulse can also be measured by the same technique by knowing how many pulses are incident on the detector. The average output, which is directly measured by the bolometer, is also a very important quantity for many kinds of applications.

We assumed that only a coherent output beam is coming out of a laser in the above discussion. However, this is not true in most practical cases, and a typical laser output includes an incoherent part of amplified spontaneous emission in addition to the coherent output. In order to remove the incoherent part of the output beam, which is spread wide over the coherent beam, the use of a spatial filter is strongly recommended. This is a device with two focusing lenses aligned on the axis of the laser beam at a distance where the two lenses share a common focal point, at which a pinhole is placed so that only the focusable component of the laser beam is passed through.

It is important in designing a laser to eliminate the incoherent part of the output. In a high-gain laser such as rare-gas halide lasers it is especially important, and sometimes the incoherent portion exceeds the coherent portion in integrated energy. However, in lasers designed for some special applications in which uniform illumination of a surface is essentially important, the lasers are designed so as to make the output incoherent while still keeping the beam divergence small.

1.3 Temporal Brightness

The temporal brightness, which is defined as the output power per unit solid angle, is a very important physical quantity that specifies the output characteristics of lasers. The higher the brightness, the higher the intensity of the focused laser beam. When the output power of a laser is rapidly changing, simultaneous measurements of the instantaneous output power, which is expressed as $f(t)$, the integrated output power W, and the beam divergence θ for a given time interval t give the brightness of the beam $B(t)$ as follows:

$$B(t)=Cf(t)/\pi\theta^2, \qquad (2)$$

where

$$C=W\Big/\int_0^t f(t)dt^{-1},$$

assuming that only the relative value of $f(t)$ is known. In the case where a time resolution of the order of 1 ns is sufficient, the instantaneous output power can be measured by detecting the relative value of the instantaneous intensity by a photodiode and displaying it on an oscilloscope. A high-speed streak camera provides a high time resolution up to the order of 1 ps. Those interested in subpicosecond resolution phenomena and their measurements should refer to the article ULTRAFAST PULSE PHENOMENA.

1.4 Spectral Brightness

Suppose that the output of a laser is stationary for the sake of simplicity. Although a laser beam is generally regarded as monochromatic light, the output energy of a laser is more or less spread in the frequency domain. By expressing the output power of a laser in a frequency domain $[\nu, \nu+d\nu]$ as

$$dI=I_0S(\nu)d\nu\,, \qquad (3)$$

where I_0 is a constant, $S(\nu)$ is defined as the normalized spectroscopic brightness. The following techniques are the most standard ones to measure the spectroscopic brightness.

1.4.1 Grating Spectrometer and Fabry–Pérot Interferometer. For the measurement of a light source having rather broad spectral width and where a resolution of only 0.01 nm (12 GHz) or lower is needed, a conventional grating spectrometer is the most suitable measuring device. A Fabry–Pérot étalon (see INTERFEROMETERS AND INTERFEROMETRY) is appropriate to measure a spectral width of 30–1 GHz. It should be noted here that both of the above instruments are made for incoherent light sources, and one should remove the spatial coherence of the beam before using them. Because a rather long time interval is normally needed in measurements with the above two instruments, inserting a ground glass or a sheet of tissue paper in front of the monochromator slit or the Fabry–Pérot étalon is sufficient.

1.4.2 Scanning Confocal Fabry–Pérot Optical Spectrum Analyzers. Units having finesse of higher than 200 are commercially available, providing a versatile way to measure the power spectrum of coherent sources having spectral widths of a few megahertz to a few gigahertz.

1.4.3 Homodyne Detector and Electronic Spectrum Analyzer. When a resolution of better than a few MHz is needed, homodyne detection by the use of a photodiode as a mixer is convenient. The key to obtaining a good signal-to-noise ratio in mixing two coherent optical beams is to align the wave fronts of the two beams with high accuracy.

1.4.4 Autocorrelation Measurements. According to the Wiener–Khinchin theorem, the Fourier transform of the autocorrelation function of a signal gives its power spectrum, and the spectral brightness of a laser beam can be obtained by measuring its autocorrelation. That is, if we describe the normalized temporal evolution of a laser output by $f(t)$, the autocorrelation function, represented by $A(\tau)$, is defined by

$$A(\tau)=\int_{-\infty}^{+\infty} f(t+\tau)f^*(t)dt. \qquad (4)$$

In principle, the spectral brightness is obtained by obtaining the experimental value of $A(\tau)$ by varying the delay time τ and then carrying out numerical Fourier transformation by a computer.

2. WAVELENGTH MEASUREMENTS

As the length standard is connected to the frequency standard by the speed of light, which is defined as $2.997\,924\,58\times10^8$ m/s, a laser wavelength measurement is equivalent to a laser frequency measurement. The wavelength of a laser relative to another is directly measured by comparing the interference fringes by the use of various types of interferometer arrangements.

2.1 Moving Michelson-Type Wavelength Meter

This most popular method essentially counts the number of fringes when one arm of a Michelson interferometer travels a known distance. The distance is normally measured by counting fringes made by another laser beam, the wavelength of which is established as a secondary standard (Hall and Lee, 1976).

A basic setup is schematically shown in Fig. 1. If the length of the fixed arm is l_1 and the movable one l_2, the difference in the optical path length for a round trip of the two arms is given by $2l_1-2l_2$, and interference of the two beams shows maxima when the following relationship holds:

$$2l_1-2l_2=n\lambda/2\ . \tag{5}$$

Therefore if two laser beams having respective wavelengths λ_1 and λ_2 are introduced to the interferometer at the same time separately and fringe counts of n_1 and n_2, respectively, are obtained for the same value of $2l_1-2l_2$, then we have

$$n_1\lambda_1/2=n_2\lambda_2/2\ . \tag{6}$$

If the wavelength of the one of the laser beams, for example, λ_1, is known, the other is given by

$$\lambda_2=(n_1/n_2)\lambda_1\ . \tag{7}$$

The technical difficulty in such a measurement is to move the end reflector M_2 smoothly without changing the direction of the reflected beam. The effect of an inevitable mechanical fluctuation of the moving reflector may be avoided by the use of a retroreflector such as a corner cube. Several wavelength standards are established on laser lines as listed in Table 1 by comparing their frequency with that of the cesium clock with a frequency chain method described below. The accuracy attained by this method ranges from a few parts in 10^7 to 10^8 for commercially available systems to a few parts in 10^{10} for the most elaborate laboratory systems.

2.2 Direct Frequency Comparison with Microwave Frequency Standard

This is not a conventional method to be applied in a routine measurement, but an elaborate way to establish secondary fre-

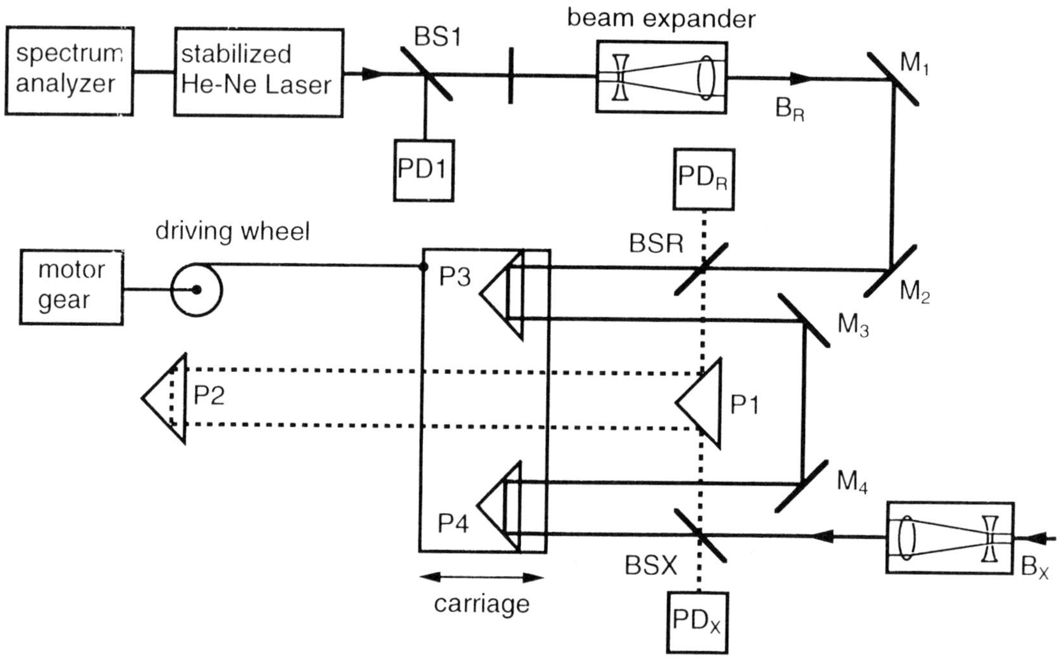

FIG. 1. A variation of the moving Michelson interferometer (Kowalski, *et al.*, 1977). The interference fringes of the reference and the sample beams are detected with PD_R and PD_X, respectively, as the carriage moves back and forth in the horizontal direction.

Table 1. Laser frequencies measured by direct comparison with calibrated microwave frequencies.

Atom or molecule	Frequency (THz)
HCN	0.890 760 550
H_2O	10.718 068 6
$^{12}C^{16}O_2$ *R*[20]	30.950 409
$^{12}C^{16}O_2$ *R*[22]	32.373 156
^{20}Ne (CH_4 *P*[7])	88.376 181 627[a]
^{20}Ne	196.780 372
^{20}Ne	260.103 264
$^{127}I_2$ [17–1 *P*(62)]	520.206 528[b]

[a]Frequency of the laser locked to the *P*[7] line of CH_4.
[b]Frequency of 17–1 *P*(62) transition of $^{127}I_2$ measured by the second harmonic of the 260-THz line of ^{20}Ne.

quency standards in optical region. Although the principle is to compare the higher harmonics of the microwave frequency standard with a laser frequency, direct comparison is impossible in the present state of art, and the frequency comparison chain must include several intermediate-frequency lasers ranging from far infrared to infrared. The frequencies of those lasers that have been determined by such a frequency comparison chain are tabulated in Table 1.

2.3 Fizeau Wavelength Meter

This is a method that can be used to determine the wavelength of a laser with pulsed output, in contrast to the two methods described above. As shown in Fig. 2, the principle is simply to compare the fringe separation observed by a Fizeau interferometer with that formed by a wavelength standard laser. The fringes can be counted by a photodiode array and the result is processed by a small computer. This is a versatile method to be used in the visible to near-infrared region, giving an accuracy of up to a few parts in 10^8.

3. MEASUREMENT OF THE CHARACTERISTIC PARAMETERS OF LASER MEDIA

The phenomenological characteristics of a laser medium are generally described by three parameters: small-signal gain, absorption coefficient, and saturation intensity, which will

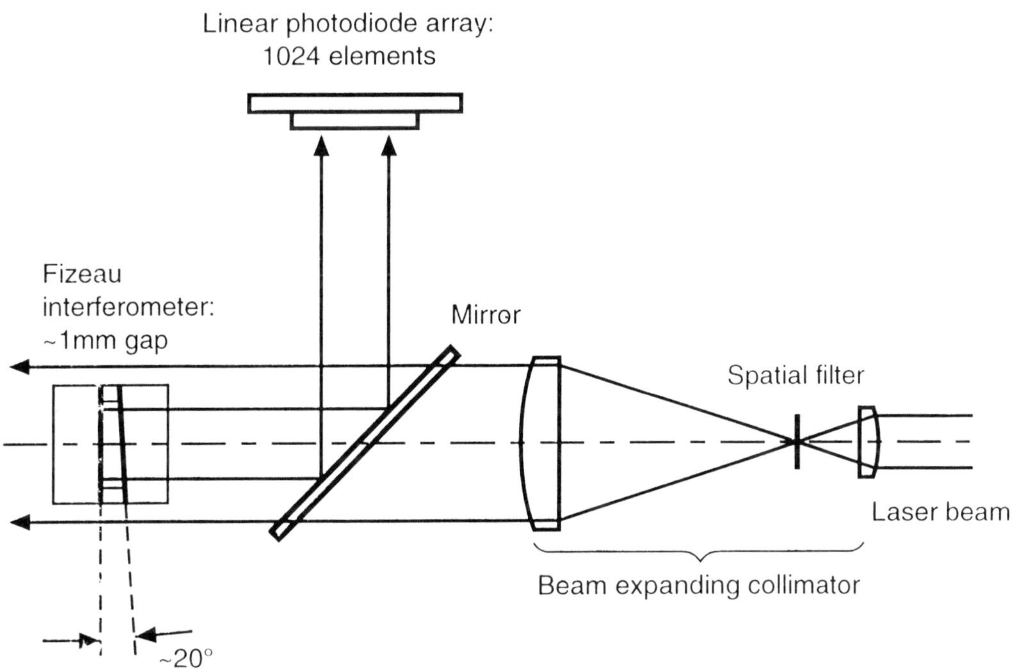

FIG. 2. Schematic diagram of the Fizeau-type wavelength meter (Snyder, *et al.*, 1977).

be designated by g_0, a, and I_S, respectively. The gain coefficient of the laser medium in operation with a laser radiation intensity of I is given by

$$g=\frac{g_0}{1+I/I_S} . \tag{8}$$

The effective gain coefficient, which is represented by g_e, is defined as

$$g_e=g-a . \tag{9}$$

Since the field intensity in the laser medium is dependent on the quality factor of the optical resonator, it is possible to determine the three constants by fitting the experimental values of the output obtained for various values of quality factor by replacing a set of reflectors that compose the resonator. However, the accuracy obtained by such measurements is rather poor, and they are not recommended if the following measurement is possible.

The most accurate determination of the above three amplifier constants can be made by realizing conditions under which the effective gain is expected to be constant in the medium. Such a condition is generally satisfied in the following three cases:

1. small-signal condition, that is, the input power is so small that saturation is negligible even after the laser beam has made a single excursion of the laser medium;
2. zero-gain condition, where the input power happens to make the saturated gain equal to the absorption coefficient, making the effective gain exactly zero; and
3. perfectly saturated gain condition, where the input gain is so large that the saturated gain is negligibly smaller than the absorption coefficient.

Figure 3 shows the output of an amplifier for a wide range of input intensity.

The small-signal condition may be checked easily by confirming the proportionality of the output and the input by making the input power level as low as possible, and g_0 is simply determined as the ratio of the output to the input power. The third condition is also defined accurately by confirming that the laser medium shows linear absorption by applying an extremely intense input power: the experimentally measured absorption coefficient under such a condition gives us the value of a with good accuracy. A good number of experimental runs with various input intensities may be needed, but in principle it is certainly possible to find the input power at which the saturated gain is exactly equal to the linear absorption and the zero effective gain condition is realized. Let us define such an intensity as I_n. Then we expect the following relationship to hold:

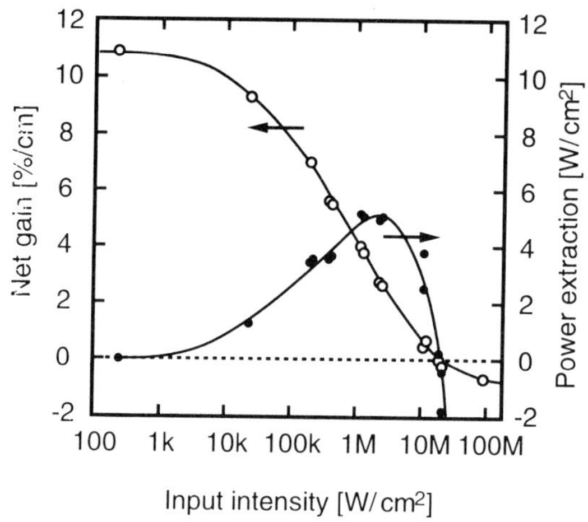

FIG. 3. An example of the output of a laser amplifier for a broad range of input intensity (Ueda and Takuma, 1987).

$$I_S=I_n/(g_0-a) , \tag{10}$$

which gives us the saturation intensity I_S from the values of g_0 and a obtained above. Thus the three key parameters can all be determined with a good accuracy.

Although the above method gives complete knowledge of the three characteristic parameters of an amplifier, it is not guaranteed that it can be applied to any laser medium. The measurement of gain under small-signal condition imposes the difficulty of obtaining a good signal-to-noise ratio in measuring the intensity of a laser beam at a very low power level. For some kinds of laser media, such a high intensity of input is needed to realize the third condition, and sometimes even the second condition, that the laser media are damaged by the laser beam, making accurate measurements impossible.

A typical example in which such measurements gave excellent results is the excimer lasers, especially the KrF laser, having high small-signal gain and rather low saturation intensity.

4. STABILIZATION OF THE LASER OUTPUT

Stabilization of the laser output is needed quite often when it is used in high-accuracy measurements. For some applications, stabilization of the output power of the laser source is needed. In many cases frequency stabilization of the laser output is needed.

4.1 Amplitude Stabilization

Amplitude stabilization can be accomplished by the use of a continuously variable attenuator and installation of a feedback loop to adjust the attenuation rate to make the output power of the attenuator constant. In practice, it can be accomplished easily when the laser beam is linearly polarized by a system as illustrated schematically in Fig. 4. The beam is passed through a Pockels cell that rotates the polarization plane of the beam in proportion to the increase of the output power. A polarizer placed behind the cell selects only the component having the same polarization as the original one. A small fraction of the output power is detected by a photodiode and the amplified detected signal is used to control the voltage applied to the Pockels cell.

The above method is the conventional way to obtain an amplitude-stabilized laser beam. However, one can expect to generate an amplitude-stabilized output directly from lasers or nonlinear devices by suppressing the amplitude noise. For example, a diode laser driven by a current-stabilized dc source shows amplitude squeezing at a pumping level much higher than the threshold value. This will be discussed in Sec. 5.3.

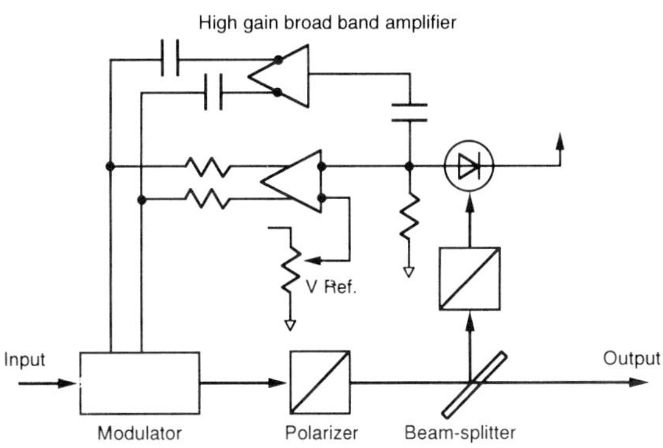

FIG. 4. Schematic diagram of an intensity stabilizer.

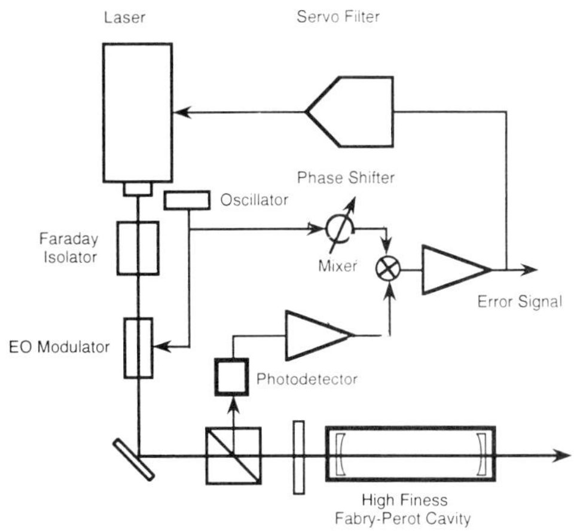

FIG. 5. Schematic diagram of a typical frequency-stabilized laser.

4.2 Frequency Stabilization

The general principle of laser frequency stabilization is to pass the laser beam through a frequency discriminator having a peak transmission or peak attenuation at a given frequency, and to control the laser frequency so that it always stays at the frequency defined by the frequency discriminator as schematically shown in Fig. 5. A Fabry–Pérot type optical resonator is widely used as the frequency discriminator, although it cannot be an absolute frequency standard unless it is calibrated with respect to some kind of absolute standard such as atomic or molecular spectral lines.

Recent progress in optical coating technology makes it possible to obtain low-loss, high-reflectivity reflectors, for example, 99.99% reflectivity and a few ppm loss, the use of which allows us to make a Fabry–Pérot resonator having a finesse of several hundred thousands by keeping a fairly high transmission rate for the beam of resonant frequency. The use of such an optical resonator gives us the possibility of controlling a laser frequency to be stabilized at a given frequency with subhertz accuracy. As a matter of fact, He-Ne lasers and laser-diode–pumped Nd:YAG lasers have been stabilized to a level of relative stability on the order of mHz, exceeding the theoretical limit of a free-running laser given by Schawlow and Townes. On the other hand, stabilization of absolute frequency involves several unknowns, such as the effect of heating of the mirror surface by the resonant radiation, the intensity of which is enhanced to an extremely high level in high-finesse resonator.

Stabilization of frequency-tunable lasers such as dye lasers, Ti sapphire lasers, or diode lasers poses difficult extra problems, such as keeping single-mode oscillation for a wide range of tuning. However, frequency stabilization up to the level of 1 MHz to 100 kHz can be realized rather easily even for those lasers.

5. SQUEEZING

According to the quantum mechanics of optical fields, uncertainties in the amplitude and phase are governed by the uncertainty relationships. That is, representing the root mean square deviations of the photon density and the phase as Δn and $\Delta\phi$, respectively, we have the relationship

$$\Delta n\ \Delta\phi \geqslant \tfrac{1}{2}\,. \qquad (11)$$

This means that photon-density noise in the laser resonator can be suppressed by allowing the phase noise to increase. A normal laser oscillator generates coherent photons, which show equal uncertainty in the photon density and the phase. The process that generates photons having less uncertainty in the photon density is called *amplitude squeezing*, and that which generates less phase noise, *phase squeezing*.

Normal laser oscillation, which is built up from spontaneous emission, generates shot-noise–limited light. However, it is possible to generate squeezed-state photons by suppressing the fluctuations. The most commonly observed example is a semiconductor diode laser driven by a current-stabilized power source at a pump level much higher than its threshold. The general scheme to generate squeezed states is to use photon generation by nonlinear effects such as degenerate optical parametric oscillation, four-photon mixing, second-harmonic generation, or the stimulated Raman effect. An experimental setup for squeezed-state generation by a degenerate parametric oscillator is shown in Fig. 6. In this mechanism, the generated photons have a strong correlation with the pumping waves, and this is the key to generation of the squeezed state. Therefore quantum noise included in the pumping wave such as spontaneous emission tends to destroy the squeezed state. This makes experimental generation of the squeezed states difficult.

The common method to detect the squeezed state is a balanced homodyne detector as shown schematically in Fig. 7. The detected signal changes periodically with the phase of the local oscillator, and squeezing appears as the minimum detected noise being much less than the noise of the coherent photons.

The use of squeezed-state photons gives us the possibility of obtaining high sensitivity or accuracy in optical measurements. In practice, a squeezed state is easily destroyed by introducing loss or spontaneous emission, and so high precaution is needed. Even so, improvement in signal-to-noise ratio has been reported in a spectroscopic measurement.

6. MANIPULATION OF LASER BEAMS

6.1 *Q* Switching

In many laser applications, only a short pulse with high peak power is needed. In such cases, *Q* switching provides a convenient way of generating a pulse of a few to a few tens of nanosecond duration by extracting a large portion of the stored energy of the laser medium. A normal laser pumped by a pulsed source exhibits pulsed oscillation, generating pulses of the laser beam reflecting the temporal behavior of the pumping source. In such an oscillation, the population inversion is limited by the induced emission. However, if the laser oscillation is suppressed by inserting a loss in the laser cavity, an inverted population much higher than the threshold value can be built up. Then the cavity loss is suddenly

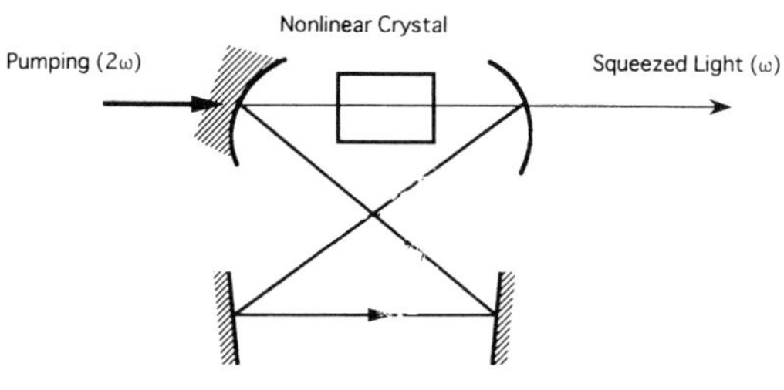

FIG. 6. Schematic diagram of a degenerate optical parametric oscillator that generates a squeezed output.

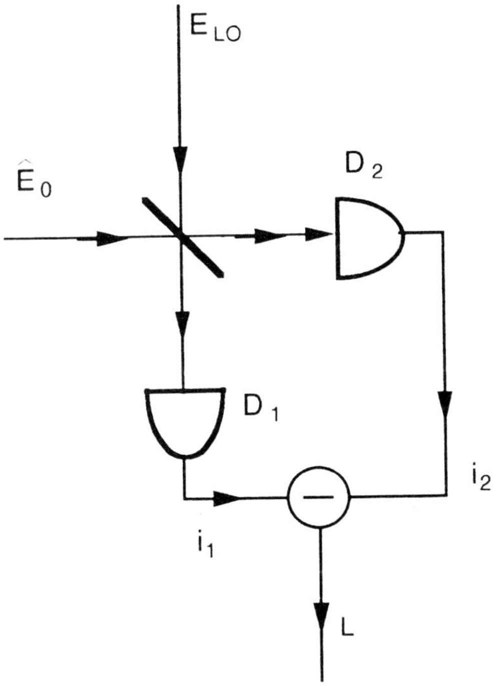

FIG. 7. Optical homodyne detector for detection of squeezed light. E_0 and E_{LO} stand for the sample beam and the reference beam, respectively. L stands for the local frequency output. D_1, D_2: detectors.

removed, and the energy stored in the medium as the inverted population can be extracted in a short time by building up the laser radiation. Such devices are often called *Q*-switched lasers and are widely used when a concentration of laser energy in a short time interval is needed.

As a *Q*-switching device, that is, a very fast shutter installed in a laser cavity, the Pockels cell is most frequently used in solid-state lasers. The laser is constructed in a way so that only linearly polarized light can be built up in the cavity by inserting a polarizer in the laser cavity, for example, and the Pockels cell rotates the direction of polarization 90° for the laser photon, making a round trip. Thus the oscillation of the laser is stopped until the population inversion reaches its maximum value by pulsed pumping and then the voltage applied on the Pockels cell is suddenly turned off. The laser radiation is built up rapidly by intense stimulated emission due to a large density of population inversion.

6.2 Modulation and Switching of Laser Beams

Many methods for laser applications require modulation or turning off and on the laser beams. There are two methods of frequency or phase modulation of laser beams. One is the acousto-optical modulator, which shows high modulation efficiency for frequencies lower than 100 MHz. However, only the electro-optical modulator is suitable for high-frequency modulation, although it is difficult to obtain high modulation efficiency.

An acousto-optical modulator consists of a transducer attached to a low-loss optical material having a high acousto-optical coefficient such as $PbMoO_4$, which is transparent to visible and near-infrared radiation. A supersonic wave is generated and propagates through the acousto-optical material in the direction perpendicular to the laser beam, and the latter is diffracted by a density grating made by the acoustic wave. The diffracted beam of the first order is frequency shifted by an amount equal to the frequency of the acoustic waves. The diffraction efficiency is approximately proportional to the power of the acoustic waves and can be higher than 50%.

For a frequency higher than 100 MHz, electro-optical modulation is the only method applicable at present. Electro-optical materials such as $LiNbO_3$ and $LiTaO_3$ have relatively high electro-optical coefficients and are most frequently used for such purpose. The electro-optical materials change their refractive indices when an electric field is applied. In most cases the electric field is applied in the direction perpendicular to the propagation direction of the laser beam and therefore the phase of the laser beam is shifted by an amount proportional to the applied electric field. The electro-optical modulator can be used as a fast intensity modulator and switching device having a high extinction rate. For more details, refer to OPTOELECTRONIC DEVICES and MODULATORS AND DEMODULATORS.

6.3 Optical Harmonic Generation

Second or higher-order harmonics of a laser beam can be generated in nonlinear optical materials. Second-harmonic generation is possible only in a medium that does not have point symmetry. Therefore piezoelectric crystals are good candidate nonlinear materials for second-harmonic generation. The intense electric field of the laser induces dipoles that oscillate at second or higher harmonics of the original frequency. The conversion efficiency is proportional to the nonlinear polarizability.

The second order nonlinear polarization, which we shall designate as $\mathbf{P}_2$, can be represented as follows when a plane wave of angular frequency ω is propagating in the z direction:

$$\mathbf{P}_2=\chi^{(2)}(\mathbf{E}_0 e^{i\omega t-ikz})^2=\chi^{(2)}\mathbf{E}_0^2 e^{2i\omega t-2ikz}, \quad (12)$$

where $\chi^{(2)}$ is the second-order nonlinear susceptibility of the medium. The above equation shows that the polarization wave of angular frequency 2ω propagates in the medium in the same direction as the fundamental laser beam with wave number $2k$. On the other hand, the wave number of the electromagnetic wave of angular frequency 2ω in the medium, which shall be represented by $k_{2\omega}$, is not equal to $2k$ because of dispersion of the medium. Thus special care must be taken in order to obtain phase matching between the polarization wave and the electromagnetic wave of the second-harmonic frequency.

The phase-matching condition can be established by the use of birefringence of anisotropic crystals. That is, in a negative uniaxial crystal, for example, one may be able to find an angle θ in refractive index space for which

$$n_e^{2\omega}(\theta)=n_o^{\omega}, \quad (13)$$

where the left-hand side represents the index of refraction of the extraordinary ray at angular frequency 2ω in the direction at an angle θ with the optical axis, and the right-hand side the index of refraction of the ordinary ray at ω. In a positive uniaxial crystal, the roles of extraordinary ray and ordinary ray should be exchanged.

The angle θ changes as a function of temperature of the crystal and the frequency of the laser beam, and it is sometimes possible to find a temperature at which the above condition is satisfied for $\theta=2\pi$. Such a case is sometimes called temperature phase matching, and is preferentially applied in the case where the phase matching is obtained at a temperature that the crystal can be easily kept at.

By choosing an appropriate nonlinear crystal, the efficiency of the second-harmonic generation can be made higher than 75% for a pulsed laser beam having high temporal spectroscopic brightness. For a continuously oscillating laser, high-efficiency generation of the second harmonic is rather difficult. The efficiency can be improved significantly in such a case by placing the nonlinear crystal in an optical cavity that is resonant at the laser frequency.

Generation of harmonics higher than the second can normally be done only with low efficiency. The efficiency can be improved greatly, but is still much lower than that of the second harmonic, by selecting a medium in which some resonance effect can be expected. In spite of the low efficiency that one expects in the high-order harmonic generation, it is useful for generating coherent radiation at extremely short wavelengths.

List of Works Cited

Hall, J. L., Lee, S. A. (1976), *Appl. Phys. Lett.* **29**, 367–340.

Jennings, D. A., Peterson, F. R., Evenson, K. M. (1979), in: H. Walther and K. W. Roth (Eds.), *Laser Spectroscopy IV*, Heidelberg: Springer, pp. 39–48.

Kowalsky, F. V., Demtröder, W., Schawlow, A. L., (1977), in: J. L. Hall and J. L. Carlsten (Eds.), *Laser Spectroscopy III*, Heidelberg: Springer, pp. 412–413.

Snyder, J. J. (1977), in: J. L. Hall and J. L. Carlsten (Eds.), *Laser Spectroscopy III*, Heidelberg: Springer, pp. 419–420.

Ueda, K., Takuma, H. (1987), in: C. Yamanaka (Ed.) *Short-Wavelength Lasers*, Heidelberg: Springer, pp. 178–185.

Further Reading

Arecchi, F. T., Schultz-Dubois, E. O. (Eds.) (1972), *Laser Handbook*, Amsterdam: North-Holland, Vols. 1 and 2.

Bass, M. Stitch, M. L. (Eds.) (1985). *Laser Handbook*, Amsterdam: North-Holland, Vols. 4 and 5.

Hecht, E. (1987), *Optics*, 2nd ed., Reading, MA: Addison-Wesley.

Mystre, P., Walls, D. F. (Eds.) (1991), *Nonclassical Effects in Quantum Optics: Key Papers in Physics*, New York: AIP.

Stitch, M. L. (Ed.) (1979), *Laser Handbook*, Amsterdam: North-Holland, Vol. 3.

Yariv, A. (1989), *Quantum Electronics*, 3rd ed., New York: Wiley.

LASERS, DYE, TECHNOLOGY AND ENGINEERING

F. J. DUARTE AND D. R. FOSTER, *Imaging Research Laboratories, Eastman Kodak Company, Rochester, New York, U.S.A.*

	Introduction	331
1.	**Laser-Pumped Pulsed Dye Lasers**	333
2.	**Flashlamp-Pumped Dye Lasers**	337
3.	**Continuous-Wave Dye Lasers**	341
4.	**Femtosecond Dye Lasers**	343
5.	**Intracavity Dispersion**	345
6.	**Applications**	347
	Glossary	349
	Works Cited	351
	Further Reading	352

INTRODUCTION

The study of dye lasers is motivated by their unparalleled versatility and intrinsic uniqueness as physical systems. For instance, dye lasers have been demonstrated to oscillate in the three states of matter, that is, in the gaseous, liquid, and solid states. However, it is largely in the liquid phase that this class of laser has shown its formidable utility and pervasiveness.

Basically, dye lasers are optically pumped, tunable, coherent light sources that can be divided into two large classes: pulsed dye lasers and continuous-wave (cw) dye lasers. Pulsed dye lasers are further categorized into the subclasses of laser-pumped pulsed dye lasers and flashlamp-pumped dye lasers. Femtosecond-pulsed dye lasers evolved from the technology of cw dye lasers. This latter class of dye lasers utilizes laser excitation.

The laser active medium consists of a solution of organic dye molecules in a liquid such as ethyl alcohol, methyl alcohol, or water (Jones, 1990). The most notable attributes of laser dye media are their broad wavelength tunability, wide spectral coverage, and practical simplicity. Hundreds of dyes have been reported to lase, and many of the more commonly used molecules for dye lasers belong to the oligophenylene, coumarin, xanthene, merocyanine, and cyanine classes. For a comprehensive listing of laser dye molecular species the reader should refer to Maeda (1984), Drexhage (1990), and Steppel (1991). As illustrated in Fig. 1, the spectral coverage provided by these dye classes extends from ~0.3 to ~1.1 μm. Further regions of the spectrum can be reached via other infrared dyes and by using dye lasers in conjunction with nonlinear techniques such as frequency doubling and Raman shifting.

Laser dyes are large organic molecules having molecular weights between 175 and 1000 amu that have strong absorption bands in the ultraviolet and visible spectral regions, and when excited by absorbed light, they exhibit intense fluorescence bands at longer wavelengths than their absorption. This red shift in wavelength maxima between absorption and fluorescence bands is called the *Stokes shift*.

One of the best known dye molecules is rhodamine 6G (a xanthene dye), depicted with its absorption and fluorescence spectra in Fig. 2. Rhodamine 6G is the most fully characterized laser dye and is often used as a standard to determine the relative performance of other laser dyes. This molecule has been reported to lase in the 550–650-nm spectral range with a variety of solvents, concentrations, and excitation sources.

The broad structure of the absorption spectrum shown in Fig. 2 corresponds to absorption transitions from the ground electronic state (S_0) to the first excited electronic state (S_1) and higher states. Each electronic state contains a multitude of vibrational and rotational levels. Because of line-broadening processes, the rotational levels are not resolved,

1-56081-067-X/94/$5.00+.50

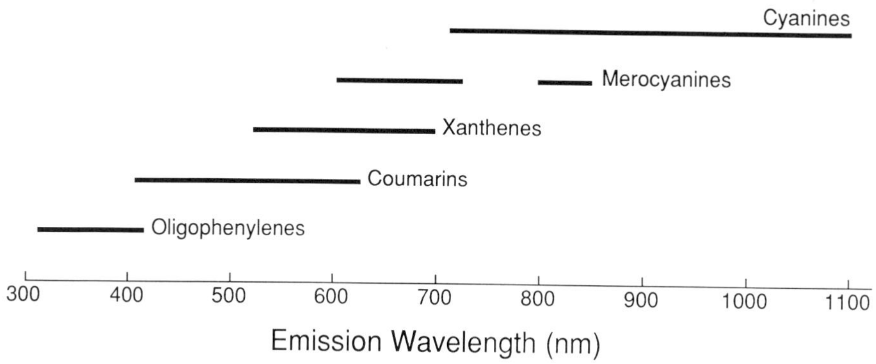

FIG. 1. Approximate spectral emission range of various dye classes. Emission from the cyanines extends beyond 1100 nm.

thus giving rise to a continuum. This property explains the continuous tunability associated with dye lasers. In addition to spectrally broad emission, the dye gain medium exhibits homogeneous broadening that enables all the excited molecules to contribute efficiently to emission in a narrow frequency band defined by intracavity dispersive elements (see Secs. 1 and 5).

The versatility of dye lasers originates in their ability to emit tunable coherent radiation. Indeed, to a large extent, the words *dye laser* have acquired synonymous status with *tunability*. Moreover, dye lasers offer tremendous operational flexibility:

1. covering the spectrum from the near ultraviolet to the near infrared,
2. emitting from femtosecond pulses to the continuous-wave regime,
3. yielding radiation from large single-pulse energies to high average powers.

Specifically, dye lasers have yielded up to 800 J in the single-pulse regime and average powers in *excess* of 2.5 kW in the high–pulse-repetition-frequency (prf) domain. Indeed, dye lasers are unrivaled in the generation of high–pulse-energy and high–average-power tunable coherent radiation in the visible. This unparalleled performance is inherently linked

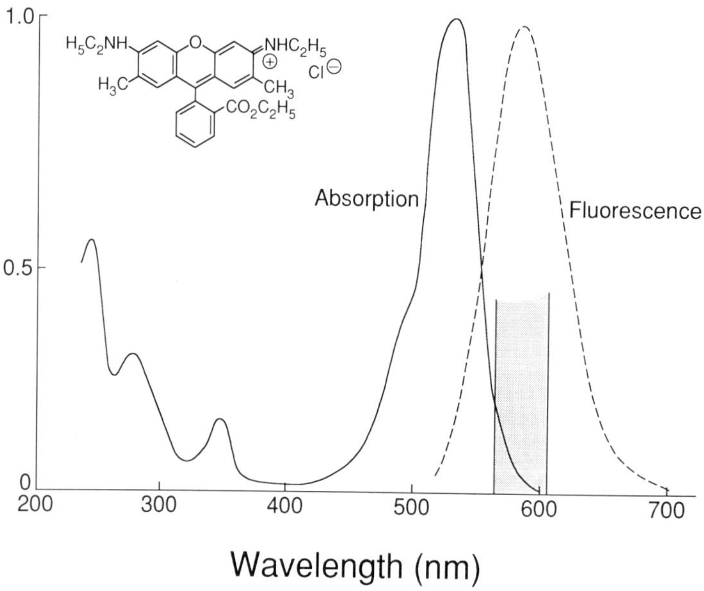

FIG. 2. Rhodamine 6G absorption and fluorescence spectra. The main absorption maximum corresponds to the $S_0 \rightarrow S_1$ transition. The shaded region is the approximate tuning range achieved with copper vapor laser (CVL) excitation. The molecular weight of this molecule is 479.

to the heat dissipation ability characteristic of liquid-phase dye molecular media. High average powers are essential to numerous applications including laser isotope separation (LIS), in industry, and in astronomy.

In general, the operational flexibility of the dye laser has yielded numerous and varied applications ranging from cancer photodynamic therapy to laser radar. The plethora of applications includes fields such as physics, chemistry, and biology. A perspective on applications is provided in Sec. 6.

The benefits of the dye laser to science extend beyond the area of applications, since dye-laser development has yielded significant contributions in the fundamental areas of amplified spontaneous emission (ASE), femtosecond laser physics, dynamical instabilities, and oscillator physics. Here, it should be emphasized that ASE refers to strong *spontaneous* emission that occurs *prior* to laser emission (see the Glossary for further information).

The dye laser has a rich literature. Since its discovery in 1966 (Sorokin and Lankard, 1966; Schäfer *et al.*, 1966) until today, there have been hundreds of papers published on dye-laser development. A perspective on the dye-laser literature and its history can be obtained from the books suggested in Further Reading.

Although we focus here mainly on liquid-phase dye lasers, alternative solid-state dye lasers using polymers (see, for example, Pacheco *et al.*, 1988) and silicate matrices are also available. Indeed, recent developments have produced very attractive organically modified silicate matrices that offer significant potential for long-lived, solid-state dye lasers for low-power applications. The oscillator and system configurations discussed here apply to liquid and/or solid-state lasers.

In this article a basic and brief survey of the elements important to dye lasers is provided. In particular, attention is given to developments in the area of intracavity physics and resonator design. This focus is justified by the numerous contributions of a *fundamental nature* made by dye lasers that are now being applied to lasers in general. Details on the science of the molecular media and other aspects of dye-laser physics are provided in the article ORGANIC DYE LASERS.

1. LASER-PUMPED PULSED DYE LASERS

Optical pumping from pulsed lasers was the first avenue of excitation demonstrated in dye lasers (Sorokin and Lankard, 1966; Schäfer *et al.*, 1966). At present there is a variety of solid-state and gas pulsed lasers that are applied to the excitation of dye lasers. These lasers offer different excitation wavelengths and energetic, temporal, and prf characteristics. Pulsed lasers suitable for dye-laser pumping include KrF, XeCl, and XeF. These lasers emit via the $B\,^2\Sigma^+_{1/2}$–$X\,^2\Sigma^+_{1/2}$ transition at $\lambda \sim 248$, 308, and 351 nm, respectively. A simple and inexpensive source of ultraviolet radiation is the nitrogen laser that emits at $\lambda \sim 337$ nm via the $C\,^3\Pi_u$–$B\,^3\Pi_g$ transition.

The copper vapor laser (CVL) emits mainly via the $^2P_{3/2}$–$^2D_{5/2}$ transition at $\lambda \sim 510$ nm and also via the secondary $^2P_{1/2}$–$^2D_{3/2}$ transition at $\lambda \sim 578$ nm. The CVL has the advantage of high efficiency, in the 1–3% range, and operates at prf's up to 30 kHz. A widely applied solid-state source is the Nd:YAG laser (YAG denotes yttrium aluminum garnet) that emits via the $^4F_{3/2}$–$^4I_{11/2}$ transition at $\lambda \sim 1064$ nm. The second, third, and fourth harmonics of this emission yield excitation at 532, 355, and 266 nm, respectively. In general, laser-pumped pulsed dye-laser emission has been shown to cover the spectrum from ~ 315 to 1100 nm.

The variety of excitation wavelengths available makes possible the design of very efficient dye-laser systems where excitation is accomplished by populating directly the first excited electronic state via the $S_0 \rightarrow S_1$ absorption transition. An important example of this type of excitation is the efficient CVL pumping of rhodamine dye lasers reported to be as high as 50% when using the transition at 510.5 nm (Webb, 1991). Furthermore, this dye under CVL excitation exhibits remarkably long lifetimes (Duarte, 1990).

The nitrogen laser is a relatively low-prf excitation source of general utility. Its ultraviolet $C\,^3\Pi_u$–$B\,^3\Pi_g$ transition is used to excite a large number of dyes whose emissions span the electromagnetic spectrum from ~ 360 to 1000 nm. Excimer lasers, such as KrF and XeCl, can be used to excite dyes further into the ultraviolet. In addition, excimer lasers provide considerably higher pulsed energies and higher prfs. Using XeCl laser excitation,

Bos (1981a) demonstrated lasing in the 335–912-nm range in an eight-cell dye-laser system. Emission at shorter wavelengths, originating at ~315 nm can be accomplished using KrF laser excitation.

Narrow-linewidth dispersive oscillator configurations are described in Fig. 3. Oscillators are important since they form the first stage of any laser system. It is the oscillator that provides the vital wavelength and linewidth information that is subsequently augmented by the amplification stages of the system. These master oscillators are usually composed of an output coupler mirror, a high–optical-quality quartz dye cell of trapezoidal geometry, a multiple-prism beam expander, and a grating either in Littrow configuration, where the angle of incidence equals the angle of diffraction [Fig. 3(a)], or near grazing-incidence configuration [Fig. 3(b)]. As illustrated in Fig. 3, the function of the beam expander is to magnify the narrow beam (w) emerging from the gain region in order to illuminate the grating completely. As explained in Sec. 5, the overall beam magnification factor is the product of the individual beam expansion coefficients provided by each prism.

Typical cavity dimensions are 1 cm for the length of the dye cell and about 10 cm for the overall cavity. In the case of low-prf operation the dye may be excited under static conditions. On the other hand, high-prf operation in the kilohertz range requires rapid flow of the dye in a direction perpendicular to the plane of propagation for heat-dissipation purposes. Linear speeds of dye flow are typically in the range 5–10 m s^{-1}. These high velocities are achieved using high-pressure pumps and a very narrow cross section at the excitation area that may be typically 10 mm in length by less than 1 mm in width. A typical excitation volume defined by the dimensions of the focused pump laser beam is 10 mm in length, with a cross-section diameter less than 200 μm.

In research prototypes optical components, such as mirrors and gratings, are usually mounted on high-resolution kinematic mounts that are in turn positioned on magnetic bases deployed over optical tables. The gratings can be rotated manually, using high-resolution Vernier micrometers, or automatically, using mechanical or piezoelectric drivers controlled electronically. Using these methods linear sub-micron resolutions can be achieved. The multiple-prism expanders are deployed over adjustable commerical optical surfaces or special platforms designed to minimize demands on intracavity space. This approach allows spatial flexibility that results in very compact cavities. In this regard, the oscillators shown in Fig. 3 can have a cavity length of 10 cm or less. Commercial cavities are often deployed along a linear Invar structure, and components are attached using traditional mechanical methods. These configurations tend to be

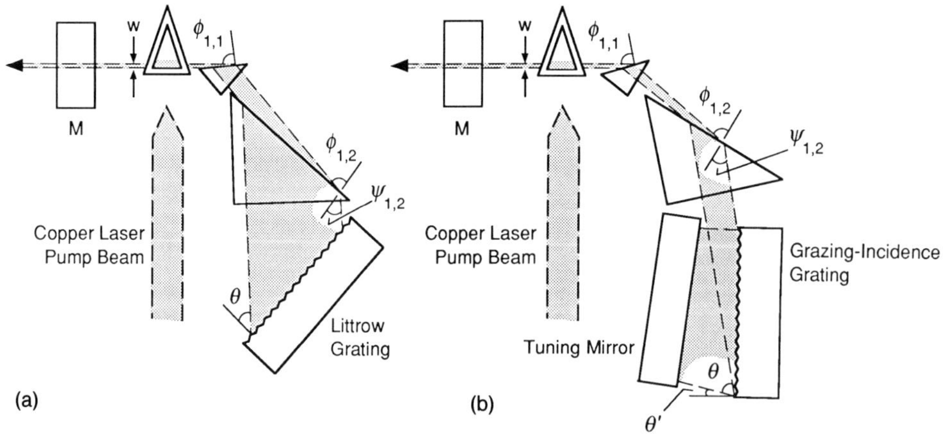

FIG. 3. Dispersive oscillator configurations showing (a) a multiple-prism Littrow (MPL) oscillator and (b) a hybrid multiple-prism grazing-incidence (HMPGI) oscillator (adapted from Duarte and Piper, 1984). These dispersive oscillator configurations apply to either liquid-phase or solid-state dye lasers.

permanent and thus provide fewer alternatives than their research counterparts.

Dispersive oscillators incorporating a multiple-prism beam expander and the grating in a Littrow configuration are classified as multiple-prism Littrow (MPL) oscillators (see, for example, Duarte, 1990). Alternatively, if a multiple-prism beam expander is deployed with a grating in a near-grazing-incidence configuration, then the cavity is classified as a hybrid multiple-prism grazing-incidence (HMPGI) oscillator.

Intracavity prismatic preexpansion in conjunction with gratings deployed in a near-grazing-incidence configuration was introduced by Duarte and Piper (1981). The term *prismatic preexpansion* refers to the beam expansion necessary to illuminate completely a grating deployed in a near-grazing-incidence configuration as illustrated in Fig. 3(b). At present, HMPGI oscillators are widely used in research and commerical cavity designs.

MPL oscillators offer higher conversion efficiencies than HMPGI oscillators given the relatively higher angle of incidence utilized to deploy the grating in the latter configuration. On the other hand, an advantage of HMPGI cavities is their intrinsic higher dispersion that yields narrower linewidths than equivalent MPL configurations. Alternative cavities include the pure grazing-incidence design (see, for example, Littman and Metcalf, 1978) and the telescopic Littrow resonator (Hänsch, 1972). For a description of distributed-feedback dye lasers see Schäfer (1990).

As listed in Table 1, MPL and HMPGI oscillators produce very narrow linewidths and nearly diffraction-limited beam quality at relatively high conversion efficiencies. In this regard, linewidth values in the sub-GHz range and characteristic of single-longitudinal-mode oscillation are readily achieved. Oscillator cavities incorporating multiple-prism expanders offer energetic advantages over pure grazing-incidence cavities and are more compact, more rugged, and easier to align than telescopic resonators. In telescopic resonators the beam-expansion function is performed by a conventional transmission or reflection telescope such as a Galilean or a Cassegrain telescope, respectively.

In addition to configurational and energetic differences, HMPGI oscillators have dispersive characteristics intrinsically distinct from those of the pure grazing-incidence cavity.

Table 1. Performance of dispersive dye laser oscillators.[a]

Cavity configuration	Laser linewidth[b]	Efficiency[c] (%)
Grazing incidence	100 MHz $\leqslant \Delta\nu \leqslant$ 2.5 GHz	2–4
MPL	60 MHz[d] $\leqslant \Delta\nu \leqslant$ 1.6 GHz	5–15
HMPGI	138 MHz[e] $\leqslant \Delta\nu \leqslant$ 1.15 GHz	4–10

[a]All values refer to closed cavity configurations.
[b]Conversion to wavelength units can be done using $\Delta\lambda = \lambda^2(\Delta\nu/c)$. For example, $\Delta\nu = 100$ MHz becomes $\Delta\lambda = 0.000112$ nm at $\lambda = 580$ nm.
[c]Efficiency values refer to laser-pumped oscillators.
[d]Oscillator incorporated an intracavity etalon (Bernhard and Rasmussen, 1981).
[e]Measured in a long-pulsed flashlamp-pumped HMPGI oscillator (Duarte *et al.*, 1991).

Wavelength tuning is inherent to dispersive oscillators, since both the prismatic beam expander and the grating provide angular dispersion. However, as outlined in Sec. 5 the prismatic beam expanders can be designed to yield zero dispersion, thus making wavelength tuning entirely dependent on the grating equation

$$m\lambda = d(\sin\theta + \sin\theta'), \qquad (1)$$

where λ is the emission wavelength, m is the diffraction order, d is the groove spacing, θ is the angle of incidence, and θ' is the angle of diffraction [see Fig. 3(b)]. For the special case of a grating deployed in the Littrow configuration where $\theta = \theta'$, the equation becomes

$$m\lambda = 2d\sin\theta. \qquad (2)$$

Thus effective wavelength tuning can be achieved by angular displacement of the grating. Using grating tuning in a MPL oscillator similar to that described in Fig. 3(a) the fluorescence spectrum from molecules such as D_2CO has been recorded as illustrated in Fig. 4. MPL and HMPGI oscillators can provide tuning ranges of up to 50 nm using a single dye.

Other means of tuning involve the use of Fabry–Pérot étalons and pressure-tuning techniques. Pressure tuning can be done by mounting the oscillator within a pressure chamber and then using gases such as O_2 and N_2. The

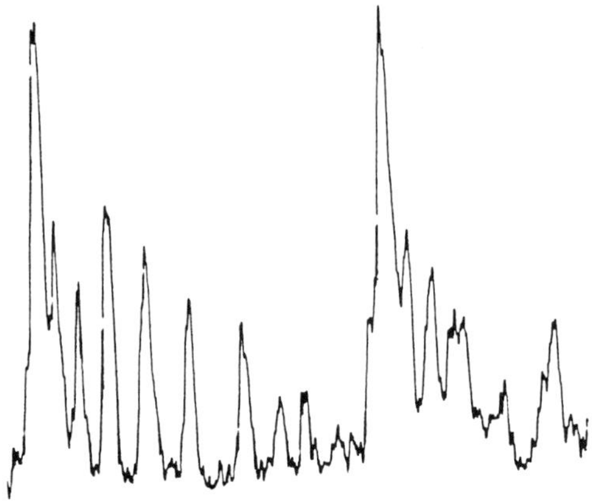

FIG. 4. Fluorescence spectrum from D_2CO due to excitation near 364 nm using a MPL oscillator ($\Delta\nu \approx 2$ GHz). The tuning wavelength range covered by the laser was 0.227 nm (~514 GHz). Tuning was achieved by rotation of the grating using a piezoelectric driver (adapted from Duarte, 1988).

method works because the index of refraction of the gas surrounding the cavity is a function of the change in pressure. In this manner, frequency can be changed almost linearly with pressure.

Dispersive oscillator configurations described here, and first developed for dye lasers, have found widespread applications in gas, solid-state, and semiconductor lasers.

Schematics of pulsed dye-laser systems are outlined in Fig. 5. As illustrated in the figure, a typical system can integrate an oscillator and several stages of amplification. Usually, research-class dye lasers incorporate one or two stages of amplification. However, larger high-power systems are composed of a master oscillator and numerous stages of amplification.

Amplification in laser-pumped dye lasers is accomplished in a single pass. In most research systems, the excitation laser beam is divided using beam splitters (see Fig. 5) with a small portion of the total energy, usually ⩽10%, used to excite the oscillator. Thus most of the energy is applied to pump the amplifiers. For a three-stage amplification system Bos (1981b) reports conversion efficiencies of ~55% and an overall gain factor approaching 229. The efficiency of the oscillator was 6% and the efficiency at the amplifiers was 60%.

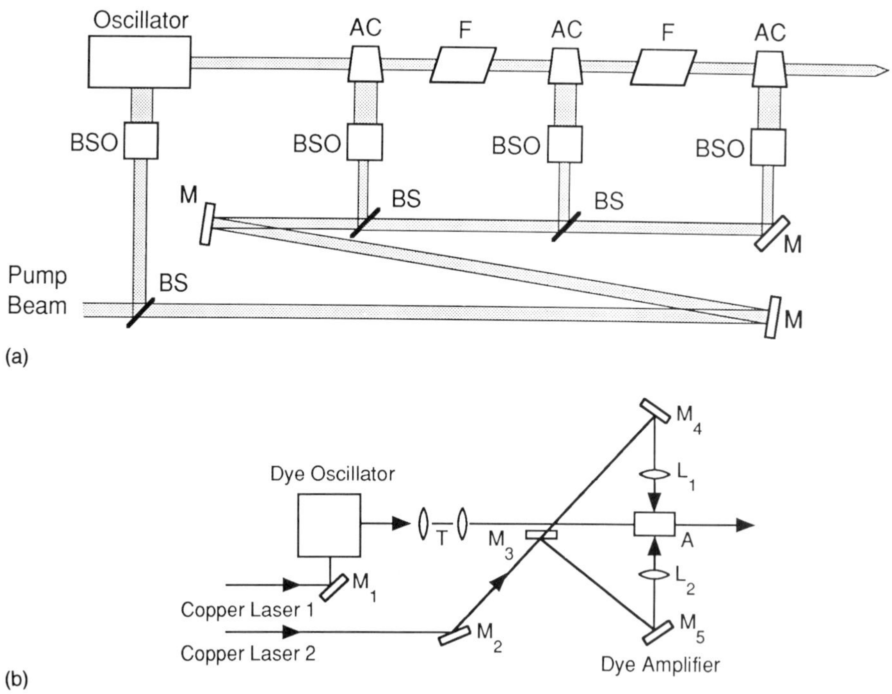

FIG. 5. (a) Transverse excitation of multistage oscillator-amplifier dye-laser system. AC, amplifier cell; BS, beam splitter; BSO, beam shaping optics; F, spectral filter; M, mirror. (b) Oscillator-amplifier system using collinear two-sided excitation of the amplifier (adapted from Hargrove and Kan, 1980,© 1980 IEEE).

This system delivered 165 mJ at a laser linewidth of $\Delta\nu \approx 320$ MHz.

Excitation of amplifiers demands careful design criteria to reduce the level of ASE. Since ASE occurs in the buildup time prior to the threshold of narrow-linewidth laser emission, an effective avenue to reduce the overall ASE of the system is to delay excitation of the amplifiers, using optical delay lines, until the narrow-linewidth emission from the oscillator arrives [see Fig. 5(a)]. Typical delay times are within the 10-ns range. An additional technique to reduce ASE consists in the deployment of spatial and spectral filters between amplification stages as demonstrated by Wallenstein and Hänsch (1975).

A useful measure of ASE can be given in terms of spectral energy density (Duarte, 1990):

$$\frac{\rho_{\mathrm{ASE}}}{\rho_l} = \left((\Delta\Lambda)^{-1} \int_{\Lambda_1}^{\Lambda_2} W(\Lambda) d\Lambda \right) \times \left((\Delta\lambda)^{-1} \int_{\lambda_1}^{\lambda_2} E(\lambda) d\lambda \right)^{-1}. \quad (3)$$

Here $W(\Lambda)$ is the ASE energy function and $\Delta\Lambda$ is the full width of the dye broadband spectral gain region. For narrow-linewidth lasing $E(\lambda)$ approaches a Gaussian distribution and $\Delta\lambda$ is in the sub-pm range. This definition takes into account the spectral brightness of the laser. For CVL-pumped dispersive dye-laser oscillators $\rho_{\mathrm{ASE}}/\rho_l \approx 5 \times 10^{-7}$ (Duarte, 1990).

In large-scale dye-laser systems the amplifier cells are usually pumped using a collinear two-sided transverse excitation geometry (Hargrove and Kan, 1980) [see Fig. 5(b)]. In systems using excitation from several pump lasers the requirements for synchronous electronic triggering can be rather demanding.

Pulsed dye lasers have demonstrated large pulsed energies and high average powers. In the area of high average power Tallman and Tennant (1991) reported on a near-ultraviolet oscillator-amplifier system yielding up to 75 W of average power using XeCl laser excitation.

The large-scale CVL-pumped pulsed dye-laser system at Lawrence Livermore National Laboratory has demonstrated reliable operation at average powers exceeding 2500 W at a prf of 13.2 kHz (Bass *et al.*, 1992). This system comprises several master-oscillator–power-amplifier (MOPA) chains operating at 50–60% photon conversion efficiencies. Each power amplifier chain is composed of three to four amplifiers. The master oscillators are of the MPL class and incorporate an intracavity étalon. A partial view of a MOPA chain is shown in Fig. 6.

2. FLASHLAMP-PUMPED DYE LASERS

An alternative source of pulsed tunable coherent radiation from the near ultraviolet to the near infrared is the flashlamp-pumped dye laser. A detailed review on this subject is given by Everett (1991).

Flashlamp-pumped dye lasers offer unique output characteristics that have found many applications. Single-pulse energies vary from a few millijoules, for compact high-performance oscillators, to several hundred joules for large devices. Pulse lengths range from the nanosecond regime to tens of microseconds. Furthermore, average powers in excess of 1 kW have been demonstrated and overall efficiencies can approach 2%.

These lasers can be excited using coaxial or linear flashlamps. The coaxial devices [Fig. 7(a)] provide uniform, inherently concentric, excitation of the gain volume and are built using a concentric arrangement with an outer reflector surrounding the flashlamp that illuminates the dye region through the cooling fluid. Coaxial flashlamp-pumped dye lasers have yielded up to 400 J per pulse and a 0.8% efficiency (Baltakov *et al.*, 1974).

The emission from flashlamps employed in dye-laser excitation can be characterized using the blackbody radiation model. A significant portion of this emission is in the 200–300-nm region. This ultraviolet radiation induces photodegradation of the dye solution via photochemical processes that cause the loss of dye molecules and the creation of photoproducts that absorb at the emission wavelength.

In addition to its thermal mission, the cooling fluid is used to absorb some of the ultraviolet radiation emitted by the lamp. Additives such as caffeine, ethanol, and $CuSO_4$ may be used as ultraviolet filters. Other filtering measures include the use of Pyrex tubing in the dye cell. These measures apply to both excitation methods.

FIG. 6. Partial view near the first amplifier of a dye-laser MOPA chain at the Lawrence Livermore National Laboratory large-scale dye-laser system. Photograph courtesy of I. L. Bass, LLNL.

A technique used to improve the conversion efficiency and lifetime of the dye solution is to employ spectral transfer dyes around the flashlamps to convert the ultraviolet radiation to longer-wavelength emission compatible with efficient excitation of the laser dye. Everett *et al.* (1986) report a 75% efficiency improvement using the dye converter stilbene 420 and coumarin 504 as laser dye.

Linear flashlamp-pumped dye lasers often employ two or more flashlamps deployed symmetrically around the dye cell. Excitation with two flashlamps may involve the use of elliptical reflectors to illuminate the active region more efficiently. Some high-energy devices have used as many as 18 linear lamps configured to yield a nearly concentric excitation flux to the dye cell [Fig. 7(b)].

Both methods of excitation employ relatively large dye volumes with active lengths in the 10–100-cm range and diameters from a few millimeters to up to 6 cm. A common feature is the use of relatively low dye concentrations that may range from 10^{-5} *M* to a few times 10^{-4} *M*.

A further successful configurational alternative is the flashlamp-pumped waveguide dye laser. These devices utilize rapid dye flow that is *transverse* to the plane of propagation. Mazzinghi *et al.* (1981) describe a planar waveguide dye-laser active volume with a 150-mm length, 55-mm height, and 0.5-mm width, excited by three flashlamps on each side [see Fig. 7(c)]. These authors report an average power of 200 W at a prf of 50 Hz. The transverse flow device described in Fig. 7(c) was designed to use either cooling fluid or a spectral transfer dye solution.

In the area of high average power Morton and Draggoo (1981) reported on a transverse-flow linear device delivering 1.2 kW at a prf of 850 Hz. Recently, Klimek *et al.* (1992) achieved ~1.4 kW at a 1.8% efficiency. The pulse energy was 140 J and the prf 10 Hz. This performance was obtained with the excitation provided by 16 linear lamps, eight on each

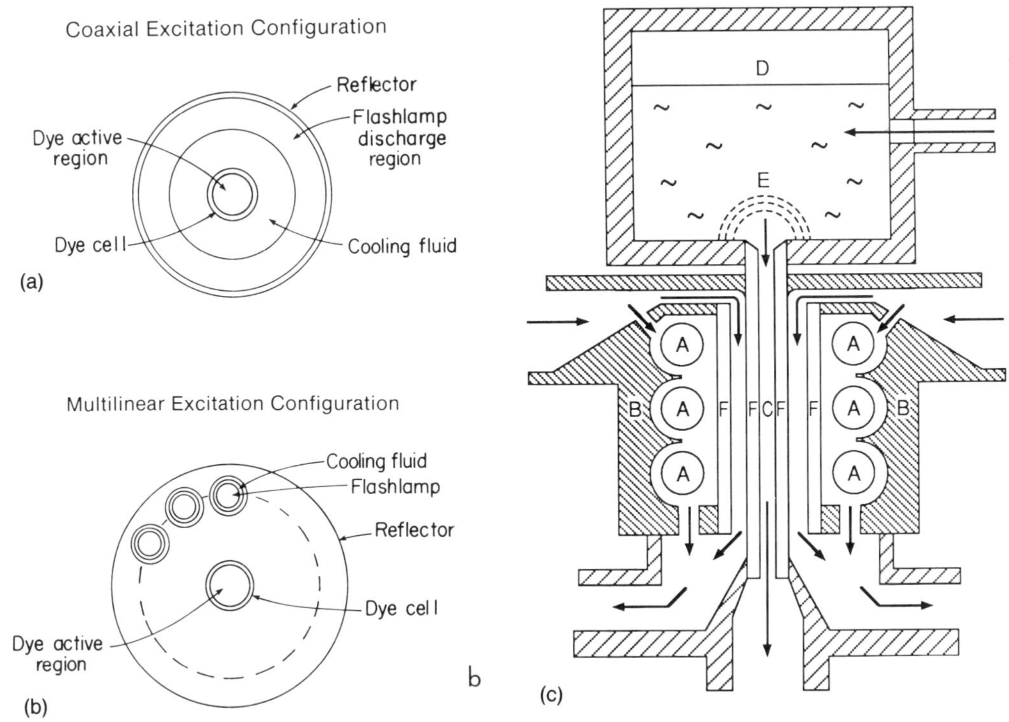

FIG. 7. (a) Cross section of a coaxial flashlamp dye laser and (b) cross section of a device using multiple-lamp linear excitation (adapted from Duarte, 1990). (c) Cross section of a transverse-flow waveguide flashlamp-pumped dye laser. Laser emission propagates normal to the page. *A*, flashlamp; *B*, flashlamp reflector; *C*, dye active region; *D*, expansion channel; *E*, filters; *F*, glass plates. The arrows indicate the flow direction of the dye and cooling fluid. (From Mazzinghi *et al*., 1981,© 1981 IEEE.)

side, in a transverse-flow device using rhodamine 590 and a spectral transfer dye. In the high–average-power transverse-flow devices the dye flows at speeds in the range 5–10 m s^{-1}.

Flashlamp-pumped dispersive dye-laser oscillators have yielded linewidths in the region $138 \leqslant \Delta\nu \leqslant 370$ MHz, at 3–5 mJ per pulse, in a nearly-diffraction-limited beam (Duarte *et al.*, 1991). Characteristic of these optical oscillators is their extremely low ASE level given by a figure of $\rho_{ASE}/\rho_l \approx 10^{-9}$. This performance is achieved by the use of low dye concentrations ($\sim 6\times10^{15}$ cm^{-3}) and the judicious application of oscillator physics to the cavity design. The MPL and HMPGI oscillators utilized in these experiments are shown schematically in Fig. 8. A temporal emission pulse from a dispersive oscillator is shown in Fig. 9. The intensity modulation is due to mode beating between two longitudinal modes separated by $\Delta\nu \approx 300$ MHz. The magnitude of the modulation shown in this case is characteristic of mode beating between a very strong longitudinal mode and a weak secondary longitudinal mode. This oscillatory behavior is due to dispersive frequency selectivity exclusively and does not result from mechanisms associated with mode locking (see Sec. 4).

These high-performance dispersive oscillators were engineered in an all-Superinvar structure and tested for mechanical and thermal stability using vehicular displacement on a rugged terrain with excellent results (Duarte *et al.*, 1991). A partial photograph of the ruggedized dispersive oscillator is shown in Fig. 10. Additional features of this laser are a stainless-steel dye reservoir and a special dye and water circulation network consisting of Teflon protected by an outer stainless-steel sleeve. In this system a single dye solution can be used for a period in *excess of one year.* A specially designed thermal control network

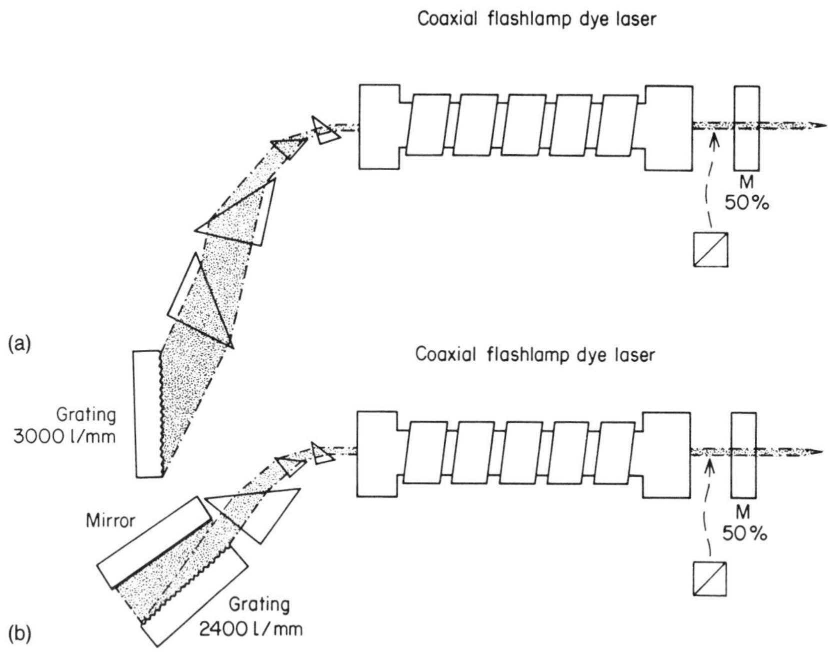

FIG. 8. High-performance flashlamp-pumped MPL and HMPGI dye-laser oscillators (from Duarte *et al.*, 1991).

allows the matching of the dye temperature to the cooling-water temperature to within 0.01 °C. This thermal control greatly reduces radial thermal gradients in the active medium, thus neutralizing an important source of active-medium inhomogeneities. The Superinvar laser chassis is attached to four pneumatic supports, each using an air pressure of 1.72 $\times 10^5$ N m^{-2}. The engineering specifications given here far exceed those of typical commerical designs and respond to requirements of long-term stable operation with narrow linewidth and diffraction-limited beam quality.

Signal amplification in flashlamp-pumped dye-laser systems is often accomplished using forced oscillators rather than single-pass amplifiers. A master-oscillator–forced-oscillator system is illustrated in Fig. 11. Here the forced oscillator uses an unstable resonator that is

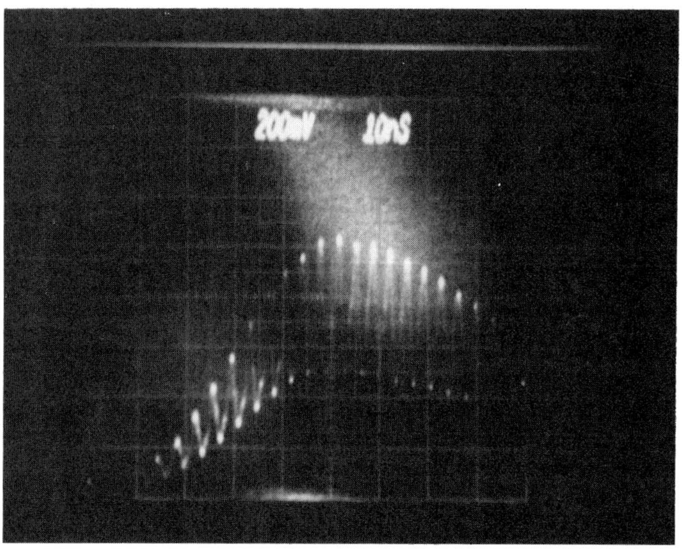

FIG. 9. Temporal pulse demonstrating double–longitudinal-mode oscillation ($\Delta\nu \approx 300$ MHz) characteristic of flashlamp-pumped dispersive dye-laser oscillators. The time scale is 10 ns div.

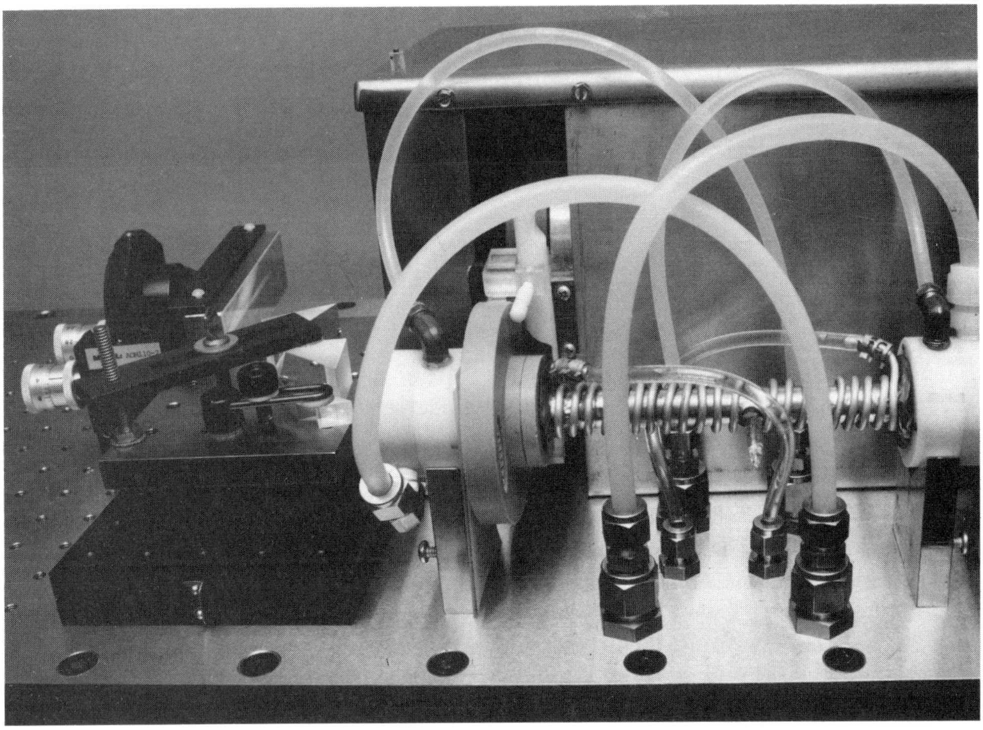

FIG. 10. Partial perspective of a ruggedized dispersive dye-laser oscillator. The material of the cavity structure is Superinvar. This oscillator has been operated with a single rhodamine 590 dye solution, at $1\times10^{-5}M$ in methanol, for *more than one year*. (From Duarte *et al.*, 1991.)

composed of a planoconcave total reflector with a concentric injection hole and a convex–concave meniscus with a high-reflectance circular coating at its center. In these systems the pulse from the master oscillator is synchronized to arrive during the buildup time of the forced-oscillator pulse.

Interesting performances from master-oscillator–forced-oscillator systems include that reported by Maeda *et al.* (1975), who quote an output energy of 4 J at $\Delta\nu\approx4$ GHz using a multiple-étalon assembly at the oscillator. The energy gain was ~267. Using a multiple-prism grating master oscillator in conjunction with a forced oscillator utilizing an unstable resonator (Fig. 11), Duarte and Conrad (1987) reported an output energy of 600 mJ at $\Delta\nu\leqslant375$ MHz and an energy gain ~60. These figures correspond to high-fidelity amplification where the interferometric characteristics of the master oscillator are preserved and the ASE level is very low. Further experiments using dispersive oscillators and a forced oscillator with an optimized unstable resonator have demonstrated high-fidelity energy amplification with gain factors in excess of 450.

3. CONTINUOUS-WAVE DYE LASERS

Continuous-wave dye lasers using Kr^+ or Ar^+ laser excitation span the electromagnetic spectrum from ~370 to ~1000 nm. Frequency-doubling techniques have been used to extend the output to the ultraviolet region from ~260 to ~390 nm. Given their unique stable single-frequency tunable operation capabilities, these lasers have made important contributions to physics, spectroscopy, chemistry, and related fields. An excellent and detailed review on the subject is provided by Hollberg (1990).

The original pumping scheme of cw dye lasers, reported by Peterson *et al.* (1970), is shown in Fig. 12. The pump beam from an Ar^+ laser is focused via a lens onto the dye gain region that is contained within the reso-

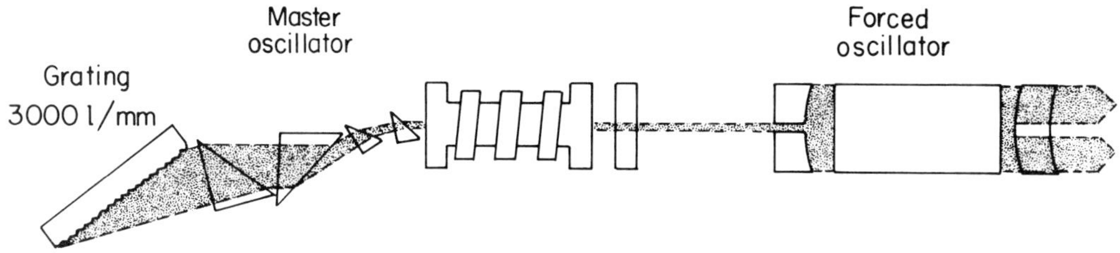

FIG. 11. Master-oscillator–forced-oscillator system. The forced oscillator utilizes a positive branch unstable resonator. (From Duarte and Conrad, 1987.)

nator. The dichroic mirrors transmit the blue-green excitation beam and reflect in the red region of the spectrum. An important feature of this design was the flowing of the dye solution at a speed of about 2 m s^{-1}. This was necessary to remove the excess heat and to neutralize the influence of the triplet state. With this device Peterson *et al.* (1970) reported an output of 30 mW with about 1 W of pump power delivered in a TM_{00} beam profile focused to a $1/e^2$ diameter of 2 mm. A refinement in this type of longitudinally excited linear configuration was the introduction of external mirrors and the integration of intracavity dispersive elements for wavelength tuning (Tuccio and Strome, 1972).

In subsequent cw dye-laser designs the dye cell was replaced by a fluid jet created by the passage of dye through a nozzle (Runge and Rosenberg, 1972). This technique coupled with the use of high-viscosity solvents, such as ethylene glycol, produces high-optical-quality surfaces at the jet and flow speeds in the 10-m s^{-1} range.

In addition to the original cavity design concepts introduced by Peterson *et al.* (1970) other cavity configurations of current use are the three-mirror folded resonator (see, for example, Peterson, 1979) and a four-mirror unidirectional ring resonator [see, for example, Hollberg (1990)]. Both these cavities are depicted in Fig. 13. Excitation radiation is usually delivered semilongitudinally at a small angle, on the propagation plane, relative to the optical axis defined by M_1 and M_2. Dispersive and frequency-selective elements (FSE) are positioned between M_2 and M_3 and between M_3 and M_4 for the folded resonator and the ring resonator, respectively. In addition, ring cavities incorporate what is called a unidirectional device (UDD), positioned next to the tuning elements, that controls the direction of the traveling wave. This optical diode selectively rotates the polarization of one of the propagating directions of the laser emission. Selective polarization rotation is accomplished by exploiting the Faraday effect on a appropriate optical material. The optical diode, or UDD, is integrated by a crystal or glass in a magnetic field that is applied parallel to the direction of propagation and a quartz plate that induces a reciprocal polarization rotation. The Faraday effect causes a rotation of the radiation of both propagating directions. Subsequent transmission by one of the propagating beams through the quartz plate neutralizes the initial rotation. On the other hand, the counterpropagating beam suffers losses induced by transmission through intracavity Brewster elements (see Hollberg, 1990, for further details).

Cavity design analysis, determination of stability conditions, and evaluation of astigmatism can be established using propagation matrices as discussed by Diels (1990). These matrices can be used to characterize the propagation of Gaussian beams. Each optical component and propagation space is represented by a particular matrix so that an overall matrix for the resonator can be derived. This allows the prediction of important parameters such as beam expansion and beam divergence.

Frequency-selective elements used in cw dye lasers include gratings, prisms, étalons, and birefringent crystals. In addition, a number of interferometers have been used to control frequency characteristics. Usually three optical elements are used to attain single-frequency operation. In a typical cavity the frequency-selective process may proceed with the use of a birefringent filter or a prism to provide the necessary bandwidth to discriminate in favor of a mode of a thin étalon, which,

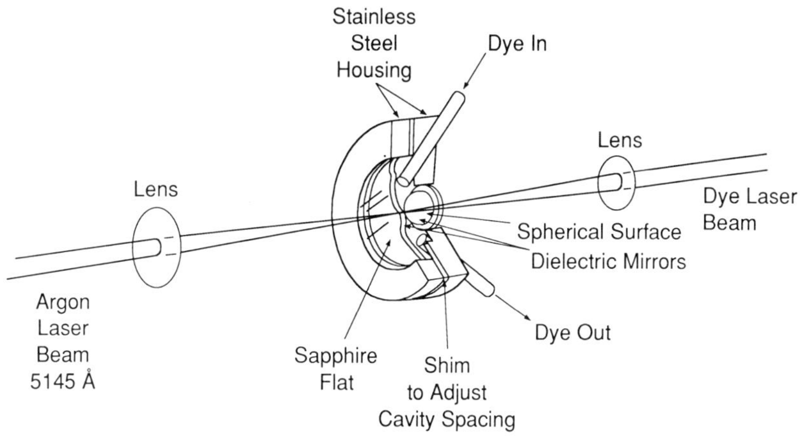

FIG. 12. Schematics of the first cw dye laser (from Peterson *et al.*, 1970). See text for details.

in turn, discriminates in favor of a mode of the second étalon with a considerably smaller free spectral range (FSR). The second étalon is designed with a FSR and finesse sufficient to select a single longitudinal mode of the cavity. In some designs single-frequency emission is achieved by replacing the second étalon by an intracavity interferometer such as a Mach–Zehnder interferometer.

Linear standing-wave cw dye-laser cavities exhibit the effect of *spatial hole burning* that causes the laser to oscillate in more than one longitudinal mode (Hollberg, 1990). This problem can be avoided in ring-cavity configurations [Fig. 13(b)] where the oscillation is in the form of a traveling wave. Furthermore, ring cavities are known to provide, with sufficient pump power, greater single-mode power. On the other hand, ring cavities involve increased optical complexity and exhibit higher pump thresholds.

As outlined earlier, a very important contribution of cw dye lasers has been the generation of very stable single-frequency oscillation. This is a topic that involves considerable technical detail such as cavity structure, materials, thermal stabilization, control of dye-flow turbulence, active servomechanisms, and a series of frequency-locking techniques. For an in-depth discussion of this topic, including post-laser stabilization methods, the reader should refer to Hollberg (1990). In this regard, Hough *et al.* (1984) demonstrated linewidths less than 750 Hz in an experiment that identified environmental noise as a principal limitation to stabilization. Post-laser stabilization using electro-optics modulators (Hall and Hänsch, 1984) can stabilize the frequency at the hertz range.

In the area of power performance Johnston *et al.* (1982) report 5.6 W of stabilized output power in a single longitudinal mode at 593 nm. This dye laser had a ring-cavity configuration and was excited by an Ar^+ laser yielding 24 W. On the other hand, Baving *et al.* (1982) report a 43-W multimode output using rhodamine 6G in a simple folded cavity and 200 W of pump power from an Ar^+ laser.

4. FEMTOSECOND DYE LASERS

The dye laser is central to the development of femtosecond laser physics and technology. Consequently, its impact in the science of femtosecond phenomena is of gigantic proportions. Here, we present an overview of the more important elements of femtosecond dye lasers. An extensive and detailed review of the subject is given by Diels (1990).

The development of ultrashort-pulse, or ultrafast, dye lasers is intimately related to the phenomenon of *mode locking*. Broadband lasers can oscillate in a number of transverse modes and a corresponding multitude of unrelated longitudinal modes whose relative phases vary randomly as a function of time. Using an intracavity aperture a single transverse mode can be selected, thus restricting oscillation to a sequence of equally spaced longitudinal modes in the frequency domain. Forcing these modes to have the same relative

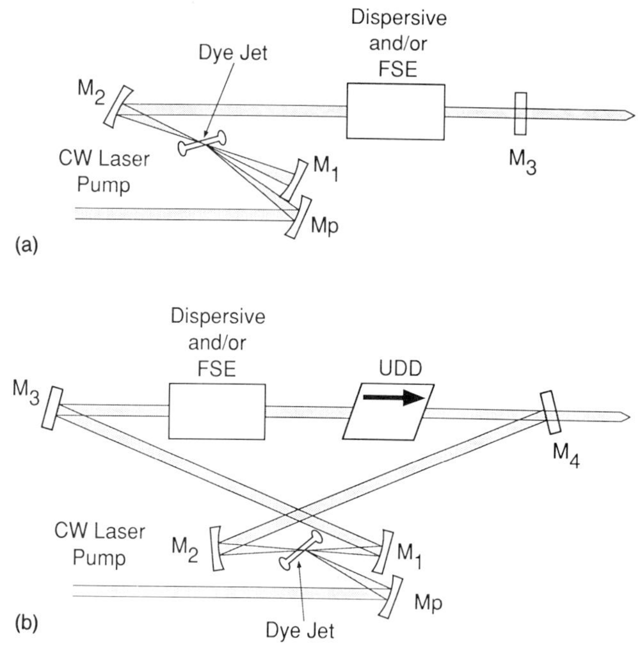

FIG. 13. Typical cw dye-laser cavity configurations. (a) The linear three-mirror folded cavity; (b) the eight-shaped ring dye-laser cavity constituted by M_1, M_2, M_3, and M_4. Here, FSE are frequency-selective elements, and UDD represents a unidirectional device (see text). (Adapted from Hollberg, 1990.)

phase results in mode locking. This yields a train of pulses in the temporal domain separated by a period inverse to the mode spacing $\delta\nu=c/2L$, where c is the speed of light and L is the cavity length. The sequence of temporal pulses corresponds to a single intracavity pulse contributing to the output on every return pass. The number of longitudinal modes allowed to oscillate is given approximately by the gain bandwidth divided by $\delta\nu$. The shortest possible pulse length is determined by the reciprocal of the dye gain bandwidth, approximately.

Mode locking in dye lasers was first demonstrated in simple flashlamp-pumped dye-laser cavities incorporating an intracavity *saturable absorber* (Schmidt and Schäfer, 1968). In this work the broadband flashlamp-pumped rhodamine 6G dye laser emitted a relatively smooth temporal pulse with a few hundred nanosecond pulse duration. Insertion of the saturable absorber 3,3′-diethyloxadicarbocyanine iodide (DODCI) caused the temporal pulse to change into a pulse train of subnanosecond pulses.

Picosecond and subpicosecond emission was achieved using simple linear cw dye-laser cavities incorporating intracavity saturable absorbers. An important development in establishing shorter pulses was the concept of colliding-pulse mode locking (CPM) introduced by Ruddock and Bradley (1976) and applied by Fork *et al.* (1981) to a ring cavity. Here, a *collision* between two counterpropagating pulses is established at the saturable absorber. This creates a transient grating due to interference at the absorber that has the effect of shortening the pulse and improving its synchronization and stability characteristics. Using the CPM technique, Fork *et al.* (1981) obtained pulses as short as 90 fs.

An additional development of importance for the femtosecond laser was the introduction of prismatic intracavity negative group-velocity dispersion (GVD) (Dietel *et al.* 1983). Pulse broadening by GVD occurs at the dye gain and dye absorber regions since the pulse transit time is affected by the dispersion of the dye refractive index. The deployment of intracavity prisms makes possible the achievement of intracavity compensation of the GVD introduced at the gain and absorber regions. This is because the GVD also depends on the second derivative of the optical path length ($d^2P/d\lambda^2$) through the intracavity prismatic array (see Fig. 14). Furthermore, the value of $d^2P/d\lambda^2$ is a function of the prismatic dispersion and its derivative (see Sec. 5) and can be made negative by adjusting the separation distance of the prisms. Thus the intracavity prismatic array is utilized to yield negative GVD. In other words, prismatic GVD compen-

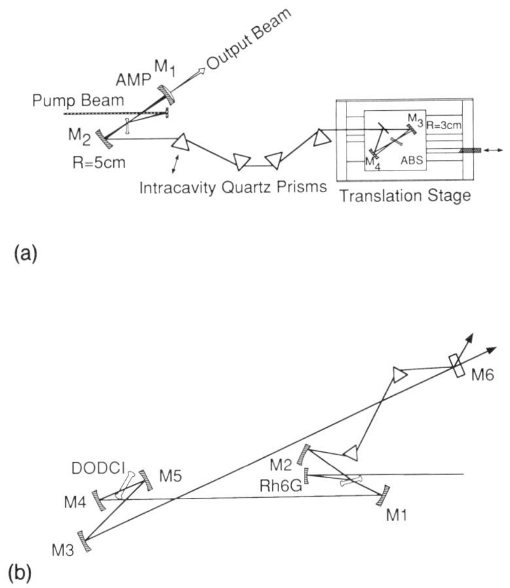

FIG. 14. (a) Linear femtosecond dye-laser cavity terminated by an antiresonant ring that serves as a broadband reflector (adapted from Jamasbi *et al.*, 1988), (b) ring femtosecond dye-laser cavity (adapted from Diels *et al.*, 1985). In both arrangements the saturable absorber is deployed to exploit the CPM effect.

sation is used to remove temporal broadening (or *chirp*) introduced at the dye regions.

Examples of linear and ring femtosecond dye lasers are given in Fig. 14. The basic elements of these lasers are the cavities established for the gain region between M_1 and M_2, the saturable absorber in its counterpropagating arrangement, and the intracavity prisms that provide dispersion compensation. Again, a propagation analysis of these cavities can be done using ray-transfer matrices (Diels, 1990). The use of GVD compensation techniques has yielded pulses shorter than 20 fs.

In addition to the passive mode locking considered so far some workers use *hybrid mode locking,* which occurs when synchronous pumping is used in conjunction with passive mode locking. Also, negative dispersion can be generated using pairs of gratings.

The shortest pulses reported to date correspond to 6 fs and were achieved by Fork *et al.* (1987) using extra-cavity pulse compression. In this experiment the authors applied a CVL to amplify, at a prf of 8 kHz, the pulses from a dye laser incorporating CPM and prismatic GVD compensation. The amplified 50-fs pulses were then propagated via grating pairs and a four-prism sequence. The optimum pulse duration was obtained by adjusting the prism and grating spacings while monitoring the compressed pulse using interferometric autocorrelation.

Measurement of femtosecond pulses involves the use of special techniques such as intensity and interferometric autocorrelations, self-convolution, asymmetric correlation, and single-shot detection. Diels (1990) provides a detailed description of these techniques. In terms of physics, femtosecond dye lasers have led to the study of various nonlinear phenomena such as the Schrödinger equation with nonlinear terms, and solitons.

5. INTRACAVITY DISPERSION

A very important subject in the physics of linewidth narrowing in high-gain pulsed lasers, and pulse compression in femtosecond lasers, is intracavity dispersion.

The return-pass dispersive linewidth in a multiple-prism grating oscillator is given by the expression

$$\Delta\lambda \approx \Delta\theta(\partial_\lambda\theta)^{-1}, \tag{4}$$

where $\partial_\lambda\theta$ is the intracavity dispersion, $\Delta\theta$ is the beam divergence, and $\partial_\lambda = \partial/\partial\lambda$. Although this expression was first adopted from classical optics, it has been shown recently that it can be derived using Dirac's notation (Duarte, 1991).

The cumulative single-pass dispersion at the mth prism (see Fig. 15) is given by Duarte (1990):

$$\partial_\lambda\phi_{2,m} = \mathscr{H}_{2,m}\partial_\lambda n_m + (k_{1,m}k_{2,m})^{-1}(\mathscr{H}_{1,m}\partial_\lambda n_m \pm \partial_\lambda\phi_{2,(m-1)}), \tag{5}$$

where $k_{1,m} = (\cos\psi_{1,m}/\cos\phi_{1,m})$, $k_{2,m} = (\cos\phi_{2,m}/\cos\psi_{2,m})$, $\mathscr{H}_{1,m} = (\tan\phi_{1,m}/n_m)$, $\mathscr{H}_{2,m} = (\tan\phi_{2,m}/n_m)$, and $\partial_\lambda n_m = \partial n_m/\partial\lambda$ is characteristic of the prism material. Here, $\phi_{1,m}$ is the angle of incidence at the mth prism and $\phi_{2,m}$ is the exit angle. Thus, $\psi_{1,m}$ and $\psi_{2,m}$ are the corresponding angles of refraction. Equation (5) is a recursive function that evaluates $\partial_\lambda\phi_{2,m}$ in ascending values of m.

The *generalized* return-pass prismatic dispersion is given by Duarte (1990):

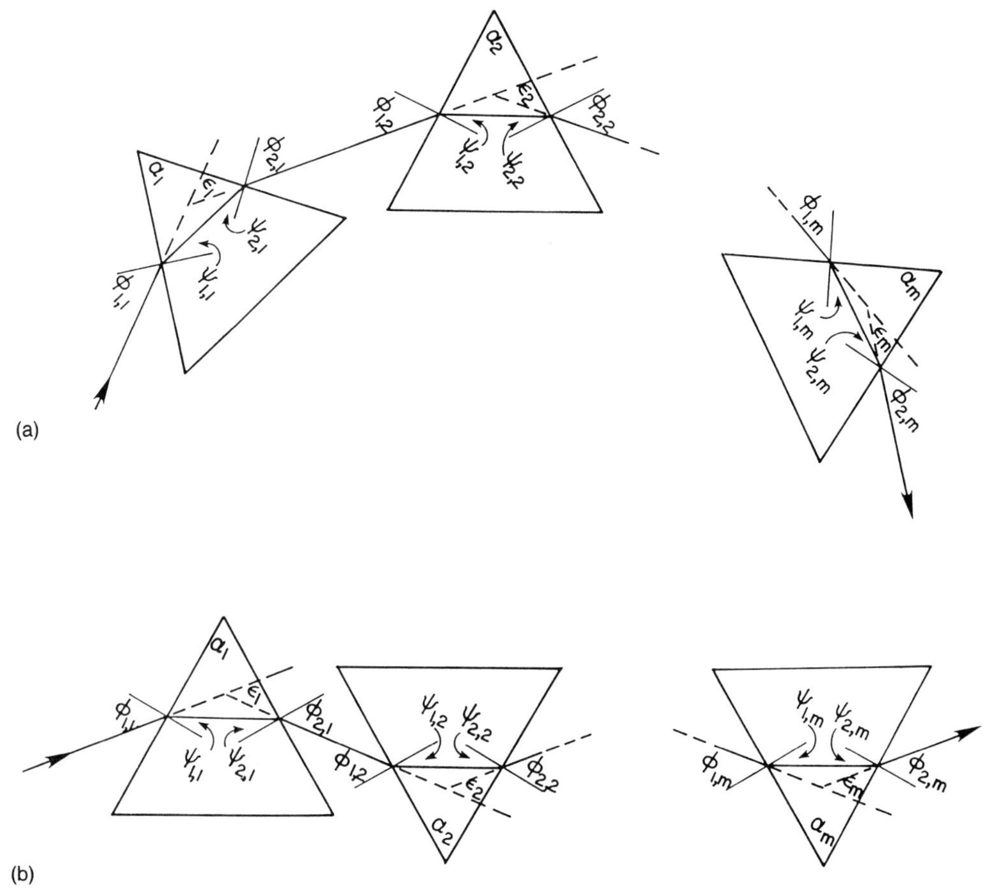

FIG. 15. Generalized multiple-prism assemblies in (a) an additive configuration and (b) a compensating configuration.

$$(\partial_\lambda \Phi)_P = 2M_1M_2 \sum_{m=1}^{r} (\pm 1)\mathscr{H}_{1,m} \times \left(\prod_{j=m}^{r} k_{1,j} \prod_{j=m}^{r} k_{2,j} \right)^{-1} \partial_\lambda n_m + 2 \sum_{m=1}^{r} (\pm 1)\mathscr{H}_{2,m} \times \left(\prod_{j=1}^{m} k_{1,j} \prod_{j=1}^{m} k_{2,j} \right) \partial_\lambda n_m, \tag{6}$$

where

$$M_1 = \prod_{m=1}^{r} k_{1,m} \tag{7a}$$

and

$$M_2 = \prod_{m=1}^{r} k_{2,m} \tag{7b}$$

are the overall beam-expansion coefficients.

For identical right-angled prisms designed for orthogonal beam exit at the Brewster angle of incidence, Eq. (6) reduces to the elegant expression

$$(\partial_\lambda \Phi)_P = 2 \sum_{m=1}^{r} (\pm 1) n^{m-1} \partial_\lambda n \tag{8}$$

and the overall beam expansion is given by $M = n^r$.

The overall return-pass intracavity dispersion is given by

$$\partial_\lambda \theta = [M_1M_2(\partial_\lambda \theta)_G + (\partial_\lambda \Phi)_P], \tag{9}$$

where $(\partial_\lambda \theta)_G$ is the grating dispersion and $(\partial_\lambda \Phi)_P$ is given by Eq. (6) or (8). In Eq. (9) the contribution from the grating dominates since the grating dispersion can be augmented by factors of up to 100 or more by intracavity beam expansion. Further, using Eq. (6) the prismatic contribution to dispersion can be made equal to zero at a given wavelength by design. This reduces beam deviations due to thermal effects and makes the tuning characteristics of the oscillator depend exclusively on the grating. In addition to dispersion equations, ray-transfer matrices are utilized to evaluate $\Delta\theta$ and other intracavity propagation parameters (see, for example, Duarte, 1990, 1991).

Equations (4)–(9) have been successfully applied to characterize and predict the dispersive linewidth in pulsed high-gain, tunable, multiple-prism grating oscillators. Agreement with measured linewidths is particularly good in high-gain oscillators where temporal and spatial dimensions allow stimulated emission to occur following a few intracavity passes. For the long-pulse regime, characteristic of flashlamp-pumped dye lasers, the dispersive linewidth value establishes the upper limit of the measured laser linewidth, which is determined by the interaction of dispersive and dynamic phenomena.

These equations transcend dye lasers and apply to pulsed lasers in general, irrespective of laser medium or emission wavelength. In this regard, the dispersive design concepts first developed for dye lasers have now been applied to gas, solid-state, and semiconductor lasers. Furthermore, the techniques of beam shaping using multiple-prism expanders have found widespread applications in electro-optics.

As mentioned previously, negative dispersion has a very important function in pulse compression, compensating for positive intracavity dispersion. Two parameters of importance in evaluating the negative dispersion effect on pulse compression are $\partial\phi_{2,m}/\partial n$ and $\partial^2\phi_{2,m}/\partial n^2$. This follows since the second derivative of the beam path length through the prismatic array is a function of these parameters (see Sec. 4). The first derivative of interest is obtained using the identity $\partial\phi_{2,m}/\partial n = \partial_\lambda \phi_{2,m}(\partial_\lambda n)^{-1}$, and the second derivative is obtained by direct differentiation (Duarte, 1990). For a single prism at Brewster's angle of incidence $\partial\phi_{2,1}/\partial n = 2$ and $\partial^2\phi_{2,1}/\partial n^2 = 4n - 2/n^3$.

6. APPLICATIONS

Dye lasers have made significant contributions in a wide variety of scientific and technological fields including physics, spectroscopy, laser isotope separation, medicine, and lidar, and in industry.

Contributions to areas of laser physics include the study of dynamic instabilities, nonlinear phenomena, oscillator physics, and femtosecond laser development.

In the area of dynamic instabilities an interesting result is the hysteresis effect observed in a cw dye laser by Hillman *et al.* (1984). These authors discovered that as the pump power was increased thresholds were reached where the output power and spectrum of their simple ring dye laser changed discontinuously. Although dynamic instabilities in homogeneously broadened lasers have been identified with strong-field interactions in the gain medium, further studies are needed for a complete understanding of the phenomena. In addition, dynamic instabilities can also originate from optical inhomogeneities in the active medium due to flow turbulence and thermal gradients. In this regard, linewidth instabilities in narrow-linewidth dispersive oscillators were found to decrease substantially as radial thermal gradients and flow turbulence were neutralized (Duarte *et al.*, 1991).

The versatility of the dye laser created a renaissance in the field of spectroscopy. In this regard, dye-laser characteristics that are important to spectroscopy are tunability, extremely narrow-linewidth emission, low ASE levels, very stable oscillation in the cw regime, and the ability to emit very short pulses (down to the femtosecond regime). Spectroscopy is the study of the wavelength or frequency dependence of any optical process where a substance absorbs or emits energy by the interaction with radiation. The tunable narrow-linewidth emission available from dye lasers allows selective excitation of atomic and molecular transitions with a high degree of specificity. The areas of applications are numerous and diverse including absorption, excitation, ionization, and saturation spectroscopy. Both cw and pulsed dye lasers have found useful roles in laser-induced fluorescence spectroscopy. On the other hand, the attainment of extremely short pulse widths has generated an explosion in the number of time-resolved and ultrafast spectroscopic measurements that have affected every branch of science and technology. For detailed descriptions on the application of dye lasers to spectroscopy the reader should refer to Klick (1990) and Stuke (1992).

In addition to spectroscopy, femtosecond dye lasers have been applied to physical measurement devices such as the femtosecond-pulse laser gyroscope developed by Diels (1991).

A significant further application of dye lasers is laser isotope separation. In atomic-vapor laser isotope separation (LIS) a metallic element such as uranium is heated and vaporized by an electron beam. The resulting vapor is then illuminated by a high-prf CVL-pumped dye-laser system (see Sec. 1) that selectively excites the ^{235}U isotope. Briefly, this is accomplished by exploiting the isotopic shift ($\delta\nu_i \approx$ few gigahertz) and the ability of narrow-linewidth dye lasers ($\Delta\nu \leqslant 1$ GHz) to tune into resonance with specific transitions. Thus, selective ^{235}U excitation can be achieved in a multistep excitation process by tuning the dye laser(s) to specific wavelengths, in the visible spectral region, compatible with a transition sequence leading to photoionization. For instance, a three-step excitation sequence may proceed by exciting the atom from the ground state to an even-parity state from where it is further excited to a higher energy odd-parity state, and subsequently to an even-parity state near the ionization potential (see Fig. 16). Then, the atom at the Rydberg state can be ionized using an infrared laser, or via collisional ionization or autoionization routes. The ion can then be extracted using electromagnetic methods. A possible alternative to the three-step excitation described here is a two-step excitation sequence involving near-ultraviolet dye lasers.

Interest in the atomic-vapor LIS process is derived from considerable economic advantages that laser enrichment of uranium is reported to have over traditional methods such as gas diffusion or centrifuge techniques.

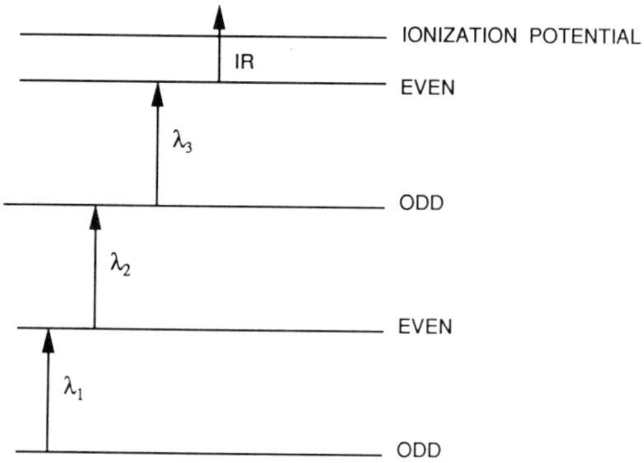

FIG. 16. Energy-level diagram illustrating a three-step excitation sequence.

For a detailed discussion of this topic and photoionization spectroscopy see Paisner and Solarz (1987).

High-prf CVL-pumped dye lasers can also be applied to material processing such as cutting and welding. In this area the high-prf characteristics of the laser yields fast processing velocities, and the good beam quality allows high spatial precision (Hargrove, 1991). For a detailed review on industrial applications of dye lasers see Klick (1990).

In addition, CVL-pumped dye lasers are suitable to applications in astronomy as a *guide star* (see Bass *et al.*, 1992). Here, the laser is tuned to the well-known 589-nm line of the atomic sodium layer that is observed 90–100 km above the earth's surface. Light from the guide star, created by the laser illumination, is collected by a Cassegrain telescope and can then by used to characterize turbulence-induced distortions. Use of this concept in conjunction with adaptive optics can lead to the correction of atmospheric distortions and thus enhance the quality of astronomical images.

An important area of research is laser radar and *lidar*. Lidar is a technique that uses laser pulses for detection and ranging of atmospheric molecular species. The measurement consists in the emission of a laser pulse and the detection of the scattered return signal. Thus, tunable dye lasers are applied to detect and map concentrations of molecular elements in the atmosphere. For instance, Kölsch *et al.* (1992) have used a mobile ground excimer-laser–pumped dye laser to map emission of nitric oxide in several European cities. Grant (1991) provides a review on differential absorption lidar (dial) measurements of water vapor profiles and considers relevant requirements on laser characteristics.

Scientists at NASA use a pair of airborne tunable dye lasers to conduct tropospheric and stratospheric ozone and aerosol measurements (see, for example, Browell, 1991). These airborne experiments have led to the measurement and mapping of O_3 depletion in the stratosphere. This depletion has been found to be particularly pronounced over Antarctica and the Artic. In the stratospheric experiments two frequency-doubled dye lasers emit at 301.5 and 311 nm for the on- and off-line dial wavelengths of O_3. In addition, part of the fundamental dye-laser emission at 603 nm is used in conjunction with a portion of 1064-nm radiation from the pump laser for aerosol measurements. All wavelengths are transmitted simultaneously. Plate X shows an O_3 distribution measured over Antarctica. These experiments are conducted to determine the dynamics and chemistry that control the distribution of O_3 in the atmosphere.

Applications including remote sensing, LIS, and spectroscopy demand tunable narrow-linewidth emission with significantly reduced levels of ASE. In this regard, the oscillator configurations described in Secs. 1 and 2 are particularly useful.

Finally, we consider medical applications of dye lasers. One of the first dye-laser applications in medicine was cancer photodynamic therapy (PDT) pioneered by Dougherty *et al.* (1978). In this technique the patient is given internally 2–3 mg/kg of a dye such as HpD (hematoporphyrin derivative). Following a given time (48–72 h) the photosensitizer is retained by the cancer tumor at an optimum concentration differential relative to healthy tissue. At that point laser radiation is delivered to the tumor, at energy densities in the 100–500-J cm^{-2} range, either directly or via an optical-fiber delivery system depending on the location of the tumor. A photosensitizer such as HpD absorbs mostly in the 400–450-nm region of the spectrum; however, since the body tissue transmits mainly in the red the dye lasers are operated at ~630 nm for cytotoxicity. Absorption of the laser light by the photosensitizer leads to a photochemical reaction that produces a reactive form of O_2 that causes loss of cellular structure and subsequently tumor destruction. The exact mechanism by which the oxygen singlet interacts with the cells continues to be subject to study.

In certain cases a particular photosensitizer may be applied locally in conjunction with an appropriate biological vehicle. Also, given the characteristics of the absorption spectrum of HpD, blue-green lasers can be used for diagnostics using fluorescence techniques. Other photosensitizers are dihematoporphyrin ether (DHE), rhodamine dyes, chlorophyll derivatives, and phthalocyanines. Types of cancer reported to have been treated with PDT include cancers of the bladder, brain, breast, lungs, skin, stomach, and vagina.

It is interesting to note that although PDT does not require narrow-linewidth laser emission, the use of systems including dispersive narrow-linewidth oscillators (as described in Secs. 1 and 2) leads to high–beam-quality emission that can be advantageous in the transmission via optical-fiber delivery systems. Also, a laser system such as the CVL-pumped dye laser can be used for the dual purpose of diagnostics and treatment.

An additional, and very successful, medical application of the dye laser is *lithotripsy* which involves the use of long-pulse flashlamp-pumped dye lasers, emitting at wavelengths such as 504 and 577 nm, to fragment ureteral calculi (see, for example, Dretler, 1987). The average success rate for this procedure can be about 80%. Further medical applications of the flashlamp-pumped dye laser are in dermatology, the treatment of vascular lessions, and thrombolysis. For a detailed account of dye lasers in medicine the reader should consult Goldman (1990). See also MEDICAL USE OF LASERS.

GLOSSARY

Amplified Spontaneous Emission (ASE): Spontaneous emission that is amplified in the gain medium. ASE exhibits some spectral narrowing and directionality. However, compared to coherent laser emission, ASE has very broadband spectral characteristics and much higher divergences. Considerable effort is devoted in the design of oscillators and amplifiers to minimize the ASE background.

Brewster Angle: Angle of incidence at which the reflection losses are a minimum. It is defined by $\tan\phi_{1,m}=n$, where n is the refractive index of the optical material.

Chirp: A temporal frequency sweep of a pulse. Typically a linear chirp will have the time-dependent frequency $\omega(t)=\omega_0+bt$.

Closed Cavity: A cavity where the laser emission exits via the output coupler mirror and not via reflection losses of dispersive elements.

Diffraction-Limited Beam: A Gaussian laser beam whose divergence is given by $\lambda/\pi w$, where w is the beam waist.

Dispersive Oscillator: A narrow-linewidth laser oscillator utilizing intracavity dispersive elements.

Fabry–Pérot Étalon: An interferometer, composed of two highly polished and parallel reflective surfaces, that is usually deployed intracavity for linewidth narrowing and fine frequency tuning.

Forced Oscillator: A multipass oscillator usually incorporating unstable resonator optics and no intracavity dispersive elements. Its function is to amplify the injected signal from a master oscillator.

Free Spectral Range (FSR): FSR of a cavity or Fabry–Pérot étalon is defined by $\text{FSR}=c/2nd$ where c is the speed of light, n the refractive index of the medium between the reflective surfaces, and d the distance between these surfaces.

Frequency Doubling: The use of a nonlinear medium such as a crystal to generate a second harmonic of the laser emission.

Homogeneous Broadening: A broadening mechanism associated with collisions and excited-state lifetimes. The broadening of the gain medium is said to be homogeneous when it can be characterized by the broadening of an individual atom or molecule in the collection. In homogeneously broadened gain media all excited atoms or molecules can emit at the same single longitudinal mode, within the frequency gain profile.

Intracavity Dispersion: The dispersion created by the use of intracavity elements such as gratings and/or prisms.

Lidar: Light detection and ranging.

Littrow Configuration: Deployment of a diffraction grating so that the angle of incidence is equal to the angle of diffraction.

Master Oscillator: A narrow-linewidth oscillator usually of the dispersive class emitting a nearly diffraction-limited beam and a very low level of ASE.

Mode Locking: The condition in which all the modes in the cavity have the same relative phase. This can be achieved with the use of an intracavity saturable absorber. The output of a mode-locked laser is a sequence, or train, of short pulses separated by $\Delta t=2L/c$, where L is the cavity length.

Photon Conversion Efficiency: Efficiency quoted in optically pumped lasers and defined as the emission laser energy over the pump laser energy.

Raman Shifting: A nonlinear optics method utilized to generate coherently a series of Stokes and anti-Stokes frequencies. Relative to the laser fundamental, Stokes radiation is of longer wavelength and anti-Stokes radiation is of shorter wavelength. A common method to generate tunable Raman-shifted radiation is to use a dye laser to excite a gas like hydrogen contained at high pressure in an optical cell.

Ray-Transfer Matrices: 2×2 and 4×4 matrices whose components are optical quantities such as beam expansion, optical path length, and dispersion. These matrices are used to characterize Gaussian beam propagation.

Saturable Absorber: A dye such as DODCI that is used intracavity. The absorption spectrum of such dye overlaps with the emission spectrum of the gain dye. The absorption of saturable absorbers decreases as the incident intracavity emission intensity increases. This feature discriminates in favor of a high-intensity pulse and in detriment of lower-intensity pulses.

Solid-State Dye Laser: A dye laser using an active medium where the dye is contained in a polymeric or a silicate matrix.

Solitons: Eigensolutions of the nonlinear Schrödinger equation $i(\partial\tilde{E}/\partial z)+\partial^2\tilde{E}/\partial t^2+|\tilde{E}|^2\tilde{E}=0$, where $\tilde{E}$ is proportional to the optical field.

Spatial Hole Burning: Refers to the spatial modulation of the gain saturation in an atomic or molecular medium. This modulation results from the interference of two propagating fields. In a linear cw dye laser cavity the interference establishes a standing-wave field at the gain region. This spatial standing wave does not use the available gain at the nodes of the field, which is then available for a secondary mode. Thus more than one lon-

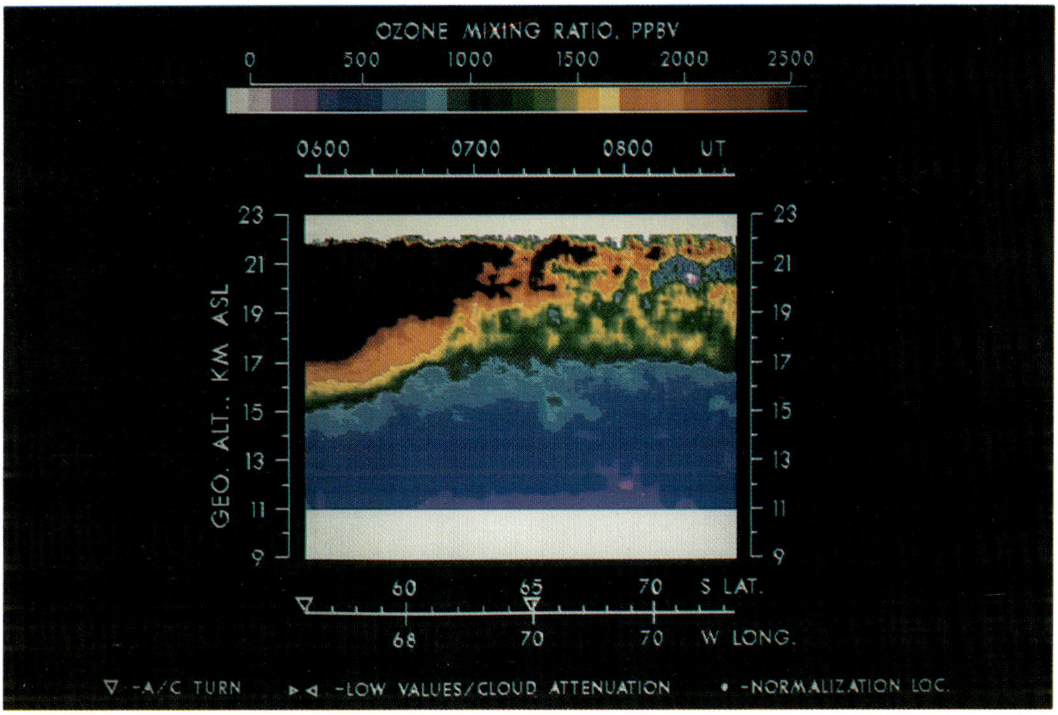

PLATE 5. Cross section of O_3 distribution obtained via airborne differential absorption lidar measurements over Antarctica in 1987. The measurement was made at an altitude of 9 km. (Courtesy of E. V. Browell, NASA.)

gitudinal mode can lase simultaneously even when the broadening is homogeneous.

Synchronous Pumping: This occurs when the gain of the dye laser is modulated by a mode-locked pump laser. The cavity mode spacing of the dye laser is adjusted to be synchronous with the mode-locking frequency of the pump laser.

Unstable Resonator: A resonator configuration often used in wide-aperture high-power pulsed lasers. Typically, the mirrors in these resonators are curved and form an intracavity reflective telescope such as a Cassegrain telescope.

List of Works Cited

Baltakov, F. N., Barikhin, B. A., Sukanov, L. V. (1974), *JETP Lett.* **19**, 174–175.

Baving, H. J., Muuss, H., Skolaut, W. (1982), *Appl Phys. B* **29**, 19–21.

Bass, I. L., Bonanno, R. E., Hackel, R. P., Hammond, P. R. (1992), *Appl. Opt.* **31**, 6993–7006.

Bernhardt, A. F., Rasmussen, P. (1981). *Appl. Phys. B* **26**, 141–146.

Bos, F. (1981a), *Appl. Opt.* **20**, 3553–3556.

Bos, F. (1981b), *Appl. Opt.* **20**, 1886–1890.

Browell, E. V. (1991), *Opt. Photon. News* **2** (10), 8–11.

Diels, J. C. (1990), in: F. J. Duarte, L. W. Hillman (Eds.). *Dye Laser Principles*, New York: Academic, pp. 41–132.

Diels, J. C. (1991), *Technical Digest, International Conference on Lasers '91*, San Diego, California, paper TA.2.

Diels, J. C., Dietel, W., Fontaine, J. J., Rudolph, W., Wilhelmi, B. (1985), *J. Opt. Soc. Am. B* **2**, 680–686.

Dietel, W. Fontaine, J. J., Diels, J. C. (1983), *Opt. Lett.* **8**, 4–6.

Dougherty, T. J., Kaufman, J. E., Goldfarb, A., Weishaupt, K. R., Boyle, D., Mittleman, A. (1978), *Cancer Res.* **38**, 2628–2635.

Dretler, S. P. (1987), *J. Endourology* **1**, 9–14.

Drexhage, K. H. (1990), in: F. P. Schäfer (Ed.), *Dye Lasers*, Berlin: Springer, pp. 155–200.

Duarte, F. J. (1988), *Lasers Optron.* **7** (2), 41–45.

Duarte, F. J. (1990), in: F. J. Duarte, L. W. Hillman (Eds.), *Dye Laser Principles*, New York: Academic, pp. 133–183; 239–285.

Duarte, F. J. (1991), in: F. J. Duarte (Ed.), *High Power Dye Lasers*, Berlin: Springer, pp. 7–43.

Duarte, F. J., Conrad, R. W. (1987), *Appl. Opt.* **26**, 2567–2571.

Duarte, F. J., Piper, J. A. (1981), *Appl. Opt.* **20**, 2113–2116.

Duarte, F. J., Piper, J. A. (1984) *Appl. Opt.* **23**, 1391–1394.

Duarte, F. J., Davenport, W. E., Ehrlich, J. J., Taylor, T. S. (1991), *Opt. Commun.* **84**, 310–316.

Everett, P. N. (1991), in: F. J. Duarte (Ed.), *High Power Dye Lasers*, Berlin: Springer, pp. 183–245.

Everett, P. N., Aldag, H. R., Ehrlich, J. J., Janes, G. S., Klimek, D. E., Landers, F. M., Pacheco, D. P. (1986), *Appl. Opt.* **25**, 2142–2147.

Fork, R. L., Greene, B. I., Shank, C. V. (1981), *Appl. Phys. Lett.* **38**, 671–672.

Fork, R. L., Brito Cruz, C. H., Becker, P. C., Shank C. V. (1987), *Opt. Lett.* **12**, 483–485.

Goldman, L. (1990), in: F. J. Duarte, L. W. Hillman (Eds.), *Dye Laser Principles*, New York: Academic, pp. 419–432.

Grant, W. B. (1991), *Opt. Eng.* **30**, 40–48.

Hall, J. L., Hänsch, T. W. (1984), *Opt. Lett.* **9**, 502–504.

Hänsch, T. W. (1972), *Appl. Opt.* **11**, 895–898.

Hargrove, R. S. (1991), *Technical Digest, International Conference on Lasers '91*, San Diego, California, paper THA.2.

Hargrove, R. S., Kan, T. (1980), *IEEE J. Quantum Electron.* **QE-16**, 1108–1113.

Hillman, L. W., Krasinski, J., Boyd, R. W., Stroud, C. R. (1984), *Phys. Rev. Lett.* **52**, 1605–1608.

Hollberg, L. (1990), in: F. J. Duarte, L. W. Hillman (Eds.), *Dye Laser Principles*, New York: Academic, pp. 185–238.

Hough, J., Hils, D., Rayman, M. D., Ma, L. S., Hollberg, L., Hall, J. L. (1984), *Appl. Phys. B* **33**, 179–185.

Jamasbi, N., Diels, J. C., Sarger, L. (1988), *J. Mod. Opt.* **35**, 1891–1906.

Johnston, T. F., Brady, R. H., Proffitt, W. (1982), *Appl. Opt.* **21**, 2307–2316.

Jones, G., II (1990), in: F. J. Duarte, L. W. Hillman (Eds.), *Dye Laser Principles*, New York: Academic, pp. 287–343.

Klick, D. (1990), in: F. J. Duarte, L. W. Hillman (Eds.), *Dye Laser Principles*, New York: Academic, pp. 345–412.

Klimek, D. E., Aldag, H. R., Russell, J. (1992), in: *Technical Digest, Conference on Lasers and Electro-Optics*, Washington, DC: Optical Society of America, p. 332.

Kölsch, H. J., Rairoux, P., Wolf, J. P., Woste, L. (1992), *Appl. Phys. B* **54**, 89–94.

Littman, M. G., Metcalf, H. J. (1978), *Appl. Opt.* **17**, 2224–2227.

Maeda, M. (1984), *Laser Dyes*, New York: Academic.

Maeda, M., Uchino, O., Okada, T., Miyazoe, Y. (1975), *Jpn. J. Appl. Phys.* **14**, 1975–1980.

Mazzinghi, P., Burlamacchi, P., Matera, M., Ranea-Sandoval, H. F., Salimbeni, R., Vanni, U. (1981), *IEEE J. Quantum Electron.* **QE-17**, 2245–2249.

Morton, R. G., Draggoo, V. G. (1981), *IEEE J. Quantum Electron.* **QE-17** (12), 222.

Pacheco, D. P., Aldag, H. R., Itzkan, I., Rostler, P. S. (1988), in: F. J. Duarte (Ed.), *Proceedings of the International Conference on Lasers '87,* McLean, VA: STS, pp. 330–337.

Paisner, J. A., Solarz R. W., (1987), in: L. J. Radziemski, R. W. Solarz, J. A. Paisner (Eds.), *Laser Spectroscopy and its Applications,* New York: Dekker, pp. 175–260.

Peterson, O. G. (1979), in: C. L. Tang (Ed.), *Methods of Experimental Physics,* New York: Academic Vol. 15, pp. 251–359.

Peterson, O. G., Tuccio, S. A., Snavely, B. B. (1970), *Appl. Phys. Lett.* **17**, 245–247.

Ruddock, I. S., Bradley, D. J. (1976), *Appl. Phys. Lett.* **29**, 296–297.

Runge, P. K., Rosenberg, R. (1972), *IEEE J. Quantum Electron.* **QE-8**, 910–911.

Schäfer, F. P. (1990), in: F. P. Schäfer (Ed.), *Dye Lasers,* Berlin: Springer, pp. 1–89.

Schäfer, F. P., Schmidt, W., Volze, J. (1966), *Appl. Phys. Lett.* **9**, 306–309.

Schmidt, W., Schäfer, F. P. (1968), *Phys. Lett.* **26A**, 558–559.

Sorokin, P. P., Lankard, J. R. (1966), *IBM J. Res. Dev.* **10**, 162–163.

Steppel, R. N. (1991), in M. J. Weber, (Ed.), *Handbook of Laser Science and Technology, Supplement* 1: *Lasers,* Boston: CRC, pp. 219–317.

Stuke, M. (Ed.) (1992), *Dye Lasers: 25 Years,* Berlin: Springer.

Tallman, C., Tennant, R. (1991), in: F. J. Duarte (Ed.), *High Power Dye Lasers,* Berlin: Springer, pp. 93–142.

Tuccio, S. A., Strome, F. C. (1972), *Appl. Opt.* **11**, 64–73.

Wallenstein, R., Hänsch, T. W. (1975), *Opt. Commun.* **14**, 353–357.

Webb, C. E. (1991), in: F. J. Duarte (Ed.), *High Power Dye Lasers,* Berlin: Springer, pp. 143–182.

Further Reading

Duarte, F. J. (Ed.) (1991), *High Power Dye Lasers,* Berlin: Springer.

Duarte, F. J. (Ed.) (1992), *Selected Papers on Dye Lasers,* SPIE Milestone Series, Bellingham, WA: SPIE, Vol. MS 45.

Duarte, F. J., Hillman, L. W. (Eds.) (1990), *Dye Laser Principles,* New York: Academic.

Maeda, M. (1984), *Laser Dyes,* New York: Academic.

Schäfer, F. P. (Ed.) (1990), *Dye Lasers,* Berlin: Springer.

Stuke, M. (Ed.) (1992), *Dye Lasers: 25 Years,* Berlin: Springer.

LASERS, FREE-ELECTRON

C. PELLEGRINI, *Department of Physics, University of California, Los Angeles, Los Angeles, California, U.S.A.*

	Introduction	353
1.	**Physical and Technical Principles**	355
1.1	Undulator Spontaneous Radiation	356
1.2	Small-Signal Gain	357
1.3	Scaling and Electron-Beam Requirements	360
1.4	Optical Klystron	360
1.5	Large-Gain, Saturation, Compton, and Raman Regimes	361
1.6	Tapered Undulator and High-Efficiency FELs	361
1.7	Optical Guiding	362
2.	**Status and Operation of FELs**	362
3.	**FEL Applications**	364
4.	**Future Developments and X-Ray FELs**	365
	Works Cited	369

INTRODUCTION

In a free-electron laser (FEL) the kinetic energy of a relativistic electron beam is transformed into electromagnetic (em) radiation. This transformation occurs when the electron beam goes through an alternating magnetic field that forces the electrons to move in an oscillatory trajectory about the axis of the system, as shown in Fig. 1. An electromagnetic wave propagates along the system axis. The magnetic field is produced by a magnet called an undulator. The electron beam is produced by a particle accelerator, which may be of many types: linear accelerator, microtron, storage ring, etc.

The undulator magnet is a periodic structure in which the field alternates between positive and negative values and has zero average value. It produces an electron trajectory having a transverse velocity perpendicular to the axis, while leaving the average velocity parallel to it unchanged. In this configuration the electron beam has a component of its velocity parallel to the electric field of the wave, allowing an energy exchange between the two. One can either transfer energy from the beam to the wave, in which case the device is a FEL, or from the wave to the beam, in which case it is an inverse FEL (IFEL). In the second case the system is acting like a particle accelerator and could be used to accelerate an electron beam to high energies.

The energy transfer can take place only if a condition of synchronism between the wave and the beam oscillations is satisfied (Sec. 1.2). This condition gives a relationship between the radiation wavelength λ, the electron velocity β_z along the undulator axis (measured in units of the light velocity c), and the undulator period λ_0:

$$\lambda=\lambda_0(1-\beta_z)/\beta_z. \qquad (1)$$

For relativistic electrons this condition can also be written in an approximate, but more convenient, form using the beam energy γ [measured in rest energy units, $E=mc^2\gamma$:

$$\lambda=(\lambda_0/2\gamma^2)(1+a_w^2), \qquad (1')$$

The dimensionless quantity

$$a_w=eB_0\,\lambda_0\,c/2\pi\,mc^2 \qquad (2)$$

is the undulator vector potential normalized to the electron rest energy mc^2; e is the electron charge and B_0 is the undulator peak magnetic field. (Here and in the rest of the paper we use mks units.) The quantity a_w is called the *undulator parameter.* Undulators can be of two types: helical and planar (Sec. 1). Equation (1′) is valid for a helical undulator. In the case of a planar undulator Eq.

1-56081-067-X/94/$5.00+.50

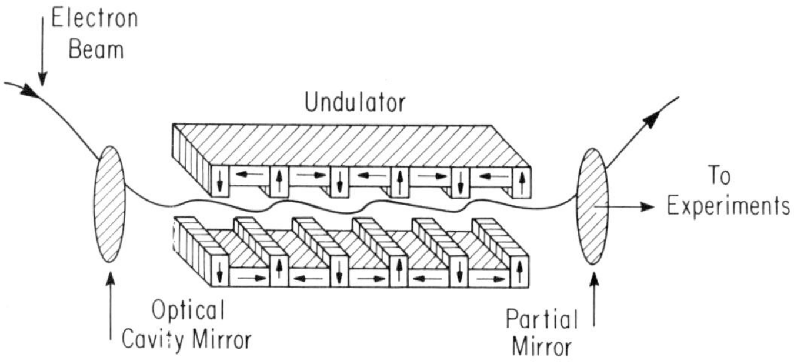

FIG. 1. Schematic representation of a FEL, showing its main components: the electron beam, undulator magnet, and optical cavity. The undulator shown is a permanent-magnet planar undulator. The arrows indicate the magnetic field direction. The beam oscillates around the undulator axis, in a plane perpendicular to the magnetic field. A series of electron bunches, synchronized with the cavity round-trip time, amplifies a radiation pulse.

(1′) is still valid if we replace the undulator parameter with its rms value $a_w/\sqrt{2}$.

Because of the dependence of the radiation wavelength on the undulator period, magnetic field, and electron-beam energy—quantities that can be easily and continuously changed—the FEL is a tunable device that can be operated over a very large frequency range. At present the range extends from the microwave to the UV. In the future it might be possible to reach the x-ray region.

The efficiency of energy transfer from the beam kinetic energy to the em wave can be quite large, up to about 40%. The beam energy not transferred to the em wave remains in the beam and can be easily taken out of the system, to be disposed of or recovered elsewhere. This fact suggests that high–average-power FELs can be designed without the problem, common in atomic and molecular lasers, of heating the lasing medium.

The time structure of the laser beam mirrors that of the electron beam. Depending on the accelerator used, one can design systems that are continuous-wave (cw) or with pulses as short as picoseconds.

Tunability, high efficiency, and time structure make the FEL a very attractive source of coherent em power. In some wavelength regions the FEL is unique. Its applications range from purely scientific research in physics, chemistry, and biology to military, medical, and industrial applications.

FELs originate in the work carried out in the 1950s and 1960s on the generation of coherent em radiation from electron beams in the microwave region (Motz, 1951; Phillips, 1960). As scientists tried to push power sources to shorter and shorter wavelengths, it became apparent that the efficiency of the microwave tubes, and the power they produced, dropped rapidly in the millimeter region. It was then realized that this problem could be overcome by using an undulator magnet to modify the beam trajectory (Madey, 1971), making it possible for the beam to interact with a wave, away from any metallic boundary. Two pioneering experiments at Stanford University (Elias *et al.*, 1976; Deacon *et al.*, 1977) proved that the FEL is a useful source of coherent radiation.

The current disadvantage to FELs is the greater complexity and cost associated with the use of a particle accelerator. For this reason the use and development of FELs are mainly oriented to the following:

1. portions of the electromagnetic spectrum, like the far infrared (FIR), or the ultraviolet (UV) below 100 nm, where atomic or molecular lasers are not readily available;
2. large–average-power, high-efficiency systems.

An order-of-magnitude comparison of the peak power obtained or expected for the FEL, compared to other sources of electromagnetic radiation, is given in Fig. 2, pointing out again the interest for FELs in the FIR and UV regions.

In Sec. 1 of this article we discuss the general physical principles of FELs and the properties and characteristics of their main components, i.e., the undulator magnets, electron beams, and optical systems. We also describe the main properties of the spontaneous radiation from an undulator, the process leading to the amplification of radiation, and the parameters that determine the FEL gain.

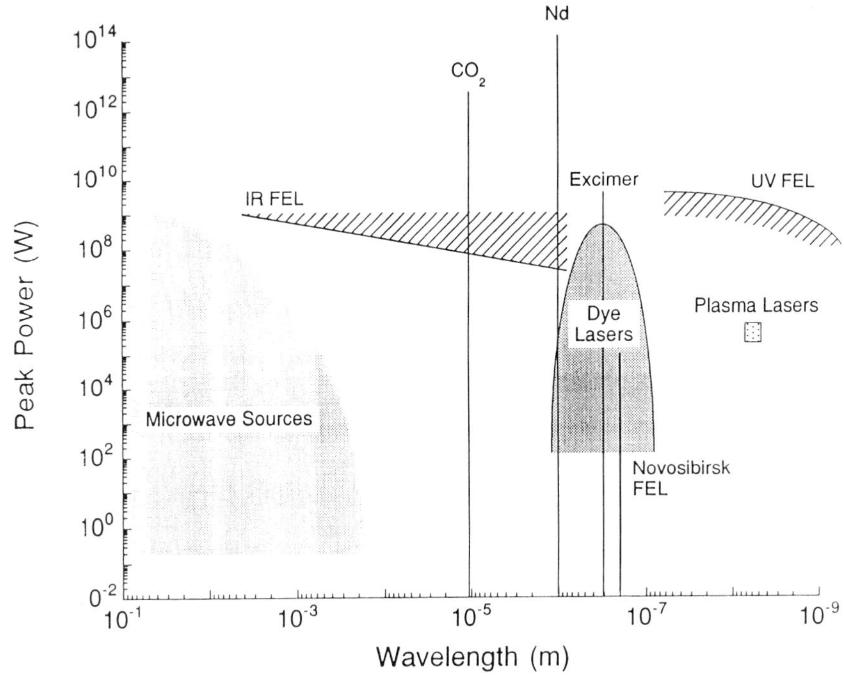

FIG. 2. Peak power vs wavelength for FELs and other coherent sources: the UV region represents expected FEL performance; the cm to mm area represents measured and expected performance; the lower boundary of the IR region is also measured performance, and the upper boundary is expected performance.

We then discuss the saturation effects, the limits on the system efficiency and other properties such as optical guiding, and the high-gain regime. A general review of FELs and a detailed discussion of their properties and applications can be found in Brau (1990); Colson *et al.* (1990); and Marshall (1985).

In Sec. 2 we give a short review of the experimental results obtained to date and ongoing development of the FEL. Sections 3 and 4 discuss FEL applications, potential FEL development, and the possibility of extending FEL operation to the x-ray spectral region.

1. PHYSICAL AND TECHNICAL PRINCIPLES

A FEL has three fundamental components: an electron beam of given energy and intensity and the associated accelerator used to produce it; the undulator magnet; and the electromagnetic wave and the associated optical components controlling its propagation. A schematic representation of a FEL is given in Fig. 1.

The accelerators used to provide the electron beam are of many types: electrostatic, induction linac, radio-frequency (rf) linac, pulsed diode, or storage rings. Some of their basic characteristics, their energy ranges, and the FEL wavelengths for which they are more commonly used are given in Table 1.

Undulator magnets are of two main types: helical or planar. In the first case the magnetic field vector rotates around the axis as a function of axial distance; in the second case its direction is fixed, and its amplitude oscillates along the axis. These magnets can be, and have been, built using a wide variety of technologies: pulsed or dc electromagnets, permanent magnets, and superconducting magnets. The field amplitude can vary from a fraction of a tesla to over 1 T, and the periods from 1 cm to many centimeters. A great effort is under way to develop undulators with periods in the millimeter range; these would allow a reduction of the beam energy for a given radiation wavelength, thus reducing the cost and complexity of the FEL. A helical undulator was used by Madey and co-workers (Elias *et al.*, 1976) in the first FEL. This undulator

Table 1. Particle accelerators for FELs.

	Energy	Peak current	Pulse length	FEL wavelength
Pulsed diode	0.1–1 MeV	1–1000 kA	10–100 ns	cm to mm
Induction linac	1–50 MeV	1–10 kA	10–100 ns	cm to μm
Electrostatic	1–10 MeV	1–5 A	1–20 μs	mm to 0.1 mm
rf linac	0.01–10 GeV	1–1000 A	1–30 ps	100 μm to nm
Storage ring	0.1–10 GeV	1–1000 A	30 ps–1 ns	1 μm to nm

was a superconducting magnet, built with two helical windings with current flowing in opposing directions.

The FEL can be operated as an oscillator, using an optical cavity to confine the radiation in the undulator region, as shown in Fig. 1, or it can be used as an amplifier. Oscillator–amplifier combinations [master oscillator, power amplifier (MOPA)], as well as systems in which one amplifies the electron spontaneous radiation emitted in traversing the undulator [self-amplified spontaneous emission (SASE)], have been also used.

The performance characteristics of some of the existing FELs have been summarized in Table 2, and are discussed in some detail in Sec. 2.

1.1 Undulator Spontaneous Radiation

For this discussion we use a reference frame with z along the undulator axis and x and y perpendicular to z. For a helical undulator, the field near the axis z is given by

$$\mathbf{B}=B_0\left[\mathbf{e}_x \cos\left(2\pi \frac{z}{\lambda_0}\right)+\mathbf{e}_y \sin\left(2\pi \frac{z}{\lambda_0}\right)\right], \quad (3)$$

where $\mathbf{e}_x$, $\mathbf{e}_y$, and $\mathbf{e}_z$ are unit vectors along the x, y, and z axes. The electron velocity in this field, in units of c, is

$$\beta=\frac{a_w}{\gamma}\left[\mathbf{e}_x \sin\left(2\pi \frac{z}{\lambda_0}\right)+\mathbf{e}_y \cos\left(2\pi \frac{z}{\lambda_0}\right)\right] + \mathbf{e}_z\beta_z. \quad (4)$$

The amplitude of the transverse velocity is a_w/γ. This quantity also represents the maximum angle between the electron trajectory and the undulator axis. For relativistic beams this quantity is much less than 1, and the longitudinal velocity is near 1. Notice that in the case of a helical undulator the axial component of the velocity, β_z, remains constant. For a planar undulator only one of the two

Table 2. Operating characteristics of some FELs.

	Wavelength	Electron energy (MeV)	Accelerator type	Micropulse duration	Peak power	Average power	Comments
LLNL-ETA	9.6 mm	4.5	Induction linac	30 ns	1 GW	N.A.[a]	Amplifier, tapered undulator
UCSB	13–338 μm	2–6	Electrostatic	1–20 μs	11 kW	0.23 W	Oscillator
Los Alamos	10–45 μm	23	rf linac	8 ps	25 MW	1 W	Oscillator
Stanford Mark III	2–8 μm	45	rf linac	1–4 ps	1 MW	3 W	Oscillator
Stanford SCA	1–5 μm	70	Superconducting rf linac	3 ps	0.85 MW	1 W	Oscillator
Vanderbilt Univ.	2.6–5.6 μm	45	rf linac	2 ps	15 MW	11 W	Oscillator
Novosibirsk	0.24–0.7 μm	350	Storage ring	75 ps	100 kW	6 mW	Oscillator, optical klystron
Boeing	0.6 μm	110	rf linac	12 ps	6 MW	50 mW	Oscillator

[a]Not available.

transverse components in Eqs. (3) and (4) is nonzero. In this case the longitudinal velocity is not constant; in fact it can be written as the sum of a constant term plus one oscillating at twice the undulator period. The electron trajectory in a helical undulator is a helix of radius $a_w\lambda_0/2\pi\gamma\beta_z$ and period equal to the undulator period. For a planar undulator it is a sinusoid in the plane perpendicular to the magnetic field.

When traversing the undulator the electrons are subject to an acceleration and radiate electromagnetic waves. This spontaneous emission is fundamental to the operation of a FEL. One relativistic electron traversing an undulator magnet emits radiation in a narrow cone, with an angular aperture of order $1/\gamma$. The radiation spectrum is a superposition of the synchrotron radiation emitted when the electron trajectory is bent in the undulator magnetic field, a narrow line introduced by the periodic nature of the motion in the same field, and its harmonics. The wavelength of the fundamental line is of the order of $\lambda_0/2\gamma^2$, which can be interpreted as the undulator period λ_0 doubly Doppler shifted by the electron motion.

The aperture $1/\gamma$ of the synchrotron radiation cone can be compared with the angle between the electron trajectory and the undulator axis, a_w/γ. If a_w is smaller than 1, the emitted radiation is contained within the synchrotron radiation cone, and the emission is predominantly in a single line. If a_w is larger than 1, the synchrotron radiation cone sweeps an angle larger than its aperture. In this case the spectrum is rich in harmonics and approaches the synchrotron radiation spectrum when a_w is very large. When $a_w \gg 1$ the magnet is usually referred to as a wiggler, reserving the name undulator for the case of a_w less than or of the order of 1. In this paper we use the generic term undulator to describe a magnet with a periodic transverse field described by Eq. (3).

The fundamental wavelength seen by an observer looking at an angle θ to the undulator axis is $\lambda = (\lambda_0/2\gamma^2)(1 + a_w^2 + \gamma^2\theta^2)$, and is the wavelength at which there is positive interference between the radiation emitted in two undulator periods. This is also, when $\theta = 0$, the wavelength corresponding to the "synchronism condition" Eq. (1).

In a planar undulator the magnetic field has only one component, say along y. Then the velocity has components in the x and z directions. The wavelength of the spontaneous radiation is the same as that for a helical undulator if the vector potential a_w is replaced by its rms value, $a_w/\sqrt{2}$.

In a helical undulator the radiation on axis contains only the fundamental harmonic given by Eq. (1). In a planar undulator the longitudinal electron velocity is modulated at twice the undulator period. This leads to a richer spectrum: all the odd harmonics of Eq. (1) appear on axis. For both type of undulators even and odd harmonics appear off axis.

The linewidth of the radiation is determined by the number N of undulator periods; the total length of the pulse emitted by one electron is $N\lambda/c$ and the corresponding linewidth of the fundamental is

$$\Delta\omega/\omega = 1/2N. \tag{5}$$

The spontaneous radiation spectrum around the fundamental is shown in Fig. 3.

When more than one electron is producing the radiation, the pulse length and angular distribution can be affected by the electron-beam characteristics. The pulse length is increased by the electron bunch length when this is larger than $N\lambda$. The radiation linewidth is increased if the beam energy spread is larger than $1/2N$, the single-electron width. Similarly, the effective source radius and angular distribution can be increased by the electron beam. These effects can reduce the FEL gain, as we will discuss in the next section.

1.2 Small-Signal Gain

Undulators are widely used in storage rings as brilliant sources of synchrotron radiation. What one is using in this case is the superposition of the spontaneous radiation of many electrons, without any phase correlation, so that the total intensity is proportional to the electron number N_e. In a FEL, as in an atomic laser, there is a phase correlation between emitting electrons. This correlation is obtained by modulating the longitudinal beam density on the scale of the radiation wavelength, a process called *bunching*. When electrons are bunched together in a distance that is short compared to the wavelength of the radiation, they emit in phase, thus increasing the intensity in the emitted line and simultaneously reducing its width.

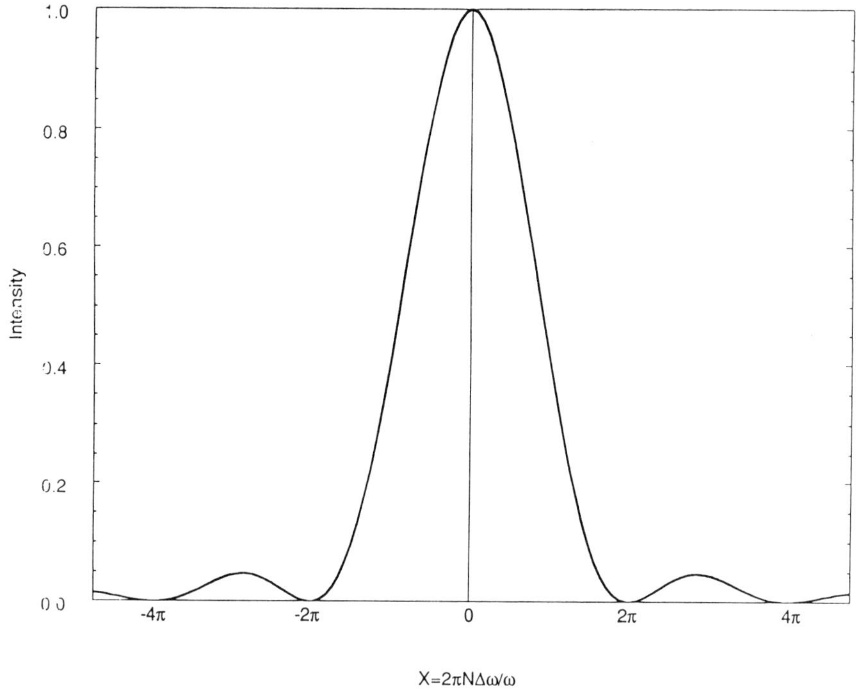

21.3p

FIG. 3. Spontaneous radiation spectrum for emission on axis near the fundamental frequency.

The process of electron bunching and increase in radiation field intensity can be described as follows. Let

$$\mathbf{E}=E_0\left[\mathbf{e}_x \cos\left(2\pi\frac{z}{\lambda}-\omega t\right)+\mathbf{e}_y \sin\left(2\pi\frac{z}{\lambda}-\omega t\right)\right] \tag{6}$$

be the electric field of a plane wave propagating along the undulator axis. This field is transverse to the undulator axis, and thus has components parallel to the transverse electron velocity, as given by Eq. (4). An energy transfer between the em field and the electron beam can take place. Using Eqs. (4) and (6) one finds

$$mc^2\frac{d\gamma}{dt}=eE_0\frac{a_w}{\gamma}\sin\Phi, \tag{7}$$

where the phase Φ is

$$\Phi=2\pi\left(\frac{1}{\lambda}+\frac{1}{\lambda_0}\right)z-\omega t+\Phi_0. \tag{8}$$

This quantity is called the *ponderomotive force phase* and plays a key role in FEL physics. The ponderomotive force phase has terms changing at the "fast" em field frequency and terms changing with the "slow" undulator period. The time-average value of the electron energy change (7), averaged over a few radiation periods, is 0 unless we satisfy a "synchronism condition" $d\Phi/dt=0$. This condition can be written as $\lambda=\lambda_w(1-\beta_z)/\beta_z$. In the case of relativistic particles, using the relationship $\beta_z=(1-1/\gamma^2-\beta_x^2-\beta_y^2)^{1/2}$, and assuming $\gamma\gg 1$, $\beta_x,\beta_y\ll 1$, this condition can be approximately written as $\lambda=\lambda_w(1+a_w^2)/2\gamma^2$. These are Eqs. (1) and (1′) given in the Introduction. If the synchronism condition is satisfied, the ponderomotive force phase is constant, and the electron beam, oscillating at the slow undulator frequency, can exchange energy with the fast oscillating em wave.

When the synchronism condition is satisfied the ponderomotive phase can be written as $\Phi=(2\pi/\lambda\beta_z)(z-\beta_z ct)+\Phi_0$. For an electron whose axial motion is described by $z=\beta_z ct+z_0$, z_0 being the initial position, the phase becomes $\Phi=2\pi(z_0/\lambda\beta_z)+\Phi_0$ and is a constant. Integration of Eq. (7) for constant phase and for a small energy change shows that, after going a distance z through the undulator, the electron energy is the initial energy plus a term proportional to sin Φ, with Φ depending on the ratio of the initial electron position within the beam to the radiation wavelength. For relativistic electrons, $\beta_z \approx 1$, a variation of z_0 by half a radiation wavelength reverses the sign of the electron energy change. Hence the interaction with the em wave modulates the energy of the electrons at a scale length equal to the em radiation wavelength.

The distribution of the initial position of the electrons within the bunch is determined by the electron source and the acceleration process. If the electron beam is produced by an accelerator using rf cavities to accelerate the beam, the distribution is characterized by two scale lengths. One is the rf wavelength; the other is the radiation wavelength. In a FEL the first is much larger than the second. The wavelengths used in rf cavities vary from a minimum of about 10 cm for an rf linac to infinity for an electrostatic accelerator. At the scale length defined by the rf system the electrons are distributed longitudinally into bunches separated by one rf wavelength. The length of these bunches is a fraction of the wavelength of the accelerating rf and varies from about 1 mm for the case of a 3-GHz ($\lambda_{\mathrm{rf}}=10$ cm) rf linac to a continuous beam for the electrostatic case.

At the scale length of the radiation wavelength the electron distribution is uniform; there is no correlation between the longitudinal electron positions at the scale of the radiation wavelength λ. As a result the average value of the energy exchange over the initial distribution of the electron longitudinal position is 0:

$$\left\langle \frac{d\gamma}{dt} \right\rangle_{\text{electrons}} \approx \langle \sin \Phi \rangle_{\text{electrons}} = 0.$$

Although the average energy exchange at this stage is 0, individual electrons have changed their energy proportionally to $\sin[\Phi(z_0/\lambda)]$. Electrons with different energies follow different trajectories in the magnetic field of the undulator. Electrons with larger energy are bent less and follow a shorter trajectory with a larger velocity; electrons with less energy are bent more and follow a longer trajectory, with a smaller velocity. As a result the electron longitudinal distribution—initially uniform at the scale of the radiation wavelength—is changed into one consisting of groups of electrons (microbunches) separated by one radiation wavelength. The microbunches have a ponderomotive force phase limited to a small fraction of 2π. The energy exchange between the electrons and the em wave, averaged over the electrons, is now nonzero. A net energy exchange takes place. The energy lost (gained) by the electrons is given to the em radiation. If the initial electron energy is slightly larger (smaller) than that defined by the synchronism condition the electrons lose (gain) energy, and the radiation field intensity increases (decreases).

This process can be summarized by saying that in the initial part of the undulator, the electrons are modulated in energy by the input em field, then the energy modulation is transformed into microbunching, and at last the microbunches radiate and the em field intensity is increased.

The energy exchange between electrons and the em field was first evaluated by Madey (Madey, 1971). The quantity of interest is the gain, defined as the relative change in the em intensity per undulator pass. When the initial em field amplitude is small—and the relative change in intensity is also small—the gain is called the small-signal gain. For an initially monoenergetic electron beam the small signal gain calculated by Madey is

$$G_s = 4\sqrt{2\pi}\,\frac{(\lambda\lambda_0)^{1/2}}{w_0^2}\,\frac{a_w^2}{(1+a_w^2)^{3/2}}\,\frac{I_p}{I_A}\,N^3 f(x), \tag{9}$$

where I_p is the beam peak current, $I_{\mathrm{A}}=ec/r_e$ is the Alfvén current (about 17 kA), r_e is the classical electron radius ($r_e \sim 2.8\times10^{-15}$ m), and πw_0^2 is the transverse cross section of the electromagnetic wave (assumed to be larger than that of the electron beam).

The function $f(x)$, with $x=2\pi N\Delta\omega/\omega$, describes the gain width and is given in Fig. 4. For a given beam energy and a given undulator, the central frequency is obtained by the synchronism condition, Eq. (1). The function $f(x)$ is simply the derivative of the spontaneous radiation line. This property has been utilized to determine the small-signal gain. The function $f(x)$ is antisymmetric. The gain is zero for $x=0$, at synchronism, and is maximum for $x \simeq 2.6$. The linewidth is of the order of $1/N$, the inverse of the number of periods in the undulator.

Since the electromagnetic radiation frequency and the electron energy are related by

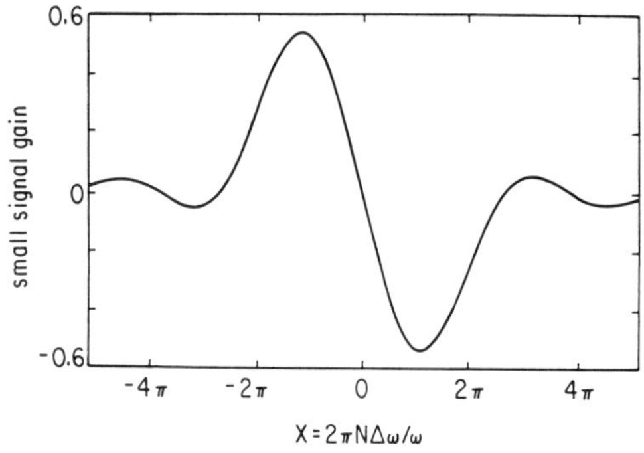

FIG. 4. FEL small signal gain vs frequency change.

the synchronism condition, Eq. (1′), we can also redefine x, for a fixed frequency, as $x=4\pi N\Delta\gamma/\gamma$. The gain linewidth then corresponds to an energy change $\Delta\gamma/\gamma$ of the order of $1/2N$. When the radiation field increases the beam kinetic energy decreases, and when the change is about $1/2N$ the gain becomes 0. Hence the maximum FEL efficiency, defined as the ratio of the laser intensity to the initial beam kinetic energy, is of the order of $1/2N$. As we will see in Sec. 1.6 the FEL efficiency can be increased to a much larger value using tapered undulators.

When the electron beam has an energy spread the gain can be reduced. Since the gain linewidth is $\Delta\gamma/\gamma=1/2N$, the gain decreases and becomes nearly 0 when the electron beam initial energy spread is of the order or larger than $1/2N$.

Although the original theoretical discussion of the FEL (Madey, 1971) used a quantum-mechanical description, it was soon realized that under most conditions a classical description is sufficient to describe the system, since the most relevant quantities, such as the gain, Eq. (9), are independent of the Planck constant.

1.3 Scaling and Electron-Beam Requirements

We can estimate the order of magnitude of the small-signal gain assuming $f(x)=0.5$, $a_w=1$. We further assume that the laser-field cross section is determined by diffraction, $\pi w_0^2=\lambda\lambda_0 N$, where $\lambda_0 N$ is the undulator length. We can now rewrite the gain as $G_s\sim\pi^2(\lambda/\lambda_0)^{1/2}(I_p/I_A)N^2$. For $\lambda=1$ μm, $\lambda_0=5$ cm, $N=100$ we have $G_s\sim0.025I_p$ (A). To get a gain of several percent the beam peak current must be several amperes at a wavelength of 1 μm. At shorter wavelengths the current must increase to keep the same value of the gain. However, the gain scales as the square root of the wavelength, a rather mild dependence.

The small-gain formula (9) is valid as long as the electron beam is monoenergetic and the beam transverse cross section is smaller than that of the radiation beam. The dependence of the emitted wavelength on the angle θ indicates that the electron angular spread also has an effect on the gain. The two conditions on the electron-beam radius and angular spread can be summarized as a single condition on the beam emittance. This quantity is the product of the beam radius and angular spread, and is an invariant when the beam is transported in a linear system. The equivalent quantity for a diffraction-limited laser beam is the radiation wavelength. The condition for a good matching of the electron beam to the radiation is that the beam emittance be smaller than the radiation wavelength.

1.4 Optical Klystron

When the small-signal gain is not large enough to overcome the optical cavity losses, perhaps because the beam current is too small, one can increase the gain by using a type of undulator called an optical klystron. This device is based on the observation that in a FEL the electrons are first modulated in energy, then this modulation is transformed into longitudinal bunching, and in the end the bunched beam produces the radiation.

In a normal undulator the energy modulation produces bunching mainly because the electron trajectory in the undulator depends on energy: particles with larger energy are bent less and have a shorter trajectory than particles with lower energy. This bunching process can be enhanced if we replace a part of the undulator with a magnet, called a dispersive magnet, designed to optimize bunching. The dispersive magnet is then followed by the remaining part of the undulator, where the beam radiates.

The dispersive magnet is designed to leave unchanged the electron trajectory inside the undulators. The system of two split undulators and the dispersive magnet is called an optical klystron, since it reproduces the process by which the microwave klystron amplifies radiation (Vinokurov *et al.*, 1977). The main difference between these two devices is that in the microwave klystron the energy modulation is produced by the interaction of a longitudinal electric field with the electron longitudinal velocity, while in the FEL the interaction is between the transverse electric field and the electron transverse velocity.

The optical klystron has been used in the Orsay and in the Novosibirsk storage-ring FELs, operating in the UV spectral region, to enhance the gain.

1.5 Large-Gain, Saturation, Compton and Raman Regimes

For sufficiently long undulators and large electron beam current the radiation field can grow exponentially in a single undulator pass (Bonifacio *et al.*, 1984). Neglecting diffraction effects and for a "good quality" beam, the exponential growth rate, called the gain length, is

$$L_G=\lambda_0/2\sqrt{3}\pi\rho. \qquad (10)$$

The FEL parameter ρ is defined as

$$\rho=\left(\frac{a_w}{4\gamma}\frac{\omega_p}{\omega_0}\right)^{2/3}, \qquad (11)$$

where $\omega_p(4\pi\, r_e c^2 n_0/\gamma)^{1/2}$ is the beam plasma frequency, n_0 is the electron-beam density, and $\omega_0=2\pi c/\lambda_0$. Typical values of ρ vary from 10^{-2} for a millimeter FEL to 10^{-3} or less for a visible or UV FEL.

The radiation intensity saturates, for an untapered undulator, when the ratio of the laser power to the electron-beam power is about equal to the FEL parameter: $P_{\text{laser}} \sim \rho P_{\text{beam}}$. At saturation the beam energy spread, $\Delta E/E$, is also of the order of ρ.

If the initial em signal for the FEL is provided by the undulator spontaneous radiation, the undulator length needed to reach saturation is about 10 gain lengths, or about $1/\rho$ undulator periods. Under these conditions the efficiency of energy transfer is again on the order of $1/N$, as for an oscillator and the small-signal–gain regime.

For low-energy, high-intensity beams, we must take into account the Coulomb forces acting between electrons in the beam. The electrons can now interact not only through and with the radiation field (the Compton regime), but also directly through space-charge forces (the Raman regime) (Sprangle and Smith, 1980). This situation corresponds to a three-wave interaction; the frequencies are the Doppler-shifted undulator frequency, the radiation frequency, and the beam plasma frequency. The synchronism condition is correspondingly modified and can be written as

$$\lambda=\frac{\lambda_0(1+a_w^2)}{2\gamma^2}\,\frac{1}{\beta_z-(\omega_p\lambda_0/2\pi c\gamma)}. \qquad (12)$$

If we compare Eq. (12) and Eq. (1′) we see that, for the same undulator, Eq. (12) gives a longer wavelength. At the low beam energies where these effects are important, the radiation wavelength is usually so large that the em signal propagates not in vacuum but in a waveguide. The presence of the waveguide modifies again the synchronism condition (Marshall, 1990).

1.6 Tapered Undulator and High-Efficiency FELs

The growth of the radiation field saturates at a level that, for a constant-period and constant-field undulator, corresponds to a relative energy transfer of the order of $1/N$, the inverse of the number of undulator periods. This saturation corresponds to the electron and radiation getting out of step, so that the synchronism condition (1) is no longer satisfied. This situation can be overcome and the energy transfer increased if we use an undulator with a variable period and/or field (Sprangle *et al.*, 1979; Kroll *et al.*, 1981), so that Eq. (1) continues to be satisfied when the electron energy changes. It has been evaluated theoretically, and confirmed experimentally, that in this way the FEL efficiency (defined as the ratio of the amount of beam kinetic energy transferred to the radiation field, to the initial beam energy), can be as large as 40% to 50%. This possibility has been first demonstrated in a high-gain amplifier, the ETA FEL, where the laser power was increased from about 100 MW to 1 GW by tapering the undulator field (Orzechowsky *et al.*, 1985).

Tapering the undulator in an oscillator to increase the efficiency produces a decrease in gain, so that the conditions of maximum gain and maximum energy transfer cannot be obtained at the same time. Systems consisting of a MOPA have been designed and built to optimize these two characteristics (Bhowmik *et al.* 1988).

1.7 Optical Guiding

The results given above on the growth rate are derived in a one-dimensional (1D) model, in which effects such as diffraction are neglected. Although the 1D model results are important and can be applied under certain conditions, three-dimensional effects must be included to have a full description of the FEL behavior.

The most important effects that appear in a 3D theory are diffraction and optical guiding (Sprangle and Tang, 1981; Moore, 1984; Scharlemann *et al.*, 1985). While the radiation tends to diffract out of the beam, thus reducing the intensity of the radiation field and the gain, optical guiding provides a mechanism to counteract diffraction and keep the FEL gain large. This mechanism is particularly important for short-wavelength FELs, where diffraction can be very large.

Optical guiding in a FEL arises from two effects: refractive guiding and gain guiding, which can be described using the beam refractive index. Refractive guiding is due to the radiation phase velocity being less than the light velocity c inside the beam; it is described by the real part of the refractive index. This effect is similar to the guiding occurring in an optical fiber.

Gain guiding is described by the imaginary part of the refractive index, which gives the rate of increase of the radiation intensity. We can describe this effect by comparing the FEL radiation Rayleigh length, $Z_R = \pi w_0^2/\lambda$, which gives the characteristic distance for diffracting light out of the beam, with the gain length. If the gain length is smaller than the Rayleigh range, the production of radiation from the beam dominates over the reduction in intensity due to diffraction. The condition for optical gain guiding can thus be written approximately as $L_G < Z_R$.

Because of gain focusing it is possible to find solutions of the FEL equation that are self-similar, i.e., undistorted in form, but exponentially growing along the axis. When the FEL reaches saturation, the gain decreases and refractive guiding becomes the dominant term.

2. STATUS AND OPERATION OF FELs

The design and construction of a FEL requires the utilization of physics and technology from the fields of optics, lasers, and particle accelerators. Situated at the crossroads of these various technologies, FEL development has required their successful integration into one system. The experience gained in building and operating FELs has confirmed the results of the theoretical analysis, showing the critical importance of the electron-beam quality, i.e., its peak current, energy spread, and emittance.

To obtain simultaneously the beam current, emittance, and energy spread required for a large gain is a difficult task. The beam characteristics required for a FEL are also different from those needed for other applications, such as high-energy and nuclear physics, synchrotron radiation sources, or medical radiation therapy. Hence particle accelerators have to be designed and optimized as FEL drivers.

As shown in Table 1, different types of accelerators have been considered and used for FELs. In fact, the choice of an optimum accelerator depends on the particular wavelength and time structure required, as one can see from an examination of Table 2, where we have listed the main characteristics of some typical FELs. Electrostatic accelerators have been shown to be well suited for the FIR and submillimeter region when long pulses, in the microsecond range, are required. rf linacs are commonly the choice for the infrared region and for pulse lengths of tens of picoseconds or picoseconds. A superconducting rf linac is being used at Stanford University and can provide a beam with extremely stable beam energy, an important characteristic to avoid wavelength drifts from pulse to pulse. Storage rings have been used at Orsay, UVSOR, Tsukuba, and Novosibirsk to obtain UV radiation down to about 0.2 μm.

We can also see from Table 2 that peak powers of tens of megawatts have been obtained in the infrared region, and that up to 1 GW has been produced in the millimeter range. The average power is up to 10 W, enough for most research applications. New systems optimized for kilowatt average power, or more, are in various stages of development.

In all accelerators considered, with the exception of storage rings, the electron-beam properties are determined by the electron source and by collective effects in the acceleration process. Much attention has been dedicated to improving the electron sources; particularly good results have been obtained recently using a photoinjector, consisting of a photocathode placed in a rf cavity (Fraser *et al.*, 1987). Figures 5 and 6 show a photograph of the photoinjector developed at Los Alamos National Laboratory, and a schematic of a photoinjector developed initially at Brookhaven National Laboratory and now being used in several other laboratories.

In a photoinjector a laser is used to extract electrons from a photocathode. The electron pulse duration and number of pulses are controlled by the laser, allowing a flexibility unique to this system. At frequencies between 1 and 3 GHz, the rf electric field on the cathode surface can be as large as 60 to 100 MV/m. Such a large field accelerates the electrons to relativistic speeds in a few centimeters, thus reducing space-charge effects. A root mean square emittance as small as 0.4 mm mrad and a peak current in excess of 100

FIG. 5. View of the photoinjector developed at the Los Alamos National Laboratory.

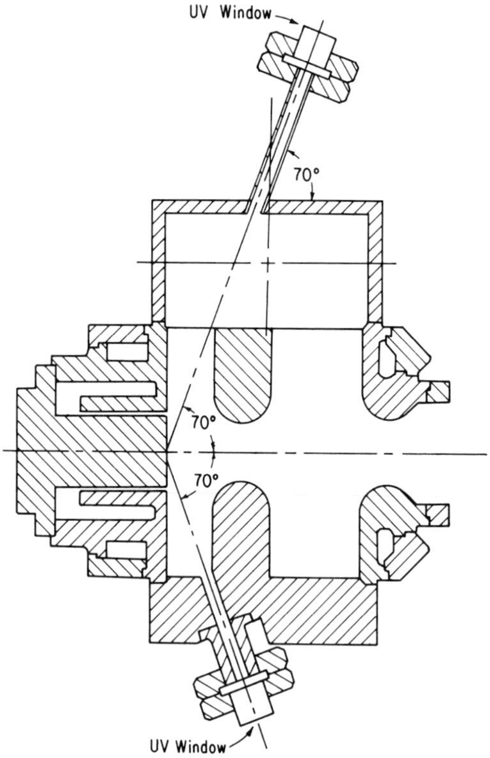

FIG. 6. Schematic of a photoinjector developed at the Brookhaven National Laboratory, consisting of $1\frac{1}{2}$ cavities operating in a π mode; the laser beam illuminating the photocathode can be injected on axis or at an angle of 70°; the accelerating field on axis can be as large as 100 MV/m; the cathode is removable.

A, at an energy of a few MeV, have been obtained with this technique.

FELs in operation, under construction, or proposed worldwide are listed in Tables 3, 4, and 5. The number of systems in this list shows the growing interest in FELs.

A more detailed description of the operating characteristics of some existing FELs is given in Table 2, showing the radiation wavelength, the accelerator type, the minimum pulse duration, the average and peak power, and whether the system is an oscillator, amplifier, or other configuration. The systems in Table 2 are typical of the present state of the art, and show performance characteristics that one can obtain and are being utilized today.

Some of the existing FELs, like the ones at Stanford University, Duke University, University of California at Santa Barbara, FOM Institute, and Vanderbilt University, are being operated as user facilities, supporting research programs in areas ranging from material science to biology and medicine. Other FELs are being developed to achieve shorter wavelengths or reach higher peak and average powers. A new generation of user facilities is also under construction or proposed, utilizing the most recent developments in FEL science and technology.

3. FEL APPLICATIONS

An examination of Fig. 2 shows that there are two regions where FELs can be superior to other sources: the infrared and millimeter- to centimeter-wave region and the short-wavelength region below 100 nm. A partial list of possible applications is given in Tables 6 and 7.

The infrared and far-infrared regions are important for research in chemistry, solid-state physics, materials science, and medicine (Yamanaka, 1992; Dlott, 1992). FELs dedicated to research in these areas and operated as user facilities are at the University of California at Santa Barbara (UCSB), Stanford Synchrotron Radiation Laboratory, Vanderbilt University, Duke University, FOM Institute, Orsay, and Osaka. The time structure of these sources varies from microsecond-long pulses for the UCSB system to picosecond pulses for the other cases. The peak power is in the megawatt range. The average power is about 11 W at the Vanderbilt University FEL (Brau, 1992).

Infrared FELs are now well established. Development work for the user facilities is aimed at improving reliability, wavelength stability, and linewidth. Near- and far-infrared FELs can provide new scientific opportunities in the study of surface and molecular physics. As an example, using the tunability and power of the FEL, one can extend the range of applicability of infrared multiphoton-excitation techniques across the entire periodic table and to practically all molecular species. Combining a FEL and synchrotron radiation, as proposed at Lawrence Berkeley Laboratory, "two-color" pump–probe experiments can be performed at essentially any wavelength (Lawrence Berkeley Laboratory, 1992). A review of infrared user facilities can be found in a paper by Smith and Schwettman (1991).

Table 3. Centimeter- to millimeter-wave FELs.

Laboratory, country, and status[a]	Wavelength (mm)	Accelerator type and energy (MeV)
LLNL, U.S., e	2	Induction linac, 7
ILE, Japan, e	1	Induction linac, 6
KEK, Japan, e	30	Induction linac, 0.75
JAERI, Japan, c	10–5	Induction linac, 1
CEA, France, e	9	Induction linac, 3
UCSB, U.S., e	1–0.06	Electrostatic, 6
Tel Aviv Univ., Israel, p	3–0.3	Electrostatic, 5
FOM, Netherlands, p	1.2	Electrostatic, 2
Rutgers Univ., U.S., e	0.25–0.16	Microtron, 19
ENEA, Italy, e	2.5–0.5	Microtron, 6
Columbia Univ., U.S., e	2–1	Pulsed diode, 1
Maryland Univ., U.S., c	3	Pulsed diode, 0.4
CEA, France, c	9	Pulsed diode, 2.5
ISAS, Japan, e	25	Pulsed diode, 0.6
RIKEN, Japan, e	30	Pulsed diode, 0.6
MIT, U.S., e	10	Pulsed diode, 0.75

[a]Status: e, existing; c, under construction; p, proposed.

Medical applications are being studied and developed in many areas. Surgery can use the region around 3 μm where water absorbs strongly. The FEL tunability can be used for diagnosis and for selective absorption in pigmented areas. The Vanderbilt University FEL has extensive facilities for medical research, including animal operating rooms, an animal care facility, and support laboratories (Brau, 1992).

FELs at Orsay, the Institute of Molecular Science in Japan, and Novosibirsk, operate in the UV. The shortest wavelength, about 240 nm, has been reached at Novosibirsk (Kuli-

Table 4. Infrared FELs.

Laboratory, country, and status[a]	Wavelength (μm)	Accelerator type and energy (MeV)
LANL, U.S., e	2.8	rf linac, 40
Boeing, U.S., c	10.6	rf linac, 34
LANL, U.S., c	3.7–0.36	rf linac, 20
Vanderbilt Univ., U.S., e	2–10	rf linac, 45
Stanford Univ., U.S., e	1–5	Superconducting rf linac, 70
Duke Univ., U.S., e	1.4–8.1	rf linac, 45
Rocketdyne, U.S., c	1.06	rf linac, 78.5
UCLA, U.S., c	5–10	rf linac, 25
JAERI, Japan, c	41.7	rf linac, 13.2
Osaka Univ., Japan, e	10–60	rf linac, 38
Tokyo Univ., Japan, e	40	rf linac, 15
ILT, Osaka, Japan, c	70–450	rf linac, 6
Orsay, France, e	2–20	rf linac, 70
CEA, France, e	15–25	rf linac, 20
FOM, Netherlands, e	8–80	rf linac, 45
Twente Univ., Netherlands, e	180–10	Microtron, 25
INFN, Italy, e	16	Superconducting rf linac, 25
Darmstadt Univ., Germany, e	2.5–6	Superconducting linac, 25
LBL, U.S., p	3–50	Superconducting rf linac, 50
Stanford Univ., U.S., c	100–1000	Superconducting rf linac, 50
ILT, Osaka, Japan, p	0.01–10	rf linac, 3.3 rf linac, 170

[a]Status: e, existing; c, under construction; p, proposed.

panov *et al.*, 1990). The next step is to reach the 100–50-nm region, particularly important for chemistry. A proposal to build a facility operating in this region has been presented recently by Brookhaven National Laboratory to perform fast spectroscopy and fast timing experiments, among others. This FEL can also be used in conjunction with a synchrotron radiation source for pump–probe type experiments (Ben-Zvi *et al.*, 1992).

At the other wavelength extreme, the millimeter- to centimeter-wave region, the FEL applications are mainly in the industrial and military areas, high-energy particle acceleration, and plasma heating. FELs at about 1 mm can be used for plasma heating in a tokamak, as an alternative to sources like gyrotrons (Urbanus *et al.*, 1992). The FEL can be used in high-energy accelerators to power high-frequency accelerating structures, at about 30 GHz, in a two-beam accelerator (Sessler, 1982; Sessler *et al.*, 1989; Amaldi and Pellegrini, 1987). The FEL can be also used to power an inverse FEL (Palmer, 1972; Courant *et al.*, 1985).

High–average-power FELs have been considered for military applications in the micron region (Schultz *et al.*, 1992). Industrial applications can be found in semiconductor manufacturing using microlithography, etching; isotope separation and reprocessing of nuclear fuel for nuclear engineering; and photoinduced chemistry (Yamanaka, 1992).

The average power needed for military and industrial applications, 10 kW to MW, has not yet been achieved. FELs for these applications need further development.

4. FUTURE DEVELOPMENTS AND X-RAY FELs

Presently there are three main lines of development for FELs:

1. user facilities or compact infrared systems;
2. short-wavelength FELs;
3. high–average-power FELs.

We now have a good understanding of the FEL physics and technology. FEL theory is well supported by experimental data in the wavelength region from centimeters to the near UV. In this region the next step in FEL development is to make the system easier to operate, more reliable, and less expensive, with a better control over the linewidth and time structure. Much of this effort is oriented toward accelerator development to improve the beam-energy stability and reproducibility, increase the peak current, and extend the flexibility in the time structure. The development of short-period, high-field undulators is also being actively pursued. The combination of an improved compact accelerator and a small-period undulator, now being developed at Los Alamos and other laboratories, can lead to FELs with lower beam energy for the same wavelength, hence FELs that are less expensive and easier to use.

FELs based on compact rf linear accelerators and using rf electron guns, in some cases with laser-driven photocathodes, are being actively developed in Europe, Japan, and the United States.

The development of an x-ray FEL with a wavelength around 10 to 1 nm, and eventually

Table 5. Visible and UV FELs.

Laboratory, country, and status[a]	Wavelength (nm)	Accelerator type and energy (MeV)
Boeing, U.S., e	600	rf linac, 110
Novosibirsk, Russia, e	240–690	Storage ring, 500
Orsay, France, e	630	Storage ring, 600
Duke Univ., U.S., c	5–400	Storage ring, 1000
IML, Japan, e	100–200	Storage ring, 750
Electrotechnical Lab., Japan, e	600	Storage ring, 230
Electrotechnical Lab., Japan, e	200–630	Storage ring, 500
Dortmund Univ., Germany, c	20–400	Storage ring, 1500
CEBAF, U.S., c	150–2000	Superconducting linac, 85
SLAC, U.S., p	4	rf linac, 7000
BNL, U.S., p	75–500	Superconducting rf linac, 250

[a]status: e, existing; c, under construction; p, proposed.

Table 6. FEL applications.

Condensed Matter	
Surface Science	
Catalysis	IR
Adsorption	IR
Selective excitation of surface layer, adsorbed molecule	IR
Semiconductors	IR
High-T_c superconductor band gap	IR
Carrier dynamics	IR
Electron-gas dynamics at metal–insulator junction	IR
Superconductors	IR
Magnetic Properties	IR
Chemistry	
Molecular vibrational excitation	IR
Reaction dynamics	IR
Photochemistry	IR
Molecular clusters, van der Waals molecule	IR
Electronic excitations	UV
Raman spectroscopy	UV
Cross photon-molecular beams	UV
Biology	
Molecular vibational excitation	IR
Microscopy, holography, cellular dynamics	1–4 nm
Crystallography	0.1 nm
Medicine	
Surgery	3μm
Diagnosis using wavelength-selective excitation	IR
Phototheraphy-selective absorption in pigmented elements	IR

in the 0.1-nm region, is very attractive and challenging. The region between 2.4 and 4.4 nm, where absorption in water is small, is particularly important for x-ray microscopy and holography. X-ray images of biological samples in an aqueous environment and *in vivo*, without dehydration or staining, can be taken in this wavelength region. If the pulse energy density in a single subpicosecond pulse is 300 mJ/mm^2, over a 0.1 mm^2 area, one can obtain an image in a single pulse, before the hydrodynamic expansion of the sample becomes important (London *et al.*, 1992). At even shorter wavelengths, of the order of 0.1 nm, a coherent source with large peak power can be used for X-ray crystallography of proteins giving amplitude and phase information. FELs in this wavelength region are still not available. The scaling laws show that the electron-beam requirements become more challenging as the wavelength is reduced. The electron energy and peak current must increase, while the emittance and energy spread must be reduced. In these wavelength regions there are no mirrors available to build low-loss optical cavities.

If the difficulties are many, the rewards can also be large, and several groups are now actively engaged in the study and development of an X-ray FEL (Cornacchia and Winick, 1992; Pellegrini, 1992). The accelerators being considered are storage rings and rf linacs (Madey and Pellegrini, 1984). The work done for the design of advanced synchrotron radiation sources in Europe, Japan, and the United States has demonstrated that storage rings can provide beams with very small emittance, a few times 10^{-3} mm mrad. When trying to obtain simultaneously a very small emittance, a small energy spread, and a large peak current, one encounters limitations due to collective instabilities. The resolution of

Table 7. High-power FEL applications.

Accelerators	
Inverse FEL	1–100 μm
Two-beam accelerator	1–10 mm
Fusion	
Inertial confinement	0.1–0.5 μm
Plasma Heating	0.1–1 mm
Industrial	
Photochemistry	IR
Nuclear engineering	IR, UV
Microlithography	UV

this problem will likely determine the shortest wavelength obtainable using a storage ring. Short-wavelength FELs based on storage rings are being built at Duke University and Dortmund University.

Another approach is based on the use of rf linacs (Newnam, 1988; Pellegrini, 1988). In this case the problems are related to reducing the beam emittance by improving the electron source, and controlling the collective effects that can blow up the emittance and energy spread during the acceleration. Recent progress in accelerator and electron source technology, related to research on linear colliders for high-energy physics and photoinjectors, make this approach very promising.

Multilayer mirrors and multifaceted metal mirrors are being developed for the x-ray FEL (Newnam, 1988). Their use would allow the construction of an oscillator at these wavelengths. The alternative to the use of mirrors is the use of the FEL as a high-gain amplifier starting from noise (self-amplified spontaneous emission). In this case the electron-beam requirements are even more stringent than in the oscillator case.

List of Works Cited

Amaldi, U., Pellegrini, C. (1987), in: M. Puglisi (Ed.), *New Techniques for Future Accelerators,* New York: Plenum.

Ben-Zvi, I., *et al.* (1992), *Nucl. Instrum. Methods A* **318**, 201–207.

Bhowmik, A., Madey, J. M. J., Benson, S. V. (1988), *Nucl. Instrum. Methods A* **272**, 183–186.

Bonifacio, R., Pellegrini, C., Narducci, L. (1984), *Opt. Commun.* **50**, 373–378.

Brau, C. (1990), *Free Electron Lasers,* New York: Academic.

Brau, C. (1992), *Nucl. Instrum. Methods A* **318**, 38–41.

Colson, W. B., Pellegrini, C., Renieri, A. (Eds.) (1990), *Laser Handbook,* Vol. 6: *Free Electron Lasers,* Amsterdam: North-Holland.

Cornacchia, M., Winick, H. (Eds.) (1992), *Proceedings of the Workshop on 4th Generation Light Sources,* Stanford Synchrotron Radiation Laboratory Report No. SSRL-92-02.

Courant, E. D., Pellegrini, C., Zakowicz, W. (1985), *Phys. Rev. A* **32**, 2813–2823.

Deacon, D. A. G., *et al.* (1977), *Phys. Rev. Lett.* **38**, 892–894.

Dlott, D. D. (1992), *Nucl. Instrum. Methods A* **318**, 26–29.

Elias, R. L., *et al.* (1976), *Phys. Rev. Lett.* **36**, 717–720.

Fraser, J. S., Sheffield, R. L., Gray, E. R., Giles, P. M. (1987), in: *Proceedings of the 1987 IEEE Particle Accelerator Conference,* IEEE87CH2387-9, p. 1705.

Kroll, N. M., Morton, P. L., Rosenbluth, M. N. (1981), *IEEE J. Quantum Electron* **QE-17**, 1436–1468.

Kulipanov, G. N., *et al.* (1990), *Nucl. Instrum. Methods A* **296**, 1–3.

Lawrence Berkeley Laboratory (1992), *An Infrared Free Electron Laser,* Design Report No. PUB-5335.

London, R., Matthews, D., Suckewer, S. (1992), in: *Proceedings of the Workshop on Applications of X-Ray Lasers,* National Technical Information Service, Virginia.

Marshall, T. C. (1985), *Free Electron Lasers,* New York: McMillan.

Marshall, T. C. (1990), in: Colson, W. B., Pellegrini, C., Renieri, A. (Eds.), *Laser Handbook,* Vol. 6: *Free Electron Lasers,* Amsterdam: North-Holland, p. 387.

Madey, J. M. J. (1971), *J. Appl. Phys.* **42**, 1906–1913.

Madey, J. M. J., Pellegrini, C. (Eds.) (1984), *Free Electron Generation of Extreme Ultraviolet Coherent Radiation,* AIP Conference Proceedings No. 118, New York: AIP.

Moore, G. T. (1984), *Opt. Commun.* **52**, 46–51.

Motz, H. (1951), *J. Appl. Phys.* **22**, 527–535.

Newnam, B. E. (1988), *Nucl. Instrum. Methods Phys. Res. B* **40/41**, 1053–1057.

Orzechowsky, T. J., *et al.* (1985), *Phys. Rev. Lett.* **54**, 889–892.

Palmer, R. B. (1972), *J. Appl. Phys.* **43**, 3014–3023.

Pellegrini, C. (1988), *Nucl. Instrum. Methods Phys. Res. A* **272**, 364–367.

Pellegrini, C. (1992), in: H. Winick, M. Cornacchia (Eds.), *Proceedings of the Workshop on IV Generation Synchrotron Radiation Sources,* Stanford Synchrotron Radiation Laboratory Report No. SSRL 92-02.

Phillips, R. M. (1960), *IRE Trans. Electron Devices* **7**, 231–241.

Scharlemann, E. T., Sessler, A. M., Wurtele, J. S. (1985), *Phys. Rev. Lett.* **54**, 1925–1928.

Schultz, J. F., *et al.* (1992), *Nucl. Instrum. Methods A* **318**, 9–15.

Sessler, A. M. (1982), in: P. Channel (Ed.), *Proceedings of the Workshop on Laser Acceleration of Particles*, AIP Conference Proceedings No. 91, New York: AIP.

Sessler, A. M., Sternbach, E., Wurtele, J. S. (1989), *Nucl. Instrum. Methods B* **40**, 1064–1068.

Smith, T. I., Schwettman, H. A. (1991), *Nucl. Instrum. Methods* A **304**, 812–821.

Sprangle, P., Tang, C. M., Manheimer, W. M. (1979), *Phys. Rev. Lett.* **43**, 1932–1936.

Sprangle, P., Smith, R. A. (1980), *Phys. Rev. A* **21**, 293–301.

Sprangle, P., Tang, C. M. (1981), *J. Appl. Phys. Lett.* **39**, 677–679.

Urbanus, W. H., *et al.* (1992), *Nucl. Instrum. Methods A* **318**, 16–19.

Vinokurov, N. A., *et al.* (1977), in *Proceedings of the 10th International Conference on High Energy Accelerators*, Serpukhov, p. 454.

Yamanaka, C. (1992), *Nucl. Instrum. Methods A* **318**, 1–8.

LASERS, GAS

J. G. EDEN, *Everitt Laboratory, University of Illinois, Urbana, Illinois, U.S.A.*

	Introduction	371
1.	**Principles of Operation**	372
2.	**Characteristics of Gas Lasers**	374
3.	**Laser Excitation Processes and Specific Gas Lasers**	375
3.1	Electron-Impact Excitation	375
3.1.1	Copper-Vapor Laser	376
3.1.2	Rare-Gas Ion Lasers	376
3.2	Excitation Transfer	378
3.2.1	He-Ne Laser	378
3.2.2	He-Cd Laser	381
3.2.3	CO_2 Laser	381
3.3	Reactive Collisions: Excimer Lasers	383
3.4	Chemical Lasers: HF	388
3.5	Other Lasers: Far-Infrared, N_2, and H_2	388
4.	**Applications**	389
4.1	Marking and Laser Evaporation	389
4.2	Microlithography	391
4.3	Photochemical Processing	392
4.3.1	Deposition and Etching of Thin Films	392
4.3.2	Synthesis of Vitamin D	393
4.4	Stereolithography	393
	Glossary	394
	Works Cited	394
	Further Reading	395

INTRODUCTION

Although the existence of stimulated emission was described by Einstein ("On the quantum theory of radiation") in 1917, and Tolman (1924) envisioned the potential of a population inversion to amplify an optical wave by stimulated emission, it was not until the 1950s that Schawlow and Townes (1958) articulated the basic features of the optical maser or laser—features such as the open resonator that would later prove to be essential to its success.

In December of 1960, Ali Javan, William R. Bennett, and Donald R. Herriott observed laser oscillation from an electric discharge in a mixture of helium and neon gases (Javan *et al.*, 1961). Following closely on the heels of Theodore Maiman's demonstration of the ruby laser in the spring of that year, this laser not only was the first gas laser but it also operated continuously. More than 10 000 different transitions in atoms and molecules in the gas phase have lased in the intervening three decades, generating radiation at wavelengths ranging from the far-infrared ($\lambda > 100\ \mu m$) to the soft–x-ray region (>1 nm). From among these have emerged several lasers that are workhorses in research as well as industrial or medical applications. Initially only a laboratory curiosity, gas lasers have been engineered to the point where they have an impact on our lives in a variety of areas. Their applications range from supermarket checkout (bar-code) scanners to welding of automobile frames to ophthalmic surgery and laser light shows. The utility of gas lasers is reflected by the prominent position they have held and continue to occupy in the laser industry. Sales in 1991 totalled approximately \$540 million—or roughly one half of the entire market—with the remainder divided among solid-state, dye (liquid), and semiconductor lasers.

This article briefly reviews the characteristics of gas lasers that are chiefly responsible for their effectiveness as both scientific and industrial tools. Since the breadth of this subject precludes an exhaustive treatment, attention is focused on those gas lasers that are widespread and have proven to be useful for research and yet are of commercial value. The basic principles common to the operation of all gas lasers and a few of their general properties are discussed in Secs. 1 and 2, respectively. Gas lasers can be broadly categorized according to the methods by which

1-56081-067-X/94/\$5.00+.50

one drives (or "pumps") the laser. Section 3 describes specific parameters of the most important laser systems in each class. Section 4 concludes, presenting selected applications of pulsed and continuous gas lasers in materials processing, photochemical synthesis, and electronics.

1. PRINCIPLES OF OPERATION

It would be well to begin by considering several of the principles that undergird the operation of all lasers—gas, solid, or liquid. The laser is an optical oscillator—that is, a device that generates coherent radiation at optical frequencies—and, viewed in its most basic form, is quite similar conceptually to a radio-frequency (rf) oscillator. Although their operating frequencies differ by several orders of magnitude (rf, 10^6–10^8 Hz; laser, 10^{13}–10^{16} Hz), both are based on identifying a medium or device that is capable of amplifying radiation of the desired frequency. As depicted in Fig. 1, the existence of amplification relies on the ability to couple power in some manner to the amplifying medium. The process or system by which this transfer of energy occurs is known as the *pump*, and several specific examples will be presented in Sec. 3. Once amplification at the desired frequency (or wavelength) is realized, oscillation is generally achieved by introducing feedback—redirecting a portion of the output power back into the amplifier. At optical frequencies, this is conveniently accomplished with mirrors.

For amplification to occur in the optical region, it is necessary that a *population inversion* be established between two energy states of the atom or molecule in question. Normally—that is, in the absence of an external source of excitation—the population of a particular excited state will be smaller than that for every state lying beneath it in energy. Said another way, at thermal equilibrium the populations of excited states diminish monotonically with increasing energy in a manner prescribed by the Boltzmann factor: $\exp(-\Delta E/kT)$, where ΔE is the energy of the excited state in question, referenced to ground, and kT is thermal energy. If this situation can be partially reversed such that the number of atoms or molecules in a particular energy level is *larger* than the population of any state

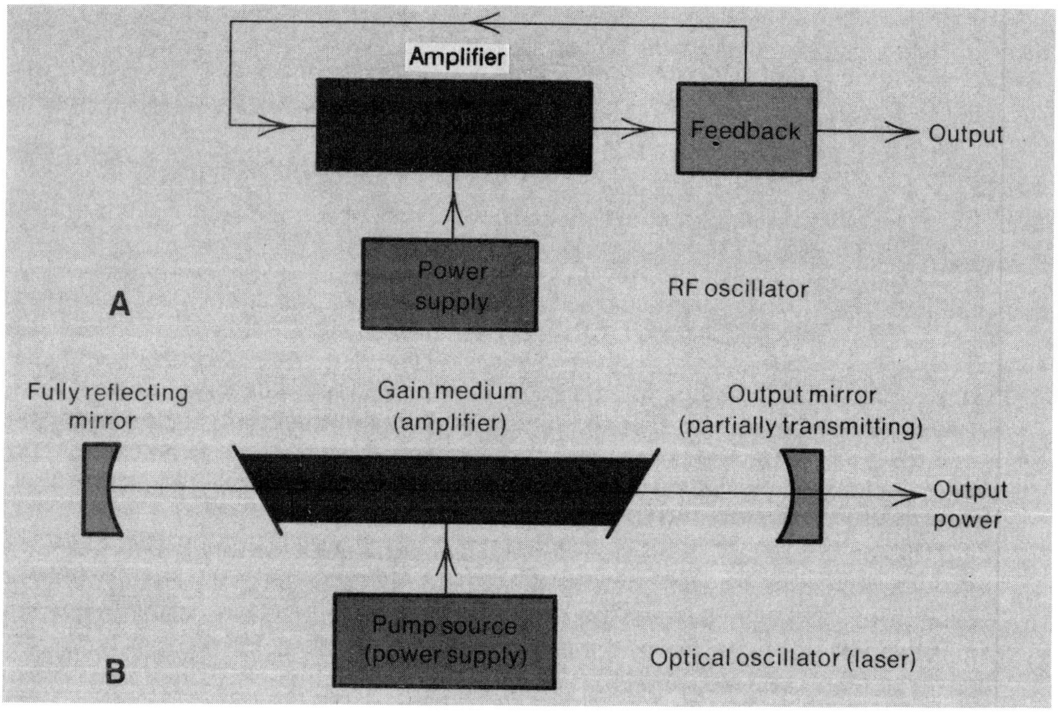

FIG. 1. Generalized diagrams for (a) radio-frequency and (b) optical (laser) oscillators. Despite their widely disparate emission frequencies, the principles underlying the operation of both are similar.

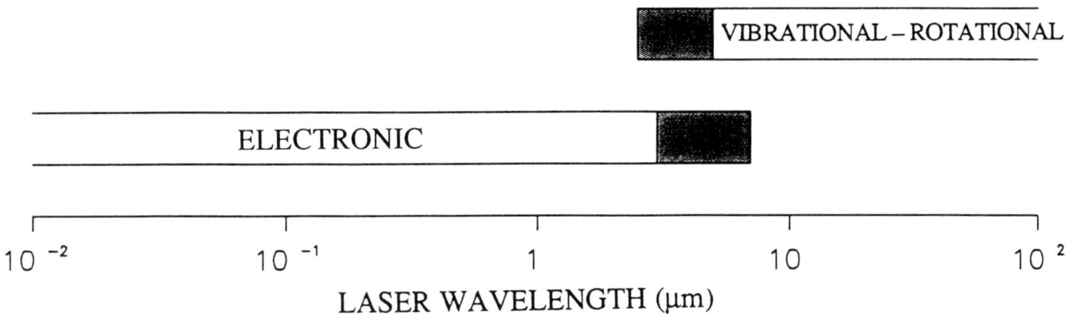

FIG. 2. Shorter-wavelength lasers involve transitions between electronic states of atoms and molecules whereas wavelengths beyond a few microns (μm) are usually the result of vibrational–rotational transitions.

below it, a population inversion and the potential for optical amplification exist.

Stimulated emission is the radiation generated when an atom or molecule undergoes a transition from one state to another lying lower in energy, but does so under the influence of an optical field whose wavelength λ coincides with the energy difference between the two states involved in the transition. To explore this further, let us consider the ramifications of having a population inversion between two states at energies E_2 and E_1 (where $E_2 > E_1$). If an optical field of wavelength

$$\lambda = hc/(E_2 - E_1) \tag{1}$$

illuminates this two-state system, then transitions from state 2 to state 1 will be induced (stimulated) by the field. Since a photon is produced in the process, this additional radiation is emitted coherently—that is, in phase with the driving optical field. Of course, absorptive transitions (which have the effect of attenuating the field) will also be taking place from state 1 to state 2, but since we started with a population inversion, stimulated emission dominates and net optical gain results. Consequently, the intensity of the optical field grows as it propagates through the inverted medium, and amplification has been realized. In the limit of weak optical fields, the intensity of the radiation increases exponentially with distance z:

$$I = I_0\, e^{\gamma_0 z}, \tag{2}$$

where I_0 is the intensity of the optical field immediately prior to entering the amplifier and γ_0, known as the small-signal gain coefficient, is proportional to

$$\lambda^2 (N_2 - N_1)/\tau. \tag{3}$$

In Eq. (3), λ is given by Eq. (1), $N_2 - N_1$ is the population inversion (expressed in cm^{-3}) between the upper and lower energy states involved in the transition, and τ is the lifetime of the emitting state.

Since the small-signal gain coefficient is proportional to the magnitude of the population inversion, much of the effort in developing new gas lasers or improving existing ones has centered on different approaches to populating specific excited states of atoms and molecules. The key is to put the species of interest into a given excited state while simultaneously minimizing the rate at which lower energy levels are populated. Several of the processes that have most often proven to be effective in pumping gas lasers are the subject of Sec. 3.

Finally, it is clear that the energy difference between the two states involved in the optical transition determines the wavelength of the laser. When attempting to develop lasers in a particular region of the spectrum, therefore, we must look for groups of atomic or molecular states whose energy differences lie in the proper range.* As illustrated qualitatively in Fig. 2, this requires transitions between *electronic* states of atoms and molecules if laser wavelengths below ~1 μm are desired. The generation of longer wavelengths (infrared–submillimeter) usually entails vibrational and rotational transitions in molecules.

*Of course, it is also necessary that quantum-mechanical selection rules allow for transitions between the desired states.

2. CHARACTERISTICS OF GAS LASERS

Although gas lasers produce a wide array of wavelengths and power levels, operate continuously or pulsed, and offer beams of differing cross-sectional profiles, they share several characteristics that distinguish them as a family. An inherent property of the gas-laser medium is its tolerance of excessive currents and voltages. That is, the gas or gas mixture is usually "self-healing" and is capable of withstanding the deposition of large amounts of power (known as specific power loading and expressed in W/cm^3). Gas lasers are also comparatively straightforward to scale in cross section—that is, to construct lasers having a large aperture. The ability to optimize quickly the composition of the gas mixture in a laser is another advantage in that it allows one to test rapidly a range of gas combinations and pressures.

For many laser applications, the high optical quality beams and pointing stability available from gas lasers represent their greatest assets. Because of the homogeneity of gaseous media and their typically low index of refraction, numerous gas lasers are noted for routinely producing beams in the fundamental spatial mode (TEM_{00}), having divergences that are close to the diffraction limit. The red He-Ne gas laser ($\lambda = 632.8$ nm), for example, typically produces a beam having a divergence angle of 1 mrad. Consequently, for 10 mW of output power, the spectral brightness of this laser—defined as intensity (W cm^{-2}) per steradian (sr)—is $\sim 10^5$ W cm^{-2} sr^{-1}, a value that is 3 orders of magnitude larger than that for the sun.

If we assume the laser beam to have a Gaussian cross-sectional intensity profile and to be operating in the TEM_{00} mode, then the *diffraction-limited spot size* for a focused beam is given by

$$W \simeq (2\lambda/\pi)(f/d) \approx 0.64\lambda(f/d), \tag{4}$$

where λ is the laser wavelength and f and d are the focal length and diameter of the focusing lens, respectively. To a first approximation, then, the smallest spot size obtainable is on the order of the wavelength λ (assuming, of course, that f/d—the inverse of the focusing system's numerical aperture—is approximately unity), and this limit is accessible *only* when the beam being focused is diverging at the diffraction limit. As a result, visible and ultraviolet gas lasers are frequently applied to those industrial and medical tasks (such as microelectronic circuit inspection) for which spatial resolution and, hence, the focusability of the optical source are of primary concern.

Entire families of gas lasers are based on the fact that producing large number densities of transient small molecules can often be accomplished in the gas phase but may be difficult or impossible to do in a solid or liquid. If lasers employing vapors (such as those of volatile metals—including Hg and Cs, metal halide salts, the halogens, etc.) as the active medium are now included, the gas-laser scientist has an enormous range of gas and vapor combinations from which to choose and an almost endless list of electronically excited species (that do not normally exist in nature) to explore.

Finally, it should be mentioned that, for wavelengths below roughly 250 nm (i.e., in the ultraviolet, $h\nu = 5$ eV), virtually all of the available laser sources are gas systems. It is important to remember that the power required to obtain oscillation scales nonlinearly with the optical frequency ν. Consequently, demonstrating lasers that operate in the short-wavelength portion of the visible, the ultraviolet, or beyond demands the deposition of a minimum of tens of kW of power per cm^3 to maintain the necessary population inversion. Such power levels frequently have catastrophic consequences for most solids and liquids, particularly for those crystals and glasses having poor thermal conductivity. As we pass the ultraviolet in moving to shorter wavelengths, the photon energies rise above those available from transitions involving outer-shell electrons in atoms and molecules. Exciting transitions involving inner-shell electrons becomes necessary, and, while this has been accomplished with gases and highly ionized plasmas, this constraint imposes increasing difficulties for solid and liquid media as the laser wavelength decreases.

The two major drawbacks of gas lasers are their size and power consumption. Although gas lasers as small as several cm in length have been demonstrated, most manufactured lasers are considerably larger (tens of cm), and pulsed, high-power commercial devices are typically 50–100 cm (or more) in length. Also, because of their low electrical-to-optical conversion efficiencies, large-frame argon-ion lasers, for example, require water cooling and more than 10 kW of electrical power to produce 20 W in the visible. As a result, gas lasers are best suited for those applications in which beam quality, peak or average power, or short wavelength (rather than size or weight) are the driving considerations.

3. LASER EXCITATION PROCESSES AND SPECIFIC GAS LASERS

The manner in which power is delivered to the upper laser level has far-reaching consequences for a gas laser, particularly insofar as its overall efficiency and engineering design are concerned. While a variety of atomic and molecular optical and collisional processes have been shown to be capable of pumping gas lasers, a handful appear repeatedly and are described here. Several of the technologically most important gas lasers that exemplify each of the pumping mechanisms also are discussed.

3.1 Electron-Impact Excitation

One of the simplest approaches to populating excited states of an atom or molecule is by electron-impact excitation. As illustrated in Fig. 3, if a fast electron has the energy necessary to excite an atom or molecule to a particular state, it is possible to transfer this energy to the heavier particle in a collision. If X represents the species of interest as the laser medium, this process can be expressed as

$$e+X\rightarrow X^{*}+e, \tag{5}$$

where the asterisk denotes an excited state and, of course, energy conservation requires that, if the electron's initial energy was E, it will emerge from the collision with energy $E-E_2$, where E_2 is the excitation energy of the X^* state. The high-energy electrons required for the process of Eq. (5) are usually produced in a low-pressure discharge. It is also clear from Fig. 3 that it is essential for both of the energy levels participating in the laser transition to be chosen carefully so that the

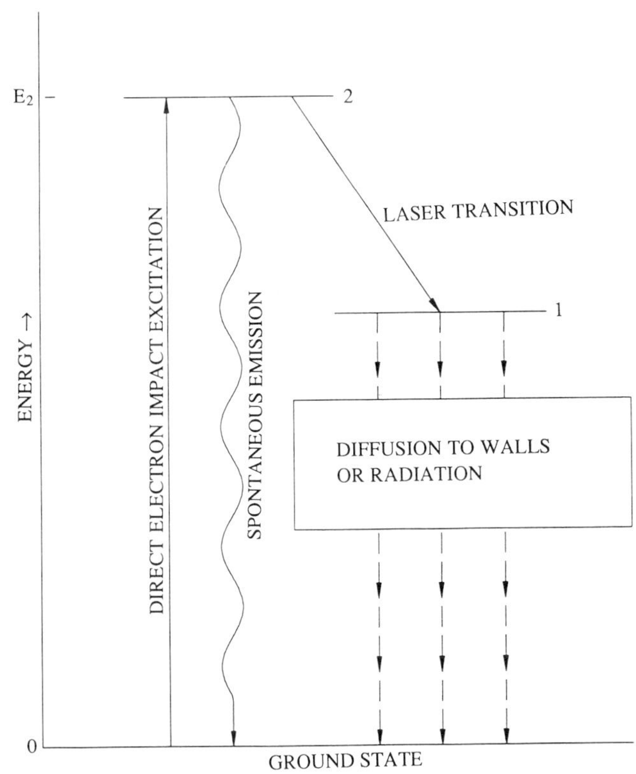

FIG. 3. Partial energy-level diagram for an atom or molecule illustrating electron-impact excitation of the upper laser level. The upper laser level can also decay by spontaneous emission but state 1 (lower laser level) may be deactivated by one of several processes, including collisions with the discharge cell wall. (After Willett, 1971.)

electron-impact excitation process populates primarily the upper laser level.

3.1.1 Copper-Vapor Laser. An immediate example of a system that is pumped by direct electron impact is the copper-vapor laser that emits at two wavelengths: 510.6 nm in the blue-green and 578.2 nm in the yellow. Figure 4 shows the states and processes underlying its operation. Collisions between high-energy ($\gtrsim 4$ eV) electrons and ground-state copper atoms preferentially populate the $4p\,^2P$ excited states that subsequently emit to one of the ^2D levels. Because of quantum-mechanical selection rules, the ^2D states are unable to radiate to the ground state and are said to be "metastable." Consequently, excited atoms in the ^2D states can be removed only by slow processes, including diffusion of the atoms to the walls of the laser cell. As a result, population of the ^2D state steadily builds, the inversion is quickly lost, and lasing stops. If the pump source is turned off, the ^2D number density will eventually decay, and the laser may once more be operated briefly. For these reasons, the copper-vapor laser is able to operate only in a pulsed mode, emitting pulses that are roughly 3 ns in duration. By operating the laser at a pulse repetition frequency (PRF) of several kHz, however, large average powers can be readily generated. Commercial devices producing up to 100 W in a beam 6 cm in diameter and operating at a PRF of 10 kHz are currently available. If gold is substituted for copper as the lasant metal, similar considerations hold except that oscillation now occurs in the red at 628 nm. The copper (or gold) vapor laser is an excellent example of the principle that the operational specifications of gas lasers (as with all lasers) are intimately tied to the details of the atomic and molecular processes occurring in the laser medium. Making *a priori* generalizations concerning the output power, repetition rate, or any other significant parameter of a group of lasers is usually not possible, and the physics of the medium in question must be closely examined for the utility of the laser to be determined.

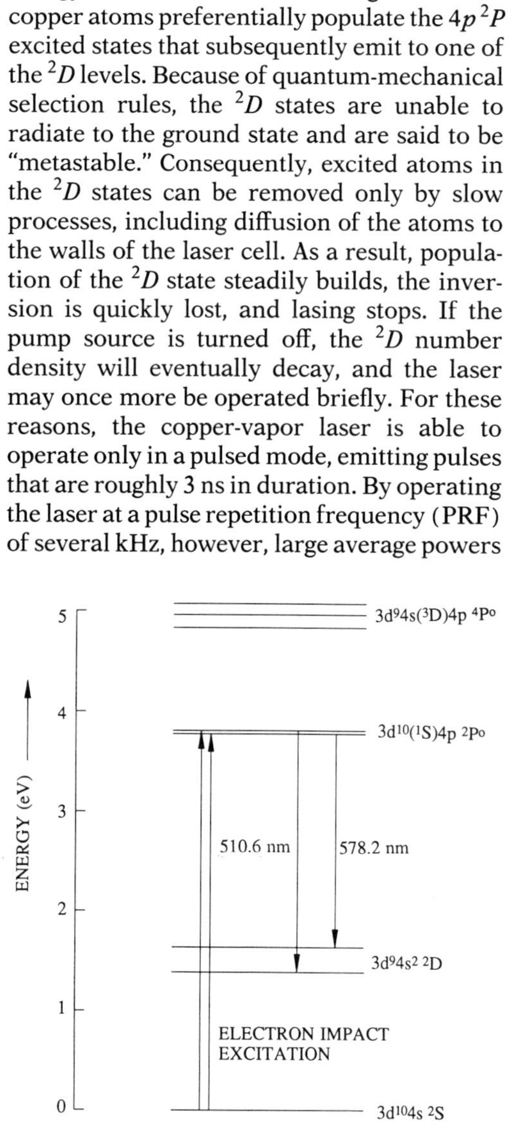

FIG. 4. Partial energy-level diagram for atomic copper showing the states relevant to the 510.6- and 578.2-nm laser lines. This laser only operates pulsed because the ^2D lower laser level is metastable to radiative decay to ground—hence, the transition "self-terminates." (After Walter *et al.*, 1966.)

3.1.2 Rare-Gas Ion Lasers. One of the major gas lasers today, from both technological and scientific standpoints, is the rare-gas ion laser. Discovered by William Bridges in 1964, the argon-ion laser is capable of continuously producing tens of watts of power (up to 30 W in commercially available systems) at several discrete wavelengths in the blue and green. Lasing occurs on several transitions of the singly or doubly charged atomic ion (Ar^+ or Ar^{2+}, respectively) that is produced from the ground state by electron impact. Because the atomic states that produce the laser lines lie above 30 eV (cf. Fig. 5) in energy, the efficiency of the laser is low, and exciting the upper laser levels requires highly energetic electrons that are produced in an intense, continuous discharge in low-pressure argon (approximately 0.2 Torr). Consider the visible Ar^+ lines at 514.5, 501.7, 496.5, 488.0, 476.5, 472.7, 465.8, 457.9, and 454.5 nm, for example. The upper states for these transitions (Ar^+ $4p\,^2S,^2P,^2D$) lie almost 20 eV above the Ar-*ion* ground state. The maximum theoretical efficiency for this laser is $<8\%$ since a total of 35 eV of energy must be invested in order to obtain 2.5–2.7-eV photons. In practice, the overall efficiencies of rare-gas ion lasers are considerably smaller ($<0.1\%$), since pumping of the upper laser levels occurs in two steps but both require electrons having energies that are well beyond the average electron energy in the dc discharge. Said another way, the energies of the electrons in a plasma are described by a distribution, and the electrons

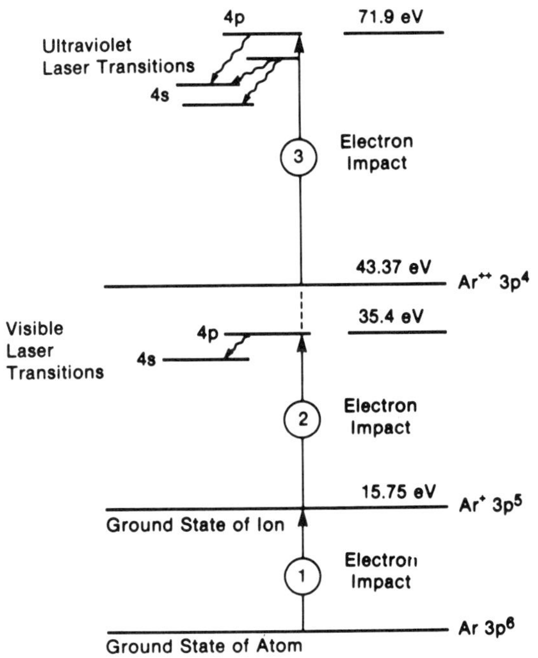

FIG. 5. Energy levels of argon that are involved in the argon-ion visible and ultraviolet lasers. The visible laser transitions range in wavelength from 454.5 nm (blue) to 514.5 nm (green) and are frequently used in laser printing, compact-disk manufacturing, and laser light shows. The ultraviolet transitions require higher plasma tube currents because of the additional electron-impact step. (Diagram courtesy of Spectra-Physics, Inc.)

required for exciting the Ar^+ laser must be in the high-energy "tail" of that distribution.

The output power of an argon-ion laser, P_0, can be expressed as

$$P_0 \propto [\gamma_0 L/(T+\alpha)]^{-1}, \qquad (6)$$

where γ_0 is again the small-signal gain coefficient, L is the length of the discharge (i.e., active length), T is the transmission of the output coupling mirror, and α represents all losses in the optical cavity other than the output coupler's transmission. In the Ar^+ laser, γ_0 is approximately linear in kT_e, the *average* electron energy, which, in turn, is proportional to the discharge current density J (A/cm^2) and is *inversely* proportional to the argon-gas fill pressure. Consequently, realizing optimal efficiencies from this laser requires large discharge densities—typically 800–1000 A cm^{-2} for a plasma-tube bore diameter of 3 mm. Because of the low electrical efficiencies for the laser, large-frame ion lasers yielding 4–30 W of output power require water cooling of the plasma tube and 5–35 kW of electrical input power. A magnetic field of ~1000 G, which serves to increase the ion density for a given discharge current, is often imposed on the plasma of the higher-power ion lasers.

Figure 6 gives a cutaway view of the structure of an Ar^+-laser plasma tube. The high discharge currents that are characteristic of the Ar^+ laser place stringent demands on the design and construction of the discharge vessel and, in particular, the structures that confine the plasma. Erosion of the walls of the tube must be minimized since an increase in the bore diameter directly translates into a lower value for J and, hence, kT_e. By using refractory materials such as beryllium oxide (BeO) and graphite, manufacturers are now able to specify tube lifetimes beyond 2000 h. Similar comments could be made for the krypton-ion laser, which emits a dozen strong lines ranging in wavelength from 337.5 to 793.1 nm.

Smaller, air-cooled versions are also commercially available that produce typically 50–100 mW of visible radiation from a much more compact package. A photograph of the head (plasma tube and cooling assembly) for a small (<50 mW), air-cooled ion laser is

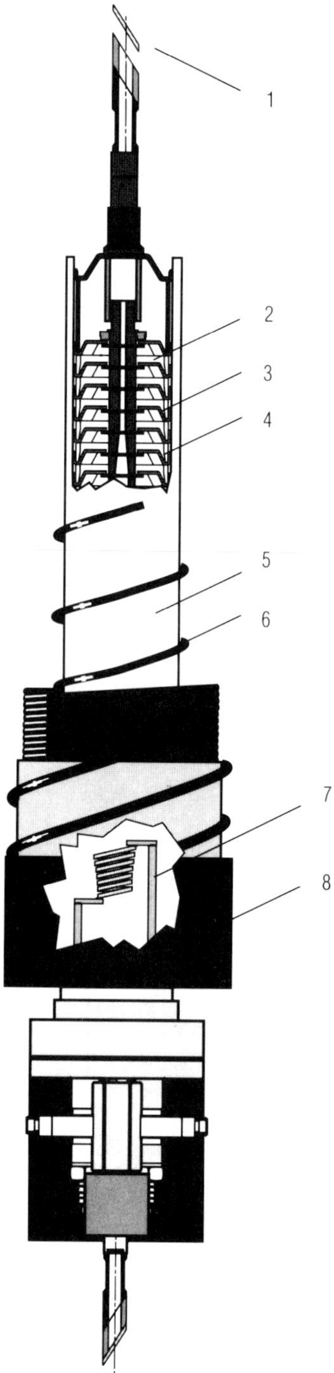

FIG. 6. Key components of the plasma tube for a large-frame (water-cooled) argon-ion laser: (1) quartz windows at Brewster's angle, (2) tungsten disks for confining the plasma and defining the gain region, (3) copper support disks for heat removal, (4) copper-to-ceramic braze, (5) ceramic cylindrical enclosure, (6) water-cooling line, (7) tungsten cathode, and (8) external magnet for plasma confinement and power enhancement. (Courtesy of Spectra-Physics, Inc.)

shown in Fig. 7. Since the plasma-tube bore is only 1 mm, however, the plasma current densities (>1000 A cm^{-2}) can actually exceed those of their larger, water-cooled counterparts. Intermediate power levels (0.3–1 W) have been generated by adopting a hybrid approach to cooling the discharge tube. The tube itself is cooled by a closed-loop water system that is interfaced to a water–air heat exchanger.

A recent development by laser manufacturers with plasma tubes capable of operating at still higher current densities (>1500 A cm^{-2}) is the demonstration of new laser lines in the deep ultraviolet (200–300 nm) from argon, krypton, and xenon. In krypton, for example, 2.5 W of total power can now be obtained from several lines oscillating in the 242–266-nm interval; similarly, a group of argon lines in the 275–306-nm range has produced 6 W.

3.2 Excitation Transfer

In many instances, it is impractical or inefficient to populate directly a particular state that is of interest as the upper level for a laser transition. If, however, the energy level in question lies close in energy to a state of another atom (or molecule) *that can be efficiently populated,* then it is frequently possible to transfer the required energy from one atom to another through a collision. The rate at which energy can be transferred from the initial to the final state is strongly dependent upon the difference in energy (denoted ΔE) between the two states. In most cases, this process is increasingly efficient as ΔE decreases, and when ΔE is comparable to or less than kT (thermal energy), the collision is said to be *resonant.* This pumping process, known as excitation transfer, is illustrated qualitatively in Fig. 8 and is at the heart of three of the best known gas lasers: He-Ne, He-Cd, and CO_2.

3.2.1 He-Ne Laser. The first gas laser, reported by Javan *et al.* (1961), emitted continuously on the 1.15-μm transition of atomic neon in the infrared. Shortly thereafter, oscillation on the red transition at 632.8 nm was also reported (White and Rigden, 1962). The critical feature that was common to both lasers was the use of helium as a vehicle to transfer energy to specific energy levels of

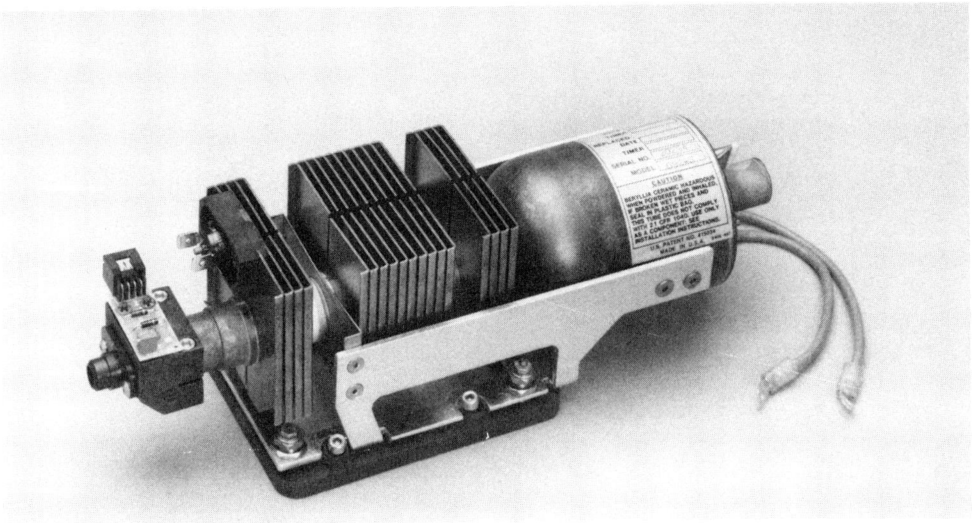

FIG. 7. Photograph of an air-cooled argon-ion laser head. (Courtesy of S. Jarrett, Spectra-Physics, Inc.)

neon. As depicted in Fig. 9, the lowest two excited states of He (2^3S and 2^1S) lie within 50 meV (or roughly twice thermal energy at 300 K) of energy levels of Ne. The He(2^1S) state, for example, rapidly transfers its energy to the $3s_2$ level of Ne by the reaction

$$\text{He}^*(2^1S) + \text{Ne} + 48 \text{ meV} \rightarrow \text{He (ground state)} + \text{Ne}^*(3s_2), \quad (7)$$

where the asterisk again denotes an excited state. A similar relation can be written for the

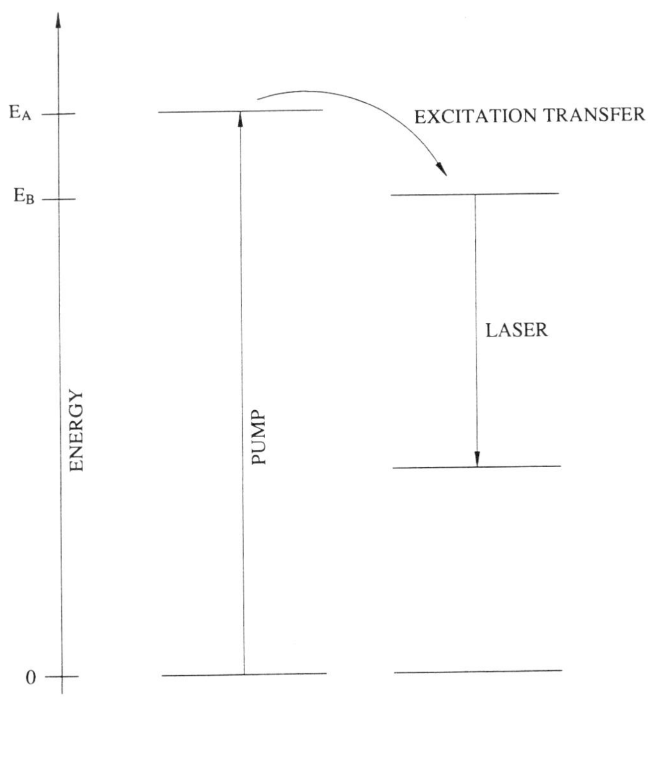

FIG. 8. Excitation transfer from atom (or molecule) *A* to *B*. Energy can generally be delivered to species *A* by a variety of processes, including electron-impact excitation (described earlier) or optical pumping.

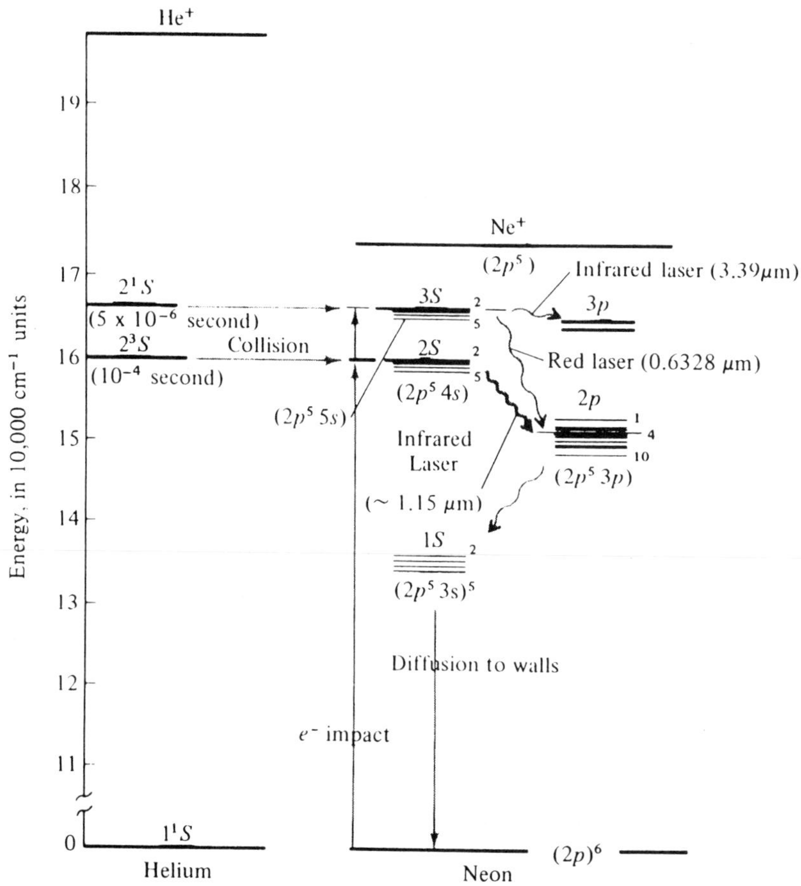

FIG. 9. Energy-level diagram for atomic He and Ne indicating excitation transfer from the 2^3S and 2^1S states of helium to energy levels of neon. (After Bennett, 1962, and Yariv, 1971.)

transfer of energy from the lowest-energy He excited state (2^3S) to one of the $2s$ levels of Ne. The He-Ne laser, then, relies on populating the 2^1S and 2^3S states of He in a discharge and transferring that energy to Ne excited states. Lasing subsequently occurs on several $3s \rightarrow 2p,3p$ and $2s \rightarrow 2p$ transitions, the strongest of which are the 1.15-μm and 632.8-nm lines mentioned earlier but also the $3s \rightarrow 3p$ transition at 3.39 μm. However, direct electron-impact excitation of the Ne levels also occurs in the He-Ne gas mixture. Since a fraction of the energy delivered to the Ne excited states will reside in the lower laser levels, a reduction in the steady-state population inversion that is attainable results, which, in turn, limits the efficiency and output power available from the laser.

The key component of the He-Ne laser is a discharge tube having an active length of 10–100 cm (commercial devices tend to the short end of this range). The bore of the discharge d (in mm) is related to the optimal gas fill pressure p (in Torr) by

$$pd \approx 3.6 \text{ Torr mm.} \tag{8}$$

So, for a discharge diameter of 4 mm, the He-Ne fill pressure should be approximately 0.9 Torr. To encourage the preferential excitation of He over Ne excited states in the discharge, the He:Ne gas ratio is typically between 5:1 and 20:1. The dependence of the output power of a He-Ne laser having a 3.0-mm bore and an active length of 55 cm on the total gas pressure is shown in Fig. 10. Under these conditions, the optimal He:Ne ratio is (5–7):1. Also, the discharge current is generally between 5 and 50 mA, which translates into current densities of 30 to 300 mA/cm² for a 4-mm-bore tube.

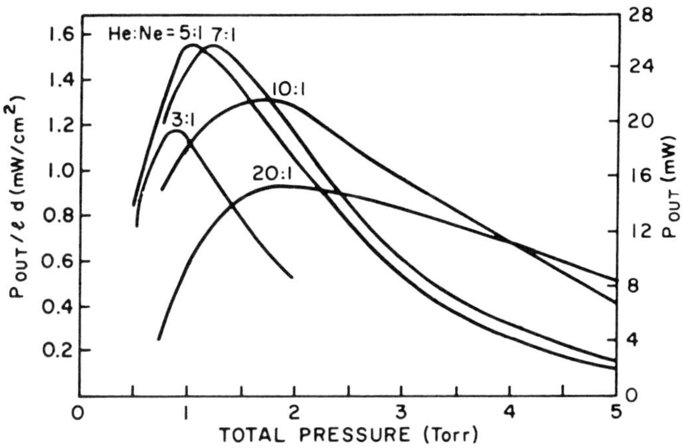

FIG. 10. Output power as a function of total pressure for a He-Ne laser having a discharge tube diameter (*d*) of 3.0 mm, and a plasma length (*l*) of 55 cm. (After Field, 1967.)

Because of the low population inversions that are inherent with continuous (cw) He-Ne lasers, the small-signal gain coefficient γ_0 is on the order of 10^{-4} cm^{-1} (or 10^{-2}% cm^{-1}). Power outputs as large as 50–100 mW are available from commercial lasers but, for low-power applications, models producing 0.5 mW in the TEM_{00} mode are also offered. Other $3s \rightarrow 2p$ transitions also lase in the visible, and several manufacturers produce a He-Ne laser having a prism in the optical cavity that enables one to select a particular (red, orange, or yellow-green) transition in the 543–633-nm region (543, 594, 612, or 633 nm).

3.2.2 He-Cd Laser. Excitation transfer from helium is also the process underlying the success of the helium-cadmium (He-Cd) laser (Silfvast *et al.*, 1966; Fowles and Hopkins, 1967). In this instance, however, the $He(2^1S,2^3S)$ states have sufficient energy to ionize cadmium atoms, thereby producing excited states of the metal *ion*. The reaction, written

$$He^*(2^1S,2^3S) + Cd\ (ground) \rightarrow (Cd^+)^* + He + e, \qquad (9)$$

is known as Penning ionization and produces an inversion on the $4d^95s^2\,{}^2D_{5/2} \rightarrow 4d^{10}5p\,{}^2P_{3/2}$ and $5s^2\,{}^2D_{3/2} \rightarrow 5p\,{}^2P_{1/2}$ transitions of the ion at 441.6 and 325.0 nm, respectively. Typically, the Cd metal vapor and helium pressures in the discharge are roughly 2 mTorr and 3–4 Torr, respectively, and to maximize excitation of the upper laser levels by electron-impact collisions requires that the current in the plasma be ~100 mA.

The He-Cd laser is a low-power laser that is usually operated continuously (cw). Commercially available models produce up to 70 mW (cw) in the blue and 25 mW in the ultraviolet (40 and 10 mW, respectively, in the TEM_{00} mode). The lifetime of the laser tube (5000–6000 h) is limited by several factors, including the deposition of cadmium metal on the optics. The short-wavelength emission of the He-Cd laser, combined with its low beam divergence (0.6–2.0 mrad) and near-diffraction-limited spot size, make it well suited for applications such as flow cytometry (cell detection and counting) or circuit inspection that require high photon energies and excellent beam quality.

A particular discharge version of the helium-cadmium laser, known as a hollow-cathode laser, will, in addition to the familiar blue visible transition, provide red and green beams as well. Figure 11 is a schematic diagram of a commercially available version of such a laser. Rather than having a longitudinal discharge (that is, the discharge axis coinciding with the optical axis), the electrode separation is much smaller, because of the transverse geometry of the discharge, and one consequence is that the plasma generates electron energies and densities higher than those typical of the longitudinal geometry. Output powers (all lines) are generally as high as 50 mW. Because of their ability to produce white light (i.e., red, blue, and green beams simultaneously), hollow-cathode He-Cd lasers are currently being applied to optical image processing systems.

3.2.3 CO_2 Laser. Most gas lasers (and particularly those operating in the visible or at

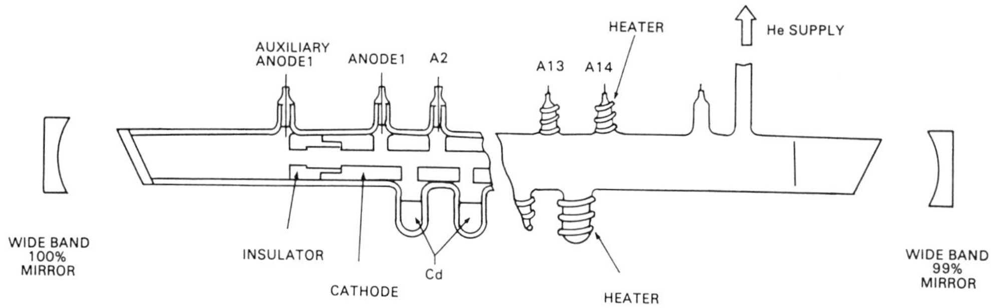

FIG. 11. Schematic diagram of a commercial hollow-cathode helium-cadmium laser that is capable of supplying the three primary colors (red: 635.5, 636.0 nm; green: 533.7, 537.8 nm; blue: 441.6 nm) in a single output beam.

shorter wavelengths) have overall efficiencies* that are typically well under 1%. In fact, few lasers of any type are as efficient as the CO_2 gas laser, which produces power at 10.6 μm in the mid-infrared with an efficiency (wall plug) of 20%. The success of this laser is attested to by its prevalence in a range of industrial applications but particularly those—such as cutting and welding metals, fabric, and wood—requiring as much as kilowatts of average power.

The unusually high efficiency of the CO_2 laser is partially attributable to a bit of serendipity that was first recognized by Patel (1964). To appreciate this, we must first consider the energy states of the CO_2 molecule. This triatomic molecule is able to vibrate in the three ways portrayed in Fig. 12. The three "modes" are known (from top to bottom) as the symmetric stretch, bending, and asymmetric stretch modes. Each mode has associated with it a set of energy levels. For the symmetric stretch, they are denoted (n00), where n is an integer. The bending-mode states are given as (0n0) and those for the asymmetric stretch as (00n). The key to the CO_2 laser's efficiency and enormous available output powers is this: as shown in Fig. 13, the first excited asymmetric stretch state (001) is almost exactly in resonance with the lowest vibrational level of nitrogen (N_2). Therefore, if we attempt to populate CO_2 (001) directly by electron collisions, little power will actually be delivered to this particular state. Rather, most of the pump energy will go into the low-lying (010) and (020) states. If, however, one takes the tack exploited so successfully for the He-Ne and He-Cd lasers and electrically excites nitrogen gas having a small amount [typically (10–20)%] of CO_2 added, then power is delivered efficiently into the first vibrational state ($v=1$) of N_2 since no intervening levels [i.e., between ground ($v=0$) and $v=1$] exist. Subsequently, energy is rapidly transferred to the CO_2 (001) state by CO_2-N_2 ($v=1$) collisions, a population inversion is created, and lasing ensues on the (001)$\rightarrow$(100) and (001)$\rightarrow$(020) transitions of CO_2 at 10.6 and 9.6 μm, respectively. This pumping process, known as *vibrational–vibrational* (V-V) energy transfer, is instrumental to the efficiency and, therefore, the utility of this laser.

Another reason for the remarkable efficiency of the CO_2 laser is the small energy ($\sim\frac{1}{4}$ eV) required to reach the upper laser level—an energy possessed at any given time by a significant fraction of the discharge electrons. Contrast this situation with the He-Ne laser, which requires almost 40 times as much energy for excitation of the upper laser level. Of course, placing such stringent demands on the electron energy relegates the He-Ne laser to the high-energy tail of the electron energy distribution and severely limits the overall efficiency of the device. For a more detailed discussion of the mechanisms critical to the operation of the CO_2 laser, the reader is referred to Verdeyen (1981).

The gas mixture for a typical CO_2 laser consists of He, N_2, and CO_2. Adding helium serves several purposes, among which is the improvement of heat removal from the laser discharge. When pumped by a longitudinal discharge that is 50 – 100 cm in length, the

*Defined as the power emitted by the laser divided by the electrical power delivered to the gain medium. If, on the other hand, the efficiency is computed by dividing the output power of the laser by the ac power consumed by the power supply, this is said to be the "wall plug" efficiency.

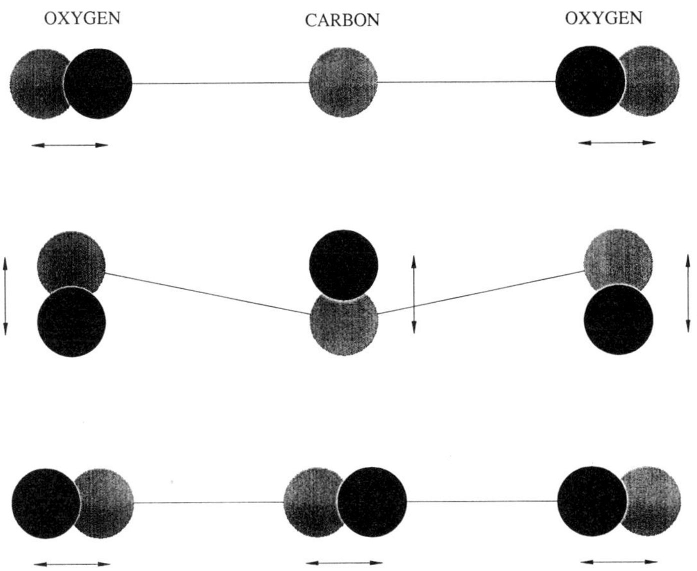

FIG. 12. CO_2 is a linear triatomic molecule that has three stretching and bending modes: (top) symmetric stretch, (middle) bending, and (bottom) asymmetric stretch. (After Yariv, 1971.)

laser will produce tens of watts of power continuously. Obtaining much higher output powers (i.e., in the kilowatt range) requires a transverse discharge such as that illustrated schematically in Fig. 14. To minimize the "residence time" of gas in the discharge, the direction of gas flow *and* the discharge are at right angles to the laser axis.

A number of other molecular lasers emitting in the infrared (such as CO, 4.9–6.66 μm) depend on vibrational–vibrational energy transfer as the dominant means for pumping the laser. A detailed account of the historical background and principal physical mechanisms for a few of these systems as well as the CO_2 laser is given by Garscadden *et al.* (1991).

3.3 Reactive Collisions: Excimer Lasers

Several of the most important lasers—particularly in the visible and ultraviolet regions of the spectrum—rely on the transient production of small molecules that do not normally exist in the gas phase but are readily formed by collisions. If A is an atom and BX is a molecule (in which B and X may consist of more than one atom each), this process is represented as

$$A^* + BX \rightarrow (AB)^* + X. \tag{10}$$

Clearly, a new species (i.e., one that was not present in the original gas mixture) is formed that is derived from the reactants. Note that A must itself be electronically excited before the reaction can proceed. A collision analogous to (10) in which all of the reactants are in their ground states may well produce no new species.

The family of lasers that best exemplifies this pumping mechanism is the excimer lasers. Excimer (a contraction for "excited dimer") is the term given to a group of molecules (usually diatomic) that are not stable in the ground state but yet have chemically stable excited states. Figure 15 is a generalized energy-level diagram for a diatomic excimer molecule. Since the ground state is dissociative (unstable), obtaining a population inversion between the excited and ground states is simplified because a molecule making a transition from $(AX)^*$ to AX finds itself on a repulsive potential curve and quickly dissociates (as AX moves down the potential hill, A and X separate).

Consequently, the number density of ground-state molecules (AX) at R_0 at any given time is quite small. If a method can be found for populating the $(AX)^*$ state, not only should lasing be obtainable but, because of the nature of the ground state, it will also be possible to tune the laser's output continuously over a modest spectral range.

Electronically excited excimer molecules can usually be produced by two types of collisions. The first involves only electrically neutral reactants and was given earlier in Eq. (10). As A^* and BX approach one another, A^* ejects one of its outermost electrons, and, in

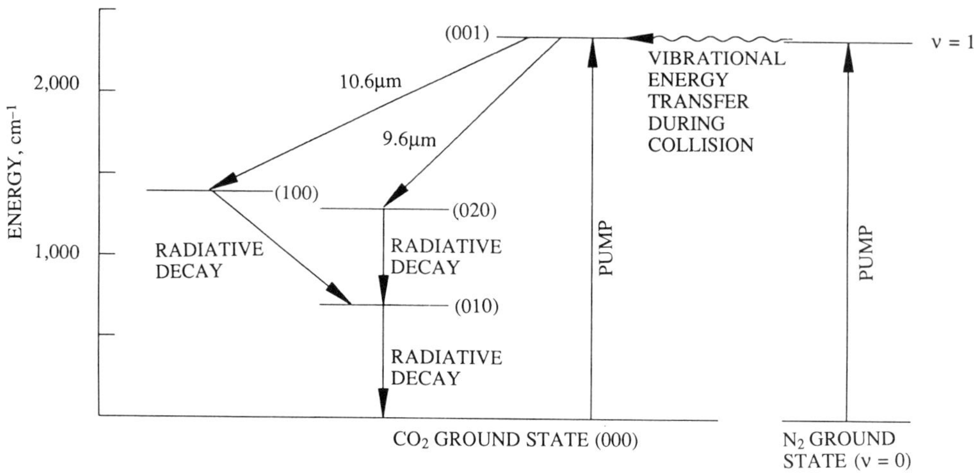

FIG. 13. Several of the energy states of the CO_2 and N_2 molecules showing the near coincidence between the first excited vibrational state ($v=1$) of N_2 and the lowest excited asymmetric stretch (001) state of CO_2. (After Yariv, 1971.)

so doing, becomes the cation A^+. The ejected, or "harpoon," electron transforms BX into the anion BX^-. As A^+ and B^- are drawn closer under the influence of the Coulombic force, the internal strain in BX^- increases until B is expelled and the A^+ and X^- ions form the stable complex AX. It is easy to see why Herschbach (1966) has applied the picturesque term *harpoon collision* to reactions of this type. A more direct approach is to begin with A^+ and X^- and form $(AX)^*$ by the *ion–ion recombination process:*

$$A^+ + X^- + M \rightarrow (AX)^* + M, \tag{11}$$

where M is any third body and is required to conserve momentum. In summary, pumping the excimer lasers usually entails delivering energy to an atomic species (excited neutral A^* or ion A^+) that is subsequently incorporated (by a collision) into an electronically excited molecule. That is, the sole purpose of the pump is to generate A^* or A^+ (generally by electron-impact excitation of ground-state A atoms), thereby initiating production of the $(AX)^*$ molecule by the harpoon or ion–ion recombination process [reaction (10) or (11), respectively].

From among those groups of excimer molecules that have lased, the rare-gas–halide lasers have emerged as the clear leader because of their efficiencies and commercial impact. Although they have been in existence only since 1975, the rare-gas–halide lasers have transformed the ultraviolet in much the same way as the CO_2 laser has dominated the infrared region.

It is well known that the chemistry of ground-state rare-gas atoms is limited since

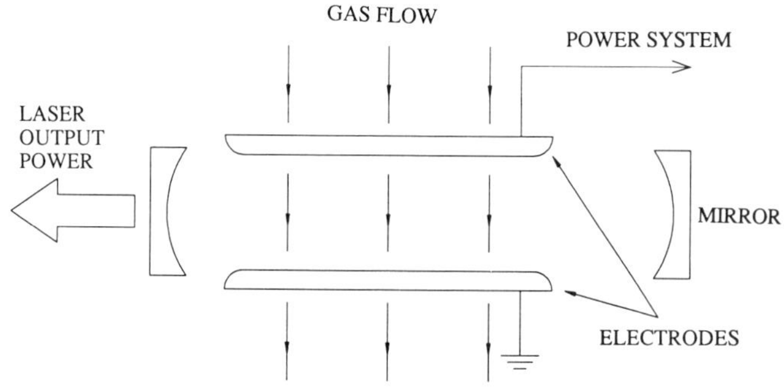

FIG. 14. Basic features of a transverse discharge-pumped CO_2 laser.

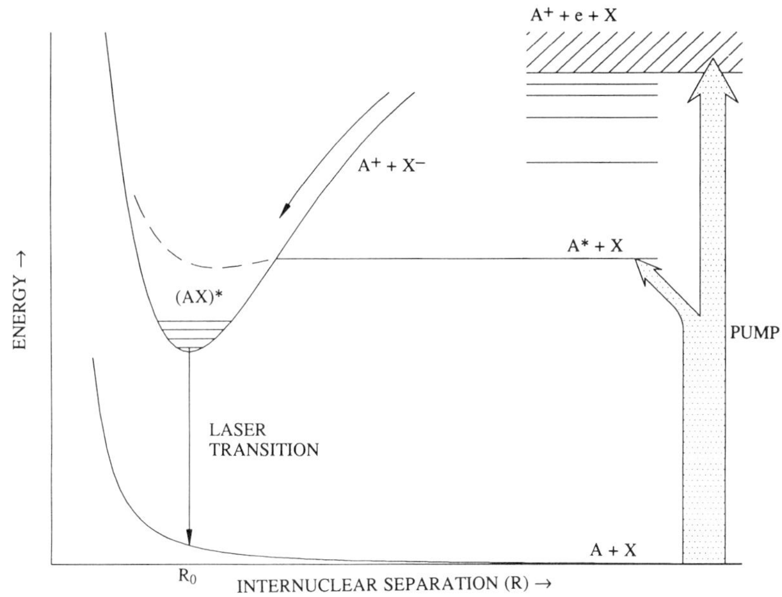

FIG. 15. Generalized energy-level diagram for the excimer molecule *AX* showing (1) its dissociative (unstable) ground state; (2) a stable, electronically excited state, $(AX)^*$; and (3) two collisional approaches to forming $(AX)^*$—from $A^* + X$ or $A^+ + X^-$ collisions.

their outer electron shells are full. However, if the outermost electron is placed in an excited state, the atom's reactivity increases markedly, and research of the past 20 years has revealed at least half a dozen families of rare-gas–containing excimer molecules that can be produced quite readily. In collisions with halogen-containing molecules, excited rare-gas atoms produce excimers with efficiencies approaching 100%. The KrF molecule, for example, is produced by the harpoon reaction

$$\mathrm{Kr}^* + \mathrm{F}_2 \rightarrow (\mathrm{KrF})^* + \mathrm{F} \quad (12)$$

[where the asterisk associated with the krypton atom denotes the lowest excited states ($5s\ ^3P_{1,2}$) of the atom], and lasing subsequently occurs in the ultraviolet at 248 nm. A comparison of the fluorescence (spontaneous emission) and laser spectra for KrF is given in Fig. 16.

Early rare-gas–halide lasers were pumped by relativistic electron beams, but today the majority are driven by pulsed, transverse dis-

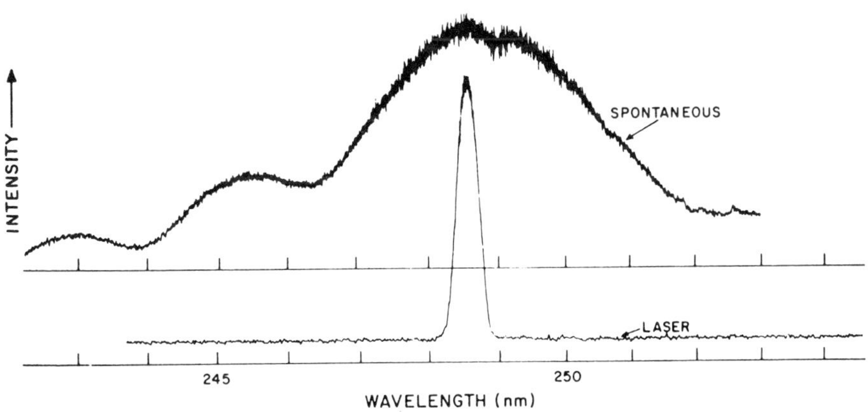

FIG. 16. Spontaneous-emission (top) and laser spectra for the KrF molecule. For the sake of comparison, the laser emission intensity has been attenuated. (After Ewing and Brau, 1975.)

charges. Several of the subsystems for a typical discharge-pumped excimer laser are illustrated in Figs. 17 and 18. Figure 17 is a diagram of the gas-flow loop for a commercially available laser. The gas mixture consists of two or more rare gases and a halogen-containing molecule known as the "donor" or "precursor." One of the rare gases is necessary as a component of the lasing molecule itself (such as Ar in ArF, Xe in XeCl, etc.) whereas the second (or third) is included for the purpose of stabilizing the discharge (by improving its heat-transport properties) or for increasing the efficiency with which rare-gas excited atoms or ions are generated. For the XeCl laser, for example, a typical gas mixture is 3–4 Torr HCl, 60 Torr Xe, and 1800 Torr He (where 1 atm$\approx$760 Torr). For long gas "lifetimes," the mixture is recirculated and periodically replenished.

Continuous purification of the laser gas is necessary because the reaction of the halogen donor molecule (or halogen atoms themselves) with the walls of the laser chamber frequently produces volatile molecules that contaminate the gas mixture and either interfere with the chemical reaction chain that pumps the laser or directly absorb the laser radiation. Most of these contaminants, such as CF_4 and other "halocarbons" (including Freons), are readily removed from the gas-flow stream by a cryogenic trap. Another factor responsible for reducing the effective lifetime of the gas mixture is the production of aerosols by the pulsed discharge. Since particulates degrade the beam quality (by scattering) and are capable of damaging the laser's optics, efforts are made to remove them with a filter.

Producing stable and uniform discharges in gas mixtures containing a halogen poses formidable engineering obstacles because of the electron affinity of the halogen atoms. As a consequence, it has proven necessary to produce a uniform background of electrons in the gas immediately prior to firing the main discharge. The subsystem responsible for doing this is known as the preionizer (see Fig. 18). It often consists simply of an array of sparks (produced by discharging a capacitor across two metal electrodes separated by a few millimeters) that emits ultraviolet and vacuum-ultraviolet ($\lambda = 100$–200 nm) photons having sufficient energy to photoionize one constituent of the gas mixture, resulting in an electron density of at least 10^9 cm^{-3}. This apparently mundane step is absolutely necessary for the uniform delivery of electrical energy into the gas and hence the proper performance of the laser. Without it, the main discharge is severely nonuniform and rapidly collapses into an arc.

Table 1 lists the critical characteristics of the rare-gas–halide lasers. Although the lowest-lying excited states of the diatomic molecule (and the upper laser level, in particular) are stable, their spontaneous-emission lifetimes τ are typically less than 20 ns, which, in turn,

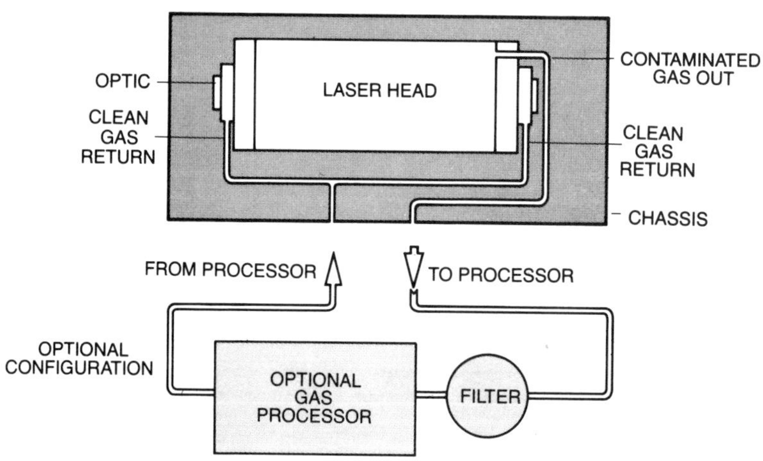

FIG. 17. Gas recirculation system for a rare-gas–halide excimer laser. (Courtesy of Questek, Inc.)

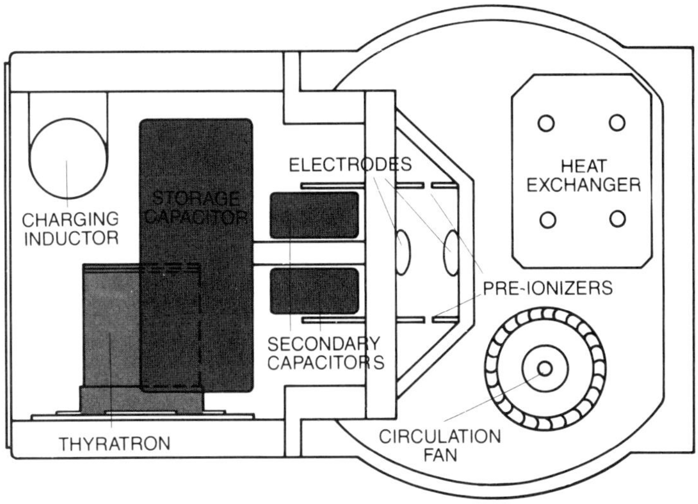

FIG. 18. Diagram of a transverse discharge system for pumping a rare-gas–halide excimer laser. The gas mixture consists of two or more rare gases and a halogen donor such as HCl or F_2, and the total gas pressure is typically 1–2 atm (absolute). The preionizer is necessary for discharge stability. Output beams are usually rectangular in cross section with an area of 1–5 cm^2. (Courtesy of Questek, Inc.)

demands that specific pumping powers exceeding (and generally much higher than) several tens of kW per cm^3 must be delivered to the gas mixture in order for reasonable optical gains to be realized. Such an enormous amount of power cannot be conveniently supplied on a continuous basis (nor long sustained) and, for this reason alone, most excimer lasers emit pulses of less than 100 ns in duration. Longer pulses have been obtained by pumping small volumes (a few to tens of cm^3) with microwave pulses or larger volumes (liters) with electron beams.

Note, too, from Table 1 that the wavelengths available from the rare-gas halides range from 353 nm (for xenon fluoride, XeF) in the near-ultraviolet to below 193 nm ($h\nu = 6.4$ eV), which lies at the threshold of the

Table 1. Key parameters of the rare-gas–halide lasers.

Lasing molecule	Wavelength λ (nm)	Excited-state lifetime τ (ns)	Peak power (MW)[a,b]	Average power (W)[a,c]
XeF	351, 353	14	17	24
XeCl	308	11	20	32
KrF	248	6.8	35	56
KrCl	222	...	5	8
ArF	193	4.2	25	25
ArCl[d]	175	...	...	...
F_2[e]	157	...	1.5	1.4

[a]The specification given is typical of that for currently available commercial models.
[b]The laser pulse widths for the rare-gas–halide lasers generally vary from 10 to 25 ns (FWHM), where FWHM is an acronym for full width at half maximum.
[c]Assumes a laser pulse repetition frequency (PRF) of 100 Hz.
[d]Although this molecule has lased, it has not yet been well characterized.
[e]This is not a rare-gas–halide molecule but is a diatomic halogen; nevertheless, this molecular laser operates in the same type of transverse discharge device that is required for the rare-gas–halide excimer lasers.

vacuum ultraviolet. As far as the average power available from these lasers is concerned, more than 100 W are now available from "off-the-shelf" devices, while the peak (instantaneous) output powers are generally 10–50 MW in 15- to 25-ns (FWHM) pulses. To achieve these results requires population inversions of 10^{14}–10^{15} cm^{-3}, which manifest themselves in a small-signal gain coefficient in the vicinity of (10–20)% cm^{-1}. The wide span in photon energies, when combined with the reasonable overall efficiencies [(1–2)%] and high peak powers of discharge devices, has prompted a virtual revolution in the applications of ultraviolet lasers to medicine, materials processing, and areas such as remote sensing. Several of these are discussed specifically in Sec. 4.

Another recently developed laser that is of increasing interest is the molecular fluorine (F_2) laser that operates at 157 nm (photon energy of 7.8 eV). Similar to the rare-gas–halide lasers in several respects—gas chemistry, design, and operation of the pulsed-power driver—this laser provides megawatts of peak power in a spectral region (vacuum ultraviolet) in which few other efficient lasers exist. It is well suited for discharge pumping, and recent devices have produced pulse energies > 100 mJ in 99.85% He–0.15% F_2 gas mixtures at a total pressure of several atmospheres. However, efficient operation of this laser requires specific power loadings of the active medium in the ~4–23-MW cm^{-3} range, which is roughly an order of magnitude larger than that required for the rare-gas–halide excimer lasers. This is a reflection of the large ionization potentials for both of the constituents of the F_2 laser gas mixture (He, F_2). As indicated in Table 1, the F_2 laser has been developed to such an extent that commercially available, discharge-pumped systems now offer watts of average output power at pulse repetition rates exceeding 50 Hz. Applications of this laser that appear to be particularly promising include microlithography and photochemistry.

3.4 Chemical Lasers: HF

Another, albeit less common, manner of producing population inversions in atoms or molecules is through highly exothermic chemical reactions involving reactants that are in their ground electronic states. Of course, for such an approach to be effective, it is necessary that the reaction culminate in the production of excited species rather than the generation of heat. The best known example of a chemically pumped laser is the hydrogen fluoride system, which is pumped by the reaction of H_2 and F_2:

$$H_2+F_2 \rightarrow 2HF \ (v \leqslant 9). \tag{13}$$

Usually, it is necessary to initiate the chemical reaction with an electrical spark, and, as noted above, the reaction leaves the HF molecule in a vibrationally excited level of its ground electronic state. Lasing occurs at 2.6–3 μm in the infrared from optical transitions occurring between inverted vibrational states of the molecule. The HF laser is capable of producing enormous pulse energies (> 1 kJ) and cw powers (1 kW) from a commercial system. Another example of a chemically pumped laser is the iodine atomic laser, which has produced more than 1 kW of power continuously at 1.3 μm.

The drawbacks associated with this pumping technique are that

1. it is usually difficult to recycle the reactants, and
2. the exothermicities available from chemical reactions severely complicate efforts to obtain lasing at wavelengths shorter than the near-infrared ($\lambda \sim 1$ μm).

Attempts to demonstrate visible chemical lasers have centered on a sequence of two or more chemical reactions to produce the required excitation energy.

3.5 Other Lasers: Far-Infrared, N_2, and H_2

No discussion of gas lasers would be complete without briefly mentioning other systems that are not currently as prominent as those discussed earlier but were nevertheless instrumental as scientific tools or in developing gas-laser science and technology. The longest-wavelength lasers rely on obtaining a population inversion between rotational states of polyatomic molecules such as HCN and H_2O to produce stimulated emission at a wide range of discrete wavelengths in the far-infrared (λ=10–100 μm) or submillimeter (100–1000 μm) spectral regions. Most are driven by a shorter-wavelength laser (such as CO_2 at 10.6 μm) and have served primarily in laboratory roles. Two areas in which these lasers have been applied are in measuring electron densities in plasmas by inverse bremsstrahlung, or in experimental simulations of planetary or astrophysical environments.

At the other wavelength extreme lie the N_2 and H_2 lasers in the ultraviolet and vacuum-ultraviolet regions, respectively. Both were discovered in the 1960s, and for more than a decade the hydrogen molecular laser (116 and ~160 nm) was the shortest-wavelength laser in existence. Over the last several years, however, that distinction has been claimed by multiply charged atomic ions (such as Se^{24+}) that lase in the soft-x-ray region when a thin metal foil of the desired material disintegrates under the influence of an intense pulse (10^{12}–10^{14} W cm^{-2}) from a Nd:YAG (YAG denotes yttrium aluminum garnet) laser (1.06 or 0.532 μm). The N_2 laser, which lases predominantly at 337 nm, served for a number of years as the optical pump for dye (liquid) lasers but in the mid-1970s was quickly displaced by the excimer and Nd:YAG (frequency-doubled or -tripled) lasers. Other gas (and vapor) lasers that have been studied extensively but have had a less pervasive impact are the pulsed rare-gas atomic lasers, various metal vapors, and the rare-gas diatomic excimers Ar_2, Kr_2, and Xe_2.

4. APPLICATIONS

Since its discovery, the laser has fired the imaginations of many, and it was early on viewed by some as a practical device with almost limitless industrial potential and by others as ushering in an era of exotic spin-offs of the "Buck Rogers" genre. Although applications were initially slow to develop, both viewpoints have since been shown to be essentially correct.

Scientific applications of the laser, including the study of atoms and molecules by laser spectroscopy, were among the earliest to be explored, and the ability of laser techniques to elucidate the structure of matter on the atomic scale must be considered to be among the most notable scientific achievements of the past 30 years. Other areas in which the laser has provided new scientific opportunities include environmental sensing (and probing of the earth's upper atmosphere, in particular) and the separation of atomic isotopes from one another in a process that relies on the narrow spectral bandwidth of the laser.

Military applications of gas lasers have been well publicized, and laser radar (LIDAR) and munitions guidance were among the first to be realized. The generally low efficiencies of gas lasers preclude their use as a weapon, but, because of excellent beam quality (near-diffraction-limited divergence), they are well suited for target identification, imaging, and tracking.

Dramatic progress in medicine, chemistry, and materials processing, among other fields, has been made possible by the advent of the gas laser, but electronics may well, in retrospect, be the area most profoundly impacted by laser technology. Inroads made by lasers into diverse areas of electronic device and system manufacturing have recently expanded significantly and in the last decade, in particular. Because of the attractive spectral and spatial properties of laser radiation, it is fair to say that lasers are well into the process of revolutionizing electronics and computer technology along two lines. The first is as a tool in processing electronic materials by trimming, drilling, and marking, for example. Less conventional applications include microlithography for semiconductor device fabrication and the deposition of films by photoablation or photochemical processes. Because of space limitations, this section reviews only a few of the emerging and established commercial applications of gas lasers in electronics manufacturing and photochemical synthesis.

4.1 Marking and Laser Evaporation

Several of the first applications to be pursued by the electronics industry—drilling, trimming, and marking of materials (see LASERS, INDUSTRIAL USES OF)—exploit the capability of coherent infrared, visible, or ultraviolet radiation to be focused to beam diameters below 100 μm. Trimming of thick-film resistors, for example, typically requires a beam spot size of ~50 μm, which is defined by the numerical aperture of the beam-delivery optics. A gas laser that is widely used in resistor trimming is the Xe atomic oscillator, which operates in the green at 537 nm.

Marking of materials is another attractive industrial application of lasers since mechanical contact with the surface to be marked is no longer required. Rather, the laser can be located well away from the workpiece or electronic component to be identified, and the laser beam is delivered to the target via an optical transmission and focusing system. Materials that are normally difficult to mark (such as ceramics and glass) with conventional approaches are quickly handled by laser systems, which are typically built around CO_2 (infrared, $\lambda=10.6$ μm) or the excimer (ultraviolet, $193 \leqslant \lambda \leqslant 351$ nm) lasers. Figure 19 is a diagram of a commercially available CO_2-laser–based marking system capable of making up to 200 000 marks per hour on paper, polyvinyl chloride (PVC), coated metals, plastic, and glass. Since the interaction of radiation with materials is a strong function of wavelength, excimer-laser marking systems yield superior results when working with certain materials including several plastics and metals. Marking glass with a CO_2 laser, for example, produces extensive microcracking in the bulk because of dissipation of heat into the material surrounding the marked region. An ArF excimer laser ($\lambda=193$ nm), however, produces clean cuts with steep side walls, as the physical mechanism responsible

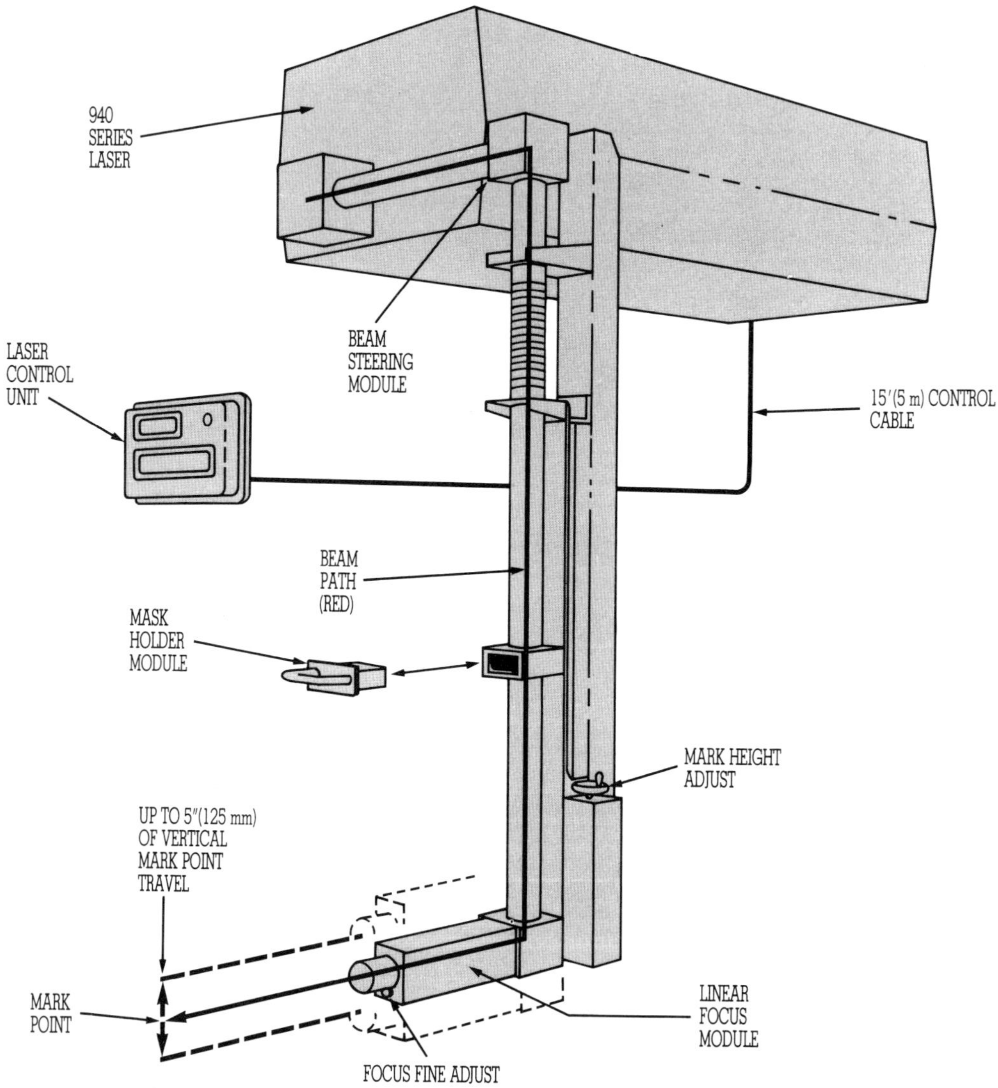

FIG. 19. Diagram of a CO_2 laser marking system, illustrating details of the beam delivery configuration. (Courtesy of Lumonics, Inc.)

for the removal of material is primarily photochemical rather than thermal in nature. Wire cladding is also readily marked with an excimer laser. A photograph of an EFTE (Tefzel)–clad wire that has been marked with characters 0.8 mm in height by an XeCl excimer laser ($\lambda=308$ nm) is shown in Fig. 20. Ultraviolet laser powers up to 75 W (250 mJ per pulse at 300 Hz) are commercially available from excimer marking systems. These industrial-duty systems are also widely employed in hermetic seam welding of electronic packages, spot welding of wire to lead frames, and integrated-circuit (IC) mask design and repair.

All of the applications discussed above depend on the rapid and localized heating of a surface that accompanies the absorption of laser radiation. This approach has also been successfully applied to depositing thin films and is known as laser evaporation or ablation. When intense laser pulses ($\gtrsim 1$ J pulse energy, 10–30 ns pulse widths) are focused onto one of a wide array of materials, the "target" is heated to the point that material is ablated from the surface. A second surface located nearby will be coated with the ejected material and studies have shown that the chemical stoichiometry of the target is, surprisingly, preserved in the deposited films. One area for which laser evaporation has provided the solution to a difficult problem is in the deposition of high-temperature superconducting films. Several research groups have deposited Y-Ba-Cu oxide films on a variety of substrates by evaporating targets with $\sim$2-J/cm^2 ultraviolet (uv) pulses from a krypton fluoride (KrF, $\lambda=248$ nm) or xenon chloride (XeCl, $\lambda=308$ nm) laser. The unique ability of laser evaporation to preserve the stoichiometry of even complex target materials has also permitted modulated structures composed of alternating epitaxial layers of II-VI compound semiconductor materials (such as zinc selenide) to be deposited. This feature, as well as the simplicity of the laser evaporation technique and the high deposition rates available (>5 Å/laser pulse), suggests that its importance to electronic device fabrication will grow in the next few years.

4.2 Microlithography

Another process critical to the electronics industry that is being markedly affected by maturing laser technology is microlithography (*q.v.*). Traditionally, the exposure of a photoresist in IC production has been accomplished with the *g* or *i* lines emitted by a mercury vapor lamp (436 or 365 nm, respectively). Shorter optical source wavelengths are desirable, however, in order to permit smaller features to be patterned onto the chip, thereby increasing the packing density. For this reason, considerable effort is being devoted worldwide to engineering development of the rare-gas–halide excimer lasers, which offer five major, discrete wavelengths between

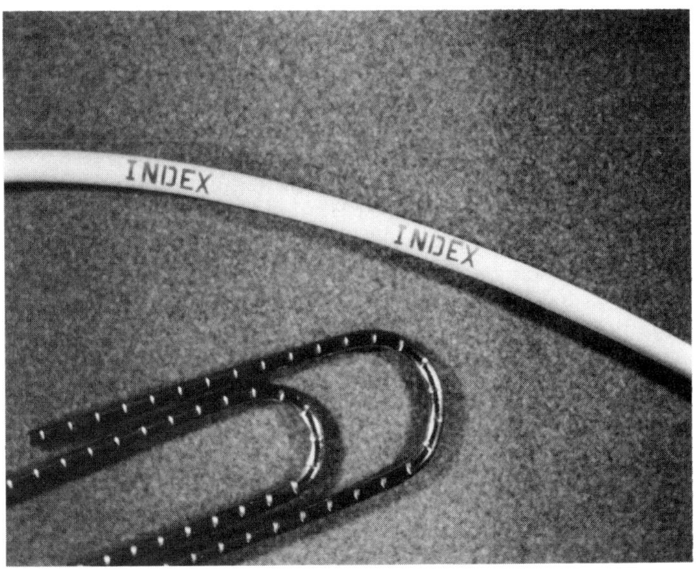

FIG. 20. Photograph of an ETFE (Tefzel)-clad wire that has been marked by a pulsed XeCl ($\lambda=308$ nm) laser. The characters are 0.8 mm in height, and similar results are obtained for Teflon cladding. (Photo courtesy of Lumonics, Inc.)

193 and 353 nm. Several members of this laser family, when operated untuned or "free-running," emit a beam having a spectral bandwidth of ~3 THz (1 part in 500) or, in terms of wavelength, ~400 pm). This breadth is a liability in microlithographic applications as it results in chromatic aberrations, which, in turn, degrade the system's spatial resolution. Consequently, firms in several countries have introduced frequency-stabilized, "injection-locked," excimer lasers having bandwidths of 50±15 GHz (3±1 pm) that are capable (in conjunction with the proper photoresist and optical system) of patterning features smaller than 0.25 μm. Both the ArF (193 nm) and KrF lasers are particularly promising sources for sub-0.5-μm lithography, and, it appears that, in the near future, Hg lamps will gradually be displaced in production by line-narrowed and stabilized KrF lasers. IBM has also recently incorporated a commercial-duty XeCl laser into the manufacturing process for multilayer microcircuits. In this instance, the laser serves in pattern-projection etching of conducting films.

4.3 Photochemical Processing

4.3.1 Deposition and Etching of Thin Films. Virtually all of the applications discussed above draw on the spatial properties of laser radiation to induce thermal processes at a surface. One emerging area that holds great promise for microelectronics device fabrication—photochemical processing—is driven by the spectral purity of a laser. The basic concept underlying the efforts in this area is that the strengths of the chemical bonds in many small molecules are less than 6 eV, an energy that corresponds to a deep uv photon. Furthermore, if a polyatomic molecule absorbs a photon of the proper wavelength, one or more bonds within the molecule are ruptured, freeing an atom or molecular radical. Exploiting such *photodissociation* processes for depositing or etching thin films has several advantages over conventional technology. A major one is that large densities of a desired atom can be produced without the need for high processing temperatures. Second, the distribution of products issuing from a photodissociation reaction changes with the photon energy (wavelength). Consequently, photochemical techniques offer considerably more control over the process than is normally available with pyrolytic (thermal) approaches.

More than 20 elemental and compound films have been deposited (or etched) by the photodissociation of polyatomic molecules. One well-studied example is aluminum. The alkyl trimethylaluminum [$Al(CH_3)_3$] is absorbing for wavelengths below ~255 nm and, for $230 \lesssim \lambda \lesssim 255$ nm, a single photon has sufficient energy to free an aluminum atom:

$$
\begin{aligned}
&Al(CH_3)_3 + h\nu(\geqslant 4.9\ \text{eV}) \\
&\quad \rightarrow Al(CH_3)_2 + CH_3 \\
&\quad \rightarrow Al + C_2H_6 + CH_3. \qquad (14)
\end{aligned}
$$

Photodissociating trimethylaluminum with a KrF excimer laser has been found to be capable of depositing Al films more than 0.5 μm in thickness at deposition rates up to 1000 Å min^{-1}. The resistivity of films deposited from both $Al(CH_3)_3$ and dimethylaluminum-hydride (DMAH) compares favorably with bulk resistivity values and shows the films to be suitable as Ohmic contacts and interconnects.

Similarly, silicon thin films have been deposited on a variety of surfaces by photodissociating the silanes SiH_4, Si_2H_6, or Si_3H_8 with an excimer laser (typically ArF, 193 nm). Deposition rates generally range from 100 to 300 Å min^{-1}, and the film properties are comparable to those for films deposited by plasma processes. With care given to surface preparation, gas-mixture flow, and substrate temperature, epitaxial (crystalline) films have also been grown at temperatures 200–400 °C lower than those normally required by existing (thermal) deposition processes. Compound films such as SiO_2 and Si_3N_4 have been deposited by irradiating mixtures of a silane (typically Si_2H_6) and either N_2O or NH_3, respectively, with uv photons from a laser or lamp. The high quality of the photochemically deposited SiO_2 films (as evidenced by pinhole densities, breakdown voltages, etc.), in particular, has led to their incorporation into commercial dynamic random access memories (DRAMs).

A similar approach has succeeded in etching semiconductor materials such as Si and GaAs except that, in this case, photodissociation is employed to free a chemically reactive species, which results in the removal of sur-

FIG. 21. Perfume-bottle "prototypes" produced by photopolymerization with an ultraviolet He-Cd gas laser. (Photograph courtesy of Spectra-Physics, Inc.)

face atoms. Photodissociating CH_3Br in the uv, for example, produces free bromine atoms that will effectively etch both metals and semiconductors. Although these processes are still in the early developmental stages, several specific reactions have already been implemented in IC production and mask repair.

4.3.2 Synthesis of Vitamin D. Vitamin D is produced naturally in humans by the interaction of sunlight with cholesterol in the skin. It has long been known that a synthetic analog to this process is the single-step photochemical conversion of 7-dehydrocholesterol (7-DHC) to pre–vitamin D, which occurs when 7-DHC absorbs a single, near-uv ($\lambda \sim 300$ nm) photon. This process has been developed commercially with $\sim$330-nm radiation from mercury arc lamps but the yield of pre–vitamin D is marginal—in the (20–50)% range. However, it has been demonstrated that irradiating 7-DHC with two separate wavelengths improves the pre–vitamin D yield considerably. When 7-DHC absorbs a 5-eV photon ($\lambda \sim 250$ nm), for example, the predominant product is tachysterol (an isomer of 7-DHC), rather than pre–vitamin D, and the tachysterol absorption spectrum peaks at longer wavelengths than do those for all the other molecules in the synthesis chain. Consequently, tachysterol can subsequently be transformed into pre–vitamin D by illuminating the molecule with $\lambda \sim 350$ nm photons. By combining the KrF (248 nm) and N_2 (337 nm) lasers in tandem, pre–vitamin D yields of $\sim$80%, or roughly double those representative of the more conventional approach, have been produced. Therefore by the use of two separate optical wavelengths, the chemistry of a synthetic chain can often be tailored so as to favor the formation of a particular product.

4.4 Stereolithography

Of the industrial applications that have been envisioned and realized for uv lasers, many of the newest and most exciting involve materials processing. Another immediate example is stereolithography, in which plastic parts and models are produced by photopolymerization with a computer-controlled uv laser. Images generated by computer-aided design (CAD) systems or obtained from x-ray or magnetic resonance imaging data, for example, can now be "printed" in three dimensions, thereby allowing accurate models to be made rapidly and inexpensively. Commercial systems utilize the focused beam from a He-Cd (325 nm) or Ar^+-ion laser to "write" objects in photopolymer contained in vats as large as $50 \times 50 \times 60$ cm^3 ($20 \times 20 \times 24$ $in.^3$). Writing speeds up to 254 cm (100 in.)/s are available, which dramatically reduce the prototype production time. Automotive parts, such as an exhaust manifold prototype, which routinely took 16 weeks to fabricate, are now produced in less than 24 h. Figure 21 is a photograph of polymer models of perfume bottles that were produced by this process. Further advances in both resolution and speed are forthcoming as several photopolymer manufacturers are developing new families of resins offering a range of mechanical properties. Other industrial materials processes that are benefitting from the properties of uv lasers include cleaning of plastic parts and the removal of polymers from electronic circuit boards.

These few examples serve to illustrate the rapid growth in the transfer of gas-laser applications into the marketplace. Other emerging applications, particularly in medical procedures, have not been discussed but have profound implications for the future. For example, the ability of ArF excimer laser radiation ($\lambda = 193$ nm; cf. Table 1) to ablate human corneal tissue in a spatially selective manner is currently being developed intensively. It is likely that, later in this decade, it will be available as an outpatient procedure for ophthalmic refractive-index corrections, thereby minimizing or entirely eliminating the need for eyeglasses. Other promising medical applications include the removal of arterial plaque (angioplasty) and the destruction of gall stones (lithotripsy) and tumors.

Therefore, while several applications are well advanced and others are yet in the embryonic stage, it is clear that many of the applications envisioned at the birth of the laser have been realized, and that possibly early in the next century its effect on our lives will be pervasive.

GLOSSARY

Efficiency: A measure of the ability of the laser to convert pump power into optical radiation. For lasers pumped electrically, efficiency is often defined as the optical power emerging from the laser divided by the electrical power delivered to the gain medium.

Electron-Impact Excitation: The process of exciting an atom or molecule (i.e., promoting it into an excited state) through a collision with an electron having the necessary energy.

Excimer: (1) A molecule that does not normally exist (that is, its ground state is unstable) but which has one or more excited states that are relatively stable; (2) a name given to a specific family of lasers—the rare-gas–halide lasers—that emit at a number of wavelengths in the 193–353-nm region in the ultraviolet.

Excitation Transfer: A common method for pumping lasers that entails delivering energy to the upper laser level from an excited state of another species (atom or molecule) that lies nearby in energy; this is usually done by a collision.

Photodissociation: Rupturing of one or more chemical bonds within a molecule by the absorption of a photon—often employed as a means of producing specific products or fragments from a precursor or "parent" molecule.

Population Inversion: A condition required for stimulated emission in which the number density of atoms or molecules in a particular excited state exceeds that for a lower-lying level.

PRF (Pulse Repetition Frequency): The number of laser pulses per second, expressed in Hz.

Pump: The process or device by which the upper laser level is populated.

Stereolithography: The process of producing three-dimensional models of objects by photopolymerization of a liquid with a computer-controlled ultraviolet laser.

Stimulated Emission: Radiation generated when an atom or molecule falls from an excited state to one lying lower in energy and does so under the influence of an optical field. For this to occur, the wavelength of the field is dictated by the energy difference between the two states involved in the transition.

Ultraviolet: Region of the electromagnetic spectrum in which the photon wavelength is between roughly 200 and 400 nm.

Vacuum Ultraviolet: 100–200-nm portion of the spectrum.

Visible: Generally taken to be the region of the electromagnetic spectrum lying between 400 (violet) and 700 nm (deep red).

List of Works Cited

Bridges, W. B. (1964), "Laser oscillation in singly ionized argon in the visible spectrum," *Appl. Phys. Lett.* **4**, 128–130.

Bennett, W. R. (1962), "Gaseous optical masers," *Appl. Opt. Suppl.* **1**, *Optical Masers*, 24–61.

Einstein, A. (1917), "On the quantum theory of radiation," *Phys. Z.* **18**, 121–128.

Ewing, J. J., Brau, C. A. (1975), "Laser action on the $^2\Sigma^+_{1/2} \rightarrow {}^2\Sigma^+_{1/2}$ bands of KrF and XeCl," *Appl. Phys. Lett.* **27**, 350–352.

Field, R. L., Jr. (1967), "Operating parameters of dc-excited He-Ne gas lasers," *Rev. Sci. Instrum.* **38**, 1720–1722.

Fowles, G. R., Hopkins, B. D. (1967), "CW laser oscillation at 4416 Å in cadmium," *IEEE J. Quantum Electron.* **QE-3**, 419.

Garscadden, A., Kushner, M. J., Eden, J. G. (1991), "Plasma physics issues in gas discharge laser development," *IEEE Trans. Plasma Sci.* **19**, 1013–1031.

Herschbach, D. R. (1966), "Reactive scattering in molecular beams," in: J. Ross (Ed.), *Advances in Chemical Physics*, Vol. 10, New York: Interscience, pp. 319–393.

Javan, A., Bennett, W. R., Jr., Herriott, D. R. (1961), "Population inversion and continuous optical maser oscillation in a gas discharge containing a He-Ne mixture," *Phys. Rev. Lett.* **6**, 106–110.

Patel, C. K. N. (1964), "Selective excitation through vibrational energy transfer and optical maser action in N_2-CO_2," *Phys. Rev. Lett.* **13**, 617–619.

Patel, C. K. N. (1968), "High power carbon dioxide lasers," *Sci. Am.* **219** (2), 22–33.

Schawlow, A. L., Townes, C. H. (1958), "Infrared and optical masers," *Phys. Rev.* **112**, 1940–1949.

Silfvast, W. T., Fowles, G. R., Hopkins, B. D. (1966), "Laser action in singly ionized Ge, Sn, Pb, In, Cd, and Zn," *Appl. Phys. Lett.* **8**, 318–319.

Tolman, R. C. (1924), "Duration of molecules in upper quantum states," *Phys. Rev.* **23**, 693–709.

Verdeyen, J. T. (1989), *Laser Electronics*, New York: Prentice-Hall, pp. 334-341, 434-444.

Walter, W. T., Solimene, N., Piltch, M., Gould, G. (1966), "Efficient pulsed gas discharge lasers," *IEEE J. Quantum Electron.* **QE-2**, 474–479.

White, A. D., Rigden, J. D. (1962), "Continuous gas maser operation in the visible," *Proc. IRE* **50**, 1697.

Willett, C. S. (1971), "Neutral gas lasers," in: R. J. Pressley (Ed.), *Handbook of Lasers*, Cleveland: CRC, pp. 183–241.

Yariv, A. (1971), *Introduction to Optical Electronics*, New York: Holt, Rinehart, and Winston, pp. 154, 157, and 158.

Further Reading

Deutsch, Thomas F. (1988), "Medical applications of lasers," *Phys. Today* **41** (10), 56–63. An overview of several medical applications of gas lasers.

Eden, J. G. (1985), "Laser technology: theory and operating principles," *IEEE Potentials* **4** (2), 7-12.

Garscadden, Alan, Kushner, Mark J., Eden, J. Gary (1991), "Plasma physics issues in gas discharge laser development," *IEEE Trans. Plasma Sci.* **19**, 1013–1031. An in-depth treatment of the underlying physical mechanisms occurring in a number of gas lasers pumped by electrical discharges.

Jain, Kanti (1990), *Excimer Laser Lithography*, Bellingham, WA: SPIE Optical Engineering. An excellent discussion of the use of excimer lasers in microlithography.

Svanberg, Sune (1991), *Atomic and Molecular Spectroscopy—Basic Aspects and Practical Applications*, Berlin: Springer. Discusses a wide array of spectroscopic applications of gas lasers.

Verdeyen, Joseph T. (1981), *Laser Electronics*, New York: Prentice-Hall. A more detailed description of lasers and laser fundamentals.

Weber, M. J. (Ed.) (1982), *Handbook of Laser Science and Technology*, Vol. 2, *Gas Lasers*, Boca Raton, FL: CRC, and more recent supplements. Contains a summary of most of the atoms and molecules that have produced stimulated emission to date and their operating wavelengths.

Weil, K. G., Weiss, A. (Eds.), (1989), *Ber. Bunsen-Ges. Phys. Chem.* **93**, No. 3. Detailed discussions concerning specific applications of gas lasers in both medicine and biology.

For a periodic update on laser products and applications, see the monthly publication *Laser Focus World*.

LASERS, INDUSTRIAL USES OF

DAVID M. ROESSLER, *Physics Department, General Motors Research and Development Center, Warren, Michigan, U.S.A.*

	Introduction	397
1.	**Basic Principles**	398
1.1	Lasers and Beam Properties	398
1.1.1	Primary Industrial Lasers	398
1.1.2	Beam Properties	399
1.2	Principles of Laser Processing	399
1.2.1	Laser–Material Interactions	399
1.2.2	Processing Regimes	400
1.2.3	Beam Optics and Delivery	401
1.2.4	General Advantages and Limitations	402
2.	**Major Applications**	403
2.1	Surface Treatment	403
2.1.1	Transformation Hardening	403
2.1.2	Surface Melting	404
2.1.3	Alloying	404
2.1.4	Cladding (Hardfacing)	405
2.1.5	Other Surface Treatments	405
2.1.6	Advantages and Limitations	405
2.2	Welding	406
2.2.1	Process	406
2.2.2	Applications	408
2.3	Drilling (Piercing)	409
2.3.1	Process	409
2.3.2	Applications	410
2.4	Cutting	412
2.4.1	Process	412
2.4.2	Applications	414
2.5	Marking	415
2.5.1	Process	415
2.5.2	Applications	416
3.	**Miscellaneous Applications**	416
3.1	Electronic Circuits	416
3.1.1	Annealing	416
3.1.2	Trimming	417
3.1.3	Scribing	417
3.1.4	Soldering	417
3.1.5	Repairs	418
3.1.6	Deposition	418
3.1.7	Oxidation and Nitridation	418
3.2	Prototype Manufacture	418
3.3	Other Processes	419
3.3.1	Ablation	420
3.3.2	Grain Refining	420
3.3.3	Curing	420
3.3.4	3D Laser Machining	421
3.3.5	Laser-Assisted Machining	421
4.	**New Developments**	421
5.	**Summary**	422
	Glossary	422
	Works Cited	423
	Further Reading	423

INTRODUCTION

The early history of the laser is well documented (Bromberg, 1991). The first man-made laser, producing pulses of red light from a synthetic ruby rod, was demonstrated in 1960. During the next few years, other materials were also found to lase, but it was ruby that was employed in 1965 for the first significant industrial machining application of lasers. The automotive industry began using lasers for materials processing in 1969 and there are now few industries that do not utilize lasers in some way. Currently, about 10% of all commercial lasers are used for materials processing, with over a third of these being used in the electrical and electronics industries.

This review excludes the many lasers used for nonprocessing applications, ranging from inspection and measurement to communication and entertainment. The average power of such lasers is usually lower than that of those used in machining, and there is no long-term change resulting from the interaction of the laser beam. Also excluded are applications for medical or military purposes. Several books exist that discuss a wide range of both machining and nonmachining uses of lasers (Ready, 1978; Duley, 1983).

The first section of this review introduces the lasers most used in materials processing

1-56081-067-X/94/$5.00+.50

and notes the relevant properties of laser beams. A very brief account is given of the physics underlying laser processing: useful references to its extensive literature can be found in the works recommended for further reading. Each application is then discussed separately, first in terms of the process itself and then by industrial examples. Understanding the former and expanding the latter are still vigorous pursuits.

1. BASIC PRINCIPLES

The role of the laser in materials processing is to produce permanent changes in a material. The laser may provide nearly all, or only a small part, of the total energy required, the remainder being supplied by mechanical or chemical forces.

1.1 Lasers and Beam Properties

1.1.1 Primary Industrial Lasers. Many materials, solid, liquid, and gas, have been made to lase, but practical requirements greatly restrict the number used industrially.

Laser action in carbon dioxide gas was discovered in 1964. The first commercial CO_2 laser appeared in 1966, with an output power of 100 W. Since then, continuous output powers of 100 kW have been obtained and 25-kW commercial lasers are already being used for manufacturing operations. Laser emission is in the 9–11-μm wavelength region, the 10.6-μm line usually having highest gain. The laser is excited by an electrical discharge in the gas and has an overall ("wall-plug") power efficiency exceeding 10%.

The most important solid-state lasers for processing applications are made from YAG (yttrium aluminum garnet, $Y_3Al_5O_{12}$) or glass, doped with a few percent of neodymium. They are usually excited by xenon or krypton flash lamps and emit near-infrared radiation of about 1-μm wavelength. The wall-plug efficiency is only about 1% for YAG and 0.1% for glass lasers. YAG is much more expensive than glass but has better thermal conductivity, allowing higher average output powers and even continuous operation. Glass can be made with much bigger dimensions but is more difficult to cool; this limits the firing rate to about one pulse per second, though each pulse may contain up to 100 J.

Over 95% of all industrial machining lasers are based upon either carbon dioxide gas or neodymium-doped solids because of their ruggedness and reliability. The remainder includes both gas (excimer, argon, etc.) and solid-state (ruby, alexandrite, etc.) lasers.

Laser action in excimers was first reported in 1975. Emission wavelengths range from 193.3 nm (ArF) to 354.0 nm (XeF). Average output powers are typically 10–100 W. Excimer lasers represent a difficult technology but their short wavelength gives them unique advantages in certain applications demanding high precision or minimal heating.

Other short-wavelength (ultraviolet and visible-range) gas lasers have low output powers (a few milliwatts for He–Cd, a few watts for argon-ion lasers). They are used where high photon energies or short wavelengths are more critical than beam power, as in processing applications like printing, lithography, or stereolithography.

The ruby laser, based on an aluminum oxide rod doped with about 0.05% Cr^{3+} by weight, has a dominant emission line at 694.3 nm. Its low efficiency limits operation to a few pulses per second, though pulse energies of several hundred joules can be obtained. While still used for drilling (and in some nonmachining applications), it has largely been supplanted by newer lasers.

Industrial lasers, like other tools, must be operated safely. In the United States, the Center for Devices and Radiologic Health (CDRH) requires lasers and laser systems to be certified and classified according to the nature of the hazards they represent, while the American National Standards Institute (ANSI) advises on their safe operation (Rockwell, 1987).

Class I lasers are sufficiently low-powered to be considered nonhazardous and require no control or warning labels. The most powerful lasers, class IV, common in processing applications, can cause biological damage to the eyes and skin, whether by direct or diffuse radiation. The potential hazards in laser processing include not only the laser beam itself but also the associated power supply (a source both of potential electric shocks and of x rays), waste products (particulates, toxic vapors), fire, and noise. In practice, beam-related accidents are less frequent than those from other hazards, the most common cause of fatalities being electric shock.

1.1.2 Beam Properties. The high spatial and temporal coherence of laser light distinguishes it from other light sources. For many processing applications, the most important characteristics are its high radiance and low beam divergence. Not only can a laser beam be efficiently propagated and steered over long distances if necessary, but it can also be focused to a very small spot, producing a high irradiance (intensity or power density).

Even moderate focusing can give power densities well above those available from other light sources. A 1-kW CO_2 laser beam, focused to a spot of 0.1 mm radius, has an intensity of 3×10^6 W/cm^2, enough to melt and vaporize most solids.

The properties of laser beams are associated with the various longitudinal and transverse resonances that occur in the laser cavity. The transverse electromagnetic (TEM) modes have characteristic spatial intensity distributions of light in a plane perpendicular to the direction of propagation. The simplest is the fundamental or lowest order mode, TEM_{00} (also loosely termed the "Gaussian" mode), for which the radial distribution of the intensity is given by

$$I(r)=I_0\exp(-2r^2/w^2)=I_0\exp(-r^2/a^2). \quad (1)$$

The intensity $I(r)$ decreases from its on-axis maximum value I_0 to a value of I_0/e at $r=a$, and I_0/e^2 at $r=w=a\sqrt{2}$. The total beam power P is $\pi a^2 I_0$, where a circle of radius a would contain 63.2% of this power while one of radius w would contain 86.5%. Higher-order or mixed modes have more complex spatial intensity distributions, but, for practical purposes, the beam size can still be defined by the area containing a given fraction of the total power.

The spreading of a laser beam as it propagates out of the resonator cavity depends on its design and on the modes that are present. It is characterized by the half-angle divergence θ (the semivertex angle of the conical envelope of the spreading beam), and increases with mode order. It may be only 0.2 mrad in He–Ne lasers, from 1 to 10 mrad for CO_2 lasers, and generally near 10 mrad for high-powered YAG lasers.

Focusing of laser beams can be accomplished by conventional optics. For a TEM_{00} mode, the radius w_0 of a diffraction-limited focal spot is given by

$$w_0 \simeq \lambda f/\pi w_1, \quad (2)$$

where w_1 is the radius of the beam at the focusing optic (of focal length f). With good optics the minimum spot size can be made comparable to the wavelength of the light. In practice, the spot size will often be larger than this but, for both low- and higher-order–mode beams, it is linearly related to the beam divergence by

$$w_0 \simeq f\theta_1. \quad (3)$$

The intensity decreases with distance from the focal plane because the beam radius increases from its minimum value of w_0. The working depth of focus, δ, is the length on either side of the focal plane for which the intensity can adequately sustain the process of interest. For a Gaussian beam, δ is about w_0^2/λ. The best position of the focal plane is not necessarily the top of the workpiece: in some applications the two may be displaced by over 1 mm.

1.2 Principles of Laser Processing

1.2.1 Laser–Material Interactions. Only laser light that is absorbed can contribute to changes in the material. This determines which laser wavelengths are appropriate for a given material. In some cases, the absorbed photons may dissociate bonds directly and be accompanied by only minor heating effects. In most industrial applications, however, absorption of the laser radiation produces significant heating. The subsequent thermal changes, either in the solid state or through melting and vaporization, determine the type of process that occurs (von Allmen, 1987). An important feature of laser irradiation is that the rates of heating and subsequent cooling can be very high.

The fraction of the total incident laser energy that is transferred to the workpiece is termed the coupling coefficient γ. It depends on numerous parameters, including the angle of incidence and polarization of the radiation, the optical properties of the material (which vary with wavelength, temperature, and surface condition), and the presence of any intervening matter. The instantaneous value of γ may change considerably during the course of the irradiation. In deep-penetration welding, for example, the initial value of γ may be much less than 0.1 before melting occurs but may be close to unity once substantial vaporization has occurred.

At low power densities, incident laser light produces negligible heating. As the power density is increased, atomic diffusion increases and, in some materials, there may be solid-state phase transformations. Further increases produce melting and vaporization. The power densities and times needed for these phase changes vary widely with the material. For example, vaporization can occur at intensities of only 10^3 W/cm^2, although 10^6 W/cm^2 is more typical.

At high irradiances, particulate matter, ions, and electrons emitted from the surface create a dense vapor. Light is absorbed and scattered by the condensing droplets and orange-white, glowing plumes of hot gas appear above the workpiece. At 10^7 W/cm^2 there is partial ionization of the vapor: it becomes more opaque, rapidly increases in temperature, and readily undergoes thermal ionization. A plasma is formed, characterized by temperatures near 20 000 K and an intense blue-white light. It may absorb nearly all of the incident laser beam and become a significant source of radiative heat transfer to the workpiece. If the plasma moves away from the surface, the laser beam may become completely decoupled from the workpiece (Ready, 1971). Figure 1 summarizes some general features of laser–surface interaction.

Few laser–material processing interactions are sufficiently simple to be described by models that yield analytic expressions for the temperature increase. Such models usually ignore the temperature dependence of the thermal and optical properties of the irradiated material and have limited value once melting or vaporization occurs. There are useful collations of many of the more interesting analytic solutions (Shkarofsky, 1975; Duley, 1976; Modest and Abakians, 1986), but most detailed analyses of laser processing rely heavily on numerical techniques. Numerous summaries of laser-processing models are available (Mazumder, 1983; Steen and Kamalu, 1983; von Allmen, 1987; Chryssolouris, 1991; Ion, 1992).

1.2.2 Processing Regimes. In general, energy densities of at least 0.1 J/cm^2 are needed to produce permanent changes in materials. Significantly higher energy densities (10^2–10^5 J/cm^2) are needed for many common applications such as cutting or welding. The energy density alone does not define what process occurs; that depends on the rate at which the energy is supplied. Figure 2 shows how different processes are controlled by the incident intensity and the interaction time. It is schematic: the limits for each regime will depend strongly on the material of interest.

Low power densities for relatively long interaction times allow significant thermal penetration prior to surface melting. This

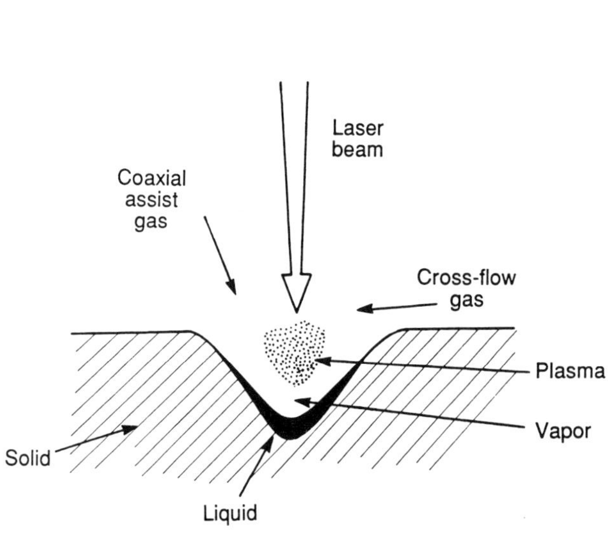

FIG. 1. Laser–surface interaction. Schematic of the primary features and phases involved in many laser machining processes. The relative contributions of melting, vaporization, and plasma formation will be determined by the intensity and duration of the irradiation.

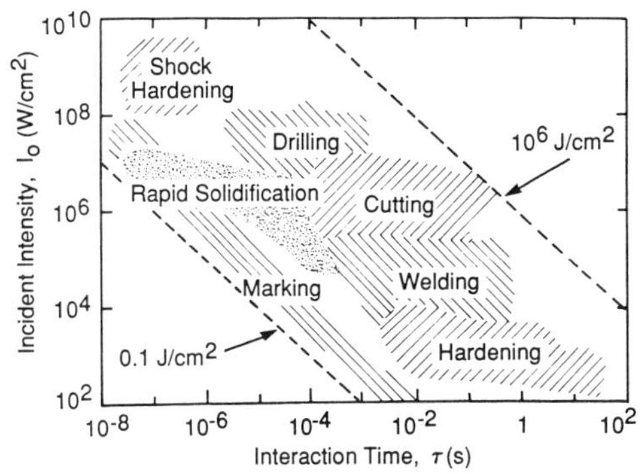

FIG. 2. Roles of incident power density (intensity) and interaction time in determining what laser processes occur. Short interaction times limit thermal penetration: high power densities increase vaporization. Energy densities used for laser processing range from a few hundred mJ/cm^2 to many thousands of J/cm^2. For a given energy density, the *rate* at which the energy is delivered controls the particular process that occurs.

regime is useful for solid-state changes. Higher intensities and shorter interaction times steepen the temperature gradient within the material. Once the surface melts, processes such as welding can occur. For many materials, intensities near 10^7 W/cm^2 and interaction times of 1 ms produce surface vaporization before there is significant thermal penetration, the conditions for ablation.

While the interaction depends strongly on the laser power density and interaction time, it is also affected by the wavelength of the radiation. Figure 3 shows scanning electron micrographs of 300-μm-diam holes made in a 75-μm-thick polyimide sheet with three different lasers: a long-pulsed CO_2, a *Q*-switched short-pulsed Nd:YAG, and a KrF excimer laser. The laser–material interaction is different for each case, varying partly with the wavelength-dependent coupling coefficient and partly with the nature of subsequent heat transfer and material removal. Thermal effects are least with the excimer laser (Znotins *et al.*, 1987).

1.2.3 Beam Optics and Delivery. Key elements include mirrors, lenses, beam splitters, polarizers, and (occasionally) fiber-optic guides. In all cases, absorption should be low to avoid thermal distortion or damage by the beam. Air or water cooling may be used to limit temperature increases and a slight overpressure of gas is often used to prevent contamination of the optics.

Lenses are sometimes antireflection coated to increase their transmittance; mirrors may have dielectric coatings to enhance their reflectance and damage threshold. Quartz is commonly used as the lens material for visible and near-IR lasers; GaAs or ZnSe is used for medium- or high-power CO_2 lasers.

Mirror substrates should have high thermal conductivity: copper is frequently used, with or without coatings. Bare copper can withstand irradiances of 10^5 W/cm^2 but is soft and oxidizes readily. A molybdenum surface improves the durability but lowers the reflectance.

Processing applications commonly entail

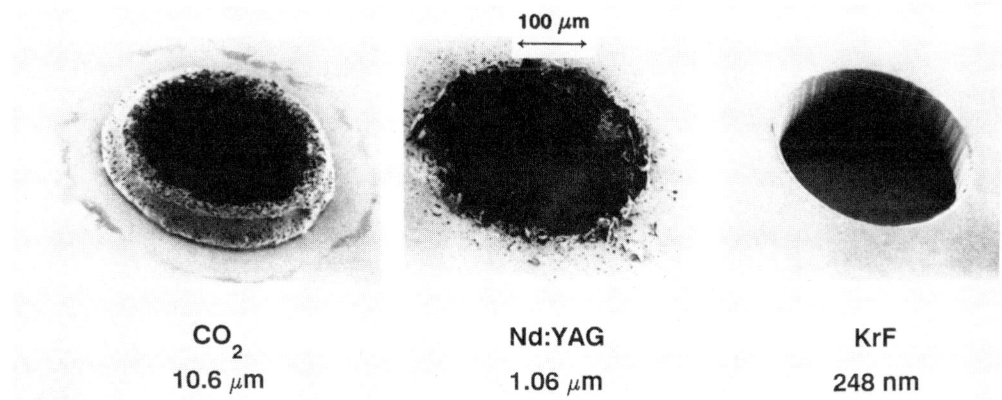

FIG. 3. Scanning electron micrographs of 300-μm-diam holes made in a 75-μm-thick polyimide sheet with three different lasers. The amount of thermal damage depends strongly on the type of laser and is least evident for the excimer (KrF). (Adapted from work of, and used by permission of, T. Znotins, Lumonics, Inc.)

relative motion between the beam and the workpiece. There are three possibilities: the beam is fixed and the part is moved; the beam is moved and the part is fixed; or both the beam and the part are moved. In low-power systems, the laser head itself, not just the beam, may be moved. Other refinements include the possibility of a single laser servicing several workstations, via beam splitting or time sharing. A further general classification occurs depending upon the laser wavelength and the availability of suitably transmissive or reflective optics for beam manipulation.

Beam manipulation via mirrors is possible for all laser wavelengths, and considerable flexibility is afforded by the use of articulated arms. The disadvantages increase with the number of mirrors. They include the cumulative loss of power when many successive reflections are involved, the need for added cooling (increasing mass and complexity) in high-powered laser systems to avoid distortion and damage to the optics, difficulties in alignment of the beam within the system, and difficulties in maintaining pointing accuracy and reproducibility over long path lengths.

Beam manipulation via flexible fibers requires the availability of low-loss optical light guides that have mechanical properties compatible with the bending and durability requirements. Currently, no materials adequate for high-power CO_2 lasers have been developed. However, quartz fibers have greatly increased the versatility of high-powered Nd:YAG lasers. Such fibers can transmit average powers exceeding 1 kW and enable the output from a single laser to be shared in power or time between several workstations or the outputs from several lasers to be combined.

Metallized wave guides, hollow channels with highly reflecting internal walls, have been used primarily with lower powered lasers. They can be flexible and may transmit over 90% of the initial power per meter of length.

Costs can be reduced, and performance enhanced, by integration of the laser with the beam delivery system. Several laser-robot devices are now available that give at least five axes of motion for the laser beam but where only four or five internal mirrors are required. Telescopic and articulated arms give flexibility, but, where reach is extensive, positioning accuracy may suffer.

1.2.4 General Advantages and Limitations. The laser provides unique processing opportunities, many of which stem from its ability to produce a precisely controlled, localized irradiance and from its lack of mechanical contact with the workpiece.

Any location on a workpiece can be processed, provided there is "line of sight" for the laser beam. This may include remote areas or situations where the workpiece must be enclosed or is accessible only through a window. Regions difficult to reach with other tools because of geometry may be marked, drilled, cut, welded, or heat-treated by a laser.

No special or low-pressure atmosphere is required (unlike most electron-beam processes, for example), although assist gases may be employed for some applications. The laser is a "clean" energy source and does not chemically contaminate the workpiece.

The laser is usually incident from a single, well-defined direction and exerts negligible mechanical pressure on the workpiece. Lack of inertial mass allows fast manipulation of the beam. Fixturing can be greatly simplified, and there is no mechanical distortion of parts that are either fragile or easily deformed. Vibration is negligible, and there is less noise than in many other machining operations.

The part of the workpiece to be machined or processed can be precisely defined by the focused laser spot. The total heat input can be regulated by processing speed, power, focusing, and pulsing parameters (width and repetition rate). Thermal damage to and distortion of sensitive components can be avoided. This allows micromachining, as in integrated electronic circuit applications. Very complex patterns, including those with fine detail, can be easily marked or cut.

Materials considered difficult to machine by traditional techniques, because of their mechanical hardness, for example, may be readily laser processed. By-products, such as scrap, particulate debris, and vapors, are usually less than those in other processes. The flexibility of the laser is particularly important where part changes are made or machining operations are modified. The relative motion of the laser beam and workpiece can be readily adjusted. Many processes are easily automated.

Costs can be saved not only in tool wear but also in the use of materials. The localized heat input and narrow width of the laser spot reduce the material lost in machining operations. In cutting, parts can be nested close together because of the narrow kerf. In welding and heat-treating, little post-process finishing may be required.

Uptime often exceeds 90% of scheduled time in manufacturing operations and productivity is high. Use of the laser can sometimes be made more efficient if the beam can be shared between several workstations or different operations. Processing quality is good and there is little scrap.

Capital costs for laser installations are high: they include both the laser and the associated apparatus (beam-delivery system, gas handling, chiller, parts handling, waste disposal, and safety features). The laser itself may account for well under a half of the total cost of some systems. In general, Nd:YAG lasers are more expensive (per output watt), and excimer lasers are much more expensive, than comparably powered CO_2 lasers.

Operating costs are moderate compared with alternative processes. Major sources are the high electrical power consumption, flash lamps (for solid-state lasers), and the amount and nature of any assist gases used.

2. MAJOR APPLICATIONS

This section discusses some of the common industrial applications of lasers by process. Useful compilations of materials processing data have now been published (Belforte and Levitt, 1986–1990).

2.1 Surface Treatment

In many applications the surface properties of a component (optical, electrical, mechanical, and so on, including cleanliness) are critical. Much industrial interest has focused on improving resistance to impact, wear, corrosion, and oxidation. Materials that intrinsically possess satisfactory properties are often expensive and may be difficult to machine by virtue of the very attributes desired (hardness, for example).

The surface properties of some common materials can be improved by laser irradiation and made comparable to those of more expensive ones. This modification may require only a thermally induced change in the material or it may involve addition of other materials to the surface. Only a thin surface region is affected and the process can be precisely controlled (Steen, 1985).

2.1.1 Transformation Hardening. It has long been known that the mechanical properties, such as hardness, of some materials may be significantly altered by heat treatment. The transformation hardening of ferrous alloys takes advantage of certain metallurgical changes that can be thermally induced while such metals remain solid.

The microstructure of many steels is dominated by a mixture of bcc ferrite (α-Fe) with carbides such as cementite (Fe_3C). When heated above a critical temperature, this material transforms into a fcc structure, austenite, which more readily dissolves carbon interstitially. The required temperature depends upon the original carbon content: it is about 912 °C for pure ferrite but 727 °C for eutectoid steel (0.77% C). The time required for the carbon to dissolve completely and become uniformly dispersed throughout the austenite also depends on the composition and microstructure of the initial alloy.

If the austenite is now cooled ("quenched") sufficiently rapidly, the carbon is retained in solution, and a hard, metastable, bcc tetragonal phase, martensite, results. This may be too hard for some applications but can be tempered to reduce brittleness; the reheating allows controlled decomposition and relieves stress.

The carbon content must usually exceed about 0.2% by weight for a mild steel to be transformation hardened. The maximum Rockwell C hardness in fully martensitic steels is less than 40 if there is only 0.1% C but over 60 when the carbon content exceeds 0.5%.

Lasers have been used for heat treatment since the early 1960s and offer several benefits over traditional furnace, flame, or induction hardening. Distortion is low because only the area of interest need be heated, the case depth (hardened thickness) can be tightly controlled, and there is often no need for forced cooling. Heat produced by optical absorption at the surface rapidly diffuses into the underlying metal in a self-quenching process.

In order to gain adequate surface coverage rates, laser powers must usually exceed several hundred watts. Typically the beam is only weakly focused, to give a spot up to 1 cm in diameter and an incident intensity of 10^4 W/cm^2 or less. The relative linear speed between the laser beam and the workpiece is typically 5–150 mm/s. Dwell times are 0.01–1 s, and hardened regions are up to several millimeters thick. If the laser heating is too rapid, the surface may melt before there has been adequate thermal penetration at lower levels, and the case depth (hardened region) will be too shallow.

Figure 4 shows the depth profile of a steel alloy that has been hardened by a 2-kW CO_2 laser scanned at 12.7 mm/s across the surface. The alloy contained about 0.4% C and, initially, was mainly pearlitic, a lamellar mixture of ferrite (α-Fe) and carbide (Fe_3C, cementite) phases. After the laser treatment, the surface zone is fully martensitic with a Vickers hardness (HV) approaching 700 [Rockwell C hardness (Rc) of 60]. The hardness remains above 600 HV (55 Rc) for a case depth of about 0.5 mm. The underlying layer, corresponding to incomplete dissolution of cementite (Fe_3C) and diffusion of carbon, shows decreasing hardness with depth. At depths greater than 0.9 mm, the temperature did not increase enough for austenite to form, and the material retains its original structure and softness.

Industrial applications are numerous, including the edges of many cutting tools, gear teeth, crankshafts, diesel engine cylinder liners, valves, piston ring grooves, and cylinder bores. The first intensive use was hardening of the cast-iron housing of automobile power-steering gears, begun in 1973 to contend with higher front-end loads on vehicles. A 1-kW cw CO_2 laser puts a stripe pattern in the housing, selectively hardening only regions known to experience high stress. Heat input is thus minimized, less than 0.1% of the total housing mass being significantly affected. Distortion is very low and there is little scrap. The rate of wear was decreased by 90% at a cost that was only 20% of that of alternative techniques, such as furnace hardening or nitriding. Furthermore, castings can be ordered on the soft side of their hardness specifications to improve machinability.

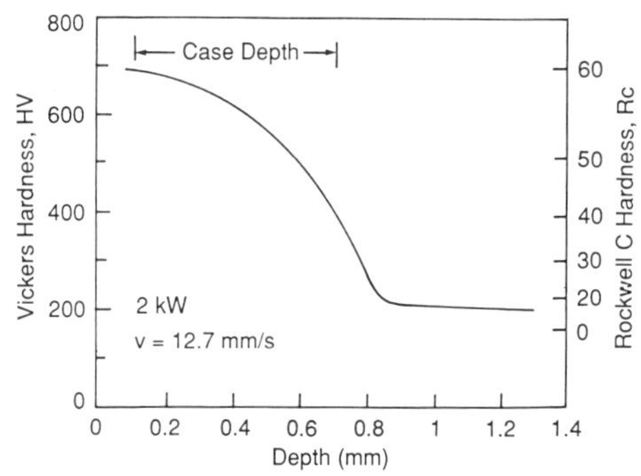

FIG. 4. Hardness vs depth below surface for a laser-hardened iron–carbon alloy. A 2-kW CO_2 laser was scanned at 12.7 mm/s across the surface, the subsequent heating and cooling rates being greatest near the surface. The temperature of material at least 0.9 mm below the surface did not reach that needed for austenitization to begin. (Adapted from work of J. R. Bradley, General Motors Research.)

2.1.2 Surface Melting. With adequate heating the surface melts, allowing homogenization of material. At high laser intensities and short interaction times, the cooling rates following irradiation are very high (10^6–10^{10} K/s), far exceeding those found in many conventional techniques, and the uniformity becomes frozen in. Metastable glassy or amorphous phases can be formed that have wear and corrosion resistance markedly different from that of the bulk material (Breinan and Kear, 1983).

The technique is relatively new and little exploited, although interest is growing for applications such as automotive camshafts and tool steels, and in the modification of coatings applied by conventional plasma and flame spray. While pre- and post-process heating can reduce surface stress cracking and warping (from the rapid heating and cooling), there is generally more distortion than in transformation hardening.

2.1.3 Alloying. Major surface modifications are possible when chemical additions are made. The desired material may be added in diverse ways: as a powder, wire, sheet, or coating prior to the laser scan, or injected or blown in during the irradiation. All of the additive flows into the surface region of the substrate when the laser traverses the workpiece. Strong convection currents in the melt pool ensure thorough mixing. The alloyed material forms a dense, fine-grained, homogeneous layer extending up to 1 mm or so below the surface (Draper and Poate, 1985).

Industrial applications commonly employ kilowatt-class CO_2 lasers and interaction times of 10–50 ms.

The technique has been used for numerous engine components (valves, piston rings) and diverse tools, such as saw blades. Inexpensive steels can be surface alloyed with elements such as nickel or chromium to become as corrosion-resistant as stainless steel. Variants of the process include "layer glazing" and particle injection (by metal carbides, for example).

2.1.4 Cladding (Hardfacing). Like alloying, laser cladding requires addition of a new material to the workpiece surface; unlike alloying, not all the additive is mixed into the substrate to form an alloy. Some remains undiluted and constitutes the new surface, which therefore has the same properties as those of the added material. The laser provides enough heat to produce a metallurgical bond between the cladding material and the base metal. The heat input must be carefully controlled: extensive melting of the base metal will result in undesirable dilution of the cladding material and possible loss of desired properties. A good bond demands some dilution but this is typically less than 10%.

Laser cladding has been investigated since 1974 and has advantages over alternatives such as plasma spraying and arc welding (Li and Mazumder, 1985). Many of the applications are similar to those for alloying, ranging from automotive valve seats to oil-drilling tools. A popular example is the use of a cobalt-based alloy to clad wear pads on jet engine turbine blades. These are made of a cast nickel-based superalloy that is optimized for creep strength. The high ambient temperatures (1600 K) and continuous blade vibration shorten blade-edge life unless wear-resistance measures are used. Manual TIG (tungsten inert gas) processes to improve wear resistance were time consuming, often diluted the nickel in the blade, and had poor reproducibility. A 2-kW CO_2 laser avoids these problems and is ten times as fast, greatly reducing costs. Consumption of the expensive cladding material, which is deposited via an inert-gas stream into the melt pool, is only half of that in TIG. The small heat-affected zone reduces the likelihood of heat cracking.

Cladding is also being used as a form of "reverse machining," where worn parts such as dies or vehicle components can be rebuilt. The aerospace industry uses this technique extensively to repair turbine blades.

2.1.5 Other Surface Treatments. The appearance and properties of surfaces are critically affected by their topography and roughness. The ability of the laser to alter surface roughness, either in degree or in nature, gives it a role in changing these characteristics.

Lasers have been used for applications from preconditioning brake linings to improving paint brightness. The latter, for example, depends both on the painting procedure and on the texture of the bare metal surface underlying the primer. A pulsed laser can be used to pattern the work roll surface at the steel mill, thereby controlling the finish of the steel sheet being produced.

Diverse surfaces, from metals and semiconductors to glasses and plastics, can be cleaned by laser irradiation. The process avoids chemical solvents and the waste disposal and safety hazards they pose. The purpose varies from improving optical or electrical characteristics to preparing the surface for subsequent bonding.

2.1.6 Advantages and Limitations. The laser can harden geometrically complex components and reach selected areas difficult to reach by conventional techniques. Inexpensive materials, rather than specialized alloys, can be used for the bulk. Forced cooling is usually unnecessary because of the self-quenching nature of the laser-induced processes. The ability to limit heat input to the region of interest minimizes distortion of the component and reduces post-treatment finishing. There is little scrap and the technique is particularly useful for applications where dimensions must be tightly controlled, such as engine cylinders.

Both the surface area to be treated and the case depth can be precisely controlled. Limiting the case depth can be useful where the bulk material has desirable properties, such as good machinability. Alloying and cladding afford rather better depth control and more diversity in material properties than does laser transformation hardening.

Disadvantages include expensive capital equipment and relatively low processing speeds. Furthermore, there is generally a need for laser beam shaping (the raw laser beam seldom has the "top-hat" uniform spatial intensity distribution ideal for surface treatment) and, for transformation hardening, the need to enhance the intrinsic laser–workpiece coupling efficiency (often by means of the preliminary addition of a highly absorbing coating).

2.2 Welding

Laser welding has been employed with many materials, including the joining of dissimilar metals. It is best suited for shallow penetrations (less than 10 mm), although much greater penetrations are possible. Speeds range from a few mm/s to well over 1 m/s.

2.2.1 Process. If a laser beam of adequate power is scanned across the surface of a metal, material melts, and the melt front propagates some distance below the surface. In most welding applications, surface melting is achieved in less than a microsecond. Figure 5 shows cross sections of a 6.5-mm-thick AISI 4140 steel plate whose surface was traversed by a 5-kW CO_2 cw laser beam at speeds of 21–85 mm/s. The new microstructures correspond to material that melted and then resolidified (the fusion zone) and to metal that was modified by the heat but did not melt (the heat-affected zone).

At the highest speed, where the laser energy absorbed per unit volume of material is least, the melt front propagates uniformly into ahemisphere. This is the conduction welding regime where the fusion zone has a somewhat semicircular cross section with an aspect ratio less than about 1.5. As the scanning speed decreases and the energy delivered per unit volume increases, there is an elongation of the melt zone directly beneath the beam. This is the "keyhole" welding regime and gives weld profiles resembling those from electron beams. For a focused cw CO_2 laser, the transition from conduction to keyhole welding occurs at powers near 1 kW and power densities of about 10^6 W/cm^2.

Numerous models of laser welding have appeared since the early 1970s (Swift-Hook and Gick, 1973; Klemens, 1976; Cline and Anthony, 1977) and have become increasingly sophisticated as they have tried to include more of the complex phenomena in the melt pool and plasma (Mazumder and Steen, 1980; Dowden *et al.*, 1987; Mazumder, 1991; Kapadia *et al.*, 1992). There are now excellent

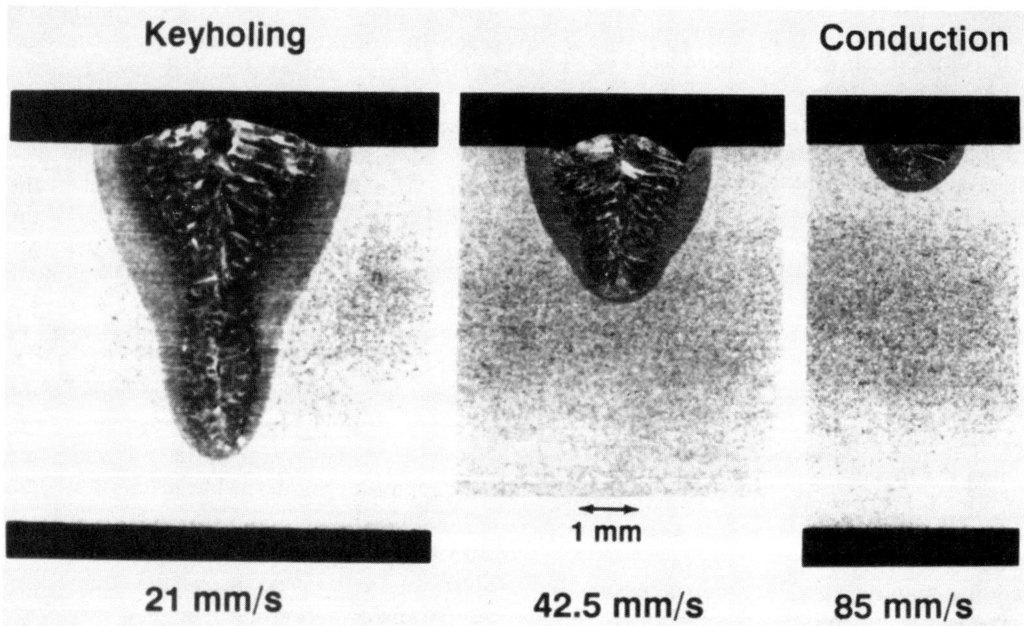

FIG. 5. Cross sections of an AISI 4140 steel plate traversed by a cw 5-kW CO_2 laser at speeds of 21–85 mm/s. The semicircular profile of the fusion zone becomes elongated as the speed is decreased and as keyhole welding takes over from conduction welding. The increased penetration occurs because a cavity is produced by vaporized material below the laser beam.

reviews of such work (Mazumder, 1983; Schuocker, 1992).

In the keyhole regime, the irradiated surface begins to evaporate during the propagation of the melt front. The vapor produces a cavity below the beam, forcing molten material aside and allowing the incident beam to penetrate deeper. The liquid flows around the keyhole, collapses back into the hole and resolidifies while the latter moves forward. The cavity is dynamically stable for a restricted range of scanning speeds. If the speed is too low, the amount of molten material is too great to be supported by the vapor pressure and no cavity can be sustained. If too fast, inadequate vaporization occurs and again no cavity is formed.

The keyhole regime has considerable practical value. The penetration is much deeper than that in the conduction mode, giving aspect ratios as high as 10:1. The heat-affected zone is narrowed to less than a quarter of the weld width, and the welding efficiency increases. Simple energy-balance considerations predict that penetration depth should increase linearly with power and inversely with speed. Empirical data for CO_2 laser powers up to 20 kW and for a wide range of speeds are in general agreement but give a slightly sublinear dependence on power. Figure 6 shows the approximately inverse relationship between penetration depth and welding speed, at a given beam power, for a cw CO_2 laser beam scanned across a thick 304 stainless-steel plate. The large bandwidth for each curve reflects the dependence upon other processing variables (focal spot size, assist gas parameters, and so on).

The thermal diffusivity of metals influences the choice of laser welding parameters. If too low, there may be inadequate penetration; if too high, much of the heat will be wasted, flowing out of the weld zone. The thermal diffusivity of alloys is often less than that of pure metals. Penetration is increased in stainless steel, for example, by use of low power but long interaction times. The high thermal diffusivity of copper mandates the use of high powers and short interaction times.

Some of the metallurgical properties of the weld, such as tensile strength, toughness, and hardness, may exceed those of the parent metal. In some cases, particularly for dissimilar metals or where the mechanical fit-up of

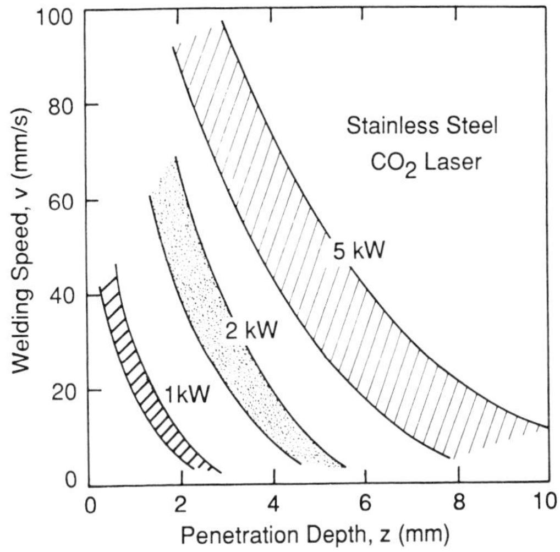

FIG. 6. Welding speed vs penetration depth of the melt zone for CO_2 laser scans across a 304 stainless-steel plate. The large bandwidth associated with each laser power reflects the dependence upon laser-beam parameters, including mode and focusing, and upon gas-assist parameters.

the parts is poor, filler metals may be added to improve the metallurgy. Benefits range from reduction of porosity to elimination of centerline cracking.

Various gases and flow geometries, from coaxial to cross flow, are utilized in laser welding. Purposes include protection of the focusing optics from contamination (especially by debris and spatter from the workpiece), protection of the weld bead from contamination, reduction in porosity, and control of the plasma or weld plume.

Inert gases are often used to blow the plasma out of the path of the incident laser beam. The direction must be controlled to avoid disruption of the weld pool. Helium reduces oxidation and plasma effects, but is expensive. Argon has a much higher molecular weight and gives excellent coverage against oxidation. However, it ionizes more readily and produces a large plasma. Results with CO_2 lasers show that the type of assist or shield gas used greatly affects the weld penetration depth. Penetration is least with argon, moderate with nitrogen or carbon dioxide, and greatest with helium. Plasma effects are generally weaker with Nd:YAG than with CO_2 laser welding.

The laser provides considerable flexibility in joint configurations, allowing significant materials and costs savings if the component can be designed to take advantage of the beam's unique properties. These offer one-sided access to the workpiece, precise positioning, and localized heat input. No mechanical contact is needed, and the material can be isolated if desired, the laser beam being introduced via a window or other optical means.

The low heat input results in narrow heat-affected zones, little distortion, and negligible thermal damage to surrounding areas, a special advantage in the welding of fragile components. Heat inputs can be reduced further by the use of a pulsed rather than continuous mode. The laser avoids some of the limitations imposed by electron beams (*e* beams): there are few x-ray hazards or potential problems from strong electric and magnetic fields, and the welding does not require a low-pressure ambient.

For many metals, weld quality is consistent and good. Laser welding is fast and easily automated. Uptime is high (90% is routine and 95% is sometimes stipulated at the time of purchase), and operating and maintenance costs can be kept below those of alternative welding techniques.

Not all metals are readily laser welded. The very rapid temperature changes induced by laser irradiation can lead to very hard, brittle weld beads in some metals and can also produce hot and cold cracking—especially in high-carbon steels. There may be unacceptably high porosity (especially in the lapwelding of zinc-coated steels), embrittlement, and cracking, or the metallurgical properties of the weld may be inadequate in general. Filler metals may not always solve such problems. Useful data have been collated on the laser weldability of many materials (Belforte and Levitt, 1986–1990).

The advent of Nd:YAG lasers with average powers at the kilowatt level means that high-powered CO_2 lasers are no longer the only option for some deep welding applications, although they continue to dominate for continuous welding at high speed. However, lasers are generally inefficient at making very deep welds. Many industrial processing lasers can make shallow penetration welds (less than a few mm) but depths above 1 cm usually require lasers with very high powers. Even though 100-kW-level CO_2 lasers have achieved weld depths of several centimeters at speeds exceeding 1 m/min, their high capital costs usually make them less attractive than electron beams.

Poor mechanical fit-up in the components to be welded may necessitate expensive fixturing of the workpiece. Furthermore, although the welding process can be automated, the laser requires special maintenance and operating skills, as well as the necessary safety precautions.

2.2.2 Applications. The laser has proved very versatile in a wide variety of welding applications. These range from low-power uses in small electronic components to the multikilowatt powers that have long been used to make seam welds in lead–acid batteries. Keyhole welding is not always preferred to conduction welding. For example, the latter is used in applications such as the hermetic sealing of electronic packages or where surface distortion must be negligible. Not all applications are metallic. The laser has proved useful in the packaging industry because of its ability to heat seal or weld a variety of thermoplastics.

If speed is critical, as in the hermetic sealing of some components, the laser must either be cw or possess a high pulse repetition rate, limiting the choice to lasers such as Nd:YAG or CO_2. However, low pulse rates are sometimes adequate for applications like the welding of wires to terminals in circuits. In such cases, a less costly Nd:glass laser can be used, despite a repetition rate of only about 1 pulse/s. In one application, a laser with a single operator replaces a dozen employees who had manually soldered the terminals: the laser welds are much faster and far more uniform.

The laser is especially useful when existing electric and magnetic properties of the workpiece must remain undisturbed. Thus a laser is preferred to TIG for the welding of stator cores.

Thin-walled or other delicate components can be laser welded because of the very low heat input, precise spot positioning, and absence of mechanical contact. Sometimes the ability of the laser to make hermetic seals is critical, as in the welding of battery cases for heart pacemakers. Another example is the use of a 50-W CO_2 laser to make a hermetic weld between the Kovar header and the Inconel body of a microdetonator can. In this application, the heating effects must be tightly controlled to prevent damage to the electrode and to avoid premature firing of the charge. Heat effects can be further reduced, if necessary, by use of a pulsed laser. It is sometimes possible to minimize distortion in wires or terminals if the pieces being welded together can be simultaneously irradiated (by a split laser beam, for example) rather than all of the heat being generated in only one part.

Lasers are widely employed to perform welding operations in microcircuits, the tightly focused beam allowing the welding of very small components, such as connecting wires on computer chips, spaced closely together. Hand soldering of wires in some circuits has now been superseded by automated pulsed laser welding, resulting in both higher-quality welds and higher processing speeds while reducing labor needs. In some electronic applications, no prestripping of the insulation is needed.

The laser butt welding of aluminum window spacers (common in double-glazed windows) has become the standard production technique, outclassing alternatives in terms of quality and productivity. Over 100 CO_2 lasers were employed for this purpose by the 1980s.

The automotive industry employs laser welding extensively. Low-power applications include the welding of components in electronic circuit boards used by the auto manufacturers. High-power (greater than 5 kW) CO_2 lasers are used extensively for welding transmission components, where there is good fit and where the low distortion and precise positioning of the weld are key benefits. Furthermore, a single laser can serve multiple workstations, the beam being shuttled between them. In Japan, *e*-beam welding still predominates, partly because uptime has been higher than in the United States.

The laser welding of automotive-body sheet metal is currently of great interest because of the savings in weight and materials costs that can be realized through joint redesign and the more efficient utilization of sheet metal. Further advantages stem from increased productivity and reductions in floor space and workers. However, many of the potential savings are offset by difficulties associated with poor fit-up and the use of coated materials.

A major use of Nd:YAG lasers with fiber-optic delivery systems is in the manufacture of disposable-cartridge razors. The razor contains two very thin floating blades, only about 1 mm wide, made of platinum-hardened stainless steel. They must be attached to very small steel support bars riding on small plastic springs. Traditional spot welding, adhesives, and mechanical stitching all proved unacceptable. Quality was poor, the process was messy, or production was slow. A 250-W laser makes 13 spot welds in each blade with negligible heat damage and is fast enough to meet production goals. Each laser completes welding a cartridge every second, so that by 1991, when over 50 lasers had been installed for this application, 5 million welds were being made every hour. Reproducibility and quality are maintained by continuous automatic monitoring of the pulse parameters. The fiber-optic guide not only simplifies fixturing but also helps to make each transmitted laser pulse more uniform.

The ability of the laser beam to be directed to normally inaccessible areas makes it particularly useful in hazardous work, such as the on-site repair of nuclear fuel rods. In such cases, a Nd:YAG laser beam can be directed to the site by a long fiber.

2.3 Drilling (Piercing)

Drilling was probably the first major industrial application of laser processing when a ruby laser was used for the initial piercing of diamond dies used to extrude wire.

2.3.1 Process. Material can be removed as solid, liquid, or vapor. Several analyses of drilling have been given (Paek and Gagliano, 1972; Minardi and Bishop, 1989), and useful references can be found in the books suggested for further reading. A simple model indicates that for a fixed irradiance H, the temperature increases roughly with the square root of the duration of the irradiation. In terms of the total energy density E delivered in a pulse of length τ,

$$\Delta T \propto H\sqrt{\tau}=E/\sqrt{\tau}. \qquad (4)$$

Thus for a given amount of laser energy, higher surface temperatures are achieved with short pulse lengths. Long pulses allow greater time for thermal diffusion into the bulk, thereby reducing drilling efficiency. A given amount of material can be removed by either a few long (several ms) pulses or many short (less than 0.5 ms) pulses. The latter may take more time but give less taper and distortion, resulting in generally better hole quality.

If peak intensities are too high, or if pulse lengths are too long, plasma formation may disrupt the process. Drilling efficiency can be improved if the time between pulses is made long enough for the plume to dissipate. High intensities (10^8 W/cm^2 in metals) can cause violent expulsion of both vapor and other material, molten and particulate (Ready, 1978; Rykalin *et al.*, 1979). Although the drilling efficiency increases initially with this added mechanism, the hole becomes deeper, making expulsion of the molten material more difficult. Some of this liquid is unable to escape from the top of the hole before resolidifying on the walls as recast. Some builds up as dross at the bottom of the hole.

Assist gases are often used to accelerate removal of loose or molten material and to flush out the hole. If the gas is inert, it supplies only mechanical energy. If reactive gases such as oxygen are used, the heated material may burn vigorously in an exothermic reaction. This energy supplements that from the laser and from mechanical processes. Sometimes the formation of oxides at the erosion front reduces cutting efficiency: this happens in the oxygen-assisted drilling (and cutting) of steels with high nickel or chromium content. There is generally poorer control of hole dimensions when reactive assist gases are used.

The lack of mechanical contact avoids problems of distortion, contamination, or tool wear. The flexibility of the laser allows holes of diverse geometry, not merely circular, to be made. Holes can be made in curved surfaces, at large angles of incidence to surfaces, in hard materials, and in regions that are difficult to reach by other means. Heat-affected zones are small and the process is both fast and reproducible.

The laser outclasses other methods for drilling holes only a few micrometers in diameter. Laser-drilled holes range from a few μm to over 1 mm in diameter. For wider holes, prohibitively high laser energies are required and the hole quality is degraded by the amount of material vaporized. Such holes are usually trepanned, the laser beam being brought off axis through a rotating lens to cut out a circular slug of material.

Holes are typically up to 25 mm deep. They can be deeper but the quality usually suffers. High aspect ratios (depth to width) are possible: values may be 10:1 in metals but even higher in ceramics. Limitations are imposed by taper, especially in high–aspect-ratio holes, though this can be reduced by the use of long–focal-length optics or by continual lowering of the focal plane. Precision holes may be possible only in thin stock. In some materials, wall roughness may be undesirable. Dross on the exit side can be removed by shotblasting or light grinding: recast on the entrance side is less easily removed but, if unacceptable, can be reduced by proper choice of processing parameters.

2.3.2 Applications. Lasers can drill or pierce holes in any material, whether hard like diamond or soft like cloth, brittle like ceramic or flexible like rubber. One of the earliest demonstrations of the ability to make holes was the piercing of razor blades. The punching of holes in diamonds used for drawing wires was the first industrial machining application of lasers. The laser-produced hole requires subsequent mechanical polishing but the process is still very much faster than mechanical drilling. It has now become the standard method but generally employs a Nd:YAG rather than a ruby laser.

Among numerous other industrial drilling applications are the holes in rubies used in watch timepiece movements, mounting holes in ceramic substrates for circuits, lubrication holes in transmission components, bleed holes in hydraulic brakes, holes in plastic medical catheters, liquid feed holes in polyethylene irrigation tubing, and the holes in the plastic upholstery covering the walls of passenger aircraft cabins (for sound deadening).

For ceramic substrates, rapidly repeated short pulses (CO_2 or Nd:YAG) are used to lessen both thermal stress cracking at the walls and the amount of resolidified material. Microcracks can be further reduced by a preheat process. The laser is faster and cheaper than diamond-tipped drill bits, and there is no wear or breakage. The latter can be reduced if the mechanical drilling is done in the "green" state, before the ceramic is fired to harden it. However, there is subsequent shrinkage and loss of precision, problems avoided in laser drilling of the fired material.

Every major supplier of cigarette paper now uses one or more CO_2 lasers to perforate the filter-tip paper with one or two lines of holes running round the circumference. These holes add dilution air to the tar and nicotine passing into the filter and are important in determining the rating of these constituents. The hole dimensions are typically 50–300 μm. The laser can be used in two ways: it can either be scanned rapidly (over 5 m/s) across the moving paper or be fired through a cylindrical mask, containing chemically etched holes, at the moving paper. Figure 7 shows scanning electron beam (SEM) micrographs of the two types of holes produced, the elongation indicating the motion of the paper during the laser scan. Current installations can produce millions of holes per minute with excellent consistency and with a quality unmatched by mechanical drilling or cutting techniques (which often deform the holes or fail to remove material completely).

The turbine blades of aircraft engines contain numerous small air bleed holes. These allow cool air from the intake to mix with the hot combustion gases and to keep the turbine blades themselves at a desired temperature. Mechanical methods are precluded because of the difficulties in drilling 1-mm-diam or smaller holes at various angles through the curved surfaces of very tough, heat-resistant alloys. Pulsed Nd:YAG lasers can drill such components at about 1 hole/s.

New aerospace applications include the drilling of millions of holes in aircraft wings for purposes such as improving air-flow characteristics and lift, and to allow a very thin sheet of oil to be sprayed out to combat ice build-up.

The laser is widely used to make the holes in the plastic nozzles of aerosol cans. The submillimeter-diameter holes are difficult to make with conventional techniques, and it is time consuming to make the modifications required when hole sizes are changed. The laser has the necessary flexibility, and can drill differently sized holes in the same com-

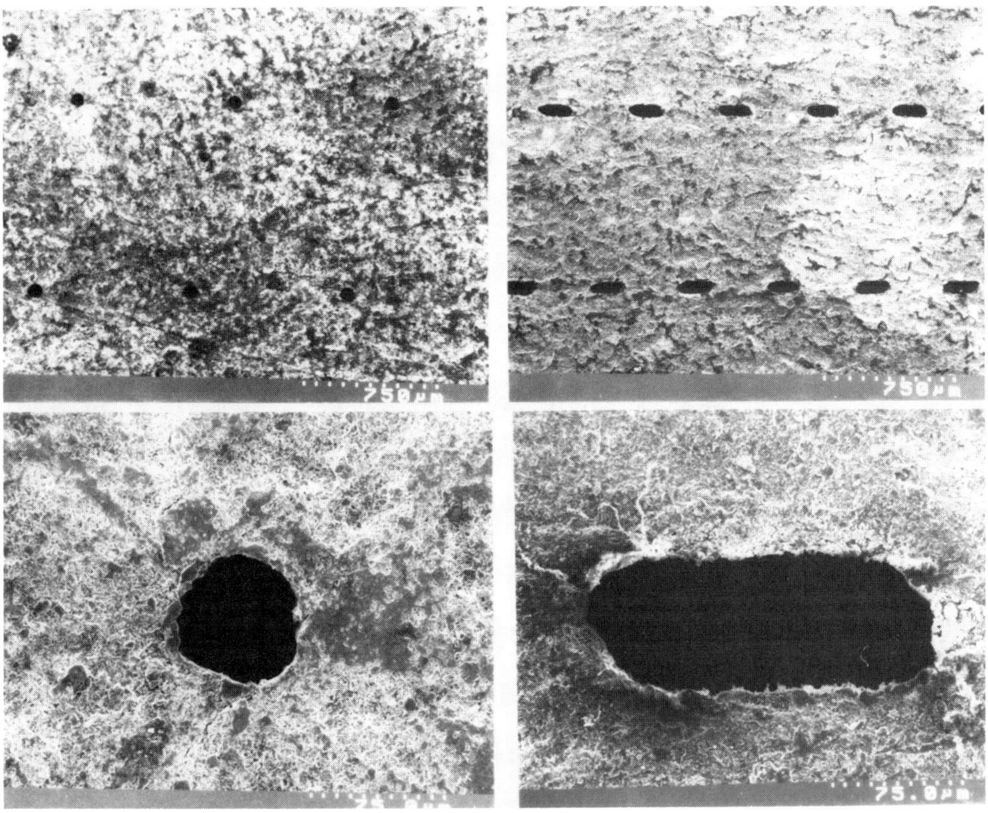

FIG. 7. Scanning electron micrographs of perforated cigarette filter paper. The holes allow air to dilute the tar and nicotine content in the filter. CO_2 laser systems can make several million holes per minute and are used by every major supplier of cigarette paper. (Photograph by C. A. Wong, General Motors Research.)

ponent if necessary by simple changes in the beam parameters.

Baby bottle nipples contain small-sized holes, typically less than 0.5 mm in diameter. The soft rubber is easily deformed and traditional techniques give clogged or nonuniform holes. These problems have been avoided since the mid-1960s by use of a pulsed CO_2 laser system that can drill several nipples per second, each nipple containing one hole for milk and two for air.

An intriguing application is the laser drilling of holes in time-release medicinal capsules. These holes permit the slow, controlled release of the contents once the capsule is ingested. The laser is uniquely suited to making the very precise holes needed in the curved surfaces of the capsules with negligible thermal or mechanical damage.

2.4 Cutting

This differs from laser drilling in that there is externally driven relative motion between the beam and the workpiece. The beam fully penetrates the material and moves parallel to its surface. The label "trimming" is sometimes used for cutting applications specifically involving the removal of flash and burrs from curved parts.

2.4.1 Process. The laser may provide only part, or all, of the energy required for material removal, the remainder being contributed by assist gases or by exothermic reactions at the cutting surface (Steen and Kamalu, 1983; Schuocker, 1987). This defines two broad classes: sublimation cutting, in which material is removed primarily by vaporization, and fusion cutting, in which molten and other material is flushed out by assist gases. The main features of a laser cutting head are shown in Fig. 8.

Sublimation cutting is suitable for materials such as plastics, which have low vaporization temperatures and small latent heats of vaporization. Fusion cutting includes the use of either inert gases, to provide the mechanical force to flush out material, or reactive gases, which additionally promote exothermic or other chemical reactions at the surface.

Assist gases are commonly directed into the kerf by a jet that is coaxial with the laser beam (Fig. 8), though off-axis flows have been used with high-pressure jets. Inert gases reduce unwanted oxidation and are imperative if burning is to be avoided in flammable materials. Furthermore, they cool the cut edges, serve to reduce backspatter, and protect the focusing optics.

Reactive gases can dramatically increase cutting rates. With oxygen, for example, the laser may be required only to raise the temperature high enough to initiate "self-burning." While the cutting speed may increase, this process can become difficult to control. Surface quality may suffer and there may also be periodic striations on the cut edges in some materials (Fig. 9). These occur when thermochemical reactions are significant and result from differences in propagation speeds between the melt front and the laser beam. In some materials, these striations can adversely affect fatigue strength, although it can usually be restored by a light abrasive sanding.

Some steels form high–melting-point oxides: these may increase the formation of dross. Some oxides are very hard and, if left on the cut edges, they may damage other materials that they may contact. Cleaner edges can be obtained with argon or nitrogen.

In many applications, the characteristically narrow cut width (kerf) provided by lasers is an important benefit. The beam quality and mode structure, which affect focusing,

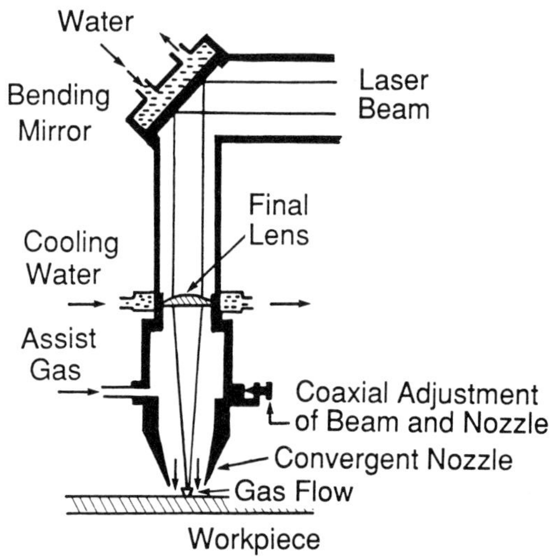

FIG. 8. Primary features of a laser cutting head, including focusing optics and assist-gas provision. Accessories to monitor or control height of the head above the workpiece are not shown.

are therefore significant. High power densities are required and are best provided by low-order–mode beams.

Often, the laser beam is not perpendicular to the material being cut, and the rate at which energy is absorbed will depend on the polarization of the beam and the angle of incidence. This can result in uneven cutting around curves, with implications that are not merely aesthetic. If edges are distorted, material being cut out may remain wedged in the hole rather than freely falling through. The problem is often remedied by use of a circular polarizer in the beam, ensuring uniform cutting independent of direction.

Focal control is critical to good cutting. If the workpiece is not flat, some type of automatic height adjustment may be needed to ensure continuous control of the focal plane position with respect to the workpiece surface. This may be provided by mechanical means, such as rollers supporting the focusing head, or via optical and electrical sensors monitoring the distance to the workpiece (not shown in Fig. 8).

Laser cutting involves complex interactions between the laser beam, the workpiece, and the various gases present. Any material, from cloth and wood to metals and ceramics, can be laser cut although the best values for the laser and gas-assist parameters will vary widely. The cutting speed v should increase with incident laser power P, but decrease with increased material thickness. Figure 10 indicates typical ranges of cutting speeds as a function of laser power normalized by the material thickness. Both CO_2 and Nd:YAG lasers have been used to cut metals well over 10 mm in thickness. The speed will vary with both material and processing parameters: the conditions that produce the highest speeds may not give acceptable cut quality.

The cutting of metals is usually slower than that of nonmetals because the laser coupling efficiency is often poorer and because the heat generated diffuses rapidly away from the interaction zone. High beam quality and high peak power densities are needed for the cutting of highly reflective and thermally conductive metals such as copper and aluminum.

The benefits offered by the laser for cutting purposes overlap those for drilling. These include the diversity of materials that can be treated, the narrow heat-affected zone and localized material damage, the lack of distortion, and the precise control possible even in complex geometries.

Significant cost savings can be achieved in time and material in short-run or prototype work, where the volume does not justify hard tooling or where rapid changes in cutting geometry are desired. The narrow kerf (comparable in width to the focused beam diame-

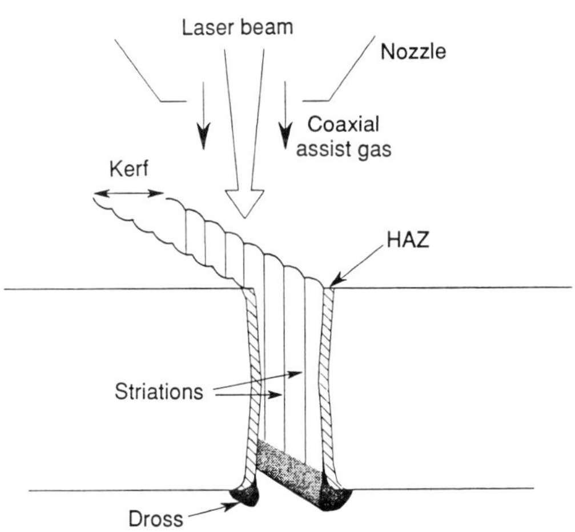

FIG. 9. Schematic of laser cutting, defining the kerf (width of cut) and showing dross on the underside and formation of striations on the walls.

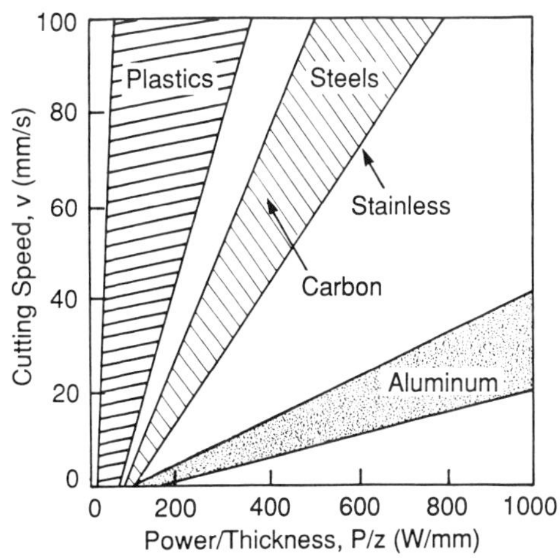

FIG. 10. Cutting speed for diverse materials as a function of power per unit thickness of workpiece. The width of the bands indicates variations due to materials and laser and gas-assist parameters. Cutting speeds used in practice will be dictated by requirements on edge quality.

ter) is much smaller than that achieved in mechanical cutting. Together with the narrow heat-affected zone, this allows close nesting of parts and efficient volume utilization of material. Even the cut-outs may be useful, and there is generally less debris and waste. Multiaxis laser cutters can handle three-dimensional parts, whereas most conventional cutting systems are restricted to cutting flat workpieces.

As in other laser processes, the same laser may be used for diverse materials and different machining patterns, even intricate shapes, by simple variation of the beam parameters or some modifications to the gas controls. The process is easily automated and, at least on thin materials, fast.

Mechanical cutting is generally several times faster than laser cutting (via vaporization), which is limited to about 5 m/s. However, the laser results in far less debris; this may offset its lower speed if waste disposal and health hazard issues are more critical in some applications.

Limitations include the high capital cost and, if large amounts of oxygen assist gas are needed, significant operating costs. Cutting speeds are low for thick samples and edge quality may also be poor. It is difficult to expel molten debris in deep kerfs: when left, it absorbs the incident beam, lowers the cutting efficiency, and degrades edge quality. Waterjet cutting is less versatile than laser cutting but is often preferred for thick materials.

2.4.2 Applications. Cutting applications constitute the main use of industrial machining lasers and range from superalloy aircraft engine parts to wooden toys. They vary widely with the particular advantage afforded by the laser over traditional techniques. The ease with which a laser can cut a material does not depend on its mechanical properties, a feature especially useful in the cutting of tough composites.

The cutting of sheet metal with lasers became significant in the early 1970s. Contour or profile cutting with a CO_2 laser remains very popular, with good economic returns. Thin-walled metal tubing is particularly suited to laser, rather than mechanical, cutting. The laser produces neither burrs nor part distortion.

The aircraft industry uses titanium extensively, despite its high cost, because it is both strong and lightweight. Lasers give much narrower kerfs and heat-affected zones than possible with techniques such as oxyacetylene or plasma arc cutting. Extensive machining had previously been necessary to remove heat-affected material from the edges, whereas little treatment is needed for laser-cut material—a chemical etch suffices for millimeter-thick stock. The materials savings offset the capital cost of the laser. Kilowatt-class CO_2 lasers give adequate speed in reactive gas cutting because titanium oxidizes so readily.

Paper cartons and boxes are stamped by flat or cylindrical plywood dieboards containing metal cutting knives or creasing blades. These are positioned in narrow slots in the board, which must be cut very precisely according to the stamping pattern. A CO_2 laser, operating at several hundred watts, can give taper-free cuts with the necessary accuracy and reproducibility in an automated process that is almost ten times as fast as the previous manual technique. This process, introduced in 1971, was one of the first industrial applications of lasers for cutting.

The speed and flexibility with which laser cutters can be programmed with different patterns has led to their widespread use in prototype and low-volume manufacture, including the manufacture of dies. They are also valuable for "late option" cutting: the laser can be used to customize a product at a late stage in its manufacture. In the automotive industry, for example, options such as roof-rack fixtures, sun roofs, left-hand versus right-hand drive features, air vents, and so on can be inserted with a laser as required. This is much less expensive than having special tooling for each option.

Trimming tools are routinely used in automotive manufacture to remove excess material from deep-drawn parts. Establishing the necessary contour to the required precision can be time consuming and expensive. The final contour can be established very quickly with a laser cutter because of the ease with which it can be programmed to accommodate modifications in the contour. The correct coordinates are then used to build the trim tool.

The lack of mechanical contact and distortion or deformation is of value in the cutting of both hard (metal pipes) and soft (plastic, rubber) materials. Rubber seals of diverse geometries can be cut by a laser without any of the mechanical distortion that would be produced by a knife.

Speed (well over 1 m/s), precision, close nesting, and the self-sealing of edges are some of the benefits provided by the CO_2 laser cutting of fabrics, whether men's suits or automotive components such as mats and seat belts. Different patterns can be easily accommodated by reprogramming of the laser-beam path. There is no thermal damage or charring, and the fraying characteristic of mechanical cuts is eliminated.

Charring may occur in some plastics, composites, and wood, but can be removed in a later process if not reduced to acceptable levels with appropriate assist gases. Wood must invariably be cut with a gas assist, and cutting speeds are strongly affected by the moisture content of the wood. In general, the laser cutting of wood is more costly than mechanical methods.

Glass has a low thermal diffusivity and may crack when laser heated. The problem can be reduced in some glasses (soda-lime and borosilicate) by the use of pulsed lasers or by a preheating process. Low-power lasers are used for cutting by controlled fracture; higher powers and assist gases are used when cutting proceeds via a significant melting phase. Quartz tubes used in the lighting industry are now cut by a laser instead of a diamond saw. There is far less dust and waste. Furthermore, the laser not only gives a much narrower kerf, it also fire-polishes the edges.

2.5 Marking

The number of lasers being used for various marking purposes will probably exceed 10 000 during the 1990s, comparable to the number used for welding and second only to the number used for cutting. The laser is superseding many traditional marking methods, such as silk screening, acid etching, labeling, stamping, and mechanical engraving.

2.5.1 Process. Marking can be achieved in numerous ways, from color change and brightness contrast to deep scoring and ablation of surfaces. Contrast can be enhanced by deliberately coloring components. Use of absorbing paints and dyes also facilitates the marking of materials normally transparent to the laser wavelength. The laser can modify chemical or physical properties of the surface, and the corresponding range of operating parameters is very broad, as indicated by Fig. 2.

Two techniques are common: mask marking (or stenciling) and beam deflection (Willis, 1983; Klimt, 1988). In the former, one or two laser pulses are fired through a reflective mask to create the whole of the desired pattern at once. It is particularly suited to standard or repeated patterns, such as decorative engravings. A focusing lens may be placed between the mask and the workpiece to reduce the image size. A TEA (transversely excited atmospheric) CO_2 laser is commonly used, each pulse delivering about 1 J of energy in about 1 μs. Penetration depths are shallow (1–25 μm) but can be increased by repetition.

The second technique uses computer-controlled galvanometer mirrors to steer the laser beam across the workpiece, tracing out the pattern as it moves. This method is appropriate when the pattern varies rapidly, as in the sequential numbering of parts. Nd:YAG lasers with average powers up to 100 W are popular. They are often Q switched, to give peak powers 1000 times greater than the average value. Penetration depths are usually several times deeper than those from mask marking. Character sizes range from 0.1 to 100 mm, with line thicknesses up to 3 mm. A dot matrix rather than a continuous mode may be used for elongated characters. Rapid dithering of the beam can be used to broaden lines.

Both CO_2 and Nd:YAG lasers are in common use, the choice depending on the absorption of the workpiece, the contrast required, the marking technique, and so on. Excimer lasers are particularly useful if thermal effects must be minimized or where optical absorption of longer-wavelength radiation is poor.

Durable marks of high quality and readability can be made on a wide range of materials. The high speed allows even rapidly moving objects to be marked on the fly. The process is easily automated and computer controlled. It is very flexible: patterns can be very complex and changes can be easily accommodated, whereas changing metal stamps is time consuming.

The finely localized and controlled heat input allows very small items to be marked as well as those susceptible to thermal damage. The lack of mechanical pressure allows delicate components to be marked (a problem with mechanical stamping), and there is no chemical contamination (a problem with inks). Selected areas difficult to reach by other means can be marked by lasers, provided there is line of sight. Operating costs are low compared to those of the consumables in ink-jet marking or adhesive labeling. The marks are more durable than those in either process, though the ease of marking and the legibility depend strongly on the material.

The primary disadvantage, as in most laser applications, is that of the high capital investment cost compared to that of some of the alternative methods (ink jet, stamping, or etching, for example).

2.5.2 Applications. Laser-marking applications span a wide range of industries, from transportation and electronics to cosmetics and food, and from vehicle identification numbers (VINs) on automobiles to labels on beverage containers. Purposes include batch or date coding, inventory control, quality assurance, safety information, government requirements, security (theft protection), personal identification, logos, and decoration.

The VIN plate in all General Motors vehicles made since 1990 is laser marked. The aluminum plate is first blackened by anodization to give good final contrast. A 100-W Nd:YAG laser then inscribes a 17-character 3-of-9 bar code and duplicate alphanumeric data, detailing serial numbers, emission standards, and manufacture. Automotive marking is rapidly expanding, in both off-line and on-line applications. One of the first industrial applications of excimer lasers was the marking of the heater symbol on the heatable outside mirrors offered on some automobiles. It is done by laser evaporation of the metallic surface. The ArF excimer laser used gives precise control and causes negligible damage to the glass, unlike the former sandblasting technique. The quality is higher and the amount of scrap is greatly reduced.

Serial numbers are engraved on handguns by a *Q*-switched Nd:YAG laser. The assembled gun cannot withstand the mechanical pressure of steel stamping, and the laser marks are both durable and difficult to duplicate. Precious stones can be marked with identification numbers as a security measure. The numbers are barely discernible to the naked eye but cannot be removed without recutting (and loss in value) of the gem.

Fragile semiconductor components, such as capacitors or silicon substrates, can be marked with manufacturing data and material properties. Excimer lasers are often used for silicon wafers. Nd:YAG lasers both hermetically weld and mark the titanium and stainless-steel cases of heart pacemakers. The very low heat input is localized and does not affect the calibration. Similarly, a pulsed TEA CO_2 laser is used to mark the sensitive thermal sensors used in automotive-emission controls. The laser has a higher capital cost, but lower running costs, than alternative marking techniques. Furthermore, the process is fully automated, giving permanent marks of high quality. Metal stamping would destroy the sensor calibration, labels would become detached in the fuel atmosphere near the engine, wax dots would melt, and electrochemical etching would be slow and difficult to automate.

3. MISCELLANEOUS APPLICATIONS

3.1 Electronic Circuits

The precisely positioned, finely focused input of the laser, together with its low heat input, has opened up a variety of applications in the electronics industry. A few of the more common ones are included here.

3.1.1 Annealing. Ion implantation (see ION-BEAM MODIFICATION OF MATERIALS) is a popular means of creating desired dopant profiles in semiconductors. However, the process damages the crystal structure, often rendering a 100-nm-thick region amorphous. Thermal annealing is then used to provide the needed electrical properties.

Lasers have been used since 1974 for annealing out damage. They provide much more localized heating than traditional oven heating and can be spatially and temporally controlled so as to anneal out damage without significantly changing the dopant profile. This is critical in integrated circuits (ICs). A rapidly scanned laser beam heats only the surface region and does not modify the dopant profile. Occasionally, a pulsed annealing technique is used, in which short laser pulses (less than 1 μs) are allowed to produce temporary surface melting.

Laser annealing has found diverse uses, from improving the conductivity of polycrystalline silicon to modifying the magnetic properties of garnet films and the superconducting properties of other metallic films (Lugomer, 1990).

3.1.2 Trimming. The electronics industry employs lasers to modify the electrical values of components such as resistors in thin- and thick-film circuits. These consist of material such as tantalum nitride (Ta_2N) on an insulating substrate such as alumina (Al_2O_3). The resistors must have very precise values, but the fabrication process commonly used (screen printing) may give variations of up to 15%.

Lasers were first used to counter this problem in 1967, by trimming the resistance to the required value. It is now standard practice to print all conductor paths wider than required (thus having resistances that are too low) and then use a laser to perform final adjustments by trimming off excess material.

Q-switched Nd:YAG lasers of only a few watts average power are frequently used for resistor trimming. The process can be monitored while the component is active in the circuit, and desired resistance values can be achieved with an accuracy of better than 0.1% within a few seconds. The low heat input prevents thermal damage to adjacent components. The precise positioning of the laser beam allows trimming to be done at a late stage, even when circuits are closely nested together. Older techniques, such as air-blasting the edges of the resistor sheet with an abrasive, were practical only if components were not too densely packed.

The resistance also depends on the chemistry of the material, such as its surface oxidation. Currently, there is increased interest in the ability of the laser to produce such chemical changes.

3.1.3 Scribing. In integrated-circuit manufacture, many identical circuit elements may be fabricated on a single large substrate. This must then be broken up to furnish the discrete elements but is often brittle, being made of materials like glass, ceramic, or silicon. These are very susceptible to thermal shock, but the laser may be operated in a pulsed rather than continuous manner in order to minimize the heat input. The laser makes either discrete or overlapping blind holes in the substrate; they form snap lines that allow well-controlled separation. The holes penetrate only about a quarter of the way into the substrate; there is negligible thermal shock and no mechanical damage.

Nd:YAG lasers are preferred for silicon substrates, CO_2 lasers for alumina. The scribing of ceramic substrates by a CO_2 laser may have been its first production application, dating back to 1967. Average powers are 10–100 W, with scribing speeds of 20–250 mm/s. This high speed, coupled with no tool wear and greater precision than given by earlier methods, has made laser scribing the standard technique in the industry. It has supplanted diamond scribing, which is expensive because the continual wear ultimately results in imprecise machining, poor-quality edges, and mechanical damage with consequent scrapping of elements.

The laser can also perform other needed micromachining operations on ceramic substrates, such as making cooling channels or assembly slots.

3.1.4 Soldering. In this variant of welding, the laser supplies the heat needed to melt solder that is already positioned on the components of interest. Where wires are insulated with plastics, a CO_2 laser can automatically burn off or melt away this coating prior to the soldering process. Power levels used in soldering, whether with CO_2 or Nd:YAG lasers, are typically in the 10–100-W range.

Electronic circuits contain fragile parts, and a low-power laser (10–20 W) restricts the heating effects to the areas needed. Traditional techniques (including infrared and vapor-phase soldering) raise the temperature of the component to that of the solder and would produce thermal damage or distortion. A CO_2 laser beam can be readily focused to spot sizes less than those of the solder pads on electronic components and can also be directed to areas normally difficult to reach. This is increasingly important as circuits continue to be miniaturized. Components must be accurately positioned, and precautions may be necessary to avoid damage from stray reflections of the laser beam.

Laser-soldering systems are expensive and give low production rates compared to wave soldering or continuous vapor-phase soldering. However, the laser offers unusually high reliability and the ability to provide access to difficult locations and intricate geometries. Quality is often superior to that in other techniques: the rapid heating and cooling rates restrict the time available for undesirable intermetallic layers to form.

3.1.5 Repairs. Defects during manufacture can often be remedied by a laser, thereby reducing waste and increasing productivity. The laser may be needed to repair broken connections or to trim off surplus material and burrs.

For example, a pulsed laser has been used to repair defective terminal connections in an electronic tube. The glass wall is transparent to the beam and the tubes can thus be salvaged rather than scrapped.

Redundancy is often built into modern integrated circuits. A pulsed Nd:YAG laser can easily vaporize defective components, without damaging adjacent parts or the substrate, and redirect channels.

Metallic masks can be repaired by selectively heating appropriate parts of the pattern (Tison, 1988). For example, various defects can occur in the fabrication of IC photomasks used for microlithography, such as too much or too little chromium deposited onto the glass. Excess material can be removed by a *Q*-switched frequency-doubled Nd:YAG laser operation at a wavelength of 0.532 μm, enabling smaller regions to be repaired than possible at 1.06 μm. Thin areas or pinholes can be repaired by vapor deposition of chromium with a cw argon-ion laser or a frequency-quadrupled Nd:YAG laser.

3.1.6 Deposition. Materials can be deposited, often as thin films, by the laser-initiated pyrolytic or photolytic decomposition of gaseous media. Many variants of chemical vapor deposition (CVD) [see CVD (CHEMICAL VAPOR DEPOSITION)] are possible, including use of the laser to ablate a solid source material or to heat a substrate. Novel materials can be produced because the laser can heat substrates to temperatures higher than usual and because nonequilibrium phases can occur during the rapid heating and cooling cycle. There are many potential applications, but many of the phenomena are only poorly understood (Boyd, 1987; Lugomer, 1990).

Primary application is to integrated circuits because of the high spatial resolution and control (although the electronics industry is not the only user of laser-assisted CVD). For example, thin gold films can be deposited over transparent defects in photomasks. Direct maskless writing applications are proliferating. Excimer lasers are of increasing interest in the fabrication of films, from high-temperature superconductors to polymers for waveguides. Whatever laser is used for film deposition, however, the localization of the effect makes the process inefficient for large-area devices.

3.1.7 Oxidation and Nitridation. The electrical characteristics of many metal and semiconductor films can be drastically changed by oxides or other surface layers. For silicon systems, the oxide passivates the surface and controls interfacial diffusion. Similarly, nitrides often provide good insulating properties and furthermore may have excellent microhardness. Such films may be grown by heating a wafer in an appropriate atmosphere.

Conventional furnace treatment requires raising the temperature of the whole wafer for a prolonged period. Lasers provide very precise localization of heating, not only for selected areas but also for controlled depths. While silicon treatments have been widely explored, many other systems are also under investigation, and several types of laser are being used (Boyd, 1987).

3.2 Prototype Manufacture

Several "desktop manufacturing" processes incorporate lasers. The part to be made is designed on a three-dimensional (3D) computer-aided design (CAD) system and then mathematically sliced into thin 2D layers. The geometry of each slice is then used to control a laser beam that traces out and builds up the model from special materials. Prototypes which formerly took weeks or months to make can now be built in a few hours. This allows more opportunity to evaluate designs and greatly reduces the lead time between concept and final production.

Savings can be effected both in time and in the costs of prototype tools. Engineering designs can be adequately proofed before changes are made to hard tooling. The new processes also avoid problems associated with part clamping and distortion. The surface may be less smooth than that produced by conventional machining, and a further finishing operation may be needed in some applications. However, the benefits generally far outweigh any such disadvantages.

Stereolithography is the best known and established of the laser-based prototyping techniques. It was first displayed publicly in 1987 and was already being implemented by industry in January 1988. There are several variants. In the most common, the CAD drawing is mathematically sliced into thin parallel layers, from 0.05 to 0.5 mm thick. The coordinates of the base layer are fed to a laser whose beam traces out the area on the surface of a photocurable liquid polymer contained in a vat. An elevator platform is positioned just below this surface. The radiation causes rapid curing of the liquid, and a solid layer is formed on top of the platform. This is then lowered to allow fresh liquid to flow over the newly formed solid, and the laser traces out the area of the next layer. The process is repeated until the full model has been obtained.

The platform is then raised out of the vat, any liquid still present on the part is removed, and a final curing occurs either by intense ultraviolet irradiation or by oven heating. Tolerances are within ± 0.1 mm. There may be further finishing processes, such as sanding and painting, but the basic prototype is completed in hours rather than days.

The limitations imposed by the early photocurable polymers and uv lasers have now been overcome. Recently developed materials require less energy to polymerize and no subsequent photocuring is required. This avoids earlier problems of shrinkage and improves dimensional accuracy. The low-powered (10-mW) He–Cd lasers (325 nm wavelength) used formerly have been supplanted by argon-ion lasers with higher powers (up to several watts), helping to increase processing speed. Part size is still limited to a maximum dimension of less than 1 m; larger components can be handled by being broken down into smaller constituents.

Applications range from the aerospace and automotive worlds to the computer and electronic industries, and from perfume bottles to medical prosthetics. In one application, the making of prototype turbine blades for jet engines, the manufacturing time has been reduced by over 95%. Automotive engineers estimate that up to 20% of the lead time between initial design and final production of vehicles can be saved.

Several modifications of the process are being investigated, including photochemical machining in which a highly focused laser beam can produce spatially selective photopolymerization. The method has inherently more dimensional accuracy than one based on layer buildup.

Other techniques employing lasers in prototype manufacture include laminated-object manufacturing (LOM) and selective laser sintering (SLS). Both use the sliced-layer concept. In LOM, the laser (25-W CO_2, for example) cuts out the layers from a metal or a paper sheet with an adhesive backing. These layers are variously brazed or pressed together under heat to give the full 3D model. Very complex parts can be made, including those with internal voids and blind holes. Further, since photocuring is not involved, the choice of laser is not as restricted as in stereolithography. Furthermore, the process is faster and larger parts can be accommodated.

SLS can currently be applied to a wider range of materials than can either stereolithography or LOM. The raw material is a powder, but can be plastic [such as ABS (acrylonitrile-butadiene-styrene), PVC (polyvinyl chloride), or nylon], ceramic, or metallic. It has usually been preheated, and the additional heat from the laser irradiation increases its temperature sufficiently to sinter it. The particles adhere because the applied heat lowers their viscosity sufficiently to overcome their surface tension: melting is not required. Unsintered powder remains to support the part as it is built up layer by layer, at a rate of about 0.5 mm/min.

The process requires fewer steps and is faster than stereolithography but the latter is better able to handle extremely fine parts. Some shrinking occurs during sintering, generating internal stresses, but can be countered by appropriate design measures.

3.3 Other Processes

Many other laser-processing applications exist, as in the following examples. Some, like 3D and laser-assisted machining, are still under development.

3.3.1 Ablation. In many laser processes, much of the material is removed by vaporization. Where, as is often the case with excimer laser irradiation, the process is not dominated by thermal energy, it is sometimes termed photoablation. This typically requires pulsed lasers with high peak intensities (10^6 W/cm^2 or greater) but produces very clean boundaries with negligible heat-affected zones.

The laser offers even more precise depth control than does chemical etching but produces no chemical contamination. It can very selectively strip off only the layer desired from a multiply layered part, such as a coating from a glass substrate or the insulation from a coaxial cable. Mechanical stripping techniques might easily damage underlying material. Provided adjacent layers differ enough in their optical and thermal properties, the laser can remove one layer while not affecting the next. For coaxial cables, the radiation from a CO_2 laser is efficiently absorbed by the insulating layers but highly reflected by the metal wire.

The electronics industry uses lasers in photoresist etching and in a host of other selective material removal applications. The excimer laser is especially useful in applications requiring high spatial resolution and no damage to adjacent material.

Lasers have been found useful in various surface-preparation applications. Good quality control in painting, or in the adhesion of two materials to be bonded, requires surface cleanliness and uniformity; these can be provided more efficiently by a laser than by alternative techniques.

High-power pulsed CO_2 lasers are used to strip paint and other surface coatings from aircraft and naval vessels. Much less waste is produced than by abrasive blasting or chemical stripping, and the laser avoids the problems introduced by the use of solvents and corrosive materials.

Laser ablation is used for the dynamic balancing of gyroscopes and other rotating components. A sensor detects any imbalance as the part rotates and triggers the firing of a Nd:YAG laser pulse when the excess material lies beneath the beam. The laser has reduced the time required to balance a gyroscope from several hours to less than a minute.

Greeting cards often exhibit filigree patterns that result from selective removal of paper by a high-power CO_2 laser directed through a photoetched template. The beam is focused onto the paper several meters from the template. The paper issues from a modified printing press and moves at high speed, together with the template, across the path of the laser beam. The process is not as fast as die-cutting but leaves no frayed or curved edges and can give high resolution. Close examination of the edges of the pattern occasionally reveals faint charring, indicating that the process was indeed laser ablation rather than mechanical cutting.

Shock hardening can occur when the surfaces of certain materials are rapidly vaporized. High-intensity ($\geqslant 10^9$ W/cm^2), pulsed laser irradiation can eject material violently, producing considerable thermal shock and recoil momentum at the surface. A high-pressure stress wave propagates into the bulk, generating dislocations and work hardening the metal (Clauer and Fairand, 1979). Sometimes an ablative coating is put on the surface to be hardened; its rapid removal by vaporization creates the required impulse.

3.3.2 Grain Refining. Cast alloys often contain undesirable inclusions, whether chemical impurities or pores. Those near the surface can degrade properties such as fatigue strength and resistance to corrosion or wear. Polycrystalline semiconductors, such as chemically vapor-deposited silicon, often exhibit low electrical conductivity because of electron scattering at grain boundaries. Laser scanning can result in the refining of grains or growth in their size and promote the desired properties.

High-power cw Nd:YAG lasers have been used to refine the magnetic-domain structure in steel used for electrical transformer cores. The high–magnetic-permeability steel contains primarily iron with about 3% silicon and has large grains. In operation, motion of domain walls in the applied alternating electric field results in significant energy losses. The laser thermally shocks the microstructure and thereby controls the magnetic domain-wall boundaries. Electrical energy losses have been reduced by up to 10%.

3.3.3 Curing. Ultraviolet light is used to cure paints and polymers in a number of industries. The process is slow, lasting days in some cases, but can be accelerated by means such as ovens. A short-wavelength laser, such as the excimer, can stimulate a fast chemical reaction in the polymer molecules, leading to shorter curing times.

Generally, the applications best suited to the laser are those that require only small areas to be treated, such as microprinting and microelectronics.

3.3.4 3D Laser Machining. Two or more laser beams can be directed onto a workpiece so that each creates an independent groove. When the grooves intersect, their geometry determines how much material falls out. Unlike cuts, the grooves do not fully penetrate the workpiece and debris can be flushed out only via the front face (Chryssolouris, 1991).

3.3.5 Laser-Assisted Machining. In most processes the laser is used as a replacement tool. It can be used to bend metals without mechanical contact, for example. In some new applications, however, lasers are used to supplement rather than to supplant traditional tools (Copley *et al.*, 1983). The laser heats the workpiece just ahead of the tool, making it easier to machine. This is of interest in the cutting of superalloys, ceramics, and other hard substances. The required tool force is reduced and so, consequently, is tool wear.

In "chip management" the laser is used to break up the long continuous spirals of thin material produced in turning operations. This simplifies the management of materials such as titanium alloys or nickel-based superalloys that are difficult to cut.

These applications of the laser, whether before, during or immediately after conventional machining, are still being developed and are less well known than welding or marking, for example.

4. NEW DEVELOPMENTS

There is little sign that the pace of laser-processing developments is lessening. Industrial lasers of all types (excimer, solid-state, CO_2) continue to be improved. Greater attention is being given to refinements such as pulse shaping and beam spatial control. There is interest both in increasing output powers and in producing more compact and reliable systems.

The role of excimer lasers, because of their short-wavelength radiation, is expected to increase significantly in applications ranging from marking to microelectronics. Their potential is being enlarged by intensive efforts to increase their output powers: several hundred watts have been obtained from XeCl (308 nm) and kilowatt levels are expected in the 1990s. The major impediments to their use will continue to be their high capital cost and maintenance requirements (compared to more common lasers).

Nd:YAG lasers with high average output powers often suffer from poor beam quality. This can be improved by means such as aperturing the beam (but thereby losing power) or changing the lasant geometry (from a rod to a slab, for example). There is also interest in the use of diode-laser arrays, rather than flash lamps, to pump Nd:YAG lasers. Flash lamps must usually be replaced after a few hundred hours of operation or a few million pulses, a significant item in the operating costs of these lasers. They also have a broad spectral output. Diode lasers both last much longer than flash lamps and have a spectral emission that is much more efficiently matched to the absorption lines of the neodymium. The capital cost of diode-laser arrays presently limits commercial interest but is expected to decrease dramatically.

Gas lasers being further developed include kW-level iodine lasers (which emit at a wavelength of 1.315 μm) and multi-kW carbon monoxide lasers ($\lambda = 5–7\ \mu$m). Even CO_2 laser technology continues to be improved: alternative types of excitation (rf and microwave) are being used; output powers are increasing even while the size of the laser head is decreasing; and the control of beam power and quality is becoming more sophisticated. Very rugged, nearly maintenance-free, compact, sealed lasers offering near-kW powers are becoming available.

Beam-delivery systems, both mirror and fiber, are becoming more efficient and versatile. Fewer mirrors have been used without loss of degrees of freedom: this gives higher optical power transmission and better stability and eases alignment procedures. Fiber systems are handling higher powers and performing a wide variety of beam-combining and -sharing operations.

Interest continues in real-time, *in situ* monitoring. The simplest type of process monitoring is to ensure that the operating parameters (laser power, mode, gas pressures, etc.) are maintained within specific ranges. It is usually more difficult to conduct real-time *in situ* monitoring of the workpiece itself, but several diagnostic techniques, notably optical, thermal and acoustic, are being assessed. They include visual appearance of the irradiated spot, the brightness and spectral distribution of the plasma, temperatures at the workpiece, noise in various frequency ranges, and so on. While this information helps and can be used in closed-loop control systems, it seldom distinguishes adequately between acceptable and unacceptable conditions, and post-process evaluations may still be needed.

Both the diversity and quality of laser applications are expected to increase. Surface treatment may expand but will probably continue to lag behind cutting, marking, and welding. Improvements in edge quality, and increases in the thicknesses that can be cut, combine with the flexibility of the laser in ensuring that cutting will remain the dominant industrial process. Although welding is already a popular application, it should become even more widespread as the potential offered by diverse joint configurations becomes realized.

5. SUMMARY

The ability of the laser to machine and control the properties of materials very precisely has opened up numerous applications. In some cases, lasers are used as direct replacement tools for existing technologies. In others, their use may allow redesign of a whole manufacturing system, with savings in materials and time by the elimination of several operations. Progress continues both in the design of industrial lasers and in the range and sophistication of processing operations.

The high capital costs of lasers limits their adoption for applications where traditional methods exist. Nonetheless, the use of lasers for industrial applications is steadily increasing; this review has included only the better known uses. The number of installed systems worldwide is expected to increase at an annual rate near 10% for the rest of the century, despite economic slowdowns.

A laser is not always the best tool for an application, even though its use is technically feasible. However, production floor experience has demonstrated that lasers can be reliable processing tools, that they can provide unique advantages in many applications, and that they can be economically viable. The laser is proving itself to be an effective tool, not an exotic toy.

GLOSSARY

Assist Gas: A gas or mixture of gases, other than those used for the laser operation itself, that are employed near the workpiece to improve processing performance.

Beam Diameter: The effective width of a laser beam, often defined on the basis of an area that encompasses a given fraction of the total beam power.

Beam Divergence θ: A measure of the angular spreading of a laser beam as it propagates.

CO_2 Laser: Gas laser in which carbon dioxide is the medium responsible for the 10-μm-wavelength radiation emitted.

Continuous Wave (cw): Continuous rather than pulsed emission from a laser.

Coupling Coefficient γ: The fraction of incident laser energy that is ultimately transferred to the workpiece and hence effective for processing.

Heat-Affected Zone (HAZ): Region adjacent to that processed that is thermally affected by the laser irradiation.

Irradiance: The optical flux per unit area falling on a surface (often but ambiguously termed "intensity").

Kerf: Width of the material removed by a cutting process.

Keyhole: Elongated hole produced in deep-penetration welding, when a dynamically stable cavity is formed amid molten material.

Mode: A standing-wave pattern generated by the radiation field in a resonator cavity.

Nd Laser: Solid-state laser, comprising a glass or YAG (yttrium aluminum garnet) rod or slab that contains a few percent of neodymium ions (Nd^{3+}).

Q Switch: A device that controls laser-cavity oscillation until a large amount of energy is available for release.

Recast: Material that, having been melted, is not removed entirely but resolidifies as a layer on the walls of a hole or cut.

Trepanning: Removal of material by the cutting out of a circular slug, usually by an off-axis motion of the laser beam with respect to the workpiece.

List of Works Cited

Belforte, D., Levitt, M. (Eds.) (1986–1990), *The Industrial Laser Annual Handbook*, Tulsa, OK: PennWell.

Boyd, I. W. (1987), *Laser Processing of Thin Films and Microstructures*, Berlin: Springer.

Breinan, E. M., Kear, B. H. (1983), in: M. Bass (Ed.), *Laser Materials Processing*, Amsterdam: North-Holland, Chap. 5.

Bromberg, J. L. (1991), *The Laser in America*, Cambridge, MA: MIT Press.

Chryssolouris, G. (1991), *Laser Machining: Theory and Practice*, New York: Springer.

Clauer, A. H., Fairand, B. P. (1979), in: E. A. Metzbower (Ed.), *Applications of Lasers in Materials Processing*, Metals Park, OH: American Society for Metals, pp. 291–315.

Cline, H. E., Anthony, T. R. (1977), *J. Appl. Phys.* **48**, 3895–3900.

Copley, S., Bass, M., Jau, B., Wallace, R. (1983), in: M. Bass (Ed.), *Laser Materials Processing*, Amsterdam: North-Holland, Chap. 6.

Dowden, J., Postacioglu, N., Davis, M., Kapadia, P. (1987), *J. Phys. D* **20**, 36–44.

Draper, C. W., Poate, J. M. (1985), *Internatl. Metals Rev.* **30**, 85–108.

Duley, W. W. (1976), *CO_2 Lasers: Effects and Applications*, New York: Academic.

Duley, W. W. (1983), *Laser Processing and Analysis of Materials*, New York: Plenum.

Ion, J. C. (1992), in: D. Belforte, M. Levitt (Eds.), *The Industrial Laser Handbook 1992-1993 Edition*, New York: Springer, pp. 39–47.

Kapadia, P., Ducharme, R., Dowden, J. (1992), in: *Proceedings of ICALEO '91* (International Congress on Applications of Lasers and Electro-Optics), Orlando, FL: Laser Institute of America, Vol. 74, pp. 53–63.

Klemens, P. G. (1976), *J. Appl. Phys.* **47**, 2165–2174.

Klimt, B. H. (1988), *Lasers Optron.* **7**, 61–67.

Li, L. J., Mazumder, J. (1985), in: K. Mukherjee, J. Mazumder (Eds.), *Laser Processing of Materials*, Warrendale, PA: Metallurgical Society of AIME, pp. 35–50.

Lugomer, S. (1990), *Laser Technology: Laser Driven Processes*, Englewood Cliffs, NJ: Prentice-Hall.

Mazumder, J. (1983), in: M. Bass (Ed.), *Laser Materials Processing*, Amsterdam: North-Holland, Chap. 3.

Mazumder, J. (1991), *Opt. Eng.* **30**, 1208–1219.

Mazumder, J., Steen, W. M. (1980), *J. Appl. Phys.* **51**, 941–947.

Minardi, A., Bishop, P. J. (1989), *Heat Transfer Manu. Mater. Process. HTD* **113**, 39–44.

Modest, M. F., Abakians, H. (1986), *Trans. ASME J. Heat Transfer*, **108**, 597–601.

Paek, U.-C., Gagliano, F. P. (1972), *IEEE J. Quantum Electron.* **QE-8**, 112–119.

Ready, J. F. (1971), *Effects of High-Power Laser Radiation*, New York: Academic.

Ready, J. F. (1978), *Industrial Applications of Lasers*, New York: Academic.

Rockwell, R. James, Jr. (1987), in: D. Belforte, M. Levitt (Eds.), *The Industrial Laser Annual Handbook*, Tulsa, OK: PennWell, p. 131.

Rykalin, N., Uglov, A., Kokora, A. (1979), *Laser Machining and Welding*, Oxford: Pergamon.

Schuocker, D. (1987), in: D. Belforte, M. Levitt (Eds.), *The Industrial Laser Annual Handbook*, Tulsa, OK: PennWell, pp. 65–79.

Schuocker, D. (1992), in: D. Belforte, M. Levitt (Eds.), *The Industrial Laser Handbook 1992–1993 Edition*, New York: Springer, pp. 67–73.

Shkarofsky, I. P. (1975), *RCA Rev.* **36**, 336–368.

Steen, W. M. (1985), *Metals Mater.* **1**, 730–736.

Steen, W. M., Kamalu, J. N. (1983), in: M. Bass (Ed.), *Laser Materials Processing*, Amsterdam: North-Holland, Chap. 2.

Swift-Hook, D. T., Gick, A. E. F. (1973), *Welding J.* **52**, 492S–499S.

Tison, J. K. (1988), *Laser Focus/Electro-Opt.* **24** (5), 88–94.

von Allmen, M. (1987), *Laser-Beam Interactions with Materials*, Berlin: Springer.

Willis, J. B. (1983), in: M. F. Kimmitt (Ed.), *Proceedings of the 1st International Conference on Lasers in Manufacturing (LIM)*, Kempston, Bedford, U. K.: IFS and North-Holland, pp. 53–62.

Znotins, T. A., Poulin, D., Reid, J. (1987), *Laser Focus* **23** (5), 54–70.

Further Reading

Bass, M. (1983), *Laser Materials Processing*, Amsterdam: North-Holland. A collection of reviews, with practical examples, of laser manufacturing processes.

Belforte, D., Levitt, M. (Eds.) (1986–1990), *The Industrial Laser Annual Handbook*, Tulsa, OK: PennWell, and (1992–1993), *The Industrial Laser Handbook*, New York: Springer. A useful resource on many aspects of laser processing.

Charschan, S. S. (Ed.) (1972), *Lasers in Industry*, Orlando, FL: Laser Institute of America. An early, but extensive and still useful, treatment of a wide variety of laser applications, including machining.

Chryssolouris, G. (1991), *Laser Machining: Theory and Practice*, New York: Springer. A graduate-level text, succinctly covering much material.

Dawes, C. (1992), *Laser Welding*, Cambridge, UK: Abington Publishing. An especially practical guide.

Duley, W. W. (1976), *CO_2 Lasers: Effects and Applications*, New York: Academic. A comprehensive account, with emphasis on the physics involved.

Lugomer, S. (1990), *Laser Technology: Laser Driven Processes*, Englewood Cliffs, NJ: Prentice Hall. A graduate-level review of diverse laser material processing, emphasizing the science involved.

Luxon, J. T., Parker, D. E., Plotkowski, P. D. (1987), *Lasers in Manufacturing: An Introduction to the Technology*, Kempton, Bedford, U.K.: IFS and Springer. A good nonspecialist introduction to both low- and high-power applications.

Ready, J. F. (1978), *Industrial Applications of Lasers*, New York: Academic. About 20% of this wide-ranging book is devoted to machining applications.

Steen, W. M. (1992), *Laser Material Processing*, London: Springer. A very readable but authoritative account that includes both principles and numerous practical applications.

LASERS, SEMICONDUCTOR

NAGAATSU OGASAWARA, *Department of Electronics Engineering, University of Electro-Communications, Tokyo, Japan*

	Introduction	425
1.	**History**	426
2.	**Structures, Materials, and Operation Principles**	427
2.1	Double Heterostructure	427
2.2	Fabry-Pérot Cavity	429
2.3	Lasing Threshold Condition	429
2.4	Stripe Geometry	429
2.5	Materials and Emission Wavelengths	430
2.5.1	III-V Compounds	430
2.5.2	IV-VI Compounds	431
2.5.3	II-VI Compounds	431
3.	**Fundamental Characteristics**	431
3.1	Light-Current Characteristics	431
3.2	Beam Profile and Polarization	432
3.3	Spectral Characteristics	434
3.4	Dynamic Characteristics	435
4.	**New Structures and Functions**	437
4.1	Quantum-Well Lasers	437
4.2	Distributed-Feedback and Distributed Bragg-Reflector Lasers	438
4.3	Semiconductor-Laser Arrays	439
4.4	Surface-Emitting Lasers	439
5.	**Summary**	440
	Glossary	440
	Works Cited	441
	Further Reading	441

INTRODUCTION

Semiconductor lasers have now grown to be key components in modern optoelectronics technology, most notably as light sources in fiber-optic telecommunications systems and optical disk systems. They have found an increasing number of applications ranging from instruments such as bar-code scanners and computer printers to solid-state laser pumps and measuring and sensoring equipment for engineering use.

Semiconductor lasers have the following features that distinguish themselves from other lasers.

1. Compactness: The typical size of a laser chip is $300 \times 200 \times 100\ \mu m^3$. The small chip contains all the ingredients of a laser structure: a resonator, a waveguide, an active medium, and a *p-n* junction to pump the active medium.
2. High efficiency: Semiconductor lasers can be driven by low electrical power [(several tens of mA)$\times$(1–2 V)]. The efficiency of converting electrical power into optical power is several tens of percent which should be compared with that of 0.1% or less for gas lasers.
3. Capability for high-speed direct modulation: Light output can be modulated at frequencies of several GHz or more simply by modulating the pumping current.
4. Wide emission spectrum: The emission wavelength is determined by the band-gap energy of the active layer material. The red–near-infrared region is covered by III-V compounds. The infrared region (3–30 μm) is covered by IV-VI compounds.
5. High reliability: Semiconductor lasers have become highly reliable devices because of the remarkable improvement in crystal quality, although rapid degradation was a great problem at the early stage of their development.
6. Sensitivity to temperature: The threshold current and the lasing wavelength strongly depend on temperature. This is a disadvantage for most applications. On the other have, wavelength tunability is useful in spectroscopic research.

This article is intended to serve as a useful reference for those who wish to grasp the

1-56081-067-X/94/$5.00+.50

basic concepts of semiconductor lasers and utilize them in engineering and research activities. Following this Introduction is a brief history of semiconductor lasers. Next, the operation principles of semiconductor lasers are described in relation to their structures and materials. The third section presents a general view of fundamental operation characteristics covering the static, spectral, and dynamic aspects. The fourth section gives the current state of the art in semiconductor laser technology by reviewing up-to-date devices incorporating new structures and new functions. Because of the limited space, citation of published papers has been kept to a minimum. Several books listed in the Further Reading section can be consulted both for further study and for access to the relevant published papers.

1. HISTORY

The advent of semiconductor injection lasers dates back to 1962, only two years later than the first achievement of laser action in ruby, when stimulated emission from forward-biased GaAs diodes was demonstrated (Hall *et al.*, 1962; Nathan *et al.*, 1962; Quist *et al.*, 1962). The stimulated emission resulted from the radiative recombination of electrons and holes in the direct-gap semiconductor GaAs. The electrons and holes were injected by forward bias into the depletion region in the vicinity of the *p-n* junction. Optical feedback was provided by polished facets perpendicular to the junction plane. The demonstration of GaAs lasers prompted the exploration of many other III-V and IV-VI compound semiconductors. Unfortunately these early homojunction devices, consisting of a single semiconductor, were not considered for serious applications since continuous operation at room temperature was not feasible because of their high threshold current densities for lasing (several tens of kA cm^{-2}).

A breakthrough was brought about in 1970 by employing a double heterostructure grown by liquid-phase epitaxy (Alferov *et al.*, 1970; Hayashi *et al.*, 1970). In the double heterostructure, stimulated emission occurred only within a thin active layer of GaAs sandwiched between *p*- and *n*-doped AlGaAs layers that have a wider band gap. The threshold current density was dramatically reduced to several kA cm^{-2} or less, and continuous operation at room temperature was eventually accomplished. The invention of the double heterostructure was obviously the most important step in the history of the development of semiconductor lasers toward practical utility.

There still remained, however, a troublesome problem of degradation; the reliability of the early double-heterostructure devices was very poor and many of them stopped their continuous operation within a few or several tens of minutes. Fortunately, intensive investigations of the degradation mechanism led to the identification of the causes of major failure: the recombination-enhanced growth of dislocations, and facet erosion (Casey and Panish, 1978, Part B, Chap. 8). These studies have revealed that a long operating life is obtained by eliminating dislocation generation during material growth and device processing and by coating the mirror facets with dielectric thin films.

Another task undertaken in parallel with the improvement of reliability was the stabilization of the optical mode in the lateral transverse direction (parallel to the junction plane). It had been recognized by the late 1970s that the lateral mode instability often observed at higher output powers was detrimental to most applications and that such an instability could be effectively suppressed by incorporating a built-in refractive-index profile in the lateral direction (Lang, 1979). A large number of laser structures have been proposed and demonstrated to introduce lateral variations of the refractive index.

As a result of these research and development efforts during the first two decades, the maturity of AlGaAs lasers reached the stage of practical use in printers and disk players early in the 1980s. Meanwhile, the rapid reduction of transmission loss in optical fibers achieved during the 1970s stimulated intensive development efforts on GaInPAs lasers, which have emission wavelengths (1.3 – 1.55 μm) in the low-loss and dispersion-free region of silica fibers. During the first half of the 1980s, GaInPAs lasers emerged as indispensable light sources in telecommunications systems using silica fibers.

Progress has continued to date in both device structures and materials. In the following, let us look at some recent major topics.

The idea of utilizing periodic gratings incorporated in a laser medium as a means of optical feedback was proposed by Kogelnik and Shank in 1971. Distributed-feedback and distributed Bragg-reflector semiconductor lasers utilizing this principle have been the subject of recent intensive studies because of the stable single–longitudinal-mode oscillation they exhibit even under high-speed modulation.

Recent advances in growth technology of semiconductor ultrathin layers by molecular-beam epitaxy and metal-organic vapor-phase epitaxy also create a new class of laser structures. Quantum-well lasers incorporating such ultrathin active layers ($\sim$ 20 nm or less) have been shown to display a variety of superior laser characteristics, such as low threshold currents and wide modulation bandwidths, which are attributed to the size quantization of electrons in the ultrathin active layers (Arakawa and Yariv, 1986).

In the visible spectrum, GaInP lasers have recently attained the level of practical use, emitting in the red spectrum (0.66 – 0.69 μm). Very recently, blue–green laser emission also was achieved in ZnCdSe lasers at cryogenic temperatures (Haase *et al.*, 1991).

2. STRUCTURES, MATERIALS, AND OPERATION PRINCIPLES

2.1 Double Heterostructure

A semiconductor laser is a diode, as shown schematically in Fig. 1. When a forward current is passed through the *p-n* junction, electrons and holes injected into the active region from the *n* and *p* regions, respectively, recombine to emit radiation. The wavelength λ of the radiation is basically determined by the relation $\lambda \sim hc/E_g$, where h is the Planck constant, c is the light velocity, and E_g is the band-gap energy of the active-region material.

For low injection currents, light is emitted through spontaneous emission. In order for lasing to take place, a sufficiently high concentration of carriers must be accumulated within the active region to induce population inversion. This is effectively accomplished by adopting a double-heterojunction (DH) structure, where a thin active layer, typically $\sim$ 0.1 μm thick, is sandwiched between *n*- and *p*- type cladding layers, which have wider band gaps than the active layer. Electrons and holes injected into the active layer through the heterojunctions are confined within the thin active layer by the potential barriers at the heteroboundaries, as illustrated in Fig. 2(a). The DH structure forms an efficient optical waveguide as well because of the refractive-index difference between the active and cladding layers shown in Figs. 2(b) and 2(c). Thus the DH structure facilitates the interaction needed for laser action between the optical field and the injected carriers.

Let us use a GaAs(active)/$Al_xGa_{1-x}As$ (cladding) DH structure as an example. The dependence of the band-gap energy of $Al_xGa_{1-x}As$ on the AlAs mole fraction x can be approximated for $x < 0.42$ (direct-gap region) by (Casey and Panish, 1978, Part A, p. 193).

$$E_g = 1.424 + 1.247x \text{ eV}. \tag{1}$$

The energy-gap difference between the active and cladding layers ΔE_g is divided into the

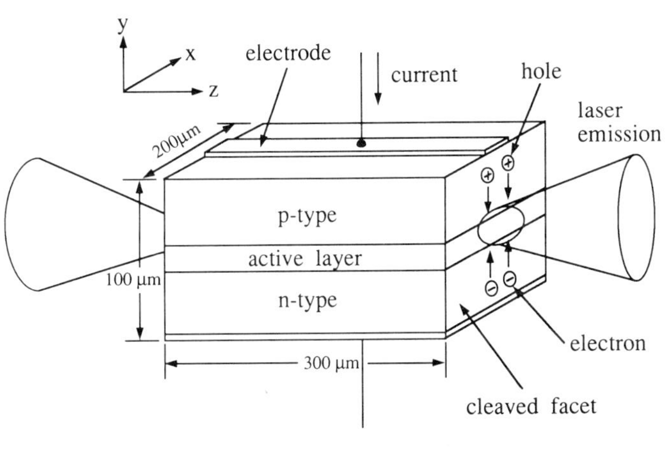

FIG. 1. Schematic Illustration of a double-heterostructure semiconductor laser.

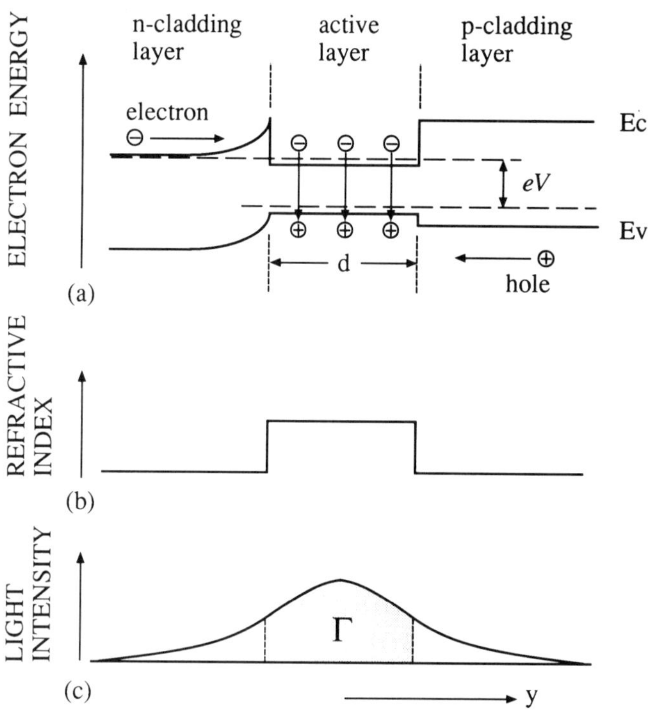

FIG. 2. Diagram illustrating carrier confinement and waveguiding in a double heterostructure. (a) Energy-band diagram at high forward bias; (b) refractive-index distribution; (c) light intensity distribution. E_c and E_v: The edges of the conduction and valence bands. V: Applied voltage. d: Active-layer thickness ($\sim$0.1 μm). Γ: Confinement factor.

band discontinuities ΔE_c in the conduction band and ΔE_v in the valence band according to (Kroemer, 1986)

$$\Delta E_c \sim 0.62 \Delta E_g \tag{2}$$

and

$$\Delta E_v \sim 0.38 \Delta E_g. \tag{3}$$

The efficiency of the carrier confinement depends strongly on the magnitude of the band discontinuity. The AlAs mole fraction x of the cladding layers is usually chosen to be greater than that of the active layer by $\sim$ 0.3. This gives rise to a ΔE_c approximately 10 times as large as the room-temperature thermal energy, which is sufficient to suppress electron diffusion over the heterobarriers. The hole leakage current is less important because of the smaller diffusion constant for holes.

The characteristics of a three-layer slab waveguide are conveniently described in terms of the normalized waveguide thickness D, defined as

$$D = (2\pi/\lambda) d \sqrt{\eta_a^2 - \eta_c^2}\,, \tag{4}$$

where η_a and η_c are the refractive indices of the active and cladding layers, respectively, and d is the active-layer thickness. For example, the condition that a waveguide support only the lowest-order fundamental mode is expressed as

$$D < \pi. \tag{5}$$

Meanwhile, the refractive index of $Al_xGa_{1-x}As$, $\eta(x)$, can be approximated for a light wavelength of $\sim$ 0.9 μm by (Casey and Panish, 1978, Part A, p. 45)

$$\eta(x) = 3.590 - 0.710x + 0.091x^2. \tag{6}$$

Thus, for a waveguide with a GaAs active layer surrounded by $Al_{0.3}Ga_{0.7}As$ layers and a wavelength λ of 0.87 μm, the condition of Eq. (5) simply becomes

$$d < 0.36\ \mu\text{m}. \tag{7}$$

Since the thickness d is typically 0.1–0.2 μm, the single, fundamental-mode condition is almost always satisfied in practical devices.

The confinement factor Γ, defined as the fraction of the electromagnetic energy of the guided mode that exists within the active layer, is an important parameter representing the extent to which the waveguide mode is confined to the active layer. Γ for a fundamental mode is approximately given by (Botez, 1978)

$$\Gamma \sim D^2/(2+D^2). \tag{8}$$

For a GaAs/$Al_{0.3}Ga_{0.7}$As waveguide with $d=0.1$ μm, $\Gamma \sim 0.27$.

2.2 Fabry-Pérot Cavity

In addition to the optical gain, optical feedback is the other ingredient for laser oscillation. This is provided by a pair of mirror facets at both ends of the devices. These facets are normally formed simply by cleaving the crystal. Since the refractive index of major semiconductor laser materials is ~ 3.6, the reflectivity of the mirror is ~ 0.3. This is very low compared with other types of lasers but is still sufficient to provide optical feedback in semiconductor lasers.

2.3 Lasing Threshold Condition

When a sufficient number of electrons and holes are accumulated to form an inverted population, the active region exhibits optical gain and can amplify light passing through it since stimulated emission overcomes interband absorption. The condition for self-sustained laser oscillation to occur is that the light makes a full round trip in the cavity without attenuation; i.e., the optical gain should equal the losses both inside the cavity and through the partially reflecting end facets. Thus, the gain coefficient at threshold g_{th} is given by the relation

$$\Gamma g_{\mathrm{th}} = \underbrace{\Gamma\alpha_a + (1-\Gamma)\alpha_c + \alpha_s}_{\alpha_i} + \frac{1}{L}\ln\frac{1}{R}. \tag{9}$$

Here, α_a and α_c denote the losses in the active and cladding layers, respectively, due to free-carrier absorption. α_s accounts for scattering loss due to heterointerfacial imperfections. The first three loss terms on the right-hand side combined are termed internal loss α_i and add up to 10–20 cm^{-1} (Casey and Panish, 1978, Part A, p. 174–176). The reflection loss $L^{-1}\ln R^{-1}$ due to output coupling ($\sim 40\ \mathrm{cm}^{-1}$ for $L \sim 300\ \mu$m, $R \sim 0.3$) is normally the largest among the loss terms. Despite the fact that both the internal and reflection losses are exceptionally large as compared with other lasers, lasing is brought about by virtue of the high optical gain in semiconductors.

It is not an easy task to analyze precisely the optical gain spectrum in semiconductor lasers. Fortunately, however, we can utilize a phenomenological linear relationship between the maximum gain g (the peak value of the gain spectrum for a given carrier density) and the injected carrier density n,

$$g(n) = \frac{\partial g}{\partial n}(n-n_t), \tag{10}$$

to a good approximation (Yamamoto *et al.*, 1983; Yamada, 1983; Ogasawara and Ito, 1988). Here $\partial g/\partial n$ is termed differential gain, and n_t denotes the carrier density required to achieve transparency where stimulated emission balances against interband absorption corresponding to the onset of population inversion. For GaAs lasers,

$$\frac{\partial g}{\partial n} \sim 3.5\times 10^{-16}\ \mathrm{cm}^2 \tag{11}$$

and

$$n_t \sim 1.5\times 10^{18}\ \mathrm{cm}^{-3}. \tag{12}$$

Substituting Eqs. (10)–(12) into Eq. (9) and assuming that $\Gamma=0.27$, $\alpha_i=10\ \mathrm{cm}^{-1}$, and $L^{-1}\ln R^{-1}=40\ \mathrm{cm}^{-1}$, we get a threshold carrier density n_{th} of $\sim 2\times 10^{18}\ \mathrm{cm}^{-3}$.

The threshold current density J_{th} is expressed as

$$J_{\mathrm{th}} = edn_{\mathrm{th}}/\tau_s, \tag{13}$$

where τ_s is the carrier lifetime due to spontaneous emission. Assuming that $\tau_s=2$–4 ns and $d=0.1\ \mu$m, we obtain a threshold current density J_{th} or $\sim 1\ \mathrm{kA\ cm}^{-2}$.

2.4 Stripe Geometry

Most modern semiconductor lasers adopt a stripe geometry, where current is injected only within a narrow region beneath a stripe contact several μm wide, in order to keep the threshold current low and to control the optical field distribution in the lateral direction. As compared with broad-area lasers, where the entire laser chip is excited, the threshold current of the stripe-geometry lasers is reduced roughly proportional to the area of the contact.

Figure 3(a) shows the simplest form of the stripe geometry (Dyment, 1967). Lasing occurs in a limited region of the active layer beneath the stripe contact where a high density of current flows. Such lasers are termed gain-guided lasers because the optical inten-

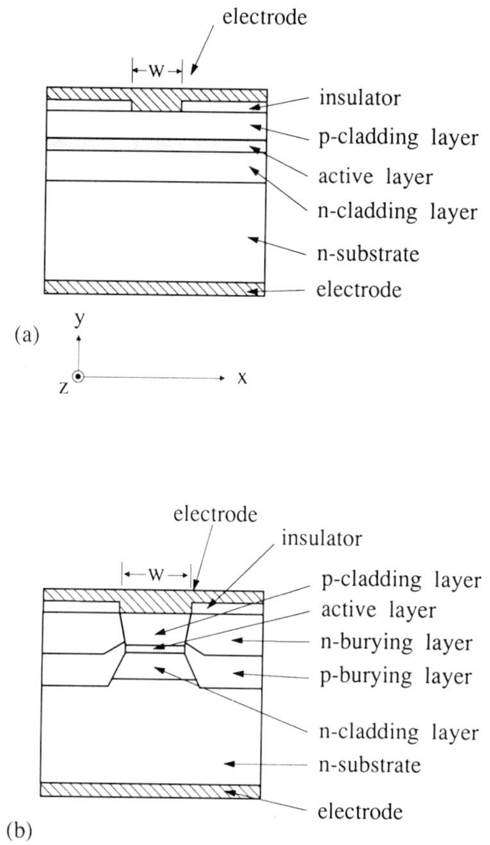

FIG. 3. Two representative examples of stripe-geometry lasers (front view of the end facet). (a) Stripe-contact laser. (b) Buried heterostructure (BH) laser. *W*: Stripe width (2–10 μm).

sity distribution in the lateral direction (x direction) is determined by the gain profile produced by carrier density distribution.

Devices incorporating a built-in refractive-index variation in the lateral direction are termed index-guided lasers. Figure 3(b) shows the structure of a buried heterostructure (BH) laser (Tsukada, 1974) as a representative example of index-guided lasers. The active region is surrounded by materials with lower refractive indices in both the vertical (y) and lateral (x) transverse directions, thus forming a waveguide structure in both directions. In the BH laser, the width of current flow is delineated by the *p-n* junction formed by the burying layers, which is reverse biased when the active region is forward biased.

2.5 Materials and Emission Wavelengths

Semiconductor laser materials that have been studied range over III-V, IV-VI, and II-VI compounds as listed in Table 1.

2.5.1 III-V Compounds. III-V compounds are the most important and popular laser materials (Casey and Panish, 1978, Part B, Chap. 5). In particular, AlGaAs(active layer)/AlGaAs(cladding layer)/GaAs(substrate) lasers emitting at 0.7–0.9 μm and GaInPAs/InP/InP lasers emitting at 1.2–1.6 μm are used in disk systems and fiber-optic communications systems, respectively. GaInP/AlGaInP/GaAs lasers emitting at 0.66–0.69 μm have recently reached the level of practical utility. All of them exhibit low-threshold, room-temperature cw operation with high reliability.

The common features among them are as follows:

1. The active layer consists of direct-gap materials.
2. Binary compounds (GaAs, InP) are used as the substrate.
3. The active and cladding layers have the same lattice constant as the substrate.

The use of direct-gap semiconductors, featuring efficient radiative recombination, as the active-region material (**1**) is essential in

Table 1. Major semiconductor-laser materials and their emission wavelengths.

Materials (active/cladding/substrate)	Emission wavelengths (μm)
III-V compounds	
AlGaAs/AlGaAs/GaAs	0.7–0.9
GaInPAs/InP/InP	1.2–1.6
GaInP/AlGaInP/GaAs	0.66–0.69
GaInPAs/GaInP/GaAs or GaPAs	0.65–0.9
GaInPAs/AlGaAs/GaAs	0.62–0.9
AlGaAsSb/AlGaAsSb/GaSb	1.1–1.7
GaInAsSb/AlGaAsSb/GaSb or InAs	2–4
InPAsSb/InPAsSb/GaSb or InAs	2–4
GaInAs(strained)/AlGaAs/GaAs	0.9–1.1
GaInPAs(strained)/GaInPAs/InP	~ 1.55
IV-VI compounds	
PbSnTe/PbSnSeTe/PbTe	6–30
PbSSe/PbS/PbS	4–7
PbEuTe/PbEuTe/PbTe	3–6
II-VI compounds	
ZnCdSe/ZnSSe/GaAs	~ 0.5

achieving laser action. Features **2** and **3** are related to the crystal quality of the hetero-interfaces. To minimize the generation of lattice defects that may impair the device reliability, the lattice parameter of the active and cladding layers should be matched to that of the high-quality binary substrate (the lattice-matching condition). The lattice parameter of $Al_xGa_{1-x}As$ is essentially independent of alloy concentration x. This is the only exception in ternary alloys, arising from the fortuity that GaAs and AlAs have virtually the same lattice parameter, the difference being as small as $\sim 0.1\%$. In quaternary alloys like $Ga_xIn_{1-x}P_yAs_{1-y}$ and $Al_xGa_yIn_{1-x-y}P$, the band-gap energy can be tuned in a certain range for a fixed lattice parameter by choosing appropriate pairs of x and y values. The design of lattice-matched DH structures is facilitated by taking advantage of this degree of freedom in quaternary alloys.

2.5.2 IV-VI Compounds. PbSnTe and PbSSe lasers emitting in the infrared region (3–30 μm) (Horikoshi, 1985) are used in high-resolution gas spectroscopy and in air-pollution monitoring. These lasers operate only at cryogenic temperatures. However, an appreciable spectral tuning is feasible by simply changing temperature and injection current, which is extremely desirable for spectroscopy.

2.5.3 II-VI Compounds. The band-gap energies of Zn and Cd chalcogenides cover the blue–green spectral region. Development of injection lasers emitting in the short-wavelength region is currently of great interest because of their feasibility of densely packing data bits onto optical disks. Intensive research efforts in recent years have opened up the prospect of success in overcoming the difficulty of p-type doping into these materials (Ohkawa *et al.*, 1991).

3. FUNDAMENTAL CHARACTERISTICS

In this section, several aspects of fundamental characteristics are described. Numerical examples are given based on AlGaAs ($\lambda \sim 0.8\ \mu$m) and GaInPAs ($\lambda \sim 1.3\ \mu$m) lasers.

3.1 Light-Current Characteristics

Figure 4 shows an example of output power (P) versus dc injection current (I) characteristics. The ordinate is the light power emitted from one end facet, and essentially the same power is emitted from the other end facet.

The threshold current at room temperature is typically several tens of mA. The dependence of the threshold current I_{th} on device temperature T is phenomenologically expressed as

$$I_{th} \propto \exp(T/T_0), \tag{14}$$

where T_0 is a constant referred to as the characteristic temperature and takes a value of 100–150 and 50–80 K for AlGaAs and GaInPAs lasers, respectively.

The light power emitted by injection above threshold is expressed as

$$P = \hbar\omega_l \frac{I - I_{th}}{e} \eta_i \frac{\frac{1}{2}L^{-1}\ln(1/R)}{\alpha_i + L^{-1}\ln(1/R)}. \tag{15}$$

Here, $(I - I_{th})/e$ is the rate of excess carrier injection beyond the threshold. η_i is the internal quantum efficiency representing the fraction of injected carriers that recombine radi-

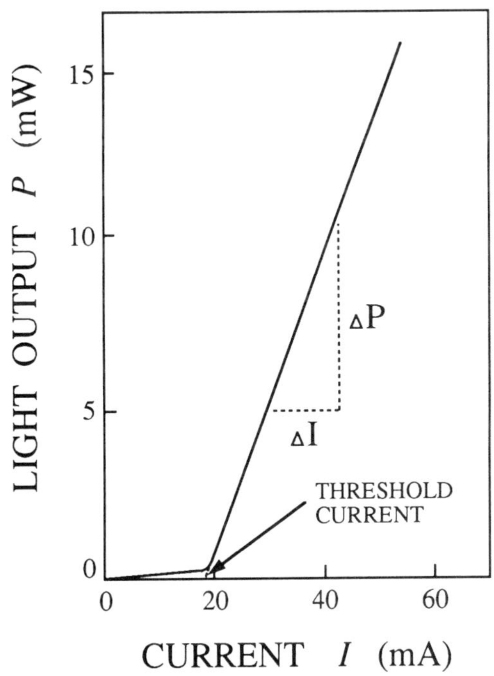

FIG. 4. Light output vs injection current curve in an AlGaAs laser.

atively and generate photons of energy $\hbar\omega_l$. The factor $\frac{1}{2}L^{-1}\ln R^{-1}/(\alpha_i+L^{-1}\ln R^{-1})$ represents the fraction of the generated photons that are coupled out of the cavity through one of the end facets. From Eq. (15), the differential efficiency η_D, defined as the slope of output power versus current curve above threshold, is written as

$$\eta_D=\frac{\Delta P}{\Delta I}=\frac{\hbar\omega_l}{e}\eta_i\frac{\frac{1}{2}L^{-1}\ln(1/R)}{\alpha_i+L^{-1}\ln(1/R)}. \quad (16)$$

The differential external quantum efficiency η_{ext}, defined as the ratio of the number of photons coupled out to the number of carriers injected, is given by

$$\eta_{ext}=\frac{\Delta P/\hbar\omega_l}{\Delta I/e}=\eta_i\frac{\frac{1}{2}L^{-1}\ln(1/R)}{\alpha_i+L^{-1}\ln(1/R)}. \quad (17)$$

Above threshold, the stimulated emission predominates over the nonradiative processes and $\eta_i\sim 1$. Typical values of $L=300\,\mu m$, $R=0.3$, and $\alpha_i=20\ cm^{-1}$ give $\eta_{ext}\sim 0.3$. Since $\hbar\omega_l\sim E_g$, $\eta_D\sim(E_g/e)\eta_{ext}$. For GaAs lasers ($E_g=1.424$ eV), we get $\eta_D\sim 0.4$ W/A.

The maximum rating for output power is typically 5–30 mW. Figure 5 shows schematically the three phenomena that determine the maximum power rating.

1. Kink: The nonlinearity in the curve of light versus current caused by lateral mode instability is referred to as a kink. Since a kink is often associated with a beam-profile shift and the generation of intensity noise, lasers cannot be used in the kink region for most applications. The lateral mode instability arises from the optical power dependence of the refractive index. By incorporating a refractive-index profile in the lateral direction to form a waveguide, the power level for kink generation can be appreciably increased.
2. Catastrophic optical damage: For optical power densities higher than several MW cm^{-2}, the end facets of semiconductor lasers may be melted. This results in a sudden decrease of output power and failure of the device. The catastrophic damage is considered to be brought about by the enhanced light absorption associated with surface states at the end facets. The importance of this process seems to depend on materials and photon energies. In GaInPAs lasers ($\lambda\sim 1.3\,\mu m$), the catastrophic optical damage practically does not occur.
3. Thermally induced saturation: Heating of the junction under high current injections raises the threshold current. This may lead to saturation of output power with an increase in injection current, especially in lasers with lower T_0 values.

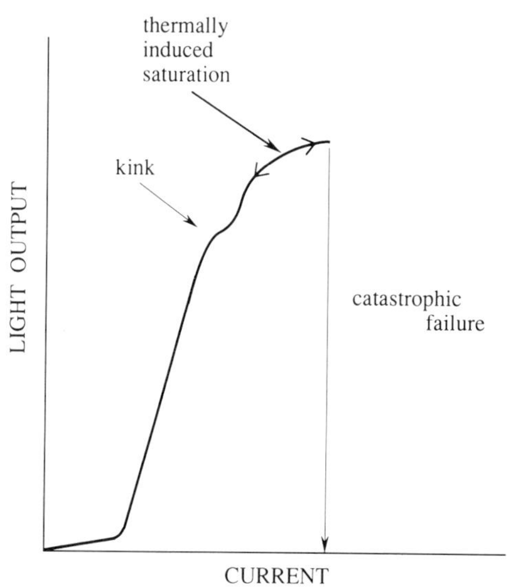

FIG. 5. Schematic illustration of the three phenomena occuring at high injection levels.

3.2 Beam Profile and Polarization

Typical radiation intensity patterns of laser diodes are shown in Figs. 6 and 7. Shown in Fig. 6 are the near-field patterns, i.e., the spatial distributions of optical intensity on the end facet in the directions (a) perpendicular and (b) parallel to the junction plane. The far-field pattern (Fig. 7) is the angular distribution of radiant intensity measured at distances several mm or more away from the facet, which is mathematically a Fourier transform of the near-field pattern. The single-lobed radiation patterns show that the transverse mode of the device is single and fundamental. Transverse-mode-stabilized lasers incorporating appropriate waveguide structures exhibit such beam profiles stably up to reasonable output power levels.

Normally, the angular width of beam divergence (the full width at half maximum) perpendicular to the junction, $\theta_\perp$ (20°–60°), is larger than that parallel to the junction $\theta_\parallel$ (10°–30°), since the near-field spot is an ellip-

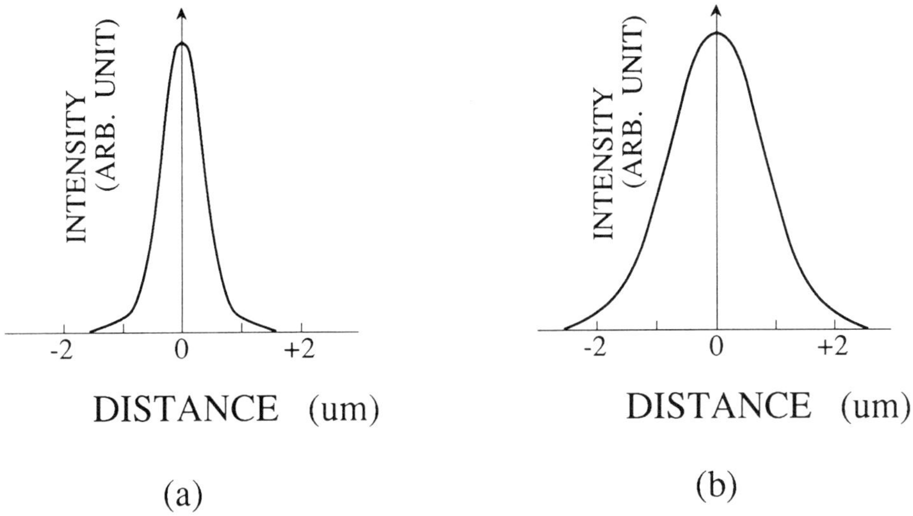

FIG. 6. Near-field optical intensity patterns (a) perpendicular and (b) parallel to the junction plane.

soid because of the large difference between the stripe width (several μm) and the active-layer thickness ($\sim$0.1 μm). $\theta_\perp$ can be reduced by adopting thinner active layers; then, most of the optical energy penetrates into the cladding layers, increasing the spot size perpendicular to the junction.

The output beam from gain-guided lasers can be considerably astigmatic; the beam waist perpendicular to the junction plane is located essentially on the end facet, while the virtual beam waist along the junction plane is formed in the cavity at a position several tens of μm behind the end facet. This is brought about by the difference of the waveguiding mechanism between the two directions. In contrast, in index-guided lasers, the beam waist, both parallel and perpendicular to the junction plane,

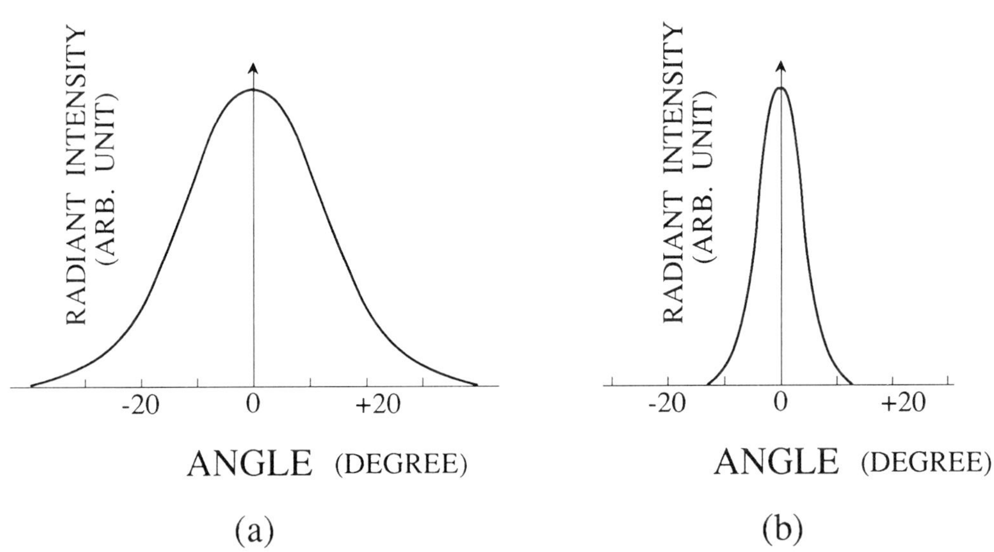

FIG. 7. Far-field radiant intensity patterns (a) perpendicular and (b) parallel to the junction plane.

is located within several μm of the facet. The astigmatism should be taken into consideration when the beam is coupled into lenses.

The laser beam is linearly polarized along the junction plane. This is because in a slab waveguide, the facet reflectivity for the TE mode is higher than that for the TM mode (Ikegami, 1972). Since the gain needed to reach threshold depends on the facet reflectivity [see Eq. (9)], the TE mode is selected for oscillation in DH lasers because of the higher reflectivity.

3.3 Spectral Characteristics

Figure 8 shows the variations of lasing spectrum with output power in an index-guided AlGaAs laser. At lower powers, the laser oscillates in several longitudinal modes. Here, the longitudinal mode wavelength λ_N is determined by the relation

$$(\tfrac{1}{2}\lambda_N/\eta)N=L \quad (N\text{: positive integers}), \tag{18}$$

where η is the effective refractive index of the waveguide and L is the cavity length. The mode spacing $\Delta\lambda$ is given by

$$\Delta\lambda=\lambda^2/2\eta_g L, \tag{19}$$

where $\eta_g=\eta-(\partial\eta/\partial\lambda)\lambda$ is the group index. $\Delta\lambda$ is typically 3 Å ($\lambda=0.8$ μm)–8 Å ($\lambda=1.3$ μm) for $L=300$ μm.

As injection is increased , the laser power tends to concentrate in a single longitudinal mode, while the power in the remaining modes saturates. This is common among index-guided lasers. Gain-guided lasers, in contrast, tend to exhibit multiple–longitudinal-mode spectra even at higher powers.

It can be seen in Fig. 8 that the lasing wavelength shifts toward longer wavelengths as the output power is increased. This is induced by the temperature rise in the active region with the increase in injection current. Shown in Fig 9 is an example of the wavelength shift due to a heat sink temperature change at a fixed injection current. Each longitudinal mode shifts at a rate of 0.5 Å K^{-1} ($\lambda\sim0.8$ μm)–0.8 Å K^{-1} ($\lambda\sim1.3$ μm) because of the temperature dependence of the refractive index. In addition, the lasing wavelength jumps toward longer wavelengths, as temperature is raised, at a rate of 2 Å K^{-1} ($\lambda\sim0.8$

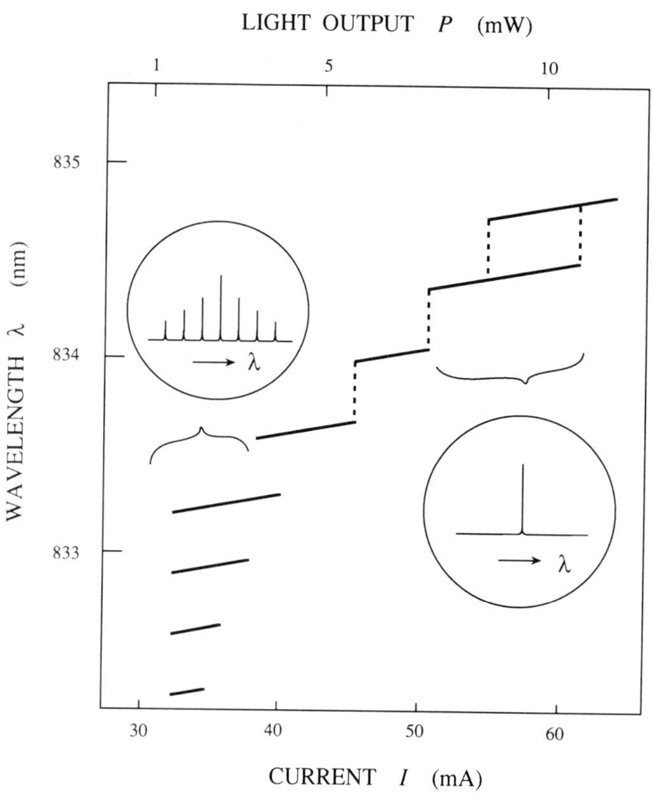

FIG. 8. Lasing wavelength vs injection current characteristics in an AlGaAs laser.

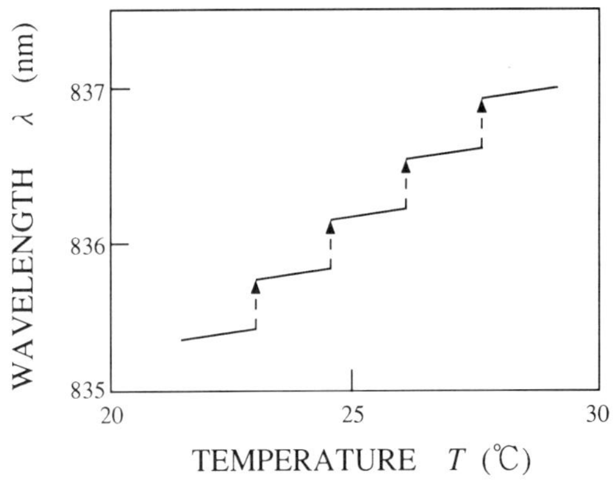

FIG. 9. Temperature dependence of the lasing wavelength.

μm)–5 Å K^{-1} ($\lambda \sim 1.3$ μm). This is caused by the temperature dependence of the band-gap energy.

At ranges of temperature and bias current where the oscillation mode jumps to a neighboring mode, it is often observed that the lasing is randomly switched between the two longitudinal modes. The random switching is associated with a large-intensity noise since the output power is different between modes by (0.1–1)%. Some devices exhibit hysteresis in the lasing wavelength versus temperature and/or injection-current characteristics. These phenomena have been interpreted in terms of nonlinear mode coupling among longitudinal modes (Ogasawara and Ito, 1988).

The spectral linewidth of a single longitudinal mode is inversely proportional to the output power to a good approximation. The product of the linewidth Δf and the output power P is typically in the range of 1–100 MHz mW. This value is 10–50 times larger than Δf_{ST} obtained from the well-known Schawlow–Townes formula. An analysis shows that the enhancement is expressed as (Henry, 1982)

$$\Delta f = \Delta f_{ST}(1+\alpha^2), \tag{20}$$

where α is the linewidth enhancement factor defined as

$$\alpha = -\frac{4\pi}{\lambda}\frac{(\partial\eta/dn)}{(\partial g/dn)} \tag{21}$$

by the derivatives of the refractive index η and the gain g with respect to the carrier density n. α represents the magnitude of the amplitude-phase coupling, inherent in semiconductor lasers, originating from the strong dependence of the refractive index on the carrier density. The linewidth is determined not only by the direct phase fluctuation caused by spontaneous emission, but there exists an additional contribution from the amplitude fluctuation since the amplitude fluctuation is coupled into the phase fluctuation through the carrier density fluctuation.

α takes a value of 2–6 depending on the active-region material, the injected-carrier density, and the lasing photon energy (Osinski and Buus, 1987), while in most gas and solid-state lasers, α can be virtually taken to be zero. This is because the gain spectrum in semiconductor lasers based on a band-to-band transition is asymmetric with respect to the gain-peak frequency (corresponding to the laser frequency) and, as a consequence, the associated refractive index in the vicinity of the lasing frequency varies appreciably with the injected-carrier density. In contrast, in ordinary lasers where the lasing transition takes place between two discrete levels, the gain spectrum is symmetric and the associated refractive-index dispersion crosses zero at the lasing frequency. The nonzero value of this parameter affects a number of semiconductor laser characteristics including linewidth broadening, lateral mode instability, and frequency chirping.

3.4 Dynamic Characteristics

The capability of direct-current modulation is one of the important advantages of semiconductor lasers. The laser response to a stepwise current pulse is schematically shown in Fig. 10. There is a delay time τ_d for the turn-on of lasing because a finite time is required before the injection carriers are accumulated to form population inversion. τ_d is approximated by (Chinone *et al.*, 1974)

$$\tau_d = \tau_s \sqrt{\frac{I_{th}}{I}} \tanh^{-1}\sqrt{\frac{I_{th}}{I}}, \tag{22}$$

where τ_s is the carrier lifetime at threshold ($\sim$2 ns), I_{th} is the threshold current, and I is the current pulse height. τ_d can be reduced by prebiasing the laser. Then τ_d is expressed by the dc bias current I_0 ($<I_{th}$) as

$$\tau_d = \tau_s \sqrt{\frac{I_{th}}{I}} \left(\tanh^{-1}\sqrt{\frac{I_{th}}{I}} - \tanh^{-1}\sqrt{\frac{I_0}{I}} \right). \tag{23}$$

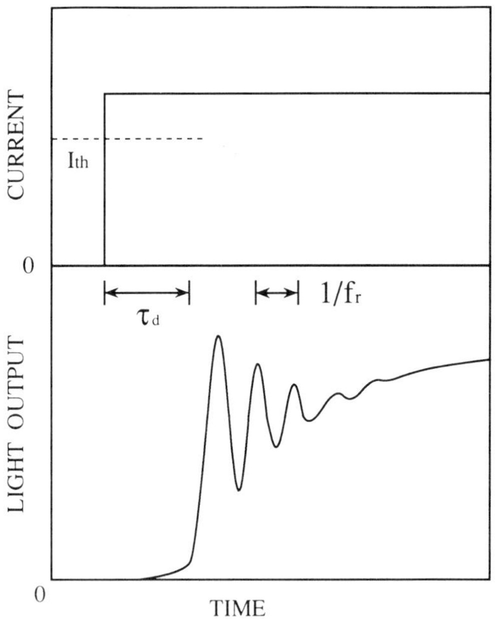

FIG. 10. Laser response to a staircase injection current. τ_d:Turn-on delay time. f_r: Relaxation oscillation frequency.

After the onset of laser oscillation, relaxation oscillations are generated for a time interval of several ns before the steady state is attained. The relaxation oscillation occurs because of the resonant interaction between photons and carriers; the energy stored in the cavity is transferred back and forth between the photon and carrier subsystems. Therefore, the periodic oscillation of the laser output is accompanied by a modulation of carrier density. The temporal variation of gain spectrum caused by the carrier-density modulation leads to a multiple–longitudinal-mode oscillation. Furthermore, the frequency of each individual mode is modulated because of the carrier-density–dependent refractive index, leading to a broadening of the time-averaged spectral linewidth. This phenomenon is referred to as frequency chirping. The magnitude of the chirping is governed by the linewidth enhancement factor α.

When a laser is biased above threshold, the bandwidth of the dynamic response of the laser is essentially determined by the relaxation oscillation frequency f_r. Let $\Delta I(\omega)e^{i\omega t}$ be a small-amplitude sinusoidal current superimposed on a dc bias I_0 ($>I_{\text{th}}$). Then the photon density inside the active region is made up of a dc component S_0 and an ac component $\Delta S(\psi)e^{i\omega t}$. The magnitude of the modulated component $\Delta S(\omega)$ is analyzed to be (Yariv, 1989)

$$\Delta S(\omega)=\frac{\Gamma\tau_p\omega_r^2}{-\omega^2+i\omega\Omega+\omega_r^2}\,\frac{\Delta I(\omega)}{eV_a}. \tag{24}$$

Here, the angular relaxation oscillation frequency ω_r is given by

$$\omega_r=2\pi f_r=\sqrt{\frac{S_0}{\tau_p}\,\frac{c}{\eta_g}\,\frac{\partial g}{\partial n}}. \tag{25}$$

Ω represents the damping of the relaxation oscillation and is given by

$$\Omega=\frac{1}{\tau_s}+S_0\frac{c}{n_g}\,\frac{\partial g}{\partial n}. \tag{26}$$

τ_p is the photon lifetime of the cavity ($\sim$2 ps) expressed as

$$\tau_p^{-1}=\frac{c}{\eta_g}\left[\alpha_i+\frac{1}{L}\ln\left(\frac{1}{R}\right)\right]. \tag{27}$$

V_a denotes the active-region volume. The magnitude of the modulated component of output power $\Delta P(\omega)$ is proportional to $\Delta S(\omega)$; i.e.,

$$\Delta P(\omega)=\frac{1}{2}\,\frac{c}{\eta_g}\,\hbar\omega_l\,\frac{1}{L}\,\ln\left(\frac{1}{R}\right)\Delta S(\omega)\,\frac{V_a}{\Gamma}. \tag{28}$$

Figure 11 shows an example of relative modulation response $|\Delta P(\omega)/\Delta P(0)|=\omega_r^2/[(\omega^2-\omega_r^2)^2-\omega^2\Omega^2]^{1/2}$. A flat response at modula-

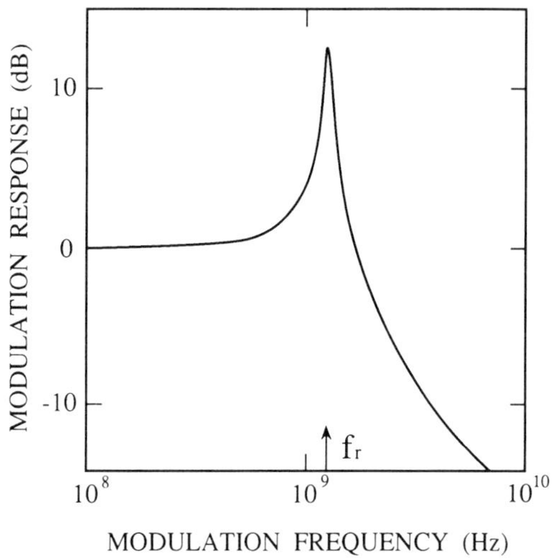

FIG. 11. Calculated small-signal modulation response. f_r is chosen to be 1.3 GHz.

tion frequencies less than f_r is followed by a peak at f_r and a sharp drop at modulation frequencies exceeding f_r. Therefore f_r is the uppermost useful modulation frequency. f_r is proportional to the square root of output power P and typically takes a value of 3–5 GHz at $P=10$ mW. In practice, however, the modulation bandwidth may be limited by the electrical parasitics associated with a specific device structure that leads to a rolloff in the modulation response. The cutoff frequency due to the parasitic rolloff varies 3–30 GHz depending on the device structure.

4. NEW STRUCTURES AND FUNCTIONS

4.1 Quantum-Well Lasers

Quantum-well (QW) lasers have an active region composed of ultrathin layers ($\sim$20 nm thick or less) forming narrow potential wells for injected carriers. Illustrated in Fig. 12 are the two representative examples of the active-region structures in AlGaAs QW lasers. Lasers comprising multiple wells [Fig. 12(a)] are termed multiple-quantum-well (MQW) lasers (Tsang, 1981a). Shown in Fig. 12(b) is the active region of a GRIN (graded index)–SCH (separate confinement heterostructure)–SQW (single-quantum-well) laser (Tsang, 1981b). In the latter structure, the injected carriers are confined in the SQW while the laser light is guided by the GRIN waveguide, thus ensuring an effective interaction between carriers and light without resort to multiple-well structure.

The quantization of electronic states in the well gives rise to a variety of superior laser characteristics over the conventional lasers with bulk active region. A major benefit is the reduction of the threshold current. Figure 13 helps us understand the low-threshold characteristics of QW lasers, where the energy distributions of the density of states and the injected carriers are compared between a bulk material and a QW structure. As the parabolic density of states in the bulk material changes into the staircase density of states in the QW structure, the energy distribution of the injected carriers narrows. The narrower energy distribution of the injected carriers leads to a narrower gain spectrum with a higher peak gain value for a given carrier density. Because of the staircase density of states, however, optical gain in QW lasers saturates for higher injection levels. Since the saturated gain value depends on the number of QW layers, SQW is preferable for high-Q cavities while MQW structure should be employed for lossy cavities. A threshold current as low as 0.55 mA has been demonstrated in a high-reflectivity coated ($R\sim0.8$) GRIN-SCH-SQW laser (Lau *et al.*, 1988). The narrowing of the gain spectrum in QW lasers as compared with bulk lasers is accompanied by an enhancement of $\partial g/\partial n$ (nearly doubled) and a reduction of α [see Eq. (21)]. This leads to an enhancement of modulation bandwidths [see

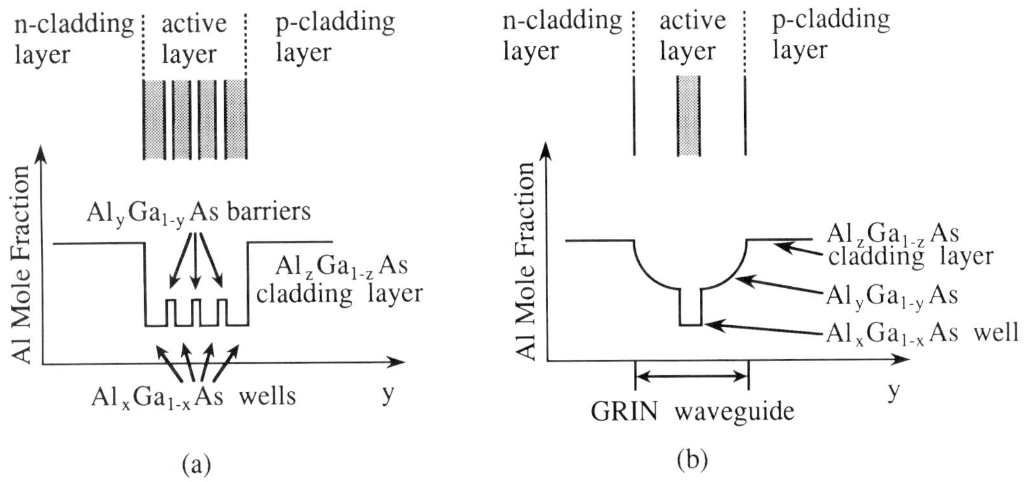

FIG. 12. Two representative examples of active-region structures in AlGaAs quantum-well (QW) lasers. (a) Multiple-quantum well (MQW) laser. (b) Graded index (GRIN)–separate confinement heterostructure (SCH)–single-quantum-well (SQW) laser.

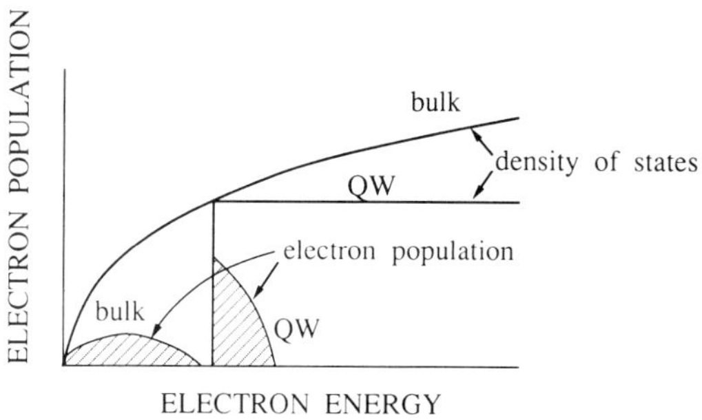

FIG. 13. Comparison of energy distribution of injected carriers between bulk and quantum-well active regions.

Eq. (25)] and a reduction of linewidths [see Eqs. (20) and (21)] under both dc and pulsed excitation (Uomi *et al.*, 1990).

The DH structure is generally formed by successive epitaxial growth of materials lattice-matched to the substrate. The lattice-matching condition imposes severe limitations on the combination of DH structure materials. However, the limitation is alleviated to a certain extent if the grown layer is thin enough to be elastically strained without any generation of dislocations. Applying this principle to the active region, we can obtain lasers whose emission wavelengths cannot be realized by lattice-matched systems. In QW lasers comprising a strained $Ga_xIn_{1-x}As$ well and AlGaAs cladding layers grown on a GaAs substrate, laser emission at 0.9–1.1 μm can be realized by varying the composition x and the well width. This wavelength region corresponds to the gap between AlGaAs and GaInPAs laser wavelengths and is fit for such applications as the excitation of Er-doped fibers and blue-green light generation by the second-harmonic generation technique.

Furthermore, it is predicted theoretically that the strain is beneficial in obtaining such good laser properties as low threshold currents, high relaxation oscillation frequencies, and low linewidth enhancement factors (Suemune *et al.*, 1988; Ohtoshi and Chinone, 1989). This is related to the strain-induced alleviation of the asymmetry in the density of states between the conduction and valence bands. The strain effects on high-speed modulation characteristics in GaInPAs lasers are now being studied experimentally.

4.2 Distributed-Feedback and Distributed Bragg-Reflector Lasers

In place of mirror facets in Fabry-Pérot cavities, periodic gratings incorporated within laser waveguides can be utilized as means of optical feedback. Integrated optical feedback from the periodic grating provides strong wavelength selectivity. Devices incorporating the grating in the pumped region are termed distributed-feedback (DFB) lasers [Fig. 14(a)], while those incorporating the grating in the passive region are termed distributed Bragg-reflector (DBR) lasers [Fig. 14(b)] (Mroziewicz *et al.*, 1991). By virtue of the strong wavelength selectivity, DFB and DBR lasers oscillate in a single longitudinal mode even under high-speed modulation, in contrast to Fabry-Pérot lasers, which exhibit multiple–longitudinal-mode oscillation when pulsed rapidly. This is an advantageous feature for optical data transmission using fibers because the spectral width determines the maximum bit rate transmitted in the presence of fiber-chromatic dispersion.

The carrier-density dependence of the refractive index can be exploited to provide

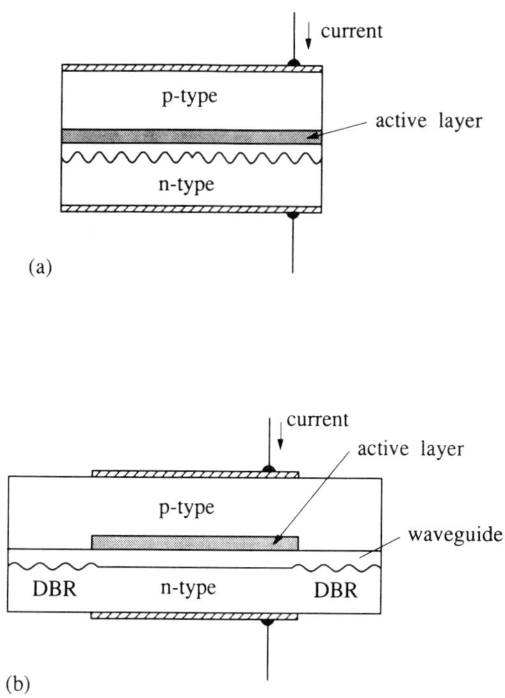

FIG. 14. (a) Distributed-feedback (DFB) and (b) distributed Bragg-reflector (DBR) lasers.

wavelength tunability to the periodic gratings (Yoshikuni, 1991). Illustrated in Fig. 15 is a tunable laser utilizing this principle, which can be tuned continuously over a wavelength range of several nm. Here, the Bragg wavelength of the DBR structure, at which the reflection loss is minimized, is tuned by the injection current I_3. The current I_2 is adjusted so that one of the resonant wavelengths of the cavity is brought into coincidence to the Bragg wavelength. Thus, by choosing appropriate pairs of I_2 and I_3, the oscillation wavelength can be varied continuously without mode jumping. Tunable lasers are expected to be used in future wavelength-division multiplexing communications systems.

4.3 Semiconductor-Laser Arrays

The major drawbacks of semiconductor lasers, the relatively low output powers and the conspicuous beam divergence, can be alleviated by integrating multiple laser stripes into an array as shown in Fig. 16. The stripes are closely spaced so that the radiation from neighboring stripes is coupled to form coherent modes of the entire array. The array modes, often refered to as supermodes, are phase-locked combinations of the individual stripe modes and are characterized by the phase relationship between the optical fields supported by adjacent stripes. If an array is properly designed so that one particular supermode, which is a uniphase superposition of the individual stripe modes (so called 0° shift mode), is excited, a single-lobed, low-divergence beam is attained; the divergence angle $\theta_{\|}$ is approximated by the relation $\theta_{\|} \sim \lambda/Ns$, where s is the center-to-center separation of adjacent stripes and N is the number of stripes.

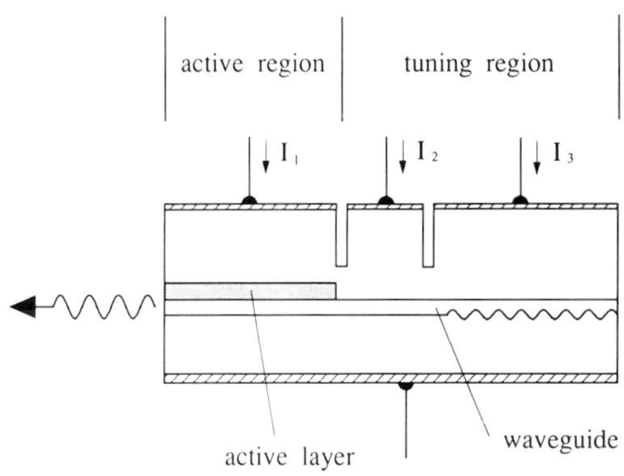

FIG. 15. Wavelength-tunable distributed Bragg-reflector (DBR) laser.

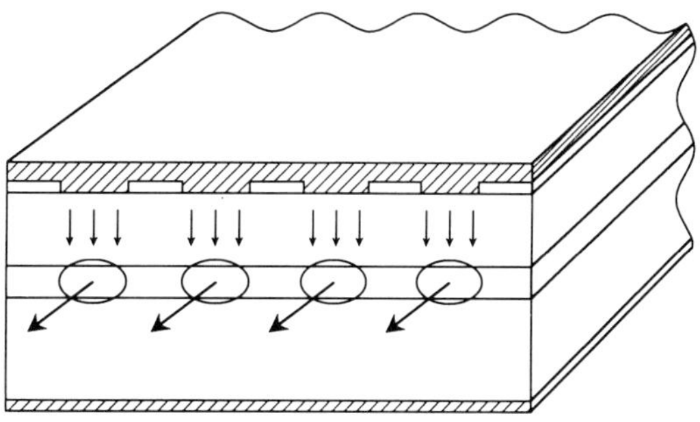

FIG. 16. Semiconductor-laser array.

A serious problem in the performance of laser arrays is their liability to multisupermode oscillation; the difference among the modal gains of the individual supermodes is so small that the lateral-mode behavior of an array is susceptible to spatial hole burning and temperature distribution in the lateral direction induced at high excitation levels. The multisupermode oscillation is generally accompanied by a multilobed output beam, which greatly reduces the utility of arrays for most potential applications. The maximum output power for the single-lobed beam operation is typically 300–500 mW.

4.4 Surface-Emitting Lasers

Conventional semiconductor lasers utilizing cleaved end facets as means of optical feedback are not fit for monolithic integration. For monolithic integration, it is desirable to have laser output normal to the wafer surface. Illustrated in Fig. 17 is the basic configuration of a surface-emitting laser incorporating two mirrors parallel to the surface to form a vertical cavity typically 5–10 μm long (Iga *et al.*, 1988). It is crucially important to increase the reflectivity of mirrors, since the reflection loss otherwise becomes very high because of the very short gain region. Bragg reflectors composed of multilayered semiconductors or dielectrics can be exploited to provide reflectivities exceeding 95%. Carrier injection is performed through a pair of electrodes in the upper and lower surfaces to excite a cylindrical active region typically of 10 μm diameter, 1–3 μm thick.

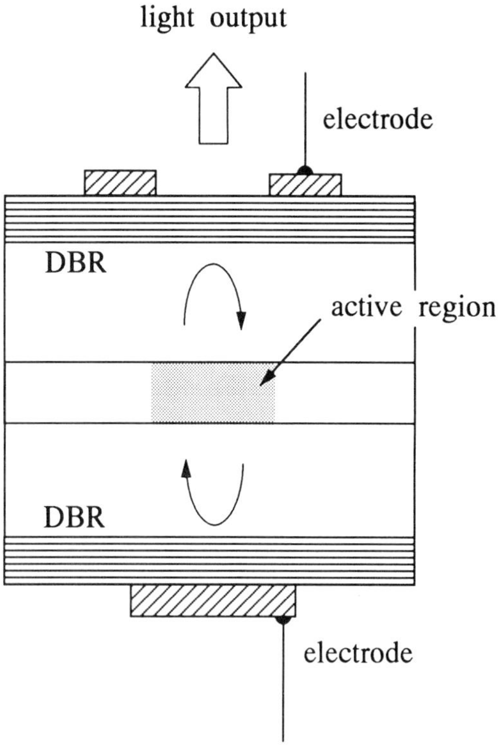

FIG. 17. Vertical-cavity surface-emitting laser.

The surface-emitting laser has several advantages over the conventional edge-emitting laser such as the capability of being integrated on a wafer to form a two-dimensional densely packed laser array, the feasibility of stable single–longitudinal-mode operation because of the large mode spacing due to the short cavity length, and the attainability of a narrow circular output beam.

5. SUMMARY

We have reviewed the history and the state of the art of semiconductor lasers. We note here that their successful development is the result of contributions of scientists and engineers from a number of fields, including semiconductor physics, quantum electronics, crystal growth, processing technologies, and system applications, as is most notably represented by the application to the fiber-optic communications systems.

Ongoing research and development activities indicate that the same will be true in the future. There is growing interest in improving laser characteristics by miniaturizing device structures. The natural extension of the QW lasers will be quantum wire and box lasers, where we can take advantage of the two- and three-dimensional quantization of electronic systems. Researchers in quantum optics are interested in the physics of microcavity semiconductor lasers, which have cavity lengths comparable to light wavelength. Such a microcavity will drastically alter the nature of spontaneous emission and can lead to ultralow-threshold ($\sim$1 μA or less), low-noise, and high-speed lasers. In order to convert the microcavity lasers as well as the quantum wire and box lasers from a laboratory curiosity to a practical light source, further progress in growth and processing technology of semiconductor microstructures is indispensable. Integration of these superior devices will find new application fields such as optical interconnection in large-scale-integration circuits and two-dimensional information processing.

GLOSSARY

Active Layer: Layer, typically $\sim$0.1 μm thick, in which stimulated emission occurs.

Cladding Layers: Layers sandwiching an active layer, made of materials that have wider band gaps than the active layer, p and n doped to facilitate carrier injection into the active layer.

Distributed Bragg Reflector Laser: Laser incorporating a periodic grating at both ends of the cavity as a means of optical feedback.

Distributed-Feedback Laser: Laser incorporating a periodic grating in the vicinity of the active layer as a means of optical feedback.

Double Heterojunction Structure: Active-region structure in which an active layer of one material is sandwiched between two cladding layers of another material.

Gain-Guided Laser: Laser in which the optical field distribution in the transverse direction is determined by the gain profile produced by carrier density distribution.

Index-Guided Laser: Laser in which the optical field distribution in the transverse direction is determined by a built-in refractive-index profile.

Linewidth Enhancement Factor: The ratio between the carrier-induced variations of the real and imaginary parts of susceptibility $\chi(n)$, i.e.,

$$\alpha=\frac{\partial[\mathrm{Re}\chi(n)]/\partial n}{\partial[Im\chi(n)]/\partial n}, \qquad (29)$$

where n is the carrier density. An equivalent expression is given in Eq. (21).

Phase-Locked Laser Array: Array of stripe lasers spaced closely so that a phase-locked oscillation among the laser stripes is attained.

Quantum-Well Laser: Laser that has ultrathin active layers forming narrow potential wells for injected carriers and giving rise to a size quantization of the electronic states.

Stripe-Geometry Laser: Laser in which current is injected only within a narrow region beneath a stripe contact several μm wide.

Surface-Emitting Laser: Laser that emits light normal to the wafer surface.

List of Works Cited

Alferov, Zh. I., Andreev, V. M., Garbuzov, D. Z., Zhilyaev, Yu. V., Morozov, E. P., Portnoi, E. L., Trofim, V. G. (1970), *Fiz. Tekh. Poluprovodn.* **4**, 1826–1829 [English translation: (1971), *Sov. Phys. Semicond.* **4**, 1573–1575.].

Arakawa, Y., Yariv, A. (1986), *IEEE J. Quantum Electron.* **QE-22**, 1887–1899.

Botez, D. (1978), *IEEE J. Quantum Electron.* **QE-14**, 230–232.

Casey, H. C., Jr., Panish, M. B. (1978), *Heterostructure Lasers*, Parts A and B. New York: Academic.

Chinone, N., Ito, R., Nakada, O. (1974), *IEEE J. Quantum Electron.* **QE-10**, 81–84.

Dyment, J. C. (1967), *Appl. Phys. Lett.* **10**, 84–86.

Haase, M. A., Qiu, J., Depuydt, J. M., Cheng, H. (1991), *Appl. Phys. Lett.* **59**, 1272–1274.

Hall, R. N., Fenner, G. E., Kingsley, J. D., Soltys, T. J., Carlson, R. O. (1962), *Phys. Rev. Lett.* **9**, 366–368.

Hayashi, I., Panish, M. B., Foy, P. W., Sumski, S. (1970), *Appl. Phys. Lett.* **17**, 109–111.

Henry, C. H. (1982), *IEEE J. Quantum Electron.* **QE-18**, 259–264.

Horikoshi, Y. (1985), in: W. T. Tsang (Ed.), *Semiconductors and Semimetals*, Vol. 22, Part C. Orlando, FL: Academic, Chap. 3.

Iga, K., Koyama, F., Kinoshita, S. (1988), *IEEE J. Quantum Electron.* **QE-24**, 1845–1855.

Ikegami, T. (1972), *IEEE J. Quantum Electron.* **QE-8**, 470–476.

Kogelnik, H., Shank, C. V. (1971), *Appl. Phys. Lett.* **18**, 152–154.

Kroemer, H. (1986), *Surf. Sci.* **174**, 299–306.

Lang, R. (1979), *IEEE J. Quantum Electron.* **QE-15**, 718–726.

Lau, K. Y., Derry, P. L., Yariv, A. (1988), *Appl. Phys. Lett.* **52**, 88–90.

Mroziewicz, B., Bugajski, M., Nakwaski, W. (1991), *Physics of Semiconductor Lasers*, Amsterdam: North-Holland, Chap. 6.

Nathan, M. I., Dumke, W. P., Burns, G., Dill, F. H., Jr., Lasher, G. (1962), *Appl. Phys. Lett.* **1**, 62–64.

Ogasawara, N., Ito, R. (1988), *Jpn. J. Appl. Phys.* **27**, 607–614 and 615–626.

Ohkawa, K., Karasawa, T., Mitsuya, T. (1991), *Jpn. J. Appl. Phys.* **30**, L152–L155.

Ohtoshi, T. Chinone, N. (1989), *IEEE Photon. Technol. Lett.* **1**, 117–119.

Osinski, M., Buus, J. (1987), *IEEE J. Quantum Electron.* **QE-23**, 9–29.

Quist, T. M., Rediker, R. H., Keyes, R. J., Krag, W. E., Lax., B., McWhorter, A. L., Zeigler, H. J. (1962), *Appl. Phys. Lett.* **1**, 91–92.

Suemune, I., Coldren, L. A., Yamanishi, M., Kan, Y. (1988), *Appl. Phys. Lett.* **53**, 1378–1380.

Tsang, W. T. (1981a), *Appl. Phys. Lett.* **39**, 786–788.

Tsang, W. T. (1981b), *Appl. Phys. Lett.* **39**, 134–137.

Tsukada, T. (1974), *J. Appl. Phys.* **45**, 4899–4906.

Uomi, K., Sasaki, S., Tsuchiya, T., Okai, M., Aoki, M., Chinone, N. (1990), *Electron. Lett.* **26**, 52–53.

Yamada, M. (1983), *IEEE J. Quantum Electron.* **QE-19**, 1365–1380.

Yamamoto, Y., Saita, S., Mukai, T. (1983), *IEEE J. Quantum Electron.* **QE-19**, 47–58.

Yariv, A. (1989), *Quantum Electronics*, New York: Wiley, Chap. 11.

Yoshikuni, Y. (1991), in: Y. Yamamoto (ed.), *Coherence, Amplification, and Quantum Effects in Semiconductor Lasers*, New York: Wiley, Chap. 4.

Further Reading

Agrawal, G. P., Dutta, N. K. (1986), *Long-Wavelength Semiconductor Lasers*, New York: Van Nostrand Reinhold.

Casey, H. C., Jr., Panish, M. B. (1978), *Heterostructure Lasers*, Parts A, and B, New York: Academic.

Kressel H., Butler J. K. (1977), *Semiconductor*

Lasers and Heterojunction LEDs New York: Academic.

Mroziewicz, B., Bugajski, M., Nakwaski, W. (1991), *Physics of Semiconductor Lasers,* Amsterdam: North-Holland.

Peterman, K. (1988), *Laser Diode Modulation and Noise,* Dordrecht: Kluwer Academic.

Thompson, G. H. B. (1980), *Physics of Semiconductor Laser Devices,* Chichester: Wiley.

Yamamoto, Y. (Ed.) (1991), *Coherence, Amplification, and Quantum Effects in Semiconductor Lasers,* New York: Wiley.

LASERS, SOLID STATE

T. Y. Fan, *Lincoln Laboratory, Massachusetts Institute of Technology, Lexington, Massachusetts, U.S.A.*

	Introduction	443
1.	**History**	444
2.	**Properties of Gain Media and Pump Sources**	445
2.1	Gain Media	445
2.1.1	Gain Media Doped with Rare-Earth Ions	446
2.1.1.1	Nd:YAG	448
2.1.1.2	Nd:Glass	449
2.1.1.3	Other Rare-Earth-Ion–Doped Lasers	449
2.1.2	Gain Media Doped with Transition-Metal Ions	450
2.1.2.1	Ruby, Alexandrite, and Other Cr^{3+}-Doped Gain Materials	450
2.1.2.2	$Ti:Al_2O_3$	452
2.1.3	Color Centers	452
2.2	Pump Sources	453
2.2.1	Lamps	453
2.2.2	Laser Sources	454
3.	**Laser Design**	454
3.1	Gain-Element Geometries	455
3.1.1	Rods	455
3.1.2	Zigzag Slabs and Disks	456
3.2	Pump Geometries	457
3.3	Laser Resonators	457
4.	**Laser Manufacture**	458
4.1	Gain-Element Fabrication	458
4.2	Optical Requirements for Solid-State Lasers	459
5.	**Applications of Solid-State Lasers**	460
5.1	Industrial Applications	460
5.2	Military Applications	460
5.3	Medical Applications	461
5.4	Communications Applications	461
5.5	Scientific Applications	462
6.	**Summary**	462
	Acknowledgment	462
	Glossary	462
	Works Cited	463
	Further Reading	463

INTRODUCTION

Solid-state lasers are a class of lasers that use an insulating dielectric solid, as opposed to a gas, liquid, semiconductor, or plasma, as the gain medium. These dielectric solids can be crystalline or amorphous, and optical gain is due to transitions between discrete electronic energy levels of dopants such as rare-earth ions or transition-metal ions with unfilled outer shells, or of defect centers, also known as color centers. Solid-state lasers have been used to generate a wide range of wavelengths from the vacuum ultraviolet to the mid-infrared. These lasers are optically excited, from either broadband sources such as flashlamps or narrow-band sources such as semiconductor or other lasers. Among the most common examples of solid-state lasers are ruby, which is sapphire doped with chromium, titanium-doped sapphire ($Ti:Al_2O_3$), neodymium-doped yttrium aluminum garnet (Nd:YAG), and neodymium-doped glasses (Nd:glass).

Solid-state lasers have a set of unique properties that set them apart from other lasers. Because they are solid, they are relatively easy to handle, are more inert, and are less susceptible to contamination compared with gas and liquid lasers. They are capable of generating high peak powers by *Q* switching, because of relatively long metastable lifetimes that allow higher energy storage compared with other classes of gain media; solid-state laser systems with $\sim 10^{14}$ W peak power have been demonstrated. They are also capable of high average power, up to ~ 1 kW, while maintaining nearly ideal output beams. Many lasers doped with transition-metal ions and color centers are capable of broadband tunability, up to 20% of the center frequency, which means that solid-state lasers are a powerful tool in optical

1-56081-067-X/94/$5.00+.50

spectroscopy. The broad gain bandwidth also allows short pulses to be generated using mode-locking and pulse-compression schemes; pulse lengths in the range of tens of femtoseconds have been generated. High-*Q* laser cavities can be built because solid-state laser materials have low optical loss. This leads to the ability to attain narrow-linewidth laser operation; subhertz laser linewidths have been demonstrated (see LASER PHYSICS).

However, solid-state lasers also have significant limitations. They are relatively inefficient; many commercial solid-state lasers have an overall power efficiency of ~1% or lower. There are also difficulties with the pump sources. Typically lamps have lifetimes of only hundreds of hours when operated continuously before they need to be replaced. In many wavelength regions, solid-state lasers either are nonexistent or perform poorly. Finally, there are significant engineering difficulties in designing and building solid-state lasers with high average power, greater than a few hundred watts, because of heat generation in the gain medium. However, there is active research ongoing to overcome all of these difficulties. Some of the limitations are beginning to be addressed by a movement to using semiconductor diode lasers as pump sources for solid-state lasers. Solid-state lasers pumped by semiconductor diode lasers have been demonstrated in the laboratory with nearly 20% overall power efficiency, and they have lifetimes of tens of thousands of hours. Active research is also under way to develop new materials for solid-state laser gain media that offer the possibility of increased performance and new wavelengths.

1. HISTORY

Solid-state lasers have been in existence longer than any other type of laser by virtue of Theodore Maiman's demonstration of a ruby laser in July 1960 at Hughes Research Laboratory (Maiman, 1960). In Maiman's experiment, ruby rods of about 10 mm diameter by 19 mm long were polished flat and parallel, and the ends were coated with evaporated silver that served as the resonator mirrors. The rod was then placed in the middle of a pulsed helical flashlamp that was in turn inserted into a reflective aluminum cylinder as shown in Fig. 1. Laser oscillation was observed on the three-level laser transition at 694.3 nm.

Following Maiman's announcement, lasers in other solid-state materials were also demonstrated. A low-threshold, four-level laser was demonstrated in uranium-doped CaF_2 at 2.5 μm (Sorokin and Stevenson, 1960). This laser was operated at liquid-helium temperature because the lower level of this transition in U^{3+} is located only ~500 cm^{-1} above the ground state; at room temperature the laser acts as a three-level laser, but at low temperature the lower laser level is depopulated. The first neodymium-doped laser was demonstrated in $CaWO_4$ in 1961; this allowed low-threshold room-temperature operation as a result of its four-level laser energy-level diagram at room temperature. Shortly thereafter the first continuous-wave solid-state laser operation was achieved in Nd:$CaWO_4$ and ruby. What followed was an explosion in the number of host materials and activator ions used for solid-state lasers with rare-earth, actinide, and transition-metal ions as dopants. The most important solid-state laser material, Nd-doped yttrium aluminum garnet ($Y_3Al_5O_{12}$ or YAG), came out of this early work (Geusic *et al.*, 1964). The first broadly tunable solid-state lasers, Ni^{2+} and Co^{2+} in MgF_2, were also demonstrated in this period (Johnson *et al.*, 1966). They operated at liquid-nitrogen temperature and were tuned with intracavity prisms.

The first glass lasers were Nd-doped fibers of ~32 μm in diameter (Snitzer, 1964), and thus these glass lasers were also the first fiber lasers. The capability of Nd-doped glasses to

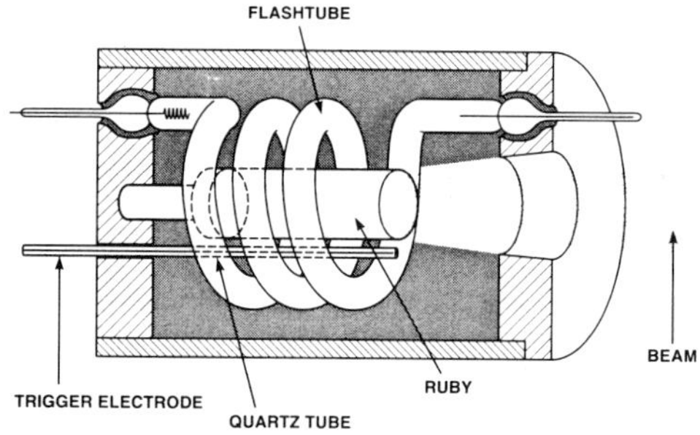

FIG. 1. Schematic diagram of the ruby laser demonstrated by Maiman. The size of the device was about 1 in. diameter by 2 in. long (adapted from Maiman *et al.*, 1961).

be used in high–energy-per-pulse laser systems was also demonstrated in a 100-J/pulse laser (Snitzer, 1964); subsequent engineering of high–energy-per-pulse Nd:glass laser systems has led to the development of a 100-kJ/pulse laser system, the Nova laser, for laser fusion experiments.

In the late 1960s and early 1970s, development of solid-state laser technology seemed to stagnate. The only practical solid-state lasers were ruby, Nd:YAG, and Nd:glass. While many other solid-state lasers had been demonstrated, these devices remained primarily laboratory curiosities because of either cryogenic temperature requirements, high threshold, or low efficiency. But two developments have helped to revive development of solid-state laser technology. The first was the demonstration of the alexandrite (Cr^{3+}-doped $BeAl_2O_4$) laser in which efficient, broadband tunable performance (~700–820 nm) was obtained (Walling *et al.*, 1980). This set off a search for other materials doped with transition-metal ions in which broadband tunability with efficient laser operation could be achieved. The second development was improvements in lifetime and efficiency of semiconductor diode lasers, in particular AlGaAs diode lasers, that allowed them to be used as practical pump sources. Higher-efficiency, longer-lifetime, and more compact solid-state lasers have been built using these diode lasers. The concept of diode pumping was not new—the use of a semiconductor diode laser as a pump source was initially demonstrated in 1964 (Keyes and Quist, 1964)—but at the time, diode lasers worked only at cryogenic temperatures.

2. PROPERTIES OF GAIN MEDIA AND PUMP SOURCES

The output wavelengths of solid-state lasers and the appropriate pump sources are determined by the positions of the energy levels of the activator ion or of the color center in a host material. Solid-state lasers have been demonstrated using many different activator ions in many host materials; however, only a few of these activator-ion–host material combinations have been found to be practical. The viability of any particular solid-state gain medium is dependent on many factors. For example, efficient, low-threshold operation is generally desired, which means that efficient pumping of the laser is needed. This requirement in turn means that there must be optical absorption in wavelength regions where practical pump sources exist.

2.1 Gain Media

Gain media for solid-state lasers can basically be broken into three categories, which are materials containing rare-earth ions, transition-metal ions, and color centers. Each of these classes of gain media has its own characteristics. Rare-earth-ion lasers have relatively narrow-band tunability and long (> 50 μs) metastable level lifetimes. In addition, the energies of the electronic states of a particular rare-earth ion change only slightly when doped into different host materials. Transition-metal-ion lasers can have relatively broad tunability and intermediate to long (> 1 μs) metastable-level lifetimes, and the energy-level diagram can be changed significantly by doping into different host materials. Finally, color-center lasers are characterized by relatively broad tunability and short metastable-level lifetimes (< 1 μs). The energy-level diagrams of these defect centers are strongly host dependent.

Host materials are chosen on the basis of many factors. First, the host material interacts with the dopant ion, and this interaction determines the position of the energy levels and the radiative and nonradiative transition strengths between energy levels. The host material must be able to incorporate the dopant ion or defect centers at sufficiently high concentrations, and good chemical stability of the dopant ion or defect center in the host is needed. Another factor is fabrication of the gain medium; it must be growable with high optical quality and sufficient size at a reasonable cost. For ease of handling, the material should be hard, have high strength, be chemically stable, and be nonhydroscopic. The thermo-optic properties of the host material are also important. Heat is dissipated in the gain medium because of nonradiative relaxation; thus high thermal conductivity is desired to allow efficient heat extraction from the gain medium, and low thermal expansion coefficient and change of refractive index with temperature are needed to reduce thermo-optic aberrations. Investigation of new ion–host material combinations is an area of active research; however, the development of useful new materials is a slow process because of the large number of factors that can affect laser performance and the long turnaround times required for bulk crystal growth.

2.1.1 Gain Media Doped with Rare-Earth Ions. The largest, and perhaps most important, class of solid-state lasers comprises materials doped with rare-earth ions. Most of the rare-earth ions are stable in the trivalent state when doped into solids, which leaves an outer shell of 4*f* electrons. The energy-level diagram of 4*f* multiplets of the trivalent rare-earth ions is shown in Fig. 2. Each multiplet contains several energy levels that are degenerate if the ion is in free space, and the energies of these multiplets are determined primarily by the Coulomb interaction between the 4*f* shell electrons and spin-orbit coupling. The effect of the host lattice, also known as the crystal field, is small because the 4*f* shell is well shielded, and splits the energy levels within a multiplet by only hundreds of inverse centimeters.

Wavelengths that have been generated in the various rare-earth ions are shown in Fig. 3. The ability to use a particular transition is determined by its spectroscopic properties in the host material. First, the host material must be transparent at the wavelength of the transition and the pump wavelengths. Most of the hosts used for solid-state lasers are oxides or fluorides; thus, the wavelength range is limited by host transparency to about 120 nm to

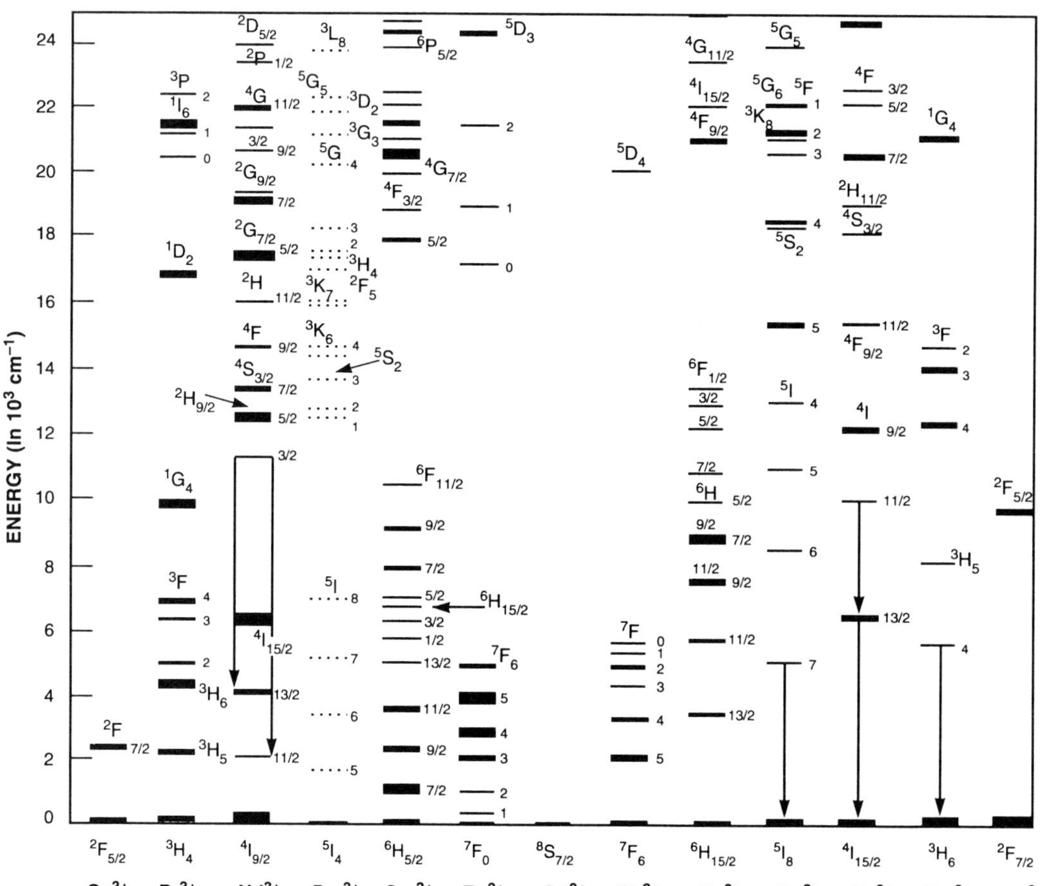

FIG. 2. Energy levels of the trivalent rear-earth ions in $LaCl_3$ (from Dieke and Crosswhite, 1963). The positions of the multiplets of the trivalent rare earths vary only slightly as the host material is changed. Pm^{3+} is shown in dotted lines because it is radioactive. While many transitions have been used for solid-state lasers, only the most important are shown with the arrows. Only a limited number of these multiplets can be upper laser levels; only those with a gap in energy of $\sim 3000\ cm^{-1}$ or more to the next lowest-lying multiplet are metastable.

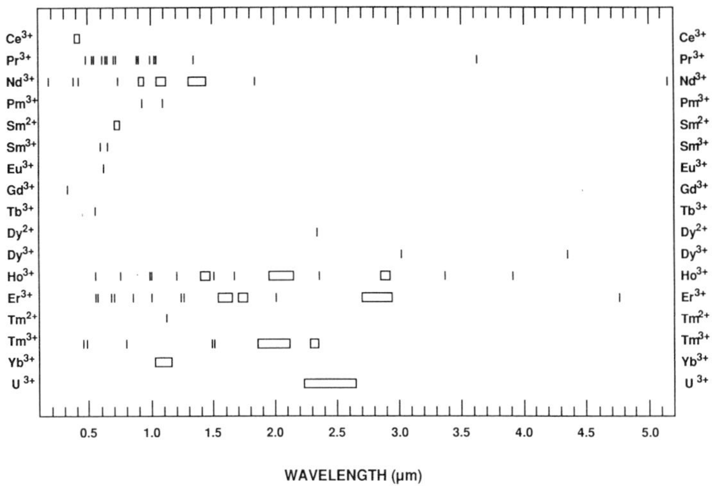

FIG. 3. Wavelengths demonstrated in rare-earth- and actinide-ion solid-state lasers.

6 μm. Second, the upper level must be metastable. Electric-dipole transitions between multiplets in rare-earth ions are only weakly allowed, which leads to long radiative lifetimes (>50 μs); however, there are also nonradiative relaxation mechanisms. The nonradiative relaxation rates are dependent on the host material and the energy gap between the multiplet in question and the next lower-lying multiplet. High nonradiative relaxation rates occur for small energy gaps and host materials with high characteristic phonon energies.

Only a few of these transitions are used in commercial devices, and these are denoted in Fig. 2. They are the $^4F_{3/2}$–$^4I_{11/2}$ transition around 1.06 μm and the $^4F_{3/2}$–$^4I_{13/2}$ transition around 1.32 μm in Nd^{3+}, the $^4I_{13/2}$–$^4I_{15/2}$ transition around 1.5 μm and the $^4I_{11/2}$–$^4I_{13/2}$ transition around 2.8 μm in Er^{3+}, and the $^5I_7$–$^5I_8$ transition around 2.1 μm in Ho^{3+}. These transitions are characterized either by low threshold and efficient operation or by output wavelengths that have important applications. The other laser transitions are for the most part limited by inefficient or currently impractical (for commercially viable products) pump schemes.

Many materials doped with rare-earth ions have been used as hosts for solid state lasers, but once again the number of host materials used in commercial products is limited. The most important set of host materials is the garnets such as YAG and $Gd_3Sc_2Ga_3O_{12}$ (gadolinium scandium gallium garnet or GSGG). The garnets have good mechanical properties, high thermal conductivity, well-understood crystal growth, and substitutional sites for rare-earth ions. Some other important host materials are $LiYF_4$ (YLF), $YAlO_3$ (YALO or YAP), YVO_4, and glasses. In addition, there are many other host materials that have been demonstrated to accommodate rare earths, and a few of these are subjects of active

Table 1. Properties of Nd:YAG, Nd:glass, and Nd:YLF.

Property	Nd:YAG	Nd:glass (LG-760)	Nd:YLF
Peak stimulated-emission cross section (10^{-19} cm^2)	3	0.4	1.8
$^4F_{3/2}$ lifetime (μs)	230	350	480
Thermal conductivity (W cm^{-1} K^{-1})	0.13	0.006	0.06
Thermal expansion (10^{-6} K^{-1})	8	12.5	8 (c axis) 13 (a axis)
Refractive index n at 1.06 μm	1.815	1.508	1.4704 (n_e) 1.4481 (n_o)
dn/dT (10^{-6} K^{-1})	10	−6.8	−4.3 (n_e) −2 (n_o)
Hardness (Knoop) (kg mm^{-2})	1350	300	300
Crystal system	Cubic	Amorphous	Tetragonal

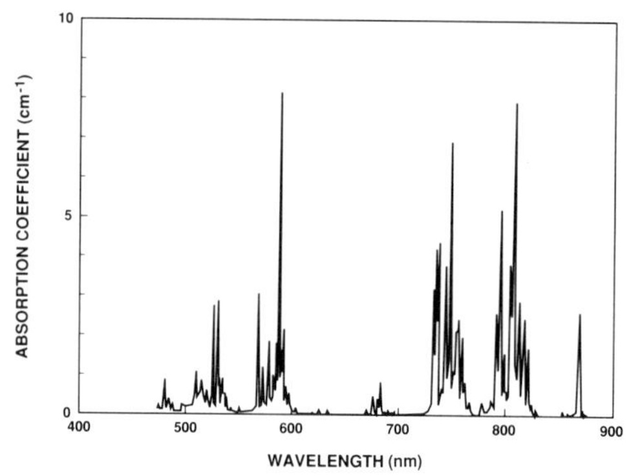

FIG. 4. Absorption spectrum in nominally 1.1-at. % Nd-doped YAG.

research to see whether they can serve as practical laser gain media.

2.1.1.1 Nd:YAG. The most widely used solid-state laser material is Nd:YAG. YAG has a combination of attractive properties for a laser host including high thermal conductivity, hardness, and strength. Some of the properties of Nd:YAG are listed in Table 1. The Nd^{3+} ion substitutes for the Y^{3+} in this material, and all sites are equivalent, which means that the spectrum of the material is homogeneously broadened. The spectroscopic properties of the Nd^{3+} ion, in particular the transition at 1.064 μm, are favorable; the gain cross section is large, $\sim 3\times 10^{-19}$ cm^2; the upper-state lifetime is long, and low laser threshold can be obtained because this is a four-level laser transition. There are also absorption bands in the near-infrared and visible portions of the spectrum so that the laser can be pumped efficiently either by AlGaAs diode lasers in the 0.81-μm wavelength range or by lamps. The absorption spectrum is shown in Fig. 4 for nominal 1.1 at. % doping of Nd. After the Nd ion is excited in these absorption bands, the ion then relaxes nonradiatively to the ${}^4F_{3/2}$ manifold, which is a metastable level; the fluorescence spectrum from this state in the 1.06- and 1.32-μm bands is shown in Fig. 5. The spectrum in Nd:YAG around 1.06 μm is composed of 12 lines (two of the lines overlap), which are transitions from two crystal-field splittings in the upper manifold to six splittings in the lower manifold. Nd:YAG lasers have been tuned to most of the fluorescence lines shown in the spectrum.

Nd:YAG crystal boules are grown by the Czochralski technique up to ~7 cm diameter and 20 cm long. The doping level of Nd is limited to ~3 at. % because of a mismatch in size of the Nd^{3+} ion and the Y^{3+} substitutional site, which leads to stress. This stress in turn leads to optical distortion and difficulties in growing high-quality crystals, and in practice, Nd:YAG crystals are grown with 1.1 at. % doping or less to keep these stress effects within reasonable bounds. Also, as the concentration is increased beyond about 1 at. %, the upper-state lifetime begins to decrease as a result of interactions between neighboring Nd^{3+} ions. This effect is known as *concentration quenching* and is another limiting factor in the concentration of Nd^{3+} in YAG. Not all of the cross-sectional area of a boule is useful for fabrication of gain elements because Nd:YAG grows with facets at the liquid–solid interface. At the intersections of these facets

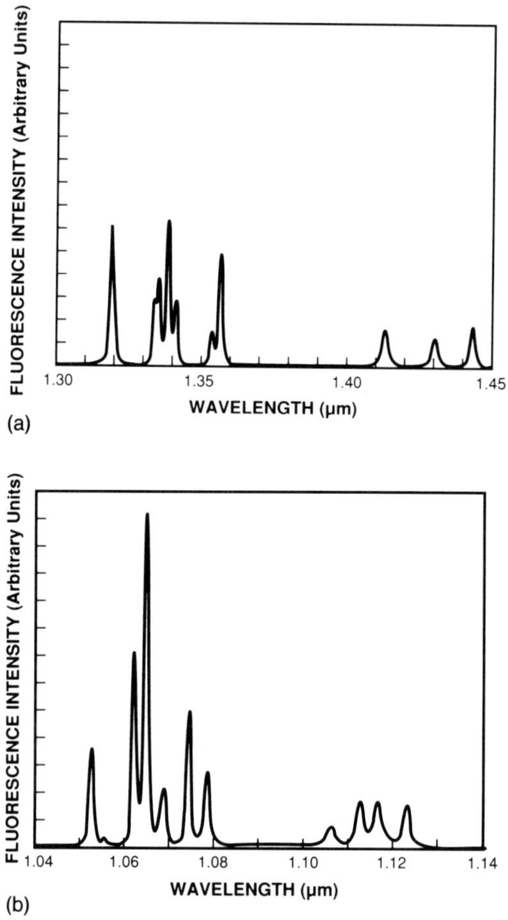

FIG. 5. Fluorescence spectrum around the two main emission bands in Nd:YAG. (a) ${}^4F_{3/2}$–${}^4I_{11/2}$, (b) ${}^4F_{3/2}$–${}^4I_{13/2}$.

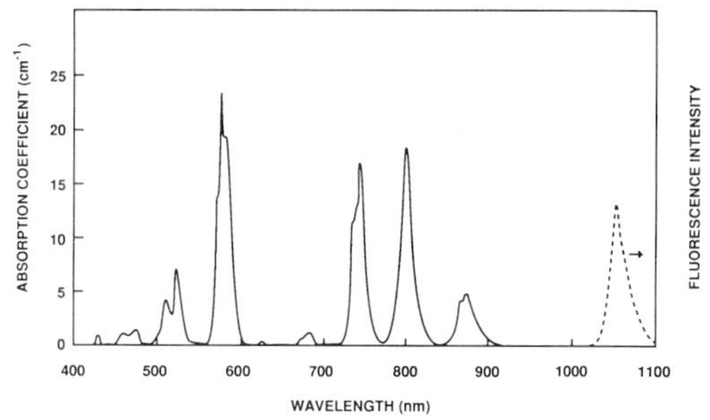

FIG. 6. Absorption spectrum (solid curve) and fluorescence spectrum around 1.06 μm (dashed curve) in Nd-doped phosphate glass. The absorption spectrum is for nominal 6 wt. % Nd_2O_3 doping.

and in the center of the boule, there is a higher incorporation of Nd and stress that leads to optical distortion. Gain elements are cut from boules with the optical axis parallel to the growth axis, and rods are limited in size to $\sim$15 mm diameter by faceting.

2.1.1.2 Nd:Glass. Nd-doped glasses constitute an important class of gain media. Glasses have several advantages as host materials. They are relatively inexpensive compared with single crystals, they can be fabricated in very large sizes, they can be optically polished to a high degree of accuracy, and their properties can be engineered to some extent by changing the glass composition. Major disadvantages relative to many single crystals are low thermal conductivity, hardness, and strength. Some of the properties of one of the more popular Nd-doped phosphate glasses are listed in Table 1.

The Nd transitions in glasses are inhomogeneously broadened by different local site symmetry of the neighboring ions around the activator ion. The fluorescence spectrum in the 1.06-μm wavelength region and the absorption spectrum in a Nd-doped phosphate glass are shown in Fig. 6. In contrast to the spectra of Nd:YAG, the Nd:glass spectrum is relatively featureless, and the increase in linewidth leads to reduced stimulated-emission cross section relative to most single crystals. In some applications this change is desirable; for example, in mode-locked lasers the wider linewidth makes it possible to attain shorter pulses, and the reduced stimulated-emission cross section makes Nd:glass lasers attractive for high-energy, short–pulse-length lasers because relatively high energy can be stored in the metastable level with reduced amplified spontaneous emission and parasitic oscillation.

Nd has been doped into numerous glass compositions; the most important of these are phosphate glasses. By slight changes in the basic glass composition, some of the properties such as change in refractive index with temperature or peak gain cross section can be altered for specific applications. In general, glasses can be doped more heavily than crystals, and in practice the upper bound on Nd^{3+} doping is established by concentration quenching, which is weaker in glasses than in YAG. This effect limits the doping to about 8 wt. % Nd_2O_3 in phosphate glass.

2.1.1.3 Other Rare-Earth-Ion–Doped Lasers. There are a large number of rare-earth-ion–host combinations that have been demonstrated, and a few of these besides Nd:YAG and Nd:glass are of some importance. One other important Nd^{3+}-doped material is Nd:YLF, which is a birefringent crystal. This material can be grown to good optical quality and has the advantages relative to Nd:YAG of a much lower change in refractive index with temperature, a two times longer upper-state lifetime, and a somewhat larger linewidth. This material has found applications in lasers where small thermally induced optical distortion is required, in continuously pumped *Q*-switched lasers because larger energy per pulse can be obtained as a result of the longer lifetime, and in mode-locked lasers where shorter pulses than in Nd:YAG lasers can be generated. The main disadvantage relative to Nd:YAG is poorer thermomechanical properties such as higher thermal expansion, reduced hardness, lower thermal conductivity, and lower yield stress; consequently, Nd:YLF is not attractive for high–average-power laser systems.

Er-doped lasers operating in the 2.8- and 1.5-μm wavelength ranges are becoming more important. The 2.8-μm wavelength is strongly absorbed by water and is of interest for use in medical applications. The most commonly used host materials for 2.8-μm Er^{3+} lasers are the garnets, in particular YAG and $Y_3Sc_2Al_3O_{12}$ (YSAG). These 2.8-μm Er^{3+} materials are sometimes codoped with Cr^{3+}, which provides additional absorption bands for lamp pumping. In this case, the Cr^{3+} dopant is known as a sensitizer because the energy absorbed by the sensitizer is transferred to the

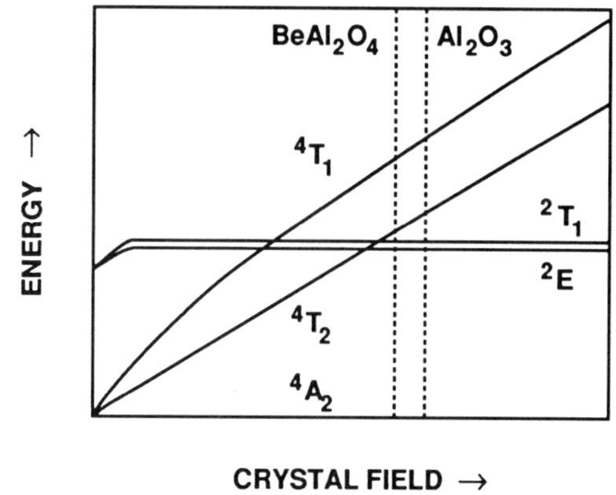

FIG. 7. Simplified crystal-field diagram for the d^3 outer shell, which represents Cr^{3+}, showing only the energy levels important for laser action. The upper laser level is ^2E for high crystal fields, $^4T_2$ for low crystal fields, and a combination of the two states for intermediate crystal fields. The vertical lines represent the crystal fields for ruby and alexandrite.

activator to increase efficiency and pump density. Er-doped glasses are used as host materials for the 1.5-μm wavelength band. The primary applications are in eyesafe laser range finders and amplifiers for 1.5-μm fiber-optic communication systems (Ainslie, 1991). In the latter application, the Er^{3+} is doped into the core of a single-mode optical fiber (see AMPLIFIERS, OPTICAL). Tm^{3+}- and Ho^{3+}-doped lasers both have outputs near 2 μm, and these lasers are of interest for applications where eyesafe wavelengths are required and in medical applications. These lasers have been demonstrated using YAG and YLF as host materials.

2.1.2 Gain Media Doped with Transition-Metal Ions. Transition-metal ions are used as activators in solid-state lasers to provide broad tunability. Lasers have been demonstrated using the first-row transition-metal ions that have 3*d*-electron outer shells with the laser transitions occurring between energy levels of the 3*d* configurations. The position of the energy levels in transition metals is affected by the host matrix more than in rare-earth ions, and an example of the dependence of the energy-level positions on the strength of the crystal field is shown in Fig. 7 for an outer-electron shell of d^3 in octahedral symmetry. The crystal field is a measure of the strength of the interaction between the dopant ion and the host matrix. By varying the crystal field by using hosts with different nearest-neighbor distances, the energy-level positions and thus the output wavelength can be changed. The linewidth of transitions can be either broadband or narrow band depending on the characteristics of the upper and lower electronic states.

The wavelength range over which laser oscillation has been demonstrated using various transition-metal ions is shown in Fig. 8. The utility of any given possible transition between two electronic states is determined by similar factors as for rare-earth ions. One of the additional complications is that the chemistry of transition-metal ion dopants is more complex than that in rare earths, since many transition-metal ions can exist in host materials in multiple valence states whereas the rare earths are almost always trivalent. Also, nonradiative relaxation rates tend to be higher than those for rare earths. Finally, excited-state absorption (ESA) is a limiting factor in many solid-state lasers doped with transition-metal ions. ESA is absorption of a photon by an ion in the upper metastable level, and this process reduces the efficiency of laser operation if the ESA occurs for either the laser or the pump wavelengths. The broadband spectra of transition-metal ions often leads to ESA at the laser wavelengths; this is a particular problem for Ni^{2+}-, V^{2+}-, and some Cr^{3+}-doped materials.

2.1.2.1 Ruby, Alexandrite, and Other Cr^{3+}-Doped Gain Materials. Cr^{3+}-doped solid-state lasers are the most studied class of transition-metal-ion lasers. They have an outer shell of $3d^3$; thus the crystal-field diagram of Fig. 7 applies. Cr^{3+} is of interest because its spectroscopic characteristics, in particular its tunability, upper-state lifetime, and gain cross

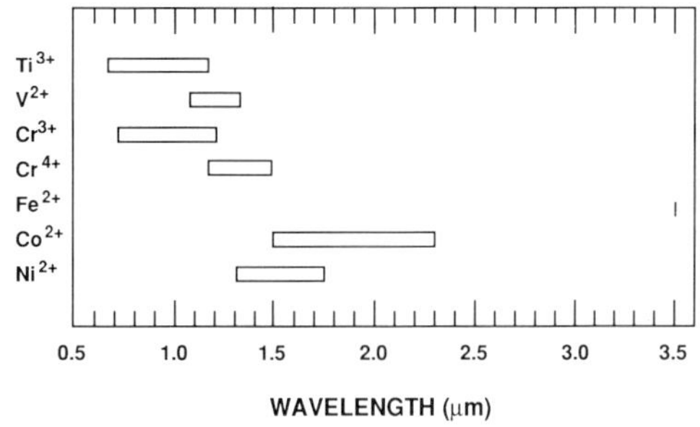

FIG. 8. Wavelengths demonstrated in transition-metal-ion lasers.

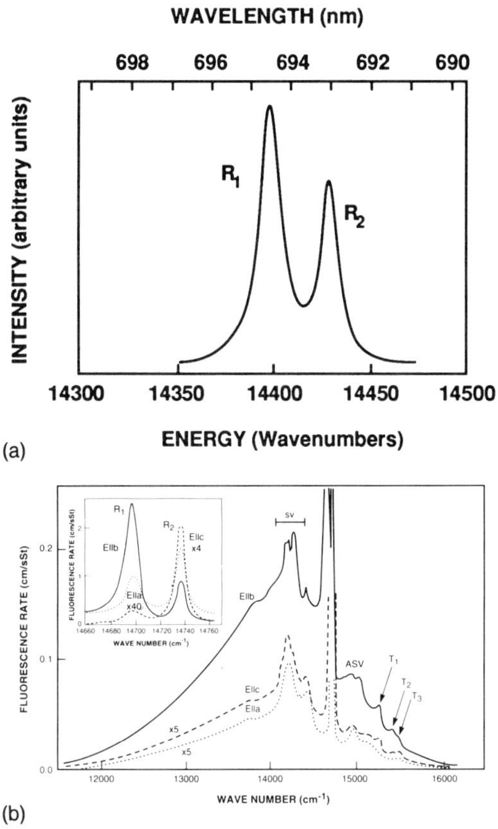

FIG. 9. Fluorescence spectra of (a) ruby and (b) alexandrite (from Walling *et al.*, 1980; © 1980 IEEE).

section, can be engineered to some degree by changing host materials. In addition, the ruby laser is one of the most important solid-state laser devices. Figure 9 shows the fluorescence spectrum of ruby and alexandrite, and some of the properties of these materials are listed in Table 2. The change in spectroscopic properties with host can be described by variations in crystal-field strength; ruby has high strength, and alexandrite has intermediate strength. Laser action has also been demonstrated in materials with low crystal-field strength such as $SrAlF_5$ and $ZnWO_4$. In high crystal fields, the ^2E energy level is the upper state; the transition between the ^2E state and the $^4A_2$ lower state is characterized by narrow linewidth and long metastable state lifetime. In intermediate crystal fields, ^2E and $^4T_2$ are simultaneously populated in the excited state, leading to both narrow-band and broadband features in the spectrum. For low crystal fields, the upper level is the $^4T_2$, and the transition between this state and the $^4A_2$ lower state is characterized by shorter lifetime and broadband gain.

Ruby lasers are useful in applications that require high energy per pulse because ruby can be effectively pumped by lamps, it has a low gain cross section that reduces amplified spontaneous emission and parasitics, and Al_2O_3 is an excellent host material as a result of its high thermal conductivity, strength, chemical stability, and ease of growth. This is a three-level laser; thus the threshold is high, but its broadband absorption and long metastable level lifetime allow efficient utilization of the pump radiation from flashlamps. The detailed energy-level diagram of ruby is shown in Fig. 10.

Alexandrite was the first broadly tunable Cr^{3+} laser, and its discovery initiated much of the work in understanding of the effect of host material on energy-level positions. Alexan-

Table 2. Properties of ruby and alexandrite.

Property	Ruby	Alexandrite
Stimulated-emission cross section at 300 K (10^{-20} cm^2)	2.5	0.7 (at 0.7 μm)
Upper-state lifetime (ms)	3	0.2
Thermal conductivity ($W\ cm^{-1}\ K^{-1}$)	0.35($\parallel c$) 0.33($\perp c$)	0.23
Thermal expansion ($10^{-6}\ K^{-1}$)	5.3($\parallel c$) 4.8($\perp c$)	6($\parallel a$) 6($\parallel b$) 7($\parallel c$)
Refractive index	1.7635 (n_o at 0.69 μm) 1.7555 (n_e)	1.7381 (n_a at 0.7 μm) 1.7436 (n_b) 1.7361 (n_c)
dn/dT ($10^{-6}\ K^{-1}$)	13 (n_o) 17 (n_e)	8
Crystal system	Trigonal	Orthorhombic

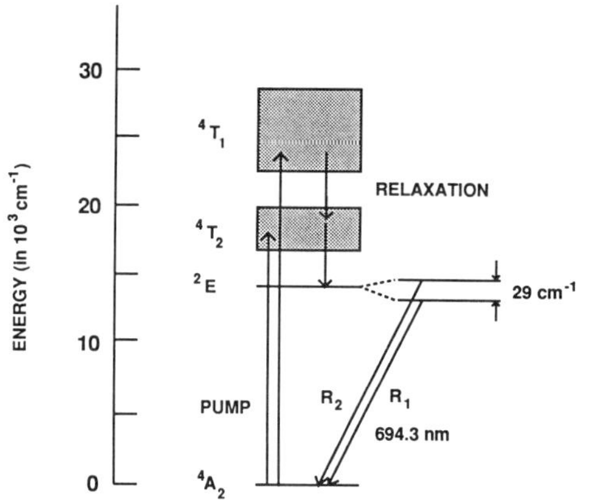

FIG. 10. Energy-level diagram of ruby.

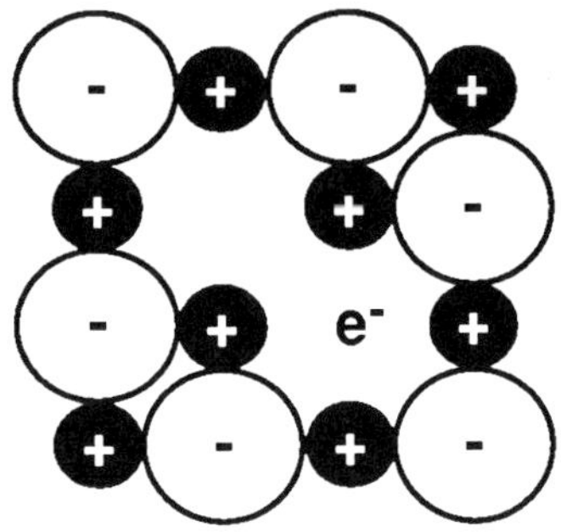

FIG. 12. Schematic of the F^{2+} color center. The small filled-in circles represent positive ions while the large unfilled circles represent negative ions. The electron is denoted by e^-.

drite lasers are useful as tunable lasers in the 700–820-nm wavelength range. Laser systems with high energy per pulse and high average power can be built using alexandrite because of its excellent thermomechanical properties. Many other Cr^{3+}-doped laser materials have been demonstrated, but only a few will ultimately prove to be useful. Excited-state absorption limits the efficiency and tuning range of many of these other host materials, particularly those with low crystal fields, and in some cases, crystal growth is difficult. Two host materials for Cr^{3+} that are becoming commercially available are $LiSrAlF_6$ and $LiCaAlF_6$, which can both be grown to good optical quality with relative ease and have only weak excited-state absorption.

2.1.2.2 Ti:Al$_2$O$_3$. The most broadly tunable solid-state laser material is $Ti:Al_2O_3$ (Moulton, 1986). The fluorescence and absorption spectra of this material are shown in Fig. 11. Titanium is trivalent in this material, which leads to a single 3*d* electron in its outer shell and thus to a simple energy-level diagram with no chance for excited-state absorption to other 3*d* energy levels. Wide tunability of this laser has been demonstrated from 670 to 1100 nm; thus this laser has found wide acceptance as a replacement for near-infrared dye lasers in scientific applications. Other host materials with Ti^{3+} have also been investigated, but $Ti:Al_2O_3$ is the only material in which efficient laser operation using this ion has been demonstrated. $Ti:Al_2O_3$ lasers can be effectively pumped by either lamps or lasers such as Cu vapor, argon ion, or frequency-doubled Nd:YAG.

2.1.3 Color Centers. Color-center lasers use electronic transitions of defect centers to provide laser gain, and the largest class of host materials for these centers comprises the al-

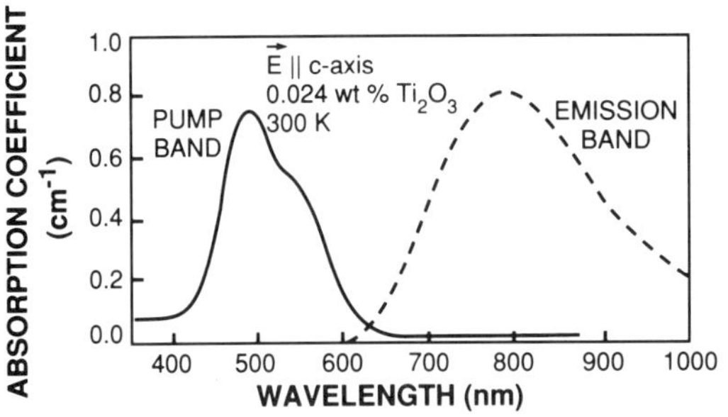

FIG. 11. Absorption and fluorescence spectra in $Ti:Al_2O_3$.

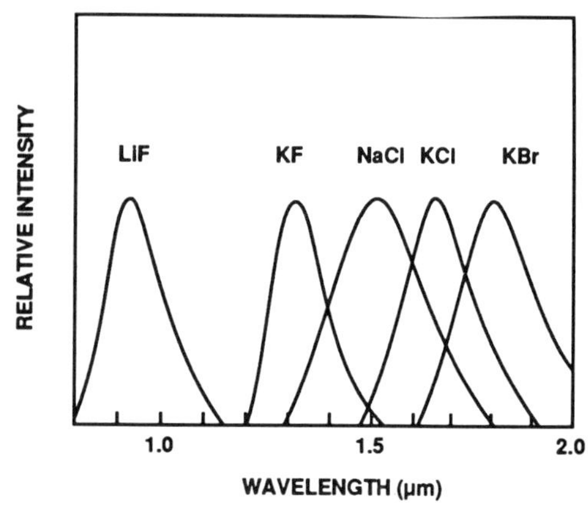

FIG. 13. Fluorescence spectra of the F^{2+} center in various alkali-halide host materials.

kali halide crystals. One example of a color center is the F^{2+} center, which consists of two adjacent halide-ion vacancies with a trapped electron and is shown in Fig. 12. This configuration can be viewed and modeled as a singly ionized H_2 molecule inside a host matrix with the vacancies acting as the nuclei. The positions of the energy levels vary with host material, and Fig. 13 illustrates the change in wavelength of the fluorescence spectra of the F^{2+} center in different host materials. Strong interaction of the defect center with the lattice in these materials leads to the broadband spectra. The laser transitions are strongly electric-dipole allowed because the upper and lower energy levels are states of opposite parity, leading to upper-state lifetimes of tens to hundreds of nanoseconds and stimulated-emission cross sections in the range of 10^{-16} cm^2.

Other color centers have also been used for lasers including the $F_A(II)$ and $Tl^0(1)$ center in which the defect centers are associated with impurities. The $F_A(II)$ center is a halide vacancy containing a trapped electron that is adjacent to an alkali impurity, and the $Tl^0(1)$ center is an alkali vacancy adjacent to a neutral Tl atom that substitutes on a halide site. Lasers based on these defect centers also work in the infrared portion of the spectrum.

Color-center lasers have found only limited use, primarily in scientific applications, because of limitations in the stability of the defect center. Most color-center lasers require cryogenic operation, and in some cases the gain element requires cryogenic storage as otherwise the color center fades. Typical fading times at room temperature are on the order of a minute or less under cw laser operation, although there are reports of pulsed operation of some color centers at room temperature. In addition, many alkali halides are not ideal laser host materials; they are soft, and many are hydroscopic.

2.2 Pump Sources

Solid-state lasers are optically pumped, either by a lamp or by another laser. The desired characteristics of these pump sources are that they be efficient converters of electrical energy to optical energy, that the wavelength of the pump source spectrally match the pump bands of the gain medium, and that the energy or power density be high. The choice of a particular pump source for a particular gain medium is dependent on how the characteristics of the pump source match the requirements for efficient, low-threshold operation of the laser and on the requirements for the application—for example, cw or pulsed operation, the average output power, and energy per pulse.

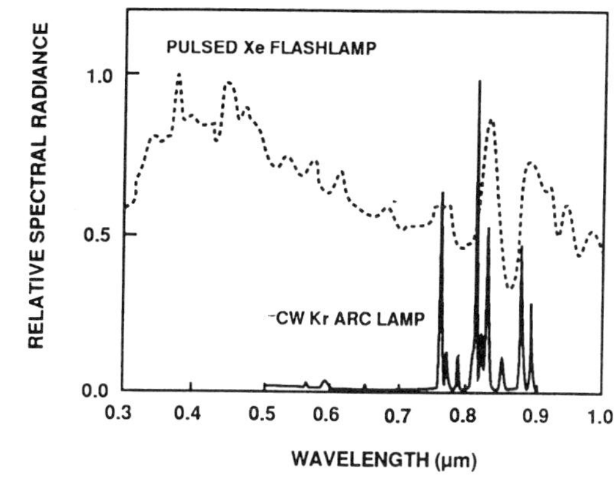

FIG. 14. Spectral power density of the outputs for cw Kr and pulsed Xe arc lamps.

2.2.1 Lamps. Lamps can be used to pump both pulsed and cw lasers. For cw lasers, linear xenon or krypton arc lamps or tungsten-halogen lamps are used. Pulsed flashlamps are gas-discharge tubes filled with xenon or krypton and come in linear or helical shapes. Helical lamps are useful for high–energy-per-pulse laser systems because of their longer arc length per length of laser rod. Other types of lamps such as Na vapor discharge lamps can be used but are less common.

The output spectra for a cw krypton arc lamp and a pulsed xenon flashlamp are shown in Fig. 14. The spectrum of a tungsten-halogen lamp closely approximates a blackbody emitter. The spectrum of a cw krypton arc lamp is dominated by line emission; these lamps are particularly useful in pumping Nd:YAG lasers because of the excellent spectral match between some of the strong lines from the lamp with absorption features in Nd:YAG. The operating temperatures and ion and electron densities are very different in pulsed flashlamps from the cw regime; the high temperatures, electron density, and ion density in pulsed flashlamps lead to a spectrum that is dominated by continuum radiation as opposed to line spectra, and the ratio of the two contributions is determined to a large extent by the current density in the flashlamp.

Lamps are available in diameters from ~2 to 20 mm and arc lengths of 25 mm to greater than 1 m and can be cooled with either air or liquids. The size of the lamp used is determined by the required spectral characteristics, size of the gain medium, energy per pulse and/or average power of lamp operation, and desired lamp lifetime. For efficient coupling of the lamp light into a gain medium, the size of the lamp should not be larger than the gain medium itself. In particular, for pumping rods, imaging techniques are often used that work efficiently only for small lamps. The area of the lamp jacket must be sufficient to remove heat to avoid excessive temperature rise, and in the pulsed mode the lamp must be large enough to avoid explosion caused by shock waves that arise from heating of the gas. The lifetime of pulsed flashlamps is a strong function of the ratio of the operating energy per pulse relative to the explosion limit. For pulsed flashlamps operating at 20% of the explosion limit, the lamp lifetime is $\sim 10^7$ pulses. In cw arc lamps, the lifetime is limited by deposition of material on the lamp wall from the electrode, which reduces the light output; the lamp lifetime is typically limited to a few hundred hours.

2.2.2 Laser Sources. Lasers are becoming increasingly common as pump sources for solid-state lasers because they have higher brightness than lamps and have directional output, which means that focusing of their beams can be used to obtain higher excitation densities than are available from lamps. Their output is also in only a narrow wavelength range; thus very efficient utilization of their output is achieved provided that the output wavelength is at an absorption feature. While it may not seem to make sense to use lasers to pump other lasers, laser-pumped lasers can provide wavelength shifts, peak-power enhancement, beam-quality improvement, and spectral-brightness improvements relative to the pump laser.

Semiconductor diode lasers are the most common pump lasers and are used primarily for pumping Nd^{3+} lasers. AlGaAs diode lasers can be fabricated to emit near 0.8 μm where there is a strong absorption in Nd^{3+}. Fabrication of large arrays of diode lasers for solid-state laser pumping has been demonstrated, and these arrays are capable of emitting >1 kW/cm^2 from areas of >10 cm^2 on a pulsed basis (Endriz *et al.*, 1992). This technology is an area of very active research that promises to lead to efficient, all–solid-state laser devices (Fan and Byer, 1988; Hughes and Barr, 1992). As pump sources, the advantages relative to lamps are low voltage requirements and longer operational lifetimes, with lifetimes of tens of thousands of hours cw and 10^9 shots pulsed. Diode lasers also reduce the thermal load for Nd^{3+} lasers. In Nd^{3+} lasers the dominant contribution to heat generation per absorbed pump photon is nonradiative relaxation, which generates heat nominally equal to the difference between the pump photon energy and the output photon energy. In pumping Nd^{3+} lasers, the average photon energy for pulsed flashlamps is significantly larger than that for AlGaAs diodes operating at 0.81 μm; the difference in thermal loading for these two pump sources in Nd:YAG is about a factor of 3.

Other examples of laser-pumped lasers are Ti:Al_2O_3, Er-doped fibers, and color-center lasers. Ti:Al_2O_3 requires pump sources in the blue-green part of the spectrum and can be pumped by Ar-ion, Cu-vapor, and frequency-doubled Nd:YAG lasers. Lasers are especially useful in pumping Ti:Al_2O_3 because of its low product of metastable-level lifetime and stimulated-emission cross section, which requires a high pump rate to achieve laser threshold (see LASER PHYSICS). Er-doped fibers are used in fiber-optic communication, and here diode lasers operating around either 0.98 or 1.48 μm are the pump sources of choice because of their long lifetime, high efficiency, and the ability to couple their output into fibers efficiently. All color-center lasers are pumped by lasers. Lamps can have output in wavelength ranges that cause a photoinduced fading of the defect center.

3. LASER DESIGN

The laser engineer is confronted with many design choices including the size and shape of the gain element, the method of coupling the pump radiation into the gain element, and the optical resonator design. The choices are driven by the characteristics of the gain medium, such as the stimulated-emission cross section, upper-state lifetime, and thermal properties;

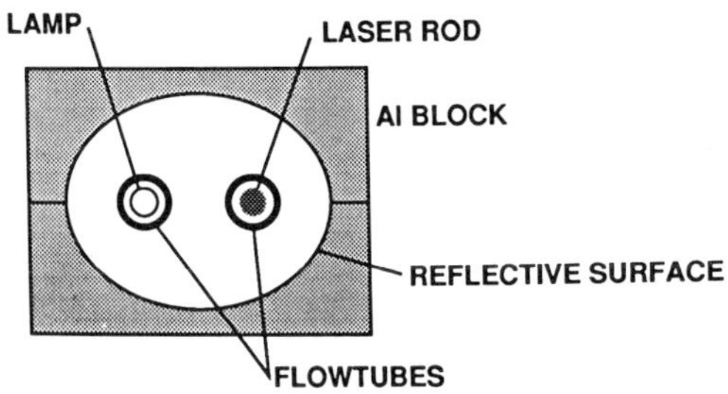

FIG. 15. Cross-sectional schematic of a pump cavity for coupling of cylindrical lamps to a laser rod. The rod and lamp are located at the foci of an ellipse so that radiation from the lamp is imaged onto the rod.

the type of pump source, lamp or laser; and the requirements of the application such as beam quality, average power, energy per pulse, and spectral output (see LASER TECHNOLOGY).

3.1 Gain-Element Geometries

Laser gain elements in solid-state lasers have been fabricated primarily in the forms of rods, slabs, and disks. The choice of gain-element geometry is governed by the average power and/or energy-per-pulse requirements, raw material (i.e., crystal boule) shape and size, and cost.

3.1.1 Rods. The most common form of solid-state gain element is a cylindrical rod in which the optical axis of the laser resonator is collinear with the axis of the cylinder. In rods the length of the cylinder is larger than the beam size of the laser. Rods are relatively inexpensive to fabricate, and the pump radiation from either linear or helical lamps can be coupled efficiently using simple pump cavities as shown in Fig. 15. Rod sizes range from ~2 mm diameter to as large as ~10 cm diameter and 2 m long. These larger-diameter rods are available only in glasses; crystalline rods are limited to ~20 mm diameter by crystal size.

A key issue in all geometries is the effect of heat generation on the optical properties of the gain element. Rods are typically cooled by flowing a coolant such as water across the barrel, which leads to a radial temperature gradient in the rod. This gradient leads to two separate effects:

1. a radially varying index of refraction due to change in refractive index with temperature, and
2. radially varying stress due to differential thermal expansion.

To first order the former effect acts as a lens with a focal length that decreases with increasing thermal load. This causes the cavity modes to vary in size with pump power, and over a sufficiently wide variation in thermal load, the cavity can change from being stable to unstable or vice versa (see LASER PHYSICS). To ameliorate this effect, laser cavities are often designed to accommodate thermal lensing at the nominal operating point of the laser.

The stress also causes a change in refractive index and causes an isotropic gain element to become birefringent, with the characteristic that the laser rod acts as a waveplate with variable retardation across its aperture and with principal axes whose directions also change across the rod aperture. If polarization elements are used inside the laser resonator, substantial loss will occur, although there are partial compensation techniques that can reduce this loss by using multiple laser rods and additional intracavity optical elements. Stress-induced birefringence can be essentially eliminated by using materials that are naturally birefringent such as YLF (as long as laser beam propagation is not along the optical axis of the crystal) since the natural birefringence is much greater than the stress-induced birefringence. At high enough thermal load, stress can eventually lead to fracture. In YAG, the stress-fracture limit is ~100 W of thermal load per linear centimeter of rod length, in YLF it is ~25 W/cm, and in phosphate glass it is ~10 W/cm. The stress-fracture limit is independent of rod diameter, and thus the average power limit of rod lasers can be increased only by lengthening the rod; how-

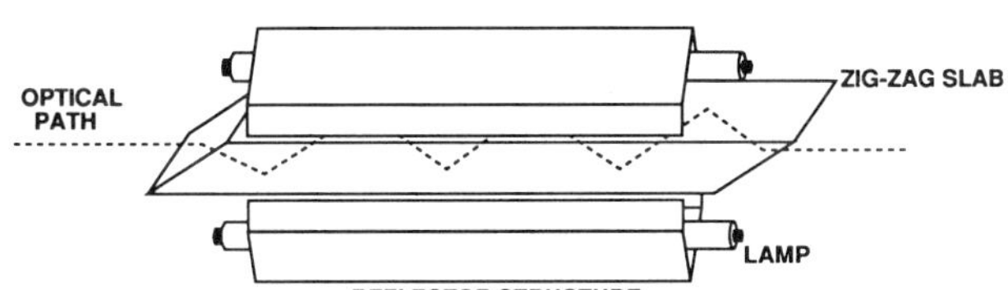

FIG. 16. Schematic of a zigzag slab laser head.

ever, rod length is ultimately limited by crystal size and lamp characteristics. High-power rod lasers (>1 kW average) have been demonstrated using multiple rods.

An optical fiber is a rod in the limit of small diameter and long length. The operation of fiber lasers and amplifiers is not limited by thermal effects because of their large surface-to-volume ratio and their long length; instead, the power obtainable from a fiber laser or amplifier is limited by damage at the ends of the fiber due to the small area and consequent high surface power density. For single-mode fiber devices, surface damage limits the average power to ~ 10 W. High gain can be achieved in fiber devices when laser pumping is employed and the pump radiation is guided by the core, because this arrangement gives high pump intensity over a long distance. In Er^{3+}-doped fiber amplifiers, unsaturated power gains in excess of 30 dB have been demonstrated.

3.1.2 Zigzag Slabs and Disks. Gain elements in the form of zigzag slabs reduce the thermal lensing and birefringence problems. In these slabs, the laser beam is incident on the slab at Brewster's angle and then propagates down the slab making total internal reflections off of two faces as shown in Fig. 16. The slab is pumped and cooled through the bounce faces while the other two faces are insulated. By cooling through only the pump faces the thermal gradient is essentially one dimensional, assuming the pump density is uniform throughout the slab, and then by zigzagging across the slab, the refractive index variation due to the thermal gradients is averaged to zero because all parts of the beam sample all parts of the slab. Thermally induced depolarization of linearly polarized beams is also reduced in slabs relative to rods. While thermally induced birefringence still exists, the principal axes of the induced birefringence are aligned across the entire slab aperture. Thus linearly polarized light with polarization along one of the principal directions will not suffer depolarization.

The stress-fracture limit in slabs scales differently from rods; it is proportional to the area of a pump face divided by the thickness of the slab. In Nd:YAG, the stress fracture limit in zigzag slabs is nominally $200A/t$ W/cm, where A is in units of cm^2 and t is in units of cm; the quantity A/t is also known as the aspect ratio. While large–aspect-ratio slabs are desirable for high average power, there are practical limits to using large–aspect-ratio slabs. A slab cannot be made too thin, otherwise it will absorb pump light inefficiently, and large–aspect-ratio slabs are difficult to fabricate. Typical Nd:YAG slab sizes are 10 cm long by 2 cm wide by 0.6 cm thick. High-power slab lasers (~ 1 kW) have been demonstrated with nearly ideal output beams and with little change in beam size as the power is varied.

Slabs do have some significant drawbacks. They are more costly to fabricate than rods because of the large area of optical surface. Another problem is coolant flow across the optical faces of the slab, because these faces can be eroded by the flow and deposits can become attached to the slab faces, leading to optical loss and/or damage. Finally, careful engineering is required to realize the benefits of zigzag slab lasers. Uniform deposition of the pump power is extremely important. In lamp-pumped systems, reflector mirror structures for directing the pump radiation are often empirically designed to achieve uniform pumping. Also, there are significant nonideal effects that occur at the ends of the slab that are difficult to model and predict. These difficulties have led to slow marketplace acceptance of slab laser technology.

Disks are used as amplifiers in laser systems with high peak power and energy per pulse, such as fusion lasers. They are typically

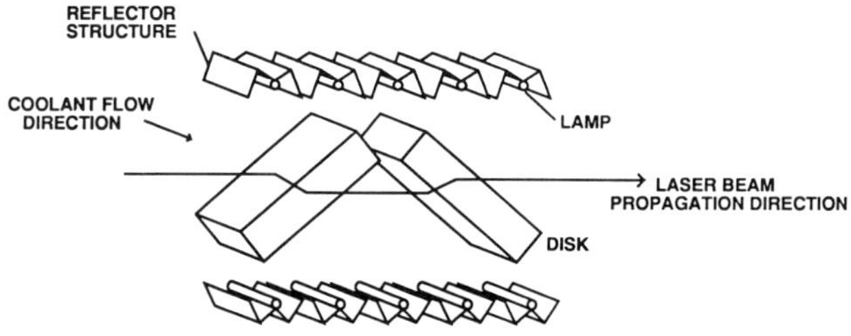

FIG. 17. Schematic of a laser disk amplifier.

inserted into the beam at Brewster's angle as shown in Fig. 17, and their large area, up to ~1 m diameter, ensures that the fluence is kept below the damage fluence. These disks are pumped by lamps through their faces and are also cooled through the faces. As in zigzag slab lasers, uniform pumping is needed to reduce thermo-optic aberrations. One difficulty with operating these lasers at high average power is that the laser beam propagates through the coolant, which can be either gas or liquid. Significant distortion of the beam can result because of thermally induced refractive-index changes in the coolant; thus disk amplifier systems have been limited primarily to low repetition rates (<1 Hz).

3.2 Pump Geometries

There are two basic pumping geometries: longitudinal (or end pumped), in which the pump radiation is directed nominally collinear to the optical axis of the laser, and transverse (side pumped), in which the pump radiation is directed nominally perpendicular to the optical axis of the laser. The choice is governed by the pump source characteristics, the gain-element geometry, and the characteristics of the gain medium. Almost all lamp-pumped solid-state lasers are side pumped, because efficient coupling of the pump radiation to rods is easily achieved since the diameter-to-length ratio of a lamp can be matched to the rod's diameter-to-length ratio. It is difficult to couple lamps to rods effectively in an end-pumped geometry because of the lack of directionality to the lamp output.

On the other hand, many laser-pumped lasers are configured in an end-pumped geometry to take advantage of the focusability of laser radiation; this geometry is preferred for efficient, low-power devices. The ability to focus laser radiation leads to high pumping rates and consequently low threshold and excellent matching to the TEM_{00} mode of the solid-state laser. However, end pumping becomes more inconvenient for higher-power devices when large numbers of pump lasers are needed and thermo-optic distortions become more of a problem. In such cases, side pumping becomes more attractive. In particular, in high–energy-per-pulse diode-pumped lasers, a large array containing many pump lasers is required, and focusing the output of such a pump source becomes difficult.

3.3 Laser Resonators

The vast majority of solid-state lasers use simple two-mirror resonators (see LASER PHYSICS). Both stable resonators and unstable resonators are used in commercial devices; the difference is that stable resonators are used in relatively low-gain lasers while unstable resonators are used in high-gain lasers with large output apertures. Stable resonators are those in which Gaussian laser modes are confined in their transverse extent upon propagation around the laser resonator; resonators not satisfying this condition are unstable. In stable resonators, the output is extracted from the cavity using a partially reflective output mirror, while in unstable resonators, the output is extracted by diffraction around a cavity mirror.

The high-gain laser systems that use unstable resonators are almost exclusively *Q*-switched Nd:YAG lasers operating at relatively low repetition rates, <100 Hz, and high energy per pulse, >100 mJ; almost all other solid-state lasers use stable resonators. Unstable resonators are appropriate for systems with high gain because they allow efficient laser operation in the fundamental transverse mode of the resonator across gain media apertures that are large. For lower-gain systems that require low-magnification unstable resonators, the mirror alignment requirements become increasingly stringent. Efficient operation in the fundamental transverse mode can be obtained with stable resonators, but only for devices in which the limiting cross-sectional aperture size is less than $\sim(\lambda L)^{1/2}$, where λ is the laser wavelength and L is the cavity length. For larger limiting apertures, the laser will run in multiple transverse modes. In fact, one challenge in engineering of solid-state lasers is the resonator design for low-gain lasers that yields laser output in the fundamental transverse mode with good extraction across a large aperture (Magni *et al.*, 1991).

Laser resonators are often designed to take into account thermal lensing in laser rods. For example, many Nd:YAG rod lasers have resonators that are unstable or at the edge of the stability region with no input power, but as the input power is increased thermal lensing of the rod is used to bring the laser cavity into stability or to its desired operating point.

However, this technique is useful only over a limited output power range of the laser.

One trend in laser resonator design is to use graded-reflectivity mirrors. Traditionally, output mirrors for stable resonators have uniform reflectivity across their surfaces, and an unstable-resonator output mirror has a region that is highly reflective over a portion of its area with the rest having no reflectivity. A typical unstable resonator for a rod laser has a highly reflective dot at its center. A graded-reflectivity mirror has a reflectivity that varies across its face with the highest reflectivity in the center of the mirror. These mirrors add an additional degree of freedom to resonator design. They have been used with stable and unstable resonators to achieve output in a single transverse mode with good beam quality and high efficiency.

More complicated resonators with a larger number of mirrors are also used in solid-state lasers. These resonators find applications in tunable lasers or single-frequency lasers. In tunable lasers, additional laser cavity elements such as birefringent tuning filters are often inserted in the cavity, and the gain element often has Brewster's-angle faces. These elements add astigmatism to the laser mode, which can be compensated by using a three- or four-mirror cavity. Single-frequency performance from solid-state lasers is often obtained by using a unidirectional ring resonator, which requires additional mirrors to form a ring path.

4. LASER MANUFACTURE

The key components in solid-state lasers are the gain element, pump source, and resonator optics. High-quality components are required for solid-state lasers if high power, good efficiency, good beam quality, and minimal environmental sensitivity are desired.

4.1 Gain-Element Fabrication

Most of the important solid-state laser crystals are grown using the Czochralski technique, including Nd:YAG, Nd:YLF, ruby, and Ti:Al_2O_3 (see CRYSTAL GROWTH). Provided that reasonable size crystals of the host material can be grown, there are additional issues that need to be addressed to determine the suitability for laser gain elements. These include the impact of dopants on the crystal-growth process, facet formation, impurity incorporation, and adequate control of crystal-growth conditions.

FIG. 18. A piece of YAG viewed along the crystal-growth axis under crossed polarizers. The light regions indicate strain. The highly strained center region is known as the core. The strained regions of the crystal cannot be used because they are optically distorted and contain stress-induced birefringence.

For example, in Nd:YAG, the Nd^{3+} ion is significantly larger than the Y^{3+} ion for which it substitutes (radius of 0.098 versus 0.090 nm). This leads to stress in the crystal, which in turn leads to refractive-index distortion and limits on the Nd concentration. The size difference also causes the distribution coefficient for neodymium to be only 0.18; that is, the ratio of the dopant concentration that solidifies to that in the melt is only 0.18. This leads to an increase in Nd concentration in the melt as the boule is grown and a resulting increase in Nd concentration along the length of the boule. Because of differences in crystal-growth rates along various crystallographic planes, the YAG solid–liquid interface is faceted. Additional dopant and stress accumulates at the intersection of the facets, leading to these parts of the boule being unusable as seen in Fig. 18. It is important that impurities be kept to a minimum, because they can lead to formation of second phases or inclusions that optically scatter, act as damage sites, or cause optical absorption at the laser wavelength. Finally, good control of mixing of the melt and of temperature needs to be maintained to achieve uniform growth rates. Poor control can lead to stress caused by nonuniform growth rates, and to striations, which are quasiperiodic composition and refractive-index variations along the length of the boule.

In glasses, there are fewer problems in gain-element fabrication. The ability to mix and cast large pieces of glass doped with rare-earth ions with the high uniformity, large size, and optical quality needed for laser gain elements is within the state of the art. Gain elements in Nd-doped phosphate glasses have been fabricated in sizes nearly as large as 1 m in diameter. Single-mode optical fibers doped with rare-earth ions have activator-ion concentration on the order of a few hundred parts per million. The doping is lower than that for most bulk gain elements because the silica glass used for most optical fibers has low solubility for rare earths. This low doping is typically not a problem in optical-fiber devices because long fiber lengths can be used to achieve efficient absorption of the pump radiation. The activator ions are introduced into the core region of a fiber preform using either a vapor-phase or aqueous-solution technique, and then fibers are drawn using standard optical-fiber fabrication methods (see FIBER OPTICS).

Color centers are created in alkali halides by two techniques, additive coloration and radiation damage. In additive coloration, the alkali halide is placed in a vapor bath of alkali metal, which then diffuses into the crystal, creating an excess of metal. To compensate for this excess of metal, vacancies with electrons also diffuse into the crystal, creating color centers. In radiation-damage techniques, high-energy radiation (x rays, γ rays, or electrons) creates electron–hole pairs, which then recombine with the recombination energy kicking a halogen atom into an interstitial space in the lattice. An electron is then left in the vacancy, creating a color center. After additive coloration or radiation damage, further processing such as annealing is often required to obtain the desired type of color center. Additive coloration is the preferred technique because the color centers are more stable.

4.2 Optical Requirements for Solid-State Lasers

The key parameters of interest for intracavity optical components are optical loss, optical distortion, surface figure, surface quality, and birefringence.

Optical loss, due to either absorption or scattering, needs to be minimized for low-threshold and high-efficiency operation; this loss is typically $\sim$0.1%/cm or less for high-quality optical components. Optical distortion and surface figure are measurements of the optical perfection of bulk optical elements and surfaces, respectively. Optical distortion and imperfect surface figure cause optical wavefronts to become distorted, which leads to additional intracavity loss and laser-beam distortion. Optical distortion in laser gain elements and on optical surfaces is typically specified to be less than $\lambda/4$ (at 633 nm) per optical element or surface. This requirement is easy to meet in glasses, but meeting this requirement in crystals can be more of a problem. Surface quality refers to residual scratches and digs in the surface of an optic after polishing. Poor surface quality leads to scatter loss, and scratches and digs are centers for optical damage under high-power operation. Optical elements that in principle are isotropic can act as birefringent elements as a result of either residual stress or stress induced by mounting; unwanted birefringence can cause loss in laser cavities with intracavity polarizing elements. In crystals such as YAG, residual stress limits the fraction of the boule that is useable. Glass elements can have low residual birefringence, but small amounts of mounting stress can lead to significant birefringence that is easily observed under crossed polarizers.

Evaporated dielectric thin-film coatings are used for mirrors, polarizers, and antireflection coatings (see also FILTERS, OPTICAL). These coatings are composed of single and multiple layers of materials such as MgF_2, ThF_4, SiO_2, CeO_2, ZrO_2, ZnS, and TiO_2 that are a quarter-wavelength thick at the wavelength of interest. Reflecting mirrors are made by alternating films of low refractive index, the first three materials, with films of high refractive index, the other materials. Simple antireflection coatings are made using a quarter-wave-thick coating of a film that has a refractive index equal to the square root of the substrate refractive index. Important characteristics for these thin-film coatings include good adhesion to the substrate, low optical absorption, good hardness, and high optical-power damage threshold. Significant effort has gone into development of high–damage-threshold optics for use with *Q*-switched and other short-pulse lasers, particularly around 1.06 μm; damage thresholds for these optics are in the range of a few GW/cm^2 peak-power density for nanosecond long pulses with an energy-density damage threshold on the order of 20 J/cm^2. Damage thresholds vary with pulse length, repetition rate, wavelength, coating technique, the coating's history, and beam size.

The support structure for the laser gain medium and for the optical resonator needs to be rigid and temperature insensitive to varying degrees depending on the type of laser and the application. Many lasers for research applications have open structures with access to all the cavity components to allow modification of the laser. Military laser systems are enclosed in very rugged structures that are compatible with large changes in environment and with the acceleration and vibration that are associated with moving platforms. Mode-locked lasers often are built with materials with low thermal-expansion coefficient, such as Invar and Zerodur, because the cavity length needs to remain fixed to within a few micrometers.

5. APPLICATIONS OF SOLID-STATE LASERS

The number of applications of solid-state lasers is increasing as these devices become more efficient, more reliable, and more flexible. Many applications rely on the solid-state laser's unique combination of high peak and average power with good beam quality. This section discusses some of the more important solid-state laser applications.

5.1 Industrial Applications

Solid-state lasers have found industrial applications ranging from cutting, welding, drilling, and marking of metals to micromachining of integrated circuits (see also LASERS, INDUSTRIAL USE OF). High-power Nd:YAG lasers, with up to 2 kW average power, are used in the metal working; in this application, the primary competition is the CO_2 laser. Nd:YAG lasers have the advantages of smaller spot sizes and somewhat higher absorption by metals, both of which are due to the shorter Nd:YAG laser wavelength. In addition, the 1-μm radiation can be delivered by optical fiber, whereas more cumbersome delivery systems are required for CO_2 lasers. On the other hand, CO_2 lasers currently offer an economic advantage of higher output power at a given price, which translates into greater workpiece speed. In practice, the choice of laser is determined by a complex tradeoff among average power, energy per pulse, wavelength, spot size, metalworking application, and material.

Solid-state lasers are also used in the electronics industry in applications such as soldering, resistor trimming, and link blowing in integrated circuits. Both CO_2 and Nd:YAG lasers are used in laser soldering, with Nd:YAG lasers having the advantage of being able to make smaller solder joints. *Q*-switched Nd lasers are the lasers of choice in thick- and thin-film resistor trimming applications, because there are fewer problems with cracking of the substrate; these thick- and thin-film hybrid circuits often use alumina substrates that weakly absorb 1-μm radiation but strongly absorb the 10-μm radiation from CO_2 lasers. Diode-pumped, frequency-doubled, *Q*-switched Nd:YAG and Nd:YLF lasers have found applications in integrated-circuit manufacturing in link blowing (i.e., cutting conductive lines) for repair of high-density random-access memory chips.

5.2 Military Applications

The two primary military uses of solid-state lasers are in rangefinders and target designators. In the first application, *Q*-switched pulses from Nd:YAG or ruby lasers are aimed at a target, and the distance is measured by the round-trip time of flight. In the second application, a Nd:YAG laser illuminates a target, and a weapon equipped with a semi-active guidance system homes in on the illuminated spot. Solid-state lasers are particularly suited for these applications because of their combination of good beam quality, high energy per pulse, short pulse length, and wavelength. The first two qualities are needed to allow operation over ranges of a few kilometers. Short pulses are needed for good range resolution in rangefinder applications, and the near-infrared wavelength of Nd:YAG lasers and the red wavelength of ruby lasers allow simple, room-temperature silicon photodetectors to be used for sensors.

Solid-state lasers are under active investigation for other military applications including electro-optic countermeasures, laser-radar systems, and space-to-underwater communications. Electro-optic countermeasures are used to jam optical detection systems such as infrared sensors on heat-seeking missiles. Laser-radar systems can be used to track and identify targets, and they have higher resolution than microwave radar because of the shorter wavelength. Laser-radar systems can also be used to remotely identify atmospheric chemicals or sense wind velocity. In space-to-underwater communications, information can be encoded on a laser beam that is transmitted through the ocean to a submarine. A narrow-band optical filter on the submarine allows detection of the narrow-band optical radiation against the background. While solid-state lasers may play a role in these applications, other laser technologies may also be suited for the task.

5.3 Medical Applications

Nd:YAG, along with CO_2 and argon ion, are the most common lasers in medicine (see MEDICAL USES OF LASERS). The 1.06-μm wavelength of the Nd:YAG laser is not strongly absorbed by water, and thus it has found use in applications that require deep heat penetration such as photocoagulation of blood and tissue welding. In surgery CO_2 lasers are the more common and are preferred in applications that require precise cutting because the 10.6-μm wavelength is strongly absorbed by water; Nd:YAG lasers are preferred when tissue ablation and control of hemorrhage are desired because the requirements of these applications match the longer absorption depth for the 1.06-μm wavelength. Nd:YAG lasers have the advantage over CO_2 lasers in ease of delivery since the 1.06-μm wavelength can be propagated down standard optical fiber. An example of one of the many Nd:YAG-laser applications is in ophthalmology; here, *Q*-switched Nd:YAG lasers are used to remove, by means of optical breakdown, opacified membranes that can develop after cataract surgery.

There is increasing interest in Ho^{3+} and Tm^{3+} lasers in the 2.1-μm wavelength range and Er^{3+} lasers in the 2.9-μm wavelength range for ablating or cutting of tissue, because these wavelengths are strongly absorbed by water, which helps to limit damage to the surrounding tissue. For 2.1-μm lasers, about 500 μm of surrounding tissue is affected, and for 2.9-μm lasers this distance is only ~ 30 μm. One advantage of the 2.1-μm relative to the 2.9-μm wavelength is in delivery of the laser radiation: 2.1-μm radiation can be propagated down currently available optical fiber. Fiber delivery systems for 2.9-μm radiation are not currently available, but there is active research in this area.

Tunable solid-state lasers are also beginning to find a role in medicine, particularly in research. The tunability allows investigation of laser–tissue interactions as a function of wavelength. Alexandrite and Ti:Al_2O_3 lasers are also being investigated as solid-state replacements for dye lasers in applications such as laser lithotripsy; one current application of alexandrite lasers is in tattoo removal.

5.4 Communications Applications

The primary solid-state laser used in communications (see OPTICAL COMMUNICATIONS) is the Er-doped single-mode optical fiber that serves as an amplifier in fiber-optic systems. Er-fiber amplifiers provide high optical gain in the 1.55-μm wavelength range and are pumped at either 0.98 or 1.48 μm by diode lasers; these amplifiers boost the signal that has been attenuated over long distances by absorption and scattering. The 1.55-μm wavelength is one of the standards used for long-distance optical communication systems. The current technique for compensating this optical loss is to install repeaters at $\sim$50-km intervals that contain a detector, signal reconditioning electronics, and a diode laser transmitter. The Er-fiber amplifier provides a simple, all-optical technique for compensating losses, and these amplifiers are expected to be installed on the next generation of transoceanic fiber-optic communication systems. Optical-fiber amplifiers doped with neodymium and praseodymium are being investigated for the other primary optical communications band at 1.3 μm; however, the performance to date is not as good as that demonstrated in the Er-doped fiber amplifiers. Nd-fiber amplifiers have been limited by excited-state absorption, and the Pr-doped fiber amplifiers are only at an early stage of development.

In practice, semiconductor diode lasers are installed as transmitters for fiber-optic communication systems, but solid-state lasers have found a role as transmitters in the development of and laboratory demonstrations of communications systems. Mode-locked color-center lasers in the 1.5-μm band are useful laboratory tools for generation of short pulses for laboratory direct-detection systems. Single-frequency Nd:YAG lasers in the 1.3-μm wavelength band have found application in demonstrations of coherent optical communications systems because of their inherently good frequency stability, narrow linewidth, and low-amplitude noise compared with diode lasers. The key disadvantage of solid-state lasers for communications applications is that external modulators must be used, whereas semiconductor diode lasers can be modulated directly by modulating their drive current.

5.5 Scientific Applications

Solid-state lasers have found wide use in the laboratory as a result of their ability to run cw, *Q* switched, or mode locked.

Q-switched Nd:YAG and Nd:YLF lasers are important scientific tools. These *Q*-switched lasers either are pumped by pulsed flashlamps and run at relatively low repetition rates (10 Hz) and at high energy per pulse, up to ~1 J, or are pumped by cw lamps and *Q* switched at high repetition rates (hundreds of hertz) and at relatively low energy per pulse, 1–2 mJ. These lasers are used for nonlinear optics studies because of their high peak power and as pump sources for other lasers. Frequency-doubled, *Q*-switched, and mode-locked Nd^{3+} lasers are used to pump $Ti:Al_2O_3$ and dye lasers, and Nd^{3+} lasers are also used to pump optical parametric oscillators that can provide tunable radiation. Nd^{3+} lasers with high peak power and high energy per pulse are used to excite plasmas efficiently by focusing the radiation on metal targets. Soft–x-ray lasers have been demonstrated using these laser-produced plasmas as the gain medium.

Tunable solid-state lasers are powerful tools in spectroscopic research. $Ti:Al_2O_3$ lasers have become the tunable laser of choice in the near infrared, replacing dye lasers in this wavelength range. The $Ti:Al_2O_3$ laser offers broader band tunability than any single laser dye, higher power, and no problems with degradation over time. Color-center lasers have found a niche as tunable sources in the 1.5-μm wavelength range.

6. SUMMARY

The solid-state laser's unique ability to generate high peak and average power with good beam quality and moderate efficiency in the near- and mid-infrared parts of the spectrum guarantees its place as an important technology in scientific, industrial, and medical applications. Current trends in solid-state laser technology include the development of efficient diode-pumped lasers, very high peak- and average-power systems, new wavelengths, optical-fiber amplifiers, and improved solid-state laser materials. Progress in these areas promises to expand significantly the utility of solid-state lasers.

Acknowledgment

This work was supported by the Department of the Air Force.

GLOSSARY

Activator ion: A dopant ion that has the energy levels used to provide optical gain.

Concentration Quenching: A decrease in the metastable-level lifetime with increase in dopant concentration that is observed in many gain media.

Host Material: A crystalline or glassy solid into which activator ions are doped.

Nonradiative Relaxation: Transitions from a higher energy level to a lower energy level by emission of phonons (as opposed to photons).

Rare-Earth Ion: Ions of elements with atomic number 57–71.

Stress-Fracture Limit: The maximum level of heat that can be generated in a gain element limited by fracture due to stresses induced by temperature gradients.

List of Works Cited

Ainslie, B. J. (1991), *J. Lightwave Tech.* **9**, 220–227.

Dieke, G. H., Crosswhite, H. M. (1963), *Appl. Opt.* **2**, 675–686.

Endriz, J. G., Vakili, M., Browder, G. S., DeVito, M. A., Haden, J. M., Harnagel, G. L., Plano, W. E., Sakomoto, M., Welch, D. F., Willing, S., Worland, D. P., Yao, H. C. (1992), *IEEE J. Quantum Electron.* **28**, 952–965.

Fan, T. Y., Byer, R. L. (1988), *IEEE J. Quantum Electron.* **24**, 895–912.

Geusic, J. E., Marcos, H. M., Van Uitert, L. G. (1964) *Appl. Phys. Lett.* **4**, 182–183.

Hughes, D. W., Barr, J. R. M. (1992), *J. Phys. D* **25**, 563–586.

Johnson, L. F., Guggenheim, H. J., Thomas, R. A. (1966), *Phys. Rev.* **149**, 179–185.

Keyes, R. J., Quist, T. M. (1964), *Appl. Phys. Lett.* **4**, 50–52.

Magni, V., Valentini, G., De Silvestri, S. (1991), *Opt. Quantum Electron.* **23**, 1105–1134.

Maiman, T. H. (1960), *Nature* **187**, 493–494.

Maiman, T. H., Hoskins, R. H., D'Haenens, I. J., Asawa, C. K., Evtuhov, V. (1961), *Phys. Rev.* **123**, 1151–1157.

Moulton, P. F. (1986), *J. Opt. Soc. Am. B* **3**, 125–133.

Snitzer, E. (1964), in: P. Grivet and N. Bloembergen (Eds.), *Quantum Electronics*, Paris: Dunod, pp. 999–1020.

Sorokin, P. P., Stevenson, M. J. (1960), *Phys. Rev. Lett.* **5**, 557–559.

Walling, J. C., Peterson, O. G., Jenssen, H. P., Morris, R. C., O'Dell, E. W. (1980), *IEEE J. Quantum Electron.* **QE-16**, 1302–1315.

Further Reading

Brown, D. C. (1981), *High-Peak-Power Nd:Glass Lasers Systems*, Berlin: Springer.

Dieke, G. M. (1968), *Spectra and Energy Levels of Rare Earth Ions in Crystals*, Wiley: New York. This covers the physics of rare earths doped into crystals.

Kaminskii, A. A. (1990), *Laser Crystals*, Berlin: Springer. An overview of the multitude of crystalline solid-state lasers that have been demonstrated.

Koechner, W. (1988), *Solid-State Laser Engineering*, Berlin: Springer. The standard book on engineering of solid-state lasers.

Moulton, P. F. (1992), *Proc. IEEE* **80**, 348–364. A good overview of transition-metal-ion lasers.

Pollack, C. R. (1991) in: R. A. Meyers (Ed.), *Encyclopedia of Lasers and Optical Technology*, San Diego: Academic. A good overview of color-center lasers.

Weber, M. J. (Ed.) (1982), *Handbook of Laser Science and Technology*, Vol. 1, Boca Raton: CRC.

Weber, M. J. (Ed.) (1991), *Handbook of Laser Science and Technology*, Supplement 1, Boca Raton: CRC. These last two handbooks contain short articles and numerous tables describing the properties of various solid-state lasers.

LATTICE VIBRATIONS, STATISTICS OF

I. P. IPATOVA, *A. F. Ioffe Physico-Technical Institute, Russian Academy of Sciences, St. Petersburg, Russia*
A. A. MARADUDIN, *Department of Physics, University of California, Irvine, California, U.S.A.*

	Introduction	465
1.	**Dynamics of a Crystal Lattice**	466
2.	**Quantization of Crystal Vibrations**	470
3.	**The Debye Specific Heat**	472
4.	**Lattice Thermal Conductivity of Insulators**	476
	Works Cited	484
	Further Reading	484

INTRODUCTION

The atoms that comprise a crystal are constantly executing small vibrations about equilibrium positions that form a structure periodic in each of three mutually perpendicular directions. At nonzero temperatures these atomic vibrations are due to thermal fluctuations in the crystal, while at the absolute zero of temperature they are due to the zero-point motion of the atoms that is associated with the quantum-mechanical uncertainty principle.

The vibrations of the atoms in a crystal play a significant role in many physical properties of the crystal. In insulating crystals, the thermodynamic, optical, elastic, and transport properties are determined by these atomic vibrations. Even in metallic and semiconducting crystals, in which the conduction electrons contribute to each of these properties, the contribution from the atomic vibrations is a significant part of the total.

The discussion of the vibrational properties of a crystal is based on the assumption that the potential energy of the crystal is a function of the instantaneous positions of the atoms in it. Because the displacements of the atoms from their equilibrium positions, as they oscillate about them, are small compared with the distance between nearest-neighbor equilibrium positions, the potential energy can be expanded in powers of these small displacements. The linear terms in this expansion are absent, since the expansion is about an equilibrium configuration. If the expansion is terminated at the terms quadratic in the atomic displacements, we have what is called the *harmonic approximation* to the potential energy. The higher-order terms in the expansion of the potential energy in powers of the atomic displacements, of third, fourth, ... order, constitute the *anharmonic* terms in the potential energy. In many cases these anharmonic terms contribute quantitative and qualitative corrections to physical properties of a crystal that are already calculated with good accuracy in the harmonic approximation, such as the thermodynamic functions. These corrections are usually small enough that they can be obtained by perturbation theory. However, as in the case of thermal expansion, they can also give rise to effects that have no counterpart in the harmonic approximation. Nevertheless, the harmonic approximation is generally a very good first approximation to the calculation of a vibrational property of a crystal, and serves as the starting point for perturbative calculations of anharmonic effects. (The exceptions to these statements are quantum crystals, e.g., solid helium, which we will not consider here.) Consequently, in what follows most of our discussion will be based on this approximation, and the effects of the anharmonic terms in the potential energy of a crystal will be discussed only briefly.

Although in principle the vibrational contribution to a physical property of a crystal can be calculated by considering the motions of the individual atoms in it, it turns out to be simpler, and physically more transparent, to exploit the special nature of the harmonic approximation—viz., that the potential energy is a positive-definite quadratic form in the

1-56081-067-X/94/$5.00+.50

atomic displacements—to reformulate the calculation in such a way that the given physical property is expressed as the sum of contributions from collective motions of *all* the atoms in a crystal. Each of these collective motions, called a *normal vibration mode* of the crystal, is a phased linear superposition of the vibrations of the individual atoms, each with the same frequency, and is independent of all the other normal modes of the crystal. The number of these normal modes is equal to the number of degrees of freedom in the crystal. The statistical properties of the normal modes are central to the way in which they contribute to a given physical property of the crystal. In this article we describe these statistical properties, and their consequences, in particular for the thermodynamic and transport properties of a crystal.

1. DYNAMICS OF A CRYSTAL LATTICE

Let us consider an infinite crystal that is periodic in three mutually perpendicular directions. The equilibrium position of the κth atom in the lth primitive unit cell is given by the vector

$$\mathbf{x}(l\kappa)=\mathbf{x}(l)+\mathbf{x}(\kappa). \tag{1}$$

In this equation the vector $\mathbf{x}(l)$ is

$$\mathbf{x}(l)=l_1\mathbf{a}_1+l_2\mathbf{a}_2+l_3\mathbf{a}_3, \tag{2}$$

where l_i $(i=1,2,3)$ is an integer, and $\mathbf{a}_i$ $(i=1,2,3)$ are the primitive translation vectors of the crystal. $\mathbf{x}(l)$ defines the equilibrium position of the lth unit cell with respect to the origin of coordinates, chosen to coincide with one of the lattice sites. The vector $\mathbf{x}(\kappa)$ $(\kappa=1,2,\ldots,r)$ gives the positions of the r atoms in the unit cell relative to the equilibrium position of the latter. The volume of a primitive unit cell is given by $v_0=\mathbf{a}_1\cdot(\mathbf{a}_2\times\mathbf{a}_3)$.

The thermal motion of the atoms of the crystal consists of small displacements of all the atoms from their equilibrium positions, and we denote the displacement of the atom $(l\kappa)$ by $\mathbf{u}(l\kappa)$. The momentum of this atom will be denoted by $\mathbf{p}(l\kappa)$.

The total kinetic energy of the crystal can be expressed in the form

$$T=\sum_{l\kappa\alpha}\frac{p_\alpha^2(l\kappa)}{2M_\kappa}, \tag{3}$$

where M_κ is the mass of the atom of type κ, and $\alpha=x,y,z$. The total potential energy Φ of the crystal is a function of the instantaneous positions of all the atoms. It can be expanded in a series of powers of the displacements $\mathbf{u}(l\kappa)$ of the atoms from their equilibrium positions,

$$\Phi=\Phi_0+\sum_{l\kappa\alpha}\Phi_\alpha(l\kappa)u_\alpha(l\kappa)+\frac{1}{2}\sum_{l\kappa\alpha}\sum_{l'\kappa'\beta}\Phi_{\alpha\beta}(l\kappa;l'\kappa')u_\alpha(l\kappa)u_\beta(l'\kappa')+\ldots \tag{4}$$

where

$$\Phi_\alpha(l\kappa)=\left(\frac{\partial\Phi}{\partial u_\alpha(l\kappa)}\right)_0,$$

$$\Phi_{\alpha\beta}(l\kappa;l'\kappa')=\left(\frac{\partial^2\Phi}{\partial u_\alpha(l\kappa)\partial u_\beta(l'\kappa')}\right)_0,\quad \text{etc.} \tag{5}$$

The approximation in which the expansion (4) contains only terms through second order in $\mathbf{u}(l\kappa)$ is called the *harmonic approximation.* The terms neglected, of higher order in $\mathbf{u}(l\kappa)$, are called *anharmonic.* They are neglected because of their smallness.

From the definition (5) for $\Phi_\alpha(l\kappa)$ we see that it has the physical meaning of the negative of the force acting in the direction α on the atom $(l\kappa)$ when all the atoms of the crystal are at their equilibrium positions. Since in equilibrium no net force can be exerted on any atom, we must have

$$\Phi_\alpha(l\kappa)\equiv 0. \tag{6}$$

The Hamiltonian of the crystal $H=T+\Phi$ in the harmonic approximation therefore has the form

$$H=\Phi_0+\sum_{l\kappa\alpha}\frac{p_\alpha^2(l\kappa)}{2M_\kappa}+\frac{1}{2}\sum_{l\kappa\alpha}\sum_{l'\kappa'\beta}\Phi_{\alpha\beta}(l\kappa;l'\kappa')u_\alpha(l\kappa)u_\beta(l'\kappa'). \tag{7}$$

The coefficients $\{\Phi_{\alpha\beta}(l\kappa;l'\kappa')\}$ are called *atomic force constants.*

From their definition, Eq. (5), the atomic force constants possess the symmetry property

$$\Phi_{\alpha\beta}(l\kappa;l'\kappa')=\Phi_{\beta\alpha}(l'\kappa';l\kappa). \tag{8}$$

In addition, they are subject to restrictions that are imposed by the invariance of the potential energy against rigid-body translations and rotations of the crystal, which are expressed by (Maradudin *et al.*, 1971, pp. 8–10)

$$\sum_{l'\kappa'} \Phi_{\alpha\beta}(l\kappa;l'\kappa')=0, \tag{9}$$

$$\sum_{l'\kappa'} \Phi_{\alpha\beta}(l\kappa;l'\kappa')x_\gamma(l'\kappa') = \sum_{l'\kappa'} \Phi_{\alpha\gamma}(l\kappa;l'\kappa')x_\beta(l'\kappa'), \tag{10}$$

respectively. Finally, the symmetry and structure of a crystal impose further restrictions on the atomic force constants. The most important of these for our purposes in this article is a consequence of the periodicity of the equilibrium crystal structure. The potential energy (4) cannot change under a displacement of this structure through any of the vectors $\mathbf{x}(l)$ given by Eq. (2). This has the consequence that the atomic force constants $\{\Phi_{\alpha\beta}(l\kappa;l'\kappa')\}$ depend on the cell indices l and l' only through their difference,

$$\Phi_{\alpha\beta}(l\kappa;l'\kappa')=\Phi_{\alpha\beta}(l-l',\kappa;0\kappa') \tag{11a}$$

$$=\Phi_{\alpha\beta}(0\kappa;l'-l,\kappa'). \tag{11b}$$

Additional restrictions on the atomic force constants due to the symmetry of a crystal are discussed on pp. 10–14 of Maradudin *et al.* (1971).

From Hamilton's equations of motion that follow from Eq. (7),

$$\dot{u}_\alpha(l\kappa)=\partial H/\partial p_\alpha(l\kappa) \tag{12a}$$

$$\dot{p}_\alpha(l\kappa)=-\partial H/\partial u_\alpha(l\kappa), \tag{12b}$$

we immediately obtain the classical equations of motion of the atoms of the crystal,

$$M_\kappa \ddot{u}_\alpha(l\kappa)=-\sum_{l'\kappa'\beta} \Phi_{\alpha\beta}(l\kappa;l'\kappa')u_\beta(l'\kappa). \tag{13}$$

The solution of the system of equations (13) can be simplified if we exploit the periodicity of the equilibrium crystal structure expressed through Eqs. (11). Since the infinite crystal is periodic, it is natural to seek a solution of the system (13) in the form of a traveling wave,

$$u_\alpha(l\kappa)=\frac{u_\alpha(\kappa)}{\sqrt{M_\kappa}} \exp[i\mathbf{k}\cdot\mathbf{x}(l)-i\omega t], \tag{14}$$

where $\mathbf{k}$ is its wave vector, ω is its frequency, and the amplitude $u_\alpha(\kappa)$ is independent of l. On substituting Eq. (14) into Eqs. (13) we obtain as the equation for the $\{u_\alpha(\kappa)\}$

$$\omega^2 u_\alpha(\kappa)=\sum_{\kappa'\beta} D_{\alpha\beta}(\kappa\kappa'|\mathbf{k})u_\beta(\kappa'), \tag{15}$$

where the matrix $D_{\alpha\beta}(\kappa\kappa'|\mathbf{k})$ does not depend on the index l, and is equal to

$$D_{\alpha\beta}(\kappa\kappa'|\mathbf{k})=\frac{1}{\sqrt{M_\kappa M_{\kappa'}}}\sum_{l'} \Phi_{\alpha\beta}(l-l',\kappa;0\kappa') \times\exp\{-i\mathbf{k}\cdot[\mathbf{x}(l)-\mathbf{x}(l')]\}. \tag{16}$$

In contrast with the system of equations (13), which is a system of an infinite number of equations in an infinite number of unknowns, the system (15) consists of $3r$ equations for each value of the wave vector $\mathbf{k}$. This result is due to the fact that the force constants $\{\Phi_{\alpha\beta}(l\kappa;l'\kappa')\}$ depend on l and l' only through their difference.

The solvability condition for the system of equations (15) is the vanishing of its determinant:

$$|D_{\alpha\beta}(\kappa\kappa'|\mathbf{k})-\omega^2\delta_{\alpha\beta}\delta_{\kappa\kappa'}|=0. \tag{17}$$

Equation (17) is an equation of degree $3r$ in ω^2. We will denote the $3r$ solutions for each value of $\mathbf{k}$ by $\omega_j^2(\mathbf{k})$, where $j=1,2,\ldots,3r$. From its definition (16) the matrix $D_{\alpha\beta}(\kappa\kappa'|\mathbf{k})$ can be shown to be Hermitian. From this it follows that all of its eigenvalues $\omega_j^2(\mathbf{k})$ are real, while the frequencies $\omega_j(\mathbf{k})$ themselves can be either purely real or purely imaginary. The stability of the crystal translates into the requirement that all the frequencies $\omega_j(\mathbf{k})$ be real. The $3r$ functions $\omega_j^2(\mathbf{k})$ for each value of $\mathbf{k}$ can be regarded as the branches of a multivalued function $\omega^2(\mathbf{k})$. Equation (15) can therefore be rewritten in a form that makes the dependence of the frequency ω and of the corresponding eigenvector $\mathbf{u}(\kappa)$ on the wave vector $\mathbf{k}$ and branch index j explicit,

$$\omega_j^2(\mathbf{k})u_\alpha(\kappa|\mathbf{k}j)=\sum_{\kappa'\beta} D_{\alpha\beta}(\kappa\kappa'|\mathbf{k})u_\beta(\kappa'|\mathbf{k}j). \tag{18}$$

The relation

$$\omega = \omega_j(\mathbf{k}) \tag{19}$$

is called the *dispersion law* for the vibrations of the crystal. Of the $3r$ solutions of Eq. (18) for each value of $\mathbf{k}$, three go to zero as $\mathbf{k}$ goes to zero. This can be seen as follows. On setting $\mathbf{k}=0$ in Eqs. (16) and (18), and dividing the latter by $\sqrt{M_{\kappa}}$, we obtain

$$\omega_j^2(\mathbf{0}) \frac{u_\alpha(\kappa|\mathbf{0}j)}{\sqrt{M_\kappa}} = \frac{1}{M_\kappa} \sum_{l'\kappa'\beta} \Phi_{\alpha\beta}(l\kappa;l'\kappa') \times \frac{u_\beta(\kappa'|\mathbf{0}j)}{\sqrt{M_{\kappa'}}}. \tag{20}$$

If for each of the three values β can assume, viz., x, y, and z, $u_\beta(\kappa'|\mathbf{0}j)/\sqrt{M_{\kappa'}}$ is independent of κ', then in view of Eq. (9) we find that the right-hand side of Eq. (20) vanishes, implying the vanishing of $\omega_j^2(\mathbf{0})$. Thus we have three solutions for $\omega_j^2(\mathbf{k})$, one for each of the coordinate axes, that vanish with vanishing $\mathbf{k}$. From the form of Eq. (14) at $\mathbf{k}=0$ we see that the condition that $u_\alpha(\kappa|\mathbf{0}j)/\sqrt{M_\kappa}$ be independent of κ means that all r atoms in a primitive unit cell move in parallel and with equal amplitudes, which corresponds to a rigid-body displacement of the entire crystal. The three branches of $\omega^2(\mathbf{k})$ that vanish as $\mathbf{k}\to\mathbf{0}$ are termed *acoustic modes,* because the same property is shared by the frequencies of a continuous elastic medium. The remaining $3r-3$ solutions, which do not go to zero as $\mathbf{k}$ goes to zero, are called the frequencies of *optical modes.* The latter name is due to the fact that an electromagnetic wave incident on the crystal can interact only with the $\mathbf{k}=\mathbf{0}$ modes whose frequencies are nonzero, and are generally in the far-infrared region of the optical spectrum.

The frequencies $\{\omega_j(\mathbf{k})\}$ belonging to the $3r$ branches of the vibrational spectrum possess the important property that they are even functions of $\mathbf{k}$,

$$\omega_j(-\mathbf{k}) = \omega_j(\mathbf{k}). \tag{21}$$

For we see from Eq. (16) that

$$D_{\alpha\beta}(\kappa\kappa'|-\mathbf{k}) = D^*_{\alpha\beta}(\kappa\kappa'|\mathbf{k}). \tag{22}$$

Consequently, if we replace $\mathbf{k}$ by $-\mathbf{k}$ in Eq. (18), and take the complex conjugate of the resulting equation, we obtain

$$\omega_j^2(-\mathbf{k})u^*_\alpha(\kappa|-\mathbf{k}j) = \sum_{\kappa'\beta} D_{\alpha\beta}(\kappa\kappa'|\mathbf{k}) \times u^*_\beta(\kappa'|-\mathbf{k}j), \tag{23}$$

where we have used the fact that $\omega_j^2(\mathbf{k})$ is real because $D_{\alpha\beta}(\kappa\kappa'|\mathbf{k})$ is a Hermitian matrix. Thus $\omega_j^2(-\mathbf{k})$ and $\omega_j^2(\mathbf{k})$ are eigenvalues of the same matrix. With no loss of generality we can equate $\omega_j^2(-\mathbf{k})$ and $\omega_j^2(\mathbf{k})$, even for values of $\mathbf{k}$ at which degenerate eigenvalues occur, which leads to Eq. (21).

To each of the $3r$ values of $\omega_j(\mathbf{k})$ corresponds a vector $u_\alpha(\kappa|\mathbf{k}j)$ whose components are the solutions of Eq. (18). The homogeneous equations (18) define $u_\alpha(\kappa|\mathbf{k}j)$ to within an arbitrary constant factor. This factor can be chosen in such a way that $u_\alpha(\kappa|\mathbf{k}j)$ satisfies the orthonormality and closure conditions

$$\sum_{\kappa,\alpha} u^*_\alpha(\kappa|\mathbf{k}j)u_\alpha(\kappa|\mathbf{k}j') = \delta_{jj'}, \tag{24a}$$

$$\sum_j u^*_\alpha(\kappa|\mathbf{k}j)u_\beta(\kappa'|\mathbf{k}j) = \delta_{\alpha\beta}\delta_{\kappa\kappa'}. \tag{24b}$$

We have not yet specified the values that the wave vector $\mathbf{k}$ introduced in Eq. (14) can assume. They are clearly determined by the boundary conditions imposed on the atomic displacements $\{\mathbf{u}(l\kappa)\}$. The most convenient boundary conditions, which preserve the simplicity of the theory developed so far, are obtained if we subdivide the infinite crystal into "macrocrystals," parallelepipeds whose edges are defined by the vectors $L\mathbf{a}_1$, $L\mathbf{a}_2$, $L\mathbf{a}_3$, where L is a large number, of the order of 10^8, and which thus contain $L^3=N$ primitive unit cells. Any one of these macrocrystals can be regarded as the physical crystal whose properties we are studying. We then postulate that the atomic displacements are periodic with the periodicity of the macrocrystals. These conditions on the displacements are called the *cyclic boundary conditions* (Born and Huang, 1954, pp. 45–46). An equivalent set of boundary conditions require that in a finite crystal the atoms at corresponding points on opposite faces of the crystal move in exactly the same way. These conditions are called *periodic boundary conditions,* and are due to Born and von Kármán (1912, 1913).

Applied to the displacements given by Eq. (14), the cyclic boundary conditions require that

$$\exp(i\mathbf{k}\cdot L\mathbf{a}_1)=\exp(i\mathbf{k}\cdot L\mathbf{a}_2)$$
$$=\exp(i\mathbf{k}\cdot L\mathbf{a}_3)=1. \quad (25)$$

To satisfy these conditions we introduce the translation vectors of the lattice reciprocal to the direct lattice defined by the vectors $\{\mathbf{x}(l)\}$:

$$\mathbf{G}(h)=h_1\mathbf{b}_1+h_2\mathbf{b}_2+h_3\mathbf{b}_3, \quad (26)$$

where h_i $(i=1,2,3)$ is an integer, and the primitive translation vectors of this *reciprocal lattice* $\mathbf{b}_i$ $(i=1,2,3)$ are defined by the set of equations

$$\mathbf{a}_i\cdot\mathbf{b}_j=2\pi\delta_{ij}. \quad (27)$$

Then from the result which follows from Eqs. (2), (26), and (27) that

$$\mathbf{x}(l)\cdot\mathbf{G}(h)=2\pi(l_1h_1+l_2h_2+l_3h_3)$$
$$=2\pi(\text{integer}), \quad (28)$$

we see that we can write the vector $\mathbf{k}$ that satisfies Eqs. (25) as

$$\mathbf{k}=\frac{1}{L}\,\mathbf{G}(h)=\frac{h_1}{L}\,\mathbf{b}_1+\frac{h_2}{L}\,\mathbf{b}_2+\frac{h_3}{L}\,\mathbf{b}_3. \quad (29)$$

The values of the integers $\{h_i\}$ are no longer unrestricted, however. From Eqs. (14) and (28) we see that adding to $\mathbf{k}$ any translation vector $\mathbf{G}(h)$ of the reciprocal lattice leaves the value of $\mathbf{u}(l\kappa)$ unchanged. This means that all the distinct solutions to Eq. (18) are obtained if we restrict $\mathbf{k}$ to lie in one unit cell of the reciprocal lattice:

$$\mathbf{k}=\frac{h_1}{L}\,\mathbf{b}_2+\frac{h_2}{L}\,\mathbf{b}_2+\frac{h_3}{L}\,\mathbf{b}_3,$$
$$h_1,h_2,h_3=0,1,2,\ldots,L-1. \quad (30)$$

Thus, there are $L^3=N$ values of $\mathbf{k}$ allowed by the cyclic boundary conditions. Since there are $3r$ solutions of Eq. (18) for each value of $\mathbf{k}$, we see that the macrocrystal has as many vibration frequencies $\{\omega_j(\mathbf{k})\}$ as it has degrees of freedom.

A related consequence of Eq. (28) is that the dynamical matrix $D_{\alpha\beta}(\kappa\kappa'|\mathbf{k})$ defined by Eq. (16) is unchanged if we add to the wave vector $\mathbf{k}$ any reciprocal-lattice vector $\mathbf{G}(h)$. It follows that the normal-mode frequency $\omega_j(\mathbf{k})$ possesses the same property: $\omega_j(\mathbf{k})=\omega_j(\mathbf{k}+\mathbf{G}(h))$. Thus, all the distinct normal-mode frequencies are obtained by restricting $\mathbf{k}$ to one unit cell of the reciprocal lattice.

In practice, it is more convenient to restrict the allowed values of the wave vector $\mathbf{k}$ to the *first Brillouin zone* of the lattice defined by the vectors $\mathbf{a}_1$, $\mathbf{a}_2$, $\mathbf{a}_3$, rather than to a unit cell of the reciprocal lattice. The first Brillouin zone is obtained by drawing vectors from the origin of the reciprocal lattice to all the points of that lattice and constructing the planes that are the perpendicular bisectors of these vectors. The smallest volume containing the origin that is enclosed by these planes is the first Brillouin zone. It has the same volume as a primitive unit cell of the reciprocal lattice, and so contains the same number N of allowed values of $\mathbf{k}$ as the latter does, and it is invariant under the operations of the crystal class to which the crystal belongs. In what follows, therefore, sums over the wave vector $\mathbf{k}$ will be understood to be over those N values allowed by the cyclic boundary conditions that are contained in the first Brillouin zone.

In most calculations the precise values that $\mathbf{k}$ is allowed to have are not important. What is important is that they are distributed uniformly and densely throughout the Brillouin zone with a density equal to $V/(2\pi)^3$, where $V=Nv_0$ is the volume of the crystal. Thus, in sums over $\mathbf{k}$ we can treat $\mathbf{k}$ as a continuous variable and replace summation by integration according to

$$\sum_k f(\mathbf{k})=\frac{V}{(2\pi)^3}\int d^3k f(\mathbf{k}), \quad (31)$$

where the integration is over the volume of the first Brillouin zone.

Since the rather artificial cyclic boundary conditions on the atomic displacements have been adopted primarily for reasons of mathematical convenience, it has to be shown that the calculation of extensive vibrational properties of the crystal is insensitive to the imposition of these or any other boundary conditions. If this were not so, it would be virtually impossible to develop a theory of crystalline vibrations that possesses any degree of generality. Such a demonstration was given by Ledermann (1944), who showed that in the limit of large N the distribution of vibration frequencies of a crystal is affected negligibly by the assumption of any particular boundary conditions in the case that the range of the interatomic forces is finite. Subsequently, Hardy (1962) provided a justification for the use of the cyclic boundary conditions in calculations of extensive properties of ionic or polar crystals, in which the constituent atoms or ions interact through long-range Coulomb forces.

2. QUANTIZATION OF CRYSTAL VIBRATIONS

The vibrational Hamiltonian of a crystal,

$$H=\sum_{l\kappa\alpha}\frac{p_\alpha^2(l\kappa)}{2M_\kappa}+\tfrac{1}{2}\sum_{l\kappa\alpha}\sum_{l'\kappa'\beta}\Phi_{\alpha\beta}(l\kappa;l'\kappa')\times u_\alpha(l\kappa)u_\beta(l'\kappa'),\quad(32)$$

is the sum of quadratic forms in the displacement components $\{u_\alpha(l\kappa)\}$ and in the conjugate momenta $\{p_\alpha(l\kappa)\}$. It can be diagonalized if we introduce the following expansions of $u_\alpha(l\kappa)$ and $p_\alpha(l\kappa)$ in traveling plane waves:

$$u_\alpha(l\kappa)=\frac{1}{\sqrt{NM_\kappa}}\sum_{\mathbf{k}j}u_\alpha(\kappa|\mathbf{k}j)\left\{\frac{1}{2}[q(\mathbf{k}j)+q(-\mathbf{k}j)]+\frac{1}{2}\left[\frac{i}{\omega_j(\mathbf{k})}\{\mathbf{p}(\mathbf{k}j)-p(-\mathbf{k}j)\}\right]\right\}e^{i\mathbf{k}\cdot\mathbf{x}(l)},\quad(33a)$$

$$p_\alpha(l\kappa)=\frac{1}{\sqrt{NM_\kappa}}\sum_{\mathbf{k}j}u_\alpha(\kappa|\mathbf{k}j)\left\{\frac{1}{2}[p(\mathbf{k}j)+p(-\mathbf{k}j)]+\frac{1}{2}\left[\frac{i}{\omega_j(\mathbf{k})}\{q(\mathbf{k}j)-q(-\mathbf{k}j)\}\right]\right\}e^{i\mathbf{k}\cdot\mathbf{x}(l)}.\quad(33b)$$

In these expansions the quantities in square brackets are real displacements and momenta, corresponding to waves propagating with the wave vector **k**. In the new variables $p(\mathbf{k}j)$ and $q(\mathbf{k}j)$ the diagonalized Hamiltonian becomes the sum of the energies of independent harmonic oscillations,

$$H=\frac{1}{2}\sum_{\mathbf{k}j}[p^2(\mathbf{k}j)+\omega_j^2(\mathbf{k})q^2(\mathbf{k}j)].\quad(34)$$

In going from Eq. (32) to Eq. (34) we have used the result that

$$\sum_l e^{i\mathbf{k}\cdot\mathbf{x}(l)}=N\Delta(\mathbf{k}),\quad(35)$$

where

$$\Delta(\mathbf{k})=\begin{cases}1 & \text{if } \mathbf{k}=\mathbf{G}(h),\\ 0 & \text{otherwise.}\end{cases}\quad(36)$$

The vibrations described by the Hamiltonian (34) are called *normal modes,* and the corresponding coordinates $q(\mathbf{k}j)$ and momenta $p(\mathbf{k}j)$ are called *normal coordinates.* They are related by $p(\mathbf{k}j)=\dot{q}(\mathbf{k}j)$.

The Hamiltonian (34) is written in a form that permits a straightforward transition to a quantum-mechanical description of the vibrations of the crystal. For this we must regard $q(\mathbf{k}j)$ and $p(\mathbf{k}j)$ as quantum-mechanical operators obeying commutation relations that contain Planck's constant,

$$[q(\mathbf{k}j),p(\mathbf{k}j)]=i\hbar\Delta(\mathbf{k}-\mathbf{k}')\delta_{jj'}.\quad(37)$$

It follows from Eq. (37) that

$$p(\mathbf{k}j)=-i\hbar\,\partial/\partial q(\mathbf{k}j).\quad(38)$$

On substituting Eq. (38) into Eq. (34) we obtain the quantum-mechanical Hamiltonian operator of the crystalline vibrations,

$$\hat{H}=\frac{1}{2}\sum_{\mathbf{k}j}\left[-\hbar^2\frac{\partial^2}{\partial q^2(\mathbf{k}j)}+\omega_j^2(\mathbf{k})q^2(\mathbf{k}j)\right].\quad(39)$$

The corresponding Schrödinger equation has the form

$$\sum_{\mathbf{k}j}\left[-\frac{\hbar^2}{2}\frac{\partial^2}{\partial q^2(\mathbf{k}j)}+\tfrac{1}{2}\omega_j^2(\mathbf{k})q^2(\mathbf{k}j)\right]\Psi=E\Psi.\quad(40)$$

Here E and Ψ are the energy and wave function of the vibrations, respectively. Since the Hamiltonian in Eq. (39) has the form of a sum of independent harmonic-oscillator Hamiltonians, one for each of the normal modes, the wave function Ψ can be represented as the product of the wave functions of the individual oscillators,

$$\Psi=\prod_{\mathbf{k}j}\psi_{n_j(\mathbf{k})}(q(\mathbf{k}j)),\quad(41)$$

while the energy E is the sum of the energies of the individual oscillators,

$$E=\sum_{\mathbf{k}j} E_{n_j(\mathbf{k})}. \tag{42}$$

In Eqs. (41) and (42) $n_j(\mathbf{k})$ is a quantum number that labels the energy level of the oscillator ($\mathbf{k}j$). By substituting Eqs. (41) and (42) into Eq. (40) we find that the Schrödinger equation for an individual normal mode has the form

$$\left[-\frac{\hbar^2}{2}\frac{\partial^2}{\partial q^2(\mathbf{k}j)}+\frac{1}{2}\omega_j^2(\mathbf{k})q^2(\mathbf{k}j)\right]\psi_{n_j(\mathbf{k})}(q(\mathbf{k}j)) = E_{n_j(\mathbf{k})}\psi_{n_j(\mathbf{k})}(q(\mathbf{k}j)). \tag{43}$$

The solution of this equation is well known from quantum mechanics (Schiff, 1949, p. 64):

$$E_{n_j(\mathbf{k})}=[n_j(\mathbf{k})+\tfrac{1}{2}]\hbar\omega_j(\mathbf{k}) \tag{44}$$

and

$$\psi_n(q)=\left(\frac{\alpha}{\pi^{1/2}2^n n!}\right)^{1/2}\exp(-\tfrac{1}{2}\alpha^2q^2)H_n(\alpha q), \tag{45}$$

where $\alpha_j^2(\mathbf{k})=\omega_j(k)/\hbar$, and $H_n(\alpha q)$ is the Hermite polynomial of nth degree, where $n=0,1,2,\ldots$. The total vibrational energy of the crystal for the state characterized by the quantum numbers $\{n_j(\mathbf{k})\}$ is then given by

$$E_{\{n_j(\mathbf{k})\}}=\sum_{\mathbf{k}j}[n_j(\mathbf{k})+\tfrac{1}{2}]\hbar\omega_j(\mathbf{k}). \tag{46}$$

The integer $n_j(\mathbf{k})$ is the number of quanta of vibrational energy $\hbar\omega_j(\mathbf{k})$ in the normal mode ($\mathbf{k}j$). This is often expressed by the statement that there are $n_j(\mathbf{k})$ *phonons* in the mode ($\mathbf{k}j$). The average number of phonons in this mode, $\bar{n}(\omega_j(\mathbf{k}))$, is temperature-dependent, and is simply obtained. Since the contribution to the total vibrational energy of the crystal is the sum of the energies of each of the vibrational modes, from the general principles of statistical physics we have that

$$\bar{n}(\omega_j(\mathbf{k}))=\frac{\Sigma_{n_j(\mathbf{k})=0}^{\infty}n_j(\mathbf{k})e^{-[n_j(\mathbf{k})+\frac{1}{2}]\hbar\omega_j(\mathbf{k})/T}}{\Sigma_{n_j(\mathbf{k})=0}^{\infty}e^{-[n_j(\mathbf{k})+\frac{1}{2}]\hbar\omega_j(\mathbf{k})/T}}, \tag{47}$$

where T is the absolute temperature in energy units, i.e., Boltzmann's constant $k_B\equiv 1$. The sums are readily evaluated, and we obtain the result

$$\bar{n}(\omega_j(\mathbf{k}))=1/(e^{\hbar\omega_j(\mathbf{k})/T}-1). \tag{48}$$

Thus we find that the equilibrium distribution of phonons in the normal mode ($\mathbf{k}j$) is described by the Bose-Einstein distribution function, in which the chemical potential μ is set equal to zero,

$$\mu=0. \tag{49}$$

The fact that the chemical potential of phonons is zero indicates that the creation and destruction of a phonon are accomplished without the expenditure of any energy.

The distribution function (48) is called the Planck distribution function. It has the following limiting forms:

$$\bar{n}(\omega_j(\mathbf{k})) = \begin{cases}\exp[-\hbar\omega_j(\mathbf{k})/T] \text{ as } T\to 0, & (50a)\\ T/\hbar\omega_j(\mathbf{k}) \text{ as } T\to\infty. & (50b)\end{cases}$$

Thus, in the limit of low temperatures, $T\ll\hbar\omega_j(\mathbf{k})$, very few phonons are present in the mode ($\mathbf{k}j$). However, in the high-temperature limit, $T\gg\hbar\omega_j(\mathbf{k})$, the mean number of phonons in this mode is proportional to the absolute temperature T.

The thermodynamic properties of the crystal can be completely described with the aid of the partition function, which in the present case has the form

$$Z=\sum_{\{n_j(\mathbf{k})\}}\exp(-E_{\{n_j(\mathbf{k})\}}/T). \tag{51}$$

If we substitute into this expression the sum (46), we obtain

$$Z=\prod_{\mathbf{k}j}\frac{\exp[-\hbar\omega_j(\mathbf{k})/T]}{1-\exp[-\hbar\omega_j(\mathbf{k})/T]}. \tag{52}$$

The Helmholtz free energy of the crystalline vibrations can be obtained from the partition function

$$F=-T\ln Z=T\sum_{\mathbf{k}j}\ln\left[2\sinh\frac{\hbar\omega_j(\mathbf{k})}{2T}\right], \tag{53}$$

as can the internal energy

$$E=F-T\left(\frac{\partial F}{\partial T}\right)_v$$

$$=\sum_{\mathbf{k}j}\left\{\frac{\hbar\omega_j(\mathbf{k})}{2}+\frac{\hbar\omega_j(\mathbf{k})}{\exp[\hbar\omega_j(\mathbf{k})/T]-1}\right\}, \qquad (54)$$

the specific heat at constant volume

$$C_v=\left(\frac{\partial E}{\partial T}\right)_v=\sum_{\mathbf{k}j}\frac{[\hbar\omega_j(\mathbf{k})/2T]^2}{\sinh^2[\hbar\omega_j(\mathbf{k})/2T]}, \qquad (55)$$

and the entropy

$$S=-\left(\frac{\partial F}{\partial T}\right)_v$$

$$=\sum_{\mathbf{k}j}\left\{\frac{\hbar\omega_j(\mathbf{k})}{2T}\coth\frac{\hbar\omega_j(\mathbf{k})}{2T}\right.$$

$$\left.-\ln\left[2\sinh\frac{\hbar\omega_j(\mathbf{k})}{2T}\right]\right\}. \qquad (56)$$

If the vibrational spectrum of the crystal $\omega_j(\mathbf{k})$ for $j=1,2,\ldots,3r$ is known, then Eqs. (53)–(56) in principle allow all four thermodynamic functions F, E, C_v, S to be calculated.

However, the number of atoms in a crystal is enormous, and the vibrational frequencies are very densely distributed in the spectrum. Therefore, it is more convenient in the calculation of these functions to use the frequency distribution function $g(\omega)$, which can be introduced in the following fashion. Let us consider an arbitrary additive function of the normal-mode frequencies,

$$S=\sum_{\mathbf{k}j}F(\omega_j(\mathbf{k})). \qquad (57)$$

It can be transformed in the following way:

$$S=\int d\omega\sum_{\mathbf{k}j}\delta(\omega-\omega_j(\mathbf{k}))F(\omega)$$

$$=\int d\omega g(\omega)F(\omega), \qquad (58)$$

where $\delta(\omega)$ is the Dirac delta function. The frequency distribution function $g(\omega)$ is therefore given by

$$g(\omega)=\sum_{\mathbf{k}j}\delta(\omega-\omega_j(\mathbf{k})). \qquad (59)$$

The function $g(\omega)$ has the meaning of the density of vibrational states, where $g(\omega)d\omega$ is the number of normal modes with frequencies in the infinitesimally narrow interval between ω and $\omega+d\omega$. The frequency distribution function $g(\omega)$ therefore satisfies the normalization condition

$$\int_0^{\omega_{\max}} d\omega\, g(\omega)=3rN, \qquad (60)$$

which indicates that the total number of normal modes of the crystal is $3rN$. In Eq. (60) $\omega_{\max}$ is the largest frequency of the vibrational spectrum.

The expressions for the free energy, the internal energy, the specific heat, and the entropy can also be expressed in terms of $g(\omega)$, and have the forms of averages over the frequency distribution function:

$$F=T\int_0^{\omega_{\max}} d\omega\, g(\omega)\ln\left[2\sinh\frac{\hbar\omega}{2T}\right], \qquad (61)$$

$$E=\int_0^{\omega_{\max}} d\omega\, g(\omega)\hbar\omega\left[\frac{1}{2}+\frac{1}{\exp(\hbar\omega/T)-1}\right], \qquad (62)$$

$$C_v=\int_0^{\omega_{\max}} d\omega\, g(\omega)\left(\frac{\hbar\omega}{2T}\right)^2\frac{1}{\sinh^2(\hbar\omega/2T)}, \qquad (63)$$

$$S=\int_0^{\omega_{\max}} d\omega\, g(\omega)\left[\frac{\hbar\omega}{2T}\coth\frac{\hbar\omega}{2T}\right.$$

$$\left.-\ln\left(2\sinh\frac{\hbar\omega}{2T}\right)\right]. \qquad (64)$$

3. THE DEBYE SPECIFIC HEAT

According to classical theory the specific heat of a solid at constant volume is determined solely by the number of degrees of freedom of the system, with each of which is associated an energy T. The resulting specific heat $C_v=\partial E/\partial T=3rN$ has a constant value, independent of temperature. This constant value was observed experimentally at elevated temperatures, and the constancy of the specific heat at such temperatures became known as the law of Dulong and Petit. At lower temperatures, however, it turned out that the specific heat is strongly temperature dependent, specifically that it decreases with decreasing temperature.

In 1893, to give a theoretical basis for the law of Dulong and Petit, Richarz (1893, 1899) was led to characterize a crystal as an assembly of atoms each of which is executing a simple harmonic motion about its mean position in the crystal. Several years later Einstein (1907), in attempting to explain the optical properties of diamond, adopted the same model of a crystalline solid, and assigned the characteristic frequency ω_0 to the motion of each atom. In this way he was led to his well known theory of lattice specific heats. The frequency distribution function $g(\omega)$ in this case can be represented by a suitably normalized delta function,

$$g(\omega)=3rN\delta(\omega-\omega_0). \tag{65}$$

On substituting Eq. (65) into Eq. (63) we obtain an exponential dependence of the specific heat on temperature of the form

$$C_v\sim\exp(-\hbar\omega_0/T), \tag{66}$$

in the low-temperature limit. However, more detailed experimental investigations of the temperature dependence of the specific heat did not confirm this strong, exponential, temperature dependence.

This discrepancy prompted several investigations of the dynamical properties of crystals, which culminated with the publication, in 1912, of fundamental papers by Debye (1912) and by Born and von Kármán (1912, 1913), which laid the basis for the modern theory of lattice dynamics. The theories of Debye and of Born and von Kármán differed from that of Einstein in their recognition that the atoms of a crystal do not all vibrate with the same frequency, but that there is a distribution of frequencies available to the atomic motions. [Even before the work of Debye and of Born and von Kármán, Einstein (1911) had suggested that this is the case to explain the disagreement between the predictions of his theory and experiment.] Debye regarded a crystal as a continuous elastic medium, whose motion was described by the equations of elasticity theory in calculating the distribution of vibrational frequencies of a crystal. It is known that the theory of elasticity describes the long-wavelength, low-frequency vibration modes of a crystal correctly [see, e.g., Maradudin *et al.* (1971), pp. 21–31]. Debye made the additional, simplifying assumption that the elastic medium is elastically isotropic. In contrast, Born and von Kármán took explicit account of the discrete and periodic nature of a real crystal. It is essentially the latter theory that has been outlined in the preceding two sections of this article. Although the theory of Born and von Kármán was more comprehensive than that of Debye, the simplicity of the latter in comparison to the former had the consequence that Debye's theory eclipsed that of Born and von Kármán for more than 20 years, and even today finds widespread application in the study of the dynamical properties of solids. Therefore, we devote the remainder of this section to a discussion of Debye's approach.

If we denote by $\mathbf{u}(\mathbf{x},t)$ the displacement of an arbitrary point $\mathbf{x}$ of an isotropic elastic medium at time t, the classical equations of motion of such a medium have the form

$$\rho\,\frac{\partial^2\mathbf{u}}{\partial t^2}=\mu\nabla^2\mathbf{u}+(\lambda+\mu)\nabla(\nabla\cdot\mathbf{u}). \tag{67}$$

Here $\rho=v_0^{-1}\,\Sigma_\kappa\,M_\kappa$ is the mass density of the medium, where v_0 is the volume of a primitive unit cell. The quantities μ and λ are called the Lamé elastic coefficients. They describe the shear and hydrostatic deformations existing in a moving isotropic continuum. The right-hand side of Eq. (67) is the force acting on an infinitesimally small volume element of the elastic medium; the left-hand side is the inertial force.

A traveling elastic wave propagating in the medium can be described by a displacement vector of the form

$$\mathbf{u}(\mathbf{x};t)=\mathbf{u}_0\,\exp(i\mathbf{k}\cdot\mathbf{x}-i\omega t). \tag{68}$$

On substituting this expression into Eq. (67) we obtain

$$\omega^2\rho\mathbf{u}_0=\mu k^2\mathbf{u}_0+(\lambda+\mu)\mathbf{k}(\mathbf{k}\cdot\mathbf{u}_0). \tag{69}$$

If the amplitude of the wave is parallel to the wave vector $\mathbf{k}$, $\mathbf{u}_{0l}\parallel\mathbf{k}$, we speak of a longitudinal wave. In this case Eq. (69) takes the form

$$[\omega_l^2\rho-(\lambda+2\mu)k^2]\mathbf{u}_{0l}=0. \tag{70}$$

The frequency of a longitudinal wave is therefore equal to

$$\omega_l(\mathbf{k})=[(\lambda+2\mu)/\rho]^{1/2}k. \tag{71}$$

For two transverse waves the amplitude is perpendicular to the wave vector $\mathbf{k}$, $\mathbf{u}_{0t}\perp\mathbf{k}$. In this case the second term on the right-hand side of Eq. (69) vanishes. The frequency of the two degenerate transverse waves is therefore obtained from the equation

$$(\omega_t^2\rho-\mu k^2)\mathbf{u}_{0t}=0, \tag{72}$$

and equals

$$\omega_t(\mathbf{k})=(\mu/\rho)^{1/2}k. \tag{73}$$

The quantities

$$c_l=[(\lambda+2\mu)/\rho]^{1/2} \quad \text{and} \quad c_t=(\mu/\rho)^{1/2} \tag{74}$$

are the speeds of one longitudinal and two transverse elastic waves, respectively.

We can now obtain the frequency distribution function $g(\omega)$ for Debye's model of a crystal. The general definition of $g(\omega)$ is given in Eq. (59). In Debye's model there is one longitudinal and two transverse modes, so that the frequency distribution function assumes the form

$$g(\omega)=\sum_{\mathbf{k}}\delta(\omega-c_lk)+2\sum_{\mathbf{k}}\delta(\omega-c_tk). \tag{75}$$

By going over from summation over **k** to integration, with the use of Eq. (31), we obtain

$$g(\omega)=\frac{V}{(2\pi)^3}\int d^3k\,\delta(\omega-c_lk)+\frac{2V}{(2\pi)^3}\int d^3k\,\delta(\omega-c_tk). \tag{76}$$

On carrying out the integrations we find

$$g(\omega)=\frac{V}{(2\pi)^3}4\pi\left(\frac{1}{c_l^3}+\frac{2}{c_t^3}\right)\omega^2. \tag{77}$$

From this result we see that $g(\omega)$ increases without limit with increasing ω. However, a real crystal possesses only $3rN$ normal mode frequencies, among which, of course, is a maximum frequency. Therefore, to impart a greater physical sense to the Debye model it was necessary to impose a normalization condition on the frequency distribution function obtained through its use, viz.,

$$\int_0^{\omega_D}\frac{V}{(2\pi)^3}4\pi\left(\frac{1}{c_l^3}+\frac{2}{c_t^3}\right)\omega^2d\omega=3Nr, \tag{78}$$

where the maximum frequency of the Debye model, ω_D, is defined by this condition. On evaluating the integral on the left-hand side of this equation we find that

$$\omega_D=2\pi\left(\frac{9Nr}{4\pi V}\right)^{1/3}\left(\frac{1}{c_l^3}+\frac{2}{c_t^3}\right)^{-1/3}. \tag{79}$$

It is convenient to rewrite $g(\omega)$ given by Eq. (77) in terms of ω_D, when it becomes

$$g(\omega)=\begin{cases}3rN(3\omega^2/\omega_D^3) & 0\leqslant\omega\leqslant\omega_D\\ 0 & \omega>\omega_D.\end{cases} \tag{80}$$

It is useful to note that ω_D is an artificial limiting frequency that is introduced into the Debye model so that we can take into account correctly the total number of degrees of freedom in the system. It may not coincide with the true maximum frequency of the vibrational spectrum of the crystal.

The frequency distribution function (80) can be used to calculate the specific heat. The result obtained describes well the temperature dependence of the low-temperature specific heat, since only the low-frequency part of the frequency distribution function is significant for the determination of the latter. On substituting Eq. (80) into Eq. (63) we obtain for the specific heat

$$C_v(T)=3rN\left(\frac{T}{\Theta_D}\right)^3\int_0^{\Theta_D/T}dx\,\frac{x^4e^x}{(e^x-1)^2}, \tag{81}$$

where we have introduced the Debye characteristic temperature

$$\Theta_D=\hbar\omega_D. \tag{82}$$

In the limiting case of low temperatures, $T\ll\Theta_D$, the upper limit on the integral in Eq. (81) can be replaced by infinity. In this case the integral can be evaluated exactly, and we obtain the following dependence of the specific heat on temperature:

$$C_v(T)=3rN\,\frac{4\pi^4}{5}\left(\frac{T}{\Theta_D}\right)^3. \tag{83}$$

This T^3 dependence of the low-temperature specific heat was in agreement with experimental results.

From Eq. (81) we see that in the Debye theory the specific heat is characterized by the single, dimensionless parameter T/Θ_D. This means that the dependence of C_v on T given by Eq. (81) is universal. This means in turn that all experimental curves of the temperature dependence of the specific heat will coincide if the appropriate value of Θ_D is assigned to each solid.

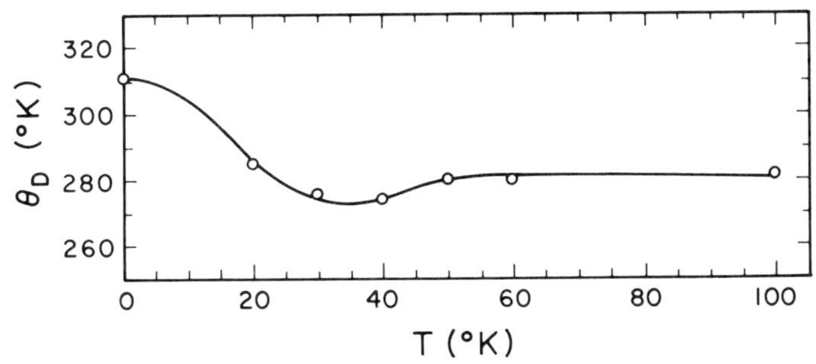

FIG. 1. The temperature dependence of the Debye characteristic temperature Θ_D for NaCl (Kellermann, 1941).

It turned out, however, that in order to obtain agreement between the Debye specific heat and an experimentally measured specific heat it was necessary to use different values of Θ_D at different temperatures. It was shown by Blackman (1934a, 1934b, 1936, 1937a, 1937b) that this temperature dependence of Θ_D can be obtained by equating the specific heat calculated numerically from the Born–von Kármán theory of lattice dynamics for a particular solid to the Debye expression for the specific heat (81). If the equation obtained in this way is regarded as an equation for Θ_D, one can establish the theoretical dependence of Θ_D on temperature. The comparison of this theoretical temperature dependence of Θ_D with that obtained from experimental data can serve as a sensitive criterion for the quality of the force-constant models used in the calculation of vibrational frequency distribution functions of crystals.

In Fig. 1 we show the temperature dependence of Θ_D for NaCl calculated by Kellermann (1941) by the method described above. From the figure we see that in the low-temperature limit Θ_D is a constant independent of temperature only at the very lowest temperatures, up to 5–7 K. Consequently, it is only at these low temperatures that the Debye T^3 law (83) is valid. The Debye temperature Θ_D is also constant over a broad temperature range at elevated temperatures. The reason for this will be discussed below.

If we compare the frequency distribution function used by Kellermann with the Debye frequency distribution function (Fig. 2), we see that they coincide only for frequencies up to the order of $\hbar\omega \sim 0.1\Theta_D$. At first glance it would seem, therefore, that the Debye expression for the specific heat, Eq. (81), should describe the temperature dependence of the specific heat in the temperature range $0 \leqslant T \leqslant 0.1\Theta_D$ ($T \lesssim 20$–50 K), in which only those normal modes of the crystal are excited that are well described by the Debye theory. As is seen from Fig. 1, the experimental data establish a different, smaller, upper limit for the applicability of the Debye theory ($T \lesssim 5$–7 K). This circumstance is related to the form of the integrand in the Debye expression (81), which is $x^4e^x/(e^x-1)^2$. For small values of x this function increases as x^2, while for large x it decreases as x^4e^{-x}. The maximum of this function occurs at $x \cong 3.922$. Consequently, the departure of the Debye frequency distribution function from the exact frequency distribution function will manifest itself in the temperature dependence of the specific heat at temperatures $T \gtrsim \frac{1}{4}(\hbar\omega/2)$. Thus, if the exact frequency distribution function coincides with the Debye frequency distribution function up to frequencies $\hbar\omega \lesssim 0.1\Theta_D$, then the Debye expression for the specific heat will describe the temperature dependence of the specific heat exactly only up to temperatures $T \lesssim 0.1\Theta_D/8 \cong 0.01\Theta_D$.

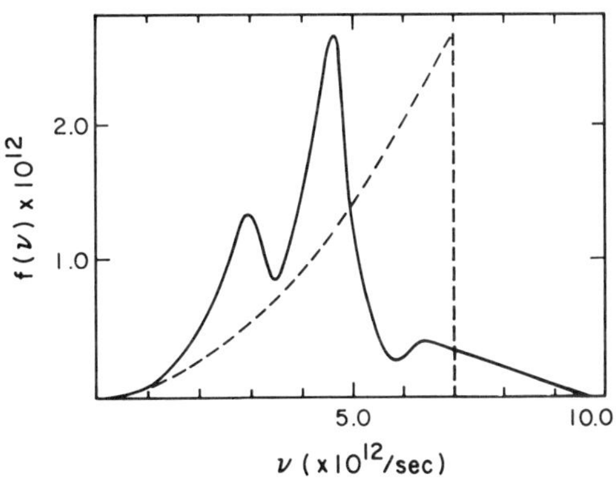

FIG. 2. The frequency distribution function $g(\omega)/3rN$ of NaCl (solid curve) compared with the corresponding Debye approximation to it (Born and Huang, 1954, p. 78).

In the high-temperature limit, $T > \hbar\omega_{max}$, the specific heat of an arbitrary crystal, in the harmonic approximation, can be obtained by expanding the integrand in Eq. (63) in powers of T^{-1} and integrating the result term by term. The result can be expressed in the form

$$C_v = 3rN\left[1 - \sum_{n=1}^{\infty} (-1)^{n+1} \frac{2n-1}{(2n)!} \times B_{2n} u_{2n} \left(\frac{\hbar\omega_{max}}{T}\right)^{2n}\right]. \quad (84)$$

In this expression, known as *Thirring's expansion* (Thirring, 1913, 1914a, 1914b), the B_{2n} are the Bernoulli numbers (Adams, 1957),

$$B_2 = \tfrac{1}{6},\quad B_4 = \tfrac{1}{30},\quad B_6 = \tfrac{1}{42},\quad B_8 = \tfrac{1}{30},$$
$$B_{10} = \tfrac{5}{66},\quad B_{12} = \tfrac{691}{2730},\quad B_{14} = \tfrac{7}{6},\ \ldots, \quad (85)$$

while u_{2n} is the dimensionless $2n$th moment of the frequency distribution function $g(\omega)$,

$$u_{2n} = \frac{1}{3rN\omega_{max}^{2n}} \int_0^{\omega_{max}} d\omega\, g(\omega)\omega^{2n} g(\omega). \quad (86)$$

The attractiveness of the expansion (84) is that the moments $\{u_{2n}\}$ can be calculated without an explicit knowledge of $g(\omega)$ from the result that

$$u_{2n} = \frac{1}{3rN\omega_{\omega_{max}}^{2n}} \sum_{\mathbf{k}j} \omega_j^{2n}(\mathbf{k}) = \frac{1}{3rN\omega_{\omega_{max}}^{2n}} \sum_{\mathbf{k}} \mathrm{Tr}D^n(\mathbf{k}), \quad (87)$$

where $D(\mathbf{k})$ is the dynamical matrix of the crystal, whose elements are given by Eq. (16). In the Debye approximation, it is found from Eqs. (80) and (86) that

$$u_{2n} = 3/(2n+3), \quad (88)$$

so that Eq. (84) in this approximation becomes

$$C_v = 3rN\left[1 - \sum_{n=1}^{\infty} (-1)^{n+1} \frac{3(2n-1)}{2n+3} \times \frac{B_{2n}}{(2n)!}\left(\frac{\Theta_D}{T}\right)^{2n}\right]. \quad (89)$$

If we equate Eqs. (84) and (89) and regard the result as an equation for Θ_D, we find that in the high-temperature limit the latter is given by

$$\Theta_D(T) = \Theta_\infty\left[1 - \frac{1}{80}\left(\frac{1}{3}\frac{u_4}{u_2} - \frac{25}{14}u_2\right)\left(\frac{\hbar\omega_{max}}{T}\right)^2 + O\left(\frac{1}{T^4}\right)\right], \quad (90a)$$

where Θ_∞, the Debye characteristic frequency at infinite temperature, is

$$\Theta_\infty = (\tfrac{5}{4}u_2)^{1/2}\hbar\omega_{max}. \quad (90b)$$

From the result given by Eq. (90a) it appears as if the near constancy of the Debye temperature of NaCl at high temperature seen in Fig. 1 is due to the smallness of the coefficient of $(\hbar\omega_{max}/T)^2$ for this crystal.

The series (84) converges only for $T > \hbar\omega_{max}/2\pi$. A rearrangement of this series, based on Euler's summation formula, that yields a series convergent for all T has been presented by Sack *et al.* (1961).

An expansion similar to Thirring's for the high-temperature specific heat was obtained for the entropy by Stern (1916). The internal energy at high temperatures also possesses an expansion of this type.

4. LATTICE THERMAL CONDUCTIVITY OF INSULATORS

Up to now in this article we have considered only the contribution of phonons to equilibrium vibrational properties of crystals. However, these vibrational excitations of a crystal play an essential role in a variety of nonequilibrium processes as well. Among these are the electrical and thermal conductivities of metals and semiconductors, in which the departures from the perfect periodicity of a crystal caused by the atomic displacements scatter the conduction electrons and give rise to nonzero electrical and thermal resistivities. In such materials the phonons themselves also contribute to the thermal conductivity, but their contribution usually is smaller than that due to the conduction electrons. The situation is quite different in the case of insulating crystals, which lack conduction electrons. The thermal conductivity of these materials is due solely to the phonons in them, and it is this case that we will use to illustrate the role of phonons in nonequilibrium processes.

In discussing the lattice thermal conductivity of an insulator it is convenient to think of the phonons in it as particles that can propagate through it and be scattered by collisions with each other (through the anharmonic terms in the potential energy of the crystal that we have neglected until now), with crystalline imperfections, and with the boundaries of the crystal. Since a phonon with wave vector $\mathbf{k}$ and branch index j is an excitation of the normal vibration mode ($\mathbf{k}j$) of the crystal that involves all the atoms in the crystal, it may seem at first glance that a particlelike description of it is inappropriate. It can be justified, however, by the use of a wave packet synthesized from phonons whose wave vectors are within a narrow range $\Delta\mathbf{k}$ about the wave vector $\mathbf{k}$. This wave packet then has a width in space of the order of $\Delta x \cong 1/\Delta k$, and can be regarded as a particle (Peierls, 1929).

Because the chemical potential of phonons is zero, Eq. (49), in phonon-mediated thermal conductivity phonons are created at the hot end of the sample and are absorbed at the cold end. The thermal conductivity tensor $\kappa_{\alpha\beta}(T)$ relates the components of the heat current in the crystal, $\mathbf{Q}$, to the components of the temperature gradient ∇T present in it:

$$Q_\alpha = -\sum_\beta \kappa_{\alpha\beta}\frac{\partial T}{\partial x_\beta}, \tag{91}$$

where the minus sign reflects the fact that heat flows from the hotter to the colder side of the crystal. It is an interesting property of this lattice thermal conduction that the heat flux, or equivalently the flux of phonons, is not accompanied by mass transport, as is the case, for example, in electrical conduction.

The thermal conductivity tensor is a symmetric, second-rank tensor $\kappa_{\alpha\beta}=\kappa_{\beta\alpha}$.

The heat current carried by all the phonons can be written as

$$\mathbf{Q}=\frac{1}{V}\sum_{\mathbf{k}j} E_{n_j(\mathbf{k})}\mathbf{v}_j(\mathbf{k}), \tag{92}$$

where $E_{n_j(\mathbf{k})}$ is the energy of the phonon ($\mathbf{k}j$) and is given by $E_{n_j(\mathbf{k})} = \hbar\omega_j(\mathbf{k})[n_j(\mathbf{k}) + \frac{1}{2}]$ [Eq. (44)], $\mathbf{v}(\mathbf{k}j)=\nabla_{\mathbf{k}}\omega_j(\mathbf{k})$ is the group velocity of this phonon, and V is the volume of the crystal. Because the frequency $\omega_j(\mathbf{k})$ is an even function of the wave vector $\mathbf{k}$, the group velocity $\mathbf{v}(\mathbf{k}j)$ is an odd function of $\mathbf{k}$, $\mathbf{v}(-\mathbf{k}j) = -\mathbf{v}(\mathbf{k}j)$. Consequently, the zero-point energy does not contribute to the heat current, which takes the form

$$\mathbf{Q}=\frac{1}{V}\sum_{\mathbf{k}j} n_j(\mathbf{k})\hbar\omega_j(\mathbf{k})\mathbf{v}_j(\mathbf{k}). \tag{93}$$

When we consider the motion of phonons of wave vector $\mathbf{k}$ and branch index j through a crystal in the presence of a temperature gradient, their number changes. This is because (Callaway, 1991, p. 169)

1. the number of phonons depends on the temperature gradient, so that $n_j(\mathbf{k})$ becomes a function of position, $n_j(\mathbf{k}) \to n_j(\mathbf{k};\mathbf{x})$, and the phonons tend to drift down the gradient; and
2. phonons collide with each other, with imperfections, and with the boundaries of the crystal.

As a result of these effects $n_j(\mathbf{k};\mathbf{x})$ deviates from its equilibrium value in the absence of the temperature gradient, $\bar{n}(\omega_j(\mathbf{k}))$. In a stationary state the total change in the phonon occupation number due to the temperature gradient, $(\partial n_j(\mathbf{k};\mathbf{x})/\partial t)_T$, and due to collisions, $(\partial n_j(\mathbf{k};\mathbf{x})/\partial t)_c$, must vanish:

$$\left(\frac{\partial n_j(\mathbf{k};\mathbf{x})}{\partial t}\right)_T + \left(\frac{\partial n_j(\mathbf{k};\mathbf{x})}{\partial t}\right)_c = 0. \tag{94}$$

This balance equation is called the *Boltzmann equation.* It expresses the fact that in a stationary state the number of phonons $n_j(\mathbf{k};\mathbf{x})d^3x$ contained in the volume elements d^3x at the point $\mathbf{x}$ in the crystal must be independent of the time.

The rate of change of $n_j(\mathbf{k};\mathbf{x})$ due to the temperature gradient can be obtained from the argument that as the phonons drift in the presence of the temperature gradient, the same number of phonons are in the volume element d^3x at $\mathbf{x}$ at the time t as were in the volume element d^3x at $\mathbf{x}-\mathbf{v}_j(\mathbf{k})t$ at the time $t-\Delta t$. The change in $n_j(\mathbf{k};\mathbf{x})$ in the time interval Δt is therefore equal to the difference of the number of phonons in these two volume elements:

$$\left(\frac{\partial n_j(\mathbf{k};\mathbf{x})}{\partial t}\right)_T \Delta t = n_j(\mathbf{k};\mathbf{x}-\mathbf{v}_j(\mathbf{k})\Delta t) - n_j(\mathbf{k};\mathbf{x}) \tag{95}$$

in the limit as $\Delta t \to 0$. On expanding the right-hand side of this equation to first order in Δt we obtain

$$\left(\frac{\partial n_j(\mathbf{k};\mathbf{x})}{\partial t}\right)_T = -\mathbf{v}_j(\mathbf{k})\cdot\nabla n_j(\mathbf{k};\mathbf{x}). \tag{96}$$

However, $n_j(\mathbf{k};\mathbf{x})$ depends on $\mathbf{x}$ only through its dependence on the temperature, which is a function of $\mathbf{x}$, so that

$$\begin{aligned}\left(\frac{\partial n_j(\mathbf{k};\mathbf{x})}{\partial t}\right)_T &= -\mathbf{v}_j(\mathbf{k})\cdot\frac{\partial n_j(\mathbf{k};\mathbf{x})}{\partial T}\nabla T \\ &\cong -\frac{\partial \bar{n}(\omega_j(\mathbf{k}))}{\partial T}\,\mathbf{v}_j(\mathbf{k})\cdot\nabla T \\ &= -\bar{n}(\omega_j(\mathbf{k}))[\bar{n}(\omega_j(\mathbf{k}))+1] \\ &\quad\times\frac{\hbar\omega_j(\mathbf{k})}{T^2}\mathbf{v}_j(\mathbf{k})\cdot\nabla T.\end{aligned} \tag{97}$$

In obtaining $(\partial n_j(\mathbf{k};\mathbf{x})/\partial t)_c$, which is due to the scattering of phonons from crystalline imperfections and from the boundaries of the crystal, and also to collisions among the phonons themselves, we have to differentiate between processes in which the total wave vector of the participating phonons is conserved and processes in which it is not. The scatterings of a phonon by crystalline imperfections (point defects, dislocations, etc.) and by the boundaries of the crystal are examples of processes in which the wave vector of the phonon scattered is not conserved. The collisions of phonons with each other are more complicated in that processes in which the total wave vectors of the initial and final phonons are the same can occur, together with processes in which they differ. This can be seen by considering the lowest-order term in the expansion of the potential energy of the crystal in powers of the atomic displacements that was neglected in obtaining the harmonic approximation, Eq. (4), viz.,

$$\Phi_3=\tfrac{1}{6}\sum_{l\kappa\alpha}\sum_{l'\kappa'\beta}\sum_{l''\kappa''\gamma}\Phi_{\alpha\beta\gamma}(l\kappa;l'\kappa';l''\kappa'')u_\alpha(l\kappa)u_\beta(l'\kappa')u_\gamma(l''\kappa''), \tag{98}$$

where $\Phi_{\alpha\beta\gamma}(l\kappa;l'\kappa';l''\kappa'')$ is the third-order atomic force constant,

$$\Phi_{\alpha\beta\gamma}(l\kappa;l'\kappa';l''\kappa'')=\left(\frac{\partial^3\Phi}{\partial u_\alpha(l\kappa)\partial u_\beta(l'\kappa')\partial u_\gamma(l''\kappa'')}\right)_0. \tag{99}$$

If we expand $u_\alpha(l\kappa)$ according to Eq. (33a),

$$u_\alpha(l\kappa)=\frac{1}{\sqrt{NM_\kappa}}\sum_{\mathbf{k}j}u_\alpha(\kappa|\mathbf{k}j)Q(\mathbf{k}j)e^{i\mathbf{k}\cdot\mathbf{x}(l)}, \tag{100}$$

where, to simplify the notation, we have defined

$$Q(\mathbf{k}j)=\tfrac{1}{2}[q(\mathbf{k}j)+q(-\mathbf{k}j)]+\frac{1}{2}\frac{i}{\omega_j(\mathbf{k})}[p(\mathbf{k}j)-p(-\mathbf{k}j)], \tag{101}$$

Eq. (98) takes the form

$$\Phi_3=\sum_{\mathbf{k}j}\sum_{\mathbf{k}'j'}\sum_{\mathbf{k}''j''}V^{(3)}(\mathbf{k}j;\mathbf{k}'j';\mathbf{k}''j'')Q(\mathbf{k}j)Q(\mathbf{k}'j')Q(\mathbf{k}''j''), \tag{102}$$

with

$$\begin{aligned}V^{(3)}(\mathbf{k}j;\mathbf{k}'j';\mathbf{k}''j'')=\frac{1}{6N^{3/2}}&\sum_{l\kappa\alpha}\sum_{l'\kappa'\beta}\sum_{l''\kappa''\gamma}\Phi_{\alpha\beta\gamma}(l\kappa;l'\kappa';l''\kappa'')\\ &\times\frac{u_\alpha(\kappa|\mathbf{k}j)}{\sqrt{M_\kappa}}\frac{u_\beta(\kappa'|\mathbf{k}'j')}{\sqrt{M_{\kappa'}}}\frac{u_\gamma(\kappa''|\mathbf{k}''j'')}{\sqrt{M_{\kappa''}}}e^{i[\mathbf{k}\cdot\mathbf{x}(l)+\mathbf{k}'\cdot\mathbf{x}(l')+\mathbf{k}''\cdot\mathbf{x}(l'')]}.\end{aligned} \tag{103}$$

The coefficient $V^{(3)}(\mathbf{k}j;\mathbf{k}'j'\mathbf{k}''j'')$ is important in phonon–phonon interactions involving three phonons because time-dependent perturbation theory tells us, for example, that the rate at which phonons $(\mathbf{k}'j')$ and $(\mathbf{k}''j'')$ collide to create a phonon $(\mathbf{k}j)$ is proportional to $|V^{(3)}(-\mathbf{k}j;\mathbf{k}'j';\mathbf{k}''j'')|^2$ (Peierls, 1929; Leibfried, 1955; Gurevich, 1986, pp. 54ff.). Now, the atomic force constant $\Phi_{\alpha\beta\gamma}(l\kappa;l'\kappa';l''\kappa'')$ appearing in the expression (103) for $V^{(3)}(\mathbf{k}j;\mathbf{k}'j';\mathbf{k}''j'')$ is invariant against the addition of the same arbitrary cell index m to each of the indices l,l',l'':

$$\Phi_{\alpha\beta\gamma}(l+m,\kappa;l'+m,\kappa';l''+m,\kappa'') = \Phi_{\alpha\beta\gamma}(l\kappa;l'\kappa';l''\kappa''), \tag{104}$$

because of the invariance of the crystal against the displacement through the translation vector $\mathbf{x}(m)$. As a consequence, $V^{(3)}(\mathbf{k}j;\mathbf{k}'j';\mathbf{k}''\mathrm{j}'')$ and $|V^{(3)}(\mathbf{k}j;\mathbf{k}'j';\mathbf{k}''\mathrm{j}'')|^2$ are proportional to $\Delta(\mathbf{k}+\mathbf{k}'+\mathbf{k}'')$, where $\Delta(\mathbf{k})$ has been defined in Eq. (36). Three-phonon scattering processes proportional to $|V^{(3)}(\mathbf{k}j;\mathbf{k}'j';\mathbf{k}''j'')|^2$ in which $\mathbf{k}+\mathbf{k}'+\mathbf{k}''=0$ have been termed *normal processes;* those in which $\mathbf{k}+\mathbf{k}'+\mathbf{k}''=\mathbf{G}(h)\neq 0$ have been termed *Umklapp (flipping-over) processes.* An Umklapp process is a scattering process in which momentum is transferred to the crystal as a whole.

The central role in lattice thermal conduction in insulators is played by the scattering processes in which the wave vector of the participating phonons is not conserved. This is particularly readily seen at very low temperatures when only very low-energy acoustic phonons are excited. Thus, in this case, if we rewrite Eq. (93) as

$$\mathbf{Q}=\frac{1}{2V}\sum_{\mathbf{k}}\sum_{j=1}^{3} n_j(\mathbf{k})\hbar\nabla_{\mathbf{k}}\omega_j^2(\mathbf{k}), \tag{105}$$

and make the simplifying assumption that the frequencies of acoustic phonons are independent of their direction of propagation,

$$\omega_j(\mathbf{k})=c_jk, \quad j=1,2,3, \tag{106}$$

where c_j is the speed of sound for the jth branch of the phonon spectrum, we obtain for the heat current

$$\mathbf{Q}=\frac{1}{V}\sum_{j=1}^{3} c_j^2 \sum_{\mathbf{k}} n_j(\mathbf{k})\hbar\mathbf{k}=\sum_{j=1}^{3} c_j^2\mathbf{P}_j, \tag{107}$$

where

$$\mathbf{P}_j=\frac{1}{V}\sum_{\mathbf{k}} n_j(\mathbf{k})\hbar\mathbf{k} \tag{108}$$

is the total wave vector of the phonons in the branch j. If we assume further that the speeds of sound are the same for all three acoustic branches, viz., that $c_j=c$ for $j=1,2,3$, then the expression for the heat current $\mathbf{Q}$ reduces to

$$\mathbf{Q}=c^2\mathbf{P}, \tag{109}$$

where

$$\mathbf{P}=\sum_j \mathbf{P}_j \tag{110}$$

is the sum of the wave vectors of all the phonons in the crystal.

If only wave-vector–conserving scattering processes existed, the total wave vector of all the phonons in the crystal, $\mathbf{P}$, could not change. Thus, in this case collisions could not lead from a state with nonzero $\mathbf{P}$ to a state with zero $\mathbf{P}$. In equilibrium we must have $\mathbf{P}=\mathbf{0}$. In a state with $\mathbf{P}\neq 0$ a heat current exists, according to Eq. (109). Therefore, in the case of wave-vector–conserving processes only, if a heat current were established it would continue to flow even in the absence of a thermal gradient, which implies zero thermal resistance or an infinite thermal conductivity. However, wave-vector–destroying processes change the total wave vector of the phonon system, and thus are essential for the existence of a finite thermal conductivity of a perfect, unbounded crystal.

Although this conclusion has been reached on the basis of several simplifying assumptions, it has a more general validity. Thus in obtaining the contribution to $(\partial n_j(\mathbf{k};\mathbf{x})/\partial t)_c$ due to wave-vector–destroying processes we exploit the fact that they cause any deviation of $n_j(\mathbf{k};\mathbf{x})$ from its equilibrium value $\bar{n}(\omega_j(\mathbf{k}))$ to disappear if the temperature gradient is turned off. A commonly made approximation that describes this is to assume that because of the scattering of phonons by processes that change the total wave vector of the crystal, $n_j(\mathbf{k};\mathbf{x})$ relaxes to its equilibrium value $\bar{n}(\omega_j(\mathbf{k}))$ according to an exponential law,

$$\left(\frac{\partial n_j(\mathbf{k};\mathbf{x})}{\partial t}\right)_U=-\frac{n_j(\mathbf{k};\mathbf{x})-\bar{n}(\omega_j(\mathbf{k}))}{\tau_j^{(u)}(\mathbf{k})}, \tag{111}$$

where $\tau_j^{(u)}(\mathbf{k})$ is the relaxation time for the mode $(\mathbf{k}j)$. It includes the contributions from all scattering processes that do not conserve the phonon wave vectors, e.g., scattering from point defects, and from sample boundaries and Umklapp processes. The result represented by Eq. (111) is called the *relaxation-time approximation.* Although Eq. (111) describes in a qualitative way how the deviation $n_j(\mathbf{k};\mathbf{x})-\bar{n}(\omega_j(\mathbf{k}))$ decays to zero as a result of wave-vector–destroying phonon-scattering processes, in general it cannot be justified from first principles. An exception to this statement, however, is provided by elastic scattering processes in which the number of phonons is conserved, such as the scattering of phonons by point defects and by the boundaries of the crystal. For such processes Eq. (111) is exact, and explicit expressions for the inverse relaxation time $\tau_j^{(u)}(\mathbf{k})$ can be obtained (Maradudin, 1966, pp. 335–338; Grimm *et al.*, 1970).

The normal processes, which conserve the total wave vector of the phonons in the crystal, cannot relax the phonon distribution $n_j(\mathbf{k};\mathbf{x})$ to the equilibrium distribution. Instead they relax it to a displaced equilibrium distribution $\bar{n}(\omega_j(\mathbf{k})-\mathbf{k}\cdot\mathbf{V})$ (Klemens, 1955):

$$\left(\frac{\partial n_j(\mathbf{k};\mathbf{x})}{\partial t}\right)_N=-\frac{n_j(\mathbf{k};\mathbf{x})-\bar{n}(\omega_j(\mathbf{k})-\mathbf{k}\cdot\mathbf{V})}{\tau_j^{(n)}(\mathbf{k})},\tag{112}$$

where

$$\bar{n}(\omega_j(\mathbf{k})-\mathbf{k}\cdot\mathbf{V})=\{\exp(\hbar[\omega_j(\mathbf{k})-\mathbf{k}\cdot\mathbf{V}]/T)-1\}^{-1},\tag{113}$$

and $\tau_j^{(n)}(\mathbf{k})$ is the corresponding relaxation time. The vector $\mathbf{V}$ has the physical interpretation of the drift velocity of the phonon gas in the temperature gradient (Gurevich, 1986, p. 120), and can be regarded as small (we will see that it is proportional to ∇T). Thus we can write to first order in $\mathbf{V}$ that

$$\bar{n}(\omega_j(\mathbf{k})-\mathbf{k}\cdot\mathbf{V})=\bar{n}(\omega_j(\mathbf{k}))+\frac{\hbar}{T}(\mathbf{V}\cdot\mathbf{k})\times\bar{n}(\omega_j(\mathbf{k}))[\bar{n}(\omega_j(\mathbf{k}))+1].\tag{114}$$

The total time rate of change of $n_j(\mathbf{k};\mathbf{x})$ due to scattering and collision processes is the sum of Eqs. (111) and (112) which, with the use of Eq. (114), is

$$\left(\frac{\partial n_j(\mathbf{k};\mathbf{x})}{\partial t}\right)_c=-\left(\frac{1}{\tau_j^{(n)}(\mathbf{k})}+\frac{1}{\tau_j^{(u)}(\mathbf{k})}\right)[n_j(\mathbf{k};\mathbf{x})-\bar{n}(\omega_j(\mathbf{k}))]+\frac{(\mathbf{V}\cdot\mathbf{k})}{\tau_j^{(n)}(\mathbf{k})}\frac{\hbar}{T}\bar{n}(\omega_j(\mathbf{k}))[\bar{n}(\omega_j(\mathbf{k}))+1].\tag{115}$$

With this result and that given by Eq. (97), the Boltzmann equation (94) takes the form

$$-\bar{n}(\omega_j(\mathbf{k}))[\bar{n}(\omega_j(\mathbf{k}))+1]\frac{\hbar\omega_j(\mathbf{k})}{T^2}\mathbf{v}_j(\mathbf{k})\cdot\nabla T-\left(\frac{1}{\tau_j^{(u)}(\mathbf{k})}+\frac{1}{\tau_j^{(n)}(\mathbf{k})}\right)[n_j(\mathbf{k};\mathbf{x})-\bar{n}(\omega_j(\mathbf{k}))]+\frac{(\mathbf{V}\cdot\mathbf{k})}{\tau_j^{(n)}(\mathbf{k})}\frac{\hbar}{T}\bar{n}(\omega_j(\mathbf{k}))[\bar{n}(\omega_j(\mathbf{k}))+1]=0,\tag{116}$$

whose solution is

$$n_j(\mathbf{k};\mathbf{x})=\bar{n}(\omega_j(\mathbf{k}))+\frac{\tau_j^{(n)}(\mathbf{k})\tau_j^{(u)}(\mathbf{k})}{\tau_j^{(n)}(\mathbf{k})+\tau_j^{(u)}(\mathbf{k})}\bar{n}(\omega_j(\mathbf{k}))[\bar{n}(\omega_j(\mathbf{k}))+1]\frac{\hbar}{T}\left\{\frac{\mathbf{V}\cdot\mathbf{k}}{\tau_j^{(n)}(\mathbf{k})}-\frac{\omega_j(\mathbf{k})}{T}\mathbf{v}_j(\mathbf{k})\cdot\nabla T\right\}.\tag{117}$$

The phonon drift velocity $\mathbf{V}$ is then obtained from the condition that the rate of change of the total phonon wave vector due to normal processes vanishes. This can be expressed by the statement

$$\sum_{\mathbf{k}j}\left(\frac{\partial n_j(\mathbf{k};\mathbf{x})}{\partial t}\right)_N\mathbf{k}=\sum_{\mathbf{k}j}\left(\frac{\bar{n}(\omega_j(\mathbf{k})-\mathbf{k}\cdot\mathbf{V})-n_j(\mathbf{k};\mathbf{x})}{\tau_j^{(n)}(\mathbf{k})}\right)\mathbf{k}=\mathbf{0},\tag{118}$$

or, with the use of Eqs. (114) and (117),

$$\sum_{\mathbf{k}j} \frac{\bar{n}(\omega_j(\mathbf{k}))[\bar{n}(\omega_j(\mathbf{k}))+1]}{\tau_j^{(n)}(\mathbf{k})+\tau_j^{(u)}(\mathbf{k})} \frac{\hbar}{T}(\mathbf{V}\cdot\mathbf{k})\mathbf{k} = -\sum_{\mathbf{k}j} \frac{\tau_j^{(u)}(\mathbf{k})\bar{n}(\omega_j(\mathbf{k}))[\bar{n}(\omega_j(\mathbf{k}))+1]}{\tau_j^{(n)}(\mathbf{k})+\tau_j^{(u)}(\mathbf{k})} \frac{\hbar\omega_j(\mathbf{k})}{T^2} \mathbf{v}_j(\mathbf{k})\cdot\nabla T\mathbf{k}. \tag{119}$$

Thus, if we write the solution of this equation formally as

$$V_\alpha = -\frac{1}{T}\sum_\beta \beta_{\alpha\beta}\frac{\partial T}{\partial x_\beta}, \tag{120}$$

then from Eqs. (91), (93), and (117) we obtain for the thermal conductivity tensor

$$\kappa_{\alpha\beta} = \frac{1}{V}\sum_{\mathbf{k}j} \tau_j^{(c)}(\mathbf{k})C_j(\mathbf{k})v_{j\alpha}(\mathbf{k})\left[v_{j\beta}(\mathbf{k}) + \frac{1}{\omega_j(\mathbf{k})\tau_j^{(n)}(\mathbf{k})}\sum_\gamma \kappa_\gamma\beta_{\gamma\beta}\right], \tag{121}$$

where $C_j(\mathbf{k}) = [\hbar\omega_j(\mathbf{k})/T]^2\bar{n}(\omega_j(\mathbf{k}))[\bar{n}(\omega_j(\mathbf{k}))+1]$ is the contribution to the specific heat of the crystal from the mode $(\mathbf{k}j)$, and $\tau_j^{(c)}(\mathbf{k})$ is the combined relaxation time defined by

$$1/\tau_j^{(c)}(\mathbf{k}) = [1/\tau_j^{(n)}(\mathbf{k})] + [1/\tau_j^{(u)}(\mathbf{k})]. \tag{122}$$

If it were not for the second term in brackets on the right-hand side of Eq. (121), this expression for $\kappa_{\alpha\beta}$ would have the form predicted by elementary kinetic theory (Kittel, 1976, pp. 143–144). The presence of the second term in brackets is a consequence of the fact that normal processes cannot relax a given phonon distribution $n_j(\mathbf{k};\mathbf{x})$ to the equilibrium distribution $\bar{n}(\omega_j(\mathbf{k}))$, but only to the displaced distribution $\bar{n}(\omega_j(\mathbf{k})-\mathbf{k}\cdot\mathbf{V})$.

It is clear that a central role in this approach to the thermal conductivity of an insulator is played by the relaxation times $\tau_j^{(n)}(\mathbf{k})$ and $\tau_j^{(u)}(\mathbf{k})$. It was shown by Herring (1954) that the relaxation time of a single low-energy mode of wave vector $\mathbf{k}$ due to three-phonon processes depends on $\mathbf{k}$ and the temperature T according to

$$1/\tau_j(\mathbf{k}) = k^a T^{5-a}, \tag{123}$$

where a is an exponent determined by crystal symmetry. The relaxation time for normal scattering processes $\tau_j^{(n)}(\mathbf{k})$ is usually written in the form (Callaway, 1959)

$$1/\tau_j^{(n)}(\mathbf{k}) = BT^3\omega_j^2(\mathbf{k}), \tag{124}$$

although an alternative form (Gurevich, 1986, p. 98),

$$1/\tau_j(\mathbf{k}) = B[T/v_j(\mathbf{k})]^5, \tag{125}$$

is also used in the limit of very low temperatures, $T \ll \Theta_D$. Both of these forms are consistent with Herring's result, Eq. (123).

For the scattering of phonons by point defects the relaxation time $\tau_j^{(d)}(\mathbf{k})$ has the temperature-independent form (Pomeranchuk, 1942; Klemens, 1951, 1955, 1957)

$$1/\tau_j^{(d)}(\mathbf{k}) = A\omega_j^4(\mathbf{k}), \tag{126}$$

where the constant coefficient A is proportional to the concentration of point defects. Equation (126) is the Rayleigh law for the scattering of long-wavelength phonons by defects whose linear dimensions are smaller than the phonon wavelength. A derivation of Eq. (126) on the basis of defect lattice dynamics for the particular case of the scattering of phonons by isotopic impurities has been given, e.g., by Maradudin (1966, pp. 335–338).

For the scattering of phonons by the boundaries of the sample, the relaxation time $\tau_j^{(b)}(\mathbf{k})$ is frequency independent and temperature independent (Casimir, 1938),

$$1/\tau_j^{(b)}(\mathbf{k}) = v_j(\mathbf{k})/L, \tag{127}$$

where L is a length that is of the order of the linear dimensions of the sample. In a microscopic derivation of Eq. (127) by Grimm *et al.* (1970), $v_j(\mathbf{k})$ is replaced by the component of the group velocity $\mathbf{v}_j(\mathbf{k})$ normal to the boundary.

Finally, the relaxation time for the scattering of phonons by three-phonon Umklapp processes has the form

$$1/\tau_j^{(\mathrm{U})}(\mathbf{k}) = C\exp(-\alpha\Theta_D/T) \tag{128}$$

at low temperatures (Gurevich, 1986, p. 123; Brüesch, 1987, p. 96), where α is a numerical factor of the order of, and smaller than, unity. The origin of this form can be understood by considering a scattering process in which two phonons with wave vectors $\mathbf{k}_1$ and $\mathbf{k}_2$ interact to create a phonon whose wave vector $\mathbf{k}_1+\mathbf{k}_2$ extends outside the first Brillouin zone. At the low temperatures we are considering, only acoustic phonons are excited, and for simplicity we can ignore the distinction between longitudinal and transverse phonons, and can suppress the branch indices of the interacting phonons. This process is illustrated for a two-dimensional square lattice of lattice constant a in Fig. 3. The addition to $\mathbf{k}_1+\mathbf{k}_2$ of a reciprocal-lattice vector $\mathbf{G}=(-2\pi/a,0,0)$, whose magnitude is equal to the length of an edge of the Brillouin zone, transforms this vector into a vector $\mathbf{k}_3$ inside the first Brillouin zone. Here k_{1x} and k_{2x} are both positive, but k_{3x} is negative. This is the origin of the name "Umklapp" for scattering processes of this type. For an Umklapp process to be possible, $|\mathbf{k}_1+\mathbf{k}_2|$ must exceed $\frac{1}{2}G_{\min}$, where $G_{\min}$ is the magnitude of the shortest reciprocal lattice vector, which in the present case is $2\pi/a$. This requires in turn that the magnitude of at least one of the two wave vectors $\mathbf{k}_1$ and $\mathbf{k}_2$ must exceed $\frac{1}{4}G_{\min}$. In the Debye approximation the Brillouin zone is replaced by a circle of radius k_{D} (in the two-dimensional case we are considering), whose area is equal to that of the Brillouin zone. Therefore k_{D} has a value $k_{\mathrm{D}}=2\sqrt{\pi}/a \gtrsim \frac{1}{2}G_{\min}$. From Fig. 3 we see that the frequency of the phonon created in this scattering process is therefore given by $\hbar\omega(\mathbf{k}_3)\lesssim\Theta_{\mathrm{D}}$, in the Debye approximation, where Θ_{D} is the Debye temperature, since $|\mathbf{k}_3|\lesssim k_{\mathrm{D}}$. The collision frequency of an Umklapp process is proportional to the product of the occupation numbers of the colliding phonons, $\bar{n}(\omega(\mathbf{k}_1))$ and $\bar{n}(\omega(\mathbf{k}_2))$, which, in the low-temperature limit, $T\ll\Theta_{\mathrm{D}}$, are given by Eq. (50):

$$\bar{n}(\omega(\mathbf{k}_1))\cong\exp[-\hbar\omega(\mathbf{k}_1)/T],$$

$$\bar{n}(\omega(\mathbf{k}_2))\cong\exp[-\hbar\omega(\mathbf{k}_2)/T], \qquad (129)$$

respectively. The product $\bar{n}(\omega(\mathbf{k}_1))\bar{n}(\omega(\mathbf{k}_2))$ is therefore given by

$$\bar{n}(\omega(\mathbf{k}_1))\bar{n}(\omega(\mathbf{k}_2)) \cong\exp\{-\hbar[\omega(\mathbf{k}_1)+\omega(\mathbf{k}_2)]/T\}. \qquad (130)$$

However, energy is conserved in the interaction, so that $\omega(\mathbf{k}_1)+\omega(\mathbf{k}_2)=\omega(\mathbf{k}_1+\mathbf{k}_2) =\omega(\mathbf{k}_1+\mathbf{k}_2+\mathbf{G})=\omega(\mathbf{k}_3)$, where we have used the result that the frequency $\omega(\mathbf{k})$ is a periodic function of $\mathbf{k}$ with period $\mathbf{G}$. It follows that the collision rate is proportional to

$$\bar{n}(\omega(\mathbf{k}_1))\bar{n}(\omega(\mathbf{k}_2))\cong\exp[-\hbar\omega(\mathbf{k}_3)/T] =\exp(-\alpha\Theta_{\mathrm{D}}/T), \qquad (131)$$

where α is a number of the order of, and smaller than, unity. This result establishes Eq. (128).

In the high-temperature limit, $T\gg\Theta_{\mathrm{D}}$, an approximate microscopic calculation (Ludwig, 1964; Gurevich, 1986, p. 96) yields the result that the relaxation time for the scattering of phonons by three-phonon Umklapp processes has the form

$$1/\tau_j^{(\mathrm{U})}(\mathbf{k})=D_j(\mathbf{k})T. \qquad (132)$$

This dependence on temperature can be understood (Kittel, 1976, p. 145) in terms of the number of phonons with which a given phonon can interact. At high temperatures, when all phonons are excited, the magnitudes of the relaxation times for normal and Umklapp processes become comparable, and we need not distinguish between these two types of scattering processes. In this limit the total number of excited phonons is proportional to the absolute temperature T [Eq. (50b)]. The collision frequency of a given phonon should be proportional to the number of phonons with which it can interact, and Eq. (132) follows from this argument.

The various scattering processes just described are assumed to be independent, so that the probabilities of their occurrence add, and consequently so do the inverse relaxation times. Thus we have that

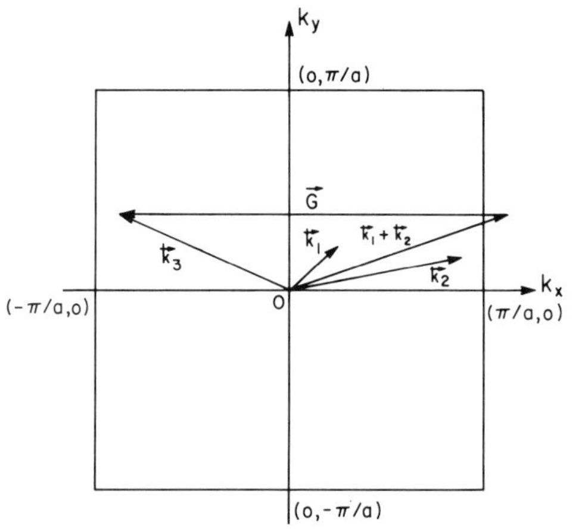

FIG. 3. The first Brillouin zone for a square lattice of lattice constant a, and the wave vectors of phonons participating in an Umklapp process.

$$\frac{1}{\tau_j^{(u)}(\mathbf{k})}=\frac{1}{\tau_j^{(d)}(\mathbf{k})}+\frac{1}{\tau_j^{(b)}(\mathbf{k})}+\frac{1}{\tau_j^{(\mathrm{U})}(\mathbf{k})}. \qquad (133)$$

The calculation of the thermal conductivity of an insulator by the procedure just described is usually carried out on the basis of the assumption of a Debye spectrum for the phonons with speeds of sound that are independent of the direction of propagation, and even that the speed of sound is the same for all three acoustic branches (Callaway, 1959). With such approximations it follows from Eq. (121) that at very low temperatures (usually below approximately 10 K), where the inverse lifetimes for the scattering of phonons by other phonons and by crystalline defects and impurities have essentially vanished, and the only scattering mechanism limiting the thermal conductivity is the scattering of phonons by the boundaries of the crystal, the thermal conductivity becomes proportional to the specific heat of the crystal, and thus increases with increasing temperature according to a T^3 law. For, in this case, if we denote by c the common speed of sound, from Eqs. (121), (127), and (133) we obtain

$$\begin{aligned}\kappa_{\alpha\beta}&=\frac{1}{V}\sum_{\mathbf{k}j}\tau_j^{(b)}(\mathbf{k})C_j(\mathbf{k})v_{j\alpha}(\mathbf{k})v_{j\beta}(\mathbf{k})\\ &=\delta_{\alpha\beta}\frac{cL}{3}\frac{1}{V}\sum_{\mathbf{k}j}C_j(\mathbf{k})\\ &=\delta_{\alpha\beta}\frac{cL}{3}\frac{C_v(T)}{V}. \qquad (134)\end{aligned}$$

At high temperatures, $T\gg\Theta_D$, the dominant scattering processes leading to a finite lattice thermal conductivity are the three-phonon Umklapp scattering processes. In this limit, we obtain from Eqs. (121), (132), (133), and the fact that $C_j(\mathbf{k})\to 1$ as $T\to\infty$, the result that

$$\begin{aligned}\kappa_{\alpha\beta}&=\frac{1}{V}\sum_{\mathbf{k}j}\tau_j^{(U)}(\mathbf{k})C_j(\mathbf{k})v_{j\alpha}(\mathbf{k})v_{j\beta}(k)\\ &\underset{T\to\infty}{\to}\frac{1}{T}\frac{1}{V}\sum_{\mathbf{k}j}\frac{v_{j\alpha}(\mathbf{k})v_{j\beta}(\mathbf{k})}{D_j(\mathbf{k})}\\ &=\delta_{\alpha\beta}\frac{1}{3}\frac{c^2}{T}\frac{1}{V}\sum_{\mathbf{k}j}\frac{1}{D_j(\mathbf{k})}. \qquad (135)\end{aligned}$$

Thus, in this limit the thermal conductivity of all insulating crystals must decrease according to a T^{-1} law (Brüesch, 1987, p. 97; Callaway, 1991, p. 175). A maximum therefore appears at intermediate temperatures in the temperature dependence of the thermal conductivity. The thermal conductivity in the temperature range where this maximum occurs is determined primarily by the scattering of the phonons by crystalline defects and impurities. In quantitative calculations of the conductivity as a function of temperature, the coefficients in the expressions for the several relaxation times entering the theory are usually obtained by fitting the resulting theoretical curves to experimental data in the limiting cases of low and high temperatures. As an example of the kinds of results that can be obtained in this way, we present in Fig. 4 plots of the lattice thermal conductivity of almost isotopically pure ^{74}Ge and of normal Ge (Callaway, 1959), calculated with the inclusion of all the scattering processes described above. The thermal conductivity of the normal Ge is seen to be smaller than that of the isotopically pure Ge because of the additional thermal resistivity of the former sample due to the scattering from the isotopic impurities in it. The agreement between theory and experiment is quite good, suggesting that the essential physics has been incorporated into the theory. The discrepancies, particularly in the vicinity of the maximum, are presumably due to the use of a relaxation-time approximation

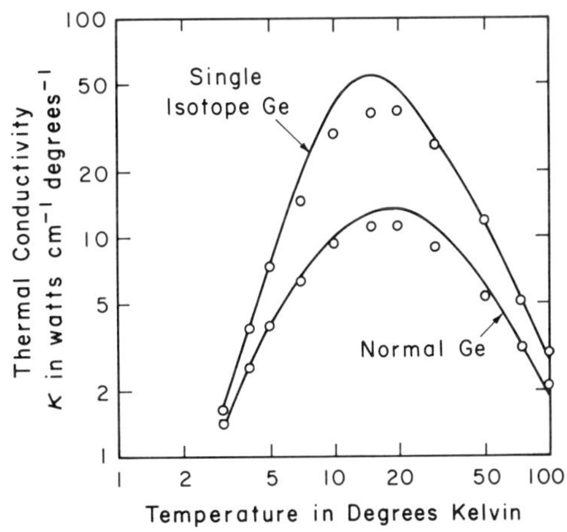

FIG. 4. The thermal conductivity of germanium. The solid curves are theoretical results calculated in the manner described in the text. The open circles are experimental data of Geballe and Hull (1958) (Callaway, 1959).

to describe three-phonon collision processes and to the neglect of the distinction between longitudinal and transverse phonons.

List of Works Cited

Adams, E. P. (1957), *Smithsonian Mathematical Formulae and Tables of Elliptic Functions*, Washington, DC: The Smithsonian Institution. Note that Adams denotes by B_n what we have called B_{2n}.

Blackman, M. (1934a), *Proc. R. Soc. London Ser. A* **148**, 365–383.

Blackman, M. (1934b), *Proc. R. Soc. London Ser. A* **148**, 384–406.

Blackman, M. (1936), *Philos. Trans. R. Soc. London Ser. A* **236**, 103–131.

Blackman, M. (1937a), *Proc. Cambridge Philos. Soc.* **33**, 94–103.

Blackman, M. (1937b), *Proc. R. Soc. London Ser. A* **159**, 416–431.

Born, M., Huang, K. (1954), *Dynamical Theory of Crystal Lattices*, Oxford: Clarendon.

Born, M., von Kármán, Th. (1912), *Phys. Z.* **13**, 297–309.

Born, M., von Kármán,Th. (1913), *Phys. Z.* **14**, 15–19.

Brüesch, P. (1987), *Phonons: Theory and Experiments III*, Berlin: Springer-Verlag.

Callaway, J. (1959), *Phys. Rev.* **113**, 1046–1051.

Callaway, J. (1991), *Quantum Theory of the Solid State*, 2nd ed., New York: Academic.

Casimir, H. B. G. (1938), *Physica* **5**, 495–500.

Debye, P. (1912), *Ann. Phys. (Leipzig)* **39**, 789–839.

Einstein, A. (1907), *Ann. Phys. (Leipzig)* **22**, 180–190.

Einstein, A. (1911), *Ann. Phys. (Leipzig)* **35**, 679–694.

Geballe, T. H., Hull, G. W. (1958), *Phys. Rev.* **110**, 773–775.

Grimm, A., Maradudin, A. A., Tong, S. Y. (1970), *J. Phys. (Paris), Colloque* **31**, C1-9–C1-19.

Gurevich, V. L. (1986), *Transport in Phonon Systems*, Amsterdam: North-Holland.

Hardy, J. R. (1962), *Philos Mag.* **7**, 315–336.

Herring, C. (1954), *Phys. Rev.* **95**, 954–965.

Kellermann, E. W. (1941), *Proc. R. Soc. London Ser. A* **178**, 17–24.

Kittel, C. (1976), *Introduction to Solid State Physics*, 5th ed., New York: Wiley.

Klemens, P. G. (1951), *Proc. R. Soc. London Ser. A* **208**, 108–133.

Klemens, P. G. (1955), *Proc. Phys. Soc. Ser. A* **68**, 1113–1128.

Klemens, P. G. (1957), *Proc. Phys. Soc. Ser. A* **70**, 833–838.

Ledermann, W. (1944), *Proc. R. Soc. London Ser. A* **182**, 362–377.

Leibfried, G. (1955), "Gittertheorie der mechanischen und thermischen Eigenschaften der Kristalle," in: S. Flügge (Ed.), *Encyclopedia of Physics*, Vol. VII, Part 1, Berlin: Springer, pp. 104–324.

Ludwig, W. (1964), "Some Aspects of Phonon-Phonon-Interaction," in: T. A. Bak (Ed.), *Phonons and Phonon Interactions*, New York: Benjamin, pp. 23–75.

Maradudin, A. A. (1966), "Theoretical and Experimental Aspects of the Effects of Point Defects and Disorder on the Vibrations of Crystals—1," in: F. Seitz, D. Turnbull (Eds.), *Solid State Physics*, Vol. 18, New York: Academic, pp. 273–420.

Maradudin, A. A., Montroll, E. W., Weiss, G. H., Ipatova, I. P. (1971), *Theory of Lattice Dynamics in the Harmonic Approximation*, New York: Academic.

Peierls, R. (1929), *Ann. Phys. (Leipzig)* **3**, 1055–1101.

Pomeranchuk, I. Ya. (1942), *J. Phys. USSR* **6**, 237–250.

Richarz, F. (1893), *Ann. Phys. (Leipzig)* **48**, 708–716.

Richarz, F. (1899), *Ann. Phys. (Leipzig)* **67**, 702–706.

Sack, R. A., Maradudin, A. A., Weiss, G. H. (1961), *Phys. Rev.* **124**, 717–722.

Schiff, L. I. (1949), *Quantum Mechanics*, New York: McGraw-Hill.

Stern, O. (1916), *Ann. Phys. (Leipzig)* **51**, 237–260.

Thirring, H. (1913), *Phys. Z.* **14**, 867–873.

Thirring, H. (1914a), *Phys. Z.* **15**, 127–133.

Thirring, H. (1914b), *Phys. Z.* **15**, 180–185.

Further Reading

Beck, H. (1975), "Second Sound and Related Thermal Conduction Phenomena," in: G. K. Horton, A. A. Maradudin (Eds.), *Dynamical Properties of Solids*, Amsterdam: North Holland, pp. 205–284.

Bilz, H., Kress, W. (1979), *Phonon Dispersion Relations in Insulators*, Berlin: Springer-Verlag.

Böttger, H. (1983), *Principles of the Theory of Lattice Dynamics*, Berlin: Akademie-Verlag.

Hardy, J. R., Karo, A. M. (1979), *The Lattice Dynamics and Statics of Alkali-Halide Crystals*, New York: Plenum.

Reissland, J. A. (1973), *The Physics of Phonons*, New York: Wiley.

LIGHT-EMITTING DIODES

M. GEORGE CRAFORD AND FRANK M. STERANKA, *Optoelectronics Division, Hewlett-Packard Company, San Jose, California, U.S.A.*

	Introduction	486
1.	**Basic Device Physics**	488
1.1	*p-n* Junction	488
1.2	Radiative and Nonradiative Recombination and Internal Quantum Efficiency	489
1.3	Direct and Indirect Energy Gap	490
1.4	Isoelectronic Impurities	490
1.5	Direct–Indirect Transition	491
1.6	Emission Wavelength versus Energy Gap and Doping	492
1.7	Device Structures	492
1.7.1	Homojunctions	492
1.7.2	Single Heterostructures	493
1.7.3	Double Heterostructures	494
1.7.4	Quantum Wells and Other Structures	494
1.8	LED Performance and Extraction Efficiency	494
1.8.1	Light Extraction	495
1.8.2	Quantum Efficiency and Power Efficiency	496
1.8.3	Luminous Performance	496
1.9	Reliability	496
1.9.1	Radiation-Enhanced Degradation	496
1.9.2	Package-Related Degradation	497
1.9.3	Typical LED Degradation Characteristics	497
2.	**Semiconductor Materials Issues**	497
2.1	Desirable Characteristics	497
2.2	Energy Gap	498
2.3	Lattice Matching	498
2.4	Substrates	499
2.5	Dopants and Impurity Incorporation	499
3.	**Visible Emitters**	500
3.1	Materials Performance Characteristics	500
3.1.1	Comparison of Different Types of LEDs	500
3.1.2	GaAsP and GaAsP:N	500
3.1.3	GaP:N, GaP:(Zn,O), and GaP	500
3.1.4	$Al_xGa_{1-x}As$	502
3.1.5	AlInGaP	502
3.1.6	GaN	504
3.1.7	ZnSe	504
3.1.8	SiC	504
3.2	Operating Characteristics	504
3.2.1	Current–Voltage Characteristics	504
3.2.2	Performance versus Current	505
3.3	Applications	506
3.3.1	Overview	506
3.3.2	Instrument and Computer	506
3.3.3	Consumer Electronics	506
3.3.4	Large-Area Display	506
3.3.5	Vehicular Lighting	507
3.3.6	Electrophotography	507
3.3.7	Trends and Future Developments	507
4.	**Infrared Emitters**	507
4.1	Materials Performance Characteristics	507
4.1.1	Comparison of Different Types of Infrared LEDs	507
4.1.2	GaAs:Zn	507
4.1.3	GaAs:Si and AlGaAs:Si	507
4.1.4	$Al_xGa_{1-x}As$	508
4.1.5	Small-Emission-Area AlGaAs and InGaAsP	509
4.2	Operating Characteristics	509
4.2.1	Current–Voltage Characteristics	509
4.3	Applications for Infrared Emitters	509
4.3.1	Remote Controls	509
4.3.2	Optocouplers	509
4.3.3	Sensors	509
4.3.4	Optical Communications	510
4.3.5	Trends and Future Developments	510
5.	**Technology**	510
5.1	Overview	510
5.2	Substrate Growth	510
5.3	Epitaxial Growth	511
5.3.1	Epitaxial Growth Overview	511
5.3.2	Liquid-Phase Epitaxy	511

Encyclopedia of Applied Physics, Vol. 8

1-56081-067-X/94/$5.00+.50

5.3.3 Vapor-Phase Epitaxy 511
5.3.4 Metal-Organic Vapor-Phase Epitaxy 511
5.3.5 Molecular-Beam Epitaxy 511
5.4 Wafer Fabrication 512
5.4.1 Junction Formation 512
5.4.2 Ohmic Contact Formation 512
5.4.3 Dicing 512
5.5 Assembly 512
5.5.1 Die Attachment............... 512
5.5.2 Wire Bonding 512
5.5.3 Encapsulation 512
Works Cited 513
Further Reading 514

INTRODUCTION

A light-emitting diode (LED) is a semiconductor device that emits visible or infrared radiation when electric current is passed through it. The basic device consists of a single *p-n* junction. The energy of the emitted photons, which determines the color, is usually approximately equal to the energy gap of the semiconductor material in the active region of the LED. Materials that are generally used for LEDs are compounds of elements from columns III and V of the periodic chart such as GaAs, GaP, AlGaAs, InGaP, GaAsP, GaAsInP, and more recently AlInGaP . The specific material is chosen on the basis of color, performance, and cost. A cross section of a LED lamp is shown in Fig. 1. The semiconductor chip is typically 250×250 μm^2 and is mounted on one of the electrical leads. A bond wire electrically connects the top of the chip to the other lead. The epoxy dome serves as a lens to focus the light and as a structural member to hold the device together. Operating currents are usually in the range of 1–50 mA, and forward voltages are about 2 V.

The external quantum efficiency (photons out per electron passed through the device), which is roughly equal to the power efficiency of visible LEDs, ranges from less than 0.1% to more than 10%. The luminous performance of visible LEDs is obtained by multiplying the power efficiency by the eye sensitivity curve as defined by the Commission Internationale de L'Eclairage, or CIE. LEDs typically have performance in the range of 1–10 l/W, but LEDs with performance as high as 20 l/W have been observed. This compares favorably with incandescent bulbs that are typically in the range of 15 l/W. LEDs usually operate at power levels of less than 0.1 W, and so they are not suitable for room illumination. However, they are widely used for large-area displays and for exterior lighting on vehicles.

The evolution of visible LEDs is shown in Fig. 2. Red GaAsP LEDs were first introduced commercially by General Electric in 1962 in very low volumes following the work of Holonyak and Bevacqua (1962). High-volume production of GaAsP emitters began in 1968 (Herzog *et al.*, 1969) and was soon followed by the introduction of GaP:(Zn,O) LEDs (Saul *et al.*, 1969). Both of these types of emitters had efficiencies of approximately 0.1 l/W and were only available in the color red. In the late 1960s and early 1970s, it was discovered that nitrogen provided an efficient recombination center in both GaP (Logan *et al.*, 1968) and GaAsP (Groves *et al.*, 1971). This led to the commercial introduction in the mid-1970s of red, orange, yellow, and green GaAsP:N and GaP:N LEDs with performance in the range of 1 l/W. Early work (Rupprecht *et al.*, 1967, Alfërov *et al.*, 1973) had indicated that AlGaAs and heterostructure AlGaAs LEDs offered potential performance advantages over GaAsP

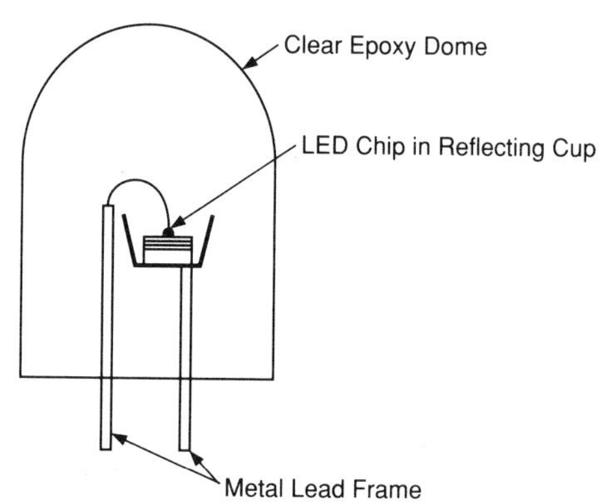

FIG. 1. Cross sectional view of a typical LED lamp. The LED chip, typically $250 \times 250 \times 250$ μm^3, is mounted in a reflective cup formed in a lead frame. Clear epoxy acts as a lens, an index-of-refraction matching medium to help extract light from the chip, and a structural member to hold the package together.

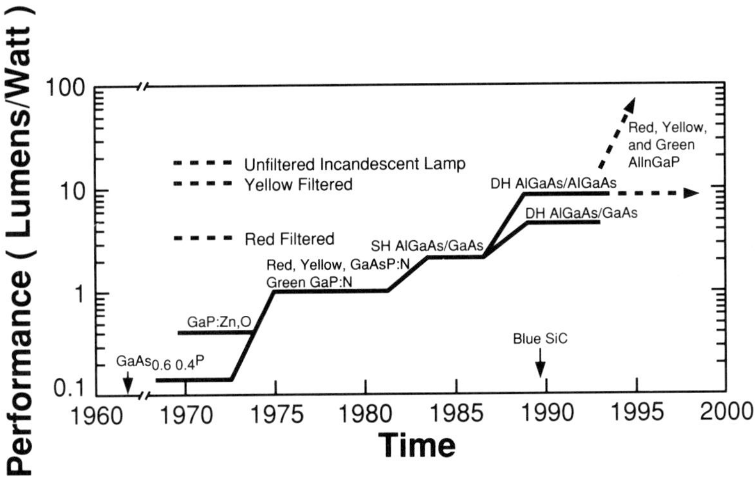

FIG. 2. Schematic illustration of the evolution of visible LEDs. Early LEDs only emitted red. The addition of nitrogen doping in GaAsP (GaAsP:N) improved red performance, and made orange and yellow possible. In GaP, nitrogen doping (GaP:N) made green available, and during the 1980s, GaP gradually improved to ~2.5 l/W (not shown). Double-heterostructure (DH) red-emitting AlGaAs devices now offer better performance than red-filtered incandescents and are used in some automobile taillights. AlInGaP has the potential to be brighter than AlGaAs in yellow and possibly red and green.

and GaP homojunctions. However, more than a decade of development was required to create high-volume liquid-phase–epitaxy (LPE) reactors capable of growing high-quality multilayer device structures (Nishizawa and Suto, 1977). It was not until the early 1980s that such LEDs became commercially available. The performance of these LEDs ranged from 2 to 10 l/W (depending upon the structure employed), and for the first time, LEDs broke the efficiency barrier of filtered incandescent bulbs, enabling them to replace light bulbs in many outdoor lighting applications.

Finally, in the early 1990s orange and yellow AlInGaP devices with efficiencies greater than 10 l/W were developed (Kuo *et al.*, 1990). The growth of such LEDs by conventional techniques such as LPE or halide-transport vapor-phase epitaxy proved intractable. The emergence of metal-organic vapor-phase epitaxy (MOVPE) as a crystal-growth technique beginning in the late 1960s (Manasevit and Simpson, 1969) and yielding high-performance AlGaAs devices in the late 1970s (Dupuis and Dapkus, 1977) later made possible the controlled growth of heterostructure AlInGaP LEDs. Their performance is expected to improve significantly over the next decade as more is learned about this crystal-growth process.

Blue SiC LEDs are also available but they are much less efficient than the other types. The development of brighter blue LEDs, possibly using ZnSe or GaN, which have shown promise in the research laboratory, is expected to be a major development area in the 1990s.

The discovery that diffused-junction GaAs diodes were efficient sources of infrared radiation (Keyes and Quist, 1962) led to the commercial introduction of infrared LEDs in the early 1960s as well. These Zn-diffused GaAs diodes had typical external quantum efficiencies of approximately 1% in air. In 1966 it was discovered (Rupprecht *et al.*, 1966) that Si-doped GaAs diodes with grown-in *p-n* junctions could provide efficiencies greater than 6%. Even higher efficiencies (up to 12%) were obtained by silicon doping a thick AlGaAs layer and then removing the original (absorbing) GaAs substrate (Dawson, 1977). Plastic-encapsulated GaAs:Si and AlGaAs:Si LEDs such as these have output powers in the 12–26-mW range when driven at 100 mA and can be reliably driven at peak currents up to 1.5 A. They have been extensively used in remote control applications for TVs and VCRs. Their switching times, however, are quite long (approximately 1 μs), and so they are unsuitable as sources for high-speed data communication. The need for fast light sources that could be used in fiber-optic links led to the development of small-area, high-radiance AlGaAs and InGaAsP LEDs in the 1970s and 1980s (see Saul *et al.*, 1985; Pearsall, 1982). They are used in medium-distance (<10 km) and low–data-rate (10 to 200 Mbit/s) applications while semiconductor lasers are used for longer-distance or higher–data-rate communications.

1. BASIC DEVICE PHYSICS

1.1 *p-n* Junction

The basic structure of an LED is among the simplest of semiconductor devices since it consists of a single *p-n* junction, where *p*-type (excess positive charge carriers called holes) and *n*-type (excess electrons or negative charge carriers) semiconductor materials are in single-crystal contact with each other as indicated in Fig. 3. When the junction is biased in the forward direction, electrons are injected into the *p*-type region, and/or holes are injected into the *n*-type region. These injected minority carriers can then recombine with majority carriers, giving up energy.

When biased in the forward direction, LED junctions begin to conduct significantly at voltages approximately equal to the semiconductor energy gap of the material involved. This can range from 0.8 to 1.5 V for infrared emitters and from 1.5 to more than 3.0 V for visible emitters (see Secs. 3.2.1 and 4.2.1). The reverse breakdown voltage is generally limited by the carrier concentration in the *n*-type region and can range from 5–10 V for heavily-doped devices to more than 50 V for many types of visible emitters.

The slope of the current-voltage characteristic after turn-on in the forward direction is known as the dynamic resistance. The dynamic resistance is determined by the area of the junction, the Ohmic-contact technology, and the conductivity of the semiconductor materials involved. The dynamic resistance can range from about 1 Ω for an IR device designed for high-current operation to 10 Ω or

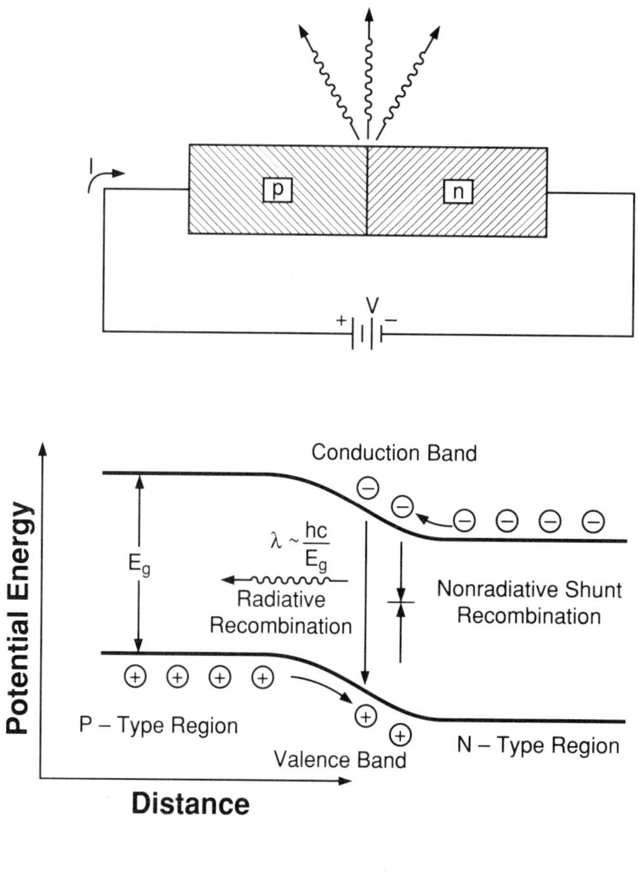

FIG. 3. Schematic of an LED and forward-biased LED junction. Radiative processes, which result in photon emission, and nonradiative processes, which generate heat, are illustrated. In indirect-band-gap material they are forbidden. Reprinted from Tannas (1985), courtesy of Van Nostrand Reinhold Company, Inc. Copyright © 1985.

more for a visible LED that is optimized for high light-output efficiency and designed to operate at 30 mA or less.

1.2 Radiative and Nonradiative Recombination and Internal Quantum Efficiency

When injected minority carriers recombine with majority carriers, the energy can be converted into either light or heat, depending upon whether the recombination is "radiative" or "nonradiative" as shown in Fig. 3. Typical radiative and nonradiative recombination processes are listed in Table 1. The internal quantum efficiency of the LED, the number of photons generated divided by the number of minority carriers injected, is determined by the relative radiative and nonradiative recombination rates. This is shown below for the case of electron injection into a *p*-type active region, which is generally the case for LEDs. The total recombination rate for electrons can be expressed as the number of electrons, n, divided by a characteristic lifetime τ:

Table 1. Examples of recombination routes.

Radiative	Nonradiative
Band to band	Phonon emission
Band to acceptor	At dislocations
Donor to acceptor	At impurities
Through isoelectronic centers	At lattice vacancies
	At interstitials

$$\text{Total recombination rate} = n/\tau = \text{radiative rate} + \text{nonradiative rate} = n/\tau_r + n/\tau_{nr}$$

$$= n(1/\tau_r + 1/\tau_{nr}),$$

where τ_r and τ_{nr} are the radiative and nonradiative decay times, respectively.

$$\text{Internal quantum efficiency} = \eta_{int} = \frac{\text{radiative rate}}{\text{total recombination rate}},$$

$$\eta_{int} = \frac{n/\tau_r}{n/\tau} = \frac{1/\tau_r}{1/\tau_r + 1/\tau_{nr}}.$$

Therefore

$$\eta_{int} = \frac{1}{1 + \tau_r/\tau_{nr}}.$$

In order to have high-performance devices, we want τ_{nr} to be large compared to τ_r The nonradiative lifetime is determined by the quality of the material, and if the nonradiative paths shown in Table 1 are minimized, τ_{nr} can be as long as a microsecond or even longer; but even in good material, τ_{nr} is more typically in the tens of nanoseconds range.

We would like to have τ_r as small as possible. At low minority-carrier injection levels, the radiative recombination rate can also be expressed as $n/\tau_r = Bnp$ or $\tau_r = 1/Bp$ where B is the recombination rate constant for a given type of semiconductor material, and n and p are the electron and hole concentrations. In order to reduce τ_r we want to have the hole concentration p as large as possible without degrading the quality of the material and hence decreasing τ_{nr}. When materials are doped above a maximum hole concentration, which is different for different materials, and can range from about 10^{17} to 10^{19} cm^{-3}, the material quality is degraded, and τ_{nr} decreases sharply.

The radiative constant is determined by the material system involved. For GaAs, B is about 2×10^{-10} cm^3/s^{-1} (Casey and Panish, 1978). Thus if $p = 10^{18}$ cm^{-3}, $\tau_r = 1/Bp = 5\times10^{-9}$ s which is a typical radiative decay time for a GaAs device. The nonradiative decay time for good-quality GaAs material is generally 10^{-7} s or longer so that the internal quantum efficiency approaches 100%. In most semiconductors, the internal quantum efficiency is much lower than 100% because of material quality problems or the fact that the recombination constant is smaller. The latter is the case for indirect materials where B can be orders of magnitude smaller as described in Sec. 1.3. The external quantum efficiency, which is the internal quantum efficiency multiplied by a light extraction factor, is lower, sometimes by more than an order of magnitude, than the internal efficiency because of the difficulty in extracting light from the interior of the LED chip. This is discussed in Sec. 1.8.

1.3 Direct and Indirect Energy Gap

The energy gap of a semiconductor is the minimum separation between the valence band, which is the continuum of allowed energy versus momentum states for the holes or *p*-type charge carriers, and the conduction band, which is the same characteristic for the electrons. This energy gap is the primary factor that determines the color of an LED, as described in Sec. 1.6. However, the separation between the conduction band and the valence band changes as a function of the momentum of the electrons and holes, and the relative momenta of the electrons and holes at the point where the energy-gap separation is a minimum is a key factor in determining the suitability of the material for the fabrication of efficient LEDs.

In a direct semiconductor, the minimum energy gap occurs at the point where both the electrons and holes have zero momentum. The electrons and holes are concentrated around this region as shown in Fig. 4. Radiative recombination occurs readily in this situation since the electrons and holes can recombine emitting a photon that has an energy

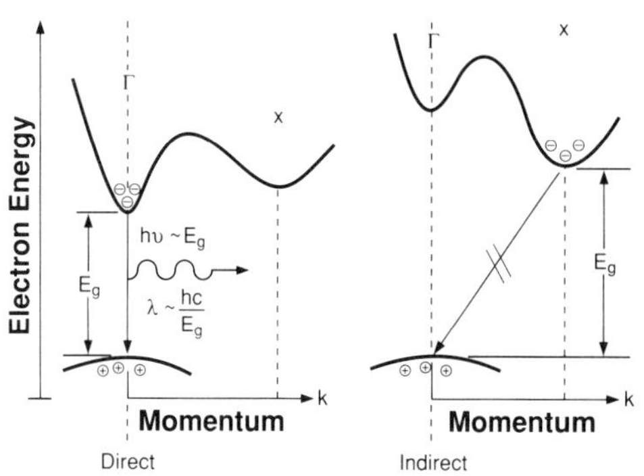

FIG. 4. Schematic diagram of conduction and valence bands in direct and indirect semiconductors. Electron energy is plotted as a function of momentum (*k*). In direct–band-gap material band-to-band radiative transitions are allowed. In indirect–band-gap material they are forbidden. Reprinted from Tannas (1985), courtesy of Van Nostrand Reinhold Company, Inc. Copyright © 1985.

equal to that given up in the recombination process—thus conserving energy. Momentum is also conserved since the photon has negligible momentum and both the electron and hole had nearly zero momentum in the beginning.

In an indirect semiconductor, on the other hand, the electrons and holes have different momenta at the point where the band separation is a minimum. The holes are still concentrated in a region with nearly zero momentum, but the electrons are concentrated in the conduction-band minimum with non-zero momentum as shown in Fig. 4. In this situation, radiative transitions can no longer readily occur. Although energy conservation can be achieved with the emission of a photon, momentum conservation cannot, since photons have nearly zero momentum and cannot make up for the momentum difference between the electron and hole.

Radiation can only occur in indirect semiconductors with the involvement of an additional interaction that allows momentum to be conserved. This interaction can occur with lattice vibrations (phonons) resulting in heat being transferred to or absorbed from the crystal lattice. This process yields an emitted photon with an energy that is different from the energy gap by an amount equal to the energy of the phonon that is simultaneously generated or absorbed. Processes of this type are relatively unlikely compared to direct recombination, and so τ_r is longer, the ratio τ_r/τ_{nr} is larger, and high quantum efficiencies are difficult to achieve in pure indirect semiconductors.

1.4 Isoelectronic Impurities

In spite of the problem presented by the band structure, indirect semiconductors are widely used for LEDs. Acceptable quantum efficiencies are achieved by intentionally adding isoelectronic impurities, which act as radiative recombination centers, to the crystal. An "isoelectronic" impurity comes from the same column of the periodic chart as the host crystal, and hence it has an electronic charge no different from that of the element it replaces. The isoelectronic impurity introduces a highly localized region of strain in the crystal that traps an electron. Since the trap is highly localized, the uncertainty in position is small and so, according to the Heisenberg

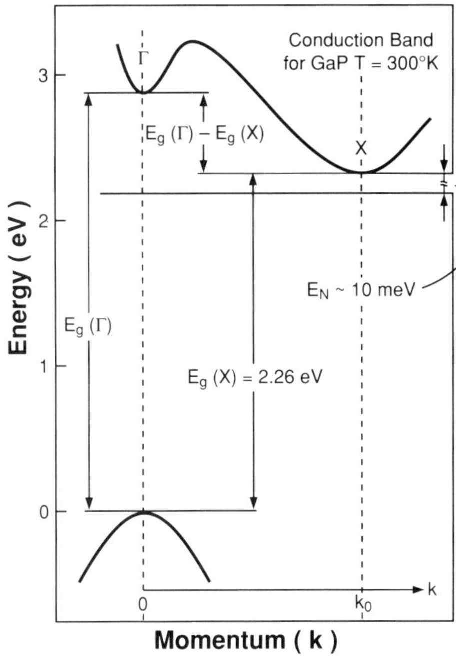

FIG. 5. Energy-band structure for an indirect semiconductor with an isoelectronic impurity (GaP with nitrogen doping is shown). The shaded region shows the probability distribution for an electron associated with the nitrogen impurity, and it can be seen that there is a significant probability that the electron will be near zero momentum so that it can recombine radiatively. Reprinted from Tannas (1985), courtesy of Van Nostrand Reinhold Company, Inc. Copyright © 1985.

uncertainty principle, the uncertainty in momentum is large. Thus, electrons trapped at isoelectronic impurities are widely distributed in momentum, and there is a significant probability that the electrons can have zero momentum and can combine radiatively. This situation is illustrated in Fig. 5 for GaP with nitrogen doping. Even with the addition of isoelectronic impurities, however, the recombination lifetime is longer than that for direct semiconductors, and the internal quantum efficiencies are typically much less than 100%.

1.5 Direct–Indirect Transition

Ternary (three-element) and quaternary (four element) alloys are used for many types of LEDs. Examples are $GaAs_{1-x}P_x$, $Al_xGa_{1-x}As$, and $(Al_xGa_{1-x})_{0.5}In_{0.5}P$. The energy gap changes with alloy composition, and, in many cases, a transition from a direct–band-gap to an indirect–band-gap semiconductor occurs at some composition (x_c). This is illustrated in Fig. 6. The quantum efficiency of the LEDs changes, typically by orders of magnitude, as this transition is crossed. The quantum efficiency as a function of alloy composition can be calculated if we assume that only electrons in the direct conduction-band minimum can recombine radiatively and that the nonradiative lifetime is the same for both direct and indirect electrons.

The total number of electrons is equal to the number of electrons in the direct minimum plus the number of electrons in the indirect minima:

$$n = n_D + n_I.$$

According to Boltzmann statistics, the ratio of the number of electrons in the indirect minima to the number in the direct minimum can be written

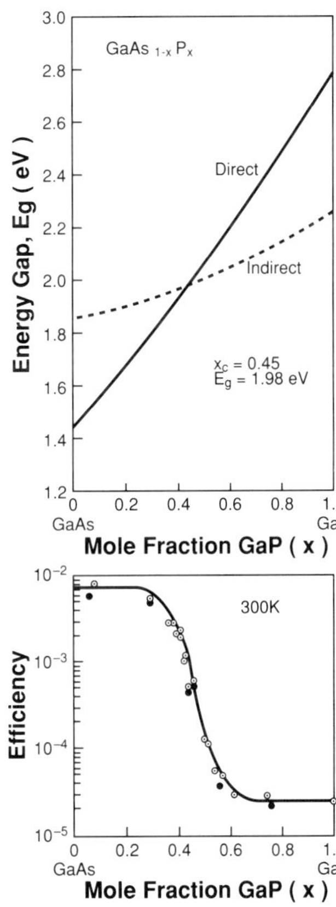

FIG. 6. Direct and indirect energy gaps plotted as functions of alloy composition. The quantum efficiency is high in the direct region to the left of the direct–indirect transition, and is orders of magnitude lower to the right. The data points are from Herzog *et al.*, 1969.

$F = n_I/n_D = Ne^{-\Delta E(x)/kT}$,

where F is the fraction of carriers in the indirect minima, N is the relative density of states in the indirect minima (typically N is > 10), and $\Delta E(x)$ is the separation between the direct and indirect minima.

Now, the radiative recombination rate $= n_D/\tau_r$, the nonradiative recombination rate $= n_I/\tau_{nr} + n_D/\tau_{nr} = n/\tau_{nr}$, and therefore

internal quantum efficiency

$$= \eta_{int} = \frac{n_D/\tau_r}{n_D/\tau_r + n/\tau_{nr}}$$

$$= \frac{1}{1 + (\tau_r/\tau_{nr})(1 + Ne^{-\Delta E/kT})}.$$

The quantum efficiency now depends on the lifetime ratio as well as the fraction of carriers in the indirect minima and drops exponentially as the separation between the minima becomes small.

1.6 Emission Wavelength versus Energy Gap and Doping

The wavelength of an LED is inversely proportional to the energy of the emitted photon, E, according to the relationship $\lambda = hc/E$, where λ is the wavelength of light, h is Planck's constant, and c is the velocity of light. The energy of the photon is equal to the energy gap if free electrons and holes are recombining, or to the energy gap minus the energy of the trap, impurity level, or phonon if these processes are involved. This can be represented as $\lambda = hc/(E - E_x)$ where E_x is the energy of the trap, etc. Typical n-type dopants have level depths of a few thousandths of an electron volt and p-type dopants have depths of a few tens of thousandths of an electron volt, both of which are small compared to the band gaps of LEDs. Therefore for most types of LEDs, the peak emission wavelength $\lambda \approx hc/E_g$. This is not necessarily the case for materials with isoelectronic traps, which can have trap depths of more than 0.1 eV.

1.7 Device Structures

1.7.1 Homojunctions. The simplest and generally least-expensive LEDs are homojunctions in which the entire epitaxial structure, and sometimes the substrate, consists of the same compound or alloy composition. Examples of common homojunction LEDs are given in Table 2, and a GaP homojunction device is shown in Fig. 7(a). The p-n junction can be formed by p-type diffusion into an n-type epitaxial film, or can be grown by changing the dopant type during the epitaxial growth process.

The optimum junction depth is a compromise between two conflicting requirements.

Table 2. Types of LEDs grouped by structure type.

Structure	Epitaxial layer/Substrate	Desirable device characteristics: Low defect density	Good injection efficiency	Good carrier confinement	Low internal absorption	Manufacturability
Homojunction	GaAsP/GaAs	No	No	No	No	Excellent
	GaAsP:N/GaP	No	No	No	Yes	Excellent
	GaP:(Zn,O)/GaP	Yes	No	No	Yes	Excellent
	GaP:N/GaP	Yes	No	No	Yes	Excellent
	GaAs:Si/GaAs	Yes	No	No	Yes	Excellent
Single heterostructure	AlGaAs/GaAs	Yes	Yes	No	No	Good
Double heterostructure	AlGaAs/GaAs	Yes	Yes	Yes	No	Good
	AlGaAs/AlGaAs	Yes	Yes	Yes	Yes	Fair but improving
	AlInGaP/GaAs	Yes	Yes	Yes	No	New (TBD)[a]

[a]TBD stands for "to be determined"

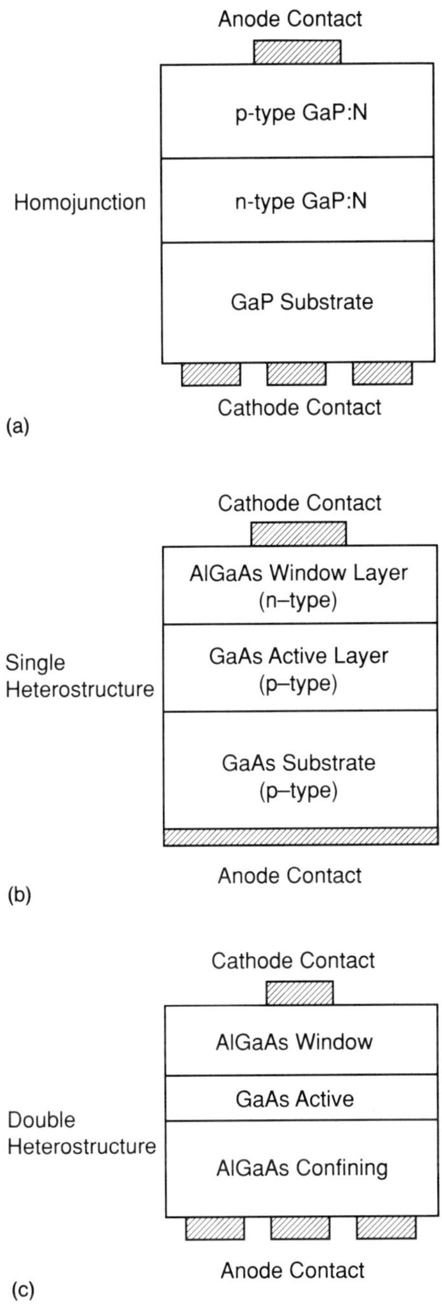

FIG. 7. Cross sections of various types of LED structures. (a) The simplest type is a homojunction in which both the *p*-type and *n*-type regions are similar material as in a GaP:N LED. (b) A single heterostructure has different material compositions on the *p*-type and *n*-type sides as in an AlGaAs/GaAs LED. (c) A double heterostructure has an additional layer on the other side of the active layer to confine the injected electrons and/or facilitate light extraction.

The junction must be deep enough so that only a small fraction of the injected carriers reaches the surface (where they generally recombine nonradiatively), and so that the sheet conductivity of the *p*-type region is high enough to avoid current crowding around the contact. On the other hand the junction should be as shallow as possible to minimize absorption. This latter requirement is particularly important for direct–band-gap devices because the absorption coefficient for near–band-gap radiation is $\sim 10^4$ cm^{-1}. Consequently, junctions in direct–band-gap homojunctions, which utilize a near-band-edge recombination processes, are generally less than 3 μm deep.

In the case of indirect semiconductors, the absorption coefficient near the band edge is only $\sim 10^3$ cm^{-1}, and the junction can be deeper. In addition, most indirect–band-gap LEDs utilize isoelectronic impurities where the transition energy is below the band edge and the absorption is substantially reduced. Consequently, in indirect–band-gap LEDs, the junction depth is often more than 10 μm, particularly for devices with grown-in junctions. The doping levels on the *p* and *n* sides of the junction are optimized both to maximize the injection of minority carriers into the preferred active region of the devices and to have the radiative lifetime in the active region be as short as possible to minimize nonradiative recombination. These are also conflicting requirements. The *p*-type region is generally the preferred active region, and it would be desirable to dope it as heavily as possible, without introducing crystal damage, in order to minimize the radiative lifetime relative to the nonradiative lifetime. It would also be desirable to dope the *n*-type region heavily to achieve effective injection. In practice it is generally found that heavy *n*-type doping results in shortened nonradiative lifetimes, and so the desired doping level cannot be achieved and device efficiency is compromised. Many of the limitations of homojunctions can be overcome with the formation of heterostructures.

1.7.2 Single Heterostructures. A single-heterostructure device has two different materials or alloy compositions on opposite sides of the junction. Devices of this type are generally not much more difficult to produce than homojunction LEDs, and they have several advantages. Consider, for example, a device with an *n*-type AlGaAs "window layer" grown on a *p*-type GaAs active layer (or AlGaAs layer of lower aluminum concentration) as shown in Fig. 7(b). The AlGaAs layer has a higher band gap and is transparent to the radiation generated in the GaAs active layer. Therefore the AlGaAs layer can be thick enough to give good current spreading and a minimum of surface recombination. Furthermore, efficient injection can be achieved since it is energetically favorable for the electrons to flow from the wider–band-gap region into the lower–energy-gap active region even if the *n*-type region is much more lightly doped than the *p*-type region. One limitation of this structure is that some of the injected electrons can penetrate quite deeply into the *p*-type region before they recombine, and much of the radiation that is generated will be absorbed before it can escape. Also, any radiation that is emitted downward will be absorbed. These limitations can be minimized with the growth of appropriate double heterostructures.

1.7.3 Double Heterostructures. The LED structure with the best efficiency is a double heterostructure in which an active region with a lower energy gap is sandwiched between two layers with higher energy gaps: an upper "window" layer and a lower "confining" layer. This structure has the advantages of a single heterostructure and, in addition, reduces the absorption of the generated radiation. The active region can be quite thin, typically 1–3 μm, since the wider band gap of the "confining" layer keeps the injected electrons from penetrating beyond the heterostructure interface. The confinement of the carriers to a narrow region results in fast and efficient recombination. The issue of fast recombination is important primarily for applications related to electronic switching or communication. Devices for these applications also often have small junction areas to minimize capacitance.

The thin active region minimizes absorption, and if the confining layer is relatively thick, some of the light that is generated can escape from the edges of the confining layer, thereby increasing the external efficiency of the device. Optimally the confining layer is grown thick enough (>100 μm) so that the substrate can be removed altogether [see Fig. 7(c)], resulting in a "transparent substrate" which gives the optimal efficiency as described in Sec. 1.8.1.

An alternative approach to growing a thick confining layer is to grow a Bragg reflector, using alternate one-quarter-wavelength-thick layers of semiconductor material with different indices of refraction in order to reflect the light back up into the chip instead of going down into the substrate. These reflectors are very effective for nearly on-axis radiation but are not effective for low-angle reflections. Therefore they work very well for surface emitting lasers but are less effective for LEDs than transparent substrates. The only disadvantage of double heterostructures compared to the structures described earlier is that they are more complicated and difficult to grow, thus adding to the cost. The thin active region and Bragg reflector layers, if utilized, may be harder to control in a production environment. Thick confining and/or transparent substrate layers are also somewhat harder to control and add time and cost. However, as production technology evolves, many of these structures will become more widely used.

1.7.4 Quantum Wells and Other Structures. LEDs with complex active regions containing one or more quantum wells have been fabricated. These structures, which are widely used for injection lasers, could give better LEDs since less absorption might be expected and injected-carrier densities in the active regions can be higher, resulting in fast radiative recombination. However, it has generally been found that the benefits gained from the added complexity are less for LEDs than lasers and have been more than offset by the anticipated cost and difficulty encountered in producing devices of this type in high volume. Consequently, devices of this type are not widely used in LED applications at this time but may become more viable in the future.

1.8 LED Performance and Extraction Efficiency

The usefulness of an LED depends upon its external performance. For visible LEDs this performance is generally expressed in terms of the luminous flux emitted divided by the power input, in units of lumens/watt. For infrared LEDs the performance is determined by the external quantum efficiency (the ratio of number of photons emitted to the number of electrons injected) expressed in percent, or as power efficiency (watts output divided by watts input), also expressed as a percentage. This section discusses the issues related to extracting radiation from the LED chip and the relationships between internal quantum efficiency, power efficiency, and luminous performance.

1.8.1 Light Extraction. The external efficiency of LEDs is much lower than the internal efficiency because of the difficulty of extracting light from the semiconductor chip. The ratio of external quantum efficiency to internal quantum efficiency can be defined as an extraction ratio R_E. The problem of internal absorption for different types of structures was discussed in Sec. 1.7, and extraction efficiencies for different types of LED structures are summarized in Table 3. Internal absorption is a major problem for LEDs because the semiconductor chip has an index of refraction of about 3.5, resulting, according to Snell's Law, in a critical angle $[\theta_c = \sin^{-1}(n_1/n_2)]$ for internal reflection of typically 17° when the chip is in air. In most cases, the LED chip is encapsulated in epoxy, with an index of about 1.5, and this critical angle increases to about 25°. Nevertheless, for the case of an LED on an absorbing substrate with a thin epitaxial layer such that edge emission is negligible, only the light that is emitted upward within a cone of 25° will escape from the chip. The extraction ratio for this upward cone can be calculated to be $(1-\cos\theta_c)/2$. One must also correct for Fresnel reflection losses $(n_1/n_2-1)^2/(n_1/n_2+1)^2$ at each interface (assuming normal incidence). The emission cone gives an extraction ratio of 0.05 and Fresnel losses are about 0.2, counting both interfaces. This results in an extraction ratio of only 0.04, or an external quantum efficiency of 4% even if the internal quantum efficiency is 100%. To make matters worse, the presence of the electrical contact on top of the chip can also block much of the light so that the extraction ratio can be substantially less than 0.04. However, in well-designed LEDs, the epitaxial layer is from several micrometers to several tens of micrometers thick, to get both good current spreading and edge emission, and the extraction ratio can be more than 0.1 even for absorbing-substrate devices.

In the case of transparent-substrate devices, much of the light that is emitted downward can bounce off of the back of the chip and escape, resulting in a (2–3)× improvement in extraction efficiency. Generally, the extraction ratio does not exceed about 0.3 even for transparent substrate devices as a result of residual absorption losses at the contacts or in the active layer, or free-carrier or other types of absorption in the layers or at the chip surface.

As much as 75% of the emitted light from LEDs can come from the edges instead of the top of the LED chip. This edge light is useful for most applications since typically the LED chip is placed in a cavity that reflects the light

Table 3. Extraction ratios for encapsulated LEDs.

Epitaxial structure type	Substrate type	Emission region	Number of cones	Calculated extraction ratio[a]	Estimated actual ratio[b]	Typical device
Thin layers (⩽10 μm)	Absorbing	Top only	1	0.04	—	GaAsP/GaAs
Thick layers (⩾40 μm)	Absorbing	Top and four sides ($\sim\frac{1}{2}$ cone each)	3	0.12	0.12	AlGaAs/GaAs
Any thickness	Transparent	Top, four sides, and reflection from back surface	6	0.24	0.3	AlGaAs/AlGaAs

[a] The calculation assumes that there are no absorption losses or randomization within the chip. If randomization occurs (due to light scattering at the chip surface) the light can make multiple passes and more can escape than predicted by the calculation.

[b] These estimates are based on measured external quantum efficiency and estimated internal quantum efficiency. The internal quantum efficiency is estimated from efficiency versus temperature measurements.

in an upward direction. In some applications, such as coupling to an optical fiber or in a direct-view display, only the top-surface emission is utilized. Different light-extraction conditions pertain in this case, and a Bragg reflector as discussed in Sec. 1.7.3 may be more effective than a transparent substrate.

1.8.2 Quantum Efficiency and Power Efficiency. The external quantum efficiency is the internal quantum efficiency multiplied by the extraction ratio,

$$\eta_{ext} = R_E \eta_{int}.$$

The power efficiency (P_E) is the power output divided by the power input:

$$P_E = \frac{(\text{photons/second}) \times (\text{emission energy/photon})}{(\text{electrons/second}) \times (\text{applied voltage})} = \eta_{ext} \frac{E_R}{V},$$

where E_R is the energy of the emitted radiation.

The applied voltage is the voltage applied directly to the junction (approximately the energy gap) plus the voltage across the series (dynamic) resistance of the LED diode:

$$V \approx E_g + IR.$$

E_R is generally approximately equal to the semiconductor energy gap for near–band-edge recombination processes. The turn-on voltage for the LED is also approximately equal to the energy gap. Therefore, for low currents, where the voltage drop due to series resistance is relatively small, we have $P_E \approx \eta_{ext}$.

This approximation is accurate to within approximately 10% for most LEDs since the series resistance is typically less than 10 Ω resulting in less than a 0.2 volt increase in voltage at a typical operating current of 20 mA, while the energy gap voltage is typically 2 V.

1.8.3 Luminous Performance. For visible LEDs, the human eye is the detector, and the luminous performance, expressed in either lumens/watt or lumens/amp, is the key performance parameter.

The luminous efficiency, or the sensitivity of the eye to radiometric energy of different wavelengths, has been established by international agreement (CIE curve). The relative eye sensitivity peaks at 1.0 at a wavelength of 555 nm. At this wavelength, 1 W of radiometric power yields 680 l. The relative eye sensitivity $V(\lambda)$ drops sharply on either side of 555 nm and is about 0.5 at either 510 or 610 nm.

The luminous performance of a visible LED in units of lumens/watt is determined by multiplying the radiometric performance by the relative eye sensitivity curve, $680V(\lambda)P_E$. In some cases it is desirable to have the luminous performance expressed in lumens/ampere, which can be obtained as follows:

$$\text{luminous performance (l/A)} = 680V(\lambda)\eta_{ext} E_R.$$

1.9 Reliability

1.9.1 Radiation-Enhanced Degradation. LEDs are inherently highly reliable devices since the recombination is a normal solid-state electrical process that does not necessarily give rise to any damage to the semiconductor. However, there is a substantial amount of energy released locally (1–2 eV) with each electron–hole recombination event, and this can cause the formation or migration of point defects or dislocations, resulting in "radiation-enhanced" light output degradation. The degradation is generally accelerated by high current densities and high temperatures (Kimerling and Lang, 1975). Under extreme drive conditions (at current densities greater than ≈ 300 A/cm^2), networks of dislocations can form and grow rapidly around crystalline defects that may be present. Carriers tend to recombine nonradiatively when they are near these dislocation networks, and "dark lines" appear in the light-emitting area. Such dark-line defects (DLDs) often occur in small–emission-area AlGaAs LEDs (Fukuda, 1991) that are used as light sources for fiber-optic communications. Because the emission areas are small, extremely large current densities are common. Such devices must typically pass a "burn-in" test that verifies that they are free of the crystalline defects that lead to DLD formation before they can be incorporated into final products.

1.9.2 Package-Related Degradation. Degradation can also be caused by the packaging methods. The encapsulating epoxy is particularly troublesome, since it has a much higher thermal coefficient of expansion than the semiconductor chip. When the LED is cooled to low temperatures, typically −20 °C or below, the epoxy contracts and substantial compressive stress is applied to the chip. When the LED is operated at low temperatures, this stress can provide a driving force for defect formation and/or migration, which results in accelerated degradation. In some cases, particularly with AlGaAs devices, DLDs can form and grow. When the LED is cooled, or temperature cycled, but not forward biased, this degradation does not occur. Another type of package-related degradation, which occurs in outdoor applications, is the discoloration of the epoxy due to ultraviolet light. The discolored epoxy absorbs some of the light emitted from the LED chip and reduces the light output of the package. Packaging-related degradation can be affected by such parameters as the proper selection of epoxy, and epoxy-curing cycles, LED chip shape, and metal reflector shape and size. Some LED materials are much more susceptible to this type of degradation than others.

1.9.3 Typical LED Degradation Characteristics. Typical LED degradation data are shown in Fig. 8. Degradation in the range of 5% to 20% is usually observed after 1000 h of operation at normal operating currents at 55 °C. The degradation generally decreases roughly linearly on a semilog plot as shown, and so the device should not reach half-brightness for more than 10^6 h. In most applications, a change in light output by less than a factor of 2 will not be noticed, and so LED reliability is generally more than adequate, particularly for indoor applications. High-performance LEDs used for outdoor applications, requiring low-temperature operation, are more susceptible to package-related degradation (Sec. 1.9.2), and must be carefully engineered to avoid nonuniform and occasionally severe degradation.

2. SEMICONDUCTOR MATERIALS ISSUES

2.1 Desirable Characteristics

Figure 9 shows the band gaps of various binary compounds and alloy systems plotted as a function of lattice parameter. In order to fabricate high-performance LEDs, it would be desirable to choose compounds and/or alloys from this set that have the following characteristics:

1. a direct energy gap corresponding to the emission wavelength desired,
2. higher–energy-gap materials that can be lattice matched to the active layer for the growth of injection, window, and confining layers for heterostructures,
3. a low-defect-density substrate that is lattice matched to the epitaxial structure and that is transparent to the emitted radiation,

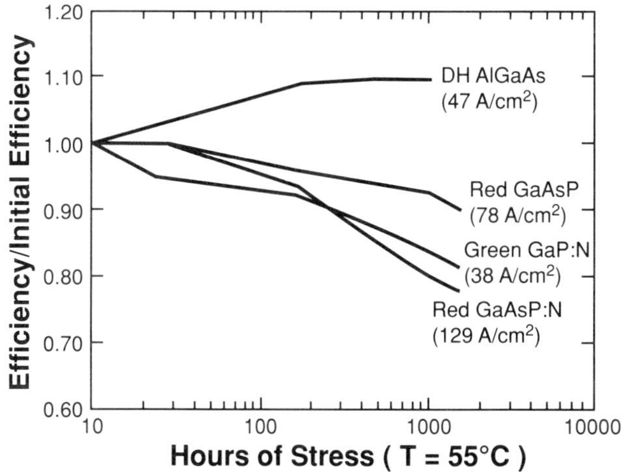

FIG. 8. Degradation data versus time are shown for several diferent types of visible LEDs at 55 °C . Most LEDs degrade by (5–20)% in light output by 1000 h, but do not fall below 50% of their initial brightness for more than 10^6 hours. The current densities shown correspond to typical drive conditions.

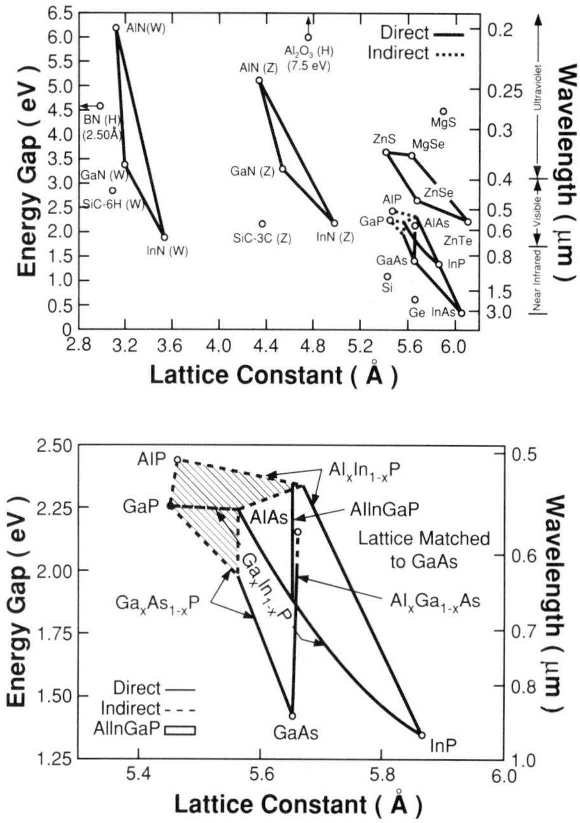

FIG. 9. (a) Energy gap vs lattice parameter for several semiconductor systems. Semiconductors that can be grown in several crystal structures are labeled Z for zincblende, W for wurtzite, and H for hexagonal. For the wurtzite and hexagonal crystals, the lattice parameter *a* is used for the lattice-constant value. (b) An enlargement of the region currently employed for most visible LEDs, showing the direct and indirect energy-gap regions.

4. the ability to add impurities to the structure to obtain a low–defect-density *p-n* junction and a low series resistance, and
5. a materials growth technology that can reproducibly and cost-effectively produce the structure as designed.

One would also like to have the lowest possible cost. Unfortunately it is not often possible to have all of these desires met simultaneously.

2.2 Energy Gap

A compound or alloy composition with an energy gap that gives the desired emission wavelength must be chosen, and, if possible, the material should have a direct energy gap. Table 4 lists energy gaps and energy-gap ranges for a variety of compounds and alloys. For infrared and red LEDs requiring energy gaps less than 2 eV, GaAs, $Al_xGa_{1-x}As$, and $GaAs_{1-x}P_x$ with appropriate compositions ($x \lesssim 0.4$) are good choices. If a compound is chosen and the energy gap is larger than desired, it may be possible to reduce the emission energy (increase the wavelength) by introducing non-band-edge recombination centers such as isoelectronic dopants. Materials with indirect band gaps and isoelectronic impurities such as GaP:N and $GaAs_{1-x}P_x$:N with $x > 0.5$ are widely used for LEDs because direct materials with energy gaps larger than about 2 eV, which are required for orange, yellow, and green LEDs, have not been available until the recent development of the AlInGaP alloy system. Materials such as ZnSe have a large enough direct band gap, but it has not been possible to form stable *p-n* junctions in ZnSe or other II-VI materials until recently.

2.3 Lattice Matching

The epitaxial layers in the device structure should, if possible, have the same lattice spacing between atoms and should match the substrate spacing as closely as possible. In other words one should try to stay on a vertical line in Fig. 9. AlGaAs and AlInGaP alloy systems grown on GaAs substrates are good examples, as is the GaInAsP system

Table 4. Materials important for LEDs.

Material system	Energy gap (eV) at 300 K	Peak emission wavelength (Å)
GaAs	1.43 direct	8670 infrared
GaP	2.26 indirect	5485 green
AlAs	2.16 indirect	5740 greenish yellow
InP	1.35 direct	9180 infrared
ZnSe	2.67 direct	4644 blue
GaN	3.39 direct	3658 ultraviolet
SiC (6*H*)	2.86 indirect	4800 blue
$GaAs_{1-x}P_x$	1.43–2.03 direct	8670–6105 infrared→red
$x_c=0.49$[a]	2.03–2.26 indirect	6105–5485 red→green
$Al_xGa_{1-x}As$	1.43–1.98 direct	8670–6525 infrared→red
$x_c=0.43$	1.98–2.14 indirect	6525–5790 red→yellow
$Ga_xIn_{1-x}P$	1.35–2.18 direct	9180–5685 infrared→yellow
$x_c=0.62$	2.18–2.26 indirect	5685–5485 yellow→green
$(Al_xGa_{1-x})_{0.5}In_{0.5}P$	1.88–2.31 direct	6596–5359 red–green
$x_c=0.70$	2.31–2.35 indirect	5359–5277 green
GaInAsP	0.724–1.35 direct	17127–9185 infrared

[a] x_c is the mole fraction at which the materials system switches from having a direct-gap band structure to an indirect-gap band structure.

grown on InP. If good lattice matching can be obtained, defects and nonradiative recombination can be minimized. It is also important to try to match the temperature coefficient of expansion because otherwise epitaxial layers that are grown lattice matched at high temperature will be strained at room temperature. If heterostructures are used, an alloy system must be chosen that allows lattice matching for alloy compositions with different energy gaps in order to permit the formation of "window" and "confining" layers without the introduction of defects at the interfaces between the layers. Some injection lasers and LEDs have been fabricated in which the active layer does not lattice match the confining layers but is so thin that it "stretches" to conform to other layers rather than creating defects. These "pseudomorphic" structures have, to date, been more effective for lasers than LEDs since the active regions must be very thin.

2.4 Substrates

Ideally the substrate should have low defect levels, be lattice matched to the epitaxial layers, and be transparent to the emitted radiation. Unfortunately, these conditions can rarely be simultaneously met. GaAs, GaP, and InP can be cost-effectively grown with reasonable levels of defects and are most commonly used. SiC and sapphire are also used in limited quantities for blue emitters. Germanium is sometimes used as a lower-cost substitute for GaAs but germanium contamination can degrade LED performance if not controlled. Silicon has also been occasionally used as a substrate for GaAs devices but the lattice match is poor and the GaAs layers are highly defective.

2.5 Dopants and Impurity Incorporation

Generally the III-V compounds, with the exception of GaN, can be easily doped either *p*-type or *n*-type so that the formation of a *p*-*n* junction is straightforward. Typical dopant impurities are listed in Table 5. The most common *p*-type dopant is zinc because it is

Table 5. Typical dopant impurities.

	p-type	*n*-type	Isoelectronic
III-V compounds	Zn	Te	N
	Be	Se	
	C	S	
	Mg	Si	
	Si	Ge	
II-VI compounds	Li	Al	
	Na	Cl	
	N	I	
IV compounds	B	P	
	Al	N	

safe and easy to handle, high doping levels can be achieved, and it can readily be diffused into most materials if photolithographically defined junctions are desired. Magnesium is often used in grown junctions instead of Zn since it diffuses less rapidly and is more stable in thin layer structures where dopant migration during operation could affect device performance. Carbon doping is increasingly used for similar reasons, but some materials cannot be doped *p*-type with carbon since it is a column-IV element and can be incorporated on either a column-III site where it would be *n*-type or a column-V site where it would be expected to be *p* type. Beryllium is also used as a *p*-type dopant. It can generally be incorporated in high concentrations, but it is a hazardous material to work with, and its use has been declining.

The impurity most commonly used for *n*-type doping is tellurium, although selenium and sulfur are also widely utilized. Silicon and germanium are also used for *n*-type dopants particularly for infrared applications.

Dopants that work well for infrared emitters in the direct–band-gap region may not be suitable for visible emitters in the vicinity of the direct–indirect transition. Deep donor levels associated with sulfur in GaAsP and silicon and other dopants in AlGaAs cause nonradiative transitions. One reason tellurium is so widely used for LEDs is that it has less often exhibited behavior of this type (Craford *et al.*, 1991).

Silicon carbide, which is used for blue LEDs, is a column-IV material and is generally doped with Al for *p*-type and N for *n*-type doping. Other wide–band-gap materials capable of blue emission such as GaN, and II-VI compounds such as ZnSe can easily be doped *n* type but are difficult to dope *p* type. It is only recently that it has been possible to form *p-n* junctions in these materials and to achieve blue-emitting devices. The performance and stability of these devices are not yet suitable for commercial LED applications.

3. VISIBLE EMITTERS

3.1 Materials Performance Characteristics

3.1.1 Comparison of Different Types of LEDs. The luminous performances for different types and colors of LEDs are summarized in Table 6. The AlGaAs and AlInGaP devices have the highest levels of performance and quantum efficiency. These devices satisfy most of the selection criteria discussed in Sec. 2, such as direct band gap, lattice-matched epitaxial structures and substrates, and double heterostructure device structures. At the present time, these devices are also among the most expensive and are primarily utilized only in applications requiring the maximum in performance. Lower-cost homojunction GaAsP and GaP devices dominate high-volume LED applications even though their performance is an order of magnitude lower.

3.1.2 GaAsP and GaAsP:N. $GaAs_{1-x}P_x$ LEDs were the first LEDs commercially produced, and they continue to be used in high volume since they are inexpensive and have adequate performance for many applications (Herzog *et al.*, 1969). $GaAs_{1-x}P_x$ has a direct energy gap, for $x \leqslant 0.45$, and emits in the infrared and red spectral regions. LEDs of this type have diffused homojunctions that can be photolithographically patterned, making them suitable for applications that require small emitting areas and/or more than one emitting region on each LED chip. The efficiency is relatively low because of internal absorption and a high defect density, due in part to the fact that GaAs and GaP have substantially different lattice parameters so that the GaAsP mixtures are not lattice matched to the GaAs (or GaP) substrates.

$GaAs_{1-x}P_x$, with $x \geqslant 0.45$, has an indirect band gap, and isoelectronic nitrogen must be added to achieve acceptable performance (Groves *et al.*, 1971; Craford *et al.*, 1972). The performance is still relatively low because of the indirect band gap and lattice mismatch with the GaP substrate. However, the GaP substrate is transparent, and so the extraction ratio is improved and GaAsP:N has higher luminous performance than direct–band-gap GaAsP on GaAs.

3.1.3 GaP:N, GaP:(Zn,O), and GaP. GaP:N is a green emitter with performance somewhat better than GaAsP:N. This improvement is due to a lower defect density since the GaP substrate and epitaxial layers are lattice matched (Logan *et al.*, 1968).

GaP:(Zn,O) emits in the red since the Zn and O atoms form a complex with a large binding energy (Saul *et al.*, 1969). The perfor-

Table 6. Characteristics of visible LEDs.

LED type	Color (peak wavelength, nm)	Typical external quantum efficiency (%)	Typical performance (l/W)	Band-gap type	Lattice matched	Structure[a]	Substrate	Epitaxial growth method[b] (*p-n* junction formation)
$GaAs_{0.60}P_{0.40}$	Red (650)	0.2	0.15	Direct	No	Homo.	GaAs	VPE (diffusion)
GaP:(Zn,O)	Red (700)	2	0.4	Indirect	Yes	Homo.	GaP	LPE (grown)
$GaAs_{0.35}P_{0.65}$:N	Red (630)	0.7	1	Indirect	No	Homo.	GaP	VPE (diffusion)
$GaAs_{0.15}P_{0.85}$:N	Yellow (585)	0.2	1	Indirect	No	Homo.	GaP	VPE (diffusion)
GaP:N	Yellow green (565)	0.4	2.5	Indirect	Yes	Homo.	GaP	LPE (grown)
GaP	Pure green (555)	0.1	0.6	Indirect	Yes	Homo.	GaP	LPE (grown)
$Al_{0.35}Ga_{0.65}As$	Red (650)	4	2	Direct	Yes	SH	GaAs	LPE (grown)
$Al_{0.35}Ga_{0.65}As$	Red (650)	8	4	Direct	Yes	DH	GaAs	LPE (grown)
$Al_{0.35}Ga_{0.65}As$	Red (650)	16	8	Direct	Yes	DH-TS	GaAs	LPE (grown)
$(Al_{0.11}Ga_{0.89})_{0.5}P$	Orange (620)	6[c]	20[c]	Direct	Yes	DH	GaAs	MOVPE (grown)
$(Al_{0.28}Ga_{0.72})_{0.5}In_{0.5}P$	Amber (595)	5[c]	20[c]	Direct	Yes	DH	GaAs	MOVPE (grown)
$(Al_{0.43}Ga_{0.57})_{0.5}In_{0.5}P$	Yellow green (570)	1[c]	6[c]	Direct	Yes	DH	GaAs	MOVPE (grown)
SiC (6H)	Blue (480)	0.02	0.04	Indirect	Yes	Homo.	SiC	CVD (grown)
GaN	Blue (430)	0.18[d]		Direct	No	Homo.	Al_2O_3	MOVPE (grown)
$ZnS_{0.07}Se_{0.93}$	Blue green (494)	0.1[e]		Direct	Yes	DH	GaAs	MBE (grown)

[a]Homo., homojunction; SH, single heterostructure; DH, double heterostructure; DH-TS, double heterostructure with transparent, epitaxially-grown substrate (original GaAs substrate removed).

[b]VPE, vapor-phase epitaxy; LPE, liquid-phase epitaxy; MOVPE, metal-organic vapor-phase epitaxy; MBE, molecular beam epitaxy (see Sec. 5.3 for a discussion of epitaxial grown technologies).

[c]Best reported results (Huang *et al.*, 1992). Typical commercial performance not yet established.

[d]Best reported results (Nakamura *et al.*, 1991a).

[e]Best reported results (Xie *et al.*, 1992).

mance of these devices is low compared to GaP:N, even though the quantum efficiency is higher, because the eye sensitivity is an order of magnitude lower for the 700-nm GaP:(Zn,O) emission than that for the green GaP:N emission.

Green-emitting LEDs can also be constructed in GaP without nitrogen doping. These devices depend upon phonons to conserve momentum (see Sec. 1.3). The absence of the nitrogen recombination center increases the radiative lifetime by about an order of magnitude. This reduces the quantum efficiency, but, since the absence of nitrogen also improves the crystal quality somewhat so that the nonradiative lifetime is also increased, the quantum efficiency is only (3–4)× lower than that for GaP:N. GaP devices also have an advantage in that the emission is greener (555 nm vs 565 nm) since the nitrogen trap is not involved.

3.1.4 $Al_xGa_{1-x}As$. The $Al_xGa_{1-x}As$ material system has a direct energy gap for compositions $x \leqslant 0.4$ and can be used to produce highly efficient red and infrared emitters. Since AlAs and GaAs have nearly the same lattice parameters, the system is well matched to GaAs substrates, so that defects can be minimized and heterostructures of various types can be grown (Nishizawa and Suto, 1977; Cook *et al.*, 1988). Three types are commercially available: single heterostructure on a GaAs substrate, double heterostructure on a GaAs substrate, and a double heterostructure with a thick confining layer and the GaAs substrate removed.

As indicated in Table 6, the performance increases as the structural complexity increases primarily because of reduced absorption as discussed in Sec. 1.7. The cost also increases with complexity, and so all three types of AlGaAs structures are utilized depending upon the price-performance trade-off desired.

Visible AlGaAs LEDs are produced by liquid-phase epitaxy (see Sec. 5.3.2). This is a mature technology that is also used for GaP LEDs. However, the growth of the multilayer AlGaAs heterostructures, particularly with thin 1–3-μm-thick active regions and ultrathick (>100 μm) confining layers, requires substantially more complex growth equipment than that used for the relatively simple GaP devices.

3.1.5 AlInGaP. The AlInGaP material system has a direct–indirect crossover in the green spectral region at an estimated energy of 2.31 eV (Cao and Stringfellow, 1990). It is the only system in which lattice-matched direct–band-gap heterostructures for colors other than infrared and red can be readily fabricated at this time. Lattice matching requires that the indium concentration be kept equal to the sum of the Al and Ga compositions since InP and GaP have lattice parameters about equal distances on opposite sides of that of the GaAs substrate (see Fig. 9). The ratio of Al to Ga can be changed to form heterostructures since Al and Ga are the same size atoms. Other wide–band-gap materials are beginning to show promise but are not yet commercially viable.

The three types of AlInGaP structures grown are shown in Fig. 10. All consist of double heterostructures on a GaAs substrate. The top window or spreading layer is grown relatively thick (~50 μm) to maximize the extraction of light from the edge of the chip (see Sec. 1.8.1).

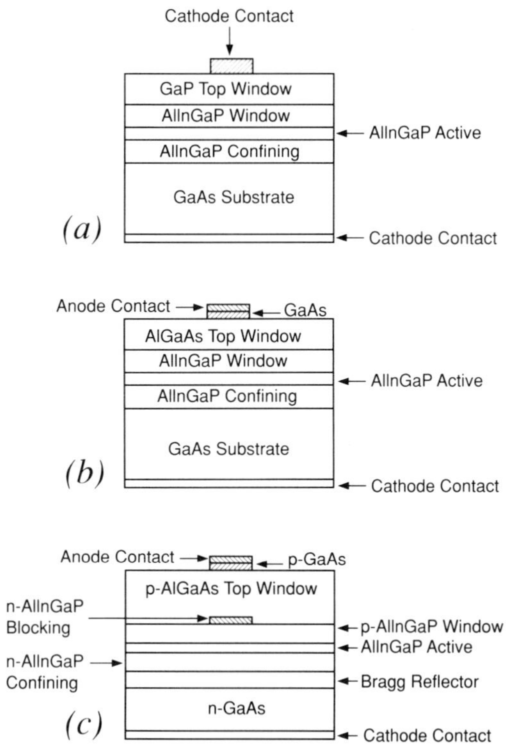

FIG. 10. Cross section of AlInGaP chip structures: (a) GaP window layer, (b) AlGaAs window layer, and (c) AlGaAs window layer with a blocking layer beneath the top contact and a Bragg reflector beneath the active layer.

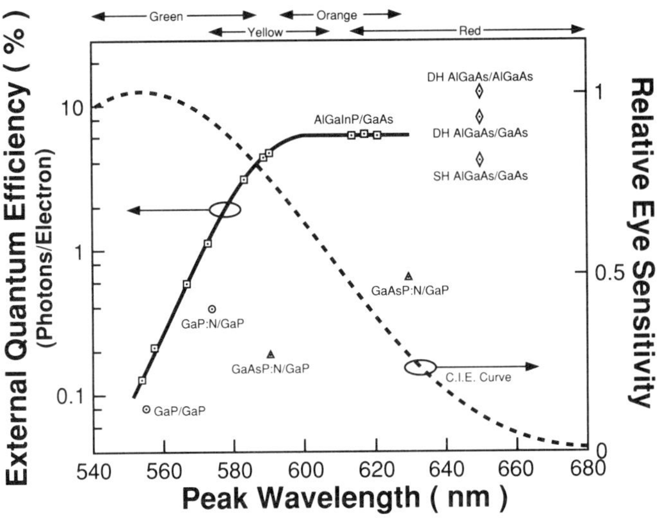

FIG. 11. AlInGaP quantum efficiency as a function of wavelength compared to other LED technologies. For red and yellow wavelengths beyond 590 nm, the devices have 5% external efficiency into air. The efficiency drops at shorter wavelengths as a result of the approaching transition from a direct to an indirect semiconductor. SH stands for single heterostructure and DH stands for double heterostructure. The CIE eye-response curve is also shown. Reprinted from Craford (1992), courtesy of the Institute of Electrical and Electronics Engineers, Inc. Copyright © 1992 IEEE.

The top window layer could, in principle, be grown of AlInGaP; but AlInGaP has a high resistivity and is costly to grow, and so AlGaAs and GaP have generally been substituted (Kuo *et al.*, 1990; Fletcher *et al.*, 1991; Sugawara *et al.*, 1991; Huang *et al.*, 1992). AlGaAs has the advantage that it is lattice matched but the disadvantage that it absorbs some of the light in the case of yellow and green emitters. GaP is transparent and inexpensive, but it is not lattice matched. This does not appear to introduce defects in the active region since the GaP layer is the last layer grown, and the best results, to date, have been obtained with a GaP window.

The quantum efficiency and luminous performance of AlInGaP devices compared to other types of LEDs are shown in Figs. 11 and 12, respectively. The external quantum efficiency for wavelengths greater than 590 nm is about 5% but falls rapidly for shorter wavelengths as the direct-indirect transition is approached. The luminous performance peaks in the region of 590 nm and falls on both sides as a result of changes in both the eye sensitivity (also shown in Fig. 11) and internal efficiency.

AlInGaP devices are clearly superior to any other technology from 590 to 620 nm and can be expected to continue to improve since the

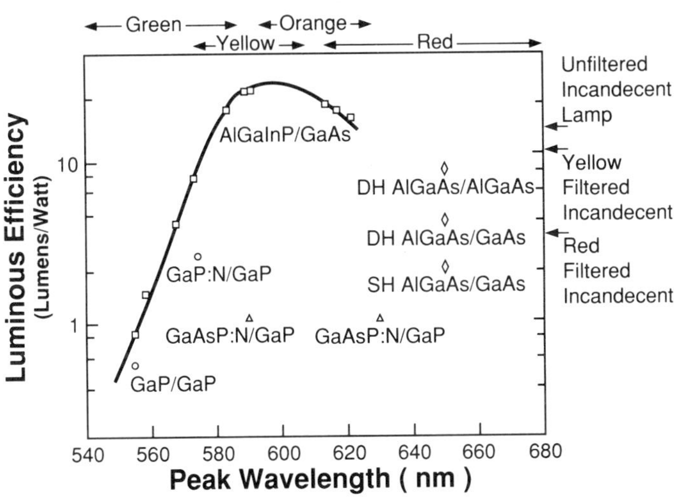

FIG. 12. LED luminous performance vs wavelength for AlInGaP compared to other LED technologies. Luminous performance is the product of power efficiency (roughly equal to quantum efficiency) and the eye's response. Reprinted from Craford (1992), courtesy of the Institute of Electrical and Electronics Engineers, Inc. Copyright © 1992 IEEE.

technology has only recently been introduced. It is reasonable to expect that AlInGaP devices will have superior performance for all wavelengths from ~550 to ~630 nm in the future and should even surpass AlGaAs at 650 nm if transparent-substrate devices can be developed.

AlInGaP is grown using the metal-organic vapor-phase epitaxy technique (see Sec. 5.3.4), which is still evolving as a production technology. As a consequence, these devices are now relatively expensive but can be expected to decrease rapidly in cost as the technology is refined.

3.1.6 GaN. GaN has a direct energy gap of 3.4 eV, which is suitable for blue LEDs, but there are two problems that have limited the application of GaN. There is no substrate that provides a good lattice match, and it is difficult to form *p-n* junctions.

The most commonly used substrate is sapphire (Al_2O_3). Sapphire is transparent, which is good, but is so poorly matched both structurally and thermally that the defect density is high. Other materials such as SiC, MgO, and ZnO are being investigated.

GaN is generally *n* type as grown, probably as a result of nitrogen vacancies. It has been extremely difficult to produce high-quality *p*-type GaN. Recently GaN doped with Mg has been converted to *p*-type through the use of electron-beam irradiation, and *p-n* junctions have been formed (Nakamura *et al.*, 1991b; Amano *et al.*, 1990). If this process can be commercialized, low-voltage GaN devices should become available. At the present time GaN devices are generally higher-voltage metal–insulator–semiconductor devices.

In the future, the AlInGaN material system can be expected to be developed. Since AlN has an energy gap of 6.2 eV, devices with wavelengths ranging into the ultraviolet should become possible.

3.1.7 ZnSe. Materials from columns II and VI of the periodic chart, such as ZnS and ZnSe mixtures, have direct energy gaps exceeding 2.5 eV and are also suitable for blue LEDs. These materials can be lattice matched to GaAs substrates, which is a major advantage compared to GaN. However, *p-n* junctions have also been difficult to form in ZnSe. Recently, nitrogen doping has been used to produce *p*-type material and to fabricate injection lasers and LEDs emitting in the blue-green spectral region (Haase *et al.*, 1991; Jeon *et al.*, 1992). These devices are promising but have had reliability problems. At the time of this writing, it is not clear whether these are fundamental limitations or if they can be improved to yield satisfactory commercial emitters.

3.1.8 SiC. Silicon carbide is a column-IV material with an indirect energy gap of 2.86 eV. It is the most widely utilized type of blue emitter at this time (Edmond *et al.*, 1992). Devices are formed by growing a SiC epitaxial film on a SiC substrate using either MOVPE or LPE. *p-n* junctions can be readily obtained by doping with Al for *p*-type and N for *n*-type material. The efficiency is relatively low because of the indirect band gap. The light output has been steadily improving as the materials growth technology is refined and, if a suitable isoelectronic impurity can be found, as in the case of GaP:N, it may be possible to improve the light output significantly.

3.2 Operating Characteristics

3.2.1 Current–Voltage Characteristics. Turn-on voltages for LEDs typically range from about 1.5 V for red devices to more than 5 V for blue devices depending upon the material system and whether or not a *p-n* junction can be formed. Some typical current–voltage characteristics are shown in Fig. 13.

The operating current for visible LEDs can range from 1 mA or less for indicator applications to 100 mA or more for illumination applications such as automotive exterior lighting. The operating voltage is the sum of the voltage at the junction plus the voltage due to the product of the forward current and the device dynamic resistance. The dynamic resistance will be higher if the LED chip is smaller in area, thicker, or made of less-conductive (more lightly doped and/or lower-mobility) material. Unfortunately, it is often desirable to use small chips to reduce cost, thick chips to get efficient light extraction from the edges, and lightly doped material to minimize nonradiative recombination. These conflicting requirements result in different optimal LED chip designs for different applications.

The reverse breakdown voltage is generally limited by avalanche breakdown in the *n*-type

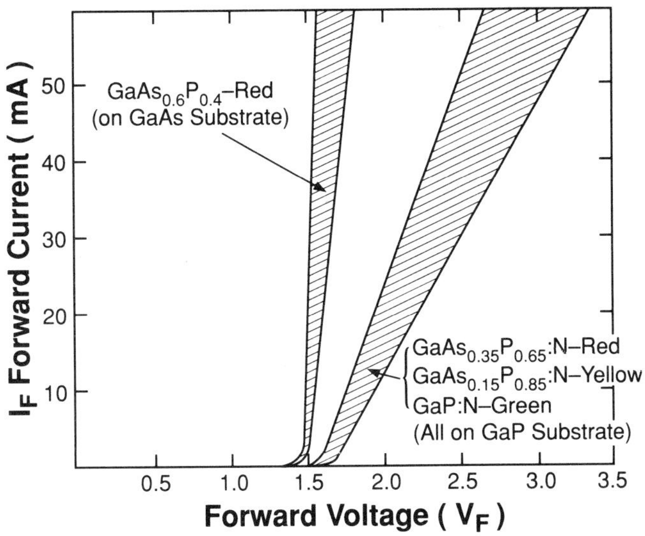

FIG. 13. Typical current-voltage characteristics for red-, yellow-, and green-emitting LEDs. The red-emitting $GaAs_{0.6}P_{0.4}$ on GaAs substrate has a typical dynamic resistance of 2–6 Ω. The nitrogen-doped devices on GaP substrates have typical dynamic resistances of 20–30 Ω. Reprinted from Tannas (1985), courtesy of Van Nostrand Reinhold Company, Inc. Copyright © 1985.

region of the LED, which is determined by the carrier concentration. In some double heterostructure devices, reverse breakdown can also occur as a result of "punch through" when the depletion region reaches an interface. In either case, the reverse breakdown voltage is generally substantially more than 10 V for visible LEDs. This is large enough for most applications, and breakdown voltage is generally not an issue as long as a high-quality *p-n* junction is utilized and reverse-current leakage is minimized.

3.2.2 Performance versus Current. The LED performance shown in Table 6 corresponds to a typical operating current in the range of 20 mA. LED performance will change with current (or current density) depending upon a variety of factors such as the concentration of nonradiative centers in the vicinity of the junction and heating due to both nonradiative recombination and Joule heating associated with the dynamic resistance. Typically LED efficiency will go through three stages—superlinear, linear, and sublinear—as a function of increasing current. This is illustrated in Fig. 14.

At low currents, the light output is superlinear, since nonradiative centers provide a shunt path that saturates as the current density increases. GaAsP devices with high defect densities typically show more superlinearity than AlGaAs devices, which are lattice matched and have low defect densities.

At moderate current densities the light output remains linear with current (and the efficiency is constant). At higher current densities, heating causes the efficiency to drop and the light output becomes sublinear with

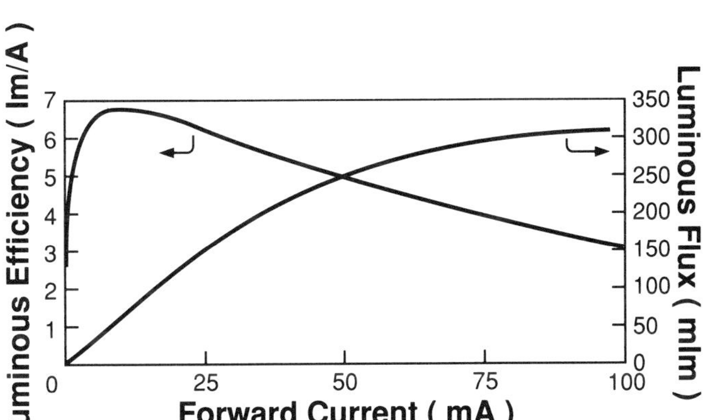

FIG. 14. Efficiency and light output as a function of d.c. current for a double-heterostructure AlGaAs LED. At low currents, the light output is superlinear. From 5 to 20 mA the light output is a linear function of current, and above 20 mA, the light output becomes sublinear because of heating effects.

current and will eventually decrease with increasing current as the heating becomes severe.

For most LEDs the light output drops at the rate of between 0.5% and 2% per degree centigrade measured at the junction. This rise in temperature of the *p-n* junction can be estimated in a given application if thermal resistance (the ratio of the steady-state rise in temperature to the power applied) for a given package is known. The thermal resistance of a typical LED lamp is 220 °C/W.

3.3 Applications

3.3.1 Overview. Traditionally, visible LEDs have been used for indicators on instruments, computers, and consumer electrical equipment. LEDs are used in the form of discrete LED lamps, numeric displays, or alphanumeric displays. These applications are particularly cost-sensitive and are dominated by lower-cost and -performance LED technologies. Emerging LED markets such as outdoor large-area displays and exterior lighting for automobiles often require LEDs with the maximum possible performance and, in addition, require devices that are robust when operated in outdoor environments. Visible LED applications are summarized in Table 7.

3.3.2 Instrument and Computer. In these markets LEDs are used both as discrete indicator lamps and as numeric and alphanumeric displays. Most of these applications are not battery powered, and traditional LED performance (~1 l/W) is often adequate. However, operation at low currents so that the LEDs can be driven directly from integrated circuits can be important, and may require higher-performance devices. Reliability is a key concern in these markets, and this gives LEDs an important advantage. Cost pressure from competitive technologies such as liquid-crystal displays is severe, particularly for numeric and alphanumeric display applications. LEDs are often used to backlight liquid-crystal displays, and for this purpose higher-performance LEDs are often utilized.

Surface-mounted LED packages have become increasingly important as LED users automate their processes in order to improve quality and reduce cost.

3.3.3 Consumer Electronics. These markets also generally use LEDs in the form of either discrete lamps or numeric displays. Applications include audio equipment, home appliances, clocks, radios, TV-channel tuning, toys, etc. This is a very-high-volume market and cost is the most critical issue. Standard lower-performance LED technologies are generally utilized. For many of these applications, the oldest and lowest-performance LED technologies [red GaAsP and GaP:(Zn,O)] are widely utilized.

3.3.4 Large-Area Display. The large-area display market has been growing rapidly. These displays can be "electronic billboards" as large as several meters high and wide, or as small as several inches high and wide. Some applications are outdoors, in which case high-performance LED technologies are required to provide sunlight viewability.

Indoor applications generally utilize the lower-performance technologies, often using red GaAsP:N and green GaP:N devices in

Table 7. Main technologies for different applications. Key: ●, primary technology; ◑, secondary technology; ○, potentially important.

Technology / Application area	Red GaAsP/GaAs	Red GaP:(Zn,O)	Red, yellow, orange GaAsP:N	Green GaP:N	Red AlGaAs	Yellow, orange AlInGaP	Blue SiC
Consumer	●	●	●	●	◑		
Industrial	◑	◑	●	●	◑	○	
Automotive							
Interior			●	●		○	
Exterior					●	○	
Large-area displays							
Interior			●	●	◑	○	○
Exterior			◑	●	●	○	
Electrophotographic	●						

pairs. By adjusting the drive current to the two LEDs, the perceived color can be changed from red to green. Blue is required to achieve full-color displays, and this is one of the major potential markets for the emerging blue technology.

The high-brightness red, orange, and yellow-amber AlInGaP and AlGaAs technologies are generally used outdoors. Electronic highway signs, where amber is generally desired, and advertising signs, where any color can be used, are among the applications.

3.3.5 Vehicular Lighting. LEDs are increasingly popular in and on vehicles because of their reliability and ruggedness as well as for styling reasons. LEDs are small and thin and do not burn out. AlGaAs and AlInGaP LEDs are incorporated in automobile trunks and "spoilers" to serve as brake lights, and they are being used on trucks and automobiles as side markers and "running lights."

LEDs are also used as indicator and display devices on the interior of vehicles. Dashboard displays generally use the GaAsP:N and GaP:N technologies, although higher-performance technologies are sometimes used to backlight liquid crystal displays.

3.3.6 Electrophotography. LEDs are generally used in electrophotographic equipment either to "write" or to "erase" a message on a revolving photoconductive drum. The LEDs are configured in a thin bar as long as the paper is wide. The drum picks up toner in selected areas determined by the LED "write" function and deposits the toner on the paper.

The LED "write bar" consists of an array of several thousand LEDs with a linear density of typically 300/in. The arrays typically consist of a series of red-emitting GaAsP monolithic chips, where each chip has 64 or 96 2–3-mil-diam junctions in a row. The drum is then "erased" by light from the LED "erase bar." The LED erase bar is a lower-resolution linear array, and discrete AlGaAs LED chips are commonly employed.

LED "write bars" compete with laser printers and at this time have a small share of the market. LEDs can provide very high-speed, high-quality printing, but the cost of LED printers is generally higher than that of laser printers.

3.3.7 Trends and Future Developments. Visible-LED markets will grow and new applications will emerge as electronics becomes increasingly pervasive. The standard LED technologies will continue to be used for applications where their light output is adequate. The higher-performance technologies will decrease in cost and will be used where high light output or low drive currents are important. Large-area displays and vehicular lighting are expected to become large markets. The development of a high-performance blue technology should occur, which would make full-color large-area displays possible and help the outdoor large-area–display market to grow rapidly.

4. INFRARED EMITTERS

4.1 Materials Performance Characteristics

4.1.1 Comparison of Different Types of Infrared LEDs. The detectors used for infrared LEDs are photodiodes or phototransistors instead of the human eye. The radiometric power output of the LED is therefore a key performance parameter. The rate at which data can be transmitted is determined in part by the LED switching speed, and this is also an important performance parameter for IR LEDs (one that is totally unimportant for most visible applications). A summary of performance characteristics for several types of IR emitters is shown in Table 8.

4.1.2 GaAs:Zn. The first IR emitters available were diffused-junction devices similar to the $GaAs_{1-x}P_x$ LEDs described in Sec. 3.1.2 (Keyes and Quist, 1962). There is a similar trade-off (Herzog *et al.*, 1972) between avoiding surface recombination problems (deep junction optimal) and reducing internal absorption (shallow junction optimal). Because of their low power output, they are no longer extensively used.

4.1.3 GaAs:Si and AlGaAs:Si. At liquid-phase epitaxial (LPE) growth temperatures >820 °C, Si incorporates as an *n*-type dopant in both GaAs and AlGaAs, while at lower temperatures it incorporates as a *p*-type dopant. By simply cooling a GaAs or AlGaAs melt on a GaAs substrate through this temperature, a *p-n* junction can be grown. The dominant

Table 8. Characteristics of IR LEDs.

LED type	Wavelength (nm)	External quantum efficiency (%)	Typical performance output power[a] (mW at 50 mA)	90% to 10% fall time (ns)	Device[b] structure	Substrate	Epitaxial growth method (*p-n* junction formation)
GaAs:Zn	900	3–4	2–3	50	Homo.	GaAs	BG[c] (diffusion)
GaAs:Si	940	9–14	6–9	1000	Homo.	GaAs	LPE (grown)
AlGaAs:Si	880	11–18	8–13	500	Homo.	GaAs	LPE (grown)
AlGaAs	780–880	4–6	3–5	20–40	SH	GaAs	LPE (grown)
AlGaAs	780–880	10–14	7–10	20–40	DH	GaAs	LPE (grown)
AlGaAs	780–880	20–27	15–20	20–40	DH-TS	GaAs	LPE (grown)
SEA[d] AlGaAs	820–880	—	—	4–15	DH	GaAs	LPE/MOVPE (grown/diffusion)
SEA[d] InGaAsP	1300–1500	—	—	$\approx$2.5	DH	InP	LPE/MOVPE (grown)

[a]Encapsulated lamp form.
[b]See Table 6 for definitions.
[c]Horizontal Bridgeman.
[d]SEA stands for small emission area.

recombination process that occurs is between donor and acceptor band-tail states, and sub-band-gap light is generated. The rest of the crystal is fairly transparent to the light generated in GaAs:Si LEDs, and thus they have high output efficiencies (Rupprecht *et al.*, 1966). In AlGaAs:Si LEDs, the GaAs substrate is typically removed (see Fig. 15), and the device is processed "*n*-side-up" to reduce internal absorption (Dawson, 1977). This donor-acceptor recombination process is, however, quite slow ($\approx 1\ \mu$s), and such LEDs cannot be used in data communication applications at frequencies greater than several hundred kilohertz. Nonetheless, because of their high output power and low cost, they continue to be used extensively as the light sources in remote controls for televisions and VCRs.

4.1.4 $Al_xGa_{1-x}As$. As was mentioned for the visible emitters (Sec. 3.1.4), the small variation in lattice parameter from GaAs to AlAs has made possible the growth of very high-quality heterostructure devices in the $Al_xGa_{1-x}As$ materials system. Single-heterostructure, double-heterostructure, and "transparent-substrate" double-heterostructure emitters are also available in the infrared portion of the spectrum. The internal efficiency of these devices is near 100% (Alfërov *et al.*,

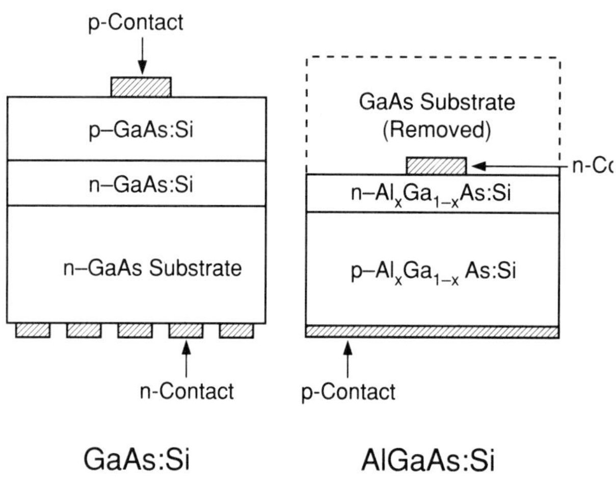

FIG. 15. Chip structures of GaAs:Si and AlGaAs:Si infrared emitters. Both the *p*-type and *n*-type epitaxial layers are grown by LPE from a single melt. In AlGaAs:Si emitters, the original GaAs substrate is removed (to reduce absorption) and the device is processed "*n*-side up".

1975) and does not vary much as the emission wavelength is changed from ~700 nm to 880 nm. High-quality devices can be grown in this wavelength region by both MOVPE and LPE. However, the thick layers necessary for efficient optical extraction are most easily grown by LPE. Double-heterostructure transparent-substrate AlGaAs emitters provide more power output than any other type of infrared LED and are roughly 20 times faster than Si-doped GaAs and AlGaAs emitters. Their use will continue to expand as more efficient manufacturing processes are developed and their cost decreases.

4.1.5 Small-Emission-Area AlGaAs and InGaAsP. Most of the IR LED emitters described above are unsuitable for fiber-optic data-communication applications as they are too slow, and it is difficult to couple the emitted light efficiently into the small (≈60 μm diameter for multimode and ≈7 μm diameter single-mode) fiber cores. To improve performance in both of these areas, the emission area is typically reduced. The resultant increased current density shortens the lifetime, and the smaller emission area is easier to focus into the fiber core. This reduction in emission area is accomplished by either introducing current-blocking layers into the structure, etching mesas, or switching to a "stripe" geometry and forming an edge emitter (see Saul *et al.*, 1985; Pearsall, 1982). The power that can be coupled into a fiber is the important performance parameter for these devices and no values are listed in Table 8 for total output power.

The wavelength of these emitters is chosen to match regions of minimum attenuation in optical fiber. The first minimum occurs at ~850 nm and AlGaAs emitters are used. Two other minima occur near 1300 and 1550 nm, and InGaAsP emitters are used.

4.2 Operating Characteristics

4.2.1 Current-Voltage Characteristics. As with the visible LEDs, the turn-on voltage of infrared LEDs varies with the band gap of the material. InGaAsP emitters have turn-on voltages of ~0.8 V while GaAs:Si and AlGaAs emitters have turn-on voltages of 1.2 V. The operating current varies dramatically depending upon the application. Some small-emission-area AlGaAs LEDs used in data-communication applications can be driven at currents as low as 10 mA, while GaAs:Si emitters used in remote-control applications are pulsed up to 1.5 A. Large chips ($400 \times 400\ \mu m^2$) are used in the high-current devices to reduce both the thermal and the dynamic resistances of the device. To keep the total forward voltage below 3.0 V (two 1.5-V batteries in series) at a current of 1.5 A, the dynamic resistance of the LED must be kept below ~1 Ω.

Again, as with the visible LEDs, the reverse breakdown voltage is limited by avalanche breakdown and is determined by the doping level on the most lightly doped side of the *p-n* junction. As the infrared emitters are usually more heavily doped than visible emitters, their breakdown voltages are typically lower, in the 7–60-V range.

4.3 Applications for Infrared Emitters

4.3.1 Remote Controls. As was mentioned above, one of the highest-volume applications for IR LEDs is for remote controls that are used in televisions, audio equipment, and VCRs. Here the information is sent from the remote control through the air in a sequence of pulses and a high-sensitivity Si *p-i-n* detector is used as the receiver. The higher the LED output power, the further one can be from the receiver and still have it work. The amount of data to be transmitted, however, is quite small, and slow GaAs:Si or AlGaAs:Si emitters are sufficient. These LEDs are driven at currents up to 1.5 A in order to maximize the transmitter–receiver distance.

4.3.2 Optocouplers. If the emitter and detector are placed opposite each other inside a light-shielded package [see Fig. 16(a)], signals can be "coupled" from one electrical circuit to another without an electrical connection. Devices of this type are called optocouplers and can be used to provide electrical isolation of up to 5 kV. They are used to isolate high- and low-voltage circuits and to eliminate interference problems. GaAs:Si, AlGaAs:Si, visible GaAsP, and AlGaAs emitters are all used in optocoupler applications depending upon the speed–power–cost trade-off desired.

4.3.3 Sensors. If there is an optically accessible path between an emitter and a detector, the presence or absence of an object in that path can be sensed. A typical example is the slot interrupter shown in Fig. 16(b) in

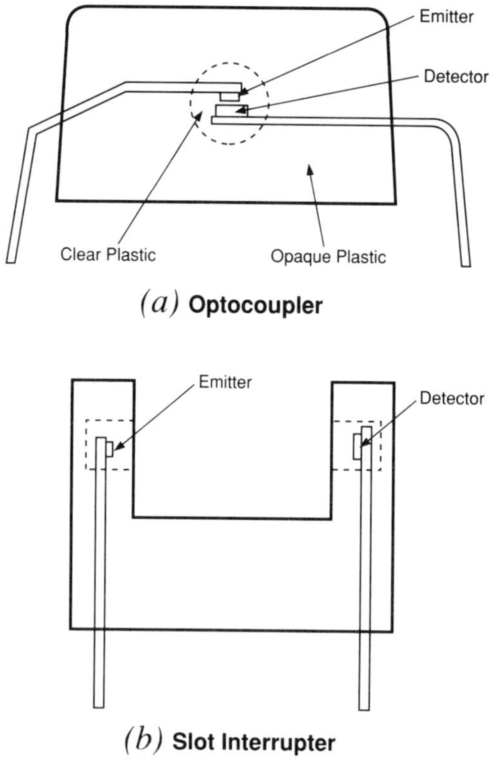

FIG. 16. (a) Optocoupler and (b) slot interrupter configurations for emitter–detector pairs. The emitter and detector in an optocoupler are shielded from external light sources by opaque plastic while the optical path is accessible in a slot interrupter. An object passing between the emitter and detector in a slot interrupter blocks the optical path and can therefore be sensed.

which an emitter and detector face each other across the opening of a U-shaped plastic holder. If an object passes between the emitter and detector, the optical path is blocked and a signal is generated. Typical applications for slot interrupters are to count objects, to sense paper in a copier, and to check for the write-protection sticker on a floppy disk.

Such sensors can also be configured to work in a reflective mode where the emitter and detector are positioned close to each other and face the same direction. If a reflective surface passes in front of them, light from the emitter is reflected to the detector and is sensed.

4.3.4 Optical Communications (*q.v.*). As was mentioned in Sec. 4.1.5, LEDs with small emission areas that can be effectively coupled into optical fibers have been developed and are the emitters of choice for data rates below 200 Mbd and distances shorter than ~8 km. For higher data rates and longer distances, laser diodes are used. The wavelength of the emitter is chosen to correspond to one of the several minima in the absorption of optical fiber. One minimum occurs at ~850 nm and AlGaAs emitters are used. Two other minima occur at 1300 and 1550 nm and InGaAsP emitters are used.

4.3.5 Trends and Future Developments. The remote-control and sensor markets are fairly mature and are not expected to grow rapidly. Data-communications applications will continue to grow as more and more Cu wire is replaced by optical fiber. Through-the-air data-communication applications such as wireless sound transmission from a television to headphones or from a personal computer to a printer are also expected to grow rapidly. These applications require both high power and high speed, and thus double-heterostructure transparent-substrate AlGaAs emitters are the best choice.

5. TECHNOLOGY

5.1 Overview

LED production technology consists of several basic steps: substrate growth, epitaxial growth, and wafer fabrication into finished components. LED technology is similar in many ways to the technology utilized to fabricate silicon integrated circuits. However, the materials used to fabricate LEDs are much more complex than elemental silicon and are more difficult to produce and process. As a consequence, the evolution of compound-semiconductor technology is far behind silicon in terms of the size of crystals grown, defect densities, volumes of device produced, and automation and cost reduction.

5.2 Substrate Growth

The most common LED substrates are GaAs, GaP, and InP. They are generally grown using the liquid-encapsulated Czochralski (LEC) technique. This technique is an extension of the Czochralski technique used to grow silicon crystals (see CRYSTAL GROWTH). A single-crystal "seed" is dipped into a molten solution of the material to be grown and, as the seed is pulled from the melt and cooled, a single crystal forms on the end. Liquid encapsulation, which is a layer of boric oxide on top of the melt, is required to keep the volatile As and P from vaporizing from the melt. The crystal growth occurs above atmospheric pressure so that the boric oxide will stay in place on top of the melt.

Typically crystals 2 and 3 in. in diameter and less than 12 in. long are grown. These crystals are then sliced into wafers 15 to 20 mils thick that are used as the substrates for the epitaxial films.

5.3 Epitaxial Growth

5.3.1 Epitaxial Growth Overview. There are four technologies used for the growth of the epitaxial films; liquid-phase epitaxy (LPE), vapor-phase epitaxy (VPE), metal-organic vapor-phase epitaxy (MOVPE), and molecular-beam epitaxy (MBE). LPE and VPE are well-established high-volume technologies used to grow GaAsP, GaP, and AlGaAs, which dominate the production of LEDs at this time. MOVPE is generally used for the growth of AlInGaP and other quaternary materials and is emerging as a LED production technology. MBE is not widely used for LED production at this time but is the only epitaxial technology that has been successfully used to produce *p-n* junctions in ZnSe. Thus, if ZnSe becomes the technology of choice for blue emitters, MBE could become an important LED growth technology.

5.3.2 Liquid-Phase Epitaxy. LPE is the highest-volume LED technology. It is used for the growth of various types of GaP, GaAs, and AlGaAs LEDs. The LPE process consists of placing a molten solution (hence the name LPE) of Ga, which is saturated with the material to be grown, in contact with the substrate wafer at high temperature (generally 600–900 °C). As the solution is cooled, an epitaxial film is formed on the substrate wafer. Successive layers with different compositions or doping types can be grown by sliding the substrate wafer from one Ga melt to another. The growth apparatus consists of graphite chambers that hold the melts and substrates and facilitate transferring the substrates from one melt to another.

The LPE technique has the advantage that it is relatively easy to grow high-quality layers. However, it becomes more complicated as the number of wafers and layers to be grown increases, and it is not suitable for the controlled growth of complex structures with layer thickness much less than a micrometer thick.

5.3.3 Vapor-Phase Epitaxy. VPE is a high-volume technology that is primarily used to grow GaAsP and GaAsP:N layers. The substrate wafers are placed in the coolest part of a heated growth chamber and the material to be grown is transported to the surface of the wafer in gaseous form (hence the name VPE) and deposited upon the relatively cool substrates [see CVD (CHEMICAL VAPOR DEPOSITION)]. The gases consist of PH_3, AsH_3, and gallium chlorides that are formed by passing HCl over Ga metal. Hydrogen and dopant gases are also present. The VPE technique has the advantages that large numbers of wafers can easily be grown upon simultaneously, and the layer composition can be readily changed as frequently as desired simply by changing the gas composition during growth. VPE has the disadvantage that AlGaAs and AlInGaP cannot be grown because the aluminum-bearing gases tend to attack the hot quartz walls of the chamber and contaminate the epitaxial films. This limitation is not present in MOVPE.

5.3.4 Metal-Organic Vapor-Phase Epitaxy. MOVPE (*q.v.*) is similar to VPE except that metal-organic gases such as trimethylgallium (TMG) are used to transport the growth metals, and the growth reaction is pyrolytic so that the substrate wafer is in hottest part of the growth chamber and the walls of the chamber can stay relatively cool. This minimizes the reactions of the growth gases with the walls and enables aluminum-containing compounds to be grown.

5.3.5 Molecular-Beam Epitaxy. MBE (*q.v.*) is a high-vacuum technology. The substrate wafers are placed in a special vacuum chamber and held at an appropriate growth temperature. The growth materials are then evaporated from controlled source ovens. MBE can grow extremely thin layers and complex structures with a high degree of control. It has not been used for cost-effective high-volume production of LEDs. If MBE is needed to produce high-performance ZnSe blue LEDs, high-volume production capability will have to be developed.

5.4 Wafer Fabrication

5.4.1 Junction Formation. Generally the *p-n* junction is formed during the epitaxial growth step. However, for GaAsP and some other types of devices this is not done. In this case, *p-n* junction diffusion is the first step in wafer fabrication. The junction is generally formed by the diffusion of zinc into the crystal. If a patterned junction or a monolithic chip with several junctions is required, a silicon nitride (Si_3N_4) film is first deposited and then holes are opened in the Si_3N_4 to determine where the Zn will diffuse. The diffusions are carried out in ampoules at temperatures from 600 to 900 °C for times ranging from minutes to hours.

5.4.2 Ohmic Contact Formation. If the *p-n* junction is formed during the epitaxial growth step, as is generally the case, wafer fabrication consists primarily of the formation of Ohmic contacts on the wafers. The Ohmic contact consists of patterned metal films, which may have several layers. Ohmic contacts are formed using evaporation, *e*-beam, or sputtering depending upon the contact metals. Typical *p*-type contacts contain alloys of Au-Zn or Au-Be. The *n*-type contacts typically use materials such as Au-Ge alloys. The contact materials are deposited on the epitaxial wafers and then heated to typically about 400 °C to form a low-resistance Ohmic contact.

Multilayer contacts are often employed because the surface of the contact must be suitable for high-speed wire bonding. This requires a relatively pure Au or Al film with a minimum of oxides that can form if Ga or other materials diffuse to the surface. Consequently a barrier layer of W or similar material is often deposited on top of the contact followed by the deposition of a final Au or Al layer. The contacts may be patterned using photolithographic techniques or by carrying out the depositions through metal masks.

5.4.3 Dicing. After the wafer fabrication is complete, the wafers are separated into chips. In some cases the wafers are tested for optical and electrical performance before dicing. Dicing is generally accomplished with a dicing saw, which uses a thin circular diamond-impregnated blade. Sometimes chips are scribed and broken, but the chip quality is better with a dicing saw. The wafer is placed on an elastomer tape before sawing, and the chips remain on the tape until they are die-attached. The tape is usually expanded, separating the chips further apart, before die attachment.

5.5 Assembly

5.5.1 Die Attachment. The LED chips are removed from the tape used in sawing or scribing and placed in position in the device assembly by high-speed die-attachment machines that operate at rates in excess of 3000 chips/h. These machines first deposit a drop of conductive epoxy and then place the chip in the epoxy. Following die attachment, the epoxy must be cured prior to wire bonding. Depending upon the final device type, the chips may be attached to a metal lead frame, a printed circuit board, or a ceramic substrate.

5.5.2 Wire Bonding. The wire bonding is also done by high-speed machines that can perform more than 3000 bonds/h. They use heat and pressure (thermocompression bonding) and/or scrubbing (ultrasonic bonding) to achieve a bond. The bonding wires are typically 1 mil thick and are gold or aluminum. Wire bonding is important in determining device reliability, and the quality of the metal bonding pad on the LED chip and process control in the bonding operation are critical. If the integrity of the bond is not good, the wire can pull off during encapsulation or when the device is in operation, particularly during temperature cycling. If the bonding process is too severe, the LED chips, which are quite fragile, can be cracked, which also causes reliability problems.

5.5.3 Encapsulation. Most LED devices are encapsulated in some type of polymer, generally epoxy. The epoxy serves to match the index of refraction better (see Sec. 1.8.1) so that more light is extracted, as a lens to focus the light in the desired radiation pattern, and as a structural medium to hold the package together and protect the fragile wire-bonded LED. However, the epoxy has a much higher coefficient of expansion than the LED chip, and so during operation and temperature cycling, it can cause stress and damage the chip instead of protecting it. In the case of integrated circuits and most other electronic devices, the encapsulating polymers are "filled" with graphite fiber or some other material that reduces the expansion coefficient. However, since LEDs must emit light, opaque filler cannot be used, and the selection and curing of the epoxy encapsulant are critical reliability issues. This is particularly true for exterior application where large temperature extremes are experienced.

The encapsulation epoxy may be completely clear so that the light can be focused in a narrow beam, or may have a small amount of diffusant, such as glass particles, added to spread the light if wide-angle viewing is desired. The epoxy may also be tinted to match the color of the LED.

The encapsulation is usually done by casting, although transfer molding is sometimes used. Injection molding has not been widely used because the polymers that can be injection molded have not been compatible with LED reliability requirements.

List of Works Cited

Alfërov, Zh. I., Chikovani, R. I., Charmakadze, R. A., Mirianashvili, G. M., Zosimov, N. K., Grigoryan, N. A. (1973), *Sov. Phys. Semicond.* **6**, 1930–1931.

Alfërov, Zh. I., Andreev, V. M., Garbuzov, D. Z., Rumyantsev, V. D. (1975), *Sov. Phys. Semicond.* **9**, 305–309.

Amano, H., Asahi, T., Akasaki, I. (1990), *Jpn. J. Appl. Phys.* **29**, L205–206.

Cao, D. S., Stringfellow, G. B. (1990), *J. Electron. Mater.* **20**, 97–101.

Casey, H. C., Jr., Panish, M. B. (1978), *Heterostructure Lasers*, Orlando, FL: Academic.

Cook, L. W., Camras, M. D., Rudaz, S. L., Steranka, F. M. (1988), *Proceedings of the 14th International Symposium on GaAs and Related Compounds*, Bristol: Institute of Physics, pp. 777–780.

Craford, M. G. (1992), *Circuits Devices* **8**, 24–29.

Craford, M. G., Shaw, R. W., Groves, W. O., Herzog, A. H. (1972), *J. Appl. Phys.* **43**, 4075–4083.

Craford, M. G., Stillman, G. E., Holonyak, N., Jr., Rossi, J. A. (1991), *J. Electron. Mater.* **20**, 3–11.

Dawson, L. R. (1977), *J. Appl. Phys.* **48**, 2485–2492.

Dupuis, R. D., Dapkus, P. D. (1977), *Appl. Phys. Lett.* **31**, 466–468.

Edmond, J. A., Kong, H.-S., Carter, C. H., Jr. (1992), in: C. Y. Yang, M. M. Rahman, G. L. Harris (Eds.), *Amorphous and Crystalline Silicon Carbide IV*, Berlin: Springer, pp. 344–351.

Fletcher, R. M., Kuo, C. P., Osentowski, T. D., Huang, K. H., Craford, M. G. (1991), *J. Electron. Mater.* **20**, 1125–1130.

Fukuda, M. (1991), *Reliability and Degradation of Semiconductor Lasers and LEDs*, Boston: Artech House.

Groves, W. O., Herzog, A. H., Craford, M. G. (1971), *Appl. Phys. Lett.* **19**, 184–186.

Haase, M. A., Q:v, J., DePoydt, J. M., Cheng, H. (1991), *Appl. Phys. Lett.* **58**, 1272–1275.

Herzog, A. H., Groves, W. O., Craford, M. G. (1969), *Appl. Phys. Lett.* **40**, 1830–1838.

Herzog, A. H., Keune, D. L., Craford, M. G. (1972), *J. Appl. Phys.* **43**, 600–608.

Holonyak, N., Jr., Bevacqua, S. F. (1962), *Appl. Phys. Lett.* **1**, 82–83.

Huang, K. H., Yu, J. G., Kuo, C. P., Fletcher, R. M., Osentowski, T. D., Stinson, L. J., Liao, A. S. H., Craford, M.G. (1992), *Appl. Phys. Lett.* **61**, 1045–1047.

Jeon, J., Ding, J., Nurmikko, A.V., Xie, W., Kobayashi, M., Gunshor, R.L. (1992), *Appl. Phys. Lett.* **60**, 892–894.

Keyes, R. J., Quist, T. M. (1962), *Proc. IRE* **50**, 1822–1823.

Kimerling, L. C., Lang, D. V. (1975), in: F. A. Huntley (Ed.), *Lattice Defects in Semiconductors 1974*, Institute of Physics Conference Proceedings No. 23, Bristol, U.K.: Institute of Physics.

Kuo, C. P., Fletcher, R. M., Osentowski, T. D., Lardizabel, M. C., Craford, M. G. (1990), *Appl. Phys. Lett.* **57**, 2937–2939.

Logan, R. A., White, H. G., Wiegmann, W. (1968), *Appl. Phys. Lett.* **13**, 139–141.

Manasevit, H. M., Simpson, W. I. (1969), *J. Electrochem. Soc.* **116**, 1725–1732.

Nakamura, S., Mukai, T., Senoh, M. (1991a), *Jpn. J. Appl. Phys.* **30**, L1998–L2001.

Nakamura, S., Senoh, M., Mukai, T. (1991b), *Jpn. J. Appl. Phys.* **30**, L1701–L1711.

Nishizawa, J., Suto, K. (1977), *J. Appl. Phys.* **48**, 3484–3445.

Pearsall, T. P. (Ed.) (1982), *GaInAsP Alloy Semiconductors*, Chichester: Wiley.

Rupprecht, H., Woodall, J. M., Konnerth, K., Pettit, D. G. (1966), *Appl. Phys. Lett.* **9**, 221–223.

Rupprecht, H., Woodall, J. M., Pettit, G. D. (1967), *Appl. Phys. Lett.* **11**, 81–83.

Saul, R. H., Armstrong, J., Hackett, W. H., Jr. (1969), *Appl. Phys. Lett.* **15**, 229–231.

Saul, R.H., Lee, T. P., Burrus, C. A. (1985), in: W.T. Tsang (Ed.), *Semiconductors and Semimetals*, Orlando: Academic, Vol. 22, 193–237.

Sugawara, H., Ishikawa, M., Hatakoshi, G. (1991), *Appl. Phys. Lett.* **58**, 1010–1012.

Tannas, L. E., Jr. (1985), *Flat Panel Displays and CRTs*, New York: Van Nostrand Reinhold.

Xie, W., Grillo, D. C., Gunshor, R. L., Kobayashi, M., Jeon, H., Ding, J., Nurmikko, A. V., Hua, G. C., Otsuka, N. (1992), *Appl. Phys. Lett.* **60**, 1999–2001.

Further Reading

Bergh, A. A., Dean, P. J. (1976), *Light Emitting Diodes*, Oxford: Clarendon.

Casey, H. C., Jr., Panish, M. B. (1978), *Heterostructure Lasers*, Orlando, FL: Academic.

Craford, M. G. (1985), in: L. E. Tannas, Jr. (Ed.), *Flat-Panel Displays and CRTs*, New York: Van Nostrand Reinhold, pp. 289–331.

Gillessen, K., Schairer, W. (1987), *Light Emitting Diodes—An Introduction*, Englewood Cliffs, NJ: Prentice Hall.

Kressel, H., Butler, J. K. (1977), New York, *Semiconductor Lasers and Heterojunction LEDs*, Academic.

Saul, R. H., Lee, T. P., Burrus, C. A. (1985), in: W. T. Tsang (Ed.), *Semiconductors and Semimetals*, Orlando, FL: Academic, Vol. 22, pp. 193–237.

Williams, E. W., Hall, R. (1978), *Luminescence and the Light Emitting Diode*, Oxford: Pergamon.

LIGHT SCATTERING

See RAMAN SCATTERING

LINEAR ACCELERATORS

See ACCELERATORS, LINEAR

LINEAR ACOUSTICS

See ACOUSTICS, LINEAR

LINEAR OPTICS

See OPTICS, LINEAR

LIQUIDS

See ACOUSTIC PROPERTIES OF LIQUIDS; DIFFUSION AND IONIC CONDUCTION IN LIQUIDS; ELECTRON STRUCTURE OF LIQUIDS; MECHANICAL PROPERTIES OF LIQUIDS

LIQUID CRYSTALS, STRUCTURE OF

J. D. LITSTER, *Department of Physics, Massachusetts Institute of Technology, Cambridge, Massachusetts, U.S.A.*
R. SHASHIDHAR, *Center for Biomolecular Science and Engineering, Naval Research Laboratory, Washington, DC, U.S.A.*

Introduction 515
1. **Nematic Liquid Crystals** 516
1.1 Optical Properties of Nematics 516
1.2 Viscoelastic Properties of Nematics 517
1.3 Device Applications of Nematics 518
1.3.1 Twisted Nematic Displays..... 518
2. **Cholesteric Liquid Crystals**... 519
2.1 Optical Properties of Cholesterics........ 519
2.2 Response of Cholesterics to Applied Fields........ 519
2.3 Applications of Cholesterics... 520
2.4 Cholesteric "Blue" Phases 520
3. **Smectic Liquid-Crystal Phases**........ 521
3.1 Smectic-*A* Phases........ 521
3.1.1 Analogy to Superconductivity 521
3.1.2 Polymorphic Smectic-*A* and Smectic-*C* Phases........ 522
3.2 Smectic-*C* Phases........ 522
3.2.1 The Absence of True Long-Range Order........ 522
3.3 Ferroelectric Smectic-*C* Phases........ 523
3.4 Bond-Orientationally Ordered Phases........ 524
3.5 Higher-Order Smectic Phases.. 525
3.6 Applications of Smectic Phases........ 525
3.6.1 Smectic-*A* Devices........ 525
3.6.2 Ferroelectric Smectic-*C* Applications........ 526
4. **Discotic Liquid Crystals**...... 526
5. **Polymer and Polymer-Dispersed Liquid Crystals** 527
5.1 Polymer Nematic and Smectic Liquid Crystals........ 527
5.1.1 Mesogenic Main-Chain and Side-Chain Polymers........ 527
5.1.2 Applications of Polymer Liquid Crystals........ 528
5.2 Polymer-Dispersed Liquid Crystals........ 528
6. **Lyotropic Phases** 528
6.1 Phases with Ordered Micelles........ 529
6.2 Lyotropic Lamellar Phases.... 529
6.3 Applications of Lyotropic Liquid Crystals........ 530
7. **Conclusions** 530
Acknowledgments 531
Glossary........ 530
Works Cited 531
Further Reading........ 532

INTRODUCTION

True liquid crystals are fluids, in the sense that they cannot support shear stress. They possess order intermediate between the high degree of order of a perfect crystal and the disorder of a common isotropic liquid. For this reason they are often called *mesophases*, and molecules that tend to form liquid crystals are said to be mesogenic. The first reported observation of the liquid crystalline state was by the Austrian botanist Friederich Reinitzer (1888). In the intervening century these materials have been investigated not only for their potential applications, but also for their intellectual interest as a window into the variety of phases that can exist in condensed matter. The latter was recognized by the award of the 1991 Nobel Prize in physics to Pierre-Gilles de Gennes of the Collège de France.

1-56081-067-X/94/$5.00+.50

In this article we shall describe the nature of the order found in the various known liquid crystal phases, their structure, and some of their optical and viscoelastic properties. As this is an encyclopedia of *applied* physics, we shall pay special attention to their applications in display devices or for other purposes.

The many liquid-crystal phases can be divided broadly into two classes. The first liquid crystals discovered, and the most studied, are those classified as *thermotropic*. These are commonly formed from a single organic compound, and the liquid-crystal phases are obtained as the temperature is changed. Many thermotropic phases are formed from "mixtures" of different compounds in order to obtain the liquid-crystal order over a desired temperature range, but these are really solutions in which the different components are distributed quite uniformly. Thermotropic liquid crystals can also be formed from polymers. These are usually classified into two types, depending on whether the polymer main-chain or attached side-chain units are mesogenic.

Most commonly, molecules exhibiting liquid-crystalline phases are rodlike, but disk-shaped molecules have also been found to have liquid-crystal phases (Chandrasekhar *et al.*, 1979). The important thing is that the molecules should have considerable shape anisotropy. For example, for mesomorphic rodlike molecules, the ratio of the length of the molecule to its breadth is typically about 5:1. Two typical thermotropic molecules are shown in Fig. 1. The thermotropic phases are broadly classified as nematic, cholesteric, and smectic; about a dozen different smectic phases have been identified.

Liquid crystals that involve molecules in solution, usually amphiphiles dissolved in water, are called *lyotropic*. As they are systems with at least two components, this provides an extra degree of freedom. The types of order that occur in lyotropic phases are analogous to those in thermotropic phases, but usually involve the ordering of aggregates of molecules. Materials that show lyotropic order may also, with somewhat different amounts of solvent or solute, show non-liquid-crystal phases, such as microemulsions, that are technologically useful.

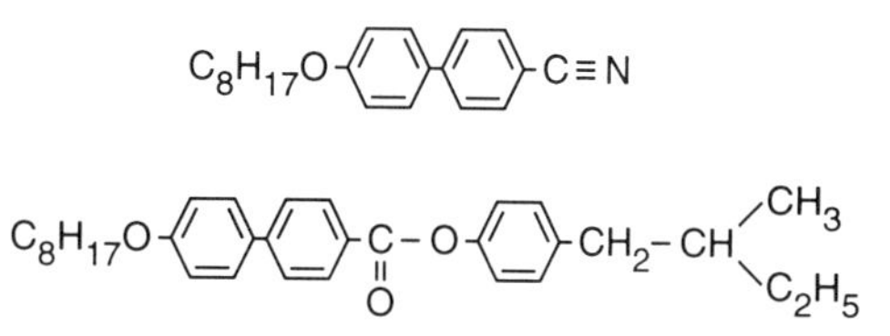

FIG. 1. The structure of two typical thermotropic liquid-crystal molecules.

1. NEMATIC LIQUID CRYSTALS

All liquid crystal phases have long-range orientational order of the molecules. The nematic phase, which is the simplest phase, has only this type of order to distinguish it from a disordered isotropic liquid. The name nematic comes from the Greek word for thread, and was given to this phase because defects in the order, somewhat analogous to the domain walls of ferromagnets, resemble threads when observed under a microscope. A schematic of the molecular order in a nematic phase is shown in Fig. 2. While the molecules are orientationally ordered, the nematic phase has continuous translational symmetry.

1.1 Optical Properties of Nematics

The optical polarizability of the nematic molecules is anisotropic, and so a phase in which molecules tend to orient parallel to one another will be birefringent. Most nematics are uniaxial; that is, there is a single optic axis that the molecules tend to align parallel to, and the material has rotational symmetry about that axis. The optic axis of the material is usually denoted by a unit vector field **n**, called the director, which is also the principal axis of the refractive-index tensor. (Recently, biaxial lyotropic nematics have been discovered; these would require two directors.) Choosing **n** to lie along the z direction, we may let θ be the angle between the symmetry axis of a molecule and the director as shown in Fig. 3. The nematic order may be described quantitatively by the order parameter

$$\mathscr{S}=\tfrac{1}{2}\langle 3\cos^2\theta-1\rangle, \quad (1)$$

where the average is taken over a small, but macroscopic, volume. In Eq. (1), $\mathscr{S}$ has been defined to be 0 for randomly oriented molecules and 1 if all were perfectly aligned. It is also possible for $\mathscr{S}$ to be negative; the minimum value $\mathscr{S}$ can have is $-\frac{1}{2}$, which occurs when all molecules are randomly oriented in

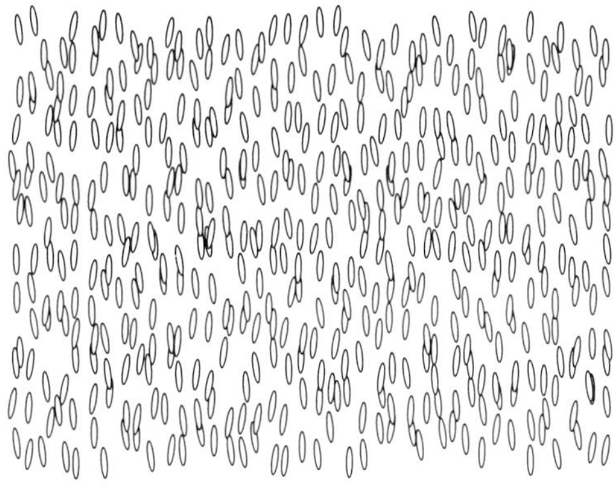

FIG. 2. Schematic of the local arrangements of molecules in a nematic liquid-crystal phase.

a plane perpendicular to the optic axis. The molecules also have an anisotropic diamagnetic susceptibility, largely due to the benzene rings most of them contain, and the diamagnetic susceptibility of the nematic will be given by

$$\chi_{\alpha\beta}=\tfrac{1}{3}\Delta\chi\mathcal{S}(3n_\alpha n_\beta-\delta_{\alpha\beta}), \qquad (2)$$

where the subscripts α and β denote Cartesian components, and $\Delta\chi$ would be the anisotropy for a perfectly ordered nematic. A similar phenomenological relation can be defined to represent the optical birefringence by expressing the dielectric constant tensor as

$$\epsilon_{\alpha\beta}=\tfrac{1}{3}\Delta\epsilon\mathcal{S}(3n_\alpha n_\beta-\delta_{\alpha\beta}). \qquad (3)$$

Equation (2) provides, in principle, a means to determine $\mathcal{S}$ from a macroscopic measurement if the fundamental diamagnetic properties of the molecule are known. The uncertainty of local field corrections makes it impossible to relate $\Delta\epsilon$ to molecular properties, and so Eq. (3) cannot be used in the same way. However, Eq. (3) does provide a quantitative description of the birefringence of nematic phases, which is large; a typical nematic has $\Delta\epsilon\simeq 1$. Because nematics are fluids, the birefringence can be modified by relatively modest applied forces, which makes them useful for displays.

1.2 Viscoelastic Properties of Nematics

The effect of electric fields, or other perturbations, on nematic liquid crystals is usually described in terms of distortions of the director field **n**. The lowest energy state of the nematic occurs when **n** is spatially uniform. Although the molecules themselves are not end-for-end symmetrical, all known nematic phases are unchanged if the direction of **n** is reversed. (If this were not so, they would be ferroelectric, and no ferroelectric nematics have been discovered.) A general distortion can be expressed in terms of three fundamental distortions commonly called splay, twist, and bend. Mathematically, these are represented by $(\nabla\cdot\mathbf{n})^2$, $(\mathbf{n}\cdot\nabla\times\mathbf{n})^2$, and $(\mathbf{n}\times\nabla\times\mathbf{n})^2$, respectively (de Gennes, 1974). The energy per unit volume of a distorted nematic in the presence of applied electric and magnetic fields is

$$\begin{aligned}\mathcal{F}_{el}=&\tfrac{1}{2}[K_1(\nabla\cdot\mathbf{n})^2+K_2(\mathbf{n}\cdot\nabla\times\mathbf{n})^2\\&+K_3(\mathbf{n}\times\nabla\times\mathbf{n})^2]+\tfrac{1}{2}\epsilon_a(\mathbf{n}\cdot\mathbf{E})^2\\&+\tfrac{1}{2}\chi_a(\mathbf{n}\cdot\mathbf{H})^2,\end{aligned} \qquad (4)$$

where ϵ_a is the anisotropy of the dielectric constant at low frequencies, $\chi_a=\chi_\parallel-\chi_\perp=\Delta\chi\mathcal{S}$ from Eq. (2), and K_1, K_2, and K_3, are called the splay, twist, and bend elastic constants. Typical values for the elastic constants are $\sim 10^{-6}$ dyn, χ_a is about 10^{-7} cgs units for molecules with benzene rings, and ϵ_a is approximately 10 for a molecule with a cyano group, like the upper one in Fig. 1. Equation (4) can be used to calculate the size of applied

FIG. 3. The coordinates used to describe the orientation of nematic liquid-crystal molecules.

electric or magnetic fields required to cause changes in the orientation of the optic axis **n** (de Gennes, 1974). The sign of χ_a for most nematics is such that the director will orient parallel to an applied magnetic field. A molecule like the upper one in Fig. 1 has a large dipole moment due to the cyano group, and will readily orient parallel to an applied electric field. There are also nematic materials whose optic axis will orient perpendicular to an electric field, but they are little used in modern devices.

The viscosity of nematic liquid crystals is also a tensor, with five independent components (de Gennes, 1974). All of the components are roughly of equal size, ~10 cP, and they, in combination with the elastic constants, determine the frequencies of the director normal modes (Martin *et al.*, 1972) and the response time of devices using nematic liquid crystals. Thermal excitation of these normal modes gives rise to substantial scattering of light, so that the mean free path of a photon can be less than 1 mm. The attenuation of light by scattering in even perfectly aligned nematic materials can be an important consideration in the design of optical devices using these materials.

1.3 Device Applications of Nematics

Most liquid-crystal display (LCD) applications use the liquid crystal to alter the transmission of light through the device. A number of effects have been investigated for use in devices. One of these is the "guest–host" display, in which strongly absorbing dye molecules are dissolved in a nematic liquid crystal that orients them; by reorienting the director variable dichroism can be obtained. However, the dominant type of liquid-crystal display in use today is the kind known as "twisted nematic" (Schadt and Helfrich, 1971), or TN, illustrated in Fig. 4. These devices have become practical thanks to the pioneering work of Gray and co-workers (Gray, 1975) at the University of Hull in England in developing materials with the required properties.

1.3.1 Twisted Nematic Displays. The display cell consists of nematic material between two transparent plates separated by a distance d, which is typically about 10 μm. The inner surface of each plate is coated with a transparent conductor such as indium-tin oxide so that an electric field may be applied across the liquid crystal. The plate surfaces are also treated to orient the director of the material in contact with them so that the optic axis will follow a particular direction in the plane of the plate. The orientation directions of the two plates are generally set at a 90° angle, which causes the director of the material between the plates to follow smoothly a 90° twist. The cell is generally used to transmit light and placed between a pair of polarizers. Light incident upon the cell will have its plane of polarization rotated to follow the twist as it passes through the cell; it will pass through if the polarizers are crossed and be blocked if they are parallel. The materials used have $\epsilon_a > 0$, so that a voltage applied between the electrodes will try to align the molecules normal to the plates. Those in contact with the plates will retain their orientation, but at a rather sharply defined threshold most of the liquid crystal between the plates will switch orientation. When this happens, the cell will no longer rotate the polarization of light passing through it. By placing the cell between crossed or parallel polarizers, one may choose to have the electric field either block or permit the transmission of light. The chief advantage of this display is that the liquid crystals are excellent electrical insulators, so that the power consumption is very small. Another advantage is that the response is determined by the voltage between the electrodes, rather than the electric field, so that TN devices are less sensitive to minor variations in the electrode spacing than those using ferroelectric liquid crystals. Several considerations are important for practical TN devices. To avoid electrochemical polarization effects at the electrodes, low-

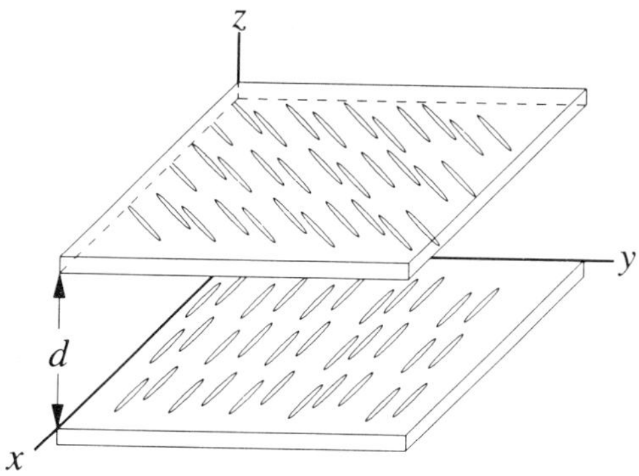

FIG. 4. Schematic of a twisted nematic display cell.

frequency ac fields must be used, rather than dc. The liquid crystal is often doped with small amounts of a chiral material to influence the choice of the twist direction. High information content displays for computer screens often employ a twist angle greater than 90°, frequently as high as 270°. Known as "supertwist," this improves the threshold response and makes possible greater levels of multiplexing and good contrast over a wider viewing angle. The best displays now use thin-film transistors at each pixel to improve performance.

2. CHOLESTERIC LIQUID CRYSTALS

The name "cholesteric" comes from the historical fact that for a long time most of the known examples of this phase were esters of cholesterol. The molecules of cholesteric phases tend to order parallel to one another as the molecules of nematics do, but they are formed from chiral molecules. The chirality of the molecules causes them to pack in a locally helical structure; in the ordered liquid crystal phase there is a macroscopic precession of the director throughout the sample. Cholesterics have a helical axis to which the director is perpendicular and around which it precesses. The distance travelled along the helical axis for one full turn of the director is called the pitch. Both right- and left-hand cholesterics exist, and the length of the pitch is often comparable to the wavelength of visible light. A cholesteric phase formed from a racemic mixture (equal numbers of molecules with right- and left-handed chirality) will be nematic, and, thermodynamically, cholesteric liquid crystals are a class of nematic.

2.1 Optical Properties of Cholesterics

The magnitude of the pitch plays an important role in determining the optical, dielectric, and magnetic properties of cholesteric phases. Even when disordered, most cholesteric materials have a high value of optical activity; polarization rotations of 300° mm^{-1} are common. Cholesteric phases have a strong circular birefringence, and propagation of light through them (Chandrasekhar, 1977) can exhibit unusual effects. When the pitch is comparable to the wavelength of light, wavelength- and polarization-selective reflections occur, so that cholesteric materials often display bright colors when reflecting white light. The color is determined by the pitch, and is used in applications we shall now discuss.

2.2 Response of Cholesterics to Applied Fields

Because cholesterics are nematics with a macroscopic twist to the director, their response to applied fields can be discussed using a modified form of Eq. (4) in which the twist term is replaced by $K_2(\mathbf{n}\cdot\nabla\times\mathbf{n}+q_c)^2$, where the pitch length $2\pi/q_c$ is long compared to molecular dimensions. The esters of cholesterol have diamagnetic anisotropies about 100 times smaller than those of common nematic materials, and the effects of magnetic fields on them are correspondingly weaker. More recently developed materials in which the chirality comes from a branched chain, as seen in the second molecule in Fig. 1, have greater response to magnetic fields. For cholesterics, we should remember that the susceptibility for fields along the helical axis is $\chi_\perp$ or $\epsilon_\perp$ of Eqs. (2) and (3). For fields perpendicular to the pitch direction, the average susceptibilities $(\chi_\parallel+\chi_\perp)/2$ and $(\epsilon_\parallel+\epsilon_\perp)/2$ will be observed. So, in applying Eq. (4) to cholesterics, $\chi_a=(\chi_\parallel-\chi_\perp)/2$ and $\epsilon_a=(\epsilon_\parallel-\epsilon_\perp)/2$ should be used. If χ_a or ϵ_a is negative, the director will align perpendicular to the appropriate field and the pitch direction will align parallel to it. When χ_a or ϵ_a is positive, the system will align with the pitch direction normal to the field, and the effect of the field will be to unwind the helix. In fact, the pitch diverges logarithmically to ∞ at a critical field that is inversely proportional to the unperturbed pitch. At this point, the cholesteric has been converted to a nematic phase by the field.

Because the pitch modulates a number of properties of the cholesteric phase as one travels along the pitch direction, cholesterics can exhibit viscoelastic behavior analogous to smectic phases. These are discussed by de Gennes (1974).

2.3 Applications of Cholesterics

Most of the applications cholesterics have found to date stem from their selective reflection of colors of light. The color reflected depends upon the pitch, which can vary with temperature through a process analogous to thermal expansion. Cholesteric materials can therefore be used as thermometers with visible readout. They have found application in measuring skin temperatures and the detection of poor contacts and hot spots on printed circuit boards. When composite materials are subjected to a step in heat flux, the resulting surface temperature distribution can reveal information about internal flaws; it has been shown (Williams, 1983) that cholesteric liquid crystals can be used in this application to provide a nondestructive evaluation of flaws in fiberglass boat hulls. By synthesizing and mixing various materials, the temperature at which visible reflections occur can be varied from roughly 0 to 50 °C, and the temperature range over which the reflected color changes from red to blue can be adjusted from about 10 to 1 °C. The materials can also be obtained commercially encapsulated in micron-sized polymer capsules, which makes them easier to handle in many applications.

2.4 Cholesteric "Blue" Phases

An interesting puzzle that has been solved during the last decade is the nature of the "blue" phases exhibited by some cholesteric materials over a narrow temperature range just below their transition to the isotropic phase. The name derives from the bluish appearance they often have; they were first observed by Reinitzer (1888). Blue phases have proved interesting to theorists seeking to understand the significance and broader implications of the chiral term in the free energy, and they have provided a challenge to experimenters seeking to elucidate their structure (Collings, 1990).

By a combination of calculations and polarization-sensitive light scattering, it has been shown that the blue phases consist of an ordered lattice of defects in the cholesteric texture (the spatial orientation field of the director) that are able to accomodate a double twist. Three blue phases are known: blue phase I has body-centered symmetry, blue phase II is simple cubic, and the structure of blue phase III, sometimes called the "fog phase," is yet unknown. It turns out that the state in which the director throughout the sample precesses about a uniform pitch axis is not the lowest elastic free-energy state for these materials. The reason is that the cholesteric would have a lower free energy of interaction between the molecules if the director could twist in two directions at once. This is not possible topologically, but it can occur through the creation of singular line defects (regions where the order parameter goes to zero), which have, however, a high free-energy cost. The tendency to form a blue phase increases as the pitch becomes shorter. The blue phases exist only over narrow temperature ranges and have yet to find applications.

3. SMECTIC LIQUID-CRYSTAL PHASES

The smectic phases represent the next step towards crystalline order beyond the nematic phase. Each smectic phase is named by a letter; these were assigned roughly in the order in which they were discovered. At this writing, smectics *A* through *L* have been identified. In the past, the structure and molecular arrangements of many smectic phases were not known. Phases were identified by their appearance under a polarizing microscope, which gave some indication of the texture (the spatial orientation field of the director), and a new phase was identified as one which did not mix with other known smectic phases. The refusal to mix was a fairly reliable indication the phases had different symmetries. As more has been learned about the structure of smectics, it has become clear that about a third of them are not truly liquid-crystal phases, because they have three-dimensional long-range crystalline order. They are crystals with especially soft elastic constants. The smectic liquid crystal phases all have some translational order in addition to the molecular orientational long-range order of the nematic and cholesteric phases. Some also have a new kind of order, called bond-orientational order, which we shall now discuss.

3.1 Smectic-*A* Phases

The simplest smectic phase is called smectic-*A*. In this phase, the molecules tend to form "layers" that are perpendicular to the director, as illustrated in Fig. 5. The term "layer" is an overstatement of the ordering. In most smectic-*A* materials the layers are not well defined, and what one really has is more accurately described as a modulation of the density

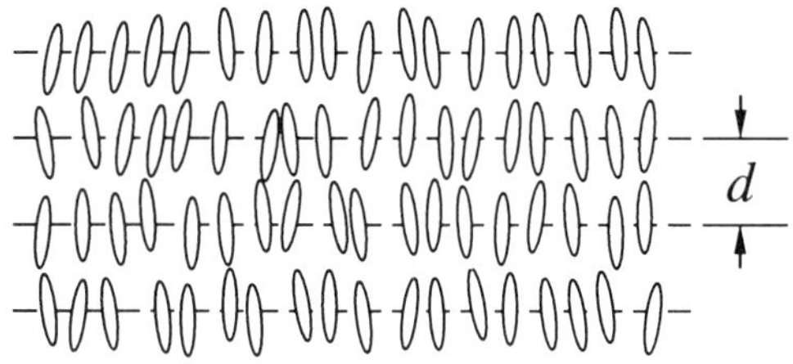

FIG. 5. Molecular arrangement in a smectic-A liquid crystal. The director is parallel to the density wave vector.

whose wave vector is along the director. In fact, the smectic-A order is commonly represented (de Gennes, 1972) by the order parameter ψ in the following expression for the density of the liquid:

$$\rho(\mathbf{r})=\rho_0[1+\text{Re }\psi(\mathbf{r})e^{iq_0z}], \qquad (5)$$

where $d = 2\pi/q_0$ is the wavelength of the density wave or the spacing of the smectic "layers." The order parameter ψ is a complex number with two degrees of freedom. The magnitude $|\psi|$ is the amplitude of the density modulation, and the phase represents a displacement of the "layers" in the wave-vector direction, which is usually taken to be z. Smectic-A order is really a one-dimensional density wave in a three-dimensional liquid, and can be said to be the simplest kind of freezing or melting to occur in nature. The smectic-A phase remains translationally invariant (liquid-like) in directions perpendicular to the density wave vector.

3.1.1 Analogy to Superconductivity. de Gennes (1972), Kobayashi (1970), and McMillan (1971) all recognized that the two-component order parameter of the smectic-A phase was similar to the order parameter used in the Ginzburg-Landau phenomenological model (Ginzburg and Landau, 1950) of superconductivity. The analogy may be seen through the following argument. Suppose, while moving through the smectic-A material from point A to point B, we compute the line integral

$$\frac{1}{d}\int_A^B \mathbf{n}\cdot d\mathbf{r}. \qquad (6)$$

This integral calculates the number of "layers" that have been crossed. If the smectic-A is free of defects, the result must be independent of the path followed, which means that the integral carried out around a closed path must yield 0. Applying Stokes's theorem, we find that $\nabla \times \mathbf{n} = 0$ in the smectic-A phase, and so, referring to Eq. (4), it is apparent that bend and twist director distortions are excluded from the smectic-A phase. This is analogous to the exclusion of magnetic fields from superconductors (the Meissner effect). In fact the phenomenological theories of the smectic-A to nematic phase transition and the superconducting to normal-state transition are mathematically similar. The liquid crystal director $\mathbf{n}$ is quite analogous to the magnetic vector potential $\mathbf{A}$ of the superconductor. As a smectic-A phase is approached in the nematic phase, the presence of smectic short-range order causes a divergence of the bend and twist elastic constants, which is the analog of fluctuation diamagnetism in superconductors. Recently a twisted grain-boundary phase has been observed (Goodby *et al.*, 1988) in smectic-A materials. This phase is analogous to a lattice of flux penetration vortices in superconductors (Toner, 1991).

3.1.2 Polymorphic Smectic-A and Smectic-C Phases. The smectic-A phase discussed above has been assumed to have a layer spacing d that is approximately equal to the length l of a molecule. However, if the molecules have a strong longitudinal dipole moment, such as the top cyanobiphenyl one in Fig. 1, there will be near-neighbor antiparallel correlations. These correlations can lead to different polymorphic forms of the basic smectic-A symmetry, which can be characterized by the ratio d/l. These have largely been elucidated through the pioneering work of the Bordeaux group (Sigaud *et al*, 1979; Hardouin *et al.*, 1983). Four forms of smectic-A are now recognized: smectic-A_1 with $d \simeq l$, smectic-A_2 with $d \simeq 2l$, smectic-A_d with $l<d<2l$, and smectic-$\tilde{A}$ which has the same local symmetry as smectic-A_2 but with a transverse dipolar modulation of the structure. The existence of these polymorphic smectic-A phases has been explained by a phenomenological model (Prost and Barois, 1983) that has two coupled order parameters, the mass-density wave and a dipolar-density wave. This approach has predicted novel critical phenomena associated with the transitions between different polymorphic types. Many of these predictions have been verified experimentally (Garland, 1990). Analogous modulated variants of the smectic-C phase, introduced below, have also been found to exist (Hardouin *et al.*, 1983; Huang *et al.*, 1984; Safinya *et al.*, 1986a).

3.2 Smectic-*C* Phases

In smectic-*C* liquid crystals the molecules are disordered within the layers as in a smectic-*A* phase, but they are also tilted with respect to the layer normal (Fig. 6); in other words, the density wave vector and the director are not parallel. This means that smectic-*C* phases are optically biaxial; often this complication can be approximated by simply referring to separate components of a director in the plane of the smectic layers and normal to them. In a well-established smectic-*C* phase enough energy is required to change the tilt angle with respect to the layers (it is coupled to the layer thickness) that it fluctuates very little. The component of the smectic-*C* director in the plane of the layers is almost as free to fluctuate in orientation as the director of a nematic phase, and these fluctuations provide a time and space dependence to the refractive indices that can scatter light strongly.

In many cases, the smectic-*C* phase will form from a smectic-*A* by a continuous tilt of the molecules. This is a second-order phase transition that has the same symmetry properties as transition to a superfluid phase in helium. It should, therefore, exhibit critical behavior. Careful experiments have shown, however, that the growth of the tilt can be described by a mean-field model (Birgeneau *et al.*, 1983). The reason for this is that a large bare coherence length reduces the effect of the thermal fluctuations responsible for critical behavior to a region very close to the transition temperature (Ginzburg, 1960). In some materials, the smectic-*C* phase forms directly from the nematic without an intervening smectic-*A* phase. In these cases the tilt angle is usually large, about 45°, and only weakly temperature dependent. The smectic-*C* to nematic transition is first order because of the effect of thermal fluctuations.

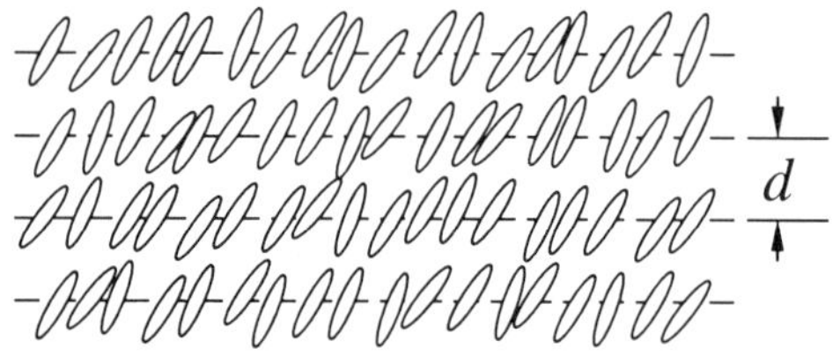

FIG. 6. Molecular arrangement in a smectic-*C* liquid crystal. The director is tilted with respect to the density-wave vector.

3.2.1 The Absence of True Long-Range Order.

As the number of spatial dimensions is reduced, ordered phases lose their order. No long-range order is believed possible in one dimension, because the entropy of the disordered state is overwhelmingly higher. Many of the common three-dimensionally ordered phases cannot have long-range order in two dimensions at finite temperature. That is because thermally excited fluctuations will destroy the order. Solids, for example, cannot exist in two or fewer dimensions because the fluctuations of the atomic positions in the crystal lattice due to thermally excited lattice vibrations become infinite in two dimensions (Landau and Lifshitz, 1980). In the language of phase transitions, we say that the lower marginal dimensionality for the existence of solid crystals is two.

Smectic-*A* and smectic-*C* liquid crystals are examples of a one-dimensional density wave in a three-dimensional liquid. The marginal dimensionality for such a material is three (Caillé, 1972), so these smectic phases can, strictly speaking, only exist in more than three dimensions. Precisely at the marginal dimensionality, the effects of the fluctuations are subtle. Whether or not long-range order exists can be discussed quantitatively by defining a correlation function

$$\mathscr{G}(\mathbf{r}) = \langle e^{i\mathbf{G}\cdot[\mathbf{u}(\mathbf{0})-\mathbf{u}(\mathbf{r})]}\rangle, \tag{7}$$

where $\mathbf{G}$ is a reciprocal-lattice vector of the crystal, and $\mathbf{u}(\mathbf{r})$ is the displacement of an atom at $\mathbf{r}$ from its equilibrium position. In the case of a smectic-*A* phase, $\mathbf{G}$ would be the smectic density wave vector q_0 and $\mathbf{u}$ the displacement of a smectic layer from its proper position. This function is useful because its Fourier transform gives the x-ray Bragg scattering from the crystal lattice or smectic density wave. If there is long-range order, $\mathscr{G}(\mathbf{r})$ will have a finite value as $r \to \infty$. If there is only short-range order, $\mathscr{G}(\mathbf{r})$ decays exponentially to zero as $e^{-\kappa r}$. At the marginal dimensionality, $\mathscr{G}(\mathbf{r})$ goes to zero at large distances, but does so algebraically as $r^{-\eta}$, where η is a small number ~ 0.1. When there is long-range order, the Bragg scattering will give a resolution-limited peak. Short-range order will give a broad peak whose width is proportional to κ. The algebraic decay at the marginal dimensionality gives a scattering peak that is a cusp and is very difficult to distinguish experimentally from the resolution-limited peak of true long-range order. The experimental trick is to use a high-resolution spectrometer with a sharply decaying resolution function (Als-Nielsen *et al.*, 1980) obtained from multiple Bragg reflections in perfect collimating crystals. Most of the properties of smectic-*A* and smectic-*C* liquid crystals are not significantly different than they would be if the order were truly long range. However, they have been of fundamental interest to demonstrate the existence of marginal dimensionality and to study in three dimensions behavior that occurs only in two dimensions in most other materials.

3.3 Ferroelectric Smectic-*C* Phases

Many liquid crystal molecules have dipole moments, and there appears to be no intrinsic reason why nematic phases could not be ferroelectric, but none has ever been observed. Meyer realized that it should be possible to obtain ferroelectricity in a smectic-*C* liquid crystal if the molecules were chiral (Meyer *et al.*, 1975). We can illustrate the idea by referring to Fig. 7. In the figure we represent the chirality by choosing molecules shaped like a hockey stick with a dot on the left side. Suppose the dot represents an electric dipole pointing from the side of the molecule. The second molecule in Fig. 1 is, in fact such a molecule. The top panel, Fig. 7(a), shows a smectic-*A* phase of these molecules. Two alternatives for a smectic-*C* phase are shown in the other two panels of the figure. Both have the end-for-end symmetry of liquid crystal phases, but the arrangement in Fig. 7(c) permits a much closer packing of the molecules and is the one that might be expected to occur once the molecules tilt. It also has the dots on the molecules visible and will have a net electric dipole moment in the plane of the smectic-*C* layers. This is the origin of the ferroelectricity that enables applied fields to control the molecular orientation and leads to the development of a device we shall discuss presently. Ferroelectric smectic-*C* phases are commonly referred to by the notation smectic-*C**. In a racemic mixture, half the molecules would have the dot on the left side, and half would have it on the right; hence there would be no net dipole moment.

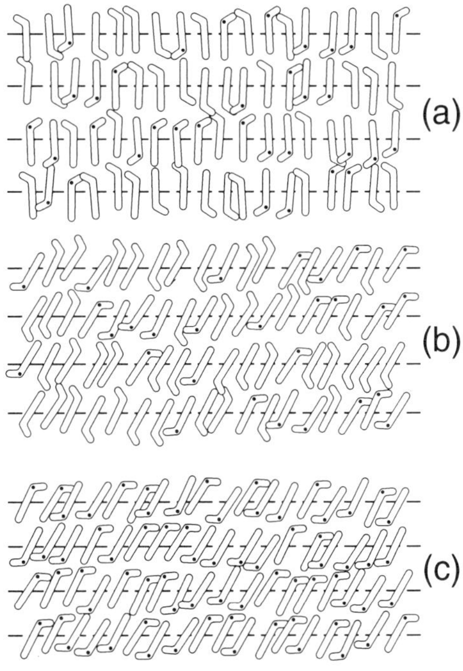

FIG. 7. Smectic phases of chiral molecules. (*a*) A smectic-*A* phase, (*b*) a tilted smectic-*C* phase with no ferroelectric order, and (*c*) a ferroelectric smectic-*C* phase.

3.4 Bond-Orientationally Ordered Phases

We mentioned earlier that the structure of many smectic liquid crystal phases was unknown until recently. The concept of bond-orientational order was the missing idea that was necessary to characterize the structure of the different smectic phases. The idea originated in theoretical investigations of melting in two dimensions. Liquids have both continuous rotational and continuous translational symmetries. In crystalline solids, both of these symmetries have been "broken" and are discrete. As we mentioned in Sec. 3.2.1 above, solids cannot have true long-range order in two dimensions. However, they can have a phase, different from a two-dimensional liquid, which has power-law decay of the correlation function given in Eq. (7). This is commonly called a Landau–Peierls state, in order to recognize earlier work of Peierls and Landau on the topic. In 1973, Kosterlitz and Thouless proposed a theory of melting in two dimensions, a dislocation-mediated transition between a two-dimensional Landau–Peierls state and a two-dimensional liquid. This was a model for breaking the continuous translational symmetry. Extending it to consider the rotational symmetry as well, Halperin, Nelson, and Young predicted the existence of an intermediate "hexatic" phase that had the discrete rotational symmetry of a crystal combined with the continuous translational symmetry of a liquid. The idea is illustrated by the sketch shown in Fig. 8. Even in a two-dimensional liquid, one might expect local arrangement of the molecules to be in hexagonal clusters whose size is a correlation length ξ, as shown in Fig. 8. There would, however, be no correlation between the positions or orientations of such clusters if they are separated by more than ξ in distance. If the material were to form a hexagonal crystal, such local clusters would be expected to have the same orientation of the imaginary lines joining near neighbors, and to be separated by an integral number of lattice vectors. In effect the translational and rotational corelation lengths have become infinite. In the hexatic phase, which has "bond orientational" order, there is no positional correlation between widely separated clusters, but lines joining near neighbors have the same orientation throughout the sample. Only the rotational correlation length is infinite. Thus the rotational symmetry is discrete although the continuous translational symmetry remains unbroken. It was recognized by Landau, and subsequently by Mermin, that it was theoretically possible for a phase with these properties to exist in nature. In effect, it would be a liquid in which one knew in which directions it was necessary to go to encounter a neighboring molecule, and these directions were the same throughout the liquid, but one did not know how far to go. Birgeneau and Litster (1978) proposed that extending these two-dimensional ideas to three dimensions might provide an explanation for some of the higher-ordered smectic phases, such as smectic-*B*, whose structure was unknown. They proposed that within the smectic layers there was bond-orientational order, but no translational order.

Although some smectic-*B* phases have three-dimensional long-range translational order (Pershan *et al.*, 1979), others have been found with only long-range bond-orientational order (Moncton and Pindak, 1979; Pindak *et al.*, 1981). These are usually called smectic-B_H to indicate the presence of hexatic order. Tilted analogs of the smectic-B_H are also known; these are the smectic-*F* and smectic-*I* phases. In the smectic-*I* the molecules tilt towards the position of their nearest neighbors, while in the case of smectic-*F* phases the tilt direction is between the nearest neighbors. There are no interlayer correlations in the smectic-B_H, but

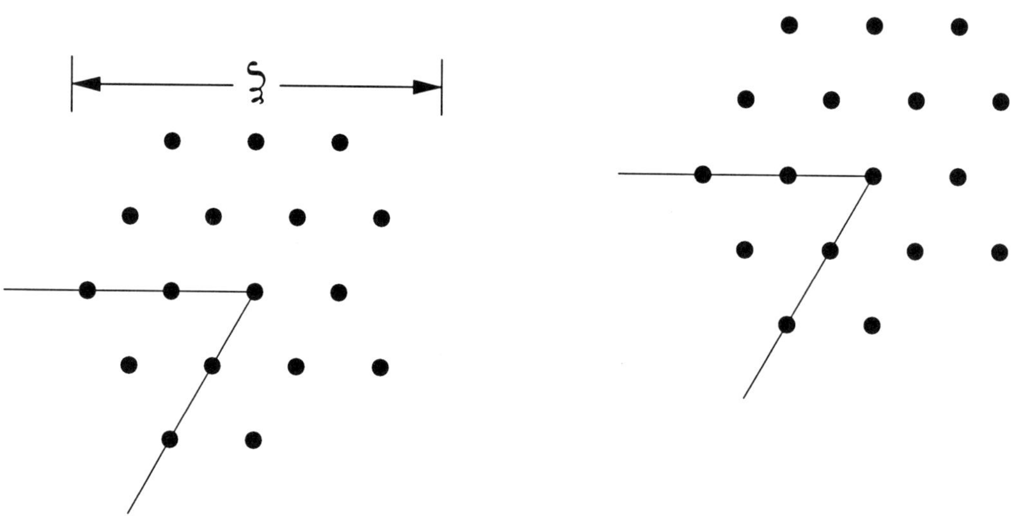

FIG. 8. Sketch to illustrate the concept of bond-orientational order. The two clusters have the same orientation but their positions are not correlated.

in the tilted phases the tilt applies an effective field to the bond-orientational order and causes interlayer correlations. In principle, the tilt will induce bond-orientational order in the smectic-*C* phase, changing it to an *F* or *I* phase. Strictly speaking, therefore, the smectic-*C* phase does not exist. This effect, so far, has been too small to detect and seems to have few practical consequences.

The smectic phases with bond-orientational order exist only over a narrow temperature range, and have not seen applications. They have, however, been studied extensively (Brock *et al.*, 1989) in order to elucidate the properties of a phase of matter whose existence would have greatly surprised most solid state physicists 15 years ago.

3.5 Higher-Order Smectic Phases

In addition to the fluid phases described so far, there are a number of smectic phases with true long-range translational order in three dimensions. These are the smectic-*B*, -*G*, -*J*, -*E*, and -*K* phases. There are differences in the molecular tilt, stacking of layers, and the extent of rotational disorder about the long axis of the molecules among these phases, but they are not true liquid crystals. They are crystalline solids, although in many cases the elastic constants can be as much as 10^6 weaker than the elastic constants of more common crystals.

3.6 Applications of Smectic Phases

3.6.1 Smectic-*A* Devices. Although the higher viscosity and greater force required to alter the texture of smectic-*A* phases renders them less suited for electro-optic displays, these are features that make them attractive for different types of memory devices (Coates, 1990). In one type of memory device the smectic-*A* is aligned "homeotropically" (molecules perpendicular to the substrate surface). This will provide a clear state for transmission of light between two glass substrates. A focused laser beam is used to heat a small area of liquid crystal into the isotropic state. When the laser is turned off, the liquid crystal cools rapidly to the smectic-*A* phase. The rapid cooling results in misalignment and the different texture scatters light, thus storing information that can be read out optically. The texture can be restored to homeotropic, and the stored information erased, either by the application of an electric field or by heating the entire liquid crystal to the isotropic state and cooling slowly. This makes possible a very high–information-content, although not particularly fast, read-and-write optical storage device.

The electroclinic effect in chiral smectic-*A* liquid crystals, such as in Fig. 7(a), can also provide device capability. In the electroclinic effect (Garoff and Meyer, 1977) an applied electric field can induce a tilt in the optic axis that is proportional to the applied voltage. This can, in principle, make possible features such as variable gray scale and analog phase modulation in a chiral smectic-*A* device.

3.6.2 Ferroelectric Smectic-*C* Applications. Ferroelectric smectic-*C** liquid crystals (FLCs) appear to have excellent potential for applications in fast electro-optic switching devices with memory. In such a device the basic configuration is the surface-stabilized ferroelectric liquid crystal (SSFLC) geometry in which the smectic-*C** material is contained between two transparent electrodes separated by about 2μm. The surfaces are treated to obtain a planar orientation of the molecules. Because they are chiral, smectic-*C** phases will have a precessing optical axis analogous to cholesteric phases. But, if the thickness of the SSFLC cell is less than the pitch of the optical axis helix, the interaction with the surfaces will unwind the helix. This produces a "bookshelf" geometry of Fig. 9 in which the smectic layers are perpendicular to the substrates and the molecular orientation is in the plane of the substrate making the tilt angle ω with respect to the layer normal. By reversing the polarity of the applied electric field, the molecules can be caused to flip to an orientation $-\omega$ with respect to the layer normal, hence rotating the optic axis by 2ω. The optimal value for ω is 22.5°, and switching times as fast as 0.1 μs have been demonstrated (Clark and Lagerwall, 1980). The switched state is also stable, and so the device provides storage, which greatly reduces the multiplexing problems of TN (twisted nematic) devices. The small electrode spacing and the fact that the response is determined by the electric field, not the applied voltage, has made these devices more difficult to manufacture. Efforts are also under way to synthesize ferroelectric liquid crystals with a significant dipole mo-

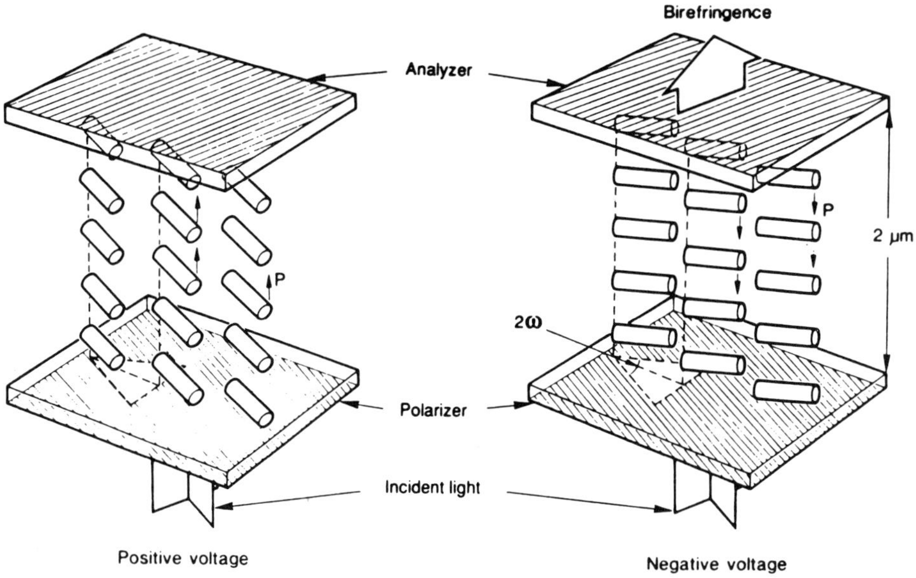

FIG. 9. The "bookshelf" geometry of a surface-stabilized ferroelectric liquid crystal display cell.

ment parallel to the long axis of the molecule; if the efforts are successful these materials could play an important role in FLC devices. Despite their fast switching times and the storage capability, SSFLC devices do not yet appear to have mounted a serious challenge to TN displays. However, this may change; at least one company (Canon Electric Company) has introduced ferroelectric liquid-crystal displays for its notebook computers.

4. DISCOTIC LIQUID CRYSTALS

Until recently it was believed that a prerequisite for the existence of a thermotropic liquid crystal phase was a rodlike shape of the molecules. However, liquid crystal phases have been demonstrated for disklike molecules (Chandrasekhar, 1979). Generally, the molecules have flat cores with long-chain substituents attached to the core. These molecules have been shown to exhibit two distinct types of mesophases: columnar and nematic. In the nematic phase, the disks are orientationally ordered without any positional order.

The columnar phases are formed by the molecules being stacked in columns, like poker chips, and then the columns are arranged in a hexagonal or rectangular fashion. A sketch of the local structure of a columnar discotic phase is shown in Fig. 10. Unlike the nematic phases of rodlike molecules, discotic nematics are usually optically negative ($\epsilon_{\parallel} < \epsilon_{\perp}$). Tilted columnar phases also exist. There is no translational order along the length of the columns. These phases are analogous to the smectic phases, except that they have a two-dimensional density wave in a three-dimensional liquid. This arrangement lacks the Landau-Peierls instability of the smectic-*A* phase and can have true long-range order.

5. POLYMER AND POLYMER-DISPERSED LIQUID CRYSTALS

5.1 Polymer Nematic and Smectic Liquid Crystals

5.1.1 Mesogenic Main-Chain and Side-Chain Polymers. Thermotropic liquid-crystal polymers evolved by researchers incorporating mesogenic monomers into polymer chains (Donald, 1992). They can be incorporated into the main polymer backbone chain, or by adding mesogenic side chains. Polymers with disklike mesogenic units are also known. The main-chain liquid crystal polymers tend to have very high melting temperatures and to be very slow in response to applied electric or

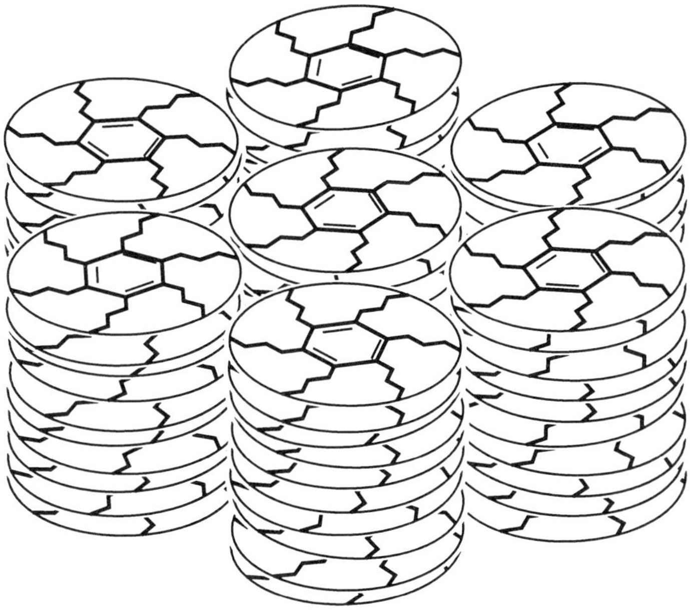

FIG. 10. Schematic of molecular arrangements in a columnar discotic liquid-crystal phase.

magnetic fields. However, one of the viscoelastic features of nematic liquid crystals is a very strong tendency to flow alignment. This can be exploited in polymer processing to produce fibers in which the polymers are oriented; such materials can have a much greater elastic modulus than those prepared conventionally from an isotropic melt. The best known example is Dupont's Kevlar®, which is spun from a nematic phase and has an elastic modulus comparable to that of steel.

The polymers with mesogenic side chains often offer response times to applied fields that are comparable to those of small-molecule liquid crystals. They can have not only nematic, but also cholesteric and smectic phases. Their viscosities and glass transition temperatures are much lower than main-chain polymers. They can be aligned easily by the same techniques (surface rubbing, skew evaporation, and surface texturing of the substrate) used for small-molecule liquid crystals. There is a potential for the polymer main chain to provide added stability to the liquid-crystal order, perhaps preserving some of the features in a solid phase, or to assist in obtaining better alignment of the liquid-crystal phase.

More recently, combined liquid crystalline polymers with mesogenic groups in both the main-chain and side groups have been synthesized. The degree of coupling of the main-chain and side-chain mesogenic groups depends on the molecular "spacer" used to attach the side groups to the main chain.

5.1.2 Applications of Polymer Liquid Crystals. Most of the work on polymeric liquid crystals has been motivated by the belief that these materials have many potential technological applications. The example of Kevlar® has inspired much of the effort. Mesogenic side-chain polymers offer potential for electro-optic displays, nonlinear optical applications, and pyroelectric applications. They may be attractive candidates for erasable holographic elements. This is an area of high current activity, and we shall have to await the results to see how valuable they turn out to be.

5.2 Polymer-Dispersed Liquid Crystals

These materials were the serendipitous result of efforts to improve the stability of liquid crystals aligned for NMR studies (Doane *et al.*, 1986). Liquid crystals can be mixed with the components of an epoxy resin; as the epoxy cures, a phase separation takes place in which much of the liquid crystal is expelled from the epoxy matrix. The result is approximately micron-sized droplets of small-molecule nematic liquid crystals dispersed in a polymer binder. Polymer-dispersed liquid crystals

(PDLCs) are now produced in several ways: solvent-induced precipitation, thermally induced precipitation, emulsion polymerization, or phase separation.

A material with a positive dielectric anisotropy can be aligned by a field applied when the PDLC is sandwiched between transparent electrodes. Normally the nematic droplets will scatter light strongly, and the material appears like a translucent shutter. If the ordinary refractive index of the liquid crystal is matched to that of the polymer matrix, the scattering will disappear when the field aligns the director. PDLCs offer the potential for high-transmission, large-area light shutters and displays with low power consumption. The absence of polarizers increases the transmission considerably, which makes them attractive for projection displays.

6. LYOTROPIC PHASES

The lyotropic liquid crystal phases contain amphiphilic or surfactant molecules. Because one end of the molecule prefers to be in contact with water and the other prefers oil, the molecules will go to the oil-water interface of a "solution" that contains both. If only oil or water is present along with the surfactant, various structures will form in order to segregate the oil-loving moiety, say, from water. The simplest type of aggregate formed by surfactant molecules in water is a micelle, shown in Fig. 11. The hydrophobic tails of the molecules can minimize contact with the water by segregating themselves in the center of the micelle. There is a price in reduced entropy as well as crowding of the hydrophobic tails to be be paid when this structure is formed. Usually double-tailed surfactants do not form micelles. In an oil solution, inverted micelles can form in which the hydrophilic moiety goes to the center of the micelle. When oil, water, and surfactants are mixed a wide variety of phases can form (Fontell, 1981). Only some of these phases have liquid crystalline order, but all are probably closely related and their existence and properties are determined by the same physical principles (Israelachvili, 1985). The surfactant phases are interesting because, as in biological systems, entropy plays a dominant role in their properties and behavior; some also provide interesting examples of two-dimensional physics. Lyotropic liquid-crystal phases are not yet as well studied or fundamentally understood as their thermotropic counterparts are.

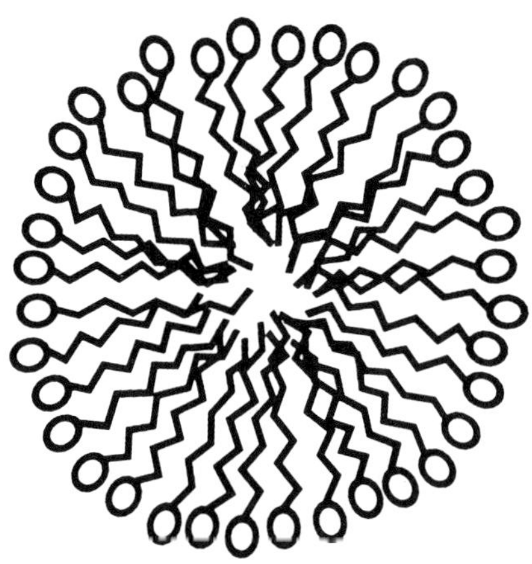

FIG. 11. Cross-sectional sketch of a micelle formed by amphiphilic molecules in water.

6.1 Phases with Ordered Micelles

Micelles are usually spherical, but under conditions that are not entirely understood, the micelle may not be spherical. Nonspherical micelles can order orientationally to form a nematic phase, and micellar nematic phases can be formed from either rodlike or disklike (Fujiwara and Reeves, 1980) micelles. As well as the usual hydrocarbon-tailed surfactants, micellar liquid crystals can be formed by surfactants with fluorocarbon tails (Tiddy, 1972; Boden *et al.*, 1979). The micellar nematic phases have viscoelastic properties (Haven *et al.*, 1981) and normal modes (Kumar *et al.*, 1983) that are closely similar to those of small-molecule thermotropic phases. In some materials the rodlike micelles can grow very long and order in a hexagonal array; these "middle soap" phases are probably the lyotropic analogs of the columnar discotic thermotropic phases. With chiral surfactants, lyotropic cholesteric phases are possible (Covello *et al.*, 1983).

6.2 Lyotropic Lamellar Phases

Many surfactants, especially biologically derived ones with double aliphatic tails, will form bilayers in aqueous solution. The phase formed depends sensitively on the spontaneous curvature of the bilayer. If it tends to be flat, then the system will form a lamellar structure as in Fig. 12; the resulting phase is closely similar to the thermotropic smectic-*A* phase; in the language of lyotropic liquid crystals (Tiddy, 1980), this phase is called L_α. If there is a tendency for the bilayer to curve, then interesting bicontinuous phases may form in which the bilayer forms a cubic lattice of intersecting curved surfaces (Gruner, 1989), sometimes referred to as the "plumber's nightmare."

The L_α phase has the same symmetry and normal-mode structure as the thermotropic smectic-*A* phase, and it also has the same lack of true long-range order in three dimensions (Safinya *et al.*, 1986b). The bilayers are more sharply defined than the smectic-*A* density wave, and the higher-order Bragg peaks are usually present in the x-ray scattering. With nonionic surfactants in water, such as the polyethylene glycol ethers (Strey *et al.*, 1990), or bilayers separated by oil (Appell *et al.*, 1989), it is possible to prepare highly swollen L_α phases in which bilayers a few nm thick are separated by many hundreds of nm. These systems are stabilized by repulsive forces between the bilayers, which can act over a very long range. In fact their properties are dominated by a previously unrecognized force predicted theoretically by Helfrich (1978). This force comes into play when any electrical forces between the bilayers are extremely weak and the bilayers are rather flexible. Its origin is entropic; the thermal fluctuations of the layers is sufficiently great that they repel through colliding with each other as they fluctuate. The highly swollen L_α phases are interesting systems in which to study the Helfrich force, as it is the dominant force stabilizing them. They are also interesting experimentally because Bragg scattering, which normally requires x rays, can be done with visible light. The energy resolution of 10^{15} available in light scattering enables a careful investigation of the dynamical behavior that one cannot obtain with x rays.

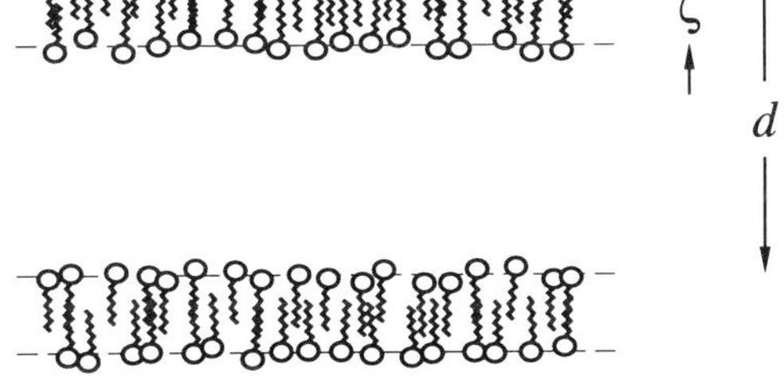

FIG. 12. A smectic-*A* lyotropic lamellar liquid-crystal phase. The surfactant bilayers of thickness ζ are separated by regions of water of thickness $d-\zeta$.

6.3 Applications of Lyotropic Liquid Crystals

Lyotropic liquid crystal phases usually occupy only a small part of the complicated phase diagrams that result when oil, water, salt, and one or more surfactants are combined. Surfactants are extraordinarily useful materials with many technological applications. These are usually connected with the ability of the surfactant molecules to reduce the interfacial energy (surface tension) between two liquids that would not ordinarily mix because of this high interfacial energy. A common example is the use of microemulsions in tertiary oil recovery. A microemulsion is a thermodynamically stable phase (Winsor, 1954; Kahlweit *et al.*, 1987) in which oil and water are separated by a random interface of surfactant. The surfactant can reduce the interfacial energy between oil and water, normally about 30 erg cm^{-2}, by 1000 times or more. This is important, for water is pumped into oil wells to force the oil out of small cracks in the rock, and reducing the oil–water surface tension greatly facilitates this process. The practice is not economical with current oil prices, but we shall undoubtedly see it at some time in the future.

The ordered liquid crystal lyotropic phases and their disordered relatives are interesting objects to study for those who seek a better understanding of classical statistical mechanics and random surfaces (Huse and Leibler, 1988), for they are systems in which thermal fluctuations and entropy play an important role. Understanding the ordered liquid-crystal phases can lead to better understanding of the forces and interactions that determine properties of other phases and materials containing biological surfactants such as phospholipids. For example, the viscoelastic properties of lyotropic liquid crystals can explain the shape of red blood cells (Deuling and Helfrich, 1976), and are believed to provide insight into biological membrane functions such as endocytosis, membrane fusion, and cell fission (Gruner, 1989) as well as the formation of vesicles and tubules (Lubensky and Prost, 1992). It also seems that studies of lyotropic liquid-crystal ordering should offer some insight into the self-assembly process (Deckman *et al.*, 1988) of Langmuir–Blodgett films.

7. CONCLUSIONS

In this article we have felt it was important to give an introduction to the structure of the various ordered phases, to indicate their role in the intellectual development of both physical chemistry and condensed matter physics, to point out special viscoelastic and optical properties of thermotropic liquid-crystal phases, and to comment upon the applications they have found. The reader should consult the literature references for more information. The journals of the Institute for Electrical and Electronic Engineers, the Society for Information Displays, and SPIE–The International Society for Optical Engineering (Jacobs, 1992) are especially good sources of information about display uses of thermotropic liquid crystals, which remains their major area of application.

ACKNOWLEDGMENTS

One of us (J.D.L.) would like to express thanks to the National Science Foundation for support from Grants No. DMR-9022933 and No. DMR-9014886 during the preparation of this article and for support of the research some of it is based on.

GLOSSARY

Amphiphile: A molecule consisting of two parts, one being soluble in water, the other in oil.

Birefringence: The condition of having two different refractive indices; the one observed depends on the polarization state of the light.

Bond-Orientational Order: A type of order in which the imaginary lines joining the centers of neighboring molecules are ordered, but the positions of the molecules are not. (See Sec. 3.4.)

Chirality: A chiral object is one which has either right- or left-handed symmetry, analogous to the difference between right- and left-handed screw threads.

Correlation Length: Sometimes called coherence length, the distance over which fluctuations in a quantity, usually an order parameter, are correlated. Commonly, the relative strength of correlations at two points separated by a distance r weakens as $e^{r/\xi}$, where ξ is the correlation length.

Director: A name given to a unit vector field that points along the local orientation of the optic axis of a liquid crystal. (See Sec. 1.1.)

Ferroelectric: A material that posesses a permanent electric polarization.

Mesogen: A term used to indicate a molecule that tends to form a liquid-crystal phase, which is sometimes called a mesophase. (Mesophase means "in-between phase.")

Micelle: Roughly sperical aggregate of amphiphilic molecules. (See Sec. 6.1.)

Optical Activity (Circular Birefringence): An optically active material is one whose refractive index differs for right and left circularly polarized light.

Order parameter: A quantity used to describe quantitatively the amount of order in a phase; see Sec. 1.1 for an example describing molecular orientational order in a nematic liquid crystal. It is defined to be zero in a disordered phase.

List of Works Cited

Als-Nielsen, J., Litster, J. D., Birgeneau, R. J., Kaplan, M., Safinya, C. R., Lindegaard-Andersen, A., Mathiesen, S. (1980), *Phys. Rev. B* **22**, 312–340.

Appell, J., Bassereau, P., Marignan, J., Porte, G. (1989), *Coll. Pol. Sci.* **267**, 600–647.

Birgeneau, R. J., Litster, J. D. (1978), *J. Phys. Lett.* **38**, 1399–1407.

Birgeneau, R. J., Garland, C. W., Kortan, A. R., Litster, J. D., Meichle, M., Ocko, B. M., Rosenblatt, C., Yu, L.-J. (1983), *Phys. Rev. A* **27**, 1251–1284.

Boden, N., Jackson, P. H., McMullen, K., Holmes, M. C. (1979), *Chem. Phys. Lett.* **65**, 476–511.

Brock, J. D., Birgeneau, R. J., Litster, J. D., Aharony, A. (1989), *Phys. Today,* **42** (7), 52–59.

Caillé, A. (1972), *C. R. Acad. Sci. Ser. B* **274**, 891–902.

Chandrasekhar, S. (1977), *Liquid Crystals,* Cambridge: Cambridge University Press.

Chandrasekhar, S., Sadashiva, B. K., Suresh, K. A., Madhusudana, N. V., Kumar, S., Shashidhar, R., Venkatesh, G. (1979), *J. Phys. (Paris) Colloq.* **40**, C3-120–3-124.

Clark, N. A., Lagerwall, S. T. (1980), *Appl. Phys. Lett.* **36**, 899–912.

Coates, D. (1990), in: B. Bahadur (Ed.), *Liquid Crystals: Applications and Uses,* Singapore: World Scientific, p. 275.

Collings, P. J. (1990), *Liquid Crystals: Nature's Delicate Phase of Matter,* Princeton: Princeton University Press, Chap. 10.

Covello, P. S., Forrest, B. J., Marcondes-Helene, M. E., Reeves, L. W., Vist, M. (1983), *J. Phys. Chem* **87**, 176–191.

Deckman, H. W., Dunsmuir, J. H., Garoff, S., McHenry, J. A., Peiffer, D. G. (1988), *J. Vac. Sci. Tech. B* **6**, 333.

Deuling, H. J., Helfrich, W. (1976), *Biophys. J.* **16**, 861–873.

Doane, J. W., Vaz, N. A., Wu, B.-G., Zumer, S. (1986), *Appl. Phys. Lett.* **48**, 269–274.

Donald, A. M. (1992), *Liquid Crystalline Polymers,* Cambridge, U.K.: Cambridge University Press.

Fontell, K. (1981), *Mol. Cryst. Liquid Cryst.* **63**, 59–72.

Fujiwara, F. Y. Reeves, L. W. (1980), *J. Phys. Chem.* **84**, 653–670.

Garland, C. W. (1990), in: J. C. Tolédano (Ed.), *Geometry Thermodynamics,* New York: Plenum.

Garoff, S., Meyer, R. B. (1977), *Phys. Rev. Lett.* **38**, 848–861.

de Gennes, P.-G. (1972), *Solid State Comm.* **10**, 753–767.

de Gennes, P.-G. (1974), *The Physics of Liquid Crystals,* London: Oxford University Press.

Ginzburg, V. L. Landau, L. D. (1950), *Sov. Phys. JETP* **2**0, 1064.

Ginzburg, V. L. (1960), *Sov. Phys. Sol. State* **2**, 1824–1834.

Goodby, J., Waugh, M. A., Stein, S. M., Chin, E., Pindak, R., Patel, J. S. (1988), *Nature* **337**, 449–452.

Gray, G. W. (1975), *J. Phys. (Paris) Colloq.* **36**, C1-337–C1-347.

Gruner, S. M. (1989), *J. Phys. Chem.* **93**, 7562.

Hardouin, F., Levelut, A. M., Achard, M. F., Sigaud, G. (1983), *J. Chim. Phys.* **80**, 65–xx.

Haven, T., Armitage, D., Saupe, A. (1981), *J. Chem. Phys.* **75**, 352–xxx.

Helfrich, W. (1978), *Z. Naturforsch.* **33a**, 305–315.

Huang, C. C., Lien, S. C. (1981), *Phys. Rev. Lett.*

Huang, C. C. Viner, J. M. (1982), *Phys. Rev. A* **25**, 3385–3396.

Huang, C. C., Lien, S. C., Dumrongrattana, S., Chiang, L. Y. (1984), *Phys. Rev. A* **30**, 965.

Huse, D. A., Leibler, S. (1988), *J. Phys. (Paris)* **49**, 605–621.

Israelachvili, J. N. (1985), *Intermolecular Surface Forces,* New York: Academic.

Jacobs S. D. (Ed.) (1992), *Selected Papers on Liquid Crystals for Optics,* Bellingham: WA: SPIE Optical Engineering.

Kahlweit, M., Strey, R., Haase, D., Kuneida, H., Schmeling, T., Faulhaber, B., Borkovec, M., Eicke, H.-F., Busse, G., Eggers, F., Funck, Th., Richmann, H., Magid, L., Söderman, O., Stilbs, P., Winkler, L., Dittrich, A., Jahn, M. (1987), *J. Coll. Int. Sci.* **118**, 436–453.

Kobayashi, K. K. (1970), *J. Phys. Soc. Jpn.* **29**, 101–109.

Kumar, S., Yu, L.-J., Litster J. D. (1983), *Phys. Rev. Lett.* **50**, 1672–1677.

Landau, L. D. Lifshitz, E. M. (1980), *Statistical Physics,* 3rd ed., Pt. 1, New York: Pergamon, Chap. 13.

Lubensky,T. C., Prost, J. (1982), *J. Phys. II France* **2**, 371–387.

Martin, P. C., Parodi, O., Pershan, P. S. (1972), *Phys. Rev. A* **6**, 2401–2413.

McMillan, W. (1971), *Phys. Rev. A* **4**, 1238.

Meyer, R. B., Liebert, L., Strzelecki, L., Keller, P. (1975), *J. Phys. (Paris) Lett.* **36**, L-69–L-71.

Moncton D. E., Pindak, R. (1979), *Phys. Rev. Lett.* **43**, 701–724.

Pershan, P. S., Aeppli, G., Litster, J. D., Birgeneau, R. J. (1981), *Mol. Cryst. Liquid Cryst.* **67**, 205.

Pindak, R., Moncton, D. E., Davey, S. C., Goodby, J. W. (1981), *Phys. Rev. Lett.* **46**, 1135–1144.

Prost, J. Barois, P. (1983), *J. Chim. Phys.* **80**, 65–72.

Reinitzer, F. (1888), *Monatsh. Chem.* **9**, 421.

Safinya, C. R., Roux, D., Smith, G. S., Sinha, S. K., Dimon, P., Clark, N. A., Bellocq, A. M. (1986a), *Phys. Rev. Lett.* **57**, 2718.

Safinya, C. R., Varaday, W. A., Chiang, L. Y., Dimon, P. (1986b), *Phys. Rev. Lett.* **57**, 432.

Schadt M., Helfrich, W. (1971), *Appl. Phys. Lett.* **18**, 127–131.

Sigaud, G., Hardouin, F., Achard, M. F., Gasparoux, H. (1979), *J. Phys. (Paris) Colloq.* C3-356–C3-368.

Strey, R., Schomäcker, R., Roux, D., Nallet, F., Olsson, U. (1990), *J. Chem. Soc. Faraday Trans.* **86**, 2253–2861.

Tiddy, G. J. T. (1972), *J. Chem. Soc. Faraday Trans. I* **68**, 608–612.

Tiddy, G. J. T. (1980), *Phys. Rep.* **57**, 1–46.

Toner, J. (1991), *Phys. Rev. B,* 8289–8296.

Williams, J. H., Jr. (1983), *Mater. Eval.* **41**, 202–210.

Winsor, P. A. (1954), *Solvent Properties of Amphiphilic Compounds*, London: Butterworths.

Further Reading

Belyakov, V. A. (1992), *Diffraction Optics of Complex-Structured Media*, New York: Springer. Discussion of the optical properties of cholesteric liquid crystals, and their relation to other scattering processes in physics.

Brock, J. D., Birgeneau, R. J., Litster, J. D., Aharony, A. (1989), *Contemp. Phys.* **30**, 321. Bond-orientational order.

Flack, J. H., Crooker, P. P., Johnson, D. L., Long, S. (1984), in: A. C. Griffin, J. F. Johnson (Eds.), *Ordered Fluids and Liquid Crystals*, New York: Plenum. Cholesteric blue phases.

Kosterlitz, J. M., Thouless, D. J. (1973), *J. Phys. C* **6**, 1181–1203. Bond-orientational order.

Landau, L. D. (1937), *Phys. Z. Sowjetunion* **11**, 545–555; republished in D. ter Haar (Ed.) (1965), *The Collected Papers of L. D. Landau*, New York: Gordon and Breach. The Landau-Peierls state of matter, which is manifested in three dimensions in smectic-*A* liquid crystals.

Meiboom, S., Sethna, J. P., Anderson, P. W., Brinkman, W. F. (1981), *Phys. Rev. Lett.* **46**, 1216;

Mermin, N. D. (1968), *Phys. Rev.* **176**, 250–254. The Landau–Peierls state of matter.

Nelson, D. R., Halperin, B. I. (1979), *Phys. Rev. B* **19**, 2457. Bond-orientational order.

Onusseit, H. Stegemeyer, H. (1981), *Z. Naturforsch.* **36a**, 1083. Cholesteric blue phases.

Pershan, P. S. (1988), *Structure of Liquid Crystal Phases*, Teaneck, NJ: World Scientific. A good overview of the structure of the various liquid crystal phases, combined with a selection of articles reprinted from the original literature.

Peierls, R. E. (1934), *Helv. Phys. Acta Suppl.* **7**, 81–xx. The Landau–Peierls state of matter.

Strandburg, K. J. (Ed.) (1992), *Bond-Orientational Order in Condensed Matter Systems*, New York: Springer. Bond-orientational order.

Wright, D. C. Mermin, N. D. (1985), *Phys. Rev.* **A 31**, 3498–3500. Cholesteric blue phases.

Young, A. P. (1979), *Phys. Rev. B* **19**, 1855–1866. Bond-orientational order.

LIQUIDS, SIMPLE, STRUCTURE OF

N. E. CUSACK, *School of Physics, University of East Anglia, Norwich, England*

	Introduction	533
1.	**What is Meant by the Structure?**	534
2.	**Distribution Functions**	534
2.1	The Pair Distribution Function	534
2.2	Long- and Short-Range Order	535
2.3	More Distribution Functions	536
2.4	Pair Distributions in Binary Systems	536
3.	**Diffraction Studies of Pair Distribution Functions**	536
3.1	The Structure Factor	537
3.2	Structure Factors and the PDF	538
3.3	Partial Structure Factors in Binary Systems	538
3.4	Neutron Diffraction and Isotopic Substitution	539
4.	**Intermolecular Forces and Liquid-State Theory**	539
4.1	Liquid-State Theory	540
4.2	Computational Aids	541
5.	**Liquid Argon and Similar Systems**	542
6.	**Metals: Some Observations**	543
7.	**Chemical Short-Range Order**	545
7.1	Chemical Thermodynamic Models	545
7.2	Chemical SRO and Structure Factors	546
8.	**Structure and the Theory of Metals**	547
8.1	The General Concept of a Metal	548
8.2	Forces and Energy	548
8.3	Structure from $\phi_{\text{eff}}(r)$ in Metals	549
8.4	A Successful Example	550
9.	**Molecular Liquids**	551
9.1	Structure Factors of Molecular Liquids	552
9.2	A Representative Experiment	553
9.3	Some General Remarks about Structure in Molecular Liquids	554
10.	**Molten Salts**	555
10.1	Halides	555
10.2	Computer Simulation and Theory	556
11.	**Liquids and Glasses**	558
12.	**Conclusion**	558
	Glossary	558
	Works Cited	559
	Further Reading	560

INTRODUCTION

Although all liquids, and solid glasses too, are structurally disordered, or noncrystalline, there are many varieties of structure, some of which may be called *simple*. What differentiates one great class of liquids from another—say, liquid metals from molten salts—is the different nature of the intermolecular forces or, one could equally well have said, the different kinds of molecule. To see which liquids might reasonably be labeled as simple, we must therefore refer to the interparticle forces. We shall adopt four criteria of simplicity which, though not logically equivalent, are such that, if one holds, it will not be surprising if the others do too.

First, the molecules should be small. Second, they should interact by force fields which, if not spherically symmetrical, depart from this by such a small amount that it may be specified by at most a few nonspherical terms. Third, it should be true to a high degree of approximation that the internal structure of the molecule is an independent entity not affected by the position of neighbors. Fourth, the molecules should act by pairwise additive forces—a well-tried assumption in the physics

1-56081-067-X/94/$5.00+.50

of simple liquids. Thus the ideal case might well be liquid argon composed of stable monatomic molecules with spherical force fields. Among the simple liquids are metals—although it was a considerable achievement of metal physics to show under what conditions the fourth criterion holds. Also simple are those molten salts in which the long-range interaction is Coulomb attraction or repulsion between ionized atoms as in NaCl. The second criterion allows us to admit small molecules like N_2 or CCl_4, but molecules with quite small numbers of atoms, such as glycerol ($CH_2OHCHOHCH_2OH$), can have shapes and dipole moments which call all four criteria into question. The remaining sections will therefore discuss liquid noble gases, a few molecular liquids, metals, and molten salts.

The criteria exclude several kinds of liquid. The structures of many nonmetallic glasses, both liquid and solid, are formed from ramifying networks of directed covalent bonds; even some elements, such as Te and Se, are excluded for this reason. Liquid crystals have rodlike or otherwise nonspherical molecules that cause structural order in one or two dimensions. Liquid polymers or solutions containing highly complex molecules such as proteins are clearly not simple. Additional complications that exclude liquids from present consideration are macroscopic quantum behavior, as in liquid He; the intermolecular bridges formed by hydrogen bonds, as in water; or large solute particles, as in colloidal solutions. Nevertheless, the diffraction techniques mentioned below for simple liquids remain important even for the complex cases we are excluding. The reader will find discussions of more complex liquids in the articles on LIQUID CRYSTALS, STRUCTURE OF; POLYMERS, STRUCTURE OF; and GLASSES.

1. WHAT IS MEANT BY THE STRUCTURE?

The structure of a pure perfect crystal is totally defined by its unit cell, and the elegant science of crystallography has solved the problem of determining unit cells. Liquids do not have unit cells, and to visualize the atomic arrangement as billiard balls jumbled in a bag is to make a good first guess. The numerous symmetries that control crystal structure mean that the environment of an atom in its unit cell is identical with that of every corresponding atom in every other unit cell. In a liquid no two atoms have the same environment or, if they happened by chance to do so, it would only be transiently, because the atoms are continually interdiffusing and changing neighbors. It follows that structure must be a statistical concept in liquids. For example, in a crystalline binary alloy of A-type and B-type atoms one can answer the question: Where exactly are the A atoms relative to the B atoms? In the liquid alloy one can answer a limited number of questions such as: What is the average number of B atoms that are nearest neighbors to an A atom? How far away are they on the average? In what follows we shall assume that our materials are homogeneous and isotropic. This not only reduces the amount of vector symbolism but is intrinsically reasonable for liquids in thermodynamic equilibrium. It might well be inadequate for strained glasses, which are not isotropic, or multiphase systems, which are not homogeneous.

2. DISTRIBUTION FUNCTIONS

2.1 The Pair Distribution Function

Let us define with some care one of the fundamental quantities for a pure liquid. Extensions of the formulas can then be made more summarily. Suppose N atoms are confined in a macroscopic volume V. Then $n_0 \equiv N/V$ is the average number density. The position of a spherical atom means the point occupied by its center. Imagine a spherical shell of radius r and thickness δr centered on any atomic position. The volume of the shell is $4\pi r^2 \delta r$, but the number of atomic centers in the shell is not necessarily $4\pi r^2 n_0 \delta r$. For example, if r is approximately the radius of an atom, the number of centers in the shell will be zero because atoms strongly resist overlap. On the other hand, if r is about the atomic diameter, the number could be up to 12 because twelve spheres can touch simultaneously a central one of equal size. In general, therefore, the number of centers in the shell is $4\pi r^2 n_0 g(r) \delta r$, where the factor $g(r)$ allows the actual number density in the shell, i.e., $n_0 g(r)$, to depart from n_0 for the geometrical packing reasons just referred to or, more generally, because of the forces of attraction and repulsion between atoms, which determine the number and distance of neighbors. $g(r)$ is to be an average. It can be looked upon as the time average of its value around any one atom, or the average at any one time of the values around all the atoms. In thermodynamic equilibrium these will be the same. $g(r)$ is called the *pair distribution function* or PDF. The quantity $4\pi r^2 g(r)$ is given the symbol $\rho(r)$ and the name *radial distribution function* or RDF: $n_0\rho(r)\delta r$ is the average number of atoms positioned in the shell of radius r.

Figure 1 shows the PDF of a simple metal. It is zero for $r \leqslant 2.4$ Å, which represents the distance of closest approach of two atoms. At larger distances the curve is *continuous*, which shows that two atoms may be at any distance apart; this is in marked contrast to crystal spacings, which are precise and discontinuous. Strictly, the crystal spacings are between abstract points about which the atoms vibrate; the liquid spacings are between the actual atomic positions. The high peak of $g(r)$ indicates the crowding of neighbors around the atom at the origin. Their number is given by the integral

$$n_1 = n_0 \int_{r_l}^{r_r} \rho(r)dr,$$

where r_l and r_r are the left- and right-hand boundaries of the peak. The values of both r_l and r_r are to some extent arbitrary, particularly if the peak has distortions or shoulders. In principle, n_1, often called the *coordination number*, is not a well-defined integer in a liquid; in a simple liquid it may be about 10. The radial distance to the first peak is the *nearest-neighbor distance*. Other peaks represent successive shells of neighbors, but not many can be seen. For $r > 10$ Å, the curve smooths out and $g(r) \rightarrow 1$.

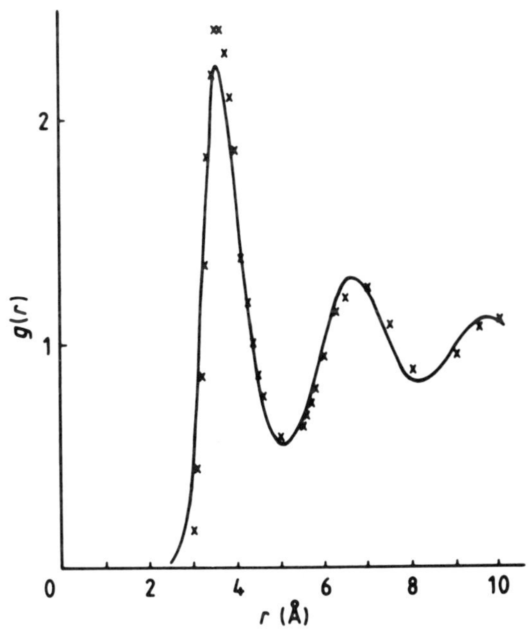

FIG. 1. The pair distribution function $g(r)$ of liquid Na. Full curve, molecular-dynamics calculation with a ϕ_{eff} (see Sec. 8.3). Crosses, experimental points from Waseda (1980).

2.2 Long- and Short-Range Order

By defining $g(r)$ we have taken one step towards saying what structure means in a liquid. It is helpful to think of $n_0 g(r) d^3r$ as the probability of finding an atom positioned at the infinitesimal volume d^3r at a distance r from another atom at the origin. Where $g(r) = 1$, this probability is just $n_0 d^3r$ and the atom may be anywhere with equal probability. Thus at large r randomness has replaced order and there is said to be no *long-range order*, (LRO). Crystals do have LRO, which in a large single crystal extends over macroscopic distances. The first high peak of $g(r)$ reveals a clear departure from randomness, which is called *short-range order* (SRO). Peaks further out indicate *intermediate-range order* (IRO), which is currently much discussed in more complex materials such as glasses (see AMORPHOUS MATERIALS, STRUCTURE OF).

If there is an atom at the origin at time $t=0$, the probability of another in the volume element d^3r at a distance r is, as we have seen, $n_0 g(r) d^3r$. If we ask the probability of the second one at time $t > 0$, a whole new field is opened up because the motion of the atoms is involved, not just their positions. In thermodynamic equilibrium this is the natural thermal motion associated with the temperature, and, to a first approximation in a simple liquid, it is an irregular oscillation about a center whose position wanders diffusively. We shall see in Sec. 3 that the PDF is explored by elastic scattering. It is *inelastic scattering* that is required to examine the *time-dependent PDF*, which is the generalization of $g(r)$ taking thermal motion into account. The articles on LATTICE VIBRATIONS, STATISTICS OF; PHONONS IN CRYSTAL LATTICES; and VIBRATIONAL STATES IN DISORDERED SYSTEMS deal with this quantity.

2.3 More Distribution Functions

$g(r)$ refers only to the distance between two atoms; it offers no directional information. It is not difficult to conceive of more informative distribution functions. For instance, the probability of finding three atomic positions in a triangle in the volume elements d^3r_1, d^3r_2, and d^3r_3 is written $n_0^3 g^{(3)}(\mathbf{r}_1,\mathbf{r}_2,\mathbf{r}_3)d^3r_1d^3r_2d^3r_3$, and $g^{(3)}$ is called the *triplet distribution*. $g^{(4)}$, $g^{(5)}$, etc., can be correspondingly defined. The more of these functions we know, the more completely defined is the structure. In fact, very little is known even about $g^{(3)}$. It is not possible to derive $g^{(3)}$ from $g(r)$; indeed, an infinite number of $g^{(3)}$'s are compatible with the same $g(r)$. Experiments yielding any information about $g^{(3)}$ are few and far between, but theory and computer simulations are used to study it. However, the vast majority of our knowledge of the structure of simple liquids and mixtures is confined to $g(r)$'s and to numbers such as n_1 that can be derived from them.

2.4 Pair Distributions in Binary Systems

Suppose we have a liquid made of atoms of two types, A and B. After the discussion of $g(r)$, the reader will probably grant that there will now be four pair distribution functions to be labeled $g_{AA}(r)$, $g_{BB}(r)$, $g_{AB}(r)$, and $g_{BA}(r)$. The first two measure the probability that two like atoms are at distance r; the last two do the same for unlike atoms. Since the number of A–B and B–A interparticle distances is the same, $g_{AB}=g_{BA}$, and only three of the four g's are independent. The g_{ij}'s are called *partial pair distribution functions*, where i, j can be A or B. From these, *partial RDFs* $\rho_{ij}(r)$ can be defined. To take an example: if there are N_B B-type atoms in the volume V, their number density is $n_B \equiv N_B/V$. The number of B atoms within a distance r_1 of an A atom is

$$n_B \int_0^{r_1} g_{AB}(r)4\pi r^2 dr = n_B \int_0^{r_1} \rho_{AB}(r)dr. \tag{1}$$

If this integral were taken over the first peak, we should have the *partial coordination number* n_{AB}, giving the number of B-type nearest neighbors to an A atom. There are four partial coordination numbers in a binary system. If g_{AB} has a large peak at small r, the $A \leftrightarrow B$ attraction favors SRO with B atoms near A atoms and vice versa. If g_{BB} has a large peak at small r, then B atoms are tending to cluster together.

In principle, triplet and higher distributions can be defined for ternary and more complex liquids. To know all these functions would be to know completely the structure of the system. Such knowledge is scarce enough even for the partial pair distributions of binary liquids, and nothing more complicated will be considered here. Therefore we now move on to consider how $g(r)$ and g_{ij} are measured.

3. DIFFRACTION STUDIES OF PAIR DISTRIBUTION FUNCTIONS

"Measured" is too strong a word; $g(r)$ is inferred by a long train of reasoning from diffraction experiments. It is true that clues to the existence of SRO may be picked up from, *inter alia*, densities, known chemical affinities, x-ray absorption, analogy with crystals of similar composition, nuclear magnetic resonance (see MAGNETIC RESONANCE AND QUADRUPOLE RESONANCE, NUCLEAR; NUCLEAR MAGNETIC RESONANCE, DETERMINATION OF STRUCTURES BY), or the dependence of electrical resistivity or volume on concentration in alloys. But most knowledge comes from measuring the scattered intensity as a function of angle when a collimated monochromatic beam is transmitted or reflected by the sample. If the beam is of x rays, the technique corresponds to x-ray crystallography (*q.v.*), though the lack of crystallinity implies the absence of Bragg peaks. To help avoid confusion we refer first to x-ray diffraction from pure monatomic liquids.

3.1 The Structure Factor

The fundamental diagram is Fig. 2. $\hbar \mathbf{k}_i$ and $\hbar \mathbf{k}_f$ are the momenta of incident and deflected photons scattered elastically by the atoms in the liquid. $\mathbf{k}$ is the wave vector of the radiation. The momentum change $\hbar \mathbf{Q}$ and its relation to the scattering angle θ are

$$\hbar \mathbf{Q} \equiv \hbar(\mathbf{k}_i - \mathbf{k}_f), \quad |\mathbf{Q}| = 2k \sin(\theta/2). \tag{2}$$

If there were only one scattering atom, the amplitude and intensity at the distant detector would be proportional to $f(\theta)$ and $|f(\theta)|^2$, respectively, where $f(\theta)$ specifies the probability with which x rays are scattered by one atom in the θ direction and is known and tabulated for all atoms. The scattered waves from all the atoms arrive at the detector with different phases and interfere with one another more or less destructively depending on θ, resulting in an angular intensity variation, $I(\theta)$. The phase differences depend on the atom-to-detector path differences and therefore on the atomic positions, which can be denoted by vectors $\mathbf{R}_1$, $\mathbf{R}_2$, ..., $\mathbf{R}_N$. Consequently the function $I(\theta)$ contains information about the positions. This fact underlies the possibility of inferring structural information from diffraction. The formulas are derived with care in numerous references [see, e.g., Waseda (1980); Cusack (1987); Guinier (1963); we shall refer to these again as "the cited texts"]. The fundamental result for the detected intensity is

$$I(Q) = \alpha N |f(\theta)|^2 S(Q). \tag{3a}$$

N and f have already been defined and f is characteristic of the element in the sample. $S(Q)$, called the *structure factor*, depends only on the positions of the atoms according to the formula

$$NS(Q) \equiv \left| \sum_{i=1}^{N} \exp(i\mathbf{Q} \cdot \mathbf{R}_i) \right|^2. \tag{3b}$$

Equation (3b) results from summing the amplitudes of all the scattered waves and then squaring to obtain the intensity. It is usual to use Q instead of θ to express the angular dependence; and, in Eq. (3b), $S(Q)$ depends only on the magnitude of Q, not the direction, because of the isotropy of liquids. Note that no units are specified for the intensity here, and the constant of proportionality, α, will depend on how the intensity is measured, e.g., by photon counting rates, and on the sample and equipment geometry and numerous experimental conditions such as absorption of the beam by the sample and its container (essential for liquids).

A great deal of subsidiary experimentation, calibration, corrections, and experience go into establishing a reliable value for α. We refer the reader to the cited texts for these details and will assume that $S(Q)$ can eventually be obtained from laboratory measurements of $I(\theta)$. The observed $S(Q)$ for liquid sodium is shown in Fig. 3, and an overall accuracy of $\sim \pm 2\%$ in $S(Q)$ results from good practice in this field. The problem is how to relate $S(Q)$ to the structure.

A crucial point is that the derivation of Eq. (3) assumes that scattering is improbable and that a photon scattered by one atom is not scattered by another. In other words, *multiple scattering* is absent. This is a reasonable assumption for x-rays; but multiple scattering always occurs to some extent and has to be ignored or subtracted in the data processing. This leaves only the distances between any two atoms to produce the path differences that determine $I(\theta)$. The latter therefore contains information only about these distances, i.e., about $g(r)$. Multiple scattering would involve $g^{(3)}$, $g^{(4)}$, etc., but as well as being a minor effect in these experiments, it is hard to relate in a usable way to the higher distribution functions. The problem therefore reduces to relating $S(Q)$ to $g(r)$.

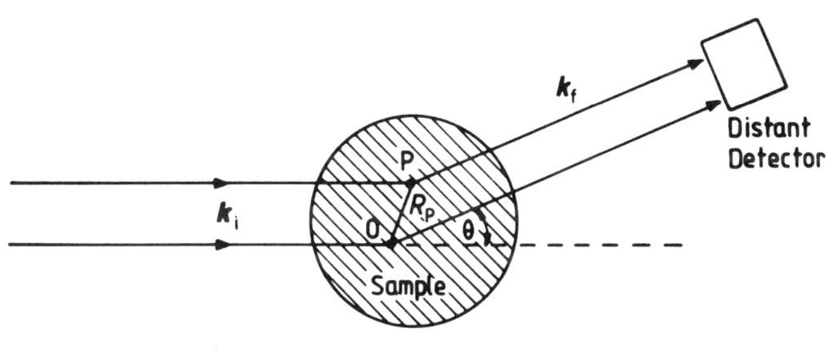

FIG. 2. Definitions of quantities in the coherent scattering calculation. O and P are two atomic positions illustrating the path difference.

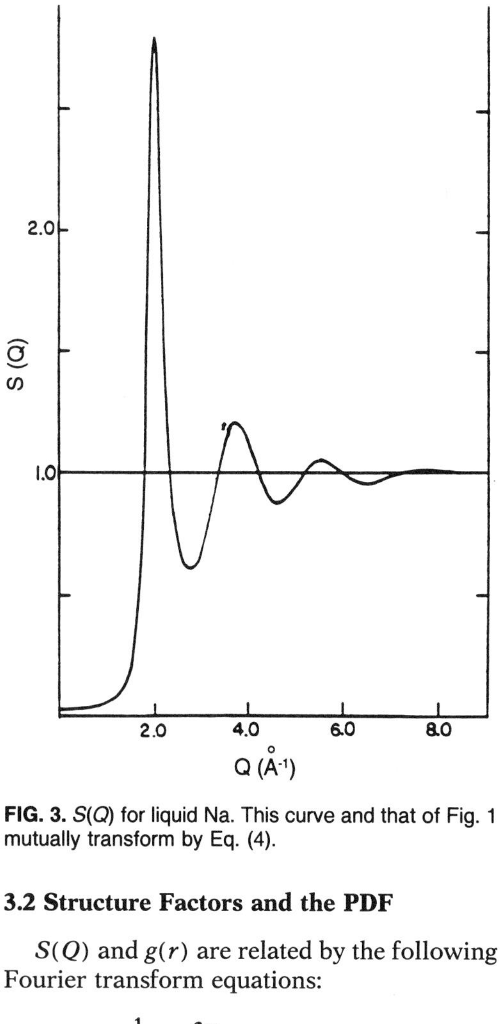

FIG. 3. $S(Q)$ for liquid Na. This curve and that of Fig. 1 mutually transform by Eq. (4).

3.2 Structure Factors and the PDF

$S(Q)$ and $g(r)$ are related by the following Fourier transform equations:

$$g(r)=1+\frac{1}{8\pi^3 n_0}\int_0^\infty [S(Q)-1] \times\frac{\sin Qr}{Qr}4\pi Q^2 dQ; \quad (4a)$$

$$S(Q)=1+n_0\int_0^\infty [g(r)-1]\frac{\sin Qr}{Qr}4\pi r^2 dr; \quad (4b)$$

$$S(Q)\to 1 \text{ as } Q\to\infty; \quad g(r)\to 1 \text{ as } r\to\infty. \quad (4c)$$

These are theoretical results demonstrated in the cited texts. Equation (4a) shows that $g(r)$ can be found by an integration over the observed $S(Q)$. Although such calculations are easily done nowadays, the most powerful computer cannot remedy a lack of experimental data, and it is clear that $S(Q)$ will not be measured as $Q\to 0$ (because observation is impeded by the strong unscattered beam at $\theta=0°$) nor as $Q\to\infty$ [because $I(\theta)$ is very small at large angles]. In many x-ray experiments $\sim 0.5 < Q\,(\text{Å}^{-1}) < 12$, but modern sources such as synchrotrons for x rays and spallation sources for neutrons extend Q to 40 Å^{-1} and more. Even so, some extrapolations or approximations in the integration, and the intrinsic errors in $S(Q)$, are unavoidable and lead to uncertainty in $g(r)$.

3.3 Partial Structure Factors in Binary Systems

A-type and B-type atoms will have different scattering amplitudes f_A and f_B. In the sum of scattered amplitudes at the detector, leading to Eq. (3), there will be terms in f_A with phases dependent on the positions of the A atoms, and corresponding terms in f_B. On squaring to get the intensity, terms in f_A^2, f_B^2, and $2f_A f_B$ will appear. As in Eq. (3), these terms will have multiplying factors depending on the A–A, B–B, and A–B distances, respectively. In other words, there will be three *partial structure factors*, S_{ij}. These are related to the $g_{ij}(r)$'s of Sec. 2.4 by relations like Eq. (4). For binary mixtures with concentrations $c_A=N_A/(N_A+N_B)$ and $c_B=N_B/(N_A+N_B)$, the detailed calculations leads to

$$I(Q)=\beta F(Q), \quad (5a)$$

$$F(Q)\equiv f_A^2[c_A c_B+c_A^2 S_{AA}(Q)] + f_B^2[c_A c_B+c_B^2 S_{BB}(Q)] + 2f_A f_B\, c_A c_B[S_{AB}(Q)-1], \quad (5b)$$

$$S_{ij}(Q)=1+n_0\int_0^\infty [g_{ij}(r)-1] \times\frac{\sin Qr}{Qr}4\pi r^2 dr, \quad (5c)$$

$$g_{ij}(r)=1+\frac{1}{8\pi^3 n_0}\int_0^\infty [S_{ij}(Q)-1] \times\frac{\sin Qr}{Qr}4\pi Q^2 dQ, \quad (5d)$$

$$S_{AB}=S_{BA}, \quad (5e)$$

where β is a calibration constant like α in Eq. (3a). This shows that, if experiments could provide the S_{ij} separately, Eq. (5d) would provide the partial pair distributions and coordination numbers.

3.4 Neutron Diffraction and Isotopic Substitution

The S_{ij} were measured separately for the first time in 1966 (Enderby *et al.*, 1966). Since the experiment relied on neutron diffraction we refer to that first.

Equations (3)–(5) apply to any kind of scattered waves if the quantities f, α, and β are reinterpreted and if the data processing is altered to accommodate different kinds of correction. Neutron beams of wavelength suited to structure studies are available at international facilities, and spectrometers are well developed. The x-ray quantities f and $|f|^2$ must be replaced by b and $|b|^2$ where b, the *scattering length*, expresses the probability of neutron scattering by the nucleus. At the wavelengths used, b is independent of θ and is usually well known. We note that $F(Q)$ contains one contribution, called *coherent scattering*, dependent on the structure factors and another, $c_A c_B(f_A - f_B)^2$, called *incoherent scattering*, dependent on the difference between the two f's. Since the system (neutron $+$nucleus) can exist in two spin states with two different b's, even a single isotope shows incoherent scattering. Moreover, a chemically pure element will usually contain several isotopes with a range of b values. In short, b_A and b_B in the formulas must be understood as averages weighted for both spin states and isotopic mix. Multiple scattering occurs with neutrons, and so does inelastic scattering, which is quite serious for low-Z samples; both must be subtracted from the data. In these and other ways, the data processing for neutrons and x rays differs, but both are aimed at correcting the measured $I(Q)$ to reveal its coherent elastic part, which gives $S(Q)$ and thence $g(r)$.

For binary liquids, separating the three unknown functions S_{ij} requires, for each concentration, three measurements of $F(Q)$, which are independent of each other on account of significantly different ratios of b_A to b_B or of f_A to f_B. Varying the concentration is not an option because this will in general change S_{ij}. Varying b is practicable nowadays because isotopic abundances can sometimes be changed at will in samples of constant chemical nature. Three mixtures A_xB_{1-x}, in which either or both A and B vary in isotopic composition, could then give three functions $F(Q)$ solvable as simultaneous equations for S_{AA}, S_{BB}, and S_{AB}. There are other possibilities. Three experiments using x-ray, neutron, and electron scattering might serve the purpose. It is also possible to use x rays alone by exploiting the fact that, for the anomalous scattering close to an x-ray absorption edge, f for a given Q depends quite strongly on wavelength. Thus measurement at three different wavelengths effectively varies the ratio of f_A to f_B.

In practice, none of these methods is without substantial difficulties, ranging from the high cost of isotopes to the enhancement of errors when three functions, each subject to experimental inaccuracy, are combined. The methods have been applied to glasses, aqueous solutions, molten salts, and alloys, but rather few partial structure factors for liquid systems have been thoroughly investigated.

The reader may wonder whether $S_x(Q)$ from x-ray diffraction should be the same as $S_n(Q)$ from neutrons. It is true that x rays are scattered by the electron density distribution and that in metals this is mainly concentrated in the ion cores. $f(Q)$ relates to spherical ions or atoms centered on the nucleus, and if this were the only scattering, the two $S(Q)$'s would be the same. But the free electrons characteristic of metals permeate the entire system including the interstices between the ions, and their density distribution will be distorted by all the electrostatic forces. Insofar as they cause scattering, they make $S_x(Q) \neq S_n(Q)$. The difference is just detectable in experiments designed to find it and can be safely neglected for the present purpose. (See Takeda *et al.*, 1990, for metals and Dore, 1990, for molecular liquids.)

4. INTERMOLECULAR FORCES AND LIQUID-STATE THEORY

Later sections will deal separately with noble-gas liquids, metals, molecular liquids, and molten salts, referring to their special characteristics. This section makes some points relevant to all of them.

The preceding sections explain what structure means in a liquid and outline the way $S(Q)$ and $g(r)$ can be found from diffraction experiments. The PDF and RDF are often referred to loosely as "the structure," but it was emphasized that they relate only to the average radial interparticle distances. Since $S(Q)$ and $g(r)$ are mutually transformable, they contain the same information. A rise in temperature increases motion and disorder, but the repulsive forces still define a nearest-neighbor distance. Heating broadens and lowers the first peak in $g(r)$ but alters its position very little. This shows that thermal expansion reorganizes the SRO and reduces n_1 without changing the nearest-neighbor distance—an important difference from the behavior of crystals.

The theory of liquids aims to derive the macroscopic properties, including the structure, from the intermolecular force. If we assume this to be pairwise and additive—not strictly true but a generally accepted approximation for simple liquids—the potential energy of two molecules i and j may be written

$$\phi_{ij}=\phi_{ij}(\mathbf{r}_{ij};\ \omega_i,\omega_j), \tag{6}$$

where $\mathbf{r}_{ij}$ is the separation of representative points or centers in the two molecules and $\boldsymbol{\omega}$ symbolizes the three angles necessary to specify a molecular orientation. For linear molecules, $\boldsymbol{\omega}$ reduces to two polar angles, and for spherical atoms or ions $\boldsymbol{\omega}$ is irrelevant.

The dependence of ϕ on $\mathbf{r}_{ij}$ contains attractive and repulsive parts. We take for granted that there is a strong short-range repulsion, though this is in fact a subtle and profound property of matter (see, e.g., Lieb, 1976, who discusses why a stone, say, is so big). The repulsion is often approximated by an infinite force, which, without an attractive force, gives us the hard-sphere (HS) model. The structure of a dense random packing of hard spheres has been much investigated by computer simulation, and its $g(r)$ is well known. It is not dissimilar from Fig. 1 and represents approximately the structure of those simple liquids whose particles have undirected attractive forces that simply pack the molecules together as closely as the repulsive forces allow. When they freeze, such liquids generally adopt one of the closer-packed crystal structures such as face-centered cubic, which are a few percent denser than their melts.

A more realistic potential represents the attractive and repulsive parts by power laws, and another much used model, named after Lennard-Jones (LJ), has two parameters characteristic of the species, viz.,

$$\phi(r)=4\epsilon\left[\left(\frac{\sigma}{r}\right)^{12}-\left(\frac{\sigma}{r}\right)^{6}\right]. \tag{7}$$

It is shown in Fig. 4. This serves well for noble gases. Many more complicated potentials have been devised for other liquids and will be exemplified below. From any chosen intermolecular potential there are two routes to the structure: statistical mechanical theory and computer simulation.

4.1 Liquid-State Theory

Statistical mechanical texts show that all thermodynamic properties, including the structure, may be obtained from the partition function, of which the classical form suits all but a few quantum liquids like He at low temperatures. For a liquid of N similar particles at temperature T, acting by a pairwise additive potential $\phi(r)$, some formulas which

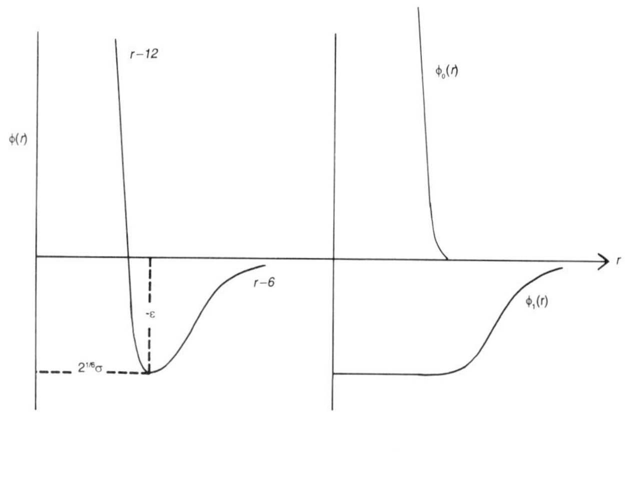

FIG. 4. The Lennard-Jones potential of Eq. (7) and its decomposition according to Eq. (11).

relate $g(r)$ to $\phi(r)$ are as follows. The total potential energy is

$$\Phi=\tfrac{1}{2}\sum_{i,j}\phi(r_{ij}). \tag{8}$$

Then

$$g(r_{12})=n_0^{-2}Q^{-1}N(N-1)\times\int\exp(-\Phi/k_BT)d^3r_3\cdots d^3r_N, \tag{9}$$

$$Q(V,T)=\int\exp(-\Phi/k_BT)d^3r^{(N)}. \tag{10}$$

Q is called the configuration integral. k_B is Boltzmann's constant and $\mathbf{r}^{(N)}$ stands for the set of N position vectors $\mathbf{r}_1,...,\mathbf{r}_N$ of the particles.

Equation (9) can be generalized by including orientational coordinates and a variety of particle species; but, even without this, its solution is a central problem of liquid-state theory, leading to hierarchies of equations, density expansions, and variation and perturbation methods for which the reader will require one of the treatises on liquid-state theory such as Hansen and McDonald (1986). This will show that there is no inherent mystery in the subject but that the technical complexity is great. The various approximate solutions give very good agreement with observed $g(r)$'s when well chosen $\phi(r)$'s are used. Specimen cases will be outlined below.

4.2 Computational Aids

Liquid studies, like the rest of condensed-matter physics, have benefited from two decades of computer simulation. *Molecular dynamics* (MD) (*q.v.*) is much used and, reduced to its barest outline, it works as follows (see also MONTE CARLO METHODS).

A computer is given the following information: coordinates and velocities for N particles of specified nature in a volume V, often cubic; the interparticle potential, $\phi(r)$; and the boundary conditions—frequently periodic to ensure that N is representative of a much larger number and that if a trajectory takes a particle out at one face, an identical one enters at the opposite face to keep n_0 constant. With use of the classical equations of motion, the computer then

1. calculates the resultant interparticle force acting on each particle at time $t=0$;
2. finds all the new positions and velocities Δt seconds later;
3. repeats for successive Δt's.

When sufficient time has elapsed for the system to be in thermodynamic equilibrium, some property, say the total kinetic energy or $g(r)$, is computed after each of many successive steps and the average taken. Thus the average equilibrium property is found. N may be $\leqslant 10^3$, $\Delta t \sim 10^{-14}$ s, and there will be many hundreds of steps. The method is of great power and versatility and there are many versions and techniques.

A new and already valuable computational aid to the interpretation of scattering data works as follows. Suppose the N particles in the box with periodic boundary conditions are allocated, not $\phi(r)$'s, but neutron scattering lengths b. Starting with an arbitrary configuration, the computer calculates $g(r)$ and Fourier transforms it to $S(Q)$. The next task is to compute a suitable statistical measure, χ^2, of the difference between $S(Q)$(calc.) and $S(Q)$(expt.) in terms of the experimental error in $S(Q)$. One particle is now displaced randomly, subject to a minimum distance of approach, and χ^2 recalculated. The new configuration is accepted in place of the old with probability $p=1$, if $\chi^2(\text{new})<\chi^2(\text{old})$, or $p=\exp[\chi^2(\text{old})-\chi^2(\text{new})]<1$ if $\chi^2(\text{new})>\chi^2(\text{old})$. Repetition of this process makes χ^2 approach a minimum and the configuration for $\chi^2(\text{min})$ is saved. A number of saved configurations could be so generated and are deemed to be statistically independent. They are all structures that reproduce the observed $S(Q)$ within experimental error. The $g(r)$'s and, in the binary case, the $g_{ij}(r)$'s can be extracted and averaged and regarded as the experimental results. Other properties of the structure, even the triplet distribution, are also available for study, and no intermolecular potential has been involved in this. The method is called *reverse Monte Carlo* (RMC) and for its implementation, including a discussion of the definition of χ^2, see, e.g., McGreevy and Pusztai (1988), McGreevy and Pusztai (1990) for molten salts, Sec. 10.2, and Howe (1990) for molecular liquids, Sec. 9.3.

5. LIQUID ARGON AND SIMILAR SYSTEMS

Liquid argon has been much studied because its liquid range is not too difficult to attain experimentally and its interatomic potential is relatively easy to envisage. At large separations the attractive force originates in fluctuations in the electron clouds, which mutually induce electric multipoles in otherwise spherical charge distributions. The dipole-dipole interaction is the strongest of these and varies as r^{-6} as in Eq. (7). At close quarters the resistance of the electron clouds to overlap creates the strong repulsion approximated by the r^{-12} term in Eq. (7) but theoretically more like $\exp(-r/r_0)$, where r_0 is a parameter. The intuitive expectation therefore is that the attractive force, which is spherically symmetrical, simply draws in the neighbors and provides the cohesion while the SRO is the result of the dense random packing of the spheres delineated by the strong repulsions. Crystalline Ar has the densest possible packing, namely face-centered cubic, but on melting near the triple point the order decreases and the density falls by about 14%. This is approximately the same as the corresponding change for hard spheres. It is now widely accepted that the shapes of curves like Figs. 1 and 5 are well approximated by hard-sphere packings for both noble-gas liquids and simple liquid metals. Structure factors have been computed for LJ and HS fluids and are almost identical (see, e.g., Hansen and McDonald, 1986). These general remarks apply to dense liquids; near the liquid-vapor critical point, the relative effect of the attractive forces on the structure increases as the distinction between the liquid and gas phases vanishes.

A good example of the application of the methods of Sec. 3 to liquid Ar is the neutron-scattering experiment of Yarnell *et al.* (1973). Figure 5 shows $g(r)$ obtained experimentally and also points calculated by computer simulation. $\phi(r)$ of LJ form is quite capable of reproducing $g(r)$ very well, and the simulation confirms that the curve is largely determined by the repulsive force.

We must omit general discussion of how to calculate $g(r)$ from $\phi(r)$ since this is a major problem of liquid-state theory. One class of approaches to it uses perturbation theory, and

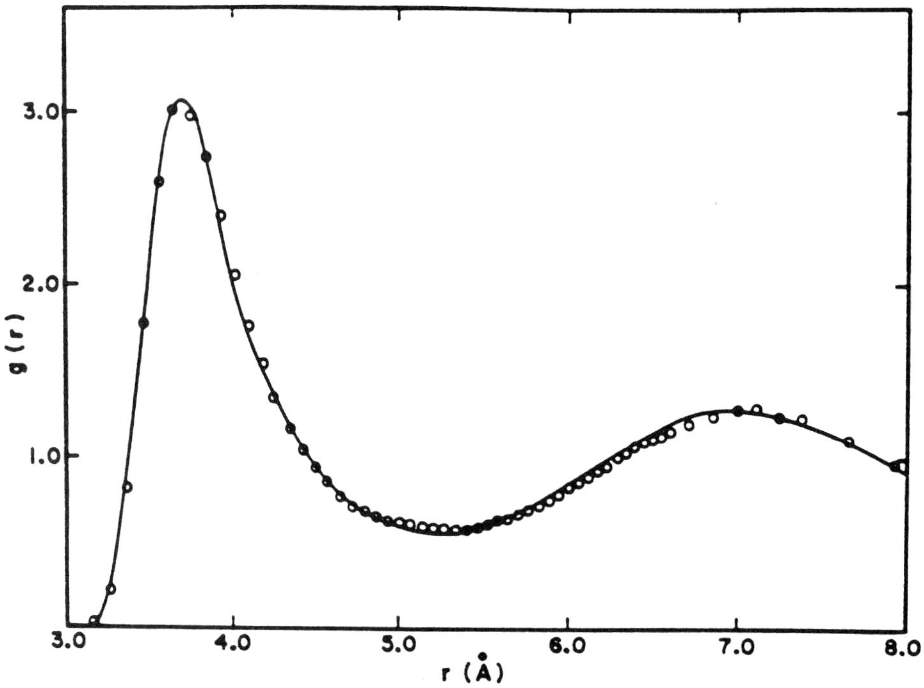

FIG. 5. $g(r)$ for liquid Ar at 85 K, $n_0=0.02125$ atom/Å^3. Full curve, experimental results; circles, computer simulation points from Eq. (7).

to give a flavor of this we refer briefly to the influential work of Weeks *et al.* (1971). The idea is to divide $\phi(r)$ into repulsive and attractive parts, ϕ_0 and ϕ_1, respectively, and to treat the hypothetical system represented by the subscript 0 as a reference system of well-known properties. This is then perturbed or corrected by the effect of ϕ_1 to describe the real system. The theory will be convenient and accurate if the perturbation corrections are small and preferably confined to one or two terms of a perturbation expansion. For the LJ potential, $\phi_{\min}=-\epsilon$ and occurs at $r=2^{1/6}\sigma$; so, for the present case (see Fig. 4),

$$\phi_0(r)=\begin{cases}\phi(r)+\epsilon, & r<2^{1/6}\sigma,\\ 0, & r\geqslant 2^{1/6}\sigma;\end{cases}$$
$$\phi_1(r)=\begin{cases}-\epsilon, & r<2^{1/6}\sigma,\\ \phi(r), & r\geqslant 2^{1/6}\sigma.\end{cases} \qquad (11)$$

Now define $y_0(r)$ by

$$g_0(r)=y_0(r)\exp[-\phi_0(r)/k_BT].$$

Since $\phi_0=0$ outside the repulsion, $y_0(r)$ is the PDF of the reference system outside the repulsion. But in precisely this region it is plausible that the PDF resembles that of a hard-sphere system, so the next step is to replace y_0 by y_d, where d refers to a system of hard spheres with diameter d; thus

$$g_0(r)\simeq y_d(r)\exp[-\phi_0(r)/k_BT]. \qquad (12)$$

Once d is chosen, $g_0(r)$ can be calculated from $\phi_0(r)$ and known properties of the HS system and can then be used as the first approximation to the theoretical $g(r)$—at least at those densities for which the structure reflects the geometry of sphere packing. In practice, Weeks *et al.* showed that, near the triple point, $g_0(r)$ represented excellently the exact result for the LJ potential given by computer simulation and therefore the experimental result too (e.g., Fig. 5). It remains, however, to see how to choose d. This is done by requiring the HS system to have the same equation of state as the reference system. Imposing this condition selects d because there is a connection between the equation of state and $g(r)$ given by statistical mechanics, viz.,

$$k_BT\left(\frac{\partial n_0}{\partial p}\right)_T=1+n_0\int[g(r)-1]d^3r, \qquad (13)$$

where p is the pressure. The equation for d is therefore

$$\int[y_d(r)\exp(-\phi_0/k_BT)-1]d^3r$$
$$=\int[y_d(r)\exp(-\phi_d/k_BT)-1]d^3r$$

in which $\phi_d=\phi_{HS}$ with diameter d.

It might be expected that $g_0(r)$ and the exact $g(r)$ agree less well when the density is lower and repulsive forces cease to dominate the structure. This is true, and corrections involving ϕ_1 become necessary. Furthermore, even at high density, thermodynamic quantities, such as the Helmholtz free energy F, require F_0 (known) to be supplemented by the first-order perturbation term $(Nn_0/2)\times\int\phi_1(r)g_0(r)d^3r$ before good results are obtained. But it is not possible to pursue thermodynamic properties here.

Similar but fewer structure studies exist for other liquified noble gases, and $\phi(r)$ functions carefully constructed from solid and gaseous state data are available for them. For example, $S(Q)$ for Kr over a range of densities was measured by Egelstaff *et al.* (1983) but does not quite agree with the computer simulation using the ϕ's just referred to. Presumably the pairwise interaction assumed for the liquid is in question here. Even if three- or many-body forces are significant, it may be possible to *adjust* the parameters in a two-body expression like Eq. (7) so that it seems sufficient by itself. But the two-body ϕ would then be to some extent fictitious, and so the interplay of the quantum theory of atomic interactions and observation of the liquid-state continues.

6. METALS: SOME OBSERVATIONS

Sections 6–8 will deal with the structures of pure metals and binary alloys. Nevertheless, such examples as there will be of chemical thermodynamics and liquid-state theory, as opposed to metal theory, are in principle applicable to nonmetals also and can be regarded as examples of general approaches.

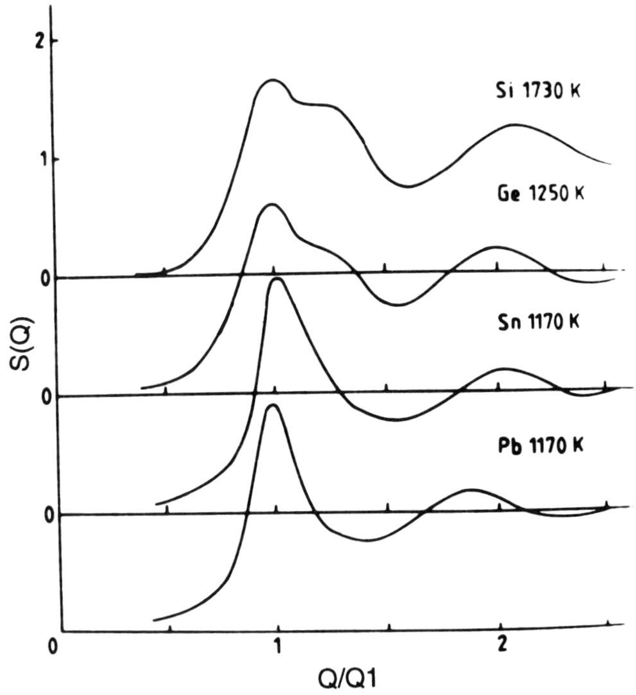

FIG. 6. Rather complicated $S(Q)$ shapes in Si and Ge; distorted first peak in Sn, tending to more close packed hard-sphere structure in Pb (see Sec. 8.4). From Hafner and Kahl (1984).

Figures 1 and 3 are typical of metals in group I of the periodic table for which $9.0 < n_1 < 10.5$. They are the simplest metals and, like noble-gas liquids, have PDFs quite close to that of a close-packed disordered system of hard spheres. The first peaks are symmetrical and the radial distances of the first and second peaks in $g(r)$ are in the ratio ~ 1.91. Be, Mg, Al, and Pb are like this also. In group-IIB metals, Zn, Cd, and Hg, the high-r and high-Q sides of the first peaks are spread out giving an asymmetrical peak though n_1 is still about 10. Although molten Si and Ge are metallic, not semiconducting like their crystals, n_1 is lower at ~ 6.6 and the ratio of the first two peak positions is higher at 2.0 to 2.3. These group-IV elements, and Ga and Sn also, have even more distorted peaks with shoulders or subsidiary maxima. Ge is discussed in detail below.

Thus there are noticeable structure trends through the periodic table and some examples are given in Fig. 6. It is plausible to see correspondence between these trends and those in crystalline elements, viz., close-packed metallic structure in groups I to III; distorted structures in group IIB and Ga, but more metallic close packing again in In and Tl; open covalent structure in group IV (Si and Ge) tending to metallic close packing at higher atomic number (Sn and Pb). Fundamental metal theory ought to account for this in both solids and liquids—see Sec. 8 below.

It is not easy to classify binary alloy structures. A random mixture of hard spheres of two different sizes would have the nearest-neighbor A–B distance between the A–A and B–B distances; the first peaks of the S_{ij} would behave similarly. The S_{AA} and S_{BB} would resemble those of pure A and B. The alloy Na-K in Fig. 7 is an example of this, and there are others, e.g., Al-Mg, Pb-Sn. There are cases, however, where the first peak of S_{AB}, but not S_{AA} or S_{BB}, has a well defined prepeak that has been interpreted as evidence of a larger size structural unit probably due to unlike-atom attraction. Figure 8 is an example with strong unlike-pair attractions.

Ordering (a preference for unlike neighbors) or *clustering* (preference for like neighbors), both departures from random mixing that should show up in S_{ij} and g_{ij}, must result from interparticle attractions and repulsions characteristic of the chemical nature of the components. The remarkable properties of some binary alloys at special compositions strongly reinforce this idea.

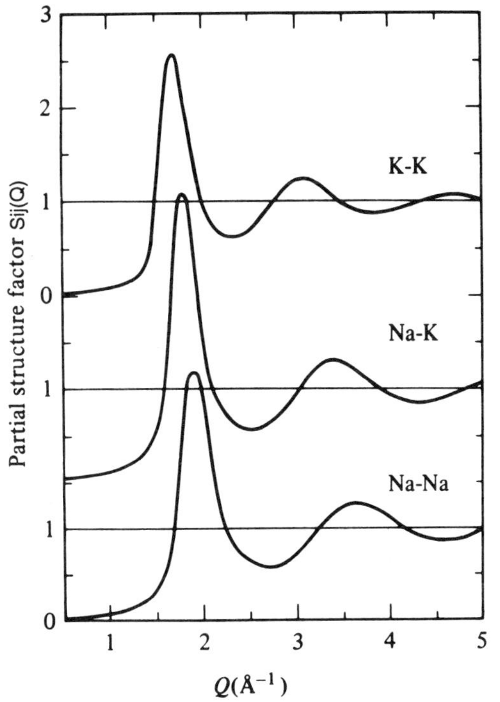

FIG. 7. Partial structure factors in Na-K; from Waseda (1980).

7. CHEMICAL SHORT-RANGE ORDER

There are surprises in the behavior of some liquid alloys that raise questions about the structure. For example, Ga and Hg form a two-phase liquid for some ranges of composition and temperature; liquid $Au_{0.5}Cs_{0.5}$ has very low electrical resistivity and behaves like a molten salt, which suggests that oppositely charged Au and Cs ions have each other as nearest neighbors (Kilic *et al.*, 1990). Anomalies such as very low electrical resistivity are often accompanied by extrema in thermodynamic properties, such as the entropy or volume of mixing, and in magnetic or thermoelectric behavior. Liquid systems well known for this include Li-*M* where *M*=Pb,Bi,Tl; Na-Sn; Cs-Pb; Mg-Bi; and Tl-Te. The significant point is that these systems have a special composition (Li_4Pb; Mg_3Bi_2; Tl_2Te, etc.), or possibly two, at which the anomalies appear most intense and that these compositions have stoichiometric proportions which suggest a chemical relation of the elements. There are usually crystalline compounds of the same composition. These liquids are often called *compound-forming alloys*. This leads to the hypothesis that SRO is caused by associations or complexes of chemically determined composition that are most in evidence when the overall concentration has the stoichiometric composition, e.g., Li_4Pb. This is called *chemical SRO*. The local associations are not conceived as stable molecules but as groupings of finite lifetime whose existence affects the $g_{ij}(r)$. This idea emphasizes the distinction between alloys of chemically attracting species and merely random mixtures.

7.1 Chemical Thermodynamic Models

On this basis many workers have tried to fit the observed thermodynamic properties by a chemical SRO model of the system. The fitted parameters enable the concentration of the hypothetical complexes to be calculated, and this knowledge allows discussion of other properties such as electrical conductivity. This is a very indirect and phenomenological way of investigating SRO, and there is considerable latitude in the fitting of numerous parameters. Nevertheless, the approach is an interesting halfway house towards a fundamental *ab initio* theory of structure, which presents profound difficulties (see below). Most workers are convinced that chemical SRO exists in the compound-forming alloys.

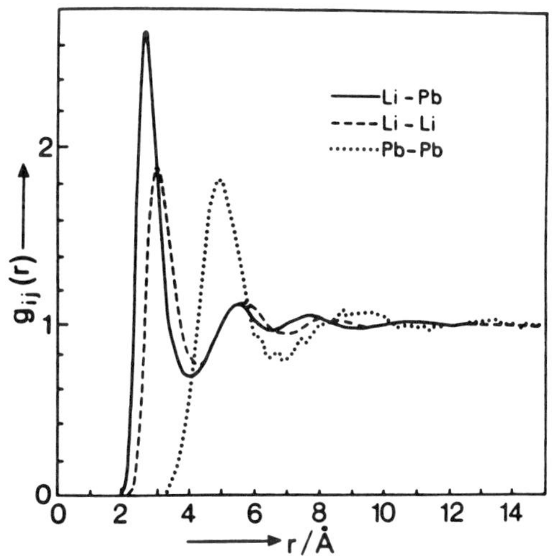

FIG. 8. Partial pair distribution functions in Li_4Pb derived by molecular dynamics. Note the preference for unlike near neighbors shown by the Li-Pb peak at 2.6 Å. From Jacucci *et al.* (1985).

Omitting detailed calculations, which can be found in Bhatia and Hargrove (1974) and Gray *et al.* (1980), the scheme is as follows. Elements A and B are attributed a tendency to form complexes $A_\mu B_\nu$, which have a Gibbs free energy of formation of $g(T)$ per mole at temperature T. The thermodynamic properties of mixing, such as the heat of mixing ΔH, can be derived from an expression for the total Gibbs free energy $G(p,T,c,N)$ of N moles of the binary mixture in volume V. This is

$$G=\sum_{i=1}^{3} n_i G_i^{(0)}+RT\sum_i n_i \ln \varphi_i$$

$$+\tfrac{1}{2}\sum_{i\neq j}\sum n_i\varphi_j\chi_{ij}, \tag{14a}$$

$$\sum_{i=1}^{3} n_i G_i^{(0)}=NcG_1^{(0)}+N(1-c)G_2^{(0)}-n_3 g, \tag{14b}$$

where the symbols have the following meanings: i or $j=1, 2, 3$ denote respectively atoms A and B and the complex $A_\mu B_\nu$; χ_{ij} is an interaction energy between i and j; $\varphi_i\equiv n_i v_i/V$, where n_i is the number of moles and v_i the partial molar volume of species i; c is the fraction of A atoms in the original binary mixture; $G^{(0)}$ is the free energy per mole of a pure constituent. The condition for chemical equilibrium is

$$\left(\frac{\partial G}{\partial n_3}\right)_{T,p,N,c}=0. \tag{15}$$

A plausible assumption based on chemical knowledge or observed anomalous composition is made for μ and ν; e.g., $\mu=4$, $\nu=1$ for Li_4Pb. Then $\Delta H(c)$, say, is computed from Eqs. (14) and (15) in terms of the parameters g and χ_{ij}, which are evaluated by fitting the observed $\Delta H(c)$. From this the three n_i can be calculated as functions of c. Having done this, Hoshino and Young (1980), for example, assumed that Li_4Pb conducted by virtue of n_1 free electrons from the n_1 Li atoms (valency 1) and $4n_2$ from the n_2 Pb atoms (valency 4) while the complexes contributed no free electrons. The thermodynamic model thus led to a curve of resistivity versus concentration which reproduced quite well the large peak observed at Li_4Pb.

7.2 Chemical SRO and Structure Factors

There is an interesting relation between the chemical SRO of compound-forming alloys and the $S_{ij}(Q)$'s of Sec. 3.3. Mathematically, a set of three new structure factors, formed by linear combinations of the $S_{ij}(Q)$'s, can be achieved in an infinite number of ways. A particular set turns out to have a useful physical interpretation and was first given a statistical mechanical basis by Bhatia and Thornton (1970). They are known as the *Bhatia-Thornton structure factors* and, omitting the argument Q, they are

$$S_{NN}=c_A^2S_{AA}+c_B^2S_{BB}+2c_Ac_BS_{AB}, \tag{16a}$$

$$S_{Nc}=c_Ac_B[c_A(S_{AA}-S_{AB})-c_B(S_{BB}-S_{AB})], \tag{16b}$$

$$S_{cc}=c_Ac_B[1+c_Ac_B(S_{AA}+S_{BB}-2S_{AB})]. \tag{16c}$$

The subscripts have the following significance. If many small equal volumes of alloy were sampled for the number of atoms in them, and for their concentration, both properties would reveal *fluctuations*, which are a fundamental property of macroscopic systems. Let ΔN and Δc stand for these fluctuations, i.e., for the departures of N and c from their average values. S_{NN} is a measure of the fluctuations in number density. S_{cc} expresses the concentration fluctuations. S_{Nc} represents the correlation between the two. In the limit $Q\rightarrow 0$, usually called the *long-wavelength limit*, $S_{NN}(0)=\langle(\Delta N)^2\rangle/\langle N\rangle$, where N is the number of atoms in the small volume in question and angular brackets indicate a statistical or ensemble average. Thus $S_{NN}(0)$ gives the mean square number density fluctuation. Similarly, $S_{Nc}(0)=\langle\Delta N\Delta c\rangle$ and $S_{cc}(0)=\langle N\rangle\langle(\Delta c)^2\rangle$. The value of these results is enhanced by the fact that for $Q\rightarrow 0$ the Bhatia-Thornton structure factors have thermodynamic expressions. The most widely used of these is

$$S_{cc}(0)=\frac{RT}{(\partial^2 G/\partial c_A^2)_{T,p,N}}=\frac{k_BT(1-c_A)}{(\partial\mu_A/\partial c_A)_{T,p,N}}. \tag{17}$$

In Eq. (17), k_B is Boltzmann's constant and μ_A is the thermodynamic quantity known as the activity of species A.

There are various thermodynamic and electrochemical ways of measuring $G(c,p,T)$ and $\mu(c,p,T)$. Equation (17) then gives $S_{cc}(0)$ as a function of c for one temperature at a time, usually for $p \sim 0$. The value of this in our present context is that the resulting curve can give strong indications about the SRO. For example, it can be shown that $S_{cc} = c_A c_B$ for a random mixture—often called an ideal solution; consequently, departures from this indicate some degree of ordering. In the extreme case $S_{cc}(0) = 0$, concentration fluctuations must have vanished—which is to say that a well-defined compound exists at the composition for which $S_{cc}(0) = 0$. At the other extreme, $S_{cc}(0) \rightarrow \infty$, the concentration fluctuations are as large as conceivable, i.e., the alloy separates into two phases. Figure 9 shows a hypothetical but not absurd case in which two tendencies to compound formation and one to phase separation are suggested. Measurements of $S_{cc}(0)$ have often stimulated the use of the chemical model of Eq. (14) by suggesting the values of μ and ν for a complex. Figure 10 shows a measured $S_{cc}(0)$ fitted by a model for a compound-forming alloy.

Since the Bhatia-Thornton structure factors are linear combinations of S_{ij} and therefore contain the same structural information, it must be possible to express the quantity $F(Q)$ of Eq. (5b) in terms of them instead of S_{ij}. Thus for neutron scattering we have

$$F(Q) = \langle\langle b \rangle\rangle^2 S_{NN} + (\Delta b)^2 S_{cc} + 2\Delta b \langle\langle b \rangle\rangle S_{Nc}, \tag{18}$$

where $\langle\langle b \rangle\rangle \equiv c_A b_A + c_B b_B$ and $\Delta b \equiv b_A - b_B$. Some elements have isotopes with both $b < 0$

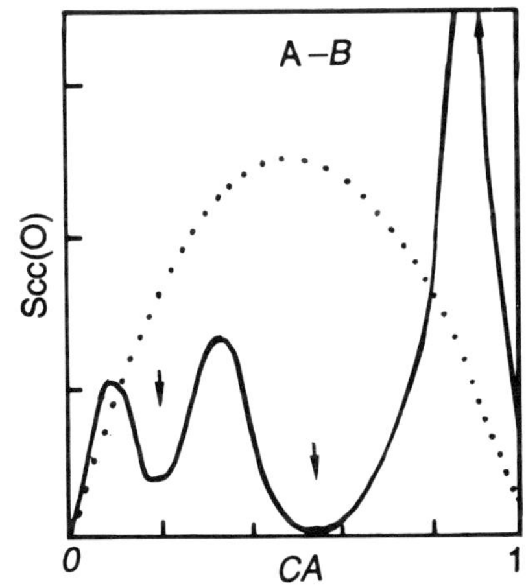

FIG. 9. A hypothetical $S_{cc}(0)$. The arrows indicate possible compositions for complex formation and the peak shows a tendency to phase separation. The dotted curve is for a random mixture or ideal solution for which $S_{cc}(0) = c_A c_B$.

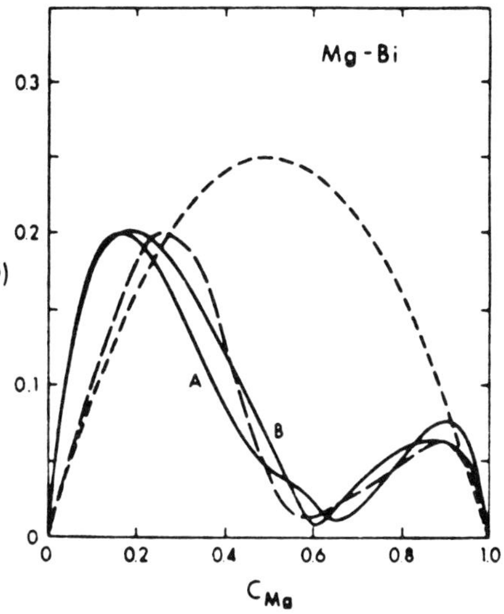

FIG. 10. A measured $S_{cc}(0)$ for Mg-Bi (long dashed curve), with $S_{cc}(0)$ ideal (short dashes) and two alternative model fittings A and B based on the ideas of Sec. 7.2. Note the tendency to compound formation at Mg_3Bi_2. From Bhatia and Hargrove (1974).

and $b > 0$. Suppose the isotopes were mixed to give $\langle\langle b \rangle\rangle = 0$. Then one measurement of $F(Q)$ gives $S_{cc}(Q)$ directly since the other terms in Eq. (18) vanish. Likewise, an isotope mixture devised to have $\Delta b = 0$ would give $S_{NN}(Q)$ directly from $F(Q)$. Naturally the possibilities of such manipulations are limited but are being tried out, particularly with metallic glasses, and are potentially powerful ways of measuring the partial structure factors separately.

8. STRUCTURE AND THE THEORY OF METALS

It is possible in principle to take the properties of an atom and deduce theoretically the structure of the condensed element. Perhaps the least exacting such exercise is to infer the structure of simple ionic crystals like NaCl from the nature of the Na^+ and Cl^- ions. Liquid metals probably present the most demanding of such tasks for three reasons: the interatomic forces are very complicated in metals; the structures are very disordered in liquids; and these two problems interact. Nevertheless, since about 1970, ways of combining liquid-state theory with metal physics have been devised, and by 1990 there was a reasonable understanding of why metallic elements, crystalline and liquid, have the structures they do.

This required a major effort of theoretical physics and the following sections therefore have the limited objective of setting out the nature of the arguments and giving a few conclusions. The following ingredients are needed for the discussion:

1. the general concept of a metal;
2. the pseudopotentials, the interionic potential, and the energy;
3. perturbation and similar techniques in liquid-state theory;
4. molecular dynamics.

8.1 The General Concept of a Metal

A pure liquid metal is a gas of electrons permeating a fluid of positively charged ions. There are about 10^{28} electrons/m^3, one per ion in Na, two in Ca, and so on according to valency z. They are loosely called *free electrons* and move through the entire system. Free electrons are responsible for the characteristic properties such as conductivity and metallic reflectivity. In a binary alloy the electron density n_e is $n_A z_A + n_B z_B$, unless complicated by compound formation as in Sec. 7.

8.2 Forces and Energy

The ions contain electrons tightly bound to nuclei. The free electrons flow through the ions under the control of two major influences:

1. the electrostatic attraction of the positive ions, and
2. a complicated response to the other electrons bound in the ions.

A powerful development of the wave-mechanical theory of solids made it possible to treat influence **2** mathematically as if it were an electrostatic force like **1** except that it is a repulsion that partially cancels the attraction **1**. The net effect of the two is quite small and is called a *weak pseudopotential.*

An electron passing through an ion may therefore be thought of as deviated by the appropriate pseudopotential. However, the ion does not suddenly exert its influence as the electron crosses a well-defined ionic boundary. As the ions are positively charged and the electron gas is mobile, the ions exert long-range influence on the gas by attracting electrons to themselves. Meanwhile the electrons are also repelling each other and a complex balance of forces sets in whose ultimate effect is to envelop each ion in a free-electron cloud. This cloud, being negative, neutralizes the effect of the positive ion rather completely at large distances, say a few atomic diameters, and less completely at smaller distances. The cloud of electrons piled up near an ion is called the ion's *screening cloud* and the total effect on an approaching electron of the ion and its cloud is called the *screened pseudopotential.*

In summary we may say that, from the point of view of one electron passing through a liquid metal, it encounters the screened pseudopotential of the ions (the nearest ones being felt more strongly) and is scattered by them. The total effect of all the ions, acting though their weak screened pseudopotentials, on all the free electrons is to produce the very electron clouds that do the screening.

Any two positive ions repel one another electrostatically, but in a metal, they have another, indirect, mutual influence because they both exert forces on the same electron gas, e.g., on each other's screening cloud. The latter contribution must depend on the density of the system because this determines the free-electron density, n_e. This contrasts with intermolecular forces in nonmetals, which are independent of density to a good approximation. However, *for a fixed density,* it turns out that the total ion-ion interaction in a metal can be treated just like the forces in nonmetallic liquids. In particular, the ions have *pairwise interaction.* The exact form of the ion-ion potential function will vary from metal to metal and differ from those of nonmetals; it depends on the screened pseudopotential. In general, it will have, as in nonmetals, a short-range repulsive core and a long-range attractive part, which, however, unlike nonmetals, exhibits maxima and minima called *Friedel oscillations* (see Fig. 11). This effective pairwise potential, $\phi_{\text{eff}}(r,n_e)$, controls the structure for any given density. It can be inserted into liquid-state theory to calculate thermodynamic properties and structure.

The expression for the energy of a metal is complicated but its structure is simple, viz.,

$$E(\text{per ion}) = E_{i,\text{ke}} + E_{\text{vol}} + E_{\text{str}}\,. \qquad (19)$$

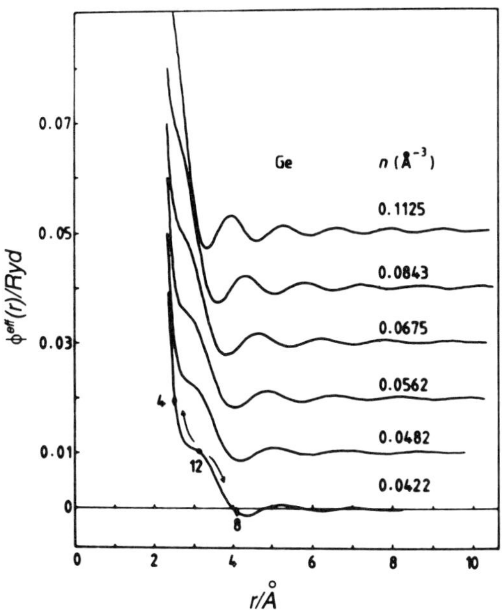

FIG. 11. ϕ_{eff} for Ge at various densities. From Arnold *et al.* (1989).

Here $E_{i,\text{ke}}$ is just the kinetic energy of the ions and is $3k_BT/2$ because their motion, unlike that of the electrons, can be treated classically. E_{vol} is independent of structure and embodies the idea that a free-electron gas has a kinetic energy that rises if the volume falls. E_{str} is due to the mutual forces of the ions. Because ϕ_{eff} is pairwise, E depends on the interparticle distances and therefore contains $g(r)$ or $S(Q)$, but also involves n_e. So far these remarks apply equally to crystals and liquids.

8.3 Structure from $\phi_{\text{eff}}(r)$ in Metals

The theory referred to in Sec. 4 assumes pairwise potentials and can therefore be taken over for liquid metals once pairwise potentials for these have been devised—provided, of course, due respect is paid to the volume-dependent aspects.

To exemplify the great progress of the 1970s (see, e.g., Ashcroft and Stroud, 1978), we first note that all the thermodynamic properties are derivable from the Helmholtz free energy, $F(\text{per atom}) = E - TS$ where S is the entropy per atom. Now there is a remarkable theorem, true in classical and quantum statistical mechanics, which works as follows. Suppose the real system, in our case a liquid metal, has its free energy conceptually divided into that of a hypothetical reference system (F_0) plus the rest (F_1) so that $F = F_0 + F_1$. Choose the reference system to consist of hard nonattracting spheres. Then the theorem says that $F \leqslant F_0 + \langle H_1 \rangle_0$, where H_1 is the difference in energy between the real liquid metal and the reference system. $\langle H_1 \rangle_0$ means the average of H_1 over all configurations of the reference system. The choice of hard-sphere reference system has advantages for this exposition, viz.,

1. the potential is zero for $r > \sigma$, the hard-sphere diameter, so that H_1 means simply the terms $E_{\text{str}} + E_{\text{vol}}$ while $E_{i,\text{ke}}$ just contributes to F_0;
2. all the properties of the hard-sphere (HS) reference system are well known and tabulated, including $S(Q)$, and depend only on σ and the density;
3. the hard-sphere $S(Q)$ is known to fit some liquid metals well, and so the reference system is structurally near to the real system.

Inserting the entropy term TS into the free energy we have

$$F \leqslant F_0 + F_1$$
$$= (3k_BT/2 - TS_{\text{HS}}) + (E_{\text{vol}} + E_{\text{str}} - TS_{\text{eg}}), \qquad (20)$$

where eg stands for electron gas. In calculating E_{str}, the $S(Q)$ for the HS system must be used because the theorem requires F_1 to be averaged over the *reference* system. The final step is a computing exercise to find the value $\sigma = \sigma_{\min}$ that minimizes $F_0 + F_1$, keeping the volume at the observed value for whatever temperature is assumed. The minimum $F_0 + F_1$ should be a good approximation to F and can be used to derive all the thermodynamic properties such as S, specific heat, etc. Furthermore, the hard-sphere $S(Q)$ for $\sigma = \sigma_{\min}$ is the theoretical prediction for the metal. Of course, the possibility of doing this depends on there being expressions for E_{str} that depend on $S(Q)$—which is exactly what the pseudopotential theory of metals provides. The same arguments apply to binary alloys, though the three S_{ij} and the three $(\phi_{ij})_{\text{eff}}$ cause technical complication.

This outlines the use of the Gibbs-Bogoliubov theorem with the hard-sphere reference system. Naturally, improvements proliferated. Better choices of reference system led to more general and reliable results at the expense of denser theory and heavier computing. But the essence remained of dividing ϕ_{eff} into an inner repulsive reference part (but not necessarily infinitely repulsive like hard spheres) and a long-range oscillatory part (like Fig. 11), together with a minimization or optimization procedure to choose the best parameters. For detailed examples see, e.g., Hafner (1977) and Hafner and Kahl (1984). The reader will have noticed that this variation method is similar in spirit to the perturbation approach of Sec. 5.

The concept of $\phi_{\text{eff}}(r)$ in metals, *for a given density*, makes molecular dynamics readily available for liquid metals. $g(r)$ and $S(Q)$ are therefore available in three ways: by diffraction experiments; by MD using $\phi_{\text{eff}}(r)$; and from liquid-state theory using $\phi_{\text{eff}}(r)$. An example comparing all three will be given below.

The problem is that $\phi_{\text{eff}}(r,n_e)$ depends on the free-electron states that are the solutions of Schrödinger's equation of motion. But these states depend on the structure, which in turn depends on ϕ_{eff}. In short, the structure and the electronic states ought to be solved in a self-consistent way. Procedures have been tried that avoid the excessive computation needed for a truly self-consistent calculation. For example, ϕ_{eff} is first obtained from a simple form of the pseudopotential theory that does not need details of the electron states. Two branches are now possible: use ϕ_{eff} with one of the forms of liquid-state theory to derive $g(r)$(theor.); or use ϕ_{eff} and MD to find $g(r)$(comp.). Either or both of these $g(r)$'s may now be compared with $g(r)$(expt.). Assuming agreement is encouraging, use the now known structure from MD to recalculate or improve knowledge of the electron states. This is a formidable program even for simple metals [see Hafner (1990)]. For transition metals (Fe, Ni, etc.) with partially occupied $3d$ shells of electrons, there is the additional rather intractable complication of incorporating the not wholly free d electrons into the expression for ϕ_{eff}. [See, e.g., Hausleiter and Hafner (1990) for an approach to this problem for the glassy transition metal alloy $Ni_{33}Y_{67}$.]

8.4 A Successful Example

Figure 11 shows $\phi_{\text{eff}}(r)$ for various densities of Ge calculated from the ideas of Sec. 8.2. The features are

1. a steep repulsive screened ion core—though not infinitely steep as in hard spheres;
2. an intermediate- and long-range oscillating part characteristic of metals and depending on the density of the electron gas;
3. an evolving relation of these as the changing density alters the position of the oscillations and spreads out the core.

If liquid Ge were controlled by the repulsive core acting alone, its structure would resemble a hard-sphere packing with $\sigma \sim 2.72$ Å$=r_1$ say, a second peak at $\sim 1.9r_1$, and $n_1 \sim 10$ to 11. But the density near the melting point is about 0.045 atom/Å^3, and Fig. 11 shows that ϕ_{eff} is distinctly structured. The observed peaks in $g(r)$ are at $1.6r_1$ and $2.25r_1$, which coincide with minima in the oscillatory part of ϕ_{eff}. Thus the atoms reduce overall energy by adjusting their positions in response to the superimposed demands of geometric sphere packing (with a periodicity fixed by σ) and an oscillatory tail (with wavelength $\lambda \propto n_e^{-1/3}$). The conflicting effects of the two different periodicities cause the peaks in $g(r)$ to peter out after ~ 7 Å. The atoms in the oscillatory troughs reduce what would have been the high peak of close-packed neighbors at $r_1 \sim 2.72$ Å and thus lower n_1 to the observed value of ~ 6.5. Using the ϕ_{eff} of Fig. 11, $g(r)$ and $S(Q)$ can be calculated by MD or liquid-state theory and compared with experimental results. Figures 12 and 13 show that

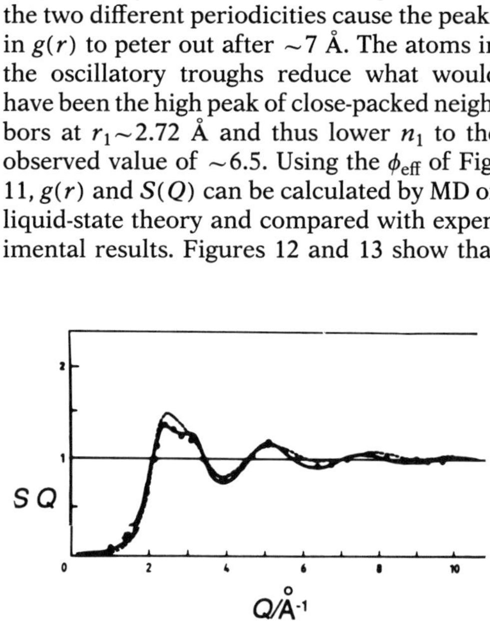

FIG. 12. $S(Q)$ for liquid Ge at ~ 1300 °C showing agreement between MD (full curve) and neutron diffraction experiments (dotted curve and circles). From Arnold *et al.* (1989).

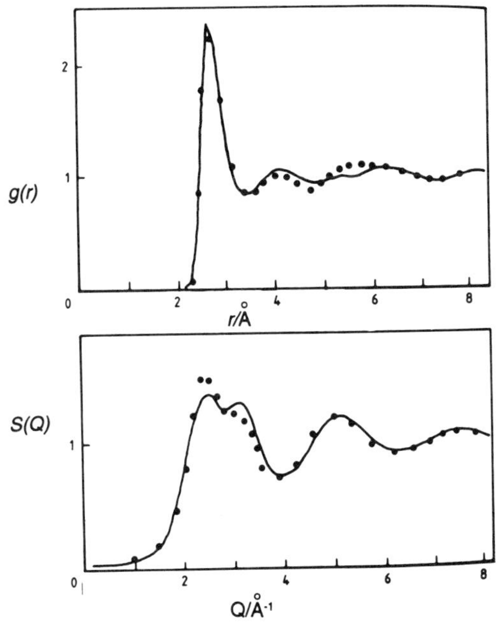

FIG. 13. *g*(*r*) and *S*(*Q*) in liquid Ge showing good but not perfect agreement between MD (full curve) and liquid-state theory (points). From Arnold *et al.* (1989).

these three approaches are beginning to agree well. [For the full theory, see Arnold *et al.* (1989).]

Figure 11 also shows that if the core radius and n_e are appropriate (e.g., the highest curve), the geometrical packing at $r \sim \sigma$ is encouraged rather than reduced by the position of the oscillatory minima: the two periodicities (σ and λ) are compatible, and structure resembling hard-sphere packing results. Calculations with suitable pseudopotentials and densities for the group I and IIA metals of the periodic table show that this is just what happens, leading to $g(r)$'s like Fig. 1. In this way detailed examination, element by element, can rationalize the trends through the periodic table referred to in Sec. 6 [see Hafner and Kahl, 1984). Both MD and liquid-state theory can be extended to alloys with the increasing complication due to three interionic potentials all dependent on the electron density, which is itself a function of composition. If chemical SRO in compound-forming alloys is due to ionic or covalent bonding forces becoming important at special compositions, then additional hypotheses for the forms of these forces are required. (For the use of exponentially screened ionic forces and application of MD to Li_4Pb, see, e.g., Jacucci *et al.*, 1985.)

9. MOLECULAR LIQUIDS

Even the simplest diatomic molecule, say N_2 or CO, involves Eq. (6) and raises the question whether the PDF is affected by the relative orientation of the molecules or whether, on the other hand, liquids of stable molecules approximate to hard-sphere packings to the extent that argon does.

Although homonuclear diatomic molecules and other special cases will not be dipoles, molecules in general will have electric dipole and quadrupole moments. The electric force fields will be nonspherical on this account and can be attractive or repulsive depending on orientation. The short-range repulsive force may also have an angular dependence. In special cases there can be a nonspherical magnetic force as in O_2. Much of the somewhat limited structural study of molecular

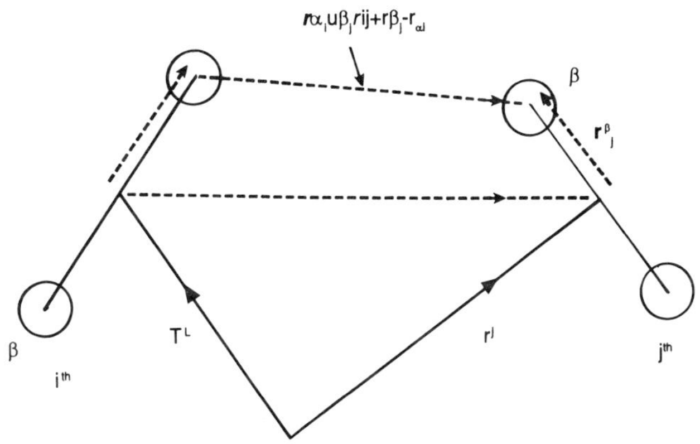

FIG. 14. The angles and distances between the *i*th and *j*th diatomic molecules composed of atoms of types α and β.

liquids has addressed the question of the orientation dependence of the pair distribution $g(\mathbf{r}_{ij};\ \omega_i,\ \omega_j)$. This is a complicated quantity, and Fig. 14 shows the angles and distances relevant to neighboring diatomic molecules to which this account will largely be restricted. Even if the three $g_{ij}(r)$ for the two kinds of atoms in, say, CCl_4 were fully determined by the methods of Sec. 3.4, they would not contain enough information to give $g(\mathbf{r}_{ij};\ \omega_i,\ \omega_j)$ uniquely, and a single diffraction experiment is even less informative.

The interaction potential can be modeled in various ways. For example, a simple and much used suggestion is

$$\phi_{ij}(\mathbf{r}_{ij};\ \omega_i,\omega_j) = \sum_{\alpha,\beta} \phi_{\alpha\beta}(\mathbf{r}_{\alpha_i\,\beta_j}) + \text{electric multipole terms.} \tag{21}$$

Here $r_{\alpha_i\,\beta_j}$ is the intermolecular distance between atoms of types α and β, and $\phi_{\alpha\beta}$ could be given the LJ form of Eq. (7). In a diatomic molecule the summation would have four terms. For comparison we quote without detailed discussion a potential used by Rodger *et al.* (1988) for Cl_2, viz.,

$$\phi(r_{ij}) = K\exp\{-\alpha[r-\sigma_0-\sigma(\Omega)]\} + C[r+\xi_0-\sigma(\Omega)]^{-6} + E_s, \tag{22}$$

where E_s is a dipole plus quadrupole term on each atom chosen to give the correct overall quadrupole for the molecule; Ω is the relative orientation of the two molecules, and $\sigma(\Omega)$ is expressed in terms of several spherical harmonic functions and contains a number of parameters which, with the other parameters in the expression, were chosen to give agreement with known properties of chlorine such as the crystal structure. This model implies orientation dependence of both attractive and repulsive forces and shows that a simple molecule can demand a quite complex potential. It is, of course, intuitively reasonable that a diatomic molecule should be dumbbell shaped, and at close quarters some relative orientations should favor closer approaches than others.

9.1 Structure Factors of Molecular Liquids

As in Sec. 3.3, the structure factor of a system of two species of atom, say CO or $GeBr_4$, is found by summing the scattered amplitudes and squaring to get the intensity. For a molecular assembly it is helpful to separate those contributions of interfering wavelets originating *within* one molecule from those coming from *different* molecules. $S_M(Q)$, now given a subscript to indicate a molecular liquid, may then be written

$$S_M(Q) = f_1(Q) + D_M(Q), \tag{23a}$$

where

$$f_1(Q) = \left(\sum_{\alpha} b_\alpha\right)^{-2} \left\langle \left| \sum_{\alpha} b_\alpha \exp(i\mathbf{Q}\cdot\mathbf{r}_\alpha) \right|^2 \right\rangle \tag{23b}$$

and

$$D_M(Q) = N_M^{-1} \left(\sum_{\alpha} b_\alpha\right)^{-2} \left\langle \sum_{i\neq j} \exp(i\mathbf{Q}\cdot\mathbf{r}_{ij}) \times \sum_{\alpha_i,\beta_j} b_\alpha b_\beta \exp[i\mathbf{Q}\cdot(\mathbf{r}_{\beta_j}-\mathbf{r}_{\alpha_i})] \right\rangle. \tag{23c}$$

In these formulas N_M is the number of molecules and α or β can be either A or B for molecules of two different kinds of atom. $f_1(Q)$, called the molecular form factor, is often calculable from the known molecular structure or available from electron diffraction studies of the gas phase. If so, a measurement of $S_M(Q)$ yields $D_M(Q)$, which includes the effect, if any, of orientation. $f_1(Q)$ can be refined by an additional factor $\exp(-\gamma_{\alpha\beta}^2 Q^2/2)$, where $\gamma_{\alpha\beta}$ is the rms thermal-vibration amplitude of the bond length, but this is often put equal to zero, thus treating the molecule as rigid with respect to internal motion; by hypothesis rigidity under intermolecular forces is already assumed. $D_M(Q)$ takes on a special form if there is no correlation of orientation at any r_{ij}. In that case, r_{ij}, r_{α_i}, and r_{β_j} are all statistically independent, and from Eq. (23c) it follows that

$$D_M(Q) = f_{2u}(Q) N_M^{-1} \left\langle \sum_{i\neq j} \exp(i\mathbf{Q}\cdot\mathbf{r}_{ij}) \right\rangle, \tag{24a}$$

where

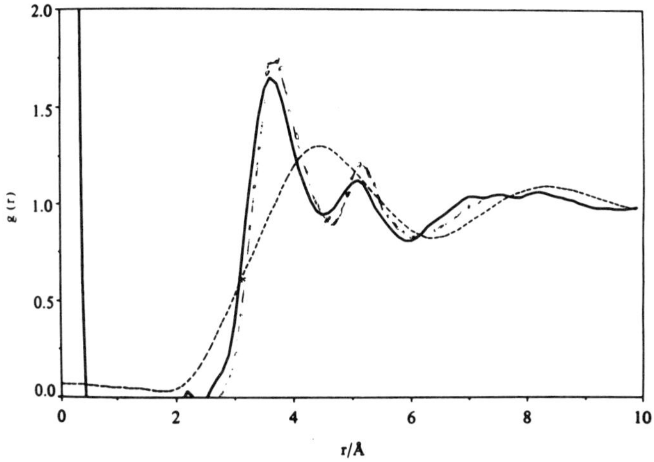

FIG. 15. Intermolecular $g(r)$ in liquid chlorine. Full line, from neutron diffraction; dashed line, calculated for an uncorrelated model; dot-dashed line, MD calculation (see text).

$$f_{2u}(Q) \equiv \left| \sum_{\alpha} b_{\alpha} \right|^{-2} \left| \sum_{\alpha} b_{\alpha} \langle \exp(i\mathbf{Q} \cdot \mathbf{r}_{\alpha}) \rangle \right|^{2}. \tag{24b}$$

Like $f_1(Q)$, $f_{2u}(Q)$ is calculable from the molecular structure: e.g., for a homonuclear diatomic molecule it is $\sin^2(Qd/2)/(Qd/2)^2$, where d is the bond length. Thus, measuring $D_M(Q)$ leads to the quantity

$$N_M^{-1} \left\langle \sum_{i \neq j} \exp(i\mathbf{Q} \cdot \mathbf{r}_{ij}) \right\rangle.$$

By direct analogy with Eq. (3b), this expression could be called $S_c(Q) - 1$, where the subscript indicates a structure factor for molecular centers and 1 is subtracted because the $i=j$ term is omitted from the summation. For uncorrelated orientation, therefore,

$$D_M(Q) = f_{2u}(Q)[S_c(Q) - 1]. \tag{25}$$

These equations have been written with neutron-scattering lengths. Similar formulas have been used for x rays, though it is a dubious approximation to treat the x-ray scattering factors for molecules as simple summations of the effect of the constituent atoms because the bonding will distort the electron distribution somewhat.

9.2 A Representative Experiment

Recent practice in this field is exemplified in a neutron experiment by Bellissent-Funel *et al.* (1990) on Cl_2. For values of $Q \leqslant 16$ Å^{-1}, at temperatures of 202 and 286 K, $S_M(Q)$ was measured and $D_M(Q)$ obtained by Eq. (23). By a Fourier transformation of $D_M(Q)$ analogous to Eq. (4a), a $g(r)$ was found, which, however, excludes the intramolecular Cl-to-Cl distance by virtue of the definition, Eq. (23c). Suppose that S_c is now calculated for a hypothetical set of molecular centers with no orientation correlation. It could then be inserted, together with f_{2u} for Cl_2, into Eq. (25) and an uncorrelated D_M calculated. The hypothetical $S_c(Q)$ was computed by MD for a monatomic liquid with an LJ potential [Eq. (7)] in thermodynamic states corresponding to those of the Cl_2 samples. The result shows that the observed $D_M(Q)$ differs markedly, having two peaks where the uncorrelated curve has one. Indeed the two peaks are thought to be a hallmark of correlated orientation in diatomic molecules. Figure 15 shows the corresponding $g(r)$'s and also one calculated by MD from the potential in Eq. (22). This potential, which builds orientation correlation into the interaction, gives a rather good fit. Furthermore, the ratio r_1/r_2 of the two peak positions in $g(r)$ is ~ 1.4, much lower than the ~ 1.95 characteristic of uncorrelated diatomic, and also monatomic, liquids. The peaks may therefore be interpreted as the effect of orientation modifying the first shell, not as two separate shells of neighbors. From these and other related arguments, it seems clear that orientational correlation exists in liquid Cl_2, but the conclusion remains rather qualitative. It is not obvious which correlations are preferred: e.g., do neighboring molecules tend to be parallel or antiparallel, or in perpendicular formation?

9.3 Some General Remarks about Structure in Molecular Liquids

It should be evident from the preceding paragraphs that even homonuclear diatomic molecules present substantial difficulties in the interpretation of data and the formulation of potentials. Hydrogen and deuterium halides evade our criteria of simplicity because hydrogen bonding plays a strong role in them; it cannot be said that their structure is well understood. Liquid N_2, CO, O_2, and the halogens have been studied by methods comparable with the Cl_2 experiment just referred to, and it appears that whereas definite orientation correlation is detectable in Cl_2, Br_2, and, especially, I_2, there is little or none in N_2, CO, and F_2. A detailed review is presented by Andreani *et al.* (1991). As indicated for Cl_2, MD simulations with comprehensive potentials are becoming available, and the understanding of molecular liquids is bound to grow as MD and experiments interact in future. In this connection an interesting development is the application of RMC (see Sec. 4.2) to halogen liquids by Howe (1990). From experimental data available up to about 1988 and RMC molecular distributions that can reproduce them, it appears that orientation correlation in the halogens persists little further than the first shell of neighbors.

Although this account has emphasized diatomic molecules for reasons of space and simplicity, the MD techniques are powerful enough to handle more complex cases, and diffraction experiments can likewise be involved with advantage, especially if isotopic substitution can be achieved. Thus there have been studies of linear triatomic molecules such as CS_2 (e.g., Tildesley and Madden, 1981), the tetrahedral molecule CCl_4 (e.g., McDonald *et al.*, 1982), and even mixtures of the two (Mittag *et al.*, 1989). In these cases the intermolecular potential had the form of Eq. (21) with LJ $\phi_{\alpha\beta}$'s and no multipole term. Thus between two CS_2 molecules there will be four S-S, four S-C, and one C-C interactions. It would be fair to say that in these cases the S_c's and $g(r)$'s from neutron and x-ray experiments, including some with isotopic substitution, are well reproduced by the computer simulations but that comparison is sometimes made difficult by discrepancies between one experiment and another and by inaccuracies in the computations. It is a measure of the difficulty of interpreting the data that firm statements about the orientation correlations cannot yet be confidently made.

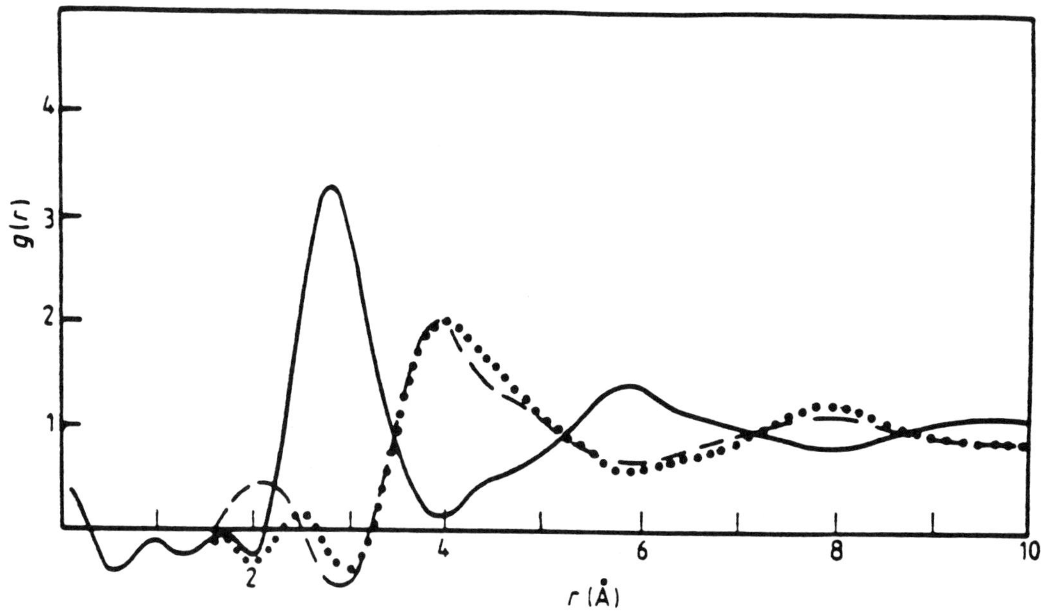

FIG. 16. Partial pair distributions in liquid NaCl; from Biggin and Enderby (1982). Full curve, g_{NaCl}; broken curve, g_{NaNa}; dotted curve, g_{ClCl}.

10. MOLTEN SALTS

The first crystalline structures determined by x-ray crystallography were those of alkali halides, and nowadays it seems natural that they should be interpenetrating lattices of positive and negative ions such that each ion has nearest neighbors of opposite sign. There seems no obvious reason why melting should alter this general picture since it does not radically change relevant properties, e.g., ionic conductivity persists. But it was not until 1975 that this was explored directly by the isotopic substitution method of Sec. 3.4 (Edwards *et al.*, 1975; Derrien and Dupuy, 1975). As in previous sections, the accumulated work in this field will be illustrated by representative examples.

10.1 Halides

Enderby and co-workers (see, e.g., Biggin and Enderby, 1982) changed the isotopic state of Cl in NaCl and obtained the three partial PDFs in Fig. 16. Using + and − to indicate cations (metal) and anions (Cl), we see that g_{++} and g_{--} are very similar with a major peak at ~4 Å; that the *nearest*-neighbor peak is in g_{+-} at 2.72 Å; and that for $r \geqslant 4$ Å, the phases of the curves conspire to cancel the charge almost completely. The similarity in g_{++} and g_{--} results from a corresponding similarity in S_{++} and S_{--}, and the first peaks in these structure factors coincide with a sharp dip in S_{+-} (see Fig. 17). There is, however, a small intrusion of like species into the first shell of unlike neighbors. The coordination number n_1, now more explicitly called n_{+-}, is 3.9. Thus the overall picture is indeed

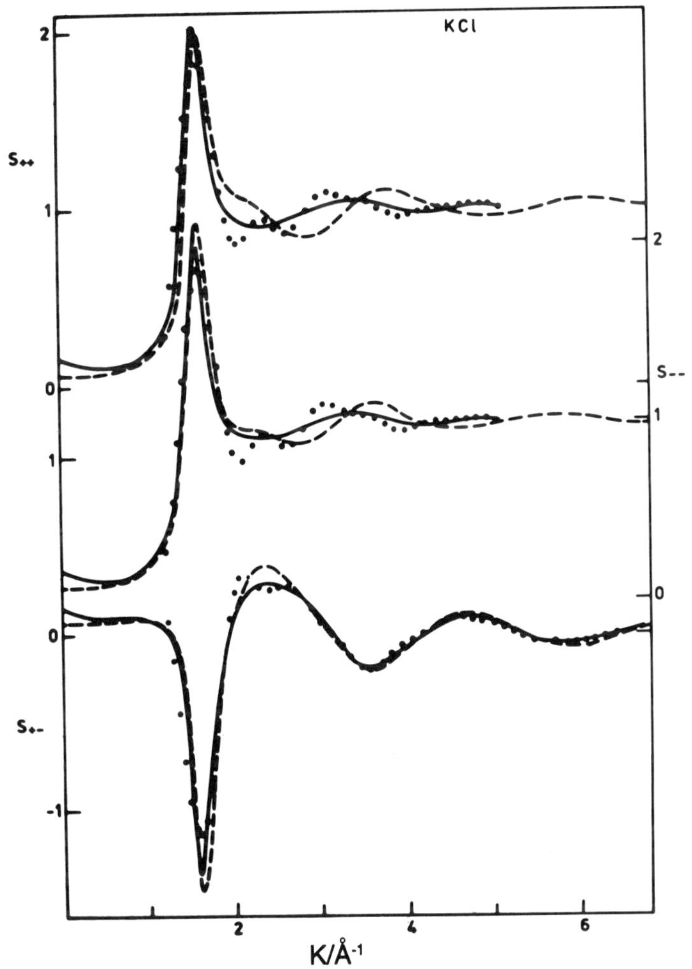

FIG. 17. $S_{ij}(Q)$ in liquid KCl. Full curve, HNC theory (see text); points, MD calculation; dashed curve, a less satisfactory theoretical result discussed by the authors. From Abernethy *et al.* (1981).

of ions surrounded by about four neighbors of opposite sign and with a next shell of mostly similar sign. However, this is only SRO, and the disordering effect of melting means that interparticle distances $\gtrsim 6$ Å are equally likely to be between like or unlike charges.

The clear interdependence of the g_{ij} in molten salts is in marked contrast to the behavior of these quantities in binary alloys. The explanation lies in the fact that the concentrations of positive and negative species in ionic melts are not independent because electrical neutrality must prevail. In binary alloys neutrality is preserved by the free electrons and the concentrations of the species can vary; there is then nothing to secure any unique relation between the peaks and troughs in g_{ij} or S_{ij}.

Group-II halides—$ZnCl_2$, $CaCl_2$, $SrCl_2$, and $BaCl_2$, etc.—are somewhat different (see, e.g., McGreevy and Mitchell, 1982). As the cation size increases from Zn to Ba there are trends of behavior believed to be associated with this size change. Let us assume a Cl^- ion at the origin; then, from Zn to Ba, while the first peak of g_{--} remains at about the same place (3.7 Å) and yields a roughly constant coordination number (3.7 to 3.8), it diminishes in height and spreads out. At the same time the penetration of Cl^- ions into the first shell of cations increases, as shown in Fig. 18. In $BaCl_2$, unlike NaCl, g_{++} and g_{--} are quite different. This is partly due to the different size of charge on Ba^{2+} and Cl^- but is also thought to be a consequence of the greater mobility of the anion. With the smallest cation, Zn^{2+}, there is a well-defined peak in g_{+-}, implying $\sim 4.3 \pm 0.3$ Cl^- ions around each Zn^{2+} ion, and the sharpness of this peak decreases progressively from Zn to Ba. These experiments yield no evidence of special Zn–Cl distances to be associated with complexes of substantial lifetime, such as Zn^+Cl or $ZnCl_3^-$ (compare Sec. 7).

10.2 Computer Simulation and Theory

The foregoing shows some of the inferences that can be drawn from neutron scattering with the isotopic method. Deriving them theoretically requires appropriate interionic potentials. As with the other types of liquid, there are model potentials of various degrees of complexity, but, for molten salts, they will be dominated by the Coulomb forces. These account for the bulk of the cohesive energy and cause the SRO of the charge distribution. A much used potential is

$$\phi_{ij}(r) = A\,\frac{z_i z_j e^2}{r} + \beta \exp[-\alpha_{ij}(\sigma_{ij} - r)] - \frac{C_{ij}}{r^6} - \frac{D_{ij}}{r^8}, \tag{26}$$

in which $z_i e$ and $z_j e$ are the ionic charges giving electrostatic attraction or repulsion $\propto r^{-2}$; the second term is the strong repulsion

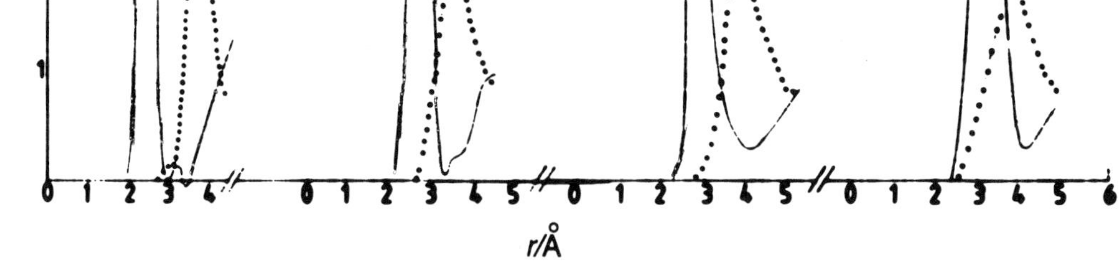

FIG. 18. Development of the penetration of like ions into the first shell of unlike neighbors as cation size increases, according to Biggin and Enderby (1981). The $SrCl_2$ curve is from molecular simulation, the others from neutron scattering. Full line, $g_{+-}(r)$; dotted line, $g_{--}(r)$.

due to overlap (an alternative to the r^{-12} term in the LJ model), and the remaining terms are the induced multipole terms (see Sec. 5) giving interactions of longer range than the repulsion but much shorter than the Coulomb field. A depends on the units, and the other symbols are parameters usually fixed by making ϕ compatible with the properties of the crystalline phases (see, e.g., Sangster and Dixon, 1976, for a detailed discussion of ϕ). With this potential, MD calculations of $g_{ij}(r)$ can be made, and the outcome is satisfactory for the alkali halides but less so for the group-II halides, suggesting that ϕ needs improvement.

Indications were given above of the perturbation method for $g(r)$ in liquid Ar and the Gibbs-Bogoliubov variation method in metals and alloys. For ionic liquids a third of the very numerous approximate methods in the theory of liquids will be invoked. The reader will appreciate that that justice can hardly be done to any of these methods short of a systematic exposition of liquid theory. What is needed is a solution of Eq. (9). Discussion is often facilitated by introducing two more correlation functions: $h(r)$, the *pair correlation function,* and $c(r)$, *the direct,* or *Ornstein-Zernike, correlation function.* These are defined for liquids by

$$h(r) \equiv g(r) - 1, \tag{27}$$

$$h(r_{12}) = c(r_{12}) + n_0 \int c(r_{13}) h(r_{23}) d^3 r_3. \tag{28}$$

$c(r_{12})$ has the effect of splitting the positional correlation $h(r_{12})$ of atoms 1 and 2 into a part depending directly on their positions and an indirect part depending through the integral on the correlated positions of all the other atoms. The range of $c(r)$ is much shorter than that of $h(r)$ or $g(r)$ and is comparable with that of $\phi(r)$. A considerable chain of reasoning, with approximations (see, e.g., Hansen and McDonald, 1986), then introduces another relation between $h(r)$, $c(r)$ and $\phi(r)$, viz.,

$$c(r) \simeq h(r) - \ln g(r) - \phi(r)/k_B T. \tag{29}$$

If this is inserted into Eq. (28), there results

$$\ln g(r_{12}) + \frac{\phi(r_{12})}{k_B T} = n_0 \int \left(h(r_{13}) - \ln g(r_{13}) - \frac{\phi(r_{13})}{k_B T} \right) h(r_{23}) d^3 r_3. \tag{30}$$

This is an integral equation soluble numerically for $g(r)$ once $\phi(r)$ is given. Equations of similar form hold for the $g_{ij}(r)$ in binary systems. This is known as the *hypernetted chain (HNC) approximation.* Unlike other approximate theories with which it is often compared in the literature, it works quite well for liquids with long-range Coulomb interactions. The $g_{ij}(r)$ and $S_{ij}(Q)$ for all the molten alkali halides have been computed by HNC and by MD, using the ϕ of Eq. (26), and the agreement, exemplified in Fig. 17, is good (Abernethy *et al.*, 1981).

With the help of the measured $S_{ij}(Q)$ and the RMC method (Sec. 4.2), it is possible to extend insight into molten salt structures because the RMC results are three-dimensional arrays consistent with the isotopic substitution experiments but not dependent on any assumed potential. For example, the first shell of neighbors can be analyzed by means of a quantity $C_{ij}(n)$, which is the fraction of ions of type i that have n ions of type j in their nearest-neighbor shell. $C_{ij}(n)$ can be extracted from the RMC structure. It turns out, for example, that ~0.4 of Na^+ ions have five Cl^- ions in the first shell in liquid NaCl; the corresponding numbers in CsCl are ~0.38 and ~6. Other Na^+ ions may have, though with lower probability, three, four, six, or seven Cl^- neighbors, giving an overall coordination number $n_{+-} = \Sigma n C_{+-}(n) = 4.85$. This emphasizes how disordered the shell is in the liquid compared with the well-defined one in the crystal and how limited the information is in a single coordination number. We also recall from Sec. 2.1 that particular n_{+-} values depend on the definition adopted by their authors. Another limitation partly removed by the RMC structures, as also by MD and MC, is the absence of angular information. If two lines are drawn from one Na^+ ion to two of its nearest-neighbor like ions, what is the distribution of the angle θ_{+++} between them? This, and the corresponding question for θ_{ijk}, where i, j, k can each be either + or −, can be answered at least approximately and begins to elucidate the triplet distribution. These angular distributions are in fact rather broad for θ_{ijk} in the alkali halides but show peaks at $\sim 48°$ for θ_{+++} and θ_{---} compared with 45° in the crystal, and near 90° and 180° for θ_{-+-} and θ_{+-+}. (For a more detailed discussion of such inferences, see McGreevy and Pusztai, 1990.)

We refer finally to $ZnCl_2$ for which the isotopic method, with or without RMC, suggests strongly that each Zn^+ ion is surrounded by a tetrahedron of four Cl^- ions. Since four anions per cation are not available, there is some vertex sharing of the tetrahedra similar to, though less orderly than, that in the crystal from which the material melts. Unlike the alkali halides, $ZnCl_2$ is very viscous just above its freezing point; its ionic conductivity is low and it readily forms a glass. These facts begin to suggest that its liquid and glassy forms depart from purely ionic interparticle forces towards ramifying networks of directional bonds more like those in glassy materials such as $GeSe_2$ or even the most well-known glass SiO_2. If so, $ZnCl_2$ would exclude itself from "simple liquids" as defined by our original criteria, and it is not surprising that, at the time of writing, it is one of several materials attracting more detailed study (Salmon, 1991.)

11. LIQUIDS AND GLASSES

One of the numerous ways of making glasses, indeed the normal way for most glasses, is to quench a liquid so rapidly that it has no time to crystallize. The result is a structurally disordered solid. All the concepts of Sec. 1–3.4 apply as well to glasses as to liquids. Much of Secs. 4.2 and 7–10 do also; e.g., glasses may be simulated by MD or discussed with pseudopotentials. The basic difference between liquids and glasses is that only the former are in thermodynamic equilibrium and therefore have their structures fully determined by the prevailing temperature and pressure. Since glasses are not in thermal equilibrium, their structure is not stable and in general changes slowly with time. Thus it is not uniquely defined: it depends on the history of the sample and especially on its mode of preparation and heat treatment. While the $g_{ij}(r)$ of a glass may well bear a general resemblance to those of a liquid of the same composition, they are rarely identical, and the glass may show more detailed features characteristic of short- or intermediate-range order. (See GLASSES.)

12. CONCLUSION

The reader may feel that the structure of simple liquids is somewhat unspectacular—all PDFs are continuous curves showing a few rather featureless peaks which soon peter out at r greater than a few atomic diameters. This just observation results from the fact that liquids are, to a first approximation, irregular packings of spheres. Size differences, the Friedel oscillation in ϕ_{eff} in metals, chemical binding forces in alloys, the orientation of nonspherical molecules, or Coulomb forces between ions all impose on the general randomness various forms of SRO and IRO that give the PDFs such features as they have. Understanding these is part of the general challenge presented by condensed matter including crystals. Determining structure from basic theory has never been easy because the energy differences between various alternative structures are small, so that quite accurate theories are necessary to establish which is the preferred structure of lowest energy. The period from 1970 to 1990 saw sufficient development in the theory of liquids and in computer simulations to justify a claim that structures are beginning to be understood from first principles even in disordered liquids. The subject is not closed, however. Apart from the problems of embracing more complex cases such as larger molecules or transition metals, there remain both experimental and theoretical difficulties in the way of understanding how structures change with density on the way to the liquid-vapor critical point. The triplet distribution and the effect on particle interactions of many-body forces also remain as challenges.

GLOSSARY

Ensemble Average: Average of a property over a large number of replicas of a system called collectively an ensemble; a basic device in statistical mechanics (*q.v.*).

Fourier Transform (FT): $F(y)$ is the FT of the function $f(x)$ if

$$F(y)=\int_{-\infty}^{\infty} f(x)\exp(-2\pi ixy)dx.$$

f is then the inverse FT of F and

$$f(x)=\int_{-\infty}^{\infty} F(y)\exp(2\pi ixy)dy.$$

There are many equivalent and related forms.

Free Energy: A fundamental thermodynamic property of any macroscopic system from which other properties may be derived. Gibbs free energy $G=E-TS+pV$; Helmholtz free energy $F=E-TS$.

Heat of Mixing: The heat given out, or taken in, when a mixture is formed from constituents.

Pairwise Interaction: Force between two particles that is unaffected by the presence of other particles.

Partition Function: The quantity $\Sigma \exp(-E_n/k_BT)$, where E_n is the energy of the nth allowed quantum state of a macroscopic system and the sum is over all states. It is fundamental for the calculation of thermodynamic properties by statistical mechanics (*q.v.*).

Stoichiometric Proportion: The proportion of constituents involved in chemical combinations.

List of Works Cited

Abernethy, G. M. Dixon, M., Gillan, M. J. (1981), *Philos. Mag.* B **43**, 1113–1123.

Andreani, C., Dore, J. C., Ricci, F. P. (1991), *Rep. Prog. Phys.* **54**, 731–788.

Arnold, A., Mauser, N., Hafner, J. (1989), *J. Phys. Condens. Matter* **1**, 965–980.

Ashcroft, N. W., Stroud, D. (1978), in: F. Seitz, H. Ehrenreich, D. Turnbull (Eds.), *Solid State Physics*, Vol. 35, New York: Academic, pp. 2–81.

Bellissent-Funel, M.-C., Buontempo, U., Petrillo, C., Ricci, F. P. (1990), *Mol. Phys.* **71**, 253–268.

Bhatia, A. B., Hargrove, W. H. (1974), *Phys. Rev. B* **10**, 3186–3196.

Bhatia, A. B., Thornton, D. E. (1970), *Phys. Rev. B* **2**, 3004–3012.

Biggin, S., Enderby, J. E. (1981), *J. Phys. C* **14**, 3577–3583.

Biggin, S., Enderby, J. E. (1982), *J. Phys. C* **15**, L305–L309.

Cusack, N. E. (1987), *The Physics of Structurally Disordered Matter*, Bristol, England: Hilger.

Derrien, J. Y., Dupuy, J. (1975), *J. Phys. (Paris)* **36**, 191–198.

Dore, J. C. (1990), in: M. C. Fairbanks, A. N. North, R. J. Newport (Eds.), *Neutron and X-Ray Scattering: Complementary Techniques*, Institute of Physics Conference Proceedings Series No. 101, Bristol, U. K.: IOP.

Edwards, F. G., Enderby, J. E., Howe, R. A., Page, D. J. (1975), *J. Phys. C* **8**, 3483–3490.

Egelstaff, P. A., Gläser, W., Litchinsky, D., Schneider, E., Suck, J.-B. (1983), *Phys. Rev. A* **27**, 1106–1165.

Enderby, J. E., North, D. M., Egelstaff, P. A. (1966), *Philos. Mag.* **14**, 961–970.

Gray, P., Cusack, N. E., Tamaki, S., Tsuchiya, Y. (1980), *Phys. Chem. Liq.* **9**, 307–322.

Guinier, A. (1963), *X-Ray Diffraction*, San Francisco: Freeman.

Hafner, J. (1977), *Phys. Rev. A* **16**, 351–364.

Hafner, J. (1990), in: H. Endo (Ed.), *Liquid and Amorphous Metals VII*, Amsterdam: North-Holland, p. 18.

Hafner, J., Kahl, G. (1984), *J. Phys. F* **14**, 2259–2278.

Hansen, J.-P., McDonald, I. R. (1986), *Theory of Simple Liquids*, London: Academic.

Hausleiter, Ch., Hafner, J. (1990), *J. Phys. Condens. Matter* **2**, 6651–6657.

Hoshino, K., Young, W. H. (1980), *J. Phys. F* **10**, 1365–1374.

Howe, M. A. (1990), *Mol. Phys.* **69**, 161–174.

Jacucci, G., Ronchetti, M., Schirmacher, W. (1985), *J. Phys. (Paris), Colloque* **46**, C8-385–C8-389.

Kilic, H. H., Pietzko, S., Schmutzler, R. W. (1990), in: H. Endo (Ed.), *Liquid and Amorphous Metals VII*, Amsterdam: North-Holland, p. 521.

Lieb, E. H. (1976), *Rev. Mod. Phys.* **48**, 553–569.

McDonald, I. R., Bounds, D. G., Klein, M. L. (1982), *Mol. Phys.* **45**, 521–542.

McGreevy, R. L., Mitchell, E. W. J. (1982), *J. Phys. C* **15**, 5537–5550.

McGreevy, R. L., Pusztai, L. (1988), *Mol. Simul.* **1**, 359–367.

McGreevy, R. L., Pusztai, L. (1990), *Proc. R. Soc. London Ser. A* **430**, 241–261.

Mittag, U., Samios, J., Dorfmuller, T., Guenster, S., Zeidler, M. D., Chieux, P. (1989), *Mol. Phys.* **67**, 1141–1151.

Rodger, P. M., Stone A. J., Tildesley, D. J. (1988), *Mol. Phys.* **63**, 173–188.

Salmon, P. S. (1992), *Proc. R. Soc. London Ser. A* **437**, 591–606.

Sangster, M. J. L., Dixon, A. (1976), *Adv. Phys.* **25**, 247–342.

Takeda, S., Harada, S., Tamaki, S., Waseda, Y. (1990), *J. Non-Cryst. Solids* **117/118**, 80–83.

Tildesley, D. J., Madden, P. A. (1981), *Mol. Phys.* **42**, 1137–1156.

Waseda, Y. (1980), *The Structure of Non-Crystalline Materials*, New York: McGraw-Hill.

Weeks, J. D., Chandler, D., Andersen, H. C. (1971), *J. Chem. Phys.* **54**, 5237–5247.

Yarnell, J. L., Katz, M. J., Wentzel, R. G., Koenig, S. H. (1973), *Phys. Rev. A* **7**, 2130–2144.

Further Reading

Cusack, N. E. (1987), *The Physics of Structurally Disordered Matter*, Bristol, England: Hilger.

Faber, T. E. (1972), *Introduction to the Theory of Liquid Metals*, Cambridge, England: Cambridge University Press.

Hafner, J. (1987), *From Hamiltonians to Phase Diagrams*, Berlin: Springer.

Hansen, J.-P. McDonald, I. R. (1986), *Theory of Simple Liquids*, London: Academic.

March, N. H. (1990), *Liquid Metals*, Cambridge, England: Cambridge University Press.

McGreevy, R. L. (1987), in: F. Seitz, H. Ehrenreich, D. Turnbull (Eds.), *Solid State Physics*, Vol. 40, New York: Academic, pp. 247–325.

Waseda, Y. (1980), *The Structure of Non-Crystalline Materials*, New York: McGraw-Hill.

LOCALIZED ELECTRON STATES

See ELECTRON STATES, LOCALIZED

LOGIC METHODS

See QUANTUM LOGIC METHODS; SEMANTIC APPROACH AND OTHER LOGIC METHODS

LOW-DIMENSIONAL PHYSICS

See ELECTRON STATES IN ZERO-, One-, and TWO-DIMENSIONAL SYSTEMS

LPCVD

See CVD (CHEMICAL VAPOR DEPOSITION)

MAGNETIC CONFINEMENT FUSION

See FUSION, MAGNETIC CONFINEMENT

MACHINES, ELECTRICAL

F. J. FRIEDLAENDER, *School of Electrical Engineering, Purdue University, West Lafayette, Indiana, U.S.A.*
D. W. NOVOTNY, *Department of Electrical and Computer Engineering, University of Wisconsin, Madison, Wisconsin, U.S.A.*

	Introduction	561
1.	**Principles**	562
2.	**Design Considerations**	563
2.1	Thermal	563
2.2	Electric Breakdown	564
2.3	Mechanical Failure	564
3.	**Direct-Current Motors and Generators**	564
3.1	General Description	564
3.2	Classification of dc Motors	566
3.3	Characteristics of dc Motors	567
3.3.1	Shunt Motors	567
3.3.2	Series Motors	567
3.3.3	Starting of Motors	568
3.4	Direct-Current Generators	568
4.	**Alternating-Current Machines**	568
4.1	Rotating Field	568
4.2	Induction Motors	569
4.2.1	Induction-Machine Equivalent Circuit and Characteristics	569
4.2.2	Single-Phase ac motors	571
4.3	Hysteresis Motors	571
4.4	Universal Motors	571
4.5	Three-Phase Synchronous Machines	572
4.5.1	Reluctance Motors	574
4.5.2	Amortisseur or Damper Windings	574
4.5.3	Transients in Synchronous Machines	574
4.6	Linear Motors	574
5.	**Permanent-Magnet Excitation**	574
6.	**Electronic Control of Electrical Machines**	575
6.1	Electronic Power Converters for Electric Drives	575
6.1.1	Controlled Rectifiers	575
6.1.2	Choppers	576
6.1.3	Phase Control of ac Voltage	576
6.1.4	Inverters	576
6.2	dc Drives	577
6.2.1	Capability Curves	578
6.2.2	High–Dynamic-Performance dc Drives (Servo Drives)	579
6.2.3	Open-Loop dc Drives	579
6.3	Variable-Speed ac Drives	580
6.3.1	Synchronous-Motor Speed Control	580
6.3.2	Induction-Motor Speed Control	581
6.3.2.1	Voltage-Controlled Induction Motor Drives	581
6.3.2.2	Adjustable-Speed, Constant-Volts/Hertz Induction-Motor Drives	581
6.4	Adjustable-Torque, Vector-Controlled ac Drives	582
6.4.1	ac Current Regulators	582
6.4.2	Field Orientation	583
6.4.3	Brushless dc Machines	584
6.5	Other ac Drives	584
	Glossary	585
	Further Reading	585

INTRODUCTION

The class of devices to be considered here are used for energy conversion with electrical energy as one component. In motors, conversion is from electrical to mechanical energy. In generators and alternators, the energy flow is reversed. Usually, the mechanical energy is supplied by means of a rotating drive, and the majority of motors supply rotational motion, but there are an increasing number of applications in which electrical energy is used to

1-56081-067-X/94/$5.00+.50

provide linear motion. In the following, motors utilizing dc, ac, and modulated power supplies are discussed, and dc generators and ac generators, usually called alternators, are considered. A final section treats the electronic control of electrical machines.

1. PRINCIPLES

All of the devices to be discussed employ a magnetic flux to produce a voltage, with the basic underlying principle being Faraday's law of magnetic induction. The calculation of the magnetic flux requires the application of Ampère's circuital law, and forces are calculated by using suitable formulations of the Lorentz force equations.

Computation of machine characteristics requires the calculation of fluxes in magnetic circuits containing ferromagnetic elements, often a difficult task complicated by the geometrical structures encountered. Hence, simplifying assumptions are usually made and magnetic circuit models employed, with circuit parameters obtained from tests on the machine. Also, computational methods are becoming increasingly available for magnetic flux computations; these are considered in the articles NUMERICAL METHODS and PHYSICS APPLICATIONS OF COMPUTERS—MODELING. Ampère's circuital law is the basic tool used for flux calculations:

$$\oint_{\mathscr{L}} \mathbf{H}\cdot d\mathbf{l} = \int_S \mathbf{J}\cdot d\mathbf{s}, \qquad (1)$$

where $\mathbf{H}$ is the field intensity vector, integrated over a closed path $\mathscr{L}$, $\mathbf{dl}$ is a length increment tangential to the path, and $\mathbf{J}$ is the current density vector, which is integrated normal to the area S enclosed by the path $\mathscr{L}$ over which the field is integrated. When $\mathbf{J}$ consists only of currents in windings (wires), i.e., there are no substantial eddy currents in the iron, the above equation is often simplified to

$$\sum Hl = \sum Ni. \qquad (2)$$

Here the individual terms Hl, known as magnetomotive forces (mmfs), are summed around a closed magnetic circuit, represented in a simplified manner by lumped magnetic circuit elements, and the current in each winding linking the magnetic circuit contributes a term Ni equal to the number of ampere turns carried by it. It should be noted that the above holds independent of the source of the current, external or induced by changing magnetic fluxes in the device.

In the magnetic circuits used in machines, ferromagnetic materials such as iron constitute the major portion of the magnetic flux path. These materials display a very nonlinear B vs H relation, where B is the flux density vector, or flux per unit area. B saturates at a flux density typical for each magnetic material (actually, not B but the magnetization M saturates—the reader is referred to the article MAGNETOSTATICS for a discussion of these concepts).

Faraday's law for lumped circuits simply states that the induced voltage V across an N-turn winding is given by

$$V = N\frac{d\phi}{dt}, \qquad (3)$$

where $d\phi/dt$ is the time rate of change of the magnetic flux ϕ linking all N turns. For sinusoidal flux variation, $\phi=\phi_m \sin\omega t$ where ϕ_m is the amplitude of the total flux and $\omega=2\pi f$ where f is the frequency of the flux variation. In the sinusoidal case, therefore,

$$V=\omega N\phi_m \cos\omega t. \qquad (4)$$

The amplitude of the voltage $V_M=\omega N\phi_m$, and

$$V_M=\sqrt{2}V_{\text{rms}}=2\pi f N\phi_m. \qquad (5)$$

With $\phi_m=B_m A$, where B_m is the amplitude of the flux density, assumed to be uniform over the area A enclosed by the coil,

$$\begin{aligned} V_{\text{rms}} &= (2\pi/\sqrt{2})fN\phi_m \\ &= 4.44fN\phi_m \\ &= 4.44fNB_mA. \end{aligned} \qquad (6)$$

In case of relative motion between flux sources and windings in which voltages are induced, it can be shown that these induced voltages can be calculated in the same manner, even if sliding contacts are involved. The literature often uses the concept of motional emf (electromotive force), which leads to the simple formula

$$V=u\,B\,l \qquad (7)$$

where V is the voltage across a straight conductor of length l moving at a velocity u perpendicular to a magnetic flux density B (i.e., u, B, and l are mutually perpendicular); in a more general case

$$V = \oint_{\mathscr{L}} \mathbf{u} \times \mathbf{B} \cdot d\mathbf{l}. \qquad (8)$$

The calculation of voltages induced in case of "general motion" requires care, but its discussion is beyond the scope of this article. In most rotating or linear-motion machines, the simple arguments mentioned in the previous paragraph are quite adequate. Equation (4) is only meaningful for a closed circuit (only a few elements may contribute to the voltage V; which ones may depend on the choice of observer in this formulation).

Forces are calculated by using the Lorentz force equation, which takes the form

$$d\mathbf{f} = I\, d\mathbf{l} \times \mathbf{B}, \qquad (9)$$

where $d\mathbf{f}$ is the force exerted on a conductor element $d\mathbf{l}$ carrying a current I, located in a field B—note vector relations. Analogously to the case of an induced voltage, the force on a straight conductor of length l carrying current I, located in a field B, is given by

$$f = I\, lB, \qquad (10)$$

with f, l, and B mutually perpendicular.

2. DESIGN CONSIDERATIONS

Limiting design considerations are those leading to premature failure of the machine. There are essentially three failure mechanisms: thermal, mechanical, and electrical breakdown (there is no equivalent magnetic breakdown, except due to excessive forces, covered under mechanical breakdown).

It must be understood that the concern here is not only with immediate failure but also "decreased lifetime" of the device. Hence, the design decision involves economic factors as well as "years of expected use." Sound scientific judgment is not always available for the last criterion.

2.1 Thermal

In a stationary device, there are copper (I^2R) losses in the conductors, magnetic losses due to cyclic magnetization of the (iron) core, and stray losses due to induced currents or alternating magnetic fields in the adjacent structure. To these must be added frictional and other losses due to motion in rotating machines. The copper losses are obtained from an integration of $J^2\rho$ (J is current density, ρ is effective resistivity) over the entire copper volume. In this calculation, proper account has to be taken of the skin effect in the conductors. Dielectric losses are usually small at power frequencies and are neglected here.

Losses in the iron path of an electromagnetic device depend on the magnetic flux density and frequency. Just as the current is not uniform over the cross section of a conductor, the flux density is largest at the surface of the magnetic material, if it is conducting. Loss calculations are much more difficult in this case and are usually based on empirical knowledge (test data) of, for instance, the particular lamination thickness of the magnetic steel used in different portions of the magnetic circuit. The situation is further complicated by the fact that the losses are anisotropic in the laminations, i.e., they depend strongly on the direction of the magnetic flux density vector $\mathbf{B}$, as do the magnetic properties, in general, in the iron parts of an electrical machine. Iron losses are described in detail in the article MAGNETIC MATERIALS. The maximum value of B that is appropriate in a given portion of the magnetic circuit is not only dictated by thermal considerations but is limited by the saturation of the magnetic flux density—too high a value of B will lead to larger magnetizing currents, which increase the no-load loss and decrease the amount of current that can be safely carried for energy conversion purposes.

The total heat generated, at given values of magnetic loading (B) and electric loading (J), depends on the volume (L^3, where L is the "linear dimension") of iron and copper. Without artificial cooling, dissipation of heat is a function of surface area, which increases only as L^2. Hence, it is clear that for optimum use of a machine, additional means of cooling must be provided as the device becomes larger. In many electric machines, the surface area is increased by the provision of fins in larger units, and fan blades are usually mounted on the armature/rotor shaft to provide additional air flow for cooling larger units. More elaborate methods of cooling may also be used.

2.2 Electric Breakdown

In the design of electrical machines, the maximum voltage must not produce electric fields that cause dielectric breakdown (failure of the insulation), corona, or sparkover. Clearly, the actual voltages must be such as to provide a margin of safety. As in other matters, the effect of aging on insulation has to be taken into account to provide trouble-free operation over the estimated lifetime of the machine. Transients are another important consideration, as the applied voltage may not be distributed evenly over all the turns during rapid voltage changes.

2.3 Mechanical Failure

Mechanical failure results from the forces that act on current-carrying conductors in a magnetic field—the same forces that provide the driving force in electric motors. Forces also arise in motors and generators as a result of their motion, specifically centrifugal forces in rotating machines. Torques on drive shafts cause torsion forces that can lead to shaft breakage. Electromagnetic forces can be calculated from the Lorentz force equation [Eqs. (9) and (10)].

Generally speaking, the important parameters in the design of a machine are the maximum flux density, determined by material properties (saturation magnetization and iron losses), and the current density J, usually dictated by thermal considerations. In the operation of the machine, the total flux and total current rather than their densities are of concern. Hence, the amount of iron and copper (dictated, in part by their cross-sectional areas) and the size of a machine are all related to the magnetic and electric "loading" of the iron and copper, at a given voltage and current rating. In ac machines, with the voltage dependent on the frequency—Eqs. (4) to (6) clearly show how voltage relates to both frequency and magnetic flux—the size of a machine at a given rating varies inversely with frequency. One can also have different ratios of copper to iron volumes, a fact emphasized in transformer design. Hence, economic factors play an important part not only in the construction but also in the operation of electric machines.

This illustrates both the economy of scale and difficulties with scale changes in a general way. An important result is the increase in power handling capability at a given size of device with frequency.

3. DIRECT-CURRENT MOTORS AND GENERATORS

3.1 General Description

A dc machine usually consists of a fixed field structure with evenly spaced salient-pole pieces (in multiples of two) with current-carrying windings to provide a magnetic field. Additional, smaller pole pieces are placed in the space between the main poles in larger dc machines (interpoles—see later discussion). Permanent magnets instead of current-carrying coils are used in many modern electric machines, especially in the smaller sizes, to produce the desired magnetic flux. Laminated pole faces are often provided in large machines to reduce eddy currents produced by magnetic field fluctuation due to armature slotting (see below).

The rotating armature consists of a steel shaft with a steel hub on which sheet-steel disks with slots on the outer periphery are placed. There is a small air gap between the field structure and the laminated armature core surface (see Fig. 1).

Windings (insulated copper conductors) are placed in the armature slots, and current is carried from the conductors through a commutator—copper conductor segments insulated from each other—to external circuits by means of brushes arranged at spacings equal to the spacing of pole pieces on the field structure. Some examples of typical armature windings are given in Figs. 2 and 3; there are many other variations.

A conductor on the rotating armature experiences a force f since it carries a current I_A in the presence of the magnetic field of density B produced by the stationary field structure. The induced voltage e and the force per conductor f, moving at a velocity u and having a

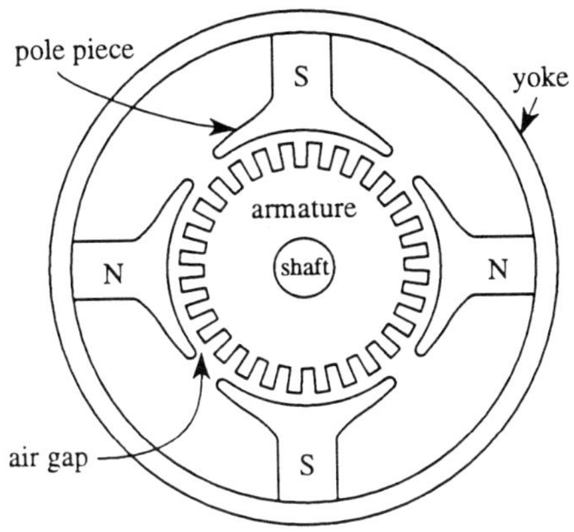

FIG. 1. dc motor. Commutator is mounted on shaft (not shown); see Fig. 2 for details of windings. Coils are placed on pole pieces to provide field excitation.

projected length l on the plane containing the **B** and **u** vectors, are given by [see Eqs. (7), (10)]

$$e = u\,B\,l \tag{11}$$

and

$$f = i\,l\,B. \tag{12}$$

More conveniently, the total generated voltage e_g for all the conductors in series between a pair of brushes is expressed by

$$e_g = K\phi\omega, \tag{13}$$

where ω is the radian velocity of the conductors (with $\omega R = u$, where R is the radius of the armature at the conductor position) and ϕ is the total air-gap flux per pole [$B\times$(pole area)]; the coefficient K is determined by geometrical factors such as armature length, the number

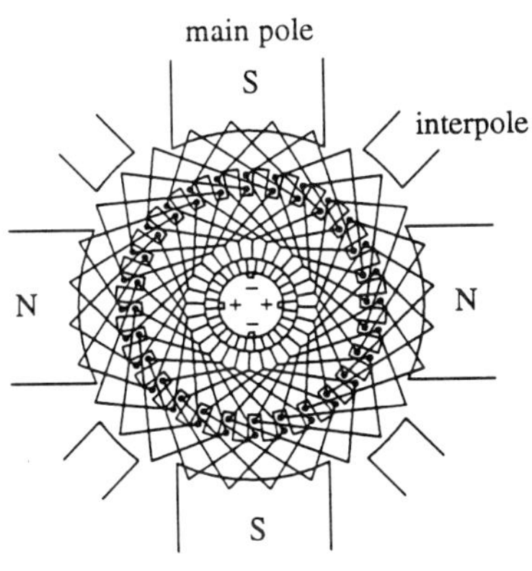

FIG. 2. dc motor with schematic of a lap armature winding shown.

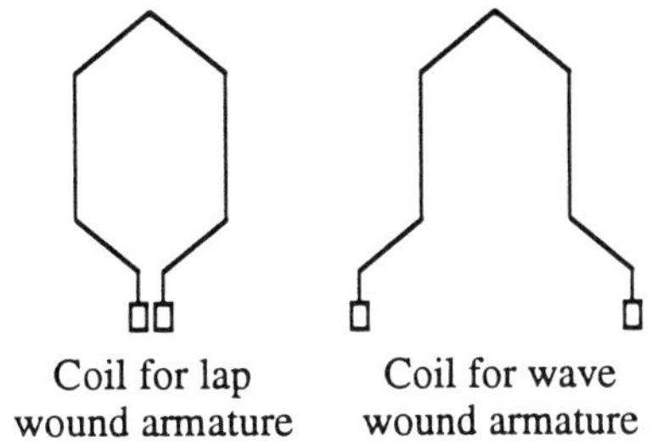

FIG. 3. Typical armature coil shapes.

of turns, and their arrangement between brushes, etc., and is a constant for a given machine. $\omega/2\pi = n$ is the speed of rotation in revolutions per second, or rpm if multiplied by 60. e_g is also referred to as back or counter emf (in motors).

The torque T_A produced by the interaction of the armature current with the field is similarly expressed as

$$T_A = K\phi I_a, \tag{14}$$

which gives the expected result

$$e_g I_A = T_A\omega \tag{15}$$

as required by energy conservation. T_A is the total torque, including friction and windage torques T_{loss} (shaft bearing friction and torques due to air motion produced by the rotating armature, often enhanced by fan blades to provide cooling). The net torque available at the shaft is $T_0 = T_A - T_{\text{loss}}$.

Basic equivalent circuits for dc machines are given in Fig. 4. From the circuit equation and torque balance (in steady state)

$$V_0 - R_A I_A = K\phi\omega, \tag{16}$$

$$T_0 + T_{\text{loss}} = T_A = K\phi I_A, \tag{17}$$

speed–torque and voltage–current relations are readily obtained for a known value of ϕ. ϕ is determined by the field excitation. ϕ and, therefore, e_g depend on the field current I_f in the nonlinear fashion shown in Fig. 5. This nonlinearity is a consequence of the magnetization characteristics of iron. A machine is considered to be a motor if $T_0 > 0$ and a generator if $I_A < 0$. In other words, energy flow from electrical energy to mechanical energy characterizes a motor, vice versa for a generator.

The magnetic flux ϕ due to the field excitation—different schemes are discussed

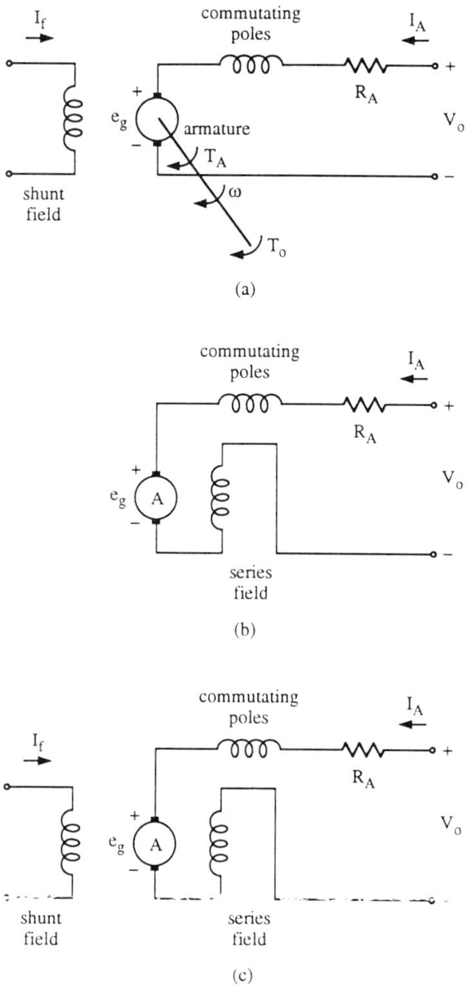

FIG. 4. Direct-current motor circuits: (a) dc shunt motor, (b) dc series motor, (c) dc compound motor.

below—can be readily obtained if the *B-H* characteristics and dimensions of each part of the magnetic path are known but are mostly determined by the air gap because of its higher reluctance (the effective permeability μ is much larger in the iron until it saturates). In practice, the experimental determination of e_g vs I_f is simple and convenient (and the results are readily adapted to all means of field excitation).

Currents flowing in the armature conductors produce an mmf that distorts the field produced by the poles. This effect is known as armature reaction. The armature current mmf essentially decreases the flux density in the trailing pole tip (in the case of a motor), and increases it in the leading pole tip. Because of saturation effects, the increase of flux density is smaller than the decrease, and the net effect of the armature reaction is to decrease the total available magnetic flux with increasing load.

Another factor that influences the operation and design of dc motors is commutation of currents as the commutator segments pass the brushes. Inductance effects cause sparking at the commutator brushes and, if not corrected, in all but the smallest motors or generators would lead to the destruction of the commutator. To counteract the effect, pole shoes (interpoles) are used in the space between the main poles (see Fig. 2), with windings connected in series with the armature in such a direction as to reduce the inductance of the armature coils under commutation to a small value. In very large machines, interpoles (also called commutating poles) do not suffice and a compensating winding, carrying armature current, is placed in slots on the main pole pieces, to oppose and hence reduce the net flux produced by the (commutating) armature conductors and avoid sparking at the commutator brushes.

3.2 Classification of dc Motors

Usually, dc motors are classified by the method used to obtain field excitation. Broadly speaking, one has to distinguish between *shunt excitation*, in which the field coils are supplied from constant or variable external dc voltage sources, and *series excitation*, in which the armature current flows through a set of field coils connected in series with the armature. Shunt coils are usually high-resistance, many-turn windings carrying a relatively small cur-

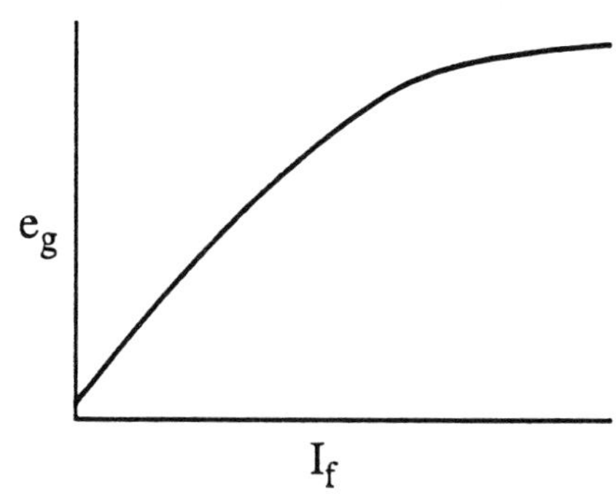

FIG. 5. Generated voltage e_g versus field current. Because of the remanent magnetic flux, e_g is not zero when I_f=0.

rent. On the other hand, series coils consist of few turns of wire, capable of carrying the total armature current at all loads.

A combination of shunt and series excitation is often used, with both a high-resistance shunt coil and a low-resistance series coil placed on each pole. Such a motor is called a *compound motor.* In a compound motor, the field excitation is given by the sum of the ampere turns of the shunt and series windings (or the difference in the case of a motor with negative compounding, which is rarely used). In Fig. 4 circuit diagrams are given for all the above.

dc motors may have fields provided by permanent magnets. Such motors behave like motors with shunt excitation, since a permanent magnet is, roughly speaking, equivalent to a winding carrying constant current.

3.3 Characteristics of dc Motors

The speed-torque characteristics of a dc motor depend on its mode of field excitation—shunt, series, or compound. Typical speed–torque curves are shown in Fig. 6.

3.3.1 Shunt Motors. As can be seen from Eq. (16),

$$\omega \approx (V_0/K\phi) \text{ if } R_A I_A \ll V_0,$$

as is usually the case. The effect of $R_A I_A$, at constant field excitation, is a small reduction in speed with increasing load. However, this effect is partly compensated by the decrease in flux ϕ due to armature reaction. A shunt motor provides, therefore, relatively constant speed; i.e., the slope of the speed–torque curve shown in Fig. 6 is very steep. Moreover, the motor speed level can be easily controlled by having the "shunt" field separately excited by means of a variable-voltage source (or a constant-voltage source with a variable series resistance or an electronic control; see Sec. 6). If the armature is also supplied from a variable-voltage source, then both V_0 and ϕ are independently controlled, resulting in a wide range of possible speeds. The easy speed control of dc shunt motors has been their traditional advantage over induction motors, offsetting their higher cost, though this advantage is now diminished by the advent of electronic control of induction motors.

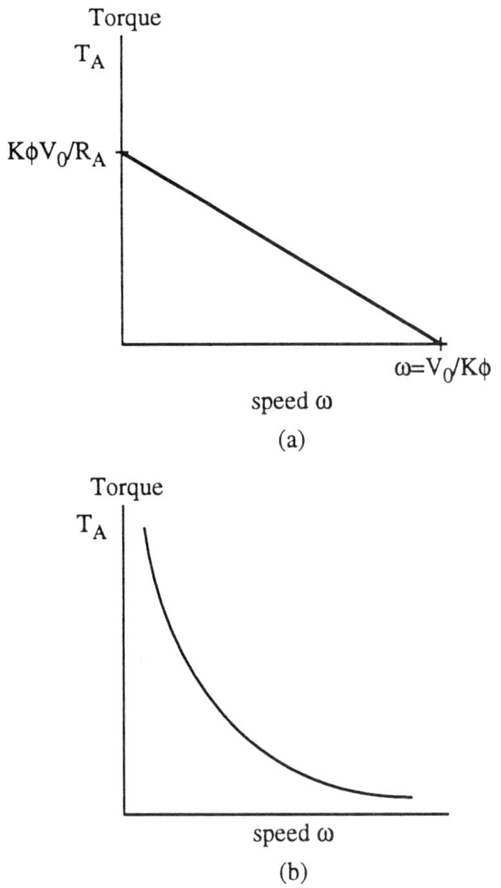

FIG. 6. Torque vs speed curves. (a) Shunt motor, (b) series motor.

3.3.2 Series Motors. When the field excitation is supplied by the armature current, ϕ becomes a function of I_A. If we neglect saturation effects, we have $\phi \propto I_A$, $T \approx K I_A^2$, and $\omega \approx (V_0/K I_A^2)$. Series motors exhibit very high torques at low speeds, at which armature current has to be limited to a safe value by means of external resistances (see Sec. 3.2.3). On the other hand, they have to be protected against excessive speeds at no load, since ϕ will drop almost to the remanent value, allowing excessive speeds. Series motors are used frequently in traction applications where the load is never removed and high starting torques are required.

The characteristics of a *compound motor* lie somewhere between those of shunt and series motors, depending on the amount of compounding (ampere turns of series winding relative to those on shunt winding, at a given load).

Operating points are readily found for any motor by plotting load-torque versus speed characteristics on the same graph as the motor torque-speed curves, and finding the intersection of the two curves, for each operating point (i.e., for each load condition and/or motor condition).

3.3.3 Starting of Motors. Transients in motors depend on mechanical and electrical time constants, with the mechanical time constants often much larger than the electrical time constants. For this reason, the starting currents of a dc motor have to be reduced by placing decreasing values of external resistance R_e in the armature circuit, as the speed ω gradually increases. I_A has to be limited to some safe value, a value that will not overheat the armature and series circuit during the starting period and will not allow excessive forces. R_e is usually decreased in steps so that I_A varies between an upper limit and the full-load value (or some similar value).

3.4 Direct-Current Generators

dc generators are essentially identical to dc motors, except that the power flow is reversed. Hence, the signs of T_A and I_A change (as well as the sign of T_0, of course). Then the applied torque T_0 must equal the sum of T_A and T_{loss}. Similarly, the output voltage V_0 is now given by the generated voltage $K\phi\omega$ minus the R_AI_A voltage drop [see Eq. (16)].

Again, generators can be classified by their field connections. The separately excited generator has its field (with a variable series resistance for voltage adjustment) connected to a constant-voltage source. Hence the generated voltage $K\phi\omega$ is a constant for a given ω (actually, armature reaction will cause some decrease of ϕ at higher load currents). A self-excited machine has its shunt field connected across the armature terminals. The resulting operating point is readily determined from the magnetic characteristic e_g vs I_f of the machine, shown in Fig. 5, and its intersection with the field load line, i.e., Ohm's law applied to e_g and I_f: $e_g = R_fI_f$ where R_f is the total resistance of the shunt field circuit. The fact that there is usually some permanent magnetic flux plays an essential role in making a self-excited generator possible (as seen in Fig. 5, e_g is not equal to zero when I_f is zero). As the armature starts rotating the remanent flux will induce a voltage in the armature, causing some current to flow, allowing the flux, and hence, voltage to "bootstrap" up to the desired value if appropriate resistance values are chosen for the field circuit.

Compounding i.e., additional field windings carrying the armature current, can be added, producing a compound dc generator. The two windings usually are aiding—both produce magnetic fluxes in the same direction. Hence, a compound generator displays a flatter V_0 vs I_A characteristic, since the series winding increases the flux at larger armature current, thus compensating for the larger R_AI_A voltage drop. Series generators are not often used, but their characteristics are readily derived from the principles given above. Typical curves for dc generator terminal voltage versus load current at a given speed are shown in Fig. 7 for various amounts of compounding.

4. ALTERNATING-CURRENT MACHINES

4.1 Rotating Field

A brief derivation of the rotating-field concept is given here, since it is the basis of most ac rotating machines. A three-phase arrangement will be considered, in which three separate sets of coils are supplied with three-phase power—three voltage sources with 120° time-phase difference between them.

We shall assume that each set of coils, arranged on the surface of a cylinder, pro-

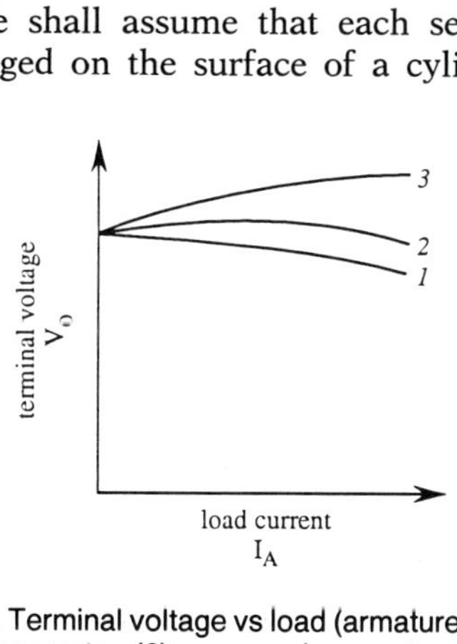

FIG. 7. Terminal voltage vs load (armature) current. (1) Shunt generator, (2)compound generator with flat compounding, (3) overcompounded generator

duces a sinusoidally distributed flux at the cylindrical surface, described by

$$B = B(t)\cos(p/2)\theta \qquad (18)$$

with

$$B(t) = B_m\cos(\omega t), \qquad (19)$$

where p is the number of poles, determined by the winding pitch (i.e., the coil span; i.e., 180° for $p=2$, 90° for $p=4$, etc.) and placement of windings and θ defines the angular position on the cylinder. Obviously p has to be an even number. One period of the pole structure (i.e., "north pole" to nearest "north pole") spans $2\pi/(p/2)$ radians. The quantity $p\theta/2 = \theta_e$ is referred to as "electrical degrees." In a typical three-phase machine, where the supply voltages differ in phase by 120°, the three identical sets of windings are displaced by $\theta_e = 120°$ from each other.

The resultant field B at an angle θ and time t is given by the addition of the three components:

$$\begin{aligned} B &= B_m\cos\omega t\ \cos(p\theta/2) + B_m\cos(\omega t + 2\pi/3) \\ &\quad \times\cos(p\theta/2 + 2\pi/3) \\ &\quad + B_m\cos(\omega t - 2\pi/3) \\ &\quad \times\cos(p\theta/2 - 2\pi/3) \\ &= (B_m/2[\cos(\omega t + p\theta/2) + \cos(\omega t - p\theta/2) \\ &\quad + \cos(\omega t + p\theta/2 + 4\pi/3) \\ &\quad + \cos(\omega t - p\theta/2) \\ &\quad + \cos(\omega t + p\theta/2 - 4\pi/3) \\ &\quad + \cos(\omega t - p\theta/2)] \\ &= \tfrac{3}{2}B_m\cos(\omega t - p\theta/2). \end{aligned} \qquad (20)$$

Thus, the resultant field is a single field of constant amplitude $\frac{3}{2}B_m$ rotating at a speed

$$\frac{d\theta}{dt} = \omega/(p/2) = 2\pi n_0, \qquad (21a)$$

where n_0 is called the synchronous speed (in revolutions per second). Expressed in rpm,

$$\frac{d\theta}{dt} = \frac{\omega}{p/2}\,\frac{60}{2\pi} = (120f/p)\ \text{rpm}. \qquad (21b)$$

Reversal of the connection to one of the three phases will reverse the direction of rotation. Note that each phase—each set of coils—produces an alternating field that can be resolved into two half-amplitude oppositely rotating fields. In this arrangement, the fields rotating in one direction have a vector sum of zero; those rotating in the opposite direction are exactly in phase and hence add to give a rotating field of amplitude $3B_m/2$. n-phase systems for $n \geqslant 3$ (n an integer), with coils displaced by $\theta_e = 2\pi/n$ radians and supplied from sources with phase differences of $2\pi/n$ radians between adjacent phases, behave similarly. $n=2$ is a single-phase system with polarity reversal.

4.2 Induction Motors

An induction motor usually has a three-phase winding arranged in slots in the laminated stator structure, producing a rotating field in the air gap, that is, the space between the stator lamination surface and that of the cylindrical rotor. There are slots symmetrically placed around the rotor armature, into which copper or aluminum bars are inserted. These bars are shorted by solid metal (copper or aluminum) rings at each end. The conductor structure, if visualized without the iron (it is often cast as a unit on the rotor and not removable), looks like a cage; hence the name "squirrel cage" induction motor is used to describe this type of motor

In an induction machine the rotating field generated by the stator windings induces voltages and, hence, currents in the usually short-circuited rotor conductors. The field caused by the rotor currents interacts with the stator field and causes the rotor to start rotating. A simple explanation: the rotor field "chases" the stator field but never catches up—if it did, there would be no induced currents in the motor and, hence, no interaction [but supplying a dc current to the rotor conductors would cause the motor to "lock" in with the rotating stator field and become a synchronous motor, rotating at $(120f/p)$rpm].

4.2.1 Induction-Machine Equivalent Circuit and Characteristics. The equivalent circuit of the "locked rotor" (rotor not turning) ac motor is essentially that of a transformer (with an air gap). As the rotor starts turning, the voltage induced in the rotor windings decreases, and goes to zero if a speed of

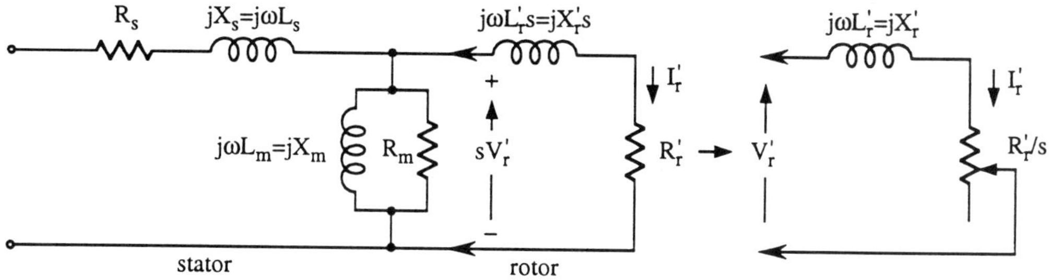

FIG. 8. Approximate equivalent circuit of induction motor showing evolution to modified rotor representation.

$2f/p=n_0$ rps is reached (i.e., the rotor moves at the speed of the rotating field). If we define slip as

$$s=(n_0-n)/n_0, \quad (22)$$

where n is the actual speed of rotation, the voltage induced in the rotor can be shown to be equal to sV_r, where V_r is the voltage induced in the "locked" rotor.

An equivalent circuit is shown in Fig. 8. It should be noted that the secondary (rotor) circuit has its values referred to the primary; hence the primed parameters (see TRANSFORMERS). As the speed of the rotor increases, and hence its speed relative to the rotating field decreases, the frequency of the voltage induced in the rotor decreases–it is always equal to $f(n_0-n)/n_0$, or sf, where f is the frequency of the stator ac supply.

It is convenient to "scale" rotor values by s, i.e.,

$$I_r'=sV_r/(R_r'+js\omega L_r')=V_r'/(R_r'/s+j\omega L_r') \quad (23)$$

without affecting the current (or the torques produced by these currents).

The evolution of the rotor circuit, referred to primary stator windings, is also shown in Fig. 8. Since $R_r'/s=R_r'+[(1-s)/s]R_r'$, Fig. 8 allows a simple interpretation of the power flow. I_r' and R_r' are the physical values of the parameters (actually R_r' may have to be adjusted for the change in the frequency of rotor currents because of skin-depth variations). The term $[(1-s)/s]R_r'$ is a fictitious resistance that models the power flow from the motor to the mechanical system via the rotating rotor shaft:

$$P=Tn_0(1-s)=I_r'^2[(1-s)/s]R_r'. \quad (24)$$

Since I_r' can be found from the equivalent circuit, torque versus speed curves for a fixed-voltage excitation of the stator are readily obtained from Eq. (24).

Typical speed–torque curves for an induction motor are shown in Fig. 9, for a series of values of rotor resistance R_r. Increasing the value of R_r shifts the point of maximum torque T_m to lower speeds, without affecting the value of maximum torque, as is readily derived from the above equations. Increasing the value of R_r consequently increases the starting torque. Squirrel-cage motors are often built to have a larger resistance upon starting than during normal operation. This is achieved by taking advantage of the fact that the depth of penetration of the currents varies with the speed: Cast aluminum or copper bars are placed in deep rotor slots, and a second winding with a higher resistance (usually cast aluminum) is placed at the top of the slots, near the air gap. As a result, the rotor resistance is higher near the surface where the currents concentrate at starting. At higher

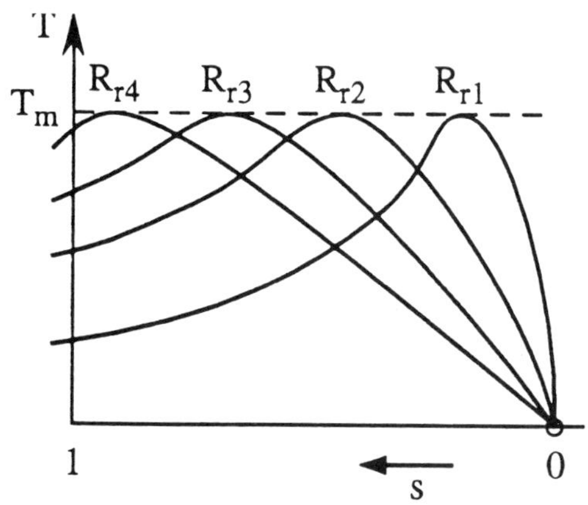

FIG. 9. Induction-motor torque-vs-slip curves for a number of values of rotor resistance R_r, with $R_{r_4}>R_{r_3}>R_{r_2}>R_{r_1}$. T_m is the maximum torque independent of value for R_r for constant voltage.

speeds, the frequency of the current decreases and, consequently, the current spreads deeper and a large portion of it is carried by the lower resistance copper bars.

As s reverses its sign (motor is driven above synchronous speed n_0) an induction motor becomes an induction alternator—it will convert mechanical energy to electrical energy. However, the stator has to be supplied by an ac source: this machine is not "self-excited" unless a bank of capacitors is connected to the terminals. Figure 10 shows torque and current versus s over a wide range of values of s defining different ranges of operation.

In some cases, instead of a squirrel-cage, separate three-phase windings are placed on the rotor and connected through slip rings to an outside circuit. This scheme allows a better control of the induction machine, but is becoming rare as more effective electronic means are being adopted to control the machine via the stator power supply.

4.2.2 Single-Phase ac Motors. The principle of operation of single-phase motors is readily explained by using an observation made in connection with the derivation of Eq. (20). Each phase produces an alternating field that can be resolved into two half-amplitude oppositely rotating fields. Superposing the torque speed curves of the two oppositely rotating fields, one finds that at the small value of slip at which induction motors operate, only the torque due to the appropriate component is significant. To get the motor started in the desired direction several schemes are used. The most common is the use of an auxiliary winding, displaced by 90° electrical from the main winding, and phase shifted by close to 90° by placing a large capacitor in series with it. A centrifugal switch may be used to disconnect this starting winding as the motor approaches the design speed. It should be noted that a single-phase motor is much noisier (pulsating rather than rotating field), larger, and less efficient than a three-phase motor of the same rating. Its main advantage is that it can be used where three-phase power is not available. A torque versus speed curve for a single-phase induction motor without an auxiliary starting winding is shown in Fig. 11.

In a *shaded-pole motor,* used for small single-phase motors with low starting torques, a short circuited copper strap is placed on a fraction of each pole (the "lagging" end of the pole). The induced currents provide a "lagging" field that, in turn, provides a small unidirectional torque, sufficient to start a fan motor, for instance.

4.3 Hysteresis Motors

Hysteresis motors (usually very small single-phase motors) derive their torque from the lagging flux produced by the motor magnetization, during subsynchronous operation. Special magnetic materials are used that are semihard (i.e., they are "permanent magnets" that can be cycled around their hysteresis loop by the stator currents). When the rotor reaches synchronous speed, the rotor flux no longer cycles and the permanent magnet rotor behaves as in a synchronous machine.

4.4 Universal Motors

Universal motors are series motors designed with relatively low field inductance, to allow them to function on either dc or ac supplies. Both stator and armature need to be laminated, and the design is usually more difficult since coils under commutation have voltages induced in them due to the ac fluxes (if used on an ac supply). Because of the higher ac impedance of a shunt field, as compared with its resistance, and because the high inductance of the shunt would lead to an undesirable phase difference between the stator and armature fields, universal motors are

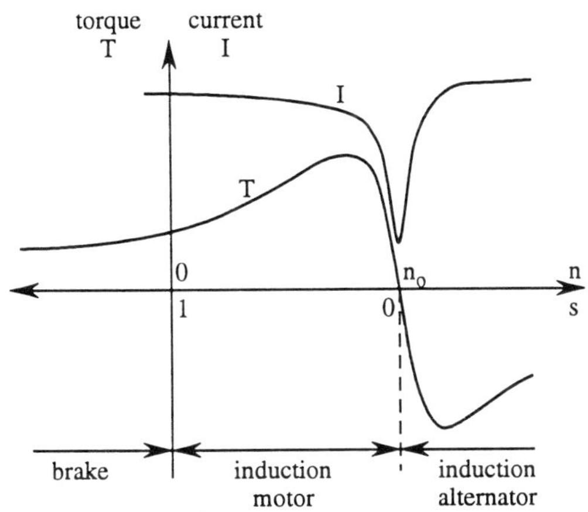

FIG. 10. Induction-motor torque and current vs slip and speed curves, showing the different regions of operation.

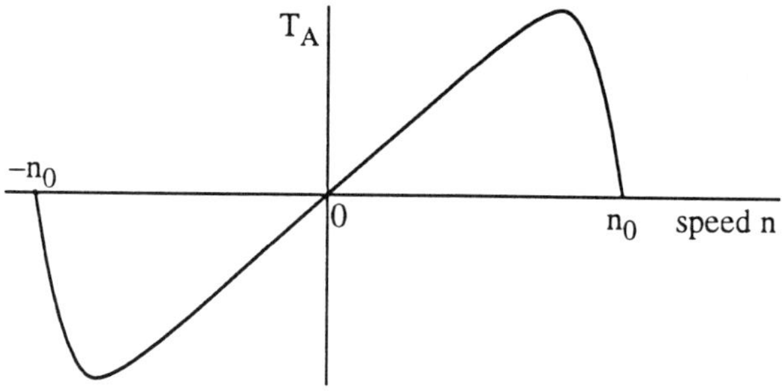

FIG. 11. Torque–speed curve for a single-phase induction motor without auxiliary starting winding.

not practical in a shunt-field configuration. Universal motors are commonly used in hand tools and similar applications.

4.5 Three-Phase Synchronous Machines

In synchronous machines, dc-excited poles rotate at a speed ω_s, where $\omega_s = 2\pi n_0/60$ rad/s, relative to stator windings similar to those described for an induction machine rotor. In a three-phase machine, there will be three windings displaced by 120° electrical from each other. The rotor may have a smooth cylindrical surface with slots to hold the dc windings (in large turbine alternators) or may consist of salient poles excited by windings, or the windings may be replaced altogether by permanent magnets. In this discussion we will explain the behavior of synchronous machines from a simple point of view without detailed considerations of harmonics—pure sine waves are only encountered in ideal machines—or of saturation effects. The reverse configuration (dc field on the stator and three-phase winding on the rotor) is possible, but rare, and can be described in a similar manner.

For symmetrical three-phase circuits, one needs to consider only one of the three phases—the other two will only differ by having their currents and voltages displaced by +120° and −120°, respectively. For smooth-rotor (cylindrical rotor) synchronous machines, the reluctance of the magnetic paths will not depend on the position of the rotating member (if one neglects the effect of the slots, which will produce harmonics but will not directly affect the fundamental component in a major way).

An understanding of the principles of operation is most easily gained by considering the magnetic fluxes produced by the field current I_f and by the currents in the three-phase windings. Suppose the two fluxes are exactly in phase so that a rotor "south" pole faces a stator "north" pole. In this situation no torque can result. The two fluxes must be out of phase, and in fact the stator windings carrying the currents that produce the mmf will ideally be situated 90° (electrical) away from the center of the rotor pole, in a region of zero flux. For the rotor field to lead the stator field (the air-gap field), a torque, that is, power, needs to be applied to the rotor shaft; the synchronous machine, then, is an alternator. If, on the other hand, the rotor field lags behind the stator (air gap) field by some angle, then a torque is applied to the rotor by the field—it is "pulled along" by the air-gap field—and the machine becomes a synchronous motor. It can be shown that the torque T (or power flow $T\omega_s$, with ω_s a constant) is proportional to the product of the stator current I_s, the rotor field ϕ_f, and $\sin\gamma$ where γ is the electrical angle between the fields produced by the rotor and the stator current I_s, respectively. A very simple model results if one neglects stator resistance (reasonable for large machines), leakage reactance, and saturation so that one can assume that the total mmf produced relates linearly to the total voltage V_t at the terminals of the ideal machine. This voltage is the sum of E_f, the voltage generated by the rotor magnetic flux, plus the "armature reaction" voltage due to the stator current. $V_s = jX_sI_s$ where X_s is the magnetizing reactance of the stator and δ is the angle between E_f and V_t (between the direction of the rotor field and the resultant field, i.e., the rotating field in the air gap due to all mmfs. One obtains

$$(V_t - E_f)/jX_s = I_s. \tag{25}$$

It follows from the above that the power P per phase of this ideal machine is

$$P = (V_t E_f \sin \delta)/X_s \tag{26}$$

with the sign of δ determining the direction of power flow. The same qualitative result was obtained from the physical reasoning above.

We can use this ideal model to understand the power flow, which has a maximum value of $V_t E_f/X_s$—if this is exceeded, the machine will fall out of synchronism (that is, the stator and rotor fields do not rotate at the same speed any more, and the average torque is zero). One can also study the use of the machine as a *synchronous condenser*, the operation when $P=0$ and hence $\delta=0$ (still neglecting losses) and V_t is fixed (supplied by a "bus," which is essentially a connection to a system that determines both V_t and the frequency f, and hence ω_s). For one value of field current I_f, $V_t=E_f$ and $I_s=0$. If I_f is increased, I_s has to increase in such a direction and phase that $\mathbf{E}_f + jX_s\mathbf{I}_s = \mathbf{V}_t$ for $E_f > V_t$ ($\delta=0$ since we are assuming a lossless, ideal machine). $\mathbf{I}_s$ has to lead $\mathbf{E}_f$ and $\mathbf{V}_t$ by 90°, i.e, $\mathbf{E}_f$ $E_f\angle 0°$ and $\mathbf{V}_t = V_t\angle 0°$ in phasor notation with $\mathbf{I}_s = I_s\angle 90°$. This machine behaves just like a capacitor, and hence this mode of operation is called that of a "synchronous condenser." A decrease in field current would cause a 90° lagging current to flow in the stator, a less desirable mode since a purely inductive load would lower the power factor of the system (assumed to be lagging), normally a poor option.

In a "real" machine, the terminal voltage V_t would be modified by additional voltage drops due to stator resistance and leakage reactance, conceptually similar to the transformer circuit model. Though complicating the circuit somewhat (see Fig. 12), the calculations needed to take account of the more realistic situation are straightforward.

Also, the nonlinear effects due to saturation in the iron have to be taken into account. The field excitation curve E_f vs I_f displays the effects of saturation—it is quite similar to that of a dc machine. Values of E_f have to be adjusted accordingly in regions of higher mmf and, hence, higher E_f values. With saturation, the value of X_s also decreases.

Experimental determination of cylindrical-rotor synchronous-machine parameters is similar to that of transformer parameters. E_f is determined by means of an open-circuit test, as a function of the field mmf. The short-circuit current is obtained, also as a function

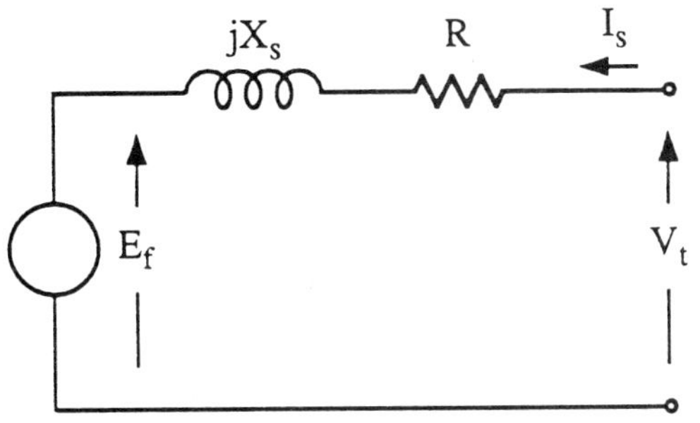

FIG. 12. Approximate equivalent circuit of a synchronous machine (per phase). X_s is the synchronous reactance and R is the total resistance of the stator circuit.

of the field mmf. The ratio of the open-circuit voltage divided by short-circuit current is the synchronous impedance at the value of the field mmf at which the ratio is taken. The synchronous impedance is the sum of the resistance plus the total reactance of the machine. The latter is also called synchronous reactance.

A salient-pole synchronous machine has separate pole pieces, with gaps between the poles. Hence, the reluctance of the magnetic path varies as the rotor rotates, with a complete period of the variation occurring as the rotor advances by 180° electrical. This situation is described most conveniently by defining two separate values of the stator reactance, the direct-axis synchronous reactance X_d for the position of the stator windings so that their mmf acts in the low-reluctance path, and a quadrature-axis synchronous reactance X_0, 90° electrical from the direct axis, when the reluctance is a maximum. The stator current is expressed in terms of its components, one in phase with the direct-axis position and the other (at 90° from the first) in phase with the quadrature-axis position of the rotor relative to the stator.

If the resistance and leakage reactance of the stator windings are neglected, one can write, approximately, for the salient-pole motor,

$$P = (V_t E_f/X_d)\sin\delta + (V_t^2/2X_d X_q) \times (X_d - X_q)\sin 2\delta. \tag{27}$$

As can be seen, the saliency adds a term called reluctance torque (or power when multiplied by ω_s) that only depends on the angle δ

between the center of the salient field structure and the peak of the air-gap flux variation of the fundamental term. Since the saliency introduces a 180° (electrical) periodicity, a "2δ" function appears.

4.5.1 Reluctance Motors. A *reluctance motor* is one that uses no dc field but relies entirely on the variation of reluctance of the structure to produce a torque. In principle, a *stepping motor* also is based on similar reluctance torques with the essential difference that, instead of three-phase excitation, two or three phase windings (arranged appropriately 90° or 120° electrical from each other) are excited successively by pulses, at a rate consistent with the speed at which the motor is to be "stepped" (and within a frequency range consistent with the capability of the system in terms of electrical and mechanical time constants).

4.5.2 Amortisseur or Damper Windings. Slots may be provided on the air-gap surface of the field structure (assumed to be the rotor here and in most of the previous discussion), and copper or aluminum bars placed in the slots and short-circuited at each end by means of a copper or aluminum ring, respectively, as in an induction (squirrel-cage) motor. The purpose is twofold. Currents are induced in these conductors when there is any "slip" between the rotating dc field structure and the synchronously rotating ac field, and will serve as a means of damping such motion. In a synchronous motor, this short-circuited winding will start the motor as an induction motor until it can be synchronized as the slip becomes small. Synchronization of synchronous machines is an important, practical topic beyond the scope of this article.

5.4.3 Transients in Synchronous Machines. These are a somewhat formidable subject. In the previous treatment, reactances have been discussed, but only for situations when all magnetic fluxes are stationary with respect to each other in a rotating frame of reference (expressed above by the term "synchronous speed"). In transients, this is no longer the case and additional "transient" and "subtransient" reactances have to be introduced, to account for the flux linkages between the dc field windings and the ac stator windings and damper windings, respectively. In salient-pole machines, direct- and quadrature-axis transients and subtransients have to be taken into account. The basic method used is based on the "conservation of flux linkages," in which one examines the rate of change of flux linkages in each circuit according to its time constant and uses the fact that flux linkages are conserved for times that are short compared to the time constant of the circuit considered.

4.6 Linear Motors

Linear motors are used to produce linear rather than rotational motion. Linear motors are built to operate as induction, synchronous, or reluctance motors. But instead of windings arranged in a cylindrical geometry, a planar geometry is used. The design may differ somewhat and larger air gaps are customary in traction devices. More sophisticated controls are needed. Only the windings in the region of the "moving part" of the motor (previously called rotor) are excited, and speed control, especially in synchronous motor design, is usually accomplished by means of variable-frequency excitation. One prospective application is the use of linear motors in high-speed transportation, with levitation integrated into the motor in some designs.

5. PERMANENT-MAGNET EXCITATION

Both dc machines and synchronous machines require a dc field structure. Conventionally, these fields were supplied by current-carrying coils placed on a pole structure (iron pole pieces) as explained above. However, with the advent of new permanent-magnet materials, and the availability of inexpensive ferrite magnets, dc fields can be supplied by substituting permanent magnets for the electromagnets in the field structure. In the automotive industry, ferrite permanent-magnet motors are commonly used in the dozens of small motors used in each automobile. Greater torque requirements are placed on some applications such as starting motors in a car or truck, so that neodymium-iron-boron permanent magnets, which provide much higher magnetic flux densities and have higher coercive forces, are used in these and other more demanding applications.

In modern practice, conventional dc motors are being replaced by so-called brushless dc motors (see Sec. 6.4.3). These are really synchronous motors with a permanent-magnet field structure. They are driven by variable ac polyphase sources, derived from solid-state supplies powered by a dc power source (or rectified ac sources). Total speed control over the entire range from starting to full speed is achieved by controlling the frequency.

6. ELECTRONIC CONTROL OF ELECTRICAL MACHINES

When supplied from conventional power sources, characterized by constant frequency and nearly constant voltage, electrical machines operate with their characteristic speed-torque curves. For all but series-connected dc machines (Sec. 4.2) and the ac equivalent, the universal motor, this results in very nearly constant-speed operation.

Adjustable-speed operation requires, as a minimum, the ability to control voltage. For efficient control of induction machines and for synchronous machines, coordinated voltage and frequency control is necessary. Because of the practical problems associated with control of frequency, traditional adjustable-speed drives utilized dc machines. However, advances in power electronics have made variable-frequency inverters quite feasible, and adjustable-speed ac drives are now readily available.

6.1 Electronic Power Converters for Electric Drives

The size and cost of all electronic power converters are strongly affected by both their continuous and short-time current capacities, which must therefore be carefully chosen and limited in the design process. An important characteristic of all electronic power converters is, therefore, very limited current capacity compared to conventional power sources. In all cases the controller must be designed to avoid the typical high inrush currents associated with normal "line starting" of motors. Current-limiting circuits and instantaneous electronic trip features are necessary to avoid damage to the electronic converter switching devices.

6.1.1 Controlled Rectifiers. Controlled rectifiers are used to provide variable dc voltage for motor control from a single-phase or three-phase ac supply. Thyristors (silicon-controlled rectifiers) are generally favored for higher power levels because of their high current and voltage ratings and low cost despite the attendant need for relatively complex timing and control circuits. Figure 13 shows a very good form of controlled bridge rectifier, called a dual bridge, capable of full four-quadrant operation. The circuit consists of two full bridges, one to handle forward (motor) current and one to handle reverse (regenerating) current. Avoiding having both bridges on simultaneously is imperative since this would result in a three-phase input fault. Reactors are often employed to limit fault current and allow recovery without shutdown. Because of control complexity and reliability considerations, the dual bridge is only used when necessitated by regeneration requirements. More typically only one bridge is used, and limited regeneration is allowed by employing a switched resistor on the dc bus. Other simplifications include semiconverters using three diodes in place of three of the thyristors, half-wave converters using only three thyristors, and single-phase versions of the various alternatives. There are also more sophisticated ac-to-dc converters (rectifiers) using gate turnoff devices such as transistors or GTOs (gate turnoff thyristors) to provide some control of the input ac wave form and power factor, but these are not currently widely used and will not be further described here.

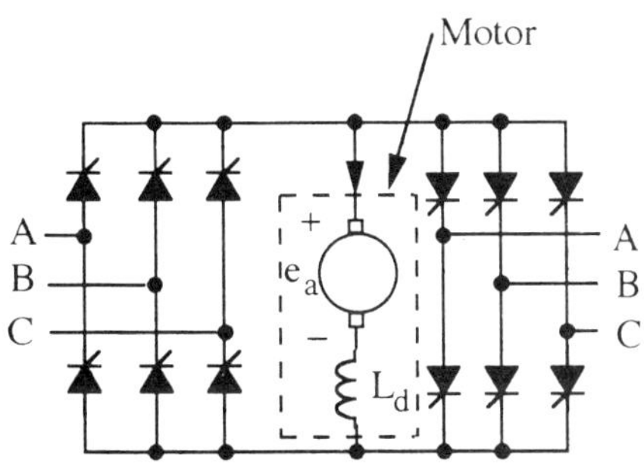

FIG. 13. Dual-bridge three-phase controlled rectifier.

When thyristors are used, the output voltage of a controlled rectifier is determined by the phase delay angle α of the thyristors. This angle is measured between the actual instant of initiation of conduction and the instant of conduction if the thyristor were replaced by a diode. Both positive and negative output voltage is available from a single bridge, and for the three-phase full bridge the voltage can be expressed as

$$V_{\mathrm{dc}}=(3\sqrt{6}/\pi)V_{\phi}\cos\alpha, \qquad (28)$$

where V_{ϕ} is the rms phase voltage. There is also an equivalent resistance, related to the ac-side line inductance, which causes a reduction in output voltage dependent on the dc load current.

6.1.2 Choppers. Power converters used to convert a constant-amplitude dc source to variable amplitude are called choppers. The *H* bridge illustrated in Fig. 14 is a four-quadrant chopper. The switches are operated in pairs, 1 and 4 yielding a positive output, 2 and 3 a negative output, and 1 and 3 or 2 and 4 a zero output. The variable voltage output is obtained by pulse-width modulation (PWM) with the output wave form consisting of variable-width voltage pulses of the desired polarity separated by segments of zero voltage. There are simpler variants that do not offer four-quadrant operation. When combined with a simple diode rectifier, choppers offer an alternative to controlled rectifiers for ac to variable-voltage dc conversion.

6.1.3 Phase Control of ac Voltage. Variable ac voltage for motor control is accomplished by utilizing back-to-back connected thyristors or TRIACS, as is done in the dimmer control of lighting. Figure 15 illustrates the scheme and a typical wave form in a three-phase system. Voltage control is obtained by varying the phase delay angle much as in a phase-controlled rectifier.

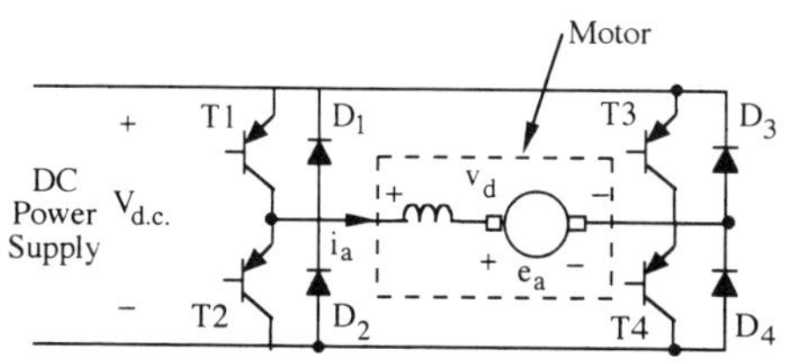

FIG. 14. Four-quadrant *H*-bridge chopper using transistors as the switching elements.

6.1.4 Inverters. Inverters are used to convert from dc to variable-frequency ac using either six-step operation (peak ac amplitude equal to the dc amplitude) or PWM operation (variable ac amplitude). At present, essentially all ac drives employ an inverter and hence require a dc supply. For ac drives operating from an ac power source, the resulting dc link system uses a simple rectifier (for PWM operation), or one of the various variable-voltage dc converters described previously, as the input converter. Figure 16 illustrates the two basic types of circuits in common use for three-phase systems.

The two circuits can be distinguished by the presence of the inverse connected diodes and the dc bus capacitor in the VSI and the dc link inductor in the CSI. Both switch blocks of the input dc to the output to create a three-phase output of rectangular wave form. The two systems differ in that the voltage source inverter (VSI) operates with three switches on at each instant and creates a direct instantaneous correspondence between dc voltage amplitude and the peak output ac voltage. In contrast, the current source inverter (CSI) only has two switches on at each instant and produces instantaneous correspondence between the dc current amplitude and the peak output ac current. The two systems are mathematical duals. Because of cost and certain technical limitations, the CSI is presently used only in very large sizes employing thyristors, and primarily with synchronous machines where leading power factor operation can be used to attain self-commutation. Some retrofit applications to large induction motors with added power factor correction capacitors also use thyristor-based CSIs.

The line-to-line output wave form of the VSI for six-step operation is a rectangular wave with two periods of zero voltage each of a duration of one-sixth of a cycle as shown in Fig. 17. The peak value is equal to the dc-side voltage resulting in a fundamental output voltage of

$$V_L=(\sqrt{6}/\pi)V_{\mathrm{dc}}, \qquad (29)$$

where V_L is the fundamental rms line voltage. The line-to-neutral wave form for the six-step VSI has a peak amplitude of two-thirds of the

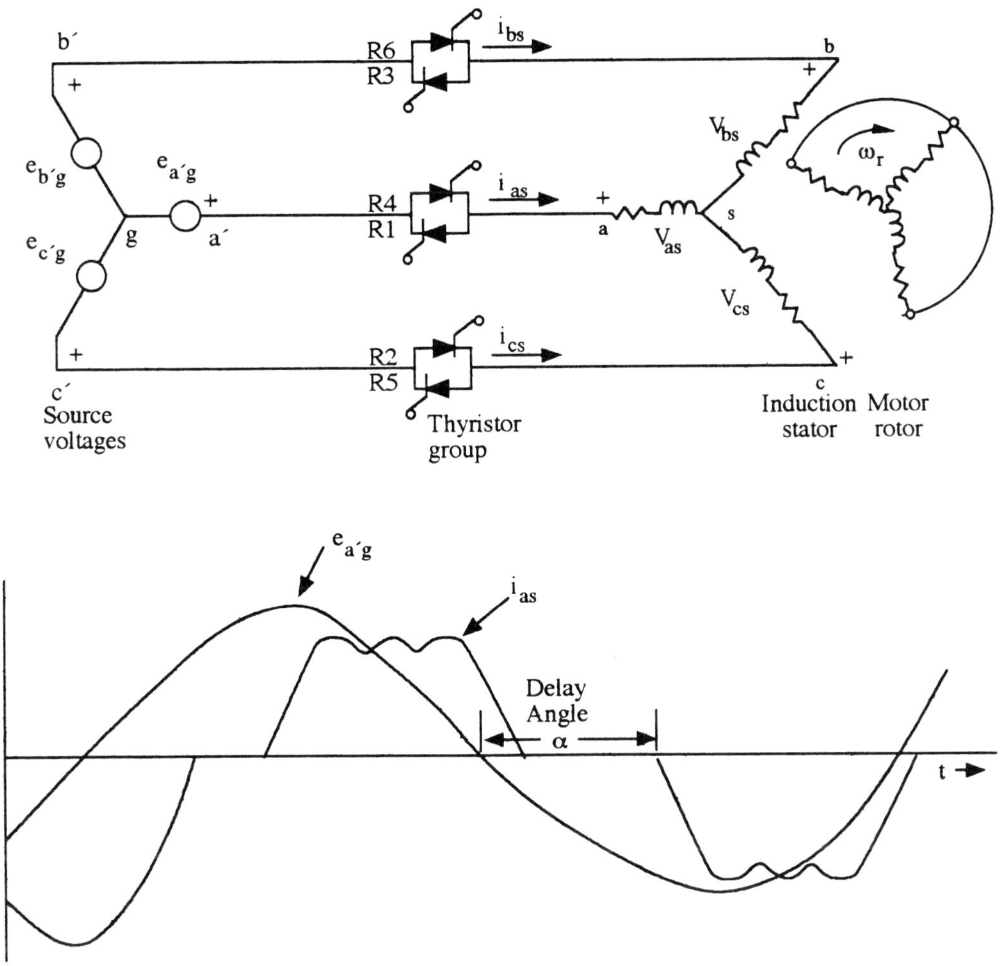

FIG. 15. Thyristor phase control of motor voltage.

dc voltage amplitude as shown in Fig. 17 and has a fundamental component of

$$V_\phi=(\sqrt{2}/\pi)V_{dc}, \tag{30}$$

where V_ϕ is the fundamental rms phase voltage. The wave forms for the CSI are also shown in Fig. 17, and the values of the fundamental components of the line and phase currents (delta connection) are given by Eqs. (29) and (30) with I replacing V. These rectangular waves have only odd harmonics without any triple multiples of the fundamentals, and the harmonic amplitudes are inversely related to harmonic order. Only the fundamental produces useful output torque; the harmonics create torque pulsations and extra heating of the motor. Control of the output frequency is obtained by varying the stepping rate of the switches. The required voltage control can be accomplished by controlling the dc-side voltage or, now more commonly, by PWM operation of the switches as in the dc chopper. Both types of inverters are fully bilateral, allowing power flow in either direction. The VSI requires reversal of the dc current and thus requires a dual converter at the input to the dc link. In contrast the CSI regenerates by reversing the dc voltage and only requires one input-controlled rectifier bridge.

6.2 dc Drives

Solving Eq. (16) for the speed ω yields

$$\omega=(V_0-R_AI_A)/K\phi\ , \tag{31}$$

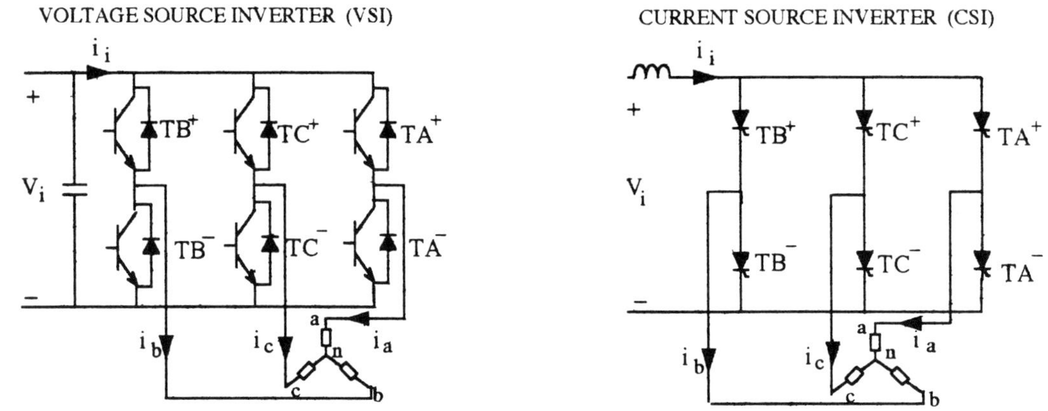

FIG. 16. Power-circuit configuration of VSI and CSI inverters.

which indicates that the dc-motor linear speed–torque characteristic can be controlled by

1. varying the armature voltage V_0, which controls the zero-speed intercept,
2. varying the armature resistance R_A, which controls the slope, or
3. varying the field flux ϕ, which controls both the slope and the zero-speed intercept.

Armature-resistance speed control is very inefficient and yields poor speed regulation and is therefore seldom used except in very small motors. Field flux control is the only way to attain operation above rated speed without exceeding rated voltage and is widely used in combination with other techniques to yield high-speed operation. It is also the only speed control method that does not require handling the full motor output power or current. It is, however, only useful at high speeds since it reduces the torque per ampere and thus the available output power unless the speed is high.

The simplest and most common electronic dc motor drive utilizes armature voltage control with a separate field supply (or permanent-magnet field). Before power electronic converters were available, the variable armature voltage was supplied by a dc generator with field control (Ward-Leonard system). Replacing the generator with an electronic converter provides much faster response and improved efficiency. However, the inherent regeneration capability of the dc generator requires a dual-bridge converter to provide equivalent performance. Also, the system isolation and input power-factor control provided by employing an ac synchronous motor to drive the dc generator are difficult to emulate in the electronic version of the system.

Power level, required control functions, and cost constraints are the primary factors that dictate selection of the type of converter used. Generally, high-power drives require thyristor-based converters and thus tend to be controlled-rectifier systems. Low-cost drives also are generally controlled-rectifier based using semiconverters or half-wave systems.

6.2.1 Capability Curves. The "capability curve" of a dc machine shows the available torque as a function of speed when the arma-

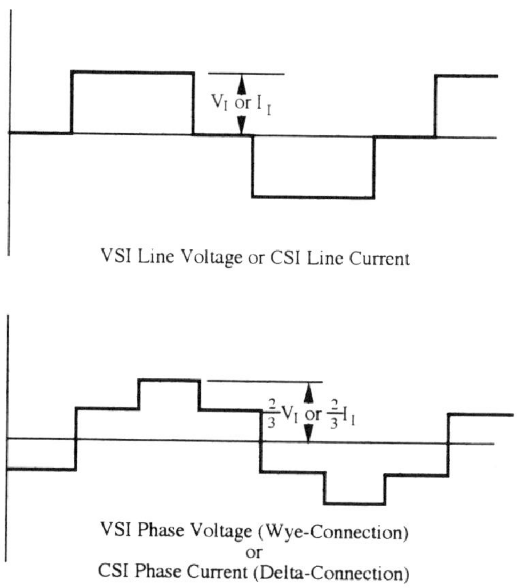

FIG. 17. Wave forms for six-step VSI and CSI inverters.

ture voltage V_0 and current I_A are limited to their rated values. As shown in Fig. 18, the total speed range is divided into two parts: the range from zero to the rated speed, where armature voltage control with full field current is employed, and the range from rated speed to the maximum allowable speed, where operation is with full-rated armature voltage and steadily reducing field current to keep the back emf e_g smaller than the terminal voltage V_0. The armature voltage control region below rated speed is called the "constant torque region" since the machine is *capable* of full-rated torque at any speed in this region. Above rated speed the torque *capability* is reduced since the field flux is smaller. This region is called the "constant (horse)power region" since with both the voltage and current held at their limit values, the power capability is constant. The term "field weakening region" is also used to describe this mode of operation at high speeds.

6.2.2 High-Dynamic-Performance dc Drives (Servo Drives). When fast response to speed or torque commands is required, a chopper-controlled system is employed except at very high power levels. This avoids the inherent dead time associated with controlled rectifiers. In high-power drives using controlled rectifiers this dead time can be reduced by employing multiple rectifier bridges phased to produce more pulses per input frequency cycle.

High-performance drives nearly always utilize an inner armature-current control loop to provide current regulation as shown in Fig. 19. With controlled current the dc motor becomes a controlled torque source since the torque and armature current are proportional when the field flux is constant. This makes the overall drive performance independent of the armature resistance, inductance, and back emf since these are all compensated by the action of the current loop.

There is an upper speed limit beyond which there is insufficient source voltage to overcome the back emf and maintain current control. Reducing the field flux (field weakening) reduces the back emf at any speed and thus extends the maximum available speed for the drive. Avoiding overcurrent in the converter is easily accomplished by simply limiting the command to the current regulator. This torque command signal is typically derived from the output of the outer loop regulator controlling speed, position, or other physical variables.

6.2.3 Open-Loop dc Drives. In applications where response time and/or precise control of speed are less important, the inner

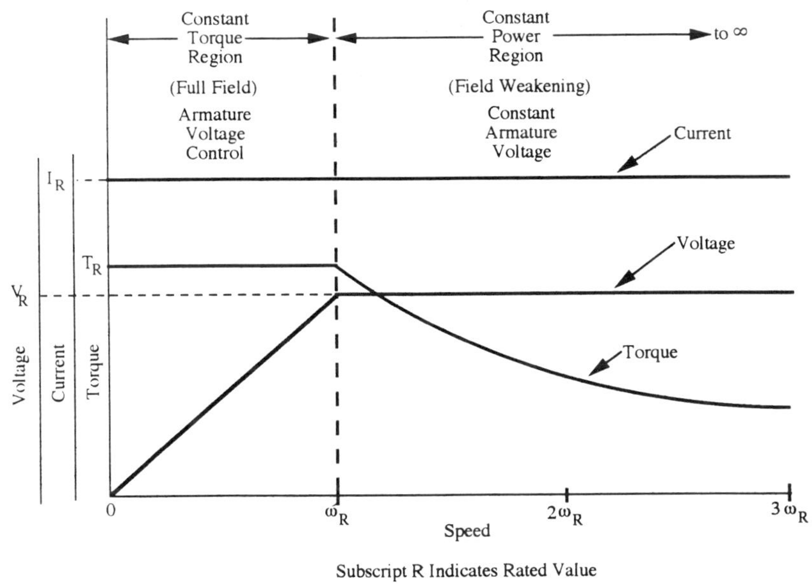

FIG. 18. Capability curves of a dc motor.

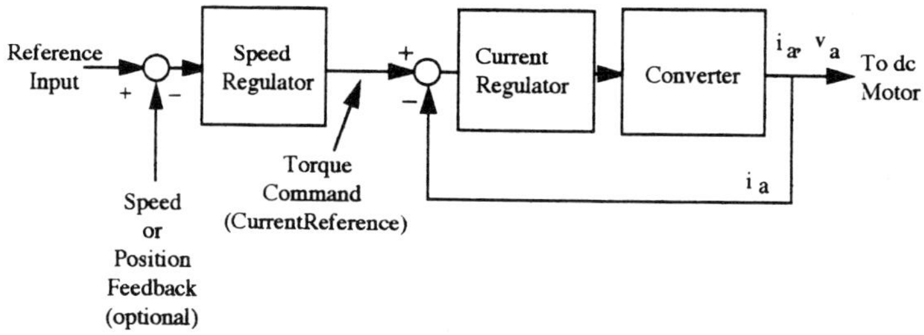

FIG. 19. High-Performance dc servo drive schematic block diagram.

current loop and sometimes even the outer speed feedback loop are eliminated. To avoid overcurrent resulting from the low dynamic impedance of the motor, an electronic current limit, as illustrated in Fig. 20, and, often, a slow ramping of the speed command that directly controls the converter output voltage are employed. The speed command can be the output of a regulator fed by speed error or simply an open-loop command. In this latter case, there is often an *IR*-drop compensator, which partially corrects for the speed droop caused by armature and converter resistance voltage drop. In most cases these simple, open-loop drives utilize some form of rectifier converter, often of the semiconverter of half-wave type.

6.3 Variable-Speed ac Drives

Unlike dc drives where there are only a few basic types, there are a large number of viable ac drive types. This variety stems from the many possible combinations of induction and synchronous machines, VSI and CSI inverters, and speed-control strategies. Only the most commonly used systems are described here.

6.3.1 Synchronous-Motor Speed Control. The only mechanism for speed control in synchronous machines is varying the frequency, which directly varies the speed of the rotating magnetic field [Eq. (21a)]. To avoid changing the flux density in the iron core, and particularly saturation, changes in frequency are typically accompanied by proportional changes in the applied voltage [Eq. (6)] leading to the term "constant-volts/hertz control" used to describe this mode of operation. When operated with constant volts/hertz and constant field current (and neglecting the armature resistance *R*; see Fig. 12), a synchronous machine develops the same torque at the same torque angle δ and the same armature current I_s at any frequency f. This very desirable result occurs because the voltage V_t, the induced voltage E_f, and the reactance drop I_sX_s in Fig. 12 are all linearly proportional to the frequency, and hence the phasor diagram relating the three quantities simply shrinks or grows as the frequency varies without altering the angle δ. Under these conditions the torque, given by the power divided by the speed, is [from Eq. (26)]

$$T=(V_tE_f/\omega X_s)\sin\delta \qquad (32)$$

and is constant. At low frequencies the armature resistance alters this simple relation and the torque is reduced unless the voltage V_t is increased above the constant-volts/hertz value (referred to as "voltage boost").

Above rated frequency the voltage is typically held constant at its rated value, and the flux is allowed to decrease [Eq. (6)]. The power capability can be maintained at the rated value for these higher frequencies if the

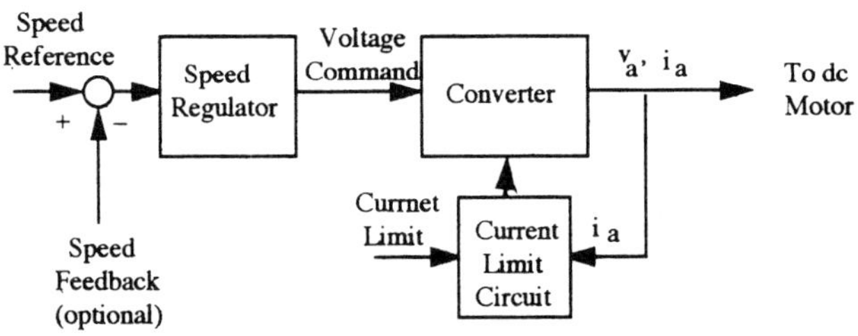

FIG. 20. Adjustable–speed dc drive schematic block diagram.

field current is appropriately reduced so as to maintain the power factor angle between V_t and I_s at a constant value. The synchronous machine thus has a "capability curve" identical to that of the dc machine illustrated in Fig. 18.

6.3.2 Induction-Motor Speed Control. Speed control in induction machines can be attained by varying the frequency and voltage as in the synchronous machine and, to a much more limited extent, by varying only the voltage.

6.3.2.1 Voltage-Controlled Induction-Motor Drives. Voltage variation alters the torque produced at any specific slip in proportion to the voltage squared and thus produces a family of torque–speed curves as illustrated in Fig. 21. Depending on the load torque–speed curve, the equilibrium point where motor torque and load torque are equal will vary with voltage, yielding speed control over a limited speed range. Since the synchronous speed is constant, speed control by voltage variation is really slip control, which results in high rotor I^2R loss as the speed is reduced.

Voltage control is therefore used only for speed control of small induction motors with fan loads in which the torque drops rapidly with speed (as speed squared). These drives are inherently inefficient since they require high-slip operation and are viable only because the rapid drop in torque makes the high

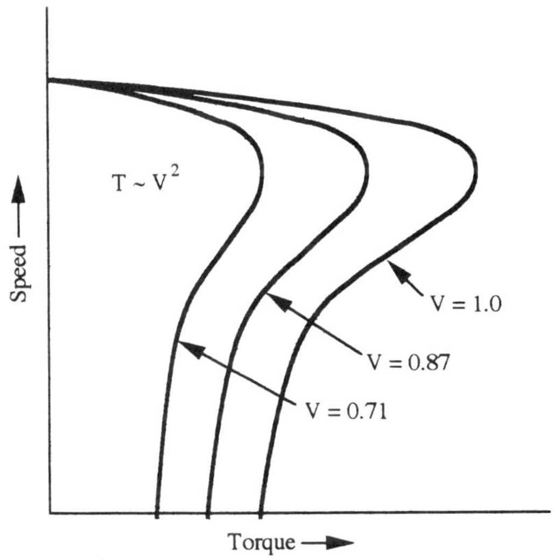

FIG. 21. Torque–speed curves of induction motor for voltage control.

percentage loss an acceptably small absolute number. Typically the voltage is controlled by varying the amount of phase delay in series connected TRIACS or back-to-back connected thyristors. This introduces additional losses as a result of the time harmonics in the voltage wave form. Phase control of voltage is also widely used in speed control of single-phase universal motors as in, for example, portable electric drills.

Voltage control, typically via series-connected thyristors in which initiation of conduction is delayed by phase control of the gate voltage, is also used to limit starting current in induction and synchronous motors and for light-load loss reduction in motors subject to variable load cycles. This last application is limited to motors that operate at very light load for long periods, since the voltage controller creates its own loss at all loads and this must be overcome by the reduction of motor core losses at light loads.

6.3.2.2 Adjustable-Speed, Constant-Volts/Hertz Induction-Motor Drives. As in the case of synchronous-machine drives, the synchronous speed of an induction motor can be varied by varying the frequency. Maintaining constant motor flux density to avoid saturation is again accomplished by keeping the ratio of voltage and frequency (volts/hertz) constant. Under this condition the induction machine exhibits essentially the same speed–torque characteristic at all frequencies. This is readily shown by rewriting the resistance R_r'/s in Fig. 8 as $R_r'/s\omega$ times ω and observing that, except for R_s, for a fixed value of $s\omega$ (slip frequency) every element in the circuit is a linear function of ω. By making the voltage vary linearly with ω, the current and torque are independent of frequency for any fixed value of $s\omega$. At very low frequencies (below 5 Hz) the stator resistance becomes significant, and voltage boost (increased volts/hertz) is employed to maintain torque capability. The result is a family of torque–speed curves, as illustrated in Fig. 22, where the various curves are simply translated up or down as the frequency is varied. Operation above base frequency at constant output voltage (and increased values of $s\omega$) allows the flux to decrease and corresponds approximately to the constant-power, field-weakening region of dc drives as illustrated in Fig. 22. The most common system at low and moderate power levels employs a PWM VSI as the variable-

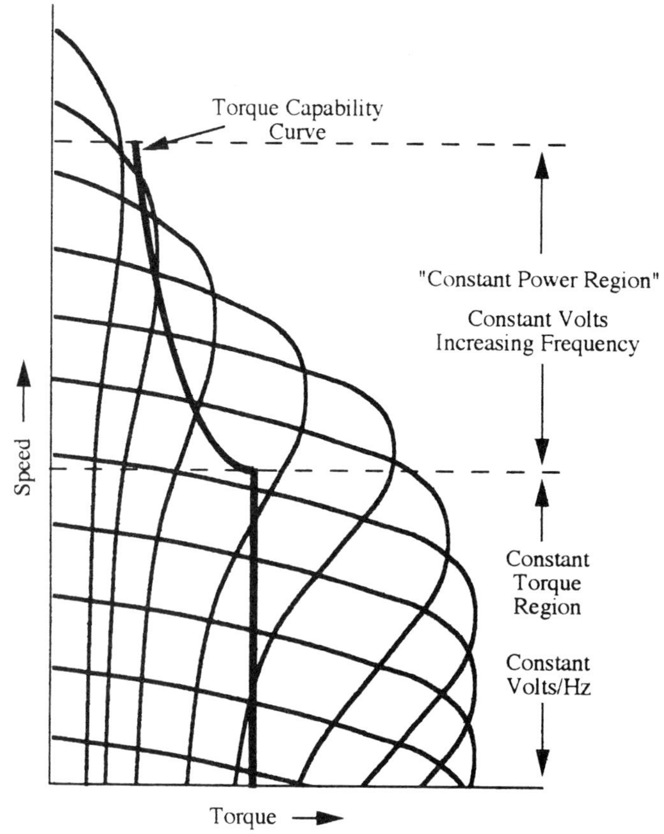

FIG. 22. Capability curves of induction motor for frequency control.

frequency supply. At higher power levels, six-step VSI or CSI inverters fed by controlled rectifiers are used. In the CSI case a voltage feedback loop is necessary to convert the CSI output to a controlled, constant-volts/hertz supply.

Motor starting is accomplished by initially applying a low frequency and slowly ramping up to the desired speed. This is necessary to keep the current within the converter rating. Unfortunately, the proper ramp rate and voltage boost depend on machine parameters and load characteristics, and effective converter utilization during starting is difficult to attain. A variety of control strategies utilizing internal converter variables such as output voltage or current, link dc current, or link power are used to assist under starting conditions or during load or speed changes. However, in the absence of feedback from the motor (speed or flux) these open-loop constant-volts/hertz drives have the fundamental problem of basically appearing as a voltage source to the machine and thus having to cope with the normal constant-voltage properties of the machine.

6.4 Adjustable-Torque, Vector-Controlled ac Drives

Torque control in ac motors is accomplished by emulating the regulated–armature-current dc motor drive (see Sec. 6.2.2). This requires controlling the spatial angle between field flux and armature mmf (leading to the name "field orientation"), which in turn requires control of both amplitude and phase of the ac excitation and leads to the name "vector control."

6.4.1 ac Current Regulators. Most torque controllers use an ac current regulator based on time-domain current feedback and a PWM inverter as shown in Fig. 23, although CSI inverters are also used at high power levels. A variety of schemes for generating the inverter voltage commands in the current regulator

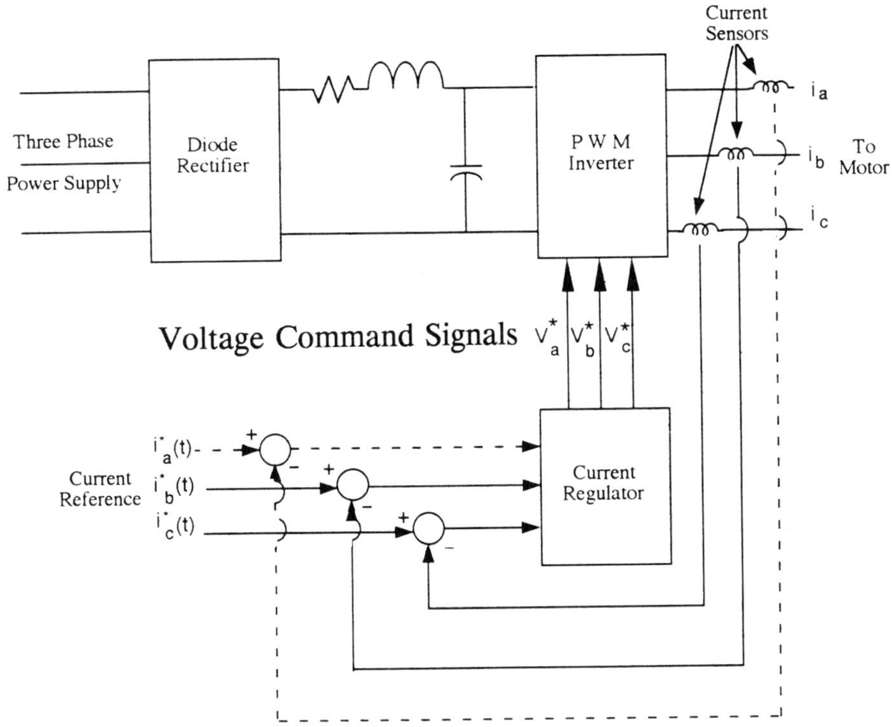

FIG. 23. Current-regulated PWM inverter for "vector control" of ac current.

block of Fig. 23 are employed, with simple on–off control with hysteresis and sine-triangle comparison (natural sampled PWM; see POWER ELECTRONICS) being the most common. The ac current-regulated inverter of Fig. 23 (or its CSI equivalent) allows positioning the machine armature (stator) mmf in space as well as eliminating the effects of armature (stator) resistance, inductance, and back emf as in the dc machine.

6.4.2 Field Orientation. The location of the field flux can be directly determined by measuring rotor position in a synchronous machine but must be estimated from indirect measurements in induction machines. The most common method in induction machines uses rotor position measurement and an estimate of slip frequency to estimate rotor flux position indirectly (indirect field orientation). As illustrated in Fig. 24, the slip estimation depends on the rotor time constant and hence is sensitive to operating conditions. The equation for the slip frequency used in the slip calculator block in Fig. 24 is

$$s\omega = \frac{R_r'}{L_r}\frac{I_T}{I_\phi} = \frac{1}{T_r}\frac{I_T}{I_\phi}, \tag{33}$$

where L_r is the total rotor self inductance, T_r the rotor (open circuit) time constant, I_T the torque command current, and I_ϕ the flux command current. Errors in the time constant cause steady-state flux and torque errors as well as noninstantaneous, usually oscillatory torque response. The sensitivity of the controller to errors in the time constant depends on the inductance-to-resistance ratio of the machine and is therefore more serious in large, high-efficiency machines. In small machines no adaptation to time-constant changes is usually necessary. For large machines a variety of adaptive controllers have been proposed, but the problem remains an area of active research.

For a synchronous machine the slip calculator in Fig. 24 is eliminated since rotor position and rotor flux position are identical (i.e., the slip frequency is zero). Direct field orientation in induction machines, in which the rotor flux position is estimated from flux

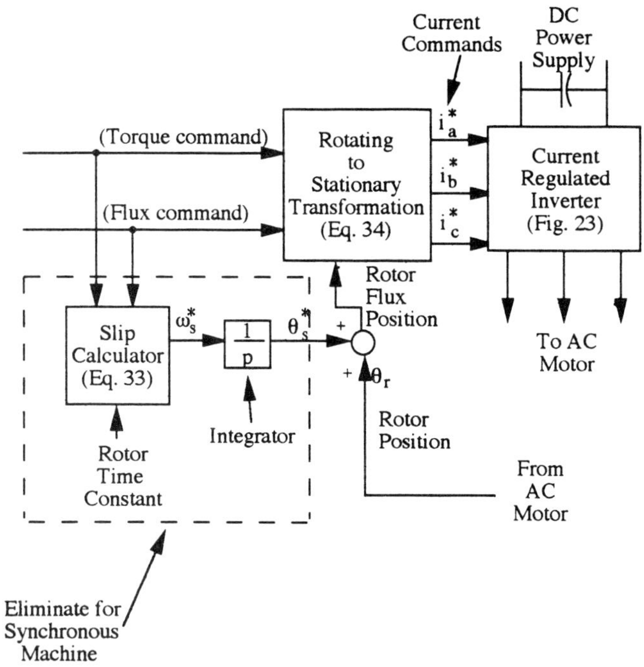

FIG. 24. Torque-controlled ac drive schematic diagram.

estimates obtained from flux coils or from armature (stator) voltage and current, is also used, but is much less common.

Torque-controlled ac drives based on field orientation are fully equivalent to torque-controlled dc servo drives. They offer the advantages of higher torque/inertia ratios, higher operating speed, and lower maintenance since there is no mechanical commutator. These drives can also be built in larger sizes than for dc machines by employing CSI inverters and synchronous motors (drives of several thousand horsepower are not uncommon) and for higher speeds since the limitations of a mechanical commutator are eliminated.

6.4.3 Brushless dc Machines. There are a large number of variants of the torque-controlled synchronous machine described in Sec. 6.4.2 and illustrated in Fig. 24 that are called brushless dc motors. They all follow the basic principle of controlling the spatial angle between flux and mmf, and all utilize some form of rotor position measurement. Most use permanent magnets for field excitation and some form of simplified position sensing using Hall-effect sensors or position estimation based on back emf. In some cases the motor back emf is intentionally caused to be trapezoidal in wave form to allow effective use of limited position-sensing accuracy and to increase the torque per ampere. The most common limitation of many of these machines is high torque ripple.

6.5 Other ac Drives

There are a number of other, less widely applied, ac drives utilizing special types of converters or special motors. For example, very high-power, low-speed applications sometimes employ cycloconverters for direct conversion of line frequency (50 or 60 Hz) to the low frequency needed to drive the machine without an intermediate dc link. Such systems are limited to frequencies below about one-third of the input frequency and utilize a large number of power-switching devices (usually thyristors) with 36 or more devices being common. Because of the complexity of controlling the switching of such systems, they are typically only used at power levels where other types of converters are unavailable because of power-switching–device limitations (for example, for large, high-voltage motors).

Another, somewhat more common, example is the use of a wound-rotor induction motor with a frequency converter feeding the rotor circuits in high-power applications requiring only a small speed-control range (20% to 40% variation). Here the motivation is to be able to use a lower power rating in the converter since it only needs to be rated to handle the slip power (i.e., 20% of the motor rating if only a ±20% speed range about synchronous speed is required), though this advantage must be balanced against the more expensive machine construction.

GLOSSARY

ac: Alternating current—a power source that alternates sinusoidally with time.

Alternator: A device to change mechanical energy to ac electrical energy.

Armature: The member of a machine in which the energy conversion takes place.

Brushless dc Motor: Synchronous motor operating on a dc supply which is converted to ac of variable frequency and voltage by means of electronic controls.

Commutator: Copper bars insulated from each other, mounted on the armature shaft, and connected to armature windings to carry currents, through brushes riding on the smooth cylindrical outer surface of the commutator, to external circuits.

Compound Motor or Generator: A dc motor or generator that has both shunt and series field coils.

Converter: A generic term for a device that alters the nature of electrical energy (for example, changes the frequency).

Diode: A power-switching device which allows current to pass in only one direction.

dc: Direct current—a power source that has a constant polarity of voltage (or current).

Four-quadrant drive: A drive capable of operating the driven machine both as a motor and as a generator and in either direction.

Generator: A device to change mechanical to electrical energy.

Hall-Effect Sensor: A magnetic-field–sensitive semiconductor device used to sense rotor position in electric drives.

Induction Motor: An ac motor in which rotor currents are set up by magnetic induction (often in a squirrel-cage conductor arrangement).

Inverter: A device for converting dc to ac.

Linear Motor: An electric motor providing linear motion rather than rotation.

Motor: A device to change electrical energy to mechanical energy.

Rectifier: A device for converting ac to dc.

Rotor: The rotating member of an ac machine.

Semiconverter: A controlled rectifier in which half of the thyristors are replaced with diodes.

Series Motor or Generator: A dc motor or generator that has the field coils connected in series with the armature.

Shunt Motor or Generator: A dc motor or generator which has the field coils connected in parallel with the armature.

Synchronous Motor or Alternator: An ac motor or alternator that runs at a constant speed dictated by the frequency of the power supply and the number of pole pairs of the machine.

Thyristor: A power-switching device which can be turned on with a small control signal but must have its voltage reversed to be turned off.

TRIAC: A bidirectional device equivalent to two thyristors connected in inverse parallel.

Universal Motor: A motor that will operate on ac or dc.

Volts/Hertz Control: A control strategy that maintains approximately constant magnetic flux in ac machines.

Wound-Rotor Induction Motor: An induction motor in which the rotor circuits are windings as in the stator and are made accessible by slip rings.

Further Reading

Bose, B. K. (1986), *Power Electronics and AC Drives*, Englewood Cliffs: Prentice-Hall.

Fitzgerald, A. E., Kingsley, C., Umans, S. D. (1990), *Electric Machinery*, 5th ed., New York: McGraw-Hill.

Jordan, Howard E. (1983), *Energy Efficient Electric Motors and Their Application*, New York: Van Nostrand-Reinhold.

Krause, P. C. (1986), *Analysis of Electric Machinery*, New York: McGraw-Hill.

Miller, T. J. E. (1989), *Brushless Permanent-Magnet and Reluctance Motor Drives*, New York: Oxford University Press.

Mohan, Ned, Undeland, Tore M., Robbins, William P. (1989), *Power Electronics Converters, Applications, and Design,* New York: Wiley.

Murphy, J. M. D. and Turnbull, F. G. (1988), *Power Electronic Control of AC Motors,* New York: Pergamon.

Nasar, S. A. (1984), *Electric Machines and Transformers,* New York: Macmillan.

Nasar, S. A. (1987), *Handbook of Electric Machines,* New York: McGraw-Hill.

Say, M. G. (1983), *Alternating Current Machines,* 5th ed., London: Pitman.

Shepherd, W., Hulley, L. N. (1987), *Power Electronics and Motor Control,* Cambridge, U.K.: Cambridge Univ. Press.

Slemon, G. R. (1953), "Equivalent Circuits for Transformers and Machines, Including Non-Linear Effects," Proc. IEE **100**, Pt. 4 (5), 129–143.

Slemon, G. R. (1992), *Electric Machines and Drives,* Reading, MA: Addison-Wesley.

MAGNETIC FILMS

See PLATES AND FILMS, MAGNETIC

MAGNETIC DEVICES

JOSEPH HEREMANS, *General Motors Research Laboratories, Warren, Michigan, U.S.A.*

	Introduction	587
1.	**Magnetic Field Sensors**	588
1.1	Galvanomagnetic Effects in Semiconductors	588
1.2	Semiconductor Magnetoresistors	590
1.3	Hall Generators	591
1.4	Magnetodiodes	593
1.5	Magnetotransistors	593
1.6	Integrated Semiconductor Magnetic Field Sensors	595
1.7	Ferromagnetic Metal Film Magnetoresistors	595
1.8	Variable-Reluctance Sensors	597
1.9	Fluxgates	598
1.10	Other Magnetometers	599
1.11	Comparison of Magnetic Field Sensors	599
2.	**Applications of Magnetic Field Sensors**	600
2.1	Magnetic Recording Heads	600
2.2	Position and Speed Sensors	601
2.3	Rotary Shaft Encoders	601
2.4	Brushless Motors	602
2.5	Shuntless dc Current Measurements	602
3.	**High-Frequency Devices**	602
3.1	General Aspects	602
3.2	Microwave Devices	604
3.2.1	Filters	604
3.2.2	Isolators	605
3.2.3	Circulators	606
3.2.4	Strip-Line, MMIC, and Thin-Film Technologies	607
3.3	Other Magnetic Devices	607
3.3.1	Antenna, Inductor, and Transformer Cores	607
3.3.2	Magnetic Memories	608
4.	**Conclusion**	609
	Acknowledgments	609
	Glossary	609
	Works Cited	610
	Further Reading	611

INTRODUCTION

Iron ferrite or magnetite was "lodestone" to the ancients. Its magnetic properties were known to the thirteenth-century Chinese, as was its property of pointing to the north (Bai Shouyi, 1982). The stone was ground and shaped into a needle, kept aloft inside a bowl of water by a floating wick, and used for guidance. The resulting compass (see Fig. 1) is the first example of a magnetic device. The torque exerted on the compass needle is the result of the action of the Earth's magnetic field on the ferrite. It exemplifies the two elements that define a "magnetic device": an element using the measurement of the intensity or direction of a magnetic field, or an element using a magnetic material.

This article is organized in the following way. The principles of operation of magnetic field-measuring devices, such as magnetoresistors, Hall generators, magnetodiodes, and magnetotransistors, are described, as well as some integrated circuits that use them. Also described are pickup coils, and fluxgates that are used in compasses. A special paragraph is included in which the performance of the different sensors is compared. A second section is devoted to applications of magnetometers. Magnetic field-measuring devices are used in compasses and attitude detectors, in magnetic recording heads (see also MAGNETIC STORAGE OF INFORMATION), in position and speed sensors, in rotary encoders, in contactless potentiometers, in dc current measurements, and in scientific instrumentation.

Microwave devices, such as filters, phase shifters, isolators, and circulators, use ferrimagnetic resonances in ferrites and garnets. Other ferromagnetic or ferrimagnetic materi-

1-56081-067-X/93/$5.00+.50

FIG. 1. A Song dynasty compass in the collection of the Museum of Chinese History, Beijing (from Bai Shouyi, 1982).

als are used in antenna cores, inductors, or even simple electromagnets. These devices are described.

Magnetic materials are also used in electrical machinery (see MACHINES, ELECTRICAL), transformers and inductors, thermometry (see MAGNETOTHERMAL DEVICES), and magnetic lenses (see MAGNETOELECTRICAL DEVICES). The reader is referred to the appropriate articles.

Recent developments in microelectronic technology have revolutionized the fabrication techniques of magnetic devices and broadened the range of useful applications. Even pickup coils can be made using thin-film technology. Magnetic field sensors with on-chip integrated amplifiers and signal conditioners are sensitive, reliable, and inexpensive, and have made magnetic position detection a practical reality. Microwave circuit elements, including gallium arsenide (GaAs) amplifiers, are integrated on semiconductor chips called monolithic microwave integrated circuits (MMICs). Soon magnetic circuit elements will be integrated (Prinz, 1990) directly in MMICs, thus lowering their cost and opening new fields of applications.

I have attempted to describe the physical principles of operation of the devices by using simplified schematic outlines. The reader can find more detailed descriptions and mathematical formulation in the references.

1. MAGNETIC FIELD SENSORS

1.1 Galvanomagnetic Effects in Semiconductors

The fundamental principles of galvanomagnetic effects are discussed in the articles ELECTRONIC TRANSPORT IN SOLIDS and TRANSPORT PHENOMENA. In this section I point out the difference between the intrinsic magnetoresistance that many electrically conductive materials exhibit and the geometrical magnetoresistance effects that are the basis of magnetoresistive sensing elements (Heremans, 1993).

Electrons or holes moving under an electric field **E** and a transverse magnetic field **H** are subject to the Lorentz force **F** proportional to the velocity **v** of the carrier and the magnetic induction **B**:

$$\mathbf{F} = \pm e(\mathbf{v} \times \mathbf{B}), \qquad (1)$$

where $e = 1.6 \times 10^{-19}$ Coulomb; the + sign is used if the carriers are holes, and the − sign if they are electrons. The Lorentz force is exerted at right angles to both **v** and **B**. As a consequence, the path of these charge carriers, or the current density vector **j**, is not collinear with the applied electric field. The angle θ between **j** and **E** is the Hall angle. Consider the case of a slab of semiconductor, of length l, width W, and thickness t along the x, y, and z directions, respectively, as shown in plan view in Fig. 2. An applied voltage V_{xx} forces a current I through the slab. A magnetic field **H** is applied normal to the plane of the slab, and the induction is **B**. The Lorentz force will distort the current lines close to the injecting contacts. Some distance along the x axis into the slab, there is an accumulation of charge carriers to one side of the slab. This results in the generation of a transverse voltage, the Hall voltage, V_{xy} in Fig. 2. In a slab of arbitrary length and width, the current lines follow the paths shown in Fig. 2. Conformal mapping techniques are used to calculate these lines (Lippmann and Kuhrt, 1958). If the slab is long enough, the transverse, or Hall, electric field balances the Lorentz force, and the current lines become parallel to the side of the sample along the x axis. In an infinitely long slab (not shown in Fig. 2) the electric fields associated with V_{xx} and V_{xy} are

$$E_x = V_{xx}/l, \qquad (2)$$

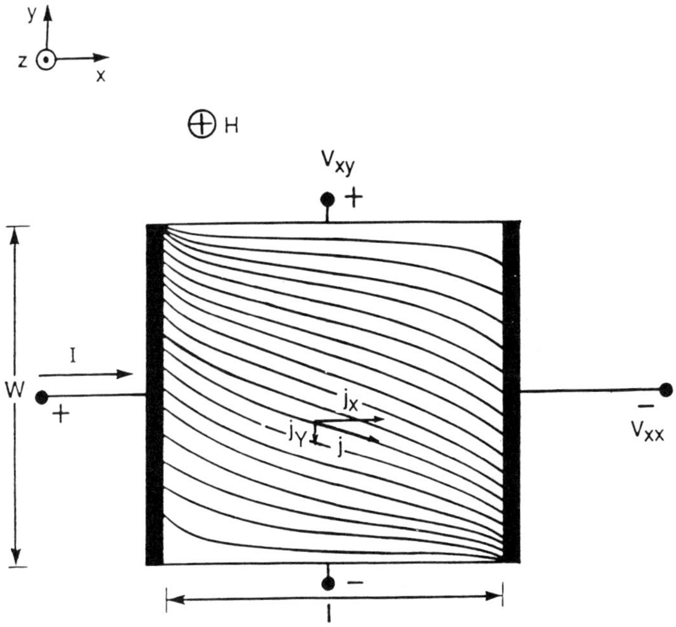

FIG. 2. Projection of a slab of semiconductor with current density (j) lines flowing between the two current electrodes (dark regions) in the presence of a magnetic field H pointing into the plane of the slab. The total current flowing is I, V_{xx} is the longitudinal voltage, and V_{xy} is the transverse, or Hall, voltage.

$$E_y = V_{xy}/W, \tag{3}$$

and the current density is mostly parallel to x:

$$j \simeq j_x \simeq I/Wt. \tag{4}$$

This geometry is commonly used to make resistivity and Hall effect measurements on materials, since these intrinsic material properties are then easily defined. The resistivity is

$$\rho_{xx} = E_x/j = (V_{xx}/I)(Wt/l). \tag{5}$$

The Hall resistivity is

$$\rho_{xy} = E_y/j = (V_{xy}/I)t \tag{6}$$

and the Hall coefficient is

$$R_{\text{H}} = \rho_{xy}/B. \tag{7}$$

In a homogeneous, nonmagnetic material, with a single type of charge carrier, R_{H} is a constant related to the density n of these carriers by

$$R_{\text{H}} = \pm 1/ne, \tag{8}$$

where the + sign is used if the carriers are holes, and the − sign if they are electrons. The resistivity ρ_{xx} is then given by

$$\rho_{xx} = 1/ne\mu, \tag{9}$$

where μ is the carrier mobility. The Hall angle is then θ, where

$$\tan\theta = \mu B. \tag{10}$$

If, furthermore, the carrier's Fermi surface is isotropic, the resistivity ρ_{xx} is independent of B. All the conditions needed for this are almost never met, and most materials have a dependence of ρ_{xx} on B that contains information about the physics of the electrical transport in that material. This is the *intrinsic magnetoresistance* and has a different origin than the geometrical magnetoresistance used in sensors.

The Hall angle can become large even in moderate magnetic fields if the mobility of a semiconductor is very large. In a rectangularly shaped slab of semiconductor, shown in Fig. 3, with $l \ll W$, there is not much length available for the current lines to become parallel to the x axis, and the current lines are similar to what they are near the contacts in Fig. 2. The current path in such a short, wide sample is considerably longer than in the same sample at zero magnetic field. This leads to a large increase in resistance $R = V_{xx}/I$. It is the *geometrical magnetoresistance* effect. The Hall voltage is not fully developed and does not counteract the Lorentz force effectively. The Corbino disk (Fig. 3) is another geometry in which the Hall voltage is shorted out and a large geometrical magnetoresistance is observed. Conformal mapping techniques can show that the Corbino disk is equivalent to an

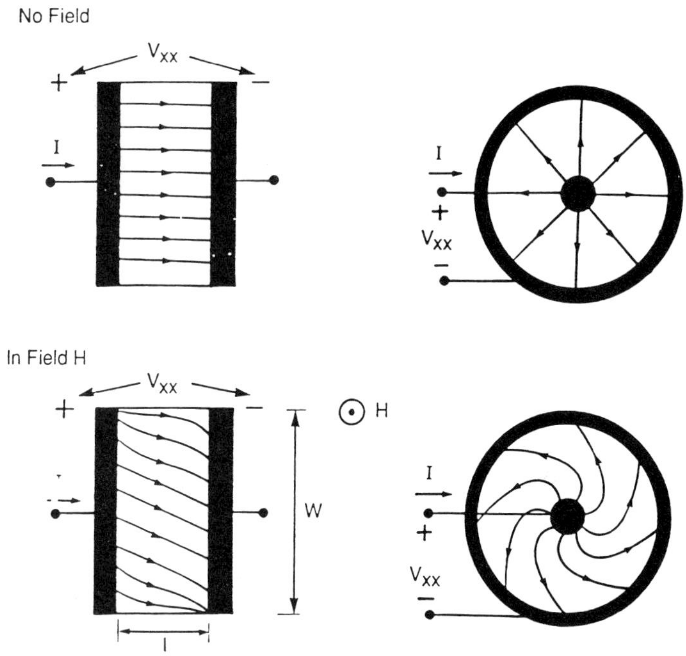

FIG. 3. Elements showing the geometrical magnetoresistance effect. A short, wide rectangle, at left, can be viewed as consisting only of the regions close to the contacts in Fig. 2. In absence of an applied magnetic field H, the current lines are normal to the contacts, but in presence of the field, they distort as shown, leading to an increase in current path length and resistance. In the Corbino disk at right, the current lines are radial in the absence of field, but lengthen as shown in the presence of field.

infinitely wide or short rectangular element, in which the Hall voltage is totally shorted out, and the geometrical magnetoresistance is maximal. The resistance of the device is then

$$R(B)=R(B=0)[1+(\rho_{xy}/\rho_{xx})^2] \times [\rho_{xx}(B)/\rho_{xx}(0)].$$

In practical devices, Corbino disks are not used, because they have a low resistance, which leads to a large power dissipation. Most devices are made from rectangular elements such as shown in Fig. 3 (left). The response of such elements is then given by

$$R(B)=R(B=0)[1+g(W/l)(\rho_{xy}/\rho_{xx})^2] \times [\rho_{xx}(B)/\rho_{xx}(0)] \quad (11)$$

or, under the conditions under which Eqs. (8) and (9) hold, and for small Hall angles $\theta<\frac{1}{8}\pi$, or $\rho_{xy}/\rho_{xx}<0.2$,

$$R(B)=R(B=0)[1+g(W/l)\mu^2B^2] \times [\rho_{xx}(B)/\rho_{xx}(0)]. \quad (12)$$

In these equations, the factor $\rho_{xx}(B)/\rho_{xx}(0)$ accounts for the intrinsic magnetoresistance, while the factor $g(W/l)$ accounts for the geometrical influence of the width-to-length ratio and is given in Fig. 4.

1.2 Semiconductor Magnetoresistors

Practical magnetoresistive devices are made out of high-mobility semiconductors, such as n-type InSb or InAs, because the device sensitivity is essentially a quadratic function of mobility [see Eq. (12)]. Unfortunately, these are also narrow-gap semiconductors, in which the carrier density is a strong function of temperature at room temperature. To compensate for this, manufacturers dope the semiconductors n-type. This technique has two disadvantages:

1. In most semiconductors, the mobility de-

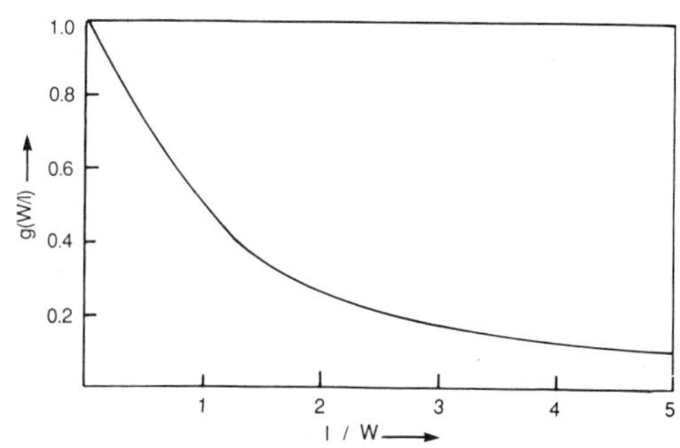

FIG. 4. The geometrical coefficient $g(W/l)$ that relates the geometrical magnetoresistance to the length/width (l/W) ratio of a rectangular element (Lippmann and Kuhrt, 1958).

creases as the doping level is increased, because of scattering of electrons by ionized impurities, and
2. when both the electron density and the electron mobility are high, the electrical conductivity is also high.

In order to keep current consumption and heat dissipation low, the ideal resistance of a magnetoresistor is in the 500- to 10 000-Ω range. Given the high conductivity of the active semiconductor used, obtaining such high device resistances requires the use of very thin slabs or films of semiconductor. Recent breakthroughs in growing thin InSb and InAs films epitaxially on insulating III-V substrates, by molecular-beam epitaxy and metal-organic chemical vapor deposition (Partin and Heremans, 1992), have made thin-film magnetoresistors very practical. Devices are made by connecting multiple rectangular elements in series (Fig. 5) to increase the resistance. The magnetoresistor shown in Fig. 5 was described by Heremans *et al.* (1993); its resistance is shown in Fig. 6 as a function of applied magnetic field at different temperatures.

The recent use of two-dimensional electron gases (see HETEROSTRUCTURES AND SUPERLATTICES, SEMICONDUCTOR) in magnetoresistors has made it possible to achieve the best compromise between high electron density and high mobility, while maintaining a reasonably large resistance per device (Heremans *et al.*, 1990). Devices based on two-dimensional electron gases have so far been limited to InAs, and work is under way to extend this to InSb-based magnetoresistors.

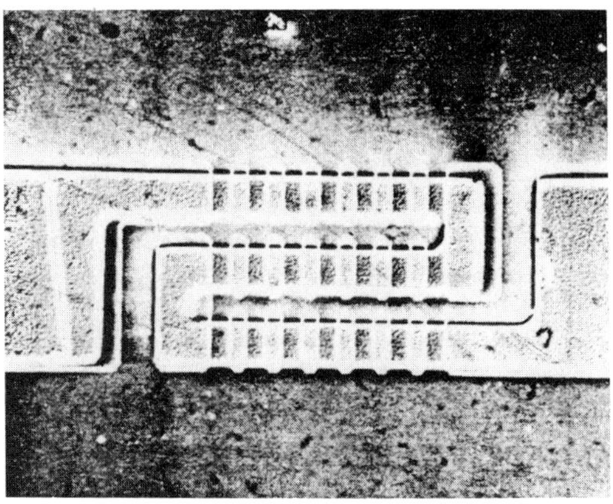

FIG. 5. A multielement magnetoresistor made from an InSb mesa etched from an epitaxial film of InSb on InP. Each rectangular element not covered by the contact metal behaves as one element described in Fig. 3.

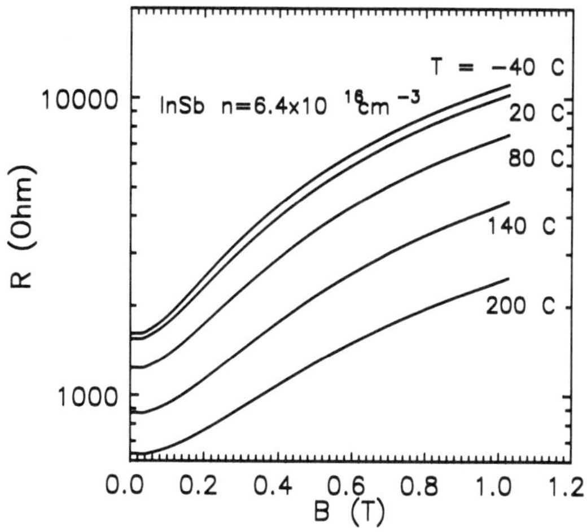

FIG. 6. The resistance of a magnetoresistor made from *n*-type InSb doped at 6.4×10^{16} cm^{-3}, as a function of applied magnetic field at the different temperatures indicated (Heremans *et al.*, 1993).

Siemens AG uses a different manufacturing process based on InSb boules that contain needlelike precipitates of metallic NiSb. The metallic needles serve the same purpose as the lateral contact stripes in the device shown in Fig. 5: they short out the Hall voltage, so that a large magnetoresistance appears. Devices are made from the boules by slicing wafers and lapping them to the desired thickness (Göpel *et al.*, 1989).

1.3 Hall Generators

The Hall voltage itself is commonly used to measure the magnetic field. Hall generators are often made in the geometry shown in Fig. 2. The Hall resistance for an infinitely long slab is defined by

$$R_{xy} = V_{xy}/I = \rho_{xy}/t. \tag{13}$$

The Hall resistance of a generator made from a material for which Eq. (7) holds is

$$R_{xy} = B/net. \tag{14}$$

A four-terminal device analysis (Heremans, 1993) of a Hall device (Fig. 2) shows that the ratio of the power output ($V_{xy}I_{out}$, where I_{out} is the current drawn from the Hall electrodes) to the power input ($V_{xx}I$) is almost proportional to the mobility of the semiconductor. Equation (14) shows that the output of a Hall generator is linear in the applied magnetic field, and that the most sensitive devices are made from thin, lightly doped slabs of material. Again, the best compromise between sensitivity, temperature stability, and moderate current consumption is obtained when Hall generators are made from thin doped slabs of high-mobility semiconductors. Two-dimensional electron gases are very well suited, but not widely used yet.

Silicon Hall generators (Popovic, 1991) can be made using an inversion layer in a metal-oxide-semiconductor (MOS) structure. Hall devices also can be made by ion-implanting of an n-type doped region of the desired geometry into p-type or semi-insulating GaAs, or p-type Si. Sometimes, n-type epitaxial Si, GaAs, or InAs grown on p-type or semi-insulating substrates is used. A mesa of the desired geometry is etched into it and equipped with current and voltage contacts. Thin lapped layers of InAs or InSb are sometimes used. Silicon is the preferred semiconductor for room-temperature applications that are very cost sensitive and do not require operation above 10 kHz, because amplifying and signal-processing electronics can be integrated on the same chip as the Hall sensor. n-type channels or implanted regions are preferred, since the electron mobility is larger than the hole mobility in Si. In applications in which a substantial current is to be drawn from the sensor's output, InAs- or even InSb-based sensors are used, because of their better power efficiency. InAs and InSb Hall sensors are also preferred in metrology applications or when high signal-to-noise ratios are desired. GaAs is used in some high-temperature applications. Figure 7 shows the output of an ion-implanted GaAs Hall generator (Siemens KSY-10). The ordinate is the Hall resistance, i.e., V_{xy}/I, the ratio of the output voltage to the input current. This device is quite stable with respect to temperature in the range from −40 to 160 °C. GaAs integrated Hall sensors for high-temperature applications are also being developed (Itakura *et al.*, 1989).

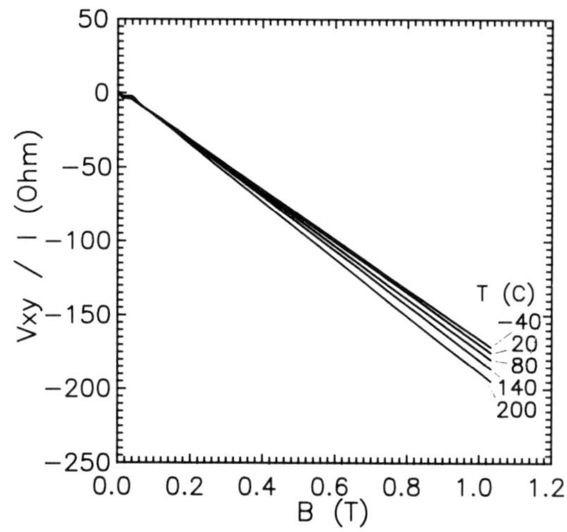

FIG. 7. Ratio of the Hall voltage to the input current, for a GaAs Hall sensor (Siemens KSY-10) made by ion implantation, shown as a function of magnetic field and temperature (after Heremans, 1993).

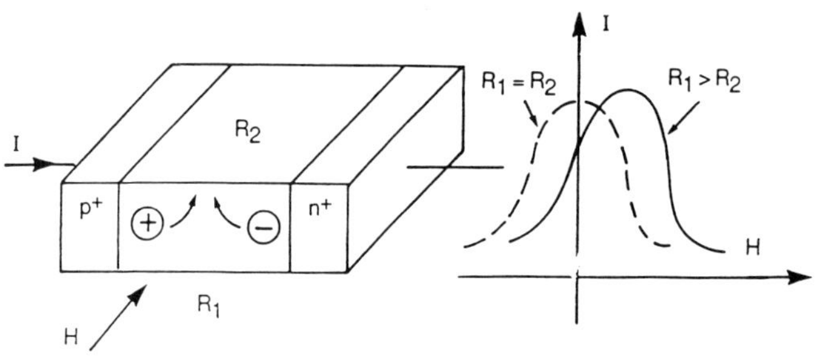

FIG. 8. A magnetodiode, made from a slab of lightly doped semiconductor sandwiched between heavily doped n and p regions. The surfaces of the lightly doped region have different recombination rates R_1 and R_2. The longitudinal voltage applied to the diode in forward biasing injects both electrons and holes in the lightly doped region. In presence of a magnetic field, the Lorentz force pushes both types of carriers in the same direction. The magnetic field H can push the carriers either toward the high–recombination-rate surface, in which case the current decreases, or toward the low–recombination-rate surface, and then the current is increased. The current versus field relation under a given bias is schematically given at right (after Cristoloveanu *et al.*, 1983).

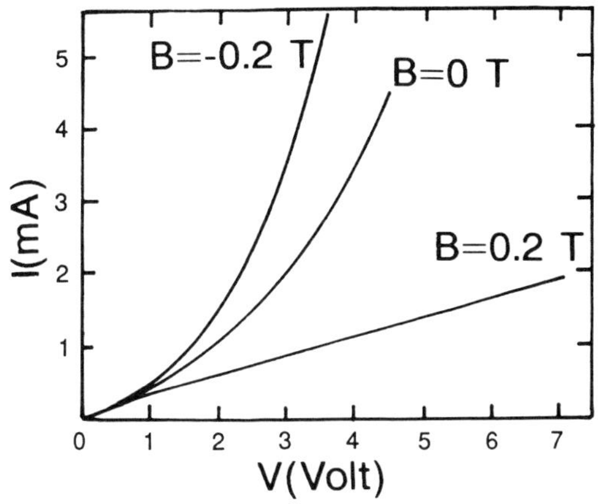

FIG. 9. Magnetic field dependence of the current-voltage characteristics of a Ge magnetodiode (after Yamada, 1968).

1.4 Magnetodiodes

Magnetodiodes (Cristoloveanu *et al.*, 1983) are *p-i-n* diodes, schematically represented in Fig. 8. They consist of an intrinsic region, contacted to a heavily *n*-type region on one side and a heavily *p*-type region on the other. The magnetodiode differs from a simple *p-i-n* diode because one of its lateral faces has been treated to form a surface at which electrons and holes have a high recombination rate R_1. This treatment can be mechanical abrasion or damaging by ion implantation. Some magnetodiodes exploit the high recombination rates at the silicon/sapphire interface. The other face is made as nearly perfect as possible, by etching for instance, to keep the recombination rate R_2 low. When the diode is forward biased, and a current I flows as shown in Fig. 8, electrons are injected from the n^+ electrode into the intrinsic region, and holes from the p^+ region. In the presence of a magnetic field H oriented as shown, the Lorentz force acting on both types of carriers pushes them in the same direction, because the Lorentz force is sensitive to the sign of the charge of the carrier; see Eq. (1). If the carriers are deflected by the magnetic field toward the face with a high recombination rate R_1, they recombine there, and the current flow is impeded. If they are deflected toward the other face, much less recombination takes place and more current flows. Current-voltage characteristics of a Ge-based *p-i-n* diode at room temperature at different field values are shown in Fig. 9. Sensitivities can be up to 1000 times higher than those of Hall generators made from similar materials. Discrete magnetodiodes have been made from Ge, Si, GaAs, silicon-on-sapphire, and CMOS technology (Baltes and Popovic, 1986).

1.5 Magnetotransistors

Many geometries have been used to exploit the influence of the Lorentz force on the flow of charge carriers in transistors. For a more complete review, the reader is referred to Baltes and Popovic (1986). Magnetotransistors can be made as FET transistors with two drains, or as bipolar transistors with two collectors. The applied magnetic field distorts the current lines and favors current flow into one of the two collectors or drains.

A dual-drain Si magnetic FET is shown in Fig. 10. A magnetic field normal to the chip plane causes a deflection of the current lines toward one of the two drains, as is shown in Fig. 11. Complementary magnetic FETs can be made using CMOS technology, and consist of a dual-drain *n*-channel device and a dual-drain *p*-channel device. The gate electrode voltage does not need to be varied during

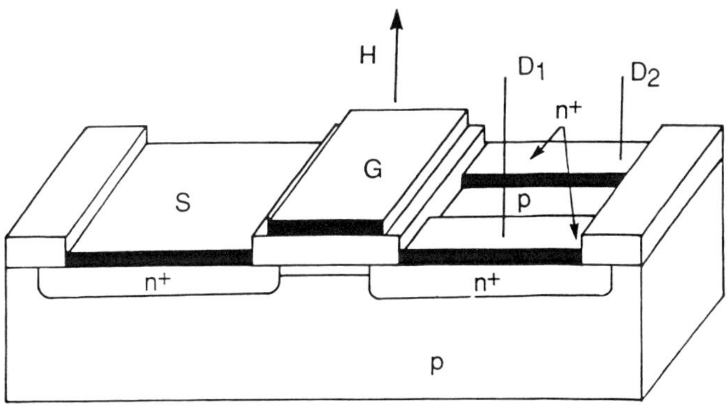

FIG. 10. Two-drain (D_1 and D_2) magnetic FET transistor. A magnetic field H exerts a Lorentz force that pushes the electrons from the source (S) preferentially toward one of the drains. The resulting current imbalance is a measure of the field H. The gate is labeled G (after Baltes and Popovic, 1986).

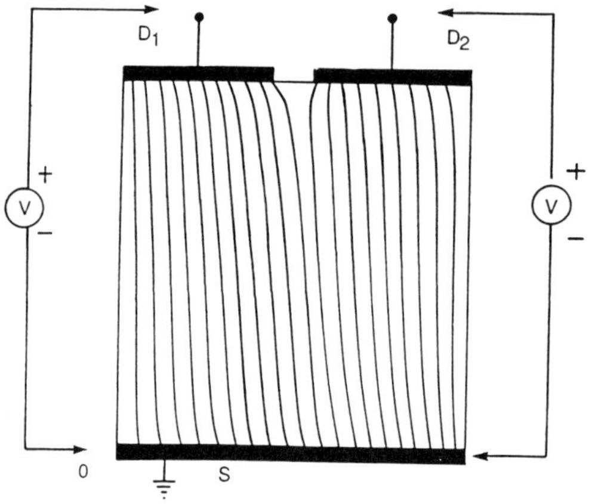

FIG. 11. Current density lines in a two-drain magnetotransistor as shown in Fig. 10 (after Baltes and Popovic, 1986).

operation; its only purpose is to create a channel of the desired surface density. In that sense, these devices are not strictly transistors. InAs tends to have a natural accumulation layer of electrons at its interface with air. Three-terminal magnetic field sensors with geometries similar to those in Fig. 10 can be etched as mesas into epitaxial layers of InAs grown on insulating GaAs, and no gate electrode is necessary (Heremans *et al.*, 1992). Similarly, GaAs/AlAs interfaces with active two-dimensional electron gases have been used by Nathan *et al.* (1990). In dual-drain magnetic FETs, the useful output voltage is measured as a voltage drop between load resistors connected to the drains; see Fig. 12.

Bipolar transistors can be designed to be sensitive to magnetic fields in different ways: some transistors have field-dependent current gains; others are designed with dual collectors. They can be made vertically or horizontally. The magnetic field can be designed to act in three different ways:

1. The Lorentz force can act on the minority carriers in the base region, or in the base-collector depletion layer, or in the low-doped collector region itself.
2. The Hall voltage can be designed to deflect the majority carriers in the base region.
3. The Lorentz force can act on both types of carriers, which then will adjust their concentration to maintain space-charge neutrality at each point.

In general, all three effects coexist to some degree. A *pnp* double-collector magnetotransistor with large sensitivity was described by Popovic and Baltes (1983). A schematic of the device is shown in Fig. 13. The output voltage is the differential voltage obtained between two load resistors connected to the two bases; see Fig. 12. The device in Fig. 13 has an emitter purposely designed with a low efficiency. The magnetic field operates on the majority-carrier current in the base (effect **2**), yielding a larger Hall field than if the Lorentz force were operating on the minority carriers alone. The base region is surrounded by a deep diffusion ring, which confines the base current to a thin slab perpendicular to the direction of the magnetic field.

Magnetotransistors can achieve sensitivities $(dI_C/dB)/I_C$ (I_C is the collector current in Fig. 12) as high as 4 T^{-1}. This seems superior to those of Si-based Hall plates, but such bipolar magnetotransistors require high base currents. When the sensitivities of existing magnetotransistors, bipolar or FET-like, are calculated including the sum of all currents, they are similar to those of Hall generators. Given the large variety of effects involved in

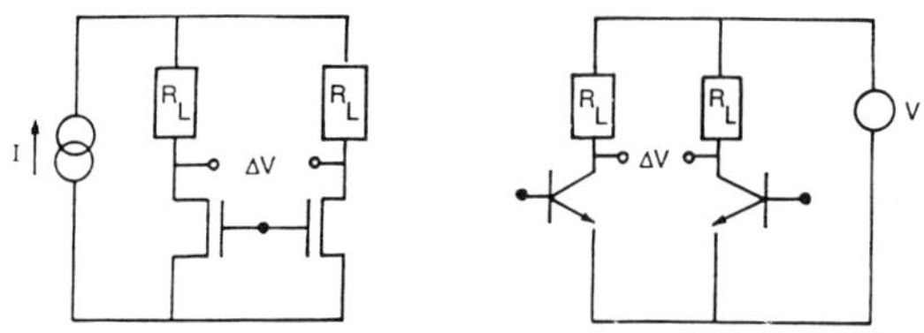

FIG. 12. The output of dual-drain FETs (left) and of bipolar magnetotransistors (right) is a current imbalance, which translates as a voltage ΔV across load resistors R_L in these diagrams.

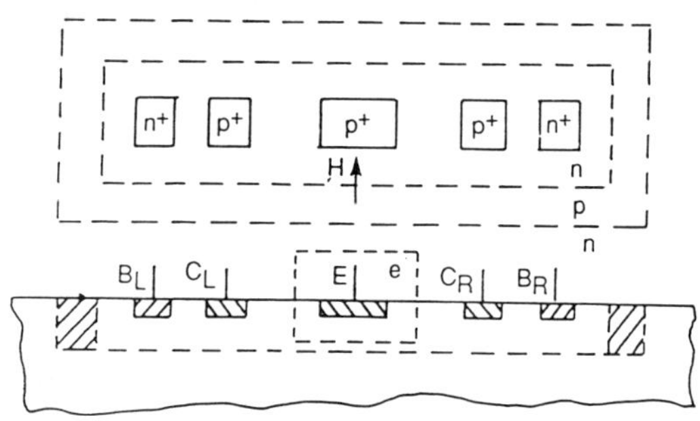

FIG. 13. Double-collector magnetotransistor. The magnetic field H modulates the ratio of the currents between the emitter (E) and each of the collectors (C_L at left and C_R at right). A diffusion ring (dashed line) confines the current. The base contacts are B_L and B_R (after Popovic and Baltes, 1983).

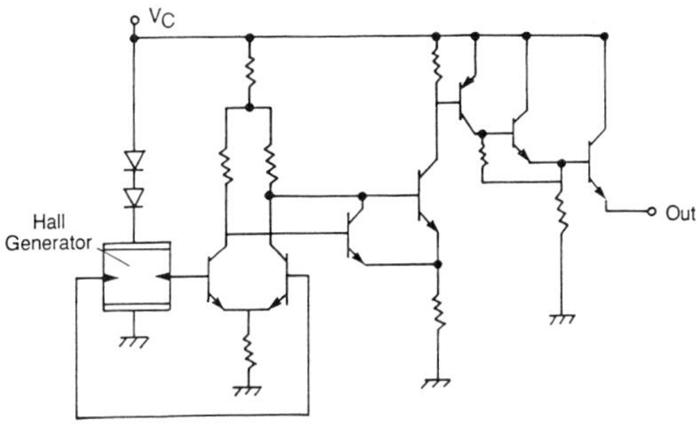

FIG. 14. Schematic diagram for an integrated silicon Hall element with amplifier (after Nakamura and Maenaka, 1990).

magnetodiode and magnetotransistor operation, these devices still have a large potential for optimization (Nathan *et al.*, 1989).

1.6 Integrated Semiconductor Magnetic Field Sensors

Silicon Hall generators are mostly used with on-chip amplification and signal-conditioning electronics. A simple circuit, used in solid-state keyboards (Nakamura and Maenaka, 1990), is shown in Fig. 14. Most integrated magnetic field sensors are rather simple combinations of the sensor with an amplifier and, when hysteresis is desired, with a latching circuit. The voltage output of a Si-based Hall generator is usually low, especially when small current consumption is also a design requirement. This implies that high-gain amplifiers must be used, with limited bandwidth. Hall generators also give a differential output, which often has to be converted to a single-ended output. For high-temperature applications, such as many automotive ones, GaAs sensors with integrated electronics are being developed (Itakura *et al.*, 1989).

InAs- or InSb-based magnetoresistors grown on an insulating GaAs substrate, in which the amplifiers and signal conditioners can be integrated, become very competitive with GaAs Hall generators, because they generate ten times higher output voltage in moderate magnetic fields (see Sec. 1.11, Comparison of Magnetic Field Sensors). Magnetoresistors give a single-ended output voltage with an offset bias.

Magnetotransistors can readily be integrated. Complementary pairs of magnetic FETs can be incorporated into a configuration reminiscent of a CMOS differential amplifier, shown in Fig. 15. Each of the usual two complementary MOS transistor pairs is replaced by the corresponding magnetic FET. Their drains are cross-coupled. When a magnetic field is applied perpendicularly to the plane of the devices, the current in one drain of one transistor increases, while the current in the drain of the other transistor connected to it decreases. The gain built into the circuit is such that small imbalances in the drain currents lead to large output-voltage variations. This circuit converts the differential output of the magnetic FET into a single-ended, amplified signal. The bipolar dual-collector magnetotransistor shown in Fig. 13 can be integrated using standard analog bipolar technology (Maenaka *et al.*, 1990).

1.7 Ferromagnetic Metal Film Magnetoresistors

The electrical resistance of certain ferromagnetic alloys is anisotropic in a magnetic field (see MAGNETIC ORDERING IN SOLIDS; MAGNETIC MATERIALS). The theoretical background

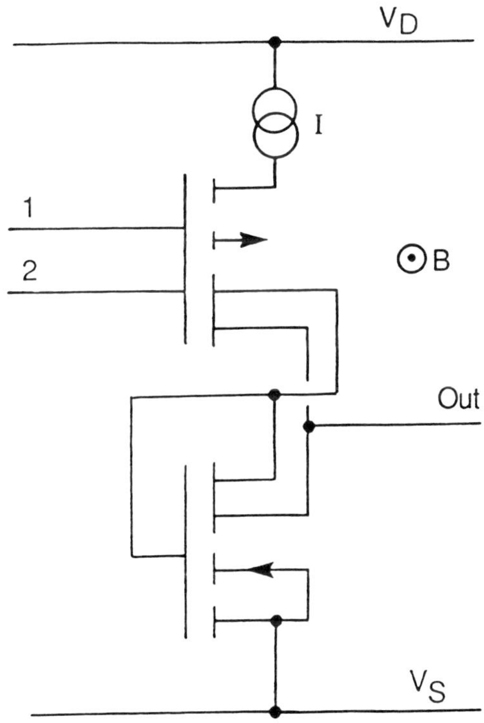

FIG. 15. Two complementary dual-drain magnetic-field effect transistors integrated with a conventional CMOS differential amplifier (after Baltes and Popovic, 1986).

for this effect is summarized by McGuire and Potter (1975). In a film of a ferromagnetic metal thin enough to be single-domain, the magnetic moment vector tends to lie in the plane because of the large demagnetization factor. The electrical resistivity measured when the current-density vector is parallel to the magnetic moment, $\rho_{\parallel}$, is different from the resistivity measured with the current perpendicular to the moment, $\rho_{\perp}$; see Fig. 16. This experimental fact has been explained by an anisotropy in the scattering of the electrons (Campbell *et al.*, 1970).

The anisotropy in the magnetoresistance of Permalloys ($Ni_{1-x}Fe_x$, $x \sim 0.2$) is used to make magnetoresistors (Dibbern, 1986). These devices are made from thin films because the electrical resistance of thin-film devices is higher than that of bulk samples, which keeps the current consumption low. A thin layer can also be made to behave like a single magnetic domain. Its magnetization with no external field applied has one single, well-defined direction in the plane, called the easy axis, also introduced during manufacture. A conceptual diagram of a device is shown in Fig. 17. There is an angle θ between the direction of current flow and the direction of the magnetization **M** along the easy axis. As an external magnetic field **H** is applied to the sample, the direction of the magnetization **M** is rotated away from the easy axis, and the angle θ changes. Since the resistivity along the magnetization direction is $\rho_{\parallel}$, different from that perpendicular to **M**, $\rho_{\perp}$, the total resistance of the film

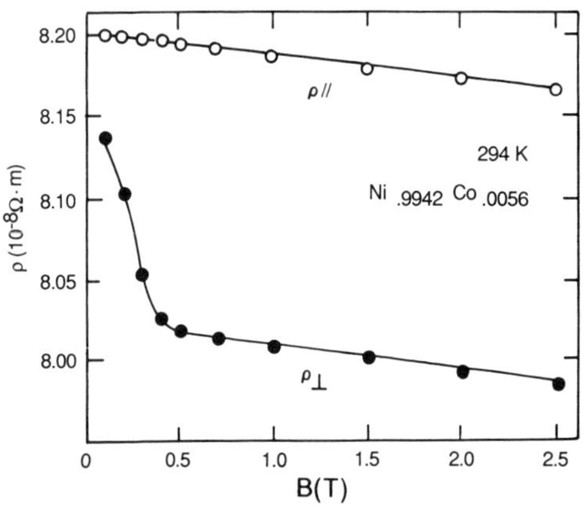

FIG. 16. Magnetoresistivity of Ni-Co at room temperature as a function of field, applied parallel ($\rho_{\parallel}$) or perpendicular ($\rho_{\perp}$) to the easy axis of magnetization (after McGuire and Potter, 1975).

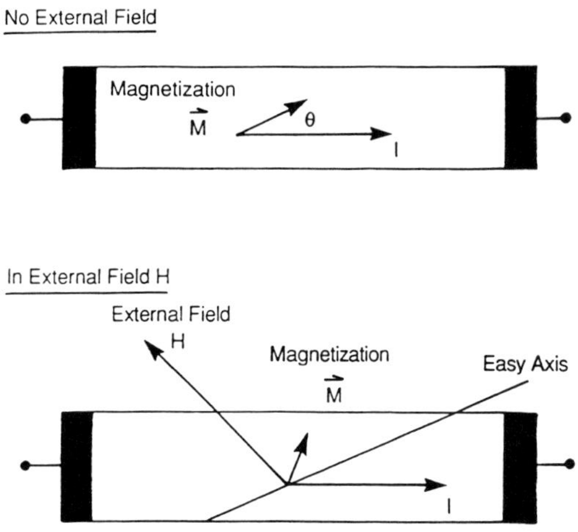

FIG. 17. Magnetoresistor made from a thin film of Permalloy. In the absence of field (top), the current makes an angle θ with respect to the magnetization **M**, which lies along the easy axis. An in-plane applied field **H** (bottom) rotates the magnetization **M**. The anisotropy in the magnetoresistivity shown in Fig. 16, and the fact that the angle between current I and moment **M** is influenced by the field, lead to a change in resistance of the strip (adapted from Lenz, 1990).

becomes a function $R(H)$ of the applied field H, via the dependence of $\theta(H)$ and $\rho_{\perp}$ on H:

$$R(H) = (l/Wt)[\rho_{\perp} + (\rho_{\parallel} - \rho_{\perp})\cos^2\theta]. \qquad (15)$$

Several techniques exist to linearize the output of these sensors; see Dibbern (1986). Ferromagnetic magnetoresistor sensors are best suited to the measurement of low fields, because they are not sensitive at fields above the saturation field of the layer.

A recent development, which may force a reassessment of the previous statement, is the discovery of giant negative magnetoresistance in Fe/Cr superlattices (Baibich *et al.*, 1988) consisting of alternating layers of magnetic and nonmagnetic metals. The magnetic moments of adjacent magnetic layers can be spontaneously aligned either ferromagnetically or antiferromagnetically, depending on the thickness of the layers. When the magnetic layers are aligned antiferromagnetically, the resistance of the structure is greater than when they are aligned ferromagnetically, because of the interplay between the spin dependence of the scattering and the magnetic orientation in the structure. An external magnetic field can change the magnetic orientation of the layers with respect to one another, and thus change the resistance of the multilayers (Barthelemy and Fert, 1991). Relative changes in resistance of 50% at applied fields of 2 T have been reported, and are shown in Fig. 18. To date, no publications have appeared describing the use of this effect in sensors, but that is likely to change. The large saturation field H_S for the Fe/Cr system is a disadvantage for sensor applications, but other systems, such as Co/Cu (Mosca *et al.*, 1991), offer better sensitivity. Also recently discovered are giant magnetoresistive effects in metal multilayers that are not magnetically coupled (Dieny *et al.*, 1991), and these too may lead to future devices.

1.8 Variable-Reluctance Sensors

This type of sensor is based on the measurement of the current induced in a coiled conductor, which is proportional to the rate of change of the magnetic flux through the coil (Göpel *et al.*, 1989, Chap. 6). A core of high-permeability magnetic material can be inserted in the coil to gather the surrounding magnetic flux. An example of such a sensor, implemented with thin-film technologies and used in a magnetic recording head, is shown in Fig. 19. The simplest output of such a device is a voltage or a current appearing between the coil end contacts as the magnetic flux through the coil is varied with time. In the example of Fig. 19, a magnetic recording medium passes over the gap (see the next section), and closes the magnetic loop that is coupled to the search coil. The sensitivity of such sensors depends on the permeability of the core material, the area of the coil, the number of turns, and the rate of change of the magnetic flux through the coil. The core material's desired properties include a high permeability for optimal sensitivity, but also a large saturation field, so that large fields can be used in the gap. A small coercivity, low hysteresis losses, and a low magnetostriction are desired to minimize noise. The core material should also have a small but nonzero uniaxial anisotropy so that the domain structure is controllable, and a high electrical resistivity to minimize eddy-current losses. The most used materials are Permalloys (Ni-rich NiFe alloys), and newer materials are being developed such as FeAlSi, Fe/C or Ni/Fe superlattices, and other materials described by Jagielinski (1990). The frequency response of search-coil sensors is limited by the ratio of the coil's inductance to its resistance, though the capacitance between the windings also can be a limiting factor. Their useful range is typically from 1 Hz to 1 MHz. The low-frequency limit implies that variable-reluctance sensors cannot be used as steady-

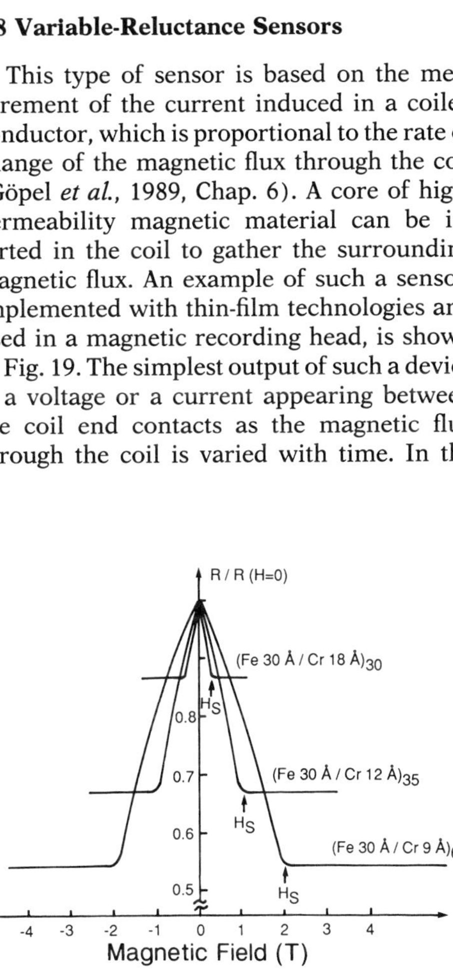

FIG. 18. Relative change in resistance of Fe/Cr multilayers as a function of magnetic field, at 4.2 K. The thicknesses of the individual layers (in Å) are given in the figure, along with the number of layers. The magnetoresistance saturates at a field H_S, but the effect persists to room temperature (after Baibich *et al.*, 1988).

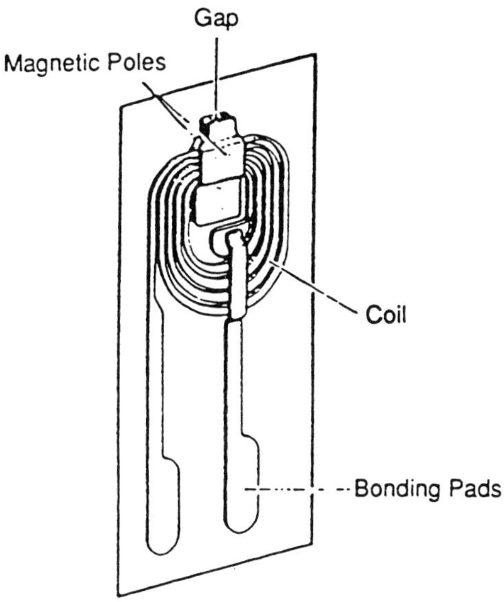

FIG. 19. Inductive head for magnetic recording/readout, made using thin-film technology (after Jagielinski, 1990).

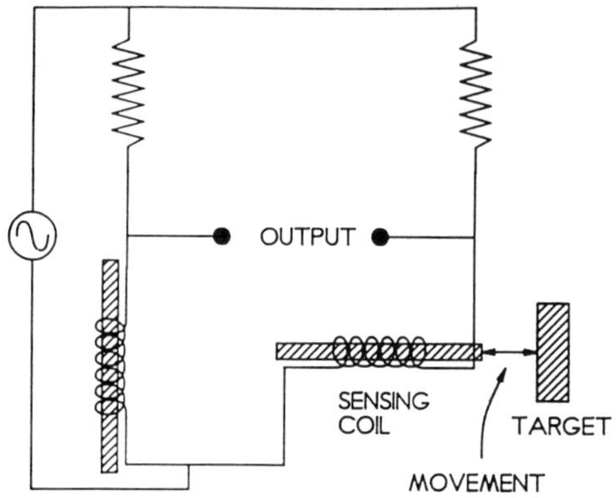

FIG. 20. Search-coil magnetometer in an inductive bridge. Movement of the ferromagnetic target changes the inductance of the coil at right, and unbalances the bridge.

state magnetometers, or in position- sensing applications. They can, however, detect field changes as small as 10^{-10} T.

Another possible way to obtain a readout from a search-coil magnetometer is to measure the coil's inductance, which will change as the magnetic circuit around the coil is changed, for instance by the proximity of a target made from ferromagnetic metal (see Fig. 20). There are two typical readout circuits used to do that. In the circuit shown in Fig. 20, the coil is one leg of a balanced inductive bridge, and the output voltage of the bridge measures the change of inductance of the coil. The coil also can be incorporated in an oscillating circuit, in such a way that a change in inductance results in a change in resonant frequency of the oscillator. These "active" search coils can be used in true position sensors.

1.9 Fluxgates

Fluxgates are often used in compasses. They exploit the magnetic hysteresis exhibited by ferromagnetic materials. They typically consist of cores of a high-permeability ferromagnet, such as Permalloy, each equipped with two coils (Geyger, 1954). There are many types of possible configurations, but only one will be described here (see Fig. 21), built around two rods of a ferromagnetic material. The long thin geometry of the rods is designed to concentrate the flux Φ from the external magnetic field **H**, which is the quantity to be measured. A sinusoidal ac current is applied to the primary windings W'_p and W''_p, which are connected in series but generate magnetic fluxes of opposing polarity in the rods. There are two secondary windings W'_s and W''_s, connected in series as shown. A third coil can eventually be added to supply a dc current I_{dc}. The primary coils are driven so that the core is saturated at each half-cycle of the current. Figure 22 shows very schematically how the voltages induced in W'_s and W''_s are not sinusoidal at frequency f, but clipped. With no external flux Φ or dc current I_{dc}, the voltages V'_s and V''_s induced in W'_s and W''_s cancel each other identically. When an external dc flux Φ or a dc current I_{dc} is applied, the voltages induced in the secondary coils become asymmetrical. A voltage consisting largely of even harmonics of f appears at the output of the two secondary coils. The voltage $V'_s + V''_s$ can be detected by an amplifier locked in at a frequency $2f$. The output can be used as a measurement of Φ, and is sensitive to the polarity of Φ. The dc coil can eventually be used to balance the external field, and the fluxgate is then used as a null detector. The

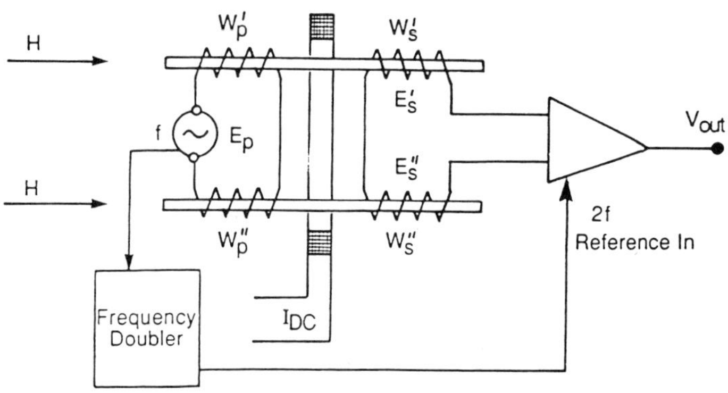

FIG. 21. A fluxgate magnetometer (after Geyger, 1954).

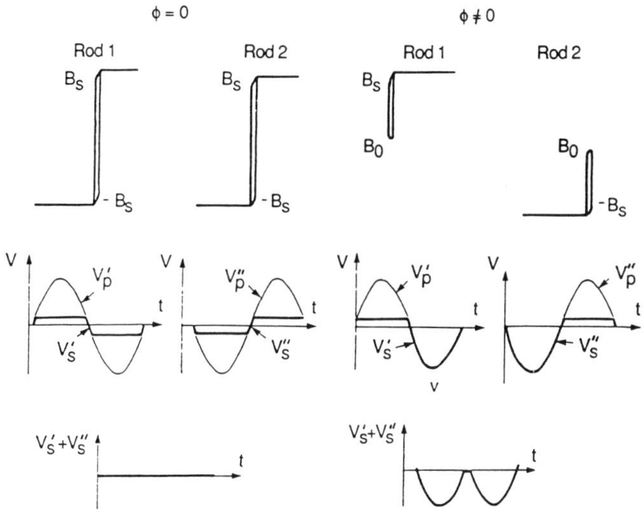

FIG. 22. Principle of operation of the fluxgate. The top figures show which part of the hysteresis curve of the core material is swept in each of the two rods at zero external flux Φ (left), and when flux Φ is present (right). The bottom shows schematically the time dependence of the voltages across the primary (V_p' and V_p'') and secondary (V_s' and V_s'') coils, again without (left) and with (right) flux Φ. The voltage at the input of the amplifier, $V_s' + V_s''$, is shown on the last line. In presence of Φ, $V_s' + V_s''$ has a large component at double the excitation frequency.

sensitivity of fluxgate magnetometers depends on the shape of the hysteresis curves, which should ideally be square. The use of a core material with low coercivity and saturation field values minimizes the power consumption of the device. Sensitivities range from 10^{-10} to 10^{-2} T, but the operation frequency is limited to about 10 kHz, because of the need to have an excitation source at a higher frequency. The power consumption of fluxgate magnetometers can be decreased by operating them on a minor hysteresis loop, because then the core does not have to be driven from saturation to saturation. Flux gates based on thin-film technology have been described by Seitz (1990).

Magnetic amplifiers are devices similar to fluxgates, in which a magnetically controlled saturation of the core of a winding is used to modulate the amount of power delivered from a source to a load. The reader is referred to descriptions by Geyger (1954) and Shea (1968).

1.10 Other Magnetometers

The devices described above are mostly small solid-state type devices. Many complex magnetometer systems exist. Scientific magnetometer systems used in the measurement of magnetic properties of materials have been reviewed by Foner (1981).

Superconductive quantum interference devices (SQUIDs) are the most sensitive instruments for measuring magnetic fields. SQUIDs are described in a separate article in this encyclopedia, and the reader is referred thence.

Optical magnetometers are based on the Zeeman effect in gases such as helium or cesium vapor; they are typically large devices with a high power consumption (Lenz, 1990). There are nuclear precession magnetometers, based on the precession of proton magnetic moments around an applied magnetic field. This type of sensor makes use of hydrocarbon fluids or ^{3}He, and the precession is sensed by a pickup coil. These magnetometers are potentially very sensitive (Lenz, 1990). Magnetooptical sensors are based on Faraday rotation of polarized light traveling through a magnetic material, for instance a terbium gallium garnet (Lenz, 1990). Unfortunately, the Faraday effect is rather small in moderate fields, except in ferromagnets. Dandridge *et al.* (1980) describe a magnetic field sensor based on a conventional single-mode optical fiber coated with a magnetostrictive layer. Under an applied field, the magnetically induced strain changes the optical path length, which can be monitored by interferometry.

1.11 Comparison of Magnetic Field Sensors

The magnetic field range to be detected determines the choice of sensor. The relative sensitivity of the sensor in its field range is the primary factor that must be considered, but other factors are the operating temperature range, the power consumption, the size, the linearity of the sensor, and whether or not sensitivity to the polarity of the field is needed.

Solid-state sensors, such as magnetoresistors, Hall generators, magnetodiodes, and magnetoresistors, have the great advantage of being integratable with amplification and signal-processing circuits, resulting in very cost-competitive systems. In Fig. 23, the sensitivity of a number of solid-state sensors is shown at 20 °C as a function of applied magnetic field. These sensors are a thin-film InSb-based magnetoresistor [*n*-doped at 6×10^{16} cm^{-3} (Partin *et al.*, 1992)], a thin-film Permalloy-based magnetoresistor (Nippondenso 076100-0202), an ion-implanted GaAs Hall generator (Siemens KSY-10), and a Si-based Hall generator with integrated on-chip amplification (Allegro 3501). In order to make a meaningful comparison, the "sensitivity" of each sensor is defined as the difference between the output voltage of the sensor at the field given in the abscissa and at zero field divided by the supply current (Heremans *et al.*, 1992). At the lowest fields, magnetoresistors based on Permalloy are most sensitive, but their sensitivity saturates at fields around 5×10^{-3} T. Hall generators give an almost linear response to field, while InSb-based magnetoresistors give a quadratic one. As a result, Hall generators are always more sensitive than magnetoresistors at low fields, and the opposite is true at high fields. The crossover field depends on the technology used: unamplified InSb-based magnetoresistors are more sensitive than the Si-based Hall generator shown with integrated amplifier above 0.03 T. The latter sensor has a cutoff frequency (3 dB) of 25 kHz, because of the large-gain amplifier used on the chip. Many other ways of conditioning the output signal can be integrated with the Si Hall generator, such as trigger circuits, latching circuits, or circuits that linearize the output and compensate for temperature over a limited range. Similar functions could be integrated on the GaAs substrates used in the InSb- or GaAs-based sensors. To summarize, Permalloy-based magnetoresistors are preferred for sensing magnetic fields below about 0.005 T, Si Hall sensors with integrated amplification for fields between 0.005 and 0.03 T, if the frequency cutoff is not a problem, and InSb-based magnetoresistors for fields above 0.03 T. These are only crude guidelines. When high-temperature (above 170 to 200 °C) operation is required, such as in automotive applications, GaAs Hall sensors can be considered along with heavily doped InSb or InAs magnetoresistors, but magnetoresistors are more sensitive.

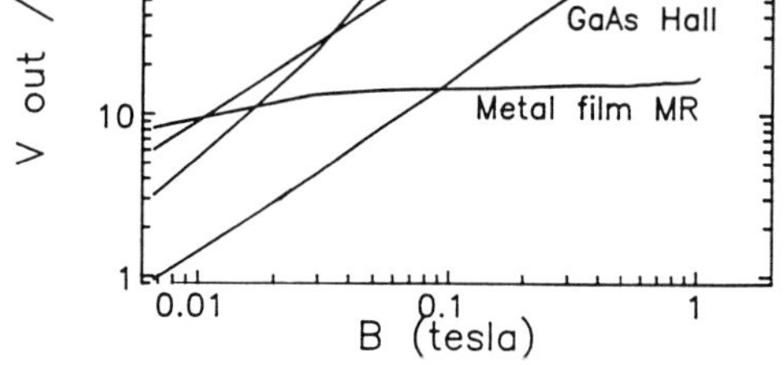

FIG. 23. Comparison of the sensitivities of different solid-state magnetometers at room temperature. The sensitivity is defined as the change in output voltage, divided by the input current, when the applied magnetic field is changed from zero to the value shown in abscissa. At very low fields, the magnetic metal film magnetoresistance (MR) is the most sensitive. It is instructive to note that when field changes above 0.03 T have to be detected, unamplified InSb-based magnetoresistors are more sensitive than Si Hall sensors with integrated amplifiers. The gain of such amplifiers is limited by the signal-to-noise ratio and the bandwidth (Heremans *et al.*, 1993).

2. APPLICATIONS OF MAGNETIC FIELD SENSORS

2.1 Magnetic Recording Heads

Magnetic recording is described elsewhere (see MAGNETIC STORAGE OF INFORMATION). In short, thin layers of magnetic recording media can be magnetized in the direction perpendicular to the film, or in the plane of the film ("longitudinal recording"). Recording heads can use search coils, magnetoresistors, or Hall generators. Figure 24 shows an example of a magnetic circuit that uses a Hall sensor to read codes magnetized into a magnetic data carrier (Siemens, 1989). Magnetoresistive heads operate on much the same principle, but use the more sensitive metal magnetoresistors (Sec. 1.7). A review of biasing methods is given by Markham and Jeffers

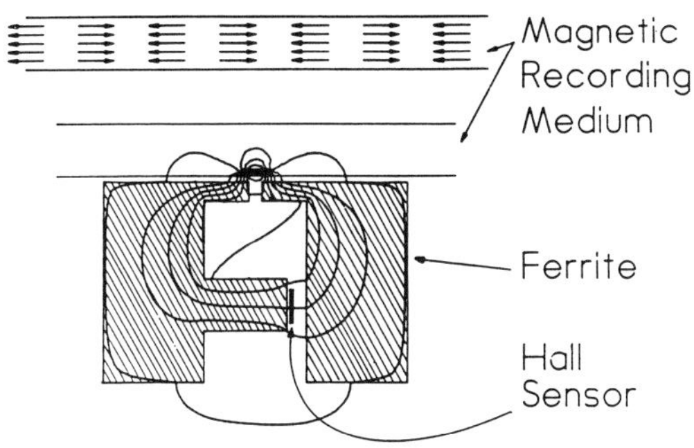

FIG. 24. Data can be encoded in a medium, the top foil, by magnetizing regions of the foil in the longitudinal direction (top). The reading head (bottom) consists of a closed ferrite flux path with two gaps: one for the magnetometer (a Hall sensor) and one close to the medium. The magnetized area (dark rectangle) in the medium generates the flux lines shown.

(1990). Inductive heads are also used, and Jagielinski (1990) shows an example of the parallel readout of ten recording tracks, using thin-film pickup devices such as shown in Fig. 19. Current technology trends are toward multitrack devices, response frequencies up to 100 MHz, and head-media spacings less than 50 nm. In longitudinal recording an increase in linear information density requires a corresponding decrease in gap width, and a higher magnetic field in the gap.

2.2 Position and Speed Sensors

With the widespread use of microprocessors in cars and airplanes, sensors that translate mechanical movement into an electrical signal are becoming very common. The simplest device that does this is the potentiometer; but the friction involved in conventional potentiometers makes them less reliable than frictionless methods. Optical techniques (a light-emitting diode and a photodetector, with or without fiber optics, for instance) are sensitive to contamination, and their applications are limited to protected environments. Magnetic position sensors are the technology of choice when reliability is a critical issue. They are based on the use of a permanent magnet whose position is changed with respect of that of a magnetic field sensor. Alternatively, the magnetic field sensor can be mounted on the magnet, and a ferromagnetic metal part moved with respect to that assembly. In either design, the mechanical movement is translated into a change in the flux through the magnetic field sensor. An example of the design of a rotary contactless potentiometer is shown in Fig. 25. Linear position sensors using a sliding magnet moving with respect to a magnetoresistor or a Hall sensor can be designed. Proximity sensors based on similar designs are also common, for instance to verify door and hatch closures (Lenz, 1990).

2.3 Rotary Shaft Encoders

Magnetoresistors or Hall sensors are widely used in rotary shaft encoders when the speed and the position of the shaft have to be detected. Pickup coils can be used when the information is not needed at zero rotation speed. A typical magnetic circuit used for sensing the position of gear teeth machined into an iron or steel wheel is shown in Fig. 26.

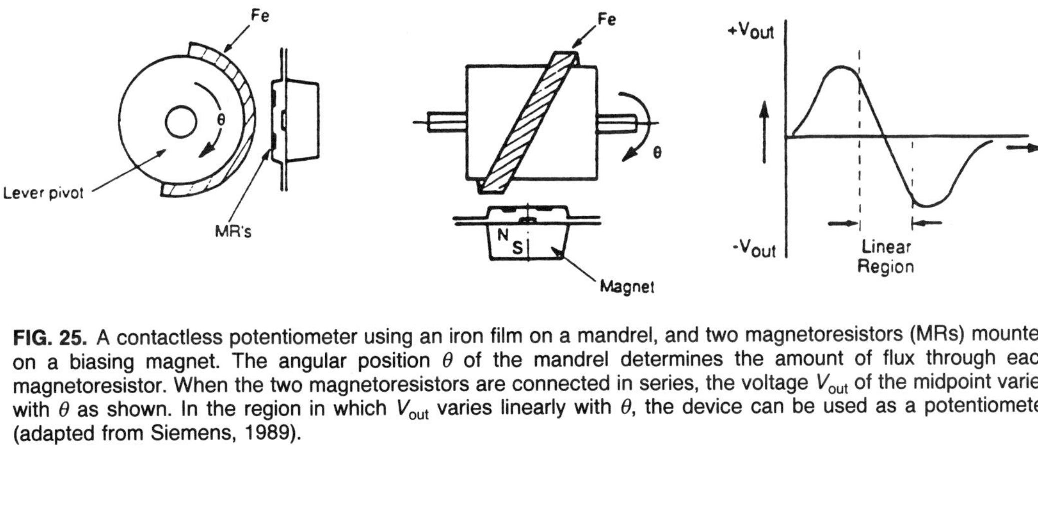

FIG. 25. A contactless potentiometer using an iron film on a mandrel, and two magnetoresistors (MRs) mounted on a biasing magnet. The angular position θ of the mandrel determines the amount of flux through each magnetoresistor. When the two magnetoresistors are connected in series, the voltage V_{out} of the midpoint varies with θ as shown. In the region in which V_{out} varies linearly with θ, the device can be used as a potentiometer (adapted from Siemens, 1989).

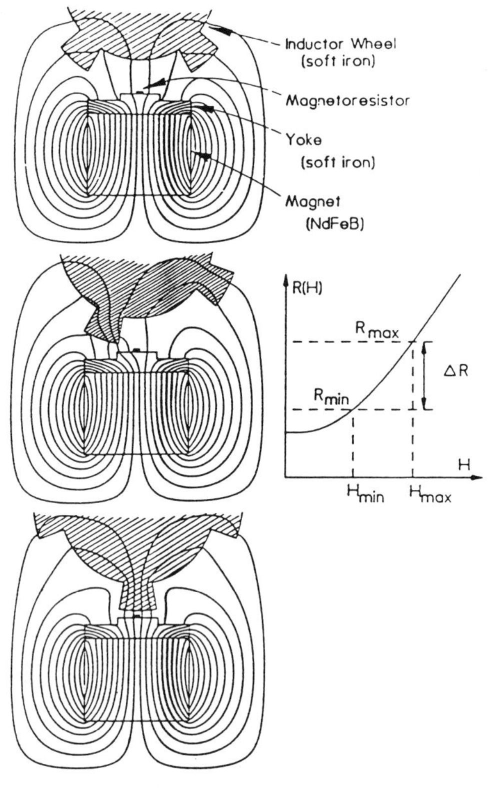

FIG. 26. Rotary shaft position sensor using a magnetoresistor mounted on an iron yoke on a bias magnet, and a cogged inductor wheel. The flux lines are shown at left, for different positions of the inductor wheel. The magnetic induction *H* at the sensor location changes with the inductor wheel angle, and so does the electrical resistance *R*(*H*) of the magnetoresistor.

The size and distance of the teeth can be used to encode absolute position sensing. This type of sensor is widely used to measure the position of the crankshaft and camshaft in a gasoline engine (see Fig. 27). Its information is fed into the engine control computer, which then determines which cylinder fires and when it does so. The ignition system shown in Fig. 27 now replaces the mechanical switches and "points" used in older car engines.

2.4 Brushless Motors

Figure 28 shows a schematic diagram of a brushless dc electric motor. The rotor is a permanent magnet (labels indicate the north and south poles). The position of the rotor is monitored by Hall sensors H_1 and H_2. These control the base currents into the bipolar transistors that control the current flow through the stator windings. Such motors have easy starting characteristics. The absence of brushes results in better reliability and the elimination of the electrical noise due to sparking. They find applications in video recorders, disk drives, turbopumps, and automotive motors. The control of the absolute rotor position can be done by a microprocessor chip, and complicated rotation patterns can be programmed in a flexible manner.

2.5 Shuntless dc Current Measurements

Shuntless dc ammeters can be made as follows: A primary coil is wound on a soft iron core that contains an air gap. The current to be measured, I_{in}, passes through this winding. The resulting flux generates a field that can be detected by the Hall sensor located in the air gap. The output of the Hall sensor is a measurement of the current I_{in}. For more precision and wider dynamic range, a secondary coil is wound around the core. A current can be made to flow through this secondary, so as to cancel the field in the gap. The Hall sensor is then used as a null detector, and the current through the secondary coil is a measure of the current through the primary.

3. HIGH-FREQUENCY DEVICES

3.1 General Aspects

The magnetic component of an electromagnetic wave couples readily to magnetic materials. Ferrites and powdered ferromagnetic materials are used as magnetic field concentrators in antenna cores, inductors and transformers. Devices operating in the microwave-frequency range also use the ferromagnetic (or ferrimagnetic) resonance. In ferromagnetic solids, the adjacent magnetic moments of the atoms interact very strongly and align parallel to each other, as long as the solid is below a critical temperature, the Curie temperature T_C. In ferrimagnetic materials, the spins are aligned antiparallel, as in antiferromagnets, but there are more spins pointing in one direction than in the other. The material thus has a net magnetic moment, but smaller than it would have if all spins were parallel. Their critical ordering temperature is called the Néel temperature, T_N. Unfortunately for device applications, solids that are strictly

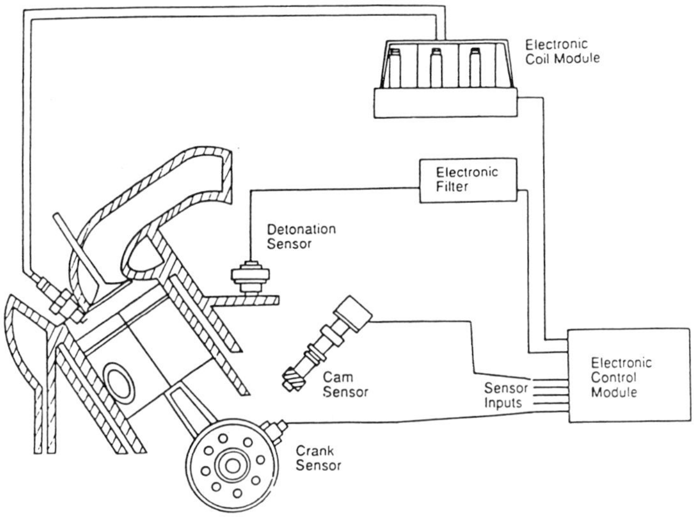

FIG. 27. Ignition system of a Buick engine. Magnetic position systems such as shown in Fig. 26 measure the position of crankshaft and camshaft, and transmit the information to a microprocessor (the electronic control module). The microprocessor, which also receives data on the air flow into the engine, the coolant temperature, the knock sensor, and the transmission, determines the spark timing and sequence. At the proper time, it switches low-voltage current to one of three coils. The secondary windings of each of these coils are connected to two spark plugs. When the current to a given coil is interrupted, both sparks fire, one in an empty cylinder, and one in a fully fueled cylinder. Such ignition systems avoid the mechanical commutation of high voltage.

ferromagnetic and with Curie temperatures above room temperature are metals. Eddy currents are generated in these electrically conducting materials at high frequencies. This leads to unacceptable energy dissipation. Magnetic materials with a high electrical resistivity, such as powdered metals, ferrites, and garnets, are preferred for high-frequency devices.

Powdered ferromagnetic materials, compacted with an insulating binder, can have reasonably low eddy-current losses. The reduction in losses in these materials usually comes at the expense of the permeability of the compact. Powdered cores are made using Permalloy, pure iron, or Sendust powder. Some properties of these powders are summarized in Table 1.

Ferrites (Smit and Wijn, 1959) are ferrimagnetic. The general chemical composition of soft ferrites is $(MO) \cdot (Fe_2O_3)$, where M is a divalent metal such as Fe, Ni, Co, Mn, Mg, Zn, Cd, or Cu. The simplest member of this class is thus iron ferrite, Fe_3O_4, or magnetite, mentioned in Fig. 1. Ferrites have high resistivities because the constituent atoms have no free 3*s* or 4*s* electrons available for conduction. Ferrites are selected on the basis of a number of properties, such as a high permeability at low field, or low losses at high frequency, or a square hysteresis loop. Impurities tend to decrease the permeability. The properties of some ferrites are given in Table 1.

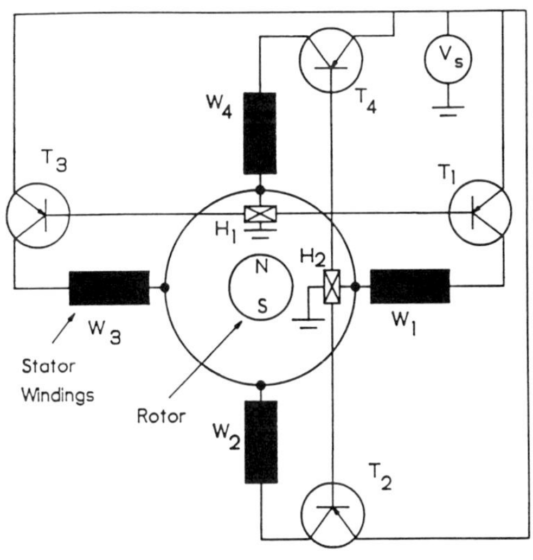

FIG. 28. Brushless electrical motor, with a permanent-magnet rotor and Hall sensors. The control currents are provided by the Hall sensors H_1 and H_2 to the transistors T which supply the windings W (adapted from Siemens, 1989).

Table 1. Magnetic properties of ferrites and pressed powders used in magnetic cores (Chikazumi, 1964). The balance for composition in the powders is Fe.

Material	Composition	Permeability, μ	Saturation magnetization (G)	Curie or Néel temp. (°C)
Pressed powders				T_C
Iron	Carbonyl Fe	20	1240	770
Mo-Permalloy	2 Mo, 81 Ni	125	560	480
Sendust	5 Al, 10 Si	80	360	500
Ferrites				T_N
Mn-Zn	50 Mn, 50 Zn	2000	200	110
Cu-Zn	40 Cu, 60 Zn	1100	—	90
Cu-Mn	40 Cu, 60 Mn	—	230	—
Ni-Zn	30 Ni, 70 Zn	80	320	130
Mg-Zn	50 Mg, 50 Zn	500	207	120
Mg-Mn	50 Mg, 50 Mn	1400	215	130

Other materials commonly used are rare-earth garnets, mostly yttrium iron garnet (YIG) or gallium-substituted YIG (GaYIG). Their general chemical formula is $(5Fe_2O_3) \cdot (3M_2O_3)$, where M represents Y or a rare-earth ion (Sm to Lu). Materials used in practice can deviate from these simple formulas. Both polycrystalline and single-crystal materials are used.

Microwave applications use the ferromagnetic resonance, whereby the spin system absorbs energy from the electromagnetic wave at a specific, resonant energy. Figure 29 shows schematically a sphere of YIG in a vertical dc field H_0. Ferromagnetic resonance can be viewed as the tendency for the spins in the garnet to precess about the vertical field H_0. For a YIG sphere, the resonance frequency of the precession is empirically given by $f_R = (28\ \mathrm{GHz/T})H_0$. The spin precession results in a rotating magnetic dipole $\mathbf{m}$ shown in Fig. 29(a). A circularly polarized microwave (rf) radiation with a magnetic vector $\mathbf{h}^+$ shown in Fig. 29(b) can excite that precession, if the radiation is at or near the frequency f_R. rf magnetic fields will build up about the sphere in the same way as the magnetic field builds up in the inductance of an *L-C* tuned circuit at resonance. Circularly polarized rf radiation with a magnetic component in the direction $\mathbf{h}^-$ shown in Fig. 29(c) does not couple to the sphere, even if the radiation is at the resonant frequency. The device is thus not "reciprocal," and will be selective with regard to the direction of propagation of a traveling rf wave. Linearly polarized fields may be constructed as the sum of two circularly polarized fields of equal strengths rotating in opposite directions. When the sphere is subjected to a linearly polarized field, it will respond to only one of the two circularly polarized components, and therefore emit only that circularly polarized field at resonance.

Some relevant properties of a few single-crystal materials used in practice are given in Table 2. The saturation magnetization limits the range of dc magnetic field available for tuning the resonance frequency. The linewidth of the resonance can be expressed as the range of dc magnetic fields ΔH over which the resonator responds (within 3 dB) at frequency f_R. This quantity depends on material parameters, and also on the shape and surface finish of the device. The resonant frequency is a function of the shape of the resonator (via the demagnetization factor; see MAGNETOSTATICS), the crystalline anisotropy (in single crystals), and interactions of the ferrite or garnet with currents in the metallic walls surrounding the device. Details are given by Soohoo (1985) and Matthaei *et al.* (1980).

3.2 Microwave Devices

3.2.1 Filters. Figure 30 shows a resonator filter built around a ferrite sphere. The incoming and outgoing waves are coupled through two loops, slightly elliptical in shape so that they can be nested perpendicularly to each other. The ferrimagnetic resonator is located in the center. The biasing dc magnetic field H_0 is parallel to the plane of both loops.

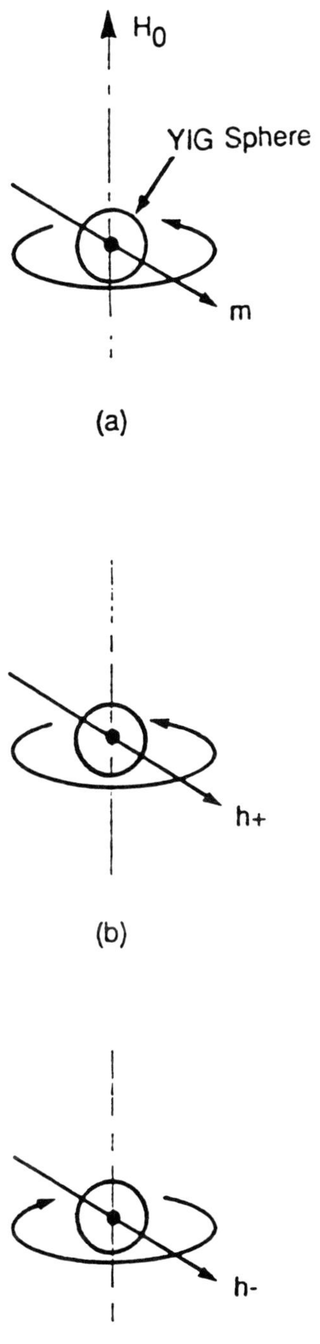

FIG. 29. (a) A YIG sphere with a circularly polarized ac moment **m**. (b) External ac fields at the resonant frequency of the sphere and circularly polarized magnetic field component $\mathbf{h}^+$ couple to the sphere. (c) If the polarization of the external field is in the wrong direction, $\mathbf{h}^-$, the wave does not couple to the YIG sphere. (After Matthaei *et al.*, 1980.)

When an rf magnetic field h_x at the resonant frequency f_R is generated by the x loop, the ferrite emits an rf moment m with components m_x and m_y. The m_y component couples to the y loop, and the wave propagates into the termination R_y. At frequencies other than f_R, no wave propagates. The frequency f_R can be tuned by changing the biasing field H_0. Such loop-coupled filters are easy to understand, but in practice there is considerable stray coupling between the loops, and the rf fields generated are not very uniform, which tends to excite higher-order modes. A much more practical, waveguide-based, device is shown in Fig. 31. Two ferrimagnetic resonating spheres are placed near the walls in the waveguide at locations such that the spheres see a strong rf magnetic field. The spheres are coupled to one another through a slot parallel to the waveguide axis; the geometry of the slot is designed to minimize coupling between the waveguides except when the spheres are excited into resonance. The spheres are placed in reduced-height waveguides to obtain adequate coupling between them and the waveguide. This also reduces the air gap for the dc biasing magnet. The steps in the waveguide height form a Tchebyscheff transformer, designed to provide impedance matching between the normal waveguide and the reduced-height portion (Matthaei *et al.*, 1980). This filter, tunable via variations of H_0, has the same attenuation and phase characteristics for both directions of transmission. Similar structures can be designed in strip-line technology. Filters can also be designed that induce a phase shift in the propagating wave, or that have different transmission coefficients for waves propagating in different directions (Matthaei *et al.*, 1980).

3.2.2 Isolators. Isolators are devices that allow wave propagation in one direction, but not in the opposite one. They are particularly useful to isolate a microwave source from the waves reflected from a load that perturbs the amplitude or frequency stability of the source. A waveguide-based ferrite isolator is shown in Fig. 32. In an empty rectangular waveguide, Fig. 32(a), the components of the rf magnetic field for the TE_{10} mode are h_x and h_z, shown in

Table 2. Properties of single-crystal ferrimagnetic materials used in microwave devices (Matthaei *et al.*, 1980).

Material	Néel temp., T_N (°C)	Saturation magnetization, $4\pi M_s$ (G)	Linewidth, ΔH (Oe)
Yttrium iron garnet (YIG)	292	1750	0.22 (@ 4 GHz)
Gallium-substituted YIGs	—	50–1750	—
(GaYIGs)	160	600	—
	206	950±50	0.7–2 (@ 4.4 GHz)
Lithium ferrite[a]	—	3550±40	3 (@ 5 GHz)
Planar ferrite, $Ba_{12}Zn_2Fe_{12}O_{22}$	—	2850	16 (8–12 GHz)

[a]Nielsen *et al.* (1962)

Fig. 32(b) as functions of x for $y=0$. Since h_x and h_z are in space and time quadrature, circular polarization is obtained at the locations where $|h_x| = |h_z|$, i.e., the points x_1 and $a-x_1$ in Fig. 32. The value of x_1 is calculated by Soohoo (1985). An isolator can be made by applying a ferrimagnetic layer as shown in Fig. 38(c) at those points, and biasing them with opposing dc magnetic fields H_0 as shown. The nonreciprocity of this structure arises from the fact that each ferrite layer couples to only one direction of polarization of the rf magnetic field, as shown in Fig. 29. Not all four strips of ferrite need to be present to make an isolator. As described, the device in Fig. 32 is an efficient isolator, giving a large attenuation for waves traveling backward (backward attenuation), and a low attenuation for forward propagating waves (low forward attenuation, or insertion loss) only at the resonant frequency. If the bias magnetic field H_0 is far away from resonance, one can obtain a phase shift at low attenuation. Such devices are "phase shifters." A phase shifter with reciprocal action can be made if the H_0 values applied to the strips at x_1 and $a-x_1$ are of the same polarity. If the bias magnets produce H_0 values of opposite polarity, as shown in Fig. 32, the phase shift is nonreciprocal.

Both the bandwidth and the reverse-to-forward absorption ratio of the isolator can be increased by adding a strip of dielectric in the waveguide, next to the ferrite strip and on the inside of it. This technique makes the fields inside the waveguide more circularly polarized throughout the cross section and over a broader wavelength range. It is described by Soohoo (1985).

3.2.3 Circulators. A circulator is a microwave-circuit component with three or more ports. In a circulator, microwave energy can propagate from one port to another only in

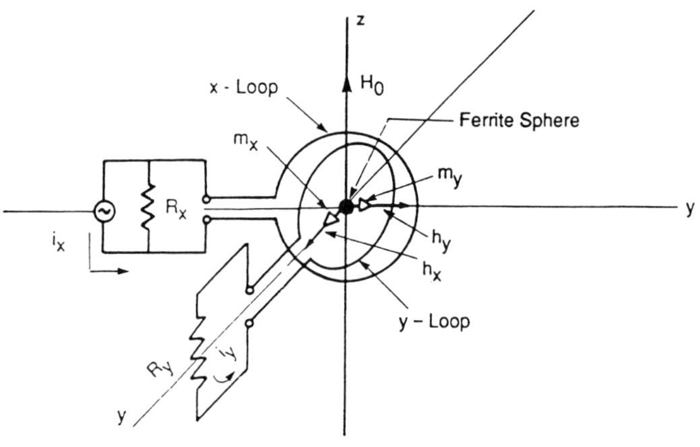

FIG. 30. A magnetically tunable filter using a single ferrite resonator in a dc biasing field H_0, and loops x and y for coupling the waves (after Matthaei *et al.*, 1980).

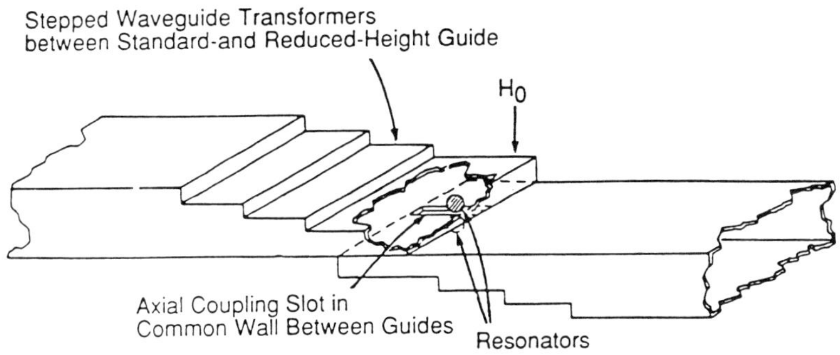

FIG. 31. A magnetically tunable waveguide filter with two resonators and an elongated axial coupling slot (after Matthaei *et al.*,et al. 1980).

one direction. In the three-port circulator shown in Fig. 33, with the dc field H_0 applied in one direction, energy can go from port 1 to port 2, from 2 to 3, and from 3 to 1, but not from 1 to 3, 3 to 2, or 2 to 1. A pedagogical example of the use of these devices is an elementary radar: attach a source to port 1, an antenna to port 2, and a detector to port 3. The wave generated at 1 propagates out through port 2 and out of the antenna. The wave reflected back from the target is received by the antenna and propagates from port 2 to port 3, where it can be detected. A Doppler radar can be made by mixing a small amount of source radiation with the wave reflected from the target. The carrier-wave frequency is filtered out at port 3 and the detector is made sensitive only to the difference frequency. This frequency is then a measure of the relative velocity of the target with respect to the radar. The circulator itself can be built as shown in Fig. 33 (top). The ferrite is attached to the top or bottom plate of the cavity at the location where the wave is circularly polarized, as in the design of the isolator, Fig. 32. A dc biasing magnetic field H_0 is applied normal to the plane of Fig. 33. Depending on the polarity of H_0, circulation is made possible either in the clockwise direction, $1 \rightarrow 2 \rightarrow 3$, or in the counterclockwise direction. This circulator is strictly a resonant device again. Dielectrics can be used, as in the design of isolators, to broaden the device's bandwidth (Fig. 33, bottom). There are many other circulator designs possible, and the reader is referred to Matthaei *et al.* (1980), Nielsen *et al.* (1962), and the article on MICROWAVE CIRCUITS.

3.2.4 Strip-Line, MMIC, and Thin-Film Technologies. The microwave-circuit elements described above used waveguide technology, for pedagogical reasons. The component elements described above are copied in strip-line technology, using ferrite or YIG thin films, and even thin-film magnets (Stancil, 1989). Ferromagnetic metals can be considered in thin-film devices, because the eddy currents are much reduced when the rf magnetic field is applied in the plane. This allows for the design of novel devices, such as the strip-line notch filter shown in Fig. 34. This filter is based on the ferromagnetic resonance in iron. The iron film shown has a resonance at $f_R \simeq 10$ GHz without external biasing. In Fig. 34, the radiation is polarized with the ac electric and magnetic fields **e** and **h** as shown. It propagates from left to right. In the iron film, which has a magnetization vector **M** as shown, radiation at the resonant frequency f_R is absorbed by the iron. Radiation at other frequencies is transmitted through the strip with much less attenuation, so that the strip serves as a notch filter.

Modern thin-film techniques can combine device-grade semiconductors with ferromagnetic films. This opens fascinating new device opportunities in monolithic microwave integrated circuits (MMICs) (Prinz, 1990).

3.3 Other Magnetic Devices

3.3.1 Antenna, Inductor, and Transformer Cores. A winding distributed about the length of a rod of high-permeability material forms an antenna in broadcast radio receivers. The core concentrates the ac magnetic flux into the antenna. Similarly, inductors and

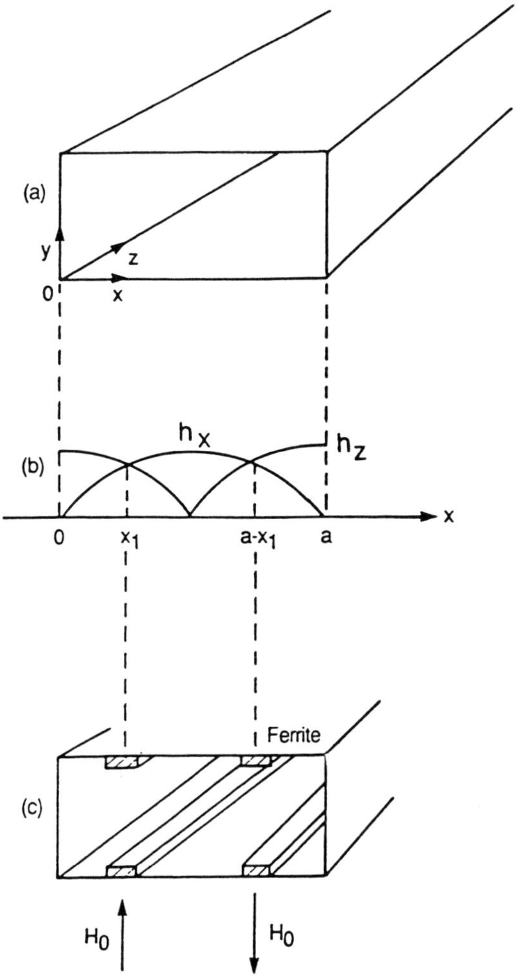

FIG. 32. A waveguide-based ferrite isolator. (a), (b) The magnetic field distribution of the TE_{10} mode in a rectangular waveguide. (c) The ferrite strips are inserted at the location where the wave is circularly polarized, i.e., $|h_z| = h_x$. The dc biasing field is H_0. (After Soohoo, 1985.)

transformers are wound around magnetic material cores. Adjustable inductors are used in many circuits. The most common form of adjustment involves changing the position of the high-permeability core material with respect to the windings. Conventional variable capacitors are difficult to seal with respect to ambient humidity; this is much less a problem in tuned inductors. The windings can easily be sealed, and moisture has little effect on the behavior of the core. Designs of antennas, inductors, and transformers use a compromise between high permeability and low losses at the intended frequency of operation. Powdered magnets and ferrites are used for cores at the higher end of the frequency spectrum. Ferromagnetic metals, such as the Ni-Fe alloys, Mumetal, Permalloy, Perminvar, and thin metglasses can be used at lower frequencies. Tape-wound cores from these alloys are available. Fe-Co, Fe-Si, Fe-Al, and pure Fe are sometimes also used in electromagnet and relay cores, since these are nearly dc applications (Fink and McKenzie, 1975).

3.3.2 Magnetic Memories. Magnetic materials are used in memories, and this type of devices is the subject of a separate article (MEMORIES). The synergistic effects of new film-growth techniques of both semiconductors and magnetic materials, already mentioned in the section on microwave thin-film techniques, also influences this field. For in-

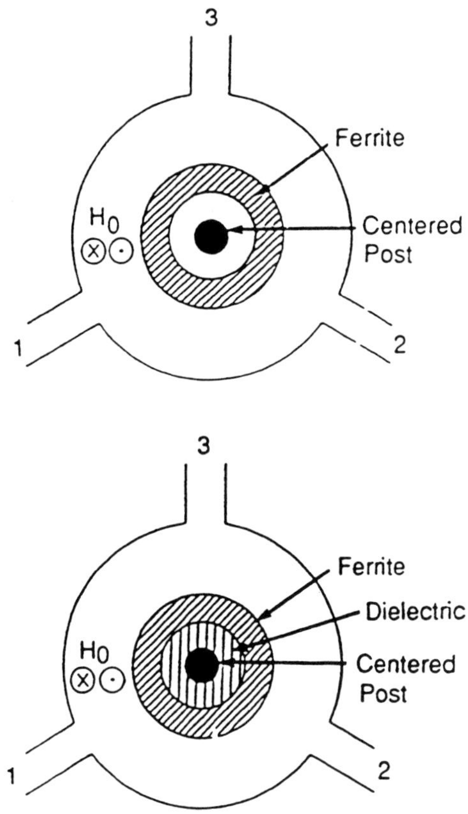

FIG. 33. (top) A waveguide circulator with three ports, 1, 2, and 3. Microwave energy can go from 1 to 2, 2 to 3, and 3 to 1, but not vice versa. The circulator is made by inserting a ferrite at the location where the wave is circularly polarized. (bottom) Dielectric loading of the circulator increases the bandwidth. (After Soohoo, 1985.)

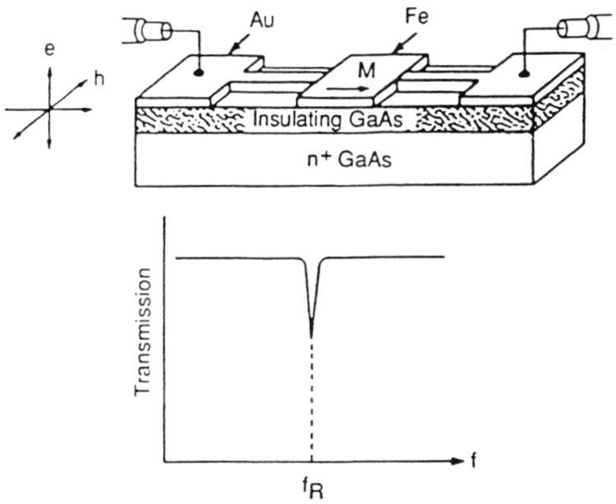

FIG. 34. Strip-line microwave notch filter, using an epitaxial iron film as resonator (after Prinz, 1990).

stance, an integrated magnetic memory element was proposed by Prinz (1990), and is shown in Fig. 35. It is the thin-film analog of the ferrite core. The element can be made from epitaxial magnetic metal deposited on a semiconductor substrate, such as Fe on Si or GaAs. In the example chosen, the iron film is patterned so that the magnetic moments align with the magnetically easy axes of Fe, here the ⟨100⟩ axes. The film can be magnetized with the moments aligned clockwise or counterclockwise, representing the 0 or 1 states of the bit. The readout of the magnetic state of the memory element can be integrated in the semiconductor, for instance as a Hall element. The magnetic loop is closed, and very little leakage flux and cross-talk are expected. The down-scaling of this memory element is only limited by semiconductor technology, because the magnetic properties of Fe are independent of size, down to maybe 200 Å.

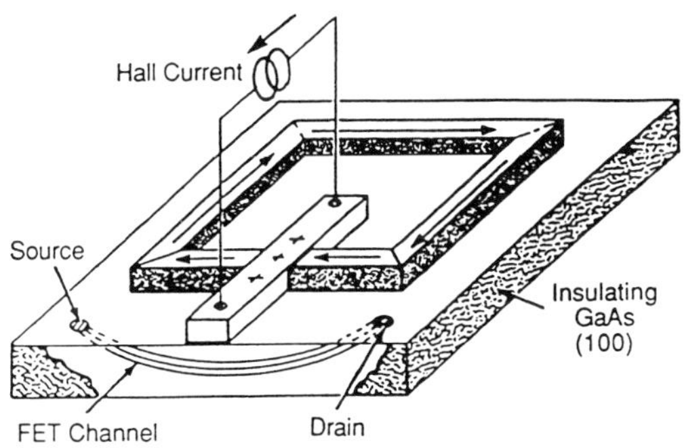

FIG. 35. Picture-frame memory element made from a thin epitaxial iron film on GaAs with a readout Hall sensor integrated in the substrate (after Prinz, 1990).

Another intergrated magnetic storage technology that uses magnetoresistors has been described by Daughton (1992), and a bipolar spin switch is described by Johnson (1993).

4. CONCLUSION

Magnetic effects and materials are widely used in electrical machines, but also in the conversion of mechanical movement into an electrical signal (position sensors) and in electronics. Magnetic device technology parallels semiconductor technology, and the synergism between the two will continue to lead to yet new devices. The ultimate synergism would be in the field of magnetic semiconductors, which were not discussed here. Only "dilute" magnetic semiconductors have been studied recently (see SEMICONDUCTORS, DILUTE MAGNETIC), and their applications are limited, because their magnetic properties are not very pronounced, especially at room temperature. Optical isolators, which perform in the visible light range the same function as their microwave equivalents described in Sec. 3.2.2, and are needed in high-data-rate optical communication systems, could be made from such materials (Furdyna *et al.*, 1991).

Acknowledgments

I want to acknowledge the help of Dr. J. Herbst and Dr. S. Foner in reading the manuscript, Dr. D. L. Partin for frequent discussions, and Mr. R. C. Dinsmore, from Dinsmore Instruments Company, for reviewing the section on fluxgates.

GLOSSARY

Circulator: Microwave device having three or more ports, labeled 1, 2, 3, etc. Microwave radiation can only propagate from port 1 to 2, 2 to 3, etc., in that order, and cannot propagate in the reverse order.

Corbino Disk: Circularly shaped slab of semiconductor, with two electrodes, one in the middle of the circle, the other on the periphery. The current lines in this element flow radially through the semiconductor. This geometry gives the largest geometrical magnetoresistance. (Can also be made from semimetals and metals.)

Fluxgate: Sensitive magnetometer using a core of magnetic material driven close to saturation, and often used in compasses.

Garnet: Ferrimagnetic mixed oxide of iron and rare-earth elements or yttrium, electrically insulating. Used as magnetic material for high-frequency applications, in which eddy-current losses have to be minimal.

Geometrical Magnetoresistance: Magnetoresistance induced in a semiconductor resistive element by choosing the geometry of the element in such a way that an applied magnetic field lengthens the current paths of the charge carriers.

Hall Generator: Device exploiting the Hall effect to generate a voltage that is used to measure the magnetic field.

Isolator (microwave or optical): Device that lets an electromagnetic wave propagate only in one direction. It is commonly placed between a source (microwave generator or laser) and its load to prevent energy reflected from the load from perturbing the source.

Magnetic Amplifier: Transformerlike device that modulates the amount of power that can be delivered from a source to a load by saturating the core of the transformer using a separate dc control current.

Magnetodiode: *p-i-n* diode in which an external applied magnetic field controls the amount of recombination in the intrinsic region.

Magnetoresistor: Resistor having a resistance that is a strong function of magnetic field, used in magnetometers. It exploits either the geometrical magnetoresistance in semiconductors, or the magnetoresistance anisotropy in magnetic metals.

Phase Shifter: Microwave device that shifts the phase of a wave propagating through it. Nonreciprocal phase shifters give different amounts of shift for waves propagating in opposing directions.

List of Works Cited

Bai Shouyi (1982), *An Outline History of China*, Beijing: Foreign Languages Press.

Baibich, M. N., Broto, J. M., Fert, A., Nguyen Van Dau, F., Petroff, F., Etienne, P., Creuzet, G., Friederich, A., Chazelas, A. (1988), *Phys. Rev. Lett.* **61**, 2472–2475.

Baltes, H. P., Popovic, R. S. (1986), *Proc. IEEE* **74**, 1107–1132.

Barthelemy, A., Fert, A. (1991), *Phys. Rev. B* **43**, 13124–13129.

Campbell, I. A., Fert, A., Jaoul, O. (1970), *J. Phys. C* **3** (Suppl. 1), S95–S101.

Chikazumi, S. (1964), *Physics of Magnetism*, New York: Wiley.

Cristoloveanu, S., Chovet, A., Malyutenko, V. K. (1983), *Sens. Actuators* **4**, 165–171.

Dandridge, A., Treten, A. B., Sipel, G. H., Jr., West, E. J., Giallorenzi, T. G. (1980), *Electron. Lett.* **16**, 408–409.

Daughton, J. M. (1992), *Thin Solid Films* **216**, 162–168.

Dibbern, U. (1986), *Sens. Actuators* **10**, 127–140.

Dieny, B., Speriosu, V. S., Parkin, S. S. P., Gurney, B. A., Wilhoit, D. R., Mauri, D. (1991), *Phys. Rev. B* **43**, 1297–1300.

Fink, D. G., McKenzie, A. A. (Eds.) (1975), *Electronic Engineers' Handbook*, New York: McGraw-Hill.

Foner, S. (1981), *IEEE Trans. Magn.* **MAG-17**, 3358–3363.

Furdyna, J., Aggarwal, R. L., Awshalom, D. D., Ehrenreich, H. H., Geschwind, S., Gunshor, R. L., Hass, K. C., Heremans, J. P., Peercy, P. S., Prinz, G. A., Ramdas, A. K., von Molnar, S., Wolff, P. A. (1991), "Diluted Magnetic Semiconductors," Panel Report, Washington, DC: National Academy Press.

Geyger, W. A. (1954), *Magnetic-Amplifier Circuits*, New York: McGraw-Hill.

Göpel, W., Hesse, J., Zemel, J. N. (Eds.) (1989), *Sensors*, Vol. 5, Weinheim, Germany: VCH.

Heremans, J. (1993), *J. Phys. D: Appl. Phys.* **26**, 1–20.

Heremans, J., Partin, D. L., Morelli, D. T., Fuller, B. K., Thrush, C. M. (1990), *Appl. Phys. Lett.* **57**, 291–293.

Heremans, J., Partin, D. L., Thrush, C. M., Green, L. (1993), *Semicond. Sci. Technol.* **8**, 424–430.

Itakura, K., Ueda, D., Hagio, M., Kazumura, M. (1989), *Electron. Lett.* **25**, 1493–1495.

Jagielinski, T. (1990), *MRS Bull.* **15** (3), 36–43.

Johnson, M. (1993), *Science* **260**, 320–323.

Lenz, J. E. (1990), *Proc. IEEE* **78**, 973–989.

Lippmann, H. J., Kuhrt, K. (1958), *Z. Naturforsch.* **13A**, 462–474.

Maenaka, K., Okada, H., Nakamura, T. (1990), *Sens. Actuators* **A21–A23**, 807–811.

Markham, D., Jeffers, F. (1990), *Proc. Am. Electrochem. Soc.* **90**, 185–204.

Matthaei, G., Young, L., Jones, E. M. T. (1980), *Microwave Filters, Impedance-Matching Networks, and Coupling Structures*, Dedham, MA: Artech House Books.

McGuire, T. R., Potter, R. I. (1975), *IEEE Trans. Magn.* **MAG-11**, 1018–1038.

Mosca, D. H., Petroff, F., Fert, A., Schroeder, P. A., Pratt, W. P., Jr., Laloee, R. (1991), *J. Magn. Magn. Mater.* **94**, L1–L5.

Nakamura, T., Maenaka, K. (1990), *Sens. Actuators* **A21–A23**, 762–769.

Nathan, A., Maenaka, K., Allegretto, W., Baltes, H. P., Nakamura, T. (1989), *IEEE Trans. Electron Devices* **36**, 108–117.

Nathan, A., Baltes, H. P., Castagnetti, R., Sugiyama, Y., Briglio, D. R. (1990), *Sens. Actuators* **A21–A23**, 776–779.

Nielsen, J. W., Lepore, D. A., Zneimer, J., Townsend, G. B. (1962), *J. Appl. Phys.* **33S**, 1379–1380.

Partin, D. L., Heremans, J. (1992), "Growth of Narow-Gap Semiconductors," in: S. Mahajan (Ed.), *Handbook on Semiconductors*, Vol. 3, 2nd ed., New York: Elsevier,

Partin, D. L., Heremans, J., Thrush, C. M., Green, L. (1992), in: *Technical Digest, IEEE Solid-State Sensor and Actuator Workshop, Hilton Head Island, SC, 22–25 June 1992*, New York: IEEE, pp. 35–38.

Popovic, R. S., Baltes, H. P. (1983), *Sensors and Actuators* **4**, 155–163.

Prinz, G. A. (1990), *Science* **250**, 1092–1097.

Seitz, T. (1990), *Sens. Actuators* **A22**, 799–802.

Shea, R. F. (1968), *Amplifier Handbook*, New York: McGraw-Hill, Chaps. 8–21.

Siemens AG (1989), *Magnetic Sensor Data Book*, München: Siemens AG.

Smit, J., Wijn, H. P. (1959), *Ferrites*, New York: Wiley.

Soohoo, R. F. (1985), *Microwave Magnetics*, New York: Harper & Row.

Stancil, D. D. (1989), *Microwave Opt. Tech. Lett.* **2**, 81–85.

Weiss, H. (1969), *Structure and Applications of Galvanomagnetic Devices*, New York: Pergamon.

Yamada, T. (1968), in: *Proceedings of the IX International Conference on the Physics of Semiconductors*, Moscow, St. Petersburg: Publishing House Nauka, pp. 672–675.

Further Reading

Baltes H. P., Popovic, R. S. (1986), *Proc. IEEE* **74**, 1107–1132.

Fink, D. G., McKenzie, A. A. (Eds.) (1975), *Electronic Engineers' Handbook*, New York: McGraw-Hill.

Furdyna, J., Aggarwal, R. L., Awshalom, D. D., Ehrenreich, H. H., Geschwind, S., Gunshor, R. L., Hass, K. C., Heremans, J., Peercy, P. S., Prinz, G. A., Ramdas, A. K., von Molnar, S., Wolff, P. A. (1991), "Diluted Magnetic Semiconductors," Panel Report, Washington, DC: National Academy Press.

Geyger, W. A. (1954), *Magnetic-Amplifier Circuits*, New York: McGraw-Hill.

Göpel, W., Hesse, J., Zemel, J. N. (Eds.) (1989), *Sensors*, Vol. 5, *Magnetic Sensors*, Weinheim, Germany: VCH.

Heremans, J. (1993), "Solid state magnetic field sensors and applications," *J. Phys. D: Appl. Phys.* **26**, 1–20.

Matthaei, G., Young, L., Jones, E. M. T. (1980), *Microwave Filters, Impedance-Matching Networks, and Coupling Structures*, Dedham, MA: Artech House Books.

Popovic, R. S. (1991), *Hall Effect Devices*, Bristol, U.K.: Hilger.

Prinz, G. A. (1990), *Science* **250**, 1092–1097.

Soohoo, R. F. (1985), *Microwave Magnetics*, New York: Harper & Row.

MAGNETIC INSTRUMENTATION

See Measurement of Magnetic Properties

LIST OF RECOMMENDED UNITS AND SYMBOLS

RECOMMENDED UNITS AND CONVERSION FACTORS

The SI system provides six basic units: meter m, kilogram kg, second s, ampere A, kelvin K, candela cd, and mole mol.

Some important derived units are also allowed and bear special names, e.g.:

1 N (newton) = 1 kg m s^{-2}
1 J (joule) = 1 N m = 1 kg m^2 s^{-2}
1 W (watt) = 1 J s^{-1} = 1 kg m^2 s^{-3}
1 Pa (pascal) = 1 N m^{-2} = 1 kg m^{-1} s^{-2}

For mass, the gram g, or the metric ton t which equals 1000 kg, may be used instead of kilogram kg.

The so-called "long ton" (UK) and "short ton" (US) have been abandoned and will not be used in the *Encyclopedia of Applied Physics.*

For pressure, the bar (name and symbol alike) may be used, and for temperature, the degree celsius °C.

From all of these, decimal multiples or fractions can be derived:

power of ten	prefix	symbol	power of ten	prefix	symbol
10	deca	da	10^{-1}	deci	d
10^2	hecto	h	10^{-2}	centi	c
10^3	kilo	k	10^{-3}	milli	m
10^6	mega	M	10^{-6}	micro	μ
10^9	giga	G	10^{-9}	nano	n
10^{12}	tera	T	10^{-12}	pico	p
10^{15}	peta	P	10^{-15}	femto	f
10^{18}	exa	E	10^{-18}	atto	a

For mass, multiples or fractions are derived from g, not from kg (since the latter already contains a prefix), e.g., mg.

Units with the prefixes are considered as one entity and can, therefore, be raised to any power, e.g., cm^3.

The liter is now considered as synonymous with dm^3 (which does not hold in the older literature!). Please use the capital letter L as unit symbol (following a recent IUPAC recommendation). The use of the Ångström unit is discouraged; it should be replaced by fractional meters (1 Å = 100 pm = 0.1 nm).

SELECTED QUANTITIES, UNITS, AND SYMBOLS

Name of quantity[a]	Symbol[b]	SI unit[c]	Name of unit	Other units
Space and time				
length*	l	m	meter[d]	
breadth, width	b	m		
height	h	m		
radius	r	m		
thickness	d	m		
area	A,S	m^2		a (are) h (hectare)
volume	V	m^3		L, l(liter)[c]
plane angle	$\alpha,\beta,\gamma,$ ϑ,φ	1, rad	radian	° (degree) ′ (minute) ″ (second)
solid angle	ω,Ω	1, sr	steradian	
wavelength	λ	m		
wave number	σ,ν	m^{-1}		
time*	t	s	second	min (minute) h (hour) d (day)
frequency	ν, f	s^{-1}		Hz (hertz)[f]
relaxation time	τ	s		
velocity	u,v	$m\ s^{-1}$		km/h
acceleration	a	$m\ s^{-2}$		
Mechanics				
mass*	m	kg	kilogram	g (gram) t (tonne)[g]
(mass) density	ρ	$kg\ m^{-3}$		g/cm^3
momentum	$\mathbf{p}$	$kg\ m\ s^{-1}$		
angular momentum	$\mathbf{L}$	$kg\ m^2\ s^{-1}$		
force	$\mathbf{F}$	N	newton[f]	
moment of force	$\mathbf{M}$	N m		
weight	G, W	N		
pressure	p	Pa	pascal	bar (bar)[h]
energy	E, W	J	joule	W h (watt hour)[i] eV (electron volt)[j]
work	W,A	J		
power	P	W	watt	J/s, V A[k]
Molecular Physics and Thermodynamics				
thermodynamic temperature*	T	K	kelvin	°C (degrees Celsius)
Celsius temperature	ϑ,t			°C
number of entities	N			
Avogadro constant[l]	N_A,L	mol^{-1}	(particles) per mole	
Boltzmann constant[m]	k,k_B	$y\ K^{-1}$		
Planck constant[n]	h	y s		
(molar) gas constant[o]	R	$y\ mol^{-1}\ K^{-1}$		
(quantity of) heat	Q	J		
entropy[p]	S	$J\ K^{-1}$		
internal energy[p]	U	J		
Helmholtz function,[p] (Helmholtz) free energy, Helmholtz energy	F,A	J		
enthalpy[p]	H	J		
Gibbs function,[p] (Gibbs) free energy, Gibbs energy	G	J		
heat capacity[p]	C_p,C_V	$J\ K^{-1}$		

Name of quantity[a]	Symbol[b]	SI unit[c]	Name of unit	Other units
Chemical Physics				
amount of substance*	n	mol	mole	
relative atomic mass	A_r	1		
relative molecular mass	M_r	1		
atomic mass constant	m_u	kg		u (atomic mass unit)[q]
mass of a portion (of substance B)	$m_B\, m(B)$	kg		g (gram)
molar mass (of substance B)	$M_B, M(B)$	$kg\ mol^{-1}$		
concentration (of substance B)	$c_B, c(B)$	$mol\ m^{-3}$		mol/L
mole fraction[r] (of substance B)	$\kappa_B, \kappa(B)$	1		
mass fraction (of substance B)	$\omega_B, \omega(B)$	1		%,%,ppm,ppb
volume fraction (of substance B)	$\varphi_B, \varphi(B)$ $\phi_B, \phi(B)$	1		%,%,ppm,ppb
mass concentration	ρ	$kg\ m^{-3}$		g/L
molality		$mol\ kg^{-1}$		mmol/kg
volume concentration[s]	σ	1		
molar volume	V_m	$m^3\ mol^{-1}$		L/mol
molar heat capacity	C_m	$J\ mol^{-1}\ K^{-1}$		
molar conductivity	Λ_m	$S\ m^2\ mol^{-1}$	(S:siemens)	
Faraday constant[t]	F	$C\ mol^{-1}$		
Electricity and Magnetism				
quantity of electricity[u]	Q	C	coulomb	
charge density	ρ	$C\ m^{-3}$		
electric potential	ϕ, V	V	volt	
electric potential difference, voltage	$U, \Delta\phi, \Delta V$	V		
electric dipole moment	$\mathbf{p}, \mathbf{p}_c$	C m		
electric current*	I	A	ampere	
electric current density	j	$A\ m^{-2}$		
electric field strength	$\mathbf{E}$	$V\ m^{-1}$		
electric displacement	$\mathbf{D}$	$C\ m^{-2}$		
capacitance	C	F	farad	
permittivity	ε	$F\ m^{-1}$		
relative permittivity	ε	1		
dielectric polarization	$\mathbf{P}$	$C\ m^{-2}$		
electric susceptibility	χ_c	1		
polarization (of a particle)	α	$m^2\ C\ V^{-1}$		
magnetic flux	Φ	Wb	weber	
magnetic flux density	$\mathbf{B}$	T	tesla	
magnetic field strength	$\mathbf{H}$	$A\ m^{-1}$		
permeability	μ	$H\ m^{-1}$, $N\ A^{-2}$	(H:henry)	
relative permeability	μ_r	1		
magnetization	M	$A\ m^{-1}$		
magnetic susceptibility	χ	1		
molar magnetic susceptibility	χ_m	$m^3\ mol^{-1}$		
(electrical) resistance	R	Ω	ohm	
(electrical) conductance	G	S	siemens	
(electrical) resistivity	ρ	Ω m		
(electrical) conductivity	κ	$S\ m^{-1}$		
self-inductance	L	H	henry	
Radiation				
radiant energy	Q, W, Q_e	J		
luminous intensity*	I	cd	candela	
radiant intensity	I_e	$W\ sr^{-1}$, W		
emissivity, emittance	ε	1		

Name of quantity[a]	Symbol[b]	SI unit[c]	Name of unit	Other units
absorptance	α	1		
reflectance	ρ, R	1		
transmittance	τ	1		
absorption coefficient:				
-linear (decadic)	a	m^{-1}		
-molar (decadic)	ε	$m^2\ mol^{-1}$		
refractive index	n	1		
molar refraction	R_m	$m^3\ mol^{-1}$		
angle of optical rotation	α	1, rad		
Transport Properties				
flux of quantity X	J_X, J	(varies)		
mass flow rate	$q_m, \dot{m}$	$kg\ s^{-1}$		
volume flow rate	$q_V, \dot{V}$	$m^3\ s^{-1}$		
heat flow rate	Φ	W		
thermal conductivity	κ, k, λ	$W\ m^{-1}\ K^{-1}$		
coefficient of heat transfer	h	$W\ m^{-2}\ K^{-1}$		
thermal diffusivity	a	$m^2\ s^{-1}$		
diffusion coefficient	D	$m^2\ s^{-1}$		
thermal diffusion coefficient	D_T	$m^2\ s^{-1}$		
viscosity	η, μ	Pa s		
kinematic viscosity	ν	$m^2\ s^{-1}$		

[a]SI base quantities are marked by asterisks (*)
[b]Recommended by IUPAC
[c]SI base units as well as derived and supplementary units are listed; all are to be used with prefixes as needed
[d]do not use "metre"
[e]do not use "litre"; $1\ L = 10^{-3}\ m^3$
[f]$1\ Hz = 1\ s^{-1}$; $1\ N = 1\ kg\ m\ s^{-2}$
[g]Formerly metric ton; $1\ t = 10^3\ kg$
[h]$1\ bar = 10^5\ Pa$
[i]$1\ W\ h = 3.6 \times 10^3\ J$
[j]$1\ eV = 1.602\ 189 \times 10^{-19}\ J$
[k]$1\ W = 1\ J/s = 1\ VA$
[l]$N_A = 6.022\ 136\ 7 \times 10^{23}\ mol^{-1}$
[m]$k = 1.380\ 658 \times 10^{-23}\ J\ K^{-1}$
[n]$h = 6.626\ 075\ 5 \times 10^{-34}\ J\ s$
[o]$R = 8.314\ 510\ J\ mol^{-1}\ K^{-1}$
[p]Molar quantities can be distinguished from the quantity of a system by adding the subscript m; e.g., molar internal energy U_m, in $J\ mol^{-1}$
[q]$1\ u = 1.660\ 565\ 5 \times 10^{-27}\ kg$
[r]A more accurate, but rather uncommon, name is "amount-of-substance fraction"
[s]σ refers to the total volume of a mixture, whereas the volume fraction φ relates the volume of a substance to the volume of several components before mixing
[t]$F = 9.648\ 530\ 9 \times 10^4\ C\ mol^{-1}$
[u]Also called electric charge